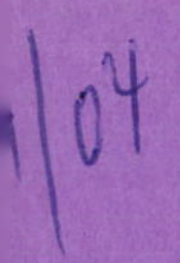

3000 800055 83305
St. Louis Community College

D0212008

The Museum of
Broadcast Communications

Encyclopedia of Radio

The Museum of Broadcast Communications

Encyclopedia of Radio

VOLUME I

A–E

Editor

CHRISTOPHER H. STERLING

Consulting Editor

MICHAEL C. KEITH

FITZROY DEARBORN
AN IMPRINT OF THE TAYLOR & FRANCIS GROUP

NEW YORK • LONDON

Published in 2004 by
Fitzroy Dearborn
An imprint of the Taylor and Francis Group
29 West 35th Street
New York, NY 10001

Published in Great Britain by
Fitzroy Dearborn
An imprint of the Taylor and Francis Group
11 New Fetter Lane
London EC4P 4EE

10 9 8 7 6 5 4 3 2 1

Library of Congress Cataloging-in-Publication Data

The Museum of Broadcast Communications encyclopedia of radio / editor,
Christopher H. Sterling ; consulting editor, Michael Keith.
p. cm.
ISBN 1-57958-249-4 (set : alk. paper) -- ISBN 1-57958-431-4 (vol. 1 :
alk. paper) -- ISBN 1-57958-432-2 (vol. 2 : alk. paper) -- ISBN
1-57958-452-7 (vol. 3 : alk. paper)
1. Radio--Encyclopedias. 2. Radio programs--Encyclopedias. I.
Sterling, Christopher H., 1943- II. Museum of Broadcast Communications.
III. Title.
TK6544.M84 2004
384.54'03--dc22

2003015683

First published in the USA and UK 2004

Typeset by Andrea Rosenberg
Printed by Edwards Brothers
Cover design by Peter Aristedes, Chicago Advertising and Design, Chicago, Illinois

Cover photos: (from upper left): Freeman Gosden and Charles Correll, *Amos 'n' Andy; Edgar Bergen and Charlie McCarthy Show;* Dick Clark; Jimmy Durante; Stan Freberg; Arthur Godfrey; Gertrude Berg of *The Goldbergs;* William Conrad (right) of *Gunsmoke;* Paul Harvey; Tom Joyner

CONTENTS

EDITOR'S INTRODUCTION

Radio means many distinct things to different people. For some, radio primarily means the "golden age" of the 1920s through the 1940s when network radio headlined the only broadcast service and provided a variety of programs for all tastes. For such listeners, radio's importance is in its programs and stars, its role as the on-the-spot recorder of history, and in its carriage of period politics, sports, and talk. (It is this period of old-time radio [OTR] that is hotly collectible—books, program premiums, recordings, magazines, and equipment of the era.) For others radio means the vibrant business of the early 21st century with huge and growing chains of stations under common ownership dependent on syndicated programs and heavy advertising. And for still others "radio" does not mean broadcasting at all, but instead refers to the transmission of voice and data, amateur or "ham" station operators, or even reception of music and talk programs on the internet. Each of these meanings, none mutually exclusive, illustrates part of radio's pervasive role in society.

Indeed radio has become part of our daily background, there but "not there" for many if not most of us. As I write this, classical music is playing from the local public radio station, for I am one of those who work better with such soothing background. And you need not look farther than the nearest teenager listening with headphones (probably while talking on the phone, monitoring television or a video, and "doing homework" on a computer) to realize how omnipresent yet invisible radio can be. And notice how often you hear radio playing in the workplace, in your car, or through ever-present headphones on public transit—and at all hours.

While most present-day media commentary focuses on television (or, increasingly, the internet), such commentary, whether praise or censure, was first directed at radio, including the fear that violent or suspenseful programs would overly excite children's imaginations, creating untold effects. Radio also established most elements of present-day electronic media industry structure. Much of what we both enjoy and bemoan today, in other words, was accomplished (or inflicted) by radio long before television arrived.

For example, that American broadcasting would depend on advertising was pretty much decided by the late 1920s, despite several concerted efforts (before and after passage of the benchmark 1934 Communications Act) to open up greater opportunities for other funding options. In turn, advertising support meant that American radio would be primarily a medium of entertainment (to attract the largest possible audience for that advertising) rather than the public or cultural service that developed in nations with other approaches to financial support. That national networks would dominate radio news and entertainment in the years before the coming of television (which would later and very quickly adopt the same pattern) was a fact by the early 1930s, with only minor modifications over the years. That government would have to selectively license broadcaster access to limited spectrum space was obvious by the early 1920s; even so, that process became fully effective only in 1927. And that government would have little to do with American radio program content (though this has again varied over time) was made clear in the laws of 1927 and 1934, reinforced by numerous court decisions in the years to follow.

On the other hand, most other nations made quite different decisions on how best to implement the promise of radio broadcasting. Certainly best known and most copied is the British Broadcasting Corporation, whose decades of invaluable public service programming were instilled from the beginning by its legendary founding director-general, John Reith. Most other industrial nations, usually copied by their colonies, also adopted a no-advertising, public-service model of radio that relied on public monies or a receiver tax to support operations. These systems became showcases of literature, music, drama, and erudite conversation—but often carrying very little light entertainment comedy, popular music, or serial drama. By 1940, after two decades of radio broadcasting development, such systems largely dominated the world of broadcasting.

When commercial television began in the United States in the 1940s, the video medium was able to develop quickly (despite its higher costs for all concerned) because radio had already established the commercial structure of the industry, the relationship between government and broadcasters, and the wealth of program formats to exploit. Television added pictures and new generations of stars, but the heavy lifting of designing an electronic media system had already been accomplished by radio. When television developed elsewhere in the 1940s and 1950s, it mirrored the public-service model of national radio broadcasting systems, though often accepting advertising to alleviate the visual medium's higher costs.

Only relatively late in the 20th century did world radio and television systems come to more closely resemble the American model of commercial support and popular-appeal programming. The reasons for this trend were and remain many and varied, though economic pressures were at their core. Because of the growing dominance of American radio program and operational methods worldwide, this encyclopedia emphasizes U.S. radio practices even as it attempts to treat the emergence and development of radio throughout the world.

What Is Included Here

Given radio's long history, how can any one reference work, even one with more than 1,500 pages and in three volumes, include all of its many roles and meanings? What should be "in" and what must be left out? How much emphasis should be given to technology, economic factors, programs, key personnel, and organizations? How best to demonstrate the many national approaches to radio, often sharply different from practice in the United States?

What you hold in your hands represents our cooperative, compromise answer to such questions. Our approach is international in scope, though with a strong emphasis on the United States and a secondary emphasis on key English-speaking nations including Britain and Canada (volumes such as these could be assembled for many other countries). Other regional entries provide at least brief comparative comment on most other systems of and approaches to radio broadcasting. We have tried to provide an ideal "first-stop" reference source for most aspects of American and English-language radio broadcasting, with sufficient coverage of other radio broadcasting traditions to provide context. Our aim has been, within some established limits of space, to be as inclusive as possible.

There are perhaps more topical or subject (as opposed to people and program) entries in this work than appear in the companion three volumes on television published in 1997 (and under revision as this is written). Partly this reflects the interests of those editing and writing both reference works, but the difference is due as well to the fact that we have had far more time to consider radio—especially that so-called "golden age" before television—and its many roles within society. Some of what is described here has been history for a half century or more, plenty of time to develop a broad appreciation of what remains important.

. . . and What May Seem Not To Be

Our focus in these volumes is entirely on radio *broadcasting*; we do not cover other applications of the technology. Aside from occasional side references, you will find little here on radio defined more broadly: as, for example, the primary means of transmission for such services as telephone, television, data, and more recent mobile services.

Even within the realm of radio broadcasting, however, we have had to make numerous and often difficult choices on what to include and how. There is a growing number of reference books (among them several other encyclopedias—see the "reference shelf" listing that follows) about radio in the United States and elsewhere; the editors, in consultation with the advisers listed on page xxix, obviously did not wish merely to parallel such works. After the decision to limit our coverage to broadcasting, we agreed to focus substantially on the United States, which has both the oldest and the most extensive radio broadcasting service in the world. But that still left a host of choices to be made: who and what to include and what to leave out or mention only in passing. Our advisory board for this project was invaluable in helping to hone the focus of these volumes. Long discussions both face-to-face and by email helped to further define our direction.

A few examples suffice to illustrate the point. Should we list and discuss a famous comedian under his name or only within his program? It was easy enough to decide in the case of Fred Allen, Jack Benny, or Edgar Bergen where a man's name (there were far fewer women for much of the period covered here than is the case today) and program title were nearly one and the same, but proved harder with Eddie Cantor or Ed Wynn where program titles varied. We have adopted what seemed the most likely way a user would seek the information, and relied on extensive cross-referencing and our comprehensive index to cover other possibilities. Some types of radio personalities—sportscasters and news commentators, for example—reflect compromises where a few (e.g., Red Barber or Edward R. Murrow) receive their own entries and others (Daniel Schorr, for example) are more briefly covered in paragraphs within larger subject entries. Likewise with radio's many different formats: we have provided entries on the most important, but some minor or splinter formats are treated within larger rubrics. Readers are encouraged to consult the index to locate such treatment.

Despite our best efforts, however, we have without a doubt missed some favorites or given insufficient emphasis to others. We welcome comments—though with a bit of trepidation after nearly five years of effort by a lot of folks—and will keep track of suggestions of new topics (and authors) for possible future revisions.

Christopher H. Sterling
Washington, D.C.

ACKNOWLEDGMENTS

As was the case with the companion *Encyclopedia of Television* (1997), many people have helped make the three volumes of the *Encyclopedia of Radio* a viable project. I am grateful first of all to **Bruce DuMont**, the founder and president of the Museum of Broadcast Communications, for entrusting the overall editorial task to me. He was most persuasive when I expressed concern about the very scope of what he proposed.

In the actual process of designing and editing this project, Michael Keith and Paul Schellinger were absolutely indispensable. **Michael C. Keith** of Boston College served as our consulting editor and was my comrade in arms on this project, serving as consulting editor throughout, or as he put it, on call 24/7/365—and he was. Mike's knowledge of all things radio is unsurpassed, and he constantly got us out of scrapes and errors and urged us on. When we needed an author or had to have an essay checked, he was there with useful input and suggestions. Mike and I co-authored several entries, and I learned more from him every time. This is a far better product for Michael's devotion to it.

Paul Schellinger, the project's final and most important editor at Fitzroy Dearborn, worked hard during the final two years or more to get the many pieces into final form, following up with authors, keeping the undersigned cool, and seeing the project through to completion. And he did it with grace and diplomacy—which was often required! Just keeping track of who was doing what—or for that matter, *not* doing what they had promised—was a huge undertaking. The project most certainly could not have been completed without Paul. And he never once had to throw himself on those bamboo bean poles under his second-story study window because of "just one more" edit or addition—though he got close.

Paul's predecessor **Steve LaRue** was another steady rock for much of this multi-year process—and had a sense of humor to boot. He handled all the correspondence and record-keeping without which no such reference work can be assembled. Likewise **Carol Burwash**, Steve's predecessor, worked with the undersigned to get the project initially designed and under way. You would not have these books in your hands without the steady efforts of these three Fitzroy Dearborn editors.

Dan Wingate, then an archivist with the MBC staff, played a central role in gathering the many photographs. This involved constant contacts with owners of photographs, obtaining permission to use the selected images and seeking photos we needed and could not find. We also appreciate the help of **Chuck Howell** and others at the Library of American Broadcasting who helped greatly with photographs and wrote several entries as well.

Our board of advisers was most helpful, especially at the opening stages of the project in 1998 as we first began to design what would be included, how long entries should be, and initial potential authors to contact. Taken as a group, the board had radio knowledge in depth and breadth, many having years of experience dating back to the golden age of

American network radio. Several members played special roles, helping to round up authors for specific categories of entries and then assisting in editing their work. **Graham Mytton** was invaluable as our man in London, gathering authors for virtually all of the British entries. Likewise, **John D. Jackson** assisted greatly with the English and French authors who contributed material on Canada. **Ed Shane** not only assisted as a board member but contributed several important entries himself—as did his spouse. And Ed came back several times offering to do more and to scare up authors—or lean on those who were late. **Michele Hilmes** was especially helpful in suggesting graduate students and other colleagues as contributors. **Peter Orlik** helped a great deal at the very beginning of the process, authoring the first entries that were used as samples for others. **Horace Newcomb** set a high standard to follow with his editorship of the *Encyclopedia of Television*—and helped to train the undersigned when I served on his editorial board.

Finally, we *all* owe thanks to the more than 240 individual authors who have written the 670 essays that make up this radio reference. They undertook all the original research, did the writing, and were generally a patient lot, accepting editorial suggestions and queries and working to tighten up their essays as we neared publication. They have waited a while to see their efforts in print, and I hope they are pleased.

Christopher H. Sterling
Washington, D.C.
September 2003

GENERAL BIBLIOGRAPHY: A RADIO REFERENCE SHELF

This listing supplements the many more specific "further reading" references appearing with individual entries. It covers those books, periodicals, and internet web sites that are broad in coverage and could be cited under many different entries. This is by no means a comprehensive listing (which would take a volume in itself), but rather points out some of the more useful reference works, monographs, and other sources relating to the history of radio, emphasizing recent titles. An asterisk (*) before a citation indicates an especially important title.

A. Books

1. Bibliographies, Statistics

*Carothers, Diane Foxhill, *Radio Broadcasting from 1920 to 1990: An Annotated Bibliography,* New York: Garland, 1991. Some 1,700 items covering radio worldwide.

Cooper, Isabella M., *Bibliography on Educational Broadcasting,* Chicago: University of Chicago Press, 1942 (reprint, New York: Arno Press, 1971). Much broader than its title suggests, this is especially useful for pre-war publications.

Greenfield, Thomas Allen, *Radio: A Reference Guide,* Westport, Connecticut: Greenwood, 1989. Combines topically arranged narrative discussion of the literature, and chapter bibliographies.

Kittross, John M., editor, *A Bibliography of Theses and Dissertations in Broadcasting, 1920–1973,* Washington, D.C.: Broadcast Education Association, 1973. Many citations concern radio. Includes both topical and key-word indexes.

Radio Marketing Guide And Fact Book, New York: Radio Advertising Bureau (annual). Statistics on radio's audience and advertising reach.

Sterling, Christopher H., *Electronic Media: A Guide to Trends in Broadcasting and Newer Technologies, 1920–1983,* New York: Praeger, 1984. More than 100 historical time-series tables and explanatory text covering radio (and other services) into the early 1980s.

Sterling, Christopher H., and George Shiers, *History of Telecommunications Technology: An Annotated Bibliography,* Lanham, Maryland: Scarecrow Press, 2000. Includes radio (chapter 9) and television and related topics—some 2,500 sources in all.

UNESCO, *Statistics on Radio and Television, 1950–1960,* Paris: UNESCO "Statistical Reports and Studies No. 8," 1963.

UNESCO, *Statistics on Radio and Television, 1960–1976,* Paris: UNESCO "Statistical Reports and Studies No. 23," 1978.

UNESCO, *Latest Statistics on Radio and Television Broadcasting.* Paris: UNESCO "Statistical Reports and Studies No. 29," 1987. Taken together, these three reports provide data on availability and use of radio transmitters, receivers, and programs, from 1950 into the 1980s.

2. *Directories and Yearbooks*

Anderton, J.T., *The Facilities of American Radio,* Cincinnati, Ohio: Duncan's American Radio, 1997. FM coverage maps for 1,300 stations in top 100 markets.

Broadcasting and Cable Yearbook, New York: Cahners/Broadcasting, 1935– (annual). One of the industry's standards, this includes a directory of all radio stations plus some statistics and listings of ancillary parts of the industry (networks, station reps, consultants).

Duncan's Radio Market Guide, Cincinnati, Ohio: Duncan's American Radio Inc. (annual). Offers a page or two on each Arbitron radio market with comparative social and business statistics, and information on operating radio stations in each.

Kneitel, Tom, *Radio Station Treasury: 1900–1946,* Commack, New York: CRB Research, 1986. Listings of stations in the U.S. and abroad for selected years—taken from contemporary sources.

M Street Radio Directory, Nashville, Tennessee: M Street Journal (annual). Sections include station listings by state and city of license, alphabetized call letters (both current and former), frequency, and market. Lists nearly 400 radio markets (including a list of the stations serving each market, and their formats and frequencies.

The Radio Power Book, New York: Billboard, 1995– (annual). One of six different Billboard directories, this includes information on major radio stations, specifically music stations, radio program syndicators, and on record labels plus Arbitron information from the top 100 markets.

Radio Yearbook, Chantilly, Virginia: BIA Financial Network (annual). Profiles all 261 Arbitron radio markets and some 10,000 stations, with full directory information on each station. Combines ratings and technical information. Includes metro, ownership, and key station contact information, listing of vendors and service providers.

R&R Directory: Ratings, Industry Directory and Program Supplier Guide, Los Angeles, California: Radio & Records (twice annually). Directory of all types of suppliers to radio stations—syndicated programs, equipment, services, etc.

3. *Radio's Technical Development*

Aitken, Hugh G.J., *Syntony and Spark: The Origins of Radio,* New York: Wiley, 1976. Classic study of Clerk Maxwell, Hertz, and Marconi.

Aitken, Hugh G.J., *The Continuous Wave: Technology and American Radio, 1900–1932,* Princeton, New Jersey: Princeton University Press, 1985. Continues the story of wireless from previous book, focusing on American developments: Fessenden and the alternator, Elwell and the arc transmitter, de Forest and his Audion, radio and cables and the national interest, the development of RCA (three chapters), and the expansion of the business based on tube technology in the 1920s.

Coe, Lewis, *Wireless Radio: A Brief History,* Jefferson, North Carolina: McFarland, 1996. Provides useful, concise survey of the work of many inventors, bringing the story up to date.

Douglas, Alan, *Radio Manufacturers of the 1920s*, 3 vols., Vestal, New York: Vestal Press, 1988–89. Combines brief text with reprinted contemporary advertising.

Dunlap, Orrin E. Jr., *Radio's 100 Men of Science,* New York: Harper, 1944. Useful mini-biographies of major inventors and engineers.

Inglis, Andrew F., *Behind The Tube: A History of Broadcasting Technology and Business,* Boston: Focal Press, 1990. Includes chapters on both AM and FM broadcasting, placing the technology of both within the larger context of industry development.

Leinwoll, Stanley, *From Spark to Satellite: A History of Radio Communication,* New York: Scribner, 1979. A wide-ranging popular history of radio into the 1970s, emphasizing the role of key inventors and developments in the expanding roles of the medium.

*Lewis, Tom, *Empire of the Air: The Men Who Made Radio,* New York: HarperCollins, 1991. Well-written analysis of the lives, work and changing relationships of Armstrong, de Forest and Sarnoff, among others.

*Maclaurin, W. Rupert, *Invention and Innovation in the Radio Industry,* New York: Macmillan, 1949 (reprint, New York: Arno Press, 1971). Thorough treatment with supporting data and much critical analysis of the process and nature of radio inventions. A study of struggles, litigation, progress, and failure of both individual inventors and industrial organizations. Includes chapters on FM and television.

*McNicol, Donald, *Radio's Conquest of Space: The Experimental Rise in Radio Communication,* New York: Murray Hill Books, 1946 (reprint, New York: Arno Press, 1974). A technical but readable record of radio progress from the mid–19th century; one of the best surveys of inventive achievement—both readable and reliable.

Mott, Robert L., *Radio Sound Effects: Who Did It, and How, in the Era of Live Broadcasting,* Jefferson, North Carolina: McFarland, 1993. Written by a radio sound-effects authority, this unique history reviews the development of sound effects in radio drama and comedy programs.

*Schiffer, Michael Brian, *The Portable Radio in American Life,* Tucson: University of Arizona Press, 1991. Valuable survey from clunky tube receivers to modern transistor models.

Shiers, George, editor, *The Development of Wireless to 1920,* New York: Arno Press, 1977. This anthology reprints 20 pioneering technical and historical papers tracing developments from the late 19th century, many by the inventors themselves including Fleming, de Forest, Fessenden, Marconi, Carl Braun, Armstrong, Elwell, and Alfred Goldsmith. Photos, diagrams, notes.

4. Histories of U.S. Radio Broadcasting

*Archer, Gleason L., *History of Radio to 1926,* New York: American Historical Society, 1938 (reprint, New York: Arno Press, 1971). While largely concerned with business aspects, this includes considerable pre-broadcast technical background and context, especially patent and related corporate rivalries. Important as the first attempt at a scholarly history. Photos, notes, appendices, index.

Archer, Gleason L., *Big Business and Radio,* New York: American Historical Company, 1939 (reprint, New York: Arno Press, 1971). Both continues previous book and fills in some of its holes, though retaining a strong RCA bias throughout. Valuable for its contemporary view of pre-war radio development.

*Barnouw, Erik, *A History of Broadcasting in the United States,* 3 vols., New York: Oxford University Press, 1966–70. Classic and well-written narrative; the first two volumes take the story to 1953 and focus most on radio. These could have been cited as "further readings" in most of the American entries in the encyclopedia.

Bergreen, Laurence, *Look Now, Pay Later: The Rise of Network Broadcasting,* Garden City, New York: Doubleday, 1980. One of the few historical treatments of networks, covering both radio and television eras.

Chase, Francis Jr., *Sound and Fury: An Informal History of Broadcasting,* New York: Harper, 1942. A well-written popular survey which emphasizes programming into the early World War II period.

*Douglas, Susan J., *Inventing American Broadcasting, 1899-1922,* Baltimore: Johns Hopkins University Press, 1987. Readable yet scholarly analysis of the combination of technological innovation, institutional development, and both visions and business realities that led to the radio broadcasting business in the early 1920s. Chapters focus on Marconi as inventor-hero, the inventors' struggles for technical distinction, wireless telegraphy in the Navy, the ups and downs of wireless as a business, the important role of amateur operators prior to World War I, initial radio regulation, the rise of military and corporate control, and the social construction of broadcasting.

Douglas, Susan J., *Listening In: Radio and the American Imagination,* New York: Times Books, 1999. More focused on programs and listener reactions than the author's earlier book.

The First 50 Years of Broadcasting: The Running Story of The Fifth Estate, Washington, D.C.: Broadcasting Publications, 1982. Based on articles that appeared in the industry weekly—one per year—thus providing something of an annotated chronology with a good deal of industry and regulatory emphasis.

*Godfrey, Donald G., and Frederic A. Leigh, editors, *Historical Dictionary of American Radio,* Westport, Connecticut: Greenwood Press, 1998. Very useful reference ranging over programs, people, organizations, and topics with dozens of contributors.

*Halper, Donna L., *Invisible Stars: A Social History of Women in American Broadcasting,* Armonk, New York: M.E. Sharpe, 2001. First attempt at an overall history of the many and varied roles of women in radio (and television), highlighting many forgotten pioneers.

Hettinger, Herman S., *A Decade of Radio Advertising,* Chicago: University of Chicago Press, 1933 (reprint, New York: Arno Press, 1971). Standard treatment of the rise of radio advertising, focusing on network developments.

Hilliard, Robert L., and Michael C. Keith, *The Broadcast Century: A Biography of American Broadcasting,* 3rd edition, Boston: Focal Press, 2001. An informal and illustrated history told in chronological fashion, with many contributed comments from important pioneering figures, and a useful chronological time-line.

*Hilmes, Michele, *Radio Voices: American Broadcasting, 1922–1952,* Minneapolis: University of Minnesota Press, 1997. Concentrating on what people heard rather than on the industry or technology, the author focuses on several key programs to illustrate the medium's appeal and success.

Hilmes, Michele, *Only Connect: A Cultural History of Broadcasting in the United States,* Belmont, California: Wadsworth, 2002. Covers the whole period of radio and television focusing on the impact of those media.

Hilmes, Michele, *Connections: A Broadcast History Reader,* Belmont, California: Wadsworth, 2003. Useful anthology of both contemporary and later research material on the whole range of American radio-television history.

*Keith, Michael C., *Talking Radio: An Oral History of American Radio in the Television Age,* Armonk, New York: M.E. Sharpe, 2000. Impressive editing job makes this readable despite about 100 contributors, and gets the reader close to decision-making in the radio business.

Landry, Robert J., *This Fascinating Radio Business,* Indianapolis: Bobbs-Merrill, 1946. Broad popular survey of the radio industry just before television entered the scene, written by the editor of the *Variety* trade paper, and including numerous references to history and technical factors.

*Lichty, Lawrence W., and Malachi C. Topping, editors, *American Broadcasting: A Source Book on the History of Radio and Television,* New York: Hastings House, 1975. Excellent anthology of articles, statistics, first-hand accounts.

Mayes, Thorn L., *Wireless Communication in the United States: The Early Development of American Radio Operating Companies,* East Greenwich, Rhode Island: New England Wireless and Steam Museum, 1989. Sorts out the host of pre-World War I wireless companies in the U.S.

Nachman, Gerald, *Raised on Radio,* New York: Pantheon, 1998. A delightful read combining considerable insight, wonderful nostalgia, and a fair bit of background information on key programs.

*Smulyan, Susan, *Selling Radio: The Commercialization of American Broadcasting 1920–1934,* Washington, D.C.: Smithsonian Institution Press, 1994. Scholarly study of the policy battles over commercial support of radio (and later television) argues that what resulted—today's commercial system—was by no means a sure thing in the early days.

*Sterling, Christopher H., and John M. Kittross, *Stay Tuned: A History of American Broadcasting,* 3rd edition, Mahwah, New Jersey: Lawrence Erlbaum, 2002. Standard history—the most comprehensive one-volume survey available—arranged by periods and then by topic. Extensive historical tables, technology glossary, and bibliography.

Taylor, Glenhall, *Before Television: The Radio Years,* Cranbury, New Jersey: A.S. Barnes, 1979. Popular history of the radio business to the late 1940s.

White, Llewellyn, *The American Radio,* Chicago: University of Chicago Press, 1947 (reprint, New York: Arno Press, 1971). Critical history that finds much wanting in the commercial radio business on the eve of television's introduction.

5. Pictorial Histories

Henderson, Amy, *On the Air: Pioneers of American Broadcasting,* Washington, D.C.: Smithsonian Institution Press, 1988. Photos and literate captions of industry leaders and key stars, based on exhibit at National Portrait Gallery.

Rhodes, B. Eric., *Blast from the Past: A Pictorial History of Radio's First 75 Years,* West Palm Beach, Florida: Streamline, 1996. Impressive collection of several hundred photos tracing the rise of commercial radio.

Settl, Irving, *A Pictorial History of Radio,* 2nd edition, New York: Grosset and Dunlap, 1967. Emphasis on programs, both local and network.

Slide, Anthony, *Great Radio Personalities in Historic Photographs,* New York: Dover, 1982. Just that, well over 100 of them, with useful captions.

6. The American Radio Industry

Albarron, Alan B, and Gregory C. Pitts, *The Radio Broadcasting Industry,* Boston: Allyn and Bacon, 2000. Useful survey of the modern business and how it operates.

Ditingo, Vincent M., *The Remaking of Radio,* Boston: Focal Press, 1994. Reviews the changing status of the radio business amidst growing competition, comparing and contrasting AM and FM and reviewing likely technology changes of the future.

Fornatale, Peter, and Joshua E. Mills, *Radio in the Television Age,* Woodstock, New York: Overlook Press, 1980. Broad description of the industry after three decades of television competition.

Frazier, Gross, and Kedlec, *Radio in 1985,* Washington, D.C.: National Association of Broadcasters, 1977. The changing economics of AM and FM radio.
*Keith, Michael C., *The Radio Station,* 5th edition, Boston: Focal Press, 2000. All aspects of operating and programming the modern commercial station sales format, news, research, promotion, traffic and billing, production, engineering and consultants. A standard, this has appeared in several languages.
McKinsey & Co., *Radio in Search of Excellence: Lessons From America's Best-run Radio Stations,* Washington, D.C.: National Association of Broadcasters, 1985. A dozen case studies of good practices.
NAB Science and Technology Department, *Station Consolidation: A Technical Planning Guide for Radio Stations,* Washington, D.C.: National Association of Broadcasters, 1997. Primary focus on how best to combine technical facilities, with a good case study.
Pease, Edward C., and Everette E. Dennis, editors, *Radio: The Forgotten Medium,* New Brunswick, New Jersey: Transaction, 1995. Twenty-two essays from *Media Studies Journal* provide a broad survey of radio today including both structure and content trends.
Radio Inc., *The Radio Book: The Complete Station Operations Manual,* 3 vols., West Palm Beach, Florida: Radio Ink, nd. Covers management, programming and promotion, and sales and marketing.
Reinsch, J. Leonard, and Elmo I. Ellis, *Radio Station Management,* 2nd edition, New York: Harper, 1960. A standard treatment by two long-time broadcasters, demonstrating the rapid changes in radio as television expanded.
Shane, Ed, *Selling Electronic Media,* Boston: Focal Press, 1999. Focuses on the selling of air time and the marketing of radio and television stations—by a top industry consultant.
Soley, Lawrence, *Free Radio,* Boulder, Colorado: Westview Press, 1999. Full account of the emergence of micro radio and the issues confronting small stations.
State of the Radio Industry, Fairfax, Virginia: Broadcast Industry Analysis (annual). Focuses on the important trends, regulatory and technological impacts concerning revenue and finance, ownership, top radio ownership groups, and technical changes.

7. American Radio Programming

For specific people, programs, or formats, see references under relevant entries; included here are only those titles that range widely.

Buxton, Frank, and Bill Owen, *The Big Broadcast, 1920–1950,* New York: Viking Press, 1972. A substantial revision of the authors' *Radio's Golden Age* (1966), this briefly lists and describes network programs.
DeLong, Thomas A., *The Mighty Music Box,* Los Angeles: Amber Crest Books, 1980. A broad history of all types of music on the air from initial classical pioneers through the various formats of the so-called golden years (into the 1950s).
*Dunning, John, *On the Air: The Encyclopedia of Old-Time Radio,* New York: Oxford University Press, 1998. Clearly *the* definitive directory of American network and major syndicated programs from the 1920s into the 1960s. Often includes full credits and, for important programs, quite lengthy discussion. This is another of the handful of invaluable books that could have been cited in "further reading" for all of the U.S. program entries.

Eberly, Phillip K., *Music in the Air: America's Changing Tastes in Popular Music, 1920–1980,* New York: Hastings House, 1982. Reviews both popular music trends and the central role of radio in spreading music's popularity.

Grams, Martin, *Radio Drama: A Comprehensive Chronicle of American Network Programs, 1932–1962,* Jefferson, North Carolina: McFarland, 2000. Listed by title with brief description and full list of episodes.

Hall, Claude, and Barbara Hall, *The Business of Radio Programs,* New York: Hastings House, 1978. Solid though now dated overview of factors in formatting of stations.

*Hilmes, Michele, and Jason Loviglio, editors, *Radio Reader: Essays in the Cultural History of Radio,* New York: Routledge, 2001. Excellent indicator of growing scholarly interest in radio past and present, offering 24 original essays.

Keith, Michael C., *Radio Programming: Consultancy and Formatics,* Boston: Focal Press, 1987. Then-current trends by one of the most prolific of writers on all aspects of radio.

Lackman, Ron, *The Encyclopedia of American Radio,* New York: Checkmark Books/Facts on File, 2000. Primarily people and programs with from one to several paragraphs on each. Updates the author's *Same Time, Same Station* (1996).

MacDonald, J. Fred, *Don't Touch That Dial! Radio Programming in American Life, 1920–1960,* Chicago: Nelson-Hall, 1979. A broad social history, focusing on network programs.

MacFarland, David T., *Future Radio Programming Strategies: Cultivating Listenership in the Digital Age,* Mahwah, New Jersey: Lawrence Erlbaum, 1997. Integrated analysis of audience needs and desires, and various approaches to both music and talk program formats in the competitive 1990s.

Maltin, Leonard, *The Great American Broadcast: A Celebration of Radio's Golden Age,* New York: Dutton, 1997. An affectionate and informal history of network radio's great years into the early 1950s discussing programs, personalities, and behind-the-scenes developments.

Norberg, Eric G., *Radio Programming: Tactics and Strategy,* Boston: Focal Press, 1996. Covers basic principles of radio programming, structuring and positioning a station, music and news as program weapons, promotion, ratings, interfacing with different parts of a station.

Pitts, Michael R., *Radio Soundtracks: A Reference Guide,* 2nd edition, Metuchen, New Jersey: Scarecrow Press, 1986. What has been recorded and how to find it.

*Shapiro, Mitchell E., *Radio Network Prime Time Programming, 1926–1967,* Jefferson, North Carolina: McFarland, 2002. Valuable charts showing changes in network program schedules over four decades. Use with Summers (below).

Sies, Luther F., *Encyclopedia of American Radio, 1920–1960,* Jefferson, North Carolina: McFarland, 2000. Individuals, programs, and stations in thousands of often very brief entries.

*Summers, Harrison B., compiler, *A Thirty-Year History of Programs Carried on National Radio Networks in the United States, 1926–1956,*Columbus: Ohio State University Department of Speech, 1958 (reprint, New York: Arno Press, 1971). Standard listing of programs with times aired, ratings, etc. Use with Shapiro (above).

Swartz, Jon D., and Robert C. Reinehr, *Handbook of Old-time Radio: A Comprehensive Guide to Golden Age Radio Listening and Collecting,* Metuchen, New Jersey: Scarecrow Press, 1993. Dated, but still useful, this offers extensive information on what program recordings are available.

Terrace, Vincent, *Radio Programs, 1920–1984: A Catalog of over 1800 Shows,* Jefferson, North Carolina: McFarland, 1998. Alphabetical directory with brief annotations.

8. American Radio Journalism

For particular radio journalists or programs, see relevant entries.

*Bliss, Edward Jr., *Now the News: The Story of Broadcast Journalism,* New York: Columbia University Press, 1991. The only overall history, roughly the first third deals with radio.

Brown, Robert J., *Manipulating The Ether: The Power of Broadcast Radio in Thirties America,* Jefferson, North Carolina: McFarland, 1998. Covers Franklin Roosevelt, *War of the Worlds* and Father Coughlin.

Charnley, Mitchell V., *News by Radio,* New York: Macmillan, 1948. Standard text with some history.

Cloud, Stanley, and Lynne Olson, *The Murrow Boys: Pioneers on the Front Lines of Broadcast Journalism,* Boston: Houghton Mifflin, 1996. Well-written assessment of the legendary CBS radio and then television news team from the 1940s to the 1980s.

Culbert, David Holbrook, *News for Everyman: Radio and Foreign Affairs in Thirties America,* Westport, Connecticut: Greenwood, 1976. Commentators and how their role developed—one of the earliest serious histories.

Hosley, David H., *As Good as Any: Foreign Correspondence on American Radio, 1930–1940,* Westport, Connecticut: Greenwood, 1984. Traces the rise of news reporting from Europe on the U.S. networks through the Battle of Britain.

Miller, Edward, *Emergency Broadcasting and 1930s American Radio,* Philadelphia: Temple University Press, 2003. Reviews coverage of the *Hindenburg* crash, Roosevelt's "Fireside Chat" broadcasts, and escapist drama on *The War of the Worlds* and *The Shadow,* among other programs, assessing the impact of each.

White, Paul W., *News on the Air,* New York: Harcourt Brace, 1947. Former head of CBS News includes a fine chapter on radio reports about D-Day.

9. American Radio Audiences and Research

Berg, Jerome S., *On the Short Waves, 1923–1945: Broadcast Listening in the Pioneer Days of Radio,* Jefferson, North Carolina: McFarland, 1999. Three main sections trace the rise of the medium in the 1920s, shortwave broadcasting and listening in the 1930s ("DXing"—the emphasis here), and wartime use of shortwave services.

*Beville, Hugh Malcolm, Jr., *Audience Ratings: Radio, Television, and Cable,* 2nd edition, Hillsdale, New Jersey: Lawrence Erlbaum, 1988. Best historical study of how methods and different audience research companies developed.

*Cantril, Hadley, and Gordon W. Allport, *The Psychology of Radio,* New York: Harper, 1935 (reprint, New York: Arno Press, 1971). Pioneering psychological study of audience impact on the then still fairly new aural medium.

Hiber, Jahn, *Winning Radio Research: Turning Research Into Ratings And Revenues,* Washington, D.C.: National Association of Broadcasters, 1987. Guidance for station managers.

*Lazarsfeld, Paul F., *Radio and the Printed Page: An Introduction to the Study of Radio and Its Role in the Communication of Ideas,* New York: Duell, Sloan and Pearce, 1940 (reprint, New York: Arno Press, 1971). Classic and pathbreaking study comparing print and radio and their varied impact.

Lazarsfeld, Paul F., and Harry Field, *The People Look at Radio,* Chapel Hill: University of North Carolina Press, 1946. See next entry.

*Lazarsfeld, Paul F., and Patricia Kendall, *Radio Listening in America: The People Look at Radio—Again,* New York: Prentice-Hall, 1948. Two national surveys of who listened to what and with what impact.

Lazarsfeld, Paul F., and Frank N. Stanton, editors, *Radio Research 1941,* New York: Duell, Sloan and Pearce, 1941. See next entry.

Lazarsfeld, Paul F., *Radio Research 1942–43,* New York: Duell, Sloan and Pearce, 1944. Two volumes of original research on different types of programs, edited by an academic and rising CBS official.

Lumley, Frederick H., *Measurement in Radio,* Columbus: Ohio State University Press, 1934 (reprint, New York: Arno Press, 1971). First book on radio audience measurement reviews a host of methods before the inception of most regular ratings research.

10. American Radio Regulation

Benjamin, Louise M., *Freedom of the Air and the Public Interest: First Amendment Rights in Broadcasting to 1935,* Carbondale: Southern Illinois University Press, 2001. Careful analysis of case law and legislation and how the basics of government's "hands off" approach to radio programming developed.

Bensman, Marvin R., *The Beginning of Broadcast Regulation in the Twentieth Century,* Jefferson, North Carolina: McFarland, 2000. The story of legislation and administrative decision-making through 1927.

Braun, Mark, *AM Stereo and the FCC: Case Study of a Marketplace Shibboleth,* Norwood, New Jersey: Ablex, 1994. Best analysis of the FCC-industry conflict over the setting of these standards and the impact of the commission's 1982 "nondecision" allowing open standards.

Davis, Stephen, *The Law of Radio Communication,* New York: McGraw-Hill, 1927. First book devoted to the then-new American radio law.

Dill, Clarence C., *Radio Law: Practice and Procedure,* Washington, D.C.: National Law Book Co., 1938. Written by the co-author of the 1934 Communications Act, this relates its early implementation.

Edelman, J. Murray, *The Licensing of Radio Services in the United States,* Urbana: University of Illinois Press, 1950. Study of FRC and FCC practices from 1927 through 1947.

*Emery, Walter B., *Broadcasting and Government,* East Lansing: Michigan State University Press, 1961; 2nd edition, 1971. Standard text with considerable historical material.

Flannery, Gerald, editor, *Commissioners of the FCC 1927–1994,* Lanham, Maryland: University Press of America, 1995. Useful short (2–3 page) mini-biographies of commissioners of both the FRC (1927–34) and FCC (since 1934).

Foust, James C., *Big Voices of The Air: The Battle Over Clear Channel Radio,* Ames: Iowa State University Press, 2000. Study of a long-lasting policy issue only finally resolved in the 1960s.

*Kahn, Frank J., editor, *Documents of American Broadcasting,* 4th edition, Englewood Cliffs, New Jersey: Prentice-Hall, 1984. Valuable anthology of key policy materials.

*Paglin, Max D., editor, *A Legislative History of the Communications Act of 1934,* New York: Oxford University Press, 1989. Includes committee hearings and floor debate.

Paglin, Max D., *The Communications Act: A Legislative History of the Major Amendments, 1934–1996,* Washington, D.C.: Pike and Fischer, 1999. Continues the previous title with the important amendments such as Communication Satellite Act of 1962, Public Broadcasting Act of 1967, various cable legislation, and the Telecommunications Act of 1996.

Robinson, Thomas Porter, *Radio Networks and the Federal Government,* New York: Columbia University Press, 1943 (reprint, New York: Arno Press, 1971). How the issues developed and the break-up of NBC's two networks.

*Rose, C.B., Jr., *National Policy for Radio Broadcasting,* New York: Harper, 1940 (reprint, New York: Arno Press, 1971). Invaluable prewar review of American radio policy concerns—many still with us today.

Rosen, Philip T., *The Modern Stentors: Radio Broadcasters and the Federal Government, 1920–1934,* Westport, Connecticut: Greenwood, 1980. The countervailing pressures between a nascent industry and nascent regulators.

*Slotten, Hugh R., *Radio and Television Regulation: Broadcast Technology in the United States, 1920–1960,* Baltimore: Johns Hopkins University Press, 2000. Three of the six case studies concern radio, including the FRC creation of radio standards and the rise of FM.

*Warner, Harry P., *Radio and Television Law* and *Radio and Television Rights,* 2 vols., Albany: Matthew Bender, 1948–53. While originally designed for practicing attorneys, these are now valuable for their very extensive historical material.

11. American Regional and State Radio Histories

For specific station histories see relevant entries.

Blanton, Parke, *Crystal Set to Satellite: The Story of California Broadcasting—the First Eighty Years,* Sacramento: California Broadcasters Association, 1987.

Brouder, Edward W., Jr., *Granite And Ether: A Chronicle of New Hampshire Broadcasting,* Bedford: New Hampshire Association of Broadcasters, 1993.

Doll, Bob, *Sparks Out of The Plowed Ground: The History of America's Small Town Radio Stations,* West Palm Beach, Florida: Streamline Press, 1996.

Dorgan, Howard, *The Airwaves of Zion: Radio And Religion in Appalachia,* Knoxville: University of Tennessee Press, 1993.

Jaker, Bill, et al., *The Airwaves of New York: Illustrated Histories of 156 AM Stations in the Metropolitan Area, 1921–1996,* Jefferson, North Carolina: McFarland, 1998. Illustrated history of all of them, some well-known and still with us and others long gone and forgotten.

Schroeder, Richard, *Texas Signs On: The Early Days of Radio and Television,* College Station: Texas A&M University Press, 1998. Illustrated history of the rise of radio and then television in the Lone Star state, relying heavily on interviews and original documents.

12. Radio Outside the United States

For radio broadcasting in specific countries or regions, see relevant entries.

Avery, Robert K., *Public Service Broadcasting in a Multichannel Environment: The History and Survival of an Ideal,* New York: Longman, 1993. Useful survey (see also Tracey, below) of how such systems are under pressure in an increasingly commercial age.

Batson, Lawrence D., *Radio Markets of the World, 1930,* Washington, D.C.: Government Printing Office, 1930 (reprint, New York: Arno Press, 1971). Aimed at pioneering advertisers, this includes profiles of radio in many nations.

Broadcasting Abroad, New York: National Advisory Council on Radio in Education "Information Series No. 7," 1932, revised 1934. Useful snapshot of very early radio developments around the world.

*Browne, Donald R., *International Radio Broadcasting: The Limits of the Limitless Medium,* New York: Praeger, 1982. The role of radio propaganda in international services—a standard history.

Codding, George A., *Broadcasting Without Barriers,* Paris: Unesco, 1959. Technology and policy concerns.

Emery, Walter B., *National and International Systems of Broadcasting,* East Lansing: Michigan State University Press, 1969. Profiles systems in a dozen nations—this was a pioneering historical and descriptive text effort.

Hale, Julian, *Radio Power: Propaganda and International Broadcasting,* Philadelphia: Temple University Press, 1975. Overall historical survey of radio's role in propaganda in war and peace.

*Head, Sydney W., *World Broadcasting Systems: A Comparative Analysis,* Belmont, California: Wadsworth, 1985. Excellent analytic survey with much on radio.

Hilliard, Robert, and Michael C. Keith, *Global Broadcasting Systems,* Boston: Focal Press, 1996. A survey of radio-television policy and structure worldwide.

Huth, Arno, *Radio Today: The Present State of Broadcasting,* Geneva, Switzerland: Geneva Research Centre "Studies XII:6," July 1942 (reprint, New York: Arno Press, 1971). Wartime analysis of radio's status.

*Katz, Elihu, and George Wedell, *Broadcasting in the Third World: Promise and Performance,* Cambridge, Massachusetts: Harvard University Press, 1977. The importance of radio in developing nations.

Schwoch, James, *The American Radio Industry and Its Latin American Activities, 1900–1939,* Champaign: University of Illinois Press, 1990. American wireless companies, growth of broadcasting, the rise of a military-industrial complex, and international conferences on radio.

Shepherd, Frank M., *Technical Development of Broadcasting in Asia-Pacific, 1964–1984,* Kuala Lumpur, Malaysia: Asia Pacific Broadcasting Union, 1984. Includes overall trends and country-by-country assessment.

Tracey, Michael, *The Decline and Fall of Public Service Broadcasting,* New York: Oxford University Press, 1998. With Avery (see above), this is a fine survey of how decades-old systems are being forced to change.

*Unesco, *Press, Film, Radio,* Paris: Unesco, 1947–51 (7 vols; reprinted in three by Arno Press, New York, "International Propaganda and Communications," 1972). Information on most countries includes extensive historical background and immediate postwar status.

Wedell, George, and Philip Crookes, et al., *Radio 2000: The Opportunities for Public and Private Radio Services in Europe,* London: John Libbey, 1991. Useful survey of developments in Europe through the 1980s.

Woods, James, *History of International Broadcasting,* 2 vols., London: IEE, 1992 and 1999. Best history of the technology behind cross-border radio broadcasts from the 1920s into the 1990s, with much on both World War II and Cold War services, with considerable information on specific nations.

B. Periodicals

Included here are only titles focused on radio that were currently publishing as of 2003.

Airplay Monitor (1993–present, weekly). Published in four different versions by the music industry weekly *Billboard*, this is aimed at music stations and offers information on trends, music rotations, and the like. The different editions cover country, Top 40, rock, and urban formats.

BE Radio (1995–present, weekly). Covers television and radio technical topics and news, from the publishers of the monthly *Broadcast Engineering*.

Broadcasting & Cable (1931–present, weekly). The principal industry trade weekly that helps to place radio in context with television, cable, and newer services. Covers advertising, programming, regulation, and people in the business. Not as comprehensive after about 1995.

Duncan's American Radio (1976–present, quarterly and annual). Provides business information and data on radio stations in the form of market guides, directories, ratings analyses, and revenue reports.

Gavin (1958–present, weekly). Focuses on radio station formats, music research, music playlists, and artists.

Inside Radio (1975–present, weekly). Includes radio industry news, station sales, management changes, and stock quotations, plus commentary (and hyperbole) by publisher Jerry DelColliano.

Journal of Radio Studies (1991–present, biannual). Covers both history and current issues in scholarly journal format—the first and only one devoted to radio.

M St. Journal (1984–present, weekly). Offers FCC news and proceedings, station sales, format changes. (*M Street Daily* is also a daily fax doing basically the same thing as *Inside Radio* without the commentary. It adds a weekly page of sales tips and a weekly update from Washington and the FCC.)

R&R: Radio & Records (1971–present, weekly). Radio programmers' main trade publication includes music charts for specific formats, industry tip sheets and interviews, and parallels for music playlist comparisons.

Radio Business Report (1983–present, weekly). Emphasis is on management and marketing issues in commercial radio business.

Radio Ink (1985–present, bi-weekly). Sales and management oriented publication containing articles by industry people who offer insights, opinions, and analysis about trends in advertising and station operations.

Radio World Magazine (1977–present, bi-weekly). Intended for station engineers, technicians, and electronics manufacturers of audio products.

C. Selected Radio Web Sites

Websites are both useful—for what they offer; and maddening—they disappear too easily. As of Spring 2003, these are some of the more interesting sites, virtually all of which offer many further references.

1. General Reference and Link Sites

Library of American Broadcasting at the University of Maryland
http://www.lib.umd.edu/LAB/
One of the better academic archives of all types of material on radio and television.

Michael Keith's Broadcast Links
http://www.michaelckeith.com/links.htm
Extensive list, heavily focused on radio both here and abroad, including domestic and international stations, services, periodicals, and radio personalities and programs.

Museum of Broadcast Communications in Chicago
http://www.museum.tv/index.shtml
The sponsor of this encyclopedia, the Museum has extensive displays and listening areas, and houses the Radio Hall of Fame.

2. *History of Radio Technology Web Sites*

Antique Radio Page by D.J. Adamson
http://members.aol.com/djadamson/arp.html
Designed for those who collect old radios, this includes books, articles, links, classified ads and more.

The Broadcast Archive by Barry Mishkind
http://www.oldradio.com/
Includes equipment and programming sections and links, plus information about the FCC, old stations, and links to other archives and organizations.

United States Early Radio History by Thomas H. White
http://earlyradiohistory.us/
A wonderfully useful site which offers full copy of a variety of pre–1920 articles and documents plus the author's valuable own research on early radio station list publications, call-letter policies, and the like.

World of Wireless
http://home.luna.nl/~arjan-muil/radio/history.html
A Dutch site (in both English and Dutch), takes the story through World War II and includes details of the owners' own collection.

3. *"Old Radio" Program Web Sites*

Virtually any radio entertainment or news program is now the subject of a site or sites of its own—listed here are some more general "master" sites that link to a host of others.

Old Time Radio
http://www.old-time.com/
Includes many logs of program series, links to other sites, information on collecting programs.

Olde Time Radio
http://www.oldetimeradio.com/
Allows one to listen to episodes of about a dozen old radio dramatic programs.

Radio Days
http://www.otr.com/index.shtml
Information on many old network radio programs (including some complete logs), OTR (old time radio) chat room and FAQs and more.

4. Modern Radio Program Sites

There are countless such sites today. What follows are some useful compendium sites allowing access to the many resources, including internet stations, by a variety of means.

Radio-Locator: Formerly MIT List of Radio Stations on the Internet by Theodric Young
http://www.radio-locator.com/
Allows user search by station call letters, format, city, state (or Canadian province), or nation. Including more than 5,000 stations and adding more weekly, this claims to be the most comprehensive such site available.

BRS Web Radio
http://www.web-radio.com/
Similar to the site above.

5. International Radio Sites

This is but a sample of useful English-language sites concerning radio broadcasting in nations other than the United States.

Asian-Pacific Broadcasting Union
http://www.abu.org.my/main.htm
Useful gateway to a variety of Asian broadcasting networks and systems, as well as regional broadcast information of all kinds.

Mike's Radio World: Live Radio on the Web by Mike Dean
http://www.mikesradioworld.com/oceana.html
Arranged by region of the country and then city. The author provides similar sites for New Zealand and some other nations as well.

British Broadcasting Corporation
http://www.bbc.co.uk/
Includes all aspects of the London-based but world-famous broadcaster, including the World Service.

Canadian Broadcasting Corporation
http://cbc.radio-canada.ca/htmen/
The English-language home-page for the Canadian CBC services, including history and other information.

Commonwealth Broadcasting Association
http://www.cba.org.uk/
Organization of public service radio and television broadcasters in the British Commonwealth, useful for a number of African nations especially.

The Offshore Radio Guide
http://www.offshore-radio.de/
Sub-titled "watery wireless websites," this includes all types of information on past and present pirate (offshore ship-based or artificial-island based) radio stations, with links to many of them.

Virgin Radio
http://www.virginradio.co.uk/
One of the many local commercial radio service providers in Britain.

United States Government International Radio Broadcasting
http://www.ibb.gov/
The opening page for the International Broadcasting Bureau, which is the controlling body for the VOA, RFE, RL, Radio Free Asia, and Radio Marti. Provides gateways to each of the individual services.

ADVISERS

Stanley R. Alten
Syracuse University

Frank J. Chorba
Washburn University

Lynn A. Christian
International Radio Consultant

Ed Cohen
Clear Channel Broadcasters

Norman Corwin
Radio Playwright

Susan J. Douglas
University of Michigan

James E. Fletcher
University of Georgia

Robert S. Fortner
Calvin College

Stan Freberg
Radio Producer

Donald G. Godfrey
Arizona State University

Marty Halperin
Pacific Pioneer Broadcasters

Gordon H. Hastings
Broadcasters' Foundation

Robert Henabery
Radio Producer

Robert L. Hilliard
Emerson College

Michele Hilmes
University of Wisconsin, Madison

Chuck Howell
Library of American Broadcasting
University of Maryland

Stanley Hubbard
Hubbard Communications

John D. Jackson
Concordia University

Jack Mitchell
University of Wisconsin, Madison

Graham Mytton
BBC World Service (ret.)

Horace Newcomb
Director, Peabody Awards
University of Georgia

Peter B. Orlik
Central Michigan University

Ed Shane
Shane Media Services

Marlin R. Taylor
XM Satellite Radio

CONTRIBUTORS

Michael H. Adams
Alan B. Albarran
Pierre Albert
Craig Allen
Steven D. Anderson
Larry Appelbaum
Edd Applegate
Sousan Arafeh
John S. Armstrong
Philip J. Auter
Robert K. Avery
Glenda R. Balas
Mary Christine Banwart
Warren Bareiss
Ray Barfield
Kyle S. Barnett
Douglas L. Battema
Mary E. Beadle
Christine Becker
Johnny Beerling
Alan Bell
Louise Benjamin
ElDean Bennett
Marvin Bensman
Jerome S. Berg
Rosemary Bergeron
William L. Bird, Jr.
Howard Blue
A. Joseph Borrell
Douglas A. Boyd
John Bradford
L. Clare Bratten
Mark Braun
Jack Brown
Michael Brown
Robert J. Brown
Donald R. Browne
John H. Bryant
Joseph G. Buchman
Karen S. Buzzard
Paul Brian Campbell
Dom Caristi
Ginger Rudeseal Carter
Dixon H. Chandler II
Frank J. Chorba
Lynn A. Christian
Claudia Clark
Kathleen Collins
Jerry Condra
Harold N. Cones
Bryan Cornell
Elizabeth Cox
Steve Craig
Tim Crook
Marie Cusson
Keri Davies
E. Alvin Davis
J.M. Dempsey
Corley Dennison
Neil Denslow
Steven Dick
John D.H. Downing
Pamela K. Doyle
Christina S. Drale
Susan Tyler Eastman
Bob Edwards
Kathryn Smoot Egan
Lyombe Eko
Sandra L. Ellis
Ralph Engelman
Erika Engstrom
Stuart L. Esrock
Charles Feldman
Michel Filion
Howard Fink
Seth Finn
Robert G. Finney
Margaret Finucane
James E. Fletcher
Corey Flintoff
Joe S. Foote
Robert C. Fordan
Robert S. Fortner
James C. Foust
Ralph Frasca
James A. Freeman
Elfriede Fürsich
Charles F. Ganzert
Ronald Garay
Philipp Gassert
Judith Gerber
Norman Gilliland

Donald G. Godfrey
Douglas Gomery
Jim Grubbs
Joanne Gula
Paul F. Gullifor
Linwood A. Hagin
Donna L. Halper
Tona J. Hangen
Margot Hardenbergh
Jeffrey D. Harman
Dorinda Hartmann
Gordon H. Hastings
Joy Elizabeth Hayes
John Allen Hendricks
Alexandra Hendriks
Ariana Hernandez-Reguant
Robert L. Hilliard
Jim Hilliker
Michele Hilmes
John Hochheimer
Jack Holgate
Herbert H. Howard
Chuck Howell
Kevin Howley
W.A. Kelly Huff
Peter E. Hunn
John D. Jackson
Randy Jacobs
Glen M. Johnson
Phylis Johnson
Sara Jones
Lynda Lee Kaid
Stephen A. Kallis, Jr.
Steve Kang
Michael C. Keith
Ari Kelman
Colum Kenny
John Michael Kittross
Frederica P. Kushner
Philip J. Lane
Matthew Lasar
Laurie Thomas Lee
Renée Legris
Frederic A. Leigh
Lawrence W. Lichty
Lucy A. Liggett
Val E. Limburg
Robert Henry Lochte
Jason Loviglio
Gregory Ferrell Lowe
Christopher Lucas
Mike Mashon
Marilyn J. Matelski
Peter E. Mayeux
Dennis W. Mazzocco
Thomas A. McCain
Jeffrey M. McCall
David McCartney
Tom McCourt
Brad McCoy
Allison McCracken
Drew O. McDaniel
Michael A. McGregor
Robert McKenzie
Elizabeth McLeod
Mike Meeske
Fritz Messere
Colin Miller
Toby Miller
Bruce Mims
Jack Minkow
Jack Mitchell
Jason Mittell
Barbara Moore
Matthew Murray
Graham Mytton
Gregory D. Newton
Greg Nielsen
D'Arcy John Oaks
William F. O'Connor
Cary O'Dell
Robert M. Ogles
Ryota Ono
Peter B. Orlik
Pierre-C. Pagé
Brian T. Pauling
Manjunath Pendakur
Douglas K. Penisten
Stephen D. Perry
Patricia Phalen
Steven Phipps
Joseph R. Piasek
Gregory G. Pitts
Mark Poindexter
Tim Pollard
Robert F. Potter
Alf Pratte
Patricia Joyner Priest
Dennis Randolph
Lawrence N. Redd
David E. Reese
Patton B. Reighard

Andre Richte
Edward A. Riedinger
Terry A. Robertson
Melinda B. Robins
América Rodríguez
Eric W. Rothenbuhler
Richard Rudin
Joseph A. Russomanno
Anne Sanderlin
Erica Scharrer
Steven R. Scherer
Karl Schmid
Clair Schulz
Ed Shane
Pam Shane
Mitchell Shapiro
Jason T. Siegel
Ron Simon
B.R. Smith
Ruth Bayard Smith
Lynn Spangler
David R. Spencer
David Spiceland
Laurie R. Squire
Michael Stamm
Christopher H. Sterling
Will Straw
Michael Streissguth
Mary Kay Switzer
Rick Sykes
Marlin R. Taylor
Matt Taylor
Herbert A. Terry
Richard Tiner
Regis Tucci
David E. Tucker
Don Rodney Vaughan
Mary Vipond
Randall Vogt
Ira Wagman
Andrew Walker
Peter Wallace
Jennifer Hyland Wang
Richard Ward
Mary Ann Watson
Brian West
Gilbert A. Williams
Sonja Williams
Wenmouth Williams, Jr.
Roger Wilmut
Stephen M. Winzenburg
Richard Wolff
Roosevelt "Rick" Wright, Jr.
Edgar B. Wycoff
Thimios Zaharopoulos

LIST OF ENTRIES

A

A.C. Nielsen Company

Developing Radio Ratings

From 1942 to 1964, the A.C. Nielsen Company was a primary provider of U.S. radio ratings. The company pioneered the commercial use of mechanical and then electronic meter devices to automatically record which stations listeners were tuning their receivers to.

Origins

Arthur C. Nielsen (1897–1980) founded his marketing company in 1923 after serving briefly as a naval officer in World War I and working with two Chicago companies. With six employees and $45,000 in capital from Nielsen's former fraternity brothers, the company specialized in performance surveys of industrial equipment. The company went bankrupt twice in its early years. A decade later, Nielsen expanded his service by launching a continuous market research service, the Nielsen Drug Index, to chart the retail flow of specific products. The Nielsen Food Index soon followed. Both were based on the same premise (which was later applied successfully to broadcast ratings): carefully develop a sample of stores and visit them periodically to measure unit sales through audits of purchase invoices and shelf stock. When projected regionally or nationally, these data provided a measure of sales that could be related directly to marketing efforts.

Nielsen entered the radio audience measurement business at the request of clients who found the food and drug indexes useful guides and desired the same assistance in purchasing radio advertising time. In 1936 Nielsen acquired the rights to a mechanical device developed by two Massachusetts Institute of Technology professors, Robert Elder and Louis Woodruff. The "Audimeter" made a graphic record on a filmstrip, providing a continuous record of radio receiver use—when it was on and to which station(s) it was tuned—over a month-long period. These early meters were both costly and cumbersome, especially as the tape had to be picked up by Nielsen personnel before the tabulation of results could begin. The tapes were then shipped to a Chicago plant where they were "read" by specially designed machines. After modifications to the meter, their use was subject to intense experimentation for four years in several Midwestern states.

Radio Ratings

The Nielsen Radio Index (NRI) ratings service, based on the meter system, was introduced commercially in December 1942 in competition with the then-dominant "Hooperatings," which used telephone surveys of sample homes. A key advantage of the Nielsen meter was that its sample (initially just 800 homes in the east-central portion of the U.S.) was not restricted to telephone-owning homes; this was important at a time when upward of a third of homes in some areas lacked the instrument. By 1946 the NRI had expanded service to some 1,100 homes over most of the country. NRI also introduced an improved meter with a mailable tape (it provided measures over two weeks) to speed delivery of the resultant ratings and to render personal staff visits to Nielsen sample homes unnecessary. The streamlined process—which could measure four separate radio receivers—allowed expansion of meter-based ratings to both FM listening and television watching.

By early 1949, the NRI sample had expanded to cover virtually all of the country except for the Mountain time zone, which was especially expensive to serve. In early 1950 Nielsen purchased the Hooper national radio and fledgling television ratings services (Hooper continued local radio market ratings for several years). By this point, Nielsen's "methodology, financial position, organization and widespread industry acceptance rendered him nearly invincible" (Beville, 1988). A year later, the NRI sample was up to 1,500 homes—and its charges to advertisers and broadcasters had nearly doubled. But these were national (network) ratings, not local-market measurements.

The Nielsen Station Index (NSI) debuted in 1954 to measure household use of both radio and television on a local-market basis. This service was not audimeter based, but rather combined the use of traditional diaries (in which audience members recorded their listening time) with a "Recordimeter" device, which signaled with light flashes and a buzzer when listeners should make a diary entry and at the same time kept a rough measure of when the receiver was on. This crude meter helped to validate the diary information provided. And the diary could provide what no meter then could—demographic information on the gender and age of the person listening. In 1959 computers were first applied to Nielsen ratings processing and analysis. By the early 1960s, NSI was measuring radio listening in more than 200 markets. But its seeming market dominance would be short-lived, for, as Hugh Beville writes,

> in 1962 Nielsen discontinued quarter-hour ratings because of declining radio listening levels and the rapidly increasing number of radio stations. This cost many client cancellations, which sparked the NSI decision to abandon radio. Not only was television seriously diminishing prime-time radio audiences, but the advent of automobile and portable receivers, plus many new independent stations, was rapidly changing basic radio listening patterns. In 1963 the local radio service was discontinued (Beville, 1988).

Contributing to the end of Nielsen radio ratings was a series of congressional hearings into the ratings for both radio and television. Nielsen became a central target in those hearings, in part because of methodological questions about some of the company's means of ratings data collection. Nielsen's system measured only home viewing, not portables. The out-of-home audience, as it became known, grew with the movement to the suburbs and the use of the automobile. In response to the changing radio audience, Nielsen created an Audimeter to be installed in automobiles. However, when his clients were unwilling to support the investment costs needed to upgrade, he decided to quit the radio business. The Nielsen Company decided to focus its investment efforts on the greater returns potential from television ratings. The end of Nielsen's radio services was a key factor leading to development of RADAR national radio ratings.

Later Years

After leaving the radio ratings business, Nielsen continued to develop its national television network and local-market ratings services. It introduced overnight ("instant") television meters in major markets in the early 1960s and slowly expanded the process to other cities and network ratings. In 1987 Nielsen introduced its still-controversial "people meter," which could measure TV receiver tuning as well as who was tuning in. With Arbitron's departure from television ratings in the late 1990s, Nielsen became the only source of both national and local-market television ratings.

When the elder Nielsen retired in 1976, A.C. Nielsen, Jr., became the company's leader. In 1984 he sold the firm to Dun and Bradstreet, which in 1998 split the marketing and media research aspects of the company. The latter was sold to Lucent Technology and then spun off to a new corporate owner, Cognizant Corporation, in mid-1998. At the end of 1999, Nielsen Media Research was purchased by a Dutch company, VNU NV.

KAREN S. BUZZARD AND CHRISTOPHER H. STERLING

See also Arbitron; Audience Research Methods; Audimeter; Cooperative Analysis of Broadcasting; Hooperatings; RADAR

Further Reading

Beville, Hugh Malcolm, Jr., *Audience Ratings: Radio, Television, and Cable,* Hillsdale, New Jersey: Erlbaum, 1985; revised edition, 1988

Buzzard, Karen S., *Chains of Gold: Marketing the Ratings and Rating the Markets,* Metuchen, New Jersey: Scarecrow, 1990

Buzzard, Karen S., "Radio Rating Pioneers: The Development of a Standardized Ratings Vocabulary," *Journal of Radio Studies* 6 no. 2 (Autumn 1999)

Nielsen, Arthur Charles, "Trends in Mechanization of Radio Advertising," *Journal of Marketing* (January 1942)

Nielsen, Arthur Charles, "Two Years of Commercial Operation of the Audimeter and NRI," *Journal of Marketing* (January 1944)

Nielsen, Arthur Charles, *How You Can Get the Ideal Radio Research Service Complete, Accurate, Impartial, Rapid, Skilfully Applied,* Chicago: Nielsen, 1946

Nielsen, Arthur Charles, *New Facts about Radio Research,* Chicago: Nielsen, 1946

Nielsen, Arthur Charles, "Audience Analysis via Audimeter Method," *Broadcasting Yearbook 1947* (1947)

Nielsen, Arthur Charles, *Greater Prosperity through Marketing Research: The First 40 Years of A.C. Nielsen Company,* New York: Newcomen Society, 1964

Rusch, H.L., "Debut of the First Nielsen Radio Index Data," *The Nielsen Researcher* (October 1939)

U.S. House of Representatives, *Evaluation of Statistical Methods Used in Obtaining Broadcast Ratings: House Report No. 193,* 87th Congress, 1st Session (23 March 1961)

U.S. House of Representatives, *Broadcast Ratings: House Report No. 1212,* 89th Congress, 2nd Session (13 January 1966)

Active Rock Format. *See* Heavy Metal/Active Rock Format

Adult Contemporary Format

Adult Contemporary (AC) music emphasizes a mixture of modern day (contemporary) and older popular hit singles.

The AC format has its origins in the arrival and eventual permanence of rock and roll as a music form in the United States during the late 1950s. According to Hyatt (1999), radio stations in the United States at that time wanted to keep airing current popular hits, a staple format that had already attracted listeners for decades. However, these stations did not want to play rock and roll and tried to find a way to keep the popular hits genre without having to play songs with a rock beat. Thus, they turned to popular songs that lacked the "heavy" sounds of rock and roll. These tunes became known by those in the radio industry as "easy listening" or "middle of the road." Hyatt refers to Adult Contemporary as being synonymous with both terms.

Beginning in the 1960s, *Billboard* magazine created a new chart listing the top records considered easy listening and middle of the road, in addition to its pop, rhythm and blues, and country lists of popular songs. *Billboard* gave several names to this chart throughout the 1960s, which listed the top 20 (and during some years the top 15 to 25) singles. These names included "Easy Listening," "Pop-Standard Singles," and "Middle-Road Singles." By 1965 the magazine had settled on "Easy Listening" to describe the chart. Artists listed on this chart included music industry veterans such as Dean Martin, Frank Sinatra, Nat King Cole, and Andy Williams. By the late 1960s, folk artists such as Peter, Paul, and Mary and Simon and Garfunkel had hits on the "Easy Listening" list. The chart also included instrumentals by Herb Alpert and the Tijuana Brass, film composer Henry Mancini, and Mason Williams during the late 1960s and early 1970s. Also during this period, Hyatt notes, "fewer people who considered themselves easy listening fans were requesting previous favorites like Frank Sinatra and Ed Ames." More notably, crossover between the "Easy Listening" and "Pop" lists started to occur at this time, with artists such as the Carpenters and Bread releasing mellow, rock-type songs that were listed on both charts.

"Easy Listening" music grew in popularity; *Billboard* expanded the chart, lengthening the list from the top 40 singles to the top 50 in 1973. Keith (1987) points out that this genre appealed to the same type of audience who listened to stations featuring soft and mellow rock music. Additionally, during the late 1970s, the number of soft and mellow rock listeners declined as the disco format grew in popularity and as the number of hit music stations increased: "It was out of this flux that the AC format emerged in earnest" (Keith, 1987). *Billboard* renamed the "Easy Listening" chart in 1979, when it became known as "Adult Contemporary." During the next 14 years, the list's length fluctuated between 40 and 50 hit singles of the genre. In 1996 the chart came to list the top 25 singles (Hyatt, 1999).

During the 1980s, the AC format became the nation's most widespread, with its target audience encompassing the 25 to 49 age group, especially women, which made it appealing to advertisers as well. By the time of AC's maturation, its audience base consisted of adults who had composed the teen listenership of Top 40 radio in the early 1970s. AC, also referred to as lite or soft rock, drew in the thirty-something listener by offering "popular, upbeat music without the harshness that often accompanies rock" (Keith, 1987). Typical artists with top AC hits during the 1980s included Lionel Richie, Billy Joel, Whitney Houston, and Phil Collins. Artists popular during the 1970s also hit the chart, such as Cher, Elton John, Barry Manilow, and Barbra Streisand.

Regarding the programming of the AC format, stations that employ the genre place greater emphasis on the music, thus minimizing disc jockey chatter. AC stations might describe themselves as "soft rockers" or as "hot, soft, lite, bright, mix or variety"; the phrase "continuous soft rock favorites of yesterday and today" serves as a common line in promotional spots (MacFarland, 1997). The music mix itself combines contemporary singles with hits from the past, though these generally do not include true "oldies." Called "recurrents," these older songs typically have just left the current chart and are usually between six months and two years old (Howard, Kievman, and Moore, 1994). Halper (1991) contends that AC music directors must keep up with the newest adult pop artists, such as those presented on VH1, the slightly more mature version of MTV, the cable music channel.

AC stations present chart toppers, both current and potential, and recurrents in blocks or sweeps, which can last as long as 28 minutes of uninterrupted music. Announcers usually follow these sweeps with recaps of song titles and artists, and commercials are limited to four or five per cluster (MacFarland, 1997). AC stations also may feature contests, all-request hours, programs that feature hits from a particular decade, and lifestyle-oriented news. As with other music-oriented formats, news takes a secondary position, although it is usually presented during drive time. Some stations feature strong on-air personalities, especially in the morning, and an upbeat delivery style similar to the Contemporary Hits Radio and Top 40 formats.

By the mid-1990s, AC came in second to country as the most popular format in the United States, even though the number of stations featuring this format dropped between 1989 and 1994. Artists with hits on the *Billboard* AC chart during the 1990s included Mariah Carey, Michael Bolton, and previous hitmakers such as Elton John and Eric Clapton. AC hits of the late 1990s exemplified the soft/lite rock, "easy listening" sounds of early AC, a key characterization of this adult-oriented radio format, as embodied in the chart-topping singles of Canadian singer Celine Dion, who headed the list of all artists with songs spending the most weeks at number one on *Billboard*'s AC chart (65, 19 of which were with one song [Hyatt, 1999]).

ERIKA ENGSTROM

See also Contemporary Hit Radio Format/Top 40; Easy Listening/Beautiful Music Format; Middle of the Road Format; Soft Rock Format

Further Reading

Halper, Donna L., *Radio Music Directing,* Boston: Focal Press, 1991

Howard, Herbert H., and Michael S. Kievman, *Radio and TV Programming,* Columbus, Ohio: Grid, 1983; 2nd edition, as *Radio, TV, and Cable Programming,* edited by Howard, Kievman, and Barbara A. Moore, Ames: Iowa State University Press, 1994

Hyatt, Wesley, *The Billboard Book of Number One Adult Contemporary Hits,* New York: Billboard Books, 1999

Keith, Michael C., *Production in Format Radio Handbook,* Lanham, Maryland: University Press of America, 1984

Keith, Michael C., and Joseph M. Krause, *The Radio Station,* Boston: Focal Press, 1986; 4th edition, by Michael C. Keith, 1997; 5th edition, Boston and Oxford: Focal Press, 2000

Keith, Michael C., *Radio Programming: Consultancy and Formatics,* Boston: Focal Press, 1987

MacFarland, David T., *Contemporary Radio Programming Strategies,* Hillsdale, New Jersey: Erlbaum, 1990; 2nd edition, as *Future Radio Programming Strategies: Cultivating Leadership in the Digital Age,* Mahwah, New Jersey: Erlbaum, 1997

Adventures in Good Music

Classical Music Program

Created by Karl Haas in 1959, *Adventures in Good Music* is one of the longest running and most widely acclaimed classical music programs in radio broadcast history.

Adventures is syndicated in more than 200 U.S. cities. The U.S. Armed Forces network beams the program to U.S. bases on all continents, and 37 Australian Broadcasting Corporation stations broadcast the show. The program is translated into Spanish in Mexico City, causing the one-hour program to run an hour and a half. And in Germany, at the request of the South German Broadcasting Corporation (Suddeutscher Rundfunk), *Adventures* is available in German under a specially formatted, select series of presentations. The program is also recorded in French for the Canadian Broadcasting Corporation (CBC).

Since 1970, Cleveland classical radio station WCLV has supervised syndication of *Adventures* through its subsidiary, Seaway Productions. Each new syndicated program—recorded by producer and host Karl Haas in his New York studio—is transmitted via Seaway Productions to a domestic satellite that beams the program to stations across the United States. Elsewhere, stations receive the program on magnetic tape reels that are duplicated at WCLV and mailed in advance of program dates.

Karl Haas began his lifetime involvement with classical music at the age of six in his hometown, Speyer-on-the-Rhine, Germany, where he studied piano under the guidance of his mother. At the age of 12, he was performing in a piano trio with friends. As a young man, Haas studied music at Germany's University of Heidelberg. At the onset of Nazi tyranny in the 1930s, Haas fled to Detroit, Michigan, where he studied at the famed Netzorg School of Music and commuted from

Detroit to New York to study with the legendary pianist Artur Schnabel.

Haas' radio career began in 1950 at Detroit station WWJ, where he was under contract to host a weekly preview of concerts performed by the Detroit Symphony. His program caught the attention of the CBC, which offered him the position of conducting a chamber orchestra and performing piano recitals for a weekly program. Based on the phenomenal popularity of his Canadian show, CBC requested that Haas incorporate a commentary about his music into the program. Following audiences' favorable response to his lively narratives, in 1959 Detroit radio station WJR hired Haas to develop his own daily one-hour music-plus-commentary program—and *Adventures in Good Music* was born.

The format of *Adventures* has remained nearly the same since its debut. Each program is fashioned around a central theme, punctuated and illustrated with musical selections and enhanced by knowledgeable and often witty commentary originated by Haas. One program may highlight the best of Bach, and yet another may challenge listeners to "Name the Composer" in a musical mystery teaser. Still other programs may seek to understand and explain the impact of humor in music or how music is relevant to current events.

In creating his *Adventures* calendar, Haas explores both the traditional and the unconventional. A sequence from a typical month commenced with a program honoring the anniversary of Chopin's birthday. By midweek, "In Every Sense of the Word" offered an exploration of the five senses and their musical equivalents. Haas scheduled a traditional St. Patrick's Day salute on March 17. Then, he finished off the month's menu with an unconventional study of "When in Rome . . ." featuring works by non-Italian composers based on Italian culture.

As an active performer on the recital concert tour circuit, Haas has held a series of biannual "live" *Adventures in Good Music* programs at New York's Metropolitan Museum of Art since 1977.

Adventures has twice been awarded the coveted George Peabody Award for excellence in broadcasting. Haas has received numerous awards in recognition of his outstanding contributions to radio and for furthering the appreciation of classical music. The French Government presented him with the Officer d'academie and Chevalier de l'ordre des arts et lettres awards. In Germany he received the prestigious First Class Order of Merit. In 1991 Haas was awarded the National Endowment for the Humanities' Charles Frankel Award. Additionally, Haas was honored with the National Telemedia Award and has received eight honorary doctorates. In March 1995 Haas was the first classical broadcaster to be nominated for induction into the Radio Hall of Fame in Chicago.

Continuing his broadcast effort to educate and entertain listeners to the joys of classical music, Haas authored the reference book *Inside Music*. Also, in 1993 and 1994 he released three compact discs, *The Romantic Piano, Story of the Bells,* and *Song and Dance,* which marked a new venue for *Adventures in Good Music.*

Karl Hass, *Adventures in Good Music*
Courtesy Karl Haas

ELIZABETH COX

See also Classical Music Format

Producer/Creator Karl Haas

Programming History

Syndicated by WCLV, Cleveland, Ohio 1959–

Further Reading

Douglas, Susan J., *Listening In: Radio and the American Imagination: From Amos 'n' Andy and Edward R. Murrow to Wolfman Jack and Howard Stern,* New York: Times Books, 1999

Haas, Karl, *Inside Music: How to Understand, Listen To, and Enjoy Good Music,* New York: Doubleday, 1984

Advertising

Advertising specifically refers to paid commercial announcements aired by a radio station. Although commercials may sometimes seem distracting to listeners, radio stations from the earliest days recognized that there had to be a way for a station to pay its operating expenses, and by the late 1920s radio stations in the United States had adopted commercial advertising.

Origins

Advertising on radio began amid controversy, as many public figures and some station operators initially felt the new medium should not depend on advertiser support. Secretary of Commerce Herbert Hoover and others believed radio should not be allowed to let advertising invade listeners' homes (although newspapers and magazines had been doing just that for decades). But as there was no other practical means of supporting operating costs, advertising on the air gradually attracted greater support.

The American Telephone and Telegraph (AT&T)-operated New York City station WEAF is generally credited with selling the first radio advertisement (what the telephone company owner termed "toll broadcasting"), although other outlets may have made similar sales at about the same time. On 28 August 1922 the Queensboro Corporation spent $100 for a 15-minute commercial message on WEAF touting a new real estate venture. The same message was repeated for five days and again a month later, resulting in many apartment sales. But despite early sales to an oil company and American Express, paid advertising on the station caught on slowly, for too little was known about radio's sales potential.

The critical turning point came in 1927–28 when several trends combined to increase acceptance of radio as an advertising medium. Among them were the development of national networks (the National Broadcasting Company [NBC] published its first pamphlet touting radio advertising in early 1927); the reduction of interference (thanks to the Federal Radio Commission [FRC]); better and less-expensive radio receivers (which led to growing audiences); the first scientific audience research on radio; the recognition by pioneering advertisers of what radio could accomplish as a sales medium; the growing interest of advertising agencies (the first book on radio advertising was published in 1927); and the general acceptance by the public of advertising as the means to pay for entertainment programming.

Radio Advertising Expands

The Depression brought about an important change in radio advertising. Commercials became more direct, intent on getting listeners to commit to a purchase and focusing on prices. Some program-length advertisements were accepted by stations hard-pressed to stay in business, as was barter advertising (exchange of station time for goods the station or its personnel could use). Advertising agencies began to develop expertise in radio, and station representative firms began to appear in the early 1930s. Radio's portion of all advertising grew from about two percent in 1928 to nearly 11 percent in 1932.

By the mid-1930s, advertising agencies were not only selling most of radio network time but were increasingly producing the programs themselves. This control continued into the early years of television. About 60 percent of all radio advertising was placed with networks (primarily NBC-Red and the Columbia Broadcasting System [CBS]) and their owned stations, with the other 40 percent going to regional and local advertising on several hundred other stations. Daytime advertising focused on soap opera audiences, whereas evening or prime-time advertising helped to support the comedy, drama, and variety programs that attracted the largest audiences. Many advertiser names appeared in program titles to emphasize their support (and control). Most advertising revenue went to the most powerful stations in larger cities.

World War II brought great prosperity to radio as advertisers flocked to buy time when newspaper and magazine advertising was limited by paper rationing. Changes in tax laws served to encourage advertising expenditure of funds that would otherwise be taxed up to 90 percent. Such "ten-cent dollars" filled radio's coffers and led to sharp declines in sustaining (not advertiser-supported) program time. Many companies producing war goods advertised to keep their names before the public, and they often supported highbrow programming with limited (but important) audiences.

Radio's post-war years were marked by a shift away from network advertising (because of television competition) and a growth of "spot" campaigns, in which advertisers would buy time on key stations in selected markets. By 1952 local radio advertising reached half of the medium's total time sales. But far more stations were sharing the advertising pie, thus sharply increasing competition. Radio also became a more direct competitor with local newspapers. Despite these trends, overall radio advertising sales increased each year, and, perhaps ironically, helped to support the expansion of television.

FM radio was a minor player in advertising sales for its first several decades. Only in the 1960s did FM outlets begin to see success in their quest for advertisers, thanks to independent programming, stereo, and a growing audience interested in quality sound. One FM station in Los Angeles experimented with an all classified-ad format but quickly failed. By about 1980, FM became the largest radio medium in terms of listeners, and soon among advertisers as well.

Still, the overall growth in radio station numbers meant that many stations were barely surviving, and a substantial proportion actually lost money in many years. Competition among stations, and between radio and other media, became tighter. Listeners noted the gradual increase in time devoted to advertising messages, and "clutter" (multiple messages played consecutively) became an issue.

Types of Announcements

In addition to entertainment programming, radio stations generally air commercials, station promotional announcements (promos), non-revenue generating announcements intended to encourage further radio listening, and public service announcements (PSAs), which air in support of not-for-profit organizations (ranging from the American Red Cross to a local civic group). All four of these categories are generally referred to as *spots* and range up to 60 seconds in length each.

Commercials are played in blocks or sets sometimes consisting of six or more announcements at a time. Depending on spot lengths, a commercial break might consume five continuous minutes of airtime. More than $19.5 billion was spent on radio advertising in the United States in 2002; about three-fourths of that total was spent on local advertising. When the advertising is sold effectively—based on the station's listening audience and program approach—a listener may benefit by receiving worthwhile consumer information.

For the potential advertiser, a radio station is in the *ear leasing* business. Just as the radio station must build listener awareness of its programming, advertising clients need listener awareness of the goods or services they sell and, most importantly, the clients need customer traffic. The job of radio advertising is to provide the ears of listeners who will hear the ad buyer's message and then visit the store or otherwise obtain the product or service advertised.

Sales Department

A sales manager or general sales manager supervises day-to-day station sales operations and helps make revenue projections for the station. The members of the sales staff are usually called account executives (AEs), although some stations may refer to them as marketing executives or marketing consultants.

It is the job of account executives to prospect for potential clients, develop client presentations, secure advertising buys, and service the account. Servicing includes ensuring that ads run when they should, updating ad copy as needed, and, in some smaller markets, collecting payment. Radio account executives are usually paid according to their sales performance. AEs may be paid a straight commission or a percentage of the sales dollars they generate. The latter compensation plan carries a strong incentive for the salesperson to produce results, but it also means the AE has little financial security.

Another approach is to pay the account executive a "draw" against commission. The draw enables the AE to receive minimum compensation based on anticipated sales. Once this minimum is reached, additional compensation is paid through sales commissions. If the AE is paid a commission based on advertising sold—rather than advertising revenue collected from clients—and later has a client who defaults on a bill, the AE may have a "charge back" to the draw and commission. In other words, the account executive must return any income earned on ads that aired but were not paid for. For this reason, many stations pay account executives based on advertising revenue collected rather than advertising sold.

As with any electronic medium, the biggest problem stations face is inventory management. For any station, "inventory" refers to the number of commercials the station has available for sale. Advertising time is a perishable commodity. Any commercial inventory not sold is lost forever. There is no effective way for the station to store, save, or warehouse the unsold commercial inventory for use at a future time when demand is higher, nor can stations effectively place additional commercials in their broadcast schedule. Airing more spots may create a short-term revenue increase, but commercial clutter is cited by listeners as one of the biggest distractions to radio listening. A decline in audience will consequently lessen the station's effectiveness in selling future advertising time.

The radio industry publication *Duncan's American Radio* estimates that radio listening in 2000 was at its lowest level in 20 years. *The Wall Street Journal* cited reasons for decreased listening: a survey of 1,071 respondents by Edison Media Research found listener perceptions of increased ad clutter on many stations. Another study found commuters who owned a cell-phone reported less listening to the radio than a year earlier.

Benefits and Disadvantages of Radio Advertising

Radio advertising, when compared with television, cable, newspaper, or magazine advertising, offers the advertiser some unique advantages. Over the course of a typical week, nearly everyone listens at least briefly. Radio reaches more than three-fourths of all consumers each day and about 95 percent of all consumers during a typical week. That exceeds the number of newspaper readers and television viewers. The typical person spends about three hours listening to radio on an average weekday, almost always while doing something else (especially driving).

There are, of course, disadvantages to advertising on radio. It is virtually impossible to buy advertising on just one or two radio stations and still meet an advertiser's marketing needs. The multitude of stations in most markets and their specialized formats (and thus relatively narrow audiences) often mean an advertiser must purchase time on multiple stations in the same market. Radio is sometimes considered a "background" medium. Listeners often tune-out commercials or, even worse, tune to another station when commercials air. Where people listen to the radio—in cars for example—often makes it difficult for consumers to benefit from such information as telephone numbers, addresses, or other product attributes. When a station's audience is perceived as being small, the client may think the ad buy will not be effective. When the station's listening audience is large, a client may think an ad campaign involves overspending for uninterested listeners.

The first job of the sales staff is to help clients understand how effective radio is when compared with competing advertising media. The second and more difficult job is to sell advertising time on a specific station. Proliferation of radio stations and continued fragmentation of audiences has made it vital for stations to market a station brand to both listeners and advertisers. Advertisers are no longer buying based solely on a station's audience. They are aware of the listener demographic profile and the station's on-air presence, which includes announcers, music, and promotional events. Listener demographics refers to listener age range, gender, ethnicity, socioeconomic background, consumer spending patterns, and a host of other qualitative variables.

Any advertiser must be concerned with both the formal and the hidden costs of purchasing radio time. The most obvious expense is the stated cost of the time, expressed either as actual dollars charged or in terms of cost per thousand listeners. Hidden cost refers to the quality or nature of the audience an advertiser is buying. How closely does this audience match the advertiser's customer profile? Significant deviation from those consumers whom the advertiser needs to reach probably indicates an inefficient advertising purchase.

Radio station owners and the Radio Advertising Bureau, an industry trade group, work to maintain radio's position as a valuable ad source. Most radio station managers acknowledge that their biggest competitors are not other radio stations in the market playing the same music and attempting to attract the same listener group. The biggest competitors for radio station time sales are usually local newspapers and, to a lesser degree, television stations, billboards, or direct mail. By the turn of the century, radio advertising was accounting for about eight percent of all advertising expenditure—an increase from the medium's low point from the 1950s into the 1980s, but far below radio's network heyday of the mid-1940s.

Radio Advertising Clients

Radio stations generally sell advertising to three distinct groups of clients: local, regional (or "national spot"), and national. The percentage of clients in each category varies with market size and the station's ratings. Small market stations air primarily local ads. Successful stations in large markets command more regional and national advertising. Nearly 80 percent of all dollars spent purchasing radio time are for local advertising.

National advertisers are often involved in local ad sales through cooperative advertising programs. These allow local retailers to share the cost of radio time with a national firm. The national company provides an advertising allowance to the local retailer, usually determined by the dollar value of the inventory purchased from the national company. This advertising allowance can be used to buy ads to promote both the national brand and the local retailer. National manufacturers may also produce radio commercials that only need the local retailer's name added as a "local tag" at the end of the ad.

Advertising Effectiveness

The effectiveness of radio advertising is gauged by measuring the reach and frequency of ad exposure. Reach refers to the number of different people who are exposed to the ad, whereas frequency refers to the number of times different people hear the ad. Even though virtually all of the population will listen to the radio at some point during the week, it will take multiple ads to ensure that all listener segments hear an ad. Also, radio ads probably won't produce the degree of effectiveness the advertiser wants if consumers are exposed to the ad only one time.

The nature of radio use suggests that consumers are often engaged in other activities while they listen to radio. To create an impression in the consumer's mind, repeated exposure to the message (frequency) is typically needed. To increase the likelihood that ads will cause the consumer to take action, frequent exposure to the message is desired. The advertiser might schedule multiple days of advertising with one or more ads per hour during a selected time period to increase frequency.

Radio advertising sales depend on quantitatively and qualitatively identifying the listeners to a particular station. Quantity is measured by radio ratings.

Research helps a radio station further quantify the listening audience—advertisers want to know how many people are listening and just who the listeners are, with respect to age, income, or gender. By collecting such listener demographic information, radio advertising effectiveness can be evaluated for specific audience segments, such as women 25 to 49 years of age.

Two of the most common calculations for comparing advertising effectiveness are "Gross Impressions" and "Cost Per Thousand" comparisons. Gross Impressions (GIs) measure the total number of people reached with a given commercial message. GIs are calculated by multiplying the AQH (average quarter hour) persons estimate for the particular daypart by the number of spots to be run in the daypart. The number of listeners or AQH persons is the number of persons listening to the station in a 15-minute period.

Cost Per Thousand provides a way to compare the cost of reaching the targeted audience either on a single station or among multiple stations. Cost per Thousand determines the cost of reaching a thousand station listeners (sometimes referred to as "Listeners Per Dollar"; in some small markets, the calculation could be cost per hundred). The simplest way to calculate Cost per Thousand is to divide the cost of the ad by the number of listeners (in thousands) who are expected to hear the ad.

It is important also to consider listener demographics. A listener profile that better matches a product or service may justify paying a higher Cost per Thousand. Another method for calculating Cost per Thousand is to divide the total cost of the ad schedule by the total number of Gross Impressions. "Reverse Cost Per Thousand" enables an account executive to determine the maximum rate per spot that a competing station can charge to remain as cost-effective as his or her own station.

It is also helpful for account executives and advertisers to know a station's "exclusive cume listeners." Rather than count listeners multiple times during the day, this calculation allows the advertiser to see how many different people listen to the station during a day. A Contemporary Hits Radio format will usually have greater listener turnover and a higher cume because there are usually several stations in a market with this format or a complementary format, and listeners are prone to change stations frequently. On the other hand, the only station in a market will have a smaller exclusive audience or cume.

Optimum Effective Scheduling is a radio ad scheduling strategy that is based on audience turnover. Optimum Effective Scheduling proposes to improve the effectiveness of a client's ad schedule by calculating the number of spots a client should run. Optimum Effective Scheduling was developed by Steve Marx and Pierre Bouvard to balance the desire for ad frequency and reach while producing an effective commercial schedule. Marx and Bouvard use station turnover or *T/O* (cume audience divided by AQH) times a constant they created, 3.29, to determine the number of spots an advertiser should schedule each week (see Marx and Bouvard, 1993).

From the standpoint of generating ad revenue for the radio station, stations with low turnover are at a disadvantage when using Optimum Effective Scheduling. Their audience listens longer and thus fewer spots are needed to produce an effective schedule of reach and frequency. Assuming ad rates per thousand listeners are reasonably comparable, these stations must attract more clients to generate the same amount of ad revenue as the station with high listener turnover.

Advertising Rates

Radio station advertising rates were once typically printed out on a rate card. Most rate cards were valid for six months to a year. Cards listed the charges for either programs or spot advertisements at different times of the day (dayparts). The card might also specify a price discount as the client purchased more ads per day or per week. This rate card is sometimes referred to as a quantity card or quantity-discount rate card. The quantity card might be an effective way to reward loyal advertising clients but is a poor technique for managing valuable advertising inventory. The radio station, with a limited inventory of commercial time, is discounting the price of its product. The discount applies, no matter what the available advertising situation is like.

Increasingly replacing formal rate cards is the grid rate card system. Using an inventory tracking software package, the grid allows a radio station to track inventory available for sale. This might mean keeping track of the number of commercial minutes sold or the total number of commercial units (spots) available for sale. The inventory management system also enables the radio station to increase or decrease its ad rate in response to customer demand. When a radio station has sold nearly all the advertising it can effectively schedule, it should be able to charge more for remaining commercial units. A grid rate card enables the station to adjust advertising rates according to the amount of inventory remaining.

Once the station's sales department has established a record with clients of pricing inventory according to demand, account executives may be more effective in pre-selling advertising time, which should decrease the likelihood of lost ad inventory. When retailers place advertising orders earlier, the station can project revenue more effectively. The longer a client waits to buy commercials, the more likely the available supply of ad time will decrease and the price of the remaining time will increase.

Radio advertising continues to be an important business for station owners. Ownership consolidation has increased sales pressures for account executives, but it has also lessened direct competition by decreasing the number of station owners. Radio's biggest challenge will be to make sure programming and advertising remain relevant to users who have at their disposal a wider range of substitute products ranging from satellite delivery audio to downloaded and home-burned CDs or MP3 audio files.

GREGORY G. PITTS AND CHRISTOPHER H. STERLING

See also Advertising Agencies; Arbitron; Commercial Load; Demographics; Market; Promotion; Radio Advertising Bureau; Station Rep Firms; WEAF

Further Reading

Albarran, Alan B., and Gregory G. Pitts, *The Radio Broadcasting Industry,* Boston: Allyn and Bacon, 2000

Arbitron Ratings Company, *Arbitron Radio Market Reference Guide: A Guide to Understanding and Using Radio Audience Estimates,* New York: Arbitron, 1987; 4th edition, 1996

Barnouw, Erik, *The Sponsor: Notes on a Modern Potentate,* New York: Oxford University Press, 1978

Dygert, Warren B., *Radio as an Advertising Medium,* New York: McGraw Hill, 1939

Felix, Edgar H., *Using Radio in Sales Promotion,* New York: McGraw Hill, 1927

Hettinger, Herman S., *A Decade of Radio Advertising,* Chicago: University of Chicago Press, 1933; reprint, New York: Arno Press, 1971

Marx, Steve, and Pierre Bouvard, *Radio Advertising's Missing Ingredient: The Optimum Effective Scheduling System,* Washington, D.C.: National Association of Broadcasters, 1991; 2nd edition, 1993

Midgley, Ned, *The Advertising and Business Side of Radio,* New York: Prentice-Hall, 1948

Radio Advertising Bureau, <www.rab.com>

Seehafer, E.F., and J.W. Laemmar, *Successful Radio and Television Advertising,* New York: McGraw Hill, 1951; 2nd edition, as *Successful Television and Radio Advertising,* 1959

Shane, Ed, *Selling Electronic Media,* Oxford and Boston: Focal Press, 1999

Streeter, Thomas, *Selling the Air: A Critique of the Policy of Commercial Broadcasting in the United States,* Chicago: University of Chicago Press, 1996

Warner, Charlie, and Joseph Buchman, *Broadcast and Cable Selling,* Belmont, California: Wadsworth, 1986; 3rd edition, 1993

Wolfe, Charles Hull, *Modern Radio Advertising,* New York: Funk and Wagnalls, 1949

Advertising Agencies

When radio broadcasting established itself in the United States and United Kingdom in the 1920s, advertising agencies were full-service organizations—planning complete advertising campaigns, producing advertising messages, and placing these messages in various media. In the United States, advertising agencies were initially reluctant to recommend radio advertising to their clients; in time, however, the agencies became supporters of radio advertising and, until the arrival of television, helped build the radio networks. In the United Kingdom, where until 1972 noncommercial radio broadcasting by the British Broadcasting Corporation (BBC) was the rule, advertising agencies lobbied for commercial radio and worked with foreign and pirate radio stations on behalf of clients; however, once commercial radio arrived, UK agencies were slow to embrace it.

Resisting Radio Advertising

The rise of radio advertising in the United States was tentative and slow. Advertising first appeared in 1922 on station WEAF in the form of sponsored time. Other stations gradually accepted sponsored programs, but many broadcasters viewed advertising agencies as competitors and were hesitant to sell them time or allow sponsorships. Anti-advertising rhetoric from listeners, critics, legislators, and regulators fueled opposition as well.

Surprisingly, advertisers and agencies distrusted the notion of radio advertising. Agencies doubted that radio advertisements would work, a sentiment shared by many advertisers. The advertising industry also believed listeners might resent radio sponsorship and, consequently, reject other forms of advertising by the same advertisers. This was of particular concern to print advertisers and their agencies.

For several years agencies warned their clients against using radio advertising. Advertisers had to produce programs themselves with assistance from station personnel. For example, in 1925 Clicquot, a soda manufacturer, worked directly with WEAF to create the *Clicquot Club Eskimos* music program because its agency did not believe in radio. There were, however, exceptions.

William H. Rankin of the Rankin advertising agency decided to test radio advertising before recommending it to clients. He bought time on WEAF for a talk about advertising but received only a small number of letters and phone calls in response. One, from a prospective client, Mineralava, led to a contract and more radio advertising. Rankin began recommending radio and another client, the Goodrich Company, sponsored a radio series.

Another early exception was the N.W. Ayer agency, which supervised *The Eveready Hour* in 1923. Ayer ensured that the show was professional and identified the sponsor in the name. The favorable attention it received attracted other sponsors to radio, with shows such as the *Bakelite Hour, The Victor Hour,* and *The Ray-O-Vac Twins.* These shows became models for later network programs.

Although opposition to radio advertising persisted into the mid-1920s, most advertising practitioners were beginning to consider its use. To win them over, the newly formed National Broadcasting Company (NBC) and Columbia Broadcasting System (CBS) radio networks hired promoters to persuade those still skeptical about the effectiveness of radio advertising.

In 1928 NBC initiated a promotional campaign to educate and encourage adoption of radio advertising. The networks targeted leading advertisers and agencies with brochures highlighting radio success stories and emphasizing radio's ability to build brand awareness and stimulate dealer goodwill. The networks also offered financial incentives by paying agencies commissions even if they were not directly involved in a client-sponsored show.

NBC loaned its employees to leading agencies to help develop radio departments. N.W. Ayer started the first full-scale radio department in 1928 and others soon followed, employing personnel who migrated from radio. The promoters urged the networks to allow agencies to sell broadcast time and produce programs. In turn, agencies recognized how lucrative program development and production could be.

Accepting Radio Advertising

The promoters' efforts were successful. By the early 1930s agencies were selling time and handling nearly all sponsored network program development and production. Agencies had gained control of prime-time radio listening and achieved great prosperity, and their radio departments became centers of power.

Sponsored radio shows of the 1920s employed "indirect advertising," simple mentions of the program's underwriter with no product description or sales pitch. The networks supported this practice with policies against direct advertising. George Washington Hill, president of the American Tobacco Company, and Albert Lasker, head of the Lord and Thomas agency, pressured the networks to allow explicit advertising messages.

Although Lasker and Hill largely conformed to the indirect advertising requirements when they launched the *Lucky Strike Dance Orchestra* in 1928, Hill, who believed strongly in intrusive radio advertising with explicit product claims, aggressively pursued this goal by forcing the issue with network executives and supporting Lucky Strike with extravagant budgets. Lord and Thomas controlled a large share of NBC's business, so Lasker had leverage as well. By 1931 women were being sold Lucky Strikes with mildness claims by opera and film stars and "slimming" messages suggesting that listeners smoke a Lucky Strike instead of eating something sweet.

The 1930s saw advertising agencies crafting selling environments for their clients in the form of elaborate comedy, variety, and dramatic series. Vaudeville came to radio as agencies began to use star talent. Young and Rubicam created *The Jack Benny Program* for General Foods' Jell-O. Lord and Thomas produced Bob Hope. J. Walter Thompson produced the *Kraft Music Hall* with Bing Crosby and *The Chase and Sanborn Hour* with Edgar Bergen and Charlie McCarthy.

After commercializing prime-time radio, the networks, with agency help, developed a daytime audience of women listeners. The networks developed 15-minute sponsored talks with recurring characters and continuing stories. Soap operas—melodramatic serials typically sponsored by manufacturers of household detergents and cleaners—were born. Most were produced by advertising agencies.

One agency, Blackett, Sample and Hummert, built a reputation for soap opera programming. Glen Sample adapted a 1920s newspaper serial into a radio show, *Betty and Bob,* sponsored by Gold Medal Flour. Sample also developed the long-running *Ma Perkins* for Procter and Gamble's Oxydol. In 1931 Frank and Anne Hummert created a daily NBC serial, *Just Plain Bill,* for Kolynos toothpaste. The Hummerts became highly prolific soap opera creators, developing nearly half the network soap operas introduced between 1932 and 1937. Soap operas were so successful that daytime radio advertising revenues doubled between 1935 and 1939.

Agencies and radio networks were determined to protect their financial success during the Depression. Indeed, their program decisions uniformly ignored economic and social problems. With the exception of *The March of Time,* produced for *Time* magazine by Batten, Barton, Durstine and Osborn, news was all but missing from sponsored programs. Radio's skilled entertainers kept Americans' minds off their despair.

The radio and advertising industries experienced continued prosperity during World War II. Agencies encouraged clients to maintain brand awareness, even if they had no products to sell, and radio benefited from such prestigious sponsorships as *General Motors' NBC Symphony Orchestra* as well as benefiting from paper shortages that limited newspaper ads. Both industries assisted the Office of War Information with insertions of war effort announcements, earning them favorable government treatment when their wartime revenues came under close scrutiny.

U.S. Postwar Changes

Envious of the power held by advertising agencies, the radio networks decided to regain control of programming. The agencies lost ground to independent producers, but the real threat to radio came from the growing medium of network television. Advertisers and their agencies shifted the system of star-studded, sponsored programs to television. Young and Rubicam found that its programs moved so easily to television that from 1949 to 1950 half of the top 10 TV shows were its productions.

Within a decade, network radio serials and soap operas had all but disappeared, taking with them substantial ad revenue. Whereas in 1931 network advertising constituted 51 percent of total radio advertising revenues, by 1960 that had fallen to just 7 percent. Radio survived by serving local listeners with format programming and attracting local advertising.

U.S. agencies became producers of commercials and buyers of spot radio time. Despite periods of renewed interest in radio and a resurgence of radio networks, for national advertisers and their agencies radio was relegated to the role of support medium.

Lobbying for Commercial Radio in Britain

The BBC's license forbids it from broadcasting advertising or sponsored programs. Other than English-language radio broadcasts from foreign and pirate stations, commercial radio did not officially exist in the United Kingdom until 1972. Still, from the start of British radio, advertising agencies lobbied for commercial broadcasting, which held out the possibility of more advertising business. Advertising on the BBC and the creation of a parallel commercial radio system were repeatedly ruled out by successive government inquiries.

In 1923 the Sykes Committee on Broadcasting heard advertising agencies' arguments, but found that radio advertising would unfairly benefit large advertisers, negatively affect the advertising revenues of the press, and lower broadcast program standards. Over a decade later the Ullswater Committee (1935) reaffirmed the Sykes Committee's conclusions.

Long before commercial radio arrived in the United Kingdom, a well-organized radio advertising industry was promoting products to a large British audience through English-language programming on foreign stations. These broadcasts emanated from many stations, the most well known of which were Radio Luxembourg and Radio Normandie. English-language broadcasting experiments in continental Europe during the 1920s attracted British listeners and sponsors. In 1929 Radio Publicity Limited started organizing English-language programs for record, food, toothpaste, and cigarette manufacturers. A year later the entrepreneurial Captain Leonard Plugge founded the International Broadcasting Company (IBC), also to arrange commercial programs.

Sponsored shows were usually produced and recorded in Britain and shipped to continental stations for transmission. The IBC established itself as a production facility, and leading British advertising agencies such as J. Walter Thompson and the London Press Exchange handled their own program production.

By 1936 radio advertising expenditures exceeded £1 million, and dozens of major advertisers, such as Lever Brothers, MacLeans, Carters Liver Pills, and Cadbury Brothers, were on Radio Luxembourg and Radio Normandie. J. Walter Thompson's major clients, including Rowntree, Horlicks, Ponds, and Kraft, were also substantially engaged in radio advertising. Between 1936 and 1939 Ponds and Horlicks spent 20 percent and 33 percent of their total advertising budgets, respectively, on radio. Radio advertising was sufficiently established by 1938 that British manufacturers spent over £1.5 million. Agency Mather and Crowther Limited compiled *Facts and Figures of Commercial Broadcasting,* and J. Walter Thompson provided prospective clients with a promotional recording showcasing their radio expertise.

After World War II Radio Luxembourg resumed its service to British advertising agencies and their clients; however, television lured listeners and advertisers away from radio. In 1946 the Institute of Practitioners in Advertising (IPA), an advertising industry association, published *Broadcasting: A Study of the Case for and against Commercial Broadcasting under State Control in the United Kingdom,* which claimed that commercial broadcasting had value to advertisers and the public. The IPA's recommendation was not to dissolve or create competition for the BBC but to allow commercial broadcasting within the BBC's existing structure.

The IPA monograph became an important piece of evidence examined by the Beveridge Committee (1949), which considered the introduction of commercial broadcasting. The advertising lobby was active in providing evidence and scrutinizing that of others, with J. Walter Thompson, for instance, helping to prepare materials presented by major advertisers, including Horlicks, Unilever, and Rowntree. The Beveridge Committee decided against commercial broadcasting, but a minority report proposed a system of national and local commercial radio that would later become a reality.

In the mid-1960s pirate radio stations, broadcasting from old forts and ships anchored just outside British territorial waters, afforded advertising agencies and their clients another opportunity to circumvent the United Kingdom's no-commercial-radio policy. The success of these stations appears to have finally led to officially sanctioned commercial radio. Advertising agencies continued to lobby for commercial radio and helped win Conservative Party support for the 1972 introduction of commercial local radio. Even after commercial radio became a reality, agencies worked to influence BBC policy.

Agencies and British Commercial Radio

The IPA joined others in 1984 to question the noncommercial future of the BBC. Two large agencies, D'Arcy MacManus Masius and Saatchi and Saatchi, issued reports indicating that advertising would allow the BBC to meet its revenue needs without raising the license fee paid by listeners. Polls indicated that the public was willing to accept this arrangement, but the Peacock Committee, which was considering the issue, rejected this option.

The BBC remains noncommercial and dependent on receiver license fees. Ironically, BBC Radio has established a commercial arm, Radio International, that allows sponsorship and advertising on the programming it markets for overseas consumption.

The 1972 Broadcasting Act established commercial Independent Local Radio in the United Kingdom, opening the door for agencies to offer radio copywriting and time-buying services; however, national advertisers and their agencies were slow to embrace commercial radio for several reasons, including incomplete geographic coverage, which precluded national reach; a lack of credible audience measurement; and restrictive advertising regulations.

In the nearly two decades that these barriers were being addressed, commercial radio struggled, developing a reputation as a "two percent medium," unable to attract more than two percent of British advertising revenues. By the early 1990s advertising time and sponsorship restrictions were lifted; coverage was essentially complete, with over 130 local broadcasters on air plus a new national station; and a new audience measurement system was in place. Nevertheless, agencies continued to ignore the medium or simply used it as a campaign extension.

A Radio Advertising Bureau marketing campaign targeting agencies and advertisers helped sell advertisers and their agencies on radio. Commercial radio started to shake its reputation in the mid-1990s when a number of blue-chip advertisers first used radio. Foote, Cone, and Belding and Ogilvy and Mather directed Lever Brothers' brands Surf and Radion, respectively, to the medium. J. Walter Thompson also encouraged Kellogg to test radio in London and Jaguar to launch a promotion for its XJ models.

The number of commercial radio services continued to grow, exceeding 250 by the close of the 1990s. Between 1992 and 2002, commercial radio revenue increased 395 percent and national radio buys were accounting for over 60 percent of radio advertising revenue. Radio was Britain's fastest-growing medium and its share of advertising revenues exceeded six percent.

RANDY JACOBS

See also Advertising; British Commercial Radio; British Pirate Radio; Radio Advertising Bureau; Radio Luxembourg; WEAF

Further Reading

Arnold, Frank Atkinson, *Broadcast Advertising: The Fourth Dimension,* New York: Wiley, and London: Chapman and Hall, 1931

Barnouw, Erik, *The Sponsor: Notes on a Modern Potentate,* New York: Oxford University Press, 1978

Baron, Mike, *Independent Radio: The Story of Commercial Radio in the United Kingdom,* Lavenham: Terrence Dalton, 1975

Burt, Frank A., *American Advertising Agencies: An Inquiry into Their Origin, Growth, Functions, and Future,* New York: Harper, 1940

Hower, Ralph M., *History of an Advertising Agency: N.W. Ayer and Son at Work, 1869–1939,* Cambridge, Massachusetts: Harvard University Press, 1939; revised edition, with subtitle dates *1869–1949,* 1949

McDonough, John, "Radio: A 75-Year Roller-Coaster Ride," *Advertising Age* (4 September 1995)

O'Malley, Tom, *Closedown? The BBC and Government Broadcasting Policy, 1979–92,* London and Boulder, Colorado: Pluto Press, 1994

Scannell, Paddy, and David Cardiff, *Social History of British Broadcasting, 1922–1939,* Oxford and Cambridge, Massachusetts: Blackwell, 1991

Seymour-Ure, Colin, *The British Press and Broadcasting since 1945,* Oxford and Cambridge, Massachusetts: Blackwell, 1991; 2nd edition, 1996

Smulyan, Susan, *Selling Radio: The Commercialization of American Broadcasting, 1920–1934,* Washington, D.C.: Smithsonian Institution Press, 1994

Wilson, H. Hubert, *Pressure Group: The Campaign for Commercial Television,* London: Secker and Warburg, 1961; with subtitle *The Campaign for Commercial Television in England,* New Brunswick, New Jersey: Rutgers University Press, 1961

Affirmative Action

Diversity in Employment, Programs, and Ownership

Affirmative action mandates equal treatment for all people regardless of gender, age, religion, sexual orientation, etc. The need for programs to assure this equal treatment depends on the amount and nature of discrimination; they are solutions to identified problems of discrimination, not processes unto themselves (Hooks, 1987). Applied to radio broadcasting, affirmative action programs have been related to discrimination in: (1) employment, (2) program content, and (3) station ownership. The rationale for affirmative action in radio was based on the desire of the Federal Communications Commission (FCC) to achieve diversity of information, defined as having many voices express opinions on many issues. The Supreme Court affirmed this goal in *Red Lion v FCC* (Honig, 1984).

Employment

Federal concern about employment diversity was initiated in the 1968 *Report of the National Advisory Commission on Civil Disorders*. The FCC, based on the public interest standard, responded with a statement about equal employment opportunity (47 CFR 73.2080, section b). The result was an examination of license renewals to determine whether the racial composition of a station's staff was similar to the demographic makeup of the community in which the station was licensed (zone of reasonableness). Short-term renewals, fines, and the threat of possible revocations could result from noncompliance. The FCC responded with a model Equal Employment Opportunity (EEO) program in 1975 to eliminate race and gender discrimination. The Supreme Court affirmed the legality of such oversight by independent regulatory agencies in *NAACP v Federal Power Commission* (1976). The FCC was committed to programming fairness and accurate representation of minority group tastes and viewpoints (FCC, 1978).

Congress, in the Cable Television Consumer Protection and Competition Act of 1992, required the FCC to monitor employment statistics for women and minorities in the cable and broadcast industries. The FCC's first report found that from 1986 to 1993 the number of women in the national workforce increased by 1.1 percent, in the broadcast industry 2.8 percent, and 3.6 percent in upper-level positions. The number of minorities increased 2.1 percent in the national workforce, 2.2 percent in the broadcast industry, and 2.4 percent in upper-level positions (FCC, 1994).

The FCC's EEO policies were overturned in *Lutheran Church v FCC* (1998). Essentially, the court found that increasing staff diversity did not necessarily lead to diversity of viewpoints in the marketplace because only a small number of station employees made programming decisions. The policy was also overbroad, much as the Supreme Court found in *Adarand v Pena* (1995). The court's response to the FCC's request for a rehearing indicated that its decision did not preclude any policies that encouraged "broad outreach" to a diverse applicant pool. The FCC has responded with a Notice of Proposed Rule Making (NPRM) suggesting that broadcasters, cable operators, and other multi-channel video programming distributors could send job announcements to recruitment organizations or to participate in job fairs, internships, etc. They could also devise their own recruitment process. Annual hiring reports would still be filed with the FCC. These rules were adopted two years later (FCC, 2000).

A portion of these rules were overturned in *DC/MD/DE Broadcasters Association v FCC* (2001). The commission responded with another NPRM suggesting that all media outlets " widely disseminate information about job openings to all segments of the community to ensure that all qualified applicants have sufficient opportunity to compete for jobs in the broadcast industry" (FCC, 2001). These rules were adopted in November 2002. What was once a requirement that media owners represent the diversity of their audiences with equal numbers of minorities on their staffs is now a program that requires them to widely distribute job opening information, attend job fairs, and offer scholarships. Statements by the commissioners decried their inability to be more forceful in this area, but stated that limitations by the courts have greatly diminished the force of regulation. Industry spokespersons were hesitant to support the new rules, saying that EEO has been over-regulated in the past (Greenberg, 2002).

Program Content

The public interest resulted in two rules requiring diversity in program content: ascertainment and the fairness doctrine. Ascertainment required stations to determine issues of public importance by surveying listeners and community leaders. The fairness doctrine required that these issues be addressed fairly. These rules, plus a decision by the Supreme Court that gave audiences the right to testify before the FCC, *United Church of Christ v FCC* (1966), resulted in increased minority participation in the 1970s until the FCC began deregulating radio in 1981 (FCC, 1981). Honig and Williams argued that deregulation was the result of a conservative FCC wishing to reduce the workload for radio stations coupled with the loss of influential groups pressuring the FCC about diversity.

Deregulation was necessary because the number of radio stations had increased from 583 in 1941 to 9,000 by the late 1980s, forcing stations to develop specialized formats to attract audiences; radio could no longer provide general services to all of its audiences. The result was the elimination of policy guidelines concerning non-entertainment programming, the ascertainment process, commercial time guidelines, and rigidly formatted program logs (FCC, 1981). The fairness doctrine was abolished in 1987. The concern for radio format changes ended in 1976 in response to the court decision in cases such as *Citizens Committee to Save WEFM v FCC* (FCC, 1976). Although the FCC was concerned with empowering broadcasters to select entertainment formats that offered the greatest commercial viability in their markets, the results of these policy decisions might have had an impact on programming oriented toward minority audiences.

The end of program content regulation for purposes of increasing diversity and the move away from numerical goals for employment after 1976 spelled the end of employment and program affirmative action policies. The FCC argued that none of these policies actually increased the diversity of information and turned to station ownership diversity as a solution.

Station Ownership

Diversity of station ownership was a goal of the FCC that assumed that who owned radio outlets would influence, if not determine, program diversity. The assumption was that increasing minority (women and ethnic minorities) owners would increase programming for such underserved audiences and thus serve the public interest. Further encouraging ownership diversity was a two-day meeting resulting from pressure from the National Black Media Coalition and the National Association of Black-Owned Broadcasters in 1977. The resulting FCC policy statement found that despite the fact that minorities comprised approximately 20 percent of the population, they controlled less than 1 percent of the over 8,500 radio stations. The FCC proposed two solutions to the lack of ownership diversity. First, tax certificates were offered to broadcasters who sold their stations to ownership teams that had a "significant minority interest." Tax certificates allowed sellers to defer capital gains taxes. Second, "distress sales" were authorized for licensees who were scheduled for revocation hearings before the FCC. The rationale was that broadcasters who would likely lose their licenses in such hearings could sell their properties at a reduced cost to minority ownership teams, producing at least some profit from the sale of the station. The market would benefit by increasing station ownership diversity. The government would also save money because costly hearings would be avoided (FCC, 1978). The result of these two solutions was the sale of 82 radio stations to minority owners between 1978 and 1982. Despite this increase, still only 2 percent of broadcast stations were minority owned (Honig, 1984). Former FCC Chair Kennard decried the lack of stations owned by minorities because only 2.5 percent of all broadcast stations had minority owners in 1997 (McConnell, 1998).

The historical basis for ownership diversity can be found in the *Policy Statement on Comparative Broadcast Hearings* (FCC, 1965). Two criteria stipulated by the FCC as integral to deciding between competing applicants for station licenses were diversification of ownership and integration of ownership/management, defined as station owners living and being active in the communities for which the license was granted. Application of these factors to diversity of station ownership was affirmed in *Citizens Communications Center v FCC* (1974). Direct application to minority owners of broadcast stations was made in *TV 9 Inc. v FCC* (1973).

The FCC was in the process of re-examining its ownership diversity procedures in the late 1980s. As more Republican members of Congress took office, along with conservative Democrats appointed during the Reagan administration, the FCC began to question its proper role in this area. Nevertheless, Congress made clear in budget resolutions that the FCC was not to make any changes.

The Supreme Court affirmed both the enhancement credits (tax certificates) and distress sales as methods for increasing minority ownership. The court's decision was twofold. First, increasing broadcast diversity was an important government goal. Second, FCC policies of diversifying ownership were determined to be reasonable means of meeting these goals. A substantial amount of data supporting this conclusion was appended to the decision (*Metro v FCC*).

Similar reasoning was used to support incentives for women to own broadcast stations, but data analyzed by the Court of Appeals failed to meet the second part of the Supreme Court's decision in Metro: no link could be established between increasing female ownership of broadcast stations and the consequent increase in programming for women. Thus, the ownership preference was held to be unconstitutional (*Lamprecht v FCC*).

Americans for Radio Diversity reported that minority ownership was up to 3.1 percent before the enactment of the Telecommunications Act of 1996. The removal of many station ownership caps has led to massive radio consolidations, however, and minority ownership has declined to 2.8 percent (2000). The decline was due in part to sharply higher station prices, which was brought about by industry consolidation.

Recent Developments

The FCC, the broadcast industry, and Congress have recently been active in exploring ways to increase diversity of radio station ownership. Then FCC Chair Reed Hundt announced a plan resulting from the standards set by the Supreme Court in *Adarand v Pena* to give preferences to women and minorities in its auction of personal communications services, originally reserved for small businesses (Jessel, 1995). More recently, FCC commissioners Kennard and Powell challenged the National Association of Broadcasters (NAB) to develop solutions. The result was the Prism Fund, funded by megamedia owners such as CBS and Disney/ABC promising $1 billion to assist minorities and women with the purchase of radio stations. The NAB has also been active, offering $10 million to encourage station purchases by minorities and women (McConnell, 1999). Fox created a vice president of diversity to increase the number of minority actors and broadcast managers (Schlosser, 2000). Congress has also been concerned with affirmative action and station ownership. Senator John McCain (R-AZ) offered a bill to reinstate tax credits for selling media properties to minorities (Albiniak, 1999).

More recently, the Quetzal/Chase Capital Partners announced the first three investments in minority owned enterprises: Blue Chip Broadcasting, Hookt.com, and Urban Box Office Networks, Inc. (J.P. Morgan Partners, 2000)

WENMOUTH WILLIAMS, JR.

See also African-Americans in Radio; Black-Oriented Radio; Deregulation of Radio; Gay and Lesbian Radio; Hispanic Radio; Native American Radio; Ownership, Mergers, and Acquisitions; Red Lion Case; Stereotypes on Radio

Further Reading

Adarand Constructors, Inc. v Pena, 515 US 20 (1995)

Albiniak, Paige, "Industry Seeds Prism Fund," *Broadcasting and Cable* (8 November 1999)

Citizens Committee to Save WEFM v Federal Communications Commission, 506 F2d 246 (1974)

Citizens Communications Center v Federal Communications Commission, 447 F2d 1201 (1971)

Federal Communications Commission, *Policy Statement on Comparative Broadcast Hearings,* 1 FCC 2d 393 (1965)

Federal Communications Commission, *Changes in Entertainment Formats of Broadcast Stations,* 60 FCC 2d 858 (1976)

Federal Communications Commission, *Statement of Policy on Minority Ownership of Broadcasting Facilities,* 68 FCC 2d 979 (1978)

Federal Communications Commission, *In the Matter of Deregulation of Radio,* 84 FCC 2d 968 (1981)

Federal Communications Commission, *In the Matter of Implementation of Commission's Equal Employment Opportunity Rules,* MM Docket 94-34 (1994)

Federal Communications Commission, *FCC Adopts New Equal Employment Opportunity Rules,* MM Dockets 98-204, 96-16, 20 (January 2000)

Federal Communications Commission, *In the Matter of 1998 Biennial Regulatory Review,* MM Docket No. 98-35 (2000)

Federal Communications Commission, *FCC Proposes New Equal Employment Opportunity (EEO) Rules for Broadcasters and Cable,* MM Docket 98-204 (2001)

Greenberg, B., "FCC's New EEO Rules Leave Most Enforcement to EEOC, Courts," *Communications Daily* (8 November 2002)

Holder, Pamela J., "A Square Peg Trying to Fit into a Round Hole: The Federal Communication Commission's Equal Employment Opportunity Regulations in *Lutheran Church Missouri-Synod v Federal Communications Commission,*" *Akron Law Journal* 32 (1999)

Honig, David, "The FCC and Its Fluctuating Commitment to Minority Ownership of Broadcast Facilities," *Howard Law Journal* 27 (1984)

Hooks, Benjamin, "Affirmative Action: A Needed Remedy," *Georgia Law Review* 12 (1987)

Jessell, H.S., "Hundt: FCC Committed to Minority Ownership," *Broadcasting and Cable* (19 June 1995)

J.P. Morgan Partners, "Quetzal/Chase Capital Partners Completes First Three Investments," *News on Quetzal* (22 May 2000)

Lutheran Church-Missouri Synod v Federal Communications Commission, 141 F3d 344 (1998)

McConnell, B., "NAB Offers $10 Million for Minority Plan," *Broadcasting and Cable* (19 February 1999)

McConnell, C., "Kennard Pushes for Women, Minorities," *Broadcasting and Cable* (8 April 1998)

Metro Broadcasting v Federal Communications Commission, 497 US 547 (1990)

Mishkin, Paul, "Symposium: Race-Based Remedies: The Making of a Turning Point in Metro and Adarand," *California Law Review* 84 (1996)

National Association of Colored People v Federal Power Commission, 425 US 662 (1976)

"Powell Joins Kennard in Pressing for TV-Radio Ownership Diversity," *Broadcasting and Cable* (23 February 1998)

Powell, M.K., *Separate Statement of Michael K. Powell: Review of the Commission's Broadcast and Cable Equal Employment Opportunity Rules and Policies and Termination of EEO Streamlining Proceeding,* MM Docket Nos. 98-204 and 96-16 (1998)

Rathbun, E.A., "Count 'em: 830," *Broadcasting and Cable* (11 October 1999)

Red Lion v Federal Communications Commission, 395 US 367 (1969)

Schlosser, J., "Wanted: VP of Diversity," *Broadcasting and Cable* (31 January 2000)

TV 9 Inc. v Federal Communications Commission, 495 F2d 929 (1973)
United Church of Christ v Federal Communications Commission, 359 F2d 996 (1966)
Williams, Wenmouth, Jr., "Impact of Commissioner Background on FCC Decisions, 1975–1990," in *Media and Public Policy,* edited by Robert J. Spitzer, Westport, Connecticut, and London: Praeger, 1993

Africa

Radio is by far the dominant and most important mass medium in Africa. Its flexibility, low cost, and oral character meet Africa's situation very well. Yet radio is less developed in Africa than it is anywhere else. There are relatively few radio stations in each of Africa's 53 nations and fewer radio sets per head of population than anywhere else in the world.

Radio remains the top medium in terms of the number of people that it reaches. Even though television has shown considerable growth (especially in the 1990s) and despite a widespread liberalization of the press over the same period, radio still outstrips both television and the press in reaching most people on the continent. The main exceptions to this are in the far south, in South Africa, where television and the press are both very strong, and in the Arab north, where television is now the dominant medium. South of the Sahara and north of the Limpopo River, radio remains dominant at the start of the 21st century. The internet is developing fast, mainly in urban areas, but its growth is slowed considerably by the very low level of development of telephone systems.

There is much variation between African countries in access to and use of radio. The weekly reach of radio ranges from about 50 percent of adults in the poorer countries to virtually everyone in the more developed ones. But even in some poor countries the reach of radio can be very high. In Tanzania, for example, nearly nine out of ten adults listen to radio in an average week. High figures for radio use contrast sharply with those for India or Pakistan, for example, where less than half the population is reached by radio.

History

There have been three distinct phases in the development of radio since the first South African broadcasts in 1924. The first phase was the colonial or settler period, when radio was primarily a medium brought in to serve the settlers and the interests of the colonial powers. Later (and in many cases not until toward the end of colonial rule) the authorities gradually introduced radio services by and for indigenous people.

The entire continent, south of the Sahara, with the exception of Liberia and Ethiopia, had been colonized by the European powers—France, Britain, Spain, Belgium, Italy, Germany, and Portugal. (At the end of World War I, Germany lost all of its African colonies, and their administration was taken over by France, Britain, and Belgium.) The domestic broadcasting systems of all European powers were at this time state (not government necessarily) monopolies such as the British independent public service model of the British Broadcasting Corporation (BBC) or the French government radio stations. The Portuguese permitted some private broadcasting by colonial settlers in their colonies, but the main picture was one of national state monopolies.

The earliest broadcasts on the continent were in South Africa. In Johannesburg, Cape Town, and Durban, three organizations—a private club, an advertising group, and a local authority—were granted licenses to broadcast. But they all soon incurred large debts and were taken over by an entrepreneur who, after some difficulty, moved the stations toward commercial viability. However, the government decided that a commercial solution would not provide the service that they sought. They looked instead at what had happened in Britain and invited John Reith, the BBC's first director-general, to come to South Africa in 1934 and help them devise a national public service form of broadcasting. The South African Broadcasting Corporation (SABC) was created in 1936 and maintained a monopoly on broadcasting there for the next 45 years.

The SABC departed from BBC's way of doing things very soon after its establishment. First, it was never far from political influence and control, both of which increased during the years of apartheid. Second, it soon began commercial services designed to make a profit to supplement license fee income for broadcasting. When neighboring Mozambique was a Portuguese colony, a successful commercial radio station there (Radio Lorenço-Marques) targeted South African audiences with popular music programs. To counter this the SABC began its own commercial service, Springbok Radio, in 1950. For most of this period, the SABC's programming was dictated by the needs and tastes of its white audiences. Until 1943, it broadcast only in Afrikaans and English, and none of its programs were directed toward African audiences. Even then, broadcasts in African languages formed only a small part of the total output. Broadcasting for Africans was expanded in the 1960s when Radio Bantu was developed during apartheid to reinforce the apartheid ideology of the separation of the races.

Elsewhere in Africa, radio was also developed first to serve European interests—in 1927 in Kenya, in 1932 in Southern Rhodesia (now Zimbabwe), in 1933 in Mozambique, and in 1935 in the French Congo. The earliest radio in British West Africa was not broadcast by wireless transmission but via wired services—subscribers had loudspeakers (linked by wire to the radio station) installed in their homes to receive the service. This was how broadcasting began in Sierra Leone in 1934, Gold Coast (now Ghana) in 1935, and Nigeria in 1936. Unlike the wireless services in Britain's other colonies, these were created with native African listeners in mind. Then in 1936 the British colonial administration decided to develop radio broadcasting throughout its African colonies as a public service for indigenous people.

In Northern Rhodesia (now Zambia), World War II provided an impetus with major consequences for the development of radio in that territory. A small radio station was established principally with the objective of carrying news (in African languages) of the war's progress to the families of soldiers fighting with the British forces in Africa and Asia. Radio also developed rapidly in other parts of Africa due to the war. The free Belgian government, exiled from German-occupied Belgium, set up a shortwave station in Léopoldville (now Kinshasa) for broadcasts to Belgium. The Free French set up their own radio stations in Cameroon and French Congo, and the French Vichy government had its own station in Dakar, Senegal.

Postwar Developments

After the war, expansion of broadcasting in most of its African colonies became official British policy. This meant that radio services would be developed principally to educate and inform African listeners. Several experts from the BBC were sent to advise on developmental issues in broadcasting, and some of them stayed to play major roles in establishing services. Most notable among these was Tom Chalmers, a successful BBC radio producer who was involved in the development of radio in Nigeria, Nyasaland (now Malawi), Northern Rhodesia (now Zambia), and Tanganyika (now Tanzania). He was the first director-general of the Tanganyika Broadcasting Corporation. Chalmers and others tried hard to separate broadcasting from government along the lines of the BBC model. But despite the establishment of public corporations in several British territories (Ghana, Nigeria, Malawi, Zambia, Uganda, and Tanganyika, and others had broadcasting corporations modeled on the BBC), the stations were all closely supervised by their respective governments and had little real independence.

The French developed a different policy. Whereas in British territories the emphasis was on broadcasting in African languages to reach the widest possible audiences, nearly all broadcasting in French territories was in the French language. Radio broadcasting was also centralized and, to a large extent, originated in France through the *Société de Radio-Diffusion de la France d'Outre-Mer* (Society for Radio Transmission to French Overseas Territories) or SORAFOM. As the society's title suggests, the prevailing philosophy was that the French territories in Africa were actually an extension of France. A series of relay stations across French Equatorial and West Africa carried the same programs. It was not until the French territories were granted independence in 1960 that separate national radio broadcasters were established in Mali, Senegal, Ivory Coast, Niger, Chad, Gabon, and other locations.

By the 1960s (the decade in which most African colonies gained their independence), all territories had radio broadcasting services. In every country they were instruments of government in much the same way as the national armed forces or the police. Broadcasters were civil servants—if not always in name, certainly in their relationships to the rest of the state apparatus. Without exception the new African governments maintained for 30 more years the monopoly of radio broadcasting established by colonial authorities. During this second phase of African broadcasting, which lasted until nearly the end of the 20th century, all 53 African countries had national broadcasting services, mostly dominated by radio. Broadcasting headquarters were generally in the capital or main city of each nation; from there, one or more national radio services were transmitted to reach the whole country. A few local and regional services were developed in Nigeria and South Africa but not many in other areas. Radio in Nigeria developed along different lines than in other African nations, reflecting that country's ethnic divisions and unique federal character. Two parallel state systems of state radio developed, often in direct competition with each other. The federal government had its own broadcasting system, and each of Nigeria's several states had its own system, as well.

Radio broadcasting in much of sub-Saharan Africa still relies heavily on shortwave (the main means of transmission for many years) to reach widely scattered populations over large areas. This is a feature of broadcasting in Africa not often seen elsewhere in the world. In Ghana, for example, all radio transmission until the 1980s was via shortwave. This means of transmission is in many respects ideal for African circumstances, although it can suffer from interference and is subject to fading and distortion. Lack of sufficient resources and infrastructure have meant that developing networks of FM or AM relays usually has not been possible, so the only way to reach an entire territory has been by shortwave. Outside of South Africa (where an FM network was quickly established in the 1960s) and the small island states, all African national broadcasters continued to use shortwave for their main national radio services at the beginning of the 21st century. So most radio receivers sold in Africa (except in South Africa) have shortwave bands on them, and virtually all radio owners outside of South Africa have ready access to international shortwave broadcasters such as the BBC, Voice of America (VOA), Radio France International, Radio Deutsche Welle, and Radio Netherlands. The South African international shortwave station, Channel Africa, is also very popular. Such international broadcasters have become popular for their African-language (Swahili, Hausa, Amharic, and Somali) transmissions and in the widely spoken languages of European origin (French, English, and Portuguese). Africa has the world's largest audiences for international shortwave radio broadcasts.

Shortwave coverage by Africa's national broadcasters is rather poor in many cases, and radio transmission remains underdeveloped on a national scale in many countries. The lack of financial resources, frequent breakdowns, power cuts, the scarcity of spare parts and other consequences of the general economic weakness in many African countries have weakened transmission capacity and performance.

Radio pluralism came late to Africa. Before 1987 there were only five or six privately owned radio stations on the entire continent—in Gambia, South Africa, Swaziland, and Liberia. In 1987 a trend to end state monopolies in almost every country began. In December, Horizon FM went on the air in Ouagadougou, the capital of Burkina Faso, launched by a local entrepreneur who announced rather defiantly that the station would have "lots of music, lots of commercials, lots of laughter, but absolutely no politics. People are tired of all that stuff." A week later the station was forced to close. The revolutionary government was clearly unhappy at this development, which it had not authorized. Horizon FM survived that early dispute and became one of several independent radio stations in Burkina Faso.

Competition

The arrival of Horizon FM was of historic significance in that it marked the beginning of the third phase of radio in Africa, one in which the national state radios continue but must compete for audiences with a growing number of independent radios. (The same trend is evident with television, which also was previously almost entirely a state monopoly.) There are important differences between state and independent radio on the continent. While the state radio services are mostly national in both reach and purpose, the new independent radio stations are mostly based in cities, and their coverage tends to be confined to the urban areas. They have also almost all been FM stations, whereas the national broadcasting stations have relied and continue to rely on a mix of transmission methods—FM, AM medium wave, and shortwave. At the turn of the century there were more than 450 independent radio stations in Africa. Most of them are the result of limited deregulation, which has invited applications for the limited coverage offered by FM. Only five or six independent radio stations existed on the entire continent 20 years earlier.

Independent radio stations in Africa can be categorized into five types. There are fully commercial stations that seek to make a profit from the sale of airtime for advertising or sponsored programs. Religious radio stations (most, but not all, Christian) use radio to communicate their faith and beliefs; some of these may carry some advertising, but most are financially supported by their sponsoring organizations and some with support from outside. The third category, comprised of community radio stations, is probably the fastest growing sector. There has been strong support in some countries for the development of very local, generally low-powered FM stations broadcasting in a community's indigenous languages or dialects. These are often staffed by volunteer helpers, are run at very low cost, and are supported by outside agencies (various non-government organizations have supported some for developmental reasons). By the year 2000 there were more than 70 community radio stations in South Africa and about 100 in West Africa, several in rural areas.

The fourth and fifth categories each emerged as the result of political and ethnic or other conflicts. Factional radio stations (some referred to as "clandestines") are used to promote a particular faction in a conflict. Somalia, a country without a government for the last decade of the 20th century, has several such stations, each supporting one of the warlords who control different parts of the country. There are similar clandestines in Sudan and Ethiopia. Some of these operate from neighboring countries rather than from within their nations of origin, for obvious reasons. Occasionally they may even broadcast from further afield. The factional radio category also includes the so-called hate radio stations. The most notorious of these was the *Radio des Mille Collines* (Radio of a Thousand Hills) in Rwanda. Broadcasting from within Rwanda (and almost certainly with the government's approval if not its backing), it was widely held to be responsible for promoting ethnic hatred and killings during the 1994 genocide.

The fifth category, humanitarian radio stations, came as a counter to the influence of factional radio. The power of radio in Africa has led various aid and relief agencies, including the United Nations, to support the establishment of humanitarian radio stations that promote peace, harmony, and democracy. Such radio stations have operated in Rwanda, Somalia, Liberia, the Democratic Republic of the Congo and the Central African Republic. One organization that has been prominent in supporting humanitarian radio has been the Fondation

Hirondelle (Swallow Foundation) in Switzerland. It has backed radio stations in areas of conflict for limited periods in countries such as Liberia, the Central African Republic, Somalia, and Rwanda.

Programs

Radio programming in Africa has suffered from the economic realities present there. It has been hampered to an even greater degree by the often heavy hand of government. Many broadcasters in state radio stations are government or civil servants, and the civil service does not lend itself to creativity, imagination, and entertainment. Much of the output of state radio stations has been dominated by government propaganda. In the early days of national independence there was a heavy emphasis on messages about nation building, with exhortations to hard work and discipline. Much of this was rather boring. A high proportion of news bulletins on radio featured heads of state visiting projects or speaking at events. Broadcasts would usually focus on what was said and who was there—sometimes remarkably uninteresting speeches would be carried in full on the radio. When the head of state visited several different projects and said more or less the same things at each appearance, subsequent broadcasts would repeat the same details. Other stories were similar—ministers and other state officials making speeches or announcements, visiting state institutions, opening schools or hospitals, and so on. Each event would be reported with the main focus on what the official said and little on any other aspect of the story.

But it would be wrong to categorize all African state radio in this way. Much of it has been entertaining and even innovative. Ad-lib or unscripted drama has often flourished, especially in Ghana, Nigeria, and Zambia; poetry and storytelling have become popular features in many countries. Local music is now a major part of the programming in many states with an emphasis on local talent in such countries as Mali, Senegal, Ghana, the two Congos, and Tanzania. Many of Africa's very successful popular music stars began their careers on radio. There have been radio stars since the early days, as the media thrive on more than mere news. Most of the time the media, radio especially, are used for entertainment. Although a cautious approach generally has been seen in news and news-related programs, this is not always the case in other creative areas.

African radio stations have been important patrons of music and, in some countries, of poetry and oral literature. In the 1970s many radio stations made regular program collection safaris into remote parts of the country to record songs, drama, poetry, and other indigenous material for later broadcast. However, in recent years these activities have been curbed by financial restrictions. Similarly, the studios of many national radio stations were once a focus for much new music, but this happens less now, largely because many state-run radio stations have stopped most payments to artists.

African radio once played a major role in popular music and still does play a role, but mainly by playing commercial records. Many African musicians find that they do better financially by marketing their own cassettes through street sellers. But copyright laws are not widely used, and few African artists are members of rights societies. Financial pressures have also slowed the growth of (and sometimes even reduced the amount of) original indigenous drama and other spoken word programs on both radio and television. The economic weakness of many states has meant that talented artists had to stop working in state radio because they were not paid adequately (sometimes not at all). The growth of successful commercial radio may change this.

Private Stations

More freedom has generally been given to the printed press in Africa than to the radio industry. Independent newspapers have been permitted to operate in most African countries, and many of them have been permitted some degree of freedom to criticize, oppose, and challenge the existing political order. The same has not been true of radio. Many African governments have been slow and reluctant to change laws and allow private broadcasting stations. Those that have legislated for independent radio have in many cases imposed restrictions on the degree to which independent stations can report news.

The reluctance to allow private radio arises in part from fear of the power of the medium. It is known that radio reaches many more people in Africa than any other single medium. Government officials may be legitimately concerned about misuse of the medium by rival political, religious, or ethnic factions, particularly when they have a shaky hold on power or rule in countries lacking in infrastructure, with weak institutions of control, and where there may be several regional, ethnic, and linguistic divisions. It is significant that there has been much greater reluctance to grant freedom to radio than to other media.

In Ethiopia, new laws to permit private broadcasters were delayed by fears that private electronic media would be critical of the government, as the private press has been. In Kenya several applications to run private radio stations were delayed for several months in 1998 and 1999, probably due to similar fears. But a major press group in the country opened its first commercial radio in Nairobi in 1999, the first example outside South Africa of major commercial press involvement in radio in Africa. In Tanzania gentle pressure has been put on the private radio stations to carry national news from the state radio station; in Zambia, the few licensed independent radio stations are not permitted to make their own news bulletins. Even after several years under new laws permitting independent broadcasts, there were still only one private commercial radio station and three private religious stations at the beginning of 2001, although the election of a new president at the end of that year led to change during 2002 and the emergence of several new independent broadcasters.

The development of independent radio should not be seen only in political terms, however. Its commercial and cultural impact and function are almost certainly of equal and perhaps greater significance. Music has always played a major part in African radio, but when the stations were almost entirely owned and paid for by the state, entertainment often took second place to other requirements. On many occasions radio schedules would be cleared for major political events. Speeches of political leaders and commentaries on national events would be given extensive coverage, with state political and administrative requirements taking precedence. With deregulation and the licensing of independent and particularly commercial stations, listeners are no longer compelled to listen to long and often tedious political broadcasts. At the same time, radio has become a much more attractive medium for advertisers, who can develop media campaigns in line with different stations' formats and content. The new and often very successful commercial stations have adopted musical policies that define their places in the market, just as their counterparts are free to do in Europe, the United States, and other parts of the world where the industry is not controlled by government. It is significant that in the African states once ruled by France, one of the most successful of the new commercial radio stations, *Radio Nostalgie*, is affiliated with a major French radio group of the same name. In 1999 it was reaching about 60 percent of all adults in Dakar, the capital of Senegal, and Abidjan, the capital of Ivory Coast.

In Ghana, the new commercial stations have been so successful that they have pushed the government's Ghana Broadcasting Corporation out of its place as one of the top eight stations in the country. Private, independent, commercial radio stations in Uganda and Nigeria, mostly broadcasting popular music, outstrip the state radio services in audience reach and share.

The reason for the success of many private stations is easy to understand. Competition from other domestic broadcasters was entirely absent in the broadcast media until their arrival, so program producers had never worried about attracting audiences or advertisers. Now this has changed in many countries (although not yet all), and there is lively competition for audiences. On the whole, radio in these countries has become livelier and more attractive. However, there has been a downside also. Whereas state radio put a strong emphasis on education and development, featuring many programs that promoted better health or provided other forms of education and improvement for the general population, competition for commercial revenue tends to push these programs out or to marginalize them. Moreover, national and state radios have broadcast programs in local minority languages for many years, which is not usually a feature of commercial stations. Community stations may increase their use of local languages and dialects in response to this shift, however.

Radio's contribution to national education and development will probably continue to be of major importance. Many developmental agencies strongly favor the use of radio in campaigns for better health, as in the campaign against AIDS and such diseases as trachoma, malaria, tuberculosis, polio, and leprosy. In the past, many broadcasts of this type were worthy but very dull. In recent years, however, there has been a welcome growth in the imaginative and entertaining use of radio to encourage development in such areas. One of the best examples is the soap opera *Twende na Wakati* (Let Us Go with the Times) in Tanzania. This regular drama features the daily lives of ordinary people, and within its entertaining story line are messages about family planning, infant nutrition, other health issues, and the changing role of women. A regular program in Senegal, *Radio Gune Yi*, made entirely by and for children, promotes the rights of children and the equal rights of girls and boys.

Technology

Radio has played a major nation-building role in Africa. This arose from an interesting and very important historical coincidence. The invention (in 1948) and commercial development of the transistor (in the 1950s and 1960s) led to very large numbers of cheap battery-operated transistor radios coming into Africa at the same time that about 40 nation-states gained their independence in the 1960s. The transistor made radios portable and cheap, liberating them from reliance on a supply of electricity, which most African homes did not have at the time. Radio rapidly became the most widespread medium in Africa, and this had important consequences for Africa's cultural and political life. It was the medium by which many, if not most, Africans gained day-to-day knowledge of their new national and international status.

At the beginning of the 21st century, new technology has arrived in the form of direct broadcasting by satellite. WorldSpace, a company based in Washington, D.C. and headed by an Ethiopian, Noah Samara, launched the first digital radio service by satellite in 1999. The technology makes very good sense in a continent where the establishment of FM relays has been so difficult due to the costs involved and problems with maintenance and security. The WorldSpace service provides several high quality radio services that can be picked up with ease and clarity anywhere on the continent. The service is being used by some African and international radio stations, and it also offers some broadcasts of its own. The main question about the satellite service is whether it will establish itself sufficiently to be commercially viable in the long term. Use of the system requires the purchase of special receivers that are currently too expensive for most African listeners. It is, however, an example of a new technology that seems to meet an African need. (It is also notable as one of the first technologies ever introduced in Africa before it became available to the rest of the world.)

Another new technology that may overtake WorldSpace is digital shortwave. The major international radio broadcasters (the BBC, VOA, Radio Deutsche Welle, and others) have joined together in a consortium, Digital Radio Mondiale, and have successfully developed a new means of shortwave transmission that employs digital coding, which vastly improves reception. If African radio stations take up this new technology (and already many are showing an active interest), it will revolutionize transmission in Africa, making high quality reception available throughout the continent.

GRAHAM MYTTON

See also Africa No. 1; Arab World Radio; Developing Nations

Further Reading

Ainslie, Rosalynde, *The Press in Africa: Communications Past and Present,* London: Gollancz, 1966; New York: Walker, 1967

Bourgault, Louise Manon, *Mass Media in Sub-Saharan Africa,* Bloomington: Indiana University Press, 1995

De Beer, A.S., editor, *Mass Media for the Nineties: The South African Handbook of Mass Communication,* Pretoria, South Africa: Van Schaik, 1993; 2nd edition, as *Mass Media, towards the Millennium: The South African Handbook of Mass Communication,* 1998

Fardon, Richard, and Graham Furniss, editors, *African Broadcast Cultures: Radio in Transition,* Oxford: Currey, and Westport, Connecticut: Praeger, 2000

Fraenkel, Peter, *Wayaleshi,* London: Weidenfeld and Nicholson, 1959

Head, Sydney, editor, *Broadcasting in Africa: A Continental Survey of Radio and Television,* Philadelphia, Pennsylvania: Temple University Press, 1974

Mytton, Graham, *Mass Communication in Africa,* London: Arnold, 1983

Tudesq, André Jean, *La radio en Afrique noire,* Paris: Editions Pédone, 1983

Unesco, *World Communications, Press, Radio, Film: Report,* Paris: Unesco, 1950; 5th edition, as *World Communications: A 200-Country Survey of Press, Radio, Television, and Film,* Paris: Unesco, Epping, Essex: Gower Press, and New York: Unipub, 1975

Unesco, *Latest Statistics on Radio and Television Broadcasting,* Paris: Unesco, 1987

Ziegler, Dhyana, and Molefi K. Asante, *Thunder and Silence: The Mass Media in Africa,* Trenton, New Jersey: Africa World Press, 1992

Africa No. 1

Africa's First Transnational Commercial Radio Station

Africa No. 1 is a private, commercial, French language radio station based in Libreville, Gabon, in Central Africa. When it was launched in 1981, Africa No. 1 was the first pan-African commercial radio station. Its main audience is in a swath of territory stretching from Congo in Central Africa to Senegal in the West. This area covers the territory of the former colonial French Equatorial Africa and today contains most of the French-speaking countries of the continent. However, with its powerful shortwave transmitters and satellite broadcasts, Africa No. 1 covers Africa, North America, the Middle East, and South America. The majority shareholders (60 percent) of Africa No. 1 are Africans, and the rest of the capital is split between a French government investment group, *Société de Financement de Radiodiffusion* (SOFIRAD), and two French companies, Havas and Sofrea.

Africa No. 1 and the Realities of Broadcasting in Africa in the 1980s

When it started service, Africa No. 1 was a breath of fresh French air over the African continent for several reasons: it was not directly controlled by any government; it offered African listeners new, exciting music and cultural programming; and it broadcast uncensored, in-depth news and analyses of African and world issues. This was a dramatic departure from the boring, doctrinal, and paternalistic programming that was broadcast on tightly controlled government radio and television stations in Africa. In effect, Africa No. 1 was the first African radio station to break with the tenets of African-style development communication—a series of African government policies that required all organs of the mass media, which were for the most part controlled by the governments, to disseminate information and messages that would help improve agricultural production, health, education, national security, and other vital areas.

When African countries gained independence in the late 1950s and early 1960s, their mass media policies originated from Western countries in general and from the United States in particular. As early as 1958, when most sub-Saharan African countries were still under colonial rule, the General Assembly of the United Nations called for the building of mass media facilities in countries that were in the process of economic and social development. This led to the United Nations' application, through its Educational Scientific and Cultural Organization (UNESCO), of the concept of development communication.

As formulated in its early stages by Western experts, the development communication perspective postulated that developing countries did not have the resources to indulge in the luxury of the liberal, watchdog journalistic model of the Western countries. Therefore, the mass media were to concentrate on the task of disseminating information and messages that would help improve agricultural production, health, education, and national security. It was believed that strategic use of information would lead to nation building and provide a "climate" for national development. Most African politicians argued that in situations of poverty, where the bare necessities were absent, having mass media that concentrated on checking government action and criticizing it was a misuse of resources.

However, the consequence of this policy was that most governments, often one-party regimes, soon had a monopoly on information dissemination. In the name of development communication, the mass media reported only news that promoted the ideological positions and interests of the governments and ruling elite of the day. Dissenting journalists were either censored, censured, or worse. The arrival of Africa No. 1 on the broadcast scene provided an alternative voice to that of the government.

Programming on Africa No. 1

For a commercial station, Africa No. 1 has a unique program format that is a mix of music, sports, and cultural documentaries. The service broke the stranglehold of African governments on news and information. The station programs a lot of African popular music, to be sure, but it also broadcasts in-depth news and information that is not heard on government radio stations. It also airs the views of dissident politicians and journalists and tells the world about the persecution of journalists and members of the opposition in African countries. Furthermore, its programming actively promotes African music, culture, and sports. The station plays mostly African music and has released compact disc compilations of the most influential music in post-independence Africa. It has also carried live broadcasts of Africa's major sporting events. Its well-researched documentaries cover African political, cultural, and historical topics.

Africa No. 1 and Liberalization of the African Airwaves

The fall of the Berlin Wall in 1989 and the end of the Cold War triggered a hurricane of instability over the African continent. Popular uprisings, strikes, sit-ins, and demands for political freedom and better standards of living broke out all over the map. In response, leader after leader, seeing that their superpower benefactors were no longer willing or able to provide unconditional support, legalized opposition parties and allowed the press more freedom. Portable transistor radios, satellites, and popular music had neutralized the power, if not the zeal, of the censor. The much-muzzled African mass media suddenly found their voices.

One of the consequences of political liberalization was the opening up of the airwaves. Press laws have been liberalized across the continent, and private radio stations broadcasting in several African and European languages are springing up in most regions of the continent. By 2000, no fewer than 21 African countries had allowed independent or alternative radio or television broadcasting facilities. This trend is quite a change from just five years earlier. Divergent voices are being heard, and many more people have access to the mass media.

All this has meant competition for Africa No. 1 from Africa and abroad. In addition to competing with African commercial stations, Africa No. 1 is experiencing increasing competition from international broadcasters such as the British Broadcasting Corporation (BBC), the Voice of America, Radio France International (RFI), Radio Deutsche Welle (Voice of Germany), and others broadcasting in English, French, Arabic, Portuguese, and several African languages. Since the liberalization of the media began in Africa in the early 1990s, several of these international broadcasters, who used to broadcast only on shortwave radio to Africa, have been allowed by a number of countries to broadcast directly to African audiences on FM frequencies. In addition, many of them have expanded their African services to include direct broadcasting by satellites and have signed rebroadcast agreements with several private African radio stations. The African audience is now splintered.

In the face of such stiff competition, Africa No. 1 has repositioned itself as the only station that is African 24 hours a day. It has also diversified its broadcast outlets and acquired FM frequencies in at least 15 African countries. Furthermore, it has obtained broadcast licenses in France, where it broadcasts on two FM frequencies from Paris. Nevertheless, Africa No. 1's expansion into the African FM market has not been all that smooth. According to the Paris-based African magazine, *Jeune Afrique,* Africa No. 1's expansion into the competitive but lucrative FM market in Abidjan was held up by the Ivorian government and by local shareholders who wanted to control the programming of the station. A compromise was reached whereby the station was not allowed to cover the local news in Abidjan. The news is broadcast from Libreville, Gabon, Africa No. 1's home station. In effect, all local news from Abidjan has to be sent to Libreville, from where it is broadcast to Abidjan and the rest of the African continent.

Besides its shortwave and FM frequencies, Africa No. 1 broadcasts on four direct broadcast satellite channels and on the internet. In addition to its broadcast activities, Africa No. 1 also serves as a rebroadcaster for such international broadcasters as RFI, the Japanese Broadcasting Corporation, and Radio Swiss International. These broadcasters send their programs to Africa No. 1's transmission center in Moyabi, Gabon, where it is rebroadcast to the rest of the continent, to South America, or to the Middle East.

Controversy

Africa No. 1 has not been without controversy. Controversies have arisen over tobacco advertising on the station and over alleged political interference in its programming. As a commercial station, Africa No. 1 survives on advertising, mostly from multinational companies doing or wishing to do business in Africa. The station made a controversial decision to accept tobacco advertising. Even as the dangers of smoking became known and Western countries banned tobacco advertising on radio and television, multinational tobacco companies were buying airtime on Africa No. 1. The station soon became known as much for the slick Marlboro and Benson and Hedges cigarette advertisements punctuating its programs as for its music and documentaries.

Despite its generally good track record, Africa No. 1, like government stations in Zaire and elsewhere, did not escape the violence that came with political liberalization in Africa in the 1990s. Africa No. 1's studios in Libreville, Gabon, were attacked by mobs who accused the station of covering up the misdeeds and corruption of the government of Gabon. Some members of the African public claimed that Africa No. 1 was not as critical of the government of Gabon as it was of other governments, because president Omar Bongo was a shareholder of the station. Because the station is private, that claim cannot be confirmed. Nevertheless, it is true that Africa No. 1 has not given as much attention to Gabonese dissidents and members of the opposition as it has given to dissidents and members of the opposition of other African countries. In addition, dissident groups from Congo (Brazzaville) claimed, in 1999, that Africa No. 1 had given in to political pressure and fired three journalists of Congolese origin who had reported on ethnic massacres in Congo's civil war.

Conclusion

Africa No. 1 is a pioneering radio station that has, in its 20 years of existence, greatly influenced Africa's political and media landscape. It has been an effective promoter of freedom of speech and expression. Its news and programming broke the monopoly of government-controlled stations over information in French-speaking Africa. Today, Africa No. 1 is the model upon which many new, independent African stations pattern their programming.

LYOMBE EKO

See also Africa; Developing Nations

Further Reading

Altbach, Philip G., and Salah M. Hassan, editors, *The Muse of Modernity: Essays on Culture and Development in Africa,* Trenton, New Jersey: Africa World Press, 1996

Eribo, Festus, and William Jong-Ebot, editors, *Press Freedom and Communication in Africa,* Trenton, New Jersey: Africa World Press, 1997

Maja-Pearce, Adewale, editor, *Directory of African Media,* Brussels: International Federation of Journalists, 1996

Monga, Célestin, *Anthropologie de la colère: Société civile et démocratie en Afrique noire,* Paris: L'Harmattan, 1994; as *The Anthropology of Anger: Civil Society and Democracy in Africa,* translated by Monga and Linda L. Fleck, Boulder, Colorado: Rienner, 1996

Ziegler, Dhyana, and Molefi Kete Asante, *Thunder and Silence: The Mass Media in Africa,* Trenton, New Jersey: Africa World Press, 1992

African-Americans in Radio

African-Americans have played an important part in American radio broadcasting from the beginnings of the medium. From early experimenters to pioneer radio performers, blacks contributed to the cultural, economic, and technical development of radio broadcasting. Though shackled by discrimination, blacks enthusiastically used their talents on radio during World War II to help America display a united domestic front. The half-century since has seen African-Americans help to change the face of American culture through radio and other media.

Origins

One of the most notable of the pioneering electricity experts was Lewis H. Latimer (1848–1928), son of an enslaved African escapee, who prepared Alexander Graham Bell's telephone patent drawings that afterwards assisted the Boston speech and hearing therapist to actually invent his device. Telephone parts were later used by others to demonstrate the wireless transmission of human speech. Talladega College, a black institution known for its solid science curriculum, provided the educational foundation that inspired Lee de Forest to become an inventor who contributed much to perfecting modern radio broadcasting.

Before World War I, some African-Americans enrolled in those YMCA radio classes that were then available to them in a number of cities, and they gained even greater access to radio technology during the war as part of the war effort. Howard University in Washington, D.C., for example, offered electrical and radio technology classes under contract with the U.S. military.

A number of African-Americans conducted radio experiments after the war by establishing amateur radio-training organizations. Members could learn how to build and repair radio transmitters and receiver sets, how to send Morse code, and how to obtain an amateur operator's license. Miles Hardy established his Pioneer Radio Society in 1921 in New York City. A year later Roland Carrington founded the Banneker Radio Club in Baltimore. One of the most active areas for black ham operators was the midwest. Operators in Ohio, Michigan, Indiana, and Kentucky can be traced to Everett Renfroe who passed his ham operator's examination in Chicago in the early 1920s. (The operators formally initiated an organization, OMIK, in the early 1950s to protect members against racism when they traveled throughout the nation.)

Early Broadcasting

Many African-American musicians took advantage of opportunities to appear on early radio broadcasts. There is evidence that the "father of the blues," W.C. Handy, performed on the Memphis ham radio station of a white amateur, Victor Laughter, as early as 1914. The "Fisk Jubilee Singers" of Fisk University and the "Hampton Singers" of Hampton College are known to have performed live on radio in the early 1920s to raise funds for their financially strapped schools. Morehouse College was repeatedly featured on radio in Atlanta. The common bond among African-American performers was a desire to display their abilities in a manner that proved black people were equal to others in society.

As American radio developed in the 1920s, the contribution of African-American musicians also increased across the country. Fletcher Henderson's music was broadcast from New Orleans in 1921. The Plantation Club in Los Angeles broadcast Kid Ory's Sunshine Orchestra in 1922. Further up the West Coast, vaudevillian George Dewey Washington made an appearance on Seattle's KFC. Clarence Jones and His Wonder Orchestra were guests on KYW in Chicago. The Symphonium Serenaders entertained KDKA listeners from Pittsburgh. On New York's WJZ, the Melrose Quartet was featured regularly, and Clarence Williams accompanied a variety of black artists for the station. Eubie Blake and Noble Sissle showcased their cast from "Shuffle Along" on Boston's WNAC during the *Boston Radio Exposition* of 1922; a year later, they were on the air at KYW in Chicago. "Shuffle Along" later became so popular on Broadway that Sissle and Noble were able to book cast appearances on New York's WJZ, WEAF, and WHN. For its opening ceremonies, WBBM in Chicago included Jimmie Wade's Moulin Rouge Orchestra. WSBC in Chicago featured remote broadcasts by Frankie "Half Pint" Jaxon, and Hughie Swift's band was heard nightly in Chicago on the same station.

Duke Ellington first performed on radio over New York's WDT and then WHN in 1923. Bessie Smith, "Empress of the Blues," was heard performing live throughout the southwest in 1923 from Memphis station WMC. In Nashville, Deford Bailey, the legendary harmonica player, was featured on WSM. He appeared regularly on radio from 1926 to 1941 and helped establish the great musical tradition of the *Grand Ole Opry*.

New York City was a beehive of African-American radio experimentation during this period. Fletcher Henderson's band, performing at the Roseland, was broadcast remotely on a regular basis. Shows from the Plantation Club in New York, another jazz center for black bands, were broadcast five nights a week. Club Alabam contracted with several different stations to remotely broadcast 47 of its shows. Remote broadcasts from the Savoy Ballroom increased to eight per week. WHN carried a program featuring the great Florence Mills in celebration of her joining the Greenwich Village Follies. A blues marathon show was programmed on WDT. Other New York stations brought in such top entertainers as Antoinette Gaines,

LeRoy Smith, Sam Wooding, Revella Hughes, and Eva "The Dixie Nightingale" Taylor. From St. George's Episcopal Church, vocalist Harry T. Burleigh was heard on WJZ and proclaimed "the leading creative genius of the Negro race."

Local Radio in the 1930s

What is known of African-Americans in and on local radio is still evolving. The paucity of information may be due to the fact that much historical attention has focused on network radio. Some researchers have found, however, that a few Northern stations had begun to observe what was then called "Negro history week," inviting guest speakers to discuss black achievements. Certain important black newsmakers occasionally spoke on a public affairs show, as labor leader A. Phillip Randolph did in 1931. By the mid-1930s, local stations in Baltimore and Philadelphia had at least one weekly program aimed at a black audience. The Museum of the City of San Francisco has discovered that Henry Starr, an African-American, was the leading pianist on KFRC's *Edna Fischer Show,* a variety program in that city during the late 1920s and early 1930s.

The Depression may have created opportunities for African-Americans in many markets during the early 1930s: African-American music was often cheaper than white music for local stations to broadcast because the licensing agency ASCAP had signed few contracts with African-American publishers and writers. A station could thus play recordings by African-Americans and not incur ASCAP fees. Although as early as 1930 there were efforts by black businessmen to purchase a radio station, such efforts would not succeed until 1949.

Strong and continuous local programs by African-Americans began with those of Jack L. Cooper in Chicago during the 1930s. Many scholars credit Cooper as being the patriarch of black radio. His newsboy experience in Cincinnati and later his professional song and dance work on stage may have helped turn his career toward communication: he became an entertainment writer for the *Chicago Defender.* While on assignment in Washington in 1925, Cooper began writing, producing, and starring in his own black vaudeville show on WCAP. Washington's racial laws, however, soon forced Cooper's return to Chicago. There, station WSBC operated successfully by brokering time to various immigrant groups and was receptive to Cooper's desire to produce and air *The All Negro Hour* in 1929. One of Cooper's live broadcasts was threatened in 1932 when the key performer abruptly quit over a pay dispute. Cooper creatively set up a phonograph, placed a microphone in front of it, and played recorded music to maintain the show's broadcast schedule and continuity thereafter. Inadvertently, he had become the first African-American disc jockey. By 1938 Cooper was brokering up to 20 hours of time on WSBC and programming church services on Sundays. Eventually he bought time on several other Chicago stations, replicated his record show, and produced news and public affairs programs that utilized his journalism skills. He also launched the first black advertising agency and radio production company.

Chicago became a focal point for broadcasting recorded black music—a mix of jazz, blues, spirituals, and hymns—when in the early 1940s Al Benson bought time on WGES and complemented Cooper's "time blocks" with his own set of programs. Cooper and Benson each organized training programs and taught young aspirants about radio. They bought and sold time, conducted market research, and wrote and produced advertisements.

The "time block" purchase method and the recorded music program also appear to have been popular among local African-American entrepreneurs in other parts of the country in the late 1930s and early 1940s. In Seattle, Bass Harris appeared on KING; in Washington, D.C., Hal Jackson was on WOOK; in Detroit, Ed Baker was featured on WJLB, while Van Douglas appeared on WJBK; in Hammond, Indiana, Eddie Honesty was on WJOB; and in New York, Joe Bostic appeared on both WCNN and WMAC. As with Cooper and Benson, most early African-American radio personalities were college educated, trained by veteran professionals, or they were experienced entrepreneurs who understood the radio business.

Network Radio and Minstrelsy

With programs such as the hugely popular *Amos 'n' Andy* featuring white performers playing black roles, one must look at the stereotypic roles played by early black performers to understand the kinds of jobs generally available to African-American actors on network radio during the 1930s. For example, Ernest Whitman was employed as Awful for *The Gibson Family.* The role of Gardenia, a humorous character, was played by Georgia Burke on the *Betty and Bob* soap opera. Even Academy Award winner Hattie McDaniel (*Gone With The Wind*) was hired to portray a mammy on the network radio series *Showboat.* McDaniel also played a more endearing role as the lead on the network series *Beulah*—but only after the role had originally been played by a white man.

Positive portrayals of blacks in network drama were rare but not absent entirely. In 1933 Juano Hernandez starred in the CBS series *John Henry, Black River Boat Giant.* Rose McClendon, Dorothy Caul, and Jack McDowell also were cast members. John Henry was portrayed as a powerful but bad ladies' man. More frequently, African-Americans were cast in stereotypical roles on network vaudeville programs including the *Eddie Cantor Show, Saturday Night Sewing Club,* and the *Rudy Vallee Show.* Cantor also hired a black female vocalist,

Thelma Carpenter, for a regular spot on his show. "Rochester," on *The Jack Benny Program,* was another black stereotype, but much of the humor came from Rochester's "bettering" his white boss.

Performers such as the Golden Gate Quartet, Southernaires, Wings Over Jordan, CBS Trumpeteers, Ethel Waters, Andy Razaf, and Paul Robeson were able to showcase their professional skills nationally at one time or another. The Mills Brothers were especially popular on radio, Duke Ellington had his own network radio show by 1936, and Marion Anderson is legendary if only for her 1939 appearance at the Lincoln Memorial in Washington, which was broadcast nationally.

World War II

With its entry into the war, however, the United States was forced to begin confronting its "Jim Crow" treatment of black people, including the paucity and negative portrayal of African-Americans on radio: the country badly needed racial tranquility at home and among its military servicemen. To help accomplish its objective, the government produced or supported network radio programs that projected African-Americans in a positive light. *Men o' War* was an all-black patriotic musical program series featuring naval personnel; it was broadcast regularly for two years over the CBS radio network. *Freedom's People*, an eight-part program on NBC, highlighted African-American achievements and featured Count Basie, Cab Callaway, George Washington Carver, W.C. Handy, Joe Louis, Jessie Owens, and other outstanding African-Americans. African-Americans portraying positive characters were also written into such network soap operas as *Our Gal Sunday* and *The Romance of Helen Trent.*

A special radio documentary, "Open Letter On Race Hatred," was broadcast on CBS in response to the 1943 Detroit, Michigan, race riot that left 35 dead. In addition, a series of discussion and public affairs programs that addressed black issues and featured black leaders aired nationally on radio networks. Ann Tanneyhill produced a show for CBS in 1943 about black women, called *Heroines in Bronze.* NBC aired programs such as *America's Town Meeting* (which, although it discussed racial issues, did so with white speakers), *The Army Hour, Too Long America,* and others. CBS broadcast *People's Platform, The Negro in the War,* and *They Call Me, Joe,* among others.

Many programs carried by the Armed Forces Radio Service (AFRS) encouraged good race relations among troops. Many programs featured black announcers, recorded music, and live bands. One AFRS station, located at the Blytheville Army Air Corps Base in Arkansas, programmed black local news, events, and recorded music in both the early morning and mid-afternoon that was listened to faithfully by African-American civilians throughout Mississippi County.

Postwar African-American Radio

After the war, African-American men and women began performing live on radio throughout America. Early Lee Wright became the South's first recognized black announcer in 1947 on WROX in Clarksdale, Mississippi. The Delta Rhythm Boys, a local gospel group from the same town, could be heard on records. The famous *King Biscuit Time,* featuring blues men Sonny Boy Williamson and Robert Lockwood, was broadcast on KFFA in Helena, Arkansas. WLAC in Nashville had begun to play a mix of black-oriented music on a nightly basis, even though they still used white announcers.

Magazine reports about the success of African-American disc jockeys began to persuade some white owners of unsuccessful radio stations to begin full-time programming of rhythm and blues music. WDIA in Memphis, Tennessee, was the first such white station to do so. It was the second new station (after KWAM in 1946), to go on the air in Memphis after the war and found survival in the new competitive market very difficult. John Pepper, one of the WDIA owners, was ready to quit. In 1948 the station's general manager and co-owner, Bert Ferguson, read a magazine article about the economic success of programming to African-Americans. Probably he was reading about Jack L. Cooper or Al Benson in Chicago. Shortly afterwards Ferguson hired Nat D. Williams to create a block of black-oriented programming in the afternoon. Williams' afternoon program soon brought financial success to WDIA and enabled Ferguson to add another time block that featured Maurice "Hot Rod" Hulbert, then A.C. "Mooha" Williams and others until WDIA's entire schedule was completely filled with various black blocks of rhythm-and-blues and gospel music programming. Fulltime black-oriented radio was born.

By the early 1950s, there were reportedly more than 500 blacks working in radio throughout the nation, playing mostly rhythm-and-blues, and working part-time on stations that otherwise provided programming for white listeners. This sudden expansion in the number of black announcers had been driven by the popularity of rhythm and blues music. The sounds conveyed joy and hope in a language that reflected the postwar vision of freedom. The men and women who played the music on radio shared the same hope for the future and spoke the same rhythmic language. The rhyme-language style popular among blacks was used in a variety of situations such as "signifying" contests or when compliments were paid on clothing styles. Al Benson used it on radio, but Jack L. Cooper did not. It was used for laudatory salutes among MCs on the black entertainment "chittlin' circuit" when introducing performers and working the audience. Some scholars trace the rhyme's origin back to Africa. It certainly found its way into the lyrics of Louis Jordan's "Beware" recorded in the mid-1940s. Nat D. Williams was associated with rhyme on Beale Street and later on WDIA. Maurice "Hot Rod" Hulbert, entertainer turned

Memphis disc jockey, moved to Baltimore radio and popularized the style. Doug "Jocko" Henderson, a Hulbert admirer, put his signature on the style in Philadelphia and syndicated his *Rocket Ship Show* to five other East Coast markets. Radio personalities such as Jack Gibson and Daddy-O Daily developed rhyme styles mutually exclusive of the Memphis linkage.

African-Americans solidified their positive presence in radio through perseverance and by promoting rhythm-and-blues music; their audience expanded because of improved education and growing wealth among African-Americans. Most major U.S. cities operated a full-time rhythm-and-blues station by the mid-1950s. As black people traveled or communicated, they spread stories about their favorite local disc jockey, among them "Frantic" Ernie, Jack "The Rapper" Gibson, Joe "Joltin Joe" Howard, "The Magnificent" Montague, "Honey Boy" Thomas, "Lucky" Cordell, Sid "The Real" McCoy, Martha Jean "The Queen," or simply, Georgie Woods.

African-American performers and rhythm-and-blues music were adopted by white people and by broadcast institutions at an increasing rate. Among those who championed the new sound were Alan Freed in Cleveland, Dick Clark in Philadelphia, George Lorenz in Buffalo, and Robert "Wolfman Jack" Smith near Del Rio, Texas. Collectively, recorded rhythm-and-blues performances and the accolades awarded black disc jockeys had made their impression. Imitation by white broadcasters was a respectful cultural compliment.

Station Ownership and Activism

Andrew "Skip" Carter bought KPRS of Kansas City, Missouri, a defunct operation, in early 1949: he became the first African-American to own a commercial radio license. Later that same year, J.B. Blayton purchased WERD in Atlanta. Dr. Haley Bell in Detroit was the first black person to construct a new radio station—WCHB first aired in 1956.

African-American disc jockeys during the 1950s found it necessary to organize professionally in order to help each other improve salaries, working conditions, equal rights, and employment. They formed the National Jazz, Rhythm-and-Blues Disc Jockey Association and attracted national attention in 1956 when members met to defend rhythm-and-blues music against its critics—those who believed that because rhythm-and-blues had black origins it was dangerous for white people to listen to. The group later changed its name to the National Association of Radio and Television Announcers (NATRA).

In the mid-1960s, managing and owning stations moved to the top of the black broadcasters' agenda. The entire black staff of disc jockeys at WVOL in Nashville went on strike in 1964 and helped head salesman, Noble Blackwell, move into the vacated general manager's job. In Chicago, Lucky Cordell was appointed manager at WVON. These moves were seen as milestone achievements within the larger civil rights movement.

Dr. Martin Luther King addressed the 1967 NATRA convention and publicly thanked its members for their valuable support of the civil rights movement. He praised the contributions of African-American announcers in general for using radio to popularize rhythm-and-blues internationally. Commentators have since praised the role played by African-American broadcasters in calming fears in the midst of the urban rebellions in the 1960s. Yet many emphasized the need for radio to open its doors even wider. African-Americans in the mid-1960s still held perhaps only one percent of the 60,000 jobs in commercial radio, and only five of 5,500 licensed radio stations were black-owned.

Pressure from activist groups and changes in federal policy brought sweeping change in the 1970s. NATRA declined as a result of growing conflict within and outside its own ranks over its priorities. Yet the organization's agenda, to promote ownership of radio stations by blacks, advanced by Del Shields and others, had won support from FCC Commissioner Nicholas Johnson and was also supported by other organizations such as Black Efforts in Soul Television, led by Bill Wright, and subsequently The National Black Media Coalition under Pluria Marshall. The predominately white National Association of Broadcasters along with the National Congressional Black Caucus also lent their efforts to addressing the ownership problem. The first black FCC Commissioner, Benjamin Hooks, was appointed by President Richard Nixon. Overall, this activism led to a modest increase in station ownership and the establishment of The National Association of Black Owned Broadcasters.

Dr. Haley Bell's acquisition of a second Detroit license in the 1960s (to operate WCHD-FM) made him the first African-American to own more than one station. James Brown's two Augusta, Georgia, stations and Percy Sutton's Inner Broadcasting, Inc., based in New York, soon followed. Dorothy Brunson's acquisition of Brown's stations made her the first African-American woman to become a station owner. Cathy Hughes of Radio, Inc. emerged as a legendary figure in radio-station financing, acquisitions, and operations. Hughes had honed her skills in advertising and sales by serving as a radio volunteer in Omaha. In 1971 she began to teach advertising at Howard University, and was soon managing WHUR-FM, owned by the university. By 1980 she had purchased Washington's WOL and added another station to her list in 1987. Deregulation helped Hughes: in the early 1990s she founded Radio One, Inc., and by 2001 her company held 48 radio station licenses and was the main reason African-Americans now owned more than 200 stations. But in many markets deregulation hurt the black goal of owning more stations. The 1996 Telecommunications Act allowed unlimited groups of radio stations (as well as multiple stations in any single market) to be owned by a single entity: this relaxation of former rules had the effect of pushing up station prices. Often unable to raise the inflated purchase prices—and with existing black-owned stations bought out by the owners of the growing radio chains—black station owners were unable to compete: the number of black-owned stations stagnated then slowly begin to decline. The only bright note was that at the turn of the century, blacks William Kennard and then Michael Powell were successive chairmen of the FCC.

Networks and Wider Distribution

The first African-American network was the short-lived National Negro Network founded by black entrepreneur W. Leonard Evans in 1954. The network signed up 40 affiliates and promised good programming: among its best were a radio drama *The Story of Ruby Valentine* and variety shows hosted by Cab Callaway and Ethel Waters. But lack of advertiser support caused the network's demise after only a few months. Two decades later two networks, the Mutual Black Network and the National Black Network, began providing news and public affairs programming to black-oriented radio stations. Together these networks reached millions of listeners.

Ronald R. Davenport, a former dean of the Duquesne University School of Law, raised enough money to purchase the struggling white-owned Mutual Black Network and assumed management of the company. In the late 1970s MBN became the Sheridan Broadcasting Company. In 1991 Sheridan Broadcasting Network and National Black Network joined forces to operate as equals in a general partnership under a new name, American Urban Radio Networks. The AURN distributes a variety of news and public affairs programming to 250 African-American-oriented radio stations. It ranks as one of broadcasting's most successful firms.

A number of highly visible existing African-American operations have made use of such newer technologies as communication satellites and the internet. For years radio personality Tom Joyner had worked in both Dallas and Chicago simultaneously. He became known as the "fly jock" for his daily commutes between the two cities. Finally he converted to uplinking his broadcast signal to satellite from his Dallas location and syndicating his program to all radio stations. His popular show is now heard coast to coast by millions of listeners daily via the ABC Radio Network's satellite system. He is not alone. Others have purchased satellite time and are thus linked to hundreds of outlets nationally. Examples include Walt "Baby" Love, for Walt "Baby" Love Productions, Lee Bailey for Bailey Broadcasting, and American Urban Radio Networks. Darnell's *Black Radio Guide* maintains a growing list of internet-only black radio stations. The heart of the human resource chain is Black College Radio whose stated purpose is to provide an annual forum for black college broadcasters, professional broadcasters, and members of the music industry to meet and discuss ways to increase minority participation in the broadcasting industry.

Although the playing field for African-Americans has seldom been level, many have persisted and mastered the skills essential for successful leadership. The number of African-Americans who worked in radio between 1920 and 2000 was comparatively small. The quality of their contribution to achieve equity is more significant to radio when culture is placed into a perspective that stretches from Latimer, Gosden, and Correll, to recorded music used in modern programming. The magnitude of African-American presence in radio through jazz, rhythm and blues (or rock and roll), blues, reggae, gospel, salsa, and rap, regardless of the performer's race, is incalculable. According to folk legend Pete Seeger, American music is Africanized music.

LAWRENCE N. REDD

See also Affirmative Action; Amos 'n' Andy; Beulah Show; Black-Oriented Radio; Black Radio Networks; Durham, Richard; Hulbert, Maurice "Hot Rod"; Joyner, Tom; KFFA; Stereotypes on Radio; WDIA; Williams, Nat D.; Wright, Early

Further Reading

Abdul-Jabbar, Kareem, and Alan Steinberg, *Black Profiles in Courage: A Legacy of African American Achievement,* New York: Morrow, 1996

Barlow, William, *Voice Over: The Making of Black Radio,* Philadelphia, Pennsylvania: Temple University Press, 1999

MacDonald, J. Fred, *Don't Touch That Dial! Radio Programming in American Life, 1920–1960,* Chicago: Nelson-Hall, 1979

Randle, William, Jr., "Black Entertainers on Radio," *Black Perspective in Music* (Spring 1977)

Redd, Lawrence N., *Rock Is Rhythm and Blues: The Impact of Mass Media,* East Lansing: Michigan State University Press, 1974

Agricultural Radio. *See* Farm/Agricultural Radio

Album-Oriented Rock Format

Album-Oriented Rock (AOR) originally referred to 33 1/3 rpm LP vinyl recordings, which distinguished it from the "single" 45 RPM recordings played on the Top-40 format.

AOR's initial popularity in the 1970s signaled the arrival of 1960s counterculture tastes into the American popular music mainstream. But far from embracing the progressive politics, lifestyle, and artistry that gave counterculture music its relevance, AOR programmers capitalized on the increasing popularity of progressive radio incubating in the antiwar, Woodstock-era FM underground by appropriating its unique characteristics. AOR programming consultants replaced the DJ-programmer with a system of cue cards and playlists, turned thematic sets into music "sweeps" designed for Arbitron's ratings methodology, and handed FM station owners a homogeneous and more manageable format. AOR radio, by stripping rock and roll of its rhythm and blues heritage and rejecting its subversive possibilities, significantly contributed to branding rock as a marketable commodity. It also helped move most radio listeners from AM to FM by the end of the 1970s. Although the AOR format prospered at the expense of the radio radicals who inspired it, some went on to run AOR stations or form broadcast consultancies, while others remained reactionary by joining in less mainstream forms of radio, typically noncommercial in nature.

The AOR playlist is comprised of selected tracks from rock albums, chosen to attract a target audience. Radio stations utilizing the AOR format skew their playlists to position the station competitively. For example, AOR stations targeting an older audience may include rock from the late 1960s and early 1970s; lighter rock tracks often attract more women; and emphasis on contemporary albums may appeal to a cosmopolitan audience. Such refinements of the AOR format have led to Classic Rock, Soft Rock, and some Alternative formats. Other Alternative formats have developed as antithetical to AOR.

JOSEPH R. PIASEK

See also Alternative Format; Arbitron; Classic Rock Format; Progressive Rock Format; Soft Rock Format; Underground Radio

Further Reading

Fornatale, Peter, and Joshua E. Mills, *Radio in the Television Age,* Woodstock, New York: Overlook Press, 1980

Keith, Michael C., *Voices in the Purple Haze: Underground Radio and the Sixties,* Westport, Connecticut: Praeger, 1997

Ladd, Jim, *Radio Waves: Life and Revolution on the FM Dial,* New York: St. Martin's Press, 1991

Neer, Richard, *FM: The Rise and Fall of Rock Radio,* New York: Villard, 2001

The Aldrich Family

Situation Comedy Program

The 20th century may not have invented teenagers, but it supplied the most memorable examples. From 1939 to 1953, the character Henry Aldrich and his imitators defined the standard crises, all poignant yet laughable. Although Henry's stage origin was unpromising, his radio personality carried on the Tom Sawyer/Penrod Schofield/Andy Hardy tradition of a good-hearted innocent who unintentionally causes mischief. Clifford Goldsmith's 1937 play *What a Life!* confined him in the principal's office, accused of stealing band instruments when he had merely cheated on a test in order to attend the dance. Henry's world grew when Rudy Vallee and Kate Smith commissioned sketches for their shows. By 1939 the vignettes expanded to a 30-minute series on the National Broadcasting Company's (NBC) Blue network. Sponsored by Jell-O and introduced by the rousing tune "This Is It," *The Aldrich Family* mingled humor, nostalgia, and complicated plots.

The opening routinely stressed the universality of Henry's zany experiences. For example, the announcer led in to the episode "The Tuxedo" by saying, "Whatever and whenever the Golden Age was, it is less important to most people than the teen age—a time of life made notable by typical American boys like Henry Aldrich and all their mishaps." Adolescence normally involves a tension between conformity and individuation, but Henry's awkward attempts to be a dutiful son/student/friend/worker as well as an independent individual seldom seem rooted in common experience. Unlike his listeners, he never matured beyond 16. Certainly he honored his parents: the show began with the memorable call from mother Alice ("Hen-ry! Hen-ry Aldrich!"), which Henry answers with an obedient, "Coming, Mother." Talking to father, lawyer Sam Aldrich, Henry resorts to elaborate but ungrammatical politeness: "Do you wish to speak to I, Father?" He tries to carry out their wishes by not leaving doors open, by cooperating with his sister Mary, and by babysitting a rambunctious tyke. Yet he inevitably upsets family order. Once Sam borrows Henry's bicycle and thoughtfully leaves a note on the rake with which his son had promised to beautify the yard. As usual, Henry forgets the leaves, doesn't see the note, and, assuming someone has stolen the bike, calls the police. By the show's end, friends, neighbors, and strangers share his teenage turmoil.

School situations likewise do little to develop Henry as a character. He once masters the Latin pluperfect subjunctive; most other times, his academic milestones mark unpredictable gaffes: when he spills glue on the shop floor and sees his teacher lose his shoe in the mess, or when he is caught on a fire escape by the principal as he tries in vain to return a teacher's grade book that his pal mistakenly picked up because it resembled a mystery story. Friendship and romance also seemed to change before Henry could understand them. A note from a girl in the next town flatters him so that he rents a costume, intending to escort her to a dance. After seeing her picture, he fobs her off on a rival who also takes his costume. The urge to make money, too, is thwarted by poor information. He buys a furnace-starting concession. The seller neglects to tell him that the owners are in Florida, a fact Henry learns from a postcard sent by the owners to his family. Unfortunately, he has already wrecked the furnace, restarted mail and milk deliveries, and burned a box of the owner's papers. Although Henry mishandles his duties, his good will eventually moderates any possibilities for serious harm.

The plots often center on Henry's quest to acquire some object: a misplaced watch given by his aunt Harriet, who wants to see it again; a tuxedo so he can attend the prom; a straw hat; an antique toy to replace the one he broke in his girlfriend Kathleen's home; a motor scooter that he will receive if he can pass his history test. His cravings for material goods parallel those of the normative middle-class citizen, but his missteps lighten greed with humor. Popular formulas for success often mislead Henry. He thinks he might get rich raising rabbits; he imitates the generosity he's seen in a film and, like the hero of Thornton Wilder's *Heaven's My Destination,* creates chaos; he writes to a Charles Atlas–type muscle developer but misplaces the letter that details his puny dimensions. Henry's ambitions do not liberate him; rather, they entangle others. His father spends some uncomfortable hours trapped in a phone booth; a friend crouches miserably in a basement cubbyhole; his chum Homer Brown, unwittingly engaged to Agnes, finds that Henry's solution is worse than commitment.

Although *The Aldrich Family* provided lasting memories, the cast changed frequently. At least three mothers, seven sisters, seven directors, and three fathers appeared. House Jameson, barely in charge as would-be patriarch Sam, had more authority as *Renfrew of the Mounted.* Ezra Stone was the best-known Henry—his reedy voice captured the nearly out-of-control mood that characterized each program. After Stone was called to military service from 1941 to 1944, Norman Tokar, Raymond Ives, Bobby Ellis, and Dickie Jones filled in until his return. Clifford Goldsmith relied on seven other writers, but his benign vision of adolescence still shaped their versions. Only Jackie Kelk remained consistent, playing Homer (a role that was a shift for Kelk, who had played the self-confident Terry on *Terry and the Pirates* and helpful Jimmy Olson on *Superman*).

Several shows copied the Aldrich formula of a well-meaning youngster who inadvertently confounds normalcy: *Archie Andrews* and *That Brewster Boy* echoed the male adolescent's turmoil; *Junior Miss, A Date with Judy, Maudie's Diary,* and *Meet Corliss Archer* presented the female version. A series of 11 Henry Aldrich movies between 1939 and 1944 made visual his arrested adolescence, but the film versions employed other actors (Jackie Cooper in the first, Jimmy Lydon in the rest). Likewise, the 1949–53 television program used five Henrys (most notably Bobby Ellis), though it retained Jameson as father, Kelk as Homer, and Leona Powers as Mrs. Brown.

JAMES A. FREEMAN

See also Comedy; Situation Comedy

Cast

Henry Aldrich	Ezra Stone (1939–42; 1945–52), Norman Tokar (1942–43), Dickie Jones (1943–44), Raymond Ives (mid 1945), Bobby Ellis (1952–53)
Sam Aldrich	House Jameson, Clyde Fillmore, Tom Shirley
Alice Aldrich	Katharine Raht, Lea Penman, Regina Wallace
Mary Aldrich	Betty Field, Patricia Peardon, Charita Bauer, Ann Lincoln, Jone Allison, Mary Mason, Mary Rolfe, Mary Shipp
Homer Brown	Jackie Kelk, Johnny Fiedler (1952–53), Jack Grimes (1952–53), Michael O'Day (1952–53)
Will Brown	Ed Begley, Arthur Vinton, Howard Smith
Homer's Mother	Agnes Moorehead, Leona Powers
Kathleen Anderson	Mary Shipp, Ethel Blume, Jean Gillespie, Ann Lincoln
Dizzy Stevens	Eddie Bracken
George Bigelow	Charles Powers
Toby Smith	Dick Van Patten
Mrs. Anderson	Alice Yourman
Willie Marshall	Norman Tokar
Aunt Harriet	Ethel Wilson
Announcers	Harry Von Zell, Dwight Weist, George Bryan, Dan Seymour, Ralph Paul

Creator/Writer

Clifford Goldsmith

Programming History

NBC	July 1939–July 1944; September 1946–April 1953
CBS	September 1944–August 1946

Further Reading

Beers, Dale, "The Aldrich Family Log Website," <www.geocities.com/Hollywood/Set/3688/aldrich.html>

Dunning, John, *Tune in Yesterday: The Ultimate Encyclopedia of Old-Time Radio, 1925–1976,* Englewood Cliffs, New Jersey: Prentice-Hall, 1976; revised edition, as *On the Air: The Encyclopedia of Old-Time Radio,* New York: Oxford University Press, 1998

Goldsmith, Clifford, "Radio Oddities," *Tune In* 4 (August 1946)

Harmon, Jim, "When Teen-Agers Were Kids," chapter six of *The Great Radio Comedians,* by Harmon, Garden City, New York: Doubleday, 1970

Swartz, Jon David, and Robert C. Reinehr, *Handbook of Old-Time Radio: A Comprehensive Guide to Golden Age Radio Listening and Collecting,* Metuchen, New Jersey: Scarecrow Press, 1993

Witham, W. Tasker, *The Adolescent in the American Novel, 1920–1960,* New York: Ungar, 1964

Alexanderson, E.F.W. 1878–1975

U.S. (Swiss-Born) Radio Engineer and Inventor

Employed for 45 years by the General Electric Company as an electrical engineer, E.F.W. Alexanderson was involved in early research and development of radio-related technology. He is associated primarily with the development of alternators that first made reliable long-distance radio transmission possible.

Ernst Alexanderson was born in Uppsala, Sweden. After graduating from the Royal Institute of Technology in Stockholm in 1900 and then studying electrical engineering for a year in Germany, Alexanderson emigrated to the United States. Shortly thereafter, in 1902, he found employment at General Electric. Alexanderson was involved in a number of projects during his first years at General Electric, until the work of Reginald Fessenden brought Alexanderson into radio research.

Fessenden was attempting to achieve transoceanic voice transmission through a radio wave. Up to that time, radio transmission had been based on Guglielmo Marconi's spark-gap system, which sent out an interrupted wave, thus creating a series of dots and dashes for Morse code. To achieve the ability to transmit a continuous wave on which a voice could be superimposed, Fessenden contracted with General Electric to construct a special alternating-current generator that could be used as the transmitter. This project was assigned to Alexanderson. After several years of design and development, in December 1906 the alternator was delivered to Fessenden, who proceeded that Christmas Eve to present the first radio voice transmission.

E.F.W. Alexanderson posing with Alexanderson alternator, ca. 1922
Courtesy Schenectady Museum Archives

In the following years, Alexanderson continued to work on and refine the alternator for General Electric, incorporating many of his own engineering ideas into its design. This General Electric alternator was commonly called the Alexanderson alternator. The design of this large machine was very similar to that of a power plant's generator, though it rotated much faster. Such high-speed rotations created complex mechanical problems that Alexanderson had to solve.

During World War I, the Alexanderson alternator began to receive much favorable attention in scientific and industrial circles when General Electric used it for transmission tests of a new vacuum tube the company had developed. Alexanderson was now concentrating on creating a 50,000-watt version of the alternator. By early 1915, developmental work on this device had reached the point where tentative plans were made to install one at an American Marconi radio transmitting station in New Brunswick, New Jersey, for field tests. While he was in the United States in 1915, Marconi visited General Electric's Schenectady, New York, laboratory to examine the alternator. He saw it as a key component in developing reliable transoceanic communication. Shortly after Marconi's visit, representatives of Marconi and of General Electric discussed a deal whereby Marconi companies would have exclusive use of the Alexanderson alternator, with General Electric retaining exclusive manufacturing rights. The agreement was not completed because Marconi was suddenly called back to Europe by the Italian government.

Although the negotiations with Marconi were on hold, General Electric proceeded with its plan to install a 50,000-watt alternator at the American Marconi New Brunswick site, and an alternator was delivered in 1917. Because of World War I, the U.S. Navy had been authorized to take over all high-powered radio transmission stations, and it soon took over American Marconi's New Brunswick facility. Using the call letters NFF, the navy used this alternator and a subsequent 200,000-watt alternator—also designed by Alexanderson and delivered to New Brunswick—to transmit propaganda and information by radio throughout Europe. In January 1918, President Woodrow Wilson's Fourteen Points message was delivered to Europe via this facility, as was a later appeal from Wilson to the German people to remove their Kaiser. For this latter message, Alexanderson himself was at the transmitter site.

After the war ended, Marconi again approached General Electric with a proposal to obtain the exclusive right to buy the Alexanderson alternator, for by that time the alternator was the best long-distance transmitter available. If General Electric had accepted Marconi's proposal, a foreign company would have gained a monopoly over American radio communications with Europe. When General Electric approached the navy for its view of the proposed contract with Marconi, navy officials expressed strong opposition. Knowing that the government preferred an American company to control the country's international radio communications, General Electric instead bought a controlling interest in American Marconi, which led to the eventual formation of the Radio Corporation of American (RCA) in 1919. Control over Alexanderson's alternator technology was one factor that led to the creation of RCA.

Alexanderson was appointed RCA's first chief engineer. General Electric, as a founder and major shareholder of RCA, loaned Alexanderson to the new company while also keeping his services at General Electric on a shared-time basis. Alexanderson stayed at RCA for four years and then returned full-time to General Electric to continue research. He concentrated on antenna design and television research, demonstrating a mechanical scanning television receiver as early as 1927.

Alexanderson retired from General Electric in 1948, but he stayed active for the next 27 years by consulting for several companies, including RCA. In these later years, Alexanderson was viewed in the scientific community as one of the pioneers of electrical engineering and early radio technology. He died in Schenectady, New York, in 1975, just two years after his final patent had been awarded.

RANDALL VOGT

See also Fessenden, Reginald; General Electric; Radio Corporation of America

E(rnst) F(redrik) W(erner) Alexanderson. Born in Uppsala, Sweden, 25 January 1878. Studied engineering at Royal Institute of Technology, Stockholm, graduated 1900; spent next year studying at Königliche Technische Hochschule, Charlottenburg, Germany; employed by General Electric Company at Schenectady, New York, from 1902 until retirement in 1948; first chief engineer, Radio Corporation of America, 1919–23. Became naturalized U.S. citizen, 1908. Awarded 344 patents, 1905–73. Died in Schenectady, New York, 14 May 1975.

Further Reading

Aitken, Hugh G.J., *The Continuous Wave: Technology and American Radio, 1900–1932,* Princeton, New Jersey: Princeton University Press, 1985

Alexanderson, E.F.W., "Trans-Oceanic Radio Communication," *Proceedings of the Institute of Radio Engineers* 8 (1920)

Brittain, James E., *Alexanderson: Pioneer in American Electrical Engineering,* Baltimore, Maryland: Johns Hopkins University Press, 1992

Kraeuter, David W., "Ernst Alexanderson," in *Radio and Television Pioneers: A Patent Bibliography,* by Kraeuter, Metuchen, New Jersey: Scarecrow Press, 1992

All India Radio

India presents huge challenges to any broadcasting institution that aspires to serve the whole nation. All India Radio (AIR), the state-run monopoly, was expected to take these challenges on and help build a modern nation state with an egalitarian social democracy. More than a billion people, nearly half of them living below the poverty line, are spread over a land mass of 1.27 million square miles. Although urbanization and industrialization are the hallmarks of postcolonial India, nearly 75 percent of the population still lives in 55,000 villages, eking out a living from farming. About 10 percent are employed in industries in urban areas. India's religious, cultural, and regional diversity is striking, with 83 percent of the population claiming Hinduism as their religion and Muslims, Christians, Sikhs, Buddhists, and Jains accounting for the rest. Fourteen officially recognized languages and hundreds of dialects coexist with English. Hindi, the official language of modern India, is slowly gaining a foothold with the masses. Uneven development characterizes India; cities such as Bangalore claim a place in the global computer industry as the "Silicon Valley of India," whereas villages have extremely bad roads and lack clean drinking water, medical facilities, and schools. Significant advances have been made in literacy rates since independence in 1947, but a mere 52 percent are functionally literate. Social inequity such as caste, class, and gender inequality can be found in urban and rural parts of the country. Untouchability is still practiced against nearly 170 million people who are cast aside in near-apartheid conditions.

Origins

Enthusiasts in India's big cities pioneered radio by organizing amateur radio clubs in the early 1920s. Their efforts, and the successful growth of radio in Europe and the United States, gave impetus to a group of Indian entrepreneurs who established the Indian Broadcasting Company on 23 July 1927. Nevertheless, by 1930 their pioneering effort to launch privately owned radio ran into trouble because of lack of revenues. Broadcasting from their two stations, located in Bombay and Calcutta, they catered to the small European community and Westernized Indians while ignoring the masses. The colonial government was faced with the rising tide of anti-imperialist sentiment in the country; being interested in the propaganda potential of broadcasting, it bought the assets of the Indian Broadcasting Company and renamed it the Indian State Broadcasting Service (ISBS).

In 1935 the colonial government took another decisive step by inviting the BBC to help develop radio; one of the BBC's senior producers, Lionel Fielden, was sent. Fielden is credited with having the name of the organization changed to All India Radio and for laying the foundations for public service broadcasting with the goal of providing information and education. He returned to England in 1940. By 1947, the year of India's independence, the AIR network had grown to 11 stations with 248,000 radio licenses.

AIR Today

AIR's growth and reach have been phenomenal in the last 50 years. There are 333 transmitters today, including 146 medium wave, 54 shortwave, and 133 FM. Some 210 radio stations cover 90 percent of India and reach 98 percent of the population. AIR claims a listenership of approximately 284 million who tune in on 111 million radio sets. Although controlled by the central government, AIR introduced advertising in 1967 and earns 808 million rupees a year (US$1 = 48 rupees). The government makes up any deficit in its operating expenses.

AIR broadcasts in 24 languages and 146 dialects for domestic audiences and in 24 languages for international audiences. Approximately 303 news bulletins are aired daily, of which 93 are intended for national listeners, whereas regional stations originate 135 news bulletins daily. In addition, there are special bulletins on sports, youth, and other major events, such as the annual Haj to Mecca by Muslims or the Kumbh Mela in Allahabad. More than 80 stations in the AIR network broadcast radio dramas in various languages. Forty percent of the broadcast time, however, is set aside for classical, light, folk, and film music. The External Service, set up to act as a cultural ambassador, airs 65 news bulletins in 16 foreign and eight Indian languages. In addition, magazine programs on sports and literature; talk shows on sociopolitical-economic issues; and classical, folk, and modern Indian music from different regions of the country are broadcast.

AIR employs well over 16,000 persons. Approximately 13,000 are regular government servants; the rest are contract employees. They are transferable every three years, and so these employees seldom come to know the community in which they work. Such a huge organization cannot escape a hierarchical structure and the formal nature of appointments, promotions, retirements, and codes of conduct. Instead of demanding commitment to listeners, the organization requires its employees to adhere to the rules and procedures of a large government department. Because the employees have very little functional freedom, creativity and innovation are sacrificed. Lethargy, apathy, and favoritism unfortunately permeate the organization.

Regulation and Autonomy

Broadcasting is a regulated monopoly of the central government. The Indian Telegraph Act of 1885 was later amended to vest the exclusive right to "establish, maintain and work" wireless apparatus in the Government of India. Consequently, AIR has functioned as an arm of the central government ever since its inception. The Ministry of Information and Broadcasting is the policy-making body for the entire broadcasting system. Generalist officers drawn from the civil service manage the ministry. The director general heads the AIR and executes policy. The government has held that any member of the elite Indian Administrative Services can function as head of AIR with equal disinterest. Hence the director general is a bureaucrat who may or may not be interested or qualified in radio.

National television grew under the umbrella of AIR and in 1976 was given a separate structure called Doordarshan, literally meaning viewing from a distance. As one would view a deity in a temple, TV audiences regularly gain a glimpse of the political establishment via Doordarshan's newscasts. With a mandate similar to radio's, television has also seen remarkable expansion and reach in the country in the last three decades. With the rise of privately controlled satellite delivery services, India now has a mixed system of public and private enterprises in television, whereas radio has clearly remained a government monopoly.

The credibility of AIR news has always been in question, however, not only because it is a government department but also because of well-reported instances of interference by the prime minister's office, irrespective of who is in power. There has been considerable pressure from private and public institutions as well as from intellectuals in the country to create an arms-length relationship between the government and the broadcast institutions ever since the National Emergency in 1975. Suspending certain articles of the constitution, the prime minister unleashed a reign of terror, which lasted almost 19 months. To silence dissent, the government engaged in mass arrests of prominent political leaders, trade unionists, human rights activists, communists, and students. There were widespread reports of torture and sterilization, especially of the poor. While the judiciary was not abolished, the ruling party in the Parliament passed certain amendments to the constitution to put the prime minister and her party loyalists above the nation's laws. The privately controlled press and cinema were subjected to intense censorship, and journalists at AIR were instructed to abandon even the pretense of journalistic fairness and balance in their coverage of events. Prominent journalists went underground to avoid arrest. The government restructured the news agencies in the country to "clean out" anyone who was not favorable to the prime minister and denied advertising in newspapers that would not change the tone of coverage of the regime. A dark moment in the history of the nation, the national emergency exemplified the extent to which the executive branch of the government could misuse its power over the media.

The debate on autonomy for broadcasting has finally resulted in Parliament passing the Prasar Bharati Act of 1990, which seeks to free radio and television from the direct control of the government and place it in the hands of an autonomous corporation that would be managed by a board. That board would be required under the law to be accountable to a broadcasting council and in turn to a statutory parliamentary committee with various powers reserved to the government. The act has not been implemented, however.

Promises Versus Reality

AIR's heavily bureaucratic ways have been the major impediment to innovation and creativity. In a highly pluralistic society with incredible linguistic, caste, and class differences, AIR has attempted not to offend any group. Controversial social and community welfare issues take a back seat while popular film music dominates. Regional language radio stations beam programs to the whole state in a formal dialect, which renders it stiff and official. As a consequence, most people find AIR boring. Radio, as a mass medium, is particularly suited to communicate in the local dialect and idiom, thereby establishing a personal connection between the broadcaster and the listener. That has not, however, been achieved in India because of the bureaucratic stranglehold on radio.

The model of a centralized national radio service with many regional and local stations intended to achieve the vision of unifying the nation was well intentioned but expensive and difficult to deliver. For development purposes, more localized micro radio operations based in community and educational institutions would have been more cost efficient and credible with audiences. The distance between the program creators and listeners would have been reduced, which in turn would have enhanced radio's credibility with the rural masses. Perhaps radio might then have met local needs better. Until recently, the government has guarded the frequencies as though they were its property and has only reluctantly allowed private program producers some space on the government-controlled stations. This may lead to licensing of private FM stations that will, in all likelihood, be urban-centered. All India Radio's local outlets around the country often are criticized for their low levels of involvement on the part of local groups. Partly in response, the Indian government began to license private radio stations in 2000, and the first of them came on air in July 2001. As of 2002 a few community radio stations had begun to appear.

AIR's long-held policy of broadcasting classical music meant that it should have developed an extensive collection of recordings by some of India's greatest performers. Lacking resources and vision, many station directors simply did not save such precious recordings. For a few rupees more, the artists would have let AIR keep the recordings and later release them in the burgeoning cassette market. Recently, AIR seems to have realized its folly, and cassette tapes of speeches by leaders such as Gandhi and Nehru are being released to the public.

Competition to radio has grown steadily from film and television. Doordarshan, with its national reach and through its regional stations, and the privately controlled satellite TV channels have stolen radio audiences. They offer similar programs of music, talk, and other shows with the power of visuals. The enormously popular film music from India's own gigantic film industry became widely available on cassettes by the mid 1970s and even penetrated rural areas in the following decade. Satellite audio services and on-line radio, operated by private companies, will be the next frontier on which AIR will have to compete.

AIR needs bold new directions in this age of the internet and FM broadcasting. What seems to be in store is an added layer of bureaucracy in the name of autonomy and higher pressure on AIR to earn more revenues and become self-sufficient. Those signs do not offer much hope for an institution with the national purpose of employing the power of the medium for social change.

MANJUNATH PENDAKUR

See also Developing World

Further Reading

Agrawal, B., "The Meaning of Hinglishness: Liberalisation and Globalisation in Indian Broadcasting," in *Programming for People: From Cultural Rights to Cultural Responsibilities,* edited by Kevin Robins, United Nations World Television Forum, 1997

Chatterji, Probhat Chandra, *Broadcasting in India,* New Delhi and Newbury Park, California: Sage, 1987; revised edition, London: Sage, 1991

Government of India, Ministry of Information and Broadcasting, *White Paper on Misuse of Mass Media during the Internal Emergency, August 1977,* Delhi: Controller of Publications, 1977

Government of India, Ministry of Information and Broadcasting, *Akash Bharati: National Broadcast Trust: Report of the Working Group on Autonomy for Akashvani and Doordarshan,* 2 volumes, New Delhi: Ministry of Information and Broadcasting, 1978

Government of India, Ministry of Information and Broadcasting, *India 1998,* New Delhi: Publications Division, 1998

Hartmann, Paul, B.R. Patil, and Anita Dighe, *The Mass Media and Village Life: An Indian Study,* New Delhi and London: Sage, 1988; Newbury Park, California: Sage, 1989

Luthra, H.R., *Indian Broadcasting,* New Delhi: Ministry of Information and Broadcasting, 1986

Manuel, Peter Lamarche, *Cassette Culture: Popular Music and Technology in North India,* Chicago: University of Chicago Press, 1993

Masani, Mehra, *Broadcasting and the People,* New Delhi: National Book Trust, India, 1976

Pasha, A.R., *Community Radio: The Voice of the People,* Bangalore, India: Voices, 1997

Pendakur, M., "Mass Media during the 1975 National Emergency in India," *Canadian Journal of Communication* 13, no. 4 (1988)

All News Format

All news is a programming format that continuously provides listeners with the latest news, sports, weather, time, and, in many cases, reports on driving conditions. This format's appeal is directed to a revolving audience continuously tuning in and out.

Radio news traces its origins to KDKA's broadcast of the 1920 presidential election returns. The station announcer requested that listeners mail postcards to the station confirming that they had heard the broadcast. By 1930 NBC and CBS were simultaneously broadcasting *Lowell Thomas and the News,* sponsored by *Literary Digest.* The unusual simulcast, in which NBC broadcast the program to the eastern half of the country and CBS to the western half, became solely the property of NBC within a year. Americans grew accustomed to turning on their radios for the latest news during the late 1930s. By the end of the decade, as social unrest increased in Europe, the voices of radio correspondents William L. Shirer, George Hicks, and Edward R. Murrow became as familiar as those of friends and neighbors. Radio was establishing itself as the leader in reporting events as they were occurring. In 1940 Americans told pollsters for the first time that they preferred radio to newspapers as their primary news source. Coverage of

World War II cemented the relationship between radio and its audience across the United States as listeners followed the progress of American troops in Europe and the Pacific.

With the end of World War II and the development of television, radio's role once again shifted. During the 1950s, radio's entertainment programming, including dramas and soap operas, moved steadily to television. Music gradually became the dominant form of programming on radio; at many stations, news was shifted to five minutes at the top of the hour.

In the 1960s radio management took another look at news programming. The trade magazine *Sponsor* described radio at that time as the "new king of the news beat" and attributed its change in status to improvements in technology, increased numbers of experienced news reporters, and recognition that newsmakers were more important than newscasters. Four radio networks were providing regular newscasts to their affiliates: ABC, NBC, and Mutual Broadcasting System (MBS) produced two newscasts hourly, and CBS, with 20 news bureaus worldwide, had nearly 50 correspondents contributing to their news programs.

The first all-news radio format in the United States was used at KFAX in San Francisco on 16 May 1960. General manager J.G. Paltridge and sales manager Ray Rhodes called the new format "newsradio." As was common at the time, KFAX (owned by Argonaut Broadcasting) sold advertisers announcements but no sponsorships of shows. Compared to today's heavy commercial schedule, spots were few, with one commercial per five minutes of news and also at station breaks. The first 25 minutes of an hour consisted of hard news followed by business news, sports, and special features. "Newsradio" died after four months due to its lack of advertisers.

In 1961 Gordon McLendon started the first commercially successful all-news radio station, X-TRA. Located in Tijuana, Mexico, it beamed its powerful signal across the border to southern California. A station promotional announcement trumpeted, "no waiting for hourly newscasts or skimpy headlines on X-TRA NEWS, the world's first and only all-news radio station. In the air everywhere over Los Angeles."

Los Angeles broadcasters were critical of the Mexican station's identification with their city, as the only address announcers mentioned was that of the sales office in Los Angeles. The station's official on-air identification, required by law, was made in Spanish and followed by Mexican music and a description of Mexican tourist attractions. The Southern California Broadcasters Association called this an unethical attempt to camouflage a Mexican station as one located in Los Angeles.

In the early days, X-TRA used "rip and read" reports (stories torn straight from wire service machines and read live on the air). The station was served by the Los Angeles City Wire Service, the Associated Press (AP), and United Press International (UPI). X-TRA also subscribed to a clipping service that provided stories from newspapers in all major U.S. cities and international capitals around the world. No one rewrote the wire copy, and there were no station reporters gathering news or conducting interviews. Newscasters alternated as anchors every 15 minutes, with a half hour in between to prepare for the next news stint. The content was somewhat repetitive, as programmers assumed that the audience would switch to a music station after hearing the most recent news. All newscasters read their copy with the sound (often recorded) of wire service teletype machines in the background, as a report in *Sponsor* magazine described it, "to suggest a newsroom setting."

McLendon brought the all-news format to WNUS (pronounced "W-news") in Chicago in 1964. He advised radio programmers considering the format not to attempt to enliven it with features and "actualities" (sound bites), suggesting that listeners wanted nothing but news. WINS, New York's first all-news radio station, ignored that advice when it switched to round-the-clock news in 1965. Owned by Westinghouse, WINS expanded the McLendon design, emphasizing on-the-scene reports and actualities. Fourteen newscasters, rotating in 30-minute shifts, anchored the newscasts. Mobile units provided live and taped reports from the five boroughs and the outskirts of the metropolitan area. A staff of more than 40 produced the news summaries, sports reports, financial news, and weather reports, plus time and traffic reports. They relied on wire service from AP and UPI and the resources of Westinghouse stations and news bureaus across the country and overseas. Despite the personnel-intensive expense of the operation, WINS started turning a profit six months into its all-news operation.

In 1966 *Broadcasting* reported that all news was a viable, profitable choice for a programming format. There were then four U.S. stations concentrating on news: WINS (New York), KYW (Philadelphia), WNUS (Chicago), and WAVA (Arlington, Virginia), and the Mexican station X-TRA in Tijuana, which also had an audience in the United States. Despite reaching the profit-making point after nine months, WNUS never gained a dominant share of the market, so management changed its program format back to music in 1968. Westinghouse, however, went on to program all news in its stations in Los Angeles and Philadelphia.

CBS's flagship station in New York, WCBS, shifted to all news in 1967, but with its own innovations and significant financial support. WCBS used helicopters for traffic reports and its own weather forecasters. It had more reporters and produced more features than other all-news stations of the period, and it had access to the resources of the respected CBS network news. Among its reporters were Ed Bradley and Charles Osgood, who would make national names for themselves and eventually shift to television news. CBS brass liked the results, and all-news formats were put into place at other CBS owned-and-operated stations in Boston, Chicago, Los Angeles, Philadelphia, and San Francisco.

NBC broke into all-news radio in 1975, introducing News and Information Service (NIS), the first national all-news service. Subscribers paid as little as $750 or as much as $15,000, depending on their market size. Stations would pay for world and national news plus sports and features, all provided by NBC anchors in New York. They broadcast for 47 minutes out of every hour with the format constructed so that subscribers could take the whole 47 minutes or as little as 23 minutes of the hourly format. Unlike traditional affiliations in which a network paid its affiliates or traded commercial availabilities, NIS had to be purchased from NBC. With the network producing the majority of the programming, the all-news format looked financially feasible for medium and small markets for the first time.

Despite the positive appearances, the all-news concept was not successful for NBC. After 18 months NBC had only 62 subscribers, significantly fewer than the projected 150 stations. Industry insiders suggested that the NBC network's unwillingness to commit its owned-and-operated stations contributed to the demise of NIS (only one adopted the all-news format). Audience numbers never reached expectations, resulting in disappointing advertising sales. All news was expensive to produce and simply was not bringing in the necessary income, so NIS went off the air two years after it began. Although it had cost $20 million, its impact on the future of all-news radio had been significant during its short life. After its demise, a number of other all-news stations continued, affiliating with other networks, including CBS. So NBC contributed to the all-news format's expansion from the top ten markets to the medium-sized markets, despite the failure of NIS.

From NBC, CBS, and Westinghouse, three basic models developed for the all-news format, all based on the "format clock." A circle divided into pie-shaped slices indicates specific times during the hour on an analog clock. News, weather, traffic reports, and features are designated on the pie slices. Today's "weather on the fours" or "traffic on the eights" are segments that appear every four and eight minutes within each hour span, based on the traditional format clock.

The Westinghouse model is based on a 20- to 22-minute cycle with short, crisp stories and repetition. This is generally a hard-news approach with headlines, weather, time checks, and traffic reports. For instance, Philadelphia's KYW is famous for announcing, "Give us 22 minutes and we'll give you the world." The CBS model tends to be more informal, with hard news, features, and commentary, having initially used dual anchors and less repetition to establish its format. NBC created a more impersonal sound with the NIS model because it was producing news for all areas of the country rather than tailored for a specific market.

All news was one of the fastest growing formats in the 1970s, but the number of all-news stations started dropping in the 1980s as the news format was combined with a less expensive format that was growing in popularity: talk. The number of news-talk stations increased as all news declined. By 1990 there were only 28 all-news stations in the United States.

In 1994 the Associated Press started a 24-hour all-news radio network to serve news and news-talk stations. AP was already providing a newswire service to 5,000 radio stations and 750 TV stations. It also provided an audio network, AP Radio, that offered stations four brief news segments per hour. Industry insiders looked back 15 years to NBC's NIS and said that new technology and changes in the radio business made AP's network likely to succeed, predicting that it would reinvigorate local radio news.

By 1995 the ubiquity of the all-news radio format was evident with the introduction of a new television sitcom, NBC's *NewsRadio,* set in an all-news radio station in Manhattan. Like previous sitcoms that had used the broadcast news business as a framework *(The Mary Tyler Moore Show, WKRP in Cincinnati,* and *Murphy Brown), NewsRadio* had little to do with news and much to do with workplace relationships. It was canceled after five seasons.

The number of genuine news-radio operations in the United States grew dramatically to 836 all-news stations in 1999: 387 AM stations and 449 FM. Of these, 451 were commercial stations and 385 were noncommercial. In Canada there were 32 all-news stations: 15 AM, 17 FM (29 commercial, and three noncommercial).

All-news formats are carefully designed for specific markets—what works in Atlanta may not be successful in Los Angeles or Minneapolis. Listening patterns, including how long a listener will stay with the station and how often, vary from market to market. The larger the population in a station's market, the more the station can afford to have frequent turnover in its audience. Smaller-market stations try to keep listeners for longer periods.

Four decades after the introduction of all-news radio, it is a stable and profitable format. Advertisers are willing to pay all-news stations top rates to reach the population segment that is attracted to news. In the United States, all-news listeners tend to be older (35-plus), better educated, and have more money than the average consumer. Because of the repetitive nature of the format, they stay with the station only a short time. But while they listen, they tend to give much closer attention to the broadcast than they give to music programming, and all-news stations tend to carry more commercials than music stations.

Despite the increased attention to news overall—international, national, business and sports—some critics and listeners in the United States have expressed concern about the reduction in local news on radio. A number of stations, recognizing that local news continues to be a labor-intensive, expensive process, are finding it economically viable to reduce their local news content and increase reliance on news services that provide more lifestyle, sports, and entertainment news. As a result, some stations that identify themselves as news-radio stations depend on syndicated services and employ few, if any, reporters to gather news. Some even label talk-show programs featuring hosts such as Don Imus, G. Gordon Liddy, Oliver North, and Howard Stern as newscasts. Other types of non-news broadcasts may also be mislabeled; for example, the magazine *Smart Money* warned its readers about "investment pros" who buy radio time to promote stocks and investment services but masquerade as business news.

At the beginning of the 21st century there are more news resources available to U.S. radio stations than ever before. AP provides audio and text to almost 4,000 stations with a variety of program formats. National syndicator Westwood One distributes CBS Radio News, Fox News Radio, and NBC Radio Network; it also owns Shadow Broadcast Services and Metro Networks, which offer a variety of news options and traffic reports. ABC provides radio service to some 3,200 affiliates. Business news has become a hot commodity with Bloomberg Business News Network and CNBC Business Radio, among others. Sports news is provided by ESPN, which has 650 affiliates in its radio network. One historical news organization dropped out of the broadcast news business in 1999, however, when UPI (which started its newswire service for broadcast stations in 1958) sold its remaining radio and TV accounts to the Associated Press. The venerable news service announced that its future plans were to focus on internet-delivered news.

By 2000 the popularity of all-news radio was spreading around the globe. Canada's largest cable company, Rogers Communications, switched its country music station in Vancouver to all news in early 1996. A spokesman described the company as "fully committed" to the format. Scandinavia's first all-news operation started in Norway in 1997. *Alltid Nyheter* (Always News) was begun by a Norwegian public service station, NRK. In addition to news gathered by its own staff, *Alltid Nyheter* uses material from CNN, Swedish Radio, Danish Radio, and BBC World Service. In France, the main all-news radio station is France-Info. After ten years without an all-news station, Montreal's 940News (English language) and Info690 (French) went on the air in December 1999. Quebec's Metromedia radio chain plans to invest $40 million in the first five years of these stations' operation. In England, listeners can tune in news from NewsDirect, London's all-news radio station.

Despite the variety of news sources available (newspapers, the internet, all-news cable networks, and local television), all-news radio continues to build a loyal following of listeners in the United States and abroad. Working people once got their news from television at home in the evening, but today's professionals often work longer hours. They listen to the news when it's most convenient—on the radio as they drive to and from work.

SANDRA L. ELLIS

See also Canadian News and Sports; KYW; News; News Agencies; Talk Radio; WCBS; Westwood One; WINS

Further Reading

Barfield, Ray E., *Listening to Radio, 1920–1950,* Westport, Connecticut: Praeger, 1996

Bliss, Edward, Jr., *Now the News: The Story of Broadcast Journalism,* New York: Columbia University Press, 1991

Charnley, Mitchell Vaughn, *News by Radio,* New York: Macmillan, 1948

Fisher, Marc, "Blackout on the Dial," *American Journalism Review* 20, no. 5 (June 1998)

Julian, Joseph, *This Was Radio: A Personal Memoir,* New York: Viking Press, 1975

Keirstead, Phillip O., *All-News Radio,* Blue Ridge Summit, Pennsylvania: Tab Books, 1980

Kotz, Pete, "The Decline of News Radio," *Des Moines Business Record* (26 October 1998)

MacFarland, David T., *Contemporary Radio Programming Strategies,* Hillsdale, New Jersey: Erlbaum, 1990; 2nd edition, as *Future Radio Programming Strategies: Cultivating Listenership in the Digital Age,* Mahwah, New Jersey: Erlbaum, 1997

Schatz, Robin D., "All-News Radio Holds Its Own in U.S. Media Markets," *International Herald Tribune* (26 June 1996)

All Night Radio

All-night radio programming has been a staple of the industry since the 1920s, when stations such as WDAF in Kansas City remained on the air far into the night to accommodate listeners who wanted to hear distant signals. Late-night programs began appearing more widely in the 1930s as the networks scheduled live big band shows, which typically played dance music into the wee small hours (often 2:00 or 3:00 A.M.). Although it is difficult to say which station first offered a regular schedule of all-night broadcasts, certainly one of the pioneers was WNEW-AM in New York. It premiered *Milkman's Matinee* (first broadcast midnight to 6 A.M. and later from 2 A.M. to 6 A.M.) on 6 August 1935, and the program remained on the air until 1992. During the same period, many major-market radio stations experimented with extended hours, with some confining their late-night programs to specific days of the week.

World War II led to a sharp increase in all-night radio programming. Feeling it their patriotic duty, many stations (among them WNAC and WEEI in Boston, WNEW and WOR in New York, WKBW in Buffalo, KDKA in Pittsburgh, WCAU in Philadelphia, and WRVA in Richmond) offered broadcasts for those legions of Americans working graveyard shifts in defense plants and factories.

After the war, all-night talk shows began to appear. Regarded by many as the father of the overnight call-in show, Barry Gray launched his program in New York City in 1945, and his "graveyard gab-a-thons" would remain among the most popular forms of this radio programming daypart. In the 1970s, all-night talk was given a significant boost with the debut of national call-in shows hosted by personalities such as Larry King and other extremely popular and sometimes controversial talkmasters. The use of toll-free 800 numbers enhanced the attraction of this format.

The primary appeal of overnight radio broadcasting lies in its companionability. A 1968 National Association of Broadcasters survey concluded that 60 percent of the all-night audience tunes in to keep from being lonely. All-night radio is also where the subcultures and countercultures tune to stay connected at an hour when the mainstream world is asleep. Insomniacs and third-shift workers, among them bakers, policemen, cab drivers, and convenience store clerks, constitute a substantial part of the loyal listenership of all-night radio. For aspiring disc jockeys and talk hosts, the overnight shift frequently serves as a training ground. It is where stations often put their most inexperienced on-air people to allow them the chance to build their skills. However, for some seasoned professionals, especially those in larger markets, the overnight shift is a preferred slot, because it is a segment of the program clock when rigid compliance to format strictures may be somewhat relaxed, thus providing them with greater opportunity to experiment and flex their creative muscles.

The program content of all-night radio tends to be eclectic. However, since the 1980s there has been a significant rise in the number of stations airing overnight talk shows, often syndicated programs, and canned programming is widely employed to fill the time slot. This period of the broadcast schedule has become famous for an often bizarre mix of program offerings. Psychics, paranormalists, conspiracy theorists, love therapists—to mention a few—are among the unique array of those who hold court over the night-time airwaves at hundreds of radio stations.

Outlets programming music over nights are frequently equally divergent in their approaches. In fact, music stations that feature a primary or single format during the day may shift gears to another, albeit complementary, form of music for their overnight hours. One quality many all-night music stations share in common is their tendency to soften or mellow their sounds to create a mood and atmosphere consonant with the nocturnal landscape. Jazz, blues, folk, and classical music are frequently given more airplay at night than during the day. Of course, not all stations vary or reconfigure their program clocks or playlists between midnight and 6 A.M. In fact, most actually mirror the programming they offer throughout the day and early evenings.

No programming ingredient has been more responsible for establishing loyal overnight followings than the radio personality; many of these have served their audiences for decades. The list of popular all-night hosts is long, if not endless. A partial list would certainly include Jean Shepherd, Norm Nathan, Franklin Hobbs, Joey Reynolds, Larry King, Henry Morgan, Jerry Williams, Jim Bohannon, Joe Franklin, Barry Farber, Larry Glick, Alison Steele, Ira Fistell, Ray Briem, Herb Jepko, Jean King, Larry Regan, Raechel Donahue, Long John Nebel, Wolfman Jack, Barry Gray, Mary Turner, Eddie Schwartz, John Luther, Mel Lindsay, Rollye James, Stan Shaw, Art Dineen, Hunter Hancock, Doug Stephan, Yvonne Daniels, Al Collins, Don Sainte John, Tom Snyder, Dave Wiken, and Art Ford. Every late-night listener has his or her favorite personality. Perhaps no all-night figure was more popular than Art Bell at the turn of the millennium. Broadcasting from a remote locale (near the infamous "Area 51") in the Nevada desert, Bell attracted an audience that consistently numbered in the millions from coast to coast.

Since 1988, Arbitron has rated overnight time slots at the behest of outlets in markets with potentially large listenerships during this segment of the broadcast schedule. Although all-night hours are not typically viewed as profit centers (in fact, they are more often thought of as "giveaway" zones), many stations have been successful enough in generating revenues to want audience statistics to help promote increased levels of advertising. In fact, the value of all-night radio as an advertising medium has risen in a world that has become increasingly 24-hour oriented.

With the predicted continuation of station consolidation and an even greater bottom-line emphasis, live and local all-night programming is expected to decline in the future. It is not unusual to find stations rebroadcasting daytime programs during their overnight hours, and the rise in bartered and syndicated shows is significant. Although the prospects for all-night radio are far from bleak, its growth will likely be influenced, if not substantially inhibited, by ongoing industry downsizing and the increasing reliance of the listening public on new forms of electronic media, such as the internet and digital audio on demand. Meanwhile, the program fare (politics and other hot-button issues) that has attracted all-night listeners for so many years appears to be on the wane, increasingly replaced by features that are far less provocative and controversial and thus more salable to advertisers.

MICHAEL C. KEITH

See also Jepko, Herb; King, Larry; Shepherd, Jean; Talk Radio; Williams, Jerry; Wolfman Jack

Further Reading

Douglas, Susan J., *Listening In: Radio and the American Imagination,* New York: Times Books, 1999

Keith, Michael C., *Sounds in the Dark: All Night Radio in American Life,* Ames: Iowa State University Press, 2001

Laufer, Peter, *Inside Talk Radio,* Secaucus, New Jersey: Carol, 1995

Munson, Wayne, *All Talk: The Talkshow in Media Culture,* Philadelphia, Pennsylvania: Temple University Press, 1993

Shepherd, Jean, *In God We Trust: All Others Pay Cash,* Garden City, New York: Doubleday, 1966

All Talk Format. *See* Talk Radio

All Things Considered

Public-Affairs Program

The seminal program of National Public Radio (NPR) first aired from an improvised studio in a run-down Washington, D.C., office building at 5:00 P.M. EDT on Monday, 3 May 1971. *All Things Considered* (*ATC*) marked the beginning of public radio as we know it. It also marked the culmination of more than a year of soul-searching about the purposes of this new enterprise. The task of defining public radio fell to the initial board of directors of NPR—a collection of managers from the largely moribund world of educational radio—and in particular to board member William Siemering, who declared NPR's first priority to be the creation of "an identifiable daily product which is consistent and reflects the highest standards of broadcast journalism." His report continued:

> This may contain some hard news, but the primary emphasis would be on interpretation, investigative reporting on public affairs, the world of ideas and the arts. The program would be well paced, flexible, and a service primarily for a general audience. It would not, however, substitute superficial blandness for genuine diversity of regions, values, and cultural and ethnic minorities which comprise American society; it would speak with many voices and many dialects.
>
> The editorial attitude would be that of inquiry, curiosity, concern for the quality of life, critical, problem-solving, and life loving. The listener should come to rely upon it as a source of information of consequence; that having listened has made a difference in his attitude toward his environment and himself.
>
> There may be regular features on consumer information, views of the world from poets, men and women of ideas and interpretive comments from scholars. Using

inputs from affiliate stations, for the first time the intellectual resources of colleges and universities will be applied to daily affairs on a national scale.

National Public Radio will not regard its audience as a "market" or in terms of its disposable income, but as curious, complex individuals who are looking for some understanding, meaning and joy in the human experience.

In the early 1970s, radio news remained a serious enterprise at commercial radio networks. The aura of Edward R. Murrow still surrounded Columbia Broadcasting System (CBS) News. CBS provided ten-minute newscasts on the top of the hour around the clock. The commercial networks maintained extensive staffs of correspondents around the world. It was against that staid but responsible incarnation of commercial radio news that NPR sought to define an "alternative." It needed to separate itself as well from the equally staid and responsible style of traditional educational radio. Although it was located in Washington, D.C., just blocks from the White House, NPR saw itself in an outsider role and took pride in not attending the news conferences to which members of the mainstream media flocked. Siemering asked, "Why do we always think that what the President did today is so important? Maybe it is more important that some unemployed person found a job today? Maybe that should lead our program?"

Siemering urged NPR to distinguish itself from commercial radio and its own past by "advancing the art of the audio medium." That mandate ultimately translated into a distinctive production style that took listeners to the scene of an event or into the lives of ordinary people instead of just talking about them. Producers would use microphones the way television reporters and documentarians used cameras. This was a new approach to radio journalism, because although listeners might recall the sounds of Murrow on the streets of London in World War II or the burning of the dirigible *Hindenburg,* these were the exceptions in radio news reporting before NPR. For the most part, radio news had meant reporters reading scripts. NPR sought to pioneer the regular use of sound in its reporting, drawing mental images more vivid than the pictures seen on television.

Even the titles of its programs suggested open-ended possibilities rather than a well-defined concept. Public radio's initial news program would be called not *The NPR Nightly News* but *All Things Considered,* suggesting both the unlimited range of its interests and the careful consideration it would give to all issues it tackled.

Siemering drafted the program's purposes as a member of the initial board. He then took on the task of implementing them as NPR's first director of programming. In the six months between his arrival at NPR's temporary facilities in Washington in November of 1970 and the program's debut the following May, Siemering took great care in hiring individuals who resonated with his vision statement. Concern with shared vision and personal compatibility outweighed more traditional standards, such as experience. In a sense, anyone with deep experience in traditional broadcasting and journalism had the wrong experience for what was to be a totally new departure. Symbolizing the priority of innovation over traditional broadcast standards, Siemering turned down an offer from the Ford Foundation to fund the salary of veteran news analyst Edward P. Morgan to anchor the new program. Siemering did not question Morgan's competence. He simply wanted something fresh.

All Things Considered's first anchor and "managing editor," then, was not Edward P. Morgan but a relatively obscure former reporter for *The New York Times* and National Broadcasting Company (NBC) Television named Robert Conley. Conley brought with him one of his former editors at *The Times,* Cleve Matthews, who would direct the news operations. Josh Darsa, formerly of CBS Television, gave NPR another experienced hand and another staff member over the age of 30. In contrast to Conley, Matthews, and Darsa, however, a motley group of young, idealistic, creative, energetic men and women more comfortable with the counterculture than conventional journalism dominated the initial staff of *ATC* and formed its distinctive personality. Men still dominated broadcasting and journalism in 1971, and affirmative action was not yet the law of the land, but Siemering hired as many women as men for the initial staff and insisted on minority representation. Only a staff representative of the whole population, he felt, could produce a program that really served the needs and interests of the whole population.

True to the spirit of the times, the character of the staff, and Siemering's own instincts, the program operated as something of a commune during its first nine months. No one was in charge. Three different individuals hosted the program during that time. The program's vision was open ended, its implementation unreliable, and the working conditions chaotic. NPR management told Siemering he would lose his job if he did not give someone operating authority over the staff and program. After consulting members of "the commune," Siemering gave the task to one of the "senior" members of the staff, 30-year-old Jack Mitchell, who imposed a structure on the previously free format and sought to develop a more consistent personality for the show. He changed the theme music from a playful little tune composed by Don Voegeli (a veteran musician at WHA in Madison, Wisconsin) to a more forceful "news" version of Voegeli's basic melody. He broke the program into three half-hour segments, fixed newscasts at the top of each hour, and organized each half hour to move from "hard news" to softer features. Commentaries by "real people" from across the country were an attempt to realize Siemering's democratic vision of radio. Commentaries by immortal broadcasters such as Goodman Ace, John Henry

(Left to Right) Robert Siegel, Linda Wertheimer, Noah Adams, *All Things Considered*
Courtesy National Public Radio

Faulk, and Henry Morgan provided a link to radio's golden age. Key to the program's evolving personality, however, would be two program hosts, a man and a woman, who would conduct their own interviews in addition to introducing produced reports. Siemering had described the hosting role as a neutral, unobtrusive "picture frame" that never called attention to itself; Mitchell moved the hosts into the picture.

ATC's first hosting team paired Mike Waters, a man with a warm voice and a natural ability to tell engaging stories, with Susan Stamberg, an enthusiastic interviewer willing to laugh out loud and reveal the emotions of a real human being. Stamberg was the first female co-anchor of a major national nightly news program, as NPR pointed out in newspaper ads when American Broadcasting Company (ABC) Television touted its appointment of Barbara Walters to co-anchor its evening network newscast as a "first." As much as her gender, Stamberg's New York accent and brash personality polarized listeners, stations, and the NPR management. When she talked on the air about her son Josh or gave her famous recipe for cranberry relish each Thanksgiving, Stamberg provided a human identity that told the world that this program was different from anything else on the air. Susan Stamberg came to personify *All Things Considered*, NPR, and public radio as a whole. Neither Waters nor Stamberg considered themselves journalists. Neither was especially interested in "news." Indeed, their ability to identify with the lay listener, posing "uninformed" questions that any reasonably intelligent non-expert might ask about complex issues, may have been part of their appeal.

In 1974 *ATC* expanded into the weekend. Waters took over the weekend assignment and was replaced on weekdays by Bob Edwards, another fine voice and excellent reader who, like Waters, provided a calming counterpoint to Stamberg's exuberance and who, unlike Waters, cared deeply about news. Many believe that Stamberg and Edwards set the standard for NPR hosting during their five years together, a period in which listeners in large numbers discovered the program and bonded with the cohosts and the institution they represented. This "perfect" combination split in 1979, when Edwards moved to host NPR's second major news effort, *Morning Edition.*

Morning Edition grew out of the example of *All Things Considered,* drawing its values, approach, and even its host from the older program. At the same time, however, *Morning Edition* changed *All Things Considered.* It broke up the team of Stamberg and Edwards, of course, but more important, *Morning Edition* turned NPR into a 24-hour-a-day news operation and forced the network to drastically increase its news-gathering capacity. Future ABC News star Cokie Roberts began her broadcasting career at NPR at about that time, joining the indomitable Nina Totenberg, who had come to NPR four years earlier, to become two of Washington's most prominent and respected news reporters. Although Siemering had written that NPR's daily magazine would include "some hard news," the program instead evolved into a primary vehicle for breaking news coverage. NPR transformed itself into a competitive news organization, eventually filling the void for quality journalism created by the decline of serious news reporting and analysis on commercial radio and the simultaneous reduction of foreign news-gathering capacity by those networks. *All Things Considered* became more serious in its approach to news, constantly raising its journalistic standards and focusing increasingly on international news as commercial radio abandoned these interests. The "alternative" to traditional broadcast journalism became the bastion of journalistic standards. The hard news squeezed, but never eliminated, the softer elements of the program that had distinguished *ATC* in its early years.

Sanford Unger, a print reporter with a journalistic résumé far stronger than that of any previous NPR host, symbolized *ATC's* evolving role as a serious player in the world of Washington journalism when he took over Edward's seat next to Stamberg. Unfortunately, Unger's on-air persona did not match his journalistic prowess. He moved on, to be replaced in 1983 by Noah Adams, who, like Mike Waters, was an "unknown" from within the NPR staff who had a soothing voice, a near-magical sense of radio, and only a marginal interest in hard news.

Tiring of the daily grind and not quite comfortable with the hardening of *ATC's* news values, both Stamberg and Adams left the program four years later in 1987—Stamberg to host the new *Weekend Edition Sunday,* a feature-oriented program that would better suit her interests, and Adams to Minnesota to take over the Saturday night slot vacated by Garrison Keillor when Keillor moved to Denmark to live with his new wife.

Thus, in a sense, *All Things Considered* started over in 1987 with a harder journalistic edge. News Vice President Robert Siegel decided to give up his big office for the chance to anchor *All Things Considered.* Reporter Rene Montaigne joined him as cohost for one year. At the end of that year, Garrison Keillor was back on Saturday night, and Noah Adams was back at NPR. The news-oriented Siegel and the feature-oriented Adams might have made an outstanding complementary team had they not both been men. NPR solved the dilemma by adding political and congressional correspondent Linda Wertheimer to the mix as a third host. Siegel, Adams, and Wertheimer would share the hosting duties, and each would have time to do some reporting as well. That arrangement served the interests of the three hosts and the philosophy of NPR's president, Doug Bennet, who chose to downplay individual personalities in favor of NPR's institutional identity as a news organization. Three interchangeable hosts on *All Things Considered* symbolized the interchangeability of reporters and other staff in an organization whose success and credibility should transcend that of any individual.

NPR's next president, Delano Lewis, expanded *All Things Considered* from 90 minutes to two hours in 1995 and moved the program forward by one hour to 4:00 P.M. EDT, to better fit peak afternoon drive-time listening. At the same time, breaks were added within each half hour, dividing the program into shorter segments similar to its companion program, *Morning Edition.* The extra time allowed *All Things Considered* to do more "soft" material among the "hard news."

Beginning with its first Peabody Award in 1973 for its "distinctive approach to broadcast journalism," *All Things Considered* has won virtually every award for broadcast journalism excellence.

JACK MITCHELL

See also Easy Aces; Edwards, Bob; Faulk, John Henry; Morgan, Henry; Morning Edition; National Public Radio; Peabody Awards; Public Radio since 1967; Siemering, William; Simon, Scott; Stamberg, Susan; Totenberg, Nina; Wertheimer, Linda

Hosts

Robert Conley (1971), Jim Russell (1971), Mike Waters (1971–74), Susan Stamberg (1972–86), Bob Edwards (1974–79), Sanford Unger (1980–83), Noah Adams (1983–86), Robert Siegel (1986–87), Rene Montaigne (1986–87), Robert Siegel (1988–), Noah Adams (1988–2002), Linda Wertheimer (1988–2002), Michelle Norris (2002–), Melissa Block (2003–)

Programming History

National Public Radio 1971–present

Further Reading

Collins, Mary, *National Public Radio: The Cast of Characters,* Washington, D.C.: Seven Locks Press, 1993

Engelman, Ralph, *Public Radio and Television in America: A Political History,* Thousand Oaks, California: Sage, 1996

Looker, Thomas, *The Sound and the Story: NPR and the Art of Radio,* Boston: Houghton Mifflin, 1995
Siegel, Robert, editor, *The NPR Interviews, 1994,* Boston: Houghton Mifflin, 1994
Stamberg, Susan, *Every Night at Five: Susan Stamberg's All Things Considered Book,* New York: Pantheon, 1982
Wertheimer, Linda, editor, *Listening to America: Twenty-Five Years in the Life of a Nation, As Heard on National Public Radio,* Boston: Houghton Mifflin, 1995

Allen, Fred 1894–1956

U.S. Radio Comedian

Fred Allen was one of the most successful radio personalities during the 1930s and 1940s. Allen's forte was topical humor. He was a satirist, drawing much of his material from the day's events. While he occasionally commented on the headlines, he more often zeroed in on the smaller "filler" items that graced the newspapers. Even though his greatest appeal was to the "thinking" audience, Allen also enjoyed enormous popular appeal. Author John Steinbeck once called Allen "unquestionably the best humorist of our time." Among his many fans were Groucho Marx, Jack Benny, and George Burns. By the end of his career Fred Allen became one of the country's leading authorities on humor.

Origins

Fred Allen was born John Florence on 31 May 1894 in Cambridge, Massachusetts. His mother died when he was very young and his father died when he was 15. After their father's death, he and his brother, Bob, lived with their aunt. At 18 he began performing on the "amateur night" circuit of vaudeville around Boston. His ability to juggle led to a hokey comedy routine. He was billed as Paul Huckle, European Entertainer, for his first appearance. He later became Fred St. James, the juggler. He soon changed his name to Freddie James and billed himself as the world's worst juggler. As time went on, the juggling became a smaller part of his act and the comedy became more prominent; he eventually abandoned juggling altogether.

In 1914 he left for New York City to try to break into big-time vaudeville. Not finding much success, in 1916 he left for a tour of Australia. While there, he polished his act and returned to the United States a year later. Upon his return he joined the New England vaudeville circuit. By 1920 he was an established vaudeville headliner, now known as Fred Allen. In 1922 Allen took his act to Broadway and there met his future wife, Portland Hoffa.

Radio

By the early 1930s radio had established itself as an important venue for many of vaudeville's top talent; by the fall of 1932 most vaudeville acts had made the switch to radio. Among the medium's new performers were singers Bing Crosby, Al Jolson, and Paul Robeson and comedians Jack Benny, Ed Wynn, the Marx Brothers, George Burns and Gracie Allen, and Fred Allen. Allen's first radio program *The Linit Bath Club Revue,* premiered on the Columbia Broadcasting System (CBS), Sunday night, 23 October 1932. Allen put his all into the weekly show, regularly working 80-hour weeks to prepare each week's program. The *Linit* show was heavily influenced by the vaudeville style—stiff stage comedy and stand-up patter backed by music. His show changed its sponsors and its name three times during the next year, first to *The Salad Bowl Revue* then to *The Sal Hepatica Revue,* then to *The Hour of Smiles.* Finally in 1934 Allen's show became famous as *Town Hall Tonight.*

Like most programs of the era, Allen's followed a formula from the beginning. In his earlier series each episode would be built around a different occupation. One week he would play a plumber, the next week a banker, then a druggist, etc. By the time *Town Hall Tonight* emerged, the formula had evolved to what is now commonly known as the "comedy variety" format. This format includes different comedic skits or monologues interspersed with music. Another feature of *Town Hall Tonight* was a regular amateur talent segment. In fact, almost half of each episode was devoted to amateur talent. Like the other elements of the program, the amateur segment was extremely popular with the audience. It was also popular with the sponsor and the network because it was very inexpensive to produce. In fact, the amateur talent segment proved so popular that it spawned a number of amateur talent programs, the most notable being *The Original Amateur Hour* of Major Bowes. Some of the amateur talent that was discovered on Allen's program included comic, and later radio and TV host,

Fred Allen
Courtesy Radio Hall of Fame

Garry Moore, actress Ann Sheridan, and a young Frank Sinatra.

It was on *Town Hall Tonight* on the evening of 30 December 1936 that Allen started his famous on-air "feud" with Jack Benny. The feud began inadvertently when Allen ad-libbed a joke about Benny's inability to play the violin well. The following week Jack Benny responded to this remark on his own program and the feud was on. The feud played on over the next three months. In March 1937 on *The Jack Benny Program* a mock fistfight, called "the battle of the century" was staged to resolve the feud once and for all. The fight was to have taken place in the ballroom of the Hotel Pierre on March 14. There was no actual winner of this contrived fistfight, and both men returned from the boxing ring tattered but friends. This was all in good fun, of course, because Allen and Benny had always been and continued to be the best of friends off the air. Although the fistfight was to have been the end of this feud, the public enjoyed it so much that Allen continued it until he left the air in 1949.

After a six-year run, *Town Hall Tonight* ended in 1939 and became simply *The Fred Allen Show*. The new weekly program contained musical numbers, a feature comedy skit by the newly created Mighty Allen Art Players (a device that was so successful it was paid homage to many years later on television by Johnny Carson in the form of the Mighty Carson Art Players), and a "News of the Day Newsreel." The newsreel was typically a satire about some obscure recent event. Harry Von Zell was the announcer and Peter Van Steeden was the musical conductor on the program.

In the fall of 1940 Allen returned to CBS and his program was renamed *The Texaco Star Theatre*, after its sponsor Texaco Gasoline. The CBS version was initially scaled down to 30 minutes; in 1941 it went back to an hour-long format for a year, but in 1942 the program returned to the 30-minute format that it would retain until its end; the show was aired on Sunday nights.

Allen's Alley

The 1940s brought change to radio. Old-style vaudeville comedy had lost its appeal. The public was becoming more sophisticated and thus expected its comedy to be more sophisticated as well. During the early 1940s "Allen's Alley" emerged. The Alley was not your typical street. In fact, it might have been the most atypical street anywhere. The actual description of the Alley was always left to the imagination of the listener. Very few details about its appearance were ever offered. The residents of "Allen's Alley" became household names, and were among the best-known characters of radio's golden age. On "Allen's Alley," Fred often visited the Brooklyn tenement of Mrs. Nussbaum (Minerva Pious). Mrs. Nussbaum was always relating her weekly problems with her husband Pierre. The network executives were concerned that Mrs. Nussbaum's thick Jewish accent might offend some listeners, but the overwhelming reaction by the audience was favorable. Also on the Alley were the farmhouse of Titus Moody (Parker Fennelly), the shack where Ajax Cassidy (Peter Donald) lived, and the antebellum mansion of southern Senator Beauregard Claghorn (Kenny Delmar). Moody was one of the funniest characters on the show, always greeting Allen with the friendly "Howdy, Bub." Cassidy, the loudmouth Irishman, was probably the least popular character on "Allen's Alley," while Senator Claghorn became the star of the Alley. His famous refrain "that's a joke, son" became part of the popular lexicon of the day.

There were other characters that became audience favorites—Senator Bloat; John Doe, the angry average citizen; and poets Falstaff Openshaw, Humphrey Titter, and, of course, Thorndyle Swinburne, the poet laureate of Boston Post Road. In addition to the appearances of his regular characters, some of Allen's most memorable moments came when he would spoof musicals. On one program in the early 1940s he and Orson Welles did a hilarious five-minute version of "Les Miserables." A few years later, his parody of Gilbert and Sullivan's "The Mikado," called "The Radio Mikado," was a smash with the audience and still remains a classic.

Throughout the 1940s Allen was frequently at odds with network officials. His battles with the networks usually involved the content of his remarks, which were often derogatory about network officials. Although ordered to stop these remarks, Allen continued. As a result he was required to submit verbatim scripts to the networks before air for their approval. The network often required him to change or delete items. Allen eventually started to include items in the script that he had no intention of using on air, so that he could use those items as bargaining chips; he would agree to cut one of those items in exchange for keeping something else. In addition Allen would often ad-lib material. As radio programs were broadcast *almost* live (there was a delay of a few seconds), the audience would sometimes hear a bleep in place of a word or phrase. On 20 April 1947 his show had 30 seconds of dead air because he refused to delete a controversial joke. Finally an angry public forced the bleeping to stop. Sometimes Allen's shows ran overtime and were cut off. Once a show ended in the middle of a skit. The following week Allen began the show with the remainder of the skit and an angry, but witty, tirade against the network for being so fussy about staying to a regular schedule.

Demise

Allen left the air for health reasons (high blood pressure) in 1944. In 1945 he returned and the title of his program returned to *The Fred Allen Show*, which it kept until its cancellation in 1949. A young Arthur Godfrey served briefly as Allen's announcer during the program's later years. Airing under different titles, *The Fred Allen Show* had a very successful 17-year run on network radio. Allen's final radio show was on 26 June 1949 and, fittingly, his last guest was Jack Benny. The program finally ended owing to a combination of competition (*Stop the Music*) and an unchanging format that had begun to wear as radio gave way to television.

Although Fred Allen was the driving force on the air and behind the scenes, many people contributed to the success of Allen's programs throughout its long run. Some of the more prominent were writers Nat Hiken, Larry Marks, and Herman Wouk; announcers Kenny Delmar, Jimmy Wallington, and Harry Von Zell; directors Vick Knight and Howard Reilly; and musical directors Al Goodman, Ferde Grofe, Lennie Hayton, Lou Katzman, and Peter Van Steeden. The most prominent sponsors during Allen's run included Bluebonnett Margarine, the Ford Motor Company, Hellmann's Mayonnaise, Ipana and Sal Hepatica, Shefford's Cheese, Tenderleaf Tea, Texaco Gasoline, and V-8 Vegetable Juice. Allen's long-standing theme song was "Smile, Darn Ya, Smile."

Like many radio personalities of the day, Fred Allen tried to make the transition to television. Unlike many of his contemporaries, the transition was difficult for Allen, because he had an inherent distrust of television. He once said of television, "They call it a medium because nothing on it is ever well done." In the same vein, he once quipped "Television is a triumph of equipment over people, and the minds that control it are so small that you could put them in the navel of a flea and still have enough room beside them for a network vice president's heart."

Allen's first venture into television was on NBC as one of the rotating hosts, along with Eddie Cantor and the team of Dean Martin and Jerry Lewis, of *The Colgate Comedy Hour.* The program debuted in September 1950 and ran for a number of years, but Allen was dropped from the rotation after only three months. His next venture was also as a rotating host (along with Bob Hope and Jerry Lester) of NBC's *Chesterfield Sound Off Time* in October 1951. The program lasted just three months. In 1952 he was to start another TV series when his first heart attack forced him into retirement. He came back in the fall of 1953 to host an NBC quiz show called *Judge For Yourself.* The program was cancelled after one season. Finally, in 1954 Allen took a panelist spot alongside Bennett Cerf, Arlene Francis, and Dorothy Kilgallen on the successful CBS quiz program *What's My Line?* Allen remained on *What's My Line?* until his death on 17 March 1956.

MITCHELL SHAPIRO

See also Benny, Jack; Comedy; Vaudeville

Fred Allen. Born John Florence in Cambridge, Massachusetts, 31 May 1894. Worked in vaudeville, performing on the amateur night circuit around Boston as a juggler and comic under the names Paul Huckle, Fred St. James, and Freddie James, 1912–14; moved to New York City trying to break into big-time vaudeville, 1914–16; left for Australia, toured the vaudeville circuit as a comic, 1916–17; joined New England vaudeville circuit as comic, 1917–20; changed name to Fred Allen, became a vaudeville headliner; 1920–22; came to Broadway as a vaudeville comic, 1922–32. Radio debut with *The Linit Bath Club Revue*, 1932, the first of several series; guest roles on radio series, 1949–50; rotating host of TV series *The Colgate Comedy Hour,* 1950; rotating host of TV series *Chesterfield's Sound Off Time*, 1951; emcee of TV series *Judge For Yourself*, 1953–54; panelist on TV series *What's My Line?*, 1954–56. Died in New York City, 17 March 1956.

Radio Series

1932–33	*The Linit Bath Club Revue*
1933	*The Salad Bowl Revue*
1934	*The Sal Hepatica Revue*
1934	*Hour of Smiles*
1934–39	*Town Hall Tonight*
1939–40	*The Fred Allen Show*
1940–44	*The Texaco Star Theatre*
1945–49	*The Fred Allen Show*

Television Series

The Colgate Comedy Hour, 1950; *Chesterfield Sound Off Time,* 1951–52; *Judge For Yourself,* 1953–54; *What's My Line?* 1954–56

Selected Publications

Treadmill To Oblivion, 1954
Much Ado About Me, 1956

Further Reading

Allen, Steve, "Fred Allen," in *The Funny Men,* by Steve Allen, New York: Simon and Schuster, 1956

Douglas, Susan J., *Listening in: Radio and the American Imagination: From Amos 'n' Andy and Edward R. Murrow to Wolfman Jack and Howard Stern,* New York: Times Books, 1999

Harmon, Jim, *The Great Radio Comedians,* Garden City, New York: Doubleday, 1970

Havig, Alan R., *Fred Allen's Radio Comedy,* Philadelphia, Pennsylvania: Temple University Press, 1990

McCarthy, Joe, editor, *Fred Allen's Letters,* Garden City, New York: Doubleday, 1965

Nachman, Gerald, *Raised on Radio: In Quest of the Lone Ranger, Jack Benny . . . ,* New York: Pantheon Books, 1998

Taylor, Robert, *Fred Allen: His Life and Wit,* Boston: Little Brown, 1989

Allen, Mel 1913–1996

U.S. Sportscaster

Mel Allen was the mellifluous "Voice of the New York Yankees" for 25 years, announcing during the era when that baseball team was regularly the World Series champion. Closely identified with the Yankees during the decades when baseball dominated U.S. sports, and a legend himself in his lifetime, Allen created heroes and dreams for radio listeners during the post-World War II period. Allen was a Yankee partisan, cheering for every hit and strikeout, and his impact was so great that he had more fans of his own than any other announcer. During the 1960s, he also represented tradition and constancy in a period of great social turmoil.

Yankee Years

Allen got his first part-time broadcasting job in 1933 as the voice of the University of Alabama's "Crimson Tide" on CBS's Birmingham affiliate. After completing law school, he was hired by CBS in 1937 and became the New York Yankee broadcaster in 1940. After army service in World War II, Allen returned to the Yankees.

The Yankees became the first professional baseball team to air all its games live, ending studio recreations of away games and giving Allen a bigger canvas to fill with stories of baseball life. For a time, as the Yankees' premiere announcer, he was joined by the equally legendary Red Barber, and as a team they were considered the best in radio. Beginning in 1941, as both a team and network broadcaster, Allen announced 20 World Series and did play-by-play for 24 All-Star Games, more than any of his contemporaries; he also called 20 college bowl games, including numerous Rose Bowls. From 1939 to 1943, Allen broadcast the New York Giants baseball games as well. One of the world's most widely recognized voices in the two decades after World War II, he narrated over 2,000 Fox Movietone newsreels as well as hundreds of short film subjects for 20th Century Fox. An enormously valued advertising spokesperson, he sold millions of dollars of Ballantine beer to Yankee baseball fans.

Allen broadcast nearly every major Yankees event from Joe DiMaggio's 56-game hitting streak in 1941 to Don Larsen's historic perfect game in the 1956 World Series to Roger Maris' record-breaking 61 home runs in 1961. Allen also handled sad farewells. On 4 July 1939, he introduced Lou Gehrig (who was then dying of what is now called Lou Gehrig's disease) to a packed Yankee Stadium before Gehrig's gripping farewell speech, "Today, I am the luckiest man in the world." In 1948 Allen also introduced a dying Babe Ruth at Yankee Stadium.

Allen was a truly gifted storyteller, sometimes carrying on long monologues that enchanted fans during slow games. He was widely admired for his comprehensive knowledge of baseball and his affection for the home team (expressed in such trademark phrases as "How about that!" and "Going, going, gone!"). Allen's authoritative words were spoken in an Alabama twang that was instantly recognizable ("Hello everybody. This is Mel Allen."). He rarely misspoke or made outright errors, and when he did, he passed them off lightly and fans forgave him. During his years with the Yankees, he created such enduring nicknames as "Joltin' Joe" DiMaggio, "Old Reliable" Tommy Henrich, and "The Scooter" Phil Rizzuto. During the Yankees' heyday, his inimitable voice was found on the radios of almost every taxi in New York City.

In an ebullient style markedly differing from Barber's detached precision, Allen peppered his broadcasts with enthusiasm for the home team, but both men were intense and focused on baseball—Allen especially so because he never married or had a family. Because he saw athletes as idols and role models for children, he felt that educating new fans about baseball was an important part of his job. This led to criticism that Allen talked too much during games, explaining rules that most fans knew quite well. But his explanations were for the new fans, he always claimed, insisting that his detailed descriptions of events viewers could see for themselves (in the television years) helped heighten the excitement. He tended to magnify the players' admirable attributes in his stories. Beyond DiMaggio's memorable home runs and outstanding batting averages, for example, Allen especially admired DiMaggio's team leadership and often drew attention to it during broadcasts. Allen was such a great favorite with both those who attended games and those who heard them over the radio that, in 1950, Yankee fans held a Mel Allen Day celebration. It raised money that he donated to the Lou Gehrig Scholarship Fund and the Babe Ruth Scholarship at the University of Alabama.

Later Career

To widespread surprise at the time, Allen was inexplicably fired by the Yankees in 1964. After a hiatus, Allen became the voice of the syndicated *This Week in Baseball* in 1977 and remained so for nearly all of the next 20 years. In 1985 he once again became the voice for Yankee games on cable television. Allen was a pioneer broadcaster with a magnetic personality and great personal charm. Those who knew him usually loved him (unless they were Yankee-haters), and thousands of fans,

Mel Allen
Courtesy CBS Photo Archive

scores of baseball players, and his broadcasting peers considered him the most unforgettable sportscaster. Hall of Fame sportscaster Lindsey Nelson called him "the best sports broadcaster of my time."

Among innumerable awards and honors, such as selection as the nation's top sportscaster for 14 consecutive years (a feat matched by none), Allen was inducted into the National Sportscasters and Sportswriters Hall of Fame in 1972, and along with Red Barber, received the first Ford C. Frick award in 1978 (placing him in the sportscaster section of the Cooperstown Baseball Hall of Fame). He was selected for the American Sportscaster Hall of Fame in 1985 and the Radio Hall of Fame in 1988. In 1992 he received the Bill Slocum Award for long and meritorious service to baseball. Allen died in 1996 at age 83.

SUSAN TYLER EASTMAN

See also Sports; Sportscasters

Mel Allen. Born Melvin Allen Israel in Johns, Alabama, 14 February 1913. Attended University of Alabama, BA, 1932; LLB, 1936; CBS Radio sportscaster, 1937; voice of New York Yankees, 1940–64; hosted *This Week in Baseball*, 1977–95; hosted Sports Channel's television coverage of New York Yankee games, 1985–96. Died in Greenwich, Connecticut, 16 June 1996.

Selected Publications

You Can't Beat the Hours: A Long, Loving Look at Big League Baseball, Including Some Yankees I Have Known (with Ed Fitzgerald), 1964

It Takes Heart (with Frank Graham, Jr.), 1959

Further Reading

Davidson, Bill, "Mel Allen: Baseball's Most Controversial Voice," *Look* 24 (27 September 1960)

Harwell, Ernie, *Tuned to Baseball,* South Bend, Indiana: Diamond Communications, 1985

Smith, Curt, *The Storytellers: From Mel Allen to Bob Costas: Sixty Years of Baseball Tales from the Broadcast Booth,* New York: Macmillan, 1995

Alternative Format

Responding to a perceived lack of inventiveness on rock music stations, some musicians and modern rock fans embraced a more experimental, less packaged, alternative sound in the 1990s. Compared to the mainstream rock primarily played on Album-Oriented Rock (AOR) stations, alternative was unpolished and unabashed; its lyrics spoke of both idealism and disenfranchisement. Radio programmers, recognizing a new trend with counter-programming potential, added alternative tracks to their playlists, developing what became known as the *alternative format.* For advertisers and record labels seeking to expand their reach, commercial alternative formats provide a highly targeted and efficient medium similar to the alternative press.

The alternative format has been implemented in a variety of ways. The hard edged modern rock version may include talk-ups, sounders, and contesting similar to Contemporary Hit Radio (CHR). Adult Alternative Album (AAA) is essentially an album-oriented rock format, but with an alternative playlist. College alternative radio, where much of the sound found its original support, often takes an eclectic approach. Other variants may include shock jocks, techno music, or music with urban appeal.

Although the alternative radio movement emerged in the 1990s, its lineage extends back through the punk rock/new wave movement of the late 1970s and progressive radio of the late 1960s to rebellious rock and roll radio of the 1950s. Each of these movements emerged from a fervent subculture demonstrating a certain disdain for what was perceived as popular music at the time. Characterized by garage bands, small venue live performances, and low budget recordings distributed by independent labels, alternative's back-to-basics approach, rejection of glitzy production, and youthful self-expression have paradoxically had popular appeal as Music Television (MTV), college radio, and rock promoters began successfully packaging and selling the new musical genre to an increasingly fragmented market. Commercial radio success, initially in the San Francisco and Seattle areas, the popularity of alternative music on college campuses, and digital distribution—including MP3 audio files on the internet—have all contributed to the vibrancy of the alternative format.

Alternative has also been known as the anti-format, associated with independent, community stations focused on political issues and social change, such as those operated in the U.S. by the Pacifica group. It may also refer to those global broadcasters with alternative worldviews and alternative means of distribution, such as the internet.

JOSEPH R. PIASEK

See also Album-Oriented Rock Format; Contemporary Hit Radio Format/Top 40; Pacifica Foundation; Progressive Rock Format

Further Reading

Campbell, Richard, *Media and Culture: An Introduction to Mass Communication,* New York: St. Martin's Press, 1998; 2nd edition, by Campbell, Christopher R. Martin, and Bettina Fabos, Boston: Bedford/St. Martin's Press, 2000

Free Speech Radio News website, <www.fsrn.org>

Lasar, Matthew, *Pacifica Radio: The Rise of an Alternative Network,* Philadelphia, Pennsylvania: Temple University Press, 1999

Vivian, John, *Media of Mass Communication,* Boston: Allyn and Bacon, 1991; 5th edition, 1998

Walker, Jesse, *Rebels on the Air: An Alternative History of Radio in America,* New York: New York University Press, 2001

Amalgamated Broadcasting System

U.S. Radio Network

The Amalgamated Broadcasting System, which survived as a corporation for 13 months but operated as an actual radio network for a mere five weeks, is better known for the myths surrounding it than for the facts of its brief existence. Despite the claims of many so-called "old-time radio" scholars, Amalgamated was never associated with station WNEW in New York (which did not exist until after the network fell into bankruptcy) and the EW in that call sign did not derive from the initials of Amalgamated's founder, comedian Ed Wynn.

Origins

Amalgamated was founded in the fall of 1932 as a program production agency. Wynn's partners in the venture were Broadway producers Arthur Hopkins and T.W. Richardson, and Hungarian-born violinist and promoter Ota Gygi. Initial press releases hinted that Irving Berlin and Daniel Frohman were interested in the project and that more than a million dollars had been committed by two nationally known agencies.

Despite these claims, nothing further was heard of the venture until January 1933 when George W. Trendle, president of the newly formed All-Michigan Network, announced his alliance with the Wynn group. The New York flagship of the network would be made up of an amalgamation of three small time-sharing stations controlled by Walter Whetstone's Standard-Cahill Corporation: WBNX, WCDA, and WMSG. This would be the first step, declared Trendle, toward building a nationwide chain of low-powered regional stations. Trendle claimed, without naming names, that five Detroit millionaires were backing the venture and that Wynn had enlisted the support of practically every theatrical man of note and 13 prominent authors.

The next five months were filled with promises but little substance. Wynn went into detail in the trade press describing the policy of the new network, declaring that it would limit advertising to indirect messages at the beginning and end of each program and that he himself planned to appear occasionally on the network once his National Broadcasting Company–Texaco contract expired. Studio space was prepared in a newly constructed building at 501 Madison Avenue in New York City, arrangements were made with Western Union for network lines, and several dates were announced for the start of broadcasting, only to be postponed at the last minute.

The industry was fast losing patience with Wynn's stalling. In the 1 June 1933 issue of *Broadcasting*, editor Sol Taishoff portrayed Amalgamated as an amateurish, slipshod operation, run entirely by show people who were decidedly naive about the realities of the broadcasting business. There was no longer any mention of Detroit millionaires and, even though the son-in-law of President Roosevelt, Curtis B. Dall, joined the company in August as chairman of the board, it was becoming evident to observers that the network's money was coming primarily out of Ed Wynn's pocket.

Operations

The Amalgamated network finally went on the air on the evening of 25 September 1933. Thirteen small Eastern stations carried the initial program—flagship WBNX; WCNW in Brooklyn, New York; WPEN in Philadelphia, Pennsylvania; WDEL in Wilmington, Delaware; WCBM in Baltimore, Maryland; WOL in Washington, D.C.; WCAP in Asbury Park, New Jersey; WHDH in Boston, Massachusetts; WCAX in Burlington, Vermont; WPRO in Providence, Rhode Island; WNBH in New Bedford, Massachusetts; WSAR in Fall River, Massachusetts; and WFAS in White Plains, New York. Despite claims that the network would soon span the continent, no additional stations were ever added.

Critic Ben Gross described the inaugural broadcast as chaotic, but a surviving recording reveals that it was in reality a dull hodgepodge of mediocre talent, the major exception being an appearance by the dynamic Broadway vocalist Jules Bledsoe. There were no commercial announcements, but on-air credits were quietly slipped into the program for the firms that provided the bar for the guests and the beer they were served. Although Gross claimed that there were hundreds of complaints from listeners unable to hear the broadcast because the noise from the rowdy studio audience drowned out the performers, the recording makes it clear that this was a fabrication. The only complaints noted in a post-broadcast article in *Broadcasting* were from technicians, who suggested that the Western Union telegraph network lines were somewhat noisier than AT&T telephone circuits. Wynn himself was not present for the inaugural, because he was occupied with motion-picture duties in Hollywood and had left Gygi in full charge of the network in his absence.

Collapse

The story of Amalgamated has a beginning and an end, but no middle. No sponsorships were ever sold. During October, a 15-hour-a-day schedule of music and talk was fed to the small eastern hookup that had taken the opening broadcast, but clearance of these sustaining programs (programs not paid for by advertising) proved difficult when affiliates insisted on carrying their own local, sponsored features.

Curtis Dall resigned as chairman of Amalgamated in early October. Ed Wynn returned to New York in mid-October and soon realized the futility of the venture. He resigned on 23 October, claiming to have spent $250,000 on the project with no hope of any return. Subsequent investigation by receivers revealed that Wynn's out-of-pocket investment was closer to $125,000.

At midnight on 1 November the network halted service to its 12 affiliates. On 3 November creditors foreclosed and Amalgamated passed into the hands of the Irving Trust Company. Liabilities totaled $38,000, with $10,000 owed in salaries to the company's 200 employees.

The assets of the network were sold at auction on 18 December, raising $10,841 toward the settlement of outstanding claims. The studio equipment was purchased for $9,800 by advertising executive Milton Biow for use in his new station in Newark, WNEW. WNEW would subsequently lease the former Amalgamated studio space at 501 Madison Avenue for its New York studio.

Ed Wynn resolved to settle all of the network's remaining debts. The stresses involved in the Amalgamated venture contributed to the failure of his marriage in 1937 and ultimately to a nervous breakdown. Ever the promoter, Ota Gygi spent much of 1934 trying to form yet another "new network" among stations in the Midwest, but he had lost all credibility. Several former Amalgamated stations became part of George Storer's American Broadcasting System.

ELIZABETH MCLEOD

See also Wynn, Ed

Further Reading

"ABS Auction Sale Raises Back Pay," *Broadcasting* (1 January 1934)
"ABS Bankrupt As Comedian Is Blamed," *Broadcasting* (15 November 1933)
"ABS Chain Makes Debut," *The Billboard* (30 September 1933)
"ABS Swan Song," *Broadcasting* (15 November 1933)
"Amalgamated Net Gets Started," *Broadcasting* (1 October 1933)
"Creative Radio Program Service Headed by Ed Wynn," *Broadcasting* (1 October 1932)
"Ed Wynn Resigns Amalgamated Post," *Broadcasting* (1 November 1933)
Gross, Ben, *I Looked and I Listened: Informal Recollections of Radio and TV*, New York: Random House, 1954
"New Third Network Embraces Old Plans," *Broadcasting* (1 February 1933)
Taishoff, Sol, "Lack of Practical Broadcaster Hampers Wynn Network Venture," *Broadcasting* (1 June 1933)
Wynn, Ed, "Why a Third Chain?" *The Billboard* (30 September 1933)
Wynn, Keenan, and James Brough, *Ed Wynn's Son*, Garden City, New York: Doubleday, 1959

Ameche, Don 1908–1993

U.S. Radio, Film, and Television Performer

As a versatile singer, actor, and host, Don Ameche was one of radio's earliest male stars and one of the medium's most popular figures in the 1930s and 1940s. He was also a relatively rare example of a star who maintained his radio career despite a period of significant Hollywood film success.

Ameche was born Dominic Amici in Kenosha, Wisconsin, in 1908. His Italian immigrant father changed the family name to Ameche shortly thereafter, and his fellow grade school students in Kenosha transformed the name Dom into Don. After a few years of grade school in Kenosha, Ameche left for a boarding school in Marion, Iowa. He remained in Iowa to attend high school, where he developed his dramatic talents under the tutelage of Father I.J. Semper, the school's drama coach. However, his parents had hoped for a lawyer in the family, not an actor, so Ameche enrolled in prelaw at Iowa's Columbia College. He had trouble staying focused on his studies, however, and as a result, he skipped around to different schools, including Marquette University and Georgetown University, before finally ending up at the University of Wisconsin, Madison, in 1928.

Acting continued to draw his interest, and he occasionally starred in campus plays with the campus's University Players. Like most successful entertainers of the period, his first steps to stardom included a mythical moment of discovery: Ameche went to see a road company play on campus one evening, but one of the leading actors had been hurt in a car accident earlier that day. Ameche arrived at the theater to purchase his ticket, and at the ticket booth, the manager of the theater recognized Ameche from his previous plays and asked him if he would fill in for the injured actor. Ameche agreed and in fact ended up staying with the stock company for the rest of the season, thereby forgoing his future career as a lawyer.

Emboldened by this success, Ameche moved to New York in 1929 and tried to foster an acting career on stage and in radio. But after an unsuccessful audition in 1930 as a singer for WMCA, he returned to Kenosha in the early 1930s. Ameche then moved to nearby Chicago at the suggestion of a friend, who told Ameche of the burgeoning opportunities for radio singers and actors in that city. After an audition with NBC Blue in 1930, Ameche was hired for a number of NBC dramas in Chicago, including *Rin Tin Tin* and *The Empire Builders*. He subsequently received a role in the show that launched him to stardom, *The First Nighter*.

The First Nighter was a 30-minute anthology drama, and the show's format fostered the illusion that listeners were hearing a Broadway play on opening night, despite its Chicago origin. As the male lead in plays for the show's first six years, Ameche was especially popular with audiences, and he quickly developed into radio's first sex symbol. In 1932 he parlayed this status into a lead role on a daily soap opera, *Betty and Bob*, the first of many daytime serials from Frank and Anne Hummert. *Betty and Bob* presented Ameche and Elizabeth Reller as a pair of seemingly incompatible newlyweds. Betty was a working-class secretary, and Bob was an urbane heir to a vast fortune. Arguments and jealousies abounded, as Bob had to accept Betty's workaday world, and she had to tolerate his dashing personality and friendliness with other women. As one can imagine, this role served to further cement Ameche's status as a radio heartthrob. Ameche also occasionally appeared on a juvenile adventure series, *Jack Armstrong, All American Boy*, with his brother Jim.

Ameche's radio success and suave persona inevitably captured Hollywood's attention. A talent scout arranged an audition with MGM, but the studio declined to sign him. After a subsequent audition with Twentieth Century Fox, Ameche signed with that studio in 1935. He then appeared in a spate of films throughout the 1930s, reaching his height of fame with *The Story of Alexander Graham Bell* in 1939. Known for the "young-man-about-town" role, Ameche was said to be second only to Shirley Temple in status at the Fox studios. Despite this Hollywood fame, Ameche continued his radio career, a choice that underscored his appreciation for the aural medium. Most notably, he starred periodically on *The Edgar Bergen–Charlie McCarthy Show*. Among other skits for this program, Ameche appeared in a regular segment called *The Bickersons*, portraying half of a quarrelsome married couple.

In the 1940s, with his film career dwindling, Ameche hosted a series of half-hour comedy-variety shows, such as *What's New?* and *The Drene Time Show*, which featured musical selections, dramatic skits, and comedy sketches, including *The Bickersons*. On his programs, Ameche offered listeners many guest stars from the film world, including Dorothy Lamour, Herbert Marshall, and Fred Astaire, illustrating the benefits to radio of his lingering connection to Hollywood. He also hosted a talent program called *Your Lucky Strike*, on which unknowns competed each week, with their talents being judged by three random housewives who voted over their telephones. In addition to his hosting duties, Ameche appeared on *Lux Radio Theater* 21 times, among the most appearances of any actor, and he guest-starred regularly on *The Jimmy Durante Show*.

In 1950 Ameche moved to New York and shifted his broadcasting career to television, beginning with hosting a quiz program, *Take a Chance* (1950). For the rest of the decade, he mainly hosted television variety shows, and he also appeared periodically on Broadway, most notably in Cole Porter's *Silk Stockings* in 1955. In 1958 Ameche returned to radio for a final time, hosting *Don Ameche's Real Life Stories*, a serial drama airing every day in half-hour installments, providing one complete narrative per week. After only occasional film and television appearances throughout the subsequent decades, Ameche made a comeback in the 1980s, winning an Academy Award for his supporting role in *Cocoon* (1985). He died in 1993, leaving behind a unique legacy as both a pioneering radio star and a successful film actor.

CHRISTINE BECKER

See also Comedy; Edgar Bergen and Charlie McCarthy Show; Jack Armstrong, All American Boy

Don Ameche. Born Dominic Amici in Kenosha, Wisconsin, 31 May 1908. Brother of Jim Ameche, radio actor. Attended Columbia College, Marquette University, Georgetown University, and University of Wisconsin, Madison; toured with Jackson Players stock company, 1928–29; starred in first major radio program, *The First Nighter*, 1930–36; often featured on *The Edgar Bergen–Charlie McCarthy Show*, 1937–48; notable film performances in *The Alexander Graham Bell Story*, 1939, and *Heaven Can Wait*, 1943. Academy Award for role in *Cocoon*, 1985. Died in Scottsdale, Arizona, 6 December 1993.

Radio Series

1930–31	*Empire Builders*; *Rin Tin Tin*
1930–36	*The First Nighter*
1932–35	*Betty and Bob*
1943–44	*What's New?*
1946–47	*The Drene Time Show*
1947–48	*The Old Gold Show*
1948–49	*Your Lucky Strike*
1958	*Don Ameche's Real Life Stories*

Television Series

Take a Chance, 1950; *The Frances Langford-Don Ameche Show*, 1951–52; *Coke Time with Eddie Fisher*, 1953; *Holiday Hotel*, 1950–51; *The Jack Carson Show*, 1954–55; *Don Ameche Theater*, 1958; *International Showtime*, 1961–65

Films

Sins of Man, 1936; *One in a Million*, 1937; *In Old Chicago*, 1938; *Alexander's Ragtime Band*, 1938; *The Three Musketeers*, 1939; *Midnight*, 1939; *The Story of Alexander Graham Bell*, 1939; *Swanee River*, 1940; *Moon Over Miami*, 1941; *Girl Trouble*, 1942; *Heaven Can Wait*, 1943; *Guest Wife*, 1945; *Sleep My Love*, 1948; *Phantom Caravan*, 1954; *The Beatniks*, 1970; *Won Ton Ton*, 1975; *Trading Places*, 1983; *Cocoon*, 1985; *Harry and the Hendersons*, 1987; *Cocoon: The Return*, 1988; *Folks!* 1992

Stage

Silk Stockings, 1955; *Holiday for Lovers*, 1957; *Goldilocks*, 1958

Further Reading

Brown, Les, "When Chi Radio Was in Bloom," *Variety* (28 March 1962)
DeLong, Thomas A., *Radio Stars: An Illustrated Biographical Dictionary of 953 Performers, 1920 through 1960*, New York: McFarland,1996
Green, Abel, and Joe Laurie Jr., *Showbiz: From Vaude to Video*, New York: Holt, 1951
Wertheim, Arthur Frank, *Radio Comedy*, New York: Oxford University Press, 1979

American Broadcasting Company

The American Broadcasting Company (ABC) came late to the radio game, appearing as an independent network only in 1945. As such, it was a weak player until the 1960s, when ABC was in the vanguard of an attempt to revive and reshape network radio in the age of television.

Origins

ABC—as a network and an owner of major radio stations—was created in the 1940s, when the Federal Communications Commission and the Department of Justice forced the National Broadcasting Company's (NBC) owner, the Radio Corporation of America (RCA), to spin off one of NBC's two radio networks. In 1943 Edward J. Noble, who had made his fortune creating, manufacturing, and selling Life Savers candy, bought NBC's Blue network and three owned and operated stations for $8 million. In 1945 Noble renamed Blue the American Broadcasting Company and began to build ABC. In 1946, for example, he acquired WXYZ-AM in Detroit from King-Trendle Broadcasting for slightly less than $3 million.

The Blue network carried a number of popular shows—including *Just Plain Bill, Easy Aces, Inner Sanctum Mystery,* and *Lum 'n' Abner.* But generally ABC shows drew last place in ratings in all of Golden Age radio's categories of programming. In the variety category, for example, ABC's *The Alan Young Show* earned but a seventh of the ratings of NBC's *Bob Hope Program,* which broadcast later the same night. The *Andrews Sisters* program drew a third of the ratings of *Your Hit Parade* on the Columbia Broadcasting System (CBS), and *Ted Mack's Original Amateur Hour* always finished far behind *Arthur Godfrey's Talent Scouts.*

Still, Drew Pearson attracted vast audiences with his reports of the goings-on in the nation's capital, and the dramatic and controversial re-creations of the *March of Time* were popular as well, helped by the movie newsreel of the same name and by the program's connection with *Time* magazine. On the prestige side, ABC's regular Saturday matinee broadcasts of the Metropolitan Opera added some class to ABC's image.

It was not that Edward Noble was not willing to acquire top talent. During the late 1940s, Noble and his managers tried to add new shows, such as *Professor Quiz, Break the Bank, This Is Your FBI, Lone Ranger, Gillette Fights,* and *Gangbusters.* For example, when ABC bought WXYZ-AM, it acquired *Lone Ranger* and *Green Hornet.* A far more temporary triumph came with the hiring of Milton Berle, for this comic appeared on the ABC radio network for only one year before, looking for a showcase better suited to his visual style, he moved to NBC television and became a national sensation.

There were two notable exceptions. In 1946 Bing Crosby moved to the ABC radio network for reasons of convenience and technical change. Crosby, who was then at the very height of his popularity as a singer and movie star, agreed to move to ABC because NBC was forcing him to broadcast his show live twice, once for the eastern and central times zones and then a second time for stations based in the mountain and pacific time zones. Crosby wanted to use audiotape to record his show at his convenience. ABC executives were more than willing to permit Crosby to use the then new audiotape technology to record his show ahead of time and then hit the links when listeners thought he was in the studio broadcasting to them.

During summer 1946, Crosby shocked the industry when he announced he was leaving NBC and long-time sponsor Kraft to sign with Philco, maker of radio sets, and appear on ABC. His weekly salary was announced at a staggering $7,500. Because Philco and ABC permitted Crosby to prerecord his *Philco Radio Time,* he was nowhere near the studio when his show debuted on Wednesday night, 16 October 1946. The Philco show proved a major ratings triumph. Because of its success, three years later, when CBS chief William S. Paley was in the midst of his celebrated "talent raids," he lured Crosby away from his three-year run on ABC. *Philco Radio Time* last ran on ABC on 1 June 1949.

The other exception to ABC's normal ratings mediocrity started in March 1948 when *Stop the Music!* premiered. Listeners quickly embraced this giant jackpot quiz show. With master of ceremonies Bert Parks as its host, musical selections were played by the Harry Salter Orchestra or sung by vocalists Kay Armen and Dick Brown. While a song was played, a telephone call was placed to a home somewhere in the United States, and when the caller answered, Parks called out "stop the music." If the person at home could name the tune, he or she won up to $20,000.

Listeners flocked to ABC on Sunday nights, and by the summer ABC truly had a hit, doubling the audience reached by Fred Allen at NBC and *Sam Spade* on CBS. With ratings high from the beginning, sponsors lined up, and ABC selected Old Gold cigarettes and Spiedel jewelry as the main advertisers. During summer 1948, demand for tickets was so high that the producers moved the show to the 4,000-seat Capitol Theater in the heart of Times Square. But ABC could not sustain the hit, and by 1952 the radio version was off the air. The fledgling ABC television network kept it on the tube—originally as just a simulcast—until 1956.

As the Golden Age in radio was ending, ABC certainly matched Mutual as a radio network, but it was rarely as successful as NBC and CBS. Building ABC as a radio network was always a struggle, yet from a network with 168 owned or affiliated stations as of the October 1943 purchase date, Noble and his managers doubled affiliations within a decade. Indeed, owning and operating radio stations and a network was lucrative enough that Noble—with the help of a $5 million loan from the Prudential Insurance Company of America—was able to launch the ABC television network. By the beginning of the 1950s, ABC not only owned a radio network and the maximum allowable number of AM and FM radio stations, but had also reached the legal limit on television stations as well—five. So successful was ABC that Noble began to attract bidders for his enterprise.

The United Paramount Takeover

In 1951, in what was up to that point the biggest transaction in broadcasting history, United Paramount (the chain of movie houses formerly owned by Paramount Pictures) paid $25 million to add ABC's five television stations, six FM radio stations, and six AM radio stations to its 644 theaters in nearly 300 cities across the United States. The FCC took two years to finally approve the deal. ABC would never have become a modern radio and television corporate powerhouse had it not been acquired by United Paramount, greatly adding to its financial resources. Leonard H. Goldenson, head of United Paramount, began to sell theaters and real estate to generate the cash necessary to build up ABC television first and ABC radio second.

On the radio side, Goldenson faced a challenge. Most of ABC's radio affiliates were lower-power stations in smaller cities. When forced to divest RCA of the Blue network a decade earlier, RCA had stacked the deck, making sure that what he transferred with Blue represented the least valuable of NBC's stations. To generate income, ABC radio management, headed by Robert Kintner, allowed advertising for products considered inappropriate by the mighty NBC and CBS, such as deodorants and laxatives. But in 1953 Goldenson felt radio would need to change as television became America's top mass medium. With AM radio now standard equipment on most new cars, and with the innovation of the inexpensive portable radio set, Goldenson reasoned that a radio market would always exist, but in a different form than had worked in the past. The question for ABC—and for all of radio in 1953—was how best to exploit the changing radio medium.

Goldenson realized that while United Paramount had gained a network with the ABC purchase, more important were the stations located in some of the nation's largest markets. The flagship station in New York City—WJZ-AM—was his most valuable radio asset, worth more than the then struggling ABC radio network. Still, whereas WNBC-AM had studios at Radio City, WJZ-AM broadcast from a modest renovated building at 7 West 66th Street, one block west of Central Park. On 1 May 1953—six and a half years after the rival network's stations were named WNBC-AM and WCBS-AM in honor of their respective parent companies—Goldenson renamed his New York City outlet WABC-AM and worked to make this 50,000-watt powerhouse a metropolitan fixture at "77" on the AM band.

Programming proved harder to change, so Goldenson stuck with what was working for the time being. In the mid 1950s that meant shows such as American Safety Razor's *Walter Winchell* on Sunday nights; Anheuser Busch's *Bill Stern's Sports Reports* at 6:30 P.M. three times a week; General Mills' *Lone Ranger* at 7:30 P.M. on Mondays, Wednesdays, and Fridays; and Mutual of Omaha's *Breakfast Club* in the mornings. Goldenson's innovation was to hire local personalities to develop followings only within the New York City metropolitan market. For WABC-AM, this meant in time Peter Lind Hayes and Mary Healy, Martin Block, Ernie Kovacs, Howard Cosell (and his sports reports), commentators John Daly and Edward P. Morgan, and rock disc jockey sensation Alan Freed.

The transition of WABC-AM to the highest-rated rock station in the United States began modestly with Martin Block's *Make Believe Ballroom,* which Goldenson bought in 1954. But it was the June 1958 hiring of Alan Freed that would signal the future of WABC-AM as a Top 40 profit-generating powerhouse. Freed would soon burn out in the payola scandals, but WABC disc jockeys "Cousin Brucie," "Big Dan Ingram," and others replaced him, and by the arrival of the Beatles in early

1964, WABC-AM had become one of the nation's most-listened-to radio stations.

Goldenson's management team rebuilt the other owned and operated ABC radio stations: WLS-AM in Chicago, WXYZ-AM in Detroit, KABC-AM (formerly KECA-AM) in Hollywood, KQV-AM in Pittsburgh, and KGO-AM in San Francisco. Each would soon take its place among the top-rated stations in its metropolitan area. Each also beefed up an FM license that had been underutilized.

For example, Chicago's clear channel WLS-AM was transformed from a major-market network affiliate to a rock and roll pioneer, beaming Top 40 hits across the Midwestern states. As the 1960s commenced, WLS-AM had joined the Top 40 elite and was being built up by a number of local disc jockey stars—none hotter, or more famous, than Larry Lujack. For a generation of listeners in the 1960s and 1970s, Lujack created and defined rock and roll.

Similar histories could be traced for all of ABC's major radio stations. In Detroit, for example, WXYZ-AM was also transformed into a radio powerhouse, and by Goldenson's own calculations it functioned as ABC's most profitable radio outlet during the 1950s and 1960s. If the selling of United Paramount's theaters and valuable real estate is properly credited with underwriting ABC television network deals with Hollywood's Walt Disney and Warner Brothers Companies, one must also credit the revenues generated by profitable radio stations such as WXYZ-AM. Indeed, the rebuilding of AM radio stations was going so well that in 1957 Goldenson separated the television side (which required fashioning alliances with Hollywood) from the radio side (which needed to transform existing properties into local hot spots, station by station) of the business. With this separation, Goldenson emphasized that television and radio management required quite different skills.

Leonard Goldenson's Radio Network Innovations

Although Leonard H. Goldenson has never been labeled as one of radio's top leaders—in the league with NBC's David Sarnoff or CBS's William S. Paley—many consider that he ought to be. Despite all the hiring and firing of radio talent during the 1950s and 1960s, ABC management at the top varied little as Goldenson and his small set of advisers built ABC radio (and television) into highly profitable media institutions. By 1985, when he stepped down, Goldenson had created a modern media conglomerate. This small-town poor boy from Pennsylvania, who managed to graduate from Harvard Law School, learned the mass entertainment business at Paramount Pictures and took over its divested theater division in 1950. He already had some experience in television from Paramount's owned and operated television station in Chicago, WBKB-TV. He had no experience in radio, but he knew of its success as an entertainment medium in cities where Paramount operated theaters.

Although most kudos for Goldenson go to his development of the ABC television network, media historians also recognize his reinvention of network radio. By refashioning a single all-things-to-all-audiences network into four—and later more—specialized radio networks in the late 1960s, Leonard Goldenson earned his place as a radio pioneer. ABC was transformed from a single radio network into the American Contemporary Network, the American Information Network, the American Entertainment Network, and the American FM Network. This specialization would set the model for network radio for the next three decades.

However, Goldenson's most significant innovation almost did not come to be. By the early 1960s Goldenson thought he had built up and milked his major-market stations for as much as he could, and he considered abandoning network radio altogether. He seriously entertained bids to sell the ABC radio network—plus all its valuable stations—to Westinghouse for a price reported to be $50 million. But once he got over the shock of the unexpected size of Westinghouse's offer, Goldenson figured that this substantial bid by Westinghouse's experienced executives did not signal the end of the Top 40 radio era; rather, new forms of radio broadcasting did have a future. He turned down Westinghouse and successfully continued to build his own radio empire as part of what was (and is) often incorrectly considered simply a television network business operation.

At the time, breaking with the mold of a single network was considered a risky proposition. The executives directly responsible for the network radio turnabout were Hal Neal and Ralph Beaudin, who had made their reputations by turning ABC-owned and -operated stations into rock and roll powerhouses. The four networks were patterned from formats of the day. The American Contemporary Network stressed middle of the road music and soft-spoken middle-aged disc jockeys. The American Information Network was all news and talk, patterned after the all-news local stations that CBS and Westinghouse were then pioneering. The American Entertainment Network was a piped-in Top 40 feed, and the FM network was a grab bag, because no one honestly knew the future of FM at that point.

In planning the four networks, Goldenson, who was already paying American Telephone and Telegraph (AT&T) for transmission by land lines, figured that four would cost only a bit more than one transmission for facilities that were being underutilized. Talent could be drawn from owned and operated stations. By late 1966 the plan for the four networks was in place, and Goldenson gave notice to all advertisers and affiliated stations that the year 1967 would be the final year for ABC as a single radio network. During summer 1967, ABC began heavy promotion of the four-network idea, and quadruple feeds commenced on the first day of 1968 to 500 affiliates.

In the first year of four-network operation, 1968, Goldensen was criticized because ABC lost $8 million. But just four years later the radio division alone was making more than $4 million annually. But by 1972 the network radio division was making $4 million profit per year. As the 1970s ended, ABC's network radio division had 1,500 affiliates and was making $17 million profit per year. In the late 1970s, ABC Contemporary had about 400 affiliates, and the American Information Network had almost 500 affiliates, as did the American Entertainment Network; however, the American FM Network never moved past 200 affiliates. The recasting and specializing of network radio worked for AM stations, but FM gradually found musical niches that would make them the leaders in radio ratings in most markets by 1980.

Goldenson continued to tinker with the format profile of both ABC's owned and operated stations and its growing number of networks. The advertising community applauded Goldenson's adaptation of focused demographics. In August 1970 ABC separated management of AM from FM owned and operated radio stations, and with the progressive rock format ABC began to remake FM outlets, which had long merely simulcast AM.

Takeovers

As the 1980s commenced, Goldenson began to slow down. In 1980 his ABC television network ranked number one, and he was able to tout ABC's radio stations as among the most popular in the nation. For example, WABC-AM in New York abandoned Top 40—after 22 years—and soon made even more money with "talk."

Goldenson needed to find a suitable successor. He wanted to pass "his" company to someone who had the skills to consistently and profitably run a mature multibillion-dollar media empire. After much looking and interviewing, Goldenson met Thomas Murphy, head of Capital Cities Broadcasting, a 30-year-old media company that as the 1980s began owned 7 television stations, 12 radio stations, an assortment of daily and weekly newspapers, and an additional assortment of magazines. Capital Cities was a Wall Street high flyer, known for its efficient management by Murphy and Dan Burke. Goldenson decided that Capital Cities was the logical successor to take over the ABC radio and television networks he had created.

At the time the deal was announced, in March 1985, it was the largest non-oil merger ever, at $3.5 billion. But although headlines warned of vast changes and ominous negative implications for news and entertainment, none ever really materialized. Murphy, Burke, and their Capital Cities executives simply merged the two media companies, sold off some duplicative properties, and then continued the process of fashioning an even more profitable, even larger media enterprise than Goldenson had created—one that encompassed forms of mass media from print to television and from film-making to radio network and station operation.

In radio, Murphy, Burke, and company changed almost nothing. They tinkered on the margins as they tried to follow (not set) trends. They smoothly and efficiently managed format makeovers as Top 40 rock and roll gave way to other formats of pop music. In general, Murphy and Burke transformed ABC's large-city AM powerhouse stations, often to middle of the road talk-format operations. Consider the example of Chicago's WLS-AM, symbol of the Top 40 era. When Murphy and Burke took over, its ratings were slipping, and so they worked to reformat WLS-AM again, even as FM was draining away listeners. By the early 1990s, WLS-AM became news-plus-talk radio 890, with no "hot jock," but instead the ramblings of Dr. Laura Schlessinger and Rush Limbaugh. Such transformations took place throughout the matrix of ABC stations, as radio continued to provide core profits to the company now known as Capital Cities/ABC.

In 1996 Murphy and Burke themselves neared retirement age, and, as Goldenson had done, they sought an alliance with a company to continue ABC. In 1996 the biggest merger in media history was announced when the Walt Disney Company acquired Capital Cities/ABC. Overnight, Disney, far more famous for its movie making and theme parks, became one of the top competitors in the world of radio.

When Disney announced its takeover of Capital Cities/ABC at the end of July 1996, the headlines blared about vast potential synergies of a Hollywood studio and a television network. Radio was considered an afterthought. Still, with New York City flagship station WABC-AM leading the list, Disney now had important radio promotional outlets in a half-dozen other top-ten media markets: in media market 2 Los Angeles (three stations), in market 3 Chicago (two stations), in market 4 San Francisco (two stations), in market 6 Detroit (three stations), in market 7 Dallas (two stations), and in market 8 Washington, D.C. (three stations).

Disney concentrated on these big cities, but its radio holdings paled in comparison to rival radio powers of the late 20th century such as CBS and Clear Channel. Yet Disney's station reach always remained vast. Disney head Michael Eisner then looked and applied synergies to these urban radio stations. He sold off Capital Cities/ABC's newspapers and other print operations but kept radio—even expanding Disney into more radio with the September 1997 launch of a new network, the ESPN radio network, with its exclusive rights for Major League Baseball for five years. Eisner also rolled out Radio Disney, a live network for families and children under age 12 with a select playlist of special music, much of it from Disney movies and television programs.

Although radio is a relatively small division at Disney, Disney management certainly recognizes radio's contribution to Disney profit accumulation. As the 21st century commenced,

ABC radio networks programmed ten services, including *American Gold, Flashback, Moneytalk, Business Week Report, Rock & Roll's Greatest,* and *Yesterday . . . Live!*—as well as ABC News Network and ABC Sports radio. Stars included canonical Paul Harvey (by the year 2000 one of the longest-running voices in radio history), the controversial "news commentator" Matt Drudge, and noted sports commentators Tony Kornheiser and Dan Patrick. ESPN offered sports news and talk, based on its companion cable television network, and ABC News expanded its long-running network services with shows spun off from television favorites such as *This Week with Sam Donaldson and Cokie Roberts.* ABC News also offered on the radio *America's Journal, World News This Week, Hal Bruno's Washington Perspective,* and *Newscall.* Disney bragged that its Radio Disney network was reaching 1.6 million children and 600,000 moms per week, while *American Gold,* starring Dick Bartley for four hours per week, offered the most-listened-to nationally syndicated oldies countdown show in the year 2000, in competition with Dick Clark, as a lure for aging baby boomers. Surely, these services—and a multitude of others—will be changed and reinvented as ABC Radio networks continue to refocus vital services.

DOUGLAS GOMERY AND CHUCK HOWELL

See also Crosby, Bing; Freed, Alan; Harvey, Paul; McNeill, Don; Network Monopoly Probe; Radio Disney; Talent Raids; WABC; WLS; WXYZ

Further Reading

ABC Radio, <www.abcradio.com>

Berle, Milton, and Haskel Frankel, *Milton Berle: An Autobiography,* New York: Delacorte Press, 1974

Buxton, Frank, and William Hugh Owen, *Radio's Golden Age: The Programs and the Personalities,* New York: Easton Valley Press, 1966; revised edition, as *The Big Broadcast, 1920–1950,* New York: Viking Press, 1972

Compaine, Benjamin M., *Who Owns the Media? Concentration of Ownership in the Mass Communications Industry,* White Plains, New York: Knowledge Industry, 1979; 3rd edition, as *Who Owns the Media? Competition and Concentration in the Mass Media Industry,* by Compaine and Douglas Gomery, Mahwah, New Jersey: Erlbaum, 2000

Federal Communications Commission, Mass Media Bureau, Policy and Rules Division, *Review of the Radio Industry, 1997,* Docket MM 98-35 (13 March 1998)

Fielding, Raymond, *The March of Time, 1935–1951,* New York: Oxford University Press, 1978

Goldenson, Leonard H., and Marvin J. Wolf, *Beating the Odds: The Untold Story behind the Rise of ABC: The Stars, Struggles, and Egos That Transformed Network Television, by the Man Who Made Them Happen,* New York: Scribner, and Toronto, Ontario: Collier Macmillan Canada, 1991

Grover, Ron, *The Disney Touch: How a Daring Management Team Revived an Entertainment Empire,* Homewood, Illinois: Business One Irwin, 1991; revised edition, as *The Disney Touch: Disney, ABC, and the Quest for the World's Greatest Media Empire,* Chicago: Irwin Professional, 1997

Hickerson, Jay, *The Ultimate History of Network Radio Programming and Guide to All Circulating Shows,* Hamden, Connecticut: Hickerson, 1992; 3rd edition, as *The New, Revised, Ultimate History of Network Radio Programming and Guide to All Circulating Shows,* 1996

Jaker, Bill, Frank Sulek, and Peter Kanze, *The Airwaves of New York: Illustrated Histories of 156 AM Stations in the Metropolitan Area, 1921–1996,* Jefferson, North Carolina: McFarland, 1998

Morgereth, Timothy A., *Bing Crosby: A Discography, Radio Program List, and Filmography,* Jefferson, North Carolina: McFarland, 1987

Quinlan, Sterling, *Inside ABC: American Broadcasting Company's Rise to Power,* New York: Hastings House, 1979

Rhoads, B. Eric, *Blast from the Past: A Pictorial History of Radio's First 75 Years,* West Palm Beach, Florida: Streamline, 1996

"Special Report on ABC's Twenty-Fifth Anniversary," *Broadcasting* (13 February 1978)

Thomas, Bob, *Winchell,* Garden City, New York: Doubleday, 1971

Whetmore, Edward Jay, *The Magic Medium: An Introduction to Radio in America,* Belmont, California: Wadsworth, 1981

White, Llewellyn, *The American Radio: A Report on the Broadcasting Industry in the United States from the Commission on Freedom of the Press,* Chicago: University of Chicago Press, 1947

Williams, Huntington, *Beyond Control: ABC and the Fate of the Networks,* New York: Atheneum, 1989

American Broadcasting Station in Europe

Office of War Information Station, 1944–1945

The American Broadcasting Station in Europe (ABSIE) was created and operated by the U.S. Office of War Information (OWI) to support the Allied invasion of Europe during World War II. The station carried news, entertainment, coded messages, propaganda, and instructions for European populations between 30 April 1944 and 4 July 1945. It broadcast from its London studios in English, French, German, Norwegian, Danish, and Dutch and offered over 42 hours of weekly programming featuring exiled statesmen, military leaders, and popular musicians. The list of prominent individuals who broadcast for the station included King Haakon of Norway; King Peter of Yugoslavia; Foreign Minister Jan Masaryk of Czechoslovakia; General Dwight D. Eisenhower; Charles DeGaulle; and entertainers Glenn Miller, Dinah Shore, and Bing Crosby.

Although the idea of an American station transmitting from Great Britain was conceived as early as the attack on Pearl Harbor in 1941, OWI did not initiate concrete planning for two and a half years. By that time, the government's widespread radio propaganda activities included shortwave broadcasts to Japan; the operation of stations in Tunis, Algiers, and Italy; and the transmission of a large number of Voice of America (VOA) programs via shortwave radio. Nevertheless, the scarcity of shortwave receivers on the continent and the Nazis' increasingly effective efforts to jam broadcasts convinced OWI officials to build a station near enough to the front that its signal could reach European listeners.

Brewster Morgan, ABSIE's first director, and Richard Condon, its engineer, started work on the station in 1942, only to have their efforts postponed when OWI assigned them to set up the Armed Forces Network to broadcast to Allied troops. By 1943 they were back on the job, ordering equipment in the summer and recruiting staff in the fall. The following period of intense activity at the station coincided with heavy air raids in London. One evening German bombs narrowly missed the headquarters that ABSIE shared with Gaumont Films and destroyed the neighboring building. Shaken, the staff continued working; by early April, they had started rehearsing programs in anticipation of the inaugural broadcast of 30 April. Listeners to that program heard OWI's European director, Robert Sherwood, caution resistance forces to avoid premature action and to await word from Allied radio before striking the enemy. Sherwood also stated that an American voice could now be heard that was committed to "telling the truth of the War" to America's friends and enemies.

To ensure that this voice was received, ABSIE broadcast using four medium wave transmitters built by the Radio Corporation of America (RCA) and six powerful shortwave transmitters leased from the British Broadcasting Corporation (BBC). Its configuration allowed listeners to tune to ABSIE on several frequencies while forcing Nazi propagandists to dedicate an increased number of transmitters to jamming. Captured German records reveal that ABSIE's efforts to counter jamming were successful enough that most of the station's broadcasts were received. A more formidable tactic than jamming was the Nazis' ban, under penalty of death, on listening to Allied radio. Postwar surveys revealed, however, that German civilians and soldiers had listened to ABSIE's programming despite the potential consequences and that stories reported on ABSIE had circulated widely by word of mouth.

ABSIE's organization mirrored that of a commercial station, an unsurprising fact given that it recruited key personnel from commercial stations and networks. Director Brewster Morgan had been a director and producer for the Columbia Broadcasting System (CBS). Station Manager Robert Saudek had worked at the National Broadcasting Company (NBC) Blue network in New York. The chief of the German language desk, Robert Bauer, had been an employee of WLWO Cincinnati prior to the war. Even William Paley, the head of CBS, helped ABSIE with planning and equipment procurement. Commercial radio's willingness to contribute staff to ABSIE reflected both its support for U.S. involvement in World War II and its conciliatory approach toward the government during the war years. Such attitudes also explain the networks' willing self-censorship at home.

OWI's U.S. offices, particularly those of its Overseas Branch in New York, provided another major source for ABSIE's staff. Phil Cohen, Morgan's successor as ABSIE's director, had been chief of OWI's Domestic Bureau in Washington. ABSIE's language desks drew heavily from the OWI's New York offices. Pierre Lazareff, previously chief of the French radio section in New York, became chief of the French desk. Jom Embretsen, chief of ABSIE's Norwegian desk, had headed OWI's Radio Program Bureau in New York. Robert Bauer, Alfred Puhan, and George Hanfmann, the three successive chiefs of the German desk, all came from New York as well.

ABSIE's language desks were central to its operation. They translated key news items while tailoring programming to specific national audiences. Lazareff made sure the French desk established close working ties with the French government in exile and aired many speeches of its officials. A popular Norwegian program prompted a deluge of audience mail when it broadcast messages from Norwegian members of the armed forces for their friends and relatives at home. The English section was specifically designed to appeal to European listeners who, saturated with overly direct propaganda in their native languages, were more trusting of English language broadcasts. Most of its programming was news, supplemented by frontline commentary.

In general, news accounted for the largest share of ABSIE's time, although special programs and rebroadcasts of BBC and VOA programs figured prominently on the daily schedule as well. Broadcasts generally avoided the most overt propaganda appeals in favor of carefully selected but usually truthful reports. This "white" propaganda contrasted with the Office of Strategic Services' use of such "black" propaganda as the invention of a revolutionary party in Germany devoted to the overthrow of the Nazis. More typical of ABSIE's approach was the news coverage in 1945, which juxtaposed stories of chaos in Nazi-occupied regions of Germany with reports of orderly conditions and fair treatment in the Allied zones. One program broadcast interviews with German prisoners of war in hopes that their families would tune in. On another broadcast, Bing Crosby performed after reading a phonetically written German script that projected a vision of the increased freedom Germans could expect after the Nazis were defeated.

ABSIE's final broadcast aired on 4 July 1945 and featured a statement by OWI director Elmer Davis asserting that ABSIE had successfully completed its mission but that the VOA and BBC would continue in its place.

BRYAN CORNELL

See also Armed Forces Radio Service; Office of War Information; Propaganda by Radio

Further Reading

Kirby, Edward Montague, and Jack W. Harris, *Star-Spangled Radio,* Chicago: Ziff-Davis, 1948

Laurie, Clayton D., *The Propaganda Warriors: America's Crusade against Nazi Germany,* Lawrence: University Press of Kansas, 1996

Pirsein, Robert William, *The Voice of America: A History of the International Broadcasting Activities of the United States Government, 1940–1962,* New York: Arno Press, 1979

Shulman, Holly Cowan, *The Voice of America: Propaganda and Democracy, 1941–1945,* Madison: University of Wisconsin Press, 1990

Winkler, Allan M., *The Politics of Propaganda: The Office of War Information, 1942–1945,* New Haven, Connecticut: Yale University Press, 1978

American Family Robinson

Soap Opera Adventure Program

The National Association of Manufacturers (NAM) raised the character concept to an art with the soap opera adventure *American Family Robinson.* Syndicated by the World Broadcasting System from late 1934 to 1940, the 15-minute transcribed episodic drama was an anomaly among the NAM's nearly exclusive investment in printed public and political relations material. The NAM's politics, like its print-oriented publicity, were underwritten by the nation's largest industrial corporations, who were *Robinson* sponsors.

Provoked by the prolabor clauses of the New Deal's National Industrial Recovery Act (NRA), in 1933 the NAM embarked on a campaign of employer opposition that forestalled the imposition of collective bargaining in the steel, chemical, and auto industries. Announcing an "active campaign of education" in September of that year, Association President Robert L. Lund explained that NRA Section 7(a) posed a special threat to employers, given the "untruthful or misleading statements about the law" made by the American Federation of Labor and "communistic groups promoting union organization." Lund concluded that the American public would become favorably disposed toward business's traditional prerogatives and institutional choices if only business leaders would "tell its story."

Drawing upon the "home service personality" expertise of its packaged goods producers, the NAM led the way in radio with the episodic adventures of the *American Family Robinson.* The program appears to have been proposed by Harry A. Bullis, General Mills vice president and chairman of NAM's public relations committee. The *American Family Robinson*'s drop-dead attacks on the New Deal reflected the print-oriented focus of the NAM's traditional publicity techniques. The interjection of editorial comment into the *Robinson*'s soap opera plot reduced series protagonist Luke Robinson, "the sanely philosophical editor of the *Centerville Herald,*" to a caricature of the factory town newspaper editor that the NAM assiduously cultivated with an open-ended supply of pro-industry preprinted mats, columns, and tracts.

Editor Robinson, the program's repository of sound thinking and common sense, is beset by social schemers and panacea peddlers. Some are threatening and even criminal, but most are simply misguided. Among the latter is Robinson's brother-in-law, William Winkle, also know as "Windy" Bill, the itinerant inventor of the "housecar." Bill's meddlesome and uninformed political ideas are as unexpected as his unannounced visits with the Robinsons. More menacing is Professor Monroe Broadbelt, the "professional organizer of the Arcadians, a group using the Depression as a lever to pry money from converts to radical economic theories" (from *American Family Robinson,* cited in MacDonald, 1979).

The story line of the *American Family Robinson* revolves around the resolution of political conflict in the home and immediate community through the application of "time-tested principles." The Robinsons are shocked when their daughter Betty falls under the oratorical spell of Professor Broadbelt, a common criminal whose turn of phrase suggests a certain Hyde Park, New York, upbringing. Complications attend Betty's engagement to the Arcadians' charismatic leader, whose first consideration is his chosen mission: "The upliftment of mankind." Broadbelt's motives, however, are neither idealistic nor romantic. In the next episode, Luke Robinson helps apprehend Broadbelt, who has skipped town with the Arcadians' treasury. Returning to Centerville, Robinson presides over the liquidation of the Arcadian movement by publicly refunding the contributions of its confused and misguided members, including his daughter's.

In certain households, interest in the *American Family Robinson* undoubtedly did exist. The program attracted an articulate audience that appreciated and responded to the NAM's send-ups of New Deal liberalism. From fan mail the NAM learned that listeners responded enthusiastically to Luke Robinson's comic foil "Windy" Bill. Written into the script as an incidental character, "Windy" soon returned to Centerville with a role expanded to include yet more meddlesome and annoying business. Other changes occurred as characters changed careers and took on new responsibilities. In 1935 Luke Robinson left the editorship of the *Centerville Herald* to become the assistant manager of the local furniture factory. Although Robinson remained the series' protagonist, a new character, "Gus Olsen," a janitor who had made the best of his lot in life, assumed Robinson's place as the managing editor and owner of the *Herald*. A tabloid "Herald" mailed to listeners from "Centerville" announced the changes and included photographs of the "Robinsons" reading their fan mail along with the paper's articles, editorials, cartoon, and crossword puzzle.

When introduced to New York City listeners in 1935, the *American Family Robinson* appeared five days a week until it changed to its regular twice-a-week schedule. In 1940, the last year of broadcast, the series appeared twice a week on 255 stations. NAM specialists considered midafternoon the optimum time for broadcasts. According to NAM Vice President for Public Relations James P. Selvage, tests showed that when scheduled between 2:00 and 3:00 P.M., the program had an excellent chance "to reach not only housewives but other members of the family." The *American Family Robinson,* Selvage wrote, presented "industry's effective answer to the Utopian promises of theorists and demagogues at present reaching such vast audiences via radio."

From the outset, Selvage had hoped to interest the National Broadcasting Company (NBC) or the Columbia Broadcasting System (CBS) in broadcasting the *American Family Robinson* on a sustaining basis. Neither was interested, and the series ended up in transcription, recorded and circulated to individual stations by the World Broadcasting System. A review of scripts submitted to NBC in October 1934 resulted in the program's banishment from the network's owned and operated stations as well. Reviewing the series' first three episodes, NBC script editor L.H. Titterton hardly knew what to make of the Robinsons, or the direction the story might take. An outline for the rest of the series and a script of the last episode received three days later confirmed Titterton's suspicion. After meeting with Selvage and Douglas Silver, the scripts' author, Titterton reported that the *American Family Robinson* proposed "to take on a definitely anti-Rooseveltian tendency." "You would probably not find in the entire series any specific sentence that could be censored," Titterton wrote to his network superiors, "but the definite intention and implication of each episode is to conduct certain propaganda against the New Deal and all its work."

WILLIAM L. BIRD JR.

Cast

Luke Robinson
The Baron
Miss Twink Pennybacker/Gloriana Day
Windy Bill
Cousin Monty, the Crooner
Professor Broadbelt
Myra
Aunt Agatha
Emmy Lou
Elsie
Mr. Popplemeyer
Letitia Holsome
Gus (Luke's assistant)
Pudgie

Producer/Creator
Harry A. Bullis

Programming History
National Industrial Council syndication, Orthacoustic transcription 1935–1940

Further Reading
Bird, William L., Jr., *Better Living: Advertising, Media, and the New Vocabulary of Business Leadership, 1935–1955,* Evanston, Illinois: Northwestern University Press, 1999
Ewen, Stuart, *PR! A Social History of Spin,* New York: Basic Books, 1996
Fones-Wolf, Elizabeth, "Creating a Favorable Business Climate: Corporations and Radio Broadcasting, 1934–1954," *Business History Review* 73 (Summer 1999)
MacDonald, J. Fred, *Don't Touch That Dial: Radio Programming in American Life, 1920–1960,* Chicago: Nelson-Hall, 1979
Marchand, Roland, *Creating the Corporate Soul: The Rise of Public Relations and Corporate Imagery in American Big Business,* Berkeley: University of California Press, 1998
Tedlow, Richard S., "The National Association of Manufacturers and Public Relations During the New Deal," *Business History Review* 50 (Spring 1976)

American Federation of Musicians

The American Federation of Musicians (AFM) represents some 150,000 members in nearly 400 local unions throughout the United States and Canada. The AFM became infamous during and after World War II under its fiery leader James Caesar Petrillo, who fought tirelessly to preserve the jobs of professional musicians at stations and networks. Petrillo defied President Roosevelt and Congress until the latter passed legislation limiting his right to pressure broadcast stations.

Origins

After several earlier attempts at organization, the AFM was founded in Indianapolis in 1896 following an invitation from Samuel Gompers, president of the American Federation of Labor (AFL), to organize and charter a musicians' trade union. Delegates from various musician organizations, representing some 3,000 musicians, created a charter stating that "any musician who receives pay for his musical services shall be considered a professional musician." The union added the phrase "of the United States and Canada" to its title in 1900. At the St. Louis World's Fair four years later, the AFM discouraged the hiring of foreign bands. It also achieved the first minimum wage scale for traveling orchestras.

The economic impact of World War I and the growing popularity of recorded music led to epic high unemployment of musicians. Prohibition was closing beer halls where musicians had worked, and by the late 1920s and early 1930s, sound-on-film technology had displaced theater orchestras. The 22,000 musicians providing in-theater musical background for silent movies were replaced with only a few hundred jobs for musicians recording sound tracks. As might be expected, the New York, Chicago, and Los Angeles AFM locals were the largest during this period, with about 25 percent of the total membership in the three cities.

AFM and Radio

AFM members initially looked to radio as a godsend, assuming that it would provide for more employment opportunities for members. And indeed larger stations did create or hire individual musicians or even orchestras. But stations in smaller markets relied on recordings or shared (networked) broadcasts of national orchestras to fill their airtime, so music was getting wider circulation, but musicians usually were not.

Development of electrical transcription around 1930 (and sound quality improvements in records sold to consumers) made the problem worse, as it was now easier for stations to produce recorded programs that sounded nearly as good as live performances. After many years of indecision, in 1937 AFM President James Petrillo originated the "standby" approach, pressuring Chicago stations to employ AFM members if recorded music was played, as backup musicians or even as "platter turners" in place of regular on-air personnel. This "featherbedding" tactic (hiring more employees than needed) was adopted by the union and expanded to other areas of the nation in the years before the U.S. entry into World War II. That the standby process originated in Chicago is central to the AFM story, for Chicago became the base of strong AFM leadership for several decades.

A onetime trumpet player ("If I was a good trumpet player I wouldn't be here. I got desperate. I hadda look for a job. I went in the union business," *New York Times*, 14 June 1956), James Caesar Petrillo joined the AFM in 1917 from a rival group. He became head of the Chicago AFM local in 1922 and kept that post for over four decades, in part due to the lack of secret ballot elections and members' fear of him. In 1928 Petrillo had demanded that radio stations in the Chicago market pay musicians for performing on the air, which ironically forced many to use recorded music. The "standby" approach followed wherein musicians were retained but often not used by broadcasters.

By 1940 Petrillo had been elected national AFM president, a post he would hold until 1958, all the while retaining his local power base (and title) with the Chicago local. He quickly expanded the union's standby tactic, requiring stations across the country that played records (as most did by that time) to hire AFM members as standby players. With strong AFM pressure, by 1944 the practice had spread across the nation, employing some 2,000 musicians.

Petrillo and the union drew negative public attention, however, by demanding that the NBC radio broadcast from the National Music Camp at Interlochen, Michigan, be canceled in 1942. (Some of the camp's final performances each season had traditionally been broadcast.) A year later, the camp's leader, Joseph E. Maddy, lost his AFM membership because he was playing with non-members. The camp proved unable to get another broadcast outlet for its concerts, though ironically most could not have joined the union in any case because they were too young.

Petrillo did not seem to be concerned with public opinion. The height of his "public be damned" mode came in August 1942, when he pulled AFM members from all recording sessions with the big record companies until they agreed to his pension and related demands. The resulting 27-month ban continued until late 1944 despite orders by the War Labor Board, pleas from President Roosevelt, and loud complaints in Congress that the AFM leader was not being supportive of the war effort. He stood his ground, and in November 1944, the last of the major recording companies (RCA Victor and Columbia) gave in to AFM demands. AFM gained the payment of 1.5–2 cents from each record sold; the money went into what became a huge performance trust fund. (The fund still helps to support popular free concerts in what is now the Petrillo Band Shell in Grant Park in Chicago.) Growth of the recording business after the war and the increasing number of jukeboxes prompted the AFM to threaten another recording musicians strike in 1948, but the parties involved settled, agreeing to continue paying AFM fees to the performance fund.

Concerned about the developing new media and what impact they might have on musicians, Petrillo in 1945 banned AFM members from performing on television or on FM dual broadcasts with AM unless standby musicians were hired. These bans were lifted only after stations again agreed to his demands for payments to musicians who were often not used at all. He also banned foreign music broadcasts except those from Canada, whose players were often AFM members. Although some of the membership grumbled, Petrillo and his supporters were all-powerful in the union and held sway. Indeed, his supporters reveled in the poor press their president achieved, publishing a booklet of negative cartoons depicting the feisty leader.

Lea Act

But pressure from broadcasters who felt blackmailed into accepting employees they did not need led Congress to take action limiting the union's power. In 1946 Clarence Lea, a Republican from California, introduced legislation to revise the Communications Act by adding a revised Section 506 concerning "coercive practices affecting broadcasting." Passed by overwhelming margins in the House and Senate and quickly signed by President Harry Truman, the Lea Act banned pressure on licensees to employ or make payments for "any person or persons in excess of the number of employees needed by such licensee to perform actual services," or "to pay or agree to pay more than once for services performed." Pre-existing contracts were allowed to stand, but renewals would have to agree with provisions of the new law.

In his usual pugnacious approach, Petrillo appealed the new legislation, using WAAF in Chicago as a test case, and promised a nationwide strike against radio if the Lea Act was found unconstitutional. The Supreme Court, however, held the act to be constitutional and thus enforceable, and the AFM lost some of its power. Petrillo remained president of the AFM for another decade, but the union gradually slipped out of news headlines.

CHRISTOPHER H. STERLING

See also Columbia Broadcasting System; Copyright; Music; Radio Corporation of America

Further Reading

American Federation of Musicians: A Brief History of the AFM <www.afm.org/about/about.htm?history>

Burlingame, Jon, *For the Record: The Struggle and Ultimate Political Rise of American Recording Musicians within Their Labor Movement*, Hollywood, California: Recording Musicians Association, 1997

Countryman, Vern, "The Organized Musicians," *University of Chicago Law Review* 16 (1948–49)

"Petrillo," *Life* (3 August 1942)
Seltzer, George, *Music Matters: The Performer and the American Federation of Musicians,* Metuchen, New Jersey: Scarecrow Press, 1989
Smith, B.R., "Is There a Case for Petrillo?" *New Republic* (15 January 1945)
U.S. Congress, House Committee on Education and Labor, *Investigation of James C. Petrillo and the American Federation of Musicians: Hearings,* 80th Congress, 1st Sess., 1947
U.S. Congress, House Committee on Education and Labor, *Restrictive Practices of the American Federation of Musicians: Hearings,* 80th Congress, 2nd Sess., 1948
U.S. v Petrillo, 332 US 1 (1947)
Warner, Harry P., *Radio and Television Rights,* Albany, New York: Bender, 1953
"What's Petrillo Up To?" *Harpers* 86 (December 1942)
White, Llewellyn, "King Canute," and "Petrillo," in *The American Radio,* by White, Chicago: University of Chicago Press, 1947

American Federation of Television and Radio Artists

The American Federation of Television and Radio Artists (AFTRA) is the national labor union or "guild" for talent in television, radio, and sound recordings. It was originally founded on 30 July 1937 as the American Federation of Radio Artists (AFRA), part of the American Federation of Labor (AFL). By 2000, AFTRA had 36 local offices throughout the United States, with a total of 80,000 members, representing performers at over 300 radio and television stations nationwide, and a workforce collectively earning over $1 billion annually under work contracted by the union. The union is still affiliated with the AFL–CIO and is headquartered in New York City.

AFTRA's membership represents four areas of broadcast employment: news and broadcasting; commercials and non-broadcast, industrial, or educational media; entertainment programming; and the recording business. Members include announcers, actors, newscasters, sportscasters, disc jockeys, talk show hosts, professional singers (including background singers and "royalty artists"), dancers, and talent working in new technologies such as CD-ROM and interactive programming. The union also franchises talent agents who represent AFTRA performers, stipulating talent agency commission fees as well as other regulations regarding the representation of performers in the union's jurisdiction.

AFTRA is party to about 400 collective bargaining agreements nationwide. These agreements generally regulate salaries and working conditions and include binding arbitration procedures for unresolved labor disputes. Union rules require AFTRA members to work only for "signatories" (employers who have signed AFTRA contracts), and members are asked to verify the signatory status of an employer before accepting a job. AFTRA was also the first industry union to establish employer-paid health insurance benefits and portable retirement plans. Any performer who has worked or plans to work in an area covered by AFTRA contracts is eligible for membership. Member dues are based on a performer's previous year's AFTRA earnings. Currently, dues and initiation fees are set by each local office, but the union has plans to implement a uniform national schedule of dues.

The union is governed by volunteer member representatives on both local and national boards of directors. National delegates are elected on a proportional basis from the locals at the union's biannual national convention. The national office publishes basic rates for the national freelance agreements for entertainment programming, commercials, sound recordings (both singing and speaking, such as for talking books), industrials (video- and audiotapes for corporate, educational, and other off-air use), and new technologies. Local AFTRA offices publish local talent guides and offer special services designed to meet local member needs, including skill development seminars, casting hot lines and bulletin boards, and credit unions and tax clinics. Local offices handle staff employment for broadcasters and newspersons at over 300 radio and television stations nationally. Because each signatory station's collective bargaining agreement is negotiated separately, it is up to the local AFTRA office to monitor and distribute information about specific station agreements. Local offices also track rates and conditions of employment for freelance work in each market, including rates for local or regional commercials and programs.

Presently, membership in AFTRA does not guarantee work or membership in other performer guilds, such as the Screen Actors Guild (SAG) or Actors Equity, although in general, both SAG and Actors Equity credit AFTRA membership and employment when evaluating applications. About 40,000 performers are members of both unions, and consequently, AFTRA and SAG have discussed a merger for several years. In 1995 the boards of directors of the two organizations approved a merger that would have created a larger union, given the combined membership of 123,000. Proponents of the merger cited the value of being able to present a united front when negotiating with an industry that was undergoing vast changes. Opponents were uneasy about the merging of the unions' health insurance and pension plans and about higher proposed dues and were said to be nervous about increasing employment pressure on the 80 percent of SAG members already earning less than $10,000 annually from acting. In early 1999 the merger got only about 50 percent of SAG voter approval, far less than the 60 percent level of approval required by the SAG constitution, although the merger was approved by two-thirds of AFTRA voters.

Following the defeat of the proposed merger between AFTRA and SAG, AFTRA's leadership continued to work toward a restructuring of the union. Between 1990 and 1993, an outside consulting agency had been commissioned to study changes affecting the industries under AFTRA jurisdiction. The study concluded with recommendations that AFTRA strengthen its national office and foster coordination among its historically strong locals. The consultants also recommended that AFTRA become more sophisticated in its use of both internal and external resources in order to match the resources of the companies with which it negotiated. Finally, it recommended that AFTRA provide more benefits and services for members and that the union find ways to involve its membership more fully in decision making and other union activity. Although the study maintained that AFTRA had "under financed" itself for many years by charging member dues that were among the lowest of any union in the broadcasting industry, the union decided it could not increase dues before first improving its services. Consequently, AFTRA reallocated its dues revenues in a series of internal changes that enabled AFTRA to add to its national staff by 36 percent to provide for better legal, financial, negotiating, and organizing services. The union also worked to enforce its existing contracts more vigorously, started a new research department, and worked to increase its lobbying presence in Washington. As of December 2002, AFTRA and SAG engaged in joint contract negotiations as an attempt to coordinate efforts in the face of continuing media (employer) consolidation.

AFTRA has a long history of supporting equal employment opportunities for women and minorities, and all AFTRA contracts include provisions for diversity and hiring fairness. AFTRA has a scholarship fund for members and dependents called the AFTRA Heller Memorial Foundation, and the union also set up the AFTRA Foundation, a tax-exempt organization funded by voluntary contributions to support educational and charitable causes.

MARK BRAUN

See also Commentators; Disk Jockeys; Female Radio Personalities; Singers on Radio; Sportscasters; Technical Organizations; Trade Associations

Further Reading

AFTRA website, <www.aftra.com> or <www.aftra.org>

Cox, Dan, "SAG, AFTRA Boards OK Merger," *Variety* (30 January 1995)

Johnson, Ted, "SAG-AFTRA's Merge Urge Hits Hurdles," *Variety* (12 February 1996)

Koenig, Allen E., *Broadcasting and Bargaining: Labor Relations in Radio and Television,* Madison: University of Wisconsin Press, 1970

Koenig, Allen E., "A History of AFTRA," *The NAEB Journal* (July-August 1965)

Leeds, Jeff, "Company Town: Screen Actors Guild Rejects AFTRA Merger," *Los Angeles Times* (29 January 1999)

Madigan, Nick, "Unions War over Marriage Proposal," *Variety* (30 November 1998)

Madigan, Nick, "SAG: Merging or Diverging?" *Variety* (25 January 1999)

American School of the Air

U.S. Educational Radio Program

First aired on 4 February 1930, with an 18-year run that ended on 30 April 1948, this Columbia Broadcasting System (CBS) half-hour educational series drew from top radio and educational talent to bring programs to U.S. and international schools and radio listeners.

The show was sponsored for a brief time by the Grisby-Grunow Company to support radio sales, and then CBS chose to retain *American School of the Air* as a sustaining "Columbia Educational Feature" overseen by the network's department of education. In 1940 the program was adapted and expanded to international educational markets in Canada and Latin America and the Philippines under the names *School of the Air of the Americas, Radio Escuela de las Americas,* and *International School of the Air.* Beginning in 1942, the *School of the Air of the Americas* was officially sponsored by the U.S. Office of War Information (OWI). In 1943 the program was deemed the "official channel for news, information, and instructions" by the OWI (CBS *Program Guide,* Winter 1943). In 1944 programs were also broadcast over the 400 stations of the Armed Forces Radio Service. The program was discontinued in 1948.

A number of educational organizations and individuals lent their names and expertise to the *American School of the Air.* Top-level national educators, such as William C. Bagley of the Teachers' College at Columbia University and U.S. Commissioner of Education John W. Studebaker, served on the national board of consultants, and educational consultants were also involved at the state and local levels. National organizations also offered conceptual and resource support to the program.

Each *American School of the Air* season ran from October through April, taking a break for the summer out-of-school months. Typically, the series offered five subseries—one for each day of the week—with titles such as *Frontiers of Democracy, The Music of America, This Living World, New Horizons, Lives between the Lines, Tales from Far and Near, Americans at Work, Wellsprings of Music, Science at Work, Music on a Holiday, Science Frontiers, Gateways to Music, Story of America, March of Science, World Neighbors, Tales of Adventure, Opinion Please,* and *Liberty Road.* Program topics included U.S. and international history and current events; music and literature; science and geography; vocational guidance and social studies; biographies; and many other topics. In 1940 CBS reported that the *American School of the Air* programs were received by more than 150,000 classrooms throughout all 48 states, reaching more than 200,000 teachers and 8 million pupils.

Some radio historians typically argue that the *American School of the Air* was part of a political strategy in early struggles over broadcast regulation. In the 1920s and early 1930s, noncommercial and citizen organizations proposed regulation, including frequency reallocation and nonprofit channel and program set-asides, to ensure that the United States' burgeoning broadcast system would remain, on some level, competitive and in the public's hands. The outcry against establishing a wholly commercial broadcast system compelled the networks to present a clear public-interest face, replete with educational, religious, and labor programming, in order to stave off binding regulation that might compromise network program time and control. CBS's *American School of the Air* was a premier effort of this type.

Educational scholars offer an alternative account of the *American School of the Air.* They focus on the program's role and function as an example of early educational technology and see the program as one of the first concerted experiments in education by radio, complete with supplemental classroom materials, teachers' manuals, and program guides.

SOUSAN ARAFEH

See also Columbia Broadcasting System; Educational Radio to 1967

Cast

Members of the New York radio pool, including Parker Fennelly, Mitzi Gould, Ray Collins, Chester Stratton; cast of *The Hamilton Family:* Gene Leonard, Betty Garde, Walter Tetley, Ruth Russell, Albert Aley, John Monks

Program Directors

Lyman Bryson, Sterling Fisher, and Leon Levine

Musical Directors

Alan Lomax, Dorothy Gordon, Channon Collinge

Writers

Hans Christian Adamson, Edward Mabley, Howard Rodman, A. Murray Dyer, Robert Aura Smith, and others

Announcers

Robert Trout, John Reed King, and others

Programming History

CBS February 1930–April 1948

Further Reading

Atkinson, Carroll, *Radio Network Contributions to Education,* Boston: Meador, 1942

Bird, Win W., *The Educational Aims and Practices of the National and Columbia Broadcasting Systems,* Seattle: University of Washington Extension Series, no 10 (August 1939)

Boemer, Marilyn Lawrence, *The Children's Hour: Radio Programs for Children, 1929–1956,* Metuchen, New Jersey: Scarecrow Press, 1989

Cuban, Larry, *How Teachers Taught: Constancy and Change in American Classrooms, 1890–1980,* New York: Longman, 1984

Smulyan, Susan, *Selling Radio: The Commercialization of American Broadcasting, 1920–1934,* Washington, D.C.: Smithsonian Institution Press, 1994

American Society of Composers, Authors, and Publishers

Established in 1914, the American Society of Composers, Authors, and Publishers (ASCAP) is the oldest music performance rights organization in the United States and the only U.S. performing rights organization whose board of directors (elected by the membership) consists entirely of member composers, songwriters, and music publishers. For almost two decades, it also was the only national organization providing copyright clearance for the broadcasting of music.

Origins

The legal foundation for ASCAP was established in the 1909 copyright law that required permission from the copyright holder in order to perform music for profit in public. With no rights clearinghouse in place, however, copyright holders faced the impossible job of individually monitoring performances of songs to which each held title. Not many years later, composer Victor Herbert was conducting performances of one of his operettas at a New York theater. At dinner one evening in a nearby restaurant, he heard the establishment's house musicians performing his composition "Sweethearts." Herbert became upset that people were paying to hear his melodies in the theater while restaurant patrons were listening to them without paying anything. He brought suit under the 1909 law. A lower court initially ruled against him because the restaurant had charged no admission fee. But the United States Supreme Court reversed the lower court in its 1917 *Herbert v Shanley* decision. In upholding the composer's claim, justice Oliver Wendell Holmes and his colleagues stated that it did not matter whether or not the performance actually resulted in a profit. The fact that it was employed as part of a profit-seeking endeavor was enough.

In 1914, before the main legal battle began, Herbert gathered eight composers, publishers' representatives, and lyricists for a meeting that ultimately would result in the establishment of ASCAP as their collection agent. In addition to Herbert, charter member composers included Irving Berlin and Rudolph Friml. Buoyed by Herbert's legal triumph three years later, ASCAP expanded its fee-seeking horizons beyond theaters and dance halls to any place where performance for profit took place. These proceeds then were distributed among ASCAP members via a sliding scale based on the number of compositions to which each held title and the musical prestige (not necessarily popularity) of each work.

ASCAP and Radio

By 1923 some radio stations had become profit-seeking (and a very few actually profit-making) enterprises that made widespread use of popular music. ASCAP therefore turned its attention to broadcasting, selecting WEAF, American Telephone and Telegraph's (AT&T) powerful New York outlet, as its test case. An aggressive protector of its own license and property rights, AT&T was not in a position to oppose ASCAP and settled on a one-year license of $500 in payment for all of the ASCAP-licensed music WEAF chose to air. This blanket license arrangement would become the industry standard. ASCAP followed this breakthrough by winning a lawsuit against station WOR in Newark, New Jersey, for the unlicensed broadcast of Francis A. McNamara's ballad "Mother Machree." Because ASCAP-affiliated composers were then the creators of virtually all popular music, stations faced the prospect of either paying up or ceasing to play the tunes that listeners expected to hear.

Perceiving themselves to be at ASCAP's mercy, major-market station owners formed the National Association of Broadcasters (NAB) in 1923 to do battle with the licensing organization. Station radio concerts were far less appealing without the melodies ASCAP controlled, but the annual license fees, which escalated upward from an initial $250, were seen as too high for many stations to pay. (Few of them had much revenue, let alone profits, at this point.) In subsequent years, ASCAP used its near-monopoly position in the music industry to charge broadcasters ever-higher rates. Continuous legal skirmishes, congressional hearings, and even frequent NAB-inspired Justice Department antitrust probes of ASCAP served only to raise the financial stakes and intensify the antagonistic relationship between NAB and ASCAP.

In 1931, for instance, ASCAP boosted its overall fees to stations by 300 percent, charging five percent of each outlet's gross income. It then broke off dealings with the NAB and began negotiations with individual broadcasters, offering three-year contracts at three percent of net income for the first year, four percent for the second, and the full five percent by the third year. By 1936 it was demanding five-year licenses.

Formation of a Competitor

When ASCAP announced yet another large increase in license fees for 1939, broadcasters took action and by the following year had established their own licensing organization, Broadcast Music Incorporated (BMI). On 1 January 1941, as BMI labored to build a catalog, most stations stopped paying their ASCAP fees and restricted their music broadcasts to airing songs with expired copyrights and folk songs that had always been in the public domain. Stephen Foster melodies, such as "Jeanie with the Light Brown Hair," became radio staples.

To ASCAP's chagrin, no groundswell of indignation arose from the radio audience. Further, singers and instrumentalists also replaced much of their repertoire with non-ASCAP material in order to keep their lucrative and visibility-enhancing radio bookings. Many performers switched from playing tunes by George Gershwin, Cole Porter, and Irving Berlin to using non-ASCAP music from South America—a key factor behind the sudden 1940s popularity of the rumba, samba, and tango. Combined with government antitrust pressure, these factors resulted in ASCAP's agreeing to offer per-program fees as well as blanket license fees and the rollback of rates to about half of what they had been collecting.

By the mid-1950s, the number of BMI tunes played over U.S. radio stations had come to parity with those licensed by ASCAP. Most of BMI's success was attributable to the explosion of rock and roll—a pulsating blend of rhythm and blues, country, and gospel music penned by songwriters outside of ASCAP's traditional constituency. BMI scooped up these composers and rode the rock and roll wave to dominance on many Top 40 format stations.

ASCAP and the Payola Scandal

In 1959 the payola scandal shook the radio industry to its core. Many disc jockeys were accused of taking unreported gifts from record promoters in exchange for "riding" (heavily playing) certain songs (payola). ASCAP added fuel to the fire when its spokespersons maintained that rock and roll, largely the creative product of BMI-affiliated composers, would never have gotten off the ground without payola. ASCAP claimed that 75 percent of the Top 50 tunes owed their success to payola—a charge meant as much to indict BMI as the practice of payola. With the subsequent passage of amendments to the 1934 Communications Act making payola a criminal offense, the radio industry moved beyond the crisis—but the resulting ASCAP-BMI animosity took a much longer time to cool.

Negotiating Music Rights

As it has for decades, ASCAP negotiates with radio stations mainly through the Radio Music License Committee (RMLC), a select group of broadcasters appointed by the NAB. Although stations technically could negotiate on their own, virtually all rely on the committee to carve out acceptable blanket and per-program license fee structures. For the period through the year 2002, blanket license fees for commercial radio stations were pegged at 1.615 percent for stations with an annual gross revenue over $150,000 or a minimum of 1 percent of adjusted gross income. For stations billing less than $150,000 per year, a flat fee schedule ranges from $450 to $1,800 depending on income. Noncommercial stations pay an annual fee determined by the U.S. Copyright Office. In 2003 this was pegged at $245 for educationally owned facilities and $460 for all other noncommercial outlets.

ASCAP determines the amount of airplay garnered by each ASCAP-member song via three methods: electronic logging information from Broadcast Data Systems (BDS), periodic logging by the radio stations themselves, and ASCAP taping of station broadcasts.

PETER B. ORLIK

See also Broadcast Music Inc.; Copyright; Licensing; National Association of Broadcasters; Payola; United States Supreme Court and Radio; WEAF

Further Reading

ASCAP website, <www.ascap.com>

Dachs, David, *Anything Goes: The World of Popular Music,* Indianapolis, Indiana: Bobbs-Merrill, 1964

Lathrop, Tad, and Jim Pettigrew, Jr., *This Business of Music Marketing and Promotion,* New York: Billboard Books, 1999

Petrozzello, Donna, "ASCAP Restructuring Rates," *Broadcasting and Cable* (12 August 1996)

Ryan, John, *The Production of Culture in the Music Industry: The ASCAP-BMI Controversy,* Lanham, Maryland: University Press of America, 1985

Smith, Wes, *The Pied Pipers of Rock 'n' Roll: Radio Deejays of the 50s and 60s,* Marietta, Georgia: Longstreet Press, 1989

American Telephone and Telegraph

American Telephone and Telegraph Company (AT&T) was a major contributor to the development of early broadcasting technology and radio networking. As a result of its refinements in vacuum tube technology and the ensuing patent disputes, AT&T became a founding shareholder in the powerful Radio Corporation of America (RCA), built the first commercial radio station (WEAF), and perfected the technology for network broadcasting. Over the years, AT&T's Bell Laboratories has pioneered many technologies used in radio.

Audion and Patent Concerns

Parent company of the Bell Telephone System, AT&T recognized the potential of de Forest's Audion tube as an amplifier for telephone circuits and secured rights to the device. Although the Audion could not be used in radio circuits due to a suit by the Marconi Company alleging patent infringements, AT&T licensed its use for telephone circuitry, quickly refining the technology and thus making transcontinental telephony a reality. AT&T also used vacuum tubes to pursue development of continuous wave transmitters necessary for voice communication as ancillary devices supporting telephone services. At the outbreak of World War I, the U.S. Navy took control of all radio patents, accelerating the development of wireless and radio receiver technology. At the end of the war, large electronic manufacturers such as AT&T, Westinghouse, and General Electric reclaimed their patents.

After the war, U.S. government officials expressed their desire to settle the patent problem quickly in order to keep key radio technology in the hands of a U.S. company. In 1919 RCA was formed as a way to pool the patents and cross-license the various technologies, making the large-scale manufacture of radio vacuum tubes possible. Under the agreement, AT&T's manufacturing arm, Western Electric, gained exclusive rights to produce long-distance transmitters and other key technology used in conjunction with wired communications. By 1922 Western Electric transmitters were powering 30 of America's pioneering radio stations, including such legendary stations as WOR in Newark, New Jersey, WHAM in Rochester, New York, and WSB in Atlanta, Georgia. AT&T soon came to realize, however, that the sale of transmitters to others conflicted with the company's strategy of beginning a nationwide commercial broadcasting service.

Birth of Radio Networking

Beginning in 1877 AT&T started experimenting with the use of telephone lines for transmission of music and entertainment. These experiments used telephones or public address systems to carry program material such as music or speeches. By 1919 AT&T had refined vacuum tube technology to the point where large, elaborate auditorium demonstrations were possible. With the advent of broadcasting, however, the need for high quality connections to bring live events to radio stations became apparent. Soon AT&T undertook experiments to test public acceptance of broadcasting.

In January 1922 AT&T began construction of its own broadcasting facilities. AT&T vice president Walter S. Gifford outlined the commercial "toll" concept of broadcasting, calling for the creation of a channel through which anyone could send out his or her own programs. AT&T originally contemplated building 38 "radiotelephone" stations linked together by the company's Long Lines division. The first two stations were constructed in New York: WBAY was erected atop the AT&T Long Lines building on Walker Street and WEAF was constructed at Western Electric's Labs on West Street. WBAY signed on to 360 meters (830 kHz) on 3 August 1922. When signal coverage from WBAY proved unsatisfactory due to the steel construction of the Walker Street building, WEAF became the company's principal transmitting facility. In 1923 WCAP, Washington, D.C., was added to AT&T's station lineup.

Both RCA and AT&T started experimenting with interconnecting stations, but RCA was limited to using Western Union telegraph lines, and these proved to be unsuitable for the transmission of high-quality voice and music. The first AT&T network experiment started on 4 January 1923, when engineers connected WEAF in New York, New York, with WNAC in Boston, Massachusetts. Soon after, Colonial H.R. Green, owner of station WMAF, convinced AT&T to provide a link from WEAF to his station in South Dartmouth, Massachusetts. Green agreed to pay AT&T $60,000 for a permanent connection, and WMAF began retransmitting WEAF programming. This arrangement gave AT&T engineers a full-time connection, which they used to experiment with transmission equipment. Other networking experiments followed. On 7 June 1923, WGY in Schenectady, New York, KDKA in Pittsburgh, Pennsylvania, and KYW in Chicago, Illinois, were connected to WEAF. AT&T used the term *chain* to refer to interconnection of radio stations. Later *chain broadcasting* became the term commonly applied to radio network broadcasting.

Using Long Lines Division's capabilities, WEAF undertook a series of spectacular remote broadcasts that generated great interest among radio listeners and gave WEAF a programming advantage over other stations. Sporting events such as the Princeton-Chicago and Harvard-Yale football games, the Dempsey-Tunney boxing match, and recitals from the Capitol Theater demonstrated that coverage of live events was of great interest to Americans. At the same time, AT&T held control over the capability to provide remote broadcasts via its telephone lines, and it began to refuse to provide hookups to other rival stations owned by RCA's radio group.

In 1924 AT&T connected radio stations in 12 major cities from Boston to Kansas City, Missouri, for special broadcasts of the Republican and Democratic national conventions. One year later, it used its circuits in the first coast-to-coast demonstration. As these experiments continued into 1925, radio network connections regularly linked WEAF and other stations in cities throughout the East and Midwest. AT&T executives began to rethink the company's involvement in broadcasting, however, as disagreements with RCA over the cross-licensing arrangements increased. Finally, in 1926 AT&T decided to discontinue broadcasting operations and sold WEAF to RCA.

RCA created the National Broadcasting Company (NBC) to operate WEAF, WJZ, and its own radio stations. Under the terms of the WEAF sale, NBC was required to lease AT&T lines for network connections whenever possible. NBC decided to form two separate networks to handle stations where there was duplication in coverage area. AT&T Long Lines engineers used red and blue pencils to trace the connection paths for NBC's new networks and NBC adopted the colors as designations for the two networks. The Red Network, with WEAF as the flagship station, was the larger and more important of the two with 25 stations; the Blue Network began with only 5 stations.

Growth of the Chains

America's growing interest in high-quality programs spurred further AT&T development of networking capabilities. As public interest in high-quality programming grew, many local stations joined one of the two NBC radio networks, but the NBC monopoly in network broadcasting was not to last for long. A small upstart, United Independent Broadcasters, was formed when Arthur Judson decided to establish a new radio network. In early 1927 Judson tried to secure telephone lines for the newly formed network, but AT&T refused to provide connections because it had signed an exclusivity agreement with RCA. By mid-1927 AT&T, under pressure from the Federal Trade Commission, agreed to provide network connections to the new network. That fall the newly named Columbia Broadcasting System (CBS) began operations with 12 affiliated stations. Soon, chain broadcasting revolutionized radio in the United States.

AT&T played a pivotal role in making network broadcasting a success. Long Lines Division developed elaborate specialized network capabilities that served both full-time networks such as NBC and CBS and specialized regional networks, such as the Don Lee Network in California and the Liberty Broadcasting System in the Southwest.

By 1928 AT&T maintained four broadcast network interconnection systems (called Red, Blue, Orange, and Purple) linking 69 radio stations together with more than 28,000 miles of wire. New York served as the central distribution point for stations in the East and South, while Chicago, Illinois, and Cincinnati, Ohio, served Midwest stations. San Francisco, California, became a switching point for the West Coast. Both telegraph and voice circuits were used to provide affiliates with networking information and programming channel feeds.

AT&T used an elaborate series of repeater stations to route high-quality audio transmissions across the nation. Special Long Lines operators provided maintenance for the system and switching for network programs. Stations that normally carried Red network programming needed to be manually switched by AT&T personnel when they wanted to carry Blue network programs. Important switching stations, such as Washington, D.C., could switch as many as 30,000 programs each year. To facilitate network quality testing, broadcasters provided musical programming for the Long Lines operators. NBC maintained a legendary jukebox at Radio City that played music whenever either of the networks was not transmitting a program. The jukebox selections provided AT&T engineers with a constant audio source to verify network quality. This practice continued through the mid-1980s, when satellites finally replaced land lines. Broadcasters worked with Bell Labs to develop equipment to interconnect broadcast stations with the telephone network. The VU meter, a visual gauge for measuring audio, was an outgrowth of that cooperation, and special terminology such as *nonemanating outputs*, (NEMO) *terminal block*, and *mults* entered broadcast parlance as a result of this relationship.

The cost of renting AT&T broadcast lines was often too high to allow local radio stations to provide live coverage of sporting events. Announcers such as Red Barber and Ronald "Dutch" Reagan—and later Gordon McLendon—made names for themselves recreating games by using sparse information provided by telegraph operators at ball games.

As broadcasting networks grew in power and size throughout the 1930s, AT&T expanded its special services. By 1939 more than 53,000 miles of special circuit wires were used to provide network services. The number of specialized networks maintained by AT&T expanded to 21 just before World War II. Network designations continued to be based on the original engineering color schemes, with NBC having its Red and Blue networks and CBS using ivory, black, pink, scarlet, and other specialized broadcast facilities. Broadcast network operations represented approximately 15 percent of AT&T's $23 million in gross revenue for Private Line Services in 1933. By 1935 broadcasters were spending more than $10 million a year for telephone lines to link their networks. Some estimate that in 1950 nearly 40 percent of AT&T's $53 million in private network gross revenues represented broadcast services. Revenues generated from broadcasting and other special private services rose throughout the golden age of radio networks, providing AT&T with substantial profits.

With the introduction of television, the decline in network radio led to a decline in AT&T's involvement in linking up network stations. Special services provided by Long Lines during this time were used to create a nationwide television network system for the growing number of television network affiliates. Long Lines continued to provide network connections for radio until the mid-1980s, when domestic communication satellites replaced land lines. The competition from satellite distribution and FCC deregulation in telecommunication services made general land line distribution of radio unprofitable.

Technical Developments

Bell Telephone Laboratories, created by AT&T in 1925, pioneered many technologies that have expanded the capability of modern radio broadcasters. In the 1930s AT&T invented stereophonic sound systems and microwave transmission, both essential technologies for today's high-quality radio programming. In the late 1940s Bell Labs invented the transistor, the forerunner of modern solid-state electronics, spawning both the transistor radio and solid-state computer era. Other key developments include the communication satellite, the light-emitting diode (LED), and the laser. Today's advanced audio technology is partially an outgrowth of the basic scientific research undertaken by AT&T.

FRITZ MESSERE

See also Columbia Broadcasting System; De Forest, Lee; National Broadcasting Company; Network Monopoly Probe; Radio Corporation of America; Stereo; WEAF

Further Reading

Aitken, Hugh G.J., *The Continuous Wave: Technology and American Radio, 1900–1932,* Princeton, New Jersey: Princeton University Press, 1985

Archer, Gleason Leonard, *History of Radio to 1926,* New York: American Historical Society, 1938

Archer, Gleason Leonard, *Big Business and Radio,* New York: American Historical Society, 1939

Banning, William Peck, *Commercial Broadcasting Pioneer: The WEAF Experiment, 1922–1926,* Cambridge, Massachusetts: Harvard University Press, 1946

Douglas, Susan, *Listening In: Radio and the American Imagination,* New York: Times Books, 1999

Sibley, Ludwell, "Program Transmission and the Early Radio Networks," *AWA Review* 3 (1988)

American Top 40

Popular Music Program

American Top 40 (*AT40*) is the longest running national music countdown broadcast on American radio during the rock era. In its 30-year history, the show has undergone a series of personnel and ownership changes. The first *AT40* show aired with veteran disc jockey Casey Kasem during the week of 4 July 1970 and was distributed in only seven U.S. markets. By 1980 the show could be heard in nearly 500 markets across the United States.

The original *AT40* program concept was created by Ron Jacobs, who with Tom Rounds founded Watermark in 1969. The program grew out of collaboration between Jacobs and K-B Productions owners Casey Kasem and Don Bustany, who sold Jacobs on the idea of a national music countdown. Despite the initial downturn in the Top 40 music format as the rapid proliferation of new FM stations popularized album-oriented rock in the early 1970s, *AT40* soon found a loyal audience.

AT40 was the first program to turn the popular local Top 40 countdown into a national syndicated show. The three-hour show was distributed weekly on records to radio stations across the United States, using the *Billboard* Top 100 as the source for the countdown. *AT40* was distributed as a boxed record set each week. Records played 30 minutes of the show per side, and the set contained cue sheets allowing stations to integrate local station breaks into the *AT40* program format. The records had to be played in the right order for the countdown to progress correctly. Program segments opened or ended with jingles identifying the program and the program host. By 1978 the general length of popular songs had increased, causing *AT40* to increase its program length from three to four hours per show. Today the show is distributed on compact discs.

One of the reasons for *AT40*'s success was the charismatic, personal voice style of Casey Kasem, the show's longest-serving host. The format initially called for a fast-paced delivery with minimal talk and a quick turnover from song to song. As the program gained momentum, Kasem's knowledge of popular music and his ability to create a sense of intimacy added interest for listeners. As the show expanded its time and found a loyal audience, special features such as the "Long Distance Dedication" became popular segments. Kasem's classic sign-off, "Keep your feet on the ground and keep reaching for the stars" became the show's trademark.

AT40's success was challenged in 1979 with the introduction of *The Weekly Top 30* hosted by Mark Elliot and in 1980 with *Dick Clark's National Music Survey*. Both of these shows were aimed at slightly different demographics than *AT40*. *The Weekly Top 30* ended in 1982, and in 1983, *Rick Dees' Weekly Top 40* aired based on the popular music chart listings in *Cashbox* magazine.

In 1988 American Broadcasting Companies (ABC) Radio Networks, which had acquired Watermark, and Kasem were unable to agree to terms for a renewal contract. By this time *AT40* had grown to become the most successful American radio program and was the sixth largest syndicated broadcasting program with an estimated 2.4 million listeners worldwide. The show boasted nearly 1,000 outlets around the world. In July 1988 ABC introduced Shadoe Stevens as the new host of *AT40*. ABC heavily promoted the transition and introduced various new features to distinguish the new show host and keep the format fresh. Stevens hosted the show until 1995. Several broadcasting companies vied for Kasem's talents, and in 1989 he signed a multimillion-dollar, multiyear contract with Westwood One to start a competing program called "*Casey's Top 40*."

The early 1990s saw a substantial change in popular music. In November 1991 *Billboard* changed the way it tabulated the Hot 100. *Billboard*'s new methodology led to a substantial increase in rap and other nontraditional pop music genres in the chart, causing many older loyal listeners to tune out. The traditional Top 40 format splintered into derivative formats. In addition, the continuing success of *Rick Dees's Weekly Top 40* and *Casey's Top 40* splintered the market for Top 40. By 1992 *AT40* had fewer than 275 stations carrying the program in the United States, although it still held the predominant position among overseas listeners.

In 1994 ABC Radio Networks acquired the Westwood One network and ownership of *AT40*. ABC now owned both the Rick Dees countdown and *AT40*. On 24 June 1994, ABC announced that it would cancel the American version of *AT40*, and the last program aired in January 1995. ABC's rights to the program terminated in 1998, and the show reverted back to Kasem and Bustany, the owners of K-B Productions. *AT40* was revived with Kasem as the host in March 1998 under the ownership of AMFM Networks. In addition to the Top 40 format, there is also an American Top 20 based on the hot adult contemporary format and another geared toward adult contemporary listeners. Even though the format of the new *AT40* is very similar to the original, the chart list is now based on the Mediabase 24/7 hit music charts.

In 2003, the distribution of the program changed to Premiere Radio Networks. American Top 40 was at that time heard on 127 U.S. stations and 14 outlets internationally.

FRITZ MESSERE

See also Kasem, Casey

Hosts

Casey Kasem (1970–88; 1998–), Shadoe Stevens (1989–95)

Creator

Ron Jacobs

Executive Producer

Tom Rounds

Programming History

Watermark Syndication	1970–88
ABC	1988–95
Radio Express	1998–2002
Premiere Radio Network	2003–

Further Reading

Durkee, Rob, *American Top 40: The Countdown of the Century*, New York: Schirmer Books, 1999

American Women in Radio and Television

Women have taken part in the business of radio broadcasting from the earliest days of the industry. Although their advancement was often slower than that of their male counterparts, they were able to contribute greatly to the development of radio. As in many professions, women organized groups to provide mutual support in their efforts for advancement and recognition. American Women in Radio and Television (AWRT) is one such organization.

Origins

On 8 April 1951, AWRT held its organizing convention at the Astor hotel in New York. According to the trade publication *Broadcasting and Telecasting,* 250 women from the fields of radio and television attended this event. They elected Edyth Meserand, assistant director of news and special features at WOR-AM-TV in New York, as the first AWRT president. Appropriately, the keynote speaker at the conference was Frieda Hennock, the first woman to be appointed a commissioner at the Federal Communications Commission (FCC).

Since 1951 AWRT has pursued its stated mission: "to advance the impact of women in the electronic media and allied fields, by educating, advocating, and acting as a resource to our members and the industry." Logically, the majority of members are women, but many men also choose to participate in this effort. The organization carries out its work through more than 30 chapters nationwide.

Function

AWRT provides professional development activities, mentoring, and job-search assistance for its members. The organization sponsors awards for excellence in the profession and publishes information resources, in print and on-line, for members and nonmembers alike. Through its lobbying efforts, AWRT has been an advocate for many causes, including better opportunities for women who aspire to own media organizations and stricter enforcement of equal employment opportunity requirements at television and radio stations. It has argued in favor of a proposal that would require broadcast organizations to keep statistics on minority and female employees, and it has taken a lead role in educating professionals about sexual harassment. At times, AWRT speaks out on behalf of general policy options that its members consider relevant to its mission. For example, the group argued against using auctions to determine spectrum ownership. The basis for this argument was the belief that auctioning broadcast frequencies would work against preserving a diversity of voices in the media marketplace.

One of the most important services AWRT provides is facilitating networking opportunities for its members. At its annual convention, professionals from all areas of electronic media discuss key issues affecting the field as well as more specific topics that are most likely to concern women. Throughout the year, AWRT serves all types of organizations by providing speakers on such topics as promoting diversity in the workplace and managing a diverse workforce.

In addition to these services, AWRT recognizes excellence in electronic media by presenting awards to individuals and companies. These awards are given for outstanding achievements in electronic media, for commitment to the issues and concerns of women, and for achievements in strengthening the role of women in the industry and contributing to the betterment of the community. Its annual Gracie awards, named in honor of broadcast pioneer Gracie Allen, recognize realistic portrayals of women in radio and television programming. Several awards are given each year to commend media contributions "by women, for women or about women." The Silver Satellite awards recognize the outstanding contributions of an individual to the broadcast industry. Among the former winners of this prestigious award are Bob Hope, who won the first Silver Satellite award in 1968, Vincent Wasilewski, former head of the National Association of Broadcasters, Mary Tyler Moore, Barbara Walters, and Pauline Frederick. Other honors given by the organization include the Star Awards, which honor individuals and companies who have shown a commitment to the concerns of women, and the Achievement Awards, which recognize a member who has both strengthened the role of women and contributed to the betterment of the community.

In 1960 the AWRT Educational Foundation was chartered to promote charitable programs, educational services, scholarships, and projects to benefit the community and the mass media. This support not only provides assistance to community organizations in need of funding, it also provides an opportunity for AWRT members and others to become actively involved in serving their communities.

PATRICIA PHALEN

See also Association for Women in Communication; Female Radio Personalities and Disk Jockeys; Women in Radio

Further Reading

Baehr, Helen, editor, *Women and Media,* Oxford and New York: Pergamon Press, 1980

Baehr, Helen, and Michele Ryan, *Shut Up and Listen! Women and Local Radio: A View from the Inside,* London: 1984

Creedon, Pamela, J., editor, *Women in Mass Communication,* Newberry Park, California, and London: Sage 1989; 2nd edition, 1993

Phalen, Patricia, "Pioneers, Girlfriends, and Wives: An Agenda for Research on Women and the Organizational Culture of Broadcasting," *The Journal of Broadcasting and Electronic Media* 44, no. 2 (Spring 2000)

Stone, Vernon A., *Let's Talk Pay in Television and Radio News,* Chicago: Bonus Books, 1993

America's Town Meeting of the Air

Public Affairs Program

For much of its 21-year run, *America's Town Meeting of the Air* (1935–56) was a Thursday evening staple in many radio homes. As part of a trend toward panel discussion shows in the 1930s, this series as well as *American Forum of the Air, People's Platform, University of Chicago Roundtable, Northwestern Reviewing Stand,* and *High School Town Meeting of the Air* were sustaining (commercial-free) programs devoted to in-depth political and social discussion. Although *America's Town Meeting of the Air* was not the first of these panel discussion programs on the air, it was the first radio program to offer debate *and* active audience participation.

The first panel discussion program, *University of Chicago Roundtable* (1931–55), was a more reserved, scholarly program featuring University of Chicago professors debating contemporary issues. *American Forum of the Air* (1934–56) developed an adversary format, with two opponents on either side of a controversial issue, which became a popular feature of later panel discussion programs. *America's Town Meeting of the Air*'s innovation was its inclusion of the live audience by using unscreened audience questions as an essential part of the discussion. Because audience members challenged guest speakers and their views, *America's Town Meeting of the Air* was an often volatile and unpredictable hour of radio programming. This serious-minded and popular program recognized the power of audience participation and influenced the format of later public-affairs programs and talk shows.

America's Town Meeting of the Air was the brainchild of George V. Denny, Jr., a former drama teacher and lecture manager. Denny was associate director of the League of Political Education, a New York–based political group founded in 1894 by suffragists that held town meetings to discuss contemporary issues. Legend has it that Denny, shocked by a neighbor's refusal to listen to President Roosevelt because he disagreed with him, sought to raise the level of political discussion in the country. He believed that a radio program could be produced that would mirror the New England town meetings of early America and promote democratic debate. Denny mentioned his idea to the director of the League of Political Education, Mrs. Richard Patterson, who brought the idea to her husband, National Broadcasting Company (NBC) Vice President Richard Patterson. Richard Patterson helped Denny develop the show and gave the hour-long program a six-week trial run in 1935 on the NBC Blue network. Inexpensive and easy to produce, *America's Town Meeting of the Air* was an efficient and effective sustaining program for NBC. From its initial airing over 18 stations on the NBC Blue network, the show reached more than 20 million listeners through more than 225 stations by 1947. The successful program found its home on the NBC Blue (later American Broadcasting Companies [ABC]) network for its entire run.

For most of its life on radio, the program refused to accept sponsors, fearing that commercial interests would interfere with the show's controversial content. For only one year, *America's Town Meeting of the Air* accepted the sponsorship of *Reader's Digest.* In its later years (1947–55), the program accepted multiple sponsors. ABC tried to simulcast the program (somewhat unsuccessfully) on television and radio in 1948–49 and again in 1952, but the program did not translate well to television. After an internal dispute, Denny, the originator of the series, was removed from the program in 1952. Despite the loss of its creator, the program lasted four more years.

America's Town Meeting of the Air welcomed listeners each week with the sound of a town crier's bell and Denny's voice calling, "Good evening, neighbors." Broadcasting from Town Hall (123 West 43rd Street in New York City), Denny assembled a live studio audience of nearly 1,800 to participate in the broadcast. Before these witnesses, the show featured two or more opponents on a controversial issue. To build suspense, each guest would have the opportunity to state his or her position and would then field unscreened questions from the live studio audience and from listeners who sent questions via telegram before the program's broadcast. The program's format was designed by Denny to present a diversity of political and social views and to bring those views into conflict before a live and often raucous studio audience.

The program's commitment to public affairs and controversial issues was established from its first broadcast on 30 May 1935. The topic for the first program was "Which Way for America—Communism, Fascism, Socialism, or Democracy?" Raymond Moley (an adviser to President Roosevelt) defended democracy, Norman Thomas made the case for socialism, A.J. Muste argued for the importance of communism, and Lawrence Dennis explained the benefits of fascism. The show was remarkable for the breadth and depth of the issues debated publicly. *America's Town Meeting of the Air* frequently addressed foreign policy or international disputes (e.g., "How Can We Advance Democracy in Asia?" or "What Kind of World Order Do We Want?" featuring a debate between H.G. Wells and Dr. Hu Shih, the Chinese ambassador to the United States) as well as domestic issues (e.g., "Does Our National Debt Imperil America's Future?" "How Essential Is Religion to Democracy?" or "Can We Depend upon Youth to Follow the American Way?"). *America's Town Meeting of the Air* also tackled the racial conflicts of the period, featuring prominent African-American scholars and writers such as Richard Wright. One of its most popular shows was the 1944 broadcast entitled "Let's Face the Race Question," with Langston Hughes, Carey McWilliams, John Temple Graves, and James Shepard. Whether discussing the detention of Japanese-Americans during World War II, debating immigration restrictions, or confronting the racial divide in the 20th century, *America's Town Meeting of the Air* offered listeners the opportunity to debate topics that were suppressed or marginalized elsewhere on radio.

In the 1930s and 1940s, prestigious panel discussion programs such as *America's Town Meeting of the Air* were sterling examples of the networks' devotion to public service. Such shows were used by the networks to fulfill their public-interest obligations to the community and to stave off government regulation in early radio. As discussed by Barbara Savage (1999), *America's Town Meeting of the Air* conducted an extensive public outreach campaign to incorporate the listening audience into the program and to increase its public profile.

According to a 1940 sales brochure, NBC supported the development of debate and discussion groups. NBC viewed the program as "a real force of public enlightenment" and an example of the network's "unexampled public service to the men and women of America." Transcripts of broadcasts were published by Columbia University Press. The program encouraged the use of transcripts in schools and sponsored editorial cartoon and essay contests on subjects such as "What Does American Democracy Mean to Me?" Despite the fact that *America's Town Meeting of the Air* originated from New York City, the program also worked carefully to promote regional interest in the program. For six months out of the year, the program traveled around the country, sponsored by local universities and civic groups.

America's Town Meeting of the Air was one of the most popular national public-affairs programs on radio. Nearly 1,000 debate groups were officially formed, and thousands more listened each week in barber shops and community centers around the country. In the 1938–39 season, nearly 250,000 program transcripts were requested; the show typically received 4,000 letters a week. The program was also critically acclaimed for its public service. *America's Town Meeting of the Air* was a multiple winner of the Peabody Award and was also recognized by the Women's National Radio Committee, the Institute for Education by Radio, and the Women's Press Bloc, among other organizations, for its educational qualities and its discussion of economic, political, and international problems.

JENNIFER HYLAND WANG

See also Public Affairs Programming

Moderator
George V. Denny, Jr.

Announcers
Howard Claney, Milton Cross, Ben Grauer, George Gunn, Ed Herlihy, Gene Kirby

Producer
Marian Carter

Directors
Wylie Adams, Leonard Blair, Richard Ritter

Programming History

NBC Blue	1935–42
NBC Blue/ABC	1942–56

Further Reading

DeLong, Thomas A., "George V. Denny," in *Radio Stars: An Illustrated Biographical Dictionary of 953 Performers, 1920 through 1960,* by DeLong, Jefferson, North Carolina: McFarland, 1996

Gregg, Robert, "America's Town Meeting of the Air, 1935–1950," Ph.D. diss., Columbia University, 1957

MacDonald, J. Fred, *Don't Touch That Dial! Radio Programming in American Life, 1920–1960,* Chicago: Nelson-Hall, 1979

Savage, Barbara Dianne, *Broadcasting Freedom: Radio, War, and the Politics of Race, 1938–1948,* Chapel Hill: University of North Carolina Press, 1999

Amos 'n' Andy

U. S. Serial (1928–1943); Situation Comedy (1943–1955); Hosted Recorded Music (1954–1960)

Amos 'n' Andy, which began as a nightly serial telling the story of Amos Jones and Andy Brown, two Georgia-born black men seeking their fortunes in the North, dominated American radio during the Depression. Combining character-driven humor with melodramatic plots, the series established the viability in broadcasting of continuing characters in a continued story and, from both a business and creative perspective, proved the most influential radio program of its era, inspiring the creation of the broadcast syndication industry and serving as the fountainhead of both the situation comedy and the soap opera. At its peak in 1930–31, the program's nightly audience exceeded 40 million people.

After 15 years and more than 4,000 episodes, the serial gave way to a weekly situation comedy and the characterizations grew more exaggerated. Today, the original *Amos 'n' Andy* is almost completely forgotten—its substance overshadowed by the unacceptability of white actors portraying African-American characters and lost to the memory of the broadly played sitcom that replaced it. Nevertheless, *Amos 'n' Andy* remains a landmark in U.S. broadcasting history.

Origins

Amos 'n' Andy grew out of *Sam 'n' Henry,* created by Freeman F. Gosden and Charles J. Correll, two former producers of home-talent revues who had begun their careers as a comic harmony team on Chicago radio in 1925. They had been asked by the management of station WGN to adapt the popular comic strip "The Gumps" for broadcasting, but were intimidated by its middle-class setting. Instead, they suggested a "radio comic strip" about two black men from the South moving to the North, characterizations that would draw on Gosden's familiarity since childhood with African-American dialect, and that would enable the performers to remain anonymous—an important consideration if the program should fail.

Sam 'n' Henry premiered on 12 January 1926 as the first nightly serial program on American radio, combining black dialect with certain character traits and storytelling themes from "The Gumps." The early episodes were often crude, but Gosden and Correll gradually learned how to tell involving stories and to create complex human characterizations.

By the spring of 1926 the performers had begun recording *Sam 'n' Henry* sketches for Victor, and the success of these records suggested to the performers that live broadcasting need not be their only course. Accordingly, the partners suggested to WGN that their programs be recorded and the recordings leased to other stations. WGN rejected the proposal, citing its ownership of the series and its characters. Gosden and Correll left WGN in December 1927, moving to station WMAQ, owned by the *Chicago Daily News,* and negotiated an agreement that included syndication rights. Arrangements were made for advance recordings of each episode on 12" 78 rpm discs that would be distributed to subscribing stations for airing in synchronization with the live broadcast from WMAQ. Correll and Gosden called this a "chainless chain" and, realizing the value of the concept, attempted to secure a patent, but were unable to do so; however, by the early 1930s their idea had formed the basis for the broadcast syndication industry.

Transition

The WMAQ series introduced Amos Jones and Andy Brown as hired hands on a farm outside Atlanta, looking ahead to their planned move to Chicago. Amos was plagued by self-doubts and worried about finding work in the North, whereas the swaggering Andy was quick to insist that he had the answers to everything.

Amos and Andy struggled until they met Sylvester, a soft-spoken, intelligent teenager patterned after the black youth who had been Gosden's closest childhood friend. Sylvester helped Amos and Andy start their own business, the Fresh Air Taxicab Company, and introduced them to a cultured, successful, middle-class businessman named William Taylor and his bright, attractive daughter Ruby, who soon became Amos' fiancée. They also met the potentate of a local fraternity, George "Kingfish" Stevens, a smooth-talking hustler who insinuated himself and his constant moneymaking schemes into their lives.

Chainless Chain to Network

Within a few months *Amos 'n' Andy* had attracted a national following and the attention of the Pepsodent Company, which negotiated to bring the serial to the coast-to-coast NBC Blue network in the summer of 1929. Amos, Andy, and the Kingfish relocated from Chicago to Harlem at the start of the network run, but otherwise the storyline continued unchanged.

At first, the program was heard at 11 P.M. Eastern time, but Pepsodent sought an earlier time slot for Eastern listeners and NBC was able to clear time at 7 P.M. As soon as the

Freeman Gosden Sr. ("Amos") left, and Charles Correll ("Andy")
Courtesy Radio Hall of Fame

change was announced, thousands of listeners in the Midwest and West wrote to complain about the move and within a week Correll and Gosden had agreed to broadcast twice nightly. This dual-broadcast plan would be widely adopted by other national sponsors as a solution to the time-zone dilemma.

The outcry over the time change offered just a hint of what was to come. By the spring of 1930 theater owners in many cities were being forced to stop the movie playing in order to present *Amos 'n' Andy* over their sound systems to hold an evening audience—dramatic evidence of an unprecedented craze that would endure for nearly two years.

Impact

As a result of its extraordinary popularity, *Amos 'n' Andy* profoundly influenced the development of dramatic radio. Working alone in a small studio, Correll and Gosden created an intimate, understated acting style—a technique requiring careful modulation of the voice, especially in the portrayal of multiple characters—that differed sharply from the broad manner of stage actors. The performers pioneered the technique of varying both the distance and the angle of their approach to the microphone to create the illusion of a group of characters. Listeners could easily imagine that that they were in the taxicab office, listening in on the conversation of close friends. The result was a uniquely absorbing experience for listeners who, in radio's short history, had never heard anything quite like *Amos 'n' Andy.*

Although minstrel-style wordplay humor was common in the formative years of the program, it was used less often as the series developed, giving way to a more sophisticated approach to characterization. Correll and Gosden were fascinated by human nature, and their approach to both comedy and drama drew from their observations of the traits and motivations that drive the actions of all people; although they often overlapped popular stereotypes of African-Americans, there was at the same time a universality to their characters that transcended race.

Central to the program was the tension between the lead characters. Amos stood as an "Everyman" figure: a sympathetic, occasionally heroic individual who combined practical intelligence and a gritty determination to succeed with deep compassion—along with a caustic sense of humor and a tendency to repress his anger until it suddenly exploded. Andy, by contrast, was a pretentious braggart, obsessed with the symbols of success but unwilling to put forth the effort required to earn them. Although Andy's overweening vanity proved his greatest weakness, he was at heart a poignant, vulnerable character, his bombast masking deep insecurity and a desperate need for approval and affection. The Kingfish was presented as a shrewd, resourceful man who might have succeeded in any career, had he applied himself, but who preferred the freedom of living by his wits.

Other characters displayed a broad range of human foibles: the rigid, hard-working Brother Crawford, the social climber Henry Van Porter, the arrogant Frederick Montgomery Gwindell, the slow-moving but honest Lightning, the flamboyant Madam Queen. Still other characters stood as bold repudiations of stereotypes: the graceful, college-educated Ruby Taylor; her quietly dignified father, the self-made millionaire Roland Weber; and the capable and effective lawyers, doctors, and bankers who advised Amos and Andy in times of crisis. Beneath the dialect and racial imagery, the series celebrated the virtues of friendship, persistence, hard work, and common sense and, as the years passed and the characterizations were refined, *Amos 'n' Andy* achieved an emotional depth rivaled by few other radio programs of the 1930s.

Above all, Correll and Gosden were gifted dramatists. Their plots flowed gradually from one into the next, with minor subplots building in importance until they took over the narrative and then receding to give way to the next major sequence; seeds for future storylines were often planted months in advance. It was this complex method of story construction that kept the program fresh and enabled Correll and Gosden to keep their audience in a constant state of suspense. The technique they developed for radio from that of the narrative comic strip endures to the present day as the standard method of storytelling in serial drama.

Storylines in *Amos 'n' Andy* usually revolved around themes of money and romance—Amos' progress toward the goal of marrying his beloved Ruby Taylor stood in contrast to Andy's romantic fumblings—with the daily challenge of making ends meet forming a constant backdrop. The taxicab company remained the foundation of Amos and Andy's enterprises, but the partners constantly explored other ventures, including a lunchroom, a hotel, a grocery, a filling station, and a 500-acre housing development. Andy invariably claimed the executive titles, while Amos shouldered the majority of the work, until Amos' temper finally blazed and Andy was forced to carry his share of the load.

The moneymaking adventures of the Kingfish moved in and out of these plotlines, and through the Depression era *Amos 'n' Andy* offered a pointed allegory for what had happened to America itself in the 1920s: Amos represented traditional economic values, believing that wealth had to be earned, whereas the Kingfish embodied the Wall Street lure of easy money, and Andy stood in the middle, the investor torn between prudence and greed.

Although *Amos 'n' Andy*'s ratings gradually declined from the peak years of the early 1930s, it remained the most popular program in its time slot until 1941. Correll and Gosden and their characters had become a seemingly permanent part of the American scene.

The early 1930s saw criticism of the dialect and lower-class characterizations in the series by some African Americans, but *Amos 'n' Andy* also had black supporters who saw the series as a humanizing influence on the portrayal of blacks in the popular media. A campaign against the program by the Pittsburgh *Courier* in mid-1931 represented the most visible black opposition the radio series would receive—and, although the paper claimed to have gathered hundreds of thousands of signatures against the series, the campaign was abruptly abandoned after six months of publicity failed to generate a clear consensus. Throughout *Amos 'n' Andy*'s run, African-American opinion remained divided on the interpretation of the complex, often contradictory racial images portrayed in the program.

A New Direction

On 19 February 1943 Correll and Gosden broadcast the final episode of the original *Amos 'n' Andy*. In a busy wartime world, the era of the early-evening comedy-drama serial was drawing to an end.

Correll and Gosden returned to the air that fall in a radically different format. The gentle, contemplative mood of the serial was replaced by a brassy Hollywoodized production, complete with studio audience, a full cast of supporting actors (most of them African-American) and a team of writers hired to translate Amos, Andy, the Kingfish, and their friends into full-fledged comedy stars. The new *Amos 'n' Andy Show* endured for the next 12 years as one of the most popular weekly programs on the air.

The sitcom initially stuck close to the flavor of the original series; with Amos having settled down to family life, the storylines in the last years of the serial had focused on Andy's romantic entanglements and on his business dealings with the Kingfish. At first the half-hour series continued in this pattern, emphasizing plots that could be wrapped up with an O. Henry-like surprise twist at the end. By 1946, however, the Kingfish had moved to the forefront, driving the plots through his eternal quest for fast money and his endless battles with his no-nonsense wife Sapphire. The subtle blend of self-importance, guilelessness, and vulnerability that had characterized Andy was gradually replaced by simple gullibility, and for the Kingfish's increasingly outlandish schemes to work, Andy had to become not just gullible but more than a little stupid. Amos receded further into the background, his presence reduced to that of a brief walk-on, in which he would tip Andy off that the Kingfish had again played him for a fool. The relaxed intimacy of the original series had been replaced by an increasing emphasis on verbal slapstick. The subtlety of the original characterizations was lost in a barrage of one-liners. At the same time, however, the new series offered African-American performers a doorway into mainstream radio, in both comedic and nondialect, nonstereotyped supporting roles.

In 1948 Correll and Gosden sold the program to the Columbia Broadcasting System (CBS), initiating a chain of events that led directly to William Paley's "Talent Raid," and the network immediately began plans to bring the series to television with an all African-American cast. The TV version of *The Amos 'n' Andy Show* was dogged by controversy as CBS took the characters even further down the path of broad comedy, culminating in a formal protest of the TV series by the National Association for the Advancement of Colored People (NAACP) in 1951. The TV series was cancelled in 1953, but remained in rerun syndication until 1966.

The radio version of *The Amos 'n' Andy Show* was not mentioned in the NAACP protest. Radio was a dying medium, however, and when the weekly show ended in May 1955 the performers had already begun their next series, *The Amos 'n' Andy Music Hall*, a nightly feature of recorded music sandwiched between prerecorded bits of dialogue. Coasting on the familiarity of the characters, this final series ran for more than six years.

On 25 November 1960 CBS aired the final broadcast of *The Amos 'n' Andy Music Hall*. After a brief comeback—in which they provided voices for the 1961–62 American Broadcast Company–TV animated series *Calvin and the Colonel*, which reworked *Amos 'n' Andy Show* plots into funny-animal stories—Correll and Gosden slipped quietly into retirement.

Although audio recordings of most of the situation comedy episodes exist, most of the serial survives only as archival scripts, stored at the University of Southern California and the Library of Congress. Modern discussions of *Amos 'n' Andy* commonly focus more on deconstruction of its racial subtext than on examination of the original program—often obscuring the seminal role Freeman Gosden and Charles Correll played in the development of American broadcasting.

ELIZABETH MCLEOD

See also African-Americans in Radio; Situation Comedy; Stereotypes on Radio; Syndication; Talent Raids

Cast

Amos Jones	Freeman Gosden
Andrew H. Brown	Charles Correll
George "Kingfish" Stevens	Freeman Gosden
John "Brother" Crawford	Freeman Gosden
Willie "Lightning" Jefferson	Freeman Gosden
Frederick Montgomery Gwindell	Freeman Gosden
Prince Ali Bendo	Freeman Gosden
Flukey Harris	Freeman Gosden
Roland Weber	Freeman Gosden
William Lewis Taylor	Freeman Gosden

Sylvester	Freeman Gosden
Madam Queen	Freeman Gosden (1931–32), Lillian Randolph (1944, 1952–53)
Henry Van Porter	Charles Correll
Pat Pending	Charles Correll
The Landlord	Charles Correll
Honest Joe the Pawnbroker	Charles Correll
Lawyer Collins	Charles Correll
Henrietta Johnson	Harriette Widmer (1935)
Ruby Taylor Jones	Elinor Harriot (1935–55)
Sapphire Stevens	Elinor Harriot (1937–38), Ernestine Wade (1939–55)
Mrs. Van Porter	Elinor Harriot (1936–38), Ernestine Wade (1939–44)
Mrs. C. F. Van DeTweezer	Elinor Harriot (1936)
Harriet Lily Crawford	Edith Davis (1935)
Pun'kin	Terry Howard (1936–37), Elinor Harriot (1937)
Arbadella Jones	Elinor Harriot (1936–39), Barbara Jean Wong (1940–54)
Genevieve Blue	Madaline Lee (1937–44)
Dorothy Blue	Madaline Lee (1937–38)
Valada Green	Ernestine Wade (1939)
Sara Fletcher	Ernestine Wade (1940–43)
Widow Armbruster	Ernestine Wade (1941–42)
Shorty Simpson	Lou Lubin (1944–1950)
Gabby Gibson	James Baskett (1944–1947)
Reverend Johnson	Ernest Whitman (1944–45)
LaGuardia Stonewall	Eddie Green (1947–49)
Algonquin J. Calhoun	Johnny Lee (1949–54)
Leroy Smith	Jester Hairston (1944–55)
Sadie Blake	Ruby Dandridge (1944)
Ramona "Mama" Smith	Amanda Randolph (1951–54)

Announcers
Bill Hay (1928–42), Del Sharbutt (1942–43), Harlow Wilcox (1943–45, 1951–55), Carleton KaDell (1945–47), Art Gilmore and John Lake (1947–48), Ken Carpenter (1949–50), Ken Niles (1950)

Writers (serial)
Freeman Gosden and Charles Correll (1928–43)

Writers (sitcom)
Bob Ross, Joe Connolly and Bob Mosher, with contributions from others

Creative Producer
Freeman Gosden

Programming History (various networks)

1928–43	*Amos 'n' Andy*
1943–55	*The Amos 'n' Andy Show*
1954–60	*Amos 'n' Andy Music Hall*

Further Reading

Alexander, H.B., "Negro Opinion and Amos and Andy," *Sociology and Social Research* 16 (March–April 1932)
"Amos 'n' Andy," *Time* (3 March 1930)
"Amos 'n' Andy: The Air's First Comic Strip," *Literary Digest* (19 April 1930)
Biel, Michael Jay, "The First Recorded Program Series—Amos 'n' Andy, 1928–29," in *The Making and Use of Recordings in Broadcasting before 1936,* Ph.D. diss., Northwestern University, 1977
Brasch, Walter M., *Black English and the Mass Media,* Amherst: University of Massachusetts Press, 1981
Clarke, A. Wellington, "If Amos and Andy Were Negroes: What Numerous Negroes in Various Walks of Life Think of the Boys," *Radio Digest* 25 (August 1930)
Correll, Charles J., and Freeman F. Gosden, *Sam 'n' Henry,* Chicago: Shrewesbury, 1926
Correll, Charles J., and Freeman F. Gosden, *All about Amos 'n' Andy and Their Creators, Correll and Gosden,* New York: Rand McNally, 1929; 2nd edition, 1930
Correll, Charles J., and Freeman F. Gosden, *Here They Are: Amos 'n' Andy,* New York: Long and Smith, 1931
Cripps, Thomas, "Amos 'n' Andy and the Debate over American Racial Integration," in *American History, American Television: Interpreting the Video Past,* edited by John E. O'Connor, New York: Ungar, 1983
Crosby, John, "Amos 'n' Andy: Ain't Dat Sumpin'!" *Colliers* (16 October 1948)
Crowell, James, "Amos 'n' Andy Tell Their Own Story in Their Own Way," *American Magazine* 109 (April 1930)
Ely, Melvin Patrick, *The Adventures of Amos 'n' Andy: A Social History of an American Phenomenon,* New York: Free Press, and Toronto: Macmillan Canada, 1991
McLeod, Elizabeth, "Amos 'n' Andy Examined," *Chuck Schaden's Nostalgia Digest and Radio Guide* 24 (June–July 1999)
"On The Air: Amos 'n' Andy," *The New Yorker* (22 March 1930)
Quest, Mark, "Amos 'n' Andy Backstage at WMAQ," *Radio Digest* 24 (March 1930)
Roberts, Harlow P., "A Key to One Sponsor's Success in Radio," *Broadcasting* (15 April 1932)
Ross, Dale Howard, "The Amos 'n' Andy Radio Program, 1928–1937: Its History, Content, and Social Significance," Ph.D. diss., University of Iowa, 1974
Wertheim, Arthur Frank, *Radio Comedy,* Oxford and New York: Oxford University Press, 1979

AM Radio

Amplitude modulated (AM) or "standard" radio broadcasting (as the Federal Communications Commission [FCC] referred to AM until 1978) was the first broadcast service. AM dominated American commercial radio through the 1970s, provided the basis for most electronic media regulatory policies, and was the medium for which programs and the programming process were first developed. After decades of growth, however, the AM business is in decline.

AM Basics

AM transmitters modulate (or vary) a carrier wave (the basic signal used to "carry" the sidebands that contain the program information) by its amplitude (loudness) rather than its frequency, and do so many thousands of times per second. Seen diagrammatically, AM waves vary in height, indicating power changes in accordance with the signal being transmitted, rather than frequency, as in FM radio. Electronic static, most of which is amplitude modulated in its natural state, cannot be separated from the desired signal, though engineers spent years attempting to do so.

In the United States, AM channels are 10 kHz wide, whereas in much of the rest of the world by the 1990s, stations were licensed to use 9 kHz (a move to do the same in the United States was defeated by industry pressure in the early 1980s). With careful monitoring, an AM station can transmit from 5,000 to 7,000 cycles per second, which is sufficient for voice and some music, but misses the overtones of true high fidelity sound. On the other hand, AM channels, being narrower, allow far more stations to be accommodated per kHz than is the case with FM.

In most countries, AM operates on the medium wave frequencies (in the United States, 535 to 1705 kHz). Such a spectrum location means that signals are propagated along and sometimes just beneath the ground (day and night), and by sky waves bouncing off the ionosphere (nighttime only). This process has its benefits and drawbacks. The former comes from the extreme distances a powerful AM signal may travel on a cold, clear, winter evening—1,000 miles and often more. Unfortunately, such transmissions can never be exactly predicted; this leads to frustration in tuning distant stations, and more importantly, interference with other outlets, even though they may be closer to the listener. Further, ground waves and sky waves arrive at the same tuning (listening) point at different times (the sky waves having traveled much greater distances), which also causes interference.

Because of the sky wave problem, more than half of all U.S. AM stations are licensed for daytime operation only (the FCC stopped issuing new daytime-only licenses in 1987). Virtually all remaining stations reduce power in the evening hours as a condition of their licenses. This greatly reduces, but does not eliminate, the sky wave problem. Because so many AM stations were crowded on the air in the half century after World War II, most now must use directional antennas to "steer" their signals away from other stations.

All of these issues make AM engineering very complex. Because of this situation and the fact that more than 500 stations were on the air when effective regulation was established in 1927, there are no allotted channels in AM, as there are with FM and television services. When applying for a license, a new station must convince the FCC that it will not cause intolerable interference to others in the same or nearby markets—a very difficult and expensive thing to prove.

In an attempt to reduce interference and add stability to licensing, the Federal Radio Commission (FRC) in 1928 established three different types of AM radio channels: clear, regional, and local (in descending order of power and coverage area). There were few clear and regional channels and hundreds of local ones. Powerful (50,000 watts) clear channel stations were designed to serve large rural regions about 750 miles across and were located at great distances from one another (at first there were no other stations on clear channels at night—hence the term "clear"). Regional stations were less powerful and covered smaller areas; these stations naturally increased as channels were reused by stations in different areas. Finally, local stations reached only 10 or 20 miles, were separated by a few hundred miles, and soon became the most common type of AM outlet, often using lower power at night, with many operating on the same channel in different areas. The system was simplified in the 1980s.

Short History

AM radio broadcasting originated from the early 20th-century radio telephony experiments of Frank Conrad, Lee de Forest, Reginald Fessenden, Charles Herrold, and others. Early radio broadcast transmitters in the 1920s were manufactured by hand, were hard to adjust and maintain, and delivered uneven and sometimes unpredictable performance. This led to considerable detail in the regulatory requirements established by the FRC after 1927. Stations required a full-time engineer for all the modifications and monitoring required. By 1941 the quality of AM technical equipment was considerably improved.

Until 1941 all U.S. commercial stations—about 600 to 700 at any one time—operated only with AM transmission and competed only with other AM outlets. The radio industry at this time was small and friendly. While by 1941 the largest cities had a dozen stations, many smaller towns had but one or two and large parts of the country had no local radio service at night. Radio networks dominated programming and advertising with relatively few stations surviving as independent operations. By 1941 there were only a handful of educational stations on the air.

AM faced its first competition when the Federal Communications Commission (FCC) approved the creation of FM stations at the beginning of 1941; television arrived by the middle of the same year. The U.S. involvement in World War II limited growth of the new services so that most people could only tune in to AM for the duration of the war. After the war, AM grew from about 900 stations in 1945 to some 2,500 by the early 1950s—a frenetic rate of growth that illustrated public interest in and demand for more local radio service. By the 1960s, the FCC initiated two different freezes on further AM licenses, steering new applicants to the FM band instead. Any town of any size had a full complement of AM stations with no room for more.

AM stations dominated the industry through these four decades: there were more of them, they earned the vast majority of radio industry income, and they reached most of the audience most of the time. Most program developments were focused on AM stations, especially the arrival in the 1950s of top-40 rock stations and their many spin-offs in the years that followed. AM station owners controlled broadcast trade associations and saw FM as merely an expansion of what they already offered.

By 1980, however, AM's competitive situation had changed dramatically. The year before, more people tuned to FM stations for the first time, and the gap continued to widen for the next two decades as FM's higher fidelity won over listeners. FM was expanding faster as well. By the turn of the century, only a quarter to a third of the radio audience regularly tuned to AM outlets. In many cases, AM stations shifted to news and talk formats, abandoning music to higher-fidelity FM. Long dominant, AM had become a minor partner in a still expanding radio business. The number of AM outlets was actually in decline by the 1990s as stations left the air unable to attract sufficient listeners and thus to make a case to advertisers. The AM business was in trouble.

Improving AM

Faced with signs of this decline, the industry manufacturers and the FCC pressed for relief by improving AM's limited technical capabilities. The first debate concerned stereo transmission, which many in the business thought held great potential for competing with FM. From 1977 to 1980 the AM industry sought to develop an agreed-upon standard for such a service that could be recommended to the FCC. Unable to make a decision among a half dozen mutually incompatible systems by as many companies, in March 1982 the FCC announced that AM broadcasters were expected to establish their own technical standard, although anti-trust laws made it impossible for the industry to overtly collaborate on such a decision. Given this confused situation, few manufacturers built receivers, few stations installed AM stereo transmission capability (about 10 percent of all AM stations on the air), and listeners were never given a reasonable opportunity to accept or reject the technology. By 1992 the AM stereo "experiment" was clearly a failure, with two systems (Kahn and Motorola) still contending to be the final choice. Under a congressional mandate to finally make a choice, the FCC in late 1993 picked the Motorola system as a de facto standard. By then it was largely too late—too few stations (and fewer listeners) cared.

That AM still needed technical improvement in order to compete remained obvious. The FCC asked the National Radio Systems Committee (NRSC), an industry group acting in an advisory capacity to the commission, to aid in the effort, admitting it was "dealing with no less an issue than the survival of the AM service." Heavily criticized for the AM stereo debacle, the FCC appeared eager to demonstrate a real commitment to AM. In a series of decisions, it adopted NRSC-1 and NRSC-2 standards, which would help to reduce AM interference and encourage manufacturers to create improved receivers.

One AM problem had always been insufficient spectrum space. Despite post-World War II growth, the AM band remained unchanged since 1952. After the International Telecommunication Union approved a recommendation for Western Hemisphere nations to add to their AM radio bands in 1979, the FCC began to shift existing services out of the affected frequencies, and in 1990 began to actively plan for AM station use of the new band (1605-1705 kHz). Most of the new band was allocated to the United States, although both Mexico and Canada received use of some frequencies as well. After deciding not to license new stations (which would merely exacerbate existing interference), in a series of decisions the FCC selected existing AM outlets (based largely on how much interference they caused or received) to shift from lower frequencies into the new space. Assignments began in 1997 and the first station, WCMZ in Miami, was operating in the new frequencies by late that year. However, the newly relocated stations often reached far smaller audiences as old receivers could not tune the new frequencies. Only time and new sets would slowly change this problem.

At the same time the United States did not adopt another proposal—to narrow AM channels from 10 kHz to 9 kHz, a decision that would have allowed hundreds of new stations on the air. Touted in the late 1970s to help bring U.S. standards in line with those of the rest of the world, the idea was shot down by industry arguments that such spacing would increase interference and would make most digital receivers obsolete, as they were calibrated for 10 kHz channels. Not as overtly stated was a strong industry belief that the last thing AM radio needed was more competing stations.

When all was said and done, however, the fact remained that AM was an older technology with an inferior sound to FM and could not accommodate developing digital formats. It could not compete on an equal basis with its newer competitors, and had not been able to for the last two decades of the 20th century, although the seeds of its decline dated back further. To a considerable degree, however, all of these technical "fixes" were band-aid approaches to a fundamental problem that could not be viably addressed without converting AM stations into digital operations.

AM Outside the U.S.

AM radio appeared in other industrial nations in the 1920s and early 1930s, and somewhat later in several of their colonies. Through World War II, AM and shortwave were the primary radio transmission choices available for broadcasting (there were some longwave stations in operation). Most nations had far fewer stations than the United States because they were smaller and didn't need as many to cover a more limited area, because they lacked advertising revenues to develop and support further stations, and because they often allocated fewer frequencies for broadcasting.

In tropical regions especially, static (often from thunderstorms) proved a substantial challenge as there was no way to carve the noise away from the desired voice or music signal. Radio stations were often clustered in the national capital and other major cities, with little radio service available in rural regions. In colonial areas, service was aimed primarily at the expatriates of the mother country rather than at indigenous peoples. Indeed, radio was used primarily to help tie the colony to the mother country.

Well into the 1970s, most foreign radio systems were typically of the public-service type, operated directly by governments or a governmentally supported independent operation (such as Britain's British Broadcasting Corporation [BBC]), and supported by fees assessed on receiver sales or ownership. Few carried advertising, and thus entertainment programming was less dominant than in the United States. On the other hand, cultural, experimental, and educational programs of all kinds were far more common on most non-U.S. radio systems.

Since the 1980s, AM's star has dimmed in Europe for the same reasons the medium became less important in the United States. The benefits of FM (called VHF in most other nations because of the spectrum it uses) radio became more widely understood and adopted, allowing foreign systems to experience greater diversity of content and far more entertainment (typically popular music) formats than had been possible with a handful of AM outlets. Expansion of FM radio and the much greater cost of television so increased overall system costs that most nations modified their AM-based public service systems to allow advertising (either on separate stations, parallel radio systems, or the formerly government-operated outlets). For all of these reasons, radio in other countries sounded far more like radio in the United States by the beginning of the 21st century than it had just a few decades earlier.

W.A. KELLY HUFF AND CHRISTOPHER H. STERLING

See also Clear Channel Stations; Federal Communications Commission; FM Radio; Frequency Allocation; International Telecommunication Union; National Radio System Committee; Stereo; United States

Further Reading

Edelman, Murray Jacob, "The Licensing of Standard Broadcast Stations," in *The Licensing of Radio Services in the United States, 1927 to 1947: A Study in Administrative Formulation of Policy,* by Edelman, Urbana: University of Illinois Press, 1950

Emery, Walter Byron, "Standard Broadcast Stations (AM)," in *Broadcasting and Government: Responsibilities and Regulations,* by Emery, East Lansing: Michigan State University Press, 1961; revised edition, 1971

Huff, W.A. Kelly, *Regulating the Future: Broadcasting Technology and Government Control,* Westport, Connecticut: Greenwood Press, 2001

Inglis, Andrew F., "AM Radio Broadcasting," in *Behind the Tube: A History of Broadcasting Technology and Business,* by Inglis, Boston: Focal Press, 1990

Special Reports on American Broadcasting, 1932–1947, New York: Arno Press, 1974 (reprints of four Federal Communications Committee reports)

Antenna

An antenna is a device designed to either radiate ("send") or intercept ("receive") radio signals in an efficient fashion. Antenna types differ according to the frequency used, the radio service involved, and the specific task at hand. Antenna size and location are important factors in their efficiency. Transmission towers for antennas that send radio signals are nearly always manufactured of steel and are either in the form of self-supporting towers or are supported with guy wires (the latter structure is less expensive but requires more ground); in the United States, they are built to national standards set in 1959 by what is now called the Electronic Industries Alliance. Tower structures are typically either square or triangular in cross-section.

Origins

Washington, D.C. dentist and wireless inventor Mahlan Loomis may have been the first user of an antenna with his experiments in the Blue Ridge Mountains west of the capital city in the late 1860s. Loomis' antenna consisted of a wire suspended beneath a high-flying kite. The real inception of modern radio antennas came decades later with Marconi's understanding of the need for a high aerial or antenna to aid in receiving wireless telegraphy signals over large distances. He and other wireless pioneers experimented with many often highly complex antenna designs before settling on the use of three or four wooden (later steel) transmission towers at a given transmitter site.

The design of antenna structures rapidly improved in the years before and after World War I. The spread of broadcasting after 1920 and the use of higher frequencies for radio services prompted development of transmission antennas with greater efficiency. Radio receivers had improved by the late 1920s to the point that they no longer always required external antennas for effective operation, thus making them easier to use and less expensive. The first transmission antennas that could propagate signals in a given direction or pattern (dubbed directional antennas) were developed in the 1930s, allowing a station to transmit its signal to a more specific area and thus help avoid interference with other stations. World War II saw the development and refinement of microwave transmission and special antennas for that service. Much was learned about how the careful location of an antenna tower could have substantial impact on its efficiency. Post-war antenna improvements included substantial refinements in methods of both transmission and reception.

AM Radio

AM radio station antennas make use of the entire transmission tower as a radiating element. To achieve the most efficient radiation, an AM antenna tower's height should equal about half the length of the radio waves being transmitted. For example, an AM station on 833 kHz, where wavelengths are about 360 meters long, should ideally use an antenna tower that is some 180 meters or about 590 feet tall (nearly the length of two football fields). Additionally, because AM makes use of ground waves to distribute its signal, the tower is usually merely the visible part of a complex system that extends under and through the ground. As good ground contact is essential for an efficient ground wave, AM antennas often feature an extensive web of copper cables radiating out from the tower location and buried just below ground level. AM antennas are sometimes located in damp or moist locales (such as swamps or marshes) to aid in ground wave conduction.

Both to help reduce interference and to focus signals where most listeners are, the vast majority of the country's AM stations transmit a directional signal. Rather than the circular coverage pattern that would result from a standard antenna (assuming flat terrain), most AM stations use two or more antenna towers to transmit signals in a pattern away from another station, or away from a body of water or some other physical feature with few listeners, and toward population centers. Multiple AM towers for different stations are often clustered together in antenna "farms" to keep potential obstructions to airplanes at a minimum. In some locations, FM antenna towers and towers for television stations are located within the same antenna farms.

FM Radio

Unlike AM antennas, where the entire tower can assist in radiating the signal, in FM broadcasting the antenna is a relatively small device mounted on top of the transmission tower. This is because the wavelengths in the VHF spectrum used by FM radio are far shorter than in the medium wave spectrum used by AM stations. An FM outlet licensed to operate at 100 MHz, where the wavelengths are about three meters long, only requires an antenna about 1.5 meters (about five feet) in length. Since FM stations rarely need a directional pattern, multiple FM antennas for a single station are uncommon. A taller transmission tower allows the antenna element to be mounted higher above average terrain, thus extending this line-of-sight service's coverage.

FM antennas can be polarized in three different ways, each affecting how the signal will be received. Horizontal polarization was long the FM standard in the United States, but most stations now employ circular polarization for better service to car radio antennas of different types. A third type, vertical polarization, is generally restricted to public FM stations seeking to avoid adjacent channel interference with television channel six (which occupies 82-88 MHz, immediately below the lowest frequencies of the FM band).

Other Types: Shortwave and Microwave

Shortwave radio station antennas are very complex because of the many different shortwave bands that may be in use during a typical broadcast period. Extensive antenna arrays of different sizes, often covering a large ground area, help to sharpen reception.

On many transmission towers across the country one can observe various ancillary dishes or horn-shaped antennas, sometimes mounted on the top of a tower, but often attached to its sides. These are nearly always for microwave transmission links, some used for voice or data links, and some for sending television signals for use by distant cable television systems.

All radio antennas in the United States must be built and operated within strict licensing and tower marking requirements. A tower's location, height, and other characteristics are specifically defined in a station's license from the Federal Communications Commission. Generally speaking, a higher antenna will help to extend a station's coverage area, although this is more true for FM than AM services. All transmission towers must be painted in wide bands of red and white to make them more visible in low light conditions, and must be electrically highlighted at night. Jurisdiction over these requirements (as well as locating towers as far away from airports as possible) is shared with the Federal Aviation Administration.

Reception Antennas

Antennas to *receive* AM or FM signals are rarely seen by consumers as they are increasingly built into radio receivers. Some home stereo systems, for example, come with a plastic-and-wire loop antenna for improved AM reception indoors. Likewise an FM antenna, usually a long wire with a "T" shaped ending, can be attached for better indoor reception.

With the inception of satellite communication in the 1960s, large dish-shaped antennas dubbed "earth stations" were built in several locations around the world. Often huge dishes up to 200 feet in diameter, they were designed to gather in weak satellite signals and boost them to an audible range. Also used to receive satellite-delivered audio signals are back-yard or building-top dishes typically called "television receive-only" ("TVRO") antennas. Direct-to-home reception of audio and video signals can now take place using small dishes of a foot or 18 inches in diameter.

CHRISTOPHER H. STERLING

See also Audio Processing; Frequency Allocation; Ground Wave, Shortwave Radio

Further Reading

"Antennas and Propagation," in *Proceedings of the Institute of Radio Engineers* (May 1962)
Carr, Joseph J., *Practical Antenna Handbook,* Blue Ridge Summit, Pennsylvania: TAB Books, 1989; 3rd edition, New York: McGraw Hill, 1998
Collin, Robert E., *Antennas and Radiowave Propagation,* New York: McGraw Hill, 1985
Hall, Gerald, editor, *The ARRL Antenna Book,* Hartford, Connecticut: American Radio Relay League, 1939; 19th edition, Newington, Connecticut, 2000
Orr, William I., *Radio Handbook,* 23rd edition, Boston: Newnes, 1997
Rudge, Alan W., et al., editors, *The Handbook of Antenna Design,* 2 vols., London: Peregrinus, 1982–83; see especially vol. 1, 1982
Setian, Leo, *Practical Communication Antennas with Wireless Applications,* Upper Saddle River, New Jersey: Prentice Hall PTR, 1998
Sterling, Christopher H., editor, *Focal Encyclopedia of Electronic Media,* Stoneham, Massachusetts: Focal Press, 1997 (CD-ROM)

Arab World Radio

There are several definitions of what makes one an Arab, but the most widely accepted one is that an Arab is someone who speaks Arabic as his/her mother tongue. The Islamic faith is also an important part of defining Arabness, but there are Christian minorities in some areas of the Levant, Egypt, and the Sudan. The Arab world stretches from Oman in the south of the Arabian Peninsula to Syria in the east, and to Morocco in the extreme west of North Africa.

The development of radio in the Arab world is almost unique. The geography, political and colonial history, language, and culture have exerted profound influences on the medium in the region.

Radio's development in the Arab world has moved through several stages. First there was the colonial period, when Italy and Britain began international broadcasting in Arabic. Also during this phase, the British and French created local stations in Palestine and in Lebanon/North Africa, respectively. Then there was a period of hostile radio propaganda, starting shortly after the Egyptian revolution in the early 1950s. Next, there was the creation in Jordan and in the Gulf states of the first radio production and transmission capabilities established by Arab nations to defend themselves against hostile Egyptian, Syrian, and later Iraqi broadcasts that called for the overthrow of governments headed by families in Jordan and the Arabian Gulf states. (For the Gulf states, the oil-rich years after the 1974 fourfold oil-price increase had made the new and expanded electronic media infrastructure affordable.) Finally, there are the contemporary radio systems.

Early Radio Broadcasting

The Arab world was targeted for the first Western broadcasts (before 1945) to a developing area with the goal of influencing its people. Italy began broadcasting across the Mediterranean in Arabic through its international radio service, Radio Bari, in 1934. The specific motivation for this effort is unclear, as there were few radio receivers in the Middle East at the time. (Receivers needed electricity, which was then available only in larger urban areas. In the 1920s and early 1930s, there were only a few low-power, privately owned stations in Egypt.) In that same year, the Egyptian government contracted with a British company to start an official radio station. At first the Italian Arabic programs had a virtual monopoly on the Middle East frequency spectrum, causing concern on the part of British diplomats serving in Egypt, Palestine, Trans-Jordan, and the Arabian Gulf sheikhdoms. The British Foreign and Dominions Office was alerted to the potential effects of such broadcasts, but reports from the field generally concluded that the programs were not effective in swaying public opinion.

Despite the measured British diplomatic response to Radio Bari broadcasts, reports circulated both unofficially and in the British press that Italy's increasingly anti-British programs had found a receptive audience and were effective in promoting Italian interests. Some of these reports came from those who understood the oral culture and the potential power of carefully crafted spoken Arabic that reached a primarily illiterate audience. Further, some well-informed Westerners in the Arab world knew about the male custom of frequenting coffee houses in the evening, drinking coffee and tea, and discussing politics. The following scene is the type of listening pattern about which some in British government grew increasingly concerned:

> When the day's work was done both the *fellaheen* (peasants) and the city dwellers would betake themselves to their favorite cafes, huddle together under a fuming oil lamp, and stolidly smoking their water pipes play game after game of backgammon until the communal loudspeakers gave forth the voice of the Bari announcer (C.J. Rolo, *Radio Goes to War,* 1941).

By 1937 the British government realized the potential danger of international armed conflict, particularly in Europe. At the time, the British Broadcasting Corporation (BBC) had a monopoly on broadcasting within the United Kingdom. Its external transmissions for British abroad, known then as the Empire Service, started in 1932 essentially as an extension of its English-only domestic service. The United Kingdom was studying the possibility of starting foreign-language broadcasts. In 1936 the British Foreign and Dominions Office asked diplomats around the world their reaction to the possibility of beginning BBC foreign-language broadcasts. The results were mixed, but posts in the Middle East were unanimous in recommending that an Arabic service be added. On 3 January 1938, the Empire Service officially started transmitting in Arabic, its first foreign-language broadcast. This event marked the beginning of the first international radio war over a developing region.

In the 1930s, the British had a strong political presence in the Arab world. BBC advisors were employed to help establish the new Egyptian radio service. Once the BBC started its Arabic service, it was decided that this should be of exceptionally high quality, as Britain had the resources, talent, and experience in the Arab world to ensure that this goal was met. The BBC hired Egyptian announcers and sought to bring to its London microphones prominent Arab leaders as well as singers and musicians, resulting in what must have been an appealing radio offering. Radio Bari's Arabic broadcasts attempted to meet the British radio challenge by increasing the strident, vituperative nature of their political commentary.

By 1938 radio ownership was becoming more common in urban homes and public coffee houses. The so-called radio war between Britain and Italy was, in fact, only a brief contest for listeners in the Arab world. Lasting from January to April 1938, verbal hostilities ended officially on 16 April 1938 with the signing of an Anglo-Italian pact. However, Nazi Germany started transmitting in Arabic for the first time in April 1938, just as the pact came into force. In 1939 both the Soviet Union and France began broadcasting in Arabic. The French, with interests then in North Africa, Lebanon, and Syria, had the advantage of either possessing medium wave facilities there for local programming or influencing domestic schedules, making possible local relays of Paris-based Arabic programming. The British had a similar advantage with their Palestine Broadcasting Service (PBS) in Jerusalem. During World War II, the main international broadcasters to the Arab world were Germany and the United Kingdom. These two countries encouraged the development of some rather strong and distinctive on-air radio personalities. The Nazi Arabic Service employed an Iraqi by the name of Yunus al-Bahri, who may have been the most gifted Arabic-language broadcaster ever to speak from Europe. However, the BBC also had popular announcers during the war. Isa Sabbagh, a broadcaster with a considerable Arab-world following, was a Palestinian who later became a U.S. citizen and worked as a foreign service officer for the U.S. Information Agency.

Modern Broadcasting

Each Arab state is unique, but from a radio broadcaster's perspective, a common language (especially the modern standard Arabic that was actually fostered by radio broadcasting), a common religion for the vast majority of Arabic listeners, a common history, and similar political interests ensure a large radio audience for international and domestic broadcasts. A large number of broadcasters from outside the area continue to transmit Arabic programming via radio to the Arab world. These organizations have attracted relatively large audiences for several reasons, in addition to those noted above: in the Arab world, only Lebanon has a system that permits private radio broadcasting, and the major international broadcasters (such as the BBC, Voice of America, and French-government-owned Radio Monte Carlo Middle East) transmit to the area via powerful medium wave facilities. Conversely, the majority of Arab states have 1- or 2-million-watt medium wave transmitters that send a domestic service or special programs throughout the Arab world. Saudi Arabia, for example, has six medium wave transmitters (2 million watts each). Every Arab country has some shortwave transmission capability. Iraq had the most powerful shortwave transmission complexes in the Middle East until Allied air forces destroyed the French-built facility in January 1991 during the brief Gulf War.

Perhaps the greatest influence on the development of Arab radio are the governments (through their Ministries of Information) that operate them. Competition—from innovative radio programmers in the Arab world, international broadcasters from the West, videocassette recorders, and traditional as well as satellite television—has convinced Arab governments that they must be responsive to listener preferences. Previously they generally refused to acknowledge these, but even the scant survey research in existence showed the need to offer listeners more than just what the government wanted them to hear, both in the areas of entertainment and Ministry of Information-mandated news. Before this realization, most radio news in the Arab world had essentially rebroadcast the government's views on a variety of social, political, and economic issues.

Increasingly, listeners want high-quality local FM stations, especially for music listening. But radio will continue to evolve from North Africa to the Gulf states as both international and local broadcasts face competition for traditional audiences, who can now turn to over 100 satellite signals receivable at home with a $200 dish and converter.

DOUGLAS A. BOYD

See also BBC World Service; Propaganda by Radio; Radio Sawa/Middle East Radio Network

Further Reading

"Arabic Broadcasts from London," *Great Britain and the East* (13 January 1938)

Bergmeier, H.J.P., and Rainer E. Lotz, *Hitler's Airwaves: The Inside Story of Nazi Radio Broadcasting and Propaganda Swing*, New Haven, Connecticut: Yale University Press, 1997

Boyd, Douglas A., *Broadcasting in the Arab World: A Survey of Radio and Television in the Middle East*, Philadelphia, Pennsylvania: Temple University Press, 1982; 3rd edition, with subtitle *A Survey of the Electronic Media in the Middle East*, Ames: Iowa State University Press, 1999

Gradin, Thomas Burnham, *The Political Use of Radio*, Geneva, Switzerland: Geneva Research Centre, 1939; reprint, New York: Arno Press, 1971

"The History of the U.A.R. Radio since Its Establishment in 1934 until Now," *Arab Broadcasts* (August 1970)

"Il Segretario di Radazione, Direzione Servizi Giornalistici e Programmi per l'Estero," *Radiotelevisione italiana* (19 October 1979)

Mansell, Gerard, *Let Truth Be Told: 50 Years of BBC External Broadcasting*, London: Weidenfeld and Nicolson, 1982

Rolo, Charles James, *Radio Goes to War: The Fourth Front*, New York: Putnam, 1942; London: Faber, 1943

Arbitron

International Media Research Firm

The Arbitron Company is a media research firm that provides information used to develop the local marketing strategies of electronic media, their advertisers, and agencies. Arbitron has three core businesses: measuring radio audiences in local markets across the United States; surveying the retail, media, and product patterns of the local market consumers; and providing survey research consulting and methodological services to the cable, telecommunication, direct broadcast satellite, on-line, and new media industries. Although begun as an audience measurement service for television, it is currently diversified into a complete service marketing firm.

As radio evolved into quite a different medium after the introduction of television, Arbitron was best able to provide a relatively inexpensive method, the personal diary, for measuring radio's listening audience. Radio stations strove to provide a continuous, distinctive sound, composed largely of music but also including news, talk, sports, and community bulletin information. As radio became more portable and available in cars, more of its audience was away from home.

Arbitron splits the field of radio measurement with Statistical Research Incorporated, whose RADAR service measures radio networks. Thomas Birch's Radio Marketing Research provided major competition for the radio marketplace before Birch left the field in 1991. Arbitron measures 276 local radio markets today by means of an open-ended mail-in personal diary developed by James Seiler (originally an improvement over C.E. Hooper's multimedia diaries). Each member of a household who is over 12 years of age receives a personal diary with a page for each day of the week without printed time divisions.

Origins

Begun by James W. Seiler in Washington, D.C., in 1949 as both a national and local television ratings service, the American Research Bureau (ARB; now known as Arbitron) succeeded because of the inexpensive method—the viewer diary—that it used to measure radio and television audiences. The diary method pioneered by Seiler met broadcasters', advertisers', and agencies' needs for more comprehensive information about the television audience, especially viewer demographics.

Seiler realized that the Federal Communications Commission freeze (1948–52) had artificially restricted television development to East Coast cities and that when the freeze was lifted, stations and the need for measurement would spread to the West Coast. On a trip to the West Coast, Seiler discovered a local service, Tele-Que, that also used a diary method to measure one-week periods. Rather than duplicating the Tele-Que service, Seiler offered to consolidate, and ARB merged with Tele-Que in 1951, which put ARB in a strong position on both coasts. Known for a time as ARB Tele-Que, by 1954 the company was known simply as the American Research Bureau.

Like all services during that period, ARB originally tailored its methods to the nuclear family, which served as the target for enormous volumes of merchandise from manufacturers and advertisers. At that time, the entire household was used as the unit of measurement. Both the diary and the meter were tailored to a lifestyle in which the family gathered around a single console radio or television and listened or watched en masse. Only one diary (called a household diary or set diary) was sent to each family, and the assumption was that the housewife would record listening for the entire family. In an era before multiple sets, only the family room console required a meter, and it generated what were known as household ratings. Diaries allowed other information desired by advertisers and broadcasters to be included, such as the number of color television sets and whether the sets had ultrahigh-frequency receivers.

ARB's first major success was with national television reports, introduced in October 1950. Without national or network ratings, Seiler would later remark, ARB would have been lost in the crowd. Only ARB, Nielsen, and a much smaller rating service called Videodex represented a national cross-section in its sampling. Although the Nielsen network service, the Nielsen Television Index, was Audimeter based, Nielsen supplemented his meter method at the local level for his Nielsen Station Index with a diary to provide demographics. The development of a meter-plus-diary service was necessitated when, in 1954, ARB introduced an electronic instantaneous meter service, called Arbitron, to supplement its diary method. Videodex, a diary-only service, was discovered to be making up numbers from a discontinued sample of warehoused diaries; this came to light during a 1963 congressional investigation into rating services and their practices. The diary provided data on both gross (or duplicated) and cumulative (or unduplicated) audiences, and data were projectable to estimates of all U.S. television homes.

The ARB diary keepers were randomly selected from telephone directories of all U.S. cities within a 50-mile radius of the television signal. Diaries, mailed to those who agreed to cooperate, were kept for a one-week period. Field personnel made two subsequent calls to ensure continued cooperation. ARB drew new samples for each one-week period.

The ARB grew rapidly throughout the 1950s. Following Hooper's death in a 1955 boating accident, ARB took over his local market reports.

Arbitron

By 1959 Seiler had added a multicity rating service that used the same method as his archrival Nielsen, an electronic household meter. He called his meter service Arbitron. This meter service offered a distinct advantage over Nielsen's mail-in Audimeter and over its own hand-tabulated diaries in use at the time, because Arbitron's data were collected instantaneously. As part of the service, each station was represented by a row of electronic lights on a display board. As viewers switched from one station to another, the lights blinked off in the row for one station and lit up in another row for a different station. Arbitron did not succeed as a national television service owing to a number of factors: stations and networks balked at the cost, Nielsen entangled Arbitron in a lengthy patent litigation suit that drained it of financial resources, and the better-capitalized Nielsen undercut the cost of ARB's station reports.

Arbitron produces estimates for three areas: the metro area, the Area of Dominant Influence (ADI), and the total survey area (TSA). The metro area is short for metropolitan area, the standard metropolitan statistical area as defined by the U.S. census, although it is occasionally more loosely employed. The ADI is an exclusive geographical area consisting of all counties in which the home-market commercial stations receive a preponderance of total viewing or listening hours. The ADI concept divided the United States into more than 200 markets, assigning each station to only the one market where it captured the largest share of audience. The ADI was the first standardized means of defining a market, because previously, media planners had used their own definitions. The ADI paved the way for demographic targeting, because with a boundary for each market, a market's performance could be related to demographics. The TSA is a nonexclusive marketing area that indicates a station's viewing or listening audience regardless of where the station is located, including areas where stations overlap. Thus, a station could be assigned to more than one TSA if listening occurred in neighboring counties or markets. Unlike television, most radio buys are based on metro areas, because radio competes primarily in the local market against such media competition as newspapers. In 1982 Arbitron switched from 4-week to 12-week measurement periods in order to reduce the influence of promotions, giveaways, and other gimmicks used by radio stations to increase listenership. Advertising rates are based on audience size.

Radio Ratings

Arbitron entered the local radio marketplace in 1963 after the Harris hearings, a congressional investigation into the ratings services that resulted in Nielsen's exit from the radio marketplace that same year. ARB's new emphasis on the field of radio audience measurement over television was also partly motivated by RKO Radio's request that Seiler conduct a study that same year of the best way to measure audiences of all media, especially radio. The outcome was the recommendation that the personal diary—a small booklet designed for each individual in the household to carry with him or her throughout the day—be used to measure radio. This was an improvement over the household or set diary previously in use, because for the first time since the mass exodus to television, radio stations were able to report a measurable audience. Radio had shifted from a mass medium where people listened around a single set to a more portable medium because of such technological innovations as the transistor. Now radio's largest audience was its out-of-home listeners, as people listened in such places as at work, in the car, and on the beach, a phenomenon that had not been recorded by the previous method, the household set diary, which conceptualized viewing as occurring at home and as a family.

Ratings Innovator

Arbitron has typically been more technologically and conceptually innovative than its competitors in responding to the marketplace. Dominant firms such as Nielsen find it more profitable to pursue a "fast second" strategy whereby they allow small pioneers a modest inroad before they respond aggressively. Arbitron's research and development, as a result, have led to a significant number of new features in product design. Seiler developed both the all-radio diary, which measured the radio audience separately rather than as part of the television sample, and the personal diary, which was sent to each member of the household.

Seiler began the practice of using a four-week ratings period (rather than the previously used one-week period). Television still uses this four-week period as the basis of its ratings, although radio now uses a 12-week period. Seiler provided the first county coverage studies, which determined station viewership on a county-by-county basis by actual measurement rather than the previously used projections.

Seiler also began the practice of "sweeps" periods, or the simultaneous measurement of all markets based on actual coverage areas. In 1966 ARB developed new survey markets, or ADI, which, in addition to the metro and total survey areas used by rating services during the period, became the industry standard for exclusively defining the geography of local markets. The ADI geography largely replaced the previously used metro areas in television. In failing to copyright the idea, however, ARB opened the door for Nielsen to introduce a somewhat similar concept, the Designated Market Areas. Much of this additional information was made possible by ARB's switch from hand-tabulated methods to use of an electronic Univac 90 computer in 1959. By 1961 ARB was measuring every U.S. television market and had twice produced national sweeps.

By the 1960s Arbitron led the other ratings services in its inclusion of age and sex demographic information; it was able to do so because of its use of the diary method, which asked specific viewers about their viewing. Arbitron's innovation offered advertisers and agencies finer and more discriminating tools to cherry-pick audiences. These innovations, together with its national sample, resulted in Arbitron's quickly emerging as a leader in local market measurement.

After merging with the Committee for Economic and Industrial Research (CEIR) in 1961, ARB expanded its computer facilities. By 1964 many members of the original ARB management, including Seiler, Roger Cooper, and John Landrith, left, citing basic differences between ARB and CEIR policy. CEIR hired new management who were not well received by major agencies. In 1967, faced with agency and station cancellations, CEIR sold Arbitron to Control Data Corporation.

By the mid-1980s, Arbitron radio audience measurement had grown to 420 markets measured four times a year, and the personal diary had become the standard tool for radio. For studies of the radio audience, Arbitron was testing a portable pocket "people meter," a passive sound-measuring methodology, for radio, television, and cable audiences. Adoption of this device was stalled because of a 1993 patent infringement suit filed by Pretesting Company of Tenafly, New Jersey, which had developed a prototype similar to Arbitron's that it had shown to Arbitron in 1994. The device is the size of a handheld beeper. It is worn by respondents and is able to detect, decode, and store signals encoded in television and radio sound transmissions. A recharging base unit collects daily data and feeds them to a central computer. Audio encoding is located at each participating radio and television station. In 1998 Arbitron moved its Portable People Meter system out of the lab and into testing in Manchester, England. By 2003, the People Meter system was being test-marketed in Philadelphia. Arbitron had negotiated an agreement with Nielsen Media Research that gave Nielsen the option to join Arbitron in future deployment of the PPM system in the United States.

KAREN S. BUZZARD

See also A.C. Nielsen Company; Audience Research Methods; Cooperative Analysis of Broadcasting; Diary; Hooperatings

Further Reading

Arbitron History, New York: Arbitron (10 February 1993)
Beville, Hugh Malcolm, Jr., *Audience Ratings: Radio, Television, and Cable,* Hillsdale, New Jersey: Erlbaum, 1985; revised edition, 1988
Buzzard, Karen S., *Chains of Gold: Marketing the Ratings and Rating the Markets,* Metuchen, New Jersey: Scarecrow Press, 1990
Layfayette, J., "Ratings Blackout: Arbitron Leave Stations Scrambling," *Electronic Media* (25 October 1993)
"Local Operations Continue: Arbitron Drops Scanamerica Ratings Service," *Communication Daily* (3 September 1992)
"Matter of Concern: Arbitron Throws in the Towel," *Communications Daily* (19 October 1993)

The Archers

British Soap Opera

The world's longest-running radio soap opera was the brainchild of Godfrey Baseley, a British Broadcasting Corporation (BBC) producer of factual agricultural and countryside programs. In the years following World War II, Britain was still subject to food rationing, and the BBC saw it as its duty to help the Ministry of Agriculture educate farmers in more efficient methods of food production. However, much of this programming was in the form of dry talks from Ministry experts, which failed to attract the farming audience.

At a meeting in 1948 between representatives of the BBC, the Ministry of Agriculture, and the National Farmers' Union, one farmer, Henry Burtt, opined that what was needed was "a farming Dick Barton." The idea seemed preposterous; *Dick Barton—Special Agent!* was an immensely popular daily thriller serial. But Baseley started to muse on the idea of using radio drama as a medium for education, and he worked doggedly to persuade his BBC masters to let him make a trial of the idea. He recruited the writers of *Dick Barton*—Geoffrey

Webb and Edward J. Mason—and obtained funding for five pilot episodes, which were aired in May 1950 on the BBC Midlands Home Service. The pilots were well received, and on 1 January 1951 the official first episode was broadcast on the BBC Light Programme.

The program was set in the fictional village of Ambridge, in the equally fictional county of Borsetshire, but its geographic location was clearly intended to be equivalent to the area somewhere to the south of Birmingham, and so typical of much of the farming country in Britain. The central characters were the Archer family of Brookfield Farm: tenant farmer Dan Archer; wife Doris; their daughter Christine; son Phil; Phil's unreliable elder brother Jack; and Jack's city-born wife, Peggy. Of the other characters, the most notable was the comic creation Walter Gabriel, a ramshackle smallholder whose cracked-voiced greeting "Hello me old pal, me old beauty" was to become one of Britain's best-known radio catch phrases.

The program was an instant success, gaining an audience of 2 million within a few weeks of its debut. It was soon moved to take over the evening slot previously occupied by *Dick Barton,* and its audience grew to 4 million by May and 6 million by the end of the year. Listeners enjoyed the mixture of comedy, as Dan helped Walter out of his regular scrapes; drama, as traditional Dan and progressive Phil clashed over plans for Brookfield; and romance, through Phil and Christine's respective love affairs.

But underlying these universal tales was the educational purpose that was the program's *raison d'être.* So when proud Phil wanted to marry Grace, the boss' daughter, he decided that he needed sufficient capital of his own to refute any suggestion that he was marrying for money. He resolved to raise £5,000 (a substantial sum then) by breeding pedigree pigs, and over the next two years listeners unconsciously learned about the finer points of pig breeding as they followed the tale of a young man's pursuit of this headstrong, well-to-do young woman.

Phil achieved his aim, and he and Grace married in 1955. But their happiness was short-lived, as the decision was made to kill Grace while she was heroically trying to save a horse from a stable fire. The fictional death—heard by a record audience of 20 million—provoked national mourning on a completely unexpected scale and in the process eclipsed the launch of independent television the same night.

The Archers has always had an air of reality unusual in a drama series. In the early days, the actors were encouraged to use a non-histrionic, understated style of delivery, which was designed to give the impression that the listener was eavesdropping on the lives of real people. Each day in Ambridge portrays the real day and date of transmission, so the program reflects the changing of the seasons that is so important to the farming community. And when the listener celebrates Easter or Christmas, the Archers do, too. The characters grow old in real time, and many cast members have been on the program for decades. The first episode featured Norman Painting as Phil and June Spencer as Peggy, both of whom were playing the same parts well after the program's 50th anniversary.

This intermingling of fact and fiction was reinforced by the inclusion of a "topical insert" in 1952, when the Archer family was heard discussing the content of the budget announcement that had been made to Parliament by the Chancellor of the Exchequer only hours earlier. Topical inserts have continued to be a feature of the program, and in more recent years they have reflected both farming matters, such as the crisis over bovine spongiform encephalopathy (or "mad cow disease"), and more general events, such as the Gulf War and the death of Princess Diana. Real-life celebrities have also been keen to appear on the program, and visitors to Ambridge have included actor Richard Todd, jazz trumpeter Humphrey Lyttleton, veteran disc jockey John Peel, and actor Britt Ekland. Most notably, the Queen's sister, Her Royal Highness Princess Margaret, appeared as herself in 1984.

Farm production eventually grew to a position of surplus, and as the original hand-in-glove relationship between *The Archers* and the government became less appropriate, so the educational function of the program was removed in 1972. However, it retained its reputation for well-researched accuracy and continued to inspire public debate on farming and non-farming topics. In 1993 the jailing of a mother of two who had aided her brother while he was on the run from the police inspired a "Free Susan Carter" campaign that provoked reaction from the Home Secretary. In 1999 the actions of Tommy Archer, who partially destroyed a crop of genetically modified oil seed rape, were soon mirrored in real life by Lord Melchett of Greenpeace. Tommy and Lord Melchett used the same defense, and both walked free from their trials.

The program moved in 1965 from the Light Programme to the mainly speech Home Service (later BBC Radio 4). Its regular pattern—five weekday episodes, each with a repeat, plus a weekly omnibus edition—was increased by the establishment of a Sunday evening episode in 1998. With weekly audiences averaging 4.5 million, it is the network's most-listened-to program after the morning news sequence *Today.* But its place in Britain's culture extends far beyond its regular listeners. The bucolic "dum-di-dum" of its signature tune "Barwick Green" still conjures up for many an image of a mythical rural heartland, the actual decline of which the program has been charting for more than 50 years.

KERI DAVIES

Cast

Dan Archer	Harry Oakes, Monte Crick, Edgar Harrison, Frank Middlemass
Doris Archer	Gwen Berryman
Christine Archer	Pamela Mant, Joyce Gibbs, Lesley Saweard
Phil Archer	Norman Painting
Jack Archer	Dennis Folwell
Peggy Archer	Thelma Rogers, June Spencer
Walter Gabriel	Robert Mawdesley, Chriss Gittins

Producer/Creator

Godfrey Baseley

Programming History

BBC Light Programme	1951–65
BBC Home Service (later BBC Radio 4)	1965–present

Further Reading

The Archers Annual, 2001, London: BBC Books, 2000
BBC Online: The Archers, <www.bbc.co.uk/radio4/archers>
Baseley, Godfrey, *The Archers: A Slice of My Life,* London: Sidgwick and Jackson, 1971
Davies, Keri, *Who's Who in The Archers 2002,* London: BBC Books, 2002 (updated annually)
Painting, Norman, *Forever Ambridge: Twenty-Five Years of 'The Archers,'* London: Joseph, 1975
Smethurst, William, *The Archers, the True Story: The History of Radio's Most Famous Programme,* London: O'Mara Books, 1996
Toye, Joanna, and Adrian Flynn, *The Archers Encyclopaedia,* London, BBC Books, 2001
Whitburn, Vanessa, *The Archers: The Official Inside Story: The Changing Face of Radio's Longest-Running Drama,* London: Virgin Books, 1996

Archives. *See* Museums and Archives of Radio; Nostalgia Radio

---

Armed Forces Radio Service

The Armed Forces Radio Service (AFRS) was the chief means of providing popular radio network programs to military forces outside the U.S. Begun during World War II, it later expanded to include television and continues to operate today.

Origins

In February 1939, shortwave radio station KGEI in San Francisco began beaming broadcasts to the Philippines for U.S. armed forces stationed there. Later these broadcasts became the responsibility of the War Department's Radio Division, which began broadcasting music and sports overseas in the spring of 1941. Late that year U.S. forces set up a 1,000-watt station on Bataan, later moved to Corregidor. They relayed news and entertainment from KGEI.

At about the same time, officers of the Panama Coast Artillery Command (PCAC), in an effort to ensure that troops would listen to widely spaced messages on a tactical radio circuit, began playing popular records in between the announcements. Early in 1940, they started broadcasting on a regular schedule. Soldiers wrote to radio stars in the United States asking for copies of programs to air. Replies were immediate. Jack Benny sent an autographed disc, and his became the first network program to be broadcast by PCAC. In September 1941, NBC sent the station 2,000 pounds of programs. PCAC signed off on the day Pearl Harbor was attacked, 7 December 1941, for fear that Japanese aircraft might home in on the signal. It resumed broadcasting in January 1943 as a part of the growing new AFRS organization.

In a very different climate, imaginative soldiers in Alaska began "bootleg" radio stations of their own, beginning with KODK on Kodiak in January 1942, and shortly thereafter, KRB in Sitka. Unlicensed, KRB was ordered off the air by the Federal Communications Commission (FCC). In February 1942, ignoring the problems of KRB, soldiers established a new station at Sitka which they identified as KGAB and later as KRAY. C.P. MacGregor, a Hollywood recording executive and radio producer, received a request from one of these Alaskan stations for any and all recordings. Quick to comply, MacGregor sought approval from the War Department to send a large shipment of glass and aluminum based recordings "out of the country" for "your radio station in Alaska." The puzzled staff members didn't know what he was talking about.

World War II

The United States Armed Forces Radio Service was established by order of the War Department on 26 May 1942. Thomas H. Lewis, vice president of the Young and Rubicam advertising agency, was commissioned as an Army major and assigned to command the new organization. Los Angeles was selected as the headquarters because of its proximity to the entertainment industry, which quickly gave its overwhelming support. The mission of the new AFRS was to provide American servicemen "a touch of home" through the broadcast of American news and entertainment. It was also intended to combat Tokyo Rose and Axis Sally, whose broadcasts to American troops from Japan and Germany were heavily laden with propaganda.

AFRS first reached its target audience via borrowed shortwave transmitters and through dissemination of "B Kits," or "Buddy Kits." These were large, so-called "portable" 16-inch turntables delivered with transcriptions of music and radio programs, produced under the supervision of Los Angeles producer and recording executive Irving Fogel.

It was Glenn Wheaton, a "dollar a year" man with the War Department's Radio Division, who first proposed a special program for the armed forces that would present entertainers requested by military personnel serving overseas. On 8 December 1941, Wheaton suggested it be called *Command Performance*. The first broadcast on 1 March 1942 featured Eddie Cantor as master of ceremonies. A special addition to the program was a recording of the Joe Louis/Buddy Baer title fight. Other Radio Division broadcasts included disc jockeys, sports roundups, and a regular section titled "News from Home." Within a week, *Command Performance* was aired on some 11 stations, and by mid-year KGEI was transmitting the show to the South Pacific.

Command Performance was moved to Hollywood, where it continued production with the enthusiastic cooperation of the entertainment industry. Stars such as Betty Hutton, Gary Cooper, Edgar Bergen, and Gene Tierney made frequent appearances. A new program, *Mail Call,* was added and early shows starred Bob Hope, Jerry Colonna, Frances Langford, and Loretta Young.

Soon other AFRS productions included *Melody Roundup,* hosted by Roy Rogers, *Personal Album,* with Bing Crosby, and *Jubilee,* which featured African-American entertainers. Easily one of the most popular programs was a musical request feature, *AEF Jukebox*, hosted by a perky "G.I. Jill" (Martha "Marty" Wilkerson).

In AFRS broadcasts, all commercials were removed for two reasons. First, neither the government nor the War Department wanted to appear to be endorsing a commercial product or service. Second, and even more compelling, performers agreed that their work could be rebroadcast overseas without payment if no additional benefit accrued to the original sponsor. The sponsors agreed to the deletion of their commercials. In the same spirit, AFRS made agreements with all broadcasting and music guilds and unions, from the American Federation of Musicians to the American Federation of Radio Artists, and these continue today. Likewise, agreements were signed with copyright agencies to permit duplication and distribution of popular music. The cooperation of the performers, producers, directors, musicians, and sponsors made possible an AFRS schedule of programs that otherwise would have been prohibitively expensive. (AFRS created special troop information and education spot announcements to replace the excised commercials.)

Shortwave broadcasts continued from both U.S. coasts to furnish troops with news and sports programming, although music, despite the vagaries of shortwave reception, was included. The War Department approved the establishment of an American radio network in England. Lloyd Sigmon, a Los Angeles radio engineer, was commissioned to establish the stations. When he arrived in England, the new Signal Corps officer discovered his mission was to oversee activation of a number of 50-watt stations with extremely limited range.

In a relatively short period there were 140 stations broadcasting AFRS programs in England, Europe, Alaska, the Caribbean, and the South Pacific. AFRS had begun shipping self-contained stations for assembly in the Pacific area, and in 1944 the "Mosquito Network" of seven stations was established in the southwest Pacific with Noumea, New Caledonia, as key station. At about the same time the "Jungle Network" was also founded. These were affiliations of stations under a single commander. By VJ Day (15 August 1945) the total number of AFRS stations worldwide had reached 154, plus 143 public address systems and hospital bedside networks. AFRS pressed its 1 millionth disc and produced its most famous *Command Performance,* "The Wedding of Dick Tracy."

Postwar Developments

At the end of the war, the first Armed Forces Radio station in Japan signed on in Kyushu in September 1945, closely followed by stations in Kure, Osaka, and Tokyo. Frankfurt became the new headquarters of the American Forces Network (AFN), Europe, and London signed-off for the last time. AFN facilities became the first American outlets to use audiotape recovered from the Germans.

As part of a worldwide reduction, in 1947 AFRS decreased production of original programming to 14 hours per week. However, stations began receiving an additional 41 hours of

Humphrey Bogart and Lauren Bacall talk to troops overseas at Jack Brown's "Showbusiness" microphone
Courtesy of Jack Brown

network programs, excluding their commercials. The Far East Network, which had reached a high of 39 stations, was gradually reduced to 16. The Berlin station served as a homing beacon for the allied Airlift in 1948–49. AFRS slowly gave up creative radio production.

In 1950, the same year that South Korea was invaded by North Korea, AFRS stations assisted in evacuation. The Seoul station became mobile as "Radio Vagabond" and eventually had to retreat to Japan. It returned to Seoul only to be driven out again, then returned permanently in May 1951. South Korea was soon well supplied with additional mobile stations known as Troubadour, Gypsy, Homesteader, Rambler, Nomad, Mercury, Meteor, and Comet.

Vietnam, Lebanon, and Desert Storm

In 1962 Armed Forces Radio and Television Service (AFRTS) Saigon signed on in Vietnam to counter propaganda broadcasts from Radio Hanoi. Da Nang's Red Beach radio transmitter was knocked off the air by enemy fire, and the Armed Forces Vietnam Network (AFVN) at Hue came under fire during the 1968 Tet offensive. The station staff was captured and spent five years in prison. Despite the fact that the network headquarters building in Saigon was nearly demolished by a car bomb, more than 500,000 troops were now receiving radio and television (which had been added in the 1950s) from AFVN stations throughout Vietnam. By 1970 the Armed Forces Thailand network was operating six manned radio and television stations and 17 relay facilities, with headquarters at Korat. In the early 1970s, as American forces pulled out, AFVN began closing stations. Saigon was the last to shut down, signing off in March 1973.

The AFRTS SATNET satellite system began program feeds from Los Angeles on a 24 hour basis in 1978. By 1982 the Los Angeles Broadcast Center was providing satellite feeds to stations from Iceland to Diego Garcia in the Indian Ocean. In 1988 satellite radio transmissions replaced the east and west coast shortwave broadcasts, providing greatly enhanced coverage and quality. A year later all satellite transmissions from the AFRTS Broadcast Center were encrypted to prevent program piracy, a move applauded by the television industry, which was still providing programs at very little cost to the Armed Forces.

In 1982 a mobile broadcasting unit was dispatched to Lebanon to support the U.S. Marine peacekeeping force. In 1983 the AFRTS station in Lebanon was knocked off the air by a bomb explosion that killed over 240 Marines. Three mobile stations were flown to Honduras to provide support for American troops operating in that country.

When Iraq invaded Kuwait in 1990, mobile stations were provided almost immediately to Dharhan and Riyadh as part of Operation Desert Shield. Two mobile stations were deployed to Saudi Arabia and in 1991 became the Armed Forces Desert Network, with headquarters in Kuwait. The stations received live news from SATNET.

AFRTS celebrated its 50th Anniversary in 1992, feted by the Pacific Pioneer Broadcasters. The organization also received the coveted George Foster Peabody Award and later a Golden Mike from the Broadcast Pioneers.

The single-channel SATNET was replaced in 1997 with multi-channel television service. These channels were identified as NewSports, Spectrum, and American Forces Network (AFN), giving overseas stations more programming choices. Later that year Direct to Sailors (DTS) service was instituted, providing ships and remote sites two live television and three live radio services, 24 hours a day. DTS also delivers a daily *Stars & Stripes* newspaper to ships and remote installations.

As of 2000, overseas stations receive seven radio music channels 24 hours a day, plus one channel of public radio and another dedicated to news, live sports, and information. Station music libraries are furnished weekly shipments of popular music on CD, keeping overseas disc jockeys abreast of their stateside counterparts. Service reaches every continent and most U.S. Navy ships through more than 300 radio and television outlets.

JACK BROWN

See also Axis Sally; British Forces Broadcasting Service; Cold War Radio; Propaganda; Radio in the American Sector; Shortwave Radio; Tokyo Rose; Voice of America; World War II and U.S. Radio

Further Reading

Brown, Jack, "A to Z of Armed Forces Radio-TV: Adak to Zaragosa," *Daily Variety* (30 October 1979)

Brown, James, "AFRTS: Good for Morale," *Los Angeles Times* (10 August 1980)

Christman, Trent, *Brass Button Broadcasters,* Paducah, Kentucky: Turner, 1992

Department of Defense, "History of AFRTS: The First 50 Years," 1992

"Radio Profile: Jack Brown," *The International Radio Report* 2, no. 32 (August 1979)

Armstrong, Edwin Howard 1890–1954

U.S. Radio Engineer and Inventor

Inventor of the regenerative circuit, the superheterodyne, and frequency modulation (FM), Edwin Howard Armstrong is remembered as perhaps the greatest radio inventor of all time.

Youth and College Years

Armstrong was born in New York City on 18 December 1890 to John Armstrong, an employee and later vice president of Oxford University Press, and Emily Smith Armstrong, a public school teacher. At the age of nine, Armstrong developed a severe case of Saint Vitus' Dance, a disorder causing involuntary contortions and contractions of muscles, especially in the face and neck, probably caused by rheumatic fever. Because of their concern over their son's illness, in 1900 the Armstrongs moved, 15 miles up the Hudson River to the relative serenity of Yonkers.

When Armstrong was 13, his father gave him a book entitled *The Boys Book of Inventions: Stories of the Wonders of Modern Science.* The next year Armstrong read *Stories of Inventors: The Adventures of Inventors and Engineers, True Incidents and Personal Experiences.* From that time on, Armstrong knew that he would be an inventor. He was particularly interested in reading about Guglielmo Marconi and his invention of the wireless telegraph. By the time he entered high school in 1905, Armstrong's inventive energies focused on wireless, and, with the help and encouragement of his family, he set up his first "laboratory" in his upstairs bedroom, from which he sent and received wireless messages with his friends. By the time he graduated from high school in 1909, Armstrong had built wireless receivers so sophisticated that he regularly received signals from as far north as Newfoundland and as far south as Key West.

Armstrong entered Columbia University in September 1909, commuting from his home in Yonkers on his high school graduation present, a red Indian motorcycle. The next year he gained admittance to the University's School of Mines, Engineering, and Chemistry, where he concentrated his studies in the electricity department. Though many of Armstrong's professors disliked him because of his willingness to question and even contradict their teachings, Armstrong found a mentor and proponent in Columbia's leading researcher in the electrical engineering department, Michael Pupin.

Because he commuted to Columbia from his home in Yonkers, Armstrong was able to continue his wireless experiments in his bedroom laboratory. In 1910, dissatisfied with his indoor antenna, Armstrong built a 125-foot tower behind his house, to which he attached his antenna wire at the top. Armstrong built the tower himself, hoisting himself skyward on a bosun's chair. This early experience demonstrated another of Armstrong's passions: the love of heights.

During his junior year at Columbia, Armstrong set out to understand the workings of the Audion tube, invented six years earlier by pioneer wireless inventor Lee de Forest. De Forest had taken the two-element "valve" (tube) of Ambrose Fleming and added another element, which he called a "grid." The grid allowed de Forest to regulate the flow of electrons through the tube and amplify them, thus providing a superior detection device for wireless signals. But de Forest never understood how or why his Audion worked. After much experimentation, Armstrong came to understand how the Audion worked and resolved to improve it by making it not only detect electromagnetic waves but amplify them as well. To do this, Armstrong fed the current back through the grid many thousands of times per second, thus amplifying it over and over again. He achieved success on 22 September 1912, noting "great amplification obtained at once." Using his new invention, Armstrong could receive signals from Ireland and Hawaii with remarkable clarity. The "feedback" or "regenerative" circuit, as it came to be known, revolutionized radio reception. Further tests of the regenerative circuit led Armstrong to discover that it could also be used to transmit continuous waves much more efficiently and powerfully than the large and expensive mechanical generators then in use. By early 1913, using modified circuitry, Armstrong successfully demonstrated the transmitting capabilities of his new invention.

Armstrong graduated from the engineering program at Columbia University in June 1913. Later that year he filed two patent applications, one in October for his wireless receiving system and one in December for his transmission circuit. At about the same time, Armstrong demonstrated his regenerative circuit to the then chief inspector of the American Marconi Wireless Company, David Sarnoff. The two quickly became close friends.

Patent Battles and World War I

Lee de Forest, believing that Armstrong had gained prominence by using his discovery, fought back by filing a patent in 1915 for an oscillating Audion, which he claimed to have discovered in 1912—a year before Armstrong filed his patents. In doing so, de Forest asserted that he was the inventor of the circuit that made the Audion work as both a receiver and a transmitter. This led to a series of patent infringement suits between the two inventors that lasted almost 20 years.

The looming patent battles between Armstrong and de Forest were delayed by the entry of the United States into World War I. During the war, the U.S. government suspended all patent cases and pooled the wireless patents in order to develop better technology for the war effort. The Navy Department controlled wireless and its development throughout the war. Armstrong joined the U.S. Army Signal Corps with the rank of captain; he was stationed in France, assigned to the division of research and inspection. Armstrong upgraded the wireless capabilities of the Expeditionary Forces on the ground and then developed a communication system for the Army Air Corps.

While Armstrong was in France, he developed his second major invention—the superheterodyne. Allied forces suspected that the Germans were using very high-frequency bands to transmit messages, anywhere from 500,000 to 3 million cycles per second. Although receivers could be built to detect such high frequencies, tuning them proved to be extremely difficult. Armstrong developed a circuit that would combine the higher frequency with lower frequencies and then amplify them so they could be heard. This new circuit permitted the precise tuning of very high frequencies. Armstrong filed a patent for his new invention in France in 1918 and in the United States in 1919 when he returned from the war. For his efforts during the war he was promoted to major. While in Paris, Armstrong learned that the Institute of Radio Engineers had awarded him the first medal of honor it ever presented.

Armstrong initially won his patent battles with de Forest and began to profit handsomely from the sale of his patents to Westinghouse and to the newly formed Radio Corporation of America (RCA). Sarnoff, then general manager and vice president of RCA, negotiated directly with Armstrong for rights to his next invention, the super-regeneration circuit. Following their agreement, which included cash and RCA stock, Armstrong became RCA's largest private shareholder. Armstrong's financial prosperity seemed secure. It was during this time that Armstrong met Marion MacInnis, Sarnoff's secretary, and began wooing her by doing tricks atop RCA's 115-foot towers in downtown New York and giving her rides in his new Hispano-Suiza automobile. They married on 1 December 1923. Armstrong's wedding gift to Marion was the first portable superheterodyne radio.

Although Armstrong had won his patent case against de Forest in 1923, the decision was never finalized because Armstrong, considering de Forest an unethical thief, refused to sign off on the judgment, which waived de Forest's court costs because he was nearly bankrupt. Nor would Armstrong agree to license his patents to de Forest. With nothing to lose, de Forest initiated another suit in federal court, this time challenging adverse decisions from the U.S. Patent Office. In 1924 the court overruled the Patent Office decisions, ignored the prior court cases, and awarded the rights to the regenerative circuit to de Forest, based on an arcane reading of the original patent applications. Armstrong lost again at the U.S. Court of Appeals and then appealed the decision to the U.S. Supreme Court. In 1928, 15 years after Armstrong filed his patent for the regenerative circuit, the Court ruled decisively in de Forest's favor. The Court did not reach the merits of the claims but made the judgment based on a technical reading of the patent applications.

Most of the community of radio engineers correctly believed that Armstrong had been wronged. Armstrong took up yet another patent suit against the owners of de Forest's regenerative circuit patent, now RCA. Armstrong lost at the trial level but won at the Court of Appeals. RCA, wanting to protect its ability to earn royalties on the patents, took the case to the Supreme Court. In 1934 the Court ruled again in favor of de Forest's claims. This time, however, the case was decided on the merits, with the Court finding that de Forest was the true inventor of the regenerative circuit. But the decision, written by Justice Cardozo, revealed serious errors in the Court's understanding of the inventions. Again the engineering community reacted with disapproval. Later that year Armstrong attended the annual meeting of the Institute of Radio Engineers, intending to return the medal of honor it had awarded him in 1918. The president of the Institute would not accept Armstrong's offer, and the assembled engineers, knowing who really invented the regenerative circuit, stood and applauded when Armstrong took the stage.

Invention of Frequency Modulation

The amplitude modulation signals used by radio stations in the 1920s shared many drawbacks; among the most serious was that they were subject to significant levels of interference. This interference led to much crackling and hissing at the radio receiver. Armstrong had studied this problem on and off since 1914, but he did not pursue it seriously until 1923. One way of eliminating static that had been discussed by some radio engineers involved a different form of modulation in which the frequency of the carrier wave was modulated instead of the amplitude. However, mathematicians considering this issue stated categorically that frequency modulation could not solve the problem. Armstrong disagreed. For the next ten years, when not distracted with patent lawsuits, Armstrong worked toward eliminating static through the use of FM.

Conventional wisdom dictated that radio should be sent via as narrow a bandwidth as possible. Widening the bandwidth, it was thought, would simply subject the signal to more interference. After several years of failure in trying to reduce interference through a narrow-band FM system, Armstrong changed course and began experimenting with wide-band FM transmission in 1931. After redesigning his transmitter and receiver to utilize *wide*-band (200 kHz) FM, Armstrong found success. In the process of designing his new system, Armstrong filed five new patents from 1930 to 1933, all of which were granted in 1933.

In December 1933 he invited Sarnoff and several RCA engineers to his laboratory, where he displayed his new invention. Skeptical of the results, Sarnoff offered RCA's transmitting space atop the Empire State Building for a field test. Armstrong conducted the first test on 9 June 1934. A receiver was placed in a house 70 miles from the transmitter. When Armstrong transmitted a signal via AM, there was significant static. When Armstrong switched over to his FM system, the static disappeared. In fact, the receiver picked up low notes from an organ that the AM signal, with its narrow bandwidth, could not even carry. In addition to high-fidelity sound, later tests with Armstrong's FM system proved the possibility of sending more than one signal simultaneously—a process known as multiplexing.

According to Sarnoff, Armstrong's invention was not an improvement, but a revolution—one that Sarnoff could not support given RCA's existing investment in AM radio and the NBC network, as well as Sarnoff's decision to spend heavily to develop television. By July 1935 Sarnoff asked Armstrong to remove his equipment so RCA could further test its television system. Sarnoff's lack of support for FM, plus RCA's recent patent suit against Armstrong, created a strain on their friendship.

Without the backing of Sarnoff and RCA, Armstrong decided to pursue the development of FM on his own using his patent-generated fortune. After securing an experimental license from the Federal Communications Commission (FCC), Armstrong constructed an FM station in Alpine, New Jersey (across the Hudson from New York City), and began testing W2XMN in 1938. By 1940 several experimental FM stations, including several built by the Yankee Network in New England, all using Armstrong's technology, were in operation. At this time, Armstrong displayed the network potential of FM by relaying FM programs from station to station over the air, over the length of the East Coast, with virtually no signal deterioration. By the end of 1940, the FCC had received over 500 applications for FM licenses and had decided that the audio portion of television signals should be transmitted by FM. Commercial FM broadcasting was authorized to begin 1 January 1941.

World War II and More Patent Battles

With the growing popularity of FM, Armstrong struck patent-licensing deals with all of the major radio manufacturers except RCA. According to the terms of these agreements, the manufacturers agreed to pay Armstrong 2 percent of all their earnings from the sale of FM receivers and related equipment. When RCA finally realized the importance of FM, it offered Armstrong $1 million for a non-exclusionary license to use the FM technology. Armstrong refused, insisting that RCA pay the same royalty as the other manufacturers. This decision by Armstrong led to fierce patent battles as well as the loss of his friendship with Sarnoff and, over the next dozen years, his fortune, his wife, and his life.

Once again, however, a war delayed the pending patent suits. When the United States entered World War II, Armstrong declined to accept royalty payments on the sale of radio equipment to the military, believing he should not profit from the war effort. He worked with military personnel to perfect FM equipment for their wireless communications links and then began working on long-range radar systems, which he continued to develop after the war.

In 1944–45, the FCC undertook a number of investigations of spectrum allocation and use, drawing on wartime research. In a hugely controversial decision, the commission decided in early 1945 to shift the FM service higher in the VHF band, to 88-108 MHz. In so doing, it made more than 50 FM radio station transmitters and half a million FM receivers obsolete after a three-year transition period. Pressured by other major broadcasters who wanted to ensure AM radio's dominance, notably William Paley at Columbia Broadcasting System (CBS), the FCC also limited the power at which FM stations could operate. Although Armstrong contested them, these actions by the FCC severely limited (and nearly terminated) FM radio broadcasting for more than a decade while the industry turned to developing television and expanding AM.

Meanwhile, unwilling to pay Armstrong the royalties he sought, RCA began developing FM circuits of its own, which its engineers claimed did not use Armstrong's inventions. By using these circuits, RCA would not have to pay Armstrong any royalties on the sale of television sets, which used FM for the audio portion of the signal. RCA convinced other set manufacturers to do the same. In July 1948 Armstrong filed suit against RCA, alleging infringement on his five basic FM patents. RCA's trial strategy was to delay the proceedings as long as possible, to a date after the expiration of Armstrong's patents. RCA's attorneys also realized that, without any royalty revenues, Armstrong would soon be broke and unable to continue prosecuting the case. The strategy worked. By 1952 Armstrong had run out of money and had to rely on credit to pay his lawyers.

In August 1953 Armstrong proposed to settle the suit against RCA, seeking $3.4 million over a ten-year period. In December RCA responded by agreeing to pay $200,000 initially, with an "option" to pay more the next year. The option meant that Armstrong was guaranteed nothing but the initial $200,000, and Armstrong rejected the offer.

The years of litigation had taken their toll. His one-time friend David Sarnoff was now his bitter enemy. His fortune was depleted. In a fit of rage in November 1953, he took his anger out on Marion, his wife of 30 years, and she fled their apartment. On the evening of 31 January 1954, Armstrong wrote a note to Marion apologizing for his actions. He then stepped outside the window of his 13th-story apartment and fell to his death.

Marion Armstrong continued the patent battles. RCA settled its case for just over $1 million in 1955. Through settlements and court decisions—the last of which came in 1967—the other equipment manufacturers began paying damages. In the end, all of Armstrong's FM patent claims were upheld.

MICHAEL A. MCGREGOR

See also de Forest, Lee; FM Radio; Radio Corporation of America; Sarnoff, David; Yankee Network

Edwin Howard Armstrong. Born in New York City, 18 December 1890. Bachelor's degree in electrical engineering, Columbia University, 1913; served in U.S. Army Signal Corps during World War I, attaining rank of major. Inventor of regenerative circuit (1912), superheterodyne (1918), superregeneration circuit (1921), and frequency modulation (FM) broadcasting (1933). Died by suicide in New York City, 31 January 1954.

Selected Publications

Operating Features of the Audion, 1917
Frequency Modulation and Its Future Uses, 1941
Nikola Tesla, 1857–1943, 1943

Further Reading

"E.H. Armstrong: The Hero as Inventor," *Harper's* (April 1956)
Lessing, Lawrence, *Man of High Fidelity: Edwin Howard Armstrong, a Biography*, Philadelphia, Pennsylvania: Lippincott, 1956
Lewis, Tom, *Empire of the Air: The Men Who Made Radio*, New York: HarperCollins, 1991
Lubell, Samuel, "Comes the Radio Revolution," *Saturday Evening Post* (6 July 1940)
Morrisey, John W., editor "The Legacies of Edwin Howard Armstrong," *Proceedings of the Radio Club of America* 63, no. 3 (November 1990)
"Revolution in Radio," *Fortune* 20 (October 1939)

Asia

The world's largest continent, Asia at the beginning of the 21st century was home to some 3.5 billion people—more than half of the world total. Although income levels vary widely, many Asians have yet to own a radio or even to make their first telephone call. Radio broadcasting systems of some Asian countries were initially modeled on the structures of their former colonial powers. For example, India, Singapore, Hong Kong, and Sri Lanka (Ceylon) were modeled on the British Broadcasting Corporation (BBC). Likewise, radio in Indochina once followed the French model.

Both medium wave (standard or AM) and shortwave have been in use for decades. But as original transmitters are becoming obsolete, use of FM is increasing. Advertising support of radio's operational costs is becoming more accepted and widespread. Asian radio's role has undergone change—including more entertainment programming (music, drama, call-in shows) in recent years to counter the growing influence and popularity of television. Experiments with digital radio have taken place in China and elsewhere, but broad introduction is years into the future.

This continental survey begins in the West and moves toward the East, briefly surveying the past and current state of radio in countries other than India and Japan, which are covered in separate entries.

Southwest and Central Asia

Turkey

Radio broadcasting in Turkey was first regulated in the Telegram and Telephone Law (enacted 1924), which granted monopoly rights to the government post, telegraph and telephone (PTT) authority. In 1926, the first radio broadcasting concession was granted to the Turkish Wireless Telephone Co., with which the PTT was a partner for a license period of ten years. Initial broadcasts followed in 1927 with installation of

5-kilowatt transmitters in both Ankara and Istanbul. The company operated these until 1936, when, on expiration of the concession, they became a state monopoly. A new transmitter of 120 kilowatts was installed in Ankara in 1938, and radio broadcast services came under the direction of the press office attached to the prime minister's office. In 1949 a new 150-kilowatt transmitter installed in Istanbul began its first broadcasts. The 1961 Constitution stipulated that an autonomous public agency should operate and supervise radio (and television) services, and the 1982 constitution continued this requirement. The Turkish Radio and Television Corporation (TRT) was established in May 1964. FM service was introduced in 1968.

By the start of the 21st century, TRT operated four national radio services. Radio 1, headquartered in Ankara (164 hours weekly), reaches 99 percent of the population and is devoted to educational and cultural programs including drama, music, and entertainment, together with news and sports. It uses 12 FM and 12 medium wave transmitters. Radio II (126 hours weekly), based in Istanbul, reaches 97 percent of the population, and its programs are also broadly cultural, including education, drama, music, entertainment, news, and sporting events. It uses three FM and three medium wave transmitters. Radio III, also based in Ankara, reaches 90 percent of the people and programs 168 weekly hours of local and foreign popular music in stereo via 94 FM transmitters. Radio IV serves 90 percent of the population with Turkish classical and folk music for 112 hours per week. TRT also provides radio service aimed at tourists in seven regions of the country, with programs composed of music, cultural topics, and news. Broadcasts are in English, German, and French, in stereo using seven FM transmitters for 61 hours a week.

In addition to the national services, regional stations operate in Ankara and Istanbul as well as six other cities with programs consisting of educational, cultural, drama, musical, and entertainment broadcasts. Expansion of this regional system continues. The Voice of Turkey Radio broadcasts internationally in 16 languages to Europe, the northeastern part of the U.S., Asia, North Africa, the Balkans, and Central and Far East Asia. The center for such broadcasts is in Ankara-Mithatpasa, and the service transmits nearly 350 hours a week.

Afghanistan

Afghan radio, government-controlled from the start, began in 1925 when the USSR donated a low-power longwave transmitter to be installed in Kabul. Three years later a German 200-kilowatt transmitter replaced the longwave unit and marked the start of regular broadcasting. Audiences were tiny—there were perhaps 1,000 receivers by 1930. Political unrest took the station off the air for two years in the early 1930s. Expansion of radio's role became a part of successive government five-year plans beginning in the 1950s. New facilities, expanded programs, and effective training programs were begun, largely with aid from Britain and Germany. By the early 1970s, Radio Afghanistan was on the air about 14 hours daily with a diversified program schedule that emphasized national culture and performers. When the monarchy was overthrown in 1973, the event was first announced over Kabul radio.

A 1978 revolution led to establishment of a pro-Soviet government, followed by a partial Soviet occupation in the 1980s, during which powerful transmitters in the USSR relayed external service. In September 1996 Talaban forces occupied Radio Afghanistan and renamed it Radio Voice of Shari'a. With the overthrow of the conservative Talaban and installation of an interim government early in 2002, the future direction of Afghan radio (which needed to virtually begin anew) was again at a crossroads.

Persia/Iran

The first broadcast in Persia was made in February 1928 from Tehran. In 1940 regular broadcasting began under the control of the Ministry of Posts, Telephones, and Telegraphs and the prime minister. By 1950 the state-controlled system included three transmitters in Tehran (on the air about six hours a day) and one at Tabriz (five hours daily) plus another 25 low-power relay stations. A limited amount of advertising helped to defray system expenses. National Iranian Radio and Television (NIRTV) operated Radio Iran by the 1960s, providing three national services 24 hours a day on both medium wave and shortwave transmitters. One service using stereo FM provided three daily hours of music for the capital city. Radio Tehran programs were also carried on 13 regional stations that provided many of their own programs. Two private stations were operated in the 1970s, one by the national oil company in Abadan and the other from a U.S. Air Force transmitter in Tehran.

With the 1979 overthrow of the Shah's government, the sole broadcaster became Islamic Republic of Iran Broadcasting, which provided five national services originating from Tehran. There are studios in 39 centers producing programs in Farsi and local languages. The external service broadcasts in 29 languages.

Central Asia

The five Central Asian republics were all part of the USSR until the early 1990s, so the first radio stations generally relayed broadcasts from Moscow Radio for much of the day, with some local programs. In later years each republic inaugurated one or more additional republican services in the local language. In all these countries, the former Soviet broadcasting apparatus has been reformed as national state broadcasting organizations of each republic. Where private media exist, these broadcasters have had difficulty in reforming themselves to meet the newly competitive environment. In most places local stations still relay some of the most popular programs from Moscow, notably key newscasts.

Although the first test transmissions in the region originated in 1922 from Tashkent in Uzbekistan, regular broadcasting began in February 1927. An external service was inaugurated in September 1947 that now broadcasts in 12 languages. Broadcasting in Uzbekistan is one of the most restricted in the region. Tashkent is the largest city and serves as the capital not only of Uzbekistan but also unofficially of the region. However, its radio and television services are exclusively government channels. There are four radio stations operating 18 hours daily. Although there are no private broadcasters, the government accepts commercials on its radio and television programs.

Broadcasting spread to Kazakhstan in 1923 when Radio Almaty was inaugurated. Private media are well developed in Kazakhstan, the largest of the Central Asian republics, and vigorous competitive media conditions have emerged, especially in the capital, Almaty. Although under restrictions and considerable informal pressure from officials, independent radio has become firmly established. Kazakh State Television and Radio Broadcasting Company is the national governmental service.

Radio Ashkhabad began broadcasting from Turkmenistan in 1927. In the 1990s, independent media have been suppressed in the country, which has the smallest population of these republics.

Radio came to Tajikistan in 1928. Broadcasting is now controlled by the official State Radio and Television Company, originating from the capital, Dushanbe. However, there are some fledgling private broadcasters struggling to get a foothold. Independent radio dates its origins from 1989, two years before establishment of the Commonwealth of Independent States. This follows a pattern familiar across the region of pirate underground stations becoming legitimate independent broadcasters after the fall of the Soviet Union.

In 1931 Radio Frunze (now Bishkek) began broadcasting from Kyrgyzstan. Pyramid Radio, the first private station, was launched in 1992. Among the five republics, only Kyrgyzstan has supported the development of comparatively free and open media. Three independent radio stations were on the air by the mid-1990s in Bishkek, Kyrgyzstan's capital, and other private stations soon developed.

South Asia

Countries in the Subcontinent were formerly British colonies, and their broadcasting systems at least initially reflected that heritage (India is covered elsewhere). Radio has always been an important medium in this region, where literacy rates are often very low and the impact of the press is limited.

Pakistan

Radio stations in what became Pakistan first went on the air in Peshawar in 1936 (an experimental station, designed with Marconi's help, under the local government) and in Lahore in 1938 (part of All India Radio). Both were used for news and propaganda during World War II by the British authorities. These relatively weak stations—covering less than ten percent of the country—formed the beginning of Radio Pakistan when the country became independent in August 1947. Announcement of the new nation was made over what was initially dubbed the Pakistan Broadcasting Service. Early operations were stymied by lack of funds and facilities. A new station aired in Karachi in 1948, and new higher-powered shortwave transmitters followed a year later. By 1950, with the addition of a station in Rawalpindi, the country had five powerful stations broadcasting in 17 languages totaling more than 100 hours a day. Domestic and international radio services expanded in the late 1950s and into the 1960s with substantial budget increases.

Radio is a service of the Pakistan Broadcasting Corporation (PBC); television is the responsibility of a wholly separate service, Pakistan Television (PTV). While PTV is a limited corporation owned by the Pakistan government, PBC is a statutory corporation. By the 1990s PBC operated nearly two dozen medium wave and shortwave stations. In 1995, the Bhutto government introduced private FM station operations in Karachi, Lahore, and Islamabad. Calling itself FM100 on the air, the new service emphasized popular music and call-in programs. BCP added its own FM service in the same three cities in 1998.

Bangladesh

Radio service was established in 1939 in Dacca (which would become capital of the modern Bangladesh) in what was then British India. From 1947 to 1971, radio served as a vital link between densely populated East Pakistan and the larger portion of that country 1,000 miles to the west. A variety of medium wave facilities were developed, and the radio service operated largely independently of the Karachi government. After considerable fighting in 1971, Bangladesh became independent of the former West Pakistan. Radio Bangladesh, a government monopoly dependent on license fees, took over the radio facilities in Dacca, although most of the transmitters had been destroyed or seriously damaged in the strife.

The radio service, now known as Bangladesh Betar (Radio Bangladesh), operates eight regional stations effectively covering the country. As the country is mainly agricultural, farm broadcasts remain an essential program feature. While a private television station was allowed to air in 1999, radio remains a government monopoly, and news broadcasts emphasize activities of the party in power. Two thirds of all programming is music and entertainment.

Ceylon/Sri Lanka

Ceylon's first broadcasts in 1923 consisted of recorded music played over a transmitter built out of parts of a radio set from a captured German submarine. A regular broadcasting service started in July 1924 in Colombo. A shortwave station was built in Ekala, 13 miles north of Colombo, for the wartime South East Asian Command's Radio SEAC, with a 100-kilowatt transmitter, used as a relay for BBC service. The facility was transferred to the Ceylon government after the war. Radio Ceylon was founded on 1 October 1949. Commercial service began a year later when Radio Ceylon added a welcome alternative of music and entertainment programs. Medium wave transmitters were expanded in the 1960s. After the country achieved independence as Sri Lanka in 1972, the station became Sri Lanka Broadcasting Corporation (SLBC).

At the turn of the century, the SLBC operated six services, broadcasting nationally in English, Sinhala, and Tamil. One focused on service to rural areas. The planned elimination of license fee revenues, however, threatened to drastically alter the funding basis of SLBC. Two community radio stations in rural Sri Lanka—Mahaweli Radio and Radio Kotmali—have attracted international attention because of their high degree of community involvement. Radio broadcasting in Sri Lanka suddenly became intensely competitive by the end of the 1990s. By 2000, some 21 private stations had been granted licenses, mostly in the capital of Colombo. Indeed, privatization proceeded so rapidly that it outpaced policy. As of 2002, a regulatory framework still had not been formulated for the country's increasingly complex broadcast situation.

Maldives

The Maldives are a group of more than 1,100 islands in widely scattered atolls in the Indian Ocean southwest of Sri Lanka. They became an independent country in 1965. Radio Maldives opened as a government monopoly on 12 March 1964, with broadcasts in English and Dhiveli. Four years later a shortwave transmitter was put into service. In 1984 this transmitter broke down, and the station now uses two AM and one FM transmitters. The service operates for about 16 hours daily and is widely listened to. News, information, and education take up nearly half of all air time, followed by entertainment. Because of the vast distances between its island groups, Radio Maldives uses satellite relays to connect outlying atolls with the national broadcasting center in Male. There are no plans for issuance of private licenses.

Nepal and Bhutan

Radio Nepal was founded on 1 April 1951, using a 250-watt shortwave transmitter in an old school building, broadcasting four-and-a-half hours a day. The station soon broadcast on medium wave (AM) and shortwave channels and by the mid-1970s was up to ten hours a day. The shortwave transmission achieves complete national coverage, but the medium wave broadcasts reach only between 80 and 90 percent of the population. Because literacy is low (54 percent according to the 2001 census) and access to television is limited to the urban elite, radio plays an important role in informing, educating, and entertaining the masses. Information and education programs make up 40 percent of the total broadcast. Entertainment programs consist mainly of Nepali, Hindi, and Western music, supplemented by traditional music. News is broadcast in Nepali and English as well as 16 other languages commonly spoken in Nepal. Radio Nepal began broadcasting recently on FM, covering Katmandu and adjoining areas with a 1-kilowatt transmitter. Nepal licensed its first independent community radio station in 1996, and Radio Sagarmatha came on air in May 1997, operated by the Nepal Forum of Environmental Journalists. Radio service from India can also be heard throughout Nepal.

Bhutan was one of the last countries in the world to initiate a radio service, which began only in 1973. At that time the National Youth Association of Bhutan, under the Ministry of Information, inaugurated a low-power shortwave transmitter. This station, known as NYAB, broadcast only on Sundays for 30 minutes of news and music. An unusual postage stamp was issued with a hole in the center, representing a phonograph record of local music. The government took over station operation in 1979 as the Bhutan Broadcasting Service (BBS), initially broadcasting three hours daily. A UNESCO-supplied shortwave transmitter in 1986 allowed the first radio service that covered the country. By 1991 the service was broadcasting 30 hours a week in four languages. With further UNESCO support as well as aid from Denmark, the system was further modernized and expanded in the early 1990s. Operation relies on government financial support. Radio receiver ownership grew from about 25 percent of households in 1989 to more than 60 percent a decade later.

In June 2000 BBS introduced an FM radio service for Western Bhutan with the main transmitter at Dobchula and one relay station at Takri in the South. Within a year the FM service was extended to central Bhutan. Plans call for total FM coverage of the country by 2007.

Southeast Asia

Burma/Myanmar

In 1926 the Radio Club of Burma was founded in Rangoon and inaugurated station 2HZ with a 40-watt medium wave transmitter. The station closed in the early 1930s. In 1938 test transmissions began from two shortwave transmitters. These were taken over by the Japanese in 1942 (there was also a Burmese resistance radio service by 1944–45) and by the South East Asia Command (SEAC) in 1945. On 15 June 1946 the Burma Broadcasting Service was founded. The military have controlled the nation since 1962 and have oppressively controlled all media. Well into the 1960s, the country still relied on the single Rangoon-based broadcast station (using both medium wave and shortwave transmitters) with a modest level of programming, the government preferring to put its media efforts into newspapers. There were no stations in other cities, nor were repeater transmitters employed, making the medium easier to control.

In 1989 the military government changed the country's name to the Union of Myanmar, and Rangoon became Yangon. The government-controlled broadcast service is known as Myanmar Radio and TV. A few opposition political radio stations operate from time to time on or near the country's borders.

Siam/Thailand

The kingdom of Siam (until 1939) was never occupied by a colonial power, unlike virtually all of its neighbors. On 25 February 1931 the first station opened in Bangkok, operated by the Office of Publicity and Armed Forces. Experimental shortwave transmissions began from Bangkok in 1937, and the following year Radio Thailand was founded, broadcasting in English and French. A second national state-run network, Tor Tor Tor, began operation in 1952. FM service was introduced in 1956. The number of radio stations grew dramatically—from less than 30 before 1960 to 250 two decades later and more than 500 by the turn of the century.

In 1992 a student uprising against the General Suchinda government drove the military from power, and in the aftermath broadcasting laws were completely rewritten. The new laws opened the system to private stations and provided relatively open and transparent licensing procedures. This led to a quick commercialization of radio as new private stations signed on the air.

National public radio is a service of the Public Relations Department (PRD) of the Government of Thailand. Private radio stations are operated commercially under long-term licenses by private firms, but under fairly tight government control. Military-operated stations provide another alternative to government broadcasts. Efforts to promote legislation that would change the nature of the PRD and make new rules for the operation of all electronic media have been considered regularly by Parliament since 2000. Radio Thailand has for more than 30 years been providing a special service to indigenous hill tribe peoples. Bangkok has the largest number of local stations in Asia, with more than 50 on the air.

Malaya/Malaysia

Radio service began in the British colony of Malaya in 1924 with the Amateur Wireless Society of Malaya in Singapore providing music from their amateur transmitters. Regular broadcasting appeared only in 1935 with the British Malaya Broadcasting Company. The company was sold to the colonial government on the eve of World War II. Under Japanese occupation (1942–45), the radio system provided news and propaganda but also offered a voice for local languages and cultures. The Japanese also established low-power stations in Penang, Malacca, and Seremban. Post-war radio (operating as Radio Malay) was again under British colonial control from 1945 until independence in 1957. In 1963 the Federation of Malaysia was formed, although Singapore withdrew two years later. Commercial radio and the country's first international radio service began about the same time as tension rose with Indonesia. The radio system was largely privatized in the early 1980s.

Radio Malaysia, operated by the government and supported by license fees and advertising, owns 21 stations in various state capitals and in East Malaysia, using medium wave, shortwave, and FM transmitters. The service provides seven national services, most of which operate on a 24-hour basis. They include national broadcasts in Malay, English, Tamil, and Mandarin (Chinese), as well as a national FM music service, special services for aboriginal peoples, and regional stations. Private commercial stations include the Time Highway Radio which operates a network, Mix FM, and Hitz FM.

Singapore

Long a British colony, Singapore became self-governing in 1959. The first radio stations appeared in mid-1936, operated by the British Malaya Broadcasting Service, a private commercial organization. The colonial government took over operations in 1940, and the Japanese occupation authorities ran the facilities from 1942 to 1945. Post-war colonial service expanded using both medium wave and shortwave transmitters to cover all of Malaya. Radio Singapore became a separate entity after Malayan independence in 1957 and was designated Radio-Television Singapore in 1965 and the Singapore Broadcasting Corporation 15 years later. Financial support came from both advertising and license fees.

The broadcasting service was privatized in 1994. The government closely controls all operations, though stations are privately held. Three private corporations operate all commercial radio stations in Singapore. In 2002 MediaCorp Radio owned the five top-rated stations plus seven other outlets, including Radio Singapore International, broadcasting in shortwave to the region. The other radio broadcasting companies are SAFRA Radio, operated by the Singapore Armed Forces Reservists' Association, and UnionWorks, jointly owned by the National Trade Unions Council. A not-for-profit arts and culture station is managed by the National Arts Council and is operated by MediaCorp.

Indonesia

Radio service makes particular sense in and for a country made up of nearly 14,000 islands, some 3,000 of them inhabited by people speaking more than 275 languages and dialects. Indeed, radio's use of Bahasa Indonesia as the official national language has helped to promote national unity.

Radio broadcasting (which grew out of earlier radio-telegraph connections with the Netherlands) began in mid-1925 with the establishment of the Batavia Radio Vereniging in what is now Jakarta. Music and entertainment programs in Dutch were aimed at the colonials. The first Indonesian-language service came in 1933. A year later the Dutch established Nederlands Indische Radio Omroep Maatschappij, which slowly developed shortwave stations in outlying areas to tie Dutch settlers with Batavia and the home country. Indonesian listeners, dissatisfied with the Dutch-sponsored radio system, began to build their own stations and by 1938 received a subsidy from the colonial authorities. From 1942 to 1945, radio came under Japanese control, focusing on news, culture, and propaganda.

With the end of Japanese occupation in 1945, Radio Republik Indonesia (RRI) was formed, and it became a voice during the struggle for independence from the Netherlands, which lasted until 1949. RRI expanded the Japanese system and by 1955 there were 25 transmitters plus some international service as well. A decade later 38 stations included the main government station in Jakarta, three more that covered most of the country, and 28 that were regional or local, though well under half the country could receive any of these signals. School broadcasts began in 1966, and a special system of rural radio forums, initiated in 1969, broadcast to farmers (often listening in community groups) in a joint venture by the government and UNESCO.

By the early 1970s the Indonesian radio system had expanded to some 50 RRI stations and another 100 regional outlets, most operated by provincial governments. By the late 1990s, after several periods of considerable political upheaval, the country enjoyed the use of nearly 1,000 radio stations including 52 RRI outlets, nearly 800 low-power (500 watt) commercial stations, and 133 stations run by local political authorities.

Brunei

Located on the island of Borneo, Brunei was under British colonial rule from 1888 until independence came in 1984. The Brunei Broadcasting Service was founded in 1957 (two years before the country became internally self-governing) using a medium wave transmitter for less than three hours daily. Radio TV Brunei operates five FM networks (National with news, religion, education and entertainment; Pelangi aimed at younger listeners with a largely musical service; Harmony directed to family listening; the Optional service of news and entertainment in English and Chinese; and the Nur Islam service of religious programs), supported by both advertising and government funds. A private FM radio station began operation in 1999.

Philippines

In June 1922 an American electrical supply company was granted temporary permits to operate three 50-watt stations in the neighbor cities of Manila and Pasay. The Radio Corporation of the Philippines subsequently acquired one of the stations and expanded radio with a station in Cebu in 1929. Radio generally followed the American pattern (and use of English) and focused on entertainment with some news in the years leading up to World War II. Four commercial stations, all owned by department stores, were based in Manila. During the Japanese occupation (1942–45), all but one were closed, and the survivor was used for news and propaganda.

After independence in 1946, radio expanded rapidly—there were 30 stations operating by 1950. By the late 1960s there were some 200 stations (40 of them in the Manila region), those numbers having risen to 350 and 55 respectively by 1972. A few chains controlled most stations and radio advertising was widely used, though political parties supported some outlets, especially during campaigns. During the period of martial law decrees by the Ferdinand Marcos government (1972–86), many nongovernment stations were closed down or severely controlled. In the years since, private commercial stations (unusual among most Asian nations) have regained considerable freedom, though all are licensed by the National Telecommunication Commission.

Philippine radio programming has always been largely entertainment centered—especially music and soap opera drama. Radio news remains of fairly low quality, often emphasizing sensationalism. Radio Veritas is an exception—a Catholic-controlled entity operating more than 50 stations with an emphasis on quality news broadcasts. By the late 1990s there were more than 500 stations (half AM and half FM) in the country, making radio by far the most important mass medium (especially in rural areas), reaching perhaps 85 percent of the population, compared to 50 percent for television and only 25 percent for the press.

Indochina

The three nations (Vietnam, Laos, and Cambodia) that now make up Indochina became French colonies in the late 19th century. Initial radio broadcasting began under colonial authorities before World War II, set up largely to serve colonial needs and create closer ties to France. Japan occupied the region from 1941 to 1945, converting radio services to models akin to its own NHK as well as for propaganda. France tried to re-establish its colonial rule in 1946 but was eventually defeated in 1954 after bitter fighting.

Vietnam

The colonial authorities established Radio Saigon in July 1930, using a 12-kilowatt shortwave transmitter. The colonial service was re-established in September 1945, adding an external service in French, English, Cantonese, and Esperanto known as *La voix du Vietnam* (Voice of Vietnam).

Vo Tuyen Viet Nam (Radio Vietnam) began in 1950 and had eliminated all French influence a few years later. South Vietnam was administered separately beginning in 1955. By 1961 there were six stations, though most programs originated from Saigon. Beginning in the early 1960s, American support helped to expand and update the radio service. By the early 1970s, Radio Vietnam provided three services to most of the country. An external service, the Voice of Freedom, was aimed at North Vietnam. A clandestine service, Giai Phong (Liberation Radio), operated from the mid-1960s (starting with less than two hours a day) under the control of North Vietnam and Radio Hanoi. By the early 1970s it had become a multilingual service providing more than 100 hours of programs per week, much of it aimed at specific ethnic groups. The Armed Forces Vietnam Network operated stations on behalf of U.S. forces from 1962 to 1973, eventually expanding to six AM-FM outlets. After the fall of Saigon in April 1975, these were closed as Vietnamese broadcasting came under control of the former North Vietnam.

In the North, radio began in Hanoi, under French colonial authorities, before World War II. The Voice of Vietnam emerged after 1955, and by the early 1960s the service had several medium wave and shortwave transmitters covering most of the Democratic Republic of Vietnam (North Vietnam). By the 1960s perhaps 20 percent of the radio audience listened to community loudspeakers connected by wired networks to government stations. Hundreds of such wired networks connected public facilities and major rural towns. During the Vietnam War, various women cited by American listeners as "Hanoi Hannah" and "Hanoi Hattie" broadcast radio propaganda from Hanoi to American military forces operating in South Vietnam.

National radio service is provided by Radio The Voice of Vietnam, with headquarters in Hanoi. Radio The Voice of Vietnam has several networks for domestic medium wave and FM broadcasting, and it is responsible for international broadcasts in a number of languages. VOV airtime expanded dramatically in the 1990s, and the organization grew to about 1,500 staff members. VOV has an internet radio service for one hour daily. Each province has its own radio station with the total now at 61, and many of the larger cities have their own municipal stations. These are loosely regulated under the Ministry of Culture, Information, and Sports. As a consequence the system is surprisingly decentralized. Even with expanding television service, radio remains an important government tool in promoting various national development campaigns.

Laos

The first regular broadcast in Laos (now the People's Democratic Republic of Laos) was inaugurated in 1951 from Vientiane by the state-owned Radiodiffusion Nationale Lao. By the 1970s the system (supported by both advertising and government funding) used a medium wave and shortwave transmitter plus regional medium wave transmitters in the royal capital city of Luang Prabang.

National Radio and National Television are separate organizations, though both report to the Ministry of Information and Culture. Radio broadcasts in five languages on two AM radio channels. One channel is exclusively Lao and another channel is programmed in foreign languages as well as Lao. An FM station in the capital Vientiane has recently been added mainly playing recorded Lao, Thai, and Western music.

Khmer Republic/Cambodia

In 1946 Radio Cambodge opened in Phnom Penh using Japanese equipment. By 1955 there were four stations in the city, though only one remained three years later. Stations were controlled by the government, and during the government of Prince Sihanouk (1960s–1970s), they devoted up to a third of their time on the air to his speeches. In the 1970s Radio Diffusion Nationale Khmère (RNK) operated the national system with 12 to 14 daily hours of programming supported in part by advertising. Various political resistance groups operated their own stations in the 1980s. In 1994 Radio FM 103 began broadcasting from Phnom Penh as a joint venture between KCS Cambodia Company and Phnom Penh Municipality. A second station opened in 1998, providing international news and music programs.

In 1994, state TV and radio were placed under the Ministry of Information and were separated into different organizations. Prior to this, all of broadcasting was united in a single organization headed by a Director-General of Radio Television who reported to the Prime Minister's Office. RNK, National Radio for Cambodia, has a staff of 560 persons. Because of competition and funding issues, staff size is expected to shrink. There are two services on AM (simultaneously transmitted on two different frequencies) and another on FM. Separate local stations—one in the north and another in the west—are presently in operation.

Government stations receive only a fraction of their budget from the government; the remainder must come from advertising. There is consequently great pressure to produce greater income from advertising. But the advertising revenues have declined due to economic conditions and the fact that more and more stations are dividing the available advertising into smaller portions.

In the main cities, especially Phnom Penh, private stations present stiff challenges to the government stations. There are 15 private radio stations in operation (and five private TV stations). Two of the private stations are said to have slightly larger cumulative audiences than the government channel.

East Asia

China

An American businessman built the first two radio stations in Shanghai in 1922 and was soon providing local newscasts. The first state-owned stations in China, in Tientsin and Beijing, were established by the Minister of Communications in May 1927. The following October, the Sun Company in Shanghai inaugurated the first private station atop its building. The number of stations in major cities in the north and east grew to total more than 70 by 1934, many of them very small, and most privately owned. Shanghai alone had 43 stations, a number of them foreign owned and serving the International Settlement. Most of China was covered by the Central Broadcasting Station, a shortwave transmitter, installed in Nanking in 1932. Cheap crystal radio receivers were in wide use. Most programs on these early radio stations consisted of lectures and talks, some news, and music. Government policy placed severe restrictions on what could be broadcast, censoring anything found to be "contrary to public peace or good morals." By 1945 what became known as the Broadcasting Corporation of China served the country through 72 medium wave transmitters.

The Chinese Communist Party established their first radio station (called "New China") in 1945 near their base at Yunan, broadcasting but two hours a day. A second soon followed, and by 1948 there were 16 transmitters operating in Communist-controlled parts of the country. With their civil war victory in 1949 the government quickly moved to take most of the 89 existing transmitters. A handful of big-city private stations were allowed to continue in operation for several years. Radio served only part of the country, being virtually unavailable in most rural regions. An international radio service began in 1950.

By the mid-1950s, a system of wired radio networks (allowing listening only to a single station) was established, with more than 2,000 transmission centers and some 13,000 community receiving centers (loudspeakers) where group listening was encouraged if not required. In addition some 60 high-power and 165 low-power stations were operating. A decade later nearly 9,000 transmission centers served some 25,000 community receiving centers, virtually one for every commune in the country. Programs focused on news, information, and political lectures with considerable education and cultural content as well. Radio propaganda was used by all sides in the contentious Cultural Revolution of the late 1960s, which set back domestic radio as staff were diverted elsewhere.

By the early 1990s there were some 1,200 stations in the country in addition to the wired networks, which taken together reached nearly all of the population (official figures placed coverage at 92.1 percent). Talk radio programs on personal and consumer topics helped the medium overcome television's inroads. At the beginning of the 21st century, China had a three-tier broadcasting system, with national, provincial, and municipal networks serving a growing number of receivers. Thus, even though there are no private stations, there is a high degree of rivalry among stations. Competition has caused radio stations to adopt imaginative and highly polished production techniques, just as is occurring in other Asian countries where many private stations are vying for listeners. China Radio International has become the largest overseas broadcaster in Asia, with programs in 44 languages.

Formosa/Taiwan

The first radio station on Formosa, then under Japanese control, appeared in Taipei in mid-1925. A network of transmitters covered the island by 1931. Only five stations remained of this system by 1945. As the Republic of China shifted to Formosa in 1949, the Broadcasting Corporation of China (BCC) became responsible for the existing stations, though initial growth was slow. Two decades later, 33 broadcasting companies, most of them private, operated 77 stations (with nearly 200 mostly medium wave transmitters), with radio reception being widely available. In 1965 the BCC's Overseas Department was established to intensify service to overseas listeners, including those on the Mainland. The BCC's first FM transmitter began operation in 1968—by 1972 four of them broadcast classical music from as many cities. The BCC offered two national services, one in Mandarin Chinese and the other in local dialects. Several stations begun to train the police became highly popular with general audiences for their traffic reports on the increasingly car-choked island.

Only after 1993 did the government loosen its tight control of radio frequencies (pushed in part by the development of unlicensed stations presenting call-in programs), allowing more local and community stations on the air—about 150 by the late 1990s. By 2000, deregulation of radio had occurred.

Some frequencies were set aside for aboriginal and minority ethnic (Hakka) broadcasts, to aid in preserving languages. With the government control loosened, radio became highly competitive. New stations such as UFO and the Voice of Taipei have become popular, exceeding audiences for the national stations. Most programs are entertainment oriented and are supported by advertising. But an educational service has existed since 1950 (with a second station added in 1966) and most stations carry a program featuring instruction in English.

Hong Kong

Radio Hong Kong was founded in the then British colony in June 1928. A shortwave transmitter was opened in 1935. Japanese occupiers ran the radio system during the war, and it took several years to rebuild the system afterwards. In 1948 Rediffusion Hong Kong introduced wired radio services in English and Chinese. Hong Kong's first commercial radio station, Hong Kong Commercial Broadcasting Company Limited, began broadcasting in 1959. By the early 1960s, there were seven stations (two FM) on the air from morning to midnight. In mid-1997 Hong Kong reverted from British to Chinese control, but as a part of the Basic Law governing the change, media remained markedly free compared to mainland standards. By 2000, there were three companies operating more than a dozen stations broadcasting in Chinese and English. One group of six stations was owned by the government.

Korea

In February 1927 Japanese authorities established a radio station at Kyongsung in Seoul which operated largely as a mouthpiece for their colonial policy. Over the next decade, additional stations were established throughout the country, broadcasting in both Japanese and Korean and depending on license fees to meet operational costs. After World War II, Korea was divided at the 38th parallel, with very different government and broadcast systems.

In the South, the U.S. military government (1945–48) took over the Japanese-built stations and helped to create the Korean Broadcasting System (KBS). American music and other formats were encouraged, as was advertising support. In 1954 the first privately owned stations (operated by Christian organizations) began broadcasting. The first commercial station, the Pusan Munwha Broadcasting Station, was established in 1959. Munhwa Broadcasting Corporation (MBC) was established on 12 December 1961 as the first public broadcaster in Korea. At various times political upheaval has affected radio operation, though never to the extent that it has with television. The first FM outlet aired in 1965, and several were on the air by 1970. MBC makes use of AM and FM with its 20 affiliates. Television competition forced radio to develop specialized formats of all kinds in an attempt to attract listeners.

By 2000, there were nearly 100 radio stations in Korea, including 42 FM outlets and one shortwave station. In addition to those run by the KBS, many are operated by religious organizations. Music and drama predominate. The American Forces Korea Network has been on the air since the beginning of the Korean War (1950–53), providing news and entertainment for U.S. military personnel stationed in Korea. Radio Korea International broadcasts overseas in ten languages.

Radio in the Democratic People's Republic of North Korea was first introduced under the Japanese and has remained under government control. The Central Broadcasting Service took over in 1945 and was a high priority for reconstruction after the Korean War. This included development of a wired system connecting Pyongyang with more than 4,000 "broadcasting booths" located in factories, farms, and other public places. It has been a key medium in the constant touting of the country's leadership cult. By the mid-1970s, radio service operated on seven medium wave and 12 shortwave transmitters. Radio receivers were regularly checked to be sure they could receive only the official domestic service and no foreign broadcasts.

Most programs are relayed from the capital city of Pyongyang and offer public affairs, culture, and some entertainment content, most of it with strong political overtones. Pyongyang FM's 14 transmitters offer music and propaganda-laced serial dramas to entice South Korean listeners. North Korean media remain the most tightly controlled on the continent—aided by widespread use of wired networks that prevent listening to foreign broadcasts.

Mongolia

In 1934 a national broadcasting service began from Ulan Bator. With substantial Soviet aid, an extensive medium wave and shortwave radio system was established that by the mid-1970s featured one national service and 20 provincial stations. Two decades later, radio served 90 percent of the country's population (television could reach only 60 percent), with some towns reached by wired networks. An overseas service was established in 1964, now called Voice of Mongolia. Until the country became a republic in 1992, all stations were owned and operated by the government and carried a strongly propagandistic program schedule.

The state-controlled system was abolished in the late 1990s. A few private stations are on the air and the government has encouraged further investment. There is little legal restriction on station ownership or operation. Three independent FM radio services now compete with Mongol Yaridz Radio, which had long operated as a monopoly and still enjoys the largest audiences. Most Russian-language programs have given way to expanded interest in learning English.

CHRISTOPHER H. STERLING AND DREW O. MCDANIEL

See also All India Radio; Arab World Radio; Developing Nations; Japan; Radio Free Asia; Russia and Soviet Union; South Pacific Islands

Further Reading

Adhikarya, Ronny, et al., *Broadcasting in Peninsular Malaysia,* London: Routledge, 1977

"Asia," in Unesco, *World Communications,* Paris: Unesco, 1950, 1951, 1956, 1964, 1975

Batson, Lawrence D., "Asia" in *Radio Markets of the World, 1932,* Washington, D.C.: Government Printing Office, 1932

Chang, W.H., *Mass Media in China: The History and the Future,* Ames: Iowa State University Press, 1989

Emery, Walter B., "Turkey: Relief from Bureaucratic Formalities," in *National and International Systems of Broadcasting: Their History, Operation and Control,* by Emery, East Lansing: Michigan State University Press, 1969

Goonasekera, Anura, and Lee Chun Wah, editors, *Asian Communication Handbook 2001,* Singapore: Asian Media Information and Communication Centre, 2001

Gunaratne, Shelton A., editor, *Handbook of the Media in Asia,* New Delhi and Thousand Oaks, California: Sage, 2000

Lent, John A., editor, *Broadcasting in Asia and the Pacific: A Continental Survey of Radio and Television,* Philadelphia, Pennsylvania: Temple University Press, 1978

McDaniel, Drew O., *Broadcasting in the Malay World: Radio, Television, and Video in Brunei, Indonesia, Malaysia, and Singapore,* Norwood, New Jersey: Ablex, 1994

McDaniel, Drew O., *Electronic Tigers of Southeast Asia: The Politics of Media, Technology, and National Development,* Ames: Iowa State University Press, 2002

Public Service Broadcasting in Asia, Singapore: Asian Media Information and Communication Centre, 1999

Shepherd, F.M., editor, *Technical Development of Broadcasting in the Asia Pacific Region, 1964–1984,* Kuala Lumpur, Malaysia: Asia Pacific Broadcasting Union, 1984

Tribolet, Leslie Bennett, *The International Aspects of Electrical Communications in the Pacific Area,* Baltimore, Maryland: Johns Hopkins Press, 1929; reprint, New York: Arno Press, 1974

Vanden Heuvel, Jon, and Everette E. Dennis, *The Unfolding Lotus: East Asia's Changing Media: A Report of the Freedom Forum Media Studies Center,* New York: Freedom Forum, 1993

Association for Women in Communications

The Association for Women in Communications (AWC) is a professional organization that supports the advancement and recognition of women in all communications fields, including journalism, advertising, public relations, radio, television, film, marketing, photography, and design. The association is also dedicated to supporting First Amendment rights and high professional standards in communications professions.

The organization consists of professional and student chapters, with a national headquarters staff and an 11-member board of directors. Since 1972 the organization membership has been open to both women and men and at the turn of the century numbered around 7,500.

Origins

AWC began in 1909 as an honorary women's journalism fraternity at the University of Washington. After collaborating on a women's edition of the school newspaper, one of the university's seven female journalism students, Georgina MacDougall, got the idea for a university organization devoted to supporting college women who wished to pursue careers as professional journalists. MacDougall enlisted the support of classmate Helen Ross, who helped formulate the mission of the organization, which would be called Theta Sigma Phi. Fellow journalism students Blanche Brace, Helen Graves, Rachel Marshall, Olive Nauermann, and Irene Somerville joined the group to form the fraternity's first chapter.

The original mission of Theta Sigma Phi was very similar to AWC's current mission. High professional standards in journalism would be encouraged, the working conditions for women in journalism would be improved, and women journalists would be recognized for superior efforts. The group took the matrix as its insignia. In printing, matrices are small brass molds used in a Linotype, a common typesetting machine of that period. In its original Latin meaning of "womb," the matrix also signified for the group a place of development and growth. *The Matrix* would become the name of Theta Sigma Phi's membership publication, a magazine for women in journalism begun in 1915.

The University of Wisconsin established the second chapter in 1910. Chapters at the Universities of Indiana, Missouri, Kansas, Oklahoma, and Oregon and at Ohio State University were also chartered during that period. For the first several years, the officers of the University of Washington chapter served as national officers and initiated the publication of *The Matrix*. By 1916 national officers were elected separately, and plans for a national convention were in the works. Although the war delayed those plans, the first Theta Sigma Phi convention was eventually held in 1918 at the University of Kansas. In 1919 the first alumnae chapter was established in Kansas City, followed by two more alumnae chapters in Des Moines and Indianapolis. Alumnae chapters would eventually become known as professional chapters.

In the years between the world wars, Theta Sigma Phi grew as a national organization, reaching 39 student and 23 alumnae chapters by 1940. The Headliner Awards were established in 1939 to recognize excellence in any communications field. Despite facing much resistance, more women were getting jobs in journalism in the 1930s and 1940s, although they were often relegated to society pages and were nearly always paid a lower wage than their male colleagues. The new medium of broadcast radio expanded job opportunities for women not only in journalism, but in entertainment and advertising as well. Theta Sigma Phi alumnae branched out into a variety of writing fields, and the organization began to broaden its scope. In 1934 the organization established a national office in New York and hired a professional director to manage its growing national affairs. That same year, First Lady Eleanor Roosevelt was given honorary membership in Theta Sigma Phi. Mrs. Roosevelt supported the cause of female journalists by closing her press conferences to men. Theta Sigma Phi president and editor of *The Matrix,* Ruby Black, covered the first lady for United Press.

During World War II, women once again found themselves filling jobs vacated by men going off to war. Several Theta Sigma Phi members became overseas correspondents, and others were promoted to editors and producers. Radio, in particular, was in great need not only of writers and correspondents, but also of engineers, directors, and other technical people. *The Matrix* published lists of radio courses and articles describing various radio jobs.

Creation of WICI

Theta Sigma Phi became more involved in political and social movements during the changing times of the 1960s and 1970s. Articles on race relations, the women's movement, and pollution appeared in *The Matrix*; several members went to Vietnam as war correspondents; and in 1973 the organization joined the National Equal Rights Amendment Coalition. Recognizing that they had outgrown the Greek letters of the original fraternity, members decided in 1972 to change the group's name to the more professional sounding Women in Communications, Incorporated (WICI). In the 1980s WICI continued to fight for equal rights; in 1980 it opened a public affairs office in Washington, D.C., to monitor legislation and to lobby on behalf of the organization. WICI also became more active in First Amendment issues, having formed the First Amendment Congress in 1979.

Membership in WICI peaked in the mid-1980s at around 13,000. By 1995, however, membership had dropped to 8,000, and the organization was deeply in debt. The members agreed that WICI needed a new strategic plan, but they disagreed on the means of attaining one. In a controversial move, the board of directors approved a recommendation from the 40-member Fundamentals for the Future Task Force to suspend elections for the open board seats, with the intention of allowing the board time to streamline and restructure the organization. In 1996 Women in Communications, Incorporated was dissolved, and the Association for Women in Communication was incorporated in Virginia. Instead of financing its own national office and staff, AWC set up a contract with Bay Media, a management firm in Arnold, Maryland, to run the national headquarters operation. In 1997 board elections were restored, and the organization stabilized with a positive cash flow.

AWC offers ten national award opportunities, including the Clarion Award competition for recognition of excellence in any communications field. The Rising Star Award is reserved for student members who demonstrate leadership potential through school and community activities.

CHRISTINA S. DRALE

See also Female Radio Personalities and Disk Jockeys; Roosevelt, Eleanor

Further Reading

The Association for Women in Communications website, <www.womcom.org>

Giobbe, Dorothy, "Democracy vs Survival," *Editor and Publisher* (30 September 1995)

The History of The Association for Women in Communication, Arnold, Maryland: The Association for Women in Communication, 1998

Audience

Over the eight decades of radio broadcasting's existence, knowledge about the medium's audience has developed and become more refined. Whereas other entries explore how radio audiences have been studied and measured, the purpose here is to characterize the audience for American radio through time.

Radio's Audience Before Television

The earliest information about radio listeners was at best anecdotal. Stations received letters from listeners (usually responding to a program), which revealed some sense of a program's geographical spread, but little else. What little research there was focused on who purchased receivers—and thus, presumably, who listened. The 1930 census gathered information on radio set ownership showing that half the urban but only 21 percent of the farm families owned a receiver. Whereas 63 percent of homes in New Jersey owned a radio, only 5 percent of Mississippi homes did.

In the early 1920s, and to some degree for several years after that, radio appealed to an upper-class audience. Manufactured receivers were often quite expensive (upwards of $1,000 in current values for better models), and only upper-income people could afford them. Programs and advertising reflected this audience. The Depression and the appearance of a variety of popular programs made radio more attractive to a wider audience.

The first concerted attempt to study patterns of the radio audience more deeply was the work of psychologist Daniel Starch, whose consulting firm conducted personal interviews with some 18,000 families across the country in 1928 and again in 1929–30 under contract to the National Broadcasting Company (NBC). The Starch researchers found that 80 percent listened daily, that radio was used about 2.5 hours per day, that listening was largely a family affair that took place in the evening, and that nearly 75 percent of the audience tuned to one or two favorite stations most of the time.

By the end of the 1930s, more than 90 percent of urban and 70 percent of rural homes owned at least one radio (half the homes in the country had two), and whereas ownership was universal in higher-income households, radios were also found in 60 percent of the poorest homes. The average receiver was on for five hours a day, and listeners developed a fierce loyalty to the characters in favorite programs (especially daily soap operas, one of the first formats whose audiences were carefully studied). Radio was also trusted, as became clear in the panic caused by the 1938 *War of the Worlds* broadcast. Research was finding, however, that as a listener's income and educational level rose, the time spent listening to radio dropped.

During the 1941–45 war, radio became the prime source of news, and listening levels reached their peak. Radio's variety of programs appealed at some time of the day to virtually everyone. Radio was available in nearly 90 percent of households and in a quarter of all cars by 1945. At the end of the 1940s—and the end of radio's monopoly of listeners—studies found that most people liked most of what they heard. Indeed, radio was ranked as doing a better job than most newspapers, churches, schools, or local government, although its reputation slipped a bit from 1945 to 1947, perhaps reflecting the end of wartime news (newspapers replaced radio as the primary news source over the same period).

Growing from related studies of the movie audience, some concern was raised about how radio affected young listeners. Programs that featured suspense and horror were said to keep children awake. Crime programs might encourage violence on the part of listeners. Considerable research was undertaken, especially at universities, but no clear results were forthcoming.

Radio Since Television

The public's growing fascination with television after 1948 initially cut down on radio listening, especially in cities with the handful of early television stations. Network audiences dropped sharply in just a few years. Radio rebounded in the 1950s, but patterns of listening were changing—radio was now largely a daytime (especially morning) medium, whereas television dominated evening time. Ironically, as radio diminished in the eyes of some of its listeners, it became the focus of more academic research. Studies began to assess the sociological and psychological reasons why people listened, but most of what was known about radio's listeners grew out of ratings and other commercial research.

As popular music formats (e.g., Top 40) appeared and as car radios became more common (half of all cars had radios in 1951, 68 percent by 1960), radio became a medium with considerable appeal to a teenage audience. Stations developed many gimmicks to keep young people listening—chiefly the use of contests and giveaways. Most parents were totally lost in this new format.

Another audience was attracted to radio, especially to the relative handful of FM stations offering classical music. These were the high-fidelity buffs who were interested in the best quality audio they could buy. They listened to AM-FM stereo broadcasts in the late 1950s and flocked to FM after stereo standards were approved in 1961. This was a relatively high-brow audience with considerable appeal to some advertisers.

By the mid-1960s, radio was in use for about 25 hours per week in the average household, with half of that from portable and car radios (in 80 percent of cars by 1965), showing radio's expanding ability to travel with its audience. Listening peaked in morning "drive time" and slowly dropped off for the rest of the day, reaching low levels in the evening. Most people turned to radio for news and weather reports and some type of music—and despite the growing number of outlets, most people still listened to only a handful of favorite stations. Radio in many cases had become background sound for other activities at work and at home. Nearly 80 percent of households listened to radio sometime during a typical week.

Until the 1970s, *radio* still meant AM stations for most people, because FM was a limited service catering primarily to an elite audience interested in fine-arts programming. However, as the number of FM stations grew and began to program independently of AM outlets, that medium's appeal increased. FM stations began to appear in ratings in major cities, and in 1979 national FM listening first exceeded that for AM. By the 1990s FM accounted for three-quarters of all radio listening.

The minority listening to AM were tuned to various talk formats, and they wanted to participate. Call-in talk shows became wildly popular, especially those with controversial hosts. Radio became almost a two-way means of expression for such listeners. Some controversy arose in the 1990s over the likely effect of some youth-appeal music lyrics that seemed to promote violent behavior.

By 2000 radio was reaching a wholly new and largely unmeasured audience—listeners tuning in via the internet. A station could now appeal to listeners well beyond its own market and even in other countries. This new mode helped to promote the splintering of radio formats—and their audiences—into more specialized categories.

CHRISTOPHER H. STERLING

See also A.C. Nielsen Company; Arbitron; Audience Research Methods; Automobile Radio; Cooperative Analysis of Broadcasting; Demographics; Hooperatings; Lazarsfeld, Paul F.; Office of Radio Research; Programming Research; Psychographics; Violence and Radio; War of the Worlds

Further Reading

Cantril, Hadley, and Gordon W. Allport, *The Psychology of Radio,* New York and London: Harper, 1935

Lazarsfeld, Paul Felix, and Patricia Kendall, *Radio Listening in America: The People Look at Radio—Again,* New York: Prentice-Hall, 1948

Lazarsfeld, Paul Felix, and Harry Hubert Field, *The People Look at Radio,* Chapel Hill: University of North Carolina Press, 1946

Mendelsohn, Harold, *Radio Today: Its Role in Contemporary Life,* Washington, D.C.: National Association of Broadcasters, 1970

Nielsen, A.C. Co., *The Radio Audience,* Chicago: Nielsen, 1955–64 (annual)

Starch, Daniel, *A Study of Radio Broadcasting Based Exclusively on Personal Interviews with Families in the United States East of the Rocky Mountains,* Cambridge, Massachusetts: Starch, 1928

Starch, Daniel, *Revised Study of Radio Broadcasting Covering the Entire United States and Including a Special Survey of the Pacific Coast,* New York: National Broadcasting Company, 1930

Sterling, Christopher H., *Electronic Media: A Guide to Trends in Broadcasting and Newer Technologies, 1920–1983,* New York: Praeger, 1984

Audience Participation Programs. *See* Quiz and Audience Participation Programs

Audience Research Methods

The necessary and sufficient condition for success in radio is an audience loyal to its favorite stations and important enough to advertisers to produce reliable levels of advertising revenue. Noncommercial radio station aims are much the same except that the advertising is called *underwriting,* and listener (audience) donations are an important source of revenue.

Research into radio audiences has been a part of radio broadcasting from its beginning. In the earliest days of radio, stations were concerned with audiences at a distance and the distant places where the signal of the station could be heard. Stations relied upon motivated listeners to send postcards and letters reporting which programs and stations they had heard.

Newspapers carried stories about the distances at which local stations had reportedly been heard; they also reported which far-distant stations had been received by readers. Consider this report in the *New York Times* (18 March 1924):

> **Pope Hears Opera on His Radio and Picks Up a London Station**
> ROME, March 17 (Associated Press)—The radio receiving set at the Vatican has been installed and Pope Pius already has been "listening in." Last night the Pontiff heard the opera "Boris Godunov" played at the Costanzi Theatre in Rome, and later picked up a London station which was broadcasting.
>
> Pope Pius expressed great pleasure at the clearness with which the sound waves were received, notwithstanding the fact that there was some static interference.
>
> The set at the Vatican is said to be powerful enough to pick up some of the stations in America, and an effort is to be made to hear KDKA (Pittsburgh). Up to the present, however, there has been no attempt made to listen in on other than Continental stations.

This interest in the reach of radio signals continues today among shortwave broadcast audiences. Many stations encourage listeners to write or e-mail them about the shortwave programs and personalities they have heard. When listeners correspond with a station, they are rewarded by receiving colorful photo cards (QSL or "distant listening" cards) featuring favorite performers.

By the late 1920s advertising on radio had grown to the point that advertisers desired to know at a quantitative level the reach of their radio commercials. Broadcasters also needed to learn whether they were charging enough for the advertising opportunities they sold. The result was systematic radio audience research.

The principal questions addressed by audience research were (1) who is likely to be in a given station's or program's audience? (2) what is the popularity of a program or station? (3) what is the success of a program or commercial announcement? and (4) what is the probable success of a program or commercial announcement that had not yet been broadcast?

The term *ratings* is often used by media professionals to refer to all measures of audience listening and sometimes to describe the commercial companies that conduct syndicated audience research. Of greatest weight in the view of broadcasters is the fact that audience research is the principal tool used to persuade advertisers that significant audiences will be delivered for their advertising. The evidence of the future value of a station or network advertising opportunity is measured by the size and composition of the audience provided by particular programming in the past, and these are reported in audience research.

The first national radio survey conducted on a systematic basis took place in 1927 when Frank Giellerup of Frank Seaman Advertising asked Archibald Crossley to study audiences for the Davis Baking Powder Company. In March and April 1928, NBC commissioned Daniel Starch, a Harvard professor and pioneer market research consultant, to conduct an extensive survey east of the Rockies. By the 1929–30 radio season a regular program rating service, Cooperative Analysis of Broadcasting (CAB) had been established, providing routine reports on the audiences of network programs. Significantly, the CAB governing board assured that the Association of National Advertisers controlled the infant rating service. Later, broadcasters rather than advertisers became the prime force in establishing standards and practices for audience research.

Methods of Data Collection

Audience research must always ask what listeners have heard (which programs, which stations). The methods of collecting this data have evolved over the history of radio ratings. They include techniques known as telephone recall, telephone coincidental, roster interviews, diaries, meters, and recorders.

In a telephone recall survey, a radio listener was called by an interviewer from the research company at some time after a program has been presented. The interviewer asked whether the respondent listened. In March 1930 the CAB interviewed regularly in 50 cities using the telephone recall method. Telephone recall interviewers asked respondents about their radio listening in the previous 24 hours, noting the time of listening, who was listening, and the programs and stations heard.

Telephone interviewers using the telephone coincidental technique asked respondents about radio listening taking place at the time of the call. Their questions revealed whether anyone was currently listening to a radio in the household, which family members were listening, and what programs and stations were being heard. In the early years of radio, this method was associated with the research firm of Clark-Hooper formed when Montgomery Clark and C.E. Hooper left the Daniel Starch organization in 1934. Later this technique became known as the Hooper ratings. As only one time of listening was researched per telephone call, this style of research was labor-intensive for the interviewer and somewhat expensive. However, it is thought by audience research authorities to be the best measure of audience activity when conducted correctly.

Roster interviews were those in which the respondents were interviewed face-to-face. At designated points in a roster interview, respondents were shown a list of programs and stations and asked to identify those that they remembered hearing within a specified period of time. This method of data collection for audience research was also known as *aided recall measurement*. The roster method was used by The Pulse during the several decades of its history and is associated with Sydney Roslow, a psychologist who formed the organization in 1941.

The diaries used in audience research are special questionnaires in booklet form in which listeners record their times of listening and the stations or programs heard. This method has the advantage of collecting many times of listening over a given period of time (typically a week). The development of this method is often associated with James Seiler, founder in 1949 of the American Research Bureau. Diaries may be kept by an individual for his/her own listening (individual diary) or by one household member for all of the household (household diary). Contemporary radio audience research in the United States and Canada asks that all individuals in selected households maintain individual diaries, a pattern called *flooding the household.*

Listening meters were devices that automatically record times and tuner settings for radio receivers. The first of these was devised by Robert Elder and Louis Woodruff of the Massachusetts Institute of Technology and first used in 1935 for CBS. As radio meters measure the potential listening in a household by monitoring which receivers are on or off at what times and to which signal each is tuned, they produce household ratings. In 1936 market researcher A.C. Nielsen attended a luncheon at which Elder spoke about the Elder-Woodruff Audimeter. He was impressed and bought out the inventors, later establishing a radio meter rating service known as the Nielsen Radio Index.

Sound recorders are potential audience research tools, and a variety of recording methods have been employed to capture listener behavior. One form of recorder, carried by the listener, tunes to a special inaudible identifying signal in the transmissions of local radio stations. The recorder logs signals received and heard by the listener at stipulated intervals throughout the day. A computer later assembles this data as a record of listener exposure to stations throughout the day. Experiments with this sort of recorder have continued for decades without producing a syndicated research service built upon this technology. The Arbitron and Nielsen companies have announced their plans to launch a research service using this method to measure audiences of both radio and TV.

Another sort of recorder captured samples of sounds heard by a listener who carried the device throughout a listening period. On a carefully controlled schedule, the recorder sampled sounds in the environment of the radio listener carrying the device. Later a computer compared the sounds recorded by the listener for their match to signals being broadcast in the market at that time. This method required a relatively unsophisticated recorder but a complex analytical system and was not used in an established system of audience research reports.

Some researchers have proposed that recorders be carried by individual listeners participating in audience research surveys. Others have proposed that stationary recorders record sounds from portable receivers operating in their vicinity, with the data then analyzed to produce audience research reports.

Samples of Listeners

In the early years of radio broadcasting, samples were typically drawn from phone listings in cities selected because they were served by radio stations affiliated with the networks sponsoring the study. During the 1940s and 1950s there were considerable efforts to produce samples that would be perceived as excellent by researchers in the broadcast and advertising industries.

Different samples are used for audience research with national and local audiences. Because the motive for much research has been to substantiate the value of advertising opportunities, the areas where surveys are conducted are called *markets.*

Networks, advertisers, and others interested in nationwide entertainment and advertising support national market audience surveys. National surveys must give weight not only to every local market and but also to listening in rural areas where national signals may reach. RADAR (Radio's All Dimension Audience Report) conducts the only regular national surveys of the radio audience.

In local market surveys the sample is drawn from three survey areas. The smallest of these is the metropolitan (metro) area, usually a core urban area as defined by the U.S. Office of Management and Budget. The largest area surveyed in a local market survey is called the total survey area (TSA), which typically is an aggregate of units of county size, including the relevant metro area or areas. The TSAs of adjacent markets may include the same counties; for example, the total survey area for City A may include Brown County because an important part of the listening in the county is to City A radio stations. At the same time, the TSA for City B may also include Brown county for the same reason. An "exclusive area" is sometimes included, in which each market consists only of those counties where the plurality of listening is to the market being surveyed. In this system of exclusive areas, any particular county can belong to only one market.

Random Sampling

In sampling for audience research, random sampling is preferred, because theoretically a random sample maximizes the probability that a sample will be very similar to the population from which it is drawn. A random sample is one in which each member of the target population is equally likely to be chosen for the survey (participants are chosen at random); each choice is also entirely independent of the others. The first criterion—equal likelihood for selection—requires that the researcher name the "sample frame" from which the sample is to be drawn (i.e., to list all members of the target population). In the early days of radio audience research, telephone directory listings were the sample frames. Listed telephones serve best as sample frames when nearly all residential telephones are listed and when nearly all residences are equipped with telephones. In the early years of audience research the first condition was typically met, but telephone penetration had yet to reach its peak.

In the past several decades residential telephone listings have become progressively poorer telephone sample frames. Although nearly all residences now have telephones, fewer and fewer have listed telephone numbers. One of the methods audience researchers have adopted to cope with this problem is random digit dialing, a method in which all possible telephone numbers within a target area are listed and the sample frame is drawn from that list. A number of variations on this procedure are used as contemporary sample telephone frames. There are sample frames for households rather than telephones; an example is a city directory. Enumerations of households within census tracts (designated by the U.S. Census Bureau for control of their surveys) also serve as household sampling frames. A quite different approach to sample frames is a frame of clusters, which are sampling units that each consist of two or more interviewing units such as residences. A city block, for example, may become a cluster in a sample frame, as it is a cluster of households or interviewing units.

In most contemporary radio audience research, sampling procedures are mixed. Thus the initial sampling frame could be a residential telephone listing, later supplemented by a second frame of telephone numbers computer-generated at random.

Producing Ratings Survey Reports

Whenever a radio audience report lists a rating (percentage of potential audience) or a listening estimate (numbers of listeners), a degree of error is also implied. This is called *sampling error,* a scientifically determined estimate of the difference between research results if the entire target population were surveyed and those obtained using a sample. The probability that such an error will occur is given by the *confidence interval* listed for any professional research study. The confidence interval standard in audience research is 95 percent, meaning that if the same study were conducted many times, the same results would be obtained at least 95 percent of the time. Commercially produced audience survey reports include descriptions of the methods used for estimating sampling error and confidence intervals.

Stations, networks, and program suppliers who use audience data in their sales and planning prefer to receive audience data at a modest cost, so the research suppliers must not spend more on audience research than their customers are willing to pay. The statistic used to compare the cost basis of competing radio audience research reports is price per "listening mention."

A listening mention is the smallest unit of reported radio listening. It consists of at least five minutes of continuous listening to a certain station or program. One listening mention, then, means that one member of the sample reported listening to a particular station/program for at least five minutes during a 15-minute interval of the time period being surveyed. The telephone coincidental method for collecting listening mentions from audiences is the most expensive method of producing routine audience reports, as any respondent can provide only one listening mention—the one in which the individual was involved at the time of the interviewer's call. The lowest cost per listening mention is nearly always a research method employing a meter or recorder, as analysis of one instrument's data can provide a train of listening mentions over months or even years.

The number of hours per week over which ratings or audience estimates are provided is another cost factor. If every hour of the week were surveyed for listening, a maximum of 672 listening mentions could be recorded. To reduce costs, audience research firms have typically limited the number of hours reported in their surveys. Thus over the years they have reported listening during prime time only, or they have excluded the hours of lowest listening (such as those between midnight and dawn).

The break-even cost of an audience survey is reached when the number of clients who will pay for the survey at a designated price meets the cost of collecting, tabulating, and printing the data they are willing to buy. When the number willing to buy increases above the break-even point, then the research company becomes profitable (at times very profitable). This explains why a number of new companies over time have entered into the radio audience research business, although relatively few have survived.

Special Research Studies

A number of research methods are used to study the desirability of using particular songs or groups of songs within the established format of a radio station or network. Each of the following music audience research methods has its advocates and detractors, but all have persisted in one form or another over the past several decades.

Telephone call-out and call-in is a method that focuses on recent musical releases and older (but still fairly recent) songs that have remained popular and are still frequently played. Each music selection being studied is prepared as a recorded *hook*—that is, a representative excerpt of a recording.

In telephone call-out research, the researcher will have previously identified a pool of qualified study participants (listeners to the station in question or to the categories of music being studied). The interviewer plays over the phone one hook at a time and asks the study participant to respond with phrases such as (1) "I've never heard of it," (2) "I dislike it strongly," (3) "I dislike it moderately," (4) "I don't care," (5) "I'm tired of it," (6) "I like it," or (7) "That's my favorite record."

Stations and networks that make use of call-out research conduct their studies weekly or semiweekly. Satellite programming services making music available to stations throughout the country conduct this research on a continuing basis in many markets.

In one form of telephone call-in, study participants receive a letter in the mail asking them to participate in the study. The letter identifies a telephone number for participants that connects them to a recording of the hooks for the study. The participant listens to the hooks when convenient, then returns the questionnaire by mail or telephones a researcher who writes down the responses read by the participant from the completed questionnaire.

Auditorium studies are surveys often used to study audience response to "oldies" (songs that were quite popular during the more distant past and that are still popular with at least part of the radio audience) and "standards" (new versions of oldies). Researchers recruit a sample of radio listeners, who assemble in an auditorium or rented meeting room. Hooks of the music are played over high quality sound systems. Participants then mark their responses to each hook in questionnaire booklets or on digital responders (keypads that summarize responses into a convenient computer file). Auditorium studies commonly include hooks for large numbers of recordings, greatly reducing the cost per hook for respondent data. Because of the effort required to assemble hooks, arrange for facilities, and recruit respondents, auditorium studies are conducted less frequently than call-out or call-in studies.

The relatively low costs of mail and internet surveys make these surveys appealing. However, mail surveys require that study participants recognize songs from written descriptions that may include the name of a song or performer or some words from a song's lyrics. This limitation often leaves researchers wondering whether the music has been correctly identified by study participants. Internet surveys permit playing hooks over a respondent's computer speakers, reducing the possibility that study participants will not recognize the music being studied. Survey questionnaires are then presented for completion on respondents' computer screens. When all answers have been provided, the data is immediately returned to the researcher's computer. As not every radio listener has access to a computer connected to the internet, study participants must be identified in a pre-survey as (1) listeners accessible by internet, and (2) listeners to the radio music being studied.

Radio Program Format Research

Radio program format research concerns the mixture of music, news, and talk programming that is best for a given station or network. Although a number of methods are used in format research, focus groups have received the greatest attention in the literature about radio.

A *focus group* is a group of research participants who are selected for their relevance to the matter being studied. Their viewpoints and opinions are collected with a guided conversation about the research topic. In the case of format studies, one strategy is to recruit a variety of groups—those who listen only to the station in question, those who listen to the station sometimes, and those who never listen to the station but by their media habits show that under some circumstances they could become listeners.

The results of station format focus group studies identify the "position" (reputation) of the station in its market. In addition, the specific language used by study participants during the focus sessions may suggest useful slogans or themes for station promotional campaigns.

Because many radio stations today are owned by corporations that own large groups of stations, station format studies may be a matter of researching format issues in several markets simultaneously. The results may lead to the choice of a station format that will function competitively in all of the group's markets.

JAMES E. FLETCHER

See also A.C. Nielsen Company; Arbitron; Auditorium Testing; Cooperative Analysis of Broadcasting; Hooperatings; Programming Research; Pulse, Inc.; RADAR

Further Reading

Beville, Hugh Malcolm, Jr., *Audience Ratings: Radio, Television, and Cable,* Hillsdale, New Jersey: Erlbaum, 1985; 2nd edition, 1988

Chappell, Matthew Napoleon, and Claude Ernest Hooper, *Radio Audience Measurement,* New York: Daye, 1944

Fletcher, James E., *Music and Program Research,* Washington, D.C.: National Association of Broadcasters, 1987

Lumley, Frederick Hillis, "Methods of Measuring Audience Reaction," *Broadcast Advertising* 5 (October 1932)

Lumley, Frederick Hillis, *Measurement in Radio,* Columbus: Ohio State University Press, 1934; reprint, New York: Arno Press, 1971

National Association of Broadcasters, *Standard Definitions of Broadcast Research Terms,* New York: National Association of Broadcasters, 1967; 3rd edition, as *Broadcast Research Definitions,* edited by James E. Fletcher, Washington, D.C.: National Association of Broadcasters, 1988

Routt, Ed, "Music Programming," in *Broadcast Programming: Strategies for Winning Television and Radio Audiences,* edited by Susan Tyler Eastman, Sydney W. Head, and Lewis Klein, Belmont, California: Wadsworth, 1981

Webster, James G., and Lawrence W. Lichty, *Ratings Analysis: Theory and Practice,* Mahwah, New Jersey: Erlbaum, 1991; 2nd edition, as *Ratings Analysis: The Theory and Practice of Audience Research,* by Webster, Lichty, and Patricia F. Phalen, Mahwah, New Jersey, and London: Erlbaum, 2000

Audimeter

The Audimeter—for audience meter—was the name of the A.C. Nielsen Company's mechanical, and later electronic, device for measuring radio and television set tuning as a way of determining a show's share of the audience, better known as its ratings.

Origins

In 1929 Claude Robinson, a student at Columbia University, applied to patent a device to "provide for scientifically measuring the broadcast listener response by making a comparative record of . . . receiving sets . . . tuned over a selected period of time." Robinson later sold his device for a few hundred dollars to the Radio Corporation of America, owner of NBC, but nothing more came of it at that time.

Many realized that the least intrusive and most accurate way to keep track of listeners' radio tuning would be to attach some kind of mechanical recorder to the set. In 1935 Frank Stanton, a social psychology student at Ohio State University, as part of his Ph.D. dissertation built and tested 10 devices to "record [radio] set operations for as long as 6 weeks." (Stanton was later research director and eventually president of the Columbia Broadcasting System.)

Others experimented with similar devices. Robert Elder of Massachusetts Institute of Technology and Louis Woodruff field-tested their device in late 1935 by measuring the audiences tuning in to Boston stations. But it was Arthur C. Nielsen, a consumer survey analyst with a degree in electrical engineering, whose wealth and fame would be made by the device. In early 1936, Nielsen heard a speech by Robert Elder, who called his device an "Audimeter." At the time the Nielsen Company, a consumer survey business, was primarily a collector of information on grocery and drug inventories.

After receiving permission to use the Robinson-RCA device and some redesign of it, in 1938 the Nielsen Company began tests in Chicago and North Carolina. In 1942 the company launched the Nielsen Radio Index based on 800 homes equipped with the Audimeter, which recorded on a paper tape the stations a radio was tuned to. In the beginning, Nielsen technicians had to visit each of the 800 homes periodically to change the tape and to gather other information from each household based on an inventory of the family's food supply. The Audimeter was usually hidden from view in a nearby closet or some other out-of-the-way place. Respondents were usually given nominal compensation for their participation, and Nielsen usually shared repair costs on any radio in which the meter had been installed. Beginning in 1949, the receiver's tuning was recorded by a small light tracing on and off on 16 millimeter motion picture film that could be removed and mailed back to the Nielsen office in Chicago for examination and tabulation by workers using microfilm readers.

Audimeter Ratings

The Nielsen Company soon supplanted the older and dominant Hooperatings, and Nielsen acquired the C.E. Hooper company in 1950. That year the Audimeter was used to record TV tuning for the Nielsen Television Index (NTI). The company also launched the Nielsen Station Index (NSI), which provided local ratings for both radio and television stations for specific market areas. In the same homes where Audimeters were in use, Nielsen obtained additional information on audience demographics by the use of diaries in which viewers were asked to record their listening and viewing of radio and television.

Throughout most of the 1950s, as television's audience grew rapidly, the measurement of radio audiences by Audimeters provided the most important information used by sponsors, advertising agencies, media buyers, and programmers. As network radio audiences declined and independent Top 40 stations rose, however, local ratings became more important. In 1941 a competitor called Pulse entered the ratings business and, with its ratings based on interviews, eventually eclipsed Nielsen.

In the late 1950s and 1960s there was much criticism of broadcasting in general, resulting from scandals involving rigged quiz shows and disk jockeys being bribed in the "payola" scheme to play specific records, and there followed lengthy congressional investigations of ratings methodologies. As a response, Arthur Nielsen tried to develop a new radio index that would be above criticism but found it would be prohibitively expensive; advertisers and stations resisted higher costs. In 1963 the Nielsen Company ended local radio measurement and the next year withdrew from national radio ratings as well. The Arbitron rating company, founded in 1949 as the American Research Bureau (ARB), continued using meters for many years to supplement its diary method of radio ratings collection.

Audimeters that merely indicate when a receiver is on, and to what station it is tuned, are now obsolete. Advertisers and station operators alike want to know *who* is listening—the listener's income, buying habits, location, level of education, etc. The Audimeter began to give way in television research (the new method was too costly to apply to radio) to the more expensive but also more useful "people meter," which can indicate who is listening by means of a remote control-type device on which each listener punches his or her key to show they are present. The people meter is connected by dedicated data lines to computers in Florida that provide overnight ratings.

Since 1999 Nielsen, with Arbitron, the largest radio ratings firm, is testing a passive, personal meter, about the size of a pager, that listeners wear to record all electronic media use. As with the rest of radio, what began as a large device—the "Audience meter"—has become much smaller and more portable.

LAWRENCE W. LICHTY

See also A.C. Nielsen Company; Arbitron

Further Reading

Banks, Mark, "A History of Broadcast Audience Research in the United States, 1920–1980, with an Emphasis on the Rating Service," Ph.D. diss., University of Tennessee, Knoxville, 1981

Beville, Hugh Malcolm, Jr., *Audience Ratings: Radio, Television, and Cable,* Hillsdale, New Jersey: Erlbaum, 1985; 2nd edition, 1988

Elder, Robert F., "Measuring Station Coverage Mechanically," *Broadcasting* (1 December 1935)

Nielsen Media Research website, <www.nielsenmedia.com>

Stanton, Frank Nicholas, "Critique of Present Methods and a New Plan for Studying Listening Behavior," Ph.D. diss., Ohio State University, 1935

Webster, James G., and Lawrence W. Lichty, *Ratings Analysis: Theory and Practice,* Hillsdale, New Jersey: Erlbaum, 1991; 2nd edition, as *Ratings Analysis: The Theory and Practice of Audience Research,* by Webster, Lichty, and Patricia F. Phalen, Mahwah, New Jersey: Erlbaum, 2000

Audio Mixer. *See* Control Board/Audio Mixer

Audio Processing

Electronic Manipulation of Sound Characteristics

Once a sound has been transduced (transformed into electrical energy for the purpose of recording or transmission), the characteristics of that sound can be electronically manipulated. These characteristics include pitch, loudness, duration, and timbre. Thus, the term *audio processing* refers to the art and science of making changes to an audio signal to improve or enhance the original sound or to create an entirely new sound based on the original. There are many technical and creative reasons for audio processing in radio.

In the early days of radio, if the audio going into the transmitter was too loud, the transmitter could be damaged. Even today, because of the potential for interference to other stations caused by overmodulation, the Federal Communications Commission (FCC) has strict rules about modulation limits.

Audio processors continuously maintain a station's compliance with these rules.

There are also many creative reasons to process audio. Consider the following examples: a commercial producer needs to transform the talent's voice into that of a space alien. In another commercial, the voices sound a little muffled; rerecording the spot through an equalizer to increase the midrange can make the voices sound louder. A third spot as recorded runs too long; it can be shortened by redubbing it through a digital signal processor using the time compression function. These are examples of problems that can easily be solved with the right audio processing in a production studio. Radio stations want their sound to be clean and crisp, bright, and distinctive. Rock-formatted stations targeting teens and young adults usually want to sound loud, regardless of the particular song being played. These are examples of the types of needs addressed by the processing equipment in the audio chain before the signal goes to the transmitter. A description of the basic characteristics of sound identifies the component parts that are manipulated during audio processing.

Characteristics of Sound

Sound is created when an object vibrates, setting into motion nearby air molecules. This motion continues as nearby air molecules are set into motion and the sound travels. This vibration can be measured and diagrammed to show the sound's waveform. The characteristics of a sound include its pitch (frequency), loudness (amplitude), tonal qualities (timbre), duration (sound envelope), and phase. A sound is described as high or low in pitch; its frequency is measured in cycles per second or hertz. Humans can hear frequencies between 20 and 20,000 hertz but usually lose the ability to hear higher frequencies as they age.

The subjective measurement of a sound's loudness is measured in decibels (dB), a relative impression. The softest sound possible to hear is measured at 0 dB; 120 dB is at the human threshold of pain. The range of difference between the softest and loudest sounds made by an object is called its dynamic range and is also measured in decibels. A live orchestra playing Tchaikovsky's *1812 Overture* complete with cannon fire will create a dynamic range well over 100 dB. The amplitude, or height, of a sound's waveform provides an electrical measure (and visual representation) of a sound's loudness. Timbre is the tonal quality of sound; each sound is made up of fundamental and harmonic tones producing complex waveforms when measured. A clarinet and flute sound different playing the same note because the timbre of the sound produced by each instrument is different. Timbre is the reason two voices in the same frequency range sound different. The sound envelope refers to the characteristics of the sound relating to its duration. The component parts of the sound envelope are the attack, decay, sustain, and release. Acoustical phase refers to the time relationship between two sounds. To say that two sounds are in phase means that the intervals of their waveforms coincide. These waves reinforce each other, and the amplitude increases. When sound waves are out of phase, the waves cancel each other out, resulting in decreased overall amplitude.

Individually or in combination, the frequency, amplitude, timbre, sound envelope, and phase of the audio used in radio can be manipulated for technical and creative reasons. The characteristics of the audio created for radio typically need adjustment and enhancement for creative reasons or to prepare the audio for more efficient transmission.

Processors Manipulate Audio Characteristics

Equipment used to process audio can generally be classified using the characteristics of sound described above. There are four general categories of audio processing: frequency, amplitude, time, and noise. Some processors work on just one of these characteristics; others combine multiple functions with a combination of factory preset and user-adjustable parameters. Some processors are circuits included in other electronic equipment, such as audio consoles, recorders, or microphones. Processing can also be included in the software written for a computer-based device such as a digital audio workstation.

An equalizer is a frequency processor; the level of specific frequencies can be increased or decreased. A filter is a specific type of equalizer and can be used to eliminate or pass through specific narrow ranges of frequencies. Low-pass, band-pass, and notch filters serve specific needs. Studio microphones often contain a processing circuit in the form of a roll-off filter. When engaged, it eliminates, or "rolls off," the bass frequencies picked up by the microphone.

Amplitude processors manipulate the dynamic range of the input audio. Three examples of amplitude processors are compressors, limiters, and expanders. A compressor evens out extreme variations in audio levels, making the quiet sections louder and the loud sections softer. A limiter is often used in conjunction with a compressor, prohibiting the loudness of an input signal from going over a predetermined level. An expander performs the opposite function of a compressor and is often used to reduce ambient noise from open microphones. Most on-air audio processing uses these types of processors to refine the audio being sent to the transmitter. Recorders often have limiter or automatic gain control circuits installed to process the input audio as it is being recorded.

A time processor manipulates the time relationships of audio signals, manipulating the time interval between a sound and its repetition. Reverberation, delay, and time compression units are examples of processors that manipulate time. Telephone talk shows depend on delay units to create a time delay to keep offensive material off the air. Commercial producers use time compression and expansion processing to meet exacting timing requirements.

Dolby and dbx noise reduction processing are methods of reducing tape noise present on analog recordings. The Dolby and dbx systems are examples of double-ended systems: a tape encoded with noise reduction must be decoded during playback. These types of processing become less important with the shift to digital audio.

Until the 1990s most processing was done using analog audio. Individual analog processors, each handling one aspect of the overall processing needs, filled the equipment racks in production and transmitter rooms. Equalizers, reverb units, compressors, limiters, and expanders all had their role. Digital processors were introduced during the 1990s. These processors converted analog audio to a digital format, processed it, and then converted the audio back to the analog form. Most processing today has moved to the digital domain. These digital signal processors allow for manipulation of multiple parameters and almost limitless fine adjustments to achieve the perfect effect. Modern on-air processors combine several different processing functions into one unit.

Audio Processing in the Audio Chain

Virtually every radio station on the air today uses some type of processing in the audio chain as the program output is sent to the transmitter. The technical reasons for processing the program audio feed date to the earliest days of radio. Engineers needed a way to keep extremely loud sounds from damaging the transmitter. The first audio processing in radio was simple dynamic range control done manually by an engineer "riding gain." The operator adjusted the level of the microphones, raising the gain for the softest sounds and lowering it during the loudest parts. During live broadcasts of classical music, the engineer was able to anticipate needed adjustments by following along on the musical score. Soon, basic electronic processors replaced manual gain riding.

Early processing in the audio chain consisted of tube automatic gain control amplifiers and peak limiters. The primary purpose of these processors was to prevent overmodulation, a critical technical issue with an amplitude-modulated signal. Operators still needed to skillfully ride gain on the program audio, because uneven audio fed to these early processors would cause artifacts, such as pumping, noise buildup, thumping, and distortion of the sound. Early processor names included the General Electric Unilevel series, the Gates Sta-Level and Level Devil, and Langevin ProGar.

Broadcast engineers generally consider the introduction of the Audimax by Columbia Broadcasting System (CBS) Laboratories to be the birth of modern radio audio processing. The Audimax, introduced by CBS in the late 1950s, was a gated wide-band compressor that successfully eliminated the noise problems of earlier compressors. The Audimax was used in tandem with the CBS Volumax, a clipper preceded by a limiter with a moderate attack time. In 1967 CBS introduced a solid-state Audimax and the FM version of the Volumax, which included a wide-band limiter and a high-frequency filter to control overload due to FM's preemphasis curve.

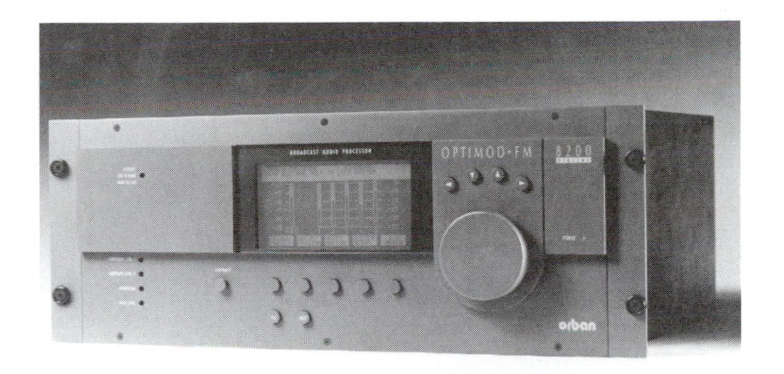

The Optimod-FM 8200 Digital Audio Processor
Courtesy of Orban

The reign of the Audimax was challenged in the early 1970s with the introduction of the Discriminate Audio Processor by Dorrough Electronics. This broadcast compressor/limiter divided the audio spectrum into three bands with gentle crossover slopes, compressing each band separately. Broadcast engineers began to make their own modifications to some of the internal adjustments, adjusting for specific program content and personal preference.

In 1975 Orban Associates introduced the Optimod-FM 8000, which combined compressor, limiter, high-frequency limiter, clipper, 15-kilohertz low-pass filters, and stereo multiplex encoder into one processor. This unit allowed for higher average modulation without interference to the 19-kilohertz stereo pilot signal. The Optimod-FM 8000 was replaced by what soon became the industry standard, the Optimod-FM 8100. A digital version, the Optimod-FM 8200, was introduced in 1992. The Optimod-AM was introduced in 1977.

The development of these processors was driven by the need for a reliable method of maintaining compliance with the FCC transmission and interference rules while allowing for creative use and adjustment of processing for competitive advantage. Along with maintaining compliance with regulatory constraints on modulation, interference, and frequency response, engineers and programmers are always looking for ways to make their stations sound better than and different from the others. Some stations have taken creative processing to extremes. During the 1960s WABC in New York was well known for the reverb used on disc jockey voices during music programs.

A station programming classical music has processing needs different from those of an urban format station. Preserving the dynamic range of an orchestral work is critical, whereas maximizing the bass frequency and loudness enhances the music aired on the urban station. Today's processors allow for this kind of flexibility in adjustment based on format and on specific goals for the sound of the station. Audio processing plays an important role in radio stations' competition for listeners. Stations targeted toward teens and young adults want to sound louder, brighter, and more noticeable than their competitors. This is where audio processing becomes something of an art. Programmers and engineers cooperate to adjust processing to attract and maintain listeners. This is a subjective process that involves trial-and-error adjustments and critical listening by station management. There is a fine line between compressing audio to boost overall loudness and creating listener fatigue. Low time-spent-listening numbers in the ratings may not be the fault of poor programming as much as of overprocessed audio.

Audio Processing in the Studios

Much of the audio sent to the on-air processor has already been processed, perhaps as it was originally recorded, dubbed in production, or mixed with other sources in the air studio to create the program output.

One of the most common forms of audio processing in the studio is equalization (EQ), which is the increase or decrease of the level of specific frequencies within the frequency spectrum of the audio being created. Many audio consoles, especially those used in the production studio, have equalization controls on each channel to allow for adjustment of the EQ of each individual audio source. At a minimum, there are controls for low-, medium-, and high-frequency ranges, but many consoles divide the frequency spectrum into more parts. The EQ controls can be used for various creative and technical purposes. Examples include matching the frequency response of different microphones so they sound the same, creating a telephone effect by decreasing the low and high frequencies of the audio from a studio microphone, adding presence to the voices in a commercial by boosting the midrange, or eliminating hum on a remote line by decreasing the low end. Equalization can also be done through an outboard equalizer; the source or console output can be routed to the equalizer for processing. These units usually divide the frequency spectrum into intervals of one-third or one-half of an octave. Each band has a slider to increase or decrease the amount of EQ on that band. Filters, a specific type of equalizer, can be used to eliminate specific narrow ranges of frequencies. Low-pass, band-pass, and notch filters are usually used to eliminate technical problems with the audio or to keep unwanted audio frequencies from getting to the transmitter.

A well-equipped production studio has a number of processing options available to producers. Until the development of digital signal processors, every effect came from a separate unit. Although many of these single-function processors are still in use and are still manufactured, digital multiple-function processors are the norm today. These are generally less expensive than the on-air multifunction processors, and a number of manufacturers provide many different models and options in their studio processor lines. Most units offer a number of factory preset effects with user-adjustable parameters. These units also allow users to create and store their own combinations of effects. The Eventide Ultra-Harmonizer, for example, provides pitch changing, time compression and expansion, delay, reverb, flanging, and sound effects as part of its inventory. The major advantage of these multifunction units is their ability to combine effects. For example, pitch change can be combined with chorusing and reverb. Flanging can be combined with stereo panning. Given the opportunity for user-created presets and parameter adjustments, the possibilities are almost limitless.

These same types of digital effects are also integrated in the software of digital audio workstations and editors. Audio processing can be added after a recording is made on a multitrack editor. The complex waveform of each track can be processed using the same type of multiple-effects options described above. An announcer can be made to sound like a group of elves through the addition of chorusing, pitch change, and reverb; each track can be processed independently. Because the changes are not made to the original sound files, any of the modifications can be easily undone and the original audio remodified.

Microphones in the production and air studios often receive special, full-time processing. An analog or digital microphone processor typically provides compression, limiting, de-essing, equalization, noise reduction, and processing functions designed specifically to enhance vocal characteristics.

JEFFREY D. HARMAN

See also Control Board/Audio Mixer; Dolby Noise Reduction; Production for Radio; Recording and Studio Equipment; WABC

Further Reading

Alten, Stanley R., *Audio in Media*, Belmont, California: Wadsworth, 1981; 6th edition, 2002

Grant, August E., and Jennifer Harman Meadows, editors, *Communication Technology Update*, 7th edition, Boston: Focal Press, 2000

Keith, Michael C., *Radio Production: Art and Science*, Boston: Focal Press, 1990

O'Donnell, Lewis B., Philip Benoit, and Carl Hausman, *Modern Radio Production*, Belmont, California: Wadsworth, 1986; 5th edition, Belmont, California, and London: Wadsworth, 2000

O'Donnell, Lewis B., Philip Benoit, and Carl Hausman, *Modern Radio Production,* Belmont, California: Wadsworth, 1986; 5th edition, by Hausman, Benoit, and O'Donnell, 2000

Reese, David E., and Lynne S. Gross, *Radio Production Worktext: Studio and Equipment,* Boston: Focal Press, 1990; 3rd edition, 1998

Audio Streaming

Carrying Sound on the Internet

Delivering audio-video (AV) content on the internet has been a long anticipated goal for the media. Audio delivery became practical in 1999 with the introduction of better streaming software and the widespread adoption of MP3 compression techniques. By 2001 major internet companies (including software, hardware, and content providers) were jockeying for positions in the new market. Streaming involves sending data but not asking the computer to record it.

Compared to other internet files, AV files are very large. At first, AV was delivered in the same manner as all other files. The user would download (receive) the entire file from a central computer. This method is still used with high quality compression (for example, MP3 files). But sending the whole file was unacceptable for two reasons. First, it took too long and required too much space on the user's computer, causing the typical home computer memory to fill up quickly. Second, after transfer, the user could keep the whole copy of the AV file. For radio it was not possible to send anything more than simple air checks or short songs, and in any case copyright holders would not allow most songs to be sent.

In 1995 Rob Glaser and company introduced RealAudio—later called RealPlayer and RealOne. The concept was simple. The software would download enough of a sound file to cover the difference between the transmission and play speed. A buffer was created on the user's computer and the file would begin to play. Only a fraction of the file needed to be transmitted before the user could enjoy it. The idea was that the file transfer would be completed shortly before the file was done playing. Continuous streaming would come later as bandwidth and compression increased.

The music industry is working to adopt both models. First, delivering whole files (download) for people to use at will—a sale model. Second, streaming content without giving it to people—a broadcast model. Even if whole files are sent, streaming is often used for more instantaneous delivery of content.

Because streaming content is expected to be one of the biggest profit centers on the web, streaming providers are engaged in a technology trade war. The system that is accepted by the majority of providers and consumers will be in a very strong position to make money. There are three levels to the competition: players, distribution, and content.

Streaming media is produced much like any other media. The target player determines how the content is stored and served. The player is the most visible element in the process. In addition to RealPlayer, Microsoft developed its own MediaPlayer and Apple added streaming capability to its QuickTime software. Finally, the MP3 compression standard allowed software vendors to create streaming media without aligning themselves with a major corporation. The main differences among these players involve cost, compression, ability, and quality.

Once produced, the digital file is encoded in one or more of the streaming formats and stored for use by the appropriate server. The server delivers files as users request them. The server may also add visual content such as advertising or graphic illustrations (such as an album cover).

Continuous programming may be delivered by a never-ending stream or by a playlist format. The latter method sends a list of files to be played rather than a never-ending stream. Playlists may make the entire file available to users, though this is a distinct disadvantage for copyright holders. On the other hand, playlists allow users to skip songs they do not want or to build playlists of their own design.

The competition in players comes down to a software choice while the competition in distribution mainly concerns hardware. Streaming files are not only very large but are also time-sensitive. If a part of the streaming content does not arrive in time, the music will stop. Distribution systems have been developed to deliver streaming files without delay. Newer server systems allow on-demand conversion from one streaming format to another. This means the producer need only store content in one format.

The first method to assure delivery was to increase the speed of delivery. Standard telephone modem delivery—the most widely used form of hooking up to the internet—has peaked at about 56,000 bits per second. Broadband delivery systems for home use are becoming increasingly available, but they are more expensive. Rollout is slow and not nearing the level required for entertainment media.

The second method is to push content closer to potential users. In 1999, companies such as Akamai Technologies and Digital Island built additional internet connections designed only to transfer streaming content. Some networks stored copies of popular media in regional servers to reduce the distance to home users. The idea was to keep the content close to users and reduce the delay caused by network transfer.

A third method is called multicasting. In traditional (unicast) distribution, a separate file was sent to every requesting receiver. When the Clinton Impeachment hearing was put on the web, thousands of people wanted the same file. If the file were three megabytes in size and 100 people wanted to get it at the same time, the network would have to accommodate a 300-megabyte transfer. In multicasting, one copy of the file is transmitted to a large number of users. Then the transmission would be a little more than the original file size. Multicasting has been hurt by the lack of standards and network compatibility.

Content providers have been working to fill streaming media. By the end of 2000, there were nearly 4,000 internet radio sites worldwide with nearly half that number in the United States. Music content led the way since it could be enjoyed before true broadband capacity could be delivered. Most were still seeking a successful means of generating revenue. Along with the search for programming came issues of copyright and payment. ASCAP and BMI created internet payment models in 1998. At about the same time, the Recording Industry Association of America (RIAA) began separate negotiation for web content payments. Their efforts resulted in successful lawsuits against MP3.com and Napster. The continuing threat of lawsuits based on the Digital Millennium Copyright Act (1996) caused most content providers to refocus their market. Pure streams of broadcast content nearly stopped in favor of copyright free content—usually new artists or in promotion of traditional media products. Eventually, a compromise will be worked out. Major media companies (especially music companies) are developing their own distribution systems so that both content producers and distributors will profit when the copyright issue can be resolved.

Content providers faced a second challenge in the early 2000s as the dot-com bubble burst. Unlike most player and distribution companies, many content providers did not have a second line of income and failed as online companies. Most players were supported by major software or hardware platforms. While distribution systems often saw corporate reorganization, broadband delivery was still seen as an acceptable investment by cable television and telephone companies. And new delivery systems show great potential including those that deliver to cellular phones, handheld players, and smart cable television boxes.

STEVEN DICK

See also American Society of Composers, Authors, and Publishers; Broadcast Music Incorporated; Internet Radio; Virtual Radio

Further Reading

CNET How-to, "Streaming Media Explained," <www.help.com/cat/2/657/337/339/808/hto/10004/1.html>

Grant, August E., and Liching Sung, editors, *Communication Technology Update,* Austin, Texas: Technology Futures, 1992; 7th edition, edited by Grant and Jennifer Harman Meadows, Boston: Focal Press, 2000

Kaye, Barbara K., and Norman J. Medoff, *The World Wide Web: A Mass Communication Perspective,* Mountain View, California: Mayfield, 1999

Miles, Peggy, and Dean Sakai, *Internet Age Broadcaster: Broadcasting, Marketing, and Business Models on the Net,* Washington, D.C.: National Association of Broadcasters, 1998

RealNetworks, "Getting Started: A Primer on Streaming Media," <www.realnetworks.com/getstarted/>

Audiotape

Introduced commercially only in the late 1940s, audio recording tape would transform radio broadcasting by removing the stigma of recorded broadcasts. Development of the technique dated back decades, to work accomplished in several countries.

Origins

Early audiotape technology drew on Danish radio pioneer Valdemar Poulsen's 1898 invention of a device called the "Telegraphone." The mechanical energy of sound was converted into a flow of electric current in a microphone and was then translated into magnetic fields, or "flux," in a small induction coil. Then, as a magnetizable steel wire or tape was drawn rapidly past the induction coil, the steel would retain a portion of that magnetic flux as a record of the original sound. This process became the basis for all later developments in magnetic recording. The Telegraphone was a grand prize winner at the 1900 International Exposition in Paris. However, it was only in the late 1930s and early 1940s that U.S. firms became interested in this technology, and even then, only two firms were actively engaged in commercializing it: the Brush Development Company of Cleveland, Ohio, and the Armour Research Foundation of the Illinois Institute of Technology in Chicago.

In contrast, the development of magnetic recording technology advanced in Europe. A turning point came in the early 1930s, when the German firm Allgemeine Elektricitäts Gesellschaft developed the Magnetophone, a high-quality broadcast recorder capable of superior sound recording and reproduction. In 1944, after almost a decade of production, the most advanced Magnetophones incorporated scores of technical innovations. The German broadcasting authority, Reichs Rundfunk Gesellschaft (RRG), became the major customer for Magnetophones, installing them in nearly all German radio studios. During World War II, the RRG took over the operations of broadcast stations in occupied countries and installed Magnetophones there as well. Thus, by the end of the war, tape recording was a standard feature in many European radio stations.

One of the chief distinguishing features of the Magnetophone system was its special recording tape. Since the late 1920s, the German chemical firm IG Farben had been developing a plastic tape base coated with a magnetic form of iron oxide that could substitute for the heavy, expensive steel recording tape used in previous designs. The particles of oxide on such a tape act as tiny individual magnets, and it was learned that it was possible to record higher frequencies at slower tape speeds than on a solid steel band or wire. The slower speed and the lower cost of materials made the coated tape much cheaper, contributing to its widespread adoption in Germany.

The structure of U.S. broadcasting militated against the creation of a demand for that technology. U.S. radio networks relied heavily on live programming distributed by telephone line. Recording represented a threat, both because the recording of a network program was piracy of the network's product and because it would then be technically possible to operate a network by distributing recorded rather than live programs. The status quo in program technology was reinforced by the oligopolistic structure of the broadcast equipment manufacturing market, which was dominated by firms such as Western Electric and the Radio Corporation of America (RCA). None of these firms would offer magnetic recording equipment until after World War II.

Postwar Innovation

In 1945 the United States enlisted the aid of its businesspeople, scientists, and engineers to collect German scientific and industrial knowledge. Some of those who became intimately familiar with Magnetophone technology while in Germany returned to the United States to play a role in the establishment of magnetic recording manufacturing there.

In 1945 Col. John T. Mullin was part of an Army Signal Corps team investigating the military applications of German electronic technology. He was told by a British officer about a tape recorder with exceptional musical quality at a Frankfurt, Germany, radio station that was being operated by the Armed Forces Radio Service (AFRS). There Mullin found German technicians working for AFRS using Magnetophone audiotape recorder/players. The technological improvements of a constant speed transport, plastic tape impregnated or coated with iron oxide, and the employment of a high AC-bias frequency mixed with the audio signal made these machines high fidelity. The first two machines acquired were turned over to the Signal Corps, and Col. Mullin disassembled two other machines and shipped them to his home in San Francisco. In 1946 Mullin designed custom record/reproduce electronics that improved the performance, rewired and reassembled the Magnetophone machines, and went into a partnership with Bill Palmer for movie soundtrack work, using those machines and the 50 reels of tape he had acquired.

In October 1946 Mullin and his partner Palmer attended the annual convention of the Society of Motion Picture Engineers, where he demonstrated the machine to the sound heads of Metro Goldwyn Mayer and Twentieth Century Fox and to the chief engineer of Altec Lansing. Mullin was then invited to an Institute of Radio Engineers meeting in May 1947 to demonstrate the German Magnetophone. It was there that employees of Ampex, a small maker of electric motors in Redwood City, California, first saw and heard the tape recorder. The U.S. government had arranged for the suspension of all German legal claims to magnetic recording technology and had sponsored its wholesale transfer to the United States. The Commerce Department released its technical reports, captured documents, and patents related to the Magnetophone, allowing any interested manufacturers access to information relating to tape-recording technology. Shortly thereafter, Ampex began its own developmental project.

In 1947 the technical staff of the *Bing Crosby Show* on American Broadcasting Companiy (ABC) arranged to have Mullin rerecord original disk recordings of the *Bing Crosby Show* onto tape and then edit them. Crosby had been with the National Broadcasting Company (NBC) until 1944, doing the *Kraft Music Hall* live, but he did not like the regimen imposed by live shows. Because NBC would not permit recorded programs, Crosby took the fall off and returned on the newly formed ABC network with a new sponsor, Philco, because ABC had agreed to let him record on electrical transcriptions as long as his ratings did not diminish. The process required cutting a record and then rerecording; what with sometimes two or three generations, the quality of sound suffered. In July 1947, after the initial demonstration of editing, Mullin was invited to give a demonstration of his equipment for Crosby's producers by taping live side by side with transcription equipment the first show for the 1947–48 season in August at the ABC-NBC studios in Hollywood. Bing Crosby Enterprises then negotiated financing for Ampex for exclusive distribution rights, and Mullin was employed to record the Crosby show on his original German equipment until the Ampex machines became available. Made with the original German tape recorders and 50 rolls of BASF tape, Mullin's first recorded demonstration show of August 1947 was broadcast over ABC on 1 October 1947.

In 1948 Alexander Poniatov and his team of engineers at Ampex introduced the first commercial audiotape recorder based on the Magnetaphone as Ampex Model 200. The first two, with serial numbers 1 and 2, were presented to John Mullin, and numbers 3–12 went into service at ABC. (To meet the contract requirements, Mullin gave his machines to ABC and later received numbers 13–14 for his contribution.)

The Minnesota Mining and Manufacturing Company (3M) of Saint Paul, Minnesota, already had experience in the manufacture of coated films from its line of adhesive tapes. Home tape machines such as the Brush Soundmirror, which used Scotch 100 paper tape supplied by 3M, had been introduced in the consumer market, but these fell far short of professional requirements. Mullin then asked 3M engineers to reverse-engineer the German product using samples of IG Farben tape and Department of Commerce technical reports. Although the Minnesota company quickly came to dominate the field, much smaller firms successfully broke into the market, competing with 3M.

The Crosby show remained tape-delayed, setting a precedent in broadcast production that remains the norm to this day. Most other network radio and recording artists quickly adopted tape to produce their shows and discs, including Burl Ives and Les Paul. Live broadcasting was soon limited mostly to local disc jockeys spinning the new long-playing 33-1/3 and 45-rpm music discs.

Mullin remained with Bing Crosby Enterprises, recording his shows and others at ABC, until 1951. As the exclusive distributor for Ampex, Bing Crosby Enterprises sold hundreds of recorders to radio stations and master recording studios. In 1951 Mullin and other engineers were spun off as the Bing Crosby Electronic Division to handle development of audio instrumentation and video recording. In 1956 the Electronic Division became the Minicom Division of 3M, where Mullin served as head of engineering and as professional recorder development manager until his retirement in 1975. He died on 24 June 1999 at age 85.

MARVIN BENSMAN

See also American Broadcasting Company; Crosby, Bing; Recording and Studio Equipment; Wire Recording

Further Reading

Angus, Robert, "History of Magnetic Recording," *Audio* 68, no. 8 (August/September 1984)

Camras, Marvin, editor, *Magnetic Tape Recording,* New York: Van Nostrand Reinhold, 1985

Hickman, E.B., "The Development of Magnetic Recording," in *American Broadcasting,* compiled by Lawrence W. Lichty and Malachi C. Topping, New York: Hastings House, 1975

Millard, Andre J., *America on Record: A History of Recorded Sound,* Cambridge and New York: Cambridge University Press, 1995

Mullin, John T., "The Birth of the Recording Industry," *Billboard* (18 November 1972)

Mullin, John T., "Creating the Craft of Tape Recording," *High Fidelity* 26, no. 4 (April 1976)

Van Praag, Phil, *The Evolution of the Audio Recorder: The "Vintage" Years, Late 1940s-Early 1970s,* Waukesha, Wisconsin: EC Designs, 1997

Auditorium Testing

Radio Market Research

Auditorium testing is a method of market research used widely in the radio industry and elsewhere to determine the effectiveness of programming and audiences' preferences in music, voice quality, commercial messages, and other program elements. Its name comes not from the room where the testing takes place, but from the "auditory" nature of the testing; that is, the subjects hear the samples being tested.

A company called ASI (now Ipsos-ASI) first used auditorium testing in the evaluation of TV programs, commercials, and movies in the 1960s. They used what they called the "Preview House" in Los Angeles as a controlled environment for such tests. Forty years later, versions of auditorium testing are still used by numerous market research firms around the world.

The basic methodology in auditorium testing starts with a careful consideration of the goals to be achieved. The client advertiser, radio station, or TV station needs to identify, in the most precise way possible, the boundaries of the testing and how the results will be used. Once the desired outcomes are known, the researchers design a testing strategy to achieve those outcomes.

With the strategy set, the research company screens and selects a group of between 75 and 200 people reflecting the demographic the client wishes to study. That demographic (a grouping according to age, gender, income, etc.) can be a random sample or one that is consistent with the station's current or desired audience, or even a subset of the audience that the client wishes to cultivate.

The assembled test group is then invited into a small auditorium and given instructions for the test. They are rarely told what is being tested or who the client is for the testing. In fact, tests often include decoy selections to keep the participants from guessing which specific radio station or product is being evaluated. The test subjects are instructed to respond to samples of music, voice, messages, images, or other content, providing some sort of rating on a scale created by the researchers. This can be accomplished with written questionnaires, a joy stick–type device that measures responses electronically, or even with a show of hands. Sometimes anecdotal comments are also solicited. Participants in auditorium testing are usually compensated for their time in order to increase the seriousness with which they approach the evaluating. The results are then tabulated and evaluated, with many variables charted, and correlations are made among the different samples tested. Ultimately the research firm can provide clients with both a review of the raw data and recommendations on how they may proceed to achieve their goals. Auditorium testing is essentially a hybrid of several market research methods, taking the group dynamic of focus groups, the larger size of diary or phone research, and the immediacy of one-on-one surveying.

Bob Goode developed a form of auditorium testing called Electronic Attitude Response System. This method uses a video readout of averaged responses of the participants correlated directly to the audio content being rated and allows researchers to determine the test audience's preferences along with their "tune-out" of program elements. It also provides researchers with a sense of which program elements are more effective if paired with others. For example, a commercial following a weather report may lead to less tune-out than if that same commercial aired after a musical selection.

Music testing is a particular strength of auditorium testing models. Whole pieces within a musical genre can be tested before they are aired on a station. More commonly, however, "hooks," short segments of songs, are tested. In markets where many stations compete for listeners within each programming genre (country, oldies, urban contemporary, etc.), the subtleties of which songs are most liked within each genre can make a major difference in the ratings successes of each station. One firm, The New Research Group, offers 600 to 1,200 musical hooks along with 100 perceptual questions, allowing the client to know not only which music is preferred, but why, in specific descriptive terms, dealing with emotions, motivations, associations, etc.

Research firms "cluster" music that appeals to test audiences in auditorium groups, because people who enjoy one song from the cluster are likely to enjoy others as well. In addition, firms use complex matrix charts to show compatibility between clusters, showing radio programmers how to broaden appeal by including more musical selections without causing tune-out by core listeners.

Auditorium testing, along with other music testing, is seen by some as limiting, in that the short hooks it tests can oversimplify otherwise interesting music that might gain acceptance upon being heard by audiences. For example, a hook from "Hey Jude" by the Beatles might not have tested well, whereas the song in its entirety was a number one hit.

Public and commercial radio stations use research, including auditorium testing, to make program decisions; for example, the Wisconsin Public Radio network has been involved in the Corporation for Public Broadcasting program research, and Denver-based Paragon Research studied public radio stations in eight markets using focus groups, surveys, and auditorium research.

As audio broadens its reach through new technologies such as satellite, Web-casting, and other distribution channels, it is likely that increasing specialization of program channels will occur, making auditorium testing more important in the precise selection of program content.

That being said, the increasing sophistication of audience behavior measurement technology imbedded in some of these new communications media may eventually render traditional auditorium testing too slow and imprecise by comparison for the emerging information needs of the industry.

PETER WALLACE

See also Audience Research Methods

Further Reading

Clemente, Mark N., *The Marketing Glossary: Key Terms, Concepts, and Applications in Marketing Management, Advertising, Sales Promotion, Public Relations, Direct Marketing, Market Research, Sales,* New York: American Management Association, 1992

Australia

Australian radio is a fascinating amalgam of the unbridled commercialism of U.S. radio and the public broadcasting ethos of Western Europe and Asia, blending the two forms into a mixed private/public system. Its history can be divided into five parts: radio as a new technology in the 1920s; the new broadcast medium of the 1930s; the emergence of a diverse production sector and debates over Australian content and wartime censorship in the 1940s; the impact of television and popular music in the 1960s; and today's changes, brought about by frequency modulation (FM), digitalization, and deregulation.

A New Technology (to 1930)

Wartime anxieties deriving from the uncoded transmission of the whereabouts of an Australian naval convoy had led to the impounding of all privately owned wireless telegraphy sets in 1915. Governments sought from very early on to exercise control over the airwaves as resources, both for military purposes and to police property, and as a source of revenue. The first public demonstration of radio took place in Sydney in 1919, an event sponsored by the Amalgamated Wireless Company of Australia (AWA). AWA and the Royal Australian Navy were in dispute over the direction the new medium should take and the framework within which it should operate. This struggle for control between AWA and the navy featured the former championing the rights of the lone user and the latter criticizing any moves toward private control.

As in the United States, when the medium's commercial and governmental potential became clearer, the private and public sectors grew increasingly antagonistic. A major conference was held in 1923 to try to sort out these differences. AWA obtained approval for a "sealed" wireless system, to operate on a competitive basis. Broadcasting companies—which often manufactured and distributed receivers—were to be licensed by the government, with audiences subscribing to particular stations. The sets were then sealed, confining listeners to the stations they had paid to hear. By the following year, cost pressures and differences within the industry led to two further conferences and the establishment of a new dual system for broadcasting: "A" licenses, funded by listeners' subscriptions—with the government retaining a proportion of the levy—and "B" licenses, financed by advertising. The A stations were required to provide a comprehensive service that would cater to all sectors of the community. The Bs, lacking the safety net provided by access to the license fee, were free of such obligations.

Much of the nation was not catered to by this fledgling industry. Concerns about rural areas, along with the legal mechanics of copyrights and patents, led to a Royal Commission into Wireless in 1927. The government was determined to maximize the capacity of the new medium to bind the equally new nation together. After failing to persuade individual license holders in the A sector to pool their resources for a nationwide grid, in 1928 the government announced that it would acquire all A class stations, in large part as a result of pressure from rural areas. The new national system would be operated by the post office, with programs provided by the private sector. The successful bidder for the contract to provide programs was called the Australian Broadcasting Company.

There was considerable innovation at the programming level: 1924 saw the first transmission of Parliament, the first radio play, and the world's inaugural broadcast from an airplane. Four years later came the first ecclesiastical opposition to beaming church services into people's homes. (Religion went on to gain a unique guarantee of airtime in Australian broadcast regulation of all sectors, because Christianity was seen as central to the moral fiber of the nation but marginal to media profitability. Such provisions were of dubious constitutional legitimacy, but they remained unchallenged for decades

[Horsfield, 1988].) By 1930 there were 290,000 sets across Australia, with 26 stations in 12 cities. The Labor Party's campaign platform included a promise to abolish license fees if elected.

A Broadcast Medium (1930s)

By the beginning of the 1930s, radio was firmly established. The critical event was the creation of the Australian Broadcasting Commission (ABC) in 1932 as a public broadcaster, still funded by licenses, in place of the three-year-old private Australian Broadcasting Company. Its enabling legislation obliged the commission to raise the educational and cultural levels of the public. One of its early chairs, W.J. Cleary, described the task of the ABC in 1934 as promoting "the finer things of life" by elevating the populace "to find interests other than material ones, to live by more than bread alone."

The first overseas transmission from Australia took place in 1933, and big increases in the sale of radio sets coincided with the ABC's descriptions of play from the 1934 cricket series between England and Australia. By the middle of the decade, the audience for radio was growing by 8,000 people each month. Complaints were made about the heavy schedule of advertisements on the commercial stations, although regulation prevented ads on Sundays. The commercial stations, now represented by the Australian Federation of Commercial Broadcasting Stations, were adjusting to the growth of the industry by networking and borrowing programs from overseas, notably *Lux Radio Theater* and radio serials. The advent of the serial and the network were linked. Multinational corporations owned the major advertising agencies, and they placed pressure on the commercial stations to deliver a big audience to clients who desired nationwide exposure for their output. They also frequently colluded with each other to rein in recalcitrant networks, assisting compliant groups and undermining others by manipulating schedules. The ABC's networking arrangements were falling into place by 1939, with two metropolitan stations established in each capital city.

In 1930 there were only a handful of commercial stations in Australia; two years later, the number had risen to 46. With more stations came greater uniformity: cutting production costs, standardizing formats, playing greater amounts of recorded music than was the case anywhere else in the world, and selling blocks of time to specific sponsors so that programs were created around the products that paid for them. Networks were established to satisfy the demands of advertisers for a national reach, with the encouragement of American-owned advertising agencies. In the 1940s, following governmental anxieties about monopolistic practice, the networks emphasized the autonomy of individual stations, which were said to rely on networks for resources rather than acting as mere conduits.

By the close of the 1930s, there were well in excess of 1 million license sales and perhaps four times that number of regular listeners. Two of every three dwellings had a set. And despite its early reputation as technologically complex, the potential of the medium to ameliorate the drudgery of domestic work even as it encouraged a habit of consumption made the female listener a target. "The men behind the microphone, in a very real sense, modulate all other sounds. To some they are folk heroes; to some women—phantom lovers" (Walker, 1973). This was also the period when the child audience was discovered: "If children were off sick, they were sometimes allowed to have the bakelite box in bed with them for the drip dramas, the afternoon children's serials and—if they were privileged—the quiz shows with tea on a tray in the evening" (Kent, 1983).

Wartime and Beyond: Diversity and Australian Content (1940–55)

World War II dramatically increased the role of the state in radio. The war brought about security restrictions on broadcast material as well as the notion that the citizenry must receive expanded coverage of global events. This expansion of service and contraction of autonomy led to both the formation of Radio Australia as an international network (and its wartime takeover by government) and the imposition of strict censorship. ABC news followed the government's line on the primacy of the Pacific theater of war, and most of these bulletins were relayed to the commercial stations. The diminution in the amount of rebroadcast British opinion and the sense that the ABC was becoming an arm of state propaganda led to serious protests from listeners. Another side effect of the war was the belated—and short-lived—opportunity for women to work as ABC news presenters. They were hired in large numbers in the absence of able-voiced men on military duty. But after the war, no woman would read the national radio news again until 1975.

In 1942 a law was passed to regulate non-ABC sectors of the industry and to provide a guarantee against political directives being issued to the ABC. The new act imposed an Australian music quota of 2.5 percent of commercial airtime. In 1949 the Australian Broadcasting Control Board (ABCB) was created as a statutory authority to regulate the industry.

One cannot draw an indelible line between the program output of the commercial stations catering to public demand and the ABC catering to public education. Both had high costs of production in comparison with today's emphasis on music or basic talk. In the years after the war, the ABC expended nearly £1 million a year on production, and the commercial stations only slightly less. Recorded material from outside Australia took up less than 5 percent of commercial airtime. Both sectors devoted a good amount of these sums to locally written plays. The ABC devoted a quarter of broadcast time to "light" and "dance" music, with much less "serious" music. This decade also saw the first episode of the ABC's popular *Blue Hills* soap opera, which was to continue until 1976. At the other end of the spectrum, the commission was required by law to broadcast Parliament from 1946, partly to raise the profile of politicians in the community and partly because of the Labor government's concerns with press bias against it.

The rationing of newsprint during the war delivered advertisers to radio on an unprecedented scale. The war also cut off the commercial stations' supply of transcriptions from North American drama, which led to the substitution of local product. A star system was created, with high-quality serials designed to capture nighttime audiences and block programming to retain interest at a particular point on the dial throughout the evening. Although commercial radio could not match the ABC's claim to having been the first network anywhere to broadcast all of Shakespeare's plays, it expended a comparable amount on drama in the 1940s. The commercial stations produced documentaries on nature, history, and medicine; devoted only about 8 percent of airtime to advertisements; broadcast cricket from overseas; and generated the preconditions for illegal off-course betting on horse races through coverage from every imaginable track.

The Challenge of Television and Pop Music (1956–70)

After 1956 television, along with the importation of Top 40 techniques from the United States and a focus on youth as desirable potential consumers, transformed radio listening. Radio shifted from a medium dominated by variety and quiz shows and drama serials to one of popular music, "talkback" (talk radio), and sports. Other enforced changes included a move toward additional use of actuality in news broadcasts, both to compete with television and as a consequence of improvements in taping facilities. Revenue and profits grew in the late 1950s to double their pre-television figures. In 1961 there were 6.5 million radio receivers and 10.54 million people in Australia.

The ornate receiver in the living-room corner of the 1930s had given way by the 1960s to an object that was on the move: the tube was replaced by the transistor. Prior to television, 40 percent of radio set sales were for console and table receivers. In 1960 these made up just 19 percent of the total. The turn toward portables (41 percent of sales) and car radios (26 percent) led the recently formed Australian Radio Advertising Bureau to characterize the trend toward "outdoor listening" and "indoor one-person audiences." Forty percent of Sunday listening was now outside the home, principally at the beach.

Overseas influence increased independently of the new medium: the ABC was importing discs for programming, and the private Macquarie network was owned by British interests from 1951 to 1965. Macquarie came to make extensive use of current-affairs material from U.S., New Zealand, British, Ceylonese, and South African sources. Programs in languages other than English were strictly limited.

The ABCB doubled the Australian music quota to 5 percent—although frequently this was not adhered to—and it also allowed advertising jingles on Sundays. A study of children's radio serials warned of the unholy effects on young people that exposure to radio could bring. Moral panics abounded with the shift toward playing rock and roll records, the turns of phrase of which provoked numerous complaints by 1960–61. About 7 percent of music on commercial radio met the ABCB's tests of local content. Talkback became possible when legislative changes in 1960 allowed stations to broadcast material using the telephone.

Expansion and Crisis (1970s–2000)

Between 1948 and 1972, Australia's urban population more than doubled, but only one new commercial radio station was added. The Labor government of 1972–75 opened up use of FM, issued additional frequencies on the AM band, and developed public access. Australia was decades behind other countries in the introduction of FM radio (despite initial trials as early as 1947) because television had been allocated the very-high-frequency (VHF) waveband. Space was found on VHF for FM signals to exist alongside those of television. By 1978 there were well over 200 commercial and ABC AM and 5 FM stations, plus 50 public access outlets. Of the commercial stations, approximately one-quarter were owned by newspaper interests.

The ABCB was succeeded in 1976 by the Australian Broadcasting Tribunal (ABT). Its first major document, *Self-Regulation for Broadcasters?* was a critical statement of the rationale for significant—albeit limited—private-sector self-determination in the industry. Nevertheless, the renamed commercial representative, the Federation of Australian Radio Broadcasters (FARB), criticized the report for its extrapolations from the model of television to radio. FARB's other protectionist activities at this time included trying to shut down the ABC's youth station in Sydney and threatening legal action to restrain the government from granting community radio licenses. The new ABT opened up the system of license renewals to public participation. It also began to show a real concern with cross-media ownership.

The ABCB had raised the Australian music quota to 10 percent of airtime in 1973, with the ABC electing to follow suit. The figure was increased to 15 percent in 1975, with major implications for both programming and the local record industry. Twenty-two million records were manufactured locally (mostly made from imported masters) in that year. By 1980, when commercial FM commenced operations, rock music and its Australian substratum were embedded in entertainment programming. A major review of Australian content on commercial radio in 1986 led to new rules, which provided that 20 percent of music broadcast between 6 A.M. and 12 midnight must be Australian. Conversely, restrictions holding advertising to 18 minutes per hour were lifted.

The ABC's movement into a more national focus took two significant steps in the mid-1970s: a national network of classical music set the seal on the higher tone to its mission, and youth station 2JJ appeared in Sydney (2JJ would later go national over the period 1989 to 1991). The ABC was reconstituted as a corporation in 1983 and charged with the responsibility to provide "innovative and comprehensive" programs "of a high standard" in order to "contribute to a sense of national identity and inform and entertain, and reflect the cultural diversity of, the Australian community." The formation in 1977 of the Special Broadcasting Service (SBS) increased ethnic broadcasting. The period also saw the advent of radio for the print-handicapped and the formation of the Central Australian Aboriginal Media Association (CAAMA). CAAMA began broadcasting by Aborigines for Aborigines halfway through 1980. The federal government allocated A$7 million to Aboriginal broadcasting in 1992, as opposed to A$65 million to ethnic broadcasting, some indication of the groups' respective political clout. By 2000 there were five licensed Aboriginal stations and special services for tribal peoples. As community radio managers moved onto an increasingly commercial footing, their stations began requiring Aboriginal groups to pay for time on the air. Thus, the amount of access available to indigenous people was strictly limited, emphasizing the importance of opening up the opportunity for Aborigines to control their own stations.

By 2000 there were four sectors of Australian radio, arching across the FM and AM bands: the ABC, the commercial stations, the SBS, and community (previously called public) stations.

The Australian Broadcasting Corporation

The ABC operates a variety of services. It offers a rural network extending across the country, in addition to metropolitan AM stations that mix talk, music, and news from local and national perspectives. Radio Australia is a single shortwave, multilingual, international service of news, music, and information that claims to be heard by 50 million people outside Australia. There are five national services: Radio National (variously, "Radio to Think By," "The Truth of Australia," or "Mind Over Chatter"); ABC Classic FM ("fine music"); JJJ-FM ("Radio That Bites"); ABC News network (alternating between political proceedings and news); and Dig Radio, an internet network. Finally, Radio Australia broadcasts to the Asia-Pacific region.

The ABC is expected by government and management to combine broad popularity (the metropolitan stations' breakfast programs) with authoritativeness (news and current affairs) and innovation (music markets neglected by commercial broadcasters but desired by cassette and compact disc manufacturers). Further, the ABC tries to ensure "that listeners across Australia hear viewpoints and perspectives not broadcast on other stations." Numerous marginal groups regard the ABC as their access to the center. At the same time, the commission continues to refer to "those original aims that saw the ABC come into being in 1932: to draw the country together by bringing radio of special quality, importance and relevance to all Australians."

For people who grew up with the ABC, their referent is never simply its actual broadcast output. It is also the *meaning* of the ABC, as Australia's foremost institution of information and culture. And for the first 25 years of its existence, it had an exclusively audio presence, a presence that continues to be enormously significant. The commission's remit is basically contradictory: a comprehensive service that should complement market-driven services, simultaneously both popular and specialist.

Commercial Stations

Commercial stations meet the needs of advertisers by attracting large audiences. The larger the number of listeners, the more the commercial stations clamor for independence from surveillance by the state: popularity, they argue, guarantees their being in step with public values and attitudes. And this in turn is their claim on the advertising dollar.

Once pilloried for their lowbrow teenage audience, commercial stations are now taken to task for ignoring this group in favor of the aging young and its taste for recycled popularity. Instead of the teenage record-buying public, 25- to 39-year-old consumers are sought by metropolitan FM stations because of their conspicuous propensity to purchase. AM stations have found their niche (considered by many to be comparatively unprofitable) in a mythic suburbia that is fond of convivial chat, of inoffensive music, or of sports radio.

The 1980s were the decade of FM. FM prided itself on the "extraordinary sophistication" of its audience research methods, targeting ever more specific categories of listener. The research is divided in five ways: focus groups for qualitative information from a few listeners; audience tracking, to find out whether listeners are loyal; callout music research, or playing music down the telephone line to gauge reaction; auditorium music tests, where respondents sit en masse and listen to hundreds of tunes; and lifestyle research, which systematizes the habits of the audience.

The commercial stations' daily cycle is differentiated through announcers defined as individual personalities rather than by a range of music. Disc jockeys are the effective markers that distinguish one service from another. This loss of diversity has been assisted by deregulatory forces. The ABT's successor, the Australian Broadcasting Authority (ABA), has replaced the single quota of Australian music with more targeted quotas specific to particular station formats (up to 15 percent for rock stations, down to 5 percent for easy listening). Regulation of offensive content has, however, continued, and stations are subject to a code of practice designed to prevent programming that offends community standards concerning violence, drugs, suicide, or hate speech.

The successor to FARB, Commercial Radio Australia (CRA) claims that 80 percent of radio listeners tune to the commercial stations. The claim represents its justification for opposing a free-market approach to the issue of licenses: room for new entrants is severely limited by the long periods needed to achieve a profit, and exclusivity is required if local commercial radio is to maintain sufficient advertising revenue to continue broadcasting. FARB opposed using the premium space on the FM band for community radio: commercial stations have the listening numbers in their favor, so they should have spectrum allocation in preference to minority interests.

Special Broadcasting Service

SBS radio is part of multiculturalism, the federal government's cultural shift in immigration settlement policy from all-out assimilation. An ethnic middle class that lacked access to media outlets and attention to their informational needs started to lobby politicians in the mid-1970s, leading to the creation of the SBS as the body responsible for stations 2EA in Sydney and 3EA in Melbourne, stations then on the air on a community basis. The SBS extended these stations' reach via a relay system to Wollongong and Newcastle.

SBS radio programming was to be along language, not community lines. As the 1980s progressed, the SBS board came to resist an overt discussion of politics by advisory groups while supporting the existence of consultative machinery: an apparent openness to public participation actually masked a very limited agenda for discussion. A further dimension emerged with the mid-1980s uptake of anti-immigration positions by the Liberal and National Parties and the 1990s push from Aboriginal and other people to forge a code of anti-discriminatory guidelines for announcers. Attention has shifted away somewhat from SBS radio since the federal government assigned further development of ethnic broadcasting to the community sectors in the mid-1980s, but the Sydney and Melbourne stations continue to attract controversy.

Community Stations

Community stations differ from the other three sectors in that their mandate is neither governmental nor commercial. They have a much more focused and limited warrant to serve local, special interest, community, or educational needs. This constituency may be highly specific, as in radio for the print-handicapped or for the residents of East Fremantle, or very broad, as in a station catering to jazz listeners from 8 to 10 o'clock and to Spanish speakers from 11 to 1. Apart from the surveillance of the ABA, which issues specific licenses for education, special interests, and geographical locations, they are also beholden to internal systems of management, often related to educational institutions or sources of program sponsorship. In addition to the reporting requirements of regulatory bodies and management committees, the sector has a peak representative body, the Community Broadcasting Association of Australia (CBAA). The association makes submissions to inquiries into the area, convenes national meetings to discuss common issues, and provides a point of articulation.

There is a significant democratic, participatory rhetoric associated with such stations. Although most have in-house training and monitor their output, the claim they frequently make is that they transform the passive audience member into an active producer of material, opening up the airwaves to alternative points of view and modes of presentation and demystifying the media. The voluntary nature of this labor also encourages a faith in the value of cooperative management, which has on occasion led to conflict over production values and the basis for making and implementing decisions on programming, finance, and personnel. Such issues can be especially awkward when paid staff seek to control the administration and quality of the station in a way that is seen as replicating the elitism of other forms of Australian media. The CBAA promulgates a "Code of Ethics" that prescribes community accountability, broad media access, participatory decision making by both presenters and listeners, quality balanced with access, no censorship other than legal requirements, and public proclamation of nondiscriminatory station policies.

The Modern Radio Industry

In 1990 Australia's 150 commercial radio stations earned A$450 million in revenue and spent over A$400 million, much of it to employ over 4,000 people. These figures mean that the sector is only marginally profitable, which pushes CRA toward contradictory postures on industry regulation. At one moment it is all for total freedom to decide what is broadcast and when, arguing that program content should be the sacrosanct domain of the implied contractual space negotiated by broadcaster and listener. But when it turns to use the airwaves, CRA wants regulation to prevent masses of new competitors. This is, however, counter to the penchant for open markets that has been very much in evidence among Commonwealth government policy makers.

At the same time, the government was engaging in a detailed investigation of metropolitan commercial radio, resulting in a 1988 National Radio Plan (NRP). It called for the conversion of ten AM stations to FM, ten additional licenses, and new services in the not-for-profit area. This was occurring against a backdrop of huge declines in ratings for AM stations. FM was the band that everybody desired to enter, and the federal government's need for cash provided a useful foil to its laissez-faire intellectual preferences: the new licenses for FM were let by competitive tender. The existing players needed to buy airspace in order to remain solvent or grow as business forces. More than half the stations in the southeastern states' capital cities changed hands between November 1986 and April 1989. In some transactions, prices paid were in excess of 25 times the value of annual profit.

The upshot of this policy innovation and supply-side disruption was that 1991 found 60 percent of the listening audience tuned to businesses run by Austereo and Hoyts. But this audience potential did not necessarily amount to profitability: Austereo was unable to find bidders for its FM and AM stations in Canberra, which had combined with its Perth license to produce a loss for 1989–90 of about A$8 million. Meanwhile, the extraordinary revival of AM ratings in Melbourne brought into question the automatic equation of FM with profits. CRA emphasized the difficulties confronting the industry, such as the aggregation of television services, the possibility of pay TV, high license fees, competition from a subsidized ABC, and too many competitors in the FM area.

Despite incurring penalties of A$50,000, 5 out of 14 metropolitan AM stations defaulted on their FM-conversion proposals by the end of 1992, because the competitive bidding system had combined with a long and deep recession to strip away the foundations of many companies. Some recompense was available, as the ABA made it possible for a licensee to run more than one station in a market, whereas the NRP had tried to acknowledge a problem with concentrated ownership. The success of the original FM stations diminished to the point of a collective loss in 1991, with the advent of new entrants and some renewal of the AM band's popularity (although most of those stations are losing money also). This should come as no surprise, because revenue from advertising is now expected to cover 23 FM stations, compared to 7 before 1990. In the first few months of its operation, the ABA issued 130 virtually free licenses for narrowcasting transmission, mostly covering tiny locations for tourist drive-through information or betting services, areas once covered by the commercial sector.

But the definition of narrow—and its implications for niche marketing by full-blown commercial services—was unclear, with 124 of the country's 150 profit-oriented stations using almost identical golden contemporary playlists! Blocks of programming, both music and news, are increasingly being purchased by many stations in order to keep a continuous service on the air without employing staff, and news bulletins are generally supplied by a small number of services. By 2002 commercial radio was making around A$745 million in revenue a year, amounting to a profit of about $140 million. The sector secured around 8 percent of all advertising revenue to media outlets.

The Future

Four interrelated factors will determine the future character of commercial radio: technology, networking, imagined audiences, and regulation. The advent of digital technology in Australia offers the prospect of enormously high-quality sound reproduction across similarly enormous distances. When combined with the centralizing drive of networking and syndication to reduce costs, this suggests a southeastern, urban broadcasting center that will swallow up both rural and metropolitan stations through a centrally delivered signal that is customized to local time, weather, and traffic conditions.

Digital audio broadcasting (DAB) has major implications for how audiences are conceived and addressed in terms of localism as well as age, income, race, values, and routine. DAB thus suggests a further homogenization of programming, but by the end of 2002 it remained a dream.

Demographic projections say that the average Australian household will increasingly be composed of the delightfully named "unoccupied person," living alone and keen to turn to casual voices in search of anomic relief. In this sense, metropolitan and suburban life become akin to rural isolation, where the regional radio presenter is a major figure and a source of delightful recognition for many people. This will encourage research into people's work and leisure activities and ideological proclivities as part of the surveillant eye of social science. It is also encouraging networks, because CD-quality commercials for national advertisers can now be transmitted from central locations to stations throughout the country. In 1999, 75 percent of commercial stations had network affiliations, and several were owned by foreign citizens, following deregulation of ownership controls. In 2000 there were 229 commercial stations serving just 19 million people—but offering a crushing sameness. Only the ABC holds out signs of difference.

TOBY MILLER

Further Reading

Allen, Yolanda, and Susan Spencer, *The Broadcasting Chronology: 1809–1980,* North Ryde, New South Wales: Australian Film and Television School, 1983

Broadcasting in Australia: The Fourth Annual Review of the Industry by the Australian Broadcasting Tribunal, Sydney: Tribunal, 1992

Collingwood, Peter, "Commercial Radio 1999: New Networks, New Technologies," *Media International Australia* 91 (1999)

Counihan, Mick, "The Formation of a Broadcasting Audience: Australian Radio in the Twenties," *Meanjin* 41, no. 2 (1982)

Counihan, Mick, "'Giving a Chance to a Youthful Muse': Radio, Records, and the First Australian Music Quota," *Media Information Australia* 64 (1992)

Higgins, C.S., and Peter D. Moss, *Sounds Real: Radio in Every Day Life,* New York and St. Lucia, Queensland: University of Queensland Press, 1982

Hodge, Errol, *Radio Wars: Truth, Propaganda, and the Struggle for Radio Australia,* Cambridge and New York: Cambridge University Press, 1995

Horsfield, Peter, "Issues in Religious Broadcasting in Australia," *Australian Journal of Communication* 14 (1988)

Jakubowicz, Andrew, "Speaking in Tongues: Multicultural Media and the Constitution of the Socially Homogeneous Australian," in *Australian Communications and the Public Sphere: Essays in Memory of Bill Bonney,* edited by Helen Wilson, South Melbourne, Victoria: Macmillan, 1989

Johnson, Lesley, *The Unseen Voice: A Cultural Study of Early Australian Radio,* London: Routledge, 1988

Jonker, Ed, "Contemporary Music and Commercial Radio," *Media Information Australia* 64 (1992)

Kent, Jacqueline, *Out of the Bakelite Box: The Heyday of Australian Radio,* Sydney: Angus and Robertson, 1983

Mackay, Ian K., *Broadcasting in Australia,* Carlton, Victoria: Melbourne University Press, 1957

Miller, J.D.B., "Radio in Our Lives," *Current Affairs Bulletin* 3, no. 12 (1949)

Mundy, Greg, "'Free-Enterprise' or 'Public Service'? The Origins of Broadcasting in the US, UK, and Australia," *Australian and New Zealand Journal of Sociology* 18, no. 3 (1982)

Potts, John, *Radio in Australia,* Sydney: New South Wales University Press, 1989

Turner, Graeme, "Who Killed the Radio Star? The Death of Teen Radio in Australia," in *Rock and Popular Music: Politics, Policies, and Institutions,* edited by Tony Bennett et al., London and New York: Routledge, 1993

Walker, R.R., *The Magic Spark: The Story of the First Fifty Years of Radio in Australia,* Melbourne: Hawthorne Press, 1973

Whitford, Irma, *Peaking on Zero: First Be Competent, Then Be a Star: A Handbook on Radio Production,* Perth, Western Australia: Murdoch University, 1991

Australian Aboriginal Radio

Over the past 15 years the Australian radio broadcast system has begun to serve the specific needs of the Aboriginal audience. By the turn of the century, progress was being made on several levels to increase the availability and variety of services offered to the nation's native minority.

Indigenous broadcasters form a unique segment of Australian Broadcasting. The Aboriginal people, through their own broadcasting services, produce programs in their own languages that enhance and preserve their culture. The Aborigines comprise only 2 percent of a total population of 17 million; half live in the coastal urban areas and the remainder in rural, more traditional communities.

ABC's Indigenous Broadcast Unit

In the early 1980s, ABC began making time available to Aborigines over some of its larger urban stations. In 1985 ABC provided three hours a day for Aboriginal groups in the northern part of Queensland. Training and support were provided by ABC, but the Aboriginal staff had full control over content. In 1986 ABC began to share a 50,000-watt shortwave transmitter with 8 KIN. This improved 8 KIN's reach across the Northern Territory, but no one knew how many Aborigines had shortwave receivers (Browne, 1990).

ABC Local Radio is committed to Aboriginal Broadcasting in three ways. First, ABC is not a funding agency but provides professional advice in the development of indigenous media organizations. Second, in 1988 ABC agreed to work toward employment equity for Aborigines and Torres Strait Islanders. An employment target of 2 percent in a variety of positions across the company has been set for ABC by an agreement with the Department of Employment Education and Training. Third, ABC Radio carries *Speaking Out* and *Awaye*, programs produced by Aboriginal and Torres Strait Islanders for a general audience. ABC also purchases programs from independent indigenous production houses and also expects to develop new programs for in-house production.

Speaking Out deals exclusively with the culture of the Aboriginal people in Australia and the politics and issues that affect them. The show airs every Sunday night live for one hour. *Awaye* airs programs that deal with the art and culture of indigenous people; the program airs on Friday and is repeated on Sunday. To hear the theme music for *Awaye*, visit their website at www.abc.net.au/message/awaye.

8 KIN Programming

By the end of 1985, the first exclusively Aboriginal station, 8 KIN, a 50,000-watt noncommercial FM station, came on the air in Alice Springs in Central Australia. Eventually, relays were added in Ntaria, Ali Curung, and St. Teresa. Until 1992, the Central Australian Aboriginal Media Association (CAAMA) operated this, the only licensed Aboriginal community broadcasting service. By June 1994, six additional licensed stations were operating for Aborigines and Torres Strait Islanders. Additional Aboriginal media associations, with community broadcasting licenses, operate in North-East Queensland, Brisbane, Perth, Darwin, and smaller stations in Western and Southern Australia. Programs on Aboriginal-owned community radio stations include news, sports, current affairs, Aboriginal music, talk back, Aboriginal oral stories, health, employment, housing, and land rights information. A special program goal is to reach both the young and the old. In addition to the seven licensed broadcasters, 11 other Aboriginal and Torres Strait Islander regional media groups produce radio programs for Australian Broadcasting Corporation (ABC) and other broadcasters. In 2001, the government provided A$3.1 million through the Community Broadcasting Foundation to facilitate radio access for indigenous Australians and ethnic communities.

Today, station 8 KIN FM broadcasts 18 hours a day and reaches a potential audience of 60,000 indigenous people from 25 major language groups. The station broadcasts in English and seven Aboriginal languages which are spoken in Central Australia. These programs account for 90 percent of the time on air or about 11 and one half hours per day. The amount of time in program categories varies from one language to another, but music accounts for 60 percent with spoken language programs comprising the remainder.

Programming includes news from wire services, local and national newspapers, and reports phoned into the station. There is a journalist coordinator and trainer and two Aboriginal journalist trainees, but much copy is translated from English. There are occasional documentaries and traditional stories. Music on the Aboriginal language program is about 75 percent Aboriginal, the rest European.

Satellite Services

Domestic satellite services became available in Australia in 1988 with the launch of AUSSAT, which made radio and television broadcasting available to remote parts of Australia for the first time. There has been debate since the early 1980s about the use of satellite by Aboriginal people as there was concern about the impact of satellite programming on traditional cultures. Two Aboriginal communities, Yuendemu in the Northern Territory (Warlpiri people) and the other in Ernabella in northern Southern Australia (the Pitjantnatjara people), were already producing video and radio. They lobbied for funding of their locally produced services, which at the time were being broadcast illegally on low power equipment (Buchtmann, 1999).

In 1984, the Department of Communication and the Department of Aboriginal Affairs recommended policies that would allow local Aboriginal people control over the new satellite service. This led to the establishment of Broadcasting in Remote Aboriginal Communities (BRACS). In many indigenous communities there were objections to programs by white people, and BRACS was considered a shield to protect their culture. Seventy-four communities were given equipment, but in many cases they did not know how to use it and were given no training. In some communities, there was no interest in learning or training and, in this situation, English programs were rebroadcast. Tribal elders in some communities were more active, and under their leadership BRACS assists in developing Aboriginal programming, especially for television.

MARY E. BEADLE

See also Developing Nations; Native American Radio

Selected Aboriginal Radio Stations as of 2002

Sydney-Koorie Radio (FM)
Brisbane-4 AAA (FM)
Port Augusta-5 UMA (FM)
Perth-6 AR (AM)
Kununurra-6 WR (AM)
Central Australia-8 KIN (FM)
Northern Australia-TEABBA (FM and Satellite)
Fitzroy Crossing-6FX (AM)

Further Reading

Browne, Donald R., "Aboriginal Radio in Australia: From Dreamtime to Prime Time?" *Journal of Communication* 40, no. 1 (1990)

Buchtmann, Lydia, "Digital Songlines: The Use of Modern Communication Technology by an Aboriginal Community in Remote Australia," <www.dca.gov.au/crf/paper99/lydia.html> (1999)

Burum, Ivo, director, *Satellite Dreaming* (video recording), Alice Springs, Northern Territory: Central Australian Aboriginal Media Association Productions, 1991

Central Australian Aboriginal Media Association website, <www.caama.com.au>

Indigenous Peoples of Australia: Media, <www.ldb.org/oz_m.htm>

Top End Aboriginal Bush Broadcasting Association website, <www.teabba.com.au>

Automation

Automation in radio refers to a method of broadcasting in which individually recorded program elements are reproduced in assigned order by equipment designed to operate with little or no human assistance. Automation systems were initially capable of performing two tasks routinely performed by disc jockeys: broadcasting music selections and commercial announcements. Refinements to technology subsequently enabled disc jockeys to record (or *voice-track*) their ad-libs in advance and to instruct systems to broadcast them at the appropriate times within the program schedule. The addition of voice-tracking capabilities assisted stations in suppressing criticism that automated broadcasting sounded "canned" in comparison with live, disc jockey supervised presentations.

Origins

Paul Schafer is credited with automating the studio operation function. After founding Schafer Electronics in 1953 to manufacture remote-control equipment for broadcast transmitters, Schafer extended the application of this technology in 1956 to program automation. Radio station KGEE used the first Schafer automation system to expand its hours of operation and offer nighttime service to its Bakersfield, California, listeners. Capable of unattended operation, Schafer's initial system consisted of two Seeburg jukeboxes for music reproduction, three Concertone open-reel tape decks for broadcasting commercial announcements, and an electromechanical switcher for storing programming instructions to control the playback sequence.

Schafer installed an automation system into a motor coach and traveled to radio stations to demonstrate its capabilities. During his visits with station owners and managers, Schafer explained how automation could reduce operating expenses while enhancing the quality of the on-air presentation. Automation systems, he asserted, not only reduced the number of personnel needed to operate the station but also executed programming instructions more reliably and with fewer mistakes than human operators could.

Schafer's identification of the economic and performance concerns were, in the late 1950s, of relatively equal importance to broadcasters. However, the economic incentive would escalate prominently in 1965 when the Federal Communications Commission (FCC) imposed the AM-FM Program Nonduplication Rule. In its effort to stimulate listener interest in the languishing FM service, the FCC reduced AM station simulcasting by their FM sister stations in the nation's 100 largest markets. A majority of the approximately 200 duopoly stations affected by the rule chose to automate their FM facilities rather than incur the expense of hiring additional announcing and engineering staffs. Contributing to the movement toward automation was management's realization that the pool of available talent had been depleted by rapid expansion in both radio and television broadcasting. Demand for qualified personnel had steadily driven up the price of labor during the 1950s. Duopolies in the smaller markets, unaffected by the nonduplication rule, were nonetheless enticed into automating when desirable local talent departed for more lucrative opportunities in metropolitan areas.

Equipment

The manufacture of automation equipment, previously a cottage industry, blossomed during the 1960s. The Radio Corporation of America (RCA) and Gates Radio Company, two prominent broadcast equipment manufacturers, entered into competition with Schafer Electronics. Unlike Schafer's initial design, which had relied upon jukeboxes for vinyl disc reproduction, next-generation systems reproduced program elements with open-reel and cartridge tape machines. This approach provided greater reliability than did the early disc-dependent systems. More important, it afforded programmers opportunities to execute formatics in a manner that more closely emulated the sound of a live, spontaneous broadcast.

As the sophistication of control switchers grew, tape-dependent systems enabled announcers to develop the technique of voice tracking. In approximately 10 to 20 minutes, announcers could record all of the comments they would normally make during the course of a four-hour, live program. By instructing the automation system to execute their voice recordings at the appropriate times, announcers were freed to pursue other creative activities during their airshifts.

Adapting a station to automated broadcasting did not, in and of itself, liberate operators from the responsibility of creating a program service; it merely facilitated the execution of programming decisions. Rather than attempting to produce tape-recorded libraries of music internally, numerous automated-station operators elected to subscribe to the services of program syndicators. In most instances, syndicators supplied stations with base libraries of recorded music, which were supplemented periodically with reels of music of current popularity.

Syndication

The syndication firm Drake-Chenault, under the supervision of legendary Top 40 programmer Bill Drake, successfully adapted the middle of the road, Top 40, country, and soul formats to automated presentation. By the mid-1970s, approximately 300 stations subscribed to the company's services. The beautiful music format was also used extensively by automated stations. Because its execution emphasized repeated segues between songs, this format was especially adaptable to automation.

Broadcaster reliance upon tape-based automation diminished during the 1980s. Among the explanations for the decline in automation's popularity was programmers' perception that listeners desired more announcer involvement within the presentation than even the most sophisticated systems could reliably provide. Although voice-tracked programming resembled live presentation, it nonetheless was incapable of emulating for listeners the spontaneity and interactivity they associated with hearing a live broadcast. An alternative approach, in which syndicators delivered live, hosted presentations simultaneously to multiple affiliates via satellite, became the preferred method of automated operation during this period.

Following the passage in 1996 of the Telecommunications Act, which sharply relaxed radio station ownership rules and led to significant consolidation of station properties, interest in automated broadcasting renewed. Owners who operated multiple stations within markets began to cluster the facilities into single studio complexes. In such situations, automation has enabled owners to share personalities among stations, thereby reducing the number of announcers required to sustain program operations. Systems now store all program elements, including voice-tracked disc jockey commentary, in hard-disk memory. These "jock-in-a-box" systems offer disc jockeys greater voice-tracking flexibility and have narrowed the gap between listeners' perceptions of live and recorded presentation. Clear Channel Communications was making use of this technique in the early 2000s, noting that a small station (say, in Boise, Idaho) could gain the use of an on-air personality for as little as $4,000 to $6,000 a year, far less than paying a real DJ on site. The DJ stays in a major market, yet appears to host local programs in multiple locations. Another trend developing among group-owned stations is to interconnect stations in multiple cities via telephone circuits to a central, or hub, production facility. A single announcing staff is thus able to provide each of the stations within the hub with individualized, market-specific commentary using the voice-tracking technique. Industry representatives estimate that approximately one in five stations now employs systems for either fully automated or announcer-assisted operation.

BRUCE MIMS

See also Drake, Bill; FM Radio; Recording and Studio Equipment; Syndication; Telecommunications Act of 1996

Further Reading

Abrams, Earl B., "Schafer Offers Stations Device for Automatic Radio Programming," *Broadcasting* (17 March 1958)

Abrams, Earl B., "Automated Radio: It's Alive and Prospering," *Broadcasting* (9 June 1969)

Antilla, Susan, "Canned Radio Means Fresh Profits," *Dun's Review* 117 (March 1981)

Bennett, Jeffrey, "RCS NT Number One for WONE," *Radio World* (25 June 1997)

Gentry, Ric, "Profile: Paul Schafer," *Broadcast Management/Engineering* 23 (April 1988)

Keith, Michael C., and Joseph M. Krause, *The Radio Station*, Boston: Focal Press, 1986; 5th edition, by Keith, 2000

Pizzi, Skip, "New Directions in Radio Automation," *Broadcast Management/Engineering* 24 (February 1989)

Routt, Edd, James B. McGrath, and Fredric A. Weiss, *The Radio Format Conundrum*, New York: Hastings House, 1978

Automobile Radios

First designed in the 1920s as separate radios to be installed optionally in automobiles, the auto or car radio eventually became a standard feature, flourishing in popularity after World War II. Today the auto radio is considered standard equipment on virtually all makes and models of cars.

Origins

Perhaps it was inevitable that two of the 20th century's most popular products—cars and radios—would unite in some way, but numerous technological barriers initially prevented such a marriage. The earliest known radio equipped car was demonstrated in St. Louis, Missouri, in 1904. In 1922 Chevrolet offered the United States' first factory unit, the Chevrolet Radio Sedan, featuring a modified Westinghouse radio with an elaborate fence-like antenna mounted on the roof. This early option alone cost one-third the price of the car itself. Few were purchased, so Chevrolet discontinued their manufacture.

Radio experimenters continued to fend for themselves. Some adapted portable, battery-powered units for use as travel radios, but interference from the automobile's engine and ignition system thwarted their widespread development. The do-it-yourselfers discovered other obstacles, including exposure to extreme heat and cold, incompatible power supplies from automobiles, and loudspeakers too weak to overcome the noise caused by driving.

Gradually these problems were resolved on several fronts and, like the technical development of the radio itself, auto radio evolved from a combination of discoveries over a period of time rather than in the immediate aftermath of a single breakthrough. Ignition interference, for example, was reduced substantially in 1927 with the invention of the spark plug suppresser, or damp resistance. Within three years, a second development, voltage conversion, allowed simplified operation of the auto radio. Enabling the auto radio to operate independently of a battery system, relying instead upon the car's electrical system, was key to the device's commercial success. By the mid-1930s widespread manufacturing was underway.

This second development, in 1930, was the result of collaborative work between Paul Galvin, Elmer Wavering, and William Lear, who later gained prominence in aviation design. At the time, Galvin manufactured battery eliminators in Chicago, products that permitted battery-operated radios to be run by the 120-volt system of household current. Galvin applied the same principle to his Studebaker automobile, demonstrating it to passersby at a meeting of the Radio Manufacturer's Association in Atlantic City, New Jersey. His first production model, the 5T71, sold for $120 (installed) and became known as a Motorola, a combination of the words *motor* and *Victrola*. That year Chrysler offered a radio as a regular option and was followed soon by other automakers. The price of an average auto radio dropped to $80 by the mid-1930s, one-eighth the cost of a typical car, as opposed to the one-third cost of 10 years earlier.

In 1933 Ford introduced auto radios compatible with specially designed dashboard panels in some of its models. By 1936 the push-button feature was added, enabling motorists to tune to a desired station safely, without glancing away from the road while driving. The adjustable, telescopic "whip" antenna, which improved reception of distant stations, was introduced in 1938.

After World War II, the popularity of the auto radio soared when automakers began to offer them as a preinstalled option. By the late 1950s, Motorola, by now the name of Galvin's company, manufactured one-third of all U.S. auto radios.

Postwar Growth

The development of the auto radio reflected the increasing mobility of the United States itself, but television's arrival in U.S. homes after World War II was swift and nearly complete by the end of the 1950s.

Many radio executives assumed the worst: their medium was dead or dying. In one sense, this was true. The amount of advertising revenue generated by network radio shows dropped by half in the five-year period ending in 1955. Such long-time popular shows as *Amos 'n' Andy*, Jack Benny, and *Ma Perkins* either moved to television or disappeared completely. Yet a paradox was emerging; despite the precipitous decline of network radio, more radio stations than ever were on the air (about 2,300 in the United States by the mid-1950s, nearly triple the number immediately before World War II) and more radios were being manufactured than ever. Surveys suggested that TV set owners were more likely to own more than one radio than were non–TV set owners and that radio listeners were tuning in more frequently, to different stations, for shorter durations of time. Television, it appeared, was threatening network radio, but not the medium of radio itself.

"Radio didn't die. It wasn't even sick," said Matthew J. Culligan, National Broadcasting Company's vice president for radio, in 1958.

> It just had to be psychoanalyzed. . . . The public just started liking [radio] in a different way, and radio went to the beach, to the park, the patio, and the automobile. . . . Radio has become a companion to the individual instead of remaining a focal point of all family entertainment. An intimacy has developed between radio and the individual.

The "disc jockey" and "drive time" had arrived. Influenced by the auto radio and its motorist listeners, radio became a predominantly music and news service after about 1955.

The new portability of radio was made possible, too, by the development of the transistor in 1948 and by the phenomenal growth of auto radios. By 1953 the first transistorized pocket-size portable radios were available, and five years later the first solid-state (tubeless) auto radios using transistors appeared on the market. In 1952 auto radios were in just over half of America's cars. By 1980 that figure reached 95 percent. The trend reflected, as J. Fred MacDonald put it, "a mobile, affluent, and commercialized America, solidly committed to television for its creative amusement, but still requiring radio for music and instantaneous information."

The proliferation of auto radios (by 1962 some 47 million cars in the United States were equipped) also coincided with the emergence of a distinct "teen culture" in the United States following World War II. The auto radio helped to promote the growth of this emerging youth culture and, like the drive-in restaurant and drive-in theater, it came to symbolize America's new level of mobility. At the same time, auto radio offered more stations. FM had been available, although not widely purchased, in car radios since the mid-1950s. With the medium's growth in popularity, sales of FM-equipped auto radios soared and prices dropped.

Radio in the years since has continually readjusted its approach to programming to meet changing audience needs, and the pervasiveness of the auto radio has remained high. Its technology has evolved in the same way consumer electronics have changed in the home. Digital audio broadcasting, introduced to the consumer in the 1990s, has spawned a new generation of auto radios. Radio continues to attract its largest audiences during commuting hours, a trend well established by the 1960s and continuing 40 years later, thanks to the near universal availability of the auto radio.

DAVID MCCARTNEY

See also Motorola

Further Reading

An Analysis of Radio-Listening in Autos, New York: Columbia Broadcasting System, 1936

Martin, Norman, "Turn Your Radio On," *Automotive Industries* 178, no. 4 (April 1998)

Master Key to Auto Radio, New York: Radio Advertising Bureau, 1962

Petrakis, Harry Mark, *The Founder's Touch: The Life of Paul Galvin of Motorola,* New York: McGraw-Hill, 1965; 3rd edition, Chicago: Motorola University Press/Ferguson, 1991

"Radio Boom," *Time* (27 April 1936)

Radio Takes to the Road, New York: National Broadcasting Company, 1936

Schiffer, Michael B., *The Portable Radio in American Life,* Tucson: University of Arizona Press, 1991

Autry, Gene 1907–1998

U.S. Radio Star and Station Owner

Born on a cattle farm near Tioga, Texas, son of a livestock dealer and horse trader, Gene Autry really was raised as a western cowboy, a role he lived on radio and in many films and television shows. Although his early dream was to be a telegraph operator and work on the railroad, Autry loved music. His preacher grandfather had taught him to sing at age 5, putting him in his church's choir, and Autry saved up to buy his first guitar at the age of 12. By the age of 15 he was singing and playing his guitar all over town. Will Rogers heard him sing as a teenager and encouraged him to pursue a radio career singing music. Autry didn't rush right into radio, but he did travel with the *Fields Brothers' Marvelous Medicine Show,* playing as the lead-in act to attract the audience so that "Professor" Fields could sell his patent medicines. He earned a hearty $15 a week for his work.

Radio Years

Autry was a teenager when radio came to Tioga. He taught himself to make his own crystal set from reading magazine articles. After he had saved up some money, he decided to take Will Rogers's advice and look into radio for his chance to sing professionally. He figured the best way would be to first cut a record, which would be his ticket through the door of radio stations. After a month in New York, he was finally given the chance to cut a test for Victor records.

He was told that he had a good voice but needed some seasoning in front of a microphone. The Victor producers recommended he spend some time on the radio and gave him a letter of recommendation to take to radio stations, something that was probably unnecessary given that many rural radio stations would let almost anyone on the air in the 1920s. Autry used the letter to land his first radio spot, a daily 15-minute show on KVOO in Tulsa, Oklahoma, in 1929. He was known as the Oklahoma Yodeling Cowboy. He worked for the railroad by day, because the radio job was unpaid, and used his on-air publicity to land singing engagements wherever he could find them. During that time, he also began to write his own music.

After six months, Autry returned to New York in the fall of 1929 and cut records for Victor; Columbia's Velvatone label; and the Conqueror label of American Record Corporation, with whom he signed an exclusive contract. As a result of these record connections, he was later invited to appear on Rudy Vallee's *Fleischmann Yeast Hour,* one of his first network engagements. He was also invited to be on *National Barn Dance* on Chicago station WLS. *National Barn Dance* was one of the most important venues for early country music. Other performers who frequented the WLS "Cornstalk Studio" from which the show originated included Pat Buttram, Lulubelle and Scotty, The Prairie Ramblers, Patsy Montana, the Cumberland Ridgerunners, and Jolly Joe Kelly. They performed what was known as "hillbilly" music in the 1920s and 1930s but which later became known as "country and western," due in part to the likes of Autry and other "singing cowboys."

These early appearances led to Autry's becoming a regular on the *National Barn Dance,* where he was paid $35 a week for performing on-air and touring with the show. The show would tour county fairs and the like during the week and return to Chicago for the Saturday night radio show, which was performed in front of a live audience of around 1,200. The cast would do two of these live radio shows in a night. One hour of the show was aired on the National Broadcasting Company (NBC). Autry had by 1931 become a WLS regular. His role included singing on the radio program and touring with the company for performances. Sears and Roebuck also sponsored him for a show using his own name, the *Gene Autry Program,* and he was a guest artist on other shows, including the *National Farm and Home Hour.*

Recording Success

Integral to Autry's network success was his recording success. One of his early recordings, "That Silver-Haired Daddy of Mine," became his first hit (he had written it with a coworker on the railroad). After it was recorded for distribution by the Sears mail-order catalogue, it sold 30,000 copies the first month. By the end of a year, it had sold half a million copies. The head of his record company and his press agent together had a gold-plated copy of the record made. When sales topped 1 million, they gave him a second gold record. In his autobiography, *Back in the Saddle Again,* Autry says that this was the start of the tradition of giving gold records for sales of 500,000 copies. The record sold over 5 million copies by 1940.

Other songs that made Autry a top-selling country star included "Tumbling Tumbleweed," "Back in the Saddle Again," "Mexicali Rose," "Peter Cottontail," "Here Comes Santa Claus," and "South of the Border." The song "Rudolph the Red-Nosed Reindeer" took Autry to the top of the pop charts for the first time when it was released in 1949. It is one of the best-selling singles of all time. Overall, Autry recorded more than 600 songs, some 200 of which he wrote or co-wrote.

Managing the Show

Keeping a band together was difficult in the early 1930s; performers regularly left to pursue other opportunities. This meant regularly searching for talent to replace those who moved on. In one instance, Autry's booking agent, J.L. Frank, was traveling in central Illinois in 1933 when he heard a man by the name of Lester "Smiley" Burnett sing on radio station WDZ out of Tuscola, Illinois. He told Autry that Burnett would be a good addition to the show for his personal appearances. Autry called Burnett at the radio station and offered him the job sight unseen. Burnett responded, "I'm getting 18 dollars a week and getting it regular," a modest salary, but one to be thankful for in the early Depression years. Autry offered him $35 dollars a week, and Burnett accepted. Burnett would go on to back up Autry on radio, co-write songs with him, and then star in 60 movies as Autry's sidekick.

Although few band members stayed with Autry through the years, similar stories show how he often went about finding them: once, Autry saw a man hitchhiking and carrying a guitar. He pulled over and asked the fellow to play for him. He didn't hire the hitchhiker, but he got the name of "the best fiddle player in the county" from the man and hired the fiddle player.

While Autry was at WLS, the station was sold to a national network, further advancing Autry's fame beyond Chicago to other cities around the country. Autry was still at WLS in 1934 when he began his film-acting career in earnest by starring in a series of westerns. After making a few films in Hollywood, he returned to Chicago during a winter storm. Suddenly, he realized that California looked much more appealing, and when another offer for a movie came in 1935, he left Chicago for good. He made eight films a year from 1935 to 1942. By 1953, when he starred in his last movie, he had been in 93 films.

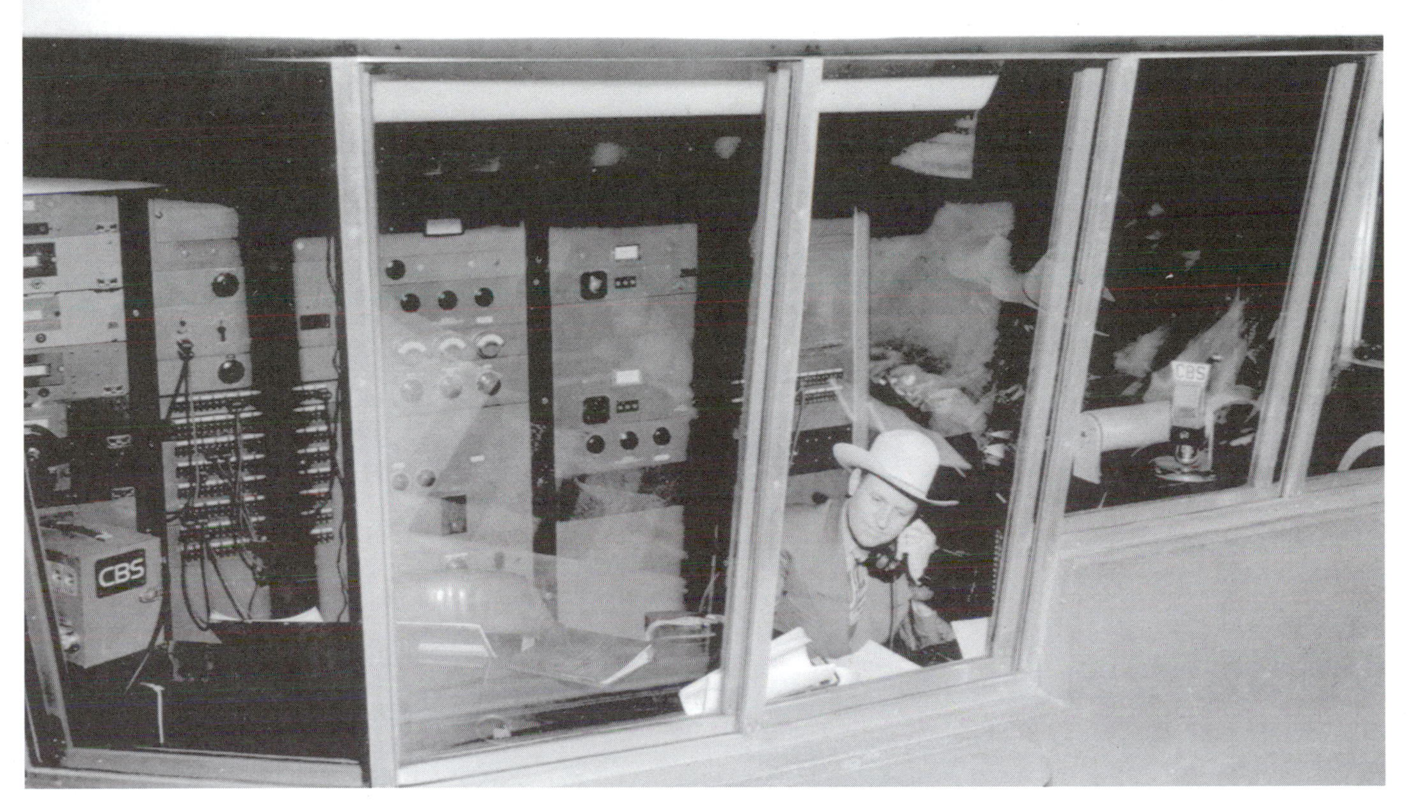

Gene Autry in control room on the *Gene Autry Show*
Courtesy CBS Photo Archive

Autry's *Melody Ranch*

In 1940 Autry made a movie called *Melody Ranch*. The Columbia Broadcasting System (CBS) then gave him a new radio show called *Gene Autry's Melody Ranch*. The program would last from 1940 to 1956, interrupted only by Autry's military service during World War II. The program included a mix of barn dance, vaudeville, Saturday matinee, and medicine show; there was even a role for his movie co-star, his horse Champion. Autry toured extensively with the show, broadcasting it from many of the small towns and rural locations where his country-style music had its greatest appeal.

In 1941, the second year of the show, *Melody Ranch* received an invitation to perform in the small town of Berwyn, Oklahoma. The town's 227 residents invited Autry because they had decided to change the town's name to Gene Autry, Oklahoma. The show performed on a flatbed railroad car to a live audience of 35,000 people who had come to the small town that day. This was in addition to the CBS radio audience and a makeshift network audience put together over several Oklahoma radio stations that carried an extended version of the broadcast.

About a month later, Autry was set to start the show at the CBS studios in Hollywood when, instead of the opening theme, a special news bulletin was announced from New York. The report gave the details of the Japanese attack on Pearl Harbor that drew the United States into World War II. The show went on as usual that night after the news bulletin, but seven months later Gene Autry enlisted in the U.S. Army Air Corps during a live *Melody Ranch* show, one of several efforts he made to promote the war effort.

Autry decided to enlist, knowing that he was ripe for the draft anyway. In July 1942 he enlisted as a G.I., cutting his pay from $600,000 the year before to $2,000 in his first year in uniform. He initially worked to entertain many of the troops and even performed many of his *Melody Ranch* shows from the base while stationed in Phoenix, Arizona. He eventually made it into flight school and became a pilot in the Air Transport Command, flying in the Pacific theater of the war. Once the war ended, he returned to the *Melody Ranch* program and to the movies.

Guests on other episodes of *Gene Autry's Melody Ranch* included First Lady Eleanor Roosevelt and a baby boy named Franklin Delano Gene Autry Johnston after both the president of the United States and the host of the radio show. One group of back-up musicians who played regularly for Autry's radio show was the Jimmy Wakeley Trio, a group that also appeared in one movie with Autry, *Heart of the Rio Grande*.

The show, sponsored by Wrigley's Doublemint gum, sometimes promoted the spearmint version of that product. The company had assured Autry that they didn't have to worry about ratings as long as the program sold gum. Nevertheless, after the show had reached the 10-year mark, Wrigley's hired additional writers, singers, actors, orchestra members, and a publicity agent—all in an effort to stem any defection of audience to the upcoming medium of television.

Because Autry saw the trends moving toward television, 1950 also marked the beginning of his own television series, *The Gene Autry Show*. The show continued for the duration of Autry's *Melody Ranch* radio program, also ending in 1956. By the time it ended, about 100 episodes of the television program had been made. Autry began to withdraw from personally performing in show business in the late 1950s. He had already made his last film, and in 1956 he ended his regular radio and television shows. He still recorded music regularly until 1962, but after that he concentrated on managing a growing entertainment corporation. He recorded his last two songs in 1964. Autry's autobiography, titled after one of his most popular films, *Back in the Saddle Again*, was published in 1978.

Ownership and Awards

By 1950 Autry was also a radio station owner, having purchased his first property a few years earlier, during World War II, when he was stationed near Phoenix. This Phoenix station, KPHO-AM, became the cornerstone of a corporate entertainment company that he called Golden West Broadcasting. Before incorporating, however, Autry ran the station with a partner, Tom Chauncey, who continued to run the station when Autry was transferred away. The partners also bought a radio station in Tucson. Soon Autry moved into television ownership also when he obtained a broadcast license from the Federal Communications Commission for KOOL in Phoenix. Autry owned KOOL-TV as well as KOOL radio interests until the early 1980s, when he sold them for $35 million.

In 1952 Autry purchased 56 percent of KMPC-AM radio in Los Angeles, which became the flagship station for Golden West Broadcasting. The corporation soon bought stations up and down the Pacific coast in San Francisco, Seattle, and Portland. The adult alternative KSCA-FM in Los Angeles was added to the group later. Golden West owned television stations as well, including KTLA in Los Angeles, bought from Paramount for $12 million in 1964. In 1985, Autry's company sold KTLA for $245 million, the highest price paid for a television station to that time.

Autry is the only entertainer to have five stars on the Hollywood walk of fame—one each for recordings, television, film, radio, and theater. Autry received many other awards, including the Songwriters Guild Life Achievement Award and the Hubert H. Humphrey Humanitarian of the Year award. He has been inducted into the National Cowboy Hall of Fame and the Western Music Association Hall of Fame, among others.

STEPHEN D. PERRY

See also Country Music Format; National Barn Dance

Gene Autrey. Born in Tioga, Texas, 29 September 1907. Served in U.S. Army Air Corp, 1942–45; started radio career as Oklahoma Yodeling Cowboy on KVOO, 1929; recorded first record, 1929; joined station WLS with *National Barn Dance*, 1931; wrote and recorded first hit song, "That Silver-Haired Daddy of Mine," 1931; first film appearance, *In Old Santa Fe*, 1934; wrote theme song, "Back in the Saddle Again," 1939; starred in own radio program, *Gene Autry's Melody Ranch*, 1940; bought first radio station, 1942; first television staring role in *The Gene Autry Show*, 1950; known for writing and recording "Here Comes Santa Claus" (1947), "Rudolph the Red-Nosed Reindeer" (1949), "Frosty the Snowman" (1950), and "Peter Cottontail" (1950); incorporated Golden West Broadcasting, 1952; ceased regular TV and radio performances, 1956; recorded last song, 1964; bought the California Angels baseball team, 1961. Died in Studio City, California, 2 October 1998.

Radio Series

1929–30	*Oklahoma Yodeling Cowboy*
1931–34	*National Barn Dance*
1930–34	*Gene Autry Program*
1940–56	*Gene Autry's Melody Ranch*

Television

The Gene Autry Show, 1950–56

Films

In Old Santa Fe, 1934; *Mystery Mountain,* 1934; *The Phantom Empire,* 1935; *Tumbling Tumbleweeds,* 1935; *Melody Trail,* 1935; *Sagebrush Troubador,* 1935; *The Singing Vagabond,* 1935; *Red River Valley,* 1936, *Comin' Round the Mountain,* 1936; *The Singing Cowboy,* 1936; *Guns and Guitars,* 1936; *Oh, Susanna!* 1936; *Ride, Ranger, Ride,* 1936; *The Big Show,* 1936; *The Old Corral,* 1936; *Git Along, Little Dogies,* 1937; *Round-Up Time in Texas,* 1937; *Yodelin' Kid from Pine Ridge,* 1937; *Manhattan Merry-Go-Round,* 1937; *Springtime in the Rockies,* 1937; *Rootin' Tootin' Rhythm,* 1937; *Public Cowboy No. 1,* 1937; *Boots and Saddles,* 1937; *Man from Music Mountain,* 1938; *Prairie Moon,* 1938; *Western Jamboree,* 1938; *Rhythm of the Saddle,* 1938; *The Old Barn Dance,* 1938; *Gold Mine in the Sky,* 1938; *Home on the Prairie,* 1939; *Mountain Rhythm,* 1939; *South of the Border,* 1939; *Rovin' Tumbleweeds,* 1939; *Mexicali Rose,* 1939; *In Old Monterey,* 1939; *Colorado Sunset,* 1939; *Blue Montana Skies,* 1939; *Men with Steel Faces,* 1940; *Gaucho Serenade,* 1940; *Ride, Tenderfoot, Ride,* 1940; *Melody Ranch,* 1940; *Rodeo Dough,* 1940; *Unusual Occupation,* 1940; *Shooting High,* 1940; *Rancho Grande,* 1940; *Carolina Moon,* 1940; *Meet Roy Rogers,* 1941; *Under Fiesta Stars,* 1941; *Sunset in Wyoming,* 1941; *The Singing Hill,* 1941; *Sierra Sue,* 1941; *Ridin' on a Rainbow,* 1941; *Down Mexico Way,* 1941; *Back in the Saddle,* 1941; *Stardust on the Sage,* 1942; *Home in Wyomin',* 1942; *Heart of the Rio Grande,* 1942; *Cowboy Serenade,* 1942; *Call of the Canyon,* 1942; *Bells of Capistrano,* 1942; *Sioux City Sue,* 1946; *Twilight on the Rio Grande,* 1947; *Trail to San Antone,* 1947; *Saddle Pals,* 1947; *Robin Hood of Texas,* 1947; *The Last Round Up,* 1947; *The Strawberry Roan,* 1948; *Loaded Pistols,* 1948; *The Cowboy and the Indians,* 1949; *Sons of New Mexico,* 1949; *Rim of the Canyon,* 1949; *Riders of the Whistling Pines,* 1949; *Riders in the Sky,* 1949; *The Big Sombrero,* 1949; *Mule Train,* 1950; *Screen Actors,* 1950; *Indian Territory,* 1950; *Cow Town,* 1950; *The Blazing Sun,* 1950; *Beyond the Purple Hills,* 1950; *Gene Autry and the Mounties,* 1951; *Texans Never Cry,* 1951; *Silver Canyon,* 1951; *Valley of Fire,* 1951; *Whirl Wind,* 1951; *The Hills of Utah,* 1951; *Night Stage to Galveston,* 1952; *Apache Country,* 1952; *Wagon Team,* 1952; *Blue Canadian Rockies,* 1952; *The Old West,* 1952; *Barbed-Wire,* 1952; *On Top of Old Smoky,* 1953; *Winning of the West,* 1953; *Saginaw Trail,* 1953; *Pack Train,* 1953; *Last of the Pony Riders,* 1953; *Goldtown Ghost Riders,* 1953; *Hollywood Bronc Busters,* 1956; *Alias Jesse James,* 1959; *Silent Treatment,* 1968; *Gene Autry, Melody of the West,* 1994

Stage

Fields Brothers Marvelous Medicine Show, 1924

Selected Publications

Rhymes of the Range, 1933
The Art of Writing Songs and How to Play a Guitar, 1933
Western Stories, 1947
Back in the Saddle Again (with Mickey Herskowitz), 1978

Further Reading

Fletcher, Steffi, *Gene Autry,* New York: Simon and Schuster, 1956

George-Warren, Holly, "Pop Music: The Singing Cowboy As Everlasting Everyman," *New York Times* (25 January 1998)

Green, Douglas B., *Singing in the Saddle: The History of the Singing Cowboy,* Nashville, Tennessee: Country Music Foundation and Vanderbilt University Press, 2002

Awards and Prizes

As with other American media, the radio broadcasting industry awards itself (and receives from others) a host of annual prizes as a prime means of recognizing key people, programs, and top performances in the business. This entry briefly describes a selection of the longer running and better known national awards available to radio broadcasters. Most are given annually, and many are open to people in both radio and television (and sometimes other media). The means of choosing winners varies tremendously across both prizes and organizations.

Armstrong Awards

Often called "Majors" in honor of inventor Edwin Howard Armstrong's World War I army rank, these awards, established in 1964, are given in recognition of excellence and originality in radio broadcasting in six content categories: news, music, news documentary, education, community service, and creative use of the medium. Additional awards may be given in the areas of technology, innovation in station programming, and outstanding service by an individual or company. Armstrong Foundation, Columbia University, New York.

Clarion Awards

Given in more than 80 categories, including radio, for people and programs concerning women in society. Association for Women in Communications, Arnold, Maryland.

Clio Awards

Given in honor of the best domestic and international advertising, these awards have been given since 1960. They include radio among other forms of advertising. More than 200 are given annually. Clio Awards, New York.

Crystal Radio Awards

Given to as many as ten radio stations per year for overall excellence in community service. National Association of Broadcasters, Washington, D.C.

Edward R. Murrow Awards

(1) Given for outstanding individual contribution to public radio. Corporation for Public Broadcasting, Washington, D.C. (2) Given for the best news department, spot news coverage, continuing coverage, investigative reporting, news series and documentary, and for overall excellence. Radio Television News Directors Association, Washington, D.C.

Freedom Foundation National Awards

Given for constructive activities on drug abuse education, ecology, patriotic programs, respect for the law, moral and spiritual values, economic education, human dignity and brotherhood. Freedom Foundation, Valley Forge, Pennsylvania.

Gabriel Awards

Given for radio programs or segments in the following categories: arts/entertainment, news/information, religion, coverage of single news story, community awareness campaigns, public service announcements, and short features. National Catholic Association of Broadcasters and Communicators, Dayton, Ohio.

George Polk Award

These vary from year to year but generally are given to honor discernment in a news story or coverage, resourcefulness in gathering information, or skill in relating the story. These are usually given for foreign, national, and local achievements. Long Island University, New York.

Gold Medal Award

Presented each year to an outstanding individual or corporate entity in radio or other electronic media. International Radio and Television Society, New York.

Golden Mike Award

Awarded to a company or an individual that has made an outstanding contribution to the art of broadcasting and the community at large. Broadcast Foundation, Greenwich, Connecticut.

International Broadcasting Award

Given to "the world's best" radio and television commercials from anywhere in the world. There are nine radio categories, and subject matter for the advertisement is open. Hollywood Radio and Television Society, Hollywood, California.

Jack R. Howard Awards

Given in honor of investigative or in-depth reporting. Scripps-Howard Foundation, Greencastle, Indiana.

Marconi Radio Awards

Given to stations or on-air personalities for excellence in and contributions to radio. National Association of Broadcasters, Washington, D.C.

Missouri Honor Medals

Given since 1930 in honor of lifetime achievement by the School of Journalism, University of Missouri, Columbia, Missouri.

National Headliner Awards

For radio network and individual station news, public service, documentary, and investigative reporting. National Headliners Club, Northfield, New Jersey.

National Radio Award

Given to an individual for significant or ongoing contributions to radio from a leadership position. National Association of Broadcasters, Washington, D.C.

New York Festivals Award

Encompasses radio advertising, programming, promotion, news, entertainment, editorials, service features, and public service announcements, among others. International Radio Program and Promotion Awards of New York, New York.

Overseas Press Club Awards

Given annually for radio spot news from abroad, or radio interpretation of foreign news. Overseas Press Club, New York.

Public Radio Program Award

Recognizes excellence in radio programming at the local or national level. Corporation for Public Broadcasting, Washington, D.C.

Radio-Mercury Awards

Given to honor creative excellence in paid radio advertising, these carry some of the largest cash prizes of any of the awards in this entry—upward of $100,000 for the top winner. Radio-Mercury, New York.

Radio Program Awards

To recognize outstanding programming for community-oriented radio, commercial or public. National Federation of Community Broadcasters, Washington, D.C.

Radio Wayne Awards

Given annually for the top account executive, sales manager, general manager, director of sales, and broadcaster of the year. They are named in honor of the late Wayne Cornils, a senior vice president of the Radio Advertising Bureau. Radio Ink, Miami, Florida.

Sigma Delta Chi Distinguished Service Award

Given for radio editorials or radio reporting. Society for Professional Journalists, Greencastle, Indiana.

Silver Baton Award

Given for outstanding work in news and public affairs during the previous year, covering both network and local radio. Can be given to an individual, program, series of programs, or a station. Alfred I. DuPont-Columbia University Awards, Graduate School of Journalism, Columbia University, New York.

CHRISTOPHER H. STERLING

See also Peabody Awards; Radio Hall of Fame

Further Reading

"Major Broadcasting and Cable Awards," *Broadcasting and Cable Yearbook* (annual listing)

Axis Sally (Mildred Gillars) 1900–1988

U.S. Shortwave Propagandist for Nazi Germany

Axis Sally's "This is Germany Calling" shortwave radio signature heralded one of the most notorious anti-Allied propaganda broadcasts of World War II. Her sexy voice and clever mix of musical entertainment with morale-shaking commentaries and dramatic sketches were familiar to GIs in every zone of the European theater of operations.

Mildred Gillars began her radio work in 1940, after efforts to secure a job as an actress in the United States failed and she moved to Berlin. With the outbreak of World War II, radio propaganda became an important instrument in Germany's psychological warfare campaign. When the Propaganda Ministry began expansion of the *Europäische Fremdsprachendienste* (European Foreign Language Service), Gillars was offered a position as an announcer and commentator for the British zone. In the summer of 1940, during the Battle of Britain, Gillars blanketed the island with anti-British propaganda from the Reich's Bremen shortwave station. In her *Bremen Sender* program, she commented on the course of the air battle (magnifying German victories) and entertained listeners with the light jazz of "Charlie and His Orchestra." On one occasion, Gillars interviewed captured British fliers at the Dulag Nord transit camp.

When British and German forces clashed in North Africa in early 1941, Gillars transferred to the *Europäische Fremdsprachendienste's* southern broadcasting zone and began transmitting her *Anzac Tattoo: From the Enemy to the Enemy* program from the 100-kilowatt medium wave unit at Graz-Doble, Austria. The Commonwealth troops of General Wavell's Eighth Army tuned in every Saturday night to this 30-minute offering, ignoring the anti-Allied remarks in preference to the musical numbers and Gillars's seductive "bedroom voice."

When the tide of battle began to shift adversely for Germany following the November 1942 Allied invasion of North Africa, the propagandistic element in Gillars's broadcasts became more prominent and more malignant. For the 65,000 U.S. First Army troops who listened to her Thursday night *Home Sweet Home* program, "Axis Sally," as the GI's now dubbed her, presented a warm picture of recently abandoned family life and added, "You'll get back to all that when the war is over, if you're still alive." Sally also made soldiers uneasy by emphasizing their pawnlike status. "How did I let myself get roped into Churchill's and Roosevelt's war business?" she had one character remark. "After all, God can save the King. Americans don't need to bother about him." Axis Sally's broadcasts constantly exploited GI lack of training and fear of imminent death. Prior to one engagement she warned, "You poor, silly, dumb lambs, on your way to be slaughtered." Personal resentments against officers, war profiteers, scheming politicians, and military discipline were also exploited. Her most effective device for undermining morale, though, was her shrewd manipulation of the GI's romantic anxieties. In a 1943 broadcast, Sally queried, "And what are your girls doing tonight, fellows? You really can't blame them for going out to have some fun, could you? It's all so empty back there now. Better to go out for some drinks with that 4-F boyfriend than sit and wait forever, doing nothing . . . especially if you boys get all mutilated and do not return in one piece." This was followed by renditions of "Can I Forget You" and "Somebody Stole My Gal."

Another disturbing feature of Axis Sally's broadcasts was her apparent ubiquity. GIs of the U.S. Sixth Corps waiting in transports off the coast of Salerno in 1943 heard her warn, "Thousands and thousands and thousands of you men going from . . . Sicily to Europe are on your *last round-up*," followed by the song of the same name. Listeners aboard one U.S. merchantman were astonished when, within 90 minutes of their sailing past Gibraltar, Axis Sally was on the air describing the convoy in detail—number and type of ships, cargo, and so on. Sally's messages were frequently beamed across the Atlantic. "You women in America waiting for the one you love. Waiting and weeping in the safety of your own rooms . . . Thinking of the son, husband, brother who is being sacrificed . . . Perishing on the fringes of Europe . . . Losing their lives, at best coming back crippled."

In mid-1943, Sally expanded her on-air operations when she and Otto Koischwitz (another influential North America service propagandist, with whom she had a long-running affair and who may have gotten her into the radio business in the first place) began originating broadcasts from Allied prisoner-of-war camps in Germany. After presenting herself as an International Red Cross representative, she allowed GIs to transmit 25-word messages to their folks back home. Although she would later claim a humanitarian motive, these *Medical Reports* were pure propaganda. Every broadcast dwelt on the idyllic character of camp life—with obliging guards and tanned and cake-eating inmates.

In the fall of 1943, Sally followed Allied forces on their invasion of the Italian mainland. Her *Midge at the Mike* program was immensely popular with GIs, despite its constant drumming of themes like homefront betrayal and violent death. Sally's sexy tones and the lively vibes featuring America's big band artists brought warmth to many remote hillside foxholes. She ended each broadcast with "That's all boys and a kiss from Sally." In February 1944, German aircraft dropped program schedules along the entire Allied line of advance.

By spring 1944, Axis Sally was the highest-paid foreign language broadcaster in the Third Reich, and her notoriety encouraged several imitators in the European theater. With the increasing artistic freedom that accompanied her rising status, she helped produce one of the best-known German propaganda dramas of World War II, *Vision of Invasion*. Broadcast on 12 May 1944 to the American homefront and to U.S. troops gathering in southeastern England for D day (three weeks before the actual operation), *Vision of Invasion* was a warning of the catastrophe awaiting Allied forces when they disembarked onto Hitler's fortified continent. The program opened with an announcer declaring "D day. D stands for Doom, Disaster, Defeat, and Death!" The action alternated between an apprehensive transport-bound GI and his grieving mother at home (played by Sally). "The invasion will be suicide," she laments, "and between 70 and 90 percent of the boys will be killed or maimed for life."

To counter Sally's growing influence, the American Armed Forces Radio Service multiplied the number of shortwave transmitters in its zones of occupation and expanded its schedule of anti-Axis programming. Morale-boosting offerings for Allied consumption were also increased, and an anti-Sally character was created in the form of "GI Jill," who presented the same mix of vibrant music and alluring commentary. In addition, U.S. broadcasters began a vigorous on-air effort to discredit Sally. At home, a federal grand jury indicted her, in absentia, for wartime treason.

In the summer of 1944, Sally was dispatched to France where, using special mobile transmitters, she tried to foment confusion and undermine morale among advancing Allied troops. After the Allied sweep out of Brittany, Sally retired to

Mildred Gillars, aka "Axis Sally," leaves court in Washington, D.C., 17 February 1949
Courtesy AP/Wide World Photos

Paris, where she endeavored to stir up French resentment against the American bombing campaign, told of increased Allied disunity, and informed soldiers of the ingratitude of "liberated" Frenchmen. After a brief stint at Hilversum studios in Holland, Sally returned to Berlin and continued her broadcasts at KWS until the end of the war. In the spring of 1945, with Allied troops closing in on Germany from all directions, Sally tried to counter the impression of German defeat by broadcasting from officer's parties and official galas. In the last weeks of the conflict, she advised GIs to shirk their duties lest they become unnecessary casualties.

After the war Sally lived in a Berlin basement until arrested by U.S. intelligence agents in March 1946. She spent 6 months at the Oberursel detention facility and 19 months in another U.S. Army prison in Germany. On 24 January 1949 she was tried in the United States for high treason and was ultimately given a 10- to 30-year sentence and a $10,000 fine. Although no firm evidence could be produced indicating her ideological commitment to Nazism (hence the relatively light sentence), enough pro-Hitler and anti-Semitic utterances could be gleaned from shortwave transcripts to condemn her. After 12 years in a West Virginia women's reformatory, she was paroled in July 1961. She returned to Ohio Wesleyan University to complete her bachelor's degree in speech communication. She died on 25 June 1988 in Columbus, Ohio.

ROBERT J. BROWN

See also Lord Haw-Haw; Propaganda; Shortwave Radio; Tokyo Rose; World War II and U.S. Radio

Axis Sally. Born Mildred Elizabeth Sisk (later Mildred Gillars) in Portland, Maine, 29 November 1900. Studied English and Drama at Ohio Wesleyan University, 1917–20; stagework in Cleveland, lived in New York as unemployed actress, 1920–28; minor role in film *Unwanted Children*, 1928; artist's model in Paris, 1929; returned to New York, 1930–33; moved to Berlin and worked at Berlitz language school, dubbed German language films, wrote film reviews, and acted as private secretary to German actress Brigitte Horney, 1934–39; propaganda broadcaster for Nazi-controlled Bremen I shortwave station, 1940; joined southern broadcasting zone of *Europäische Fremdsprachendienste* (European Foreign Language Service) and hosted anti-Allied propaganda broadcasts, 1941–44; hosted propaganda broadcasts from Paris, the Netherlands, and Berlin, late 1944–45; in hiding in Berlin, 1945–46; arrested, March 1946; in U.S. Army detention center at Oberursel with other U.S. radio propagandists; released December 1946; rearrested by U.S. Army, January 1947; in U.S. Army prison, 1947–49; on trial for treason, found guilty, and received 10–30 year prison sentence and $10,000 fine, 1949; women's reformatory, in Alderson, West Virginia, 1949–61; paroled, 1961; taught French and German at Catholic school, Columbus, Ohio, 1961–72; returned to Ohio Wesleyan to complete bachelor's degree in speech communication, 1972. Died in Columbus, Ohio, 25 June 1988.

Radio Series

1940	*Bremen Sender*
1941–42	*Anzac Tattoo: From the Enemy to the Enemy*
1942–45	*Home Sweet Home*
1943–45	*Medical Reports*
1943–45	*Midge at the Mike*
1944	*Vision of Invasion*

Further Reading

Bergmeier, Horst J.P., and Rainer E. Lotz, *Hitler's Airwaves: The Inside Story of Nazi Radio Broadcasting and Propaganda Swing,* New Haven, Connecticut: Yale University Press, 1997

Edwards, John Carver, *Berlin Calling: American Broadcasters in Service to the Third Reich,* New York: Praeger, 1991

Ettlinger, Harold, *The Axis on the Air,* Indianapolis, Indiana: Bobbs-Merrill, 1943

Kris, Ernst, and Hans Speier, *German Radio Propaganda,* London and New York: Oxford University Press, 1944

Van Dyne, Edward, "No Other Gal Like Axis Sal," *The Saturday Evening Post* (15 January 1944)

B

Barber, Red 1908–1992

U.S. Radio and TV Sportscaster

Respected for his integrity and admired for his folksy, literary style, Red Barber ranks as one of the greatest sportscasters in broadcasting history.

"The Ol' Redhead" was born in Columbus, Mississippi, and grew up in Sanford, Florida. Barber inherited the storytelling ability of his father, a railroad engineer. From his mother, an English teacher, he absorbed a love of language. While studying English at the University of Florida in 1930, Barber joined the staff of WRUF, the campus radio station. Among his duties were play-by-play broadcasts of the University of Florida football games.

In 1934, at WSAI in Cincinnati, Ohio, Barber began a 33-season career of broadcasting major-league baseball. Thanks to the innovations of Cincinnati Reds general manager Larry MacPhail, Barber was at the microphone for the first night game in the big leagues, and he broadcast from an airplane when the Reds became the first team to travel by air.

Following MacPhail to Brooklyn in 1939, Barber enjoyed more "firsts," including the first televised major-league game on 23 August 1939. That NBC game between the Dodgers and Reds also featured Barber doing the first TV commercials. Later that year, Barber did play-by-play announcement of the first football games ever televised.

Barber called the first integrated game in 1947 when the Dodgers' Jackie Robinson became the first black player in the major leagues. Like many Southerners of his generation, Barber opposed integration and considered leaving the Dodgers. He knew listeners would focus on how he handled the situation and said he had "the hottest microphone in broadcasting" at the time. Robinson's enormous courage in the face of threats on his life made Barber a convert to civil rights. Robinson and the black players who followed him into the majors praised Barber for treating them like other ballplayers and never referring to their color. They said that was exactly what they wanted. Robinson led the Dodgers to the 1947 World Series, the first to be televised, and Red Barber called the games.

New York listeners and viewers were charmed by Barber's folksy expressions. From his "catbird seat," Barber described an outfielder "movin' as easy as a bank of fog," a pitcher who was "no slouch with the willow," and an infielder trying to handle a ball "slicker than oiled okra." A close game was "tighter than a new pair of shoes on a rainy day," but a one-sided game was "tied up in a croker sack." A bad game was "full of fleas." A pitcher might be "as wild as a hungry chicken hawk on a frosty morning," but another might be so good "he could toss a lamb chop past a hungry wolf." A fight was a "rhubarb," and a game with a lot of hits was an example of "tearin' up the pea patch."

Barber was no rube, however, and he never played the clown. His expressions came as naturally to him as his use of words such as *concomitant* and *penultimate*. He surely was alone among play-by-play announcers in quoting Byron and Coleridge to describe action on a baseball diamond.

Calling the games as a detached journalist, Barber refused to root for the teams that paid his salary. In the days when road games were sometimes described by announcers reading telegraph reports, Barber refused to pretend he was actually watching a game. He also rejected the superstition of ignoring a no-hitter in progress. He believed the pitcher's no-hit performance was the obvious lead story of the game.

Hired by Edward R. Murrow in 1946, Barber served as Director of Sports at CBS for nine years. Among Barber's innovations was the football "round-up," in which the network would broadcast bits of several games on the same Saturday afternoon, switching to a more interesting match when a game became dull or one-sided. Out of one of these round-ups emerged Barber's protégé, Vin Scully, who joined Barber and Connie Desmond on the Dodger broadcasts in 1950.

Red Barber
Courtesy Radio Hall of Fame

Barber could also be heard on *The Old Gold Hour* with bandleader Sammy Kaye and later with Woody Herman. His work for CBS included many celebrity interviews with people not connected to sports, including Fred Astaire, Francis Cardinal Spellman, and former president Herbert Hoover.

Leaving the Dodgers after the 1953 season, Barber joined his old rival Mel Allen in the broadcast booth at Yankee Stadium. Barber and Allen, the two great stars of their era, remain linked in the memories of many baseball fans. When the Baseball Hall of Fame began honoring broadcasters with the Ford C. Frick Award in 1968, it could not decide between Barber and Allen, so it gave the award to both.

Barber's independent streak was costly. Named to broadcast the 1953 World Series, he insisted on negotiating his fee. He was removed from the broadcast and was never asked to call another World Series. When CBS took control of the Yankees, it gave Barber three former ballplayers as broadcast partners. Barber believed broadcasting was a professional career, not a lark for ex-athletes who could no longer play the game. The 1966 Yankees were a bad team that drew just 413 fans to a game in September. Barber's reference to the sparse attendance did not endear him to management. He was fired after the 1966 season.

Barber retired from daily broadcasting and returned to Florida, where he wrote five books. From 1981 until his death in 1992, Barber was a commentator on National Public Radio's *Morning Edition*. Speaking from his home in Tallahassee, Barber, with his four-minute chat each Friday, was the most popular feature on NPR. Listeners loved hearing his stories about sports and the early days of broadcasting, but they especially enjoyed hearing him talk about his garden, the adventures of his pet cat, and the opera he had enjoyed that week on public television. He spoke with pride of his on-air support of blood drives during World War II, his USO tours, and his decades as a lay reader in the Episcopal Church.

BOB EDWARDS

See also Sports on Radio; Sportscasters

Red Barber. Born Walter Lanier Barber in Columbus, Mississippi, 17 February 1908. Joined staff of WRUF, Gainesville, Florida, 1930, and broadcast play-by-play of University of Florida football; staff announcer for WLW/WSAI, Cincinnati, Ohio, 1934–38, broadcasting Cincinnati Reds baseball, University of Cincinnati football, and some Ohio State football games; broadcast voice of Brooklyn Dodgers, 1939–53; play-by-play broadcaster for New York Yankees, 1954–66; director of sports, CBS, 1946–55; worked 13 World Series, five All-Star games, several NFL Championship Games, and 13 college football bowl games; newspaper columnist for the *New York Journal-American, Christian Science Monitor, Miami Herald,* and *Tallahassee Democrat;* commentator for National Public Radio's *Morning Edition,* 1981–92. Died in Tallahassee, Florida, 22 October 1992.

Selected Publications

The Rhubarb Patch (with pictures by Barry Stein), 1954
Rhubarb in the Catbird Seat (with Robert Creamer), 1968
Walk in the Spirit, 1969
The Broadcasters, 1970
Show Me the Way to Go Home, 1971
1947, When All Hell Broke Loose in Baseball, 1982

Further Reading

Barber, Lylah, *Lylah: A Memoir,* Chapel Hill, North Carolina: Algonquin Books, 1985
Edwards, Bob, *Fridays with Red: A Radio Friendship,* New York: Simon and Schuster, 1993
Smith, Curt, *Voices of the Game: The First Full-Scale Overview of Baseball Broadcasting, 1921 to the Present,* South Bend, Indiana: Diamond, 1987; 2nd edition, New York: Simon and Schuster, 1992

Barnouw, Erik 1908–2001

U.S. (Dutch-Born) Broadcast Writer, Producer, and Historian

Renowned as a media scholar and practitioner, Erik Barnouw had a long career that involved work with commercial networks, the federal government, documentary filmmakers, and academe. He is best known, however, as American broadcasting's premier historian.

Origins

Barnouw was born in the Netherlands and spent his first decade there, moving with his family to New York after World War I. He became very active in theater while attending Princeton

University from 1925 to 1929, and wrote and acted in several plays and musicals, including *Open Collars* and the *Princeton Triangle Show.* On graduation he turned down a chance to stay on for graduate degrees and become a faculty member, opting instead for the theater world. The theater was one of three offers he considered; in the end, Barnouw was luckily able to enjoy all three of them. The theater stint lasted only a few months, and he then accepted a position as a writer for the new business monthly magazine *Fortune.* When the Depression began to cut into that option (first his hours were reduced, then he lost his job), he turned to the third—a fellowship to travel overseas. The fellowship included a few months of his studying drama with Max Reinhardt in Vienna, plus travel to other spots in Europe and North Africa. On Barnouw's return in 1931, he found an America that had plunged even deeper into the Depression and that offered few job opportunities of any sort.

Radio Years

A chance meeting led to a position with the Erwin Wasey ad agency, which had just won the Camel Cigarettes radio advertising account. Barnouw was assigned to direct the *Camel Quarter Hour* program, both the program and its advertising. This variety show was broadcast six nights a week. As he later recalled: at that point "I had never had a radio. I went out and bought a radio . . . [and] listened to it all weekend, so I could come to work Monday knowing a lot about it." He remained with the Erwin Wasey agency until 1935, then spent two years with the Arthur Kudner agency. By mid-decade he was directing six or seven network (CBS and NBC) radio dramas every week, among them *Bobby Benson's Adventures*, a 15-minute children's program. Though he had sought a salary of only $30 a week for this job, he was started at $65, an indication of national radio's economic health (despite the perilous times) and growing importance.

By 1937 Barnouw felt burned-out from the long and hectic weeks dealing with multiple radio programs. A writing professor at Columbia University contacted him while he relaxed in Maine for the summer, and offered him a chance to teach radio writing during the coming academic year. As Barnouw later noted, this was fortuitous timing; the networks were increasing their sustaining (i.e., non-advertiser supported) programs, many of them dramas, in the face of press and government criticism of radio's direction. The networks needed and were willing to help support ways of training writers for the medium; they helped pay tuition for some of the people attending Barnouw's course at Columbia. Among those attending his first class were a young Bernard Malamud and (under an assumed name) novelist Pearl Buck, who had just won a Nobel Prize. Barnouw's first book, *Handbook of Radio Writing,* was the result of his experiences in this class, it also included scripts from his work with both CBS and NBC.

Teaching about how to write for radio (something he had deplored before he tried it) increased his interest and activity in writing for radio. He served as script editor for a CBS radio series in 1939–40. He also wrote many *Cavalcade of America* and *Theater Guild on the Air* scripts during and after World War II. As many radio employees were being called up for wartime work, Barnouw found himself working full-time again in network radio, becoming a script editor for NBC's public service program writers. As a result of advertisers increasing their expenditures during the war, the number of public service programs also greatly increased; there was plenty for Barnouw to do. The NBC position led to his being appointed the head of educational programs for the Armed Forces Radio Service in Washington, D.C., in 1944–45.

Television and Film

After the war, Barnouw returned to New York and Columbia University, this time as a full-time faculty member in the School of Arts. Among his first roles was that of bridging academe and network: the university made use of NBC television studios to train writers and production people. He expanded the courses in his department to include television and film as well as radio. He would chair this department until 1968, and he remained with the university until his retirement in 1973.

What many perceive to have been Barnouw's signature contribution to radio began in 1959 when he was approached by Oxford University Press to prepare a three-volume history of American broadcasting to parallel the history of the BBC that Oxford was planning to publish in England. Barnouw had already scheduled a year (1961) in India on a Fulbright fellowship (out of which came a history of Indian film), but on his return he devoted himself to this project. The first volume appeared in 1966 to laudatory reviews; it was followed by the two further volumes in 1968 and 1970. The trilogy almost overnight became the standard history of U.S. radio and television. He followed up with a one-volume version focused on television in 1975, and revised it twice before his death.

Barnouw spent the final years of his career in Washington. He became a fellow at the Woodrow Wilson Center of the Smithsonian Institution in 1976, from which residency came his history of the unique role of American broadcast advertising, *The Sponsor,* in 1978. From 1978 until his retirement in 1981, Barnouw served as founding chief or director of the Library of Congress's Motion Picture, Broadcasting and Recorded Sound Division. Even after he retired, and up until the time of his death in mid-2001, Barnouw continued to write insightful analyses of the state of American media.

CHRISTOPHER H. STERLING

See also Education about Radio

Erik Barnouw. Born in The Hague, Netherlands, 23 June 1908, son of Adriaan Barnouw, a noted linguist and later Columbia University professor of Dutch, and Anne Midgely. Immigrated with his family to the U.S., 1919. Graduated in English, Princeton University, 1929. Briefly joined Cukor-Kondolf stock theater company in Rochester, New York, as assistant stage manager. Writer for *Fortune* magazine, 1929–30. Radio program director and radio writer for two different advertising agencies, 1931–37; writer for CBS, 1939–40; editor, NBC Script Division, 1942–44; supervisor, Education Division, Armed Forces Radio Service, 1944–45. President, radio writer's guild, 1947–49; at Columbia University as part-time instructor from 1937; full-time faculty 1946–73, organizing film division and serving as its chair until 1968. Co-founder (in 1954) and head of Writers Guild of America, 1957–59. Peabody Award, 1944; Fulbright scholar in India, 1961; Guggenheim Fellowship, 1969; Bancroft Prize in American History, 1971; George Polk Award, 1971; Frank Luther Mott Journalism Award, 1971; Woodrow Wilson Center Fellow, 1976. Chief of Motion Picture, Broadcasting and Recorded Sound Division, Library of Congress, 1978–81. Died 19 July 2001 at Fair Haven, Vermont.

Selected Credits

Radio

(years indicate program span, not necessarily years Barnouw was involved)

Producer, *The Camel Quarter-Hour*, CBS, 1931–32

Director, *Bobby Benson's Adventures*, CBS, 1932–36

Director or writer, *The Cavalcade of America*, CBS, 1935–39

Director or writer, *Theater Guild on the Air*, CBS, 1943–44; 1945–49, ABC; 1949–53, NBC

Producer, *Words at War*, NBC, 1943–45

Writer, *The Conspiracy of Silence*, 1948

Television

The Ignorant, Ignorant Cowboy, public health campaign.

Motion Pictures

Producer, *Freedom to Read*, 1954

Producer, writer (with Herbert Wechsler), *Decision: The Constitution In Action*, 1959

Producer, *Hiroshima-Nagasaki, August 1945*, 1970 (Atlanta Film Festival Award winner)

Selected Publications

Handbook of Radio Writing, 1938; 2nd edition, 1947

Radio Drama in Action: 25 Plays of a Changing World, 1942

Handbook of Radio Production, 1949

Mass Communication: Television, Radio, Film, Press, 1956

The Television Writer, 1962

Indian Film (with S. Krishnaswamy), 1963; 2nd edition, 1980

A History of Broadcasting in the United States, 3 vols.

- *A Tower in Babel* [to 1933], 1966
- *Golden Web* [1933–53], 1968
- *The Image Empire* [from 1953] 1970

Documentary: A History of the Nonfiction Film, 1974; revised editions, 1983, 1993

Tube of Plenty: The Evolution of American Television, 1975; revised editions, 1982, 1990

The Sponsor: Notes on a Modern Potentate, 1978

The Magician and the Cinema, 1981

International Encyclopedia of Communications (editor), 1989

Media Marathon: A Twentieth-Century Memoir, 1996

Conglomerates and the Media, 1997

Media Lost and Found, 2001

Further Reading

"A Festschrift in Honor of Erik Barnouw On the Occasion of His Ninetieth Birthday," *Wide Angle* 20, no. 2 (April 1998)

"Papers Presented at the 'Future of Media Historical Research,' a conference at the Media Studies Project, Woodrow Wilson Center, September 27, 1991," *Film & History* 22, nos. 2-3 (May/September 1991)

Beautiful Music Format. *See* Easy Listening/Beautiful Music Format

---

Bell Telephone Laboratories

At the forefront of communications research, Bell Laboratories for decades was regarded as the largest and most successful private research organization in the world. Created as the research and development division of American Telephone and Telegraph (AT&T), Bell Labs is now a part of Lucent Technologies. Best known for its development of the transistor, laser technology, and information theory, the organization boasts more than 27,000 patents and 11 Nobel laureates. Headquartered in Murray Hill, New Jersey, Bell Labs consists of a global community of some 16,000 people in 16 countries.

It was nearly 50 years after the invention of the telephone when AT&T formed Bell Telephone Laboratories as a subsidiary in 1925. The unit was created to merge and centralize the research and engineering work of AT&T and its manufacturing and supply arm, Western Electric. Frank B. Jewett was its first president.

Bell Labs can be credited with major roles in the development of computer technology, the microelectronics industry, and a host of modern communications technologies. In its first three years of operation, researchers demonstrated long-distance television transmission, sound/motion pictures, the artificial larynx, and the negative feedback amplifier (used to reduce distortion in radio and telephone transmissions). During the 1930s, researchers developed an electrical digital computer, a radio altimeter (a new means of radio transmission), and radio astronomy. In addition, they conducted research on stereophonic sound. During World War II, the U.S. military profited from research at Bell Labs on radar and wireless communications.

One of the most notable Bell Labs inventions was the transistor, a small electronic component with a semiconductor, which is now found in virtually every electronic device and led to solid-state communications and transformation of the electronics industry. The transistor was invented in 1947 when a team of scientists took initial semiconductor research and improved upon it in order to amplify signals in the same way as the vacuum tube did but with more reliability and much less power and space consumption. Bell Labs developed techniques to make the transistor practical, and it eventually became a fundamental and essential component of radio, television, telephone, and entertainment equipment. The transistor radio, one of the first mass-produced products based on this invention, appeared in 1954, quickly becoming the best seller in consumer product history and significantly influencing popular culture. The transistor paved the way for portability, miniaturization, and better car radios. Later, the invention of the integrated circuit, which organizes numerous transistors and other electronic components on a silicon wafer, took the transistor innovation to a new level and sparked the Information Age.

Other key contributions of Bell Labs include the publication of "information theory" in 1948, the invention of laser technology in 1958, and the development of the solar battery and communications satellite by the early 1960s. Bell Labs introduced software-controlled telephone switches long before personal computers, in addition to the first electric microphone for hands-free telephone conversations. Other developments in which Bell Labs researchers assisted included cellular mobile radio, fiber optics, light-emitting diodes, charge-coupled devices (used in cameras), the UNIX operating system, C and C++ programming languages, and High-Definition Television (HDTV).

For more than 50 years, Bell Telephone Laboratories enjoyed financial support from its parent company AT&T, a regulated monopoly that could pass along research costs to customers. The research approach was primarily academic, with a corporate philosophy of giving topflight researchers freedom and autonomy. Bell Labs focused on being the first or best in such areas as publishing papers, setting transmission records, and building the most powerful laser diode. Eleven of its researchers have been awarded the Nobel Prize, nine have received the National Medal of Science, and seven have received the National Medal of Technology. An Emmy Award was earned for the institution's work on HDTV. Nobel Prize winners include Clinton J. Davisson, who demonstrated the wave nature of matter, a foundation for much of today's solid-state electronics. John Bardeen, Walter H. Brattain, and William Shockley were honored with a Nobel Prize in 1956 for inventing the transistor, and Arno Penzias and Robert Wilson received the Nobel Prize for detecting background radiation supporting the Big Bang theory.

By the 1980s competitive pressures and then divestiture led AT&T to refocus its research efforts. The terms of a 1956 agreement with the U.S. Department of Justice restricted what use AT&T could make of technologies unrelated to its core telephone business. Bell Labs had to license others to utilize many of its patents—especially the valuable transistor. The court-ordered 1984 break-up of AT&T, however, freed Bell labs to engage in direct marketing as well as licensing of its innovations. Research Director Arno Penzias pushed to bring wide-ranging research projects more in line with the company's telecommunications business. Although there was concern that Bell Labs would suffer without the former cross-subsidies, AT&T pledged to continue its tradition of both primary and applied research. Critics noted, however, the slow but steady decline of fundamental scientific research in favor of more applied work to develop products to meet customer needs.

Bell Labs experienced a dramatic change in 1996 when AT&T underwent another major restructuring (a trivestiture), giving up 75 percent of its Bell Laboratories staff to its new offspring, Lucent Technologies. The remaining 25 percent, made up of computer scientists, mathematicians, and other information scientists remained to form the new AT&T Laboratories, supporting the telecommunications provider's businesses in long distance and other services. Some physical science positions were terminated due to the corporate reorganization.

Bell Labs initially experienced a resurgence under its new parent company, Lucent Technologies, a high-tech company that develops and manufactures telecommunications technology and equipment for AT&T and others. Lucent also gave Bell Labs a higher profile, featuring the research and development unit in the company slogan: "Lucent Technologies. Bell Labs Innovations." About 80 percent of Bell Labs employees are part of and integrated into Lucent's business units. The other 20 percent work for what is called the Central Labs, which consist of three major technical divisions: Advanced Technologies, Technology Officer Division, and Research.

During the 1990s, Bell Labs led the world with more citations than, for example, IBM and top academic institutions. Even today, despite serious cut-backs and reorientation of research priorities, the Labs average better than three patent applications every business day. A number of technologies for the digital audio broadcasting (DAB) market have been patented, including the perceptual audio coding (PAC) algorithm. The PAC encoder converts AM or FM radio signals into high-quality digital signals and enables the transmission of digital audio over a variety of wireless and wireline channels, including in-band, on-channel (IBOC) systems and the internet.

At the dawn of the 21st century, Lucent, like most of the telecommunications industry, experienced financial troubles. As a result, Bell Labs lost funding and people, particularly in the physical sciences. Restructuring and streamlining shifted the focus to addressing the needs of communications service providers. There are now two primary operating units, focusing on wireline networks and mobile networks. Web-based customer solutions and service intelligence are emphasized. Bell Labs continues to be Lucent's innovation engine.

LAURIE THOMAS LEE

See also American Telephone and Telegraph; Transistor Radios

Further Reading

Bell Labs Innovations, <www.bell-labs.com>

Bell Telephone Laboratories, *Impact: A Compilation of Bell System Innovations in Science and Engineering,* 2nd edition, Murray Hill, New Jersey: Bell Laboratories, 1981

Buderi, Robert, "Bell Labs Is Dead: Long Live Bell Labs," *Technology Review* 101, no. 5 (September/October 1998)

Feder, Toni, "Bell Labs Research Regroups as Parent Lucent Shrinks," *Physics Today* 54, no. 10 (October 2001)

Mabon, Prescott C., *Mission Communications: The Story of Bell Laboratories,* Murray Hill, New Jersey: Bell Telephone Laboratories, 1975

Mueser, Roland, editor, *Bell Laboratories Innovation in Telecommunications, 1925–1977,* Murray Hill, New Jersey: Bell Laboratories, 1979

O'Neill, E.F., editor, *A History of Engineering and Science in the Bell System: Transmission Technology (1925–1975),* Indianapolis: AT&T Bell Laboratories, 1985

Service, Robert S., "Relaunching Bell Labs," *Science* (3 May 1996)

Southworth, George C., *Forty Years of Radio Research,* New York: Gordon and Breach, 1962

Benny, Jack 1894–1974

U.S. Radio Comedian

Jack Benny was one of network radio's top comedy stars for several decades, and one of the few who successfully made the transition from vaudeville to radio, and later from radio to television.

Vaudeville Origins

Jack Benny was born Benjamin (Benny) Kubelsky in Chicago on 14 February 1894, but he claimed Waukegan, Illinois, as his birthplace because his parents lived there. Benny's father bought him a violin and paid for lessons, and after graduating from grammar school Benny played with a children's orchestra in stores, at parties, and at bar mitzvahs. He entered Central High School in Waukegan in 1909 and worked in vaudeville as a violinist in the orchestra pit at the Barrison Theater, also in Waukegan. Matinee performances required time away from school and led to his flunking out. His disappointed father found him a haberdashery job and enrolled him in Waukegan

Business College. Benny was so uninterested that his father let him return to the Barrison, which closed when Benny was 16. Benny then formed the act "Salisbury and Benny—From Grand Opera to Ragtime" with Cora Salisbury, a mature pianist. They played the Midwest until 1913, when Benny joined pianist Lyman Woods to form "Bennie and Woods." Their crowning moment came when they played New York's Palace, the leading vaudeville theater. In November 1917 Benny's mother died, and for the rest of his life Benny feared the cancer that had claimed her life. He also regretted that he had not become a serious violinist as she had hoped he would.

In 1918 Benny joined the navy at Great Lakes Training Station and never saw Woods again. He joined the *Great Lakes Review,* a navy theatrical show that performed across the Midwest for Navy Relief. Benny had never talked onstage in vaudeville. In the navy show, he had two lines that he delivered in a flat voice the audience loved. Upon his discharge, Benny performed solo to modest reviews as "Ben Benny—Fiddleology and Fun" and "Ben Benny—Aristocrat of Humor." He also had some success composing piano solos. He soon decided, however, to put all his emphasis on being a comedian in his act "A Few Minutes with Ben Benny." Ben Bernie, bandleader and comedian, was unhappy with the name similarity, and so Jack Benny was created. In "A Few Minutes with Jack Benny," music disappeared from his act. By 1926 Benny was well established as a vaudeville comedian when he met Sadie Marks, a hosiery salesgirl who in 1927 would become his wife. They adopted a daughter, Joan, in 1934.

In 1926 Benny landed a role in *The Great Temptations,* a musical revue. After that, Metro-Goldwyn-Mayer signed him to a film contract. He was used primarily to emcee industry dinners and premieres, but he did play himself in *The Hollywood Revue of 1929.* He asked to be released from his contract and returned to the stage in *The Earl Carroll Vanities* in 1930.

Network Radio

Benny's first radio appearance was on an interview show hosted by *New York Daily News* columnist Ed Sullivan in 1932. Benny said: "This is Jack Benny. There will be a slight pause while everyone says, 'Who cares?'" Benny was given the *Canada Dry Program,* which aired on National Broadcasting Company (NBC) Blue (2 May 1932 to 26 October 1932) and on the Columbia Broadcasting System (CBS; 30 October 1932 to 26 January 1933). Benny tried new ideas, such as humorously working the advertisement into the show. Canada Dry was at first unimpressed, but thousands of letters changed their minds. During the second year of Benny's work for NBC, he was given the *Chevrolet Program* on NBC Red (3 March 1933 to 23 June 1933; 1 October 1933 to 1 April 1934), but network President William F. Knudsen dropped the show until protests from dealers changed his mind. However, NBC Red had already begun to broadcast Benny's *General Tire Revue* (6 April 1934 to 28 September 1934). Sadie joined the cast as Mary Livingstone, a fictitious character from Plainfield, New Jersey. There was no real Mary Livingstone, but fans who remembered her sent letters to her for the rest of Sadie's life.

In 1934 Benny signed to do the *Jell-O Program Starring Jack Benny* on NBC Blue in New York (14 October 1934 to 14 July 1935). The show moved to Hollywood on NBC Blue (29 September 1935 to 21 June 1936) and then on NBC Red (4 October 1936 to 31 May 1942). By 1936 the *Jell-O Program Starring Jack Benny* was rated number one. Benny introduced many running bits (including his famous "feud" with fellow radio comedian Fred Allen), which continued on television. In his basement was the famous vault and Carmichael the polar bear. He satirized motion pictures, always maintained that he was 39, drove a very old Maxwell car, and cultivated his stingy image. He demonstrated frustration, with chin in hand, by stating, "Well!" Sometimes he showed aggravation by saying, "Now cut that out!" His most famous line reflecting stinginess came during a scene in which he was held up by a robber who exclaimed, "Your money or your life!" After a lengthy delay, the robber repeated the demand. The audience laughed profusely and at great length until Benny responded, "I'm thinking!" Benny also employed sound effects, perhaps more successfully than anyone to that point in radio. Although he was a proficient violinist in real life, Benny continued to use it as a funny prop and sometimes played a sloppy version of his theme song, *Love in Bloom.* Benny and his writers did not employ any particular gag on a weekly basis but rotated and mixed old jokes with new ideas to keep material fresh. No bit was overplayed, but the audience always knew that the vault, Carmichael, and the Maxwell were there, even though they may have been part of the script only a few times per season.

Benny's radio program featured regular cast members, such as Mary Livingstone, Dennis Day, Phil Harris, and others. Perhaps the most beloved character was Eddie Anderson's "Rochester." The show returned to New York on occasion, and Rochester was introduced as a Pullman porter in a skit about one of the train trips back to Los Angeles. Gravel-voiced Rochester was so well received he was written into the program permanently under the premise that Benny had hired him away from the railroad. Rochester used affectionate insults to put Benny in his place, and the relationship worked perfectly into Benny's trademark of self-deprecating humor.

After moving to California, Benny signed a long-term contract with Paramount to appear in a number of movies, including *College Holiday* (1936) with George Burns and Gracie Allen and *Love Thy Neighbor* (1940) with Fred Allen. *Buck Benny Rides Again,* playing on one of the running radio gags, satirized Western films with film actor Andy Devine portraying

Jack Benny
Courtesy Radio Hall of Fame

his sidekick. Benny had success in his films, but one would not know it based on his treatment of film in his radio act. One recurring joke focused on the 1946 film *The Horn Blows at Midnight.* Although it made some money, it was his least successful movie effort. His agent, Irving Fein, said that although Benny later had other parts in movies, "the record books will state that *The Horn Blows at Midnight* really did blow taps to his motion-picture career."

While Bob Hope rightfully received tremendous credit for entertaining American troops abroad, Jack Benny deserves credit in this respect as well. Many jokes focused on his experiences as a World War I sailor. Unlike other entertainers, Benny and his accompanying performers wore civilian clothes, because he realized that military personnel were tired of seeing nothing but uniforms. On one tour, Benny discovered GI Jack Paar; three years later, Benny gave Paar his show business start by signing Paar as his summer radio replacement. Benny also performed in Korea (1950–53) but returned so exhausted that he was ordered to a week's bed rest. Although he wanted to entertain in Vietnam, he was in his 70s by then, and the memory of Korea prevented his participation.

The *Jack Benny Program for Grape-Nuts and Grape-Nuts Flakes* on NBC Red (4 October 1942 to 4 June 1944) dropped from first in the ratings to trail Bob Hope and Edgar Bergen and Charlie McCarthy. Benny hired new writers and launched the *Lucky Strike Program Starring Jack Benny* on NBC (1 October 1944 to 26 December 1948), but by 1945 the program had dropped from the top five. It was decided to have a contest in which listeners would write in 25 words or less on the subject, "I can't stand Jack Benny because . . . ." Great publicity was generated, and the show returned to number one in 1947. In 1948 the *Lucky Strike Program Starring Jack Benny* left NBC for CBS (2 January 1949 to 22 May 1955). William Paley, CBS chairman, had seen television's future and raided NBC for almost all of its top performers and programs in order to move them all to television.

Television

The Jack Benny Program on CBS-TV (1950–64) began with a series of live specials from New York on 28 October 1950. Benny's first television line was, "I'd give a million dollars to know what I look like on television." For five years Benny continued to do radio even as his TV show became more popular. Indeed, much of the time the two were simulcast, often leaving radio listeners in the dark as audiences (or laugh tracks) indicated some sight gag. His only concession was to reduce the number of radio shows from 39 to 35 programs per year. The number of TV programs steadily increased to 9 in the third year, then to 13 and 16. In 1960–61 the number rose to 26. Benny's TV show rotated with other programs on the schedule, including *Private Secretary, Bachelor Father,* and *The George Gobel Show.* Many of his television shows were adapted from radio. Initially television made Benny nervous, but he eventually began to like it more than radio. In 1955 Benny left radio behind to concentrate solely on TV. Lucky Strike sponsored the TV show from 1950 to 1959, and other sponsors took over until 1965. The entire TV series was done in black and white. The show's last year was broadcast on NBC (1964–65). *The Jack Benny Program* won eight Emmy awards for Best Comedian and Best Comedy Program.

In 1957 Benny won the first trustees award of the Academy of Television Arts and Sciences. The award read: "Jack Benny—For his significant contribution to the television industry as a showman. For the high standard for all to emulate, set by his personal skill and excellence as a performer. For the consistency, quality and good taste of his program through many years and many media." Until his death, Benny appeared sporadically as a guest star and in his own specials, featuring themes and all-star casts. The specials included *Carnival Nights, Jack Benny's Bag, Jack Benny's Birthday Party, Jack Benny's Twentieth Anniversary Special, Everything You Always Wanted to Know about Jack Benny but Were Afraid to Ask, Jack Benny's* First *Farewell Special,* and *Jack Benny's* Second *Farewell Special,* his last show.

W.A. KELLY HUFF

See also African Americans in Radio; Allen, Fred; Comedy; Stereotypes on Radio; Talent Raids; Vaudeville

Jack Benny. Born Benjamin Kubelsky in Chicago, Illinois, 14 February 1894. Worked in vaudeville as violinist in orchestra pit and then with "Salisbury and Benny" and "Bennie and Woods," 1909–18; served in U.S. Navy, World War I; returned to vaudeville as comic and musician Ben Benny; small-part Broadway actor during 1920s; radio debut, *The Ed Sullivan Show,* 1932. Recipient of eight Emmys. Died in Beverly Hills, California, 26 December 1974.

Radio Series
1932–55 *The Jack Benny Program*

Television Series
The Jack Benny Program, 1950–64, 1964–65

Films
Bright Moments (short), 1928; *The Hollywood Revue of 1929,* 1929; *Chasing Rainbows,* 1929; *The Medicine Man,* 1930; *Mr. Broadway,* 1933; *Transatlantic Merry-Go-Round,* 1934; *Broadway Melody of 1936,* 1935; *It's in the Air,* 1935; *The Big Broadcast of 1937,* 1936; *College Holiday,* 1936;

Artists and Models, 1937; *Manhattan Merry-Go-Round,* 1937; *Artists and Models Abroad,* 1938; *Man about Town,* 1939; *Buck Benny Rides Again,* 1940; *Love Thy Neighbor,* 1940; *Charley's Aunt,* 1941; *To Be or Not to Be,* 1942; *George Washington Slept Here,* 1942; *The Meanest Man in the World,* 1943; *Hollywood Canteen,* 1944; *It's in the Bag!* 1945; *The Horn Blows at Midnight,* 1945; *Without Reservations,* 1946; *The Lucky Stiff,* 1949; *Somebody Loves Me,* 1952; *It's a Mad Mad Mad Mad World,* 1963; *A Guide for the Married Man,* 1967; *The Man,* 1972

Stage

The Great Temptations, 1927; *The Earl Carroll Vanities,* 1930

Publication

Sunday Nights at Seven: The Jack Benny Story (with Joan Benny), 1990

Further Reading

Allen, Steve, "Jack Benny," in *The Funny Men,* by Allen, New York: Simon and Schuster, 1956

Benny, Mary Livingstone, Hilliard Marks, and Marcia Borie, *Jack Benny,* Garden City, New York: Doubleday, and London: Robson, 1978

Fein, Irving A., *Jack Benny: An Intimate Biography,* New York: Putnam, 1976

Harmon, Jim, "The Thirty-Nine Year-Old Skinflint," in *The Great Radio Comedians,* by Harmon, Garden City, New York: Doubleday, 1970

Josefsberg, Milt, *The Jack Benny Show,* New Rochelle, New York: Arlington House, 1977

Museum of Television and Radio, New York, *Jack Benny: The Radio and Television Work,* New York: HarperPerennial, 1991

Wertheim, Arthur Frank, *Radio Comedy,* New York: Oxford University Press, 1979

The Beulah Show

Situation Comedy

Based on a character that first appeared in *Fibber McGee and Molly,* this spin-off program marked an important transition. A black character, Beulah was, for the show's first two seasons, portrayed by white men. Only in later seasons did black women play the black characters—including the title character of *The Beulah Show.* For its era, however, the program helped to break racial barriers by introducing blacks to on-air roles.

The Beulah character, merely the latest black domestic in a radio program (Rochester on the Jack Benny program was probably the best known), had first appeared on *Fibber McGee and Molly* in early 1944 and became an instant hit. Portrayed by Marlin Hurt (who had himself been raised by a black maid and had thus picked up some of the "right sound" in childhood, developing a reputation as a good portrayer of blacks on radio), the character soon was delivering lines that became widely popular catch phrases across the country—"Looove dat man!" and the regular stand-by, "Somebody bawl fo' Beulah?"

Beulah was played as a central part of the white middle-class family that employed her. She was good natured and respectful, but not subservient. Indeed she was often sarcastic, though rarely directly to her employers. She ran the household and solved problems–the core of program stories. Her radio friends included a shiftless boyfriend and the next door domestic, among others.

The weekly series seemed on its way to a long run when Hurt died at age 40 of a heart attack and, lacking its key actor, the program left the air. In the spring of 1947 it returned, with yet another white man (Bob Corley) playing the black domestic. Only that fall, when the program switched to CBS as a 15-minute program every weekday, did a black woman (Hatie McDaniel) begin to play the title part.

The program moved to television for four seasons beginning in 1950, though lacking some of the comic bite of the radio original. Black characters were played by black actors from the beginning—the first network television series where this was so.

CHRISTOPHER H. STERLING

See also African-Americans in Radio; Amos 'n' Andy; Fibber McGee and Molly; Stereotypes on Radio

Cast

Beulah	Marlin Hurt (1945–46); Bob Corley (1947); Hatie McDaniel (1947–52); Lillian or Amanda Randolph (1952–53)
Bill Jackson	Marlin Hurt (1945–46), Ernie Whitman (1947–53)
Harry Henderson	Hugh Studebaker
Alice Henderson	Mary Jane Croft
Donnie Henderson	Henry Blair
Announcer	Ken Niles (1947), Marvin Miller (1947–53), Johnny Jacobs (1954)

Programming History

CBS (as *The Marlin Hurt and Beulah Show*)	1945–46
ABC	1947
CBS (15 minutes weekdays)	1947–54
ABC-TV	1950–53

Further Reading

Bogroghkozy, Aniko, "Beulah," in volume 1 of *Encyclopedia of Television,* 3 vols., edited by Horace Newcomb, Chicago and London: Fitzroy Dearborn, 1997

Nachman, Gerald, "No WASPs Need Apply," in *Raised on Radio,* by Nachman, New York: Pantheon Books, 1998

Beville, Hugh Malcolm 1908–1988

U.S. Audience Research Executive

The economics of commercial broadcasting are driven by estimates of audience size and type, and organizations such as the Arbitron Company and SRI (originally Stanford Research Institute) provide measurements of listenership to radio stations and networks. These measurements determine prices the media can charge for commercial time. The methods of audience research have changed greatly in the years since 1930, when the first system of ratings was introduced. Along the way, media researchers such as Hugh Beville influenced the direction and quality of the audience research process, helping to shape today's business of broadcasting.

Beville went to work for the National Broadcasting Company (NBC) at age 22 and spent the greater part of his career working in research for the company's radio and television networks, only leaving for a brief period during World War II to serve in the army. NBC had been in existence for only three years in 1930, when Beville was hired as a statistician. That same year, the Cooperative Analysis of Broadcasting released its first radio rating report. Beville's 38-year career coincided with the development of the field of audience research, and he made significant contributions to its development. Although he is perhaps best known for his insights into the ratings collection and analysis process, he is also credited with improving both quantitative and qualitative audience research methodologies.

As early as the mid 1930s, Beville was involved with the Joint Committee on Radio Research to study radio audiences. He developed the first nationwide study of daytime audiences and was one of the first researchers to identify and track seasonal listening patterns. In his later years he was closely involved in decisions about crediting videotape recorder usage in television audience ratings, and he contributed to the planning and development of the People Meter system of measurement. Concerned with differences in population estimates among rating services, he helped develop a more compatible system of estimation using data from Market Statistics, Inc. He also worked on a system for estimating station circulation—something most industry practitioners take for granted now that the process has been refined.

Beville rose through the ranks at NBC during his 38 years there, from statistician to manager of research, director of research, and finally to vice president for planning and research in 1964. After retiring from NBC in 1968, he became a professor of business administration at Long Island College in Southampton, New York, teaching marketing and management. From 1972 to 1982 he was the executive director of the Electronic Media Ratings Council in New York and later served that organization as an independent consultant. Throughout his career, Beville enjoyed widespread respect in the media industries. Leo Bogart, a friend and colleague, described him as "a very earnest, soft-spoken gentleman, with sound professional judgment and a good eye for picking talented associates."

Beville fought many battles to ensure that audience ratings were estimated fairly and interpreted correctly. During his tenure at NBC, he served as the networks' representative to the

Advertising Research Foundation for 15 years. In 1963 he helped to establish the Broadcast Ratings Council (BRC), which later became the Electronic Media Research Council (EMRC) and eventually the Media Research Council (MRC), an organization formed in response to congressional hearings that questioned the integrity of the audience measurement process. Its mission is to ensure that ratings firms follow research protocols so that users of the data are able to correctly interpret the information.

One of Beville's major concerns was that companies supplying audience research data were too secretive about what they were doing. He argued that their research should be transparent—that clients should know exactly what they were getting. On several occasions he admonished the users of audience research data to recognize the drawbacks and limitations of statistical information as well as its usefulness. In a 1981 NAB newsletter he was quoted as saying: ". . . every measure of everything has limitations. Knowing this is part of the requirement of becoming a professional working to make these limitations known and clear—and as limited as possible." He believed that there could be no perfect audience measurement system and that users of audience data had to live with trade-offs between cost, accuracy, and/or speed of delivery. At the same time, he was always concerned with studying the measurement process to improve research methods. He was a founding member of the American Association of Public Opinion Research (AAPOR), an organization that today continues the quest for better research.

The broadcasting industry often turned to Beville for his opinions on current issues. He wrote analyses for trade journals, spoke at industry events, and was frequently consulted about issues that affected audience measurement. His *Audience Ratings: Radio, Television, Cable* became an indispensable text in the field, explaining to lay and professional readers what the ratings process is all about.

In 1986 the Market Research Council (MRC), an organization that Beville had formerly served as president, selected him for its Research Hall of Fame. In 1989 the National Association of Broadcasters (NAB) and the Broadcast Education Association (BEA) announced the creation of the Hugh Malcolm Beville award to "recognize noted researchers who have made major contributions to the advancement of audience research in the broadcasting industry." The first award was given to William Rubens, retired vice president of research for NBC.

PATRICIA PHALEN

See also Advertising; Audience Research Methods; Cooperative Analysis of Broadcasting; Electronic Media Rating Council; National Broadcasting Company

Hugh Malcolm Beville. Born 18 April 1908. Attended Syracuse University, B.S., 1930; New York University, M.B.A., 1966; statistician, manager of research, director of research, vice president for planning and research, NBC, 1930–68; established Broadcast Ratings Council, 1963; member of American Association of Public Opinion Research; elected by Market Research Council to Research Hall of Fame, 1986; National Association of Broadcasters and Broadcast Education Association created Hugh Malcolm Beville award in his honor, 1989. Died in Port Washington, New York, 25 March 1988.

Selected Publications

The Market Research Council, 1958–1989, 1989
Audience Ratings: Radio, Television, Cable, 1985; revised edition, 1988

Further Reading

Banks, Mark James, *A History of Broadcast Audience Research in the United States, 1920–1980, with an Emphasis on the Rating Services*, Ph.D. diss., University of Tennessee, 1981
Buzzard, Karen S., *Chains of Gold: Marketing the Ratings and Rating the Markets*, Metuchen, New Jersey: Scarecrow Press, 1990
Buzzard, Karen S. "Radio Ratings Pioneers: The Development of a Standardized Ratings Vocabulary," *Journal of Radio Studies* 6, no. 2, (Autumn 1999)
Webster, James, Patricia Phalen, and Lawrence Lichty, *Ratings Analysis: The Theory and Practice of Audience Research*, 2nd edition, Mahwah, New Jersey: Erlbaum, 2000

Big D Jamboree

Country Music Radio Show

Although the *Big D Jamboree* never rivaled the influence of the mighty *Grand Ole Opry* or even small *Opry* cousins such as the *Louisiana Hayride* and *Wheeling Jamboree,* it was nonetheless a potent regional force that helped raise a number of country music artists, and later rock and roll artists, to national prominence. The radio barn dance, which broadcast on 50,000-watt KRLD in Dallas, Texas, also acted as an important stage for local talent.

The *Jamboree* had its roots in the *Texas Barn Dance,* a live country music show first staged at Dallas' Sportatorium in 1946. The *Texas Barn Dance* became the *Lone Star Jamboree* when it found its first radio home on WFAA in Dallas a year later. However, WFAA already featured a country music stage show (the *Saturday Night Shindig*), so in 1948 the show put down more permanent stakes on the airwaves of KRLD. Rechristened the *Big D Jamboree,* the barn dance debuted over KRLD on 16 October 1948.

KRLD, named for the Radio Laboratories of Dallas, had begun broadcasting on 31 October 1926 and achieved its 50,000-watt designation in 1938. The station's power allowed it to cover a 100-mile radius during daytime hours, and to reach more than 30 states during the nighttime hours. The *Big D Jamboree* indeed had a powerful conduit through which to reach its radio audience. In the early 1950s, the *Jamboree* gained wider distribution when the Columbia Broadcasting System (CBS) radio network agreed to feature the show on its *Saturday Night Country Style,* a program that featured various country music barn dances around the United States.

The show that would become the *Big D Jamboree* was the brainchild of Sportatorium owner Ed McLemore (who also staged wrestling matches in his venue), Dallas nightclub proprietor Slim McDonald, and KLIF radio (Dallas) disc jockey Big Al Turner. By the time the show appeared on KRLD, only McLemore still had a hand in producing the show. Turner, the show's host going back to the *Texas Barn Dance* days, emceed the *Big D Jamboree* at KRLD for a short time, but that role soon went to KRLD personality Johnny Hicks, who would be the on-air voice most associated with the program during its run. KRLD's Johnny Harper was the *Jamboree*'s announcer, and he also shared producing credit with McLemore and Hicks.

Cast members of the *Big D Jamboree* who achieved national success in country music included Billy Walker, Sonny James, Ray Price, Lefty Frizzell, Hank Locklin, and Charline Arthur. Many other country music performers visited the *Big D Jamboree* frequently and found it a career-amplifying stage; among them were Jim Reeves, Hank Snow, Ferlin Husky, Hank Thompson, Johnny Cash, and Hank Williams. Local acts who never enjoyed much national fame but who nonetheless expanded their profile in Texas included the Callahan Brothers, Romana Reeves, Sid King and the Five Strings, Jimmie Heap, Gene O'Quin, Riley Crabtree, and Okie Jones.

The rise of rock and roll in the mid- to late 1950s was one of the factors that ultimately silenced the *Big D Jamboree* and other barn dances featured on radio, but ironically the *Jamboree* had a role in fueling the dissemination of the rock sound during the genre's early days. Elvis Presley, who toured frequently in Texas at the outset of his career, appeared often on the program before he became nationally known, as did other important figures such as Carl Perkins and Gene Vincent (who was managed by Ed McLemore). Other notable purveyors of the new sound who also appeared regularly on the *Jamboree* were the Belew Twins, Wanda Jackson, Johnny Carroll, and Werly Fairburn.

By the late 1950s, the *Big D Jamboree*'s way had become uneven as rock and roll increasingly overshadowed country music, the primary staple of the program. The once-vital show limped along into the mid-1960s before fading. The *Jamboree* was briefly revived in 1970, but it failed to recapture the glory that inspired historian Kevin Coffey (2000) to call it "an enviable presence on the Southwestern music scene."

MICHAEL STREISSGUTH

See also Country Music Format

Hosts
Big Al Turner, Johnny Hicks

Producers
Big Al Turner, Johnny Hicks, Ed McLemore, Johnny Harper

Programming History
KRLD 16 October 1948–early 1960s

Further Reading
The Big D Jamboree Live, Volumes 1 and 2 (compact disc), Dallas, Texas: Dragon Street Records, 2000; see especially the liner notes by Kevin Coffey

Cooper, Daniel, *Lefty Frizzell: The Honky-Tonk Life of Country Music's Greatest Singer,* Boston: Little Brown, 1995

Gals of the Big D Jamboree (compact disc), Dallas, Texas: Dragon Street Records, 2001 (see especially liner notes by Kevin Coffey)
Guys of Big D Jamboree (compact disc), Dallas, Texas: Dragon Street Records, 2002 (see especially liner notes by Kevin Coffey)
Kingsbury, Paul, editor, *The Encyclopedia of Country Music,* New York: Oxford University Press, 1998
Malone, Bill C., *Country Music U.S.A.,* Austin: University of Texas Press, 1968; revised edition, 1985
Wilonsky, Robert, "Big D Jamboree," *Dallas Observer* (20–26 May 1993)

Biondi, Dick 1933–

U.S. Top 40 Radio Personality

As Top 40 radio grew and was embraced by the public, Dick Biondi served as a driving force in its development. Along the way, he earned a reputation as one of the bad boys of radio. It is a title he wears proudly, along with his claim that he has been fired 23 times during his career. Biondi has collected a variety of aliases and nicknames over the years that personify his reputation: the Wild Itralian [*sic*], the Screamer, Daddy-O Substitute, the Supersonic Spaghetti Slurper, the Big Noise from Buffalo, and the Limp Linguini. In 2001 Biondi celebrated his 50th year in broadcasting, still on the air, still performing at remote broadcasts, and still capturing the ears and hearts of fans in the Chicago area radio market with his shows on Oldies 104.3, WJMK.

Early Days

In 1941, as a child of eight, Biondi, the son of a firefighter and a short-order cook from Endicott, New York, moved from playing disc jockey using a wooden spoon as a microphone to his first on-air appearance at WNDO in Auburn reading a commercial. Later, he helped out at WINR in Binghamton. By 1951 he had his own shows, working a split shift in Corning, New York.

The next year, 1952, after being fired by a new manager, Biondi landed in Alexandria, Louisiana, working primarily as a utility or substitute disc jockey, but also hosting an all-black show called *Jammin Jive* on KYSO. Another change in management found Biondi on the road to York, Pennsylvania, in 1954.

From 1954 to 1958, Biondi held court in Youngstown, Ohio, at WHOT-AM. Literally working from the ground up, Biondi helped put down the tiles on the floor before the station went on the air. Biondi honed his talent for picking hit records while working in Youngstown by observing the reactions of his audiences as they watched performers at record hops hosted by Biondi.

WKBW in Buffalo, New York, hosted Biondi from 1958 to 1960. It was the departure of George "Hound Dog" Lorenz, one of radio's legendary disc jockeys and the man who introduced rock and roll to WKBW's evening listeners, that made Biondi's move to Buffalo possible. Whereas Lorenz had programmed his show without restrictions, new management in 1958 established a Top 40 format they expected Biondi to follow. Biondi, although not thrilled about the restrictions, took on the assignment in what he has described as something of a guerilla warfare role.

Chicago in the 1960s

In 1960 Chicago's WLS, one of the clear channel stations, became a premier Top 40 radio station. On 2 May 1960, Biondi unleashed his special brand of patter on the nine-to-midnight shift at WLS. For the next three years, he consistently ranked as one of the most highly rated air personalities in Chicago, regularly attracting well over half of the listeners in the area. Nationwide, Biondi led the country with a Pulse rating that showed an average 60 share of the national audience.

In mid-1963 Biondi left WLS because of a dispute over the number of commercials during his program and took his act to KRLA Pasadena/Los Angeles. Here, Biondi worked with such notable entertainers as Casey Kasem and Bob Eubanks. In addition to his radio work, Biondi was involved in promoting and hosting rock and roll acts throughout southern California through his *Dick Biondi Road Show.*

In October 1967 Biondi returned to Chicago, but this time to WLS's archrival, WCFL. Biondi describes working for Ken Draper, the program director, as being far different than his earlier experiences. Biondi and the other on-air personalities were referred to as "talent" rather than "jocks." The atmosphere was generally supportive, rather than combative.

1970s and Beyond

After a final disagreement with WCFL management in 1973 over his "abrasive style," Biondi moved briefly to an on-air position in Cincinnati, en route to North Myrtle Beach, South Carolina. At the time, Biondi intended, if not to retire, at least to take it easy for a time. In 1977, faced with a dwindling bank account, he presented himself at local radio station WNMB looking for an on-air position. Without the general manager's realizing his legendary status, Biondi was hired (at a salary of about $100 per week) and remained in the area until spring 1983.

A six-figure salary offer from WBBM-FM to host an oldies show brought Biondi back to Chicago as morning disc jockey. The expense to the station was well justified by a complete and virtually instant sellout of commercial time during Biondi's show. By August 1984, Biondi was firmly entrenched at Oldies 104.3, WJMK in Chicago, where he remains at this writing, easily his longest tenure in any post. He still makes a number of personal appearances each week and delights his audience with interviews and personalized greetings from the many legends of rock and roll he counts among his friends.

Stunts and Memories

Taking to the recording studio, Biondi had a minor novelty song hit with his original "On Top of a Pizza," which sold more than 11,000 copies. Biondi delighted in ordering off-the-wall food for delivery to the station, including peanut butter and sauerkraut pizza. During his time at KRLA in Los Angeles, he spent ten days in a cage with a chimpanzee and a typewriter. The idea was that eventually the chimpanzee would peck out the station's call letters (something the chimp never did manage). He stayed atop a flagpole at Idora Park in Youngstown for three days and nights.

Biondi delighted his audiences with a seemingly never-ending stream of the worst possible knock-knock jokes ("Knock-knock. Who's There? Biondi. Biondi who? Beyon-di blue horizon"). While airing Gillette shaving razor commercials, Biondi captivated his young male listeners by teaching them, on the air, the finer points of shaving. When he appeared at local high schools, he often dyed his beard to match the school colors.

Black slacks are a Biondi trademark. The tradition began when Biondi met Elvis Presley during a Cleveland concert. Elvis wore a kelly green jacket and black slacks. During a trip to Memphis, Biondi collected fallen leaves from the grounds of Presley's Graceland mansion, took them home, and then awarded them to his listeners. Biondi once wore a shirt that Elvis had autographed and then flung himself, wearing the shirt, into the audience. Fans went home with both a piece of Elvis and Biondi, while Biondi ended up in the hospital with multiple cuts and bruises.

Biondi's Goal

The only hobby Biondi indulges in is golf, and he categorizes himself as a duffer. He also says he likes to write. Biondi lists his one goal in life as wanting to be "the oldest, active, working rock-n-roll disc jockey in the United States." He's proud that he's maintained his skinny physique and that he still sports a full head of hair.

JIM GRUBBS

See also Contemporary Hit Radio Format/Top 40; Disk Jockeys; KRLA; Radio Hall of Fame; WBBM; WCFL; WLS

Dick Biondi. Born in Endicott, New York, 11 September 1932. First radio appearance in Auburn, New York, WNDO, 1941; disc jockey, Corning, New York, 1951; hosted "Jammin Jive," KYSO, Alexandria, Louisiana, 1952; disc jockey and original staff member, WHOT-AM, Youngstown, Ohio, 1954–58; disc jockey, WKBW Buffalo, New York, 1958–60; member of the original Top 40 announcing staff at WLS, Chicago, Illinois, 1960–63; held position as the number one nationally rated disc jockey with a Pulse average share of sixty, 1961–62; deejay at KRLA (Los Angeles, California) and entertainment entrepreneur (Dick Biondi Road Show) from 1963–65; introduced the Beatles in Los Angeles, 1964; disc jockey, WCFL, Chicago, 1967–73; disc jockey, WNMB, North Myrtle Beach, South Carolina, 1977–83; morning show host, WBBM-FM, Chicago, 1983–84; oldies disc jockey, WJMK, Chicago, 1984–present; inducted into the Rock and Roll Hall of Fame and Museum, 1998; inducted into the Radio Hall of Fame, 1998.

Selected Recordings

The Pizza Song (On Top of A Pizza), 1961; *Crusin' 1960*, 1970

Further Reading

Fong-Torres, Ben, *The Hits Just Keep on Coming: The History of Top 40 Radio*, San Francisco: Miller Freeman Books, 1998

Gray, Andy, "Hall of Fame's New Wing Is Big, Bright, and Brash," *Warren (Ohio) Tribune Chronicle* (3 April 1998)

Salholz, Eloise, "The Return of the 'Wild Italian,'" *Newsweek* (26 September 1983)

Smith, Wes, *The Pied Pipers of Rock 'n' Roll: Radio Deejays of the 50s and 60s*, Marietta, Georgia: Longstreet Press, 1989

Birch Scarborough Research

Birch Scarborough Research was a radio research firm in competition with Arbitron throughout the 1980s and early 1990s. The measurement of local radio audiences provided by the Birch reports and the qualitative research provided by the Scarborough service helped to fill the void left in syndicated local radio audience measurement after the Pulse organization closed down its operation in April 1978. Perhaps no competitor challenged Arbitron with as much consistency and industry support as did Birch Scarborough in the 1980s.

Birch Scarborough Research began in 1979 when Thomas Birch first started a radio ratings service called Radio Marketing Research. Birch had tested a system for measuring market shares and surveying music audiences by phone that met with success during test runs. Birch's research was first used to help determine programming, and his monthly service grew to include 18 markets by 1980. By March 1982 Birch was able to compete with Media Statistic, taking many Mediatrend subscribers from the other service. This gave him some major market subscribers and lent further credibility to his service. By 1984 Birch served 93 markets with the standard report format and he had hired two former Arbitron executives to work for his organization, making the service an even greater challenger in the radio research business. The executives were Richard Weinstein, who served as president and would later go on to become executive director of the Electronic Media Rating Council (now known as the Media Rating Council), and William Livek, who served as vice president.

The Birch system relied on telephone interviews that asked respondents to report ("recall") their listening pattern during the past 24 hours. Only one designated person per household was used in the survey, with the phone interviewer asking to speak to the person with the last or most recent birthday, an approach commonly known as the "last birthday" method. The selected respondent was asked about stations heard on the radio during the previous 24 hours, the location(s) (such as at home, at work, or in the car) where the listening took place, and in which time periods the listening took place. (Time periods or "dayparts" included 6 A.M. to 10 A.M. that day, 10 A.M. to 3 P.M. that day, 3 P.M. to 7 P.M. the previous day, and 7 P.M. to midnight the previous day.) The Birch system used pre-designated households in its sample; households were randomly selected from the Total Telephone Frame listing developed by A.C. Nielsen for the Nielsen Station Index. Each household was called during the evening hours, with three attempts made to reach the household and the designated respondent.

Birch served as a major competitor to Arbitron in the radio research market for a number of reasons. Its service was fairly inexpensive; it provided monthly data that was delivered every two weeks; and its research included qualitative components. Data collection was done in interview centers that allowed for oversight of interviewers and presumably ensured the quality of responses. And finally, Birch Scarborough laid claim to a response rate of over 60 percent, higher than that of Arbitron.

The basic Birch radio report was provided both monthly and quarterly. Monthly reports provided a combination of findings from the two most recent months of interviewing. Quarterly reports included average quarter-hour listening habits, daily listening habits, and weekly cumulative measures that were broken down by daypart, demographic group, and location of listening. Measures for ratings, shares, and cumulative audience were reported on an average quarter-hour and/or daypart basis.

Birch made efforts to weight its findings and thus balance the results obtained by compiling data from households of different sizes. Weighting was also used with data recorded for different days of the week, as not all respondents could be reached on the same day and their interviews often reflected listening on different days of the week (with different program schedules). Balancing methods were also used to account for disparities in ethnicity, age, sex, and county location factors.

Birch Scarborough conducted a number of studies on its own methodological research throughout the late 1980s and in 1990; for instance, in 1988 it conducted an analysis of those telephone calls placed via random digit dialing that resulted in no answer and busy signals. The results of this study were utilized by Birch to calculate response rates.

In 1989 the firm analyzed the number of attempts that were being made to reach specific demographic groups with its surveys. This analysis found no significant differences by gender or age group in the number of completed interviews. A separate study in 1989 examined the "seven day methods test." It compared the weekly cumulative numbers obtained via single-day interviews that examined listening on the current day and previous day with cumulative numbers based on successive daily interviews throughout the week, finding no significant differences between the two groups of results.

In 1990 Birch Scarborough completed a case study of the Hispanic market to evaluate how Hispanic respondents viewed the Birch interview. Personal interviews were conducted in San Antonio, Texas, and Miami, Florida, to determine (1) how well the respondents understood the Birch interview; (2) how they recalled their listening habits from the previous day; (3) which language they preferred to use during the interview; and (4) how the language used affected their responses. This case study was also used to gather other comments about the Birch survey.

Despite its status at the time as the nation's second-largest market research company, the Birch Radio Ratings Service was discontinued in 1992. Birch Scarborough attributed the move to financial losses. Scarborough Consumer Media and Retail Services continued, and a marketing arrangement made Arbitron the exclusive provider of Scarborough's qualitative data to radio and television stations. This precluded any further competition between the two companies.

MATT TAYLOR

See also A.C. Nielsen Company; Arbitron; Audience Research Methods; Electronic Media Rating Council

Further Reading

Beville, Hugh Malcolm, Jr., *Audience Ratings: Radio, Television, and Cable,* Hillsdale, New Jersey: Erlbaum, 1985; 2nd edition, 1988

Webster, James G., and Lawrence W. Lichty, *Ratings Analysis: Theory and Practice,* Hillsdale, New Jersey: Erlbaum, 1991; 2nd edition, as *Ratings Analysis: The Theory and Practice of Audience Research,* by Webster, Lichty, and Patricia F. Phalen, Mahwah, New Jersey: Erlbaum, 2000

Blacklisting

Blacklisting was a highly organized, institutionalized effort to deny employment to individuals assumed to be members of the Communist Party or to have communist sympathies. Begun shortly after the end of World War II, blacklisting in radio was a by-product of the larger hunt for communists led by Senator Joseph McCarthy and others. The senator was not, however, a major figure in the radio version of this witch-hunt. The entertainment industry had its own inspired group.

Some researchers trace the beginnings of blacklisting in radio to the founding in 1947 of *Counterattack: The Newsletter of Facts on Communism.* Three former Federal Bureau of Investigation agents, Theodore Kirkpatrick, Kenneth Bierly, and John Keenan, founded American Business Consultants and began publishing the aforementioned newsletter. They sent copies to advertising agencies, broadcasting executives, and sponsors along with offers to do special investigations. The newsletter listed entertainers of all types along with their supposed communist activities.

Others date the beginnings of blacklisting a little earlier, just after the end of World War II. Evidently a list of between 80 and 100 "undesirables" was circulated among broadcasting executives and shown to directors. At one network, the list came with a memo advising, "For Your Information: Keep these names in mind when casting."

Context

In order to understand blacklisting and why it worked, it is first necessary to understand the political climate of the post–World War II world. The war ended in 1945 with the Soviet Union in control of Eastern Europe. Within four years, communists came to power in China. The atomic bomb became a shared weapon. Alger Hiss and the Rosenbergs became front-page news when Senator Joseph McCarthy of Wisconsin first proclaimed that there were spies in the State Department in 1950, the same year the Korean War started. In short, it was a time ripe for demagoguery and exploitation. Into this climate of fear stepped the blacklisters.

Blacklisters needed the help of both the general public and the broadcasting industry to succeed. They needed the general public to be afraid of communism as of a mortal enemy. This meant that any method used to defeat such an enemy was allowable. If the public could be convinced of the danger, then the firing of the occasional innocent actor, writer, and so on would be understandable and permissible. This was, after all, a life-and-death struggle with an enemy who would use any means at his disposal to succeed. Therefore, one must be willing to use any means available to defeat him—including the sacrificing of some civil rights. This "end justifies the means" argument was a relatively strong one, considering the political state of the world. Finally, one also needed the public to believe that the entertainment industry was a prime target of an international communist conspiracy.

Blacklisters needed the broadcast industry, including advertisers and advertising agencies, to believe something else entirely. They needed the industry to believe that they, the blacklisters, could institute product boycotts and that such boycotts could ruin an advertiser, agency, or product. From the late 1940s through the middle 1950s, broadcasters, advertisers, and agencies all acted as if this were possible. From a distance of half a century, it is possible to wonder why broadcasters failed to truly question such assumptions (although some did, at great personal risk), but it is always necessary to remember time and place when discussing blacklisting. What the acceptance of such assumptions meant, however, was that instead of discussing whether blacklisting itself was morally correct, people argued over whether a particular individual should or should not be included on one of the many lists being circulated. Few asked whether the lists should be published to begin with.

Although the impression is often of a large, corporate force instituting blacklisting throughout the entertainment industry, the opposite is closer to the truth. There were Bierly, Keenen, and Kirkpatrick, who founded American Business Consultants and published the regular newsletter, *Counterattack*. In 1950 they would also be responsible for publishing *Red Channels*. There was Lawrence Johnson, a Syracuse supermarket owner; Vincent Hartnett, a talent agent associated with AWARE; Daniel T. O'Shea at the Columbia Broadcasting System (CBS); Jack Wren at Batton, Barton, Durston, and Osborne; and George Sokolsky of the Hearst papers. In addition to the publications of American Business Consultants, there was the American Legion's *Firing Line*, the *Brooklyn Tablet*, and the *American Mercury*. Although not all-inclusive, this list includes many of the major groups and individuals involved in the process.

The avowed goal of blacklisters was to root out communists in the entertainment industry. The following is from the September 1947 issue of *Counterattack*:

> The most important thing of all is to base your whole policy on a firmly moral foundation. Space should not be rented to the Communist Party or to any Communist front. Supplies should not be sold to them. They should not be allowed to participate in meetings or to have time on the air or to advertise in the press. No concession should ever be made to them for any business reason. Communist actors, announcers, directors, writers, producers, etc., whether in radio, theater, or movies, should be barred to the extent permissible by law and union contracts.

How It Worked

The way blacklisting worked was relatively simple. Entertainers of all types were listed along with their supposed communist affiliations. Networks and advertising agencies then used the lists when deciding whom to hire. The names were gathered from a number of different sources. Old editions of the *Daily Worker* were searched for incriminating references. Office stationery, letters, publicity, and the like from groups labeled communist fronts by the U.S. Attorney General's Office or by the blacklisters themselves were also scoured for names. Names were also supplied by friendly witnesses to governmental agencies that were supposedly searching for communist infiltration of the entertainment industry. Prominent among these groups were the Tenney Committee in California and the House Un-American Activities Committee of the United States House of Representatives. Sometimes, as was eventually shown in court, the listings contained half-truths. Sometimes they were outright lies. Always, the blacklisters were after "names."

Once an entertainer was "listed" in one of the blacklisters' publications, it became almost impossible to find work. The problem was in the way radio was supported. In theater or film, there is a direct correlation between success and people attending the event, but that was not true in radio. Advertisers placed commercials, often in programs they themselves produced, with the idea that people would hear the advertisement and buy the product. Any negative publicity surrounding the program was thought to reflect on the product itself. Advertisers feared listener boycotts of their products if the programs they produced used actors listed in blacklisting publications.

Lawrence Johnson, a Syracuse supermarket owner and active supporter of blacklisting, was excellent at instilling the fear of a boycott into an advertiser's mind. When notified that a program was using actors listed in *Counterattack* or by AWARE, Johnson would write a letter to the program's sponsor. Johnson would offer to hold a test in his supermarkets. A sign in front of a competitor's brand would say that it sponsored programs that used only pro-American artists and shunned "Stalin's Little Creatures"—a phrase for which Johnson was famous. The sign in front of the sponsor's product was to explain why its maker chose to use communist fronters on its program. Johnson then said he would hold the letter for a few days awaiting a reply. He threatened that if he received no reply, he would send a copy to the United States Chamber of Commerce, the Sons of the American Revolution, the Catholic War Veterans, the Super Market Institute in Chicago, and others. The goal was to scare the sponsor into believing that blacklisters could really create a meaningful product boycott. No one at the network level ever called the bluff.

The year 1950 became a high-water mark for blacklisting in the United States when American Business Consultants published what may be the most successful blacklist. Appearing just before the outbreak of the Korean War, *Red Channels* listed 151 artists in the entertainment field and their communist affiliations. Although many broadcast executives claimed to be appalled by the names included on the list, the book nevertheless became known as the "Bible of Madison Avenue." Two examples should prove the book's effectiveness.

Irene Wicker was one of those listed in *Red Channels*. Kellogg's had sponsored her *Singing Lady* program. The sponsor dropped the program after the publication showed she had one listing. Her sole citation was that she had sponsored a petition for the reelection of Benjamin J. Davis to Congress. The citation was based on an item in the *Daily Worker*. Wicker claimed she had never heard of Davis and went to great lengths to prove she had never sponsored a petition for his reelection. Her lawyer even got a court order to examine all 30,000 names on Davis's petitions. Wicker's was not among them. Although this "cleared" her, it still did not make her employable, for now she had become controversial.

Another of those listed by *Red Channels* was Jean Muir. A former movie actress, she had been hired by the National Broadcasting Company (NBC) to play the role of the mother in *The Aldrich Family*—a former radio drama being developed for television. *Red Channels* included nine listings for her. When Kirkpatrick was informed that NBC was going to use an actor listed in *Red Channels,* he organized a protest over her hiring. The end result was that Muir was paid in full for her contract but was never seen on the program. The industry never attempted to determine how widespread the protest was. The concern was that there was a protest at all. General Foods, the program's sponsor, issued a press release stating, "The use of controversial personalities or the discussion of controversial subjects in our advertising may provide unfavorable criticism and even antagonism among sizeable groups of customers." The sponsor did not want its product placed in any negative light at all. The publicity surrounding the Muir case had made the whole issue of blacklisting much more public than either the networks or the advertising agencies wanted it to be. As a result of the Muir case, blacklisting became institutionalized. The networks and agencies developed a system whereby all those involved in programs were screened ahead of time. Those found to have some sort of "communist affiliation" were simply never offered employment, rather than fired later. This cut back on some of the negative publicity.

Like Wicker, Muir also tried to clear herself, but she remained unemployable. The real point of the Muir case is that by the end of 1950, the industry had accepted the blacklisters' standard on employability. If someone was listed, the networks would not employ him or her. NBC had tried with Muir, but when confronted with a token protest, NBC caved in, thus allowing *Red Channels* and publications like it to set the standards by which performers would be judged. Understand that Muir was not a part of a communist plot, nor was she a member of the Communist Party. She had merely participated in liberal political activities in the 1930s. For those actions, her career was destroyed.

To speak out against the blacklisters was to put one's own career in jeopardy. Raymond Swing was chief commentator for Voice of America when he was invited to debate blacklisting with Kirkpatrick before the Radio Executives Club of New York. While vigorously defending the American system of government, he attacked those in charge of the radio industry:

> If, by some bleak and dreadful tragedy, American radio should come under the control of persons intent on producing a single conformity of thinking in America, it will not be the pressure groups or the blacklisters who will be to blame, but those now in charge of radio. They have it in their keeping, and what happens to it will be their doing.

After his appearance, Swing found himself under attack by both the blacklisters at *Counterattack* and by Senator Joseph McCarthy. Although it was reported that industry executives applauded Swing's role, they did nothing to change blacklisting in radio. That would be left to a single radio personality—John Henry Faulk.

Blacklisters, and to some degree sponsors and networks, believed themselves to be immune from prosecution. After all, blacklisters merely transcribed their information from other publications. If there was a mistake, it was the fault of the publication from which the information had been gathered. Those listed should sue those publications. Sponsors and networks felt immune because they certainly had a right to hire those people they felt best suited the job. This was to change.

John Henry Faulk

In 1956 John Henry Faulk was elected second vice president of the New York Chapter of the American Federation of Television and Radio Artists (AFTRA). Charles Collingwood was elected President and Orsen Bean first vice president. All had made their anti-blacklisting beliefs known. AWARE was particularly upset and sought to get both Bean and Faulk fired. Bean became unemployed almost immediately but was able to rely on his club work as a stand-up comedian. Faulk, however, was vulnerable. He was an employee of CBS Radio and had been doing some television at the time of his election. Shortly after his election, his name appeared in the publications of AWARE, which accused him of collusion and fellow-traveling. His radio sponsors deserted him, and CBS fired him. Faulk then took the unusual step of suing AWARE and Vincent Hartnett. It took six years before the case came to trial. Hartnett eventually admitted, "I was sold a barrel of false information," when questioned about the listings next to Faulk's name. Many of the citations on Faulk were incorrect, and others were intentionally misleading. In 1962 the jury awarded Faulk $3,500,000. The award was later reduced to approximately half a million dollars, and Faulk saw little of the money. Lawrence Johnson, who had avoided testifying, died of an overdose of barbiturates the day the verdict was announced. Thanks to the publicity from this case, blacklisting was, more or less, officially dead.

It was not just a single lawsuit that ended blacklisting but several things that came together in the late 1950s. First, the Faulk lawsuit placed advertisers and networks alike on notice that to maintain an official blacklist was to court financial disaster. Second, advertisers were moving away from the program production end of broadcasting. This removed them from the day-to-day hiring of entertainment personnel and made them less vulnerable to threats. Third, it became widely known that many blacklisted individuals had continued to work under assumed identities with no negative consequences for networks or advertisers. By the late 1950s blacklisters were no longer able to raise the same level of response from the public over their allegations. Combined, these factors helped end the effective reign of blacklisters.

Looking back, the real goal of the blacklisters seems to have been publicity. They were constantly after names. No plot was ever uncovered to use radio to convert the masses to communism. No evidence was ever found of a left-wing conspiracy to blacklist anticommunist actors. No spies were found in the radio industry. Blacklisting became a self-perpetuating effort at continued publicity. It destroyed careers and, in some cases, lives.

The real issue was whether there should have been a blacklist at all. Unfortunately, that particular issue was rarely raised. Yes, there was some editorializing during and after the Muir case, but the bottom line is that *Red Channels* was both accepted and used by networks and advertising agencies alike. Some, such as Edward R. Murrow, raised the issue, but they had difficulty sustaining it. Although on the one hand, Murrow was allowed to fight Senator Joseph McCarthy on his CBS program *See It Now,* on the other hand, the network was running its own in-house blacklisting organization headed by Daniel T. O'Shea.

It is also true that the relative number affected was really quite small when compared with the number of people employed in radio and television. This was of no consolation, however, to the Muirs, Wickers, and others who were ruined by the process. The real conclusion is that the broadcasting industry lacked the will to fight blacklisting. Although it is possible to find exceptions to this pattern, they are most notable because they are exceptions. The industry as a whole allowed both itself and the First Amendment to be battered at the hands of blacklisters.

DAVID E. TUCKER

See also Aldrich Family; Cold War Radio; Faulk, John Henry; Red Channels; Swing, Raymond Graham

Further Reading

Buckley, William F., Jr., and L. Brent Bozell, *McCarthy and His Enemies: The Record and Its Meaning,* Chicago: Regnery, 1954

Cogley, John, *Report on Blacklisting,* 2 vols., New York: Fund for the Republic, 1956; see especially vol. 2, *Radio-Television*

Dunne, Philip, *Take Two: A Life in the Movies and Politics,* New York: McGraw-Hill, 1980; updated edition, New York: Limelight Editions, 1992

Faulk, John Henry, *Fear on Trial,* New York: Simon and Schuster, 1964

Foley, Karen Sue, *The Political Blacklist in the Broadcast Industry: The Decade of the 1950s,* New York: Arno Press, 1979

Kanfer, Stefan, *A Journal of the Plague Years,* New York: Atheneum, 1973

Miller, Merle, *The Judges and the Judged,* Garden City, New York: Doubleday, 1952

Red Channels: The Report of Communist Influence in Radio and Television, New York: Counterattack, 1950

Black-Oriented Radio

Although African-Americans have participated in radio since its inception in the early 1920s, specific programs directed to blacks did not develop in any appreciable form until the late 1940s and early 1950s. Historically, black-oriented radio first provided music and comedy. Later, public affairs, news, and programming for the entire community made their way to the airwaves.

Early Black Radio

The first African-American to have a commercially sustained radio program was Jack Leroy Cooper, a former vaudevillian and entrepreneur who began announcing on Washington, D.C.'s WCAP radio station in 1925. Later, in Chicago, Cooper worked at WSBC, where he started the *All Negro Hour* in 1929. Among Cooper's many accomplishments were hiring African-Americans to work as announcers and salespeople, playing gospel music, broadcasting sports, and developing a missing persons program to help individuals find loved ones. Moreover, Cooper created the concept of the disc jockey when his studio musician walked out: Cooper began playing records and talking between them when a local musician's union demanded that his pianist go on strike. In 1947 *Ebony Magazine* called Cooper the "Dean of African-American Disc Jockeys." By that time he was responsible for more than 50 programs broadcast on four Chicago radio stations.

Early network radio developed programs that included black characters; however, these were often in stereotypical roles. On some shows, such as *Amos 'n' Andy,* whites portrayed blacks on the air in stereotypical fashion, and in other programs African-Americans portrayed themselves in this manner. Most often, African-Americans were featured as maids, butlers, and gardeners and in other vocational or domestic-helper roles.

Amos 'n' Andy became one of the most popular radio programs of all time. African-Americans listened to the show and probably laughed at the antics of its characters, even though the program often portrayed blacks in an unfavorable light. Nevertheless, the National Association for the Advancement of Colored People and some other organizations believed that the program demeaned African-Americans and urged the Columbia Broadcasting System (CBS) to cancel the show.

In the many comedies broadcast on network radio, African-Americans played key roles in the success of the programs. Eddie Anderson earned fame as "Rochester" on the *Jack Benny Show.* Hattie McDaniel played "mammy" roles on the *Optimistic Doughnut Hour* and in the radio series *Showboat.* Later, McDaniel played the lead role in *Beulah.*

During the early years of radio, black music such as jazz and blues was often heard on network radio. Bessie Smith's live blues performance was broadcast from a Memphis radio station in 1924. In addition, groups such as the Hampton Singers performed on radio in that same year. Jazz, especially, received a great deal of airplay through the 1940s. Bandleaders such as Cab Calloway, Duke Ellington, and Thomas "Fats" Waller were among the many African-American musicians who made regular broadcasts on early radio. Marian Anderson, Roland Hayes, and Paul Robeson were among the African-American vocalists heard on early radio broadcasts.

Early non-entertainment programming such as public-affairs programs reported on the status of the black family, educational activities in the black community, farming techniques, and occasionally racial issues. In 1933 CBS broadcast *John Henry, Black River Boat Giant,* a positive drama featuring African-Americans. By the 1940s the National Broadcasting Company's (NBC) *Freedom's People,* an eight-part series featuring dramatic vignettes by African-Americans such as Paul Robeson, Joe Louis, and A. Philip Randolph, was broadcast. Non-network radio broadcasts included Roy Ottley's *New World a'Comin'* and Richard Durham's *Destination Freedom,* which focused on historical treatments of African-American experiences. These programs provided positive portrayals of African-Americans to radio audiences. In addition, radio stations began gospel music and church-service broadcasts in the 1940s.

NBC's Blue network broadcast *America's Negro Soldiers* in 1941. Sponsored by the U.S. Department of War, *America's Negro Soldiers* included patriotic vignettes that highlighted the historical contributions black soldiers made to the U.S. Army. Other program components included music, singing, and tap dancing, but the program omitted references to racial discrimination in American society. CBS's *Open Letter on Race Hatred,* however, examined the causes and consequences of the Detroit Race Riot of 1943. Other radio programs developed for African-American listeners during World War II were "Judgment Day" (1942), "Beyond the Call of Duty" (1943), "Fighting Men" (1943), "Gallant Black Eagle" (1943), and "The Negro in War" (1945).

Postwar Rise of Black-Oriented Radio

By the late 1940s radio began to broadcast programs targeted directly to predominantly African-American audiences. For example, in 1946 CBS and NBC produced specials that highlighted significant events in the African-American community. The CBS program shed a spotlight on "National Negro Newspaper Week," and the NBC program focused on Nat King Cole, the famous singer. Two major factors had an impact on the networks' attempts to reach African-American listeners. First, national advertisers recognized African-American economic power. Thus, companies that produced products such as canned milk, flour, and lard directed their advertising messages directly to black consumers via radio.

Second, television began to siphon off advertising dollars, audiences, and top-name performers from network radio. Aside from the money involved, radio performers soon realized that they would also receive greater exposure to larger audiences on television than on radio. Station owners responded to these developments by changing their formats and playing jazz, rhythm and blues, blues, rock and roll, and other black musical forms to appeal to African-American listeners. They also hired disc jockeys whose words, personalities, and music dramatically increased the number of black listeners. When first introduced, radio stations broadcast rhythm and blues, blues, and other black music formats in segments. A few hours during the day was set aside for these broadcasts. Eventually, stations began to build their entire formats around "black-appeal" programming.

In 1948, for example, WDIA in Memphis, Tennessee, broadcast its first program to black audiences. Nat D. Williams, pioneer black disc jockey at WDIA, hosted this show and many others for years to come. Soon after the initial broadcast, WDIA began an all-black programming format. Prior to WDIA's efforts, few black-appeal radio stations or programs existed. Notable exceptions included Cooper's *All Negro Hour* and Chicago disc jockey Al Benson's programs. WDIA's programming included public affairs, news, public service announcements, and other community service announcements and promotions.

Radio stations across the United States quickly imitated WDIA, which also became known as the "Mid-South Giant," because of its broadcast signal. The station reached audiences in Mississippi, Tennessee, and Arkansas. Other black-oriented stations hired flamboyant black disc jockeys such as "Jockey" Jack Gibson, also known as "Jack the Rapper"; Maurice "Hot Rod" Hulbert; and Peggy Mitchell Beckwith to play music, advertise and promote products, and especially to communicate with African-American listeners. Their unique personalities and knowledge of black music and recording artists catapulted black-appeal radio stations to unprecedented popularity among listeners.

Black disc jockeys in northern urban areas performed a number of other functions at the radio station. They often provided useful public service advice and served as counselors to many of the newly arrived migrants from the South, informing them about where best to shop and how to avoid the dangers in their new urban environments.

The popularity of their radio presentations, sometimes referred to as "personality" radio, began to decline in the late 1950s, mainly as a consequence of the payola scandal and the movement toward formatting in radio. Payola, or the payment of unreported money to play records, was legal but became rampant in the industry. The U.S. government outlawed the practice in the late 1950s. Thus, disc jockeys lost the opportunity to play records they deemed popular or attractive to listeners. Instead, that role eventually became one for program directors and other managers to take over.

Moreover, black-oriented radio stations began using a more tightly controlled music format, which did not allow disc jockeys to express themselves as they had in the past. Instead, black-oriented radio stations began to promote call letters, dial positions, and themes, such as "The Quiet Storm" and "the Black experience in sound."

In addition to the disc jockeys who worked for black-oriented radio stations during these years, African-American news reporters and public-affairs announcers also found jobs. Eddie Castleberry and Roy Woods, Sr., became well known for their announcing and reporting skills.

Throughout the civil rights movement, black-oriented radio stations assisted in the struggle for African-American human rights. Broadcasts from these stations provided listeners with accounts of newsworthy events, such as marches, boycotts, and voter registration drives. Additionally, black-oriented stations were often in the forefront in bringing attention to societal ills suffered by African-Americans, including police brutality and violence directed at them. The Reverend Martin Luther King, Jr.'s sermons and speeches were widely broadcast on black radio stations. Message music from black recording artists such as Curtis Mayfield and Little Milton found airplay on black-oriented radio stations.

Black Music and Black Ownership

Although hundreds of radio stations played black music, hired African-American announcers, and used promotions that appealed directly to black listeners, few of these stations were actually owned by African-Americans. J.B. Blayton bought WERD in Atlanta, Georgia, in 1948, becoming the first African-American to own a commercial radio station. WERD played black music and employed disc jockeys such as Jack Gibson, "Joltin' Joe Howard," and Helen Lawrence to appeal to African-American listeners.

By 1970 African-Americans owned only 16 stations out of more than 7,000 commercially operated facilities. Throughout the 1970s, the number of black-owned radio stations increased to 140. During the late 1980s and 1990s, the number of black-owned radio stations further increased but then started to decline to levels approximating those of the late 1970s.

Several factors contributed to this decline, among them greater consolidation in the radio industry, broadcast deregulation, and advertising practices that had a negative impact on the overall dollars generated by black-owned stations. The combination of consolidation in the radio industry and group owners' greater control over the advertising dollars in local markets left the often poorly financed black-owned radio stations unable to compete economically in today's marketplace, forcing many owners to sell.

In addition, some studies have cited the lack of access to investment capital and the lack of policies and incentives that promote African-American ownership of radio stations. One change adversely affecting black ownership was a Federal Communications Commission (FCC) decision not to "extend enhancement credits" for African-American ownership. Enhancement credits helped make African-American applications more competitive in comparative hearings. Other deregulatory actions adversely affecting black radio station ownership included the relaxing of ownership caps. In 1992, for instance, the FCC relaxed national ownership limits, allowing a broadcaster to own up to 18 AM and FM stations nationally.

Moreover, in 1995 Congress repealed the FCC's tax certificate program. This industry incentive had provided tax benefits to the seller of a media property that was sold to a minority broadcaster. Finally, the passage of the Telecommunications Act of 1996 further deregulated the industry. The act removed all national caps on radio station ownership. On the local level, ownership restrictions were considerably liberalized, allowing increased ownership of stations, up to 50 percent of stations in a market, up to a maximum of eight, depending on market size.

The combined impact of these changes has made it difficult for African-American owners to generate revenues to compete successfully with group-owned stations. These changes, however, do not necessarily affect advertising revenues for all black-oriented radio stations, because many are owned by conglomerates and use scale economies to achieve efficiencies and revenue generation.

Finally, African-American ownership, or the lack thereof, does not affect the number of black formats available to listeners. Black-oriented radio stations play music that African-Americans expect to hear. Many of these stations developed into outlets that emphasized music programming and used promotional slogans such as "Soul Music" stations, "The Total Black experience in Sound," and "The Quiet Storm" to appeal to listeners. Non-entertainment programming on some of these stations, however, suffers when local ownership disappears.

Black-Oriented Radio Formats Today

There are nearly 500 black-oriented radio stations operating in the United States today. The most popular format on these radio stations is urban contemporary. This format plays music from several genres, including rhythm and blues, urban adult contemporary, dance, urban gospel, rap music, and jazz. Fifty-seven percent of all African-Americans aged 12 and older listen to urban contemporary formatted radio stations. Moreover, black-oriented radio has popular appeal among other ethnic groups, including Asians, whites, and Hispanics. The majority of its listeners, however, are African-Americans—indeed, 90 percent of the listeners to black-oriented stations are minorities. Eighty percent of listeners to the urban contemporary format are minority group members. Other black-oriented formats include "black talk" and blues. General market radio stations attract a 21 percent minority audience.

Whether black-owned or not, most black-oriented radio stations attempt to establish ties with the local communities in which they are licensed. In addition to the music these stations play, they also offer public-affairs programming, news, and public service announcements.

Black-oriented noncommercial radio stations also play an important role in entertaining and informing the African-American community. Sometimes called "community stations," these stations schedule programs with the idea of helping communities create strong identities. Most of these stations allow the public access to the airwaves, especially those who may not otherwise have an opportunity to play their styles of music, such as jazz, reggae, or other Caribbean sounds. These stations also give individuals an opportunity to voice controversial and unpopular opinions.

For example, noncommercial stations provide forums to discuss issues such as police brutality, racism, disparities in incarceration rates, poverty, and other forms of racial discrimination. Moreover, black-oriented noncommercial radio stations are often staffed and operated by "activists" who argue that "mainstream" commercial black-oriented radio stations fail to adequately educate and inform African-Americans on such issues as AIDS, the U.S. legal system, and racial discrimination. Thus, black-oriented noncommercial radio stations provide news, public affairs, and other information often excluded from black-oriented commercial radio stations.

GILBERT A. WILLIAMS

See also African Americans in Radio; Amos 'n' Andy; Black Radio Networks; Blues Format; Community Radio; Cooper, Jack L.; Durham, Richard; Hulbert, Maurice "Hot Rod"; Jazz Format; Joyner, Tom; Payola; Urban Contemporary Format; Williams, Nat; WDIA

Further Reading

Alexander, Keith, "Riding the Signal: Cathy Hughes Takes Command of Radio Airwaves," *Emerge* (1 September 1999)

"Annual Negro Radio Issue," *Sponsor* (1952–56)

Berry, Venise T., and Carmen L. Manning-Miller, editors, *Mediated Messages and African-American Culture,* Thousand Oaks, California: Sage, 1996

Kernan, Michael, "Around the Mall and Beyond," *Smithsonian* 27 (April 1996)

MacDonald, J. Fred, *Don't Touch That Dial! Radio Programming in American Life, 1920–1960,* Chicago: Nelson-Hall, 1979

Newman, Mark, *Entrepreneurs of Profit and Pride: From Black-Appeal to Radio Soul,* New York: Praeger, 1988

Spaulding, Norman W., "History of Black Oriented Radio in Chicago, 1929–1963," Ph.D. diss., University of Illinois, 1981

Williams, Gilbert Anthony, "The Black Disc Jockey As a Cultural Hero," *Popular Music and Society* 10 (Summer 1986)

Williams, Gilbert Anthony, *Legendary Pioneers of Black Radio,* Westport, Connecticut: Praeger, 1998

Black Radio Networks

Although the major national radio networks got their start in the 1920s, a network dedicated to African-American listeners did not make its debut until 1954, when the National Negro network (NNN) went on the air. It differed from the older networks in that it did not own any radio stations, but it resembled them in providing programs to affiliate stations.

Origins

Two driving forces helped launch the NNN. The first impetus had to do with finding a way of reaching African-American consumers. National advertisers, seeking ways to increase market share, decided that by using "Negro-appeal" radio stations they could better achieve their goals. For many national companies, sales to African-American consumers often represented the difference between breaking even and increased sales, and finding a way to reach them with advertising was thus an important goal.

Another driving force for the creation and development of a national black radio network came from a desire of African-American entertainers to reach a national black audience with their programs. Toward those two ends, Leonard Evans, publisher of a black radio trade magazine, organized the NNN in 1954 in order to distribute programming to affiliates. The NNN, for example, produced and distributed *The Story of Ruby Valentine,* a soap opera starring Ruby Dee and Juanita Hall. African-American entertainers Cab Calloway and Ethel Waters also produced NNN programs. Moreover, other programming on the network represented a range: there was high-brow fare, such as symphony concerts broadcast from black colleges, for example a concert at North Carolina College in Durham hosted by African-American disc jockey Norfley Whitted. There was also personality radio, featuring the latest rhythm and blues, blues, and jazz music. These music programs were hosted by African-American disc jockeys at the various NNN affiliates around the country.

Ruby Valentine was broadcast on 45 radio stations. Pulse ratings indicated that the show received a 2.0 rating among African-Americans in 1954. The show and its network lasted three years. National advertisers pulled away from the network, realizing that local black DJs had probably more appeal than national DJs.

After the demise of NNN, other attempts to create a black radio network were led by Chicago disc jockey Sid McCoy, whose syndicated programs were heard in 61 markets in 1957. McCoy's programs featured interviews with well-known personalities from the world of music. In 1958 McCoy's *Showcase,* a talk program aimed at African-American listeners, became a staple on radio stations in 32 markets.

Norman Spaulding organized Feature Broadcasting Company in 1960. Feature produced radio programs that covered sports, domestic issues, and black history; it also produced a program moderated by Ethel Waters called *Advice to the Housewife.* These syndicated efforts laid the foundation for more ambitious developments in black radio networking.

Developments in the 1970s

The Mutual Broadcasting System helped launch the first black all-news radio network in 1972, the Mutual Black network (MBN). The MBN had two principal bases of operation, New York and Washington, D.C. The network produced five-minute news and sports reports and distributed them to its affiliates daily. The New York office was led by veteran broadcast journalist Sheldon Lewis, and long-time news reporter Ed Castleberry headed up MBN's Washington, D.C., office. MBN distributed programming to approximately 90 affiliates, using telephone lines subleased from the Mutual Broadcasting Service. Later, as the number and types of programs increased and changed, MBN used leased satellite transmission facilities to distribute its programming. MBN employed approximately 50 people, with about half working in each of its two main offices.

Another black all-news radio network got its start just a few months after the MBN operation began. The Sheridan Broadcasting network (SBN) in Pittsburgh was developed as part of the Sheridan Broadcasting Corporation. Ron Davenport, Philadelphia native and entrepreneur, along with other investors purchased four radio stations—WAMO AM/FM (Pittsburgh), WUFO-AM (Buffalo, New York), and WILD-AM (Boston). These stations formed the initial media investments of the Sheridan Broadcasting Corporation.

In 1976 MBN, which had been struggling financially, merged with Davenport's Sheridan Broadcasting Corporation when the latter purchased 49 percent of MBN. In 1979 Sheridan bought the remaining 51 percent of the shares, and it became part of SBN. Upon gaining control of MBN, Sheridan expanded its programming offerings to affiliates to include *Money Smarts,* a financial report broadcast daily, and *Coming Soon,* a movie review program. In addition, the network produced *Major League Baseball Notebook, NFL Playbook,* and the *NBA Report* to provide listeners with coverage of the nation's professional athletes in those sports. In addition to sports coverage, the SBN also broadcast Lou Rawls's *Parade of Stars* telethon, an annual fund-raiser for the United Negro College Fund. By 1990 SBN boasted more than 150 affiliates and grossed more than $15 million in annual revenues.

A third black-owned radio news network was established in 1973 in New York City. The National Black Network (NBN) employed 50 people. NBN used a combination of telephone lines, satellite interconnection, and microwave relays to distribute its programming nationally. NBN broadcast its news to affiliates in Los Angeles, New York, and five other major U.S. cities. Eugene D. Jackson became NBN's first president. Sidney Small played a significant role in securing financing for the organization, and Del Racee, another founding member, brought radio station operations and know-how to the group of founding members.

NBN's initial offerings included *Black Issues and the Black Press,* a weekly 30-minute news show, and *One Black Man's Opinion,* which featured the commentaries of veteran newsman Roy Wood, Sr., and aired five days a week. Also in its lineup of programs was the *Ossie Davis and Ruby Dee Story Hour,* a one-hour weekly series hosted by this husband-and-wife team, featuring poetry, historical anecdotes, interviews, and music. By 1977 NBN served 80 affiliate stations.

Aside from the information and entertainment NBN provided to affiliates and listeners, the network also increased the available options of national advertisers to reach African-American consumers. NBN's demographic profile indicated that it had a 54 percent audience share among African-American women, a 47 percent share of African-American men, and a 73 percent share of the listening audience of African-American teens aged 12–17.

Modern Black Radio Networks

By the end of the 1980s, NBN served 94 affiliates, reaching nearly 20 million African-American listeners each week with news, sports, and information programming. NBN had gross revenues exceeding $10 million by the late 1980s. Its expanded programming services included such shows as *Energy Insight,* a consumers' program, and *Short Cuts.* In addition, the network added a late-night talk show, *Night Talk,* hosted by Bob Law. Its parent company, Unity Broadcasting, continued to expand and purchased two radio stations, WDAS AM/FM.

In 1991 SBN bought NBN, creating the American Urban Radio network (AURN). By 2000 this network had more than 250 affiliates and reached nearly 90 percent of African-American listeners. The network was headquartered in Pittsburgh and was the only black radio network in America at the beginning of the 21st century. AURN offers affiliates news; public-affairs programs; and syndicated features on finance and money, health, and minority business ventures. In addition, AURN produces and distributes sports features, entertainment, and cultural offerings, including programs that focus on black music, comedy, media, and women.

By the late 1990s, AURN had become the third largest radio network operating in the United States. Its five divisions—entertainment, marketing and promotion, news, public affairs, and sports—produce programs especially designed for African-American listeners. For example, its STRZ Entertainment network offers programs on black music (*USA Music Magazine*), media, comedy (*STRZ Funline*), and shows for women (*Cameos of Black Women*). The news division, American Urban News (AUN)/SBN News, distributes two separate news reports. AUN news is a three-and-a-half-minute news summary, delivered on the hour from 6 A.M. to 10 P.M. each day. The SBN newscast is broadcast in five-minute segments at half past the hour. These satellite-delivered newscasts reach approximately 8 million listeners each week.

The Urban Public Affairs network (UPAN) is responsible for special programs, such as those developed for Black History Month, and for regular features covering consumer issues, health, minority business activities, and financial matters. Special programs on UPAN include, for example, memorials created for the Dr. Martin Luther King Jr. birthday holiday and for the national elections (*Election Day: America*).

The Sports network on SBN offers the same programming that it offered prior to the merger of the two black radio groups. The SBN Urban network programs are new, however. This AURN network distributes to affiliates marketing and promotional materials and services, including direct-mail campaigns and sweepstakes promotions.

GILBERT A. WILLIAMS

See also African-Americans in Radio; Black-Oriented Radio; Mutual Broadcasting System

Further Reading

Barlow, William, *Voice Over: The Making of Black Radio,* Philadelphia, Pennsylvania: Temple University Press, 1999

Edmerson, Estelle, "A Descriptive Study of the American Negro in the United States Professional Radio, 1922–1953," Master's thesis, University of California, Los Angeles, 1954

Newman, Mark, *Entrepreneurs of Profit and Pride: From Black Appeal to Radio Soul,* New York: Praeger, 1988

"NNN: Negro Radio's Network," SPONSOR (20 September 1954)

Spaulding, Norman, "History of Black-Oriented Radio in Chicago, 1929–1963," Ph.D. diss., University of Illinois, 1981

Williams, Gilbert Anthony, *Legendary Pioneers of Black Radio,* Westport, Connecticut: Praeger, 1998

Block, Martin 1903–1967

U.S. Radio Announcer and Disk Jockey

The show most often credited with initiating the modern disc jockey movement was Martin Block's *Make Believe Ballroom* on WNEW in New York City. Block's *Ballroom* began in February 1935, during the trial of Bruno Hauptmann for the kidnapping of the Lindbergh baby. Block, who had started working at WNEW in December 1934 after moving to New York from California, was a $20-a-week staff announcer engineering the station's broadcasts of the trial.

Block convinced station manager Bernice Judis to let him use phonograph records to fill some of the gaps between trial segments. When the next recess arrived, Block introduced and played several records. His manner of presentation, however, was unique: Block pretended that the show was a live broadcast from a giant dance hall with a glittering chandelier. He played a master of ceremonies role, introducing songs as if the bands were actually there. The public was intensely interested in the Hauptmann trial, and Block found a ready-made audience. He located a sponsor, Retardo Weight Reducing Pills, and convinced the station to add the show to its regular schedule.

Martin Block
Courtesy Library of American Broadcasting

In time, Block's *Make Believe Ballroom* became the nation's preeminent disc jockey show, and during various periods it was broadcast over the Mutual and American Broadcasting Companies (ABC) networks. It was even syndicated over the Voice of America briefly. Thus Block became the first popular icon of the disc jockey genre and also its first millionaire.

Two important elements in Block's success were his voice, the tonal qualities of which evoked a physical sensation in some listeners, and his on-air personality. Because he had held various jobs as a salesperson prior to his radio career, Block could be quite persuasive. One New York bakery credited him with increasing their sales by 144,000 doughnuts in one week, and an appliance dealer in New Jersey claimed that his show helped the store sell 109 refrigerators during one of the state's worst snowstorms. Others have suggested that Block helped establish the careers of both Spike Jones and Dinah Shore.

Although the press labeled him "Block the Jock" and "the Lord High Admiral of the Whirling Disk," some radio personalities challenge Block's claim to being the first platter pilot. Al Jarvis was one. Jarvis was in radio on the West Coast when Block worked at XEFD in Tijuana and at KMPC in Beverly Hills, and Jarvis claimed to have had a record program on KFWB that premiered in March 1932 called *The World's Largest Make Believe Ballroom.*

Although others may have originally pioneered disc jockey programs, Block was the first record spinner to gain national recognition. By the end of World War II, he was making $100,000 a year at WNEW in New York. *Variety* celebrated the 15th anniversary of *Make Believe Ballroom* in 1950 with a number of specially dedicated articles, including tributes from Perry Como and Guy Lombardo. Block's work presaged the programming specialization, clock segmentation, and reliance on recorded music that would prevail in radio following television's arrival. Block's son Joel followed his father's footsteps in part: he co-hosts the *Earth and Sky* feature carried on public radio stations.

CHARLES F. GANZERT

Martin Block. Born in Los Angeles, California, 1903. Moved with his mother to Baltimore and New York after his father's death; began working at age 13 as an office boy for General Electric Vice President Own Young; worked as traveling salesman and pitchman selling razor blades, candy bars, and vacuum cleaners as a young man; moved back to the West Coast in 1930; began working in radio at XEFD in Tijuana and KMPC in Beverly Hills; hired at WNEW in New York City, 1934; later broadcast for WABC and WOR. Died in New York City, 19 September 1967.

Radio Series

1935–54 *Make Believe Ballroom*
1944–50 *Chesterfield Supper Club*
1954–61 *The Martin Block Show*

Further Reading

"Martin Block Parlays a $20 a Week Idea into a New Radio Pattern," *Variety* (8 February 1950)
Passman, Arnold, *The Deejays,* New York: Macmillan, 1971

Blore, Chuck 1930–

U.S. Advertising Agency Executive and Innovator of Radio Promotion

Chuck Blore was hooked by radio early in life. As an adult, he applied his infatuation with the medium to programming, promotion, and advertising for radio stations. His advertising expertise was later used in campaigns for a variety of products and services. One of the commercial concepts he created to advertise radio stations on television was still in use 25 years after he first introduced it. Another campaign ran a record 11 years in one city.

Blore claimed he was raking leaves at his East Los Angeles home at age 11 when his mother asked, "What do you want to do when you grow up?" At the time he hadn't given the idea any thought. "But Al Jarvis was on the radio while I was raking," he said, "and I realized that's what I wanted to do. I told my mother, 'I want to talk on the radio.'" Radio became so much a part of his teenage life that Blore quit high school at age 17 to join the Navy to learn more about the medium. The Navy led him to El Paso, Texas, where he programmed KELP radio and achieved a 74 share of the listening audience.

Blore became well known in the radio industry for his innovative work as program manager for KFWB in Los Angeles, where he introduced "Color Radio" in 1958. The sound included elaborate singing logos and songs about the station, all mixed with Top 40 music. KFWB was one of the first stations to adopt a consistent, 24-hour-a-day sound that made it easily identifiable, especially against the "block programming" of the era. As "Color Channel 98," KFWB earned audience shares in the 30 range in southern California from the inception of the concept through the mid-1960s.

KFWB earned Blore a place among the originators of the Top 40 format and won him the "Man of the Year" awards from three trade publications in 1961—*The Gavin Report*, *Billboard* magazine, and *Broadcasting* magazine. Blore says he lists the award from *The Gavin Report* first because it was the most respected music publication of its time. Not long after those awards, Blore was named National Program Director for the Crowell-Collier radio group, overseeing the programming and branding concepts for KDWB in Minneapolis and KEWB in San Francisco, as well as for the Los Angeles station.

He held that position until mid-1964 when he stepped away from day to day radio to form Chuck Blore Creative Services. In partnership with writer Don Richman, Blore applied his ideas to other radio stations. Blore and Richman produced jingles and singing logos for radio stations and advertising and marketing campaigns for consumer products and services. They enlisted the Johnny Mann Singers to provide the vocals, and Johnny Mann wrote the musical arrangements. Blore had first worked with Mann and his singers on music for "Color Radio" at KFWB. By the time the new venture began, Mann had achieved a national hit record, "Love Me With All Your Heart," and that added value to his performance on the Blore jingles.

Among Blore's first work for radio were jingles for WCAR in Detroit and WKYC in Cincinnati using the theme, "Here is Love in Your Ear." Other packages were developed for KRLA in Los Angeles, KYA in San Francisco, and for CBS Radio. Blore's most famous jingles, however, are the ones he conceived for WCFL in Chicago, using lyrics that can only be termed hyperbolic:

> Chicago is saved! Hurray and hallelujah—Chicago is saved! WCFL!

Another ran:

> Never was your radio so radiant, never has your set had so much sun,
> What we have for your ears is ear-resistible, sounds of love and sounds of fun,
> You can tell it's CFL . . . WCFL!

Blore also included in the package a song for St. Swithin's Day, giving WCFL music that was noticeably out of the ordinary. The WCFL jingle packages were first produced in 1966 and updated each of the following two years. Radio memorabilia collectors still buy, sell, and trade copies of Blore's WCFL productions at several internet sites.

For a brief period in the early 1970s, Blore returned to day-to-day radio to develop a concept called "Entertainment Radio" for KIIS-AM in Los Angeles. Instead of relying on disc jockies and announcers, Entertainment Radio used the talents of the KIIS creative department to produce "mini-dramas" that picked up the theme of a song as an introduction—dialog of a couple falling in love, for instance, followed by the Carpenters' "We've Only Just Begun." KIIS also allowed listeners to become involved on the air by doing tasks usually reserved for disc jockies. A typical KIIS weather forecast would have a listener commenting, "The smog looks like pea soup."

In 1975, Blore created a television commercial called "The Remarkable Mouth"—a close up of an attractive woman's mouth from which emanates the music, announcers, jingles, and logos of a radio station. At the end of the commercial, an announcer says, "You have a remarkable mouth." The woman replies, "You have a remarkable radio station."

The first Remarkable Mouth commercial was produced for WTAE Radio in Pittsburgh, Pennsylvania. Because WTAE was co-owned with a television station, the radio operation received free airtime. A WTAE executive had previously worked with Blore in Los Angeles and asked him for ideas to promote radio on TV.

Blore said he could not believe a radio station wanted to advertise on TV because radio and TV were so competitive at the time; such cross promotion just was not done. He leapt at the opportunity to create a commercial to advertise his favorite medium. Blore's company had just completed a series of commercials for the Hollywood Bowl concert venue in which a male actor appeared to have the music and the cheering audiences coming from his mouth as he spoke. Blore and Richman adapted the idea to radio and hired a female model they had used in a billboard campaign in Los Angeles.

Since its introduction, the "Remarkable Mouth" commercial has been on the air somewhere in the world. Blore produced versions of the commercial for broadcasters in Lithuania, Russia, Ireland, England, and Venezuela, and a long list of other countries. In early 2001, "Remarkable Mouth" was revived in Los Angeles for KCBS-FM and its Classic Rock format called "The Arrow."

Chuck Blore
Courtesy The Chuck Blore Company

Two other Blore commercials also became radio promotion classics. "The Janitor," an adaptation of "Remarkable Mouth," showed a late-night janitor interrupting his work at a radio station to step into a studio and pretend to be on the air. From his mouth came all the programming broadcast during the previous day on that station. The commercial, first produced for KABC, Los Angeles, was syndicated to news and talk stations across the U.S.

Blore's "Deborah" commercial was also a simple concept: A woman's face was on screen as she acted as spokesmodel for the station. She recited the benefits of the radio station while quick video cuts attracted the viewer's eye to the screen. "Deborah" was also syndicated nationally and achieved one of longest runs of any single commercial for any product when it was played for 11 years on Atlanta television to promote WKHX, the Country station known as "Kicks."

The success of his advertising campaigns for radio led to Blore's induction into the PROMAX Hall of Fame, established by the organization of radio and television promotion and marketing managers.

Blore companies have produced branding, imaging, and advertising for broadcast and cable television networks including ABC, CBS, Fox, NBC, CNBC, The Discovery Channel, and The Learning Channel. The company also provides imaging and advertising for internet sites such as MTV's SonicNet, HitComedy.com, and WB's Entertaindom. Now known as the Chuck Blore Company, Blore's organization has won more than 400 major advertising awards. *Adweek* magazine called the firm "the most honored company in broadcast advertising history."

ED SHANE

See also DJs; Promotion on Radio; WCFL

Chuck Blore. Born in Los Angeles, California, 10 April 1930. Joined the U.S. Navy in 1948. After discharge from the Navy in 1950, joined KGAN, Kingman, Arizona; left after three months to join KTKT, Tucson, Arizona. In 1955, recruited by Gordon McLendon to be Program Director for KELP, El Paso, Texas. Introduced "Color Radio" on KFWB, Los Angeles, 1958. Named "Man of the Year" in 1961 by *The Gavin Report*, *Billboard* and *Broadcasting*. National Program Director, Crowell-Collier Stations, 1962–64. Founded Chuck Blore Creative Services, 1964. Inducted into PROMAX Hall of Fame, 1995.

Further Reading

Hall, Claude, and Barbara Hall, *This Business of Radio Programming: A Comprehensive Look at Modern Programming Techniques Used throughout the Radio World,* New York: Billboard Books, 1977

Blue Book

Broadcast Policy Statement

More formally titled *Public Service Responsibility of Broadcast Licensees*, this 1946 Federal Communications Commission (FCC) report on radio's program and advertising shortcomings gave rise to a lasting controversy concerning the agency's supervisory role over broadcasting's practices.

Background

FCC concerns about radio advertising and programming were anything but new—they had been a part of commission discussion and some legal cases since the commission's creation in 1934. With the approaching end of World War II, the FCC was better able to focus on domestic issues, and incoming chairman Paul Porter proposed a study of radio program practices on which the commission might base overall policy guidelines that could assist in its station licensing decisions. What several commissioners felt was needed was a comprehensive analysis of program and advertising promises stations made in applying for licenses versus their actual performance three years later when that license came up for renewal.

In mid-1945 former British Broadcasting Company (BBC) official Charles A. Siepmann was hired to work with attorney Elinor Bonteque and the FCC staff to develop a workable study of "promise versus performance" measures, including such measures as the amount of advertising a station carried per hour or week, the proportion of locally produced programs provided, and the proportion of sustaining (non-sponsored) programs offered. Because it was too costly and time consuming to survey all 900 AM radio stations then on the air, a few sample cases would have to be relied on to provide a picture of current industry practices. Even before the study got under way, the commission began to hold up once-routine license renewals in cases where there was evidence of serious promise-versus-performance problems. By early 1946 more than 300 stations—nearly a third of all those on the air—were in license limbo.

What It Said

On 7 March 1946, the FCC released a 149-page mimeographed report in light blue covers titled *Public Service Responsibilities of Broadcast Licensees.* Demand for copies led to a printed version of 59 pages, and it is these that are usually found in libraries and archives today. The "Blue Book" (as it was quickly dubbed by all parties) was divided into five parts: (1) a discussion of the commission's concern with program service (which presented five case studies of specific stations found wanting); (2) the FCC's legal jurisdiction with respect to program service; (3) four specific aspects of the public interest in program service; (4) a review of relevant economic issues; and (5) a summary and conclusion including proposals for future commission policy.

The five case studies in Part 1 each pinpointed a different problem. KIEV in Glendale, California, was found to have promised considerable local cultural and public service programming and limited advertising—but instead to have provided a largely sponsored recorded music service, meeting almost none of its original promises. WSNY in Schenectady, New York, had been granted a license in a comparative hearing based on programming promises that, at renewal time some years later, had not been fulfilled. Station WTOL in Toledo, Ohio, had obtained a full-time authorization (it had been a daytime-only operation), again based on certain promises concerning local public service programs, which were found "conspicuous by their absence" four years later. Baltimore station WBAL changed ownership in the mid-1930s, and a decade later it was found to be providing a service largely bereft of promised local sustaining programs. And finally, station KHMO in Hannibal, Missouri, obtained a license in a court action in the mid-1930s, based in part on programming commitments that it was not fulfilling by early 1945.

The second part of the Blue Book, concerning the "commission jurisdiction with respect to program service," focused on legal issues raised at the time of the FCC's creation from the former Federal Radio Commission. Written by Bonteque, this section concluded that the FCC "is under an affirmative duty, in its public interest determinations, to give full consideration to program service."

The specifics of that determination were spelled out in Part 3. The Blue Book defined the public interest to include four specific requirements of all radio stations. The first was to carry sustaining programs—those not paid for by commercial sponsors—because such programs provided a vital balance to advertiser-supported programs, especially for minority audiences and program experimentation. Including several full-page charts illustrating station practices, this was the longest single part (nearly 24 pages in the printed version) of the Blue Book. The second requirement was to carry local and live programs to reflect local community concerns and interests. Excessive reliance on national commercial programs was held to be an example of poor practice. Carrying discussion of public issues was the third requirement. Another lengthy section of eight pages was devoted to the fourth requirement—not carrying too much advertising.

The fourth portion of the Blue Book focused on economic aspects—essentially the profits made by the industry. Here 14 tables demonstrated the substantial returns stations had made during the war, suggesting that a profitable business like radio broadcasting could easily support a larger public service role.

Finally, the Blue Book turned to the role of the public and government with some specific procedural proposals for future regulation (some of this section was written by Siepmann). Among these proposals were creation of uniform definitions of program types, segments of the broadcast day, selection of a composite week on which program reports would be based, some revisions in license and renewal application forms, and procedures on renewal actions. All of these proposals were designed to allow ready comparison of practice across stations. The same section also called for more radio criticism, self-regulation, radio listener councils, and education about radio in colleges and universities.

Impact

Publication of the Blue Book brought forth an instant negative radio industry response, including rhetoric that the government was trying to take control of radio or censor broadcasters. At the least, industry figures argued, they should have had a chance to comment upon the cases and methods used and the findings reached before the report was released. Along with other critics, they also held that the FCC had no authority to regulate as it seemed to intend; at the same time, the report was criticized for emphasizing a few bad actors in an otherwise well-meaning and effective industry. Even some of those sympathetic to the report's intent felt the distinction concerning benefits of sustaining and commercial programming was overstated. And the financial section probably overstated the industry's profits, because the war years, in retrospect, were clearly an unusual period (given wartime limits on newspaper advertising to save paper and tax provisions making it beneficial for companies making war products to keep their names in the public eye with radio advertising).

Did the publication have any lasting effect? In the end, no station lost its license for the kind of transgressions described in the Blue Book. Virtually all the licensees designated for renewal hearings because of Blue Book issues were eventually renewed—and hundreds of new stations took to the air as well. Yet the FCC never withdrew or replaced the document, which remained in place as a statement of policy thinking for years to come. Still, a decade later, radio broadcasters were carrying even fewer sustaining programs in what had become a "local," although heavily commercialized, music service. By 1959 the trade weekly *Broadcasting* noted that the report was long out of print and was "now something of a collector's item."

CHRISTOPHER H. STERLING

See also Federal Communications Commission; Regulation; Siepmann, Charles A.

Further Reading

"FCC's Blue Book," *Freedom of Information Center Publication* 90 (1961) (major portions of the text are excerpted here)

Meyer, Richard J., "The Blue Book," *Journal of Broadcasting* 6 (Summer 1962)

Miller, Justin, *The Blue Book: An Analysis,* Washington, D.C.: National Association of Broadcasters, 1947

Siepmann, Charles A., *Radio's Second Chance,* Boston: Little Brown, 1946

Siepmann, Charles A., "Storm in the Radio World," *The American Mercury* (August 1946)

Siepmann, Charles A., *The Radio Listener's Bill of Rights: Democracy, Radio, and You,* New York: Anti-Defamation League of B'nai B'rith, 1948

U.S. Federal Communications Commission, *Public Service Responsibilities of Broadcast Licensees,* Washington, D.C.: FCC, 1946; reprint, New York: Arno Press, 1974

White, Llewellyn, "The Blue Book," in *The American Radio: A Report on the Broadcasting Industry in the United States from the Commission on Freedom of the Press,* by White, Chicago: University of Chicago Press, 1947

Blue Network

The Blue network was one of two radio networks operated by the National Broadcasting Company (NBC) from 1927 until 1943. After its sale in 1943, this network continued using the Blue network name for a year, until it was renamed the American Broadcasting Companies (ABC).

Origins

The Blue network, predecessor of the ABC Radio network, traces its roots to the early 1920s, when two informal networks linked a few radio stations in the U.S. Northeast to carry broadcasts from New York. The American Telephone and Telegraph (AT&T) network was the stronger of the two, feeding sponsored programs and special events from the company's New York station, WEAF (later WNBC, now WFAN). Starting in 1923, the second network fed programs from New York station WJZ (now WABC) to other Northeastern stations of the "Radio Group" operated by the Radio Corporation of America (RCA), General Electric, and Westinghouse Electric.

In 1926 a patent agreement reached between the Radio Group and AT&T heralded the beginning of serious network broadcasting in the United States. The agreement provided that the Radio Group would operate radio stations and networks, and AT&T would provide telephone lines to connect stations for network broadcasting. RCA established the National Broadcasting Company, a new corporation, in 1926 to operate local stations and radio networks. In turn, NBC bought WEAF and the telephone company's network from AT&T for $1 million to complete the settlement. NBC then announced that it would provide the best programs available for broadcasting in the United States and that it would provide these programs to other stations throughout the country. NBC was launched with a gala inaugural broadcast from New York on 15 November 1926.

At the time of its establishment, NBC had two stations in both New York (WEAF and WJZ) and Washington, D.C. (WRC and WMAL) as well as two affiliates in several other cities. Instead of duplicating the same program on both stations in the same community, NBC devised a plan starting in early 1927 for two semi-independent networks that would carry separate programs most of the time. These two networks, known as the NBC Red and NBC Blue networks, were originated by NBC's two New York flagship stations, WEAF, the former AT&T station, and RCA's station WJZ. On 23 December 1928, NBC linked together its eastern and Pacific coastal stations, known as the "Orange" network, establishing the first transcontinental network service.

As 1927 began, a number of lavish sponsored programs were on the air. Concerts, classical or semi-classical, were presented by several orchestras. Live radio drama was attempted as early as 1928 by the *Eveready Hour.* Remote pickups of dance bands from New York's hotel ballrooms continued to be a prominent feature of both the Red and Blue networks during late night.

When NBC began in 1927, there were 10 stations on each network. At the end of six months of operation, NBC's Red network had a chain of 15 stations, including WEAF in the East, and NBC Blue had 10 stations including WJZ. Eight additional stations were affiliated with both networks. In January 1928, one year after the network began regular daily service, NBC had 48 affiliates. Ten years later, in 1938, there were 154 NBC affiliates, including 23 on the Red network and 24 on the Blue, with the remainder choosing programs from both. However, NBC Red had considerably more of the high-power clear channel stations, making it the stronger competitor.

Mode of Operation

From the start, the Red network outstripped the Blue network in terms of popular programming. The NBC Red network enjoyed the heritage of the AT&T chain, whose pre-merger advertisers paid performing talent well, whereas the Radio Group's WJZ had largely used free talent. With the Red network's lineup of powerful stations and strong popular programming, many sponsors insisted on placing their programs on NBC Red. Furthermore, to placate the government during the rapid growth of commercialism on radio, NBC deliberately programmed NBC Blue as a complementary service to the Red network, providing extensive news, public service, and cultural programming. Although NBC Blue had some popular sponsored shows, its schedules consisted largely of sustaining (non-sponsored) public-affairs talk programs, concert music, classic drama, and late-night dance bands. New programs often made their debut on NBC Blue and were moved to the Red network when they became popular. Because the Red network stations carried about three-fourths of NBC's commercial programs, industry observers commented that NBC, from 1927 until 1943, used the Blue network more as a foil than as an all-out competitor with the Columbia Broadcasting System (CBS).

Despite its secondary role, the NBC Blue network launched what was to become radio's first sensationally popular serial drama. In 1929 NBC Blue began carrying the nightly *Amos 'n' Andy* comedy show, which depicted the activities of a group of affable black characters living in Harlem. *Amos 'n' Andy* soon dominated all radio listening in the early evening hour of 7:00 Eastern time. The Cooperative Analysis of Broadcasting reported that more than half of all radio homes in the nation regularly tuned to this program during the 1930–31 season.

During the 1930s, NBC Blue also carried additional five- and six-day-a-week serialized dramas, including *Little Orphan Annie, Lum 'n' Abner, Vic and Sade, Clara Lu and Em,* and *Betty and Bob*. Several news commentators, including Lowell Thomas, also were heard five nights a week on the network. Other regular NBC Blue network programs included concerts by the NBC Symphony Orchestra, *Sherlock Holmes* and other mystery dramas, and the popular *Quiz Kids* program featuring gifted youngsters. NBC Blue served rural audiences with its *National Farm and Home Hour,* offered adaptations of classic drama on *Radio Guild,* and provided the *Walter Damrosch Music Appreciation Hour* for students.

NBC's main competitor during the 1930s was CBS, which was founded in 1927. NBC and CBS together controlled almost all of the most powerful clear channel and regional stations—so much so that a third rival, the Mutual Broadcasting System (MBS), found it extremely difficult to obtain competitive station affiliations after its founding in 1934. Mutual's complaints to the Federal Communications Commission (FCC) resulted in an investigation of radio network practices beginning in 1938. The FCC concluded that the extent of control exercised by NBC and CBS over the radio network industry was not in the public interest; in 1941 the Commission issued a new set of "Chain Broadcasting Regulations" that made it illegal for one company to operate more than one national radio network.

Separation from the National Broadcasting Company and Network Sale

In January 1942, NBC officially split the operation of the two networks, making the Blue network a separate subsidiary of RCA. After bitter litigation, the U.S. Supreme Court upheld the FCC's action, forcing NBC to sell one of its networks. In October 1943, the FCC approved the $8 million purchase of the Blue network by Edward J. Noble, whose fortune was derived from Life Savers candy. The new company was named the Blue Network, Incorporated. One year later, the network was renamed the American Broadcasting Company (ABC).

In the 1940s, the Blue Network/ABC became a more aggressive competitor of NBC and CBS but continued the public service traditions of NBC Blue. ABC hired conductor Paul Whiteman as its musical director and substituted the Boston Symphony for the NBC Symphony Orchestra and the *Blue Theatre Players* for the *Radio Guild*. The Blue network began carrying the Saturday matinee performances of the Metropolitan Opera early in the 1940s. It also carried a Sunday night blues/jazz show called the *Chamber Music Society of Lower Basin Street*. Both the Opera and *Basin Street* were hosted by famous opera announcer Milton J. Cross. The Blue network was also known for its stable of mystery programs, including *Sherlock Holmes, Gangbusters,* and *Counterspy,* as well as for its series of children's adventure shows in the late afternoons, including *Jack Armstrong, the All-American Boy*. During the daytime hours, the Blue network also counter-programmed the NBC and CBS soap operas with variety shows, the most famous of which was *Don McNeal's Breakfast Club,* a long-running morning show originating from Chicago.

During the war years of the 1940s, the Blue network was heavily engaged in news broadcasting. However, lacking the resources to maintain a worldwide news operation, the Blue network instead hired a number of commentators who presented a spectrum of views on current events. In this unique arrangement, the network's newsmen ranged from ultraconservative to ultraliberal. However, none was more controversial, nor more sensationally popular than columnist-commentator Walter Winchell, who attracted a huge audience for his Sunday night news and gossip programs. Serious public-affairs programming included the weekly *America's Town Meeting of the Air,* which featured speakers both for and against issues of the day. In another business innovation, the Blue network (and Mutual) offered some of its news programs to local advertisers in an effort to broaden the network's sponsorship and revenue base.

The separation of the Blue network from NBC in 1943 introduced a new and more competitive era for the radio networks. NBC and CBS continued to be the strongest rivals, but the Blue network, no longer subsidized by NBC, had to struggle (with Mutual) for third place in the network industry. Both had growing strength in programming but limited resources for competing in the radio and the soon-to-come television network field. Eventually, in 1953, ABC merged with Paramount Theatres and became a much stronger organization in preparation for the coming of television.

HERBERT H. HOWARD

See also American Broadcasting Company; American Telephone and Telegraph; McNeal, Don; Mutual Broadcasting System; National Broadcasting Company; Network Monopoly Probe; Radio Corporation of America; WEAF

Further Reading

Archer, Gleason L., *Big Business and Radio,* New York: American Historical Company, 1939; reprint, New York: Arno Press, 1971

Kisseloff, Jeff, *The Box: An Oral History of Television, 1920–1961,* New York: Viking, 1995

Lichty, Lawrence W., and Malachi C. Topping, compilers, *American Broadcasting: A Source Book on the History of Radio and Television,* New York: Hastings House, 1975

Robinson, Thomas Porter, *Radio Networks and the Federal Government,* New York: Columbia University Press, 1943; reprint, New York: Arno Press, 1979

Summers, Harrison B., editor, *A Thirty-Year History of Programs Carried on National Radio Networks in the United States, 1926–1956,* Columbus: Ohio State University, 1958

Blues Format

The blues radio format is defined most eloquently by blues music itself. Blues songwriters often explore subjects that deal with real-life situations, and it is not uncommon for listeners to contact a blues host between selections to share their testimony after hearing a certain blues selection. Says renowned *King Biscuit Time* disc jockey Sonny Payne, it is the "history of the African-American people" surviving enslavement, post-reconstruction, and legal segregation, songs of human beings just dealing with life. The unsugarcoated "facts of life" themes often found in the lyrics can be beneficial, nonetheless. The music helps people forget their problems, and it imbues the human spirit with strength. Like other musical genres, the blues format can serve as a cathartic experience. "The blues is the truth," according to the late legendary record promoter Dave Clark.

Radio Blues and Disc Jockeys

Bessie Smith sang the blues live on WMC, a Memphis, Tennessee, radio station, as early as 1924. The regular remote broadcasts from The Palace on Beal Street appear to have continued until sometime in the 1930s. The legacy of blues presence on Memphis radio programming eventually influenced the owners of WDIA radio, the shape of black radio, and lives of legendary listeners such as B.B. King, Rufus Thomas, and Elvis Presley, whose first commercial success was the recording of Arthur Crudup's "That's Alright Mama."

In the early 2000s, WMPR-FM in Jackson, Mississippi, devoted 11 hours per day to blues. Most blues programs are limited to certain time blocks during a radio station's weekly air schedule. One exception is WAVN-AM in Memphis, which in 2003 devoted its entire program schedule to blues. Many noncommercial radio stations (public, community, and college) have increasingly programmed blues for the past 30 years. At least one radio station in many major markets and college communities can be found devoting selected block schedules to blues. National Public Radio downlinks via satellite a blues program, *Portraits in Blue,* to its affiliates each week. The Handy Foundation in Memphis circles the globe to record live blues concerts and syndicates the performances in a magazine format called *Beal St. Caravan.* Blues programming can be heard on the internet, and the trend is growing rapidly. Emerging satellite services such as Sirius and XM had begun to provide continuous blues programming by the early 21st century.

Disc jockeys who work in the radio blues format often travel to blues festivals around the country to keep up with current trends and developments. They exchange ideas, conduct interviews with historical and leading artists, and then broadcast them on their local blues programs back home. Such periodicals as *Living Blues* and *Big City Blues* can provide invaluable cultural information for the program producer. It is fair to say that most men and women who join the still loose network of blues programmers take that step seriously. In essence, they become part of a respected culture that was pioneered by men and women who struggled valiantly to regain their human dignity and make life better for everyone. A serious blues disc jockey will know—and play—the music of Sonny Boy Williamson, Robert Lockwood, Muddy Waters, or B.B. King. And the blues enthusiast—whether disc jockey or listener—might consider revisiting or discovering the rich origin of the blues radio format, which began in the Mississippi Delta "On the Arkansas Side."

Chicago: Al Benson

During the early 1940s in Chicago, Al Benson (following the precedent of Jack L. Cooper, another Chicago entrepreneur) began purchasing blocks of time on several different radio stations to program black music, much of which was blues. An important key to Benson's success was the format he designed, which permitted him and his hired announcers to speak the language of many transplanted Southerners and to promote the products of sponsors. His use of recorded blues music and his training of young broadcasters such as Vivian Carter and Sid McCoy appears to have accompanied the rise in popularity of black disc jockeys and blues programming. Carter later cofounded Vee Jay Records and helped develop the legendary Jimmy Reed. She launched the Beatles' first recordings in the United States. Benson's block programs, broadcast on various stations, remain a major contribution to the blues radio format. By 1947 there were at least 17 blues-oriented radio programs being broadcast in the United States. Several programs aired on various stations in Los Angeles, and Leroy White and others were very popular in Detroit.

Helena, Arkansas: *King Biscuit Time*

Helena, Arkansas, located on the west bank of the Mississippi River, is a small city that became home to the longest-running blues program on radio, *King Biscuit Time*. Shortly after KFFA Radio was established in 1941, bluesmen Sonny Boy Williamson and Robert Lockwood, Jr., met with their white childhood friend, Sonny Payne, who worked at the station and helped get them on the air. Sam Anderson, the station manager and part owner, agreed to sell Williamson and Lockwood a block of airtime, but the blues duo had no money. Anderson referred them to a potential sponsor, Max Moore, a wholesale grocer who needed to sell a huge backlog of flour from his warehouse. A financial deal was struck, and a tight program structure was agreed upon.

Williamson and Lockwood opened their 15-minute show Monday through Friday with a theme song that was followed by an Anderson voice-over announcement: "Pass the biscuits boys, it's King Biscuit Time." Mixing performances of blues songs with casual conversation about where the duo would be performing in the area, Williamson and Lockwood were a success. Listeners in a 100-mile radius of KFFA's transmitter embraced the blues program and quickly purchased all of Moore's existing supply of King Biscuit Flour.

KFFA Radio has continued broadcasting *King Biscuit Time,* uninterrupted, for six decades and had logged nearly 14,000 blues shows by the turn of the century. Robert Lockwood, Jr., and the late Sonny Boy Williamson have grown into legends in both the blues and radio programming history. The show made Max Moore wealthy and the late Sam Anderson's KFFA world famous. Sonny Payne now hosts *King Biscuit Time* in a half-hour disc jockey format. Visitors from around the world frequently stop in at the Delta Cultural Center in Helena to catch the program, 12:00 to 12:30 P.M. Some guests even get a chance to be interviewed live by Payne. Each year up to 90,000 blues lovers from around the world flock to Helena, Arkansas, to attend a blues festival in honor of *King Biscuit Time* and the return of Robert Lockwood, Jr., to center stage.

Nashville: WLAC Radio

Francis Hill, a white woman, sang the blues live on WLAC in the late 1930s. Then, sometime in the mid-1940s two black record promoters were welcomed into the WLAC studios by Gene Nobles. One of the promoters is believed to have been Dave Clark. Nobles, white and handicapped, held down the night shift for WLAC's 50,000-watt clear channel signal, which blanketed the South, Midwest, parts of Canada, and the Caribbean. After Nobles began playing a few of the promoters' black records several nights a week, listeners began writing from as far away as Detroit, Michigan, and the Bahamas for more blues and boogie. Nobles came to the attention of Randy Wood, a white businessman in Gallatin, Tennessee, about 40 miles away. Wood bought some advertising spots to promote the sale of several thousand records by black artists that he discovered after purchasing an appliance store. Again, the audience responded and bought out Wood's phonograph stock.

Gene Nobles was soon hosting a blues-oriented program on a radio station that many African-Americans referred to simply as "Randy's" (WLAC). The disc jockey-run show focused on promoting a C.O.D. mail-order system operated by Randy's Record Shop in Gallatin, Tennessee. The primary pitch involved promoting sets of phonograph records made up of five or six unrelated 78-rpm singles. To promote sales, one or two records were played each night from various sets called "specials" (e.g., "The Treasure of Love Special" or "The Old Time Gospel Special").

Ernie's Record Mart and Buckley's Record Shop, both in Nashville, soon imitated the successful Randy Wood format. Each store bought time blocks, which were spread among WLAC's additional blues-oriented programming with traditional spots and per-inquiry advertisements. By the early 1950s WLAC Radio's entire night-time schedule was bought out. John Richburg, Bill Allen, and Herman Grizzard joined Nobles to formulate a powerful programming block from 9:00 P.M. to early morning, Monday through Sunday. All of the disc jockeys were white, but they addressed the audience fairly, respected the culture, and won acceptance and trust from a largely, though not exclusively, black audience. Don Whitehead, an African-American, joined the news staff in the 1960s.

Memphis: WDIA Radio

John Pepper and Bert Ferguson, two white businessmen, found themselves unable to attract white listeners or money to their newly built WDIA Radio just as Randy Wood was gaining success. While on a trip to New Orleans, Ferguson encountered a copy of *Negro Digest* and read a success story about Al Benson. The magazine caused him to recall the 1930s live radio broadcasts from Beale Street featuring the skillful Nat D. Williams. When he returned to Memphis, he sought the assistance of Williams, a black educator, journalist, and Beal Street impresario. In an afternoon block of time, Monday through Friday, Williams developed and hosted a blues-oriented show, and the radio audience bonded with his style, laughter, and cultural knowledge. Williams' success led to the hiring of other black announcers until WDIA's entire programming schedule consisted of blues, rhythm and blues, and gospel. It was the birth of full-time radio devoted to these genres.

WDIA Radio intermingled its music with several public service announcements, called "Goodwill Announcements" by the station, to help educate and inform African-Americans living in the mid-South's tristate region: Tennessee, Arkansas, and Mississippi. A.C. Williams, another educator turned WDIA radio announcer, maintains that the foundation of black political achievement in Memphis, which is now very organized, began with public-affairs programming on WDIA Radio. The station's 1950s programming model remains at the pinnacle of blues radio formats. WDIA's programming philosophy served as a model for other radio legends who continued to promote or program the blues wherever their career paths led them: Maurice "Hot Rod" Hulbert in Baltimore; Martha Jean Steinberg in Detroit; and Rufus Thomas and B.B. King as performers around the world.

Blues Radio Format Diffused

The blues format was still strong in 1953 when more than 500 black disc jockeys were reported to be working in radio, mostly in block formats or part-time situations. A few years later, black military veterans returning home from service brought reports that Europeans loved the "real blues." They cited John Lee Hooker, Howlin' Wolf, Muddy Waters, Sonny Boy Williamson, and others as being revered. Indeed, the Animals, the Rolling Stones, and Canned Heat advanced blues programming on white commercial radio stations in the 1960s after they included blues songs by the great African-American masters on their early albums. Curious fans who studied the origins of English rock performers became more aware of the blues. In addition, 1960s FM radio, in need of program material and open to experimentation, also began playing blues. Many young white soul radio station listeners who became attracted to rhythm and blues made additional cultural explorations and discovered the blues. The blues format increasingly made its way onto the programming schedules of noncommercial radio as the number of FM public, college, and community radio stations expanded.

LAWRENCE N. REDD

See also Black-Oriented Radio; Black Radio Networks; Hulbert, Maurice "Hot Rod"; KFFA; King Biscuit Flower Hour; Thomas, Rufus; WDIA; Williams, Nat D.; WLAC

Further Reading

Redd, Lawrence N., *Rock Is Rhythm and Blues: The Impact of Mass Media,* East Lansing: Michigan State University Press, 1974

Routt, Edd, James McGrath, and Frederic Weiss, *The Radio Format Conundrum,* New York: Hastings House, 1978

Williams, Gilbert A., *Legendary Pioneers of Black Radio,* Westport, Connecticut: Praeger, 1998

Board for International Broadcasting

Directing Radio Free Europe and Radio Liberty

The Board for International Broadcasting (BIB) was created in 1973 to oversee and fund Radio Free Europe (RFE) and Radio Liberty (RL), two surrogate radio stations that broadcast into countries behind the Iron Curtain. In 1994 the BIB's oversight responsibilities were turned over to the Broadcasting Board of Governors when the United States' international broadcasting operations were reorganized and all nonmilitary government-financed international operations were consolidated.

RFE had been established in 1949 as a nonprofit private corporation to broadcast news and current-affairs programs to Central and Eastern European countries in the Soviet political and military orbit (Albania, Bulgaria, Czechoslovakia, East Germany, Hungary, Poland, Romania, and Yugoslavia, as well as three Baltic countries that had been absorbed into the Union of Soviet Socialist Republics [USSR] in 1940 but that the United States did not recognize as part of the USSR—Estonia, Latvia, and Lithuania). RL was created in 1951 to broadcast the same type of programs into what was then the USSR. These two operations were called "surrogate" stations because they broadcast news and public-affairs programs about the target countries themselves and considered themselves competitors of the domestic services in their target areas, rather than programs primarily about the United States and the West, which was the responsibility of the Voice of America.

Originally both RFE and RL had been funded principally and covertly by the U.S. Congress through the Central Intelligence Agency (CIA), but they also received some private funding, thanks to publicity campaigns that made it seem that private money was all that kept the stations afloat. Suspicions of CIA involvement in the activities of the two stations grew over time, and in 1972 CIA involvement in their operations was acknowledged and ended, and the two stations were put under the direction of the Department of State. But the State Department did not want to oversee their operations.

In 1973 the Presidential Study Commission on International Radio Broadcasting, headed by Milton Eisenhower, officially recognized the prior role of the CIA and the fiction that all of RFE/RL's funding had come from private sources and recommended that a separate board be established to oversee the two services' operations. This recommendation resulted in the Act for the Board for International Broadcasting in 1973. This act declared that the purpose of RFE and RL would be to provide "an independent broadcast media operating in a manner not inconsistent with the broad foreign policy objectives of the United States and in accordance with high professional standards" and that their operations were "in the national interest." The BIB was authorized to make grants to RFE/RL, to review their mission and operations, to evaluate their effectiveness, to encourage efficient use of resources, to conduct audits, and to make sure that their operations were in no way inconsistent with the foreign policy objectives of the United States. The board was to make an annual report to the president and to Congress through the foreign relations committees of the House and Senate.

The BIB was founded solely to oversee and serve as the conduit for funding RFE and RL. It was composed of nine bipartisan members appointed by the president of the United States and confirmed by the U.S. Senate. Terms were three years in length, with one-third of the BIB members changing each year; no more than five members could be of the same political party. The BIB continued to function in this capacity when the two broadcasting organizations were merged into RFE/RL in 1975. Struggles between the BIB and the RFE/RL board of directors resulted in continuing conflict. In 1982 new congressional legislation under the Pell Amendment eliminated the private corporate board and made the members of the BIB also the board of directors of RFE/RL: there were essentially two separate and parallel boards but with the same members.

The collapse of the Soviet Union following the so-called Velvet Revolution in Eastern Europe, which resulted in the collapse of Soviet hegemony and the destruction of the Berlin Wall in 1989, led to a new reassessment of the extent of American international broadcasting activities and to a questioning of whether or not surrogate radio stations were still necessary. With the ardent support of many leaders in the newly democratic states of Eastern and Central Europe, RFE/RL survived, but another reorganization occurred, and a new oversight agency was created. The BIB's duties were transferred to this new agency, the Broadcasting Board of Governors, in 1994.

ROBERT S. FORTNER

See also Broadcasting Board of Governors; Radio Free Europe/Radio Liberty

Further Reading

Board for International Broadcasting, *Annual Report,* Washington, D.C.: GPO, 1974–95

Buell, William A., "Radio Free Europe/Radio Liberty in the Mid 1980s," in *Western Broadcasting over the Iron Curtain,* edited by Kenneth R.M. Short, New York: St. Martin's Press, and London: Croom Helm, 1986

Fortner, Robert S., *International Communication: History, Conflict, and Control of the Global Metropolis,* Belmont, California: Wadsworth, 1993

Tyson, James L., *U.S. International Broadcasting and National Security,* New York: Ramapo Press, 1983

United States Presidential Study Commission on International Radio Broadcasting, *The Right to Know: Report,* Washington, D.C.: GPO, 1973

Bob and Ray

Bob Elliott (1923–)
Ray Goulding (1922–1990)

U.S. Radio Comedians

From their base in New York City, Bob and Ray affirmed that radio comedy was alive and well, despite the emergence of television, in the late 20th century. Bob Elliott (1923–) and Ray Goulding (1922–1990) were both born in Massachusetts, where they worked at local radio stations before World War II. Coincidentally, after the war they ended up at WHDH, Boston. Elliott was a morning disc jockey, and Goulding did the news. They developed an instant comedic chemistry. "I began staying in the studio," Goulding said, "and bailing him out with some chatter, what with all the awful records he had to play." Soon, the program director asked them to do a 25-minute comedy show before baseball games called *Matinée with Bob and Ray.* Elliott recalled, "They had to have that rhyme, and it's the only reason we're Bob and Ray and not Ray and Bob."

In 1951 Bob and Ray sat in for Goulding's older brother Phil and Morey Amsterdam on WMGM, New York. That stint led to a successful audition for the National Broadcasting Company (NBC). The network offered them a 15-minute show each evening, and Bob and Ray moved to New York, where they remained for nearly 40 years. Soon they had a two-and-a-half-hour morning show, a half-hour evening show, and a 15-minute live television program in addition to their original show. Early on the duo decided to call their program *The Bob and Ray Show,* a simple title that they continued to use throughout their career at NBC, which lasted until 1973, and on other stations and networks. The only exception was a TV game show, *The Name's the Same,* which they hosted for ABC television briefly in 1955.

In 1953 Bob and Ray moved their television show to American Broadcasting Companies (ABC) and jumped to WINS radio for the next three years. They began a regular feature on the NBC Radio network program *Monitor* and developed an afternoon show for the Mutual Radio network. By 1956 they had landed at WOR, where they stayed off and on for more than 20 years. They also had a show on the Columbia Broadcasting System (CBS) Radio network in the late 1950s.

Although radio was their primary medium, Bob and Ray also starred as "Bert and Harry" in a long-running series of television commercials for Piels Beer and became regular guests on both the *Today* and *Tonight* shows on NBC television. In 1970 they opened on Broadway with *Bob and Ray: The Two and Only,* a revue based on characters from the radio skits; they subsequently took the show on tour and released a live album of the performance. They appeared in two movies, *Cold Turkey* (1971) and *Author, Author* (1982), and published three books of scripts from their radio shows. They returned to the stage briefly in 1984 with sold-out performances at Carnegie Hall.

In 1981 Bob and Ray were inducted into the National Association of Broadcasters Hall of Fame and were named "Men of the Year" by the Broadcast Pioneers. The next year, the Museum of Broadcasting in New York presented a retrospective of Bob and Ray's career that set attendance records and was held over for four months. During the 1980s more than 200 National Public Radio (NPR) affiliates carried *The Bob and Ray Public Radio Show.* They continued regular radio broadcasts until Goulding's death in 1990. Elliott still plays occasional bit parts on television, often in productions written by and starring his son Chris. Many classic Bob and Ray performances were recorded and remain in circulation.

Humorists as diverse as Bob Newhart, Phil Proctor, and Roy Blount Jr. have paid homage to Bob and Ray. Their comedy has been described as wry, low-key, elegant, restrained, and seductive. The *New York Times* once called them "a couple of master comedians who live in a large, comfortable, friendly house right next door to reality." *Kirkus Review* applauded their ability "to take the stupid words right out of our mouths and, with sweet innocence, toss them in our faces."

Bob Elliott and Ray Goulding, "Bob and Ray"
Courtesy of "Bob and Ray" (Bob Elliott-Ray Goulding)

Typical Bob and Ray routines feature normal people who do bizarre things—the lighthouse keeper whose lighthouse is 40 miles inland, the professor of penmanship who teaches executives to write *il*legibly, a world champion low jumper, or the editor of *Wasting Time* magazine. Often the skit involves an interview in which the comedy hinges on one absurdity—the frustration of waiting for answers from the president of the Slow Talkers of America, or the misprint on a script that causes the oblivious host to ask questions that his guest just answered. Recurring characters include the casts of the soap opera "Mary Backstayge, Noble Wife" and of the adventure serial "Tippy, the Wonder Dog"; pompous sportscaster Biff Burns; and the intrepid reporter on the beat, Wally Ballou, who always upcut his cue and thus began each feature:

—ly Ballou standing here with a gentleman my staff tells me is one of the most unusual and interesting interviews we've ever lined up. I wonder if you'd tell us your name, sir?

Man: No, I'm afraid I can't do that . . .

Ballou: You hiding from the police or something like that?

Man: No, I can't tell you my name because I am one of the very few people in America with a name that is completely unpronounceable.

Ballou: Well . . . could you spell it for us?

Man: That's all you can do with it. It's spelled: W-W-Q-L-C-W.

Ballou: W-W-Q-L-C-W. Are you sure that's a name and not the call letters of some radio station?

Man: No, it's my name all right. But there's no way to pronounce it. I've been trying for years and it's got me beat.

Ballou: I certainly never heard it before. What nationality is it?

Man: Well, my grandfather came from Iraq, originally. And I've got a hunch that when he changed the letters from the Arabic alphabet into English, he goofed something awful.

Ballou: I guess that could be. Do you still have relatives back in the old country?

Man: Oh yeah. Cousins . . . and things like that.

Ballou: And how do they pronounce the name?

Man: They pronounce it Abernathy. (Elliott and Goulding, 1985)

ROBERT HENRY LOCHTE

See also Comedy; WOR

Robert B. (Bob) Elliott. Born in Boston, Massachusetts, 26 May 1923. Studied at the Feagin School of Drama and Radio, New York City; served in the U.S. Army in Europe, 1943–1946; worked at WHDH Boston, 1946–1951, where he met longtime partner Ray Goulding and began writing and performing radio comedy routines. Elliott and Goulding moved to New York in 1951, and their daily comedy programs became a staple of network and local radio for nearly 40 years. During the period, *The Bob and Ray Show* and the duo's sketches were heard on the NBC, ABC, Mutual, and CBS radio networks, as well as National Public Radio. Elliott and Goulding won George Foster Peabody Awards for their work in 1952 and 1957, and were nominated for Grammy Awards for recorded comedy in 1987 and 1988.

Raymond Chester (Ray) Goulding. Born in Lowell, Massachusetts, 20 March 1922. Served in the U.S. Army from 1942–1946, rising to the rank of 1st Lieutenant. His career was closely connected to that of Bob Elliott, above. Died 24 March 1990, in Manhasset, New York.

Radio Series

1946–51 *Matinee with Bob and Ray*
1951–77 *The Bob and Ray Show*
1983–90 *The Bob and Ray Public Radio Show*

Television Series

The Bob and Ray Show, 1952–53; *The Name's the Same*, 1955

Selected Recordings

Bob and Ray, the Two and Only, 1970; *Vintage Bob and Ray*, 1974; *A Night of Two Stars Recorded Live at Carnegie Hall*, 1987; *The Best of Bob and Ray, Vol. 1*, 1988

Selected Publications

Write if You Get Work: The Best of Bob and Ray, 1975
From Approximately Coast to Coast . . . It's the Bob and Ray Show, 1983
The New! Improved! Bob and Ray Book, 1985

Further Reading

Balliett, Whitney, "Profiles: Their Own Gravity," *New Yorker* (24 September 1973)

Border Radio

Mexican-Based Stations Aimed at the United States

Mexico-based radio stations, located in cities near the United States border and often beaming signals of great wattage, offered programs and advertising not always found on U.S. radio stations licensed by the Federal Communications Commission (FCC). At various times in radio history, these "border blasters" temporarily filled programming gaps and advertiser needs that stations licensed in the United States could not or would not provide. But United States–based stations always adapted, and border radio stations would fade into obscurity until the next time that they could successfully counter-program.

Origins

Border radio stations, located in Mexican cities bordering the United States from California to Texas, came into being in the 1930s, when broadcasting became big business and U.S. network programming defined itself through specific genres of programming and advertising. Border stations could transmit more powerful signals than U.S. law permitted, could and did advertise products considered fraudulent under U.S. law, and could and did offer programming—particularly "hillbilly" music—that U.S. networks failed to offer.

Although the U.S. government officially worked through a dominant U.S. network—the National Broadcasting Company (NBC)—and its powerful owner, the Radio Corporation of America (RCA), to expand global markets, "border blasters" looked to fill unserved market niches. Although actual audience comparisons are impossible to make since border stations did not subscribe to ratings services such as Hooper, the border stations' own records of selling products in the 1930s are indeed impressive. The official authorities on both sides of the border never liked these clever entrepreneurs but often could do little about directly shutting them down. Border stations such as XED-AM—located across the border from Laredo, Texas—successfully sold Mexican lottery tickets by mail to listeners in the United States, who could also listen to XED-AM for the results. Lottery promotion was at that time strictly forbidden under U.S. radio law.

The Mexican authorities accommodated "outlaw" radio entrepreneurs—some of whom, such as Dr. John Brinkley, had been denied broadcasting licenses in the United States—because it seemed to them that the United States and Canada had divided up all the long-range frequencies between themselves, allocating none for Mexico. In 1931 Dr. Brinkley opened XER-AM (called XERA-AM by 1935) in Villa Acuna, Mexico; later in the 1930s, Brinkley also bought XED-AM, changing its name to XEAW-AM. Indeed, these constant changes were one of the key traits of border radio, because entrepreneurs knew that they risked prosecution if and when Mexican and U.S. authorities came to some agreement.

Brinkley used border radio and its hillbilly music to make money by selling "medical miracles" that the American Medical Association (AMA) deemed fraudulent. (The AMA had pressured the Federal Radio Commission to get Brinkley off the air.) He built a transmitter with 300-foot towers. Out of the range of American restriction, station XER-AM started broadcasting with a power of 75,000 watts, with a remote studio linked by phone lines to the Rosewell Hotel in Brinkley's new headquarters in Del Rio, Texas. The station started operating in October 1931, with gala celebrations in both towns. XER-AM offered more than just hours of pseudoscientific lectures from Dr. Brinkley: it also featured the stars of country music of the day—singing cowboys, fiddlers, a Mexican studio orchestra, and many guests.

Thanks to XER-AM's amazing power, Brinkley could be heard as far away as Chicago. His busy Mexican lobbyists succeeded in allowing him to boost power, which made XER-AM for a time the most powerful radio station in the world at a shattering 1 million watts, a signal that for a short time smashed everything in its path and could be heard in New York and Philadelphia—sometimes to the exclusion of all other channels.

Even broadcasting at *only* 100,000 watts (twice the power of the largest American stations), Brinkley was able to reach his potential customers. Significant in radio history, he pioneered the use of electrical transcription discs, even as NBC and the Columbia Broadcasting System (CBS) were insisting that listeners preferred live broadcasts. Brinkley also deserves a place in the history of country music, because he kept alive the career of the fabled Carter Family. But in time both U.S. and Mexican authorities took away his radio stations under the provisions of the North American Regional Broadcasting Agreement treaty, which mandated which country got to use which frequencies. Brinkley died in 1942 before the U.S. Internal Revenue Service could finish suing him for failure to pay taxes.

Although Brinkley was surely the most extreme case, border stations' power generally ranged from 50,000 to 500,000 watts. Listeners reported hearing Mexico-based signals in all parts of the United States and even in Canada. Network affiliates located near a border signal on the AM dial were often drowned out, or at least interfered with, as border stations overwhelmed them. And since border stations were beyond any code of good conduct that network radio or U.S. law required, they could sell and say almost anything they wanted; indeed, border stations hawked items and made claims that would have been disallowed and even prosecuted in the United States, such as pitches for miracle medicines and sexual stimulants and the hawking of donations for phony religious institutions.

Consider the case of Crazy Water Crystals, owned by Carr P. Collins, entrepreneur and political adviser to Texas politician W. Lee "Pappy" O'Daniel (elected governor in 1938 and 1940). Crazy Water Crystals promised to revive a sluggish system; the crystals were produced in Mineral Wells, Texas, by evaporation of the town's fabled "Crazy" water. In 1941, when the United States and Mexico began to cooperate as part of the efforts surrounding World War II, Mexican authorities confiscated Collins' station.

Country Music

Border radio fulfilled the needs of the audience for hillbilly music, needs that the networks only partially met with *The Grand Ole Opry* and *The National Barn Dance*. In the 1930s there were many local hillbilly radio shows, but the supply never matched the demand, so border stations often blanketed the United States with songs by the Carter Family, Cowboy Slim Rinehart, Patsy Montana, and others.

The greatest beneficiaries were the Carter Family. To call the Carter Family—A.P. Carter, Sara Carter, and Maybelle Carter—the first family of country music is a historical truth, because their famed Bristol, Tennessee, recording sessions in 1928 established country music as a recording, and later as a radio, musical genre. But by the mid-1930s, their style had been supplanted by that of singing cowboys such as *The National Barn Dance*'s Gene Autry and the rising stars of the *Grand Ole Opry*, such as Roy Acuff and Ernest Tubb. Thus, few were surprised that the Carters were tempted by the lucrative contract offered by XERA-AM from 1938 to 1942 to work for Brinkley. They needed the money, and Brinkley gave them unparalleled exposure. Jimmie Rodgers, a nascent country star, helped inaugurate XED-AM in Reynosa, Mexico, for similar reasons.

More obscure hillbilly stars benefited as well. Nolan "Cowboy Slim" Rinehart, often called the "king of border radio," was border radio's answer to Gene Autry and the other singing cowboys. Rinehart began his singing career just as border radio was beginning, and although he first appeared on KSKY-AM from Dallas, he gravitated to XEPN-AM in Piedras Negras, Mexico, across the Rio Grande from Eagle Pass, Texas. After his initial appearances on XEPN-AM, the station was deluged with mail, and soon Rinehart was being electrically transcribed and then played on all border stations from Tijuana east to Reynosa. Rinehart had no contract with a U.S. record label, and so he made his additional monies on tour selling songbooks. This was a marginal existence, with few of the paths to fame and fortune enjoyed by those on the *Grand Ole Opry*.

Listeners were dedicated, and some even became country music stars. The case of Hank Thompson is instructive. Born in 1925, Thompson loved these border stations while he was growing up in Waco, Texas: they alone played and programmed country music nearly all day. Border radio should be remembered not only for creating stars, but also as an inspiration for future stars, who as children had access to inspiration around the clock from border stations. Webb Pierce, Jim Reeves, and other stars of the 1950s appeared live on XERF-AM with country disc jockey Paul Kallinger partially as a payback. Border stations helped develop the music that would later become known as "country and western," which would by the year 2000 be simply known as country, the most popular format on radio.

Rock and Roll

Top-40 pioneered rock music on U.S. radio. But since U.S. stations avoided playing rock's raunchier records, border stations in Mexico filled the gaps. This phenomenon is exemplified by the career of disc jockey Wolfman Jack, who, in the late 1950s, after a series of disc jockey jobs in the United States, appeared on XERF-AM, across from Del Rio, Texas, and sold collections of hit records while "spinning rock" in his own unique style. Although Wolfman Jack's broadcasts hardly constituted anything new in format radio, other than their utter outrageousness, they became far more famous after the fact as a result of the hit movie *American Graffiti*, a tribute song by the Guess Who, and a nationally syndicated radio program in the United States.

All-News Format

But border radio should not be remembered solely for fostering interest in country and rock music. The first commercially successful all-news radio operation in North America went on the air in May 1961 from XETRA-AM (pronounced "x-tra") from Tijuana, Mexico, and was aimed at southern California, not at Mexican audiences. This 50,000-watt AM station was operated by radio pioneer Gordon McLendon. Before McLendon took over, it was border station XEAK-AM, which played rock music aimed at southern California teenagers. By 1961 there was a glut of rock format stations in southern California, so McLendon tried an all-news format instead. XETRA-AM was a headline service, with a 15-minute rotation that was later stretched to 30 minutes when McLendon discovered that Los Angeles commuters were trapped in their cars for far more than a quarter hour. McLendon went to great lengths to disguise XETRA-AM's Mexican base and tried to make it seem like just another Los Angeles AM radio station. Jingles repeated over and over: "The world's first and only all-news radio station. In the air everywhere over Los Angeles." The only address announced was that of the Los Angeles sales office. The station was required to give its call letters and location every hour, so McLendon ran a tape spoken in Spanish in a soft, feminine voice that was backed by Hispanic music, followed in English by a description of Mexico's tourist attractions, suggesting to listeners that XETRA-AM was running an advertisement for vacations in Mexico rather than the required call letters and station location.

Los Angeles radio competitors complained to the FCC, contending that such masking was certainly unethical and possibly illegal. At first, because of Gordon McLendon's reputation as a radio pioneer, XETRA-AM was able to draw even national advertisers. By 1962 the station was making a profit, in part because it was strictly a "rip-and-read" station employing no actual reporters, only a dozen announcers who rewrote wire and newspaper copy and who frequently rotated shifts so as to make the broadcasts seem fresh and new. In the background the teletype's tick-tick-ticking was ever-present. XETRA-AM sounded as though its announcers were sitting in a busy, active newspaper office. But in the end, like the rock format, this format proved too easy to copy, and with competition came lower profits. Eventually McLendon turned to other, more profitable ventures.

Later Incarnations

In the 1980s, the United States and Mexico reached an international agreement that allowed shared use of clear channel stations. U.S. radio owners now cared less, however, because FM's limited-distance signals had become audience favorites, and AM's long-range radio was less valuable. The border stations went the way of the clear channel AM stations that had once blanketed much of the United States, and with common U.S. owners and all-recorded sounds, the niche programming of FM radio fulfilled the needs of the marketplace far better than the limited number of AM stations that broadcast from the 1930s through the 1960s. There are still border stations, but now nearly all of them create programs in Spanish for audiences in nearby U.S. communities and compete in the major radio markets with dozens of other stations.

DOUGLAS GOMERY

See also All News Format; Brinkley, John R.; Country Radio Format; Grand Ole Opry; Music; National Barn Dance; North American Regional Broadcast Agreement

Further Reading

Carson, Gerald, *The Roguish World of Doctor Brinkley,* New York: Rinehart, 1960

Fowler, Gene, and Bill Crawford, *Border Radio: Quacks, Yodelers, Pitchmen, Psychics, and Other Amazing Broadcasters of the American Airwaves,* Austin: Texas Monthly Press, 1987

Garay, Ronald, *Gordon McLendon: The Maverick of Radio,* New York: Greenwood Press, 1992

Landry, Robert J., *This Fascinating Radio Business,* Indianapolis, Indiana, and New York: Bobbs-Merrill, 1946

Malone, Bill C., *Country Music U.S.A.,* Austin: University of Texas Press, 1968; revised edition, 1985

Porterfield, Nolan, *Jimmie Rodgers: The Life and Times of America's Blue Yodeler,* Urbana: University of Illinois Press, 1979

Routt, Edd, James B. McGrath, and Fredric A. Weiss, *The Radio Format Conundrum,* New York: Hastings House, 1978

Schwoch, James, *The American Radio Industry and Its Latin American Activities, 1900–1939,* Urbana: University of Illinois Press, 1990

Whetmore, Edward Jay, *The Magic Medium: An Introduction to Radio in America,* Belmont, California: Wadsworth, 1981

Bose, Amar G. 1929–

U.S. Inventor and Acoustic Researcher

The world listens to audio differently because an M.I.T. graduate student was disappointed with what he heard in stereo systems. While pursuing a graduate degree in the 1950s, Amar Bose shopped for stereo speakers, but found none that could reproduce the realism of a live concert hall performance. Not even speakers with impressive technical specifications satisfied Bose. That fact led him to conduct extensive research into speaker design and acoustics, to pursue the field of psychoacoustics—the human perception of sound—and, ultimately, to found the company synonymous with quality audio systems.

As a teenager, Bose earned money by repairing model trains, gaining practical experience that helped prepare him for the Massachusetts Institute of Technology, where he studied electrical engineering. Bose later worked on a series of consulting projects for a variety of companies including Standard Oil of New Jersey; Edgerton, Germeshausen and Grier; and the Epsco Corporation. From 1962 to 1964 he was in charge of research on prosthetics in a project for Liberty Mutual Insurance, Harvard University and Massachusetts General Hospital. The work led to development of the Boston Arm for amputees. Bose holds numerous patents in the fields of acoustics, electronics, nonlinear systems and communications theory.

Bose's research and patents led to the formation of the Bose Corporation in 1964. The company introduced significantly new design concepts that helped to recreate the impact of live music by reflecting an audio speaker's sound off walls and ceilings, creating a "bigger" sound. Among the products that Bose Corporation points to with pride are the 901® Direct/Reflecting® speaker system, introduced in 1968; it brought international acclaim to the company and to Bose himself. Then in 1972 Bose entered the professional ranks, producing a loudspeaker system designed exclusively for professional musicians. A direct descendant of the 901® speaker system was introduced in 1975. Known as the 301® Direct/Reflecting® speaker system, it became one of the world's best-selling speakers. In 1982 Bose introduced a custom-designed, factory-installed audio system specifically for automobiles. The introduction of Acoustimass® speaker technology in 1986 changed conventional thinking about the relationship between speaker size and sound. Speakers small enough to fit in the palm of a hand produced sound quality previously thought impossible. At the other end of the size spectrum, Bose's Auditioner® audio demonstration technology allowed builders, architects, and facility managers to hear precisely what an audio system will sound like in their arenas and other large-scale venues as early as the blueprint stage.

Fourteen years of research led to the development of acoustic waveguide speaker technology, found in the Bose Wave® radio, introduced in 1993. The Wave radio was launched with an advertising campaign that depended upon network radio commercials and print advertising in national magazines. The campaign made the product well known and introduced the public to Amar Bose, thanks primarily to personal endorsements by radio commentator Paul Harvey who spoke of Bose as an old friend. The advertising was so pervasive that *Monitoring Times* magazine claimed, "You would have to be Amish not to have been bombarded by TV and print ads trumpeting the extraordinary sound of the Bose Wave radio. It's depicted as transforming a room into a concert hall and rising up into a looming entertainment presence by simply turning it on." The comment was followed by a positive review of the new technology, as *Monitoring Times* reported that the Wave radio "astounded audiophiles and set a new standard for the nearly forgotten table radio."

Since most consumers don't know what to do with typical stereo receiver functions such as equalization, tone control, and balance, Bose dispensed with them entirely. Instead, automatic signal processing and active electronic equalization were performed by special circuitry in the radio.

Bose became the world's number one speaker manufacturer, commanding nearly 25% of the market. The company makes its products in North America and Ireland and has nearly 90 stores worldwide.

Amar Bose is Chairman of the Board and Technical Director of the Bose Corporation and is the company's largest shareholder. He continues as a Professor at M.I.T. Bose is a fellow of the Institute of Electrical and Electronic Engineers and a member of the Audio Engineering Society. He holds honorary doctorates from Berkley College of Music and from Framingham State College. In 1987, Intellectual Property Owners named him Inventor of the Year; in 1991 he was elected to the American Academy of Arts and Sciences; and in 2000 he was inducted into the Radio Hall of Fame. Bose's personal wealth, estimated at more than half a billion dollars, earned him a ranking in *Forbes* magazine's list of the wealthiest people in the United States. The Bose Foundation, which he also runs, has donated more than $6 million to M.I.T.

ED SHANE

See also High Fidelity; Receivers

Amar G. Bose. Born 2 November 1929 in Philadelphia, Pennsylvania, son of Noni Gopal, a political dissident who had immigrated from Calcutta, India, and Charlotte Mechlin Bose.

B.S. and B.M., 1952; Doctor of Science, Electrical Engineering 1956, all MIT. MIT faculty (assistant professor, 1957; associate professor, 1960; professor, 1966). Fulbright scholar, India, 1957. Founded Bose Corporation, 1964. Honorary doctorates from Berkeley College of Music, Framingham State College. Fellow, Institute of Electrical and Electronic Engineers; member, Audio Engineering Society; elected to the American Academy of Arts and Sciences, 1991; elected to Radio Hall of Fame, 2000. Holds patents in acoustics, nonlinear systems, and communication.

Selected Publications

Introductory Network Theory (with Kenneth N. Stevens), 1965

"Sound Recording and Reproduction: Part One: Devices, Measurements, and Perception," *Technology Review* (June 1973)

"Sound Recording and Reproduction: Part Two: Spatial and Temporal Dimensions," *Technology Review* (July/August 1973)

Further Reading

Ammeson, Jane, "Sound is Golden for Dr. Bose," *Compass* (February 1991)

Bulkeley, William M., "How an MIT Professor Came to Dominate Stereo Speaker Sales," *The Wall Street Journal* (31 December 1996)

"Vox Populi," *The Economist* 10, no. 20 (15 January 2000)

Bowman, Robert. ***See*** **Far East Broadcasting Company**

Brazil

Pioneer radio experimentation occurred in Brazil during the last decade of the 19th and the first decade of the 20th centuries. From the early 1920s to the mid-1930s, radio's technical, legal, and commercial foundations were laid. From the end of that period until the early 1960s, radio reigned as a major media power and enjoyed its golden age. Since the 1960s, it has become one among many different kinds of media in Brazil, dwarfed now by television and the internet.

Beginnings

The "Marconi of Brazil" was a priest, Fr. Roberto Landell de Moura (1861–1928), who experimented extensively and incessantly with wireless communication. By the early 1890s, he had already anticipated or accompanied several European and American inventions for wireless sound transmission. For a time, he lived in New York City, and in October 1904, he obtained U.S. Patent No. 771,917 for a "wave-transmitter." Quite unfortunately, however, not only was there no interest in his work in Brazil, there was even suspicion of it. With his technical genius spurned by the people he sought to help, he died a disappointed man.

During the rubber boom in the Amazon at the beginning of the 20th century, an American company, Amazon Wireless, attempted to set up radio service similar to others that had been successful in Central America. The endeavor failed because of legal complications and poorly understood equatorial conditions for radio operations. Nevertheless, by the following decade, Brazilian ships and coastal stations had wireless communication.

The first Brazilian radio broadcast station was the *Rádio Sociedade do Rio de Janeiro* (Radio Society of Rio de Janeiro), with the call letters PRA-2. The station was founded on 20 April 1923 by Professor Edgard Roquette Pinto of the Brazilian Academy of Science, an anthropologist who knew the value of radio from his participation in pioneering expeditions to indigenous regions. The station programmed news and sedate music.

Because of strictures remaining from World War I, Brazil, like other countries, legally prohibited ownership of radio equipment. These restrictions were loosened as radio became a national craze, one that started in 1922, during the international exhibition commemorating the centennial of Brazilian independence. On Independence Day (7 September), the

Westinghouse company mounted a 500-watt transmitter atop Corcovado Mountain, the high peak near the capital, Rio de Janeiro. The equipment broadcast a presidential address down into the city's few but "magical" receivers.

The craze saw radio clubs spread throughout the country, into the northeastern states of Pernambuco and Bahia, into the Amazon region, and into the southern states of Rio Grande do Sul, Paraná, and São Paulo. The club format allowed financing of broadcasting stations through membership fees. Such membership authorized individuals to purchase a receiver. Additional costs included government fees and taxes. By 1927 twelve stations broadcast daily in the country.

By the end of the 1920s Brazil was second to Argentina as the largest importer of radio equipment from the United States, the main provider. The volume of business was worth more than half a million dollars annually. So that they could withstand tropical conditions, American manufacturers built sets with reinforced metal parts. U.S. diplomatic personnel aided not only in the sale of U.S. equipment but also in the promotion of U.S. program style. Before World War II, U.S. radio faced serious competition from British (Marconi) and German (Telefunken) interests. By the early 1930s Brazil itself began manufacturing radios.

Although nominally prohibited, advertising increasingly became an issue for station owners as their audiences grew faster than their income, which was limited to members' fees. On-air commercials, along with conditions and standards for equipment, training of personnel, control of technical operations, and responsibility for program content, were addressed in the establishment by federal decree in 1932 of the national *Comissão Técnica de Rádio* (Radio Technical Commission).

The dictates of this agency required some radio clubs to close, because they could not afford to meet the new equipment standards. This was true of Brazil's first radio station, which was donated to the newly established Ministry of Education and Culture. It would soon become one of the most renowned broadcasters in the country.

Most important was that the new legislation allowed for advertising. By the mid-1930s growing commercial revenue allowed radio to become an important media power, competing with newspapers and later with the mass magazines that were to emerge the following decade. For the next two decades, this income would finance a golden age of radio, creating a renaissance of popular culture in sports, music, drama, and comedy. Advertising changed the very purpose of radio as perceived by its Brazilian founders. They, like others around the world, envisioned radio as a sober, refined vehicle for the communication of education and culture, the "British Broadcasting Corporation (BBC) model." The competition among stations for listeners (to generate the highest possible advertising revenue) produced a more raucous model, which came to prevail in Brazil.

Radio's Golden Age

In 1930 the rural planter aristocracy was removed from power by President Getúlio Vargas. He would remain in power, using increasingly authoritarian tactics, until he was ousted in 1945. Radio was required by the government to carry a daily program of government news, the *Hora do Brazil* (*Hour of Brazil*). Vargas installed a propaganda and censorship bureau that controlled radio, film, and newspapers. Prohibited from engaging in politics, the new media amply compensated by producing the hallmarks of modern Brazilian popular culture.

Thanks to radio, soccer and samba became definitive marks of Brazilian culture. Soccer achieved its first international fame with the broadcast in 1938 of the World Cup match in France. The programming of music and the rise of the recording industry encouraged the popularization of the samba along with other types of Brazilian song. From this radio music environment emerged singers such as Carmen Miranda and samba composers such as Noel Rosa and Ary Barroso. Radio made the samba schools and carnival of Rio de Janeiro popular throughout the country, enhancing the city's identity as the center of national popular culture.

Audiences were subsequently attracted to radio for drama and comedy programs, which began to be broadcast in the late 1930s. Stage actors and comedians now had opportunities not only in the emerging cinema industry but also in radio. Serial dramas, known as *novelas* in Portuguese, produced a generation of actors who contributed to the renaissance of the Brazilian theater after World War II and later provided the talent for television drama.

Comedy programs, especially satire, which were played before live audiences, became a vital part of radio during the 1940s in the development of variety programs. Radio also affected the Portuguese language, helping to create a popular national style for expressing narrative (news reports), hyperbole (advertising), and deeply felt sentiments (soccer). In a country in which slavery continued until almost 1900, radio was both a vehicle and promotor of modernization. Its advertisements unveiled the glittering products of the modern age: movies, records, cars, electrical appliances, and more. It projected a Brazil that was industrializing, urbanizing, and modernizing.

The competition for advertising revenue provoked by radio caused newspapers to begin acquiring radio stations and forming national networks. The prime example of this kind of consolidation was *O Globo* newspaper, established in the 1920s: it created *Radio Globo* a decade later and then founded today's huge Brazilian-European television conglomerate, *TV Globo*. The radio station most fondly remembered from the golden age, however, the one that attracted the largest audiences and produced some of the most innovative and elaborate programming, was the Rio station *Radio Nacional*.

Modern Radio

The advent of television in the 1950s began the decline of radio, a decline that accelerated with the inauguration of color TV in the early 1970s. Nonetheless, because almost all Brazilians could afford to own a transistor radio, the number of radio stations continued to increase, and radio remained a stable part of everyday Brazilian life. Offering more economical advertising than television, it competed quite well.

An armed forces coup in 1964 inaugurated a military regime that endured until 1985 and that established a number of organizational changes in telecommunications. The Ministry of Communications was established in 1967. Under its auspices was created the state radio broadcasting company, the *Empresa Brasileira de Radiodifusão* (RADIOBRÁS), which was given responsibility for maintaining the technical quality and national coordination of the radio system. During the 1990s, Brazil launched several telecommunications satellites.

By the turn of the century, half of the Brazilian population of 170 million (the fifth largest population in the world) owned a radio. Brazil has nearly 3,000 radio stations—40 percent FM, 60 percent AM. Some stations broadcast using what in Brazil is termed an *onda tropical* (tropical wave) at 2,300 to 5,060 kilohertz. Radio programming involves mainly music, news, and sports. Since 1985, "phone-in" programs allowing listeners free expression have become very popular.

Radio has been a fundamental contributor to the modern technical and cultural development of Brazil. The development of Brazilian radio laid the technical foundations for Brazil to create its television industry, the largest in South America. That development in turn formed the basis for Brazil's achieving the largest computer industry on the continent.

EDWARD A. RIEDINGER

See also Landell de Moura, Roberto; South America

Further Reading

Federico, Maria Elvira Bonavita, *História da comunicação: Rádio e TV no Brasil,* Petrópolis, Brazil: Vozes, 1982
Fornari, Ernani, *O "incrível" Padre Landell de Moura,* Rio de Janeiro and Pôrto Alegre, Brazil: Globo, 1960; 2nd edition, Rio de Janeiro: Biblioteca do Exército Editora, 1984
Fox, Elizabeth, *Latin American Broadcasting: From Tango to Telenovela,* Luton, Bedfordshire: University of Luton Press, 1997
Schwoch, James, "The United States and the Global Growth of Radio, 1900–1930, in Brazil and in the Third World," Ph.D. diss., Northwestern University, 1985
Schwoch, James, *The American Radio Industry and Its Latin American Activities, 1900–1939,* Champaign: University of Illinois Press, 1990
Shaw, Lisa, *The Social History of the Brazilian Samba,* Aldershot, Hampshire, and Brookfield, Vermont: Ashgate, 1999
Straubhaar, Joseph Dean, "Mass Communication and the Elites," in *Modern Brazil: Elites and Masses in Historical Perspective,* edited by Michael L. Conniff and Frank D. McCann, Lincoln: University of Nebraska Press, 1989

Brice, Fanny 1891–1951

U.S. Stage and Radio Comedian

Fanny Brice played a key role in the history of radio. Not only was she one of the earliest women to headline a major prime-time show, but her portrayal of the inquisitive, mischievous Baby Snooks innovated the child-centered situation comedy, later to be developed by radio and television shows such as *The Aldrich Family* and *Dennis the Menace.*

Brice was born Fannie Borach on 29 October 1891 on New York's Lower East Side to Jewish immigrant parents, Charles Borach and Rose Stern. In the lively local vaudeville scene, where Fanny earned a small but steady income appearing in amateur nights from the age of 14, she learned the "Yiddish" accent so common to the ethnic humor of the time. Her first professional successes occurred in the field of burlesque, in humorous singing and dancing acts. One of her earliest hits was the song "Sadie Salome," composed by Irving Berlin, which Florenz Ziegfeld brought to a larger venue in his *Ziegfeld Follies of 1910.* Brice appeared in the *Follies* steadily from 1910 through 1923, diversifying also into vaudeville on the Orpheum and RKO circuits and into light comedy both in the United States and in Europe. The song "My Man," first performed in the 1921 *Follies,* became her biggest hit and trademark vehicle. Brice's stage popularity peaked with two Broadway vehicles produced by her soon-to-be ex-husband, Billy Rose, *Sweet and Low* and *Crazy*

Quilt, which toured across the United States from 1930 through early 1932.

Though Brice developed many comic personae and acts, she is best known (aside from the perennial "My Man") for songs such as "Sadie Salome," "Second Hand Rose," and "I'm an Indian" and for routines such as "Mrs. Cohen at the Beach," which drew from the vaudeville tradition of Yiddish dialect and humor. Her function in the *Ziegfeld Follies* may have been to embody, under a "disguise" of ethnicity, the working-class elements of burlesque that Ziegfeld had so carefully excised from his elevated "celebration of the American girl." Brice was adept at negotiating the double-edged weapon of ethnic humor, at once taking possession of the "othered" characterization while simultaneously disavowing or disarming it, notably through the physical, almost slapstick quality of her performance. Impersonating head-injured ballet dancers ("Becky Is Back in the Ballet"), Jewish evangelists ("Soul-Saving Sadie," a takeoff on Aimee Semple McPherson), graceless fan dancers, or the hypersexual movie vamp ("I'm Bad"), Brice's genius for physical satire both took the edge off her ethnic and sexual humor and marked out a performance space unique to Brice—a daring one for a female comedian. Often earthy and sometimes slightly bawdy, full of double entendres and subtle references, Brice's combination of ethnic humor and physical satire earned her the highest acclaim from stage critics and audiences alike.

Brice's fortunes in radio, however, would be another matter. Brice first ventured onto the air on the *Philco Hour* in February 1930, a stint that proved unsuccessful. She returned to the stage, but in 1933 she agreed to appear as a guest on J. Walter Thompson's *Fleischmann Yeast Hour* (*The Rudy Vallee Show*). This led to a contract with Standard Brands for the prestigious *Chase and Sanborn Hour* with host Eddie Cantor. Here her brash, bawdy humor, emanating as it did from a woman, proved troublesome to the National Broadcasting Company's (NBC) Continuity Acceptance Department. After a flurry of censoring memos and script deletions, Brice came up with a new strategy designed to allow her the comic freedom that her persona as a mature, sexual woman could not: the precocious child, Baby Snooks. Snooks had been introduced earlier, as "Babykins" in the stage show *Corned Beef and Roses,* but Brice built her character into a continuing role on radio. In the *Ziegfeld Follies of 1934* she brought Snooks onstage into the spotlight of national fame, which then carried back onto radio in the *Ziegfeld Follies of the Air,* sponsored by Colgate-Palmolive on the Columbia Broadcasting System (CBS) in 1936. Brice subsequently took Snooks to MGM/Maxwell House's *Good News of 1938* on NBC with Hanley Stafford as Daddy. Finally, in 1949, after a few more shifts in program and a brief return to the stage, the Dancer Fitzgerald Sample agency brought the *Baby Snooks Show* to NBC, where it ran to great popular acclaim until Brice's untimely death in May 1951.

"Schnooks," as Brice always called her, offered listeners a good-natured critique of adult hypocrisies, an exploration of the pitfalls of the English language, and an inverted world seen through the eyes of a precociously resistant and troublesome child, whose deflation of some of the most respected traditions, concepts, and institutions of American culture provoked sympathetic laughter while remaining safely contained and corrected by her age and innocence. In a 1939 skit, Daddy, upset by Snooks' terrible grades in school, decides to hire a tutor. In the opening lines, Snooks parodies a common attribution of schooling for women:

> Daddy: I'm hiring a private teacher to *make* you work.
> Snooks: Waaahhh! . . . I don't want no private teacher.
> Daddy: Oh now listen, dear, it's for your own good. She'll make you a little lady.
> Snooks: I don't wanna be a lady . . . and I don't wanna go to school.
> Daddy (voice rising): Well, what *do* you want to do!
> Snooks (smugly): I want to get married.

When the teacher arrives, speaking with a pretentious upper-class accent, Snooks refuses to be intimidated or to cooperate.

> Teacher: Now come here and kiss me, little one.
> Snooks: What for? I ain't done nothin'.
> Daddy: Now kiss your teacher, Snooksie.
> Snooks: *You* kiss her, Daddy . . .
> Teacher: Come here.
> Snooks: Leave me alone.
> Teacher (threateningly): When I beckon like this, it means I want you to come.
> Snooks: When I stick out my tongue like this, it means I ain't comin'!

Finally the teacher, at the end of her patience, turns Snooks over her knee and spanks her.

> Teacher: There! That'll impress it on your mind.
> Snooks: That ain't where my mind is!

This last line is a reworking of one that Brice attempted to use in 1933 on the *Chase and Sanborn* show:

> Fanny: Abe, why do you spank the boy like that?
> Abe: I spanked him to impress it on his mind.
> Fanny: Where do you think his mind is?

That time, network continuity acceptance editors objected and deleted the line immediately. Now, from the mouth of a child, it could be uttered over the air without repercussions.

Fanny Brice as "Baby Snooks"
Courtesy CBS Photo Archive

Having not quite made it to television—and it is doubtful whether the 59-year-old actress could have carried Snooks to television without the grotesque overwhelming the humorous—Brice faded slowly from memory, her contributions to the development of broadcast program forms recalled in brief asides but never deeply assessed. However, her child-centered situation comedy would soon become the staple of television. And Snooks is the character with whom Brice spent more of her life and on whom she expended more of her comic energy than any of her justly famous stage routines—despite what later semi-biographical works such as Barbra Streisand's two films choose to remember. She deserves a more prominent place in the history of broadcasting than past accounts have permitted. Brice's Snooks marks a significant moment in the movement of women's humor from the private sphere to the public arena. Once heard, Baby Snooks speaks in a voice that is hard to forget.

MICHELE HILMES

See also Vaudeville; Women in Radio

Fanny Brice. Born Fannie Borach in New York City, 29 October 1891. Appeared on vaudeville stage starting at age 14, with regular employment starting at 15; played in New York burlesque; headlined *Ziegfeld Follies of 1910* and several other years, 1911–34; played in light theater in London, 1914; back to New York stage, 1913–18; numerous film roles, 1928–46; married Billy Rose, Broadway producer, in 1929 (divorced 1938); starred in two Billy Rose-produced revues, *Sweet and Low,* 1930, and *Crazy Quilt,* 1931; radio debut on *The Fleischmann Yeast Hour,* 1930; appeared on *Ziegfeld Follies of the Air,* 1936; toured on stage in *The New Ziegfeld Follies of 1936–37;* returned to radio as Baby Snooks, 1938–51. Died in Los Angeles, California, 29 May 1951.

Radio Series

1930	*The Fleischmann Yeast Hour*
1933	*The Royal Vagabonds*
1933–34	*The Chase and Sanborn Hour*
1936	*Ziegfeld Follies of the Air*
1938–40	*Good News of 1938, 1939, 1940*
1940–44	*Maxwell House Coffee Time*
1944–45	*Toasties Time*
1945–48; 1949–51	*The Baby Snooks Show*

Films

My Man, 1928; *Night Club,* 1929; *Be Yourself!* 1930; *Crime without Passion,* 1934; *The Great Ziegfeld,* 1936; *Everybody Sing,* 1938; and *Ziegfeld Follies,* 1946

Stage

Ziegfeld Follies of 1910, 1911, 1916, 1917, 1920, 1921, and *1923; The Honeymoon Express,* 1913; *Why Worry?* 1918; *Music Box Revue,* 1924; *Fanny,* 1926; *Fioretta,* 1929; *Sweet and Low,* 1930; *Crazy Quilt,* 1931; *Ziegfeld Follies of 1934,* 1934; *The New Ziegfeld Follies of 1936–1937,* 1936–37

Further Reading

Goldman, Herbert G., *Fanny Brice: The Original Funny Girl,* New York: Oxford University Press, 1992
Grossman, Barbara Wallace, *Funny Woman: The Life and Times of Fanny Brice,* Bloomington: Indiana University Press, 1991
Hilmes, Michele, "Fanny Brice and the 'Schnooks' Strategy," in *The Life of the Party: Comediennes in Hollywood,* edited by Kristine Brunovska Karnick, New York: New York University Press, 2003
Unterbrink, Mary, *Funny Women: American Comediennes, 1860–1985,* Jefferson, North Carolina: McFarland, 1987

Brinkley, John R. 1885–1942

U.S. Radio Broadcaster

John Brinkley's broadcasts in the 1920s provided one of the first benchmark tests of the power of the federal government to control radio. A medical quack who used radio to promote his fraudulent products, Brinkley was forced off the air in the United States, only to continue his popular programs from a Mexican station for another decade.

Origins

Born in North Carolina, John Richard Brinkley came to Milford, Kansas, a town of several hundred persons, after World War I in response to an ad for a town doctor. He had graduated from the Eclectic Medical University of Kansas City and the Kansas City College of Medicine and Surgery. Dr. Morris Fishbein, head of the American Medical Association at the time, labeled both schools "diploma mills."

Brinkley began broadcasting in 1923 and quickly made his KFKB (Kansas First, Kansas Best) one of the most popular stations in Kansas and the Midwest. In addition to the station, Brinkley owned the Brinkley Hospital and the Brinkley Pharmaceutical Association. Three times each day for an hour or more, Brinkley hosted a program, *Medical Question Box,* over KFKB. During the program he answered letters from listeners and prescribed cures for their ailments, generally advising them to use Brinkley pharmaceutical medicines. These talks included his now-notorious claims to restore potency in men by grafting live tissue from goats.

Over-the-air prescriptions brought Brinkley between $15,000 and $20,000 each month. Brinkley and his staff performed thousands of "goat gland" operations at fees ranging from $250 to $1,500. Brinkley made himself popular by distributing large amounts of money to the town and financing building projects that employed residents. Among his projects were a large sanitarium and the Brinkley Methodist Memorial Church.

Landmark Legal Case

Complaints from the medical establishment against "radio quacks" reached the Federal Radio Commission (FRC) in 1929 and focused attention on the hazards to public health of these broadcasts. Although the FRC initially ruled that it had no authority to act under the Radio Act of 1927, by early 1930 commissioners changed their minds. Fearing negative consequences of widespread dispersal of questionable medical information, the commission decided it could determine whether programs were in the public interest, convenience, or necessity.

In hearings in 1930, the commissioners noted that KFKB carried Brinkley's talks and Brinkley Hospital programming about 3 hours out of each 15-hour broadcast day. Other station programs included music, lectures, weather broadcasts, political discussions, baseball reports, and coverage of national events. They also noted that Brinkley's *Medical Question Box* contained material of a possibly sexual nature, bordering on indecency. Under the Radio Act, the FRC could revoke a license for obscene, indecent, or profane broadcasts. After the hearings, the FRC refused to renew Brinkley's license on the ground that the station was being operated for his personal and financial interest and not in the public interest. Brinkley appealed.

The Court of Appeals upheld the FRC on 2 February 1931. The court ruled that KFKB was operated solely for Brinkley's personal interest and that the *Medical Question Box* was "inimical" to public health and safety, as it was devoted to "diagnosing and prescribing treatments of cases from symptoms given in letters" written by patients he had never seen. The court held that in license renewal, an important consideration was past performance. The court noted that censorship was not involved, because the FRC did not subject Brinkley's broadcasts to scrutiny prior to release, but the court added that the commission had a right to note a station's past conduct.

The Kansas Medical Board revoked Brinkley's medical license for "unprofessional conduct." Afterward, he ran for governor of Kansas in both 1930 and 1932 and nearly won as a write-in candidate. Although interested in the governorship itself, Brinkley undoubtedly also wanted the right to choose members of the State Board of Medical Examiners, which had revoked his license. Meanwhile, he headed to Texas and Mexico, where he continued his broadcasts.

Broadcasting from Mexico

In 1931 Mexican authorities banned his physical entry, but by using Mexican citizens as a front Brinkley was able to erect a station in Villa Acuna, Coahila, Mexico, opposite his new hometown of Del Rio, Texas. Using telephone hookups from his home in Del Rio, he began broadcasting from station XER in October. Again, he gave medical advice over the airwaves and answered letters sent to his *Medical Question Box*. He directed people to contact his hospital in Milford, Kansas, for further treatment. Broadcasting at 50,000 watts on 735 kilocycles, XER drowned out U.S. stations close to it on the dial and interfered with CKAC in Montreal, Canada.

Protests came from Mexican citizens and focused on station employees, who were all U.S. citizens, and broadcasts, which were rarely in Spanish. Criticism of this "Yankee imperialism" grew, as did complaints by U.S. officials, until the Mexican government forced Brinkley to close XER by mid-1934. Brinkley then continued his broadcasts by purchasing broadcast time over XEPN in Piedras Negras, Mexico, and using telephone hookups to both Mexico and Abilene, Kansas, from Del Rio, Texas. Other broadcasts were carried through a studio at Eagle Pass, Texas, just across the Rio Grande from Piedras Negras.

Mexico, meanwhile, promulgated a set of regulations for Mexican broadcasters that eliminated a number of conflicts with U.S. broadcasters. Commercial stations could be licensed only to Mexicans, had to be operated by Mexicans, and had to employ a staff comprising no less than 80 percent Mexican citizens. All advertising rates had to be approved by Mexico's Department of Commerce, and all medical advertising had to receive approval of the Minister of Health. All programs were to be in Spanish, unless the station received special permission from the government. These laws were aimed not only at Brinkley but also at the radio mystics, astrologers, and fortune tellers who transferred their activities to stations south of the border when FRC pressure halted such broadcasts in the United States.

In January 1935 Brinkley was charged with violating Section 325(b) of the newly passed Communications Act of 1934. This provision made it illegal to maintain, use, or locate a studio or apparatus in the United States for the purpose of transmitting sound waves electrically to a radio station in a foreign country for rebroadcast back to the United States without first securing a permit from the Federal Communications Commission. Brinkley, however, continued to sidestep the law for five more years until the Mexican government finally forced him off the air in 1940 with the signing of the North American Regional Broadcasting Agreement.

Demise

In 1941 Brinkley suffered a series of heart attacks, and one of his legs was amputated. By then, his fortune had disappeared, and he was forced to declare bankruptcy. He was also charged with mail fraud, but he never came to trial, as he died in San Antonio, Texas, on 26 May 1942.

Brinkley's widespread popularity was abundantly evident in his two campaigns to become governor of Kansas, as well as in the large audiences for his Kansas and later Mexican radio stations. His importance to radio history hinges on the FRC and Court of Appeals decisions denying KFKB's license renewal on the basis of his past programming record and how it compared to the public interest, convenience, or necessity measure of the 1927 Radio Act. The court decision was the first judicial affirmation of the FRC's right to make such decisions.

LOUISE BENJAMIN

See also Border Radio; Federal Radio Commission; First Amendment and Radio; North American Regional Broadcasting Agreement

John R. Brinkley. Born in Beta, North Carolina, 8 July 1885. Graduated from Eclectic Medical University of Kansas City, May 1914, and Kansas City College of Medicine and Surgery (honorary degree), 1919; began broadcasting on KFKB, 1923; famous for promoting "goat gland" surgery to aid male virility; FRC ordered him off air in United States, 1931; moved broadcast operations to Mexico; forced off air in 1940 with signing of North American Regional Broadcasting Agreement. Died in San Antonio, Texas, 26 May 1942.

Further Reading

Benjamin, Louise M., *Freedom of Speech and the Public Interest,* Carbondale: University of Southern Illinois Press, 2001

"The Brinkley Case," in *Documents of American Broadcasting,* edited by Frank J. Kahn, 4th edition, Englewood Cliffs, New Jersey: Prentice-Hall, 1984

Carson, Gerald, *The Roguish World of Doctor Brinkley,* New York: Rinehart, 1960

Juhnke, Eric S., *Quacks and Crusaders: The Fabulous Careers of John Brinkley,* Norman Baker, and Harry Hoxsey, Lawrence: University Press of Kansas, 2002

Lee, R. Alton, *The Bizarre Careers of John R. Brinkley,* Lexington: University Press of Kentucky, 2002

Resler, Ansel Harlan, "The Impact of John R. Brinkley on Broadcasting in the United States," Ph.D. diss., Northwestern University, 1958

Shelby, Maurice E., Jr., "John R. Brinkley: His Contribution to Broadcasting," in *American Broadcasting: A Source Book on the History of Radio and Television,* edited by Lawrence W. Lichty and Malachi C. Topping, New York: Hastings House, 1975

Wood, Clement, *The Life of a Man: A Biography of John R. Brinkley,* Kansas City, Missouri: Goshorn Publishing, 1934

British Broadcasting Corporation

The British Broadcasting Corporation (BBC) has dominated the history of U.K. radio, operating as a monopoly until the early 1970s. Since 1973, when commercial radio stations were licensed for the first time, independent local, regional, and national radio expanded to the point that by 1997 more people listened to commercial radio in an average week than to the BBC. Nevertheless, BBC influence and presence remains pervasive and compelling, although more direct competition has been an agent for reform, more rapid evolution, and diversification. Certainly competition from illegal "pirate" services in the 1960s was a major factor in transforming the national networks and brought into being a national popular music network responding to the needs of postwar youth culture.

Character and Issues

The BBC was an important prototype of a public corporation operating as the mechanism for state broadcasting (often as a monopoly), a model since replicated in many countries. Radio broadcasting emerged in the early 1920s when Britain was still a global imperial power. Radio technology was initially utilized by amateurs as an enthusiastic hobby, and inventors, engineers, and commercial entrepreneurs saw the potential to make money. Government intervention determined the future and development of radio, but other factors shaping the BBC included influences from abroad, individual people, and existing forms of mass media. Government saw radio as a potential tool or even a weapon. The British mass media audience was literate, gaining in disposable income, and lived within a society where political and social pressures sought a more equal distribution of wealth. The notion of "equality" had finally won the vote for women over thirty.

BBC history has been heavily colored by powerful cultural and political myths. For one thing the British Broadcasting *Company* began in 1922 as a private entity before it became (in 1927) a public corporation and with a Royal Charter and statutory license agreement. It is wrongly assumed that there was political and social consensus that the one-company private monopoly concept was "the right way forward." It is also wrongly assumed that this regulated one company monopoly was a "civilized" reaction to the capitalist and market driven chaos of American radio in the early 1920s and that the planning and introduction of approved radio in Britain was based on the public interest. In fact the trademark centralism of BBC radio, with power vested in London, was only imposed after the BBC became a public corporation. Until 1929 BBC Radio was something of a network of local stations with culturally autonomous production centers and a lively and creative partnership of local broadcasters and loyal listeners.

One of the remarkable aspects of BBC history is that the 1927 model remained the status quo until 1973, while the structure and constitution of BBC funding has remained remarkably consistent. This stability has laid the ground for the BBC's transformation into a powerful cultural and political force in British society.

The BBC as a public corporation can be described in many ways:

Funding: While the government decides the cost of the annual listener license, the BBC has maximum independence concerning the spending of the resulting revenue. Initially its sole income was derived from the collection of the license fee, which has become a legally enforceable taxation on any U.K. citizen who wishes to listen to the radio or watch television using a receiver. The separate license for radio was abolished in 1971. The BBC has always operated commercial "hybrid" activities such as selling programs abroad, merchandising products, and publishing in other media such as magazines, cassettes, CDs, the internet, books, videos, DVDs, and satellite television. From the outset the BBC cross-promoted the media linked products such as *The Radio Times*, *The Listener,* and other publishing products that carried considerable amounts of advertising.

Governance: The Chairman and Board of Governors constitute the legal personality of the BBC. They are supposed by convention to be chosen by the government not as representatives of sectional political interests but on the basis of their experience and standing. Although political parties in office are supposed to avoid political prejudice in these appointments, it can be argued that this convention has often been undermined. The problem has been to achieve reform of the BBC in line with reforms in the public sector without making the BBC subject at any time to the political policies of the party in power.

For example, BBC Director-General Alisdair Milne officially resigned in 1987, but in subsequent interviews it became clear he was dismissed on the initiative of a new chairman (Marmaduke Hussey) who had been deliberately selected by the Conservative government to "sort out the BBC." The trigger for the dismissal had been a public controversy over government anger and opposition to BBC programming policy on the Northern Ireland troubles. It can also be argued that he had an inability to see the need for root and branch management reforms to make the BBC more efficient.

Government is allowed to intervene only in BBC decision-making during a national emergency or if it is clearly shown that the BBC has not abided by the Charter and terms of the license granted by Parliament. The Corporation has complete editorial independence in the production and scheduling of its programs, although its history shows frequent incidents and periods of self-censorship and compromising political influence.

These basic descriptors raise any number of issues. For example, why did British radio begin as a monopoly devoid of competition and regulated by the state? Why was Britain behind the U.S. in the development of radio as a business? Despite this, why was radio so successful in building audiences given the lack of listener choice? And whose interests did BBC broadcasting serve—the government, the BBC, or the audience? Finally, how has the development of the BBC matched broader changes and developments in British society?

Origins

In 1912 the Marconi Company's energetic managing director, Godfrey Isaacs, persuaded the Postmaster General (the minister responsible for posts and telecommunications in the Liberal Government) to present a plan to link the British Empire with a network of 18 radio stations. Marconi was granted the contract. Isaacs pulled off a stroke of commercial opportunism by buying the troubled American United Wireless Company, which held the rights to Lee de Forest's vital triode vacuum tube. However, allegations of government corruption undermined an imaginative and developmental scheme for global expansion of radio by British interests. Meanwhile, other countries accelerated both government and private radio development. These included not only the United States but such European countries as France, the Netherlands, and Belgium. In response, the British Postmaster General dispatched a senior civil servant, F.J Brown, to the United States to study the American scene and make appropriate recommendations for the best course of licensing and regulating the growth of radio in Britain.

There is a wide discrepancy between what Brown observed and what he actually reported. He was in the U.S. during the winter of 1921–22. American stations had not yet discovered the potential of radio advertising (the WEAF experiments selling real estate did not air until August 1922). The radio Brown observed was then dominated by educational objectives as the 400 stations on the air were largely run by public, civic, and religious institutions. Their funding depended on donations, selling radio sets, or other funding *not* including the sale of advertising. Radio was widely perceived as a public democratic medium, not as a commercial free for all. But Brown downplayed this American radio diversity and emphasized the "dangers" of "a large number of firms broadcasting." He concluded it was impossible to have a variety of broadcasting stations in Britain because "it would result only in a sort of chaos, only in a much more aggravated form than that which arises in the United States."

While it can be argued that the political imperative in Britain in the early 1920s was social and political control, Brown seems to have ignored that American radio was characterized by an explosion of freedom of expression in the arts, entertainment, in education, opinion and even the potential for radio journalism. Thus British civil servants and commissions of enquiry for decades to come also overlooked this reality. The government and the BBC had the advantage that few people then traveled abroad to experience alternative approaches to radio.

Brown and his Post Office superiors invited proposals from interested parties to develop one or two private stations to provide national broadcasting, although it was made very clear they preferred the idea of only one. Despite a subsequent battle between Marconi and the American-controlled Metropolitan Vickers company (Westinghouse was a major shareholder) a single station company soon emerged out of a coalition of interests and government inspired/cajoled compromises. While it seems bizarre in hindsight, the cultural imperative of British radio broadcasting was set by one civil servant, F.J. Brown, who defined the following statutory brief for the BBC: "to educate, inform, and entertain the British public, but with no newsgathering, advertising or controversial content to be originated by the company."

Reith's Influence

The BBC's first director general, John C.W. Reith, was appointed in 1922 and two years later produced a book, *Broadcast Over Britain,* which became an influential blueprint for public corporatism. Reith wrote: "in these days, when efforts are being made towards the nationalization of the public services and of certain essential industries of the country, the progress of broadcasting has been cited as the most outstanding example of the potentiality of a combination of private enterprise and of public control." As Asa Briggs later recognized, Reith argued that the BBC should exercise the "brute force of monopoly" which would reinforce the other three fundamentals of broadcasting: public service, a sense of moral obligation, and assured finance. Although Reith and his supporters would never recognize it at the time, what was being proposed for the BBC was a totalitarian institution constructed out of a reform of a socialist concept and serving within a capitalist economy.

Reith quickly established four objectives in public service broadcasting: information, education, entertainment, and high standards. He wrote in *Broadcast Over Britain,* "It is occasionally indicated to us that we are apparently setting out to give the public what we think they need—and not what they want—but few know what they want and very few what they need. . . . In any case it is better to over-estimate the mentality of the public than to under-estimate it." A year later the Crawford Parliamentary Committee was appointed to investigate the future of British broadcasting. If Reith could persuade them to adopt public corporatism as a monopoly, he would have achieved his aims. By maneuvering the broadcasting organization away from private commerce into the public sphere, Reith argued that the BBC would be insulated from the ravages of competitive market economy which would undermine his determination to give the public what he believed would be good for them.

But could government and Parliament trust the BBC to serve the public interest? The British General Strike of 1926 became the supreme test. In order to win the confidence of the political establishment, Reith ensured the BBC was the voice of the British state. He did so by retaining the dignity of acquiescent autonomy and pretending that it was the voice of the British Constitution. His eye was on the prize of public corporation and not on the right of any other voice of moderation to broadcast its point of view. The next year saw the creation of the British Broadcasting Corporation, answerable ultimately to Parliament but with day-to-day control left to the judgment of a Board of Governors that were supposed to be appointed on the basis of their standing and experience.

Still, and despite Reith's best efforts, the BBC was continually attacked in Parliament and in newspapers such as the *Daily Mail* and *The Times* for disseminating "left-wing propaganda." These papers remain today among the strongest critics of the BBC. It was claimed that "pink Bolshevism" was prevalent in the BBC's interpretation of news, in talks, addresses to school children, and entertainment programs. The pressure of this newspaper criticism generated self-censorship. (For example, a storm over a plan by Filson Young to write a play called *Titanic* led to the BBC withdrawing the commission in 1932 without the author having put one word to paper.) Accusations of bias were often combined with a threat that the BBC's funding should be remodeled, its Charter reconstituted, and competition introduced to break up the existing monopoly.

In January 1924 newspapers first reported on the need to crack down on illegal broadcasters. It was clear that the maintenance of the monopoly in radio broadcasting demanded investigative and enforcement machinery. The Radio Society of Great Britain assisted the Post Office by forming a broadcasting "flying squad" which sought to track down the owners of the unlicensed transmitting sets in the North Surrey area who had been "interrupting the official broadcast programs by howling and making other disturbances expressive of disapproval of certain items sent out." This effort was clearly a precursor to the Post Office's radio detector vans, which from 1932 roamed the streets of Britain in search of households that had not paid the required license fee. By October 1932 criminal prosecutions for non-license payments had begun. The hunt for "radio pirates" was undertaken by the Post Office. The threat of detection and prosecution resulted in a record 154,000 wireless licenses being taken out in the first 11 days of that month, with an average of 14,000 a day (compared with September's figure of only 2,800 licenses being issued daily). The new war on "pirates" was planned on the assumption that there had been at least 5 million wireless sets in the country, and the number of unlicensed sets had even been placed as high as 2 million.

Despite the 1930s Depression, the BBC grew steadily larger in the prewar years. The BBC employed nearly 800 people when it became a Corporation in 1927, and had expanded to almost 5,000 by 1939, one year after Reith left. As occurred with many other industries, World War II accelerated investment and activity in the BBC. By 1945 the staff had grown to 11,500.

Programs

During the first decade of BBC development, the organization was essentially a middle-class institution with a commitment to education as an agent of social advancement. Supplementing Reith's early call for the on-air use of a standardized "received pronunciation," the BBC began to present regional accents in features, documentary programs and continuity in the 1930s as it began to intersect with the British working class and demonstrate the potential of the developing meritocracy. BBC radio offered a wider canvas for the rhetoric of politics and during election campaigns connected a growing electorate with the body politic. For the 1929 General Election the BBC had begun to broadcast views of party political leaders and reported election results faster than newspapers could.

The BBC became the voice of cultural and political consensus as well as a location for those voices. This paradox was evident in 1929 when *The Radio Times*, the BBC's weekly listing magazine, which now lists all radio and TV services available in the U.K., promoted the racist stereotyping of black Americans using such language as "Hear Dem Darkies Singin" (to promote a program of "minstrel plantation singing") yet at the same time celebrated Paul Robeson with a live broadcast of his singing at a concert in Bournemouth, and published a critique of African-American classical music by the Harlem Renaissance poet Countee Cullen. A similar ambiguity was evident in the irrelevant identification of alleged and convicted criminals as Jews. Arthur Koestler's brother changed his name to Green and pursued an on-air career at the BBC with all traces of his Jewish heritage expunged.

Despite silencing far-left political voices during the 1930s, the BBC has tended to adopt a more moderate position during periods of political and cultural paranoia. During the 1920s and 1930s it had to negotiate an ideological struggle between communism and fascism. It took its cue from government policy so that during the Spanish Civil War, General Franco's forces were described as "insurgents" rather than "rebels." In 1927 the BBC censored a commissioned play entitled *Machines* by Reginald Berkeley because it was deemed politically controversial in its criticism of the social and spiritual impact of capitalism on the individual. During and after World War II, BBC radio would become a rich and lively location for a wide range of artistic and documentary programming that articulated political, cultural, and social dissent. In the later Cold War, left-wing American writers would participate in BBC arts and cultural programming after moving to a more liberal Europe. BBC radio commissioned, produced, and broadcast to large audiences material that was restricted on the stage in terms of language and content.

The BBC has also responded to the gradual, though slow development of sexual equality through the acceptance of women in various areas of work. In the 1920s and 1930s women began to have a noticeable presence in broadcasting. U.S. born Elizabeth Welch had her own radio program in the 1930s and became Britain's first black television presenter at Alexandria Palace in 1936. Joan Littlewood and Olive Shapely participated in the pioneering BBC radio features movement at Manchester in the 1930s. Audrey Russell became the BBC's first roving radio correspondent during World War II. The war accelerated further the participation of women in male dominated arenas of work and professional culture. By the 1950s women were taking leading editorial roles in prestigious radio programs, and two decades later they were being appointed to senior BBC executive positions in BBC radio.

From 1955, live programs began to abandon their reliance on the scripting of every word. Radio performance began to sound more natural and no longer depended on the acting ability of program participants. The BBC became a location for the origination and dissemination of popular culture through comedy and satire such as *The Goon Show* and the subversive utterances of disc jockeys such as Kenny Everett. More cost-effective recording processes introduced in the late 1950s meant that a live production culture with half of all programs providing a live audio theatre gave way to a schedule with 90 percent pre-recorded programming on BBC national networks in 1975.

Changing Networks

Development of a "General Forces Programme" during World War II, where entertainment was more evident than information and education, helped to set the stage for the creation of "The Light Programme" after the war. The presence of over a million U.S. service personnel, including a substantial number of African Americans, and exposure to the programs of the American Forces Radio Network resulted in American styles of radio presentation and formats being adopted by the BBC. Demand for a greater variety of music and popular entertainment formats was evident in the 1945 inception of "The Light Programme." A year later "The Third Programme" provided a sound stage for highbrow programming, while the "Home Service" continued to offer mainstream news, drama, and cultural programming.

The changing identity of national regions in Northern Ireland, Scotland, and Wales was maintained by separate services for these areas. Their existence could be regarded as an influential factor in the establishment of a parliament for Scotland, and national assemblies for Northern Ireland and Wales between 1997 and 1999.

In 1967 the networks were added to and redrawn as BBC Radio1, BBC Radio 2, BBC Radio 3 and BBC Radio 4. BBC Radio 1 was set up to cater to the national youth culture audience for music and Radio 2 tended to serve the evolving demand for popular music for people aged over 40. As the state began to recognize the democratic nature of popular culture and have more confidence in diverse voices participating in the public sphere of media debate, the BBC changed its rules on program preparation and production. In 1994, following the success of a temporary radio news service during the 1991 Gulf War, the BBC recognized the growing demand for more concentrated news and sports programming with the establishment of the national *Radio Five Live* service making use of vacant BBC AM frequencies.

By the early 2000s there were five BBC national radio networks, three national BBC stations for Wales, Scotland, and Northern Ireland, and a network of 39 BBC local radio services in England. By 2003 the BBC had established a portfolio of digital radio services serving more targeted "narrowcast" markets in music, light entertainment, speech, and ethnic minority programming. Two years earlier the BBC had positive public feedback for projected national digital stations that included a service focusing on black music, news, and speech aimed at a young audience (1Xtra), a station focusing on the music that helped shape the generation from the 1970s to the 1990s (6 Music), a new speech-based service mixing old and new comedy, drama, stories, and features and also focusing on storytelling for children (BBC 7), a niche sports service widening the radio coverage of live sport (*Five Live Sports Plus)*, the delivery of the BBC World Service to the domestic audience, and an upgrading of an embryonic Asian Network into a national station focusing on news and sports for British Asians.

Tuning In

The need for BBC listener research developed out of commercial competition from abroad in the 1930s, which soon professionalized BBC links between programming and audience. The audience research section, led by Robert Silvey after 1936, highlighted the impact of overseas commercial programming from Luxembourg and Normandy in northern France. The dour BBC symphony of religious services, talks, and classical music on Sundays was rapidly abandoned by U.K. listeners who preferred the light band music and bright lights of consumer advertising on the English speaking commercial radio services from the continent targeted at British audiences. The dry biscuit of Holy Communion on the BBC Sabbath offering was a poor second to the Cadburys Chocolate sponsorship on Radio Luxembourg by a ratio of 2:8. Audience research soon contributed to the adoption of more "fixed point" scheduling. The BBC realized its programs needed to follow the social habits of its audience.

The evolution of domestic radio technology aided this process. By the late 1920s there was a change from using crystal sets that required the use of headphones (and which tended to be home-made and largely used by males) to manufactured tube-powered receivers that enabled housebound women to listen during the day. Later, car radios served the demand for programs that followed commuting habits at breakfast and drive-time. The 1950s development of the portable transistor radio became a fashionable tool of reception by young people and underlined the need for the BBC to respond to their interest in popular musical formats.

Competitive Pressure

BBC Radio began to lose its monopoly position through the 1950s as audiences tuned to commercial music services from Europe and various offshore pirate stations. This process was hastened after the beginning of independent television in 1956, although domestic commercial local radio began only in 1973, and a commercial national radio service two decades later. The explosion of consumer youth culture and popular rock and roll music meant that radio offered a new platform for retaining a mass audience.

The BBC more recently came under pressure from changes in the political-economic consensus as government control and direction of the economy gave way to market economics. The BBC was urged to generate some of its own income rather than depend on license fee funding. Staff levels were reduced by almost a third. However, despite the ferocity of political criticism, mainly from the right, and regular Parliamentary enquiries into BBC funding and operations, the BBC's constitution today remains very similar to that of 1927, surely an unprecedented record in the world's public service radio.

By 2003 most BBC radio services had internet dimensions by way of live audio streaming or lateral levels of text, audio, and video accompaniment that was downloadable on demand. In addition, many established programs on networks such as Radio 4, Radio 3 and BBC Five Live were being archived with a back catalogue available to listeners from the BBC website. BBC Radio's share of total listening was 51.6 percent, an increase of 5.6 percent from 1998. Radio had overtaken television as the most consumed medium in the U.K.

Radio 3 continued its public service role in supporting he Promenade concerts, World Music, and five orchestras. Radio 4 was the largest commissioner of new writing in the world. Radio 1 was reaching over half of the country's 15 to 24 year olds. BBC radio comedy was feeding the television medium with the successful transfer from sound to vision of series such as *Dead Ringers*, *Alan Partridge*, *Goodness Gracious Me*, and *The League of Gentlemen.*

Out of a total license fee income in the year 2001–2002 of nearly £2.6 billion, £302 million was being spent on domestic radio services. The Foreign and Commonwealth Office funded the BBC World Service through Grant-in-Aid with £205 million. The number of people being jailed for defaulting on license fee prosecution fines had dropped to 17 compared to 148 in 1998–99. These figures indicate that BBC Radio is at the time of writing the most generously public funded radio organization in the world. The only comparable level of funding was NHK in Japan and the regional German public radio stations.

In 2003 the U.K. government was seeking to consolidate its regulation of communications under one body known as Ofcom. The BBC had successfully lobbied to resist its absorption into this framework. However, Ofcom will take over the Broadcasting Standards Commission role of adjudicating on complaints about taste, invasion of privacy, and fairness. The BBC has also been under pressure to submit its financial auditing to greater external scrutiny.

TIM CROOK

See also Archers; Cooke, Alistair; Gillard, Frank; Promenade Concerts; Reith, John

Further Reading

Barnard, Stephen, *On the Radio: Music Radio in Britain,* Milton Keynes: Open University Press, 1989

Briggs, Asa, *The History of Broadcasting in the United Kingdom,* 5 vols., Oxford and New York: Oxford University Press, 1961–95

Briggs, Asa, *The BBC: The First Fifty Years,* Oxford and New York: Oxford University Press, 1985

British Broadcasting Corporation, *Annual Report and Accounts,* London: BBC Information Services (annual)

Burns, Tom, *The BBC: Public Institution and Private World,* London: Macmillan, and New York: Holmes and Meier, 1977

Cathcart, Rex, *The Most Contrary Region: The BBC in Northern Ireland 1924–84,* Belfast: Blackstaff Press, 1984

Crisell, Andrew, *Understanding Radio,* London and New York: Methuen, 1986; 2nd edition, London and New York: Routledge, 1994

Donovan, Paul, *The Radio Companion*, London: HarperCollins, 1991

Great Britain National Audit Office, *Management of the BBC World Service: Report by the Comptroller and Auditor General,* London: HMSO, 1992

McDowell, W.H., *The History of BBC Broadcasting in Scotland, 1923–83,* Edinburgh: Edinburgh University Press, 1992

Miall, Leonard, *Inside the BBC,* London: Weidenfeld and Nicolson, 1994

Paulu, Burton, *British Broadcasting: Radio and Television in the United Kingdom,* Minneapolis: University of Minnesota Press, 1956

Paulu, Burton, *British Broadcasting in Transition,* Minneapolis: University of Minnesota Press, and London: Macmillan, 1961

Pawley, Edward, *BBC Engineering 1922–72,* London: BBC, 1972

Reith, John, *Broadcast Over Britain,* London: Hodder and Stoughton, 1924

Seymour-Ure, Colin, *The British Press and Broadcasting since 1945,* Oxford and Cambridge, Massachusetts: Blackwell, 1991; 2nd edition, 1996

Street, Seàn, *A Concise History of British Radio 1922–2002,* Devon: Kelly Publications, 2002

British Broadcasting Corporation: Broadcasting House

BBC London Headquarters

Broadcasting House, also known as BH, is the headquarters of the British Broadcasting Corporation. The BH is a 1932 building in Portland Place, London. It originally cost £350,000 and was regarded by many critics as formal, cold, and pretentious. The shiplike shape was the second choice, after a "top hat" design. *Architectural Review*'s description of "the labyrinthine pokiness of the interior" added to the general condemnation of the building's human and aesthetic qualities. Even the "Rhapsody to Broadcasting House" specially composed to celebrate its inauguration was panned by the music critics.

Lord Asa Briggs, in his monumental history of the BBC, says that the BBC's 1932 move from Savoy Hill (behind the famous Savoy Hotel), to the new custom-built Broadcasting House represented the myth that, before, all was "intimacy and harmony" and that everything after was "bureaucracy and conflict."

The architect Lieutenant Colonel G. Val Mayer was given the task of replacing Savoy Hill, a charming Georgian house and garden designed by John Nash, with a modern building that would provide office accommodation and a complex of sound studios for production and broadcast. Although BH contained 12 floors and 350 offices for 700 people, within its first year of using the facility the BBC found itself too big for its new headquarters. The original BH was expanded by a large extension, projected before the war, but only opened in 1961. It contains both offices and studios and interconnects with the original 1932 building.

Val Mayer created something that the press at the time compared to a beached ocean liner. Mayer had to respect the unusual shape of the plot and the function of broadcasting. A central "boiler-house" of soundproofed studios was wrapped in an exterior of concrete offices. Three floors were below ground level, and some of the studios were bomb-proof. BH had the fastest elevators in London and became the first London building to install central air conditioning, because the central tower of silent sound studios in the middle had no natural ventilation.

From the beginning, nearly every aspect of the building's existence was associated with controversy. The staff did not like their new home, and BBC Director General John Reith asked his colleagues to "ring to the inconveniences a good heart." Reith himself expressed his dislike of the BBC's new home and was embarrassed when his name was included in a confident Latin dedication in the entrance hall: "Deo Omnipotenti."

Sculptor Eric Gill's external carvings and reliefs led to a complaint in the House of Commons. In March 1933 a Member of Parliament asked the home secretary if he would make the police compel the BBC to remove immediately the statue placed over the front entrance, as it was "objectionable to public morals and decency." He was referring to "Prospero and Ariel," and in particular to the display of Ariel's genitalia, which, according to oral history, had been reduced in size after the BBC's governors had inspected the work by looking up from the pavement. It is claimed that one of the governors climbed the scaffolding with a notebook and tape measure and informed Eric Gill, "In my view this young man is uncommonly well hung." Gill was the celebrity artist of his time; he was known as "the married monk" and carved the statue in situ wearing medieval dress; he refused to accept a free wireless set.

A major eccentricity of the building is that there appears to be no logical connection between the stairs and the number of floors. This is because the original design had to be compromised in a "rights to light" dispute with neighbors (meaning the new BH could not cut off all the natural light to the existing buildings nearby), so that from the sixth floor upwards, the top of the building was narrowed by a mansard roof. Staff complaints led to an internal inquiry, and the report presented on 1 January 1934 savaged the building's functional features, complaining of noise leakage between studios, not enough lighting to read scripts and scores, doors that were not wide enough for pianos, and the fact that the Bakerloo Line underground could be heard in the studios that were below ground level. The concert hall was supposed to be a superstudio that would accommodate 100 musicians. Acoustic problems meant that it could only cope with 30 to 35.

The only appreciation for the pioneering modernism of Australian Raymond MacGrath's interior design was to be found in *Architectural Review*, the editor of which described the art deco and jazz age colors and curves as the "New Tower of London." Apart from the council chamber, an auditorium, and the entrance hall, with Gill's symbolic figure of the Sower scattering seeds as a man might broadcast ideas that grow wherever they are heard, virtually nothing of the original interior has survived—not even the mock chapel, with an altar and cross projected electronically for religious broadcasts. Other than several contemporaneous publications, the 1930s film "Death at Broadcasting House" in black and white is one of the few records available of the building's original interior.

The first broadcast from the building by Henry Hall's new BBC Dance Orchestra on 15 March 1932 was also captured on newsreel. The BH survived the Blitz, even though a 500-pound delayed-action bomb killed seven people on 15 October 1940. The sound was recorded during the nine o'clock news, but newsreader Bruce Belfrage paused and had to continue for security reasons. More damage was caused by a land mine exploding in Portland Place on 8 December 1940.

The steady increase in space needs led the BBC to take over the old Langham Hotel, across Portland Place from BH, for offices. In the 1980s, the BBC held an architectural competition, won by Foster Associates, to build a modern Radio Centre on the same site. It was never built because of financial restrictions, and plans to expand BBC facilities near the BBC's television centre at White City. In 1988 the BBC was prosecuted and fined because inadequate maintenance of a cooling tower on top of the building had caused the deaths from Legionnaires' disease of three people.

BH has been made a "listed" (historic) building. BBC radio news and current affairs programs including the network *BBC Five Live* were moved to a purpose built bi-media news center at White City in West London in 1995. But the move was unpopular. The radio dimension of programming felt dominated by the size, ego, and costs of television. The BBC's Director of Radio and Music at the time of writing, Jenny Abramsky, observed that the radio program teams did not agree that synergies would emerge and would ensure that the BBC spoke with one voice.

White City was also an unpopular location for journalism. It was far from Central London and the seat of power and decision-making. The lease on Bush House, the headquarters of BBC External Services was also due to expire. In 2001, the BBC decided to redevelop BH into a new home for BBC Radio & Music, BBC News, and the BBC World Service. The structure of the original 1932 building and surviving features such as the entrance hall and council chamber would be preserved in an integrated complex of 140 studios, a central atrium, and a newsroom half the size of a football pitch. Two streets have been closed to the public as four adjacent buildings are demolished and excavated to within three meters of a London underground tunnel. The scheme seeks to transform BH into the world's largest broadcasting news hub accommodating over 5,000 members of staff.

The plan has been designed by British architect Sir Richard MacCormac and his team and is described by the BBC as "a remarkable combination of the old and the new." The British *Observer* newspaper on 3 January 2003 described it as "a glass and marble palace," expected to be completed by 2008. The studios of old BH have been redesigned for the digital age. The BBC claims the new complex will include public spaces and amenities "to make the BBC more accessible and welcoming to visitors." The new "BBC Palace" is being equipped for national and international radio, television, and online services. During the late 1990s it accommodated an interactive "BBC Experience" exhibition, which purported to inform paying customers about the history of the corporation.

TIM CROOK

Further Reading

Architectural Review (August 1932) (special issue on Broadcasting House)

British Broadcasting Corporation, *A Technical Description of "Broadcasting House,"* London: British Broadcasting Corporation, 1932

Broadcasting House, London: British Broadcasting Corporation, 1932

Donovan, Paul, *The Radio Companion,* London: HarperCollins, 1991

Mayer, Lt-Col. G. Val, *Notes on the Building* in *BBC Yearbook 1932,* London: British Broadcasting Corporation, 1932

Reid, Colin, *Action Stations: A History of "Broadcasting House,"* London: Robson House, 1987

Walsh, Conal, *BBC's News Palace in Peril,* London: *The Observer,* 5 January 2003

British Broadcasting Corporation: BBC Local Radio

It is not widely known that the BBC in its early years (to 1929) was really a network of local radio stations. The original private company had been the result of a merger of stations in London, Manchester, and Birmingham, some of which had been owned originally by American companies such as Westinghouse and Western Electric.

Origins

The limited reach of early transmitters meant that it was only feasible to broadcast locally produced programming. In 1923 a further six main stations were added to the network, Newcastle, Cardiff, Glasgow, Aberdeen, Bournemouth, and Belfast.

Between 1924 and 1925 more "relay stations" were set up at Sheffield, Plymouth, Edinburgh, Liverpool, Leeds-Bradford, Hull, Nottingham, Dundee, Stoke-on-Trent, and Swansea. Telephone lines linked the stations, allowing a pool of national programming to be received and relayed from London. In 1925 London programs would be provided on two complete evenings every week. Frequently stations would pool their resources and stage simultaneous broadcasts.

There do not appear to be any sound archives that would enable the modern listener to appreciate the local flavor of the broadcasting from this period. But early issues of the *Radio Times* and other magazines for radio enthusiasts present a picture of a warm relationship between early broadcasters and their audiences and of programming that was both popular and cultural. Scannell and Cardiff (1991) report that local stations originated programs ranging from early quiz shows and phone-in programs to live interactive dramas with listeners winning prizes for supplying the best endings. Writer Patrick Campbell recalled that his first introduction to radio was in 1924 when he went to the Bournemouth studio every Wednesday to talk in *Children's Hour* on behalf of a young person's charity group. The edition of the *Radio Times* for 30 August 1929 reveals the preparations for a play called *The Pennillion Singer,* which "deals with an exciting time in the 'Hungry Forties' when the small farmers and yeoman of South Wales were up against adverse fate in many ways, and specially against the tyranny of the toll-gate." This play about an armed rebellion with farmers disguising themselves as "Rebecca's Daughters" touched local political sensitivities and was originated and broadcast from the BBC's Cardiff station 5WA. In the same week listeners could hear Professor Patchett from Bournemouth's 6BM on his summer holiday experiences in the "New Germany," and Captain H. La Chard at Plymouth's 5PY on "various aspects of life in Borneo." A lively local radio culture generated a rich exchange of local music, storytelling, entertainment, and information, but Scannell and Cardiff (1991) argue that between 1927 and 1930 this would be "quite deliberately eradicated by the policy of centralization."

By 1929 the BBC's Director-General Sir John Reith and his executives at the London station 2LO had already embarked on a process of centralization and control. It took the form of regular meetings between London and local station directors, touring by London inspectors, and the submission of program schedules to London in advance of broadcast. Soon the BBC eradicated the high cost of local radio repertories and orchestras through centralization. The savings enabled the concentration of budgets in London, which was thought better suited to producing expensive concerts, variety shows, and extravagant drama productions. Head of Talks Charles Siepmann and Director of Drama Productions Val Gielgud were given full authority over their regional strands of programming to maintain "London standards" throughout the country. Local radio gave way to "regionalization," with production centers at Glasgow, Manchester, Birmingham, Cardiff, and London. The call signs of the smaller stations had disappeared by 1930, and the spontaneity of their programs and loyalty of their audiences were quickly forgotten.

U.S. Radio Inspires a Revival

A proposal for reviving local radio appeared in the report of the Beveridge Broadcasting Committee in 1949. VHF (FM) radio extended the spectrum of wavelengths. The former BBC war correspondent Frank Gillard was now climbing up the ladder of BBC management, and in 1954 he had submitted a report on his observations of U.S. radio. He was impressed by WVPO Stroudsburg, in the Pocono Mountains of Pennsylvania, which operated in daylight hours only, served a community of 15,000 people with a staff of 13, and, according to Gillard, "spoke to its listeners as a familiar friend and neighbour." He noticed that the whole operation was conducted "with the utmost informality." When he became responsible for BBC radio programming, Gillard campaigned passionately for a local dimension to BBC broadcasting. He was responsible for the BBC promise to the Pilkington Committee's enquiry into broadcasting in 1961 that "BBC Local Radio will be friendly, reliable and in touch with people's lives." In the same year Gillard had arranged 16 closed-circuit radio experiments in towns such as Dumfries, Dundee, the Isle of Wight, and at Bournemouth on the South Coast, which had been the location of the BBC's first local stations started in 1923. When he realized that nobody on the Pilkington Committee seemed sympathetic to the idea of local radio, Gillard impressed some of the members by playing recordings of the closed circuit experiment.

By 1966 the government had approved the start of nine BBC local stations as a two-year experiment. Initially some of the funding came from local authorities. The first of the new BBC local radio stations started in Leicester because the local City Council was prepared to contribute £104,000 toward its costs. The notion of "radio on the rates," or tax-supported radio, would not last, however, and local radio is now wholly funded by a share of the BBC's license fee income. Unlike public radio in the USA and Australia, there is no listener subscription, and any pledges for money from listeners are reserved for charitable projects such as the annual "Children in Need."

BBC Radio Leicester first aired on 8 November 1967 with a pledge from a government minister that the station "should never forget that it is home town radio with its own Leicester individuality and that it must always be bright and attractive." The first audience figures disclosed that 25 percent of local listeners were tuning in.

It is important to emphasize that the BBC's commitment to local radio was not simply a romantic celebration of the diversity of local cultures. Gillard's ambitious plan for a network of more than 90 BBC local stations was a political tactic to head off the clamor for legalizing commercial radio. It enabled the BBC to justify its monopolistic control of all license fee funds. Leicester was followed by seven other local services in Brighton, Durham, Leeds, Merseyside, Nottingham, Sheffield, and Stoke-on-Trent. The BBC's Chief Correspondent Kate Adie was then one of the first reporters working at the station in Durham. Local radio pioneered live and unscripted programming. It tended to be a mixture of news, local current affairs, sport, and record request programs. The first charter for local radio, written by Frank Gillard, declared: "Station managers will be free to provide programmes, which in their judgement best meet the needs of their communities."

A second wave of BBC local stations was opened in 1969, in Birmingham, Blackburn, Bristol, Derby, Humberside, London, Manchester, Medway, Newcastle, Oxford, Solent, and Teesside. By 1978, 22 local stations were broadcasting at least six hours of programming every day with news on the hour, talk at breakfast, and sports coverage on Saturdays. Some stations were able to originate drama. BBC Merseyside in Liverpool became a platform for local dramatists. Merseyside and other stations such as BBC Stoke originated soap operas, which were funded by police and health authorities. Although BBC services are not permitted to receive commercial sponsorship or sell air time for advertising, the BBC Charter does permit the receipt of funding from government/public bodies for educational and public information services. Using entertainment and dramatic formats to educate listeners became an imaginative mechanism to secure funding from outside the license fee.

Competition with Commercial Local Radio

Britain's first licensed independent station, the news and speech service LBC in London, began broadcasting in October 1973. A week later the music format station Capital opened. British independent radio was legislated on a local basis with a public service broadcasting remit. A proportion of profits was "taxed" and reinvested in training, engineering, and prestigious programming projects. In its early days Capital was able to include a range of program styles such as news features at breakfast, a soap opera, and a daily social action campaign. The BBC stations found themselves in a head-to-head competition even though the program schedules of Independent Local Radio stations were primarily music based.

The impact on BBC Radio London was catastrophic, and since the inception and expansion of commercial radio the BBC's local London service has never been able to establish any significant market share. It has repeatedly reinvented itself. Radio London became GLR (Greater London Radio, "London Live") and more recently "BBC London." In the provinces, however, BBC local services were able to sustain a commanding share of listeners.

By 1997 the weekly audience for BBC local stations was in the region of 7 million listeners. By 2003 the weekly reach had increased to 10.7 million listeners per week—over 20 percent of the population. It saw itself as providing a service to those over the age of 55 with a speech-based programming format, which provided space for local comment and opinion. In view of the fact that BBC local radio's 39 stations compete with more than 250 local commercial stations, the desire to target people over 55 might be seen as a decision to cater to marketing categories not prioritized by Independent Local Radio. The early ambition to establish a network of more than 90 stations had to be tempered by the periodical financial crises affecting the BBC in the 1980s and 1990s. The relatively unfashionable arena of local radio was an easy target for staff cutbacks and cancelled capital projects. There were occasional aberrations of management judgment that had to be reversed, such as merging BBC Berkshire with BBC Oxford to form the station BBC Thames Valley. The fusion of these two stations liquidated the reflection of the separate cultural dimensions and social identities of a significant city and town in southern England. It caused resentment. It was the equivalent of expecting the respective soccer teams for Oxford and Reading to merge under a different name. The current policy of respecting "localness" has resulted in stations with regional reaches such as Southern Counties, Wiltshire Sound, and Radio Devon introducing separate output at key times of the day from the major towns and cities of their catchment areas. Merging BBC Guildford and BBC Brighton created BBC Southern Counties. The effective closure of the station in Brighton, which is a thriving town on the South Coast and in a different county, and running the programming from Guildford, which is in the county of Surrey, was the antithesis of the spirit of "localness."

By the 1980s the U.K. Home Office had given approval for the establishment of 38 local stations that could broadcast to 90 percent of the population. As independent commercial radio became both regional and national, with a substantial diversification into niche musical formats, it could be argued that BBC local radio found itself being marginalized into providing programming for marketing categories in the upper middle-age range that did not interest advertisers.

Deregulation of commercial radio and the reduction of newsrooms by independent stations increased the importance of BBC local radio journalism. BBC local stations perfected the technique of "snowline" services during weather crises, and they have been given official roles in future emergency plans. The BBC saw its local services as "a helping hand in times of trial, a smiling friend in times of success, and a questioning intelligence in times of controversy. And at all times it was to be ambassador, inquisitor, commentator and reporter."

By 1990 a new charter, primarily written by Ronald Neil, the then Managing Director of Regional Broadcasting, committed BBC local radio to a minimum speech/music ratio of

60/40 between 6 AM and 6 PM, with speech rising to 100 percent at the peak periods of breakfast and afternoon commute—wide-ranging speech, addressing local concerns and interests, and journalism as the bedrock that should be respected in local radio as it is on the programs *World at One* and *Panorama.*

Social and Cultural Objectives

By 1997 commercial radio had overtaken the BBC with a 50.6 percent share of all listeners. By 2002 BBC radio's share had bounced back to 51.6 percent, and radio had overtaken television as the most used medium in the U.K. The loyalty of many BBC local listeners has now been challenged by the rich offerings of formats elsewhere. Audiences in most metropolitan markets have shrunk, although the Birmingham station BBC Radio WM has been a notable exception. With the commercial stations abandoning their public service speech-based programming, and concentrating on music from the pop charts, contemporary, and "gold" nostalgia hits, the BBC local stations have catered to an older audience that prefers a speech format. Simulcast "regional" programs share costs, and a sustaining overnight service of the national network BBC Radio Five Live has provided for 24-hour broadcasting. The BBC seems to see those over 55 as its main audience for local radio; 30 percent of its audience under 45 tunes in for shorter periods, to cherry pick specific programs and information services such as news, sports, travel, and weather. The remit is therefore overwhelmingly public service. A survey by the BBC has discovered that 91 percent of the population think it is important to continue with public service radio and over 80 percent thought this genre of radio should be informative, educational, entertaining, and catering to all age groups and tastes. The BBC's Head of Radio at the time of writing, Jenny Abramsky, believes public service radio should have range, ambition, and "a duty to contribute to culture—both popular and high." This may be the case with the national networks such as Radio 4 where programming is budgeted on the basis of £11,000 an hour. BBC local radio costs £300 an hour.

In 2002 all of the BBC's local stations were underwritten by a budget of £108 million. This means that the costs and resources are not available to deliver significant educational objectives. However, the local stations have made some contribution to the collection of oral history. In 1999 BBC local staff interviewed over 6,000 people about their lives over the 20th century. In conclusion, the profile of the audience and the very style of broadcasting do not encourage the expression of non-mainstream views in either the arts or journalism.

A BBC poll of listeners between 1994 and 1997 sought to determine local radio's strengths and weaknesses. BBC local stations were often seen as "a phone-in Citizens Advice Bureau." Listeners' loyalty was based on the absence of advertising, their perception of better presented, more in-depth news, presenters asking more intelligent questions, a calmer texture of programming, and a feeling that BBC local stations offered better coverage of local affairs. Criticisms were that BBC local radio was too parochial, sounded like institutional radio at times, was amateurish, and transmitted badly researched interviews and uninteresting chat.

The BBC has encouraged its local stations to be more upbeat, to broadcast more interesting and varied content, as it tries to meet the challenge of increasing competition in the 21st century. A key objective is radio programming with an element of fun and professional presenters who seem to be more accessible than their national counterparts. Its public service role is also centered on serving specialist subject areas such as religion output, serving ethnic communities, and presenting sports and social action broadcasting. BBC local stations are firmly committed to digital radio and multi-media provision through the internet, and many BBC local radio services are now establishing and exploring their presence on the internet.

Many commentators feel that BBC local radio is well placed to meet the social demands of contemporary Britain. Demographic trends indicate an increasing older population, one with a continuing appetite for local radio services, for news about "where I live," which is consistent with sociological research that reveals a stability in communities and a sense of the importance of local identity and attachment.

Although Britain has become a much more multi-racial, multi-cultural and multi-religious society, the BBC was shaken by 1994 research indicating that listeners from ethnic communities found the license fee poor value for money. The BBC responded with improved ethnic representation in BBC local stations, and the BBC local service in London has since then certainly maintained a better record for employing non-white broadcasters and journalists. The BBC Asian Network in the Midlands is regarded as a success story: it tries to meet the needs of a substantial British Asian population in Central England and connects them with BBC World Service broadcasts in English and Asian languages.

In October 2002 the BBC Asian Network went nationwide when it was launched as a digital station dedicated to broadcasting a mix of speech and music to second and third generation British Asians. Vijay Sharma said it would become "a one stop shop for Asian communities where they can get daily national news, top international stories, big consumer stories, and music ranging from the latest in British Asian sounds to old favourites." This development indicates that the BBC's interpretation of "localness" is coming to terms with social associations that go beyond geographical locations and recognizes the diversity of social and cultural communities.

TIM CROOK

See also Gillard, Frank; Reith, John C.W.

Further Reading

Adie, Kate, *The Kindness of Strangers,* London: Headline, 2002

Baehr, Helen, and Michele Ryan, *Shut up and Listen! Women and Local Radio: A View from the Inside,* London and New York: Comedia, 1984

Barnard, Stephen, *On the Radio: Music Radio in Britain,* Milton Keynes, Buckinghamshire, and Philadelphia, Pennsylvania: Open University Press, 1989

Barnard, Stephen, *Studying Radio,* London: Arnold, and New York: Oxford University Press, 2000

Briggs, Asa, *The History of Broadcasting in the United Kingdom,* 5 vols., Oxford and New York: Oxford University Press, 1961–1995; see especially vol. 5, *Competition,* 1995

Chantler, Paul, and Sim Harris, *Local Radio Journalism,* Oxford: Focal Press, 1992; 2nd edition, Oxford and Boston, 1997

Chapman, N., editor, *BBC Local Radio 2000: A Review for a New Era,* London: BBC, 1999

Crisell, Andrew, *Understanding Radio,* London and New York: Methuen, 1986; 2nd edition, London and New York: Routledge, 1994

Crisell, Andrew, *An Introductory History of British Broadcasting,* London and New York: Routledge, 1997

Crook, Tim, *International Radio Journalism: History, Theory, and Practice,* London and New York: Routledge, 1998

Donovan, Paul, *The Radio Companion: The A–Z Guide to Radio from Its Inception to the Present Day,* London: HarperCollins, 1991

Eckersley, Myles, *Prospero's Wireless: A Biography of Peter Pendleton Eckersley,* Romsey, Hampshire: Myles Books, 1997

Hartley, Ian, *2ZY to NBH: An Informal History of the BBC in Manchester and the North West,* Altrincham: Willow Publishing, 1987

Hendy, David, *Radio in the Global Age,* Cambridge: Polity Press, and Malden, Massachusetts: Blackwell, 2000

Herbert, John, *Journalism in the Digital Age: Theory and Practice for Broadcast, Print, and On-Line Media,* Oxford and New York: Focal Press, 2000

Herbert, John, *Practising Global Journalism: Exploring Reporting Issues Worldwide,* Oxford and Boston: Focal Press, 2001

Lewis, Peter M., and Jerry Booth, *The Invisible Medium: Public, Commercial, and Community Radio,* London: Macmillan Education, 1989; Washington, D.C.: Howard University Press, 1990

Local Radio Workshop, *Local Radio in London,* London: Local Radio Workshop, 1982; 2nd edition, as *Nothing Local about It: London's Local Radio,* London: Comedia, 1983

Scannell, Paddy, and David Cardiff, *A Social History of British Broadcasting,* Oxford and Cambridge, Massachusetts: Blackwell, 1991

Shingler, Martin, and Cindy Wieringa, *On Air: Methods and Meanings of Radio,* London and New York: Arnold, 1998

Wilby, Pete, and Andy Conroy, *The Radio Handbook,* London and New York: Routledge, 1994

British Broadcasting Corporation: BBC Monitoring Service

The British Broadcasting Corporation Monitoring Service was formed just before World War II in order to listen to enemy radio stations and report their contents to the government and to other departments of the BBC. By the end of the century, 60 years later, the service was monitoring, in conjunction with its American partner the Foreign Broadcast Information Service (FBIS), 100 languages from over 150 countries and serving a variety of customers. Its main sources are radio, television, news agencies, the foreign press, and, increasingly, the internet. The service has often brought the first news of important developments and has become in effect a news agency, but one that simply reports words without any gloss.

The need for such a service became apparent in the 1930s. As fascist powers Germany and Italy broadcast violent anti-British propaganda, both the Foreign Office and the BBC did some listening in, using shorthand typists wearing headphones. In the summer of 1939, with war approaching, the new Ministry of Information formally asked the BBC to undertake the task on a wider, more professional basis and agreed to cover the cost.

Because it was feared that London might be heavily bombed, the service was at first based in the Midlands, in the town of Evesham in Worcestershire. In 1943 it was moved to Caversham Park, near Reading, west of London, where it remains.

The first surviving report, dated 27 August 1939, said that the highlight of German news programs in German was "the new ration decree." News in English for Africa the same day "suggested that Poland was in a nervous, anxious and bellicose state and that the army chiefs were losing control over the lower army groups."

Monitoring in Wartime

A few days later, the German army invaded Poland, and Britain and France declared war. The early wartime experiences of monitors were described later by retired members of staff. One of them said

> I was sent almost immediately . . . to the hut which housed the engineers. There was a long table down the centre of the room and on it stood a row of big black boxes. They turned out to be radio sets, most delicate instruments, each with about half a dozen coils representing a range of frequencies (a term then still unknown to most of us). Housed in movable boxes, these were tuned by large calibrated dials. The engineers were rightly very proud of these sets and also protective of them; it was months before they allowed us to operate them ourselves.

During the war, monitoring became a 24-hour, seven-days-a-week operation. Verbatim texts of speeches by Nazi leaders and lists of prisoners of war were provided. One notable coup was the "eavesdropping" on the weekly article written by the Nazi propaganda chief, Josef Goebbels, for the periodical *Das Reich*. The article was transmitted in advance to the German authorities in Norway on the Hellschrieber, the German tape machine. The Monitoring Service picked it up and transmitted it to the BBC German Service, which wrote a commentary on the article before it appeared—much to the amazement of the Nazis.

In 1942, shortly after the United States had entered the war, its own monitoring service, the FBIS, began a collaboration with the BBC service, which was formalized in 1947 and lasts to this day. The two services divide the world between them, which allows them to operate more economically than if they both tried to cover everything. The BBC concentrates on Russia, Central Asia, Iran, parts of Europe and Africa, and a few other areas, and the FBIS concentrates on the rest of the world; there is some overlapping. The arrangement has been described as a shining example of international cooperation.

Cold War

After the war, the service began to look farther afield, particularly to the communist world. Because so many official pronouncements are first made on radio, it picked up many developments that it reported to the world either through the BBC World Service or one of the news agencies that subscribe to the Monitoring Service; these include the death of Stalin in 1953, the nationalization of the Suez Canal in 1956, and the crushing of the Hungarian revolution by Soviet tanks the same year. In 1962 the Monitoring Service played a part in bringing the Cuban missile crisis to an end. At a dangerous stage in the crisis, it monitored a Moscow Radio broadcast in which the Soviet leader, Nikita Khruschev, replying to a message from President Kennedy, said, "The Soviet government has ordered the dismantling of the bases and the despatch of the equipment to the USSR. I appreciate your assurance that the United States will not invade Cuba."

The message was flashed to Washington via the FBIS. Kennedy replied immediately, even though he had not received the official text, because, he said, "I attach tremendous significance to acting quickly with a view to solving the Cuban crisis."

Sometimes the behavior of radio stations gave a hint that an important announcement was on its way. On the evening of 26 August 1968, for example, it seemed that momentous events were in the offing in Czechoslovakia. The Czech monitor on evening duty stayed on to listen to the first bulletin after midnight, which normally repeated old news. At 12:50 A.M. he heard this: "In a short while the Czechoslovak radio will be broadcasting an extremely important news item. Stay at your receivers, wake all your fellow citizens." This was followed by an interval of music, then an announcement that Warsaw Pact troops had invaded the country to put an end to the experiment of "communism with a human face," the so-called Prague spring.

The Monitoring Service received a foreshadowing of the overthrow of President Ceauçescu of Romania in 1989. Listening to a live relay of a rally Ceauçescu was addressing, the monitor noticed what sounded like a scream. The relay went silent, to be resumed later. The recording was played over and over again for checking, and finally the BBC correspondent in Central Europe was alerted. The president was overthrown, but the scream was never identified.

The attempted coup against Russian leader Mikhail Gorbachev in 1991 was heralded by radio. Just before three o'clock in the morning on 19 August, all the separate radio channels were merged into one, a break with the normal pattern. Ten minutes later, the Soviet news agency announced the formation of a state emergency committee to run the country because, according to the agency, of Mr. Gorbachev's state of health. Once the Monitoring Service heard this, the BBC broke the news.

The Gulf War in the same year provided the Monitoring Service with a severe test. There was a proliferation of new radio stations, such as the "Mother of Battles," "The Voice of Free Iraq," "Holy Mecca," "Voice of Peace," and "Voice of the Gulf," as well as Baghdad Radio and other stations in the area.

The Monitoring Service became the main source of news from within Iraq, and Caversham Park was visited by journalists and TV crews from many parts of the world, including Japan, Canada, the United States, the Netherlands, and Britain itself.

Monitoring Today

The tasks of the Monitoring Service and its working arrangements altered considerably over the first 60 years of its existence. Caversham Park itself remains a pleasant environment, with extensive grounds and graceful vistas over the Thames valley. The building dates from the 19th century but has been enlarged since then. The receiving aerials are in an old deer park a few miles away (Crowsley Park), an electrically quiet area in which the local electricity company has agreed not to station any overhead power cables. Here there are aerials suited to the whole range of radio frequencies and a number of satellite receiving dishes. There are more dishes in Caversham Park itself. The first dish went into operation in 1981, covering one of the Soviet television channels. Since then, there has been a huge increase in the number of satellite TV and radio channels, and they now account for over 70 percent of the services received.

The monitors themselves are foreign nationals living in Britain or at Monitoring's overseas units or British graduates with a good grasp of languages. They listen on headphones, but transmissions are recorded, and it is from the recordings that translations are made, which can be checked for accuracy and archived. The monitors have been described as sedentary correspondents, gathering news by sitting and listening.

In a booklet written in 1979, a former member of the service made the point that, in addition to good hearing, a monitor needs "a wide knowledge of current affairs, politics, economics, history, world geography, a knowledge in depth of the language or languages being monitored, and a fluent and idiomatic command of English." Accuracy and speed, he said, were the keynotes; he continues: "The monitor wages a constant struggle against the unreliability of sound. In that struggle background knowledge and an intuitive gift for mental association are major allies."

The use of satellites has made transmissions easier to hear since those days, but there are still problems. Mishearing can lead to a mistranslation, and even if the words are clear, their meaning may be obscure. Television presents its own problems: pictures can be open to subjective interpretations.

In 1994 the Foreign Office undertook a review of the whole operation, which changed its direction and the arrangements for funding. The service now has "stakeholders," customers who specify their requirements and who are serviced in various ways. Stakeholders are major government departments concerned with foreign policy and international security and the BBC itself. British diplomatic missions abroad that are on-line

Listeners in Eritrea with clockwork radio
Courtesy BBC World Press Office

can receive the service direct. Material is also available to the public by commercial subscription to hard-copy publications and tailored on-line services. Customers include the media, major corporations and investment houses, consultants, charities, academic institutions, and freelance researchers. The stakeholders require that the service receive as much commercial income as possible.

The Monitoring Service and the FBIS between them select about 150,000 words a day from the millions they receive. The selection process still depends on the knowledge, skills, and judgment of the individual men and women who make up the service, but it is conveyed to the customers in radically different ways from the past. At one time, information was conveyed by teleprinter or the printed word.

Printed information from the Monitoring Service continues in several forms. There is the *Summary of World Broadcasts*, a regular publication that developed from the wartime *Digest*. The *Summary* is divided geographically—the former Soviet Union and Baltic states; Central Europe and the Balkans; Asia and the Pacific; the Middle East and North Africa; and Africa, Latin America, and the Caribbean. A weekly publication called *World Media* gives a complete picture of developments in radio, television, satellite communications, and news agencies. Another weekly publication, *Inside Central Asia*, is in the form of a briefing document covering the various states in the region. However, the trend is for an increasing amount of material to be supplied in soft-copy form. It can be accessed either by searching the worldwide web database (with a password, or "pull" service) or by means of a profiled e-mail service from the database (a "push" service). This service enables customers to extract only information relevant to them, meaning that material can be passed much more quickly.

The Monitoring Service has always been able to provide up-to-the-minute news, and this continues with the Newsfile. This is a 24-hour-a-day service that now provides a concise headline for each story, a summary of the main points, acknowledgment of the source, and relevant quotes. Audio and video material from a wide range of countries is available for actuality inserts—a useful service for broadcasters, in particular the BBC World Service, one of Monitoring's main customers.

There is constant consultation with customers to ascertain their requirements. There is still high interest in a number of geographical areas, such as the former Soviet Union, but there is also increasing interest in specific subjects, including developments in the foreign media, reaction to British policies, energy, human rights, crime, telecommunications, and terrorism. Some services are targeted for customers according to criteria they provide and are sent by fax or e-mail. Some are packaged to meet particular requirements on an ad hoc basis. The service also provides reference material, such as cabinet lists and biographies, consultations and radio and TV interviews with in-house specialists, and monitoring of the use made of radio frequencies.

There has been a major retraining program to raise standards of English among the monitors in order to improve their understanding of customer needs and their basic skills, including use of the internet. This is to enable them to release material directly to the newsroom or to external customers instead of having to pass it through a process of editing.

The end of communism in Europe and technical developments have meant that instead of dealing with state-controlled media with a single voice, the Monitoring Service now deals with a huge proliferation of media. The challenge now is to select the most authoritative, authentic, and representative of these many different voices. Caversham Park remains the headquarters, but the Monitoring Service has set up a number of regional sites to gain access to the increasing range of local sources. This began in 1961, when a unit was established in Nairobi to cover East and Central Africa. Much later came others—Moscow and Tashkent in 1994, Baku in 1997, Kiev in 1998, and Rabat in 2000.

ANDREW WALKER

See also Foreign Broadcast Information Service

Further Reading

Kris, Ernst, and Hans Speier, *German Radio Propaganda,* London and New York: Oxford University Press, 1944

Renier, Olive, and Vladimir Rubinstein, *Assigned to Listen: The Evesham Experience, 1939–43,* London: British Broadcasting Corporation, 1986

Rolo, Charles J., *Radio Goes to War,* New York: Putman, 1942; London: Faber, 1943

Walker, Andrew, *A Skyful of Freedom: 60 Years of BBC World Service,* London: Broadside Books, 1992

British Broadcasting Corporation: BBC Orchestras

For all of its history, the British Broadcasting Corporation has operated under a mandate to provide information, entertainment, and culture for citizens of the United Kingdom and beyond. As the BBC's programming has not been driven exclusively by ratings, its original programming has been oriented toward cultural quality rather than commercial appeal. This focus on cultural quality has often been given expression in the presentation of classical and symphonic musical performances, and the stars of those performances have most often been the BBC Orchestras. Several different orchestras provide original music programming for the BBC, including the BBC Symphony Orchestra, the BBC Philharmonic Orchestra, the BBC Concert Orchestra, the BBC Scottish Symphony Orchestra, and the BBC National Orchestra of Wales.

The BBC Symphony Orchestra is considered the flagship orchestra of the BBC. Based in London, it was formed in 1930 under Sir Adrian Boult, only three years after the incorporation of the BBC. Its primary purpose was to serve as a permanent, full-time orchestra performing music for broadcast. The orchestra developed a reputation for performing new music and has provided premier performances of hundreds of musical pieces, many of which were specifically written for broadcast on the BBC. One very significant charge given to the orchestra was to provide much of the music for the world-famous Henry Woods Promenade Concerts (the "Proms"), a two-month festival held each summer in London's Royal Albert Hall and broadcast over the BBC's Radio 3. The orchestra also performs another festival each January, recognizing and celebrating the work of a 20th-century composer. In addition to these performances for broadcast, the BBC Symphony Orchestra stays busy with concert appearances, recording sessions, and international tours.

The BBC Philharmonic Orchestra, based in Manchester, originated as a small ensemble of 12 players assembled to perform music for Manchester's first radio station, which began broadcasting in 1922 and was known by the call sign 2ZY. Consequently, the orchestra was first called the 2ZY Orchestra. Typically, the 2ZY Orchestra performed lighter music because of its smaller size. On occasion, however, extra players were brought in and more elaborate concertos and even symphonies were offered. After a few years, the radio station 2ZY became part of what is now the BBC. 2ZY's station manager, Dan Godfrey, Jr., had so impressed the BBC that he was moved to London to oversee the orchestra there. Shortly thereafter, under a new leader, T.H. Morrison, the 2ZY Orchestra was renamed the Northern Wireless Orchestra (NWO).

The BBC established the BBC Symphony Orchestra in 1930 to serve the entire BBC system, a decision that led to a reduction in support of regional orchestras such as the Northern Wireless Orchestra. To the public's dismay, the NWO became a nine-piece ensemble, the Northern Studio Orchestra. But it soon became clear that one orchestra could not meet the level of demand for broadcast music, and the regional orchestras were restored. The Manchester orchestra was re-established as the BBC Northern Orchestra, and in spite of continuing threats to its existence, the orchestra endured. During the second half of the 20th century, it achieved symphonic strength, found new prestige leading to invitations to perform at events such as the Proms, and embraced its new name, the BBC Philharmonic Orchestra. The philharmonic performs for a number of festivals in the United Kingdom and internationally. It also makes studio recordings that are available to the public.

The BBC Concert Orchestra, based in North London, was established in 1952, in part to supplement the broadcast work of the symphony orchestra and the regional groups. Its repertoire combined traditional classical music with light, popular works. The concert orchestra performs regularly for BBC Radio 2, as it is featured each Friday evening on the program *Friday Night Is Music Night.* Other performances include broadcast music for BBC Radio 3 and BBC Television, studio recordings, and annual appearances at the Proms. In addition, the concert orchestra hosts its own festival of popular classics at the Royal Festival Hall in London and makes other concert appearances throughout the United Kingdom and the world.

The BBC Scottish Symphony Orchestra is another of the BBC's regional performing groups. Established in 1935 as the BBC Scottish Orchestra, it was Scotland's first full-time orchestra. Its primary role was performing music for broadcast over BBC Scotland. Early on, however, the BBC Scottish Orchestra also became involved with the Edinburgh Festival, which allowed the public to enjoy the orchestra in person and offered players an opportunity to perform with many internationally acclaimed musicians and artists. As the orchestra grew in fame and stature, its name was changed in 1967 to the BBC Scottish Symphony Orchestra. Since the 1960s, the BBC Scottish Symphony Orchestra has regularly performed at the Proms in London, toured internationally, and—more recently—recorded a number of popular, award-winning compact discs.

The BBC National Orchestra of Wales is another regional symphony orchestra that was created to perform music for broadcast over the BBC. Its music is heard frequently on BBC Radio 3, BBC Radio Cymru/Wales, and BBC Television. The National Orchestra of Wales performs several times each year at the Proms, as the other BBC orchestras do; it also tours internationally and makes recordings.

A sixth "orchestra," the BBC Big Band, features jazz music and is the only remaining survivor of several popular and light music orchestras that were discontinued during the 1960s and 1970s.

Each of the BBC orchestras continues to pursue its primary purpose of providing music for broadcast over the BBC, especially for BBC Radio 3, the classical music channel. But each orchestra also has broadened its output to include international tours, concert appearances, studio recordings, and even educational efforts to teach young people about fine music. In every case, the BBC through its orchestras fulfills its mission of informing, entertaining, and providing cultural enrichment to citizens of the United Kingdom and many other parts of the world.

RICHARD TINER

See also Promenade Concerts

Further Reading

BBC Online: The BBC Orchestras <www.bbc.co.uk/orchestras>

Boult, Adrian, *My Own Trumpet,* London: Hamish Hamilton, 1973

Briggs, Asa, *The BBC: The First Fifty Years,* Oxford and New York: Oxford University Press, 1985

Cain, John, *The BBC: 70 Years of Broadcasting,* London: British Broadcasting Corporation, 1992

Carpenter, Humphrey, *The Envy of the World: Fifty Years of the BBC Third Programme and Radio 3, 1946–96,* London: Weidenfeld and Nicolson, 1996

Doctor, Jennifer R., *The BBC and Ultra-Modern Music, 1922–36: Shaping a Nation's Tastes,* Cambridge and New York: Cambridge University Press, 1999

Kenyon, Nicholas, *The BBC Symphony Orchestra: The First Fifty Years, 1930–1980,* London: British Broadcasting Corporation, 1981

Miall, Leonard, *Inside the BBC: British Broadcasting Characters,* London: Weidenfeld and Nicolson, 1994

Orchestra Net, <www.orchestranet.co.uk>

British Broadcasting Corporation: BBC Radio Programming

BBC radio programming presents a rich and complex mixture of old-style and modern formats. Sequential format stations based on the personality of presenters and reflecting changing fashions in music, such as Radio Five Live and Radio 1, were established to compete with commercial radio services. Radio 1 was created in 1967 as a response to the success of pirate music services. Radio Five Live, concentrating on news and sports, was established as a response to the success of the London Broadcasting Company and to preempt the expected development of national talk commercial services. BBC Radio 4 and Radio 3 and to a lesser extent Radio 2 have preserved an output based on individual and separate programs, some of which have a heritage going back several decades.

The key changes in the content of the BBC's national radio stations and the development of BBC local radio came about as a result of implementing the policy document *Broadcasting in the Seventies,* published in 1969. The idea was to establish a coherency of programming. Radio 4 was transformed from its old Home Service style of mixed speech and music into a wholly speech-oriented station where journalism would be allowed to bloom, current affairs would have a central role, and drama, comedy, science, and coverage of the arts would have a national platform. The current affairs and discussion programs *The World Tonight*, *PM*, and *Start The Week* all began in 1970. Radio 3 tried to abandon some of the aspects of its previous incarnation in 1946 as the Third Programme. It would become a national station concentrating on music and the arts. In 1969 BBC Radio 1 and BBC Radio 2 simulcast programming during many periods of the day. From 1970 the stations developed separate identities for different audiences. Radio 1 targeted the youth culture audience in popular music. Radio 2 tried to keep in step with the musical tastes of people in their forties and fifties. Stations in Scotland, Wales, and Northern Ireland have maintained separate centers for radio programming that seek to reflect the politics and concerns of their communities.

At the time of writing the development of digital radio has resulted in the addition of five more national stations. 1Xtra, launched in August 2002, is dedicated to playing the very best in contemporary black music for a young audience. 6 Music launched in March 2002 with a remit to draw on the best music from the past through to today and look for the best music of tomorrow. The station claimed it played contemporary songs that other networks would deem too daring. Five Live Sport Extra launched in January 2002 as a supplement to Radio Five Live and offers sports fans the option of listening to more football, rugby, cricket, tennis, and Formula 1 racing. The Asian Network was launched as a national digital channel in October 2002 providing speech and music to Britain's Asian communities that are concentrated mainly in the Midlands and South East. BBC 7 launched in December 2002 and specialized in a mix of BBC comedy, drama, book readings, and children's radio programs.

In its 1949 *BBC Year Book,* an attempt was made to set out the objectives for originating and commissioning BBC radio drama. This manifesto or "cultural agenda" provides a useful framework to define, evaluate, and categorize the history and contemporary nature of many aspects of BBC radio programming. First, the BBC sought "to maintain whatever the basic quality, interest, or importance of the individual production may be, a generally high professional level both of acting and technical interpretation." Second, it sought to provide a balanced schedule of plays and drama programming. This includes classical plays of established international repute, which are "susceptible to microphone treatment," and have entertainment as well as cultural value. Third, BBC radio has sought to encourage interested authors to write plays conceived specially in terms of the broadcasting medium. Fourth, BBC radio has striven to fulfill the demand of the listening public for "popular dramatic entertainment," and this has been developed through Listener Research and the continued investment and maintenance of soaps, series, and serials as well as adaptations of successful stage plays and films. BBC radio has also been committed to presenting dramatized serials or single productions of novels and short stories for radio "without unreasonable distortion either of form or of spirit." And fifth, the BBC expressed the desire to give to "the English listener some of the more outstanding examples of contemporary dramatic work from the Continent of Europe."

These worthy aims do not give us much of a flavor of what it was like to make programs and how they sounded, particularly in the decades before systematic archiving. Most of the established histories of the BBC depend on the dry and rather stuffy content of the official written archives. The human side to the program making and listening is more likely to be found in private papers, manuscripts, and autobiographies. George Orwell worked as a producer in the BBC during World War II and described his time as trying to exist in a cross between a public (privately funded) girl's school and lunatic asylum. The actor Maurice Gorham in *Sound & Fury: Twenty-One Years in the BBC* (1948) said the BBC was blighted by petty hierarchies and suffocating bureaucracy. He described the 1930s as "the era of the stuffed shirts." The rather pompous and imperious image of John Reith and his colleagues is balanced by Gorham's moving observation that many of these survivors of World War I still suffered from shell-shock and tried to cope through excess drinking and clandestine affairs with their secretaries.

Drama and Literature

The 1949 five-streamed approach to drama can be observed in other categories of BBC programming. The tensions of establishing these aims, or failing to achieve them, can also account for controversies over the development of BBC radio programming since 1922. The BBC has also extended the presentation of alternative cultures by producing plays and literature from Africa, Asia, South America and others parts of the world. BBC Drama director Val Gielgud established a regular drama slot, *World Theatre,* in the 1950s which is representative of the role the BBC played in fostering a creative reception of dramatic literature from overseas.

Radio programming developed parallel to the modern movement in both art and literature. Authors such as H G Wells, James Joyce, and Virginia Woolf not only set up a new story telling vision and rhythm, but also were dramatized and vocalized by the new radio medium. Joyce was exploring the phonetic qualities and resonance of written language in an era which was rediscovering the acoustic space of the oral tradition through radio. Socialist authors including Jack London and George Orwell introduced an increasingly iconoclastic and political style of writing. Modern poetry and prose were debated and introduced to a wider cultural constituency through BBC radio programming.

Radio was found to be a natural medium for exploring the inner world of humans. The radio plays of Samuel Beckett on the Third Programme in 1957 formed a new genre of psychological subjectivity in audio drama. The director of such early Beckett radio plays as *All That Fall* and *Embers*, Donald McWhinnie, deliberately eschewed the production fashion of realism or pseudo-naturalism to paint a soundscape that used new electronic techniques, including the stylistic use of actors' voices to make animal and mechanical sounds. Thus the texture of the program's content articulated the abstract qualities of Beckett's absurd theater of the mind.

BBC cultural programs were influenced by the way Bertolt Brecht and his Berliner Ensemble revolutionized the relationship between performance and audience in theater. This shift in the philosophical center of gravity could be identified in the development of more experimental, irreverent, and satirical forms and content in some BBC programming.

Radio drama programming was initially derived from live theater and novels. A sketch from *Cyrano de Bergerac* and a telescoping of Shakespeare's *Twelfth Night* in 1922 are among the first radio drama events. The first British play especially written for the microphone was *Danger* by Richard Hughes, first aired in 1924. The first head of radio drama R.E. Jeffrey developed the genre at the BBC until early 1929, when 28-year-old Val Gielgud took over from him as Director of Productions. Jeffrey may have been a casualty of a public row over the commissioning of Reginald Berkeley's play *Machines,* which was regarded as a breach of the BBC's then statutory prohibition on broadcasting matters of "political controversy." *Machines* had been commissioned by the BBC, but in 1927 they were reluctant to produce a drama that blatantly criticized industrial assembly line capitalism. The BBC clumsily tried to reject the work by implying it was badly written. Berkeley responded by publishing his play and the correspondence he had had with BBC executives. The row was a devastating embarrassment for the BBC. Questions were raised in Parliament. Berkeley fired a ringing condemnation of BBC cowardice: "A great instrument of intellectual development is being blunted and misused for want of courage. It is no good replying that British broadcasting is better than any other. It ought to be. And it ought to be better than it is."

The dramatist George Bernard Shaw created another storm when he took advantage of his role in chairing a live BBC radio debate in 1927 and declared: "If you find, then, an energetic force of military and police breaking into this hall, destroying the microphone and leading me away in custody, I must ask you not to offer any resistance. (Laughter.) Your remedy is a constitutional one. You must vote against the Government at the next election. (Laughter and cheers.)" The BBC tried to suppress reporting of his remarks, but the American journalist César Saerchinger tracked down a shorthand note and published Shaw's witty attack on BBC censorship in *Hello America!: Radio Adventures in Europe* (1938).

Landmarks in script development for the microphone play before World War II included Berkeley's *The White Chateau,* which dramatized the suffering and tragedy of World War I and became somewhat iconic for the veterans and their families when broadcast on Armistice Night 1925. It also became the first full-length radio play published in book form and promoted by the BBC. Tyrone Guthrie was another early radio drama pioneer. His play *The Squirrel's Cage,* broadcast in 1929, was about childhood fear and adult monotony with suburbanites getting no further than the animal rotating on its wheel. The lead actors were Mabel Constanduros and Michael Hogan, themselves accomplished radio dramatic writers. They had devised and acted all the parts in the BBC's first radio soap based on a London cockney family called *The Buggins.* Mabel Constanduros has been somewhat neglected by broadcast historians. Her ability to deliver improvised stand-up and scripted comedy and mimic a gallery of characters became evident during her first broadcast at the BBC's Savoy Hill studios in 1925. *The Buggins* series began in 1928 and was so successful it generated spin-offs in book and record sales. Constanduros helped found the British tradition of soap opera and situation comedy in both radio and television.

More sound-based styles of dramatization and production also developed through adaptation of novels such as Compton Mackenzie's *Carnival* and Joseph Conrad's *Lord Jim.* A faster paced, short scene-based structure of juxtaposing realistic sound backdrops and locations also contributed to advances in audio drama writing and performance. The BBC also supported the production of epic and literary stage classics. The 12-part *Great Play* series of 1928–29 introduced traditional texts from foreign cultures.

During the 1920s and 1930s, Peter Cresswell, Howard Rose, Val Gielgud, and Mary Hope Allen nurtured the development of radio drama as an art form. Gielgud transcended the period between 1929 and 1963. Donald MacWinnie, Barbara Bray, Nesta Pain, and Douglas Cleverdon are but a few examples of significant figures who combined with the talents of Martin Esslin, John Tydeman, William Ash, Jeremy Mortimer, Kate Rowland, and many others to populate a powerful culture of director/editors in BBC radio drama over the 50-year period between 1950 and 2000.

Despite Val Gielgud's prejudices against popular soap operas, many series and serials became established during World War II, including *Frontline Family,* which became *The Robinsons,* and *Mrs Dale's Diary,* which became *The Dales. The Archers,* which began in 1951, has become the world's longest running soap opera. The program is a remarkable phenomenon in radio history. It attracts Radio 4's largest audiences, and through a constant process of reinventing itself with new characters and plot-lines it sustains substantial audiences and listener loyalty.

The popular genre has also been represented in the adventure serial as *Dick Barton, Special Agent* (1946), *Send For Paul Temple* (1938), and in the comic science-fiction *The Hitchhiker's Guide To the Galaxy* (1978). The latter, created by the late Douglas Adams, became an instant cult classic after its first broadcast in 1978. Radio aficionados claimed the transfer to television was never as good. The series will be remembered for ironic and melancholic characters such as Marvin, the Paranoid Android, and Zaphod Beeblebrox.

From 1939 onwards BBC radio drama was often described as "A National Theatre of the Air." During World War II, Dorothy L. Sayers' *The Man Born To Be King* had a major impact as a radio drama event. In twelve parts between December 1941 and October 1942, Sayers dramatized the story of Jesus Christ in modern language. It was the first time Christ had been portrayed in a publicly broadcast serial. Sayers used colloquialisms to make sense of the New Testament. The Elders of the Synagogue were described as like those "found in every parish council, always highly respectable, often quarrelsome and sometimes in a crucifying mood." Matthew the Publican was "a contemptible little Quisling official fleecing his own countrymen in the name of the occupying power and enriching himself in the process till something came to change his heart, and not presumably his social status or his pronunciation."

The postwar period was also rich in the number and variety of distinctive and original radio playwrights who were recognized as making a literary contribution through their radio plays. The poet and classical scholar Louis MacNeice specialized in writing and producing poetic features. *Christopher Columbus,* starring Laurence Olivier as Columbus, celebrated the 450th anniversary of the transatlantic voyage in 1942. MacNeice said he used vocal music to concentrate on "the emotional truth of the legend rather than let it dissolve into a maelstrom of historical details." *The Dark Tower* in 1946 was an imaginative fantasy about suffering and salvation with music composed by Benjamin Britten. McNeice described his work as a parable of spiritual quest "concerned with real questions of faith and doubt, of doom and free will, of temptation and self-sacrifice."

Dylan Thomas' *Under Milk Wood,* first broadcast in January 1954, is probably Britain's most celebrated piece of radio. Described as "a play for voices," it captures the thoughts, emotions, and dreams of the inhabitants of a small Welsh village, Llareggub. Many radio listeners are familiar with the rich and mellifluous voice of Richard Burton as the narrator purring the opening lines: "To begin at the beginning: It is spring, moonless night in the small town, starless and Bible-black, the cobblestreets silent and the hunched, courters' and rabbits' wood limping invisible down to the sloeback, slow, black, crowblack, fishing boat bobbing sea."

The radio plays of Angela Carter (1940–1992), directed by Glyn Dearman, have begun to attract considerable critical attention. Carter said she wrote for radio because it retains "the atavistic lure, the atavistic power, of voices in the dark, and the writer who gives the word to those voices retains some of the authority of the most antique tellers of tales." *Vampirella* (1976) creates the nightmarish world of a female vampire. *Come unto These Yellow Sands* (1979) is a creative drama-documentary on the Victorian painter Richard Dadd. *The Company of Wolves* (1980) is a surreal reworking of the Red Riding Hood folk-tale, and *Puss in Boots* (1982) is an old comedy for radio that features a multiple orgasm accompaniment to the *1812 Overture* (the program had to be approved by the controller of Radio 4 for transmission). *A Self-Made Man* (1984) is a fake radio documentary on the life of Edwardian novelist Ronald Firbank. Carter described herself as a child of the radio age and she believed "it is *par excellence,* the medium for the depiction of madness; for the exploration of the private worlds of the old, the alienated, the lonely." While her scripts are readily available, the sound of her radio plays can only be heard by appointment at the National Sound Archive in London.

The originality of voice in radio drama is certainly represented in the contemporary work of Tom Stoppard, David Pownall, and Lee Hall. Stoppard's successful stage play *Indian Ink* was an original radio drama commission and broadcast by the BBC as *In The Native State* in 1991. *In the Native State* explores the relationship between an Indian painter and the European woman who poses naked for him. The story switches between England in the present day and India in 1930. Lee Hall's radio play *Spoonface Steinberg* became a best-selling audiocassette and has been successfully adapted in film and theater. It presents the monologue of a 7-year-old autistic girl dying from cancer.

BBC radio programming has also introduced literature and storytelling to the mass audience through such programs as *A Book at Bedtime,* which began on the Light Programme in January 1949 with a reading of John Buchan's *The Three Hostages.* The 15-minute episodic readings now broadcast at night on BBC Radio 4 have also provided a platform for popular and prestigious writers. Eight continuous hours of BBC Radio 4 on FM were given over to a reading by Stephen Fry of one of J.K Rowling's *Harry Potter* novels on a December day in 2000.

The Feature

The idea of the radio feature was pioneered by BBC radio through the 1930s as a hybrid of drama and documentary, equivocating between reality and fantasy. In Manchester a production center including D.G. Bridson, Archie Harding, Joan Littlewood, and Olive Shapely extended an iconoclastic approach so that the feature form adapted itself to the social stresses experienced by the BBC audience during times of mass unemployment, insecurity, and depression.

More mobile recording techniques enabled radio to reproduce phonically the dynamics of the documentary photography movement. Poetic features wove drama and music to provide creative representations of historical events such as The *March of the '45* (1936). Dramatic features also engaged with contemporary crises such as *Crisis in Spain* (1931) on the abdication of the Spanish monarchy, the six part *Shadow of the Swastika* (1940) on German anti-Semitism, and in historical commemoration such as *Bomber* (1995), an eight-part series blending documentary interview with a dramatization of Len Deighton's novel about the final mission of an RAF Lancaster bomber.

A separate features department headed by Laurence Gilliam continued to cultivate avant-garde creativity up until its closure in 1965. The work of Charles Parker and Dennis Mitchell was recognized in international festivals such as Prix Italia. Parker's *Radio Ballads* were inspired by the work of the U.S. radio documentarian and dramatist Norman Corwin. Human dignity and sympathetic characterization are the hallmarks of radio features produced by Piers Plowright and John Theocaris. Editors such as Peter Everett and Richard Bannerman have originated new approaches to radio documentary form and content such as the *Soundtrack* series for BBC Radio 4 in 1983 and *Take The Plunge* in the 1990s, which empowered documentary subjects with audio-diary technology. Brian King and Sarah Rowlands have produced a genre of audio-vérité that sought to tell the stories of institutions and professions through "microphone on the wall" montage, revealing secrets within those institutions. Human voice and musicology have been blended in the work of Alan Hall, which again has flown the BBC banner in international awards and festivals.

Talks

Talks and cultural magazine programs have in the context of public service broadcasting made BBC radio a "university of the air." Beginning in 1941, *The Brain's Trust* explored complex philosophical and scientific concepts with brightness, spontaneity, and entertainment. The program was built around listeners' questions and with a panel that started with Dr Julian Huxley, Commander A.B. Campbell, and C.E.M. Joad. Although the radio version was last broadcast in May 1949, it has left its legacy in discussion programs such as *Night Waves* on BBC Radio 3 and *Front Row* on BBC Radio 4. The dramatist Arnold Wesker once said that for a working class boy without qualifications BBC Radio "was my university education."

Music: Classical

The BBC's early commitment to "serving as a standard of excellence" meant that music programming policy was dominated in the early days through the celebration and promotion of classical music and opera. In 1928–29, the BBC mounted a special season of 12 Great Operas, accompanied by the sale of the score and libretto in published booklets. Before centralization of programming in 1927, the several centers of local production spawned a variety of orchestras and chamber groups that performed a rich and varied schedule of live concerts and series. The BBC assumed sponsorship of Sir Henry Wood's annual Promenade Concerts at the Albert Hall in 1927 (known as "The Proms") and founded its own symphony orchestra in 1930.

The BBC became the most significant patron of classical music in Great Britain. The intensity of orchestral performance in programming was combined with explication and criticism through "Talks" series given by Percy Scholes and Sir Walford Davies. The 15-minute series *The Foundations of Music* ran between 1927 and 1937. The early foundation of a policy on "World Music" can be identified in 1929 with the inception of programs celebrating "The Negro Spiritual" and Paul Robeson's BBC debut in a live concert from Bournemouth. BBC radio used documentary to investigate and introduce new developments in modern music. BBC Radio 3 is now a platform for covering the pioneering and adventurous promotion of world music through programs such as *Late Junction* and the enthusiastic presentation of Andy Kershaw.

The postwar establishment of the Third Programme guaranteed a location for high culture in musical expression and so enabled the BBC to serve the growing demand for separate formats of popular music on the Light Programme network as well as maintaining its cultural commitment to what was described as "the urbane and cosmopolitan." There was an undoubted overlapping in music genre and programming formats. In one set of circumstances the classical could become very popular and in another, popular styles of music could be contextualized with classical legitimacy. The BBC Concert Orchestra founded in 1952 maintained a repertoire of classical, light opera, light music, and popular song. For example in 1989 it performed the entire Gilbert and Sullivan canon.

Music: Popular

BBC Radio 1 in 1967 was a response to the cultural assault of pirate offshore broadcasters, which powerfully served the baby boom youth generation that had expanded the record market for rock and roll. The Light Programme (later BBC Radio 2) was a response to the World War II-era influence of Jazz and Big Band Swing music broadcast by the American Services Network in Britain and Europe.

Most of the leading pirate DJs of the early 1960s were eventually paraded as BBC Radio 1's first line-up of presenters. They included Tony Blackburn, Keith Skues, Dave Cash, Kenny Everett, and John Peel. The pirate dance music station Kiss 100 in London became legitimized by regulation and licensing and its success led to an acquisition of much of its DJ talent and music format in a reinvention of BBC Radio 1 in the middle 1990s. Advances in recording technology and reproduction and agreements on needle-time became a disincentive for the BBC to maintain a large range of live orchestras, and many BBC "light" orchestras were shut down in 1980.

The writer Compton Mackenzie is sometimes referred to as the first DJ because he selected music discs and put together a live program from Savoy Hill in 1924. In 1927 the BBC's first regularly scheduled disc jockey, Christopher Stone, would present recitals rather than "gigs" or "jam sessions." His programs were a mix of recorded music and talk.

A key musical program in BBC history is *Music While You Work,* which began in June 1940 as a non-stop medley of popular tunes played by a different live band each day. The objective was to provide light musical entertainment to a round-the-clock sequence of shifts in the factories. Since many male workers had been called up during the war, many of the listeners were women; newsreels from the time depict them singing along to the hits heard in the program. *Music While You Work* continued on the Light Programme until 1967.

BBC popular music programmers would battle with the music publishing and record industries over rights and royalties. This would lead to conflict and boycotts, but eventually they cooperated for mutual benefit. As in the U.S., there were payola scandals and legal disputes. But this would not stop the broadcasting of popular music from becoming a launch pad for the modern celebrity. Early bandleader Jack Payne was appointed Director of BBC Dance Music in 1926 and formed the BBC Dance Orchestra in 1928, two years before the creation of the BBC's first classical orchestra. When Payne left in 1932 to earn greater riches through touring and record sales, Henry Hall took over the role of BBC heart-throb and musical celebrity.

Sports

Some of the early attempts to provide sports by means of radio were such events as the Derby for flat horse racing in 1921 and a prize boxing fight between Kid Lewis and George Carpentier at Olympia in May 1922. Lobbying by newspapers against the coverage by radio of sports meant that many attempts to organize such coverage during the early years of the BBC were frustrated. However, Royal Charter and incorporation in 1927 was followed by a breakthrough in outside broadcasts. Credit for this should go to producer Lance Sieveking, who successfully recruited Captain H.B.T. Wakelam. Wakelam brought spontaneity and excitement to ad-libbed sports commentary. This was evident at the first live coverage of a Rugby International between England and Wales at Twickenham in January 1927.

Most media histories have a tendency to neglect the contribution of BBC sports broadcasting in enhancing the popularity of the medium and emblematizing local, regional, and national identity through sports. *Sports Report* became a broadcasting institution. This hour-long roundup of sports news and results started on the Light Programme in January 1948, moved to Radio 2, to Radio 5, and to Radio Five Live. It retained its popular signature tune *Out of the Blue* for more than 40 years. One of the BBC's Directors-General, John Birt, once said: "The jaunty signature tune of *Sports Report* would summon the unpunctual from all parts of the home to hear the Everton and Liverpool scores before tea." Sports has also formed the basis for a wide range of popular quiz programs such as *Sporting Chance,* refereed by the popular cricket commentator Brian Johnston; the program aired Saturday afternoons on the Light Programme during the 1950s and 60s. The commentator John Arlott became known as "the voice of English summer" because of his cricket commentaries for more than 30 years. One of his producers once said: "You could smell the grass when he was talking."

Arlott and Brian Johnston became associated with *Test Match Special*, the live, ball-by-ball BBC radio coverage of cricket Test matches. The BBC would assign an entire frequency from one of its national networks to accommodate an "institution in the sporting world." The competitive bidding for sporting rights in a more fragmented media landscape in the 1990s has made it difficult for the BBC to hold onto its exclusive commentary presence in a number of sporting events.

Competition increases the costs, but at the time of writing, BBC Radio Five Live still dominated radio coverage of British soccer.

News and Public Affairs

There is no doubt that World War II was the catalyst for building substantial audience loyalty for news programs. Prior to 1939 the BBC had to break away from its dependence on the national news agencies and the successful lobbying by newspapers to block radio news broadcasting in the morning and during the day. The BBC had to indulge in independent newsgathering during the General Strike of 1926 because the national newspapers could not be printed. Gradually the BBC acquired "observers" to act as reporters. During the Munich crisis of 1938, the BBC was allowed to run bulletins during the morning and throughout the day.

The crisis of world war enabled news to acquire a social value because it was a source of information. *The Nine O'clock News* in the evening, followed by *War Report,* built up record audience ratings. The BBC's reputation in journalism in Britain has been established around key programs established after the war on the Home Service, which became BBC Radio 4 in 1967. The Radio 4 schedule for news is punctuated by the strength and quality of the *News Briefing* at 6 A.M. followed by the *Today* breakfast program, *The World at One* at lunchtime, *PM* for drivetime, *The Six O'clock News* for evening listeners, *The World Tonight* at ten P.M., and *News at Midnight,* which has become a half-hour program of reports from correspondents and reporters on the day's events. Authority, variety, and quality in radio journalism were further extended when the Radio Five network became BBC Radio Five Live, a dedicated news and sports service, in 1994. Its first controller, Jenny Abramsky, successfully established popularity and reputation for a more rapid response to news events within Britain and abroad. Radio Five Live harnessed the global resources of BBC news gathering through the use of foreign correspondents and news bureaus.

The Week in Westminster was established in November 1929 to meet the expanded franchise of women voters and the first generation of women MPs in Parliament. *From Our Own Correspondent* began in 1955 as a forum for BBC reporters to give more personal firsthand accounts of their experiences of crisis and developing news stories. Alistair Cooke's *Letter From America* is in some respects a permanent *From Our Own Correspondent* despatch from the U.S. *Woman's Hour* has developed a journalistic agenda for women since its inception in 1946. The first edition had a male presenter and contained a talk on *Mother's Midday Meal* and *Putting Your Best Face Forward.* However, with a growing confidence in women's rights and as feminist issues began to take center stage in mainstream media, *Woman's Hour* was able to bring a range of taboo subjects into the open. The magazine format has been a vehicle for establishing the broadcasting reputations of Jean Metcalfe, Marjorie Anderson, Sue Macgregor, and Jenni Murray.

Vaudeville, Variety, and Light Entertainment

Light entertainment is the BBC's phrase that broadly defines show business, quiz, and entertainment programs. In BBC radio it owed its traditions and much of its early programming to theatrical vaudeville. In the early years the theater industry and vaudeville artists were wary of the new medium, just as they were in the U.S. Sketches, jokes, and routines that could be performed repeatedly on stage would be exhausted in the moment of only one broadcast. The racist stereotyping inherent in the genre of black faced minstrelling transferred to radio through the series *Kentucky Minstrels,* which ran from 1933 to 1950 and was the precursor for the TV *Black and White Minstrel Show.*

Landmarks in BBC radio variety programming have been identified as *In Town Tonight* and *Band Waggon,* featuring comedians Arthur Askey and Richard Murdoch. Between 1938 and 1939 Askey and Murdoch developed the world of an imaginary top floor apartment at Broadcasting House and built comic routines around sound motifs such as the Greenwich Time Signal. During World War II, *ITMA* ("It's That Man Again") became a vehicle for ideological entertainment. The comedian Tommy Handley was a master of ceremonies for an ensemble of surreal characters such as Mrs. Mopp, the cleaning woman with the bottomless bucket, Colonel Chinstrap, Funf the German spy, Ali Oop and others. Mrs Mopp's catchphrase was 'Can I do yer now, Sir?' *ITMA* attracted a peak audience of 15 million listeners a week and each show captured the attention of 40 percent of the population.

The Goon Show, which began airing in May 1951, marked an exquisite extension of radio's potential for surreal comedy. *Round The Horne* written mainly by Barry Took and Marty Feldman and featuring performances by Kenneth Horne and Kenneth Williams, would be an example of a successor to the *Goon Show* tradition. Other postwar light entertainment programs that attracted large followings include *Much Binding In The Marsh*, *The Glums, Take it From Here, Educating Archie, The Navy Lark, Beyond Our Ken,* and *Tony Hancock.*

Most of the successful contemporary artists in British television comedy were first established on BBC radio through weekly satire programs such as *Weekending,* which ceased airing in 1997. Chris Morris collaborated with Peter Cook on improvised sequences on BBC Radio 3, co-wrote a series which satirized the clichés and conventions of news coverage on BBC Radio 4, and at the time of writing has established a large following of younger listeners with a surreal ambient world of dysfunctional characters in the series *Blue Jam*, broadcast on BBC Radio 1 after midnight.

TIM CROOK

See also Archers; Cooke, Alistair; Cooper, Giles; Desert Island Discs; Drama, Worldwide; Goon Show

Further Reading

Bridson, D.G., *Prospero and Ariel: The Rise and Fall of Radio,* London: Gollancz, 1971
Crook, Tim, *Radio Drama: Theory and Practice,* London and New York: Routledge, 1999
Evans, Elwyn, *Radio: A Guide to Broadcasting Techniques,* London: Barrie and Jenkins, 1977
Gielgud, Val Henry, *British Radio Drama, 1922–1956: A Survey,* London: Harrap, 1957
McDonnell, James, *Public Service Broadcasting: A Reader,* New York and London: Routledge, 1991
Midwinter, Eric C., *Out of Focus: Old Age, the Press, and Broadcasting,* London: Centre for Policy on Ageing, 1991
Palmer, Richard, *School Broadcasting in Britain,* London: BBC, 1947
Parker, Derek, *Radio: The Great Years,* Newton Abbot, Devon, and North Pomfret, Vermont: David and Charles, 1977
Reville, Nicholas, *Broadcasting: The New Law,* London: Butterworths, and Austin, Texas: Butterworth Legal, 1991
Robinson, John, *Learning over the Air: 60 Years of Partnership in Adult Learning,* London: BBC, 1982
Shelton, Edward, *UK Radio: The New Era,* London and New York: Economist Intelligence Unit, 1991
Snagge, John, and Michael Barsley, *Those Vintage Years of Radio,* London: Pitman, 1972
Wilby, Pete, and Andy Conroy, *The Radio Handbook,* New York and London: Routledge, 1994

British Broadcasting Corporation: BBC World Service

In 2003 the British Broadcasting Corporation World Service broadcast for 24 hours a day in English and for varying periods in 43 other languages. Founded in 1932, it has been known by different names over time. It gained a reputation for integrity during World War II, when it put heart into the peoples of Nazi-occupied Europe. It claims the largest audience of any international broadcaster, with 150 million regular listeners in 2003, and it was described by the United Nations Secretary-General Kofi Annan (in October 1998) as "perhaps Britain's greatest gift to the world this century."

The mission statement of the World Service says that its main aims are "to deliver objective information and reflect the values of a free and democratic society; to help meet the need for education and English-language teaching; and to give access to the best of British culture and entertainment."

Origins

The origins of the World Service were modest enough. The development of radio in Britain came later than in the United States and took a different form. BBC Director General John Reith was interested in broadcasting as a way of linking the British Empire. Technical developments in shortwave transmission over long distances made this possible in the mid-1920s, but the BBC did not immediately take advantage of them. This was partly due to the conservatism of its engineers but mainly attributable to disputes over funding. The government was in favor of the idea but not of paying for it. Reith argued at first that the British license payers should not be asked to cover the expense. However, in the financial crisis of 1931, when sacrifices were called for all around, he changed his mind, citing the national interest. A shortwave transmitter was built, and the Empire Service, as it was called, opened on 19 December 1932. A few days later it carried a historic broadcast by King George V, speaking to his empire for the first time:

> Through one of the marvels of modern science I am enabled this Christmas Day to speak to all my peoples throughout the empire. . . . I speak now from my home and from my heart to you all, to men and women so cut off by the snows and the deserts or the seas that only voices out of the air can reach them.

In the beginning, the Empire Service had five separate transmissions, each lasting two hours, that were directed to areas of the world where it was evening, peak listening time. In the following years transmissions increased, until by the outbreak of war in 1939 the Empire Service was broadcasting for 18 hours a day.

Much of the output was taken from the domestic BBC service, but there were some specially produced programs, and in 1934 the Empire Service established its own news section. The programs were not universally popular: "flabby and uninspiring" was one description. The BBC had a monopoly at home, but not abroad, and fascist Italy and Germany saw radio as an ideal medium for propaganda, using it in an effective and innovative way. The Nazis concentrated their efforts at first on German immigrants in America, both North and South, hoping to convert them to the cause.

The Italians broadcast mainly in Arabic to the Middle East, where Britain had considerable political and economic interests, including a League of Nations mandate over Palestine. The Italian radio mixed entertainment with tales of alleged British atrocities and such choice items of invective as "The empire of the British is decadent" and "Eden [the Foreign Secretary] is a clown in the hands of the Freemasons."

The Nazis later turned their attention to the Middle East in similar terms; faced with such a barrage of hostility, the British government considered how best to counter it and determined not to reply in kind but rather to put forth the British view. After toying with the idea of a government radio station based in Cyprus, the government turned in the end to the BBC. Senior executives in the Empire Service were not enthusiastic about broadcasting in foreign languages; they saw it as a form of propaganda, alien to the traditionally objective tone of the BBC.

Reith overcame management objections and extracted conditions from the government. Services in foreign languages would have to be paid for with government funds, and, crucially, the BBC would have to have the same editorial freedom as it did with services for home listeners. Prestige, Reith said, depended on broadcasting that was both truthful and comprehensive, in other words, not leaving out items that might be embarrassing or critical of government policy. On these terms, the BBC began broadcasting in foreign languages, first in Arabic in January 1938, then a few months later in Spanish and Portuguese for Latin America.

Later the same year, the Munich crisis served to increase the number of languages used. At one point, the government decided in a last-minute policy decision to broadcast a speech by Prime Minister Neville Chamberlain in German, Italian, and French translations. The BBC undertook the broadcasts. It is unlikely that the broadcasts were heard by the intended recipients, but from then on, those three languages were added to the list.

Wartime Expansion

By the time Britain and France went to war in September 1939, the BBC had added Afrikaans language broadcasting to try to influence the people of South Africa, and the BBC was broadcasting in Spanish and Portuguese to Europe as well as Latin America. When the war ended six years later, it was broadcasting in more than 40 languages, ranging from Albanian to Welsh (for Welsh-speaking inhabitants of Patagonia in Argentina).

The broadcasts had been a source of hope and inspiration to millions of people in Nazi-occupied Europe and had actively helped the resistance movement. The Empire Service had acquired a reputation for truth and was acknowledged as the foremost international broadcaster in the world. This reputation did not come easily. From the beginning, the external service followed a policy of not trying to conceal military defeats, on the argument that it would thus be more readily believed when there were victories to report. The course of the war proved the case, but the first three years produced virtually nothing but defeats, and the BBC was accused of lowering morale—"an enemy within the gates," as Winston Churchill described it once. Europe was occupied from the north of Norway to the south of France, and Hitler's propaganda machine was able to make use of all the radio stations in this large area.

The BBC responded with a huge expansion of broadcasting abroad, decreed by the government. New transmitters were ordered—some from the United States—and new staff were recruited; personnel numbers went up by more than 500 percent in the first 18 months of the war. To accommodate the extra staff, the BBC rented offices in Bush House, a building in central London erected in the 1920s by Irving T. Bush of New York and dedicated to the friendship of the English-speaking peoples, as an inscription on the top of the building still proclaims. Bush House continued to be used after the war as the headquarters of all BBC services directed abroad.

London at that time was host to a large number of governments in exile, and several of them used BBC facilities to broadcast to their own people. General Charles de Gaulle arrived in June 1940, an obscure junior minister in the French government. After France surrendered, he broadcast to the country's armed forces, calling on them to continue the fight. At first his words had little effect, but as the war continued his voice became well-known in France; his reputation was founded by radio. Winston Churchill himself spoke in French over the BBC. "Français, prenez garde, c'est moi, Churchill, qui vous parle !" [French, be on guard, it is I, Churchill, who speaks to you!] he growled in his distinctive voice, which elderly French people still remembered with emotion decades later.

Much of the wartime work of the BBC went unreported at the time, but one initiative that was widely publicized was the "V for Victory" campaign, which began in 1941. It was the brainchild of the man producing programs for Belgium, who noted that *V* was the initial letter of *Victoire* in French and *Vrijheid* (freedom) in Flemish, thus encompassing both languages of his country. He encouraged people to chalk the letter on walls, doors, and other suitable surfaces, and its use spread to other countries. The Morse code for *V*, three dots and dash, corresponds to the opening notes of Beethoven's Fifth Symphony, and this rhythm also became part of the act: people were encouraged to simulate it in ordinary life—a teacher clapping her hands to summon schoolchildren, for example, or customers calling a waiter.

Winston Churchill encouraged the campaign by adopting the V sign with two upraised fingers. But the campaign lasted little more than a year: it was criticized then and later for having risked lives for no obvious end—there was no strategic follow-up. The Nazis adopted it for their own ends by claiming that *V* stood for victory against Bolshevism and the German word *Viktoria* and by pointing out that Beethoven was German anyway.

More practical methods of encouraging resistance included broadcasting news at dictation speed or by Morse code for the clandestine newspapers that sprang up all over occupied Europe and passing coded messages to resistance groups. These messages gave notice of impending operations, including D day, or the arrival of agents or documents, and they were always repeated. For example, "Le diable jongle avec les âmes, nous disons le diable jongle avec les âmes" [The devil juggles with souls, we say the devil juggles with souls] had a precise meaning for somebody crouched in a cellar with headphones clamped to his ears.

During the war, the BBC also expanded its transmissions outside Europe. The Empire Service vanished to become the General Overseas Service in English. It developed regional offshoots for the Pacific, the Caribbean, and North America. The North American Service began in 1940 when the United States was still neutral, and it was intended to be an important medium for putting forward the British case. Presentation and contents were adapted to meet the requirements of a North American audience. Canadian presenters were hired, and leading British literary and artistic figures were used as speakers.

A new type of program, *Radio Newsreel,* was devised, which consisted of eyewitness accounts by reporters and ordinary men and women, bringing the sounds of war home to the American people. *Radio Newsreel* was later taken up by other parts of the BBC, and it continued in the World Service until the 1990s, when the cinema newsreels from which it took its name were long forgotten.

The Persian Service also began in 1940, and in the following year British troops moved into the country to forestall German expansion. The Shah abdicated, and an excitable journalist claimed that this was the first time a ruler had been toppled by radio. Broadcasting also began to India in both English and Hindi. The programs in English were intended to show that, whatever the evils of British rule, it was preferable to that of the Nazis. T.S. Eliot, E.M. Forster, and George Orwell (under his real name of Eric Blair) were among the speakers hoping to appeal to the Indian intelligentsia. After much government prodding, transmissions in Japanese were started in 1943. Since the Japanese people were forbidden to own shortwave sets, however, nobody heard them.

Postwar Activities

With the end of the war, the new Labour government decided to continue with broadcasting abroad and put the various language services, which had grown up haphazardly, on a regular footing. The government stipulated that the Foreign Office should determine the languages and the amount of time devoted to each one (and pay for them) but that the BBC should be entirely responsible for the contents. A new director general, Sir William Haley, set out guidelines that have been followed ever since. The External Services, as they were called then, should provide "an accurate, dispassionate and impartial" flow of news, seen through British eyes but international in scope. In matters of international controversy, the official British view would be given due prominence, but opposing foreign views were to be carefully explained, and conflicting opinions with serious backing in Britain itself were to be given due weight.

One of the fruits of peace was the opening of a service in Russian. The BBC had wanted to start such a service in 1941, when Hitler's invasion turned the Soviet Union into an ally, but the Soviet response was that it would be pointless because all private radio sets that could receive the service had been confiscated. After the war, the restriction was removed, and the service began in March 1946. It was intended to express the friendship of the British people for the Russian people after the great victory of 1945, but before long the Cold War changed the nature of the dialogue. Critics in Britain accused the service of "moral compromise and appeasement" in its transmissions—not being tough enough on the Soviets. But Moscow attacked the service in violent terms—as "mad agitators and disruptionists," for example, and "a crying radio crocodile"—and jammed it from 1949 to 1987, with occasional breaks in periods of détente.

The immediate postwar years were not happy ones for the External Services. Economic problems led to cuts in government spending, particularly during the Korean War of 1950–53. Services in English and other languages were slashed, and some were abolished altogether. Important capital expenditure projects were postponed. At the same time, other countries were increasing their efforts in international broadcasting: the United States, the Soviet Union, and China all overtook the BBC in hours of broadcasting, and Egypt and West Germany were not far behind.

Relations with the government were soured temporarily over the Suez crisis of 1956. President Nasser of Egypt nationalized the Suez Canal. Some months later, Britain and France invaded his country, ostensibly to separate Egyptian and Israeli forces that had attacked across the Sinai desert—a so-called police action. There was an international outcry and vocal opposition in Britain itself, all of which the BBC reported. Prime Minister Anthony Eden felt that since Britain was effectively at war, a radio station financed by the government should not publicize antigovernment sentiments. He talked of taking the BBC over, and there was strong criticism of the corporation in Parliament. However, the crisis was quickly resolved. British and French troops called a cease-fire, and Eden resigned. The BBC argued that if it had failed to report the criticism of the government's action, which was publicized everywhere else, it would have lost all credibility abroad. It has also argued since then that the episode shows its independence of the government.

However, there have been occasions when the World Service has acceded to requests—not orders—from the government. When Rhodesia (now Zimbabwe) declared illegal independence in 1965, the government asked the BBC to mount a special program for white Rhodesians to bring home to them their isolation and the consequences of their government's action. A transmitter was set up near the border, but transmissions were jammed, and the few Rhodesians who did hear the program dismissed it as propaganda.

On the eve of the Arab-Israeli war in 1967, the Soviet Union threatened to break off talks with the British Foreign Secretary if the Russian Service went ahead with plans to broadcast excerpts from a book by Stalin's daughter, Svetlana, who had fled to the West. Since the talks had been set up to try to avert the war, the BBC agreed at the highest level to postpone the broadcast. However, the talks proved unsuccessful, the Foreign Secretary returned home, and the broadcast went out 48 hours later.

In 1975, when Idi Amin was in power in Uganda, the World Service postponed a review of a book about him by a British expatriate living in Uganda. This was at the request of the Foreign Office, which said that it would infuriate Amin and so endanger the lives of Britons living there.

New Horizons

After the Suez episode, the BBC expanded its Arabic Service and began broadcasting in African languages—Hausa for West Africa, Swahili and Somali for East Africa. Transmissions in Afrikaans were dropped. This was the prelude to a change in attitudes and priorities in a world that was itself changing. Britain shed most of its remaining colonial possessions in the 1960s. Many of these new nations—in Africa and the Caribbean—set up their own radio networks with the help of people seconded from the BBC.

The radio audience was increasing enormously thanks to the invention of the transistor. This tiny device revolutionized radio by making possible small, lightweight portable sets. Before the transistor appeared on the scene in the 1950s, most radios were in Europe and North America. In the 20 or so years after 1956, the number of radios in sub-Saharan Africa grew from under half a million to over 22 million, in China from 1 million to about 50 million, and in India from 1 million to 18 million.

New listeners meant new types of programs. The General Overseas Service was no longer seen as aimed at the expatriate Briton but at anyone who could hear it. It became the World Service in 1965, and the short news program *Home News from Britain* was renamed *News about Britain*. The title of *World Service* was given to all the External Services in 1988.

The BBC was one of a number of Western services broadcasting to the Soviet bloc during the Cold War—collectively known in Russia as "The Voices" and widely listened to. Their efforts clearly helped in bringing about the fall of communism there by showing that a more attractive alternative existed in the West and by reporting events ignored by Soviet official media. The biggest and most popular stations were American: Radio Liberty (for the Soviet Union), Radio Free Europe (for the satellite countries), and Voice of America. The U.S. approach was harder than that of the BBC; the latter adopted a "Give them the facts and let them make up their own minds" attitude, which some Russian exiles, such as Alexander Solzhenitsyn, criticized as "wishy washy." However, when Solzhenitsyn was expelled from the Soviet Union, the first interview he gave was to a member of the BBC Russian Service whose voice he recognized.

The World Service was singled out for praise by former Soviet President Mikhail Gorbachev at a press conference on his return to Moscow after an attempted coup against him in 1991. He was told that nobody from the BBC was present. "Never mind," said Gorbachev with a smile, "The BBC knows everything already." The breakup of the Soviet Union, with its monolithic structure, created problems in broadcasting terms. For example, with the creation of a separate Ukrainian state, it became necessary to add that language, because the people there could not be expected to listen to programs in Russian any longer. Other languages added since then have included Azeri, Kazakh, and Uzbek for Central Asia.

Similarly, the fragmentation of Yugoslavia meant that Serbo-Croatian was no longer an acceptable language, as it had been since the war; the BBC had to broadcast separately in Serbian and Croatian, and it also added Macedonian. The regimes that succeeded communism in Europe are of variable quality, and not all are wedded to the idea of free expression in the media; it has remained necessary to continue broadcasts to them. There are still Third World countries, too, with state-controlled media, whose people rely on outside broadcasters such as the BBC to inform them about what is going on inside their own boundaries as well as the rest of the world.

This has often led to complaints from the regimes affected. For example, in the late 1970s, the Shah of Iran, remembering the fate of his father, convinced himself and others that the Persian Service was encouraging revolution by reporting the increasing opposition to his regime. A campaign was launched against the service, with British businessmen, politicians, and others being provided with fake transcripts and encouraged to put pressure on the BBC. There was even a proposal to sabotage the transmitters (in Cyprus and Oman), which was fortunately vetoed by the Shah. Independent investigation showed there was no evidence of bias against the Shah. Reports of opposition were balanced by statements of support for him. The British ambassador at the time, who had been critical of the BBC, agreed years later that its only fault had been telling the people what their own media were concealing.

The government of Burma went so far as to produce a book in 1988 called *A Skyful of Lies,* outlining what it said was the misinformation disseminated by the BBC and the Voice of America. "That the BBC is particularly trying to subvert Burma is especially clear," the book says. The country was run by a military regime that cracked down on any dissent, and the Burmese Service was very popular. In the following year, it received 98,000 letters, the highest number received by any language service; one of the letter writers wrote, "Tuning into the BBC is like sharing a bit of its freedom as our own."

Television and Restructuring

The World Service began television services in 1991. It had been planning the move for several years, but the government refused its request for a subsidy to start the service, and in the end the BBC went ahead on its own. BBC World, as it is now called, is sent by satellite and is received either directly or through cable companies; it is financed by local advertising.

The BBC underwent a restructuring in 1996 that involved the merging of all news and program output in English, whether for domestic or overseas broadcasting. A number of prominent people expressed fears about the continued distinctiveness of the World Service and the link between the foreign language departments and the news operation under these conditions. A campaign, "Save the World Service," was launched with three former managing directors of the service taking part. In the end, a compromise was reached and a number of safeguards put in place. The World Service now commissions programs in English from other departments of the BBC but retains its own newsroom in Bush House, preparing bulletins in English and foreign languages.

Reception Difficulties

Since the earliest days, radio reception has been a constant concern. Direct shortwave broadcasting has limitations: it is overcrowded, it is affected by such uncontrollable phenomena as sunspots, it weakens over long distances, and it is subject to jamming. The Soviet Union, China, Libya, Iraq, and Argentina are among the countries that have jammed BBC transmissions at various times. The original Empire Service sought a solution to reception difficulties by transcribing programs onto discs and mailing them for local rebroadcasting.

However, this method is not suitable for topical material, particularly news. Even before World War II, it was recognized that the signal needed to be boosted by relay stations in various parts of the world. The war held up progress, but the first relay was set up in Tebrau, Malaya, in 1949; it later moved to Singapore. Delays in the provision of funds for capital expenditure, already noted, held up the work, but other relays opened in the 1950s and 1960s in Cyprus, Ascension Island in the Atlantic, and the Omani island of Masirah.

However, the BBC still lagged behind the expansion plans of other international broadcasters. In 1981 a program costing £100 million was undertaken to improve audibility worldwide, although the government insisted that some language services should be closed to help pay for it. Transmitters were modernized, including some in Britain itself that dated from the war, and new relays were opened in, for example, Hong Kong and the Seychelles, and transmissions to them were conveyed by satellite.

Satellites have also made it possible for World Service programs to be broadcast on FM stations throughout the world. The 1990s saw a huge increase in this development, starting with a handful and ending with over 1,000 rebroadcasters. Some are on the air for 24 hours a day, others for only an hour or two. The broadcasts are in English or the local language. In some cases, they involve joint programming with the local broadcaster. The rapid growth of this form of broadcasting now in 130 capital cities all over the world has helped to boost the audience for the World Service.

The World Service first went on-line in 1995, in Polish, followed by English. This has been increased to all 43 languages, with nine of them having an update every 24 hours.

Since the war, the World Service has taught English by radio and then by television as well as audio- and videocassettes. The service began training broadcasters in other parts of the world in 1989 and has since set up a Training Trust, which undertakes training in more than 30 countries, including three schools of broadcast journalism in Eastern Europe.

The World Service has cooperated with the International Red Cross to help reunite refugee families in Rwanda, Burundi, and Kosovo. These programs are in the local languages, Kinyarwanda and Albanian, and are transmitted by both FM and shortwave. Training and humanitarian activities are funded separately from broadcasting, which continues to be paid for by a government grant. For 2003–04, this amounted to £201 million, to increase to £239 in 2005–06.

ANDREW WALKER

See also Cold War Radio; International Radio Broadcasting; Jamming; Propaganda by Radio; Shortwave Radio; World War II and U.S. Radio

Further Reading

BBC World Service website, <www.bbc.co.uk/worldservice/index.shtml>

Beachcroft, Thomas Owen, *Calling All Nations,* Wembley, Middlesex: British Broadcasting Corporation, 1942

Bennett, Jeremy, *British Broadcasting and the Danish Resistance Movement 1940–45: A Study of the Wartime Broadcasts of the BBC Danish Service,* Cambridge: Cambridge University Press, 1966

British Broadcasting Corporation External Services Publicity Unit, *Voice for the World: 50 Years of Broadcasting to the World: 1932–1982: BBC,* London: BBC External Services Publicity Unit, 1982

Cohen, Yoel, *Media Diplomacy: The Foreign Office in the Mass Communication Age,* London: Cass, 1986

Hale, Julian, *Radio Power: Propaganda and International Broadcasting,* London: Elek, and Philadelphia, Pennsylvania: Temple University Press, 1975

Hills, George, *Broadcasting beyond One's Frontiers,* London: British Broadcasting Corporation, 1971

Mansell, Gerard, *Let Truth Be Told: 50 Years of BBC External Broadcasting,* London: Weidenfeld and Nicolson, 1982

Partner, Peter, *Arab Voices: The BBC Arabic Service 1938–88,* London: BBC, 1988

Tusa, John, *Conversations with the World,* London: BBC Books, 1990

Walker, Andrew, *A Skyful of Freedom: 60 Years of the BBC World Service,* London: Broadside Books, 1992

British Commercial Radio

In the second half of the 1990s, commercial radio was the fastest growing advertising medium in the United Kingdom. Only a quarter of a century earlier, however, commercial radio did not exist, and for its first two decades it had to struggle to convince advertisers and agencies that radio advertising works. The dramatic change in its fortunes was brought about by a happy confluence of events: regulation was substantially relaxed, the number of commercial services rapidly increased, and the industry at last found a way of winning the confidence of the advertising world.

Some countries have had commercial radio systems for as long as they have had radio, but for over half a century the only legitimate radio services in the United Kingdom were those of the British Broadcasting Corporation (BBC), which began sound broadcasting in 1922. Even commercial television in the United Kingdom predated commercial radio by some 18 years. In these facts are found some of the reasons why, when it belatedly arrived in 1973, commercial radio struggled to become established. But the first stations also had other severe handicaps: heavy-handed regulation, high costs, and a repressive music copyright regime.

Origins

During the long period when British radio consisted solely of BBC public service radio, commercial radio was being introduced and developed in some other countries, but others had to wait even longer than the U.K. for the public ownership monopoly to be broken. The BBC monopoly of sound broadcasting might have gone on even longer, but two developments paved the way for the introduction of commercial radio. One was the emergence of the offshore pirate stations in the 1960s, which not only spurred the BBC into creating its first pop music service, Radio 1 (1967), but demonstrated that there was an audience for commercial services. The second development was an increasingly insistent lobby for commercial radio to be introduced.

Commercial television had begun in 1955, regulated by the Independent Television Authority (ITA). After the 1972 Broadcasting Act performed a neat (as the government of the day saw it) double shuffle, the ITA changed one initial and added an embryo "radio division" to become the IBA (Independent Broadcasting Authority). As with commercial television, the

IBA was to be technically "the broadcaster"—building, owning, and operating the transmitters—and the independent radio stations, as they were known, would be holders of franchises to supply program services. Regulation of radio mirrored that of commercial television in many respects: it was detailed, severe, and costly. The commercial television contractors lived with this system because they had what the first Lord Thomson of Fleet once described as "a licence to print money." But it was to be almost two decades before government accepted that commercial radio needed to be much more lightly regulated than television if it was to prosper.

The first radio franchises were offered in 1972, and the first stations began broadcasting in the autumn of 1973. The first two were in London: Capital Radio, designated a "general entertainment" station, and London Broadcasting (LBC), designated "news and information." More stations came on the air over the next few months in Glasgow (appropriately on New Year's Eve), Birmingham, and Manchester, and within a year others in Newcastle, Swansea, Sheffield, and Liverpool brought the total to nine. Over the next three years the system grew to 19 stations, each separately owned and locally funded, for these were the franchise conditions.

Then there was a gap of four years while the government commissioned and then digested the report of a royal commission under the chairmanship of Lord Annan, which looked at the future of all broadcasting. The franchising of commercial radio stations was not resumed until 1980 and then proceeded at the rate of four or five a year until 1985, by which time there were 49 stations.

Independent radio, like independent television 18 years earlier, was conceived as public service broadcasting, to be funded by advertising instead of the license fee that sustains the BBC. Thus many obligations were imposed on the companies that won franchises. They had to produce schedules that offered a wide range of programming designed to appeal to all tastes and age groups. They had to carry a news service approved by the IBA. They had to produce programs on religion and for children. And they had to pay "rentals" for their franchises that averaged 10 percent of their incomes.

Independent radio struggled in its early years. With bright and breezy programming and close identification with the life of the areas they served, independent radio stations soon began winning large numbers of listeners from both the BBC local services (which had begun in 1967) and from the BBC national networks, but they found it very difficult to attract sufficient advertising revenue to meet their substantial costs. In addition to the rentals paid to the IBA, copyright royalties were the highest in the world, averaging 12 to 13 percent, and the IBA (pressured by the Musicians' Union) insisted that stations spend another 3 percent of their revenue on providing employment for musicians. So no less than 25 percent of a station's income was committed before it had paid one employee or met any of the usual costs of running a business.

Remarkably, only one of the early radio companies went bankrupt, Centre Radio (Leicester), which closed in 1983 with debts of more than £1 million, but others were in dire straits. In fact, in the early 1980s managing a commercial radio station in Britain was regarded as being about as secure as managing a soccer team; in the years 1981–83, when there were just over 30 stations, there were 20 changes of managing directors.

Turnaround

In 1984, by which time there were 33 commercial radio companies of which more than half were losing money, the managements of the companies decided that something had to be done. All were members of the Association of Independent Radio Contractors (AIRC; now the Commercial Radio Companies Association), and on 23 June 1984, AIRC called a "council of war." On that hot summer afternoon, the chairmen and managing directors of all 33 companies locked themselves in a hotel room near Heathrow Airport and made a series of decisions that were to change the whole thrust of radio development in the United Kingdom. In the shorter term—which was of more immediate concern to those companies clinging to existence—they stopped an industry from collapse.

The AIRC made three key decisions: to force the IBA to reduce rentals substantially and to remove some of its more irksome rules; to press the government for revision of the Broadcasting Act; and to commission an independent report demonstrating the scope for deregulating radio. All three initiatives succeeded: within 18 months, the IBA had cut rentals by 35 percent and had dropped many of its most petty rules. Within two years the Home Office (then the government department responsible for broadcasting) was actively reviewing the broadcasting legislation, and the report, by the Economist Informatics Ltd., showed clearly and authoritatively that radio could be largely deregulated.

The reform process is never a swift one. AIRC made its first formal submission to the government on what a new Broadcasting Act should contain in January 1986. The AIRC made another submission in October, and in February 1987, the government published a Green Paper, "Radio: Choices and Opportunities," which was widely welcomed and contained nearly all the industry's proposals. However, it was a further three years before new broadcasting legislation was enacted. The delay—caused largely by indecision over structural changes for television—was frustrating for a radio industry that was totally clear about its needs; fortunately, business improved in the late 1980s, and over the four years to 1989, revenue increased by 85 percent.

Then there was another downturn, not because radio had lost its way, but because the general economic situation had deteriorated. Retailing was an inevitable early casualty, and all media soon felt the trickle-down effect. In fact, radio was somewhat better placed to withstand recession than it had been a few years earlier, and although national sales were hit hard in 1990, local revenues held up well for most stations thanks to the success story radio had created in the preceding four years and the increased professionalism of station managements and sales teams.

In parallel with the somewhat slow-moving legislative developments and the financial adventures, much was happening to the radio services during this period. British radio—both the BBC and the commercial sector—had always simulcast, transmitting each service on at least two frequencies, one on medium wave (now more widely known as AM, or amplitude modulation) and one on VHF (FM or frequency modulation). The commercial services had been conceived as FM stations (the BBC already had FM frequencies for all its medium wave services), but in the early 1970s, the bulk of radio listening was still AM, so to give the fledgling services a better chance they were also allocated AM frequencies.

Simulcasting continued until 1988. By then the Green Paper had been published and the government's intention to create many more commercial radio services had been announced. A long-standing excuse for the limited number of radio services in the United Kingdom (as compared, say, to the United States and to most European countries) was shortage of radio spectrum. Under international agreements, Britain was due to get more FM spectrum in the 1990s, but to hasten the government's objective of widening listener choice simulcasting had to go. The then-home secretary, Douglas Hurd, coined the injunction, "Use it or lose it," and said that the commercial radio companies could each effectively become two stations by having separate programming on their AM and FM frequencies—"splitting," as it became known. If they didn't do so, they would run the risk at the next franchise renewal of losing one of them.

Between 1988 and 1990, all the larger stations and a number of the smaller ones began broadcasting two services where previously there had been one. The pattern was fairly general. The station's existing service, perhaps slightly modified to appeal even more to younger listeners, was retained on FM (which by this time was being used by about two-thirds of all listeners), and a new service, broadly labeled "Gold" and aimed at those over age 35, was started on AM. There were one or two exceptions to the general trend: LBC was required by the IBA to stay with speech, although they would have preferred to make one of their services music oriented. Piccadilly (Manchester) put its existing output on AM and sought a yuppie audience on FM (which didn't work and was eventually abandoned), and Radio City (Liverpool) initially went for a talk service on AM, but within six months the station had dropped it for the Gold music format.

In under two years, "splitting" dramatically increased the number of commercial services, from 60 in 1988 to over 100 in the autumn of 1990. At the same time, several new services were launched. The pattern of development of the first 15 years had been for all new stations to be in "white space"—areas not previously allocated a commercial station—with the object of eventually covering the whole country. In 1988, although it was known that new legislation was on its way that would provide for many more stations, such was the pressure from would-be new broadcasters on ministers that the Home Office asked the IBA, "Is there anything you can do in the meantime?"

The IBA came up with "incremental" contracts—additional franchises within areas that had a commercial service (or perhaps two, if the original station had "split"). The IBA claimed that these incremental contracts would widen listener choice ahead of the new act, because they would be awarded to groups whose programming plans offered services markedly different from those already available.

The IBA eventually offered 26 such franchises. One was never applied for, one was awarded but then surrendered when the winners failed to raise enough cash, one had its license withdrawn, and 23 went on air. For the most part, the incrementals struggled, much as the first stations had 16 years earlier, but for different reasons. The incremental franchises were supposed to bring new blood into the commercial radio business (existing contractors were told they could not apply, and some of the new entrants actually described themselves as "the new wave"), but within a short time the established stations had to come to the rescue with cash injections and management know-how. Many of the new stations struggled, some because they came into existence when the economy was entering a downturn and, others because they underestimated the cost and complexity of launching even a small station. Happily, the majority survived, aided in some cases by refinancing and changes of ownership.

The largest number of incremental franchises was created in London, where there had always been the greatest clamor from entrepreneurs for the chance to broadcast. In 1988 London had just two commercial stations; two years later it had 12, and by 1999 the figure was 23. Although this number may seem small compared with the major U.S. markets, London also receives the three national commercial stations and the five BBC national networks, and it has its own BBC local station. The commercial services now have 61 percent of the total radio audience in the capital, to the BBC's 36 percent.

By the end of February 2003, there were 266 analogue commercial radio services broadcasting in the U.K., comprised of three national stations, 16 regional stations, and 247 local stations. A majority of these stations also broadcast digitally, and there were also digital-only services.

There was one national digital multiplex (carrying eight radio services) and 41 local digital multiplexes which between them carried over 250 services. All these were licensed by the Radio Authority and were, of course, additional to the BBC's digital services. There were also 78 satellite and 13 cable radio services.

Shares of listening for the survey period ending in December 2002, were: BBC 52.5 percent; commercial 45.5 percent. Commercial's share had been marginally ahead for a time at the end of the 1990s, but BBC Radio has come back strongly in recent years.

Radio Authority

The new legislation, for which the industry had fought so hard and so long, was the 1990 Broadcasting Act, which saw the end of the IBA and the introduction of separate regulators for commercial radio and commercial television. Since 1 January 1991, commercial radio has been the responsibility of the Radio Authority. Although it has had its detractors, the Radio Authority has operated with a minimal level of interference with stations' operations and has steadily created many more opportunities to broadcast. The 1990 act swept away the concept of the regulator as technically the broadcaster; stations now hold licenses instead of franchises and are now responsible for their own transmission arrangements.

New ownership rules were introduced in the 1990 act and modified in the 1996 Broadcasting Act, and these have allowed groups to develop, although there is still a ceiling on the proportion of the industry that any one company can control. This is based on a somewhat complex formula of ownership "points." Every license has a point score that depends on the potential audience, that is, the number of adults in its predicted coverage area. No group can hold licenses that add up to more than 15 percent of the total points in issue at any one time, but of course, as the Radio Authority issues more and more licenses, the total, and the 15 percent ceiling, rises.

Unlike the U.S. system, there is still not the freedom for owners to determine what formats their stations will follow or to change them at will. When a license is awarded, it is for a specific format. Subsequent minor adjustments to the content can be made with the Radio Authority's approval, but total change is not possible.

The major groupings of stations that have emerged in recent years include GWR, which holds 36 local station licenses and also owns the national station Classic FM; EMAP Radio, which owns 18 stations, including market leaders in five major cities; Capital Radio (14 stations); and Scottish Radio Holdings, which among its 15 licenses owns the leading stations in all the main cities and towns in Scotland and Northern Ireland. There are more than a dozen smaller groups, each owning a handful of stations, of which the most significant is Chrysalis Radio, which owns 6 of the regional stations.

Music copyright—both the level of royalties and the restrictions imposed by the record companies—was a major problem for the industry throughout its first two decades. Even before the first commercial stations came on air toward the end of 1973, the regulator, the IBA (technically the broadcaster) made deals with the Performing Rights Society (PRS; a composers' organization) and Phonographic Performance Limited (PPL; record companies) to which the stations were bound for the first five years. These agreements delivered very high royalties (much higher than the BBC had ever paid) to both PRS and PPL and also allowed the latter to limit stations to playing eight hours of records a day, the infamous "needle-time" restriction.

The moment those first five-year contracts expired, the radio trade association, AIRC, referred both agreements to the Performing Rights Tribunal (now the Copyright Tribunal). There followed no fewer than 15 years of litigation—costing the industry millions of pounds in legal fees—and on-off negotiation before, in 1993, AIRC won a significant victory over the record companies at the Copyright Tribunal and peace broke out. PPL had wanted up to 20 percent of a station's revenue on a sliding scale that would rise with both usage and revenue. The tribunal decision broadly gave PPL a flat rate of 5 percent of a station's net advertising and sponsorship revenue (still high by other countries' standards and, of course, the United States has no such right) and gave the stations the unfettered right to play as many records as they wished. Two years later AIRC struck a deal with PRS at a slightly higher figure.

Of course, with royalties linked to revenue and with the industry expanding so rapidly in the second half of the 1990s, both in terms of number of stations and of revenues earned, the copyright organizations can have no complaints: they have seen their incomes from commercial radio increase spectacularly with no extra effort on their parts.

Digital Radio

After a difficult start and a long period of struggle against restrictive regulation, punitive royalties, and advertiser indifference, commercial radio in the United Kingdom enters the new century as an undoubted success story. However, in the modern world every industry faces new challenges, usually of a technological nature, and radio is no exception. The new challenge is digital radio.

Current transmissions, whether in AM or FM, are analog, and the signals pass through the ether from transmitter to receiver in wave forms that are all too easily affected by topography, especially by high buildings in towns and cities. Digital transmissions turn the sounds from the studio into data, which is not subject to such interference and which, when turned back into sound by the receiver, is almost as clear as the sound from a compact disc. Not only this, but digital radio makes much better use of spectrum, with six or more high-quality channels able to be radiated from the one transmitter.

These "advantages" of digital radio have encouraged broadcasters across Europe, and especially the public broadcasters, to develop a single European standard, known as Eureka 147, which all countries have adopted, and to press forward with the building of transmitters and the testing of new program services. All the countries have had to find new blocks of spectrum for digital radio, and there is no overall uniformity: some place it in Band III, and others choose the much higher frequencies of the L-Band. In the United Kingdom, spectrum in the L-Band will not be available until about 2007, and so the limited allocation made so far in Band III means that at present not all local commercial radio services will be able to begin digital transmission even if they wish to. The intention is that once the bulk of radio listeners have switched to digital, the spectrum currently occupied by analog FM services will be released—either for more digital radio or for other uses, such as mobile phones. However, even the most bullish advocates of digital radio accept that such a switch-over is a decade or more away.

A major delaying factor is that new receivers are required to be able to receive digital broadcasts, but unfortunately, the receiver manufacturers have not moved as fast as the broadcasters. They decided that initially they would cater primarily to the more expensive end of the market and produced only receivers for automobiles or hi-fi systems that sold for many hundreds of pounds. As a result of this, public interest in digital radio was slow to develop, but when a cheaper digital tuner, which could be linked to a home PC, became available interest rapidly picked up, and by the end of 2002 portable digital receivers costing around £100 were on sale, but not in huge quantities.

In the United Kingdom, the prime mover into digital radio was the BBC, and by 1999 it had built a digital transmitter network covering 60 percent of the population at a cost of £14 million. Its existing five national networks and several new services, including a sports channel and Parliamentary channel, are being transmitted digitally.

The commercial sector, inevitably, has had to be more cautious, because its revenues are not guaranteed as are those of the public broadcaster, and it has to make profits to satisfy shareholders. For a commercial station to embrace the Eureka 147 system of digital radio, it has to bite the bullet of doubled transmission costs for at least ten years with no immediate prospect of increasing its audience and therefore its revenue.

A national digital network has been launched by the commercial sector—with the largest radio group, GWR, the prime mover—and in 1999 the Radio Authority began offering licenses to run local digital radio systems, but managements are decidedly nervous about the implications of the digital revolution. For example, there was only one application to run the digital multiplex for England's second city, Birmingham, and that was a combined effort by two of the largest radio groups, Capital and EMAP. However, by the beginning of 2003 there was a national commercial multiplex carrying eight stations, and the Radio Authority had licensed 41 local digital multiplexes carrying over 250 services. The overwhelming majority of digital services, both locally and nationally, are simulcasts of established analogue services.

Digital is unquestionably a bigger advance for radio than the introduction of FM broadcasting in Britain in the 1960s. In addition to vastly improved reception (especially in cars), the Eureka 147 system offers exciting opportunities for the operators to add value to their sound broadcasts with text and to make money from separate and additional data services. But for commercial radio, there are the two major headaches of long-term additional transmission costs and getting listeners to buy new receivers. The digital era will come for radio—as it has for all other forms of electronic communications—and commercial radio stations will eventually be beneficiaries. Only the timing is in doubt at the beginning of the 21st century.

BRIAN WEST

See also Capital Radio; Digital Audio Broadcasting; London Broadcasting Company

Websites

Capital Radio: Britain's Leading Commercial Radio Group, <www.capitalradio.plc.uk>
Classic FM, <www.classicfm.co.uk>
Commercial Radio Companies Association, <www.crca.co.uk>
GWR Group: The UK's Leading Commercial Radio Broadcasting Company, <www.gwrgroup.musicradio.com>
The Department for Culture, Media and Sport, <www.culture.gov.uk>
Radio Advertising Bureau OnLine, <www.rab.co.uk>
The Radio Authority Online, <www.radioauthority.org.uk>
The Radiocommunications Agency, <www.radio.gov.uk>
Virgin Radio: The Home of Ten Great Songs in A Row, <www.virginradio.co.uk>

Further Reading

Baron, Mike, *Independent Radio: The Story of Independent Radio in the United Kingdom,* Lavenham: Dalton, 1975

Crook, Tim, *International Radio Journalism: History, Theory and Practice,* London and New York: Routledge, 1998
Donovan, Paul, *The Radio Companion,* London: HarperCollins, 1991; London: Grafton, 1992
Garfield, Simon, *The Nation's Favourite: The True Adventures of Radio 1,* London: Faber, 1998
Mitchell, Caroline, *Women and Radio: Airing Differences,* London and New York: Routledge, 2000

British Disk Jockeys

It would be easy but incorrect to assume that British disc jockeys were essentially pale imitations of a style of radio presenter most often associated with the United States. Certainly, many British disc jockeys—including some of the best known and most enduring in the profession—consciously copied attributes heard from across the Atlantic. Because British commercial radio modeled its programming largely on what had been heard in the United States—promoted by station identification jingles created and recorded in the United States—and played much of the same recorded material, it would be surprising if the people who linked the program material did not also sound similar to those in North America.

Origins

One of the key distinctions between the British and U.S. systems is the dominance—made real by an official monopoly for more than 50 years—of the noncommercial British Broadcasting Corporation (BBC). The distinction was made even greater by the strict limits on the amount of play of "commercial gramophone records" imposed by the copyright authorities and the Musicians' Union—the so-called needletime agreements. For decades, this limited the scope for presentation of true "disc" programs: popular music shows either were all live or contained a mixture of "live" or specially recorded tracks interspersed with occasional gramophone discs. It was the active determination of the BBC's senior managers not to follow the U.S. style of commercial programming, which, until the mid-1960s, pushed the informal and youth-oriented style of music presentation to the margins. In the pre–World War II era, the main outlet for disc jockey-type programs on the BBC were those that specialized in playing American jazz and blues records (music genres that, like rock and roll some 20 years later, were viewed by the corporation's hierarchy with deep suspicion and even hostility; some subgenres of music were regarded as entirely beyond the pale—scat singing was officially banned in 1936). When such records were played, they were introduced with typical BBC solemnity by BBC announcers, the best known from this period being Christopher Stone.

On continental Europe, though, unhindered by the BBC's cultural and stylistic attitudes, a number of entrepreneurs set up commercial radio services in the 1930s targeted at the British audience. Record request programs became established in this period, and some of the programs and stations were very popular indeed, especially on Sundays, when the BBC, reacting to religious sensibilities (not least those of its first director-general, John Reith), broadcast "serious" music and talk programs. Such was the dominance of the BBC's approach that even the continental commercial stations often pretended that music on discs was being played by live singers and dance band orchestras.

As with so much else of British life, World War II had a major effect on the public's attitude and forced the Corporation to encompass more record programs—notably *Family Favourites* (which developed from the wartime *Forces' Favourites*) and the weekday *Housewives' Choice.* Many of the presenters on the latter were taken from the world of variety or were even recording stars themselves: programs were scrupulously scripted and rehearsed, and the records were played by a team of technical operators and engineers—the concept of the record presenter operating his own "board" was fiercely resisted on the BBC, as it was for the first 30 years or so of the postwar Radio Luxembourg. Therefore, the concept of the "art" and technique of the disc jockey—which is generally thought to include technical competence as well as broadcasting ability and appeal—remained a "foreign" concept in both senses of the word. Nevertheless, it was Radio Luxembourg that again introduced the U.K. audience to rock and roll—the exciting new youth-appeal music from the United States—and presented it in a way that can certainly be identified with the term *disc jockey.*

The most influential and most idiosyncratic disc jockey from the late 1950s to early 1960s era on Luxembourg was Jimmy Savile. Like his rather more conservative colleagues Pete Murray, David Jacobs, and Jimmy Young, he was to find a home on the BBC networks. The latter is thought to have presented the BBC's first unscripted record show—in 1963.

Another figure who had an enormous influence on the "education" and emerging music tastes of British youth was

Brian Matthew, who, from 1958, presented *Saturday Club,* a program that emerged from *Skiffle Club* and featured a mixture of recorded and live music—notable and regular guests were the Beatles. The show became required listening for a whole generation of young Brits, the first to come of age after postwar austerity. Matthew—who, over 45 years later, continues to present a Saturday show on the Light Programme's successor, Radio 2—had an avuncular style, which provided a comforting context for BBC bosses who, like their predecessors in the 1920s and 1930s, were extremely perturbed by the new musical "fad" from the United States. Another important broadcaster from this period was Jack Jackson, who established a style that has been much imitated—a carefully constructed program linking "pop" records with comedy clips. Jackson provided much of the inspiration for the country's most innovative and admired disc jockey, Kenny Everett.

However, the offshore "pirate" stations—which established Everett and numerous other disc jockeys—were what really exposed British audiences to the true concept of the disc jockey and established the disc jockey as a distinct part of the entertainment business: as a role model, an arbiter of taste, a spokesperson for young people, and a mediator between the music industry and the listener. The good disc jockey, it was recognized, though neither comedian, commentator, nor journalist as such, often utilized the attributes of these to create something unique. The test of this uniqueness was that when a disc jockey was on vacation, the substitute, although playing the same records and broadcasting the same features, nevertheless would sound very different from the usual host. Not only did these presenters operate their own equipment, but they also consciously adopted many of the mannerisms and styles—and even, like Johnnie Walker, Dave Cash, and many others—the names of their counterparts in the United States. Although most of the stations adopted—and to some extent adapted—the Top 40 radio model from America, there were also "beautiful music" stations and, by 1967, the adoption of "underground" rock radio. The best-known and enduring figures of the Top 40 and "progressive rock" styles, both of whom continue to broadcast into the 21st century, are, respectively, Tony Blackburn and John Peel. For a while both broadcast on the same "pirate" station—the enormously successful Radio London, which was backed and programmed by Texans. Blackburn is a particularly interesting example of the disc jockey's art, because, although he has presented programs with very different musical styles—Top 40, soul, "oldies," jazz, and blues (he has been a consistent champion of soul music and continues to present a weekly soul show for "Jazz FM")—his style—bright, upbeat, and interlaced with the corny jokes for which he is notorious—has remained virtually unchanged in nearly 40 years of broadcasting. Peel also, after a brief imposed flirtation with an upbeat Top 40 style when he worked in Texas in the early 1960s, has also maintained the same slow, ironic, and somewhat lugubrious style. Peel is perhaps the great British disc jockey survivor—he is the only disc jockey from the original lineup when the BBC launched its pop and rock network Radio 1 in 1967 who was still broadcasting on the station in 2003.

Opposition to Radio 1's Dominance

If the cult of the disc jockey needed any further entrenchment in the United Kingdom, Radio 1 established the profession in the consciousness of the British public. For over six years after the "pirate" stations were effectively made illegal, in 1967 Radio 1 had a national monopoly, and its disc jockeys became every bit as famous—indeed in many cases more so—than the recording artists they were playing. For the first time in Britain, it became legitimate to aspire to the role of radio disc jockey in its own right, rather than being a disc jockey as an adjunct to a career based in show business or journalism.

A variety of broadcasting styles were heard, from the frantic, fast-talking, wise-cracking Top 40 style to more contemplative and "credible" but still mainstream approaches to the music, with the latter approach perhaps personified by Johnnie Walker, who stayed on the pirate Radio Caroline after the government's new Antipirate legislation came into effect but who, after a brief hiatus, established himself on daytime Radio 1. In 1976, however, frustrated by being forced to play Top 40 "teenybop" music such as the Bay City Rollers, he quit the United Kingdom for the United States and secured a place on San Francisco's KSAN. He returned to the United Kingdom and in the late 1990s began presenting the *Drivetime* show on BBC Radio 2. Walker was also notable for criticizing the attitude of many disc jockeys before and since of regarding the music they play as being an almost irritating irrelevance to their shows—most were more than happy to let their producers decide on their playlists and indeed had no engagement with the music industry and certainly rarely went to live "gigs." Walker also rejected the desired common career path of many Radio 1 and commercial disc jockeys, who saw their radio work as merely a stepping-stone to the more glamorous and better-paid world of television presenting. Walker, like John Peel, had no such ambitions and indeed found his obligatory appearances on the BBC's hugely popular television chart show *Top of the Pops* to be an embarrassing ordeal.

Commercial Radio DJs

In the mid-1970s local commercial radio, legally established in 1972, sprouted up in conurbations across the United Kingdom, and the disc jockey became a familiar sound and sight to the public—disc jockeys were now accessible as well as famous. The regulatory requirements of the new commercial system required these disc jockeys to do more than spin discs and spout trivial chat: they had to be able to be part-journalists and community activists as well. The dominance of the standard BBC speech patterns and accent was also undermined by the local stations. Regional accents—and to some extent attitudes—were often not only tolerated but positively encouraged as station managements sought to capitalize on their "localness" in contrast to the seeming remoteness of their national BBC rivals.

A good example of this type of disc jockey—who in fact began his career on his local BBC station in 1970 but moved to the commercial service BRMB a few years later—is Les Ross, who maintains a pronounced Birmingham accent and continues to broadcast in the city in 2003, switching to Saga FM, which targets the over-50 demographic. Ross' style has remained unchanged in its fundamentals over this 33-year period, yet when exposed to U.S. radio in the mid-1970s, he introduced more scripted gags and produced comedy along with his trademark sharp, ad-libbed wit. In recent years, along with most local commercial FM stations—and again something that has been copied directly from the United States—he has developed a "zoo" format on his show.

The deregulation of British commercial radio led to more diverse music formats: disc jockeys who specialized in particular music genres often were able to break out from the "ghetto" evening slots to which they had been confined on mainstream stations, to niche format services. Nevertheless, one of the main criticisms of British commercial radio—and the disc jockeys who present on it—is the blandness and similarity of stations in both style and content. Many critics have pointed to a "mid-Atlantic" sound that owes nothing to the locality—or even the nation—in which the station is situated. Certainly the vast majority of commercial radio disc jockeys have little or no control over the music they present, which is usually selected using computer software, supervised by a head of music or programme controller, who is sometimes based many miles away.

Career Patterns

The career origins of British disc jockeys have changed significantly. The early disc jockeys were mostly drawn from either an announcing or general entertainment background and, occasionally, from journalism; from the late 1970s an increasingly common background was the club scene, although the approach needed in a noisy discotheque where the audience can be seen and its reactions gauged is clearly very different from that required in the intimate, personal medium of radio, with its invisible audience. An increasingly common source of talent has been children's television, and many television "youth" programs have a style, structure, and attitude that owe a lot to personality disc jockey programs.

A new phenomenon emerged in the early 1990s that had its echo in the old variety background of disc jockeys: the burgeoning "alternative comedy" scene in Britain (it was sometimes said that comedy was "the new rock and roll") launched several disc jockeys, especially in the high-profile breakfast shows on big-city stations, as management sought to find something different and marketable for their services. Coupled with this trend has been the development of what might be called the "postmodern/ironic" style of disc jockey, especially on Radio 1 and the larger commercial FM stations: disc jockeys who, like John Peel and Johnnie Walker, eschew the traditional terminally cheerful, positive, showbiz-obsessed, glamorous lifestyle disc jockey and adopt, if not a sullen, then certainly a downbeat and often cynical "real" approach—being a disc jockey now means you can audibly have a "bad day."

A Male-Dominated Profession

One enduring characteristic of British disc jockeys is that the vast majority continue to be male. Although the situation is now slightly more balanced than it was in the pre-1970s days, when less than a handful of women earned their livings as radio disc jockeys, a survey carried out by the University of Sunderland in 1999 suggested that only 11 percent of disc jockeys in England, Scotland, and Wales were women—and nearly two-fifths of commercial stations had no female disc jockeys. Antiquated attitudes of management formed in the days when women stayed at home and therefore constituted the main daytime audience—and, it was presumed, would on the whole rather listen to a male, with his vicarious seductive approach, than another woman—may also be partly responsible for this disparity.

In her autobiography, Anne Nightingale (who after John Peel has probably presented for more years on Radio 1 than any other disc jockey) describes how the network's early controllers regarded disc jockeys as "substitute husbands." Most of the female presenters who have made it to the corporation's pop and rock network share a common characteristic: they are "ladettes"—that is, they have the same attitudes, style, and approach as their male counterparts. As Anne Nightingale puts it, " I wanted to *be* a DJ, *be* one of them, *be* one of the boys" (emphasis in original). It is certainly the case that many more men than women seek work as radio disc jockeys: audition tapes from men typically outnumber those from women by a factor of at least 25 to 1. The BBC can be fairly praised in this area for allowing two women in succession to host its key breakfast show slot on Radio 1. Zoe Ball (solo from 1998, after a year cohosting with a male disc jockey) and Sara Cox (from 2000) followed a previously unbroken line of male disc jockeys from the start of the station in 1967.

A role model for the aspiring British disc jockey in the new century might be Chris Evans. He began as "gofer" in Manchester's Piccadilly Radio, went on to BBC local radio and quickly moved over to Radio 1, where he eventually won the coveted breakfast show produced by his own company while simultaneously establishing himself as a major television presenter and producer. After quitting Radio 1, he switched to national commercial rival Virgin Radio, which his company then bought—then sold, at an enormous profit—while Evans continued to present the breakfast show there.

RICHARD RUDIN

See also BBC Local Radio; Capital Radio; Everett, Kenny; London Broadcasting Company; Radio Luxembourg

Further Reading

Barnard, Stephen, *On the Radio: Music Radio in Britain,* Milton Keynes, Buckinghamshire, and Philadelphia, Pennsylvania: Open University Press, 1989
Baron, Mike, *Independent Radio: The Story of Independent Radio in the United Kingdom,* Lavenham, Suffolk: Dalton, 1975
Chapman, Robert, *Selling the Sixties: The Pirates and Pop Music Radio,* London and New York: Routledge, 1992
Crisell, Andrew, *An Introductory History of British Broadcasting,* London and New York: Routledge, 1997
Fong-Torres, Ben, *The Hits Just Keep on Coming: The History of Top 40 Radio,* San Francisco: Miller Freeman Books, 1998
Garay, Ronald, *Gordon McLendon: The Maverick of Radio,* New York: Greenwood Press, 1992
Garfield, Simon, *The Nation's Favourite: The True Adventures of Radio 1,* London: Faber, 1998
Lewis, Peter M., and Jerry Booth, *The Invisible Medium: Public, Commercial, and Community Radio,* London: Macmillan, 1989; Washington, D.C.: Howard University Press, 1990
Nightingale, Annie, *Wicked Speed,* London: Sidgwick and Jackson, 1999

British Forces Broadcasting Service

The British Forces Broadcasting Service (BFBS) has its roots in the establishment of the British Forces' Experimental Service in Algiers, which began broadcasting on 1 January 1944. The first transmitter was a German model captured in Tunisia. By the end of 1944, 74 officers and people of other ranks were divided between five stations, often broadcasting from positions that had been occupied by the retreating German army just a few days previously. In Rome, station B5 even claimed to broadcast the first phone-in request show.

On 10 May 1945—just 48 hours after Germany's surrender—the words "The British Forces network" (BFN) were heard for the first time, and a studio center—a true radio *station*—was quickly established at the famous Musikhalle in Hamburg. The programs became increasingly sophisticated and varied—the BFN dance orchestra made its debut in May 1946—and original drama productions became a regular feature. Some of the best-known British postwar actors, musicians, scriptwriters, and singers gained their first experience with BFN Hamburg.

In the same period, the All Forces Programme was established in India; the British Pacific Programme broadcast over the transmitter of Radio Australia, which could also be heard in Singapore; and a Forces Broadcasting Service continued to develop in the Middle East and in several European locations, including Trieste and Austria. By the late 1940s and early 1950s, major stations had been established in Kenya, Malta, Cyprus, Libya, and Gibraltar. In short, the precedent was established that wherever in the world British troops were to be found in any quantity, a Forces radio service would be established to serve them, providing a mixture of vital information, education, morale-boosting entertainment, and a "link with home."

For millions of civilians "back home," the existence of BFN was indelibly linked in the mind with Sunday lunchtimes through an enormously popular record request program, which was cohosted by a British Broadcasting Corporation (BBC) presenter in London and a BFN announcer in Hamburg (later Cologne) and broadcast simultaneously over both the BBC's Light Programme and BFN. *Family Favourites* began in October 1945, and at its peak the show had a U.K. audience of some 16 million. The program survived, in slightly varied formats and time slots, until January 1980, by which time direct communications between troops and their families and friends back home had largely become easy and routine. The most famous of the on-air partnerships, from 1949, was between Jean Metcalfe for the BBC and the then Royal Air Force Squadron Leader

Cliff Michelmore, who was to become one of the best-known broadcasters in the United Kingdom. The two met for the first time during Michelmore's visit to London a couple of months after their on-air relationship began, and, much to the fascination of the British public and popular press, this quickly turned into a "radio romance," and the couple married.

In 1953 BFN was forced to give up its AM (mediumwave) frequency of 274 meters—which had been "commandeered" toward the end of the war but which was now reallocated under the Copenhagen Plan—and share the 247-meter frequency with the BBC's Light Programme. Because AM radio waves travel further at night, programming output was reduced to just a couple of hours a day in the winter months so as not to interfere with reception of the BBC's programs in the United Kingdom. The farsighted and technically innovative solution to this problem was to move transmissions to a new waveband and transmission standard. Thus, in February 1956, BFN became the first English-speaking network to move wholly to the very-high-frequency (VHF) band, using frequency modulation (FM) transmission. Two years before this, the main broadcasting studios had moved from Hamburg to a modern studio complex in two renovated villas in a high-class district of Cologne. By this time it was estimated that several million German civilians were tuning in to the BFN.

At the beginning of the 1960s, a standard name for the multiple services across the globe—BFBS—was mandated by headquarters. This period also saw the end of military conscription ("the draft") in the United Kingdom, and, increasingly, broadcasting staff were recruited directly from civilian life in the United Kingdom rather than being "seconded" from their military duties. Throughout the rapidly changing background of both the United Kingdom's military commitments and broadcasting styles in the 1960s and 1970s, BFBS continued to serve its special audience wherever they were stationed. Major locations provided stations with local output, backed by network programming taped at the BFBS London studios that featured some of the best-known U.K. presenters and journalists. Where radio transmissions were impractical—such as at very small military outposts and on board navy ships and submarines—programs were recorded on cassette and posted out for local relay.

In September 1975 the radio network in Germany was augmented by a television service, although it was a further seven years before TV programs could be broadcast "live" from the United Kingdom.

The biggest shake-up of the organization since its inception came about in 1982, when BFBS, which had been a branch of the Ministry of Defence (with U.K. staff treated as civil servants), became part of the new Services Sound and Vision Corporation (SSVC), a self-supporting registered charity formed by a merger between BFBS and the Services Kinema Corporation (SKC). Income is derived from a mixture of grants from the U.K. government and commercial activities. Any surpluses are donated to welfare support for the armed forces.

With the "peace dividend" following the end of the Cold War in the early 1990s, British troops were massively cut back in West Germany, which had always been the largest BFBS overseas operation. Even before these political and military upheavals, however, it had been decided to move BFBS Germany's main studios from Cologne—which for years had been some distance from the main garrisons—to Herford, supported by a number of smaller contribution and "opt out" studios. The Berlin station closed in 1994 after 33 years of operation, as the World War II Allies withdrew from the formerly divided city, which was once more to be capital of a united Germany. During the Cold War, a substantial and loyal audience—at considerable risk—had listened to the service from behind the Iron Curtain.

Today, two BFBS radio stations and the television service are available to British Forces personnel in Germany, the Balkans, Cyprus, Gibraltar, Brunei, the Falklands, and Belize. Temporary stations were also set up in Afghanistan in 2001 and Kuwait in 2003—the latter also gaining a large and appreciative audience of U.S. service personnel—in response to the U.K.'s military involvement in the conflicts in Afghanistan and Iraq. Northern Ireland now also has its own radio service, and in 1999 local radio stations were set up at a number of army garrisons in England, using the old name of BFN. The major overseas BFBS radio services broadcast a mixture of locally originated programming of information and entertainment combined with network programs, both from the BBC and those specially made for a services audience, produced at state-of-the-art digital studios in Buckinghamshire, England, and transmitted by satellite around the clock.

RICHARD RUDIN

See also Armed Forces Radio Service

Further Reading

Barnard, Stephen, *On the Radio: Music Radio in Britain*, Milton Keynes, Buckinghamshire, and Philadelphia, Pennsylvania: Open University Press, 1989

British Forces Broadcasting Service, <www.ssvc.com/bfbs>

Crisell, Andrew, *An Introductory History of British Broadcasting*, London and New York: Routledge, 1997

Grace, Alan, *This Is the British Forces Network: The Story of Forces Broadcasting in Germany*, Stroud, Gloucestershire: Sutton, 1996

Lewis, Peter M., and Jerry Booth, *The Invisible Medium: Public, Commercial, and Community Radio*, London: Macmillan Education, 1989; Washington, D.C.: Howard University Press, 1990

Taylor-Wilkie, Doreen, *A Microphone and a Frequency: Forty Years of Forces Broadcasting*, London: Heinemann, 1983

British Pirate Radio

Operating Without a License

Pirate radio is a phrase used to describe broadcasts from stations operating without government licenses. Offshore commercial stations lacking licenses sought from the mid-1950s through the 1960s to break the British Broadcasting Corporation's (BBC) monopoly of radio broadcasting. Similar pirate stations operated offshore from other nations in Europe and elsewhere. The British pirate stations helped to galvanize change in BBC radio, paving the way for British commercial radio and constituting a milestone in the international spread of commercialized broadcasting, the result of Britain's leadership in world broadcasting during that period.

Factors Leading to Pirate Radio

Four elements were important in the development of the British pirate stations. First was the continued monopoly of domestic British radio by the BBC (though Radio Luxembourg was widely tuned as well), despite the availability of commercial television since 1955. Although commercial radio was not licensed nationally until 1973, the logical disparity between the availability of commercial options for one medium and not the other presented a wedge for proponents of commercial radio.

The second significant factor was the territorial jurisdiction of Britain (and other European nations) at the time, which was defined as ending three miles offshore. A ship anchored just a few miles off the coast was not subject to British laws, and a radio transmitter could be set up in such a ship or on any of a handful of derelict offshore forts without formally contravening the BBC's monopoly.

Third was the ongoing struggle in British broadcasting between the elitists and the popularizers, or—depending on one's viewpoint—the public educators and the crass commercializers. World War II had forced concessions to the musical tastes of enlisted troops in the shape of the General Forces Programme, a BBC service later retitled the Light Programme. However, these concessions were fought every step of the way within the BBC and by other traditionalists, so that by the 1950s the Light Programme was losing the younger, postwar-generation audience.

The grimly staid public atmosphere of the 1950s was the fourth major contributor to the development of pirate stations. This social climate resulted from two wars in which the British had taken a tremendous pounding, from the intervening Depression years of the 1930s, and also from the years of severe austerity that followed World War II. The British people had faced the necessity of sacrifice and had then gone on to make a virtue of it. For the younger generation, this public culture seemed impossibly stifling. Dramatic signals that a turning point had been reached included *The Goon Show*, the "Angry Young Man" theater and "Northern social realist" cinema of that decade, the victorious Conservative Party's slogan for its third election victory in 1959 ("You never had it so good!"), and the thrilling music of such American rock and roll stars as Bill Haley and the Comets and Elvis Presley. Popular music, and therefore a different kind of radio, were at the heart of the transition. Yet the BBC's agreement with the Musicians' Union considerably limited the needle time for this music.

Pirate Radio Stations

Pirate stations broadcasting principally to other European countries preceded the development of pirate operations directed at British audiences. In 1958 Radio Mercur began broadcasting a few miles off the Danish coast. In 1960 Radio Nord went on the air off the Swedish coast, and in the same year Radio Veronica started transmitting off the Dutch coast. Veronica, which lasted until 1974, began an English-language service in 1961. These ventures broadcast popular music overwhelmingly targeting the youth audience, playing to that generation's growing budget for leisure products and its impatience at older styles of music. Rock and roll music from the United States had great appeal, partly because of its energy and partly because of the images it projected of a consumption-oriented culture with plenty of space for pleasurable activities.

In Britain as elsewhere in Europe, the pirate stations' distinctive programming feature was their use of the U.S. Top 40 format, then quite foreign to European radio. Inevitably there was considerable interest in these stations on the part of U.S. record company executives, who recognized a new avenue for getting their products heard and thus sold. Some of these stations were directly backed by U.S. entrepreneurs such as Gordon McLendon, one of those originally responsible for the Top 40 format. Radio London, one of the major British pirate stations, was backed by Texas automobile dealers and an oil baron. These stations were not exclusively musical outlets, nor did they feature just rock and roll. But rock and roll was the element of their programming that drew the most attention, both from those who loved them and those who were alarmed at their potential to influence the morals of the younger generation.

Ronan O'Rahilly, an Irish citizen and somewhat flamboyant music entrepreneur at home in very smart and trendy Chelsea circles, spearheaded the most prominent of all the British pirate stations, Radio Caroline. Caroline began transmitting in

1964 and was named after President Kennedy's daughter. In Britain, the youthful-looking president had symbolically represented a refreshing new age, and his daughter's name was symptomatic of the fresh tide in British life of which O'Rahilly himself was part. Radio Caroline was emblematic of London's "Swinging Sixties."

Radio Caroline's sponsors quickly established two ship-based stations, Caroline North and Caroline South. Caroline North was more original in its programming, attuned both to the Merseyside (Liverpool) beat then made famous by the Beatles and to new trends in African-American music such as the hits from Motown. Caroline South found itself in increasing financial trouble for a variety of reasons and had to be rescued in 1966 by Tom Lodge, one of the key figures from Caroline North. His most important contribution to the development of British radio and popular music was the establishment of disc jockeys as the pivotal cultural entrepreneurs of the stations, with their musical intuition providing the stations' heartbeat.

The other leading British pirate station of that period was Radio London, also established in 1964. Its format was considerably less freewheeling than Radio Caroline's and served as a model for Radio One, the BBC pop music channel begun by the BBC in July 1967 in direct response to the pirates' popularity (and also due to some internal pressure to develop a more audience-responsive programming policy). Radio London's programming was always much more culturally conservative than Radio Caroline's, and thus it attracted considerably more support from elite circles.

A number of problematic conditions challenged the pirate station developers. The North Sea, where most of the pirate station ships were positioned, is subject to very powerful gales and stormy weather. In a number of cases there was a sharp disparity between the disc jockeys' spartan and dangerous working conditions and the ritzy administration offices in central London. A number of investors saw the pirate station ventures as instant cash cows, with predictable effects on financial policies and stability. They often wildly inflated their audience size. Only about a half-dozen syndicates owned all 21 pirate stations available to British listeners.

Decline

A number of other troubling events shadowed the stations. Harry Featherbee, director and one of the three founders of Radio Invicta in June 1964, drowned in circumstances that some found suspicious. Radio City, previously Radio Sutch, became the target of a control battle between Reginald Calvert and Oliver Smedley. The former launched a military-style boarding party to repossess his transmitter and was later shot dead in Smedley's home.

These and other factors resulted in a loss of popularity of the pirate stations that enabled the British government to reassert its authority and monopoly of the airwaves. Beginning in August 1966, the Marine Offences Act effectively choked off the pirates' revenue stream by outlawing the use of their channels by British advertisers.

Despite their relatively short term of operation and shadowy dimensions, these rebels against British radio authorities had a lasting impact on British broadcasting and culture. In addition to the change in focus from traditional music to a format that appealed to younger listeners, the language used in music radio changed from a carefully articulated southern English accent and vocabulary to a mid-Atlantic intonation and patter. In Britain, the debate continues (mostly along generational lines) as to whether this change signified the reverse cultural colonization of the United Kingdom or its welcome introduction to the dynamism of U.S.—and especially African-American—popular culture.

British pirate stations continued to operate in the 21st century—mainly urban and carrying ethnic minority content. Such stations appear and disappear rapidly and are difficult to track down. Many operate only on the weekends.

JOHN D.H. DOWNING

Further Reading

Barnard, Stephen, *On The Radio: Music Radio in Britain,* Milton Keynes, Buckinghamshire, and Philadelphia, Pennsylvania: Open University Press, 1989
Chapman, Robert, *Selling the Sixties: The Pirates and Pop Music Radio,* London and New York: Routledge, 1992
Crisell, Andrew, *An Introductory History of British Broadcasting,* London and New York: Routledge, 1997
Harris, Paul, *When Pirates Ruled the Waves,* London: Impulse, 1968; 4th edition, 1970
Henry, Stuart, and Mike von Joel, *Pirate Radio: Then and Now,* Poole, Dorset: Blandford Press, 1984
Hind, John, and Stephen Mosco, *Rebel Radio: The Full Story Of British Pirate Radio,* London: Pluto Press, 1985

British Radio Journalism

Changing Styles of Radio News

Radio offered British journalists an opportunity to develop a new genre of journalism to supplement and eventually compete with the traditional press. Reporting opportunities associated with economic, social, and military crises such as the general strike of 1926, Mussolini's invasion of Abyssinia and the outbreak of the Spanish Civil War in 1936, the Munich crisis in 1938, World War II (1939–45), the Suez Crisis of 1956, and the Falklands War of 1982 have marked highlights in the development of British radio journalism.

On the other hand, British radio news was slower to evolve than American broadcast journalism. Elegance and journalistic edge through microphone reporting did not emerge so quickly as in the U.S. because of the hostility of the established newspaper media, which successfully lobbied the government to restrict early British radio news to operate as a mere replication of news agency copy. Radio news transmissions were restricted to other than peak listening periods to avoid competition with either morning or evening newspapers. Because it depended on Parliament for its royal charter to be renewed, the British Broadcasting Corporation (BBC) was also limited by its perceived need to avoid political controversy and maintain a neutral stance. The lack of competition between public and private broadcasters (until 1955 for television, and 1973 for radio) may well have been an additional reason for the slow development of radio news.

Given this historical situation, the selection of journalists made in this entry is based primarily on the contributions they made in originating and developing radio journalistic practice. Attention is also paid to the social, cultural, and political impact of their journalism. Evaluation of the importance of an individual radio journalist is partially determined by peer recognition as well as whether his or her work has been seriously analyzed by either academic or professional critics.

Apart from the multi-volume BBC history by Lord Asa Briggs (1961–95), *A Social History of British Broadcasting* by Scannell and Cardiff (1991), and several autobiographies and biographies of major broadcasting figures, there is little published information on the subject. In addition to reporters, editorial figures that advanced the form and content of radio journalism also merit consideration. Apart from significant reporters such as Richard Dimbleby and Audrey Russell, important news editors included R.T. Clark, William Hardcastle, and Isa Benzie. Radio journalism has also been present in some dramatic programs. For example, the 1931 feature *Crisis in Spain* used actors and story-telling techniques. Produced and written by Lance Sieveking and Archie Harding, the drama was journalistic in both its approach and its impact on political and public opinion.

Founders and Pioneers

Apart from some early experiments sponsored by the *Daily Mail,* the first significant official, licensed news broadcast was broadcast by the then-private British Broadcasting Company on 14 November 1922. Presented by the first director of programs, Arthur Burrows, the bulletin was based on news agency copy and was somewhat self-consciously read twice to give listeners the chance to make notes. Scannell and Cardiff credit Burrows with establishing general BBC principles of taste and editorial policy on the basis of letters he wrote to Reuters that sought to distinguish a different "socio-psychology of reception" on the part of listeners compared with newspaper readers. He argued that BBC news copy from agencies should eliminate those crimes and tragedies that did not have national and international importance. In perhaps the first indication of sensitivity to racial representation, he also held there was no need to mention the Jewish origin of people in the news.

During the British general strike of 1926, the absence of most newspapers led to the broadcasting of news bulletins throughout the day. Some BBC staff began their own independent "newsgathering." The BBC's managing director, John Reith, helped to establish radio's journalistic role as a newscaster. His was the voice that announced the inception of the strike by interrupting normal programming—and also announced its end. Although Reith believed that the BBC should aspire to be a neutral integrator, in reality BBC coverage was biased on the government side and Reith conceded that the Corporation was "for the government in the crisis." Cautious judgment in news selection was demonstrated on 12 May 1926 while Reith was reading the 1 P.M. news and Stuart Hibberd crept into the studio with agency tape announcing the end of the strike. On it Reith scribbled, "Get this confirmed from No. 10 [Downing Street]." He also lobbied hard for an end of a statutory BBC ban concerning the broadcast of "matters of controversy." The BBC had been censored by a minister called the Postmaster General who prevented the broadcasting of matters of political, industrial, or religious controversy. The campaign by Reith succeeded when the ban was suspended in 1928 as an experiment. The policy of leaving this to the discretion of the director-general and the governors eventually became an established convention that made the broadcasting of political news possible.

Although not a journalist by training, Hilda Matheson established a small independent "News Section" in 1927 while she was Head of Talks. She also commissioned a former newspaper journalist, Philip Macer-Wright, to carry out a feasibility study on whether the BBC could become a major provider of news. Macer-Wright's report, produced in 1928, advocated accredited journalistic experts on finance, sports, law, and science. To make an independent BBC news service attractive he also urged human-interest news that was simply and attractively conveyed. He set out the idea of "radio news values" with a consecutive flow of home, overseas, and sports news. He agreed with Matheson that the news needed to be written specifically for listeners' ears as opposed to merely using stories written for newspaper readers.

Crises of the 1930s

Events of the 1930s were significant in the development of news form, content, and style. Reith negotiated the BBC's gradual development of its own independent newsgathering as well as more flexibility in providing a greater number of radio news broadcasts. In so doing, Reith gradually eased the BBC into a position whereby it could readily report both domestic and international news stories. He accomplished this, however, by what today would be regarded as fatally compromising BBC journalistic integrity. If an unemployment march was reported, for example, he would assure the government that only its statements would be broadcast. The concept of journalistic independence and "integrity" had yet to evolve. There was no room for journalism of conviction. News content followed the conventions of the established news agencies. Political events had to represent a balance of mainstream opinion. Journalists who overstepped the mark into editorializing would be criticized and find that their contracts would not be renewed.

By 1933 Vernon Bartlett was operating as the BBC's first foreign correspondent. Following German withdrawal from a League of Nations Disarmament conference, he broadcast an analysis arguing that the German decision flowed from the injustices of the Versailles Treaty. This was condemned as editorializing. As Scannell and Cardiff state, "The BBC quietly dispensed with Bartlett's services, and he was not asked to talk again for several years."

In 1930 John Watt, a producer in the Talks Department, originated the idea of a newsreel program that would include commentary and dramatization of news events. Lionel Fielden produced such a program in 1933. Two years later the first separate News Department was established under the editorship of John Coatman. He recruited such journalists as Kenneth Adam, R.T. Clark, Michael Balkwill, Ralph Murray, Tony Wigan, Richard Dimbleby, Charles Gardner, and David Howarth. R.T. Clark succeeded Coatman in 1937, though on taking a firm stand against political pressure about how the BBC was covering the Spanish Civil War, he became involved in a row with the director-general, was dismissed, and later reinstated thanks to a successful petition by concerned BBC staff.

Ralph Murray succeeded in covering some key stories in Europe as a BBC correspondent/observer, but the cultural and political shackles of trying to mesh with the government's appeasement policy prevented the developing news service from sending reporters to Abyssinia or Spain, or matching the on-the-spot, dramatic Columbia Broadcasting System coverage of the German *Anschluss* of Austria.

Richard Dimbleby's live and unscripted reports from the French-Spanish border in 1939 were emotionally moving and an indication of progress in style. Although not acknowledged at the time, Dimbleby demonstrated the advantages of combining emotion, verbal pictures, dramatic sound, and authoritative command of spoken English in radio journalism. His enthusiastic letter of application to the BBC, often quoted in media history publications, pays homage to American methods of radio journalism. Likewise, his telephone report from the scene of the Crystal Palace fire in 1936 was markedly superior in style and confidence to the self-conscious and halting performances of other BBC broadcasters.

Several academics have emphasized coverage of the 1938 Munich crisis as a crucial event in BBC radio journalism. There was a relaxation of restrictive rules on bulletin timings plus a series of live remote broadcasts of Prime Minister Neville Chamberlain's statements from Heston airport and Downing Street in an unfolding and developing story. On the other hand, the BBC did not accord equal coverage to critical voices. Its failure to warn the country of the inevitability of war led John Coatman to write a lengthy memorandum, "The BBC and National Defence," which was highly critical of the lack of balance in news coverage and the exclusion of anti-appeasement voices. Coatman wrote: "I say, with a full sense of responsibility and, since I was for over three years Chief News Editor, with a certain authority, that in the past we have not played the part which our duty to the people of this country called us to play. We have, in fact, taken part in a conspiracy of silence." Winston Churchill's eloquent calls for rearmament were heard on U.S. radio networks but not over the BBC. Instead, the BBC presented a solidly governmental slant on events.

World War II

World War II marked another turning point in the development of the technology, style, and importance of radio news. The war saw an acceleration in the use of portable technology, from cumbersome mobile recording vans to "midget" recorders, which were introduced during the D-Day invasion of June 1944.

The pool of BBC radio journalists expanded and a number of individuals developed distinctive styles of vivid broadcasting, including the construction of word pictures in dramatic contexts. They included Frank Gillard, Audrey Russell, Edward Ward, Wynford Vaughan-Thomas, Godfrey Talbot, Colin Wills, Doug Willis, Thomas Cadette, and Patrick Gordon-Walker and correspondents such as Richard Dimbleby recruited just before the war.

A decision by the BBC to provide greater accuracy than German or Italian broadcasters resulted in an increase in audiences for radio journalism programs, such as *War Report*, beyond all previous measurements. The war correspondents became a fundamental link between the home population and service people overseas. The resulting journalism was still primarily patriotic cheerleading. There was rigorous censorship of reporters (at the same time, Audrey Russell challenged—albeit unsuccessfully—male dominance of this field and her inability to report from the front).

Richard Dimbleby's report from the Belsen concentration camp in April 1945 carries historical significance for the poetic and humanistic quality of its writing and performance. He had to challenge an attempt to censor it from editors who feared the report was too shocking and would be disbelieved. Dimbleby threatened resignation. Unlike Edward R. Murrow's famous report on the liberation of the Buchanwald concentration camp, Dimbledy reported that some of the inmates were Jewish and, more than other journalists, made clear the reality of the Nazi's Final Solution against European Jewry.

Postwar News

The postwar period is something of a black hole in BBC programming history and is substantially under-researched. British radio journalism played a significant role in disseminating news of key world events from the late 1940s into the 1970s. As in World War II, some journalists, such as John Nixen during the Palestine Mandate emergency of 1949, paid with their lives. A BBC radio journalist witnessed and reported the assassination of Mahatma Ghandi. James Cameron's and Rene Cutforth's broadcasts during the 1950–53 Korean War were distinctive in their highlighting of the injustice of war and the suffering of civilian noncombatants. Foreign correspondents served both BBC news and current affairs programs, which retained their authority and cultural resonance even with growing competition from television. The BBC'S inflexible editorial conservatism gave way to a more relaxed style of using actuality sound and informal language as modern youth culture and counter-culture movements emerged amid postwar prosperity.

The New Zealand–born head of BBC news, Tahu Hole, is held up as a symbol of conservatism and unpopular editorial judgment during the 1950s. He earned his nickname, "Hole and Perfect," because he had a reputation for maintaining a policy of safety first. In an unflattering profile of Hole, the BBC Foreign Correspondent Leonard Miall said that his reign was characterized by insecure and uncertain news judgment. His fear of making mistakes led to slow and pedestrian bulletins where all items broadcast had to be supported by at least two sources. Under him, some limited progress was made recruiting women. After Audrey Russell became a freelance commentator, the BBC advertised for a trained, experienced female journalist and in 1951 Sally Holloway was selected from among nearly 400 applicants. Briggs evaluates the BBC's radio journalistic coverage of the 1956 Suez crisis as a mark of greater editorial independence and resistance to government pressure.

Postwar Competition

Postwar expansion of television in Britain did not cause a diminution in the resources and output of BBC radio journalism. Continuity of funding and arm's length regulation of the BBC (through its royal charter and a board of governors) may account for the political and cultural stability that has led to a continuity in quality radio journalism through long-running programs. The monopoly of BBC radio until 1973 was financed by a license fee—a compulsory taxation on owning a radio or television receiver. The separate license for radio was abolished in April 1971. Now BBC radio journalism is funded by a share-out from television license fee revenue. For over four decades the government took a diminishing cut from license fee income. Eventually all license fee income went directly to the BBC and payment was enforced through criminal prosecution. Failure to pay fines could lead to imprisonment.

For example, *From Our Own Correspondent*, started in 1955, has given space to a more personal and creative expression of reporter opinions than regular BBC newscasts. A series of published volumes of correspondents' scripts from this program has given this genre of radio a literary textual value, since the scripts can be read by the listeners as a permanent record. Other news programs, such as *The World At One*, *Today*, BBC Radio 4's breakfast news and current affairs program, *P.M.*, *The World Tonight*, and BBC World Service programs such as *News Hour* and *Outlook*, are also examples of longevity and continuity creating influential environments for radio reporting. The *Today* program was originated and editorially pioneered by women journalists including Isa Benzie, Janet Quigley, and Elisabeth Rowley.

Andrew Boyle and William Hardcastle, former editor of the *Daily Mail* newspaper, originated *The World At One* in 1965. Many of the leading broadcast journalists of the last 40 years had associations with this program, including both Margaret Howard and Sue Macgregor.

Some radio journalistic traditions at the BBC are linked to much earlier programs such as *The Week in Westminster,* which began in November 1929 and was launched by producer Marjorie Wace. Other BBC programs are characterized by the individual associated with them, such as Alistair Cooke's long-running and highly popular commentary *Letter from America* (begun in 1946), and Roger Cook's *Checkpoint,* which was a vigorous investigative program championing the victims of swindles and social injustice. Other successful formats supporting investigative radio journalism include *File on Four,* which nurtured the editorial talents of Helen Boaden, who was appointed BBC Radio 4 controller in 2000.

By 1973 the BBC found itself competing with licensed commercial radio for the first time and this generated expansion and experimentation in style, formats, and the number of radio news programs. Despite the realities of market economics, radio news from the United Kingdom's first independent station (LBC in London), as well as the independent radio news agency IRN encouraged the greater use of actuality (on-the-spot sound) on the air and a return to greater reporter spontaneity as had been exemplified by such pioneers as Richard Dimbleby. LBC was inspired by the New York–based all-news station WINS. The more flexible programming response to crises such as the 1982 Falklands War by LBC/IRN journalists prompted the BBC to explore its own presentation of longer radio journalistic formats such as the national network *Radio Five Live* initiated in 1994.

The ITN multimedia group now produces most independent British radio journalism, and a large proportion of ITN broadcasters emerged from the generation of reporters who worked at LBC/IRN during the 1970s and 1980s and were engaged in a lively competition with the BBC. They include Jo Andrews, Jon Snow, Paul Davies, Mark Easton, Julian Rush, Lindsay Taylor, and Simon Israel.

British radio journalism has been slow to represent the changing nature of the communities it serves. Early black and ethnic programming tended to be ghettoized in terms of token programs for blacks and Asians. Only Choice FM and Sunrise in London and the BBC's Asian Network, based in the Midlands, could be said to reflect the diversity and depth of coverage evident in the ethnic press. The craft of radio journalism has been an entry point for iconic figures in British broadcasting such as the ITN newscaster Trevor MacDonald, but at the time of writing, Britain's nonwhite communities were substantially underrepresented in radio news.

British radio journalism is a continuing story without a conclusion. The ability of radio news to spawn individual reporting and writing that has profound cultural resonance is demonstrated by the work of BBC journalist Fergal Keane, whose published volume of foreign correspondent dispatches, *Letter to Daniel,* was an international bestseller. Editorial figures such as Jenny Abramsky have brought about significant changes in the way British radio journalism is consumed and communicated. Abramsky presided over the launch of the national news and sports channel Radio Five Live, the launch of the global television channel BBC News 24, and as Director of Radio and Music has unraveled the subjugation of radio news from bi-media fusion.

Extensive use of the internet by both the BBC and ITN has extended the social and cultural reach of the radio medium. Digitalization has both accelerated and expanded the transmission of radio news programs. Radio journalists now work in an inter-media environment and are more engaged with their listeners within a global medium that has greater speed and distribution than ever before.

In 2003 the BBC's Head of Radio Jenny Abramsky asserted that "Radio paints pictures, conveys images, gets inside your head, stimulating your imagination. And it takes time to acquire those skills. And great radio reporting uses sound to convey the sense of place." She set about dismantling the bi-media production culture in 2000 because "put simply radio is about painting pictures, television is about shooting them." It can be argued that as a result BBC Radio 4's breakfast news and current affairs *Today* program now has more listeners than any radio format in Greater London. The political and cultural importance of BBC radio journalism was exemplified in 1989 when the new U.S. Ambassador to London Henry Catto was advised: "In the States the most important program you must appear on is on television, in Britain it's on radio."

TIM CROOK

See also British Broadcasting Corporation; Cooke, Alistair; Gillard, Frank; Reith, John C.W.

Further Reading

Crisell, Andrew, *An Introductory History of British Broadcasting,* London and New York: Routledge, 1997
Crook, Tim, *International Radio Journalism: History, Theory, and Practice,* London and New York: Routledge, 1998
Dimbleby, Jonathan, *Richard Dimbleby: A Biography,* London: Hodder and Stoughton, 1975
Donovan, Paul, *All Our Todays: Forty Years of Radio 4's "Today" Programme,* London: Jonathan Cape, 1997
Grant, Ted, editor, *The Best of "From Our Own Correspondent" 5,* London and New York: Tauris, 1994
Hawkins, Desmond, and Donald Boyd, editors, *War Report: A Record of Dispatches Broadcast by the BBC's War Correspondents with the Allied Expeditionary Force, 6 June 1944–5 May 1945,* London and New York: Cumberlege and Oxford University Press, 1946; reprint, as *War Report, D-Day to VE-Day: Radio Reports from the Western Front, 1944–45,* edited by Hawkins, London: BBC Books, 1994

Haworth, B., "The British Broadcasting Corporation, Nazi Germany, and the Foreign Office, 1933–1936," *Historical Journal of Film, Radio, and Television* 1, no. 1 (1981)
Hickman, Tom, *What Did You Do in the War, Auntie?* London: BBC Books, 1995
Holloway, S., "That Was the BBC News," *The Oldie* (November 1997)
Hunter, F., "Hilda Matheson and the BBC, 1926–1940," in *This Working-Day World: Women's Lives and Culture(s) in Britain, 1914–1945,* edited by Sybil Oldfield, London and Bristol, Pennsylvania: Taylor and Francis, 1994
Keane, Fergal, *Letter to Daniel: Despatches from the Heart,* edited by Tony Grant, London: BBC Books, and New York: Penguin, 1996
Lazar, Roger, editor, *From Our Own Correspondent,* London: British Broadcasting Corporation, 1980
Luckhurst, Tim, *This is Today: A Biography of the Today Programme,* London: Aurum Press, 2001
Miall, Leonard, editor, *Richard Dimbleby, Broadcaster: By His Colleagues,* London: British Broadcasting Company, 1966
Miall, Leonard, *Inside the BBC: British Broadcasting Characters,* London: Weidenfeld and Nicolson, 1994
Morrison, David E., and Howard Tumber, *Journalists at War: The Dynamics of News Reporting during the Falklands Conflict,* London: Sage, 1988
Rawnsley, Gary D., "Cold War Radio in Crisis: The BBC Overseas Services, the Suez Crisis, and the 1956 Hungarian Uprising," *Historical Journal of Film, Radio, and Television* 16, no. 2 (1996)
Scannell, Paddy, and David Cardiff, *A Social History of British Broadcasting,* Oxford and Cambridge, Massachusetts: Blackwell, 1991
Short, Kenneth R.M., "A Note on BBC Television News and the Munich Crisis, 1938," *Historical Journal of Film, Radio, and Television* 9, no. 2 (1989)
Talbot, Godfrey, *Ten Seconds From Now: A Broadcaster's Story,* London: Hutchinson, 1973
Talbot, Godfrey, *Permission to Speak,* London: Hutchinson, 1976
West, W.J., *Truth Betrayed,* London: Duckworth, 1987

Broadcast Education Association

Serving College and University Faculty

The Broadcast Education Association (BEA), located in Washington, D.C., is a U.S. organization for professors, students, and electronic media professionals who prepare college students to learn more about, and possibly enter, the broadcasting, electronic media, and emerging technologies industries when they graduate. By 2000, the BEA had more than 1,450 individual members (professors, students, and professionals), 250 institutional members (colleges and universities), and more than 85 associate members (associations and companies), as well as several important corporate contributors. The BEA is a 501 (c) 3 not-for-profit higher education association and is primarily funded through membership dues, corporate contributions, industry grants, and publications.

Origins

The association traces its lineage back to 1948, when the University Association for Professional Radio Education (UAPRE) was established with members representing ten colleges and universities. The organization was dissolved in 1955, and a new organization, the Association for Professional Broadcasting Education (APBE), was created. At the APBE's first annual meeting in Chicago in 1956, the organization established the *Journal of Broadcasting,* the first scholarly research periodical about radio and television, which produced its first quarterly issue the following winter. The APBE was established with close ties to the professional broadcasting community through the National Association of Broadcasters (NAB). Membership consisted of academic institutions and NAB member broadcasting stations. The Association's connection to NAB remained very close in the following years. NAB provided an executive secretary, office space, and a substantial yearly cash grant to maintain APBE's operation. The APBE became the Broadcast Education Association in 1973. In 1985 the *Journal of Broadcasting* was renamed the *Journal of Broadcasting and Electronic Media*. The BEA has grown over the years and today includes members from all around the world; it publishes two scholarly journals (it added the *Journal of Radio Studies* in 1998), a quarterly membership magazine, *Feedback,* and issues a variety of student scholarships and holds a vibrant annual national convention.

In its initial years, UAPRE and APBE focused substantially on radio issues and training, just as college and university academic departments did. Published research was largely descriptive and historical, with little focus on the audience and less on research methodology. The emphasis was on educating students for professional careers. The growing focus on television in the 1950s left radio concerns behind. Early annual conventions—well into the 1960s—attracted about 100 faculty and student participants for a day of educational sessions.

As the field matured, so did its research output. NAB and APBE cooperated in a series of annual research grants beginning in 1966. In 1968 the annual convention expanded to two days, and research paper sessions made their appearance, attracting more attendees. By the mid-1970s, articles in the *Journal of Broadcasting and Electronic Media* increasingly reflected more social science research into audience patterns and uses of both radio and television.

The Broadcast Education Association Today

The BEA publishes two respected journals. The *Journal of Broadcasting and Electronic Media* is a quarterly research journal considered to be one of the leading publications in the communication field, with articles about new developments, trends, and research in electronic media. The *Journal of Radio Studies,* officially adopted by BEA in 1998, is published biannually and is the first and only publication exclusively dedicated to industry and academic radio research. Additionally, *Feedback,* a membership publication, appears quarterly with articles on pedagogy and industry analysis and reviews of books and instructional materials.

The Association has a paid staff of two (its first part-time executive secretary, Dr. Harold Niven, began work in 1963; he became a full-time paid president in 1984): an executive director and an assistant to the executive director. BEA is governed by a board of directors comprising mainly electronic media faculty and industry professionals. The BEA holds an annual convention in Las Vegas each spring that spans three days and is attended by more than 1,000 people. The convention is held in the days immediately preceding the NAB convention, and on the last day there are sessions cosponsored by both BEA and NAB.

The Association is made up of divisions representing various areas of interest to members. The BEA administers scholarships, a new faculty research grant, and a dissertation award. The Distinguished Education Service Award recognizes someone who has made a significant and lasting contribution to the American system of electronic media education by virtue of a singular achievement or through continuing service on behalf of electronic media education.

The Association serves as a repository for information about teaching and research through its website. Among the resources available is the "BEA Syllabus Project," in which professors can access sample syllabi, course outlines, and textbook choices for a variety of classes in radio, television, and new media. Another popular feature is the website's listing of academic job openings.

STEVEN D. ANDERSON

See also College Radio; Education about Radio; Intercollegiate Broadcasting System; National Association of Educational Broadcasters

Further Reading

Broadcast Education Association website, <www.beaweb.org>

Journal of Broadcasting and Electronic Media, "35th Anniversary Issue Symposium: Founder and Editor Comments," 35, no. 1 (Winter 1991)

Kittross, John M., "Six Decades of Education for Broadcasting and Counting," *Feedback* 31, no. 3 (Fall 1989)

Kittross, John M., "A History of the BEA," *Feedback* 40, no. 2 (Spring 1999)

Niven, Harold, *Broadcast Education,* Washington, D.C.: APBE.BEA, 1965–86 (7 reports)

Niven, Harold, "Milestones in Broadcast Education," *Feedback* 26 (Summer 1985)

Broadcast Music Incorporated

Broadcast Music Incorporated (BMI) brought competition to the business of music performance rights licensing in the United States. Established in reaction to what was perceived by radio broadcasters as predatory pricing by the American Society of Composers, Authors and Publishers (ASCAP), BMI gradually rose to parity with ASCAP and its songs now dominate the playlists of most contemporary music formats.

Origins

In the years following the 1923 negotiation of its first broadcast performance rights license, ASCAP demanded higher and higher copyright fees from stations for airing the music the public expected to hear. As ASCAP controlled the performance rights to virtually all songs being played by U.S. radio stations, broadcasters believed that they had no choice but to pay the rates ASCAP demanded. But in 1939, faced with the onset of yet another price increase, the broadcasting industry rebelled. Sidney M. Kaye, a young CBS copyright attorney, designed the blueprint for a new licensing agency to be called Broadcast Music Incorporated. As presented to key radio executives in Chicago in the autumn of 1939, broadcasters would, under Kaye's plan, pledge sums equal to 50 percent of their 1937 ASCAP copyright payments as seed money to launch the new organization. In exchange for these payments, participating broadcasters received non-dividend-paying BMI stock (most of which they or their successor companies still hold). On 14 October 1939, BMI's charter as a nonprofit venture was filed, and the agency's offices opened in New York on 15 February 1940.

ASCAP did not take the new effort seriously and soon announced a 100-percent rate increase for 1941 (which would amount to five to ten percent of a station's advertising revenues). In response, 650 broadcasters signed BMI licenses by the end of the year, with only 200 primarily small stations re-signing with ASCAP. Broadcasters who were anxious about what the loss of ASCAP material would do to their programming were encouraged to buy BMI stock by a BMI pamphlet that observed, "The public selects its favorites from the music which it hears and does not miss what it does not hear." On 1 January 1941, the broadcasters' boycott of ASCAP officially began.

Setting up a new rights agency was one thing; acquiring music for it to license was quite another. BMI began life with only eight songs, all of which had been commissioned specifically for its catalog from non-ASCAP composers. While BMI sought to find and sign nonaffiliated writers, radio stations that had turned in their ASCAP licenses had no music to program except these eight tunes and songs with expired copyrights. American radio thus entered the "Jeanie with the Light Brown Hair" era, so named for an incessantly aired public domain tune by 19th-century composer Stephen Foster.

As Foster and folk songs filled the ether, BMI looked for new sources of material to license. The popular works of George Gershwin, Cole Porter, Irving Berlin, and scores of others were all ASCAP-licensed. Music from Britain and the rest of Europe could not be used because foreign composers were members of rights organizations that had signed reciprocal agreements with ASCAP. ASCAP had not entered the South American market in any significant way, however; consequently, the music of Latin America soon came to dominate radio program schedules. The sudden and widespread popularity of sambas, tangos, and rumbas during the early 1940s was thus the result of legal necessity rather than of intrinsic musical merit. Faced with a growing competitive threat from BMI, ASCAP agreed to roll back its rates late in 1941, but it was too late to repair the damage.

ASCAP v. BMI

BMI was now firmly established as a licensing rival. Over the next 15 years, BMI rose to parity with ASCAP principally by signing songwriters that ASCAP had ignored: young mainstream composers rebelling against ASCAP's royalty payout system, which favored more established writers; country-and-western composers from the hinterlands; and later, rock-and-roll songsters who combined black blues and white country stylings into a new, rhythmically pulsating phenomenon. Soon, a number of major publishers such as E.B. Marks and M.M. Cole affiliated with BMI. The organization also advanced seed money to new publishers who agreed to be represented by it. BMI prospered under Kaye, who rose from vice president and general counsel to chairman of the board. He was assisted by Carl Haverlin, a former vice president of the Mutual Broadcasting System who began his BMI career as director of station relations and became its president in 1947.

ASCAP and its select members counterattacked with charges that BMI and the broadcasters were conspiring to promote musical trash. Broadway legend Oscar Hammerstein charged that "BMI songs have been rammed down the public's ears," and other detractors asserted that BMI stood for "Bad Music, Inc." Nevertheless, buoyed by broadcasters' resentment of past ASCAP arrogance and the growing 1950s appeal of the rock-and-roll songwriters whom BMI discovered and nurtured, the new organization came to dominate the radio pop charts.

In 1959 when the payola scandal (illegal payment for record promotion) was fully disclosed, ASCAP sought to make it a BMI issue by maintaining that BMI-dominated rock-and-roll music would never have become popular without under-the-table bribes. With a few high-profile disk jockey firings and the passage of federal anti-payola legislation, the radio industry weathered the storm and so did BMI. The organization further insulated itself against future attacks on the quality of its catalog by broadening its musical base. Within a few years, BMI had signed affiliation agreements with jazz composers such as Thelonius Monk, folk writers such as Pete Seeger, classical icons such as William Schuman, and Broadway mainstays Sheldon Harnick and Jerry Bock.

BMI Today

Nevertheless, as a primarily broadcaster-owned-and-directed enterprise, BMI remains vulnerable to the undocumented charge that it is more sympathetic to broadcaster interests than to those of its affiliated composers and publishers. BMI's 2002 rates, however, were very close to those assessed by ASCAP: 1.605 percent of adjusted net revenue for stations billing more than $150,000 and 1.445 percent for stations billing less than that figure. BMI also offers stations both blanket and per-program license options, as does ASCAP, and negotiates with the radio industry through the Radio Music License Committee (RMLC), whose members are appointed by the National Association of Broadcasters (NAB).

Under their BMI license agreements, radio stations periodically fill out BMI logs listing the music played during a given week. Outlets logging at any particular time are selected as part of a sample designed to reflect all sizes, formats, and geographic locales. This sample is then used to project national usage of individual BMI-licensed tunes, with license fee payments accordingly divided among member composers and publishers.

PETER B. ORLIK

See also American Society of Composers, Authors and Publishers; Copyright

Further Reading

BMI Fiftieth Anniversary History Book, available on the BMI website, <http:www.bmi.com/library/brochures/historybook/index.asp>

Dachs, David, *Anything Goes: The World of Popular Music,* Indianapolis, Indiana: Bobbs-Merrill, 1964

"Fifth Estater—Francis Williams Preston," *Broadcasting* (23 March 1987)

Lathrop, Tad, and Jim Pettigrew, Jr., *This Business of Music Marketing and Promotion,* New York: Billboard Books, 1999

Ryan, John, *The Production of Culture in the Music Industry: The ASCAP-BMI Controversy,* Lanham, Maryland: University Press of America, 1985

Broadcasting Board of Governors

Oversees U.S. International Radio Services

By the early 21st century, more than 100 million listeners, viewers, and internet users around the world tuned to U.S. international broadcasting programs on a weekly basis. Since 1995, the Broadcasting Board of Governors has been the federal entity supervising all these international services. The board developed out of a series of government reorganizations, brought about in part by the end of the Cold War, although listeners to the various radio services probably noticed little change.

Origin

The inception of the Broadcasting Board of Governors came with the International Broadcasting Act (Public Law 103-236), which President Bill Clinton signed on 30 April 1994. The new law established an International Broadcasting Bureau (IBB) within the U.S. Information Agency (USIA). The IBB was designed to administer the formerly separate Voice of America, Worldnet TV and film services, the Office of Cuba Broadcasting (operating Radio and TV Martí), and a supporting Office of Engineering and Technical Services. The formation of the IBB was intended to generate economic savings through greater administrative efficiency. The same act created a president-appointed Broadcasting Board of Governors (BBG), also within the U.S. Information Agency, to exercise jurisdiction over all U.S. government international broadcasting efforts, radio and television. The BBG held its first organizational meeting in early September 1995.

The BBG was designed to oversee IBB operations (such as appointing its director) as well as to supervise the two separate international radio organizations receiving federal funding: Radio Free Europe/Radio Liberty (RFE/RL), and the new Radio Free Asia (RFA), authorized in the same legislation. RFE/RL had been supervised by the Board for International Broadcasting (BIB) for the previous two decades. The IBB was dissolved and plans were made to privatize all aspects of both RFE and RL by the turn of the century (though that did not, in the end, take place).

Operations

The final organizational step came three years later (21 October 1998), when President Clinton approved the Foreign Affairs Reform and Restructuring Act (Public Law 105-277), said by many observers to be the single most important legislation affecting U. S. government international broadcasting in nearly a half century. Under its provisions the USIA was dissolved and the BBG became a fully independent federal agency operation on 1 October 1999. The BBG's eight bipartisan members are appointed by the president (and confirmed by the Senate) and the secretary of state serves as a ninth ex officio member. Early in 2002 the BBG created a wholly new radio service to serve the Middle East, Radio Sawa.

The BBG is intended to act as a "firewall" to protect the professional independence and integrity of the several broadcast services from the political process. It is also authorized to evaluate the mission, operation, and quality of each of the broadcasting activities; to allocate funds among the various broadcast services; to ensure compliance with broadcasting standards (especially with regard to news and public affairs); to determine addition and deletion of language services; and to submit annual reports on its activities (and those of the individual broadcast services) to the president and Congress.

CHRISTOPHER H. STERLING

See also Board for International Broadcasting; International Radio Broadcasting; Radio Free Asia; Radio Free Europe/Radio Liberty; Radio Martí; Radio Sawa/Middle East Radio Network; Voice of America

Further Reading

Board for International Broadcasting, *1995 Annual Report on Radio Free Europe/Radio Liberty, Inc.*, Washington: Government Printing Office, 1995

Broadcasting Board of Governors website, <http://www.bbg.gov/index.cfm> (provides link to annual reports)

Broadcasting House. *See* British Broadcasting Corporation: Broadcasting House

Broadcasting Rating Council. *See* Media Rating Council

Brokerage in Radio

Buying and Selling Stations

Radio station brokers specialize in the buying and selling of radio stations, representing one side or the other in such transactions. As more stations change hands each year, especially in recent years, the role of the broker is an increasingly important one. Commercial radio station licenses in the United States are issued for a finite period, but after each license term there is an expectation of license renewal. Because the expiration of a radio station's license does not usually correspond to the timing of a station's sale, the Federal Communications Commission (FCC) will readily grant a license transfer from a current licensee to a prospective owner, provided that the prospective owner is an acceptable licensee under the FCC ownership rules. The licensee and the prospective owner must submit a request to the FCC for a license transfer.

The assets associated with a station are sold or transferred to another entity either through a conventional sale or through an exchange of assets commonly called a "swap." Just as is true with the sale of any other business, a variety of external or internal events can cause an owner to consider the sale of a station. Externally, a radio station's geographic market or audience may change in a manner that is incompatible with a particular owner's goals. Internally, the particular financial structure that supports a given station may require that the station be "refinanced" in a manner so comprehensive as to require a sale. Other factors that commonly trigger the decision by a station owner to sell include death and consequent estate issues for shareholders, as well as disagreement among principal owners.

Role of Brokers

Radio station sales can be handled by the owners themselves, their attorneys, accountants, small business brokers, investment banking firms, or specialists such as radio station brokers. As the name implies, radio station brokers are industry-specific agents, and, as such, these brokers specialize in representing buyers or sellers of radio stations. After years of specializing in these kinds of transactions, radio station brokers are often also able to assist their clients in refining the future economic projections for a station's operation. One of a broker's main tasks is to properly guide and manage the expectations of his or her clients. Because station brokers are especially familiar with the radio industry, they can often spot unrealistic economic assumptions made by their clients. When a radio station broker is working for a seller, it is his or her responsibility to coordinate efforts with the station's owner, lawyers, and/or accountants to help ensure a desired economic or strategic result. In those instances when radio station brokers work for buyers, the broker's responsibility is to assist the buying principals and their financial advisers in locating and purchasing radio stations that fit the buyers' criteria.

After a definitive agreement is reached between station buyers and sellers, all radio station license transfers must be approved by the FCC. Radio station brokers will usually encourage owners to obtain legal advice from attorneys who are familiar with the execution and submission of the proper forms required by the FCC. Following correct FCC procedure is imperative, because failure to do so can result in severe fines or even in license revocation by the FCC.

Professional radio station brokers attempt the marketing and sale of radio stations so as to create minimum disruption to a station's personnel, revenue, and profitability. This challenge can be difficult to meet. In order for a station to benefit from being sold at the highest price, it is in the seller's best interest that the greatest number of potential buyers be approached; however, the larger the number of buyers contacted, the more likely it is that the employees of the station will learn that the station is being offered for sale. This awareness can create unpleasant instability among the station's staff. Similarly, station advertisers may also learn that the station's ownership is expected to change, and, as a result, the advertisers may be inclined to limit or change the plans for their advertising expenditures in a manner adverse to the station's economic well-being. Radio station brokers are paid to navigate this difficult road.

How Sales Are Made

The normal procedure followed by a station owner who anticipates selling his or her station first includes the choice of a radio station broker or others experienced in the selling of businesses similar to radio stations. Most radio station brokers are known to station owners and are listed in various radio industry publications. Once a broker is selected, a fee structure is negotiated. Fee structures vary depending on the nature and anticipated price for the property being sold. Most frequently, brokerage fees range from 6 percent to as low as 1 percent of the sale price. The resulting percentage is related to the size of the transaction, with the larger transactions paying lower percentages to the brokers. The seller should confer with the broker and with various advisers in setting an asking price for the radio station, because a wide array of factors must be considered in the price-setting process. Pricing considerations should include data from comparable sales, past economic performance of the specific assets being sold, and the anticipated future earnings performance for the assets.

Most sellers instruct their brokers to secure assurances of confidentiality from the prospects being approached during the sales process. Such assurances are often contained within a confidentiality agreement that is signed by potential buyers before they are given specific information with respect to a purchase opportunity. Potential buyers are furnished with certain information by the radio station broker about the station being offered for sale, commonly referred to as a "book." The book usually contains general information about the station, economic facts pertaining to the market being served, the station's competition, its audience, and its historical financial performance. The prospective buyer will review the book and based on its information will prepare various financial projections with respect to what the buyer feels the station may earn for its owners over a future period of time. Such future projections are called "pro forma estimates," and each may contain a different set of assumptions with regard to items such as competition, ratings, and revenue. Each buyer typically has his or her own set of pro forma objectives and will measure the relative attractiveness of each acquisition opportunity against these objectives.

Once a buyer becomes relatively comfortable with the material he or she has reviewed, the buyer may seek to enter into a written agreement with the licensee. This document is typically called a "letter of intent." The letter of intent usually sets forth various terms and conditions under which the buyer will proceed. This agreement also sets forth the intent of the buyer with respect to confidentiality, pricing, and timing of the contemplated transaction. The letter of intent typically also includes agreement on the procedure and responsibility for the preparation and negotiation of a definitive purchase and sale agreement to be used in the sale. The letter of intent will frequently provide the buyer with an *exclusive* period of time during which time only this buyer or his or her agents can conduct a thorough investigation of the various factors influencing the station's operation. This period is commonly referred to as the buyer's opportunity to conduct "due diligence." Either at the conclusion of such investigation or at the same time such investigation is progressing, the buyer and seller frequently agree to move cooperatively toward the formulation of a definitive purchase and sale agreement. Sometimes, for various reasons, the seller and buyer eliminate the step that involves a letter of intent and instead move directly to a definitive purchase and sale agreement.

There are a number of factors influencing a buyer's and a seller's decision on whether or not to include a letter of intent in the purchase process. Among the consideration for sellers is whether or not they wish to "encumber" their flexibility in negotiating the sale of the station with other potential buyers during the time a letter of intent is in force. Sellers are also frequently concerned that, notwithstanding an agreement as to confidentiality, word of the possible transaction might "leak" during the period that the station is under a letter of intent. Included in the decision process regarding letters of intent for buyers is whether or not a buyer wishes to expend the money and effort to perform due diligence and to continue contemplating the purchase of a specific station, without any firm rights to actually compel a sale of the station to this particular buyer.

Once a definitive purchase and sale agreement is executed, it is filed for consideration with the FCC. The FCC review process includes an opportunity for the public, the FCC, and other governmental agencies to register any objections to the license transfer. If there are no objections, the FCC will typically render its "preliminary" approval within a generally predictable number of days. Thereafter, there is an additional period of time before the FCC approval automatically becomes a "Final Order." The closing on a station's sale transaction usually takes place within a reasonably short period of time following issuance of the Final Order.

There are many strategies that drive the desire to purchase or sell a particular radio station. The radio station broker becomes conversant with the client's plans with respect to economic goals. FCC legal limitations on station ownership, as well as Department of Justice considerations with respect to market dominance leading to unfair competition, are among the factors that constrain buying and selling strategies. Informed radio station brokers assist their clients in conceptualizing and implementing their acquisition or exit strategies.

Sales Trends

In the 38 years from 1954 through 1992, FCC files indicate that nearly 20,000 radio stations changed ownership—some of them several times. The volume of radio station sales exploded with passage of the Telecommunications Reform Act of 1996. The sudden heated demand for the ownership of radio station "clusters" occurred simultaneous with an extremely robust public stock market, which provided large amounts of investment capital to those companies that were able to take advantage of an unprecedented opportunity to rapidly amass a large number of radio stations. In the four-year period following passage of the act, ownership of 7,839 radio stations changed hands, with well-capitalized radio companies emerging as highly acquisitive in markets of all sizes. As these clustered acquisitions continued, the single- or two-station owners came under increased competitive pressure. In 1996 nearly 21 percent, or 2,157 stations, changed hands. In 1997 a similar number, 2,250 stations, were sold. In 1998 the number of stations changing hands went down only slightly, to 1,740 stations. Consolidation of radio station ownership extended into even the smaller markets as the beneficial economics of consolidation were clearly established. In 1997 the average station sold for about $8 million; this price dropped to about $5 million in 1998, as stations in the smaller markets began to represent more of the sales.

JACK MINKOW

See also Consultants; Licensing; Ownership, Mergers, and Acquisition

Further Reading

Krasnow, Erwin G., *The Politics of Broadcast Regulation*, 3rd edition, New York: St. Martin's Press, 1982

Vogel, Harold L., *Entertainment Industry Economics: A Guide for Financial Analysis*, 5th edition, Cambridge and New York: Cambridge University Press, 2001

Brown, Himan 1910–

U.S. Radio Producer

Himan Brown directed and produced some of the most memorable dramas in radio history. His opening signatures are classics: the steam train of *Grand Central Station,* the urgent "calling all cars" and sirens of *Dick Tracy,* and the creaking door of *Inner Sanctum* and *CBS Radio Mystery Theater.* The latter sound effect was the first sound to be trademarked. Brown's longevity in the business is remarkable: his program creation and directing credits span eight decades (1920s–90s), and he has directed more than 30,000 shows.

Brown was born in Brooklyn, New York, in 1910 to poor parents who had immigrated from Odessa (in what is now Ukraine). Brown could not speak English when he started school; he remained fluent in Yiddish throughout his life. He became involved in theater as a teenager, performing at the Brooklyn Jewish Center under the direction of Moss Hart. He also assisted with his father's dress contracting business throughout his school years, which culminated in a law degree from Brooklyn College in 1931 when he was just 21.

Like many who became successful in the embryonic days of radio, Brown's chutzpah played a role. In 1927 he convinced the licensee of WRNY at the Roosevelt Hotel to let him read poetry (billed as *Hi Brow Readings*). He then landed an audition at the National Broadcasting Company (NBC), where his Jewish dialect characterizations, honed on the Borscht Circuit and drawn from Milt Gross' cartoons, resulted in a month-long stint on the network. First, however, he had to secure, through determined, repeated visits, Gross' permission to use the work.

Gertrude Berg, creator of *The Goldbergs,* heard Brown's performance and asked for help in selling her concept. He sold the series, first called *The Rise of the Goldbergs,* to NBC and played the part of Jake Goldberg in early episodes. Brown later claimed that Berg pushed him out of the show, abrogating their joint partnership agreement. The experience soured Brown but taught him that his forte was program packaging and sales.

Functioning as an independent packager during an era in which advertising agencies predominated, Brown matched sponsors with program ideas. He acted in, cast, and directed his earliest serials, *Little Italy* and *Bronx Marriage Bureau,* which were seasonal runs for sponsors Blue Coal and Goodman's Matzos.

Brown began working in the early 1930s with Anne Ashenhurst, who, with Frank Hummert—later her husband—produced nearly half of the woman-oriented serials in the mid-1930s. Brown produced and directed the Hummert serials *Marie, the Little French Princess; David Harum; John's Other Wife;* and *Way Down East.* He often directed as many as four productions daily.

Some of Brown's shows became well known for the premiums they offered. *David Harum* gave away a horse every week during one promotion. Listeners were also asked to suggest names for Harum's horse; 400,000 suggestions poured in—all attached to sponsor Bab-O's labels. The program was also notable because it was one of the first in which the main character, rather than the announcer, pitched the product.

Brown gave many talented people their first break in radio: the actress Agnes Morehead, writer Irwin Shaw (who wrote *Dick Tracy* and *The Gumps* for two years), lyricist and composer Frank Loesser, and others.

His works were sometimes criticized for being too sexy (*The Thin Man*), too violent (*Dick Tracy*), or too scary (*Inner Sanctum*). *Inner Sanctum* was especially popular, ranking in the top 20 shows for more than 10 years.

During World War II, he worked with the Office of War Information and the Writers War Board. He integrated conservation and war bond appeals into his programs (such as the children's pledge to save paper on *Terry and the Pirates:* "Turn in every scrap you can, to lick the Nazis and Japan") and produced patriotic home-front serials such as *Green Valley, U.S.A.* Before America even entered the war, Brown coproduced a series called *Main Street, U.S.A.* to dramatize the fascist threat.

Brown was blacklisted in *Red Channels* in 1950, along with many notables in the broadcast industry. Years later, he testified in John Henry Faulk's libel trial that he himself had been pressured to drop a blacklisted cast member.

Brown produced public service films (such as *A Morning for Jimmy* for the Urban League), televised mystery shows (the syndicated *Inner Sanctum Mysteries*), and a handful of movies in the 1950 and 1960s. He also produced several star-studded televised specials called *The Stars Salute* to raise money for the Federation of Jewish Philanthropies. He directed spectacular Hanukkah festivals held in Madison Square Garden for 18 years running, raising $500 million in bonds for the new nation of Israel.

His signature series, *CBS Radio Mystery Theater,* debuted in 1974. Brown received more than 100,000 letters when he called for early signs of support for the show. Recognized in 1975 with a Peabody Award, the program ran seven days a week for nine years—an amazing record of almost 1,500 original episodes. At its apex, it aired on 350 stations and drew an audience of 5 million—in an era in which radio drama was thought to be extinct.

Himan Brown directing
Courtesy CBS Photo Archive

In 1977, Brown created the children's program *The General Mills Adventure Theater,* dramas commended by the National Education Association for invigorating student interest in literature. Brown's later works (for example, *Americans All*) dramatized biographies of famous Americans for Voice of America. He donates his time to such projects, and his foundation, the Radio Drama Network, often finances production costs, working in conjunction with universities and groups such as the Freedom Forum.

Brown is a generous philanthropist and an avid art collector. Still going strong at the beginning of the new millennium, he participated in radio drama workshops around the country, directed live dramas, and worked tirelessly to revive radio drama.

PATRICIA JOYNER PRIEST

See also Blacklisting; Faulk, John Henry; Goldbergs; Inner Sanctum Mysteries; Jewish Radio; Red Channels; Stereotypes

Himan Brown. Born in Brooklyn, New York, 21 July 1910. B.A. in Liberal Arts, Brooklyn College, and L.L.B., Brooklyn Law School, jointly awarded in 1931; radio debut in *Hi Brow Readings,* 1927. Recipient: Mystery Writers of America Raven Award for *Inner Sanctum,* 1948; Radio and Television Editors' Director of the Year Award, 1958; American Federation of Television and Radio Artists commendation, Writers Guild Award, Mystery Writers of America Raven Award, and Peabody Award, 1974, for *CBS Radio Mystery Theater*; Public Relations Society of America/New York's Big Apple Award for work on behalf of the United Jewish Appeal/Federation of New York, 1994; inducted into Emerson Radio Hall of Fame, 1988, and Museum of Broadcast Communication's Radio Hall of Fame, 1995; designated American Broadcast Pioneer, 1997.

Radio Series

1927	*The Hi Brow Readings; Milt Gross' Nize Baby Readings*
1929–34, 1936–45, 1949–50	*The Rise of the Goldbergs* (later titled *The Goldbergs*)
1931, 1934–37	*The Gumps*
1932–34	*The Bronx Marriage Bureau*
1933–34	*Little Italy*
1933	*Jack Dempsey's Gymnasium*
1933–35	*Marie, the Little French Princess*
1934–35	*Peggy's Doctor*
1934–39, 1943–48	*Dick Tracy*
1935–36	*Flash Gordon*
1935	*Captain Tom's Log*
1936–37	*Way Down East*
1936–42	*John's Other Wife*
1936–51	*David Harum*
1936–38	*Thatcher Colt Mysteries*
1937–39, 1941–48	*Terry and the Pirates*
1937–42, 1944–54	*Grand Central Station*
1937	*Dr. Friendly*
1938	*Main Street, U.S.A.*
1938–48, 1951–52, 1955	*Joyce Jordan, Girl Interne* (became *Joyce Jordan, M.D.*, in 1942)
1938–40	*Your Family and Mine*
1939–40	*Hilda Hope, M.D.*
1941	*City Desk*
1941–50	*The Adventures of the Thin Man*
1941–52	*Inner Sanctum Mysteries*
1941–49, 1954	*Bulldog Drummond*
1941–44, 1948–49, 1951	*Philip Morris Playhouse*
1942–44	*Green Valley, U.S.A.*
1943–44, 1946, 1950–51	*The Adventures of Nero Wolfe*
1947	*International Airport*

1949–53	*The Affairs of Peter Salem*
1951	*The Private Files of Rex Saunders*
1951–55	*Barrie Craig, Confidential Investigator*
1955–56, 1959–60	*The NBC Radio Theater (Morning Matinee)*
1974–82	*CBS Radio Mystery Theater*
1977–78	*The General Mills Adventure Theater*
1984	*Americans All*
1988	*We, the People*
1990	*A More Perfect Union*
1998–99	*They Were Giants*
1999	*They Made Headlines*

Television
Lights Out, 1949–52; *Inner Sanctum Mysteries,* 1954; *His Honor, Homer Bell,* 1955; *The Chevy Mystery Show,* 1960; *The Stars Salute,* 1960–66

Films
That Night, 1957; *The Violators,* 1957

Selected Publications
Strange Tales from CBS Radio Mystery Theater (editor), 1973

Further Reading

Dunning, John, *On the Air: The Encyclopedia of Old-Time Radio,* New York: Oxford University Press, 1998

Hickerson, Jay, *The Ultimate History of Network Radio Programming and Guide to All Circulating Shows,* Hamden, Connecticut: Hickerson, 1992; 3rd edition, as *The New, Revised, Ultimate History of Network Radio Programming and Guide to All Circulating Shows,* 1996

Nachman, Gerald, *Raised on Radio,* New York: Pantheon Books, 1998

Payton, Gordon, and Martin Grams, Jr., *The CBS Radio Mystery Theater: An Episode Guide and Handbook to Nine Years of Broadcasting, 1974–1982,* Jefferson, North Carolina: McFarland, 1999

Skutch, Ira, editor, *Five Directors: The Golden Years of Radio,* Lanham, Maryland: Scarecrow Press, and Los Angeles: Directors Guild of America, 1998

Stedman, Raymond William, *The Serials: Suspense and Drama by Installment,* Norman: University of Oklahoma Press, 1971; 2nd edition, 1977

"Three Witnesses to Faulk Suit Link Store Operator to 'Blacklisting,'" *New York Times* (18 May 1962)

C

Cable Radio

Cable radio is a program service offered by a cable television system. Usually providing many different talk and music program types, the service is typically offered as an extra feature to television cable subscribers.

Technology

Technically, cable radio is relatively easy to offer. At a cable system head-end, one or more FM antennae are aimed toward stations that can be received by the cable operator. In the past, most such systems used an all-band approach. That is, local stations across the entire FM radio spectrum (88 to 108 MHz) are received, amplified, and carried on the cable system.

To receive cable radio, a signal splitter at the subscriber's location provides a second connection for the FM tuner. Although many systems charge an extra fee for the service (most are low—$1 or $2 per month), seldom are security measures taken that would require payment before the cable radio feed could be used.

With the introduction of multiple channels on such premium services as Home Box Office (HBO), cable operators often use cable channel space to carry the stereo audio signal of such services to subscribers' homes. Audio from the satellite is fed to an FM modulator located on a locally unused FM channel. Before availability of television sets that supported stereo audio, such schemes were popular among television fans.

Economics

Just as television superstations developed with a national cable television audience in mind, so have a few radio superstations. Classical station WFMT in Chicago is a notable pioneer in this area. Some cable systems even allow local FM signal origination. This is usually done in conjunction with a nearby college or university. For example, WDBS, a long-time "closed carrier current" station on the campus of the University of Illinois, (Champaign-Urbana) has a spot on the local cable system. Additionally, such local origination is often carried on the audio carrier of locally originated television channels.

The relative ease of hooking home receivers into cable radio service has discouraged its active promotion by cable systems. Subscribers soon learn that they can hook up their cable through an easily purchased splitter and not have to pay a monthly subscriber fee. A few cable systems attempt to eliminate this theft of services either by trapping the range of frequencies through a filter or by using a cable audio converter to shift the service first to an unused frequency range and then to convert it back once the subscriber pays to rent the necessary equipment.

Policy

As cable delivery grew—from only 70 U.S. communities in 1950 to more than 32,000 communities in 1995—over-the-air broadcasters came to believe that they were being denied potential revenue from cable operators. Provisions of the Communications Act of 1934 require that a station that desires to rebroadcast the signal of another outlet must first obtain permission from the originating station. As amended, the Communications Act now prohibits cable operators (and other multichannel video program distributors) from retransmitting commercial television, low-power television, or radio broadcast signals without first obtaining consent.

In mid-1993, faced with the daunting task under new rules of obtaining permission from every FM station within a 57-mile radius of their receiving antennae, many cable systems curtailed their cable radio offerings. Only the locally originated channels, generally stereo audio for a few premium services, remain on the systems.

In 1982 National Public Radio (NPR) commissioned a study on the future of cable audio as a possible revenue stream to support other NPR operations. The report concluded that the future of cable audio, although bright, would only be profitable if such services generated revenue for both the cable operator and, of course, NPR. Three models were proposed. The first relied on advertiser support, very much like traditional over-the-air radio stations. Another model suggested that cable operators might be willing to pay for audio services as long as they could sell them as premium services with an appropriate profit margin built in. The model used for public television, where services are supported by corporate sponsorships and individual donations, was also suggested. In the NPR report, an important element for the success of cable audio was the restriction of access to services through secure channels. The report also noted the superior quality of the processed FM signals it proposed to deliver, as compared to the signals of the all-band FM approach. At the time of the NPR report, digital audio, although technically feasible, was not in wide use because it was cost-prohibitive. And, of course, audio streaming on the internet was years into the future.

The Future

Although NPR provided the vision, it took commercial interests and a breakthrough in technology to actually capitalize on the concept of cable audio services. Especially with the introduction of digital, multichannel, CD-quality audio streams such as those in the Digital Music Express (DMX) and Music Choice (formerly Digital Cable Radio) services—services not available over the air—a small but eager audience signed up for service. The set-top digital converter box is similar to that used for pay-per-view video events, and it is addressable. In recent years, American Telephone and Telegraph (AT&T) cable services, among others, have included a variety of such audio services as part of their tier of digitally transmitted services.

Although it appears that some form of cable radio will continue into the future, according to Dwight Brooks, a contributing author to textbooks on broadcast programming, cable operators are skeptical about growth for this medium, citing a figure of only 15 percent penetration among basic cable subscribers for audio services. With proper copyright clearance, such services are being effectively marketed to business locations to provide background music services.

The internet has the ability to translate the essence of cable radio into a viable service. However, broadband connections are required to achieve similar fidelity, and the introduction of the Digital Millennium Copyright Act (DMCA) has had a chilling effect on internet radio by imposing significant fees for internet transmission of copyrighted material.

A variation on the pay cable radio approach proposed by NPR is now available to consumers via direct satellite connection. The XM Satellite Radio Service and Sirius Satellite Radio Network deliver dozens of channels of audio programming directly to consumers who subscribe to their services and who have purchased a proprietary receiver. Some XM services include commercials; Sirius programming is entirely commercial free and carries a slightly higher subscription fee.

JIM GRUBBS

See also Digital Audio Broadcasting; Digital Satellite Radio; Internet Radio

Further Reading

Bartlett, Eugene R., *Cable Television Handbook,* New York: McGraw Hill, 2000

Crotts, G. Gail, Joshua Noah Koenig, Richard Moss, and Ann Stookey, *Listening to the Future: Cable Audio in the 80s,* Washington, D.C.: National Public Radio, 1982

Eastman, Susan Tyler, Sydney W. Head, and Lewis Klein, *Broadcast Programming, Strategies for Winning Television and Radio Audiences,* Belmont, California: Wadsworth, 1981; 5th edition, as *Broadcast/Cable Programming: Strategies and Practices,* by Eastman and Douglas A. Ferguson, 1997

Hollowell, Mary Louise, editor, *The Cable/Broadband Communications Book, vol. 2, 1980–1981,* White Plains, New York: Knowledge Industry, 1980

Vane, Edwin T., and Lynne S. Gross, *Programming for TV, Radio, and Cable,* Boston: Focal Press, 1994

Call Letters

WJCU. KCBS. WRR. Unique combinations of alphabetic letters such as these, known as *call letters,* are used to identify individual radio (and television) stations. In addition to when they sign on or off, broadcast stations must give an identification announcement each hour—near the top of the hour and during a natural break in their programming. Radio stations give an aural identification, usually an announcer voicing the information, but sometimes a station jingle or musical identification.

According to FCC regulations, legal station identification consists of the station's call letters followed by the location of the station. Nothing can be placed between the call letters and the city of license, with the exception of the name of the licensee and/or the station's frequency or channel number. Station identification regulations (Section 73.1201) are found in the *Code of Federal Regulations,* Title 47, Part 73, Subpart H—"Rules Applicable to All Broadcast Stations."

Current policy assigns call letters east of the Mississippi River with a beginning *W* and those west of the Mississippi with a *K*. All modern call signs consist of the appropriate beginning letter plus three additional letters, and they can have a suffix, such as *-AM* or *-FM,* to denote the actual type of radio station. At one time, the FCC would not release objectionable call letter combinations; even the "mild" *SEX* combination was withheld. However, during the deregulatory 1980s, the FCC became less concerned about this and deferred to the courts in disputes regarding call signs that might be objectionable or too similar to another station's.

In the late 1990s the procedures regarding the designation of call letters were altered when the FCC replaced the existing manual system with an on-line system for electronic submission of requests for new or modified call signs. Through the FCC's website (www.fcc.gov), stations can determine the availability of call letters, request specific call letters or modify an existing call sign, and determine and submit the appropriate fees.

Historical Origins

The concept of radio station identification has its roots in the maritime industry, for which an International Code of Signals noted in the 1850s that signal flags, which included letters, were to be used to identify vessels. As radio, or rather wireless, developed in the late 1890s and early 1900s, telegraph operators used informal, one- or two-letter call signs as a condensed way to identify their stations. The 1906 Berlin International Wireless Telegraph Convention attempted to formalize a system of three-letter call signs, but at the time there was little cooperation. Individual wireless operators or wireless companies merely chose their own identification, which often consisted of one or two letters or a combination of letters and numbers with little consideration for duplicate calls.

The 1912 London International Radiotelegraphic Convention continued to formalize a system of station identification that was the beginning of the *K* and *W* series assigned to U.S. ships; other letters were assigned to vessels from other nations. The Radio Act of 1912 gave responsibility for licensing of U.S. ships and shore radio stations to the Bureau of Navigation in the Department of Commerce. Call signs were designated as a three-letter random sequence, with *K* calls for the west and *W* calls for the east. What would become early "radio stations" actually fell under the status of Amateur and Special Land Stations, which had a different call-sign system. Nine Radio Inspection Districts were established, and call letters were assigned with the District Number plus two alphabetic letters, such as 6XE, 9XM, or 8MK.

As more and more stations went on the air, the international agreements of 1912 were employed for all stations, and many pioneer radio stations were assigned three-letter *K* or *W* call letters. The dividing line for *K* and *W* stations was originally the eastern state boundaries of New Mexico, Colorado, Wyoming, and Montana; however, this was moved to the Mississippi River in early 1923. Existing stations were allowed to keep their previously assigned call letters. Because of this change and a few quirky assignments, some pioneer stations do not follow the current *K/W* demarcation, notably KDKA in Pittsburgh, KYW in Philadelphia, and WOW in Omaha. The move to four-letter call signs took place in the early 1920s as the number of radio stations coming on the air escalated rapidly and additional call letters were needed.

Call Letters Used to Promote Station Image

Although early call letter combinations were merely random assignments, many modern call signs have been carefully chosen and have a specific context for the particular station. In fact, many stations trademark their call signs. In addition to being the legal identification for a radio station, call letters have become an important artistic or imaging statement used to help market the station. From WAAA (Winston-Salem, North Carolina) to KZZZ (Bullhead City, Arizona), stations have tried to dream up memorable call signs. Even a casual examination of radio call letters will reveal several categories that these station identifiers fall into.

Many stations use their call letters to recognize a current or past station owner or licensee. KABC (Los Angeles) and WCBS (New York) denote the network organization associated with each station. Chicago's WGN stands for "World's Greatest Newspaper," which in turn refers to *The Chicago Tribune* and the station owner, the Tribune Company. KLBJ (AM) and (FM) licensee, The LBJS Broadcasting Company in Austin, Texas, recognizes owner and former first lady Lady Bird Johnson.

Besides the station licensee, a station's format offers a logical reason to request a certain set of call letters. For example, WINS (New York) stands for the basic programming elements of "information, news, and sports." Just as WJZZ (Roswell, Georgia; Smooth Jazz) and WHTZ (Newark, New Jersey; Top 40/Hits) readily describe music formats, WFAN (New York) is the monogram for an all-sports station. WGOD (Charlotte Amalie, Virgin Islands) makes it pretty clear it's a religious station; however, you need to know that WBFC (Stanton, Kentucky) stands for "We Broadcast For Christ." And if you really just don't want to bother with a format description, you could be like WGR (Buffalo, New York) and be the "World's Greatest Radio" station.

From dogs (WDOG, Allendale, South Carolina) and cats (KCAT, Pine Bluff, Arkansas) to frogs (WFRG, Utica, New York) and pigs (KPIG, Freedom, California), station call letters that denote animals are quite common. Even less ordinary beasts make an appearance with WFOX (Gainsville, Georgia); Chandler, Arizona's camel, KMLE; and KEGL, the Eagle, in Fort Worth, Texas. Animal-based call signs are not only memorable, but they make it easy for the station to add an appropriate mascot to their marketing efforts. Even the lowly WORM (Savannah, Tennessee) is accounted for, and the human species isn't left out either, with KMAN (Manhattan, Kansas); KBOY (Medford, Oregon), and WGRL (Noblesville, Indiana).

A station's location—either its city of license or its frequency—has been a prevalent theme for clever call letters. WARE—found in Ware, Massachusetts—is the only current set of call letters that is exactly the same as the city of license. For a number of years WACO in Waco, Texas, was another, but radio station WACO is now KKTK (although there is still a WACO-FM in Waco). Stations in cities of more than four letters have had to settle for using just the first few letters, so we find WPRO in Providence, Rhode Island; WORC in Worcester, Massachusetts; KSTP in St. Paul, Minnesota; and KSL in Salt Lake City, Utah. AM stations using frequency-based call signs, especially at the upper end of the band, include WTOP (Washington, D.C.) near the "top" of the dial at 1500 and WXVI in Montgomery, Alabama, at 1600. KIOI (San Francisco) is found at 101.3 FM, and near the end of the FM band at 106.5 is KEND in Roswell, New Mexico.

Many radio stations request call letters that help define a characteristic of the locale where the station is found. Pioneer station WSB in Atlanta stands for "Welcome South, Brother." KABL refers to San Francisco's cable car; KSPD to Boise, Idaho's potato or "spud"; and in what better market than Detroit would you find station WCAR? Cow country territory gives us KATL (Miles City, Montana), WCOW (Sparta, Wisconsin), and KMOO (Mineola, Texas). You could also do a weather forecast with call signs—from WSUN (Tampa, Florida) and WSNO (Barre, Vermont) to KICY (Nome, Alaska) and KFOG (San Francisco). There's also WWET (Valdosta, Georgia), KDRY (Alamo Heights, Texas), and WIND (Chicago).

Finally, there is another group of call signs that are colorful because the sound or spelling of the letter combination is memorable. For example, there is a WHAK (Rogers City, Michigan), a WHAM (Rochester, New York), and a WOMP (Bellaire, Ohio), as well as a KRAK (Hesperia, California), a KICK (Palmyra, Missouri), and a KPOW (Powell Wyoming). Broadcast journalists will be pleased to learn there is a WHO (Des Moines, Iowa), a WHAT (Philadelphia), a WHEN (Syracuse, New York), a WHER (Heidelberg, Mississippi), and a WHYY (Philadelphia).

Maybe all this call letter image information is making you think WOW (Omaha) and WWEE (McMinnville, Tennessee), but there are many more creative call signs yet to be devised. With a *K* or *W* combined with three other alphabetic letters, there are over 35,000 unique call letter combinations possible, which is almost three times as many as there are current radio stations.

DAVID E. REESE

See also Frequency Allocation; Licensing

Further Reading

Archer, Gleason Leonard, *History of Radio to 1926,* New York: American Historical Society, 1938; reprint, New York: Arno Press, 1971

Kahn, Frank J., editor, *Documents of American Broadcasting,* New York: Appleton-Century-Crofts, 1968; 4th edition, Englewood Cliffs, New Jersey: Prentice Hall, 1984

Mishkind, Barry, "A Pause for Station Identification," *Radio World* (30 September 1998 and 28 October 1998)

Peterson, Alan, "WILD, WAKY, KRZY Call Letter Combos," *Radio World* (27 December 1995)

Stark, Phyllis, "Stations Spell Out Tradition," *Billboard* (13 March 1993)

United States Callsign Policies, <www.earlyradiohistory.us/recap.htm>

Can You Top This?

U.S. Comedy Panel Program

Perhaps not believable in an era of fast-changing television program tastes, this simple half-hour (15 minutes in its final NBC season) panel program of three men telling jokes lasted nearly 15 years on network radio. The title came from the attempts of the joke tellers to "top" the previous joke and get a louder measured laugh from a studio audience.

Known as the "Knights of the Clown Table," the program's three starring personalities all shared great joke-telling memories and abilities. Ed Ford had been given the title of "Senator" at a political gathering some years previous (Ford also produced and owned the program); Harry Hershfield was already a well-known cartoonist and after-dinner speaker; and Joe Laurie, Jr., had knocked around vaudeville and other jobs before eventually migrating to radio. Ford was said to be the hardest man to get to crack a smile. Radio program authority John Dunning reports that between the three of them, they probably knew something like 15,000 jokes. And all three (plus joke teller Peter Donald) could and did employ a variety of funny dialects and odd-ball characters.

And indeed, the program did not thrive on originality; many of the jokes used were old. To tie the program to its listeners, the audience was encouraged to send in their best jokes (for which they received $10 for each one used on the air) to be told on the air by joke teller Peter Donald. These were followed by the panelists telling their own jokes in the same vein. Audience applause was judged on a score of from one to a thousand by a "laugh/applause meter" displayed so the panel and studio audience could see it. The joke getting the loudest response (the most decibels on the meter) won. Listeners could win up to $25 if their joke was not successfully topped by the panel.

The series later transferred to television, for five months on ABC (1950–51) and then as a syndicated series two decades later, hosted by Wink Martindale and later Dennis James. The radio series was inducted into the Radio Hall of Fame in 1989.

CHRISTOPHER H. STERLING

See also Comedy; Radio Hall of Fame

Cast

Jokesters	"Senator" Ed Ford, Harry Hershfield, Joe Laurie, Jr.
Host	Ward Wilson
Joke-Teller	Peter Donald
Announcer	Charles Stark

Programming History

WOR, New York	1940–1945
NBC	1942–48
Mutual	1948–50
ABC	1950–51
NBC	1953–54

Further Reading

Dunning, John, *Tune in Yesterday: The Ultimate Encyclopedia of Old-Time Radio, 1925–1976,* Englewood Cliffs, New Jersey: Prentice-Hall, 1976; revised edition, as *On the Air: The Encyclopedia of Old-Time Radio,* New York: Oxford University Press, 1998

Ford, Edward Hastings, Harry Hershfield, and Joe Laurie, Jr., *Can You Top This?* New York: Grosset and Dunlap, 1945

Ford, Edward Hastings, Harry Hershfield, and Joe Laurie, Jr., *Cream of the Crop: The New Can You Top This? Laugh Roundup,* New York: Grosset and Dunlap, 1947

Canada

Canadian radio history offers a case study of the development of a communications medium in a country with a very large territory but a relatively small (and linguistically divided) population. That the Canadian federal government has always been interested in fostering the growth of a technology with such potential to draw Canadians together is not surprising. But the question of how best to accomplish that goal has been constantly debated, particularly given the enormous challenge posed by Canada's contiguity to the United States, a world leader in broadcasting from the earliest days. The history of

Canadian radio is the story of the creation of a mixed public/private system mandated to fulfill certain national goals, constantly struggling against both economic and social forces favoring continentalism.

Early Growth and Regulation

As in other countries, wireless telegraphy and telephony (radio) developed in Canada in the early 20th century as an experimental technology for ship-to-shore and other point-to-point communications. Guglielmo Marconi's first wireless transmission across the Atlantic Ocean was received in St. John's, Newfoundland, in late 1901. The first broadcast of music by wireless was originated by Canadian-born Reginald Fessenden, an employee of the National Electric Signalling Company, from Brant Rock, Massachusetts, in December 1906. As these milestones indicate, the evolution of radio in Canada was part of an international undertaking, instituted mainly by electrical companies, that progressed rapidly in the first two decades of the 20th century. By the beginning of World War I, there were also in Canada, as in other countries, a number of amateurs ("hams"), mainly boys and young men, experimenting with home-built crystal sets and transmitters. During the war, the technology was further developed for military purposes, and many soldiers and aviators, in Canada as elsewhere, learned how to use radio equipment.

The first broadcast in Canada, on 20 May 1920, was an experiment conducted by the Marconi Company of Canada, sending a concert by a soprano soloist from the company's Montreal laboratory to a listening audience of distinguished members of the Royal Society of Canada over 100 miles away in Ottawa. Subsequently, Marconi engineers demonstrated their equipment at exhibitions and trade shows, and by the winter of 1920–21, the company was broadcasting two hours a week of musical programming under their experimental license XWA (later CFCF) Montreal. As interest mounted and other companies began marketing their radio equipment this way, more experimental licenses were issued, until in April 1922 the federal government began licensing stations dedicated specifically to private commercial broadcasting.

Canadian radio was regulated from 1905 by the Radio Branch of the Department of Marine and Fisheries, for reasons of national security and to prevent interference between transmitters. Once broadcasting began, the Branch's policy was to encourage private enterprises to build radio stations, in the belief that this was the only way to provide service quickly to Canada's large territory and scattered population. Moreover, as in the United States, other predecessor technologies such as the telegraph, telephone, and undersea cable systems were privately owned. Canada thus opted for competition in radio development, choosing not to follow the British model of monopoly.

About 80 radio stations were set up in Canada in the 1920s, mainly by electrical companies, newspapers, and retailers. Most of the stations were small and struggled to survive on the still-meager income from advertising (which was allowed after 1923). Two national networks created in the late 1920s, one operated by the government-owned Canadian National Railways and the other privately owned (the Trans-Canada Broadcasting Company), also struggled to make a go of providing service to such a far-flung territory.

The financial difficulties of the early broadcasters were exacerbated by the fact that Canada lacked large electrical manufacturing companies; its electrical industry since the late 19th century had largely been a branch-plant operation of American firms. Companies such as General Electric and Westinghouse set up factories in Canada to circumvent Canadian tariff walls, but they did not set up radio stations because Canadians could hear their powerful flagship stations in the northern United States. As Canadian listeners became more demanding of high-quality programming (partly because of their familiarity with American offerings), Canadian stations fell further behind in their ability to offer comparable service profitably.

By 1931 approximately one in three Canadian homes had radio receivers. Radio ownership tended to cluster in cities and in the more prosperous areas. The largely French-speaking province of Quebec lagged in radio ownership, partly because of poverty but more importantly because there were few French-language stations of quality (the exception being CKAC Montreal, owned by *La Presse* newspaper). Although musical programs from English-language and American stations held some appeal, the reluctance of the Quebecois to purchase radios is understandable from this perspective.

Like broadcasting stations, radio receivers were also licensed by the Radio Branch. Owners were required to purchase a license annually, the cost of which ranged from $1 in the early years to $2.50 in the 1940s. This policy, instituted to enable officials to track down sets causing interference, also provided the Radio Branch with most of its income. Many radio owners evaded paying the license fee, however, complaining that it was unfair that they had to pay this tariff when their American neighbors got their radio "free."

Creating a Public Broadcasting Body

By 1930 four Canadian stations had become affiliates of the U.S. networks, because affiliation provided them with popular programs that attracted lucrative advertising. This situation raised alarm bells in some nationalist cultural and political circles in Canada. Although Canadians have for the most part accepted the liberal and free-speech assumptions on which private-enterprise media ownership is based, by the end of the 1920s the case of radio began to be considered unique. Not only did radio penetrate into homes, but its coverage, simultaneity, and emotional impact offered important opportunities for nation building.

The worries of the English-Canadian cultural nationalists were given an airing when in 1928 the Liberal government of Mackenzie King set up a Royal Commission on Broadcasting headed by retired banker Sir John Aird. Although the origins of the Commission lay in a controversy over the nonrenewal of the licenses of some religious stations owned by the Jehovah's Witnesses, its deliberations opened up for the public the debate about the "Canadianness" of radio. The Aird Commission's principal recommendation was that *all* broadcasting stations in Canada should come under the ownership of the federal government. This idea of course met with opposition from many of the private station owners, but they did admit to wanting the government to subsidize their operations, either by supplying them with programs or by financing network hookups.

Although the recommendations of many Canadian Royal Commissions languish unfulfilled, in this case intense lobbying by a group of cultural activists called the Canadian Radio League kept the issue of public radio alive. The two young leaders of the Radio League, Alan Plaunt and Graham Spry, organized individuals and voluntary associations across the country to support the concept of a Canadian radio network that would be national, noncommercial, and public service-oriented. In Spry's famous phrase, Canada's choice lay between "the State and the United States." Finally, in 1932 the government of Conservative Prime Minister R.B. Bennett acceded to the argument that Canadian radio must not be allowed to fall into American hands, as it was likely to do if it remained underfinanced by Canadian private enterprise. The Bennett government passed the Radio Broadcasting Act establishing the government-owned Canadian Radio Broadcasting Commission (CRBC) in May 1932. The Act was a compromise: although the CRBC was to be the government broadcasting body, it would not be a monopoly. Private stations would continue to exist, at least in the medium term. Nor would the CRBC be completely noncommercial; to supplement its income and provide outlets for Canadian businesses, a certain amount of advertising was allowed on local CRBC stations.

The CRBC had two main functions. First, it was a network (the *only* national network) financed by the fees collected from radio owners, providing programming in both English and French and network connections for a few government-owned and many affiliated private stations from coast to coast. (Affiliated stations either reserved some of their time for CRBC programs or aired them on a discretionary ad hoc basis. This arrangement provided inexpensive programming for needy private stations and coverage for the CRBC.) Second, the CRBC became the regulatory body for all Canadian broadcasters.

In its brief existence, the CRBC ran into many organizational, financial, and political problems. But it was successful enough in regulating and enhancing the national distribution of Canadian programming that in 1936, rather than killing it, the newly elected Liberal government of Mackenzie King replaced it with a structurally sounder successor, the Canadian Broadcasting Corporation (CBC). The CBC remains Canada's public broadcaster, still operating alongside a strong private sector.

Although Canadian radio began exclusively in the private sector, the system set up in 1932, which continues to this day, is a mixed public/private one. Through the affiliate system, the public and private elements have been intertwined and interdependent. They have also, inevitably perhaps, been competitors, both for advertising dollars and for audiences.

The Balance Shifts

Just before World War II, an internal reorganization of the CBC separated the French and English language services, and the *Société Radio-Canada* grew alongside the English language network, providing significant indigenous French-language programming to Quebec and, to some extent, to French speakers outside Quebec. During the war, both the CBC and the private broadcasters came into their own, and receiving set ownership became almost universal. The CBC established a news service in 1941 that played a very important role in informing Canadians about the war effort; the CBC also established many successful and celebrated talks, drama, and musical programs. Network advertising time on the CBC (about 15 percent of its income came from advertising) was in such demand that the Corporation set up a second national English-language network in 1944.

Meanwhile, as advertising income enriched the private broadcasters, especially in large cities such as Toronto, Montreal, and Vancouver, they became more aggressive in seeking increases in power and better frequencies and began to challenge the CBC's dual role as their competitor and regulator. By 1948 total operating revenues of the private stations were twice those of the CBC. As of 1956 there were 20 CBC-owned and -operated stations in Canada, almost 100 privately owned stations affiliated with the CBC, and another 70 private independent (nonaffiliated) stations, the latter group including some of the wealthiest and most powerful stations in the country.

In 1949 the federal government appointed a Royal Commission (known as the Massey Commission after its chairman, Vincent Massey) to examine the state of Canadian cultural and intellectual life. Broadcasting was one of the Commission's major focuses, especially given the imminent arrival of television. The commissioners concluded that the CBC played an essential role in Canadian society, that it needed greater financial support, and that it should be given priority jurisdiction over the development of television in Canada. However, the 1950s brought changes that in the end moved Canadian broadcasting in almost the opposite direction. The Conservative government that was elected in the late 1950s, much less sympathetic to the CBC than previous Liberal administrations had been, acceded to various long-standing demands of the private broadcasters, most importantly the ending of the CBC's regulatory authority in favor of an independent body, first the Board of Broadcast Governors and now the Canadian Radio–Television and Telecommunications Commission (CRTC). The "balance" built into the mixed broadcasting system in the early 1930s, a balance that gave predominance to the CBC, began to shift toward the private broadcasters, and the CBC gradually became a smaller player in the system. Today CBC radio attracts only 10 percent of the listening audience.

During the 1950s, as television wooed away sponsors and programs, all of Canadian radio changed dramatically. As in the United States, the principal dramas, comedies, and variety shows moved to television, and radio quickly lost much of its evening audience. Radio stations changed their formats, first to music and then in some cases to talk, with prime time becoming the drive times in the morning and late afternoon.

Historically, both private and public radio in Canada have aired a mixture of Canadian and American material. Responding to listener demand, popular American shows such as *Fibber McGee and Molly* were picked up from the U.S. networks and broadcast across Canada on the CBC national network in the late 1930s and 1940s. Much of the popular music on Canadian stations has been American, although content rules introduced in the 1970s have increased the percentage of Canadian music. Today, a growing amount of other kinds of American content is heard on Canadian private radio, especially syndicated talk shows. CBC radio's content is, on the other hand, mostly Canadian in origin.

In the early 1950s the government decided to cancel the unpopular receiving-license fee, partly because enforcement for television set owners was deemed too difficult. The early years of television development by the CBC were financed by a special tax on the purchase of TV sets, but subsequently the CBC has been dependent on annual parliamentary grants supplemented by advertising income from the TV network (CBC radio has been commercial-free since 1975). This dependence has made the public broadcaster peculiarly vulnerable to political criticism and to the effects of budgetary ups and downs. In the 1990s the budget for the radio service of the CBC and *Radio-Canada* has been slashed dramatically.

Meanwhile, private radio stations, especially AM stations, have been exposed to economic fluctuations as well. The competition for advertising dollars has been intense, and many stations have experimented with innumerable format changes in desperate survival ploys. In the late 1990s the CRTC eased some ownership and advertising rules and allowed more private station networks in order to enable private stations to regain their profitability.

Probably the most significant factor in explaining the development of Canadian radio is the country's proximity to the United States, the world leader in broadcasting technology and programming. There is little doubt, for example, that the principal motive for the creation of the public broadcaster in the 1930s was to provide an alternative to the previously purely private commercial broadcasting system that was destined for financial reasons to become increasingly American owned and Americanized. Similarly, whether listening directly to United States stations beaming across the border or to American network programs picked up for transmission by Canadian outlets, Canadian audiences have always been captivated by American radio. Canadian governments would not—indeed could not—prevent Canadians from listening to their favorite American shows. They have, however, attempted by various positive measures, such as the creation of the CRBC/CBC and the introduction of content regulations, to make room for some Canadian programming on Canada's airwaves. Thus, although the historical development of Canadian radio paralleled and imitated that of the United States in many ways, it also possessed a number of unique features, the most important of which has been the nationalistically inspired desire to use the medium of radio to foster Canadian cultural identity, and the consequent activism of the federal government in the radio field. Whether this unique mixed system can prevail in a climate of government retreat and commercialized globalization remains a key question for the future.

MARY VIPOND

See also CFCF; CHED; CHUM; CKAC; CKLW; Fessenden, Reginald; *as well as essays immediately following this essay treating specific aspects of Canadian radio broadcasting*

Further Reading

Bird, Roger, editor, *Documents of Canadian Broadcasting,* Ottawa: Carleton University Press, 1988

Eaman, Ross Allan, *Channels of Influence: CBC Audience Research and the Canadian Public,* Toronto and Buffalo, New York: University of Toronto Press, 1994

Filion, Michel, *Radiodiffusion et société distincte: Des origines de la radio jusqu'à la révolution tranquille au Québec,* Laval, Quebec: Méridien, 1994

Peers, Frank W., *The Politics of Canadian Broadcasting, 1920–1951,* Toronto: University of Toronto Press, 1969

Raboy, Marc, *Missed Opportunities: The Story of Canada's Broadcasting Policy,* Montreal and Buffalo, New York: McGill-Queen's University Press, 1990

Vipond, Mary, *Listening In: The First Decade of Canadian Broadcasting, 1922–1932,* Montreal and Buffalo, New York: McGill-Queen's University Press, 1992
Vipond, Mary, "Financing Canadian Public Broadcasting: Licence Fees and the 'Culture of Caution,'" *Historical Journal of Film, Radio, and Television* 15, no. 2 (1995)
Vipond, Mary, "The Continental Marketplace: Authority, Advertisers, and Audiences in Canadian News Broadcasting, 1932–1936," *Journal of Radio Studies* 6 (1999)
Weir, Earnest Austin, *The Struggle for National Broadcasting in Canada,* Toronto: McClelland and Stewart, 1965

Canadian Radio Archives

The largest radio archives in Canada are the production archives of the state-run Canadian Broadcasting Corporation (CBC). The CBC maintains archives at its English network production headquarters in Toronto and at its French network production headquarters in Montreal, as well as in regional production centers and local stations across Canada. The CBC's extensive collection of programs is primarily for rebroadcast and research by CBC staff. In order to make this substantial cache of Canada's radio broadcasting heritage more accessible to the public, the CBC has donated copies to public archives across Canada.

The first formal radio archives at the CBC began in Toronto in 1959. The English and French networks subsequently gathered, organized, and cataloged disc recordings from the late 1930s to the 1960s as well as the magnetic tape masters that gradually replaced discs as the production and archiving format. The CBC has also been actively acquiring a selection of contemporary programs. The high cost of archiving broadcast recordings means that the CBC is not able to keep tapes of everything it broadcasts. The emphasis is on news and current affairs, drama and other arts programming. In the case of music broadcasts, the archives maintain a smaller selection of recordings, restricted to those of Canadian content.

The largest Canadian public archives collection of radio broadcasts is at the National Archives of Canada. Most of the radio recordings are of CBC radio programs on disc and tape. The CBC has donated almost all of its discs to the Archives but retains tape copies of a large selection. The earliest broadcast recording at the Archives is coverage of the official celebration of Canada's 60th anniversary in 1927, broadcast by the radio network of the Canadian National Railway, a predecessor to the CBC. Another highlight is an excerpt of the round-the-clock coverage of the rescue of three men trapped in a mine in Moose River, Nova Scotia, in April 1936, which was broadcast by the Canadian Radio Broadcasting Commission several months before that entity was replaced by the CBC. The Moose River Mine broadcast captured the attention of radio listeners across Canada and the United States, demonstrating that radio had an immediacy that newspapers and newsreels could not equal.

Among the most historically valuable programs in the CBC disc collection are the thousands of reports in French and English by war correspondents for the CBC Overseas Broadcast Unit, among them Matthew Halton and Marcel Ouimet. The war recordings cover the contribution of the Canadian military overseas and in some postwar events. Many reports are enlivened by actual battle sounds, recorded by the cumbersome portable disc recording equipment of the era.

The CBC disc collection also provides a cross-section of the range of program types available to Canadian listeners, especially from the 1940s to the 1960s. By the 1950s, when radio faced increasing competition from television, news and current affairs held onto their place on the schedule, but many drama and entertainment programs gradually disappeared. It is possible to trace this shift in the radio schedule through the types of programs represented in the CBC disc collection.

The CBC collection at the National Archives of Canada also includes some more recent programming, but the best source of network programs from the 1970s onward are the CBC's own archives. The National Archives of Canada has disc and tape copies of spoken-word CBC shortwave broadcasts, in English, French, German, and other languages.

Recordings of broadcasts by privately owned Canadian radio stations are considerably rarer, not only at the National Archives of Canada but also at other archives across the country. This reflects the fact that private radio broadcasters have never had financial resources on the scale of those enjoyed by the publicly funded CBC to create and maintain archives. Radio stations have, however, made recordings of programs for their own research and rebroadcast and to preserve noteworthy broadcasts as part of a station's history. The archival record of private radio, especially, exists thanks to the efforts of historically-minded behind-the-scenes individuals who, without prodding by any official station archival policy, recorded an often eclectic selection of private radio broadcasts, saved them, and donated them to public archives. Despite the growing appreciation in recent decades of the historical value of radio broadcasting, most contemporary private-sector radio broadcasting in Canada is still not being saved. The risk to the broadcasting heritage is compounded by frequent changes in station ownership and personnel as radio struggles to compete with television, the internet, and other means of mass communication.

Among the holdings of private radio at the National Archives of Canada are discs from Toronto station CFRB from the 1940s and 1950s and discs dating from 1938 to 1956 from Montreal station CFCF, the oldest radio station in Canada. There are some examples of day-to-day programming originated by the stations, but many recordings are of speeches by major public figures, often transmissions picked up from British and American sources, of which the CBC and private stations tended to save disc copies. The U.S. network affiliations of major Canadian radio stations are reflected in the occasional copies of American shows from the 1940s and 1950s. The Harry E. Foster collection consists of discs, tapes, and scripts for hundreds of radio programs broadcast on private radio stations in the 1940s and 1950s, from such series as *Men in Scarlet* and *The Adventures of Jimmy Dale;* from a news highlights program entitled *Headliners;* and from *The Northern Electric Hour,* a program of orchestral music. The hundreds of discs and tapes donated by Joseph Cardin, who worked at radio station CJSO in Sorel, Quebec, provide a rich resource for the study of French private radio from the 1940s to the 1980s.

Another source of private radio recordings are tapes donated by public figures, notably politicians. Interviews and speeches predominate, but examples of radio's democratic impact are illustrated by radio phone-in shows in which listeners were able to express their opinions and sometimes to vent at a hapless politician. The National Archives of Canada collection also includes examples from the 1970s and 1980s of news reports fed by private-sector news services, such as Standard Broadcast News, to member radio stations. There are also copies of radio programs submitted by stations across Canada to awards competitions such as those held by the Canadian Association of Broadcasters.

The National Library of Canada acquires published recordings, including an extensive collection of the transcription discs of music performances issued by the CBC shortwave service.

Other public archives across the country have smaller but nonetheless significant collections of radio broadcast recordings, often primarily from the CBC. There are collections of recordings in provincial and territorial archives, university archives, local historical societies, and other organizations. The Folklore and Language Archives at Memorial University of Newfoundland in St. John's, Newfoundland, has hundreds of hours of programs broadcast by Newfoundland radio stations from the late 1930s to the early 1960s. Included are programs in the series *The Doyle Bulletin,* which relayed personal messages from people to their friends and relatives in an era when transportation was more difficult and many people did not have telephones. Political events, music, some local drama, and even sound effects of fishing boats are also in the collection. The university also acquires contemporary recordings of local CBC arts programs. The provincial archives in New Brunswick, Nova Scotia, and Alberta are among the government-run regional archives with collections of CBC broadcasts produced in their geographical areas. Most regional public archives in Canada also have small collections from private radio stations, often providing a unique archival record of local news and music. The Saskatchewan Archives Board actively does what many archives talk about doing but seldom carry out: they regularly contact radio stations to record one full day of programming, thus documenting the flow of the broadcast day as a listener experiences it. The Centre for Broadcasting Studies at Concordia University in Montreal has an extensive collection of CBC radio drama scripts.

Archives in Canada still have considerable work to do to make it easier for researchers to find out what still exists for particular radio stations and networks. Slowly, information about radio archives is becoming more accessible through in-house databases and on-line catalogs. These automated tools provide a mix of general and detailed information and also alert researchers to a wide range of other archival records necessary for the study of broadcasting history. They include oral history interviews with people who worked in Canadian radio from its earliest days, photographs, and administrative files. Access to copies of the actual broadcasts is more limited. In most cases, the broadcaster has retained copyright to the recordings, and researchers wishing to obtain copies must first obtain the permission of the copyright owner.

ROSEMARY BERGERON

See also Canadian Broadcasting Corporation; CFCF; Museums and Archives of Radio

Further Reading

Dick, Ernest J., "An Archival Acquisition Strategy for the Broadcast Records of the Canadian Broadcasting Corporation," *Historical Journal of Film, Radio, and Television* 11, no. 3 (1991)

Langham, Josephine, "Tuning In: Canadian Radio Resources," *Archivaria* 9 (Winter 1979–80)

Canadian Radio Policy

Canadian radio policy is among the world's most complex. This peculiar situation may be explained in a number of ways. Historically, Canadian radio has encountered numerous challenges in the context of vast geography and low-density demography. On the economic front, it had to survive hardship and to distinguish itself from the powerful U.S. broadcasting system. Culturally, Canadian radio has been, and still is, challenged by the constant attraction foreign productions hold for most Canadian citizens. Furthermore, the federal government's involvement in the radio industry has resulted in a particular broadcasting structure in which the public element has to coexist with the ever-growing private element. Therefore, regulation has various results: the Canadian radio industry is viable, but the creation of a truly Canadian culture remains an objective to be attained. The challenges Canadian radio faced 80 years ago are still very evident.

Development

Preceded by the Wireless Telegraphy Act of 1905, the Radiotelegraph Act of 1913 gave the Canadian government power to license the use of airwaves—considered public property just like other natural resources. During the 1920s, stations popped up everywhere, but radio broadcasting remained without strong direction from the federal government, which was only mandated to issue licenses and to manage the frequency spectrum allocated to Canada under the terms of international agreements. Canadian radio was soon threatened with being integrated into the U.S. broadcasting system, in spite of a 1923 amendment to the Radiotelegraph Act that gave only British citizens (this included Canadians) the right to obtain broadcasting licenses. Radio program content, however, remained nearly free of any constraints. This situation resulted in an abundant use of foreign (i.e., U.S.) programs, mostly in English-speaking Canada. Canadian content nevertheless became part of the debate over the future of Canadian broadcasting, and commercial stations' freedom became an issue of national debate.

Laws and Commissions

The report of the 1929 Royal Commission on Radio Broadcasting, formed under the chairmanship of John Aird, directly supported the centralizing movement that characterized the formation of the Canadian state, and it recommended a governmental takeover of private radio. The Canadian Association of Broadcasters, which had been created in 1926 to foster and protect the interests of existing radio stations, strongly opposed this recommendation. Nevertheless, the Broadcasting Act in 1932 created the Canadian Radio Broadcasting Commission (CRBC), which was empowered to decide upon the numbers and locations of radio stations in Canada. The CRBC was also mandated to establish a national radio service by creating a Canada-wide network. National private radio networks were forbidden, but private regional chain broadcasting was tolerated.

The Broadcasting Act of 1936 replaced the CRBC with the Canadian Broadcasting Corporation (CBC) and mandated it to establish a national radio service and to control the broadcasting system as a whole. During the next two decades, the CBC was successful with the first assignment, but less so with the second: private stations grew and became financially secure, although they were forbidden to amalgamate into private networks. Always facing potential conflicts of interest, the CBC—as both the national broadcaster and system regulator—imposed little regulation on its affiliates (or on private independent stations).

The Canadian broadcasting system took form as a compromise between nationalistic and commercial objectives. This mixed system of public and private ownership was to cease when the CBC would be capable of becoming the dominant actor in the national radio service, but this too was not an easy challenge to meet. Before long, installation and production costs exceeded the CBC's resources. In order to expand broadcasting coverage, the corporation issued many licenses to private interests.

The main objectives of the Canadian system were in fact hardly realizable. This was already evident in reports of the Royal Commission on National Development in the Arts, Letters, and Sciences (led by Massey and released in 1951) and of the Royal Commission on Broadcasting (led by Fowler and released in 1957), which examined the problem of radio (and television) commercialization, a problem closely linked with the so-called Americanization of private radio's cultural content. Their reports also revealed that public broadcasting was overloaded with responsibilities, such as the promotion of national unity and Canadian identity.

The Broadcasting Act of 1958 was a very liberal interpretation of the Fowler report; it placed the private sector—henceforth allowed to create its own networks—and the public sector on equal grounds in terms of legal recognition. To manage this major change in public policy, both sectors were to be refereed by an independent regulatory institution, the Board of Broadcast Governors (BBG), which was responsible for the application of the act. This decision demonstrates a substantial evolution of the Canadian broadcasting system: the 1932 act foresaw the nationalization of private undertakings, the 1936 act made them complementary to the national service, and the

1958 act gave commercial stations enough autonomy to compete openly with the CBC.

To satisfy popular demand, notably for television coverage after 1952, private enterprises became inevitable. The creation of commercial networks, allowed by the 1958 act, was definitely authorized by the Broadcasting Act of 1968. Its concept of a *single broadcasting system* was a paradox, because the system in reality comprised two distinct components, with somehow different—if not contradictory—objectives.

On a regular basis, the regulatory body, the government, or private stations (if not all three) have been blamed for the failure to provide Canadians with a truly Canadian system: the 1965 Advisory Committee on Broadcasting, the 1969 Special Senate Committee on the Mass Media, the 1979 Consultative Committee on the Implications of Telecommunications for Canadian Sovereignty, the 1982 Federal Cultural Policy Review Committee, and the 1986 Task Force on Broadcasting Policy all deplored in one way or another the inefficiency of Canadian content regulation, the lack of goodwill by private stations toward the spirit of the Canadian policy, and the abuse of broadcasting as a profit-making instrument. And they all agreed on the absolute necessity for Canada to preserve a distinctive broadcasting system. Since 1968 the foreign ownership of broadcasting undertakings, limited to 20 percent of voting shares, has continued to guarantee Canadian economic interests without necessarily supporting Canadian cultural content.

The Regulatory Body

Control over broadcasting in Canada is carried out by a single authority: like the BBG before 1968, the Canadian Radio Television Commission (CRTC), which became the Canadian Radio–Television and Telecommunications Commission in 1976, exerts its power through the licensing process, because the license terms and conditions are, in principle, compulsory requirements assigned to each station or network. As an administrative tribunal, the CRTC must also regulate the whole radio industry according to the Broadcasting Act, which, after many unsuccessful attempts, was reformulated in 1991 without changing its structure and main orientations. Interestingly enough, the 1991 act states that "the Commission [CRTC] shall regulate and supervise all aspects of the Canadian broadcasting system set [by the act]," but also that "[t]he Canadian broadcasting system should be regulated and supervised in a flexible manner that . . . facilitates the provision of Canadian programs to Canadians."

Although the CRTC plays an important role, because it regulates every station by its licensing process, it has been preoccupied with deregulation since the mid-1980s. In this context, the very essence of the rationale for government intervention—a national cultural identity—is jeopardized. Many federal cultural institutions (including the CBC) have declined because of budget constraints and the relative failure of governmental involvement. Notwithstanding the recent conjuncture, the government is limited by commercial realities that render imports of foreign cultural products more profitable than the production of comparable domestic goods. Such a situation arises from the severe reaction brought about by most attempts to exert additional control. The CRTC, like many similar institutions, must manage the clash between national cultural objectives and pragmatic economic concerns.

Public Radio

Public service radio belongs to the CBC. Canadian public radio is in reality government broadcasting with a specific agenda, as shown in some provisions of the Broadcasting Act, such as subsection 3(m)(vi), which states that the programming of the CBC should "contribute to shared national consciousness and identity." Since 1975 commercial advertising has almost completely vanished from CBC radio as a voluntary measure ratified by the CRTC. Therefore, the CBC can present a more original (some would say a better-quality) program, which does not have to compete with the private sector's program for the large and popular audience shares, an otherwise vital precondition for generating revenues. Nevertheless, its relative marginal situation with the Canadian audience (and income tax payers) renders the CBC vulnerable to budget cuts in an era of governmental downsizing. The CBC possesses and operates its own stations, but to ensure its total coverage it also depends on private independent stations, which as affiliates broadcast a mix of national and local programs.

Even more marginal than public radio, community radio is collective property and is mandated to serve a defined clientele: campus, native, and ethnic stations are among this group. Community radio is not profitable and is financed mainly by governmental subsidies.

Private Broadcasters

In raw numbers, as well as in financial terms, the private independent stations together make up by far the most important group in the Canadian radio industry. Private radio is essentially a commercial undertaking whose mandate is to serve local markets. Private radio is more flexible than the CBC, but the broadcasting legislation nevertheless encourages private radio's participation in the main national objectives, such as the creation and presentation of Canadian programming. Nevertheless, the CRTC, as the regulatory body, must take into account the financial viability of private undertakings.

Since the end of the 1980s, the radio industry has suffered from low profitability, an overall phenomenon especially obvious in the AM sector. A long-term decline in radio's share of advertising and increased competition among stations are the predominant factors here. In the mid-1980s, marketing expenditures shifted from media advertising to promotional activities on local markets, which made up 75 percent of the total advertising revenue base for radio stations; at the national level, among several factors responsible for the loss of advertising revenue is the overflow of U.S. advertising into Canada and the scant use of media advertising by emerging large retail stores of U.S. origin. At the same time, the net increase of new licenses significantly reduced the advertising revenue per station.

Traditionally, the private stations complained about the so-called unfair competition generated by direct federal involvement, but, unlike television, CBC radio has not impinged on the advertising market since the mid-1970s. As revenues cannot increase, it has been suggested that expenses, such as the payment of license fees, might be reduced. This solution is hardly acceptable for the CRTC, which is a financially autonomous institution. The CRTC is nevertheless preoccupied with private stations' financial situation, particularly that of the AM radio: since 1986 AM stations have been free to advertise as much as they wish. Under previous regulations, AM stations were restricted to 250 minutes of commercial content per day and 1,500 minutes per week. Noting that such regulation did not increase advertisements to an unacceptable level in AM radio, the CRTC in 1993 removed the limit of 150 minutes of commercial message during the broadcast day imposed on FM licensees.

Canadian Content Regulation

Canadian nationalists have long been preoccupied with preserving and enhancing the country's political, economic, and cultural sovereignty. Since the 1930s, the challenge has been presented as the protection of Canadian mass media against the powerful U.S. system by means of a cultural boundary. Therefore, the federal government's intervention has been aimed at creating cultural identity as a mean of national defense and national unity.

Encouraging Canadian content is undoubtedly the ultimate objective of the broadcasting system. Surprisingly, the first regulation imposing foreign import quotas on AM stations was not introduced until the 1970s. The situation was then critical. For example, in 1968, Canadian musical selections accounted for between 4 and 7 percent of all music played. The 1970 regulation imposed a 30 percent minimum of Canadian content. In 1986, new regulations were introduced covering both AM and FM radio, although different requirements applied. For example, in spite of common objectives pertaining to political broadcasts, AM radio had to respect the 30 percent Canadian content rule, whereas, for the rising FM sector, requirements varied from 10 percent for the "easy listening" format to 30 percent for "country" stations. Subjected to a daily limit on commercial advertising (150 minutes), FM radio on the other hand benefited from an incentive to encourage Canadian content, because the CRTC's radio regulation stated that commercial messages broadcast during Canadian feature segments would not be considered as commercial messages. Since 1991 every popular FM station must program a minimum of 30 percent of Canadian selections scheduled in a reasonable manner throughout each broadcast day. In 1998 this radio regulation was redefined: every commercial station, AM or FM, must devote at least 35 percent of its musical content to Canadian selections, especially between 6:00 A.M. and 6:00 P.M. from Monday to Friday. Any musical selection that meets two of the following conditions is deemed to be "Canadian": the music or lyrics are performed by a Canadian, the music is composed by a Canadian, the lyrics are written by a Canadian, or the musical selection is a live performance recorded or broadcast in Canada.

This regulation aims to expose Canadian audiences to Canadian musical performers and thus to strengthen the Canadian music industry. Cultural and industrial objectives are closely interrelated. In addition, stations licensed to operate in the French language, either on the AM or FM bands, must devote at least 65 percent of their weekly vocal selections (at least 55 percent from Monday to Friday between 6:00 A.M. and 6:00 P.M.) to musical selections in French. This regulation aims to preserve language diversity, particularly in metropolitan markets such as Montreal, where English language stations are over-represented. Furthermore, French language popular music broadcasters are encouraged to strive for a Canadian content level of 50 percent. Conversely, English language stations are invited to promote and financially assist Canadian talent.

A Canadian Culture?

Throughout the history of Canadian radio, U.S. influence has been considered a major threat. On the one hand, economic imperialism jeopardized the emerging Canadian radio system as major American networks could have easily overwhelmed stations north of the border before the Broadcasting Act of 1932 (and its subsequent versions) put an end to this movement. On the other hand, cultural imperialism—a more subtle form of influence—is probably as strong as ever today, notwithstanding the Canadian content regulation and the incentives to Canadian talent development.

Has radio in fact succeeded in nurturing Canadian culture? If the Canadianization of radio broadcasting has been a success in terms of ownership, because only Canadian citizens or corporations (in which at least four-fifths of owners or controlling persons are Canadians) are issued licenses, it has been less so in terms of Canadian content. The approximately 30 percent quota hardly meets the spirit and the letter of the act. Although Canadian content quotas provide only a minimum limit, very few stations broadcast more Canadian music than the imposed minimum. It should be stressed that CBC targets of 50 percent Canadian content for popular music and 20 percent for classical music are sometimes attained, but these outstanding results must be tempered by considering the CBC audience share, which includes only about one-tenth of the listening population.

Furthermore, subsection 3(s) of the Broadcasting Act remains ambiguous, as it states that "private networks and programming undertakings should, to an extent consistent with the financial and other resources available to them, (i) contribute significantly to the creation and presentation of Canadian programming, and (ii) be responsive to the evolving demands of the public." Yet it still has to be demonstrated that Canadians prefer Canadian cultural products. The case of prime-time television and the success of private radio in terms of audience share are obvious enough to suggest that many Canadian consumers (or listeners) are not reluctant to buy (or listen to) foreign products. In this sense, their role in whether or not the creation of a distinct Canadian society remains an objective to be pursued is still far from over. Market globalization and new technologies such as audio digital transmission might very well revolutionize the entire radio environment early in the 21st century. Canadian nationalists can still fear for the country's cultural sovereignty, because there is little reason to believe that the permeability of Canada's southern boundary will not increase.

MICHEL FILION

Further Reading

Audley, Paul, *Canada's Cultural Industries: Broadcasting, Publishing, Records, and Film,* Toronto: Lorimer, 1983

Ellis, David, *Evolution of the Canadian Broadcasting System: Objectives and Realities, 1928–1968,* Ottawa: Canadian Department of Communications, 1979

Filion, Michel, *Radiodiffusion et société distincte: Des origines de la radio jusqu'à la révolution tranquille au Québec,* Laval, Québec: Méridien, 1994

Filion, Michel, "Broadcasting and Cultural Identity: The Canadian Experience since 1920," *Media, Culture, and Society* 18, no. 3 (1996)

Filion, Michel, "Radio," in *The Cultural Industries in Canada: Problems, Policies, and Prospects,* edited by Michael Dorland, Toronto: Lorimer, 1996

Grant, Peter S., Anthony H.A. Keenleyside, and Michel Racicot, *Canadian Broadcast and Cable Regulatory Handbook,* Vancouver, British Columbia: McCarthy Tétrault, 1983

Peers, Frank W., *The Politics of Canadian Broadcasting, 1920–1951,* Toronto: University of Toronto Press, 1969

Raboy, Marc, *Missed Opportunities: The Story of Canada's Broadcasting Policy,* Montreal and Buffalo, New York: McGill-Queen's University Press, 1990

Task Force on the Introduction of Digital Radio, *Digital Radio: The Sound of the Future: The Canadian Vision,* Ottawa: Minister of Supply and Services, 1993

Vipond, Mary, *Listening In: The First Decade of Canadian Broadcasting, 1922–1932,* Montreal and Buffalo, New York: McGill-Queen's University Press, 1992

Vipond, Mary, "The Beginnings of Public Broadcasting in Canada: The CRBC, 1932–1936," *Canadian Journal of Communication* 19 (1994)

Weir, Earnest Austin, *The Struggle for National Broadcasting in Canada,* Toronto: McClelland and Stewart, 1965

Canadian Radio Programming

Continuing French and English Traditions

Much of the development of Canada's radio programming may be seen in light of the country's wish to avoid total dominance by U.S. radio. Although Canada's French tradition in Quebec made distinct programming easier, English-language programming faced a stiff challenge from the beginning.

French-Language Programs

The Beginning Years

Radio broadcasting as a medium of social communication appeared in Quebec and Canada in 1922. The federal Minister of Maritime Affairs and Fisheries established a new code of communications and created a new category of licenses, know as "private commercial broadcasting licenses." Many enterprises in Quebec and Canada applied for and obtained broadcasting licenses. During the month of April, the federal Minister of Maritime Affairs granted about 20 broadcasting licenses, including the French-language station CKAC (*La Presse*) as well as CFCF (Marconi, Montreal), CFCA (*Toronto Star*), CHCB (Marconi, Toronto), CJSC (*Evening Telegram)*, CJCG (*Winnipeg Free Press*), CJCA (*Edmonton Journal)*, and CJCE (*Vancouver Sun)*.

The creation of radio station CKAC in 1922 by the major French-language newspaper *La Presse* created the link between the technological dimension and journalistic expertise. This harmonious fusion was made possible by the founder of CKAC, Jacques-Narcisse Cartier, an expert technician as well as a seasoned journalist with experience in Montreal and at British and American newspapers. Cartier—pilot in the Royal Air Force, collaborator with Guglielmo Marconi and technician at Marconi stations in Nova Scotia and New York, personal friend of the Radio Corporation of America's (RCA) David Sarnoff, technician at Telefunken, and a businessman and corporate leader who would later take the helm of two major Montreal newspapers—represented the spirit of initiative that a French Quebecker could bring to the avant-garde technologies of his time.

Radio's early use of live music led to links between radio and all the musical groups, vocalists, composers, teachers, and hosts who formed the core of Montreal's very active cultural life. Cartier was a musician himself, a pianist and organist; his social milieu included musicians and contacts with the designers at the Casavant organ manufacturing company. This is why, in December 1922, Cartier had a Casavant organ installed in the CKAC studio. In spring 1923 Raoul Vennat, musicologist and importer of music, began a weekly series of concerts in which musicians and singers performed the most recent works of French music. In 1923 an operetta, "Les Cloches de Corneville," was aired with an orchestra of 25 musicians and a choir of 38 singers. In 1925 CKAC offered a series of live piano lessons given by Emiliano Renaud, a teacher of international reputation back from a career in New York. In 1929 J.-Arthur Dupont negotiated an association with the American Columbia Broadcasting System (CBS) network. This was done to connect CKAC into a circuit of concerts aired live by a group of American stations, which CKAC then joined in 1930 when it created its own symphony orchestra. Creation of the "Quatuor Alouette" allowed for original harmonizations of international folkloric music to be broadcast on the radio. In 1933 Dupont, now the program director at the Canadian Radio Broadcasting Commission, arranged for the airing in Canada of live opera from the Metropolitan Opera in New York, which had already been broadcasting on American stations for two years. Quebec-crafted songs were introduced to radio by Lionel Daunais, who had a regular broadcast of French-language songs at the CCR and who regularly hosted *L'Heure provinciale* at CKAC, a cultural and educational biweekly magazine financially supported by the Quebec government. It became the model for educational radio; over ten years, it broadcast close to a thousand scientific and cultural lecture programs. But *L'Heure provinciale* was also a privileged forum for introducing musicians, singers, and other Quebec artists who became regularly featured artists on radio and in the theater in the decades that followed. All the great artists of this period got their start on radio. The group that best symbolized this sociocultural dynamic was without a doubt the operetta troupe of the "Variétés lyriques," founded by Charles Goulet and Lionel Daunais, whose artistic activity would span 30 years, integrating artists from Quebec as well as singers from Europe into their productions.

But Quebec radio also wanted to be an information service from its very beginnings, as expressed by Cartier in 1925 before a committee of the federal parliament in a statement on "the true role of radio in the life of a people." He was thus developing a strong position that *La Presse* had presented in an editorial as early as May 1922. For Cartier, radio had a triple mission of information, education, and entertainment.

From the beginning, CKAC systematically integrated information programs, news bulletins, and entertainment, or "magazines." By 1925 Cartier established a tradition of live reports from the scene of major political and sports events: hockey (broadcast from Boston, because the Montreal Forum refused to give its permission) and, more importantly, the federal electoral campaign of October 1925, during which Cartier and Dupont, with the collaboration of the Marconi Company, developed a mobile unit for live reports and covered electoral assemblies all over Montreal and the medium-sized towns across the region. News programs became more elaborate by the beginning of the 1930s, despite the prohibitions of the Canadian Press agency. As an alternative, Dupont negotiated an association with CBS whereby he received news bulletins that were then translated and aired in French.

In June 1938, in time for the Fête Nationale in Quebec, CKAC launched the first important radio-journal, *Les nouvelles de chez-nous* (Local News), which was hosted until 1954 by one of the best actors and communicators of the period, Albert Duquesne. This was a coup for CKAC, which took a commanding lead over the public broadcaster, Radio-Canada, where the news (mostly foreign and from distant places) was only broadcast late in the evenings. There was a major difference between the translated news items from American or English-Canadian agencies and the network of reporters that CKAC had put in place to cover news across Quebec: important information could also be about the goings-on in Quebec.

Radio Development in Quebec

By the 1940s the population of Quebec, some 3 million, was served by 30 radio stations, the most powerful of which, CKAC, covered a large portion of the territory as well as the French-Canadian populations of New England. Radio became an essential service for public culture, independent of both political and religious influences. Closely associated with the professional milieus of music and theater, radio brought society a multifaceted and original discourse and an openness to information, values, and models circulating everywhere in the Western world.

Radio of the 1940s and 1950s was primarily a concert hall as well as a music-hall stage, and it served as an intimate venue for original Quebec singer-songwriters. Radio-Canada took over and became the producer of concerts for large and smaller ensembles. These broadcasts were, of course, aired live. In certain radio music halls, where sketches and songs, humor and editorial comments alternated, the background environment was always that of music played by an orchestra. Even late-morning variety programs such as *Les joyeux troubadours* were accompanied by a small orchestra in the studio.

Radio in this period was also an editorial room for daily news. It was at the beginning of the 1940s that radio became a major source of information. CKAC's Information Service had been created in 1939, station CHRC in Quebec City had specialized in news broadcasts, and in 1941 Radio-Canada created its own Information Service. At the outbreak of World War II in September 1939 Radio-Canada stayed on the air for a month in order to ensure that the news was continually updated, 24 hours a day. In 1941 the Canadian Press agency created an affiliate for radio news, *Press News Ltd.* Journalist Jovette Bernier created a daily humorous program, *Quelles nouvelles,* as an antidote to the anxieties generated by the daily news.

Around 1960 the presence of parliamentary correspondents became the trademark of various stations, with slogans such as "the news as it happens" or "news on the hour."

Radio also recognized its educational mission. Following the British model, and undoubtedly inspired by educational stations in the United States, Quebec radio proposed a programming schedule offering a broad range of programs geared toward adult education and the development of more advanced students. The resulting university of the airwaves was called "Radio Collège." However, school broadcasts for students at the elementary and high school levels were not developed as quickly and would only appear in the 1960s.

Quebec radio was also the proving ground for an important cultural innovation: the invention of an original form of radio-based literature. Authors of great talent produced original works for the radio medium as a means of experimentation and of reaching a large audience. Unlike other countries (France in particular), where literary broadcasts on the radio consisted mostly of adaptations of published works, in Quebec a body of original works was created. The most diverse forms were used—theater, dramatic serials, historical works, humorous sketches, tales, memoirs, monologues, essays, and poetry. Between 1930 and 1970, half a million pages of literary texts were written for radio by about 1,000 authors. The serial or radio-novel, because of its structure, based on chapters broadcast over many years—often 10 or 20—became a major building block of the collective imagination. Works such as *Un homme et son péché, Métropole, Faubourg à m'lasse, Le survenant, Jeunesse dorée, Le curé de village, Madeleine et Pierre, La rue des pignons, La famille Plouffe, Le ciel par-dessus les toits, Nazaire et Barnabé,* and many others became daily meeting places for the general public. These radio-novels were works by *auteurs,* in the full literary sense of the term within the European tradition, rather than industrial productions on the American model.

Long-running and important series transformed radio into a creative laboratory or a theatrical repertory company for the benefit of the general public. One must note the cultural influence of the following series in terms of their contribution to artistic creativity: *Le théâtre de chez-nous, Radio-théâtre, Radio-théâtre miniature, Radio-théâtre Ford, Théâtre dans un fauteuil, Le radio-théâtre de Radio-Canada,* and *Nouveautés dramatiques.* Providing an opening toward the international classical and modern repertories, the very long-running series *Sur toutes les scènes du monde* introduced the public for over 30 years to European authors writing in many languages (in French translations), to American authors, and to major classical and modern French authors.

Three humorous works stood out for their literary value: *Carte blanche,* a satire on society and its traditional culture; *Chez Miville,* a parody of the ideologies of change of the 1960s; and *D'une certaine manière,* an ironic look at the new currents of thought of the period.

Since 1970: More Stations and Missions

In the last quarter of the 20th century, the number of Quebec French-language radio stations more than doubled, from 70 to nearly 150. Radio broadcasting was no longer considered by the Canadian Radio Television Commission (CRTC, the federal regulation agency) as a service but as a business to be opened up to free-market competition. The concept of serving the needs of a territory to ensure communication has given way to the notion of markets to conquer, develop, and consolidate. For all commercial stations, radio is first and foremost an enterprise that has to provide a return on investments comparable to that of any other business. Only public radio has escaped this definition, but it is also under attack, forced to restrict its field of action so as not to interfere with the private sector, and every successful program that it broadcasts is perceived by the private sector as unfair competition. The very legitimacy of the public sector meets with such opposition that in Quebec and elsewhere, associations have been formed to bolster public opinion in favor of governmental financing of public radio.

In Quebec, community radio stations appeared in the mid-1970s. With limited means and relying on financing from associations and on voluntary involvement, they came to reach out to diverse audiences, specialized in major cities and generalized in many of the more remote regions.

Therefore, during this period, Quebec radio went through not only a fragmentation of its audience, but also a profound modification of the social consensus on the role of radio in society. Private stations, rooted in their networks, closed their news offices, no longer hired journalists, and rebroadcast news bulletins written by a central agency. The content on private radio stations, in Canada as in the United States, became polarized into talk or music-based radio. In both cases, the music aired was imposed by the record companies and by the commercial circuits of the distributors. Radio was no longer an experimental stage for young Quebec artists, but a link in the distribution chain of the major producers.

Within the domain of talk radio, however, certain stations maintained a format that continued, in a manner acceptable to today's popular culture, missions geared toward information and education. Whether it be through in-depth interviews, exchanges between hosts and audience members, the formula of on-air telephone polls, or commentary on daily events, certain stations kept elements of what had defined the originality of Quebec radio.

On public radio, this culture of analysis of and commentary on current affairs occupied a privileged niche. Public radio remained, especially in drive time, the most practical medium as well as the most economical.

Community radio stations reached out to more specific audiences. They also allowed for the broadcast of international news related to countries of origin, especially news not circulated by the major news agencies or commercial stations. Encouraging solidarity within or among specific groups, discussions on social questions, and the expression of more traditional values were the characteristics of many community radio stations. Also, in many regions, community radio stations were the instrument for the dissemination of a regional identity, as well as an experimental stage for nonprofessional creators and communicators.

During the 1980s a broad range of missions was offered by radio to the variety of Quebec audiences. Young audiences interested in pop culture, young adults at the university level, retired persons concerned by issues related to volunteer work, working audiences seeking to travel efficiently through the city, exhausted listeners seeking calmness—all these audiences could, at any time of the day, find their own type of radio. They could find rock music, classical music, ballads, American music from the Anglophone stations, familiar conversations with the audience, hosts specializing in provoking strong reactions by commenting on current affairs of major and minor importance, religious programs, university courses, and sports—all kinds of sports, live and in rebroadcasts, described and commented upon. Within all these types of content, advertising is in control of the selection of major time slots and their content. It is this very busy landscape, a type of untamed wilderness, that has become the popular cultural environment of Canadian French-language radio for the last few decades.

PIERRE-C. PAGÉ

English-Language Programming

On the evening of 20 May 1920, singer Dorothy Lutton strolled to a microphone in the Chateau Laurier Hotel in Canada's capital city of Ottawa. Her audience consisted of Prime Minister Sir Robert Borden, the future prime minister William Lyon MacKenzie King, the Duke of Devonshire, and Arctic explorer Vilhjalmur Steffansson. The four dignitaries and assorted guests had been part of an audience listening to a lecture on war inventions. Now they were to be treated to Miss Lutton's songs, as were a number of others in and around the Montreal area. The singer's voice was being carried to radio station XWA by telegraph wires. Those fortunate enough to have access to receiving sets were about to hear the first organized radio program to be broadcast in Canada.

Radio penetration and the consequent appetite for programming characterized much of the 1920s. Regulations were few; facilities ranged from small studios in the back of retail stores to state-of-the-art facilities in big cities. Programs were experimental and erratic. Until the early 1930s, radio licenses remained exclusively in private hands, and the private sector determined program tastes.

Nearly nine out of ten Canadians lived within ninety miles of the U.S. border, much as they do today. Large American corporations such as RCA and CBS took an early interest in broadcasting, with the consequence that they had a head start in developing popular and marketable programming. American tastes in programming soon became Canadian tastes. Canadians quickly demonstrated their preferences for popular American programs. Throughout the late 1920s and early 1930s, NBC's *Amos 'n' Andy* was the most popular radio program in Canada. In 1925 the Toronto newspaper the *Telegram* asked readers to report on their favorite radio stations. The first 17 places were filled by stations in Pittsburgh, Schenectady, Buffalo, and New York City. U.S. programs dominated Canadian radio sets, which were seldom tuned to the Canadian National Railways (CNR) network, which began producing dramatic programs, high-quality musical shows, children's programs, and some information programming in 1925.

Canadian stations were more than willing to affiliate with American networks. CFCF joined NBC—the program schedule was top-heavy with musical programming, and CFCF broadcast no news. The station carried only one Canadian network show, *Melody Mike's Music Shop*. Together, CFRB and CKGW in Toronto, along with CFCF and CKAC in Montreal, turned over one-third of their cumulative broadcast day to American-produced programming. CFRB, affiliated with CBS, joined the network for half its broadcast day. By 1931 nearly all radio comedy and drama was being produced by advertising agencies with business offices in both Canada and the United States. There were virtually no dramatic programs produced in Canada. Only one serious dramatic program appeared on the airwaves during the early 1930s. The CNR historical series *Romance of Canada,* written by Merrill Denison and directed by Tyrone Guthrie, was a harbinger of things to come in Canadian Broadcasting Corporation (CBC) radio drama.

The late 1920s witnessed a number of significant events in program development. When Canada celebrated the Diamond Jubilee of Confederation on 1 July 1927, a number of stations hooked up in the country's first network to broadcast the celebration program. In March 1928 the Department of Marine and Fisheries turned its attention to what Canadians were hearing on Canadian radio and canceled the licenses of a number of stations operated by the International Bible Students Association (Jehovah's Witnesses).

After its creation in 1936, the CBC enjoyed legal protection as the only national radio network. As a consequence, programming became divided: CBC programming was more national in character, and the private stations designed programs for local tastes. Whereas the CBC broadcast hockey games and live drama from coast to coast, local stations developed talk shows, some local drama, and live dance music programs. In spite of their subservient place in the broadcasting universe, private stations continued to attract huge audiences, built mainly on American imports, such as the *Jack Benny Show, Gangbusters,* and *Green Hornet*. After many of these programs had either ceased to exist or moved into television, Toronto's CFRB continued to broadcast *Yours Truly, Johnny Dollar* into the early 1960s.

When World War II broke out, CBC Radio joined the war effort with a stream of carefully composed propaganda programs intended to keep up spirits at home. The war effort permeated information, entertainment, and musical programming. The CBC established a second network in 1944: the founding network became the Trans-Canada and the new operation the Dominion network. Most serious programming remained with Trans-Canada stations. In 1947 CBC Radio launched *CBC Wednesday Night,* an eclectic mixture of music and information aimed at a high-end audience. The Dominion network was established to act as a commercially based programming source for Canadian productions of a lighter nature. It was hoped that privately owned stations would join the Dominion network. The network was a marginal success. With the increasing reach of television, it closed in 1962 and converted its Toronto flagship station to a French language operation.

Following World War II, the CBC continued in its dual role of broadcaster and regulator. Private broadcasters never accepted their subservient role and finally, in 1958, the newly elected federal government of John Diefenbaker revised the Broadcasting Act and removed the regulatory powers from the CBC, turning them over to a new agency, the Board of Broadcast Governors. When the Board began to move into programming questions, the private broadcasters resisted.

The conflict pointed out a number of weaknesses in the Broadcasting Act, which in turn led to revisions in 1968, when the Board of Broadcast Governors was replaced by the CRTC with a much stronger nationalist mandate. The agency imposed Canadian content requirements on not only television but AM radio as well. Although FM radio did not have to meet specific targets during its formative years, the CRTC dictated how much spoken-word programming had to be carried, as well as news and sports content. Application forms for licenses and license renewals contained pages on which broadcasters had to calculate down to the second each kind of broadcasting endeavor that would be undertaken during a week of programming. Once accepted, the CRTC treated this Promise of Performance as a contract, not a guideline.

The CRTC was created when broadcasting in Canada was entering the first phases of extensive expansion. FM radio was beginning to erode the base that AM radio had enjoyed since the 1920s. Private radio defined a survival agenda based on local news, popular music, and in some cases phone-in shows.

The CBC was one of the first organizations to respond to the new environment. Faced with collapsing audiences, the radio service underwent radical programming changes. The new CBC programming was to be based on information on the AM side (now Radio One) and music on the FM side (now Radio Two). Shows such as *Morningside* (now *This Morning*), *As It Happens, Metro Morning,* and others focused on current affairs. The new stereo service offered an eclectic mix of what it now calls "Classics and Beyond." In the early 1990s CBC radio began to abandon many of its AM stations across the country, opting to place Radio One programs on FM channels as well. Private AM stations started to abandon music programming near the end of the 1980s, opting instead for a variety of talk formats, including phone-in shows and all-news or sports formats. Many AM stations followed the CBC lead and abandoned their channels for FM alternatives.

DAVID R. SPENCER

Further Reading

Bird, Roger Anthony, editor, *Documents of Canadian Broadcasting,* Ottawa: Carleton University Press, 1988

MacMillan, Keith, "Radiodiffusion," in *Encyclopédie de la musique au Canada,* vol. 3, edited by Helmut Kallmann, Gilles Potvin, and Kenneth Winters, Montreal: Fides, 1993

Pagé, Pierre, *Répertoire des oeuvres de la littérature radiophonique québécoise, 1930–1970,* Montreal: Fides, 1975

Pagé, Pierre, and Renée Legris, *Le comique et l'humour à la radio québécoise: Aperçus historiques et textes choisis, 1930–1970,* vol. 1, Montreal: La Presse, 1976, and vol. 2, Montreal: Fides, 1979

Romanow, Walter I., and Walter C. Soderlund, *Media Canada: An Introductory Analysis,* Mississauga, Ontario: Copp Clark Pitman, 1992; 2nd edition, Toronto: Copp Clark, 1996

Rutherford, Paul, *The Making of the Canadian Media,* Toronto: McGraw-Hill Ryerson, 1978

Rutherford, Paul, *When Television Was Young: Prime Time Canada, 1952–1967,* Toronto and Buffalo, New York: University of Toronto Press, 1990

Vipond, Mary, *Listening In: The First Decade of Canadian Broadcasting, 1922–1932,* Montreal and Buffalo, New York: McGill-Queen's University Press, 1992

Canadian Broadcasting Corporation

Canada's Public Broadcaster

The Canadian Broadcasting Corporation (CBC) is a public resource that operates under the 1991 Broadcasting Act and is accountable to the Parliament of Canada through the Minister of Canadian Heritage, to whom it reports annually. Financing comes mainly through public funds supplied by advertising revenue on television, as well as various other revenue sources.

Origins

The CBC was created by an Act of Parliament on 2 November 1936. It took over the facilities and staff of the Canadian Radio Broadcasting Commission, which had been the first attempt to provide a national broadcasting service. The CBC offered programs of mainly Canadian content in English and French. A few programs were relayed from the British Broadcasting Corporation (BBC), which had introduced its Empire Service in 1932.

Shortly after the CBC's formation, a study of coverage and reception conditions across Canada showed that only 49 percent of listeners, mostly in the large cities, could receive the CBC. A start was therefore made on network expansion. In 1937 two high-power stations were installed—CBL at Hornby near Toronto and CBF at Verchères near Montreal—which increased coverage to 76 percent of the population. These stations were followed by CBK Watrous, Saskatchewan, and CBA Sackville, New Brunswick. These and subsequent high-power stations were publicly owned, with private stations used as supplementary outlets.

In 1947 the Corporation broadcast the first *CBC Wednesday Night,* a program concept that was new to North America—a full evening of ambitious and more serious programming. The idea was borrowed from the BBC, and the program included operas and classical drama. But since most listeners preferred the popular music offered by private commercial stations, the program was short-lived. In the Montreal–Windsor corridor where most Canadians live, American stations offered a greater variety of program formats and choices. This is still true today.

During World War II, listeners heard some of the most spectacular and successful news reporting in Canadian radio history. CBC was the first broadcaster to use broadcast vans, known as "Big Betsy," to make disc recordings on location. Some American correspondents also made use of these facilities for their own news reports. Art Holmes made the best recordings of the sounds of battle from the London Blitz, as well as from remote and dangerous warfronts. Listeners were glued to their radios each evening to hear news from correspondents such as Matthew Halton, Bob Bowman, Peter Stursberg, and Marcel Ouimet.

After World War II, many private stations began to press for networks of their own, for better facilities, and for increased coverage by using high-power frequencies originally allocated to the CBC. They also believed that regulatory authority on licensing should not belong to a body that was itself engaged in broadcasting. Between 1936 and 1958 the CBC had also regulated private broadcasting, and therefore the Broadcasting Act of 1958 established the Board of Broadcast Governors. A further Act in 1968 established what is now known as the Canadian Radio-television and Telecommunications Commission (CRTC) to "regulate and supervise all aspects of the Canadian broadcasting system." The CRTC has wide regulatory powers and also has the authority to issue broadcasting licenses and prescribe their terms. It is located in Gatineau, Quebec, adjacent to Ottawa.

After television was launched in 1952, CBC Radio gradually lost much of its audience and popularity. In the 1970s CBC Vice President Laurent Picard launched a campaign to revive radio. Local news and feature programming were increased, and longer network shows were introduced. This strategy recaptured much of the audience. Nevertheless, during the last decade the CBC, in common with other public broadcasters such as the BBC and the Australian Broadcasting Corporation, has had to downsize some of its operations. Television and radio production facilities in smaller centers have either closed or reduced their output.

The CBC operates four radio services, two in English and two in French. The French division is known as *Société Radio-Canada* and has its head office in Montreal. The English networks originate from Toronto and are known as Radio One and Radio Two. Radio One provides a full service of general interest and information programs, whereas Radio Two offers mainly classical music and fine arts. In the far north there are periodic broadcasts in the various aboriginal languages throughout the day.

News and Current Affairs

From its foundation, the CBC had made use of the entire Canadian Press news service, because it was free. Following the outbreak of World War II, the Corporation felt the need to have its own independent news service. This was inaugurated on 1 January 1941, with Lorne Greene as its first anchor. Greene joined the CBC in 1939 and became known as the "voice of doom" because of his deep, ominous voice. He was best known for his role as Ben Cartwright in the American television series *Bonanza*.

Today, CBC Radio broadcasts news bulletins every hour. These are mostly five minutes in duration, but at peak listening times they run to ten minutes (e.g. *World Report* and *Canada at Five*). Each evening, the *World at Six* covers news stories in greater depth. Canadian politics is covered each weekend on *The House*.

Comedy and Variety

The Happy Gang was one of Canada's longest-running radio shows (1937–59). Hugh Bartlett was the program's first host. The show's music and comedy were a source of lighthearted entertainment during the dark years of World War II, and the program caught the imagination of Canadian listeners over the years.

In 1946 the comedy team of *Wayne and Shuster* joined CBC Radio. Johnny Wayne and Frank Shuster were probably Canada's best-known comedians. They had already been performing for 15 years, initially with CFRB in Toronto and later touring Canada and entertaining Canadian troops during the war. The show was first called *Johnny Home,* and it ran for 52 weeks. The name was soon changed to *Wayne and Shuster,* the first time the Corporation had ever used performers' names in the title of a network show. The program was truly Canadian. One broadcast featured sportscaster Foster Hewitt and included a skit of a hockey game between the Toronto Maple Leafs and the "Mimico Mice." Wayne and Shuster also did several sketches from Shakespeare's plays. The program switched to television in 1954 and was often featured on American TV as well. Wayne died in 1990 and Shuster in 2002. The show may still be seen in reruns.

Royal Canadian Airfarce was the CBC's premier comedy series, airing on radio from 1973 until 1997. Based on current news headlines, it is now carried only on television and stars Roger Abbott, John Morgan, Don Ferguson, and Luba Goy. In 1996 the CBC announced budget cuts that would seriously affect the show. Ivan Fecan, the former CBC executive who helped launch the television series, had left the Corporation and was an executive at Baton Broadcasting, one of Canada's major broadcasting groups. He offered the cast a lucrative contract with Baton. Realizing that this would cause a programming calamity, CBC decided to rescind the cuts. The show now has the highest rating on CBC television.

Telephone Call-in Programs

Cross Country Checkup, a national call-in program broadcast on Sundays, began in 1965 from Montreal. It is something of a tradition, having been broadcast almost without a break since 1965. It is a vehicle for listeners to air their views on various Canadian issues. The program got its name from the topic of the first broadcast, which asked whether there should be a national publicly funded health care system. People liked it, so the name was kept. Since 1995 Newfoundlander Rex Murphy has hosted the program. Nearly half a million listeners tune in each week to hear or take part in a lively discussion with invited guests. The program now originates from Toronto.

As It Happens debuted on 18 November 1968, initially as an experiment. The weekly program originated in Toronto and was broadcast "live" for two hours in each time zone across Canada. The first hosts were Harry Brown and William Ronald. They held a dialogue by telephone with politicians and newsmakers from across Canada and around the world. In 1971 Barbara Frum and Cy Strange joined the team. For the next ten years, Frum attracted a large and loyal audience before moving on to television. She was one of Canada's best-known journalists and media personalities until her death in 1992 at the age of 54.

In 1973 *As It Happens* changed to its current 90-minute format each evening, still with two hosts. Some of the events covered in the 1980s and 1990s included the formation of new Canadian political parties and the Quebec separation referendum. Among those on the team over the years have been Alan Maitland, Elizabeth Gray, Dennis Trudeau, and Michael Enright. The present hosts are Mary Lou Finlay, one of Canada's most respected broadcast journalists, and Barbara Budd, broadcaster and actress.

Feature Programs

Ideas, originally called *The Best Ideas You'll Hear Tonight*, began in 1965 and is a program of contemporary thought with a listening audience of 350,000. It is presented by Paul Kennedy each weekday evening. The program includes the Massey Lectures, which are sponsored by CBC Radio in cooperation with Massey College in the University of Toronto and which were created to honor former Governor-General Vincent Massey.

Richardson's Roundup debuted in 1997 and is broadcast on weekday afternoons on Radio One. It is hosted by Bill Richardson and originates in Vancouver. Richardson is much loved by CBC Radio audiences for his work as a host on both Radio One and Radio Two. The heart of the *Roundup* is its listeners, with their stories, music requests, letters, and phone calls. The storytellers are entertaining and sometimes outrageous. The music can be anything from the latest pop melody to songs of long ago.

This Morning, hosted by Jennifer Westaway, is a show about Canada—a showcase of Canadian writers, musicians, and artists. It is a lively mix of drama, comedy, satire, and music. In the fall of 2002 the CBC revised its program schedule, and *This Morning* was reduced from a three-hour to a one-hour show.

This Morning is followed by *Sounds Like Canada*, a 2-hour mix of in-studio and on-location production, where former *This Morning* host Shelagh Rogers seeks to fill the airwaves with the voices and sounds from all over Canada and bring them into the listener's home. The highlights of each day's program are rebroadcast in *Sounds Like Canada Tonight*.

Science and the Environment

Quirks and Quarks, CBC's award-winning science program, has been on the air for more than 20 years. Each week, science journalist Bob McDonald features information about new scientific discoveries. He also looks at political, social, environmental, and ethical issues. The program has an estimated audience of half a million in Canada, as well as an international audience on shortwave via Radio Canada International (RCI). It continues to be the most enjoyed program on CBC Radio and was recently chosen as one of the top ten programs in the world by an international shortwave radio journal.

Drama

CBC English Radio is a major producer of drama on Canadian radio. In 1998–99, close to 100 original radio drama episodes were broadcast, most of them by Canadian writers and all of them with Canadian casts. *(For full discussion of Canadian drama programs, see the Programming and Drama sections of this survey.)*

Sports

Live sports events are mainly featured on television; however, reports on games are covered throughout the day in the regular newscasts. *Inside Track* began in 1985 and includes interviews with Canadian and overseas sports personalities. The program covers topics ranging from the Olympics to national sports and has won several awards over the years for its documentaries.

Cross-Cultural Programs

Because Canada is a bilingual country, the CBC attempts to build bridges between the two official language communities by providing a number of cross-cultural programs. *C'est la vie* is a weekly current-affairs program on francophone issues and life in French-speaking Canada, including short vignettes and interviews with people in the news. A popular feature of the show is "Word of the Week," where listeners can improve their French vocabulary. The program is hosted by Bernard St.-Laurent, a former host on *Cross Country Checkup*. Another cross-cultural program is *À propos*, featuring francophone music for English-speaking audiences.

Other Programs

CBC Overnight, inaugurated in 1995, airs between 1:00 and 6:00 A.M. in each time zone. It includes information highlights from other public broadcasters around the world, such as the BBC, Deutsche Welle, Radio Australia, and South Africa's Channel Africa. There are CBC newscasts every hour. The program has become very popular, especially with night-shift workers.

Music and the Arts

In 1998 English Radio presented approximately 80 orchestral broadcasts and 120 chamber music concerts. Canadian choirs are also heard regularly on English Radio.

Gilmour's Albums was the longest-running CBC program hosted by one person. Every weekend for nearly half a century, the gentle, knowledgeable, and impeccable Clyde Gilmour presented a variety of music or spoken-word recordings, anything from Bob Newhart and Rosemary Clooney to the classics. Every record he featured on his show was from his own collection. Born in Alberta, Gilmour began his broadcasting and journalistic career as a movie critic, interviewing many famous personalities and reviewing movies. He died in 1997 at the age of 85, and his record collection now forms part of the CBC Music Library.

Music for A While is a Radio Two early-evening program of chamber and orchestral classical music, hosted by Danielle Charbonneau, a well-known French radio announcer from Montreal.

The *Vinyl Cafe* is an imaginary record store where the eccentric owner Dave lends program host Stuart McLean compact discs and vinyl albums for use on the show. A regular feature is the "Concert Series," where McLean visits concert halls across Canada and records material for broadcast each month.

Regional Broadcasting

Each CBC region has local programming during the morning and afternoon drive times and at lunchtime. One long-running weekend morning show in Ontario and Quebec is *Fresh Air,* currently hosted by Jeff Goodes. It features a variety of music and interviews. A popular segment, especially with seniors, is "Adrian's Music," in which CBC music archivist Adrian Shuman researches songs listeners have requested. The program lives up to its name, providing a breath of fresh air.

External Broadcasting

RCI is the overseas service of the CBC, originating from Montreal. It broadcasts mainly on shortwave, using high-powered transmitters from a 316-acre site at Sackville, New Brunswick, on Canada's Atlantic coast. This location provides an excellent transmission path to Europe and Africa. In parts of the world where reception is difficult, airtime is leased on other international stations as part of a reciprocal arrangement. The first test transmissions in English and French were made on 25 December 1944 for Canadian forces in Europe, and the station officially opened on 25 February 1945.

RCI's programming includes live relays of some domestic shows, such as the *World at Six* and *As It Happens*. Others, such as *Quirks and Quarks* and the *Vinyl Cafe,* are broadcast at prime-time hours in the listening area. RCI also produces its own programs, catering to an international audience. In addition to news, there is a weekly show, *Maple Leaf Mailbag,* in which listeners' letters and questions about Canada are answered. *Business Sense* takes an in-depth look at Canadian companies that are making their mark in the global economy. In *Media Zone,* Canadian journalists express their ideas about topical issues facing Canadians. *Spotlight* focuses on all facets of artistic and cultural life in Canada. *Canada in the World* looks at Canadian initiatives around the world and considers how Canada deals with other countries on a multilateral basis.

RCI has an estimated audience of 5 million listeners each week and is heard in the Americas, Europe, the Middle East, Africa, and the Far East. It is on the air in seven languages: English, French, Spanish, Russian, Ukrainian, Chinese, and Arabic. Music and feature programs are also sent to about 200 radio stations in other countries.

RCI is funded by a grant from the Canadian government. In recent years it has faced a number of potential closures due to budget cuts, but it has managed to survive.

Strategic Directions

In recent years, many of the AM stations have moved to FM, which is cheaper to run and provides better audio quality. In 2002 the CBC's program services were distributed through satellite in combination with microwave and land lines, feeding 103 CBC-owned stations, 1,164 CBC rebroadcasters, 26 privately affiliated stations, and 282 affiliated or community rebroadcasters and stations.

As mentioned above, during 2002 CBC Radio made several changes to its program schedules in order to attract more listeners to Radio One and Radio Two. Some programs were reduced in length, others were discontinued and new programs introduced. *This Morning* and *Sounds Like Canada* have already been mentioned. Another example is *Dispatches,* a weekly program in which veteran reporter Rick McInnes-Rae takes listeners to places they may never see, hear voices of people from these places, and provides insights and issues that they might never have confronted before.

Radio 3 (note that the numeral is used here, unlike the other English networks) is unique. It was originally planned as the CBC's third off-air FM radio network, targeted at young people. In December 1999 it was felt that the timing was not favorable for the launch of a third off-air network, so in June 2000 Radio 3 was launched, producing on-air programming for 33 hours a week on Radio Two, and on the internet primarily as a small portal to several other core websites. Radio 3 programs carried on CBC Radio Two include weekend programs *RadioSonic, JustConcerts, NewMusicCanada* and *Radio On,* along with the weekday program *BraveNewWaves.* Radio 3 is also exploring digital audio broadcasting in Vancouver from the existing DAB experimental transmitter.

COLIN MILLER

Further Reading

Ellis, David, *Evolution of the Canadian Broadcasting System, 1928–1968*, Ottawa: Department of Communications, 1979

Frum, Barbara, *As It Happened*, Toronto: McClelland and Stewart, 1976

Gzowski, Peter, *Peter Gzowski's Book about This Country in the Morning*, Edmonton, Alberta: Hurtig, 1974

Hall, James L., *Radio Canada International: Voice of a Middle Power*, East Lansing: Michigan State University Press, 1997

McNeil, Bill, and Morris Wolfe, *Signing On: The Birth of Radio in Canada*, Toronto: Doubleday Canada, and Garden City, New York: Doubleday, 1982

Nash, Knowlton, *The Microphone Wars: A History of Triumph and Betrayal at the CBC*, Toronto: McClelland and Stewart, 1994

Nash, Knowlton, *Cue the Elephant! Backstage Tales at the CBC*, Toronto: McClelland and Stewart, 1996

Peers, Frank W., *The Politics of Canadian Broadcasting, 1920–1951*, Toronto: University of Toronto Press, 1969

Skene, Wayne, *Fade to Black: A Requiem for the CBC*, Vancouver, British Columbia: Douglas and McIntyre, 1993

Stewart, Sandy, *A Pictorial History of Radio in Canada*, Toronto: Gage, 1975; revised edition, as *From Coast to Coast: A Personal History of Radio in Canada*, Montreal and New York: CBC, 1985

Troyer, Warner, *The Sound and the Fury: An Anecdotal History of Canadian Broadcasting*, Toronto: Wiley, 1980

Canadian Radio and Multiculturalism

Over the last three decades radio, no less than other entertainment and information services, has responded to the fact of multicultural communities. Gender, sexual orientation, physical and mental disabilities, language, race, and ethnicity—each label signifying a category of people—have become the elements of a discourse focusing on equality, recognition, and identity. Since the inauguration of the Royal Commission on Bilingualism and Biculturalism in Canada during the early 1960s and the United States Congress' passage of the Ethnic Heritage Studies Program Act in 1974, broadcasting in general—and radio in particular—has had to recognize diversity in its content, in the composition of its workforce, and in its remuneration practices.

The Concept

As an adjective, *multicultural* may be used simply to describe the demographic realities of a state, region, or community. As a noun, *multiculturalism* may be used to refer to a particular ideology, wherein racial and ethnic diversity is highly valued in and of itself. The term *Multiculturalism* may also refer to social policies that recognize demographic realities, promote equality, and combat racism and discrimination.

A major difference between Canadian and American usage of the concept is that in Canada, the idea of multiculturalism has become a foundation for federal policy, whereas in the United States it remains a topic of heated debate in civil society. Multiculturalism in Canada assumed yet another nuance of meaning when multiculturalism within a bilingual (French/English) framework became a policy of the federal government. Thus, the emergence of multicultural policies in Canada was rooted in the fact of French/English biculturalism. English and French language private radio stations were on air at the beginning of broadcasting in Canada. The national public broadcaster (CRBC, now the CBC) began programming in 1932 in both languages; by 1938 separate English and French networks were in place. The story of Canadian radio and multiculturalism is situated in the context of this duality.

A subsequent set of policies and constitutional changes occurred during the 1970s and the early 1980s. The Official Languages Act of 1969, the Multiculturalism Act of 1971, and the Canadian Human Rights Act of 1978 set the groundwork for the Canadian Charter of Rights and Freedoms, which accompanied the Constitutional Act of 1982. Each of these acts added legitimacy to the application of multicultural policies in broadcasting.

In order to trace the path of multicultural policies in broadcasting, it is necessary to look at the interplay of four institutions: (1) the Broadcasting Act and the Canadian Broadcasting System, (2) the Canadian Radio-television and Telecommunications Commission, (3) the Canadian Association of Broadcasters, and (4) the Canadian Broadcasting Standards Council. Each of these institutions is the locus of policies and regulations governing the recognition of diversity in broadcasting.

Although there have been several versions of the Broadcasting Act since 1932, references here are to the most recent version, which was passed in 1991. It is significant that the act considers Canadian broadcasters, public and private alike, as a single system—the Canadian Broadcasting System—to which the act and all subsequent regulations apply. Furthermore, the intrusion of the act into the private sector is based on the premise that the airwaves are public property. It is within the general mandate of the "enhancement of national identity and cultural sovereignty . . . operating primarily in the English and French languages" that multicultural policies and regulations must find their niche. The act accents both employment equity and program content with respect to cultural diversity, noting gender, age, aboriginal status, language, and cultural and racial diversity.

The task of implementing this position was given to the Canadian Radio-television and Telecommunications Commission (CRTC). The CRTC's mandate was originally provided for in the 1968 Broadcasting Act.

Guidelines for the development of employment equity are set out in the commission's regulations, and licensees are required to report on progress. As for programming, the commission requires that it reflect the linguistic duality, cultural diversity, and social values of Canada, as well as national, regional, and community voices.

A good deal of the monitoring and vetting of complaints defers to the Canadian Association of Broadcasters (CAB) and its creation, the Canadian Broadcast Standards Council (CBSC). CAB members include 402 radio stations, 78 television stations, 1 network, and 15 specialty services. By February 1998, the CAB had two codes in place, one specific to violence on television and the other a general code of ethics. The human rights clause in the code stipulates that,

> every person has a right to full and equal recognition and to enjoy certain fundamental rights and freedoms, broadcasters shall endeavor to ensure, to the best of their ability, that their programming contains no abusive or discriminatory material or comment that is based on matters of race, national or ethnic origin, color, religion, age, sex, marital status, or physical or mental handicap . . . [and, further, that] television and radio programming shall portray the wide spectrum of Canadian life. Women and men shall be portrayed with fair and equitable demographic diversity, taking into account age, civil status, race, ethnocultural origin, physical appearance, sexual orientation, background, religion, occupation, socioeconomic condition, and leisure activities, while actively pursuing a wide range of interests. Portrayals should also take into account the roles and contributions of the mentally, physically and socially challenged.

The articles of this code are the basis upon which complaints are received by the CBSC. The CBSC, an independent, nonprofit organization, was established by the CAB. Its membership includes 387 private-sector radio and television stations and networks programming in English, French, and other languages. With the approval of the CRTC, the council plays an intermediate regulatory role, free of government formalities and sanctions.

The Howard Stern Affair: A Case in Point

The response of the CBSC to complaints regarding Howard Stern's appearance on Canadian radio and the response of the radio stations to the council provide a recent illustration of how the system operates within the regulatory field surrounding cultural diversity. The *Howard Stern Show* was first syndicated to Canada on 2 September 1997 and aired over CHOM-FM in Montreal and CILQ-FM in Toronto. Over the two weeks following the first broadcast, the CBSC received over 1,000 signed complaints in addition to complaints directed to the CRTC, which were forwarded to the CBSC. The CBSC, in turn, forwarded all complaints to the broadcasters, who, according to established procedures, were required to respond in writing to each of the complainants.

The complaints opened at least three lines of inquiry. First, the national origin of the broadcast brought into play sections of the Broadcasting Act limiting foreign content and talent. Second, the show appeared to violate the CAB Code of Ethics with respect to offensive statements directed to cultural groups, and third, it appeared to violate the code regarding sex-role stereotyping. The stations' responses to the complainants noted that the content of the show did not reflect the views or opinions of the broadcasters and that on-air advisories were given; furthermore, the responses defended the content as comedy, not intended as serious commentary on social or political issues. The rationale presented by the broadcasters was insufficient to prevent an investigation by the CBSC's regional councils in Quebec and Ontario. It is important to note that such an exercise emerges from voluntarily established codes drawn up by the CAB, an industry-wide trade association. The CBSC condemned the show. The sanctions were twofold: first, the stations were required to announce the decisions during prime time and within 30 days of notification and to provide confirmation of the airing of the decisions to the CBSC and to each of the complainants. The CBSC can do no more. Second, the response of the stations will influence license renewal hearings before the CRTC, which has the authority to refuse renewal. CHOM-FM in Montreal canceled the show, and CILQ-FM in Toronto continued to carry it.

Radio Practices

Apart from employment equity, an improving situation in Canadian radio, and the curbing of bluntly offensive material, multicultural sensitivity in radio entails the broadcasting of programs of special interest to minorities, of minority languages, and of popular music programming that meets minority needs, in other words, programming that permits members of minority groups to "hear themselves speak." Mainstream French- and English-language stations are weak in this respect. A few, very few, in metropolitan areas will broadcast alternative music and music specialized to the interests of various ethnic groups, and some will, in off hours, broadcast local community events in a language other than French or English. But for the most part, "ethnic" programming is to be found on alternative radio—community and internet radio.

Radio stations owned and operated by universities and colleges are the most notable in this respect. Most such stations broadcast directly to identified minorities in their region. For example, CJSW-FM, broadcasting from the University of Calgary, airs programs in German, Serbian, Croatian, and Chinese. CHSR-FM, operated by the University of New Brunswick in Fredericton, broadcasts multilingual programs sponsored by their South Asian, Muslim, and Chinese Student Associations. CISM-FM, owned by the University of Montreal and broadcasting in French, provides Afro-Haitian coverage. CHUO-FM, the station of the University of Ottawa, broadcasts in French and English, with coverage in several other languages, including German, Korean, and Cantonese.

In addition, independently owned and membership-supported community radio stations provide programming for a variety of minorities. CHIR-FM, Toronto, broadcasts regular programming in Greek and English over cable and the internet. CIBL-FM in Montreal, broadcasting principally in French, covers a wide variety of music meeting the interests of several local minority groups. CFRO-FM out of Vancouver presents programs in Spanish, Amharic, Farsi, Armenian, and Salish, a First Nations language. CKWR-FM of Waterloo, Ontario, broadcasts a newsmagazine on Monday evening featuring gay, lesbian, and bisexual issues. Several stations broadcast programs to the Mohawk people of Ontario, Quebec, and New York. These include, among others, CKON-FM, which straddles the three borders in Akwesasne; KWE-FM out of Tyendinaga, near Belleville, Ontario; and CKRK-FM out of Kahnawake, near Montreal. CFWE, "The Aboriginal Voice of Alberta," broadcasts via satellite to 200 communities across Canada. These are but a few of the several private and community stations broadcasting local programs to First Nation communities.

In addition, the CBC inaugurated broadcasting in the Arctic territories in 1960. Presently, CBC Nunavut broadcasts across the eastern Arctic and northern Quebec in Inuktitut and English. In the west, CBC Northern Territories serves its listeners in six different languages with 48 hours of locally produced programming, 17 hours in aboriginal languages. CFRT, Radio Iqaluit, a French language station, serves its listeners in Nunavut and northern Quebec in French, Inuktitut, and English.

JOHN D. JACKSON

See also Native American Radio

Further Reading

Canadian Broadcast Standards Council, <www.cbsc.ca>

Canadian Department of Justice, *Broadcasting Act* (1991, c.11), <www.canada.justice.gc.ca/STABLE/EN.LAWS/Chap/B/B-9.01.html>

Clement, E.U., "Women's Resistence to Paternalism: An Analysis of Selected CBC Radio Drama," Master's thesis, Concordia University, Montreal, 1995

Conseil de la radiodiffusion et des télécommunications canadiennes; Canada Radio-Television and Telecommunications Commission, "Implementation of an Employment Equity Policy," Public Notice CRTC 1992-59, Ottawa, 1 September 1992," <http://www.crtc.gc.ca/archive/ENG/Notices/1992/PB92-59.HTM>

Keith, Michael C., *Signals in the Air: Native Broadcasting in America,* Westport, Connecticut: Praeger, 1995

Kymlicka, Will, *Multicultural Citizenship: A Liberal Theory of Minority Rights,* New York: Oxford University Press, 1995

Li, Peter S., editor, *Race and Ethnic Relations in Canada,* Toronto and New York: Oxford University Press, 1990; 2nd edition, Don Mills, Ontario, and New York: Oxford University Press, 1999

Royal Commission on Bilingualism and Biculturalism, *Report of the Royal Commission on Bilingualism and Biculturalism,* 5 vols., Ottawa: Queen's Printer, 1967; see especially vols.1 and 4

Smelser, Neil J., and Jeffrey C. Alexander, *Diversity and Its Discontents: Cultural Conflict and Common Ground in Contemporary American Society,* Princeton, New Jersey: Princeton University Press, 1999

Stiles, J. Mark, and William Litwack, *Native Broadcasting in the North of Canada: A New and Potent Force,* Ottawa: Canadian Commission for UNESCO, 1988

Canadian Radio and the Music Industry

Preserving Canadian Culture

As is the case in many countries, relations between the radio and recording industries in Canada have been marked by ongoing tensions. Many of these, such as the desire by private broadcasters to reach audiences older than the adolescent population of active record buyers, are not unique to Canada. Others are rooted in the differences between the ownership and regulatory frameworks under which these two industries operate in the Canadian context. Although all radio stations in Canada are Canadian owned (as required by law), over 80 percent of all recordings sold in Canada are released by the Canadian subsidiaries of multinational music companies. Whereas the radio industry is subject to a high level of regulation, multinational record companies within Canada operate under few restrictions. Although radio broadcasters have no economic interest in the success of Canadian musicians and recordings, they have often taken the blame for the lack of such success and have been the vehicle for policies intended to remedy that lack.

The situation is further complicated in Canada by a broadcasting system that includes four public networks (all operated by the Canadian Broadcasting Corporation [CBC]) and some 700 private radio stations. Although the CBC is committed, in its mandate and ongoing programming strategies, to the promotion of Canadian culture, music occupies a much less important place within its schedule than is the case for private broadcasters. Indeed, the prominence accorded to such commercially marginal forms as classical, jazz, and blues music within the programming of the CBC has, throughout much of its history, reduced its usefulness to a domestic recording industry that is oriented toward rock and other commercial forms. Although private radio has a much greater impact on the dissemination of music, it has no commitment to stimulating the sales of Canadian recordings beyond those measures imposed upon the industry by government regulators (discussed later).

Throughout most of the 20th century, the growth of the private radio industry in Canada has been intimately bound up with developments occurring in the United States. From the 1920s through the 1940s, connections between radio in the two countries were formal, as the most important English Canadian private radio stations (such as CFCF in Montreal and CFRB in Toronto) affiliated themselves with the Columbia Broadcasting System (CBS) or the National Broadcasting Company (NBC) and carried their programming. Although French-language radio, during this same period, broadcast concerts by francophone folksingers or country musicians (such as La Bolduc), music programming on English-language radio included large numbers of music and variety shows originating in the United States. In the 1950s, Canadian radio programming was transformed along lines similar to those observable in the United States, away from network-based block programming and toward local station formats based on recorded music and on-air announcers. Although this development ended the direct affiliation of Canadian radio stations with U.S. networks, it also led to their adoption of music programming formats that were often identical to those developed in the United States and to playlists dominated by music from elsewhere.

In the 1960s private radio broadcasting in Canada came under increased criticism as musical performers and owners of Canadian-based recording companies blamed radio for the flight of Canadian musicians to the United States. Informal surveys made public during that decade showed alarmingly low levels of Canadian recordings on the playlists of Canadian radio stations. Among the journalists urging the Canadian government to take action to support Canadian music was Walter Greely, editor of the music industry trade magazine *RPM,* who pushed for the establishment of Canadian quotas for radio station playlists. Although many radio stations undertook voluntary campaigns to promote Canadian music during the late 1960s and early 1970s, the pressure to enact regulations to support a domestic recording industry mounted.

When implemented in 1971 by the Canadian Radio and Television Commission, Canadian content regulations required of radio stations that a significant percentage of their musical selections be Canadian. The Canadian character of recordings is determined by a point system, which assigns up to four points to selections depending on whether the artist, music composer, lyricist, or producer is Canadian. To qualify under Canadian content regulations, a musical selection must be accredited two points. Although the percentages of Canadian music programming required of broadcasters have increased several times since 1971 (by 1999 the required level for commercial broadcasters was 35 percent), these regulations have generated ongoing controversy over their fairness and effectiveness. For example, although certain recordings by the Canadian artist Bryan Adams have been disqualified (because their producer and composer or lyricist were not Canadian), others, such as U.S. singer Jennifer Warnes' renditions of Leonard Cohen compositions, have been deemed to have Canadian content (because the lyrics and music were by Cohen, a Canadian.) To many observers, this seems an unjust fixation on technical criteria at the expense of the obviously greater "Canadian" quality of one national celebrity over another.

Throughout the 1970s, radio stations complained that a shortage of high-quality Canadian recordings limited their ability to meet these quotas while maintaining the interest of listeners. Nevertheless, this same period was one that saw a significant growth in the Canadian-owned sector of the domestic recording industry, as important record companies (such as Attic and Anthem) produced hit recordings and developed back catalogs. Although a causal link between radio quotas and the growth of a domestic recording industry has never been demonstrated in convincing statistical fashion, both are seen as having laid the groundwork for the enormous worldwide success of Canadian music in the 1990s, when Celine Dion, Shania Twain, Bryan Adams, and Alanis Morissette became among the world's best-selling recording artists.

As in the United States, relationships between the recording and radio industries have been marked by tension over conflicting objectives. Whereas record companies typically wish for radio stations to play new, youth-oriented music, Canadian radio stations since the 1970s have moved away from such music in an effort to reach older, more affluent audiences. A 1998 survey by Statistics Canada suggested that the most popular format on Canadian private radio stations was adult contemporary (at 25 percent of the total audience), followed by oldies, all-talk, and country formats, at approximately 12 percent each. Indeed, adult contemporary is even more popular in the predominantly French-speaking province of Quebec, where record companies often complain that the only forms of French-language music likely to receive airplay are those involving solo vocalists. Although long-term analyses of the Quebec music industries have stressed the need to develop a greater variety of musical genres in order to ensure that younger generations maintain an interest in francophone music, the homogeneity of radio formats in the province is seen as a significant barrier to such variety.

More specific tensions between the radio and recording industries have to do with the effects of repeated airplay on the sales of records. Here, there are significant differences between language groups. French-language record companies frequently complain that their recordings are overexposed on radio stations, played in high rotations that dissuade potential buyers from purchasing them. This overexposure is the result of government regulations that insist on a high level of French-language content for stations designated as French, and of audience tastes that favor locally produced French-language music over music from France or other francophone companies. English-language record companies, in contrast, complain about low levels of rotation for new recordings, making it difficult for songs to "take off" and acquire hit status.

Because the allocation of frequencies to new radio stations in Canada is determined by a federal regulatory agency, following competitive public hearings, public debate over the musical needs of particular communities is often heated. A well-known (and unsuccessful) attempt to license a dance music station in Toronto, for example, involved lobbying on the part of African-Canadian community groups, associations of nightclub disc jockeys, and small record companies. Government regulators awarded the frequency to those proposing a country music format, citing the greater commercial viability of a country station and responding to a well-organized campaign on the part of the local country music industry. Nevertheless, radio stations, once licensed for a particular format, may usually shift formats in response to changing market conditions with relative ease. Since the late 1990s, the Toronto area has had dance-oriented formats as a result of strategic format changes by stations licensed for other purposes. This tension between a public service conception of radio broadcasting and market competition will likely become more acute with the introduction of digital radio. As radio-based music services become more specialized, it is difficult to imagine how formats devoted to tightly defined niches (such as Broadway musicals or cool jazz) will meet Canadian content quotas or other regulatory objectives intended to support local music industries.

WILL STRAW

Further Reading

Canadian Communications Foundation, "The History of Canadian Broadcasting," <www.rcc.ryerson.ca/ccf/index2.html>

Filion, Michel, "Radio," in *The Cultural Industries in Canada: Problems, Policies, and Prospects,* edited by Michael Dorland, Toronto: Lorimer, 1996

Government of Canada, "Statistics Canada," *The Daily* (22 July 1999)

Straw, Will, "Sound Recording," in *The Cultural Industries in Canada: Problems, Policies, and Prospects,* edited by Michael Dorland, Toronto: Lorimer, 1996

Vincent, Pierre, "Les grandes compagnies de disques: Que font-elles avec notre argent," *La presse* (8 July 1971)

Canadian Radio Drama

Reflecting the country's largely two-language culture, Canadian radio drama exhibits a distinctly two-sided appearance. Quite different cadres of writers, actors, and producers developed often strikingly different dramatic traditions.

French Radio Drama

It is within the context of private radio, and more specifically at CKAC—the first French-language radio station in North America, launched in 1922—that the first experiments in radio drama writing were developed. There have been close to 100 French radio drama series aired since 1923, lasting from one to ten years, and nearly 300 radio drama serials, which were aired by various Quebec radio stations, particularly by CKAC (1923 to 1955), CKVL (1948 to 1968), and CBF (1937 to 1972). Nearly 500 authors and more than 5,000 of their Quebec creations had been broadcast up to 2000.

Origins of Radio Drama at CKAC

On 5 April 1923, less than one year after its inauguration, CKAC broadcast its first radio play, "*Félix Poutré*"—an 1871 classic of Quebec literature written by Louis Fréchette that was produced by Jacques-Narcisse Cartier, the president and founder of CKAC. The production used 13 actors placed around a microphone and presented eventful moments from the trial of a *Patriote* in the Canadian Revolution of 1837, who was condemned to death but was saved from hanging thanks to his convincing mimicry of madness.

The next step was to create a general cultural program, *L'Heure provinciale* (1929–39), funded by the Quebec government, which broadcast, aside from poetry and opera, excerpts from classical French and Quebec repertories. The producer of this program, Henri Letondal, drew from various theatrical publications, such as *La Petite Illustration,* for his choice of contemporary plays. *La Demi-heure théâtrale du Docteur J.O. Lambert* (1933–37) also presented adaptations of European works and laid the ground for yet another phase, the development of a program dedicated to works from Quebec, *Le Théâtre de chez-nous* (1938–49), created by Henri Letondal. These radio plays explored typical Quebec situations, drawing from the comic, dramatic, psychological, and social essences of the culture and using a language marked by the vocabulary and accents of the 1930s.

Radio productions from Quebec can be grouped into three types of dramatic writing, each of which left its mark on various eras: (1) radio dramas, (2) radio serials, and (3) comic sketches. Radio dramas stemming from stage productions were established at the end of the 1930s with the series "*Le Théâtre de chez-nous*"*;* the radio play *Un coucher de soleil* (1942) by Henri Letondal is a good example of this genre. The radio drama was defined more as a style of radio writing during the 1950s, and it became a more distinctly experimental and symbolic language within the scope of the radio serial. *Le Coureur de marathon* (1951) by Claude Gauvreau, *L'Homme qui regardait couler l'eau* (1951) by Yvette Naubert, and *Confession d'un héros* (1961) by Hubert Aquin, as well as the works of Louis Pelland, *Le Véridique procès de Barbe-bleue* (1954) and *Voltaire s'en va-t-en Canada* (1971), are models of this trend.

However, the first major works of radio were created within the genre of the serial: *Le Curé de village* (1935–38) and *La Pension Velder* (1937–42) by Robert Choquette, and *Un homme et son péché* (1939–65) by Claude-Henri Grignon. Then, during World War II, production increased, and works of note such as *Jeunesse dorée* (1940–65) by Jean Desprez and *La Fiancée du commando* (1942–47) by Paul Gury were aired. During the 1950s and 1960s, themes became more diverse; a work of anticommunist propaganda during the Cold War stands out—*Béni fut son berceau* (1951) by Françoise Loranger—as does a portrait of regional and maritime life in Quebec's Gaspé region—*Je vous ai tant aimé* (1951–54) by Jovette Bernier. The cycle of radio serials that developed from 1934 to 1974 included an original work based on criminal intrigue, *Marie Tellier, avocate* (1964–69) by Maurice Gagnon; crime shows were a genre rarely practiced at Radio-Canada. A characteristic of these radio serials was the use of short scenes that fragmented the plots and sectioned them into five 15-minute programs per week. The content was built around known character types, evolving within complex situations, which in turn were interlaced within several plots. A narrator explained the essential elements necessary for understanding the plot. The historical serial was generally built around a series of programs in which the plot was completed within each episode, but each of these smaller plots also fit into a longer chronological narrative. Each program was a 30- or 60-minute weekly event. *Le Ciel par-dessus les toits* (1947–55) by Guy Dufresne, *L'Histoire du Canada* (1957–60) by Jean Laforest, *Histoire de Montréal* (1967–68) by Yves Thériault, and *Les Visages de l'Amour* (1955–70) by Charlotte Savary are important works in this genre. Finally, comic sketches filled in certain daily niches, such as *Quelles nouvelles?* (1939–58) by Jovette Bernier or *Chez Miville* (1956–70) with the authors Albert Brie, Louis-Martin Tard, Louis Landry, and Michel Dudragne. These authors developed a characteristic language and dialogue structure, making good use of witticisms, puns, irony, and situational comedy revolving around archetypal characters. The themes explored were sociopolitical and served as a focal point for a critique of Quebec society in the 1960s, the period of the Quiet Revolution in Quebec.

From Private Radio at CKAC to Public Radio at CRCM and CBF

By the end of the 1930s, the status of the principal radio genre was sufficiently strengthened that public radio in Canada could propose a programming schedule that contained the principal models already being used at CKAC. Between 1932 and 1936, the Canadian Broadcasting Commission (CBC) set in place an important programming schedule while Canada waited for the law that would create public radio, CBC/Société Radio-Canada (1936). Two series broadcast on CRCM are worth mentioning: *Promenades en Nouvelle-France* (1933–34) by Robert Choquette, which presented historical portraits, and *Le Fabuliste La Fontaine à Montréal* (1934) by the same author, which consisted of 15 half-hour comedies. These programs projected a humorous and ironic image of Montreal's bourgeois society. At CBF, launched on 3 November 1937, many years would go by before original Quebec creations would be broadcast within specially designed programs such as the series *Radio-théâtre canadien* (1951–53), even though radio serials were already being broadcast at CBF as early as 1938. During CBF's first 15 years, the station offered works principally by classical and contemporary authors from the international literary scene. These important radio drama series included *Le Radiothéâtre de Radio-Canada* (1943–56), the drama segment of *Radio-Collège* (1941–56), which in 1951 became *Sur toutes les scènes du monde* (1951–75).

Influences of the Massey Report on French Canadian Radio Drama at CBF

In the wake of the *Massey Report* (1951) on Canadian arts and culture, a program was developed, specifically dedicated to young authors, in order to encourage Quebec creations. The program *Nouveautés dramatiques* (1950–62), produced by Guy Beaulne, proposed works from numerous authors; some of these works were, from a dramatic perspective, of very high quality. Other works of note include *Zone interdite* (1950), a series written by Pierre Dagenais; *Flagrant délit, Billet de faveur* (1955), and *Les Ineffables* (1956), series produced by Hubert Aquin; and *Le Théâtre canadien* (1955). All of these series presented works by young authors such as Jacques Languirand, Luan Asllani, Marcel Blouin, and Georges Cartier, who explored, among other genres, the tradition of the Theatre of the Absurd at the urging of producer Aquin. During the 1960s, two series stood out: *Le Petit théâtre de poche* (1965) and *Studio d'essai* (1968), but one must note that their authors wrote more frequently for television. The 1970s ushered in a revival of radio drama broadcasting. *Premières* (1971–86) became the major series, airing the plays of many Quebec authors whose inventiveness, research, and thematic development rejuvenated radio writing. The series *Escale* (1978–83) offered a variety of literary genres: radiophonic short stories, introspective radio dramas, and tales of fantasy or whimsy, as well as monologues and comedic dialogues. *La Feuillaison* (1972–87), produced by Jean-Pierre Saulnier, continued in the tradition of experimental theater. Each of these series submitted its best scripts to the international contest organized by *La Communauté radiophonique des Programmes de langue française,* which awards Le Prix Paul Gilson.

Rupture in Programming and Revival of Radio Drama

If the end of the 1980s marked a rupture in the programming of radio drama at CBF-FM, because the important series had ended by 1987, producers Claude Godin and Line Meloche, beginning in 1991, nevertheless returned to the airwaves with two experimental series: *Videoclip* and *Atelier de création radiophonique*. Jean-Pierre Saulnier proposed a historical series for the occasion of the 500th anniversary of the discovery of America; Saulnier produced a work by Yves Sioui-Durand and Catherine Joncas, *La Redécouverte de l'Amérique* (1992), a wonderfully poetic drama. The next year, Line Meloche produced another historical serial, *Les Fils de la liberté* (1993) by Louis Caron (scripted by Annie Piérard). By the mid-1990s, "radiofiction" opened up toward new aural and thematic aesthetics. Line Meloche was the originator of this change. Radio language in the 1990s was in a state of transformation: explorations included fantasy, tales of crime and espionage, and the new techniques of communication. Concurrently, another original experiment was undertaken by producer and author Cynthia Dubois on the theme of eroticism. The series *Je vais et je viens entre les mots* (1995–96) demonstrated the author's outstanding creativity. Dubois also wrote *L'Arbre de vie* (1997–), a very postmodern work, with its splintered structure incorporating diverse forms of dialogue within a theme in which the couple and the family are the two polar extremes. Between the acerbic dialogue of the conflicting lovers, an interview, an open-line show, or even a monologue could be interjected, all of which converge to create a new dynamic between the author, the characters, the actors, and the listeners.

In radio drama writing, particularly since the end of the 1980s, the treatment of sound has in many ways amplified the construction of meaning. The themes are increasingly supported by the sound effects and the musical selections, the role of which is just as important as scenery and costumes are for stage productions. Furthermore, it has been discovered that textual structure on the radio can support quoted aural inserts, flashbacks, superimposed sophisticated sound effects in the foreground, and very powerful musical sound environments that at times can drown out the voices of the actors or bring about a reduction of the narrative function. Introspective narration has been used extensively in order to convey a particular sense of time and an interior or exterior space different from that of the main narrative voice—a traditional technique that has been given a new contextual function in recent productions. The statements become more complex, supported by more sophisticated technical processes (filters, vocal interplay, distances from the microphone) and by the use of nonlinear and ever more diversified dramatic structures.

Producers

After Jacques-Narcisse Cartier, the first radio producer, Henri Letondal, Robert Choquette, and Fred Barry followed; all of these laid the foundation for the first models of dramatic productions at CKAC. A few names should be added, because at CBF, Jacques Auger, Guy Mauffette, Lucien Thériault, Armand Plante, and Florent Forget all played determining roles in the success of certain productions between 1930 and 1950. In the 1950s and 1960s, producers continued a tradition of defining radio as an art form and leaning toward a more diverse aural aesthetic. Their experiments were all milestones on the road leading to an artistic vision of dramatic writing in Quebec. Their work was an exploration of the new technical means that opened the way to research into the modernity of the language of radio. The early 1970s were a watershed period in the history of Quebec radio theater, where a new interrogation of the specificity of sound began with Madeleine Gérôme, Jean-Pierre Saulnier, and Gérard Binet. One of Jacques Languirand's works, "Feed back" (1971), was significant in this respect because it used the resources of sound recording in order to give structure to the idea of the failure of communication—as it became obsession following a nuclear holocaust—by the repetition of the same aural and narrative motifs. The radio play by Monique Bosco, "Le Cri de la folle enfouie dans l'asile de la mort" (1978), produced by Madeleine Gérôme, presented a sound creation by musician Gabriel Carpentier and was proposed for the Italia Prize. "Belles de nuit" (1983) by Yolande Villemaire, produced by Jean-Pierre Saulnier, won an award, the Prix du concours des oeuvres radiophoniques de Radio-Canada. It must also be noted that producers experimented with new styles in a context of the production of meaning in which the symbolic system of sound participated as an equal partner. During the past decade, producers have explored freer forms and new musical and sound effect codes, which have served as support for new themes and acting techniques.

Since the autumn of 1996, and each season since, a dramatic series, *Radiofictions en direct,* has been broadcast live on the French network, CBF-FM, of La Société Radio-Canada. These radio events, played live in a concert hall, call upon the talents of musicians, a sound effects technician, and actors whose experience ensures an exceptional level of quality to the broadcasts. The producer, Line Meloche, has succeeded in rekindling interest at the end of the century for an art form in which radio theater and stage theater find a common ground.

RENÉE LEGRIS

Further Reading

Hamel, Réginald, editor, *Panorama de la Littérature Québécoise Contemporaine,* Montreal: Guérin, 1997

Legris, Renée, *Robert Choquette: Romancier et Dramaturge de la Radio-Télévision,* Montreal: Fides, 1977

Legris, Renée, *Dictionnaire des Auteurs du Radio-Feuilleton Québécois,* Montreal: Fides, 1981

Legris, Renée, et al., *Propagande de Guerre et Nationalismes dans le Radio-Feuilleton 1939–1955,* Montreal: Fides, 1981

Legris, Renée, "Radio Drama in Quebec," in *The Oxford Companion to Canadian Theatre,* edited by Eugene Benson and Leonard W. Conolly, Toronto: Oxford University Press, 1989

Pagé, Pierre, Renée Legris, and Louise Blouin, *Répertoire des Oeuvres de la Littérature Radiophonique Québécoise: 1930–1970,* Montreal: Fides, 1975

Pagé, Pierre, and Renée Legris, *Le Comique et l'Humour à la Radio Québécoise: Aperçus Historiques et Textes Choisis, 1930–1970,* vol.1, Montreal: La Presse, 1976; vol. 2, Montreal: Fides, 1979

Pagé, Pierre, "La Radiodiffusion/Broadcasting 1922–1997," *Fréquence/Frequency,* 7–8 (1997)

English Radio Drama

Early English-Canadian radio drama was much influenced by radio drama in the United States as practiced by the pioneer networks the National Broadcasting Company (NBC) and the Columbia Broadcasting System (CBS) from the late 1920s. By the mid-1930s drama was American radio's most popular form: from the many popular dramatic series—mystery and adventure, soap opera, variety, and comedy—to the prestigious "sustaining" (unsponsored) anthology dramatic series, serious and experimental. The earliest of these were NBC's *Radio Guild,* which started in 1929, and the CBS *Columbia Workshop,* which began in 1931. Their experimental techniques were aimed at adapting the universal dramatic mode to the limitations and strengths of this new sound-based medium. By 1938, of the 26 leading American evening radio programs, 20 were dramatic. Radio drama was gaining recognition as a distinct creative and technical mode of theater.

Although there were some early radio-drama experiments in Canada starting in the mid-1920s, the most popular radio-drama programs among English-speaking Canadians through the 1930s were the American dramatic series. These were received by Canadians from powerful distant American stations or were broadcast by Canadian affiliates of the major American radio networks. Some were even broadcast by the first Canadian network, the Canadian National Railways Radio Department (CNR Radio) from the late 1920s, and by the nationalized English-Canadian networks that followed it: the Canadian Radio Broadcasting Commission (CRBC) from 1933 until 1936 and the Canadian Broadcasting Corporation (CBC) thereafter. These pioneer American radio dramas had a strong influence on early Canadian dramas in this medium.

The golden age of American radio drama was the decade of the 1930s. Although there were some important American radio-drama achievements even until the mid-1950s, American television soon captured sponsors, budgets, and audiences, relegating American radio to a secondary role. This U.S. cultural and commercial transformation from radio to television was the opportunity for nationalized Canadian radio to complete its own network of radio-drama production centers in the 1940s and to achieve its own golden age. Unlike French-Canadian radio drama, important productions of which were broadcast on several private stations as well as on the French-language CBC (*Radio-Canada*), most significant English-language Canadian radio drama was produced on only the CBC. Although CBC drama "producers" (each one both produced and directed) learned many basic technical and creative lessons from American popular and serious radio dramas, CBC radio drama did develop original creative styles and techniques in the 1940s and 1950s, particularly under its senior producers, Andrew Allan and Esse W. Ljungh. The Canadian golden age of radio drama lasted until well into the 1960s.

The Beginnings of English-Canadian Radio Drama: CNR Radio, CKUA, and the CRBC

The earliest English-Canadian radio-drama network broadcasts were produced from mid-1925 at the Moncton, New Brunswick, station of CNR Radio. They were mainly popular post-Victorian stage plays, transposed to radio without much understanding of the need for adaptation to the sound medium, especially to its lack of visual dramatic cues. The first regular weekly anthology drama series in Canada, called the *CNRV Players,* began broadcasts over CNR Radio's national network in 1926; it was written by Jack Gillmore and produced by him over station CNRV in Vancouver. These broadcasts included adaptations of Shakespeare and other classical plays and fictions, adaptations of many standard modern stage plays, and a few original radio plays commissioned by Gillmore. He grasped the distinctive nature of radio drama, and his radio adaptations for this new sound medium accommodated its strengths and limitations, even before the 1929 start of the NBC *Radio Guild.* Gillmore's series lasted until the 1932 nationalization of the CNR Network, which became the CRBC.

Another pioneer radio-drama series, the *CKUA Players,* was broadcast throughout the 1930s over CKUA, the independent radio station of the University of Alberta. Produced mainly by Sheila Marryat, it included some original plays by such Canadian writers as Gwen Pharis Ringwood and Elsie Park Gowan. From the late 1930s, this series was also broadcast over an informal western Canadian radio network and even over the CBC's regional and national networks.

An ambitious series of dramatizations of Canadian history called *Romance of Canada* had been commissioned by CNR Radio in 1930. The dramatist chosen was Merrill Denison, a well-known Canadian stage writer. The producer was Tyrone Guthrie, a London theater director who had begun his career as a writer and producer of British Broadcasting Corporation (BBC) radio dramas (and who would return 20 years later to found the Stratford, Ontario, Shakespeare Festival). Guthrie produced the first 14-play season of *Romance of Canada.* The plays of the second season were directed by his protégé Rupert Caplan, also a stage professional, who had just returned from acting at the Province Town Playhouse in New York. Caplan went on to a long career as a senior producer of radio drama—first for CNR Radio; then, after it was nationalized, for the CRBC; and finally for its successor, the CBC.

Between its founding in 1933 and its transformation in 1936, the CRBC increased the number of its weekly English-language radio-drama series to as many as 17. Its best-known national series was Rupert Caplan's *Radio Theatre Guild* (a name echoing NBC's *Radio Guild*), which broadcast original Canadian, American, and European plays. Also very popular were *The Youngbloods of Beaver Bend* and a series produced by Don Henshaw, *Forgotten Footsteps.*

CBC Radio Drama's Golden Age

When the CRBC became the CBC in 1936 (mainly a political change of administration), its first national supervisor of drama, Rupert Lucas, further expanded the radio-drama offerings of the network, including not only adaptations from Shakespeare and the theater classics and from classical fiction, but also some original plays and documentaries written primarily for radio. Lucas established a national radio-drama series at CBC Toronto and set up parallel regional production units in Montreal, Winnipeg, and Vancouver, each with its own major regional series.

When Canada entered World War II in 1939, the CBC began to make educational and propaganda programs for its nationwide audiences, as the American networks were to do starting in 1941. On the other hand, the CBC never abandoned its practice of ambitious original radio-drama productions, which survived the war to become, in effect, the National Canadian Theatre. The visionary who accomplished this was Andrew Allan, who had worked in Toronto from the mid-1930s as a radio producer of popular plays and variety shows; he worked in England from 1937 and returned to Canada in September 1939 to become regional drama producer in Vancouver. There he earned an impressive reputation as a producer of original play series, particularly *Vancouver Theatre,* and as the organizer of an excellent repertory company of writers and actors. With the departure of Rupert Lucas for NBC in New York in 1943, Allan was appointed national drama supervisor in his place, a position he held until 1955. He invited many of his Vancouver repertory team to work with him in Toronto, particularly the actor John Drainie and the writers Fletcher Markle, Len Peterson, and Lister Sinclair (who was also an excellent actor). Allan added to his CBC Toronto repertory company a number of other Canadian dramatists and actors as well as the composer-conductor Lucio Augustini. He also invited Esse W. Ljungh to Toronto to help produce the growing number of national series.

One of Allan's first acts as CBC drama supervisor in 1944 was to create an hour-long weekly national anthology series of original Canadian radio plays out of Toronto, called *Stage* (*Stage '44, Stage '45,* etc., later *CBC Stage*). The year *Stage* was founded—1944—marks the beginning of the Canadian golden age of radio drama. This was made possible partly because (as noted) American radio's golden age largely ended with the postwar introduction of television. Allan's major goal for the *Stage* series was to create a professional Canadian theater. Although there were many experienced Canadian actors and dramatists and an excellent semiprofessional Dominion Drama Festival, until 1944 there was no professional stage-theater institution in Canada, as existed in many other countries. This was mainly because professional American and British theater companies dominated (and often physically owned) the major stage theaters in Canadian cities, and these companies mainly offered popular American and British stage plays. Allan believed that the CBC offered a unique opportunity to lay the foundation for a professional Canadian theatre drawing on Canadian plays, producers, and actors. He conceived the idea that the CBC Radio Drama Department could become Canada's first professional national theater, and thus the subtitle he gave to *Stage*: *Canada's National Theatre on the Air.* Allan produced virtually every play in this premier national anthology series until he retired as national drama supervisor in 1955, and he produced other plays in this series until 1960. All were real-time, live-to-air productions. Among Allan's best playwrights were Peterson, Sinclair, and Markle, but other notable playwrights included W.O. Mitchell, Patricia Joudry, Gerald Nixon, Reuben Ship, Joseph Schull, and Tommy Tweed.

In 1947 Harry Boyle, CBC's program supervisor, created a second ambitious weekly national anthology series, called *CBC Wednesday Night,* which broadcast a schedule of original radio plays and important classical and modern European and American dramas and adaptations. Each lengthy *Wednesday Night* program also included related talks or documentaries and often music, providing a whole cultural evening. The plays in this series were produced by the four senior CBC producers, Allan, Esse Ljungh, J. Frank Willis (also CBC head of features), all out of Toronto, and Rupert Caplan (of *Romance of Canada* fame) from Montreal. That same year, the physical network of CBC regional drama production centers was completed, with facilities at Halifax and Calgary. By 1947, then, CBC was broadcasting weekly: two anthology series of full-length plays, plus a major half-hour drama series from each of its six regional studios. From the early 1940s to the early 1960s, some 600 serious dramas were produced by the CBC, at least half of them original dramas for radio. During the 1950s the CBC broadcast in total some 20 weekly CBC English-language radio-drama series, including also the whole panoply of popular dramatic forms lost to American radio with the coming of television to the United States in the mid-1940s.

Mature and Shrinking CBC Radio Drama after 1960

By the mid-1950s conditions became less ideal for Canadian radio drama. CBC Television's English network began in 1952, soon to be joined by a private television network, CTV, each with several drama series. As in the United States, television gradually began to steal away both drama professionals from CBC Radio and radio's audiences for drama. The Shakespeare Festival at Stratford, Ontario, one of the first professional Canadian stage companies, was also founded in the early 1950s by Tyrone Guthrie (the creator of *Romance of Canada*); it also lured away CBC Radio's drama-trained theater professionals. The Stratford Festival was soon joined by a growing number of other professional stage companies, beginning the movement toward a mature Canadian professional stage-theater institution, which burgeoned in the 1970s. Allan's vision was being realized, and, ironically but inevitably, it was weakening his original radio-drama creation. Nevertheless, Allan's successor, Esse W. Ljungh, produced and sponsored many notable plays in the mature CBC sound medium from 1955 to the mid-1970s.

By the late 1960s, the CBC Radio Drama Department, having lost its function as the only professional medium for Canadian drama, gradually reduced production. Being outside the spotlight was nevertheless an opportunity for the next generation of national producers, John Reeves in Toronto and Gerald Newman in Vancouver, to experiment with new dramatic forms and techniques. These experiments were aided by the 1960s move from live-to-air performances to taped productions, which came to resemble the out-of-sequence recording practiced in film production; however, radio plays also lost the edge of the previous live productions. With the growth of television as the major entertainment medium, Canadian radio (like American radio almost a generation before) was gradually changing into a medium for mainly music, news, and talk. Although *Stage* and *Wednesday Night* (later known as *Tuesday Night*) continued until the mid-1970s, the institution of CBC radio drama built up in the first decades was slowly disappearing.

There was a revival of cultural radio in the 1980s, and the number of national anthology drama series grew again. A subtle change, though, was taking place in the way listeners were being addressed. As a policy decision, CBC English radio drama became more populist in form and content and was aimed at a more general audience. A new flagship series of original Canadian radio plays, *Sunday Matinee,* was founded, but there was also the more populist *Vanishing Point* and *Stereo Theatre* and the dramas on *Morningside.* By the early 1990s, partly because of drastic cuts in government funding, none of the above series had survived in its original forms, and CBC radio drama was once more shrinking. The 1990s witnessed the disappearance of the previous forms and functions of CBC's former "Senior Service," the Radio Drama Department. Nevertheless, exemplary radio theater, serious and popular, is still being produced by the CBC in the new millennium. Canadian audiences can continue to experience the particular sound theater offered by radio and can learn how to use their imaginations to apprehend the sound images and the strong emotional communication of this unique dramatic form.

HOWARD FINK

See also Drama, U.S.; Drama, Worldwide; Playwrights on Radio

Further Reading

Drainie, Bronwyn, *Living the Part: John Drainie and the Dilemma of Canadian Stardom*, Toronto: Macmillan, 1988
Fink, Howard, "The Sponsor's v. the Nation's Choice: North American Radio Drama," in *Radio Drama*, edited by Peter Lewis, London and New York: Longman, 1981
Fink, Howard, and John Jackson, editors, *All the Bright Company: Radio Drama Produced by Andrew Allan*, Toronto and Kingston, Ontario: CBC Enterprises, 1987
Frick, N. Alice, *Image in the Mind: CBC Radio Drama 1944–54*, Toronto, Ontario: Canadian Stage and Arts Publications, 1987

Canadian Radio Satire

Radio satire is a continuation of an ancient tradition of humor that converges or collides the serious with the comic; ironically reverses social, linguistic, and bodily hierarchies; ridicules the traditional from the point of view of the contemporary; and addresses aspects of the human condition that range from the darkest, most cynical, and acerbic to the most lighthearted, mindless, and silly. Radio satire, like prose and poetry, needs to be understood in terms of its place within the scale of possible comic expression, which ranges from the serious to the light and the reception of which often crosses the boundaries of the scholarly and the popular.

Two Cultures

Radio satire provides countries around the world with entertainment in a familiar voice and accent that expresses cultures of laughter in their local communities and languages and that does so in ways that confirm or transgress complex political and moral issues. Canadian radio satire needs to be understood as one example of how two distinct societies, English-speaking Canada and French-speaking Quebec, laugh at themselves and at each other (Nielsen, 1999). Most of the time, the two audiences are not aware of exactly what it is the other is laughing about, because the vast majority of English audiences have no knowledge of French programs and vice versa. Although French-language satire on the private networks and on *La Société de Radio-Canada* (the French public network) addresses the small number of French-speaking minorities across Canada, its primary audience lives in the province of Quebec, where the majority of French speakers reside. The distinctness of Quebec society is defined in terms of the French language and culture and its differences from the rest of Canada and North America. On the other hand, satire on the English-language private networks and on the publicly funded Canadian Broadcasting Corporation (CBC) addresses a distinctly English-Canadian society that is typically defined in terms of its differences from the United States rather than differences from French Quebec. As Nielsen suggests in his many writings on this topic, when a French or English accent appears in radio satire on either network, it is almost always about laughing at the other or at differences with America.

A further difference in the way the two radio cultures within Canada have developed is readily identifiable. Although satire in English Canada has its origins in private radio networks in the 1930s, the public radio network has traditionally produced the majority of programs in this genre. This is explained by the possibility of filling the airwaves with programming from the United States. Although many early American satires were translated into French, the fact that most programs were not translated meant that a demand for local programming was stronger in French-speaking Quebec than in the rest of Canada (de la Guarde, 1991). In Quebec, private networks have pursued commercial programs, whereas the public networks have tended to produce more seriously engaged cultural and social material.

English-Language Satire

The earliest examples in English-speaking Canada that mix serious and light radio satire themes were Jack Bawdry's Vancouver production of *Millie and Lizzy* (1930–35) and Art MacGregor and Frank Deaville's Calgary production of *Woodhouse and Hawkins* (1933–44). Bawdry's series was a political satire on the Great Depression from the point of view of two working-class women, and MacGregor and Deaville's series was a lighter satire on the theme of the "country bumpkin." The former celebrated working-class values and determination, and the latter used a mixture of accent and vernacular to poke fun at rural traditions from a contemporary urban viewpoint. Two other figures at the light end of the satirical scale that would go on to dominate English-Canadian comedy until the 1960s also began their careers in this period. Johnny Wayne and Frank Schuster began in private radio in the late 1930s before eventually hosting a regular satirical series that would run from 1947 to 1950 before moving on to a career on television that would end in 1989. Wayne and Schuster's satires followed the burlesque format. The first act featured light stand-up comedy routines, followed by word plays and songs and then a parody of a contemporary play or musical that often included sexual (and, by today's standards, sexist) references.

Toward the end of the 1930s, the publicly funded CBC came to dominate program production in almost all fields. From the outset, the CBC sought to bring together the best artistic talents from the various regions—Halifax, Vancouver, Winnipeg, Montreal—to create production teams for the national network based in Toronto. Between 1936 and 1961, the CBC produced more than 300 radio theater series, including more than 8,000 individual plays, of which half were original productions. Around 70 satiric radio theater plays were broadcast between 1940 and 1952, of which 50 were written for the prestigious *Stage* series, directed by Andrew Allan. Allan produced over 450 shows during the first 12 years of the series. During the same period, more than 70 writers and over 150 actors and actresses were employed. A reading of the themes of all the plays produced by Allan suggests that the questioning of the social order was more evident at the beginning of the series (1943–48) than in the final period of production (1948–55).

The influence of western Canadian satirical writers in the production teams of the *Stage* series was especially marked in the first period. Len Peterson (Saskatchewan), W.O. Mitchell (Saskatchewan), and Tommy Tweed (Manitoba), along with Fletcher Markle, Lister Sinclair, Bernard Braden, Andrew Allan (the director who brought the others from Vancouver), and, somewhat later, Alan King, made up the key writers whose texts severely criticized society. The writers of *Stage* who were famous for their critical satirical spirit came from western Canada, and those who were associated mainly with light drama and comedy that were generally non-controversial originated in central Canada (Ontario and Quebec). Generally, critiques of social class and the economy—that is, the expression of an active opposition to the social order—were most evident between 1944 and 1948. After 1948 social criticism became more introspective and focused on the questioning of such cultural norms as traditional family values or gender roles, rather than on social classes or the economic system.

The distinction between serious and popular radio comedy has its origins in ancient forms, as was mentioned previously. However, it should be pointed out that the carnival origins of satire are heavily concentrated on grotesque elements and on reference to the lower bodily stratum. As one critic has remarked, "Vaudeville and music hall humour had been centered in the groin and heart. Radio humour located above the neckline" (Clark, 1997). In its first decades, radio satire presented a reified version of carnival laughter in the sense that there remained words that could not be uttered, comic reversals that could not be achieved, and levels of laughter that could never be expressed. In the golden age of radio, satire was sanitized. Nonetheless, certain of *Stage*'s ironic satires did carry out a hierarchical inversion, one of the most fundamental conditions of seriocomic satire. In principle, the inversion is based on the carnivalesque logic of opposition, the simultaneous process of negation and synthesis that links the worlds of the "serious" and the "comic" rather than substituting one for the other or replacing higher strata with lower ones. *Stage*'s radio literature is mediated by the moral horizon of the era and hence offered little or no swearing or grotesque realism.

A key early figure who would challenge the moral and political horizon of his day was Max Ferguson, whose comic stylings in *Rawhide* began on the English-language CBC in

1958. Ferguson blended impersonation, satiric political comedy, and music in a morning radio show. His goal exceeded the lighter version of seriocomedy and purposefully pushed the limits of the genre. He is quoted as having said, "My goal is to be taken off the air." In 1961 his wish came true when *Rawhide* was canceled following a particularly acerbic attack on a member of parliament.

Since 1960 CBC radio has produced over 80 satires in the form of one-hour radio plays and irregular mini-satires series. An example of the latter is the 15-minute weekly comedy series on the three-hour *Morning* program over the last 20 years. The *Morning* series also carried the Charley Farquhusan character, a send-up of the rural-urban theme played by Don Harron, one of the program's early hosts. By 1971 the main satirical program for the CBC became the weekly half-hour series the *Royal Canadian Air Farce,* created by Roger Abott and Don Ferguson. In its early years, the *Farce* was a marginal series that broadcast the studio radio performance of four comedians. In its fifth season it shifted toward a more vaudevillian style and took its show on the road to perform live broadcasts on location. It continued to develop the old vaudevillian technique, and in 1996—its final season in radio—it remained one of the only live traveling radio comedy series in North America. Its style grew from a wide mixture of the short sketch, stand-up comedy, English music hall, and theater of the absurd. The *Farce* is probably one of the most important programs in Canadian radio history, given its pioneering role in stretching the possibilities of what could be said or presented in the genre on a public medium.

In the 1990s the CBC produced a series of similar studio comedies such as the *Frantics* and *Double Exposure* and, more recently, live programs such as *Radio Free Vestibule* and *Madly off in All Directions.* All these series built on the farce's political satire, but none have retained the traveling live broadcast format in quite the same way. After 20 years on radio, the *Farce* transferred its production to the television studio, where it has enjoyed a successful run as a mainstream light and popular seriocomedy in the late 1990s.

Since the 1980s, English-Canadian seriocomedy has had a whole cycle of popular successes both nationally and in the United States. Historically, this flight of talent has been from the visual arts and not radio. The exodus to the United States entertainment industry began in the early days of cinema. The vaudeville-style physical comedians have had the most success in the American industry, and many of them did get their start in radio satire—from Mary Pickford to Allen Young, Leslie Nielsen, and Jim Carrey. Young was the first to leave the CBC to star in the second biggest budgeted television program to come out of Hollywood in 1949: *The Alan Young Show.* Wayne and Schuster also did American television in the 1950s, and since the 1960s a disproportionate Canadian influence has been clearly observable in seriocomedy in the United States—from the late John Candy to Martin Short, Eugene Levy, Andrea Martin, Mike Myers, Howie Mandal, and many others (Pevere and Diamond, 1996).

French-Language Satire

French-language radio satire also has its origins in private radio during the 1930s. The earliest examples of mixing serious and light radio satire themes for popular audiences are Eduard Baudry's *Par le trou de la serrure* (1932–33) and Alfred Rousseau's *Les Amours de Ti-Jos et les mémoires de Max Potvin* (1938–45). Baudry's series was one of the first light cultural satires on day-to-day family life in Montreal, whereas Rousseau worked on one of the first burlesque-style variety programs, *Radio-Divertissement Molson* (1935–38), and developed it further in *Les Amours de Ti-Jos* (1938–45). Rousseau was the first to innovate through a burlesque and vaudevillian style that mixed songs, monologues, and character sketches. The celebrated Quebec radio dramatist Robert Choquette also began his career writing satirical series for private radio, as did Gatien Gélinas, who is considered to be one of Quebec's first indigenous playwrights. Choquette's first works, *La fabuliste La Fontaine à Montréal* (1934) and *Vacances d'artistes* (1935)—like those of Gélinas, *Le Caroussel de la gaieté* (1937–38) and *Le train de Plaisir* (1938–40)—were weekly social satires that played with the reversal of upper and lower social strata by satirizing the poverty of French Canadians while celebrating their ability to create and express themselves in vernacular language. The most popular radio satire in the history of Quebec, which satirized social themes in the rural-urban context by using vernacular language and elements from the theater of the absurd, was *Nazaire and Barnabie* (1939–58) by Olivier Légaré.

A standard theme across the history of social satires on Quebec radio established in these early series relates to language. Social satires draw from the deep tension between traditional and modern culture through the ironic use of subdialects, local oral traditions, and regional accents. Language is stratified from top to bottom and is defined through a struggle between the peripheral forces of popular speech and the centralizing pull of literary correctness. Language stratification plays a key role in establishing the scale of satire, which ranges from the serious to the light and which addresses audiences that are potentially both popular and scholarly. Historically in Quebec, private radio produced more of the lighter, "popular" entertainment and less of the more serious or "scholarly" radio plays. Radio-Canada does not produce as much popular entertainment as the private networks do, even though the public network often addresses a popular audience. Although private radio allows certain popular voices to speak in their own slang, Radio-Canada tends to treat the vernacular voice as something that can be innovated (as on private radio) but also, and more frequently, as something negative that should be corrected. Like the reader or the spectator, the addressee of satirical works on Radio-Canada is most often a listener from the middle class. He or she can be part of the popular or the scholarly audience, but when a popular addressee appears in a role, he or she is often the object of satirical ridicule, parody, or irony.

Once we understand that the scale of narratives ranges from serious to light while the audience ranges from scholarly to popular, we can better situate the variety, burlesque, and satirical magazines that developed social satire from the 1940s to 1970. Among the best examples are the variety show *Radio-Carabin* (1944–53) by Émilien Labelle, Laurent Jodoin, and Paul Leduc; the cabaret show *Chez Miville* (1956–70) by Paul Legendre; and the satirical magazine *Carte blanche* (1951–53) by Fernand Seguin, André Roche, and Roger Rolland.

The variety show is composed of a mixture of songs, music, and light humorous skits. Each *Radio-Carabin* show lasted 30 minutes. The satirical skits often conveyed such serious social issues as housing problems and poverty and mocked "high society." The cabaret series *Chez Miville,* which aired every morning between 8:00 and 9:00 A.M., differed from the variety show because it interspersed serious journalistic interviews or editorials on moral or political themes with periods of music and light skits. Comic stereotypes of the time were created through parody rather than political irony. Although the show avoided the most extreme versions of political satire and theater of the absurd, and although it was perhaps the most popular morning radio program throughout the 1960s, it ended very soon after the famous "October Crisis" in 1970. The federal government suspended civil liberties and sent orders for mass arrests of artists and intellectuals suspected of collaboration with the terrorist group le Front de la Libération de Québec, which had kidnapped and ransomed a federal politician and a British diplomat. Radio-Canada producers were reportedly very nervous about any political or moral satire, however light, that might be directed against the government of the day (Pagé and Legris, 1979).

In contrast to the variety show or cabaret, the satirical magazine *Carte blanche* was designed especially to be critical. It was composed of distinct sections oriented toward a totalizing satire of Quebec culture and society. Although it remained faithful to the entertainment principle, its aims were more serious than those of the variety show or the cabaret. The series as a whole satirized the predominant worldview of the early 1950s. The narrative scenes concerned paradoxes of the institutions of Quebec society. Theater, the novel, poetry, art, and music, as well as political parties, mass media, and educational and bureaucratic institutions, were all treated with irony and satire. The series addressed itself to a scholarly audience of Quebec intellectuals and celebrities. Before the end of its third season, the writers (Seguin, Roche, and Rolland) decided to abandon the show rather than bow to the pressure of censorship.

Satire on public radio after 1970 became increasingly intertwined with information, sports, and music. These programs replaced the cycle of dramatic and satirical programs that flourished in the pre-television era. Unlike English Canada, where the *Royal Canadian Air Farce* dominated public radio satire from 1972 to 1996, Quebec radio satire has been much more broadly distributed across a variety of programs that extend elements of the genres discussed previously. A surprising amount of satire continues to be produced on Quebec radio, but it is no longer sustained in a series format. On the private networks, satire is most typically used to enhance the morning shows (*Y'é trop de bonne heure,* hosted by Norman Brathwaite) and the afternoon drive-time programs (*Y'é pas trop tard,* hosted by Patrice Lécuyer). The topics of discussion in these programs range from news to sports, weather, the arts, and entertainment. The radio announcer who discusses the issues of the day often slips into a satirical, lighthearted dialogue or comic improvisation with his sidekick or with the regular specialist who comes on air to talk about traffic or weather.

Radio satire on the private networks tends to have a secondary role, in the sense that it is "sprinkled" into the show to lighten it up and is therefore only a small part of a larger program. On the public networks, satirical slots or capsules are introduced rather than "sprinkled" into daily cultural magazines. The main difference from the programs broadcast on the private networks is that the satires are animated by comedians and have their own well-blocked slots within the programs. For example, satire can be heard on the program *Indicatif présent* in the sketch *"Si j'étais premier ministre,"* in which well-known comedians such as Yvon Duchamps are asked what they would do if they were elected to political office. The comedians don't miss the opportunity to mock politicians and their institutions. Other programs, such as *En direct,* satirize news clips from television and parody journalists and other "serious" professions.

Three of the most important seriocomedy radio satires from the 1970s into the 1990s were *Rock et belles oreilles* (1991–), *Le festival de l'humour* (1974–88), and *Les insolences d'un téléphone* (1968–96). All three were produced by Quebec's private radio stations. *Rock et belles oreilles,* a weekly one-hour satirical magazine in the tradition of *Carte blanche,* developed both light and serious parodies of language and of social conventions, advertisements, popular music, and television programs. *Le festival de l'humour* was a one-hour live satirical magazine that parodied the main news events each week. *Les insolences d'un téléphone* presented a new kind of direct satire. The key segment in the program has the comedian Tex Lecors telephone people and pretend to be someone else in order to get a response and to engage the person in a mock dialogue. The show was a huge success and opened new ground, inspiring direct satire in various television programs, both French and English.

Contemporary French-language radio satires draw from a long tradition of the light variety of cabaret and burlesque forms as well as the serious magazine and social comedies of the 1930s and 1940s. The nihilism and theater of the absurd that entered French-language radio in the 1950s often informs contemporary direct satires. The early social comedies established a tradition of social satire around the stratification of the French language in Quebec. Many of Quebec's most famous writers began their careers writing social satires. The best generic example of social satire on the French side, one that pushed the limits of critique of its own society, is the series *Carte blanche* (1950–53).

Satire has a long history in private and public radio in Canada and Quebec. It is durable partly because of its capacity to adapt itself to any context and partly because its basic ingredients—critiquing tradition from the perspective of emerging contemporary values, reversing hierarchies, and stratifying language—have remained intact. Given radio's extraordinary durability as a means of communication, it seems reasonable to conclude that satire will develop new boundaries, which will in turn be challenged by new satires in response to new value orientations, generational contexts, and ever more innovative forms of experimentation in the local cultures of laughter around the world.

MARIE CUSSON AND GREG NIELSEN

Further Reading

Clark, Andrew Beatty, *Stand and Deliver: Inside Canadian Comedy,* Toronto, Ontario: Doubleday Canada, 1997

De la Guarde, Roger, "Y a-t-il un public dans la salle?" in *Communication, publique et société: Repères pour la réflexion et l'action,* edited by Michel Beauchamp and Bernard Dagenais, Boucherville, Quebec: Morin, 1991

Fink, Howard, and Brian Morrison, *Canadian National Theatre on the Air, 1925–1961: CBC—CRBC—CNR Radio Drama in English: A Descriptive Bibliography and Union List,* Toronto: University of Toronto Press, 1983

Nielsen, Greg Marc, *Le Canada de Radio-Canada : Sociologie critique et dialogisme culturel,* Toronto, Ontario: Gref, 1994

Nielsen, Greg Marc, "The CBC and Canadian Society," *Canadart: Revista do Núcleo de Estudos Canadenses da Universidade do Estado da Bahia, Associação Brasileira de Estudos Canadenses* 2 (1995)

Nielsen, Greg Marc, "Quebec's Case against Canadian Broadcasting Policy," *Fréquence; Frequency* 3–4 (1995)

Nielsen, Greg Marc, "Culture and the Politics of Being Québécois: Identity and Communication," in *Quebec Society: Critical Issues,* edited by Marcel Fournier, M. Michael Rosenberg, and Deena White, Scarborough, Ontario: Prentice Hall Canada, 1997

Nielsen, Greg, "Two Countries, One State, Two Social Imaginations: A Comparison of CBC and Radio-Canada Seriocomedy," *Journal of Radio Studies* 6, no. 1 (1999)

Pagé, Pierre, and Renée Legris, *Le comique et l'humour à la radio québécoise: Aperçu historique et textes choisis, 1930–1970,* vol. 1, Montreal: La Presse, 1976, and vol. 2, Montreal: Fides, 1979

Pavelich, Joan E., "Irony in English- and French-Canadian Radio Drama (1940–1960)," Ph.D. diss., Université de Sherbrooke, 1984

Pevere, Geoff, and Greig Dymond, *Mondo Canuck: A Canadian Pop Culture Odyssey,* Scarborough, Ontario: Prentice Hall Canada, 1996

Canadian News and Sports Broadcasting

News-gathering organizations did not ignore the introduction of radio in the 1920s. In August 1922 the Radio branch of the Department of Marine and Fisheries revealed that 14 Canadian newspapers held radio licenses. Many insightful newspaper owners saw radio as a potentially profitable addendum to the business of supplying news and information. Others saw it purely as a medium of entertainment. No matter which view they took, the newspaper owners recognized early in the game that radio, if held by other hands, had the potential to undermine their bottom lines. In a fashion similar to internet development today, the owners felt it better to be on the inside should the medium prosper, rather than watching from beyond. However, when promised profits failed to emerge, many newspapers, including the affluent *Toronto Daily Star,* abandoned their broadcasting activities, with the consequence that news and information suffered a decade-long setback.

The country's first major current-affairs information program was carried by the Canadian Radio Broadcasting Commission's (CRBC) predecessor, the Canadian National Railways network. Gratton O'Leary, editor of the *Ottawa Journal,* broadcast a 15-minute weekly program called *Canada Today,* in which he discussed major issues that had been reported the previous week in his newspaper. In spite of his ties to the federal Conservative Party, O'Leary promised to be impartial on the air. However, O'Leary never kept his promise. When the United States began turning up the heat to get Germany, France, and Italy to pay their war debts and reparation payments, O'Leary claimed that loans made to the United States for southern reconstruction after the Civil War by several European states had never been repaid. The American consul in Montreal, Wesley Frost, called for O'Leary to be forced to desist. In spite of the pressure, the program was not canceled.

Early Canadian radio broadcasters had developed an uneasy relationship with the country's largest news- and sports-gathering organization, the Canadian Press, by the early 1930s. In the 1920s, few if any stations carried significant news and sports programming. Reluctantly, the Canadian Press offered to allow radio stations access to its wire services free of charge on the provision that the stations would not sell newscasts to advertisers. The wire service itself had been constituted as a nonprofit cooperative with the precise mandate to serve Canadian newspapers with national and international news and sports. When the CRBC, Canada's first public broadcaster, took to the airwaves in 1933, it announced that it would sell newscasts to prospective advertisers. Much to the chagrin of Canadian Press, private stations soon followed suit.

In spite of the ongoing battles between broadcasters and the Canadian Press, the emergence of the publicly owned Canadian Broadcasting Corporation (CBC) did much to advance information programming on radio, mainly in the area of what we now call current affairs. During the late 1930s, the CBC launched farm programs, women's programs, political broadcasts, and extensive coverage of major events such as the 1939 royal tour of Canada by King George VI and his wife Elizabeth just before the outbreak of World War II. That same year Leonard Brockington, chairman of the CBC, asked the Board of Governors to approve a policy of nonpartisanship for what the CBC deemed controversial programming. This fairness doctrine still guides CBC news and current affairs to this day. However, spot news and sports coverage as a regular feature lagged well behind. That would change with the outbreak of World War II.

The Canadian Press realized that its battle with broadcasters seemed to be endless. As a consequence, the agency established Press News in 1941 for broadcasters. Young journalists such as Scott Young (father of singer Neil Young) and Jim Coleman (later an icon in the sports reporting community) were hired to reduce the Canadian Press' wordy newspaper copy to broadcast format. Sam Ross, the first manager of Press News, was given a mandate to sell the service to any Canadian broadcaster willing to pay the fee. The days of free Canadian Press copy in the broadcast newsroom had ended. That same year, the CBC founded its first national news service by establishing five newsrooms across the country under the direction of Daniel McArthur. McArthur remained convinced throughout his career that reporting spot news and interpreting and analyzing current affairs, although related, were two separate activities. McArthur had a mandate to expand CBC news coverage, which in 1939 constituted only 9.4 percent of the national network's programming. By 1941 he managed to increase this amount to 20 percent. The bombing of Pearl Harbor was a turning point in news coverage in both Canada and the United States. The CBC broadcast its own bulletins adjacent to feeds it carried from the United States. In the week following the disaster, CBC Radio News broadcast bulletins every hour on the hour, establishing a pattern that would soon be copied by private radio.

In an ironic turn of fate, the war proved to be a boon to the goings-on in CBC news. Two CBC newsmen accompanied the first Canadian contingent to Europe. They followed the soldiers and reported from battles throughout the course of the war. A CBC reporter was assigned to the British Broadcasting Corporation in London to help develop shortwave broadcasts to North America. The CBC bought a six-ton van, which it converted into a mobile war reporting studio. In the winter of 1940, more than 1,000 reports were recorded on soft-cut discs during the six-month period. As CBC historian Austin Weir reported, three half-hour war programs were sent back to Canada weekly featuring interviews with service personnel, rides in war planes and tanks, and numerous notes of human interest.

The war coverage spawned other current-affairs shows at the CBC. The Talks Department produced several new programs covering a myriad of topics. By the end of the war, news and information were an essential part of radio programming on both public and private stations. In 1953 the Canadian Press severed its Press News service from the newspaper cooperative and launched Broadcast News. However, the publishers appointed one of their own, Roy Thomson, as the first president of the new entity. Gordon Love, a television executive from Calgary, was appointed vice president, and Charles Edwards was named manager. Edwards was well aware of the potential held by Broadcast News. Shortly after his appointment, he connected 27 Canadian stations with prerecorded news items. The system was called Tapex News and eventually evolved into the voice service of Broadcast News.

In spite of its bumpy start, information programming took on a life of its own at the CBC. Evening newscasts became a regular part of the schedule, and by the mid-1950s lively and sometimes controversial current-affairs programming began to appear on CBC stations. Most privately owned affiliates broadcast short local newscasts during the supper hour and in the late evening following the network national news. In Toronto, producer Ross McLean launched *Close Up*. The local station CBLT produced the somewhat racy *Tabloid* program. A quiz show based on the weekly newspaper headlines was launched named *Front Page Challenge*. The CBC had begun to establish its reputation as a reliable and consistent purveyor of information programming. It certainly was aided by the fact that the Corporation had a monopoly on national network programming. Until the all news service CKO (which no longer exists) came on the air in the mid 1970s, CBC had a monopoly on radio networks. It also had a monopoly on television networks from 1952 until 1960.

In 1971 CBC Radio split its AM and FM services and revamped its program schedules. New shows such as *This Country in the Morning, Later That Same Day, Radio Noon, Metro Morning,* and *As It Happens,* all based on the delivery of news and information, became the mainstays of the AM network. *As It Happens,* which continues today, can also be heard on the shortwave service of the CBC Radio-Canada International and on selected National Public Radio stations in the United States. Its format of interview and call-out has been a leader in the international broadcast journalism field. It set a trend that more and more AM radio stations in Canada, faced with stiff competition with FM stations with superior sound quality, followed by turning to sports and information programming. However, the Canadian Radio and Television and Telecommunications Commission made one serious licensing mistake in the 1970s. It approved a coast-to-coast network of 12 all-news and -sports radio stations called CKO All Canada News Radio. With the exception of its Montreal license, all stations broadcast on FM. The network never turned a profit and closed its doors in late 1989.

CBC Radio provided an excellent model that was later used when television came to Canada. Initially the newsrooms, especially at the reporter level, tended to integrate television and radio personnel and facilities. As CBC budgets increased, so did the separation between the two media, but as finances declined in the mid 1980s, once again CBC reporters faced double duty.

Today, the CBC continues to be the leader in news and information programming. It offers an evening one-hour news and current-affairs program at 10 P.M. on the national network, one hour earlier on NewsWorld. Its competitors, CTV and Global, also offer evening newscasts. The CBC broadcasts an investigative journalism program called *The Fifth Estate,* a business program called *Venture,* a consumer-oriented program called *Market Place,* and a documentary series entitled *Witness.* It operates a 24-hour all-news channel, CBC NewsWorld, and its French language affiliate, RDI. CTV operates a 24-hour news headline service, CTV News1, as well as CTV SportsNet. The network is also attempting a merger with TSN, the country's first all-sports television specialty channel. All-news radio made a major comeback when CFTR Radio 680 in Toronto dropped its pop music format and opted for news and information. It was followed by Canada's largest English-speaking private station, CFRB Toronto, with a mixed format of talk, news, and sports. Virtually every major city in the country now has access to broadcast news and information on a 24-hour, seven-day-a-week basis.

DAVID R. SPENCER

See also All News Format

Further Reading

Bird, Roger Anthony, editor, *Documents of Canadian Broadcasting,* Ottawa: Carleton University Press, 1988

Peers, Frank W., *The Public Eye: Television and the Politics of Canadian Broadcasting, 1952–1968,* Toronto and Buffalo, New York: University of Toronto Press, 1979

Romanow, Walter I., and Walter C. Soderlund, *Media Canada: An Introductory Analysis,* Mississauga, Ontario: Copp Clark Pitman, 1992; 2nd edition, Toronto: Copp Clark, 1996

Rutherford, Paul, *The Making of the Canadian Media,* Toronto and New York: McGraw Hill Ryerson, 1978

Rutherford, Paul, *When Television Was Young: Prime Time Canada, 1952–1967,* Toronto and Buffalo, New York: University of Toronto Press, 1990

Vipond, Mary, *Listening In: The First Decade of Canadian Broadcasting, 1922–1932,* Montreal and Buffalo, New York: McGill-Queen's University Press, 1992

Vipond, Mary, "The Continental Marketplace: Authority, Advertisers, and Audiences in Canadian News Broadcasting, 1932–1936," *Journal of Radio Studies* 6 (1999)

Weir, Earnest Austin, *The Struggle for National Broadcasting in Canada,* Toronto: McClelland and Stewart, 1965

Canadian Talk Radio

Canadian radio stations provide listeners with a wide variety of spoken-word programs, from news and documentaries to talk radio formats consisting of discussions between hosts, guests, and listeners who telephone the show. Canadian talk radio reflects the two main influences of radio in Canada: the public broadcasting model of the Canadian Broadcasting Corporation (CBC) and the U.S. roots of private-sector radio formats.

Talk programs figure in the schedules of most private radio stations, but the amount of talk has varied from station to station over time. This shuffling has been especially active since the 1980s as AM and FM stations have competed for increasingly fragmented listener segments; AM has struggled to regain profitability, and radio overall has tried to fend off competition from other media. Talk programs sometimes form part of the programming mix of stations specializing in music. There are also all-talk/news stations in Toronto, Vancouver, Montreal, and other major cities.

Origins

Older forms of spoken-word programs conditioned Canadian radio listeners to expect an emphasis on information and education. In the face of frequent format change by private radio, the CBC has been the most consistent source of spoken-word programs. The CBC provides a varied schedule of music, drama, news, interviews, and discussions. It has been easier for the CBC to maintain a relatively stable percentage of spoken-word programs than it has been for the private sector because the government-owned CBC has a more constant source of funding than privately owned stations. The CBC has thus had the freedom to broadcast programs that often attract fewer listeners than would be acceptable for private stations, which are dependent on audience size and the resultant advertising revenues.

In 1937 the CBC appointed a director of talks to develop a series on issues of contemporary public concern, with programs done by experts skilled in speaking on the radio. By 1941 the CBC calculated that it had broadcast approximately 1,250 different speakers. Commentaries by individuals and discussions between several speakers were typical of the forms of talk on the CBC at that time. For listeners tired of serious talk about current affairs, the public network aired talks on other subjects such as consumer information, cooking, and literature.

Private radio also broadcast a range of spoken-word programs, especially before the 1950s. In the mid-1950s the popularity of disc jockeys and rock and roll began to push recorded music to the forefront. Some stations, then as now, felt that programming based heavily on recorded music was cheaper to produce than news and other spoken-word programs. Talk programs were rejuvenated, however, as technology became better able to provide broadcast-quality reproduction of telephone calls. Borrowing a new talk format popular in the United States, private radio stations in Canada began broadcasting open-line call-in shows in the late 1950s. By the mid-1960s these call-in shows had become a fixture on stations across the country. The programs gave listeners a sense of participation, even though only some wanted to speak on the air and even fewer actually made it on to the air. Unlike the somewhat patronizing one-way lecture or in-studio interviews of experts, open-line shows gave ordinary people a chance to express their opinions. The call-in shows rapidly became a key ratings weapon.

In Vancouver, radio stations CJOR and CKNW competed for listeners by pitting abrasive open-line host Pat Burns against the equally controversial Jack Webster. Both Burns and Webster considered themselves muckraking reporters rather than mere entertainers. Burns also amused and irritated listeners with his phone-out format. Callers heard him telephone such major figures as former U.S. President Harry S. Truman and grill him about the bombing of Hiroshima. Burns' ambush-style telephone calls raised questions about ethics, but it made for compelling radio. Few Canadian radio talk show hosts then or since have managed to gain the fame of Burns and Webster, but most medium and large markets developed their own local hotline shows with loyal followings. The list of other longtime open-line hosts in Canada includes Rafe Mair, Gilles Proulx, Lucien Jarraud, Lowell Green, John Gilbert, Lorne Harasen, Peter Warren, Roger Delorme, Paul Arcand, and Tom Cherington.

In addition to the general-subject hotlines, where the topic for discussion changes from show to show, talk series devoted to specific subjects have also been a fixture of Canadian talk radio. In the late 1950s, for example, Montreal broadcaster Reine Charrier pioneered a show about love and sex, broadcasting under the name Madame X. As society became more open about sex, call-in shows became more explicit and hosts no longer felt the need to use pseudonyms. By the late 1990s Vancouver sex therapist Rhona Raskin's call-in show was syndicated on stations across Canada and the United States. Listeners in Canada can also tune in to shows specializing in gardening, computers, personal finance, health, home renovation, sports, car repair, and a wealth of other subjects.

Recent Trends

In October 1965 the CBC began a coast-to-coast network open-line show, *Cross County Checkup*. At first, CBC officials were reluctant to adopt the popular phone-in format because open-line shows already had a reputation for being too sensational. *Cross Country Checkup* hosts avoided the confrontational style adopted by Burns, Webster, and many of their fellow open-line hosts on the private stations. The live broadcast quickly became a national forum for serious discussion of issues in the public sphere, although it has occasionally allowed lighter discussions about such subjects as favorite books. Hosts over the years have included Betty Shapiro, Elizabeth Gray, and Rex Murphy. The CBC also produces regional call-in shows in French and English. For CBC listeners in northern Canada, talk programs in native languages feature news, interviews, and phone-ins, an important service, particularly in sparsely populated areas.

It has become common for politicians to be guests on call-in shows, particularly during election campaigns. Their comments make headlines, as Prime Minister Kim Campbell found out in 1993 when she underestimated the price of milk. Politicians run the risk of being caught in mistakes, but they also gain an opportunity to be on the air unedited. Talk radio has taken on a uniquely Canadian flavor at several crisis points in the nation's history. In 1964, for example, English-language station CHUM in Toronto and French-language station CJMS in Montreal used a bilingual call-in show to discuss the emotionally charged subjects of bilingualism and the growing independence movement in Quebec. In 1990 radio stations in Toronto and Montreal again shared an open-line show on which people discussed the Canadian government's controversial constitutional reforms and the resulting polarization of views between Quebec and the rest of Canada. A station in St. John's, Newfoundland, where the provincial government voted against the reforms, and a station in Vancouver, a part of Canada that usually feels excluded from such debates, shared a similar show.

The power of talk radio to bring strong opinion to the airwaves frequently attracts complaints by listeners angered by the opinions expressed or the host's treatment of guests and callers. The federal broadcast regulator, the Canadian Radio–Television and Telecommunications Commission (CRTC), has guidelines for talk radio. The Canadian Broadcast Standards Council, an organization set up by Canada's private broadcasters, uses codes developed by the private broadcasters themselves to judge the validity of complaints. The CRTC guidelines and the private broadcasters' own codes both aim to ensure that broadcasters can continue to air spontaneous, entertaining, and informative forums for differing points of view, as long as those points of view do not convey racist, sexist, inaccurate, or other harmful commentary.

Satellite and internet technologies have made it easier to expand the broadcast reach of radio, nudging Canadian talk radio beyond the local to a national and international audience. Canadian radio stations have been quick to jump into internet broadcasting without losing sight of the essentially local appeal of much talk radio. Technology has also made it easier and cheaper for Canadian radio stations to carry programs originating outside Canada. Talk shows hosted by Dr. Laura Schlessinger, Dr. Joy Browne, and Mike Siegel are among the American imports heard in recent years on English-language private stations, alongside programs produced in Canada. The most publicized and controversial talk radio import has been *The Howard Stern Show,* which was picked up by radio stations in Montreal and Toronto in 1997, immediately attracting high ratings. In 1998 the Montreal station dropped the program amid complaints that Stern pushed talk radio beyond standards acceptable to Canadians, but this did not stop the Toronto station from continuing with the program until 2001.

Canadian talk radio has exhibited a U.S. influence principally by imitating the basic U.S. format. This is especially true of French radio, which, because of the language barrier, does not import U.S. programs. The durability of Canadian talk radio demonstrates that it succeeds as entertainment, but talk shows of Canadian origin tend to focus on information and not solely entertainment, a reflection to some extent on the CBC's traditional role as a national forum for current affairs.

ROSEMARY BERGERON

Further Reading

Becker, Jane, "Radio's Public Confessors: The Air Is Full of Secrets," *Maclean's* 74 (September 1961)

Davis, Ted, "Safeguards or Censorship? Proposed Guidelines May Restrict Talk Show Content," *Broadcaster* 47 (October 1988)

Dexter, Susan, "The Mouth That Roars," *Maclean's* 79 (October 1966)

Elliott, Tim, "Open Line Shows: The People Talk Back," *Broadcaster* 32 (January 1973)

Levin, Mark, "Will the Spoken Word Save AM Radio?" *Media* 3 (Spring 1996)

Roberts, Avril D., "Hold That Open-Line or Move to a More Manageable Format?" *Broadcaster* 37 (March 1978)

Webster, Jack, *Webster! An Autobiography,* Vancouver, British Columbia: Douglas and McIntyre, 1990

Cantor, Eddie 1892–1964

U.S. Radio Comedian

Although few people are familiar with him today, Eddie Cantor was one of radio's most popular performers in the 1930s and 1940s. In those days he made between $5,000 and $10,000 a week and had consistently high ratings.

Origins

Born Isidor Iskowitz (there are several variant spellings) in 1892, Cantor was raised in poverty by his grandmother. He was attracted to performing at a very young age and was discovered by comedian Gus Edwards while working as a singing waiter in Coney Island, New York. By 1912 he was performing in vaudeville houses, touring with Edwards and George Jessel. Known for his bulging eyes—his nickname was "Banjo Eyes"—and his frenetic energy, at first Cantor performed in blackface, a common convention in vaudeville at that time. He was a close friend of Bert Williams, one of the highest-paid black performers in vaudeville, who was also expected to perform in blackface.

By 1917 Cantor was appearing in the Ziegfeld Follies of 1917, along with such famous performers as Fannie Brice, Will Rogers, W.C. Fields, and Bert Williams. His vaudeville performances won him top billing and critical acclaim; he set box office records with his starring role in *Make it Snappy* in 1923, and by 1926 he was featured in his first movie, a film version of his successful 1924 Broadway show *Kid Boots*.

Radio

Cantor had begun making occasional radio appearances as early as 1921 and would sometimes perform for a charitable event that was being broadcast (throughout his life, he was known for his philanthropy), but his radio success really began with a popular variety program—his first network show—on the National Broadcasting Company (NBC), in September 1931. *The Chase and Sanborn Hour,* sponsored by Chase and Sanborn coffee, lasted until late 1934.

In his program Cantor sang and told jokes and had a cast of talented performers who also took part. He especially enjoyed having the studio audience interact with the cast. According to Dunning (1976; 1998), this was quite unusual in 1931: "Before Cantor, audiences were sternly warned to make no noise . . . while the shows were on the air. [Not even] laughter was permitted." But Cantor changed that: he wanted the audience to have a good time, and that attitude certainly contributed to his popularity. Not only did Cantor have a cast of regular performers, he also used his show to introduce new talent. Among the stars first heard on his program in the 1930s were comedienne Gracie Allen and dialect comic Harry Einstein, whose character "Nick Parkyakarkas" went over very well at a time when ethnic humor was popular. During the early 1940s Cantor helped launch the career of vocalist Dinah Shore. He also helped several black performers, such as singer Thelma Carpenter, at a time when black vocalists were not usually in the regular cast of the predominantly white variety shows.

The versatile Cantor was also successful in his own career as a singer; he had several hit songs, among them "Ida," "(Potatoes are Cheaper, Tomatoes are Cheaper) Now's the Time to Fall in Love," and "If You Knew Susie." He also continued appearing in movies; in 1934 *Kid Millions* grossed more than $2 million even though America was in the midst of the Depression. He became so famous internationally that his testimonial advertisements could be seen in European magazines, and he was frequently on the cover of U.S. fan publications such as *Radio Stars*. "Looney Tunes and Merrie Melodies" used him as a character in a 1933 cartoon, a parody of the hit song "Shuffle Off to Buffalo," in which the cartoon characters all chanted "We Want Cantor!" until the cartoon version of Eddie appeared.

In 1935 and again in 1940 and 1946 he had a new radio show with a new sponsor, but his style remained unchanged over the years; he was still the genial and energetic host with the clever one-liners and the topical humor. When not singing or performing skits about the events of the day, he made frequent jokes about his wife Ida and their five daughters; although today some critics find this misogynistic, back then the audience felt as if they were included in Eddie's life, and people felt they really knew his family. His radio career lasted till 1949, at which time, along with many others, he moved to television.

In addition to his success as a performer, Eddie Cantor had the respect of his colleagues in the entertainment business. He was a founder of the Screen Actors Guild and served as its president from 1933 to 1935. In 1937 he became the first president of the American Federation of Radio Artists. He helped to start the March of Dimes to fight Infantile Paralysis, and he often gave benefit concerts for orphans' homes and hospitals. He contributed time and money to help Jewish refugees during World War II. He was a guest on such talk shows as *America's Town Meeting of the Air* and, although the stock market crash of 1929 cost him most of his fortune, he continued to help the poor. In a 1936 interview with *Radio Stars* magazine, he stressed the importance of giving to charity and feeding those

Eddie Cantor
Courtesy Radio Hall of Fame

who are hungry. At times, some of his critics said his perfectionism made him difficult to work with, but no one could ever dispute his dedication to charitable causes

Although Cantor's humor has not aged well, he deserves to be remembered as one of the most popular and influential performers during radio's golden age, a man who brought laughter to millions of devoted fans. A year before his death, he issued a book of essays, *As I Remember Them,* a retrospective about some of the celebrities he had worked with during his 50-year career. He had first written about his life in a 1928 autobiography, *My Life Is in Your Hands,* and in 1959 he wrote about his philosophy of living in *The Way I See It.* He died in October 1964; only a few months earlier he had received a medal from President Lyndon Johnson for his years of humanitarian work.

DONNA L. HALPER

See also Comedy; Vaudeville

Eddie Cantor. Born Isidor Iskowitz, New York City, 31 January 1892. Stage debut at the Clinton Music Hall, 1907; won his first vaudeville talent show and a $5 prize, at Miner's Bowery Theater, 1908; toured with comedian Gus Edwards' *Kid Kabaret,* 1912–14; joined Ziegfeld's Follies 1917–19; made first radio appearance, 1921; first film appearance in the silent movie version of *Kid Boots,* 1926; wrote first edition of his autobiography, *My Life Is in Your Hands,* 1927; elected president of National Vaudeville Artists, 1928; had hit record, "Makin' Whoopee," from his Broadway show, 1929; suffered serious financial setback during the Depression which prompted his book, *Caught Short: A Saga of Wailing Wall Street,* 1929; starred in own radio program, 1931; first introduced comedienne Gracie Allen on his radio show, 1932; named first national president, Screen Actors' Guild, 1933–35; elected first national president of American Federation of Radio Artists, 1937; helped to create the March of Dimes, 1937–38; one-man show, "My Forty Years in Show Business," at Carnegie Hall. Recipient: Honorary Doctorate of Humane Letters, Temple University, 1951; special Academy Award for distinguished service to the motion picture industry, 1956; Medallion of Valor, State of Israel, 1962. Died in Hollywood, California, 10 October 1964.

Radio Series

1931–34 *The Chase and Sanborn Hour*
1935–54 *The Eddie Cantor Show*
1945 *Arch Oboler's Plays*
1948 *The Comedy Writers Show*
1949–50 *Take It or Leave It*
1950–52 *The Big Show*

Films

Kid Boots, 1926; *The Speed Hound,* 1927; *Follies,* 1927; *Special Delivery,* 1927; *Glorifying the American Girl,* 1929; *That Party in Person,* 1929; *Getting a Ticket,* 1929; *Whoopee!,* 1930; *Insurance,* 1930; *Mr. Lemon of Orange,* 1931; *Palmy Days,* 1931; *The Kid from Spain,* 1932; *Roman Scandals,* 1933; *Kid Millions,* 1934; *Hollywood Cavalcade,* 1934; *Screen Snapshots No. 11,* 1934; *Strike Me Pink,* 1936; *Ali Baba Goes to Town,* 1937; *Forty Little Mothers,* 1940; *Thank Your Lucky Stars,* 1943; *Hollywood Canteen,* 1944; *Show Business,* 1944; *Rhapsody in Blue,* 1945; *If You Knew Susie,* 1948; *The Story of Will Rogers,* 1952; *The Eddie Cantor Story,* 1953

Television

The Colgate Comedy Hour, 1950–54; *The Eddie Cantor Comedy Theatre,* 1955; *Seidman and Son,* 1956

Stage

Kid Kabaret, 1912–14; *Canary Cottage,* 1916; *Midnight Frolic,* 1917; *Ziegfeld's Follies,* 1917–19; *Midnight Rounders,* 1920–21; *Make It Snappy,* 1922; *Kid Boots,* 1923–26; *Ziegfeld's Follies,* 1927; *Ziegfeld's Whoopee,* 1928–30; *Banjo Eyes,* 1942

Selected Publications

My Life Is in Your Hands (with David Freedman), 1928; reissued, *With a New Chapter Bringing the Story Up to 1932,* 1932
Caught Short: A Saga of Wailing Wall Street, 1929
Between the Acts, 1930
Yoo-Hoo, Prosperity! The Eddie Cantor Five-year Plan (with David Freedman), 1931
Take My Life (with Janes Kesner Moris Ardmore), revised edition, 1957
The Way I See It (edited by Phyllis Rosenteur), 1959

Further Reading

Dunning, John, *Tune in Yesterday: The Ultimate Encyclopedia of Old-Time Radio, 1925–1976,* Englewood Cliffs, New Jersey: Prentice-Hall, 1976; revised edition, as *On the Air: The Encyclopedia of Old-Time Radio,* New York: Oxford University Press, 1998
"Eddie Cantor Dead; Comedy Star Was 72," *New York Times* (11 October 1963)
Goldman, Herbert G., *Banjo Eyes: Eddie Cantor and the Birth of Modern Stardom,* New York: Oxford University Press, 1997
Fisher, James, *Eddie Cantor: A Bio-Bibliography,* Westport, Connecticut: Greenwood Press, 1997
Variety (14 October 1964) (numerous essays and tributes to Cantor, published the week he died)

Capehart Corporation

The history of the Capehart Corporation in Fort Wayne, Indiana, dates back to the late 1920s, when entrepreneur Homer Earl Capehart (1897–1970) established the foundations for the enterprise. Capehart was known for producing quality high-end phonographs, radios, radio-console combinations, and jukeboxes.

Homer E. Capehart was born 6 June 1897 in Algiers, Indiana, and he grew up on a farm. After high school he enlisted in the U.S. Army from 1917 to 1919 and advanced to the rank of sergeant. He joined the J.I. Case Corporation as a salesman and soon earned a reputation as a man who could sell anything. He moved from sales to entrepreneurship, at first manufacturing and selling popcorn poppers. In 1928 he established the Automatic Phonograph Corporation; by 1929 the company was manufacturing "talking machines" and was known as the Capehart Automatic Phonograph Corporation. Capehart served as founder and president from 1927 to 1932. During the 1930s Depression era, when other companies such as Philco and the Radio Corporation of America (RCA) were developing low-priced consumer radio sets to encourage sales, Capehart stood stubbornly behind the company's high-quality, expensive receivers. This decision led the company to the brink of bankruptcy. In the early 1930s, at the height of the Depression, Capehart joined Wurlitzer, a producer of jukeboxes, and as a result the Capehart Corporation was saved. Capehart himself served as vice president of the Wurlitzer Company from 1933 to 1938. The joining of the two companies was a complementary success: Wurlitzer sold jukeboxes, which in turn sold records, which in turn created a demand for the Capehart phonograph. The investment helped make Capehart a wealthy man. Despite success with Wurlitzer, Homer Capehart was forever the adventurer and entrepreneur, and by the end of the 1930s he was ready to move into real estate.

In 1938 the Capehart Company and all its "real estate, plants, factories . . . all patents, patent licenses and patent application rights, and trade marks" were sold to the Farnsworth Television and Radio Corporation. Farnsworth kept the name Capehart because of its reputation for quality radio and phonograph manufacturing. The Capehart manufacturing entities were retooled to manufacture both Farnsworth and Capehart brand-name radio and television receivers intended for consumer sale. The Farnsworth Corporation was banking on the Capehart organization's reputation for quality to launch its entrance into the manufacturing business. However, World War II intervened, and the plants were converted a second time, this time for the manufacturing of armed forces communication equipment.

Following the war, the name Capehart surfaced again. By 1949 the International Telephone and Telegraph Corporation

(ITT) had purchased the Farnsworth Television and Radio Corporation, and the Capehart-Farnsworth division of the company was returned to consumer manufacturing. However, even with the financial backing of ITT, the Capehart-Farnsworth sets were never able to capture a significant share of the radio and television manufacturing market. They were competing against the giants of radio manufacturing at the time—RCA, General Electric, Philco, and Westinghouse. By 1954 the Capehart-Farnsworth division of ITT was split. The Farnsworth Electronic division continued as a wholly owned subsidiary of ITT, but the Capehart manufacturing was sold in 1956 to the Ben Gross Corporation, a holding company. The manufacturing properties in Fort Wayne were retained by ITT, the remaining assets were sold, and the Capehart name disappeared from the history of radio and television.

DONALD G. GODFREY

See also High Fidelity; Receivers

Further Reading

Godfrey, Donald G., *Philo T. Farnsworth: The Father of Television,* Salt Lake City: University of Utah Press, 2001

Pickett, William B., *Homer E. Capehart: A Senator's Life, 1897–1979,* Indianapolis: Indiana Historical Society, 1990

Sampson, Anthony, *The Sovereign State of ITT,* New York: Stein and Day, 1973

Capital Radio

London Commercial Station

Capital Radio in London was the second authorized commercial radio station in the United Kingdom, going on air 16 October 1973, just eight days after the first such service, the London Broadcasting Company (LBC). Until then the British Broadcasting Corporation (BBC) had been the only organization allowed to operate radio stations within Britain for more than 50 years. London was unique in the first phase of development of what was officially called Independent Local Radio (ILR), in that two franchises were awarded by the Independent Broadcasting Authority (IBA) to cover a defined geographical area, rather than the usual one. The "general and entertainment" franchise was—to some surprise—awarded to Capital, after a hard-fought contest against seven other consortia, to run what was expected to be the most lucrative and prestigious station in the ILR system. The chairman was the internationally renowned actor and film director Richard Attenborough.

Capital promised "quality pop," making a virtue of the fact that, unlike the BBC's pop network Radio 1, it was in FM stereo as well as on AM. Capital initially played a more adventurous and sophisticated selection of tracks, many of which were album cuts rather than Top 40 singles. The overall station style was largely modeled on adult contemporary FM stations in the United States. The station had some familiar voices, because many of the disc jockeys had previously worked for the BBC and for "pirate" radio; others had gained experience in commercial radio overseas. Nor was the station a continuous diet of popular music. In line with the IBA's demands and expectations, Capital's early programming contained an ambitious schedule of specialist music shows, original drama, arts magazines, children's shows, community action slots, and much more, as well as its own news service and a commitment to public-affairs programming, including phone-ins and documentaries.

Unfortunately, the station's debut coincided with an economic recession that would have hit advertising revenues for even a successful station. Capital's initial audience figures were disappointing: the first ratings, around Christmas 1973, indicated a weekly "reach" of about 1 million—about a tenth of the potential audience—and this created a financial crisis at the station. Richard Attenborough later admitted that the station had faced closure within a year of its launch and that he was only able to stave off bankruptcy by offering paintings from his private art collection as collateral against the company's mounting debts. The shareholding structure of the company was also radically changed: in February 1975 the Canada-based Standard Broadcasting Company became the station's largest shareholder, increasing its share of the company's stock to just under 25 percent.

The music policy was switched to more or less pure Top 40; within a year the news service was scrapped (although it was restored some years later), to be replaced by the Independent Radio News (IRN) bulletin produced by London rival LBC, and drama was abandoned. The well-known BBC television and radio presenter Michael Aspel was hired to present the 9 A.M. to noon weekday slot aimed at housewives, and Kenny

Everett and Dave Cash re-created their successful on-air partnership of pirate radio days during the vital breakfast time period. Gradually these changes built a strong and loyal listener base. Although programming certainly became more populist, the station did not renege on commitments to its public service broadcasting and charitable commitments, which have endured through major changes in the company's development. By 2002 the Eastertime "Help a London Child" radiothon had raised some £14 million.

The Capital Radio group has three other charities of its own operated by its stations in different parts of the U.K. From its very early days, Capital Radio has run community social action initiatives, heavily promoted on-air but often providing confidential, personal, advice off-air. "Helpline" and "Jobfinder" were two of the services that helped justify the early "Cuddly Capital" promotional line. In the summer of 2002 the Capital FM network launched "Call a Course" aimed at helping students find appropriate college courses.

Public service commitments, it appeared, could co-exist happily with a profitable media business, because in 1987 Capital became the first U.K. radio company to float shares on the London Stock Exchange. The public company was in a position to fully exploit the changes in the media and business environment in the late 1980s caused by the government's decision to fundamentally change the licensing and regulatory structure of commercial radio: stations were allowed to target their program content to specific audiences and to drop much of their public service and "minority listening" obligations; many more stations were licensed, creating commercial competition in each area; and services were encouraged—then compelled—to end simulcasting and operate different services on their AM and FM transmitters. Accordingly, in 1988 the company split its services into Capital Gold—featuring oldies from the 1960s through the 1980s, with sport, comedy, and personality presenters—and Capital FM, aimed at a young adult audience and featuring a contemporary hit radio format and big-prize contests. This change had the effect of increasing the overall audience: Capital FM established a seemingly unassailable position as the most listened-to radio service in London (over all BBC and commercial network and local rivals), with Capital Gold often rated number two.

However, the new century saw this dominance compromised both by increasing ratings success of BBC network services and by other London commercial stations—including the Capital-owned rock service XFM. A new Head of Music for Capital FM was appointed in early 2003 and a more diverse music policy implemented. Press reports noted some spectacular losses to the FM service, especially in the key breakfast slot. This program, hosted since 1987 by Chris Tarrant (also a national TV personality through his hosting of such shows as *Who Wants To Be a Millionaire?*), became the company flagship and was responsible for around 15 percent of the group's entire revenues. Increasingly lengthy and frequent vacations by Tarrant and rumors that he was about to quit the show contributed to a major fall in the company's share price. During this period, official listening figures showed Capital FM in third overall place in the London market—behind the BBC's speech network Radio 4 and music and personality service Radio 2—with the AM service, Capital Gold, failing to make even the top 10 in market share. Some media pundits even compared the downturn in Capital's position to that of the once dominant Independent Television (ITV) commercial TV network, which had been squeezed by newer commercial rivals and by a more ratings-conscious BBC television. Nevertheless, Capital FM's share was still around 50 percent greater than its nearest commercial rival, and the breakfast show—even though it was surveyed only in the greater London area—had a larger audience than was achieved across the whole country by the equivalent shows on two of the three national commercial stations.

Deregulation, among other things, meant that commercial radio groups were allowed to acquire other stations. In 1993 Capital bought the Birmingham stations BRMB FM and Xtra AM; a year later Capital added seven stations on the English south coast, followed by Fox FM in Oxford and Red Dragon FM and Touch Radio in South Wales. In 1997 Capital proposed to buy the national commercial rock station Virgin Radio from Richard Branson. However, the deal was referred to the Monopolies and Mergers Commission, and before this investigation could be completed, Virgin was bought by Chris Evans' Ginger Media Group. One of Capital's most controversial acquisitions came in 1999: XFM in London had been awarded a license as a "new music" station, and there were street demonstrations from fans of the original station sound when Capital made major changes to the music output and presenter lineup.

Capital has also been a major investor in digital radio. By the spring of 2003 it partly or wholly owned the licenses for eleven multiplexes, and provided more than 30 program services, including a simulcast of XFM, and a new children's station (in partnership with the Disney corporation), Capital Disney, to multiplex license-holders, as well as an Adult Contemporary service on the commercial Digital One network.

Nor are the company's business interests confined to radio services: in 1996 it acquired the My Kinda Town restaurant chain in London, which it renamed Capital Radio Restaurants. However, Capital's radio success was not matched in this venture, and the restaurants were abandoned less than three years later.

RICHARD RUDIN

See also British Commercial Radio; London Broadcasting Company

Further Reading

Barnard, Stephen, *On the Radio: Music Radio in Britain,* Milton Keynes, Buckinghamshire, and Philadelphia, Pennsylvania: Open University Press, 1989

Crisell, Andrew, *An Introductory History of British Broadcasting,* London and New York: Routledge, 1997

Crook, Tim, *International Radio Journalism: History, Theory, and Practice,* London and New York: Routledge, 1998

Lewis, Peter M., and Jerry Booth, *The Invisible Medium: Public, Commercial, and Community Radio,* London: Macmillan Education, 1989; Washington, D.C.: Howard University Press, 1990

Shingler, Martin, and Cindy Wieringa, *On Air: Methods and Meanings of Radio,* London and New York: Arnold, 1998

Captain Midnight

Adventure Program

Among the many syndicated and network daily serials aimed at younger listeners was this aviation-related program of adventure that involved code-breaking and worldwide travels. The debut of *Captain Midnight* is generally given as 17 October 1938; however, since the show was originally syndicated under the sponsorship of Skelly Oil, it is possible that different stations first aired *Captain Midnight* on different start dates. The initial sponsor owned Spartan Aircraft, and had previously sponsored *The Air Adventures of Jimmie Allen,* another aviation oriented radio serial. The writers from *Jimmie Allen,* Robert Burtt and Wilfred Moore, both World War I pilots, were assigned to create the new show. With pilots scripting the show, the aviation content was accurate.

The initial adventures of the show involved the title character, Charles J. ("Jim" or "Red") Albright, who was referred to primarily by his alias "Captain Midnight," as an independent pilot who acted altruistically to fight wrongdoing, along with his ward, Chuck Ramsay, a girl sidekick, Patsy Donovan, and various others, including a mechanic, Ichabod Mudd. His chief adversary was a criminal, Ivan Shark, who led a gang with his daughter Fury, two aides, Fang and Gardo. Stories involved adventures in the western United States, Mexico, and Canada.

In 1940, the program changed sponsors and first aired on a national network, Mutual. Ovaltine had previously sponsored *Little Orphan Annie,* a 15-minute adventure serial based on the newspaper comic strip, since 1930. Ovaltine dropped sponsorship of that show in favor of *Captain Midnight,* possibly because the international tensions of the era required a mature hero in the eyes of the sponsor. The initial program under the new sponsor provided the hero with an "origin" story (he earned the code name Captain Midnight because of an exploit during World War I) and a secret organization to head. For *Orphan Annie,* Ovaltine had developed a club (Radio Orphan Annie's Secret Society) and a "Decoder Pin"; and these were concepts carried over to *Captain Midnight.* The hero headed a paramilitary organization, the Secret Squadron, which was supposedly set up by a high U.S. government official. Its identifying badge was a cipher device, the Code-O-Graph, which was used, like its *Orphan Annie* predecessors, to decrypt "secret messages" provided at the close of some episodes, to provide a hint of the next day's broadcast.

Most of the main characters from the Skelly show were retained, including Chuck Ramsay, Ichabod Mudd, Ivan Shark, Fury Shark, Fang, and Gardo. One exception was that Patsy Donovan was dropped, and a new girl sidekick, Joyce Ryan, was added. The nature of the Secret Squadron, which was supposed to fight sabotage and espionage, enabled the program to have adventures around the world, including the Caribbean, Central and South America, and China, as well as in the United States. New villains were introduced: the Barracuda shortly before the U.S. entry into World War II, and Baron von Karp, Admiral Himakito, and Señor Schrecker during the war. (An interesting sidelight: well before the 1941 Japanese attack, Captain Midnight found plans for Pearl Harbor in The Barracuda's headquarters in Japanese-occupied China.) After the war, the program continued to be set in locales across the world, and the major villain, Ivan Shark, became prominent again.

The program retained its 15-minute serial format through June of 1949. In September of that year, it changed format to half-hour, complete-in-one-program stories. These alternated with *Tom Mix Ralston Straight Shooters,* running on Tuesdays and Thursdays, until the middle of December, when it went off the air.

The first sponsor, Skelly Oil, aimed most of its products to adults—gasoline, bottled gas, and motor oil. The program attracted a fairly large minority of adult listeners, despite its scheduled spot in the middle of the hour devoted to juvenile programs. This audience carried over to Ovaltine sponsorship. As a result, the vocabulary, dialog, and concepts were more mature than those normally found in a children's adventure show.

One notable aspect of *Captain Midnight* was that women were not relegated to stereotypical roles of the time. Joyce Ryan, a teenage Secret Squadron member, routinely faced the same dangers as her male counterparts, including going on commando raids and participating in aerial dogfights. Likewise, Fury Shark was as courageous as her father, and as scheming. Neither expected special treatment because of their gender. This was reflected in the handbooks that came with the Code-O-Graph premiums, where both genders were encouraged to go after exciting careers.

A television version of *Captain Midnight* was aired on CBS (1953–57), sponsored by Ovaltine, but it differed significantly from the radio program. When it was rereleased as a syndicated show, the hero's name was changed to Jet Jackson, and the new name was spliced into the sound track.

STEPHEN A. KALLIS, JR.

See also Premiums

Cast

Captain Midnight	Ed Prentiss (1938–39), Bill Bouchey (1939–40), Ed Prentiss (1940–49), Paul Barnes (1949)
Chuck Ramsay	Billy Rose (1938–41), Jack Bivans (1941–44), Johnny Coons (1944–46), Jack Bivans (1946–49)
Joyce Ryan	Marilou Neumayer (1940–46), Angeline Orr (1946–49)
Ivan Shark	Boris Aplon
Ichabod Mudd	Hugh Studebaker (1940–46), Sherman Marks (1946–48), Art Hern (1948–49)
Fury Shark	Rene Rodier (1938–40), Sharon Grainger (1940–1949)
Patsy Donovan	Alice Sherry Gootkin
Kelly, SS-11	Olan Soulé

Creators/Writers

Robert Burtt and Wilfred Moore

Programming History

Syndicated	October 1938–March 1940
Mutual	September 1940–July 1942
NBC Blue	September 1942–June 1945
Mutual	September 1945–December 1949

Further Reading

Kallis, Stephen A., Jr., "The Code-O-Graph Cipher Disks," *Cryptologia* 5, no. 2 (April 1981)

Kallis, Stephen A., Jr., "Flying with the Secret Squadron," *Yesterday's Magazette* 25, no. 1 (January/February 1997)

Kallis, Stephen A., Jr., *Radio's Captain Midnight: The Wartime Biography,* London and Jefferson, North Carolina: McFarland, 2000

Tumbusch, Tom, *Tomart's Price Guide to Radio Premium and Cereal Box Collectibles: Including Comic Character, Pulp Hero, TV, and Other Premiums,* Radnor, Pennsylvania: Wallace-Homestead, 1991

Widner, James F., "Cast Photos for *Captain Midnight,*" <www.otr.com/cm_cast.html>

Winterbotham, Russell R., *Joyce of the Secret Squadron: A Captain Midnight Adventure,* Racine, Wisconsin: Whitman, 1942

Car Radio. *See* Automobile Radio

Car Talk

Advice and Humor Call-in Program

Few radio programs can deliver on promotional announcements that promise advice on both car repair and human relationships. But then few programs blend the serious and the sophomoric into an hour-long show that is both funny and helpful. National Public Radio's (NPR) *Car Talk* is one.

The hosts, Tom and Ray Magliozzi, better known to their fans as "Click" and "Clack," are brothers who opened a do-it-yourself counterculture garage in Cambridge, Massachusetts, in 1973. As hippies evolved into people with real jobs and cars became more complicated, the brothers offered more conventional car repair.

In 1977 the Magliozzis were invited to appear on a talk show on WBUR-FM with other area mechanics. Tom accepted, and when he returned the following week, he brought Ray along. Later they were given their own WBUR talk show in which they gave advice and tried to drum up business for their garage. The show is produced by Dewey, Cheetham, and Howe—a company the Magliozzis named with the same self-deprecation that drives the program—and it still originates at WBUR.

In January 1987 NPR host Susan Stamberg invited the brothers to be weekly contributors to NPR's *Weekend Edition*. On October 31 of that same year, *Car Talk* premiered as a national program. After more than ten years on the air, NPR broadcasts *Car Talk* on more than 550 NPR stations nationwide to over 3.8 million laughing fans. The program received the George Foster Peabody Award in 1992. In 1998 the Museum of Broadcast Communications in Chicago inducted Tom and Ray Magliozzi into the Radio Hall of Fame.

Although cars and human responses to cars are the foundations of the show, *Car Talk* is about laughing. Tom and Ray are ready to laugh out loud at themselves, at cars, and at callers. Often the show begins with a humorous piece about a serious issue such as global warming or politics. Sometimes one of the brothers begins with a tirade against oversized automobile engines or people who drive while talking on cell phones.

Both brothers are graduates of the Massachusetts Institute of Technology; Tom has a doctorate in marketing and has taught at Boston and Suffolk Universities. Ray still runs the garage and is a consultant to the Consumer Affairs Division of the Massachusetts attorney general's office. Their education and work backgrounds provide fertile areas for them to make fun of each other, their schools, and all areas of higher education. They also make fun of each other's expertise or lack thereof. When one takes a caller's question, he will often say that the other doesn't know what he's talking about and can't possibly give a good answer. Callers come in for their share of jibes, too, mostly in the form of gentle teasing.

Literary references, puns, and joking references to NPR news reporters and hosts spark the show: for example, reading the standard NPR underwriter line, the Magliozzis remind listeners that "Support for *Car Talk* comes from bogus parking tickets we put on cars all over the NPR parking lot" or that "Support for *Car Talk* comes from the small but regular deductions we make from Carl Kassell's retirement account." Callers are encouraged to banter and allowed to star as story tellers. In the midst of the fun and zany comments, real questions about spark plugs, used cars, problems with mechanics or dealer service shops, and personal issues with cars do get answered.

On the air, Ray is the one who actually tries to answer the automotive questions. He's the director of the show and keeps it moving. Tom make jokes, insults Ray and callers, and laughs

Tom and Ray Magliozzi, *Car Talk*
Courtesy National Public Radio

the most. Both men are honest to the point of bluntness when it comes to how to deal with bad mechanics or auto manufacturers. Ray says that they started cracking jokes the first time they were unable to answer a caller's questions. The more they laughed, the better they enjoyed the show. And the bigger the audience became. Producer Doug Berman, who has worked with the Magliozzis for 12 years, said in an interview in *Brill's Content,* "They're like the kids in the back of the class that used to joke and make you laugh, and you didn't want to laugh because you'd get in trouble" (Greenstein).

Car Talk is a tightly structured show: it begins with a thought piece, usually humorous. There are three segments that feature phone calls. Most weeks a puzzle is featured in the "third half" of the program. Music from many genres, as long as the lyrics mention something automotive, is used as audio bumpers to separate the segments.

"Stump the Chumps" is an irregular feature in which callers are brought back to reveal whether Click and Clack gave the correct answer to their automotive questions. This feature gives rise to much self-deprecating humor. It also establishes credibility, because most of the time the answer was correct and saved the caller both time and money.

During the show's closing credits, puns reign. From research statistician Marge Inoverra to pseudonym consultant Norm Deplume, the end of *Car Talk* is a high point. The names are accompanied by the appropriate accent when required. Although the basic names repeat week after week, there's always a new one to catch the ear of the faithful and keep Tom laughing.

Car Talk closes with an underwriting statement that offers a final opportunity to make a joke about another NPR host: "And even though Scott Simon sends his resume to MTV every time he hears us say it, this is NPR, National Public Radio."

PAM SHANE

See also Comedy; National Public Radio

Programming History

WBUR-FM	1977–87
NPR (550 stations)	1987–present

Further Reading

Car Talk website, <www.cartalk.cars.com>
Greenstein, Jennifer, "The Car Talk Guys Just Want to Have Fun," *Brill's Content* (October 1999)
Magliozzi, Tom, and Ray Magliozzi, "The Mechanics of Buying a Great Used Car," *Friendly Exchange* (Summer 1999)
Magliozzi, Tom, and Ray Magliozzi, *In Our Humble Opinion,* New York: Penguin, 2000

Cavalcade of America

U.S. Radio Drama

Sponsored by E.I. du Pont de Nemours and Company, *Cavalcade of America* established the dramatic anthology program format among a generation of public relations and advertising specialists, as well as its reluctant sponsor, in a period when continuous institutional promotion by radio was not generally practiced and when the value of radio in prosecuting even short-term public relations campaigns was not fully appreciated. Because the DuPont Company's previous radio use had been limited to the efforts of company officials who personally helped underwrite the anti–New Deal talks of the American Liberty League, the National Association of Manufacturers, and other pro-business groups, the debut of *Cavalcade* was a signal event in the conservative seedtime of modern broadcast entertainment. What became the longest-running radio program of its kind debuted 9 October 1935 and ran until 1953 with only two brief lapses. In 1952 *Cavalcade* moved to television, where it remained until 1955. Although *Cavalcade*'s sponsor never relinquished its editorial prerogative, by 1940 DuPont acceded to their specialists' attempts to bury the program's more troublesome aspects in the dramatic subtext of "Better Things for Better Living."

A positive expression of corporate social leadership supervised by the advertising and public relations specialists of Batten, Barton, Durstine, and Osborn (BBD&O), the *Cavalcade* exemplified the higher concepts of corporate public affairs, far removed from the give and take of American party politics, which by 1935 had become manifest in a daily cycle of reaction and attack. By the early 1950s, company advertising and public relations specialists proudly pointed to increasingly favorable opinion polling data associating DuPont with "Better Things for Better Living." Reflecting on BBD&O's long and successful relationship with DuPont, Bruce Barton attributed the turnaround in part to two factors: women's nylons and the *Cavalcade of America.*

BBD&O's aggressive merchandising of *Cavalcade* involved celebrated authors, dramatists, actors, actresses, educators, and historians. From 1935 to 1938 the *Cavalcade*'s historical advisers included Dixon Ryan Fox—the president of Union College and the New York Historical Association—and Professor Arthur M. Schlesinger of Harvard. The arrangement enabled BBD&O to merchandise the program as a contribution to the "new social history" with which Fox and Schlesinger had become identified as co-editors of the 12-volume *A History of American Life*. Suspended between the liberal sensibilities of the new social history, represented by the collaboration of Fox and Schlesinger, and the sponsor's predilection for rhetorical attacks upon Franklin Roosevelt's New Deal, the *Cavalcade* offered a counter-subversive drama of self-reliance, resourcefulness, and defiance animated by the misfortunes of typically natural phenomena: grasshopper plagues, flash floods, fire, drought, dust storms, blizzards, ice floes, and log jams. Successful resolution demanded heroic acts of voluntarism, community spirit, and the sterner stuff that defined a heritage. As one flinty character explained while he helped extinguish a forest fire threatening his town, "What we struggled to get, we fight to keep."

The dramatization of the personal meaning of business enterprise played a role in the *Cavalcade*'s striking use of female protagonists. In its first season, the *Cavalcade* presented a hierarchical schedule of broadcasts beginning with "Women's Emancipation," the story of Elizabeth Cady Stanton, Lucretia Mott, William Lloyd Garrison, and Susan B. Anthony; "Women in Public Service," the story of Jane Addams and Hull House; "Loyalty to Family," the story of frontier widow Ann Harper; and "Self-Reliance," the story of planter Eliza Lucas' efforts to establish indigo in Carolina. Many *Cavalcade* women turned up as agents of production. For example, "The Search for Iron," broadcast in 1938, dramatized the story of the Merrit family's discovery of a massive iron ore deposit in Minnesota's Mesabi Range. The search, spanning three generations, featured matriarch Hepzabeth Merrit, log-hewn home life, and a frontier quest for resources. The concluding "story of chemistry" explained how miners used DuPont dynamite to excavate iron ore from "mother Earth," an example of the modern world's extraordinary engineering feats and of dynamite's use for constructive projects. Not without lighter moments, the "Search for Iron" began, as did many early *Cavalcade* broadcasts, with a medley of popular show tunes, in this case "Someday My Prince Will Come" and "Heigh Ho" from Walt Disney's *Snow White and the Seven Dwarfs*.

By the early 1940s, the *Cavalcade of America* had become the commitment to well-merchandised institutional entertainment that its specialists had long sought. Specialists attributed the *Cavalcade*'s success to its capacity to assimilate the functions of broadcast education and entertainment, with each adjusted to fit the circumstances of the changing leadership of the DuPont Company; the inroads of middle management using positivist audience research; and the onset of World War II, which made possible and even desirable the expression of democratic sensibilities.

After 1940 the *Cavalcade* featured a new mixture of amateur and academic historians who assumed greater program responsibilities. Professor Frank Monaghan of Yale delivered on-air story introductions. A memorable broadcast performance by poet and Lincoln biographer Carl Sandburg and the performance of poet Stephen Vincent Benét's *The People, Yes* signaled the relaxation of the program sponsor's editorial outlook. Thereafter, the formulaic dramatization of the American past culminating in "better living" distanced itself from the crisis of Depression-era business leadership that had called the *Cavalcade* into being. Ever so slowly, the *Cavalcade* decamped from the usable past for the intimate terrain of "more," "new," and "better living" merchandised in a build-up of stars and stories.

In concert with program producer BBD&O, the National Broadcasting Company (NBC) took the *Cavalcade* on the road for timely broadcast performances before the network's "pressure groups." The first of three remote broadcasts originated from the Chicago Civic Opera House, starring Raymond Massey in Robert Sherwood's adaptation of Sandburg's *Abraham Lincoln: The War Years*. Another starred Helen Hayes in "Jane Addams of Hull House," broadcast from the Milwaukee convention of the General Federation of Women's Clubs. A third program, attended by DuPont's Richmond, Virginia, employees, featured Philip Merivale in "Robert E. Lee," based on historian Douglas Southall Freeman's biography of the general.

Program specialists acknowledged the advantage of featuring characters already familiar to listeners, many of whom regarded historical figures as voices of authority. The ideal protagonist was heroic yet humble. Of the 750 *Cavalcade* radio programs broadcast from 1935 to 1953, biographical treatments of George Washington and Abraham Lincoln led the list (15 programs each), followed by Benjamin Franklin (9 programs) and Thomas Jefferson (8). Washington personified a recurring *Cavalcade* metaphor cementing America's revolutionary struggle for freedom with business' modern-day struggle to escape the regulatory tyranny of the New Deal. In a dramatization of the first inauguration entitled "Plain Mr. President," for example, the *Cavalcade*'s Washington invoked the "sacred fire of liberty and the destiny of the republican form of government . . . staked on the experiment entrusted to the American people." Washington prayed that "the invisible hand of the almighty being guide the people of the United States to wise measures, for our free government must win the affection of its citizens and command the respect of the world." The weekly "story of chemistry," entitled "news of

Cavalcade of America, 1938
Courtesy CBS Photo Archive

chemistry's work in our world," noted that "Washington, the practical economist, would no doubt have been pleased with modern house paints that actually clean themselves."

Gaining the confidence of their sponsor, who at last warmed to the idea of entertainment, the *Cavalcade*'s producers found themselves able to take advantage of a wider range of story material. This new range of material expanded the program's original basis in the historical past and the world of letters to feature adaptations of Hollywood screenplays and original works for radio that dramatized democratic sensibilities. In fall 1940, the *Cavalcade* presented the story of "Wild Bill Hickok" woven around a ballad composed and performed by Woodie Guthrie; "Town Crier" Alexander Woolcott, on loan from the Columbia Broadcasting System (CBS), who performed his "word picture" of "The Battle Hymn of the Republic"; and a special Christmas night broadcast of Marc Connelly's "The Green Pastures" featuring the Hall Johnson Choir. The adaptation of popular screenplays the following season enlarged upon the plan. In November 1941 the *Cavalcade* presented Henry Fonda in "Drums along the Mohawk" and Errol Flynn in "They Died with Their Boots On"; in the weeks following Pearl Harbor, the program featured Orson Welles in "The Great Man Votes" and James Cagney in "Captains of the Clouds." The appearance of stars who volunteered personal feelings about the company at the conclusion of select broadcasts spoke volumes for the program sponsor's growing confidence in a corporate public relations strategy inconceivable in the early years of the program.

The *Cavalcade* signaled an appreciation among specialists and business leaders alike that carefully scripted investments in dramatic anthology programming could, in the long run, reestablish a political climate conducive to the autonomous expansion of corporate enterprise. Business' contest with the administration for social and political leadership would continue, specialists hoped, divorced from rhetorical reaction and counterproductive short-term effects.

After World War II, the dramatic anthology became the preferred vehicle of corporate public relations among the clients of BBD&O, with tremendous significance for the television of the 1950s. BBD&O-produced programs included *Cavalcade of America, General Electric Theater, U.S. Steel Hour* (*Theater Guild on the Air*), and *Armstrong Circle Theater.* As the prototype of well-merchandised institutional entertainment, the *Cavalcade* set the precedent for them all, including the merchandising of programs undertaken by *General Electric Theater* host and program supervisor Ronald Reagan.

Ever responsive to the need of the moment, the free enterprise subtext of radio's *Cavalcade* continued unabated. At times, company public relations and advertising specialists seemed incapable of any other than dramaturgical expression. When the DuPont Company became entangled in an antitrust suit in 1949, for example, the *Cavalcade* dramatized the benefits of large-scale monopoly in "Wire to the West," the story of Western Union's consolidation of rival telegraph companies; in "Beyond Cheyenne," a story about "how the packing industry started as small business and became big business"; and in "The Immortal Blacksmith," "a story of the invention of the electric motor by Tom Davenport, . . . which never amounted to anything until big companies took hold of it and converted its power into conveniences for the millions." The *Cavalcade's* sponsor's reluctance to broadcast a more explicit defense spoke for a certain dramatic success.

WILLIAM L. BIRD, JR.

Narrator/Host
Walter Huston

Announcers
Frank Singiser, Gabriel Heatter, Basil Ruysdael, Clayton "Bud" Collyer, Gayne Whitman, Ted Pearson

Actors
John McIntire, Jeanette Nolan, Agnes Moorehead, Kenny Delmar, Edwin Jerome, Ray Collins, Orson Welles, Karl Swenson, Ted Jewett, Jack Smart, Paul Stewart, Bill Johnstone, Frank Readick, Raymon Edward Johnson, Ted de Corsia, Everett Sloane, Luis Van Rooten, Mickey Rooney, Cary Grant, Tyrone Power, and Ronald Reagan

Producer/Directors
Homer Fickett, Roger Pryor, Jack Zoller, Paul Stewart, and Bill Sweets

Writers
Arthur Miller, Norman Rosten, Robert Tallman, Peter Lyon, Robert Richards, Stuart Hawkins, Arthur Arent, Edith Sommer, Halsted Welles, Henry Denker, Priscilla Kent, Virginia Radcliffe, Frank Gabrielson, Margaret Lewerth, Morton Wishengrad, George Faulkner, Irv Tunick

Programming History
CBS 1935–39
NBC Blue January 1940–June 1940
NBC Red 1940–53

Further Reading

Barnouw, Erik, editor, *Radio Drama in Action: Twenty-Five Plays of a Changing World,* New York: Rinehart, 1945

Barnouw, Erik, *Media Marathon: A Twentieth-Century Memoir,* Durham, North Carolina: Duke University Press, 1996

Bird, William L., Jr., *Better Living: Advertising, Media, and the New Vocabulary of Business Leadership, 1935–1955,* Evanston, Illinois: Northwestern University Press, 1999

Burk, Robert F, *The Corporate State and the Broker State: The Du Ponts and American National Politics, 1925–1940,* Cambridge, Massachusetts: Harvard University Press, 1990

Fones-Wolf, Elizabeth, "Creating a Favorable Business Climate: Corporations and Radio Broadcasting, 1934–1954," *Business History Review* 73 (Summer 1999)

Fox, Dixon Ryan, and Arthur Meier Schlesinger, editors, *The Cavalcade of America,* Springfield, Massachusetts: Milton Bradley, 1937

Fox, Dixon Ryan, and Arthur Meier Schlesinger, editors, *The Cavalcade of America, Series 2,* Springfield, Massachusetts: Milton Bradley, 1938

Golden, L.L.L., *Only by Public Consent: American Corporations Search for Favorable Opinion,* New York: Hawthorn Books, 1968

Marchand, Roland, *Creating the Corporate Soul: The Rise of Public Relations and Corporate Imagery in American Big Business,* Berkeley: University of California Press, 1998

Miller, Arthur, *Timebends: A Life,* New York: Grove Press, and London: Methuen, 1987

Walker, Strother Holland, and Paul Sklar, *Business Finds Its Voice: Management's Effort to Sell the Business Idea to the Public,* New York and London: Harper, 1938

Wolfskill, George, *The Revolt of the Conservatives: A History of the American Liberty League, 1934–1940,* Boston: Houghton Mifflin, 1962

Wolfskill, George, and John A. Hudson, *All but the People: Franklin D. Roosevelt and His Critics, 1933–39,* New York: Macmillan, 1969

Censorship

Censorship means prior restraint—stopping something from being published or broadcast before it can appear. Radio censorship often determines who gets to broadcast and what is broadcast. It can take many forms: state monopoly of radio facilities and political expression; program monitoring by military or civilian bodies; "private" censorship of controversial topics by station authorities; specific stipulations of what constitutes acceptable quality and good taste in radio programming; the denial of the right to broadcast to minority groups, religions, races, and ethnicities; the list goes on and on. If censorship is understood more broadly as the regulation of the transmission and reception of representations and opinions, it could be argued to transpire at all levels of the radio communication process—through the actions of governments, networks, stations, advertisers, producers, performers, parents, and listeners themselves.

Radio Censorship in Europe

From the 1920s to the present day, radio has been mobilized as a tool for the purposes of authoritarian governments and colonial authorities around the world. Internationally, radio censorship has most significantly and powerfully taken the form of its classic definition: to suppress unofficial and oppositional political voices before they can be heard and to prohibit unauthorized material and information.

Radio broadcasting in the former Yugoslavia, for example, was heavily controlled by the government prior to World War II. After the ascendance of a socialist regime in the 1940s, radio was coordinated by Jugoslavenska Radiotelevizija (JRT), although it was a highly federalized arrangement, with regional broadcasting networks located in the various republics serving the multilingual and culturally and ethnically pluralist populations included under the state. Yugoslavia's relative autonomy from the Soviet Union resulted in a heavier infiltration of Western news and entertainment media and a more open news broadcasting policy. Local radio stations operated independently of JRT, but a basic censorship was exercised over all Yugoslav radio. Criticism of the basic communist system was prohibited, as was any personal attack directed toward Chief of State Marshall Tito. Nothing could be broadcast that might "exacerbate the troublesome animosities dividing the various Yugoslav nationalities" (Paulu, 1974). Negative viewpoints regarding the Soviet Union were not permitted, for fear that such views might antagonize Soviet leaders into military intervention.

Clandestine *samizdat* radio stations helped to propel the cultural and political reforms that swept across Eastern Europe in the early 1990s. But although most of those nations subsequently embraced privatized radio as a symbol of newfound democratization, Serbia's troubled late 1990s history resulted in heavy-handed reassertions of state control. The Milosevic regime revoked operating licenses and physically dismantled independent radio stations during times of anti-government protests and North Atlantic Treaty Organization (NATO) military actions. Likewise, radio is closely controlled (censored, as need be) in most Third World countries, where the media are either a voice of the state or are held by those close to the party in power.

Great Britain

The censorship situation in the United Kingdom has been much different, but similarly complex. The British Broadcasting Corporation (BBC) was formed as an autonomous public monopoly in Britain in 1926 after four years of commercial operation. Although outright instances of radio censorship have been few, the organization has routinely encountered various pressures from the party and prime minister in power. Historically, the relations between Broadcasting House and Whitehall have been rather too cordial for many critics, raising questions about the political and cultural neutrality of the corporation. Although BBC News has earned a much-vaunted reputation for impartiality, during the General Strike of 1926 and over the course of World War II, "the BBC became an integral part of the state's information machinery" (Schlesinger, 1978). Close government oversight of reporting was likewise maintained during the Falklands/Malvinas conflict with Argentina in the early 1980s.

Once established in 1955, commercial television was subject to the self-regulatory authority of the Independent Television Authority. Cat-and-mouse adventures between licensing authorities and pirate radio ships—unauthorized music stations transmitting from vessels in the North Sea—enlivened the regulatory scene in the 1970s and forced the BBC to adopt more popular programming. Despite the recent proliferation of independent and community radio stations in the United Kingdom, pirate outfits catering to fringe musical tastes are still common in metropolitan areas.

The primary government influence over U.K. radio has involved the coverage of events in Northern Ireland. Irish Republicans have long criticized the BBC and the Independent Broadcasting Authority (IBA) stations for their biased or partial coverage of the "Troubles." Unofficial agreements and tacit understandings between government and broadcasting establishments were the most common cause for this, although the relationship had clearly become strained in many areas by the late 1980s. The Conservative government responded with extraordinary provisions, including prohibiting the direct reporting of Sinn Fein members—a measure that backfired, antagonizing journalists and garnering widespread condemnation as unwarranted government interference.

The "Absence" of Official Radio Censorship in the United States

Officially, no government censorship of regular radio programming has ever existed in the United States. The First Amendment's prohibition of laws concerning speech and the press are the primary barrier to such activity. Under the Radio Act of 1912, the secretary of commerce and labor was obliged to issue radio licenses to all applicants. Section 29 of the Radio Act of 1927 stipulated that:

> Nothing in this act shall be understood or construed to give the Commission the power of censorship over the radio communications or signals transmitted by any radio station, and no regulation or condition shall be promulgated or fixed by the Commission which shall interfere with the right of free speech by means of radio communication.

This clause was reproduced as Section 326 of the Communications Act of 1934, and it has been applied to the Federal Communications Commission's (FCC) oversight of broadcasting ever since.

The real history of radio censorship in the United States, however, has been far less clear-cut than this official situation would suggest. The "no-censorship" clause was designed to allay fears that a government agency might impose its political will against the First Amendment rights of the press. The assumption that censorship was an act of "prior restraint" by the government on a private citizen, company, or organization was reiterated in this legislation. But if we understand censorship to constitute a more diverse set of limitations and restrictions, patterns of censorship have existed throughout the history of American radio broadcasting.

The no-censorship clause was crucial because it differentiated the "democratic" American system of broadcasting from the state-controlled or state-affiliated systems adopted by most other nations (such as Yugoslavia and Britain). The specter of political control loomed over early discussions about radio regulation and justified the adoption of a system that endorsed private commercial development of the airwaves. In the process, censorship was frequently regarded as a black-and-white issue: the presence or absence of government control. The broadcasting industry was highly successful in soliciting support for its two philosophies on the subject: (1) the argument that government censorship was a slippery slope (i.e., once established in any measure, it would tend toward the kind of political despotism present in authoritarian media systems); and (2) the idea that the government should not be allowed to impose its elitist standards of taste and culture on the American public by determining program content (hence, a paternalist radio model was unacceptable).

In the early 1930s, the federal courts recognized the Federal Radio Commission's (and subsequently the FCC's) right to consider past programming performance when deciding whether to renew or revoke a broadcasting license. Because no radio station can legally broadcast without a license, this "subsequent review" power has long been recognized as an indirect form of censorship, producing the "chilling effect" on broadcasters of avoiding controversial material that might antagonize the commission. In truth, the commission has rarely revoked or failed to renew licenses.

The Emergence of Self-Censorship in the United States

Most censorship in American radio has consisted of self-regulation by networks, stations, advertisers, and performers. Except for certain situations involving political candidates, broadcasters can refuse anybody access to their facilities. This "editorial control" has inspired well-founded criticisms that particular political opinions, news items, and entertainment forms have been routinely excluded from the radio airwaves. *Market censorship*, where the commercial basis of the industry discourages the airing of certain "unpopular" topics or minority perspectives, is often responsible for these restrictions. Allegations have also surfaced that networks and station owners—controlled by wealthier, politically conservative individuals—have prohibited left-wing viewpoints and protests against the broadcasting industry from reaching the microphone. Conversely, throughout the 1930s, Republican congressmen and conservative commentators such as Boake Carter objected that stations and networks, fearing or favoring the Democratic administration, refused them equitable opportunities to air their perspectives.

Private censorship refers to the various program (or advertising) prohibitions undertaken by radio stations and networks. The most commonly restricted subjects during radio's golden age were labor unrest, socialist politics, pacifism, political "radicalism," birth control advocacy, criticism of advertising, anti-Prohibition speeches, unorthodox medical practices, unorthodox religious opinions, excessive excitement in children's shows, "offensive" words, and suggestive situations. Private censorship often stemmed from stations' unwillingness to offend advertisers or listeners (based upon feedback or the assumed preferences of their audience). Such actions were not always unfounded or irresponsible. Popular radio priest Father Charles Coughlin's anti-Semitic remarks resulted in his program's cancellation by a number of stations in the late 1930s and led to his eventual removal from the air. Significantly, however, most networks and many stations responded to Coughlin by formalizing policies refusing to accept paid programming that addressed "controversial" issues. As a result, the limits of radio discourse were further circumscribed.

Obscenity and Indecency in Radio

In response to an outbreak of "radio vandalism," in 1914 the Department of Commerce stipulated that amateur licensees must refrain from profane or obscene words. This preoccupation with maintaining standards of good taste and upholding the moral order continued into the broadcasting era. The one exception to the no-censorship clause of the Radio Act and the Communications Act is the following addendum: "No person within the jurisdiction of the United States shall utter any obscene, indecent, or profane language by means of radio communication."

In other words, this was the one legislated area in which prior restraint was permissible: broadcasters airing obscene, indecent, or profane material could expect license revocation or nonrenewal. During its tenure, the Federal Radio Commission interpreted this clause broadly, arguing that because radio entered the home and was accessible by children, indiscretions in this area were unacceptable. Several licenses for smaller stations were revoked following "vulgar" and "offensive" broadcasts, encouraging a higher degree of caution among other broadcasters. Any mention of "sex" was avoided, leading to the widespread cancellation of academic lectures on venereal diseases and birth control methods.

Self-Regulation

In fact, the larger stations and the radio networks justified their dominance within the industry based upon their ability to uphold "good taste" in programming. The commission supported the notion that "quality radio service" was best represented by vigilant self-monitoring of programs and performers. The corroboration between official government regulation and industry self-regulation solidified in the 1930s. The National Association of Broadcasters (NAB) emerged as the primary industry lobbying group and developed continuing working relations with the FCC and Congress. Dominated by the larger commercial entities throughout its history, the NAB encouraged its members to more aggressively self-censor.

Without formal government outlines of what was permissible over the airwaves, program producers personally took on the obligations of unofficial censorship. Most stations codified their censorship policies, justifying them in terms of universal community interests. KSD, St. Louis, prided itself on its ability to exercise "an inflexible censorship over all programs offered for broadcasting . . . [to protect] listeners and advertisers against association with the unworthy." In the early 1930s, the trade magazine *Variety* described how the current policy, "somewhat along the lines of an honor system, makes a censor of everybody in the studio, from actors to control room engineers. Nobody has been taught what to avoid or bar and the material washing is left to personal discretion" (quoted in Rorty, 1934).

This gatekeeping function was formalized as the decade progressed, especially as the networks asserted their oversight functions. The Columbia Broadcasting System's (CBS) 1935 policies focused on children's programming, listing themes that would not be permitted:

> The exalting, as modern heroes, of gangsters, criminals and racketeers will not be allowed. Disrespect for either parental or other proper authority must not be glorified or encouraged. Recklessness and abandon must not be falsely identified with a healthy spirit of adventure.

The National Broadcasting Company (NBC) likewise institutionalized restrictions, prohibiting such subjects as "off-color" songs and jokes, astrology and fortune-telling, irreverent references to the deity, and "questionable statements." The NAB followed suit, issuing in 1939 a more stringent code of "accepted standards of good taste" for its members. CBS and NBC established "Standards and Practices" and "Continuity Acceptance" departments to enforce "courtesy and good taste" and to guarantee programming appropriate for "homes . . . of all types . . . and all members of the family."

The self-censorship system was similar to that of the motion picture industry, but it differed in certain respects. The Hollywood movie studios submitted scripts and films to a semi-independent body to preview and approve. Radio censorship was less centralized; most radio programs were created (and self-censored) by sponsors and their advertising agencies. Networks and stations were therefore usually dealing with third parties, not their own productions. The sheer volume of radio programming meant that continuous monitoring of all stations' output was impractical. Radio guidelines were also harder to enforce, because most broadcasts were transmitted live. The radio networks and stations required all programs and speakers to submit scripts in advance and forbade ad-libbing, but this cumbersome "blue-penciling" review process was never comprehensively enforced, and it failed to account for misinterpretations or unscheduled deviations from the script during broadcast. Writers and performers frequently challenged the networks' censorship provisions, slipping in double entendres or tiptoeing on the brink of "tastelessness" with their gags and dramas. Nevertheless, major infractions of the self-regulatory codes were few and far between, and the FCC wholeheartedly supported the application of private censorship as a preferential alternative to official program supervision.

The close cooperation between government and big industry objectives in radio that had developed during the 1930s was indicated by the formal alliances forged during World War II. The administration's faith in the ideological integrity of radio business interests was confirmed when President Roosevelt appointed top radio journalists and executives to posts in the Office of War Information (OWI) and the Office of Censorship. Networks, stations, and sponsors obliged the OWI by providing hours of free airtime to government programs and bond drives. Dramatic scripts were rewritten to encourage patriotism, enlistment, and home front support for the war effort. The Office of Censorship issued guidelines of prohibited topics such as weather reports and troop movements, and it required the downplaying of racial antagonisms—all of which broadcasters followed willingly.

Radio Censorship in the United States after World War II

As the networks shifted their interests to television in the postwar period, the tight mechanisms of self-regulation that had developed in the 1930s and 1940s began to break down. Radio stations shifted away from a mass-appeal broadcasting model to a format-based system that targeted particular localities and audience groups. In the process, minority tastes, unorthodox political opinions, and non-mainstream moralities were serviced. In increasingly competitive urban radio markets, commercial broadcasters began to "push the envelope" and schedule controversial and sensational programming. A rise in noncommercial community radio stations resulted in programming that resonated with more politically and aesthetically progressive audiences. The NAB Code of Program Standards was abolished in the 1980s. Censorship reemerged around the fringes of the electromagnetic spectrum.

In the 1970s, the FCC reprimanded several "indecent" radio broadcasters. The trend toward "topless radio"—call-in talk shows inviting sexual anecdotes from listeners—resulted in fines and warnings from the Commission. Various stations associated with the Pacifica Foundation (a listener-supported organization serving avant-garde tastes and addressing political subjects) were chastised for their indiscretions. Most significantly, Pacifica member WBAI, New York, broadcast an unexpurgated sketch called "Seven Dirty Words" by comedian George Carlin. This resulted in a U.S. Supreme Court decision declaring that, although the sketch was constitutionally protected speech, the FCC had the right to restrict indecent expression over the airwaves. The FCC more aggressively reasserted this right in the 1990s, in response to the daytime scheduling of "shock jocks." The shock jocks were largely the employees of ratings-hungry radio networks and station conglomerates, who revamped their criticisms of the FCC's intervention into program content as First Amendment infringements.

A burgeoning "microradio" movement, which broadcasts to immediate localities using cheap, portable, low-power transmitters, more recently flustered the FCC. Advocates of microradio argue that it allows greater access to the airwaves for marginalized voices. The FCC long refused to license broadcasters under 100 watts and considered such microbroadcasting illegal. In 2000 the FCC began to license low-power FM transmitters; microradio proponents consider this an attempt by the commission to commercialize the movement and extend its authority over radio content.

A significant censorship issue for the future concerns internet radio, which many forecast will supersede broadcast radio if issues of listener access and portability can be resolved. The international implications are massive, because the delivery of audio over the internet renders discrepancies in signal strength and frequency allocation irrelevant. Censorship based upon geographical factors disappears as a result. Internet radio seemingly offers a solution to national/state censorship, representing a technological means to circumvent authoritarian attempts to prohibit or limit broadcast transmissions.

MATTHEW MURRAY

See also Communications Act of 1934; Controversial Issues; Equal Time Rule; Fairness Doctrine; Federal Communications Commission; First Amendment and Radio; Internet Radio; Licensing; Low-Power Radio; Obscenity and Indecency on Radio; Propaganda by Radio; Seven Dirty Words Case; Shock Jocks; Topless Radio; United States Supreme Court and Radio; Wireless Acts of 1910 and 1912/Radio Acts of 1912 and 1927

Further Reading

Brindze, Ruth, *Not to Be Broadcast: The Truth about the Radio,* New York: Vanguard Press, 1937

Chase, Francis Seabury, Jr., *Sound and Fury: An Informal History of Broadcasting,* New York and London: Harper, 1942

Coons, John E., editor, *Freedom and Responsibility in Broadcasting,* Evanston, Illinois: Northwestern University Press, 1961

Hilmes, Michele, *Radio Voices: American Broadcasting, 1922–1952,* Minneapolis: University of Minnesota Press, 1997

Krattenmaker, Thomas G., and Lucas A. Powe, Jr., *Regulating Broadcast Programming,* Cambridge, Massachusetts: MIT Press, 1994

Lipschultz, Jeremy Harris, *Broadcast Indecency: F.C.C. Regulation and the First Amendment,* Boston: Focal Press, 1997

Murray, Matthew, "'The Tendency to Deprave and Corrupt Morals': Regulation and Irregular Sexuality in Golden Age Radio Comedy," in *The Radio Reader: Essays in the Cultural History of Radio,* edited by Michele Hilmes and Jason Loviglio, New York and London: Routledge, 2002

Paulu, Burton, *Radio and Television Broadcasting in Eastern Europe,* Minneapolis: University of Minnesota Press, 1974
Rivera-Sanchez, Milagros, "Developing an Indecency Standard: The Federal Communications Commission and the Regulation of Offensive Speech, 1927–1964," *Journalism History* 20, no. 1 (Spring 1994)
Rorty, James, *Order on the Air!* New York: Day, 1934
Schlesinger, Philip, *Putting "Reality" Together: BBC News,* London: Constable, 1978; Beverly Hills, California: Sage, 1979
Smead, Elmer E., *Freedom of Speech by Radio and Television,* Washington, D.C.: Public Affairs Press, 1959
Summers, Harrison Boyd, compiler, *Radio Censorship,* New York: Wilson, 1939
Summers, Robert Edward, compiler, *Wartime Censorship of Press and Radio,* New York: Wilson, 1942

CFCF

Montreal, Quebec Station

Based in Montreal, CFCF ("Canada's First, Canada's Finest") holds the distinction of being the country's oldest radio station and is arguably the first radio station in North America.

Origins

The foundations for CFCF were laid through a series of experiments with the transmission of electromagnetic signals. The Marconi Wireless Telegraph Company of Canada (hereafter referred to as Canadian Marconi) received a license from the Canadian government to conduct experiments with radio, including the erection and operation of wireless telegraph stations on ships for navigational and commercial purposes. *Report of the Department of the Naval Service* (the Canadian Department of Marine and Fisheries oversaw radio services at this time), dated 31 March 1915, lists CFCF's predecessor, XWA (Experimental Wireless Apparatus), as the sole experimental "wireless telephony" station.

Although many place the distinction of "oldest radio station" on KDKA Pittsburgh, an equally strong case can be made for the precedence of XWA. Both stations experimented with broadcasts to local ham operators in 1919, but there are no existing records indicating that signals were received. If the date when scheduled broadcast began is used as a starting point, XWA's broadcast of musical programming on 20 May 1920 occurred six months before KDKA's broadcast of the Harding-Cox election returns on 2 November 1920. XWA's inaugural program featured an orchestra and soloist Dorothy Lutton as part of a special meeting of the Royal Society of Canada at the Chateau Laurier Hotel in Ottawa. It was possible to receive the broadcast as far as Ottawa, more than 100 miles away. XWA was christened CFCF on 4 November 1920. Early programming from XWA/CFCF consisted mainly of weather reports and the playing of gramophone records on a wind-up Victrola.

CFCF moved into its first real broadcast studios at the Canada Cement Building in Phillips Square in 1922. Performers heard regularly from the Phillips Square location included the dance bands of Joseph Smith from the Mount-Royal Hotel, Andy Tipaldi from the Ritz-Carlton, and Harold Leonard from the Windsor. In 1923 a yacht race from Lake St. Louis was described using a portable hand-held transmitter. Catering to the station's wealthier radio set owners, CFCF made arrangements with *The Financial Times* to provide bulletins from the Montreal Stock Exchange to be broadcast during the noon hour. By 1923, livestock and financial market reports also appeared.

In 1927, CFCF participated in the first coast-to-coast network radio experiment in Canada. A number of privately run stations were linked together using the telegraph lines from the Canadian National and Canadian Pacific railways along with local and provincially operated telephone lines. The network was arranged to cover a number of celebrations held in Ottawa to mark Canada's Diamond Jubilee. The broadcast also aired worldwide on CFCF's shortwave station, VE9DR (later CFCX).

By 1930 Canadian private radio stations began forming affiliate relationships with the Columbia Broadcasting System (CBS) and the National Broadcasting Company (NBC). Although the American networks used the Canadian stations on a station-by-station and program-by-program basis, the links became more frequent and regular. By 1932 CFCF had permanent affiliate status with NBC. The station's program lineup mixed American programming, such as *Miracles of*

Magnolia, Gloom Chasers, Hotel New Yorker Concert Ensemble, and *Amos 'n' Andy* with locally produced, sometimes bilingual programming. With the dissolution of the Canadian National Railway's radio service during the 1930s, CFCF's manager (and former CNR employee) Vic George negotiated a deal to use CNR telegraph circuits in the evenings so that stations from as far as London, Ontario and Halifax, Nova Scotia could exchange programming. By the end of World War II, CFCF switched its affiliation to the American Broadcasting Companies (ABC). One of the first ABC broadcasts heard on CFCF was the Saturday afternoon live broadcast from the Metropolitan Opera.

Between 1922 and 1935, the station moved its studios three times, from Phillips Square to the penthouse of the Mount Royal Hotel and finally to the King's Hall Building on St. Catherine Street East. However, their stay at the King's Hall location was short-lived, as an explosion and fire destroyed the building in January 1948. The station moved to a temporary studio on Cote des Neiges, and variety programs such as *Little Players on the Air* and *The Good Neighbour Club* were back on the air in a matter of weeks. The station was forced to move again in 1957 because of another fire. Within days, temporary studios were established in the penthouse of the Dominion Square building until a complete broadcast complex could be built on the 6th floor.

Ownership and Program Changes

In 1968 the federal government passed new legislation limiting foreign ownership of broadcasting entities. Because the ownership of Canadian Marconi was held by the English Electric Company of Britain, the company had to divest its holdings and find a new owner. In 1972 the Canadian Radio-Television and Telecommunications Commission (CRTC) approved the ownership of the station by Multiple Access, owned by the sons and daughters of Samuel Bronfman. CFCF remained in the broadcast division of Multiple Access until 31 August 1979, at which time Jean Adelard Pouillot purchased CFCF and all of its broadcasting assets. In 1985 CFCF radio moved to the Park Street Extension district in the city's north end, where the television station was located.

In 1988 the station was purchased by Mount Royal Broadcasting, operated by Pierre Berland and Pierre Arcand. On 1 May 1989 the station relocated downtown to 1200 McGill College Avenue. As a condition of sale, the station had to abandon its call letters in order to differentiate itself from the television broadcaster CFCF-12. The condition was met on 9 September 1991, when CFCF became radio station CIQC. Upon the expiration of their lease in 1996, CIQC moved its studios to Gordon Street in the Montreal suburb of Verdun.

The change of call letters also resulted in a change of programming format, as the station moved from adult standards to country music. In 1993, CIQC switched formats again, towards a talk radio and community affairs format. The result was an eclectic mix of "Ask the Experts" programs (ranging in topics from automobiles to holistic medicine), phone-in programs featuring local celebrities (including Jim Duff, Ted Tevan, Mitch Melnick and Howard Galganov), and syndicated American programming. The station also broadcast Montreal Canadiens and Montreal Expos games. However, a dwindling Anglophone population, a weak frequency, and increased competition from rival English language stations eroded CFCF/CIQC's status within the Montreal radio environment. In late 1998 the station successfully applied to the CRTC to shift its frequency to the vacant, and stronger, 940 AM. The CRTC allowed the station to simulcast its broadcasts on both the 600 and 940 AM frequencies for a period of six months. In December 1999, the station became "News 940" and CIQC's call letters changed to CINW. In August 2000, the company was purchased by the Corus Entertainment company and has retained its all-news format.

IRA WAGMAN

See also CKAC; KDKA

Further Reading

Allard, Thomas James, *Straight Up: Private Broadcasting in Canada, 1918–1958,* Ottawa, Ontario: Canadian Communications Foundation, 1979

Stewart, Sandy, *A Pictorial History of Radio in Canada,* Toronto, Ontario: Gage, 1975; as *From Coast to Coast: A Personal History of Radio in Canada,* Montreal, Quebec, and New York: CBC Enterprises, 1985

Troyer, Warner, *The Sound and the Fury: An Anecdotal History of Canadian Broadcasting,* Toronto, Ontario: Wiley, 1980

Vipond, Mary, *Listening In: The First Decade of Canadian Broadcasting, 1922–1932,* Montreal and Buffalo, New York: McGill-Queen's University Press, 1992

Weir, Earnest Austin, *The Struggle for National Broadcasting in Canada,* Toronto, Ontario: McClelland and Stewart, 1965

Chain Broadcasting. *See* Network Monopoly Probe

See Network Monopoly Probe

CHED

Edmonton, Alberta Station

In spite of its northern location, radio station CHED in Edmonton, Alberta, was an important force in the Canadian radio industry as one of the leading rock music stations in the country before the supremacy of FM radio. In January 1953 the Board of Broadcast Governors (BBG), the Canadian broadcasting regulator, granted a license to the station, which was to be owned and operated at 1080 kilohertz, to local businessman Hugh Sibbald and to Lloyd Moffatt, owner of a number of radio stations in the Canadian West. The station's application was vigorously opposed by the Alberta government, which had financed a competitor, station CKUA. The granting of the license by the BBG is indicative of the lack of authority by the provincial government in broadcasting—an area considered to be a matter of national interest.

On 13 July 1953 the station adopted the call letters CHED and shortly thereafter named Don McKay, a former chairman of the Edmonton Chamber of Commerce, as the general manager of the station. The station then hired local commentators Guy Vaughan (a former newsreader at the Canadian Broadcasting Corporation [CBC]), Ron Chase, and Bob McGavin as newsreaders. This was in part the station's attempt to provide, in the words of McKay, "a happy blend of familiar and new voices." Other members of the CHED staff included Warner Troyer, Stu Phillips, and Phil Floyd. The news director of the station was Allan Slaight, the future president of the Standard Broadcasting Network. During this time, the station moved into its first broadcasting location, at 10006 107th Street, in the city's downtown. The station would remain in that location until it moved into its present building at 5204 84th Street in 1982.

CHED first aired on 3 March 1954, the 50th anniversary of the city's founding. The emphasis of the early programming was on more familiar concert and show tunes, from Broadway musical scores to popular movie sound tracks, as well as music from swing bands of the era. This material was complemented by news every hour on the half hour and weather information on the hour, as well as periodic sports news and on-the-spot features announced by Bart Gibb. Full local, national, and international news came from wire services, including the British United Press and Broadcast News, as well as from CHED's city reporting staff. In 1955 then programmer and disc jockey and future station manager Murray McIntyre "Jerry" Forbes was responsible for spearheading the "Santas Anonymous" toy-drive campaign, which has become one of the largest Christmas toy drives in North America, currently delivering more than 26,000 toys to needy families in the Edmonton area.

On 1 November 1962, CHED underwent an important technical change, moving its frequency to 630 kHz and increasing its power to 10,000 watts day and night (originally, the station had only 1,000 watts of nighttime power). This gave the station substantial coverage throughout the northern region of Alberta and into Saskatchewan.

The change of frequency and subsequent increase in coverage marked the beginning of a substantial period of prosperity for the station, beginning in the late 1960s. The station had adopted a Top 50 rock format in 1957, but it achieved audience supremacy with the rise of rock and the development of "personality radio," which saw the development of old-time "boss jocks"—well-known local personalities—in prominent roles as disc jockeys who hosted a variety of local events, introduced concerts and other promotional opportunities for the station, and participated in community initiatives. An example of this occurred in 1967, when the local newspaper, the *Edmonton Journal,* temporarily discontinued the comic pages because of a newsprint shortage. In response to this development, CHED announcers read and described the comic strips to audiences over the air. Newsreader Bob McCord read *Peanuts*, and other local personalities appeared at the station to read other areas, including Mayor Ivor Dent's reading of the *Wizard of Id.* The rationale, according to McCord, was that "with all of this world's troubles, it's no time to be without the funnies."

The success of personality radio was most prominent among CHED's roster of disc jockeys. With a team that included Len Theusen, Bruce Bowie, Chuck Chandler, Wes Montgomery, and Keith James, CHED dominated the AM market with a morning show that occasionally posted a 50 percent share of the audience. Theusen's radio show, which aired evenings between 9 P.M. and midnight, was among the most popular shows on the dial. Theusen himself was not popular with everyone: the station was once forced off the air when he was attacked on the air by a former disc jockey. The station's local popularity did not go unnoticed by the music industry, and CHED was featured in a 1973 issue of *Billboard* magazine's spotlight on Canadian radio.

By the 1980s, changes in ownership, programming format, and the local broadcast landscape resulted in the substantial reorganization of the station. In 1989 Moffat's Edmonton division (630/CHED) formed a partnership with Maclean Hunter station CKNG-FM (Power-92). The CHED management team was responsible for the operation of both radio stations until their sale to Western International Communications Ltd. (WIC) on 1 September 1992. More than a year later, the station changed its format to news/talk with the slogan "Alberta's Information Superstation" on 1 December 1993. By 1995 CHED had secured a virtual monopoly on the broadcasting rights for major professional sports; the station now broadcasts Edmonton Oilers hockey and the Edmonton Eskimos of the Canadian Football League. In late 1999, WIC's broadcasting (radio and television) holdings were divided up, and Corus Entertainment (a division of Shaw Communications) purchased the radio properties, including the operation of CHED.

IRA WAGMAN

See also Canadian Radio and the Music Industry

Further Reading

Allard, Thomas James, *Straight Up: Private Broadcasting in Canada, 1918–1958,* Ottawa, Ontario: Canadian Communications Foundation, 1979
"Attacker Forces CHED off the Air," *Edmonton Journal* (26 October 1972)
Finlayson, Dave, "The Chucker's Back and Glad to Be Here," *Edmonton Journal* (10 August 1997)
"Radio Station CHED Airs Journal Comics," *Edmonton Journal* (31 August 1967)
Troyer, Warner, *The Sound and the Fury: An Anecdotal History of Canadian Broadcasting,* Toronto, Ontario: Wiley, 1980

Chenault, Gene 1919–

U.S. Station Owner and Manager, Top 40 Format Pioneer, Consultant, and Syndicator

Gene Chenault's name is less famous than that of his partner, programmer Bill Drake. Yet behind the impact of Drake's contributions to the Top 40 format were Chenault's management skills and his sales and marketing concepts. The two men altered U.S. radio and American popular culture in the 1960s. From a single radio station, Chenault and Drake spread their sales, programming, and promotion philosophies across California to stations in other major U.S. markets. The company they formed together, Drake-Chenault Enterprises, parlayed their success into one of the largest tape syndication services.

Early Years

Born Lester Eugene Chenault in 1919, Chenault spent his high school days as a radio actor in Los Angeles, including performances on national broadcasts via the Mutual network. He fell in love with radio, he said, when he and his father listened together to the broadcast of the Jack Dempsey-Gene Tunney prize fight in 1926. It was then that he knew he wanted to pursue radio as a career. After graduation, he landed a job in Fresno at KFRE but interrupted that with service in the army in World War II.

At the end of the war, Chenault joined an engineer friend who was planning an application for a new station in Fresno. In October 1947 the new station, KYNO at 1300 kHz—Fresno's fourth signal—was on the air with Chenault as general manager and managing partner. Over time he would acquire sole ownership.

KYNO was affiliated with the Don Lee/Mutual Network, and into the 1950s the station relied heavily on network programming. When the arrival of television challenged radio, especially network affiliates, Chenault explored his options for KYNO and decided on rock and roll. He phased in the new sound during 1955 and 1956 as Mutual's programs grew fewer. The Top 40 format, coupled with big money giveaways, captured the market: KYNO became dominant, achieving 60 shares in the Hooper Ratings.

By 1962 KYNO's huge shares had attracted a competitor. Fresno's KMAK was bought by a station group that owned KMEN in San Bernardino and a station in Hawaii, and that company installed a format called "Circus Radio" at KMAK, relying on outrageous disc jockey stunts to get attention.

Opening the checkbook was Chenault's strategy for defending KYNO. If KMAK staged a contest with a $1,500 prize, KYNO upped the ante to $2,000. The ultimate weapon at Chenault's disposal was programming. After a few short-term program directors at KYNO, Chenault imported the tall, soft-spoken Drake from San Francisco. Drake had made a name for himself at WAKE in Atlanta and had moved to WAKE's sister station, KYA in San Francisco. Chenault heard Drake's work and set up a meeting through a mutual friend.

As program director, Drake tightened the music policy to a short playlist with heavy repetition. Talk by disc jockeys was pared to a minimum, and the commercial load was reduced. Within a year KYNO had scored a decisive win in the battle, prompting challenger KMAK to change to country music.

Drake-Chenault

Chenault and Drake formed American Independent Radio in 1964 to provide consultation to stations that wished to model their success on the Fresno victory at KYNO. Chenault advised on sales and marketing while Drake advised on programming. Their first client was KGB in San Diego, owned by Willet Brown, a founder of the Mutual Broadcasting System. KGB was San Diego's oldest radio station. At the time Brown signed with Drake and Chenault, KGB was also last in the ratings.

As he had done in Fresno, Drake introduced a tightly formatted fast turnover of hit songs. For their staff Chenault and Drake relied on air talents they knew well: Les Turpin and K.O. Bayley from KYNO and Robert W. Morgan, who had competed with them as KMAK's morning man. Within a year there was another ratings and revenue victory. KGB rose to number one in the city, soundly defeating KCBQ and KDEO, the Top 40 competitors.

The success in San Diego proved that Chenault's and Drake's policies worked beyond Fresno. They were ready to take the next challenge, and Willet Brown made the introduction that would cause it to happen. Brown knew Thomas F. O'Neil, chairman of General Tire, owner of RKO Radio. That company's KHJ in Los Angeles was faring poorly in the market. O'Neil agreed to sign Chenault and Drake, and they moved the concept—and their team—to Los Angeles in 1965.

Drake called the approach at KHJ an offshoot of Fresno and San Diego with very few minor adjustments. A key difference in Los Angeles was the introduction of a new name, "Boss Radio." It gave the station an easily remembered verbal reference. It also resonated with the station's potential audience: the word *boss* came from California surfer slang for *good,* as in "That's a *boss* wave."

KHJ reached number one in the Los Angeles ratings within one year, and the feat provided decisive proof of the Drake and Chenault formula. As a result, RKO hired the team as consultants to KFRC in San Francisco. Further success led to a contract to add Boss Radio to additional RKO stations, including CKLW in Windsor, Ontario (serving Detroit); WRKO in Boston; and WHBQ in Memphis.

Stations outside the RKO chain sought the ratings and revenue gains they read about in trade publications, and Chenault accommodated them. The first non-RKO client was KAKC in Tulsa, which achieved the same positive results as Drake's and Chenault's first California stations. Not so for WUBE in Cincinnati, an inferior facility with poor coverage of the market. Competitor WSAI defeated WUBE soundly, and Drake and Chenault pulled out within a year. Observers said that Chenault's natural sales ability and his exuberance for making the deal caused him to overlook the station's deficiencies.

What came to be known as "the Drake format" soon included the number-one radio market in the United States: New York. RKO assigned Drake and Chenault the consulting job at WOR-FM to go against Top 40 powerhouses WABC and WMCA. In 1967 WABC remained the leading New York station and WOR-FM was number two.

WOR-FM proved that FM was not a secondary medium and that the Drake and Chenault policies could be effective on FM. Capitalizing on this success and on the knowledge that KHJ in Los Angeles was buying taped music services for its FM sister station, Chenault proposed that the consultancy begin creating the taped programming, directed by Drake. The two men created a syndication service called Drake-Chenault Enterprises (DCE).

Late in 1968 three RKO FM stations—KHJ-FM in Los Angeles, KFRC-FM in San Francisco, and WROR-FM in Boston—were programmed with the automated tape service "Hit Parade '68" from DCE. Tapes were recorded in Fresno at the facilities of KYNO. Robert W. Morgan and other KHJ personalities traveled to Fresno to record voice tracks.

DCE also debuted a taped service called "Stereo Rock," an attempt to capitalize on the trend to play longer, rock-oriented cuts. In essence, it was a "progressive rock" service, and one of DCE's few failures. "Bill [Drake] was never hot about it," said RKO Radio president Bruce Johnson. The idea was attributed to Chenault, who was looking for a new product to sell.

Johnson, who became president of RKO Radio in 1972, told an interviewer that he was upset that "some of the work that was being done by the music director [at KHJ] was going to Fresno and other places." He felt that Drake's programming team was working more for DCE than they were for RKO—60 percent DCE, 40 percent RKO, he said. In 1973 Johnson canceled the consultation agreement.

Later that same year Chenault and Drake entered a five-year contract to manage and program KIQQ-FM in Los Angeles, known as "K-100." They brought personalities Robert W. Morgan and "The Real" Don Steele from KHJ and attempted to replicate the Boss Radio formula. On the air they promoted the station as "the dawn of a new radio day," but there was little new about the sound. RKO's Johnson called the move "revenge" against KHJ and RKO. The project never reached the level of success that Drake and Chenault had enjoyed previously. In a 1976 interview Chenault characterized KIQQ as "behind target."

While KIQQ foundered, Drake-Chenault syndication flourished. Broadcast automation of the late 1960s and early 1970s was typically used by easy-listening formats because the hardware was not capable of the tight-paced cue needed for up-tempo formats. DCE engineers devised a way to put cue tones on their tapes one second earlier to start the next reel of tape in time for the segue to be tight. With that technology, DCE could supply a variety of contemporary-sounding formats.

The Fresno syndication operation moved in 1973 to Canoga Park, California. At that time DCE had two studios and 15 employees. The following year, the firm grew to 100 radio station clients. By 1976 clients increased to 350, and DCE's staff was at 50. The company built its own tape duplication facility in Canoga Park to maintain its strict audio standards.

In 1979 DCE produced *The History of Rock n Roll,* an ambitious 50-hour radio documentary. The program met with phenomenal success as radio stations clamored to schedule it, first as a blockbuster weekend special, then in repeat broadcasts of shorter segments. The response to the documentary caused Drake and Chenault to create a new division within DCE called "The History of Rock n Roll, Inc.," which produced additional programs for syndication, among them *The Motown Story,* a long-form history of Detroit's Motown Records.

Later Years

The introduction of satellite delivery and the compact disc combined to make reel-to-reel automation outmoded. Drake Chenault Enterprises felt the pinch with a gradual loss of radio station clients to services using newer technology. In 1986 DCE was sold to Wagontrain Enterprises. The Canoga Park studios and duplication facilities were dismantled and moved to Wagontrain's Albuquerque headquarters in 1987.

A year or so before the Wagontrain purchase, Bill Drake elected to retire from DCE. Chenault stayed with DCE until the transfer to Wagontrain was complete. He retired in 1986 to the home he bought in Encino, California, in 1967.

Wagontrain licensed the name "Drake-Chenault" and continued to produce tapes. It also purchased the reel-to-reel operation of TM Productions in Dallas and added satellite distribution in a partnership with Jones Intercable of Denver. The combined operations were later sold to Broadcast Programming, Inc., of Seattle, which ultimately merged with Jones to form Jones Radio Networks.

In 1996 both Chenault and Drake were inducted into the California Broadcasters Hall of Fame.

ED SHANE

See also Consultants; Contemporary Hit Radio Format/Top 40; Drake, Bill; Morgan, Robert W.

Lester Eugene Chenault. Born 1919. Radio actor in Los Angeles; general manager and managing partner, KYNO, Fresno, California, 1947–64; established consulting company, American Independent Radio, 1964, with partner Bill Drake; successful stations included KGB, San Diego; KHJ, Los Angeles; CKLW, Windsor, Ontario; WOR-FM, New York. With Drake, created Drake-Chenault Enterprises syndication service in 1967 to provide music programming to radio stations. Produced the documentary, *The History of Rock n Roll,* 1979. Retired to Encino, California, 1986. Inducted into California Broadcasters Hall of Fame, 1996.

Further Reading

"Editorial: More For Better?" *Billboard* (22 July 1967)
Goulart, Elwood F., "The Mystique and the Mass Persuasion: A Rhetorical Analysis of the Drake-Chenault Radio Programming, 1965–1976," <www.bossradioforever.com>
Hall, Claude, *This Business of Radio Programming: A Comprehensive Look at Modern Programming Techniques Used Throughout the Radio World,* New York: Billboard Publications, 1977
Larson, Lanny, "The Fruit of Imagination," *Fresno Bee* (1 January 2000)

Children's Novels and Radio

Much of the excitement over the introduction of wireless communication and radio broadcasting is reflected in a host of books aimed at younger readers, several series of which appeared before 1930, most of them during the early days of broadcasting in the 1920s. Mistakenly referred to as "dime novels," a phrase more fitting to 19th-century magazine fiction, at least 160 separate volumes focusing on wireless or radio appeared from 1908 to as late as the 1960s. Written for an audience of young boys and girls, there were single titles and series, all promoting the use of wireless and radio technology to save lives, solve problems, win friends, and punish the lawless. The earliest books concentrated on shipboard wireless themes; the youthful characters employed the new invention to warn of storms, pirates, and smugglers. By the 1920s, stories of broadcasting to an audience appeared in a few titles, but mostly plots centered on the hobby of radio-set construction and sending and receiving messages—often in the service of law enforcement, with the moral of the story demonstrating that young people use radio for the greater good.

The earliest such volume, John Trowbridge's *Story of a Wireless Telegraph Boy* (1908) appears to be the first children's book focusing on radio. This was followed by Harrie Irving Hancock's *The Motor Boat Club and the Wireless; or, The Don, Dash and Dare Cruise* (1909); James Otis' *The Wireless Station at Silver Fox Farm* (1910); and one of the famous Tom Swift series—Victor Appleton's (pseudonym for Howard Garis) *Tom Swift and His Wireless Message; or, The Castaways of Earthquake Island* (1911). After 1912 the pace picked up, with several other wireless and radio-related boys titles, including the first multivolume radio series—the six-title *The Ocean Wireless Boys,* in which high school boy Jack Ready and friends use wireless to find a lost ocean liner, warn ships away from icebergs, and fight in World War I. Likely influenced by the *Titanic* disaster, many of the later wireless stories show how radio can help ensure the safety of ships at sea.

By far the best-known and most hotly collected titles today are the two *Radio Boys* series that appeared in the 1920s under the names of Gerald Breckenridge and Alan Chapman (the latter a pseudonym) for a number of different syndicate authors. Unlike the earlier wireless tales, most of the Chapman series took place on dry land (though each featured a different short preface by *S.S. Republic* wireless hero Jack Binns). The first volume had the boys building a radio in order to win a cash prize, and in others, the boys share their hobby by taking their radio equipment to homes for the aged, hospitals, and other venues where the less fortunate would not otherwise have access to this wondrous new device. Chapman's radio boys were of working-class background, and many of their adventures took place near their small town in New York. In all 13 books of the Chapman series, the same two story elements are repeated and resolved: a local group of bullies is thwarted using nonviolent methods, and a criminal is brought to justice using radio, resulting in accolades from the community. During the eight-year run of the Chapman series, his boys remained the same age and in the same year in school.

The Breckenridge radio boys series is quite different. Though written during the same period, these boys are of upper-middle-class background: their fathers are doctors, lawyers, and bankers. The Breckenridge radio boys grow up, age, and progress in school during the series. And whereas the Chapman boys spend some time away from their small town, the Breckenridge youth travel the world, go to Yale, and become embroiled in foreign adventure. In a story very atypical of a juvenile series, *The Radio Boys as Soldiers of Fortune* (1925), radio boy-now-man Jack, having graduated from college, gets married and moves to Mexico to learn an aspect of the family business. Jack befriends local citizens who are trying to oust the current dictator and return a popular democracy to power. In a bizarre twist, Jack invents television, called a "televisor," and uses it to spy on the dictator and thus ensure his removal and replacement by democratic forces. This early portrayal of television is quite realistic.

What unites both series is the hobby of radio and its use in keeping communities safe while promoting law and order. What differentiates them is their focus and the career paths depicted in the stories. Whereas the Chapman boys stay in high school and experiment with radio locally, the Breckenridge boys graduate from college and go into banking, medicine, and the law. Not a career in radio for these savvy lads, but the hobby of radio as entertainment, as a way to spend their newfound wealth. Another difference may have been in the focus of the series publisher: Chapman was a pseudonym used by the Edward Stratemeyer Syndicate, the major owner of juvenile series in the first third of the 20th century. Stratemeyer would develop a story synopsis and then hire writers to do the book under the pseudonym, and so volumes in the same series were often written by different writers. Gerald Breckenridge was not part of the syndicate; he used his own name, and it is possible that he enjoyed writing about his boys' aging and progressing through life. Unlike the syndicate writers, Breckenridge likely had more artistic freedom, as long as he sold books.

At least two other *Radio Boys* series were published, and by 1922 a *Radio Girls* series, though of only four titles, appeared. By the next decade, when broadcasting was more

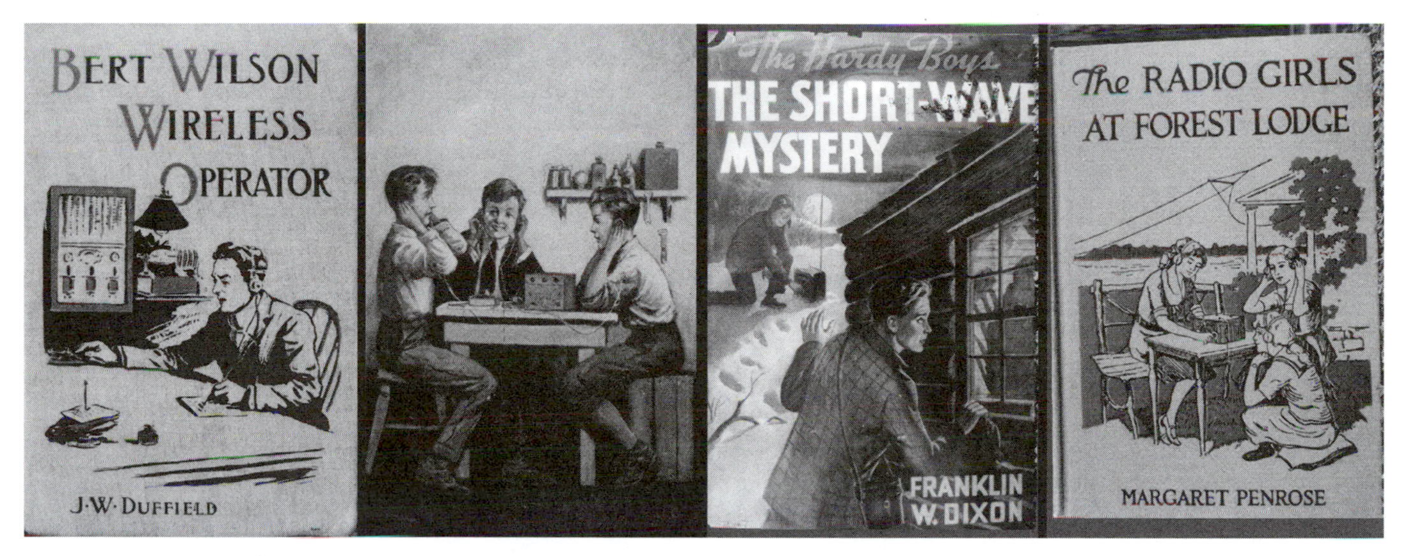

Several examples of juvenile children's stories involving radio
Courtesy of Michael A. Adams

fully formed, Ruthe S. Wheeler's *Janet Hardy in Radio City* (1935) featured a high school performer who gets the lead in a film, writes a radio script, and ends up in Radio City for its premiere. Betty Baxter Anderson's *Four Girls and a Radio* (1944) included broadcast entertainment consisting of accordion and vocal, and Julie Campbell's *Ginny Gordon and the Broadcast Mystery* (1956) featured a young woman who discovers she has talent as a radio interviewer and a solver of mysteries.

Radio broadcasting was commonplace when Franklin W. Dixon's *The Hardy Boys and the Short Wave Mystery* (1944) was introduced, but there was still hobby interest. Tracking down the source of an illegal transmission used in a smuggling operation, the boys aid law enforcement and further post–World War II interest in amateur radio. *The Hardy Boys* was but one of several longer juvenile fiction series that featured at least one story centering on radio; others included *The Bobbsey Twins, The Navy Boys, The Brighton Boys,* and *Bert Wilson.* Series that continued into the 1950s, such as *Tom Swift Jr.* and *Rick Brant Electronic Adventures,* used radio and combined it, actually overwhelmed it, with undersea adventure, microelectronics, space science, and robotics, but radio transmission was always at the basis of their experiments, and saving the day—for local law enforcement, the government, the armed services, and the community—was always the overriding use portrayed for radio.

Nearly all such books followed a basic pattern. Running 250 or more pages in most cases, with a frontispiece drawing or painting featuring the heroes at a key point in the story, the books offered fast-moving adventure stories wrapped in gaudy dust wrappers (few of which survive today). Their titles were often formatted with a main title followed by *or* and a subtitle hinting at the excitement within. The writing was often exaggerated and certainly old-fashioned by today's standards, and stories sometimes ended with a cliff-hanging reference to the next book in the series. The heroes or heroines rarely aged, though a series might appear over nearly a decade. Using cheap paper and inexpensively bound, these books were intended to be enjoyed and discarded, and few survive today in good condition.

More than entertainment, the depiction of radio in early juvenile novels may have influenced and reinforced some of the cultural and social convictions held by young people. More than merely a technical device, radio in these novels was almost always related to the ideals and preservation of community and family, career choice, patriotism, and attitudes about law and crime, and the stories may have encouraged their young readers toward discovery and invention as adults. There is a clear "right and wrong" point of view in these books, and the books champion the radio hero who uses wireless and later broadcasting to do good for communities, to preserve a way of life, to promote a common good, and to save lives.

What was the real significance of radio as depicted in such novels? First, the simple stories mirrored the public fantasy and its knowledge and sometimes misunderstanding of communication and later entertainment using this 20th-century invention. The stories traced the evolution of radio from a spark-gapped, Morse-coded curiosity into a powerful medium to which everyone listened. These stories of young men and women of high school age mostly provided escape, but in the process, both the hobby and the business of radio and broadcasting were portrayed as a force for public service, for the good of community, and a way to reinforce our view of ourselves. Overwritten though they are, these books remain a wonderful window on the excitement created first by wireless and later by radio broadcasting. The "gee-whiz" nature of the stories and the central role of radio in each is a good indicator of the general public fascination with cat's whiskers, DXing, silent nights, and crystal sets.

MICHAEL H. ADAMS AND CHRISTOPHER H. STERLING

Further Reading

Adams, Mike, "Will the Real Radio Boys Please Stand Up?" *Antique Wireless Classified* 8–9 (September 1991, December 1991, May 1992, and December 1992)

Dizer, John T., Jr., *Tom Swift and Company: "Boys Books" by Stratemeyer and Others,* Jefferson, North Carolina: McFarland, 1982

Lisle, Larry, "Reading Radio Fiction," *QST* (April 1994)

McGrath, J.J., "Radio Boys Revisited," *QST* (July 1976)

Prager, Arthur, *Rascals at Large; or, The Clue in the Old Nostalgia,* New York: Doubleday, 1971

Sterling, Christopher H., "Dime Novel Radio," *Old Timer's Bulletin: Official Journal of the Antique Wireless Association* 13 (September 1973)

Sterling, Christopher H., *Children's Novels Devoted to or Including Telegraph, Telephone, Wireless, Radio or Television, 1879–1995,* 5th edition, Washington, D.C.: George Washington University School of Media and Public Affairs, 1996

Children's Programs

Children and teenagers have always been particular fans of radio (and targets for radio's advertisers). From children's programs in the "golden age" before television to the advent of rock and roll music programming through today's diet of educational and entertainment options, radio has been a consistent element of youth culture.

Radio Before Television

Programs

The early days of radio offered a host of programs specially designed for young audiences, as well as several family-friendly options that encouraged parents and children alike to gather in front of the centrally located radio receiver. In the 1930s and 1940s, children everywhere rushed home from after-school activities and errands to listen closely to a series of three to five consecutively run quarter-hour serials largely sponsored by breakfast food companies, collectively called the *Children's Hour.* In the ever-popular category of children's adventure programs, young listeners followed the interplanetary exploits of *Buck Rogers in the Twenty-fifth Century,* the aviation adventures of *Captain Midnight,* the high school heroics of *Jack Armstrong, the All-American Boy,* or the daunting detective work of *Little Orphan Annie.* Westerns such as *Tom Mix, Sky King,* and the wildly popular *Lone Ranger* sparked the imaginations of children across the country and made "Kemo Sabe" a household phrase.

Science fiction was also a favorite genre for young radio listeners. Adapted from comic books was *Buck Rogers in the Twenty-fifth Century,* on which the hero Buck and his faithful assistant and love interest Wilma fought the evil powers of Killer Kane on the planet Niagara. Other early sci-fi favorites of children included *Flash Gordon, Tom Corbett, Space Patrol,* and *Space Cadet.*

Themes and Messages

Typical themes and messages in the wartime and immediate postwar programs included defending "good"—freedom, justice, honor, and other "American values"—from "evil." A series of episodes, for instance, had Tom Mix fighting an enemy balloon perched over the plains or had Superman rushing off to rescue a captured Lois Lane. The Lone Ranger and Tonto battled the wilds of the desert, and Dick Tracy solved crimes and foiled villains in a number of contexts. Perils and pitfalls plagued characters but were expertly averted by the stars of the shows. The hero concept was omnipresent, with viewers left hoping that they would be as strong, as brave, and as principled as those largely male characters who saved the day time and time again.

In fact, during the period to 1948, the potential for violent and aggressive messages to influence child listeners was a prominent and controversial concern. Newspapers and magazines of the 1930s and 1940s invited experts to comment on the issue, with most suggesting there was little ground for concern, citing psychoanalytic and cathartic theories. The discourse was very similar to the controversy that would later arise about television violence, yet it occurred well before the scholarly attention of psychologists and other researchers to the influence of the media and prior to the research and theories that were advanced to explain and demonstrate the influence of media violence on aggression.

When the country was at war, enemies on such programs as *Terry and the Pirates* and *Chandu the Magician* were from the Far East; in *Little Orphan Annie* or *Jack Armstrong, the All-American Boy,* villains had German, Russian, or Italian accents. Some were blatantly referred to as Nazis or obliquely called spies. The "damsel in distress" was a frequent theme, and often highly stereotypical gender representations prevailed. One exception was Little Orphan Annie, who solved crimes à la Nancy Drew and was even more adventurous and plucky than her comic strip character.

Program Openings and Sound Effects

Attention-getting introductions perked the ears and piqued the interest of youngsters across the country. "Look! Up in the sky . . . it's a bird . . . it's a plane . . . it's *Superman*!" heralded the beginning of a child favorite, as did the sing-along openings of *Little Orphan Annie* ("Who's that little chatter-box . . . The one with the pretty auburn locks . . . Who can it be? . . . It's Little Orphan Annie") or *Jack Armstrong* ("Raise the flag for Hudson High, boys . . . Show them how we stand"). *Terry and the Pirates* opened to the waterfront sounds and hearty bellows of a busy harbor.

The artful use of sound characterized much of the radio offerings of the day, encouraging the "theater of the mind" that fed off young imaginations. The radio tower communications of *Hop Harrigan* added to the realism and authentic feel of the program. The sirens and screeching brakes of police squad cars in *Dick Tracy* sent many a pulse racing. And of course, the gun sound effect after "Faster than a speeding bullet" and the train sound effect following "More powerful than a locomotive" reinforced the power and strength of the visual image of Superman aroused in children's minds.

Sponsors, Premiums, Announcers

Almost equally entertaining were the commercial jingles and premium offers intertwined with children's adventure plots by such advertisers as Ralston Purina (sponsor of *Tom Mix*), General Mills' Wheaties (*Jack Armstrong*), Kellogg's Pep cereal (*Superman*), and Ovaltine (*Little Orphan Annie, Captain Midnight*). Children saved numerous box tops and anxiously awaited the arrival of program-associated toys and gadgets. Successful receipt of a nail from the shoe of *Tom Mix* horse Tony, a *Dick Tracy* badge, or a secret decoder ring that made sense of encrypted messages in *Little Orphan Annie* instantly elevated a child to envied status in the neighborhood. Millions of boxes of hot or cold cereal and tons of hot cocoa and peanut butter were consumed with the ulterior motive of a *Captain Midnight* Key-O-Matic Code-A-Graph, a *Little Orphan Annie* Shake-Up Mug, a *Green Hornet* ring, or a *Tom Mix* Straight-Shooter Medal. Premiums also encouraged further listening and careful attention to the program, because secret messages were woven into the fabric of plots and called for premium decoders for deciphering. Many children's radio programs also offered club memberships to avid followers and follow-up premiums that complemented past offers.

The sponsorship of an entire radio program by a single advertiser allowed for a blending of program and ad copy that heightened the ability of the ad to entertain. Episodes of *Tom Mix* both began and ended with a musical message promoting the qualities of Ralston cereals. At the beginning of the show, the "Tom Mix Straight-Shooters" asked kids (in a Texas accent) to "Start the mornin' with Hot Ralston" amid the sounds of horse hooves and cowboy yips and yells. At the end of the show, the singers urged listeners to "Take a tip from Tom . . . go and tell your mom . . . Hot Ralston can't be beat." The characters, the music, the sound effects, and the sponsors' messages all contributed to the overall theme, in this case a Western, a strategy successfully pursued by many radio programs.

The announcer/host was often used to create a transition from the message of the sponsor to the program and back or even to speak the commercial message, as well as to introduce the program or unveil the latest premium offer. *The Lone Ranger* and Tonto were introduced by Fred Foy, Brace Beemer, and others, who occupied the announcer role by inviting audience members to "return with us now to those thrilling days of yesteryear. From out of the past come the thundering hoofbeats." Before his *60 Minutes* tenure, Mike Wallace was a convincing voice for *Sky King* and sponsor Peter Pan Peanut Butter on National Broadcasting Company (NBC) and Mutual in the mid-1940s. Announcer Pierre André read copy that persuaded young listeners to fully experience the escapades of *Captain Midnight* with membership into the Secret Squadron or the purchase of a decoder badge.

Weekend Programming

The programs for children during weekend hours generally involved storytelling, singing, and playing educational games. The target age group was often lower on the weekends as well, with many shows appealing to pre-kindergarten and early elementary school children.

Saturday mornings were populated by Archie, Jughead, Veronica, and friends on *Archie Andrews,* replete with story lines featuring the harebrained hijinks of Jughead and the budding romances between the more suave characters. There was also *No School Today,* in which characters Big Jon and Sparkie would tell various adventure tales to very young listeners. Boston's WCOP ran the *Children's Song Bag* on Saturday mornings, while Mutual affiliates carried a show in which birds sang the notes of popular songs, called the *Hartz Mountain Canary Hour, Canary Pet Show,* and *American Radio Warblers* at various times during its 15-year history (late 1930s to early 1950s). Beginning on WGN in Chicago in 1932 was the *Singing Story Lady,* hosted by Ireene Wicker who, true to her title, sang songs and told stories.

Sunday afternoon's gem was *Quiz Kids,* which offered brain-teasing knowledge questions asked of panelists under 16 years of age (aired from 1940–54). The series began on NBC Blue and ended on Columbia Broadcasting System (CBS): time slots changed, but it was generally broadcast on Sunday in late afternoon or early evening. Hosted by Joe Kelly, *Quiz Kids* featured very intelligent young panelists who defined difficult words and performed other mentally challenging tasks. Sponsors of the show included Alka Seltzer and One-a-Day vitamins.

In many markets, radio personalities and other prominent people (including New York City Mayor Fiorello LaGuardia during a newspaper strike in 1945) read the comics to children on Sunday mornings. Among the most beloved children's radio programs was *Let's Pretend,* which was hosted by Nila Mack and enjoyed a 20-year stint (1934–54) on CBS. In the program, a group of child actors (the "Pretenders") acted out a half-hour story built around the names of three objects sent in by young audience members. The actors also played roles in other charming fairy tales written by Mack. The precursor to *Let's Pretend,* a show called *Adventures of Helen and Mary,* had run for five years before Mack changed the name and made the concept more popular with children by involving more young actors. The enthusiastic opening included a theme song ("Hello, hello . . . Come on, let's go . . . It's time for *Let's Pretend*") and a rousing "Hello, Pretenders!" shouted by recurring host Uncle Bill Adams, to which the chorus of child actors would respond "Hello, Uncle Bill!"

Radio After Television (Since 1948)

Music Programming

After television arrived and radio came to be primarily devoted to music, teenage audiences remained among the most loyal fans to radio's new programming. Music is the primary force drawing young people to radio. The lure increases with age: 85 to 95 percent of those aged 2 to 18 who listen to the radio are listening to music programs. From rock and roll in the 1950s through alternative and rap in the 1990s and into the new millennium, teens and music have gone hand in hand. Teenagers use music for relaxation and entertainment as well as to keep up to date on popular trends. Another gratification teens derive from music is its social utility; it is something to talk about with their friends. But perhaps what is unique about teenagers and music is the relationship of music to self-identity. Young people use musical styles and favorite performing artists as ways of defining and expressing themselves in a manner that has no equivalent at any other time in life. Evidence for the relationship between music and identity is not limited to teenagers' uses of radio. It is also apparent in concert T-shirts and other styles of dress, posters and sometimes even lyrics displayed by teens in their bedrooms and school lockers, and their treasured compact disc collections. Research suggests that as teens move toward adulthood, their preferred means of listening to music will not be radio but their own sets of compact discs, tapes, and records.

Music programs such as *American Top 40* (launched in 1970) and *Rick Dees' Weekly Top 40* (launched in 1983) have long drawn faithful young audiences on weekends. The contemporary hit format (Top 40) is highly favored by children. When they become teens, also like album-oriented/classic rock, country, rap, and alternative.

In addition to Top 40 favorites, children's radio offers musical options not likely to be found on stations targeting an older audience, with a heavy emphasis on songs from Disney or other kids' movies and novelty tunes. A list of the top 30 songs on Radio Disney affiliate WSDZ in St. Louis for 20 January 2003 featured the songs of pop stars Britney Spears, Jennifer Lopez, and 'N Sync, as well as the child-friendly tunes "Hampsterdance Song" by Hampton and the Hampsters and "Who Let the Dogs Out" by Baha Men.

Talk, News, and Educational Programs

Young people also have a wide variety of non-music radio programming options. Statistics show that 11- to 18-year-olds in the United States spend an average of five to seven minutes with talk radio per day, five minutes with radio news, and six minutes with other types of radio programs. Many local markets offer their own version of WXPN Philadelphia's *Kid's Corner,* an entertaining combination of talk, games, and novelty songs for 8- to 14-year-olds. One of the most popular features of *Kid's Corner* is the news segment, in which 10- to 17-year-olds who have participated in reporter training workshops present stories they've gathered on such topics as women's issues and politics. Somewhat similarly, National Public Radio (NPR) features *Teenage Diaries,* a program in which teens act as newscasters reporting on themselves. They conduct interviews with friends and family and for several months chronicle many aspects of their lives on audiotape, providing reflections on what it means to be a teenager in contemporary society. The older children become, the more often they tune into such non-music radio programs.

Countless stations provide call-in or talk shows tackling youth-oriented issues and subjects of concern, including such diverse topics as cheating in school, romance and young crushes, and struggles with self-esteem and "fitting in." WXPN host Kathy O'Connell has encountered callers interested in talking about light-hearted issues such as their computers or their pets as well as those eager to discuss more weighty topics such as animal rights and AIDS. Other youngsters call in not to discuss issues but to sing a song, play an instrument, or tell a joke. The program has boasted some 400 attempted calls per night.

Educational children's radio enjoys a niche-market position. Endeavors such as the *Kinetic City Super Crew,* sponsored by the American Association for the Advancement of Science, have proven successful at stimulating children's interest in scientific concepts as well as provoking them to work on science projects at home. This particular 30-minute program was launched in 1993 on several public radio stations across the nation and targeted third to fifth graders. The program showcased four students' attempts at solving problems of the world using science, with the goal of encouraging interest in science as well as advancing knowledge and honing critical thinking skills. The Super Crew, comprising Annalee, Joaquin, Chantel, and Alvin, aged 12 to 16, solved mysteries by "traveling" to different locations around the world and gathering information. A scientific study of over 250 fourth graders found that there were educational benefits for both boys and girls in listening to *Kinetic City Super Crew.*

Radio can also be a source of information regarding politics. Studies have found that children turn to television and radio much more often than to newspapers or magazines when gathering information about politics and political campaigns. The national civics education program *Kids Voting USA* found that children overwhelmingly turn to broadcast media for current events and civics information gathering to complete their school assignments. Children and teenagers are often exposed to the news media when their parents tune in, and they are present in the broadcast news audience in surprisingly large numbers. They use radio and television to provide them with information about public figures, policies, and events. In turn, this knowledge helps shape their opinions, values, and behaviors.

As with adults, radio and other media can also set the agenda of youngsters by highlighting some occurrences as newsworthy while shunning others. Through news bias, be it intentional or inadvertent, radio can also influence young people's interpretations of events. Because children often lack first-hand information to counter media messages, these influences can be stronger and more dramatic for child than for adult audiences.

Cultural Programming and the Arts

Radio has been an outlet as well as a vehicle for appreciation of the arts, including theater and literature. Many radio plays are written for young performers or written by young playwrights. All encourage expression, imagination, and creativity. *Children's Radio Theater,* for example, began in 1977 at WPFW in Washington, D.C., and was picked up by more than 100 public radio affiliates by the mid-1980s. The series featured 30-minute plays written by children aged 5 to 17 and performed by professional actors of all ages. The scripts were chosen by teachers, librarians, and actors affiliated with the nationwide Henny Penny Playwriting Contest. The troupe broadcast its last performance in December 1997 on WPFW in New York City.

1994 saw the launch of *Rabbit Ears Radio,* a radio program founded by Rabbit Ears, the Connecticut-based publishers of video and book-and-tapes. It enjoyed a two-year run distributed by Public Radio International. It featured stories narrated by celebrities and was hosted by Mel Gibson and Meg Ryan, who described the history of the story and the music chosen to accompany it and also introduced the narrator. Examples included *John Henry,* narrated by Denzel Washington with music by B.B. King, and *The Velveteen Rabbit,* narrated by Meryl Streep with music by George Winston. Many local radio stations also offer programs in which stories are read to children over the air, often created through collaborations between stations and local libraries or colleges. Examples include Northeast Indiana Public Radio's "Folk Tales from the Briar Patch" offered on two weekday afternoons and "Magic Hat Storytime" broadcast on Sundays on the campus station at Rowan University in Glassboro, New Jersey.

Children's Radio Networks

In addition to individual children's programs, there have also been attempts—some successful, some not—at establishing children's radio networks. In 1990 an early entry into the all-children's radio scene was launched: Orlando-based Kid's Choice Broadcasting network boasted an advisory board at the time that included Peggy Charren of Action for Children's Television fame as well as Peter Yarrow of the musical group Peter, Paul, and Mary, who wrote original music and the theme song for the network. Among the programming alternatives offered by Kid's Choice were a 6 to 10 A.M. segment called *New Day Highway* (a family show with news, music, and special features), *Curbside Carnival* from 2 to 6 P.M. (offering story and exercise segments and music from around the globe), and a 10 A.M. to 2 P.M. slot targeted to preschoolers featuring music and information on numbers, colors, and shapes.

Foremost among obstacles preventing adoption of a kids' radio format cited by executives was the difficulty in providing potential advertisers with viable listenership data, because Arbitron didn't measure the presence of young children in the radio audience. Station owners and managers have noted that many advertisers, both local and national, would have welcomed the opportunity to address child audiences with their advertising messages. The potential of another outlet, in addition to television, for peddling the breakfast cereals, snack foods, clothing, and other products favored by a young audience was appealing to many advertisers. The tradition in the industry to base pricing, placement, and other crucial advertising decisions on ratings data that were lacking for children's radio was, however, a real impediment to the growth of children's radio at that time.

Therefore, the decision of Arbitron in 1993 to gather data on children aged 2 to 11 years had the potential to be groundbreaking. Arbitron contracted with the Children's Satellite network's Radio Aahs, the nation's only 24-hour children's network at the time. At the request of advertisers, then-president of the Children's Satellite network Bill Barnett arrived at an agreement with Arbitron whereby they would target households with children aged 2 to 11 in return for a subscription fee. To address the needs of the young audience, Arbitron adopted new strategies of training and assistance: older children filled out their own diaries, and Arbitron also conducted sessions to inform parents about completing surveys for younger children. For Radio Aahs, a network that had already been successful at reaching 15 percent of the nation just three years after its debut in Minneapolis, the ratings information offered the potential of additional growth and strengthening of advertising base for the future. Yet the Arbitron agreement turned out to be short-lived and Radio Aahs resumed the practice of relying on call-in statistics to woo advertisers.

Modern Networks and Programs

Despite the breaking of new ground by Arbitron and Radio Aahs, measuring the 2- to 11-year-old audience is still exceptionally rare today. Nevertheless, the 1990s spawned a remarkable growth spurt in children's radio. The growth was due in part to the mere possibility of obtaining ratings data for young child audiences. It was also attributed to the many pioneers and trailblazers whose occasional forays into children's radio had been largely successful. There was also a commonly held view that the nation's youngsters had been an underserved radio audience for too long. For all of these reasons, a resurgence of additional programs and networks geared toward children occurred in the mid-1990s. Some of these fledgling networks are presently enjoying growing financial success and increasing patronage by young audiences, while others have gone belly up.

One of the clear success stories is American Broadcasting Companies' (ABC) Radio Disney, launched in 1997 and targeting 2- to 11-year-olds, boasted 45 subscribing stations and an audience of 1.6 million nationwide in the year 2000, offering pop, soundtrack, and novelty songs as well as safety and education tips and news. Music programming accounts for the vast majority of Radio Disney's offerings, and the network showcases such kid-friendly acts as Weird Al Yankovic, 'N Sync, Backstreet Boys, and teen heartthrobs Christina Aguilera and Britney Spears. The network uses focus groups with parents to determine whether music is appropriate for its audience and has often aired edited or alternate versions of songs to ensure that the family-friendly principle is met. Reactions of audience members are monitored carefully through e-mail and phone calls to the station. Radio Disney also gives away prizes that range from the somewhat typical—compact discs, Pokémon cards, and concert tickets—to the downright luxurious, such as snowboarding trips, visits to space camp, and even a day with a recording artist. Audience research data is gathered by Statistical Research, Inc. and has helped attract sponsors.

Fox Kids network's weekly *Fox Kids Countdown* program had garnered well over 200 affiliates by 1997 and reached an estimated 3 million listeners, mostly in the 8- to 14-year-old age group. The host is Los Angeles disc jockey Chris Leary, who helps draw in the Sunday morning audience with movie promotions and spots advertising Fox television shows. Songs requested by kids, call-in shows, and guest celebrities help make the program, now called *Fox All Access Countdown*, appealing to its affiliates, 95 percent of which are FM outlets.

In the mid 1990s, the Children's Broadcasting Corporation delivered 24-hour programming through Radio Aahs via satellite to over 40 percent of the country, allowing children aged 2 to 11 to listen to young disc jockeys, games, contests, news, and educational or self-esteem messages. A glimpse at the Radio Aahs programming schedule revealed such educationally entertaining fare as *All-American Alarm Clock* (6 to 9 A.M.), *Alphabet Soup* (noon to 1 P.M.), and *Evening Theater* (8 to 9 P.M.). The network also featured weekly live broadcasts from Universal Studios Hollywood and Universal Studios Florida. Later nightly slots were designed for adults tackling parenting issues.

Yet the Children's Broadcasting Corporation soon faced stiff competition from the growing Radio Disney network. Indeed, the two companies were in litigation from 1996 to 2002 as Children's Broadcasting Corporation filed suit against Radio Disney for breach of contract and use of trade secrets, claiming that after the Corporation hired ABC as a consultant, ABC used that information to design Radio Disney. The Children's Broadcasting Corporation ultimately received an award of $9.5 million, yet the judgment was too little, too late for the struggling network, which had gone off the air in 1998.

Despite new growth for more fortunate kid-friendly options, the late 1990s also saw the demise of a pioneer force in children's radio, KidStar Radio. The Seattle-based KidStar Interactive Media organization had distributed 24-hour children's programming from its home at KKDZ (AM) to AM stations in such major markets as San Francisco, Boston, Houston, and Detroit. The network made its debut in May 1993 and used 45 different interactive phone lines to record kids' input on songs, elicit their views on social issues, and even allow them to leave a message for their state senators. Sponsored programs included *Virtual Safari,* in which kids encountered adventures with animals (sponsored by GapKids), and *Zack and Zoey's Survival Guide,* in which the title characters were eighth graders who passed on words of wisdom from their experiences at school (sponsored by the Disney Channel). Citing the loss of a crucial investment deal, KidStar folded in 1997, leaving its affiliates scrambling to replace the child-oriented music (from disc jockeys ranging in age from 9 to 14), news, sports, and entertainment programming that KidStar Radio had provided.

Today, two- to seven-year-olds in the United States listen to the radio an average of 24 minutes per day. Eight to ten-year-olds listen for 26 minutes. From age 11 to 13, average radio use is 44 minutes per day, whereas for 14- to 18-year-olds it's 65 minutes per day. The most common times for radio listening among youngsters are after school on weekdays (3 to 7 P.M.) and on weekends from late morning through the afternoon (10 A.M. to 3 P.M.).

Radio is still an important and consistent presence in the lives of young people. Many fundamental characteristics of radio remain unchanged. In addition to musical programming, modern radio programs for children are still populated by heroes and villains, main characters and sidekicks, and the occasional presentations of aggression and gender stereotypes. The adventures and trials and tribulations deemed interesting or relevant to children and teens are still the focus of many shows, much like in the days of old-time radio. Radio is still used to inform young people of topics, events, and central figures in public life. It remains an outlet or a vehicle for creative expression and appreciation of the arts. And radio is still a source of distraction, a means of escape, and a companion. Frequently entertaining and often educational, radio remains a means of exercising the imagination of America's youth.

ERICA SCHARRER

See also American Top 40; Captain Midnight; The Green Hornet; Jack Armstrong, All American Boy; Let's Pretend; Little Orphan Annie; The Lone Ranger; Radio Disney; Science Fiction Programs; Westerns

Further Reading

Anderson, Arthur, *Let's Pretend: A History of Radio's Best Loved Children's Show by a Longtime Cast Member,* Jefferson, North Carolina: McFarland, 1994

Boemer, Marilyn Lawrence, *The Children's Hour: Radio Programs for Children 1929–56,* Metuchen, New Jersey: Scarecrow, 1989

Clark, W.R., "Radio Listening Habits of Children," *Journal of Social Psychology* 12 (1940)

Cooper, Jim, "Arbitron to Measure Children's Radio," *Broadcasting and Cable* (23 August 1993)

Eisenberg, Azriel L., *Children and Radio Programs,* New York: Columbia University Press, 1936

Garfinkel, Simson, "Children's Science Radio," *Technology Review* 96, no. 7 (1993)

Gordon, Dorothy Lerner, *All Children Listen,* New York: G.W. Stewart, 1942

Gruenberg, Sidonie Matsner, "Radio and the Child," *Annals of the American Academy of Political and Social Science* 177 (January 1935)

Lyness, Paul I., "Radio Habits of the Young Audience," *Broadcasting* (25 September 1950)

McCormick, Moira, "Children's Radio Continues to Liven Up Airwaves," *Billboard* (9 November 1996)

Nachman, Gerald, *Raised on Radio: In Quest of the Lone Ranger, Jack Benny . . . ,* New York: Pantheon Books, 1998

Petrozzello, Donna, "Children's Radio: A Format Whose Time Has Come?" *Broadcasting and Cable* (7 October 1996)

Petrozzello, Donna, "KidStar Calls It Quits," *Broadcasting and Cable* (24 February 1997)

"Radio in the Life of the Child," in *Educational Broadcasting,* Chicago: University of Chicago Press, 1936 (symposium)

Roberts, Donald F., et al., *Kids and Media at the New Millennium,* 2 vols., Menlo Park, California: Henry J. Kaiser Family Foundation, 1999

Stark, Phyllis, "Slow Growth for Youth-Oriented Talk Shows: Finding Engaging Topics, Vocal Callers among Challenges," *Billboard* (7 March 1992)

Taylor, Chuck, "Radio Disney Tunes In Young Listeners and Turns Youth Pop Craze on Its Ear," *Billboard* (25 March 2000)

CHUM

Toronto, Ontario Station

For people growing up in Toronto and Southern Ontario during the 1950s, 1960s, and 1970s, listening to 1050 CHUM-AM was something of a religion, making it quite possibly Canada's most influential radio station at that time. It was also Canada's first "hit parade" outlet.

The station began broadcasting during the daylight hours from the Hermant Building at 21 Dundas Square on 28 October 1945 on 1050 kilohertz with 1,000 watts of power. Operated by York Broadcasters, the station was owned by entrepreneur Jack Q'Part, who saw it as providing an excellent avenue for marketing his patent medicine products. However, many feel that the station really didn't find its niche until ten years later when Allan Waters, president of a pharmaceutical company owned by Part, purchased the radio station in December 1954. His first task was to change the station's format to rock and roll. Waters was inspired by the success of stations owned by Todd Storz in the United States that were successfully reaching younger audiences through the repetitious play of new records. Waters altered the programming day to broadcast 24 hours a day, and he moved the station's transmitter, increasing power to 10,000 watts. On 27 May 1957, the station unveiled its Top 40 music format. It took two or three years before advertisers became interested in the station, as the Top 40 format was unproven in the Canadian marketplace. During this time the station moved twice, to 225 Mutual Street and 250 Adelaide Street West, before settling in 1959 at 1331 Yonge Street, where it remained at the beginning of the 21st century.

Among the most significant developments occurring during the 1950s was the establishment of the CHUM chart, a Top 50 list of hit singles issued weekly. The first CHUM chart appeared on 27 May 1957, with Elvis Presley's "All Shook Up" assuming the number one position, ahead of Pat Boone's "Love Letters in the Sand." CHUM's chart remained a Top 50 list until 1968 when it was reduced to 30 songs. It remained in that format until 1986. In 1984 the chart was converted to a television format with a program that aired weekly on CITY-TV, a local Toronto station owned by the CHUM radio group. The TV program incorporated (newly emerging) music videos. Available for listeners at local record stores and published in newspapers, the CHUM chart served as an important marketing tool to boost record sales, particularly of domestic acts such as Lighthouse and Blood, Sweat and Tears. (A complete list of CHUM charts is available at the station's website, www.1050chum.com.)

In the 1960s the station's prominence as Canada's strongest radio station continued. This was due largely to the emergence of rock and roll, the station's relentless promotional activities, and its roster of popular disc jockeys. Before heading off on his long stint with American Bandstand, Dick Clark hosted a series of shows broadcast on CHUM. Other popular disc jockeys of the era include Dave Johnson, Mike Darrow, Bob Laine, Bob McAdorey, and Duff Roman. One of the biggest celebrities was "Jungle" Jay Nelson, who, after working as the host of an eclectic children's program on WKBW in Buffalo, New York, was recruited by CHUM in 1963 to replace the popular morning host Al Boliska.

On 1 September 1963 CHUM-FM opened at 104.5 megahertz with 18,000 watts of power, broadcast from the same Yonge Street location. Its program content was different from that of the AM station, as CHUM-FM broadcast classical and fine arts programming. In 1964 the station upgraded to 50,000 watts and started transmitting from Clarkston, Ontario. This transformed CHUM from a Toronto area radio station to an outlet that reached most of the population of Ontario and some U.S. cities on the Eastern seaboard, as well. By 1967 the station traded under the name of CHUM Limited on the Toronto Stock Exchange, and CHUM-FM began broadcasting 24 hours a day.

Owing to failing ratings and the superiority of FM radio by the 1980s, CHUM-AM underwent a number of format changes, moving away from the Top 40 format (Top 40 was moved, along with the CHUM chart, to the FM station) toward a format of "favorites of yesterday and today" in 1986. (That same year, the station was knocked off the air when the transmitter tower was sabotaged. There were rumors that this act was motivated by the unpopular change in musical format, but in fact one of the perpetrators had an ongoing grudge with the station because the tower interfered with his television reception.) The 1986 format change was followed by a change to soft rock in 1988. This strategy was also short-lived, and in 1989 the station adopted an all-oldies broadcast format. Previously, CHUM broadcast Toronto Argonaut football in the Canadian Football League, as well as the Toronto Blue Jays baseball games.

The station also featured a number of prominent personalities over the years. John Roberts, later an anchorman working with CBS television, began his career as night-time disc jockey J.D. Roberts on CHUM before moving to work on CITY-TV. Other prominent personalities formerly with CHUM include Bob McGee, Dick Smyth, Tom Rivers, and Allan Slaight, a CHUM program director and general manager in the 1960s and later president of the Standard Broadcasting Network, which owns a number of radio stations across Canada.

On 7 May 2000, 1050 CHUM signed off its popular oldies format and began an experiment in all-sports radio using a number of CHUM-owned radio stations across Canada. 1050 CHUM became the anchor station in the fledgling TEAM sports network. On 27 August 2002, the station abruptly dropped the all-sports format, returning to a mix of hits from the 1950s, 1960s, and 1970s. On 29 November 2002, Allan Waters stepped down as chairman and president of the CHUM group, which now owns a number of television and radio outlets across Canada. His two sons, Jim and Ron, continued to hold other executive posts with the company.

IRA WAGMAN

See also Contemporary Hit Radio Format/Top 40

Further Reading

Canadian Communications Foundation: The History of Canadian Broadcasting, <www.rcc.ryerson.ca/ccf/index2.html>

Farrell, Allen, *The Chum Story: From the Charts to Your Hearts,* Toronto, Ontario: Stoddart, 2001

Fortner, Brad, M.G. Stevens, and Sonya Felix, "1050 CHUM 40th Anniversary Issue," *Broadcaster* 56, no. 7 (Summer 1997)

Hall, Roy, *The First 27 Years of the "Chum Chart": A Complete Listing of Every Record to Make the "Chum Chart" from Its Beginning on the 27th May, 1957 through 31st December, 1983,* Rexdale, Ontario: Stardust, 1984

Jennings, Nicholas, *Before the Gold Rush: Flashbacks to the Dawn of the Canadian Sound,* Toronto, Ontario: Penguin, 1997

Miller, Cal, and Hanoch Bordan, "Yonge Street Blocked As CHUM Tower Is Sabotaged," *Toronto Star* (11 Aug 1986)

Churchill, Winston Spencer 1874–1965

British Statesman and Political Broadcaster

Winston S. Churchill used the worldwide facilities of the British Broadcasting Corporation (BBC) to deliver some of the most memorable speeches of the World War II era to the largest audience ever reached by a British politician. Between 1926 and 1954, his evocative phrases and magnificent epigrams helped encourage and inspire the enemies of militant fascism and postwar communism and proved to be the most powerful force for overcoming U.S. isolationism. Using Burke, Gibbon, and Cromwell as models, his "great flights of oratory" and "sudden swoops into the intimate and conversational" made him the most eloquent statesman of his day and were instrumental in ensuring his long political success. According to *Time,* "Few orators since Demosthenes have evoked the emotional quality of the prime minister's exhortations."

Early Use of Radio

Churchill first manifested an interest in the political uses of broadcasting when, as Chancellor of the Exchequer in the Baldwin government (1924–29), he attempted to use the BBC to end the General Strike of 1926 by presenting the government's position directly to the nation. Political motives also underlay his highly partisan annual budget speeches and his vigorous campaign for state-controlled radio, both of which brought him into constant conflict with the BBC's director general, John Reith.

The Prime Minister's official residence, No. 10 Downing Street in London, was first wired for broadcasting in 1926, although equipment to allow broadcasts to originate from No. 10 was not put in place until late 1938. The nearby underground Cabinet War Rooms (now a museum) also had broadcast capability by the time the war began in September 1939. From one of these sites (or from Chequers, the country house of Prime Ministers, which received broadcasting equipment during the war), Churchill broadcast 24 times between June 1940 and May 1945.

The 1930s were Churchill's "wilderness years." He took to the air only four times between 1934 and 1939, generally as an opponent of Indian self-government and of German re-armament (both unpopular positions). After 1936 his broadcasts stressed the danger of appeasing Hitlerian expansionism. On 16 October 1938, after Prime Minister Chamberlain's failure at Munich to thwart German designs on Czechoslovakia, Churchill delivered a stinging rebuke of Hitler over the BBC and called attention to Britain's military unpreparedness. When the Nazi blitz into Poland ignited World War II, Churchill's warnings seemed vindicated, and he was rewarded with a return to his old post as First Lord of the Admiralty.

As First Lord, Churchill regularly broadcast to the nation on developments in the war at sea. In December 1939 he reported jubilantly on the sinking of the German pocket battleship *Graf Spee* in Montevideo harbor. In order to sustain morale during the bleak period of military inactivity (Phony War), he routinely exaggerated Germany's U-boat losses.

Finest Hour: 1940–45

In May 1940 Churchill replaced Chamberlain as prime minister. It was a dark hour for Britain. Allied forces had recently been driven from Norway and were in headlong retreat before the victorious Wehrmacht in France. Strong leadership and inspired guidance were necessary if Britain was to surmount the challenges of total war. During his premiership, Churchill would deliver 33 major wartime speeches by radio. All were carried by the BBC within Britain and shortwaved to North America and throughout the Empire. Many were translated into Danish, Dutch, Serbo-Croatian, and several other languages. As Britain confronted the gravest national crisis in its long history, Churchill employed his speeches for three purposes: to keep the nation abreast of the progress of Allied operations against the Axis; to rally and reassure Britons at times of disappointment and defeat; and to satisfy world opinion, especially in the United States, of Britain's resolve to persevere until victory was achieved. In the first few years of conflict, when British forces were on the defensive in all theaters, Churchill's "fighting words" seemed to be the most formidable weapon in the national arsenal. Edward R. Murrow marveled at the way in which Churchill "mobilized the English language and sent it into battle," and parliamentarian Josiah Wedgewood believed his speeches to be "worth a thousand guns." Most of Churchill's major addresses were broadcast, and the few that were not carried live from either the House of Commons, the BBC's Broadcast House, or Chequers were conveyed in summary form immediately afterwards. Complete transcripts were published in morning newspapers and collected in annual bound anthologies. Murrow reported on the prime minister's activities in his nightly transmissions from London, and Raymond Swing and H.V. Kaltenborn routinely analyzed his speeches on their programs. Columbia Broadcasting System (CBS) correspondent Eric Severeid believed Churchill's voice was as familiar in the United States as it was in Britain, where it became an eagerly awaited wartime ritual to cluster around the living-room or pub set and hear Big Ben's nine chimes and the defiant and rousing words that followed.

In his first broadcast as prime minister on 19 May 1940, Churchill told the nation, "I have nothing to offer but blood, toil, tears and sweat." With Britain suffering grievous losses in France, he urged his countrymen to "be men of valor" and accept the sacrifices the developing war would soon require of them. The next day German forces reached the Channel coast at Abbeville, effectively cutting the Allied armies in two. With his shaky French, Churchill took to the shortwaves and tried to rally General Weygand's beleaguered troops. By the beginning of June, Britain's expeditionary force was surrounded and lost 30,000 (and saved on the order of 300,000) men while evacuating the port of Dunkirk. In a 4 June 1940 speech, Churchill accepted this ignominious setback but confidently pledged that Britain would continue the struggle against Hitler "whatever the cost may be." "We shall fight on the beaches," he proclaimed, "we shall fight on the landing grounds, we shall fight in the fields and in the streets . . . we shall never surrender." The House of Commons was deeply moved by this speech, and Vita Sackville-West, listening at home, was "stirred by his Elizabethan phrases" and the "massive backing of power and resolve behind them."

On 17 June 1940, France sued for peace, and Britain found herself alone against the overwhelming might of Nazi Germany. The next evening, Churchill informed the world of the tragedy and once again asserted Britain's "inflexible resolve to continue the war." In what is often regarded as his greatest rhetorical performance, the prime minister defiantly stated:

> The whole fury and might of the enemy must very soon be turned on us. Hitler knows that he will have to break us in this island or lose the war. If we can stand up to him, all Europe may be free and the life of the world may move forward into broad, sunlit uplands. But if we fail, then the whole world . . . will sink into the abyss of a new Dark Age . . . Let us therefore brace ourselves to our duties and so bear ourselves that, if the British Empire and its Commonwealth last for a thousand years, men will still say, "This was their finest hour."

In the summer of 1940, Hitler launched a massive air campaign against Britain and began preparations for "Operation Sea-Lion"—the cross-Channel invasion of the Home Islands. On 14 July, as the Luftwaffe attempted to bomb British cities into submission, Churchill told the 65 percent of the entire population listening to him that he "would rather see London laid in ruins and ashes than tamely and abjectly enslaved." In June and July, the Royal Air Force lost over 500 pilots successfully defending British airspace against the marauders. On 20 August Churchill expressed the sentiments of a grateful nation. "Never in the field of human conflict," he declared, "has so much been owed by so many to so few." Violet Bonham-Carter said of the prime minister's hour-long broadcast, "Nothing so simple, so majestic, so true has been said in so great a moment of human history."

The air engagements of the Battle of Britain and the Blitz strained Britain's resources to the breaking point and rendered her strategic situation perilous. If she was not to succumb, the economic and military assistance of the neutral United States was imperative. Many of Churchill's 1940–41 broadcasts were geared toward securing greater U.S. involvement in the European war. In his 4 June 1940 broadcast, the prime minister spoke confidently of the "New World liberating the Old." On 20 August he likened Anglo-American unity to the irresistible flow of the Mississippi River, and on 11 September he urged the two nations to "join hands [and] rebuild the temples of man's freedom." On 9 February 1941 he transmitted a direct appeal to Franklin Roosevelt over BBC shortwave. "Put your confidence in us," he pleaded. "Give us the tools and we will finish the job." Churchill's rhetorical assault on American isolationism bore fruit, and within a month the U.S. Senate passed the Lend-Lease Act, guaranteeing Britain a regular supply of all required war material.

Despite America's increased material and moral commitment, Britain's war fortunes continued to deteriorate in the spring of 1941. By April, the Royal Army had been evicted from Greece, and German U-boats were inflicting appalling losses on Allied merchantmen in the Atlantic. On 27 April 1941 Churchill went on the air to overcome the pessimism that recent events had generated. "When we face with a steady eye the difficulties which lie before us," he assured, "we may derive new confidence from remembering those we have already overcome." Over 77 percent of the adult population heard this speech (the largest audience achieved by a British premier to that date), and most were heartened. When Churchill asked her if she had tuned in to the broadcast, his long-time friend Violet Bonham-Carter replied, "Of course. Everyone in England listens when you speak."

Churchill was considerably encouraged by diplomatic events in the second half of 1941. On 21 June, Hitler's "Operation Barbarossa" brought the Soviet Union into the anti-Nazi coalition. The following evening, Churchill broadcast to the world the news that Britain was no longer alone. During the same period, the United States was increasingly assuming the role of a cobelligerent. Churchill used the occasion of a 16 June broadcast accepting an honorary doctorate from the University of Rochester to praise Roosevelt's pro-Allied tendencies, and on 14 August he reported on the Anglo-American agreement on war aims achieved in the Atlantic Charter. *Time* magazine remarked of this latter broadcast: "Churchill told the world and the world hung on his words. . . . His timing was matchless [and] he lashed Britain's enemies with the splendor of Elizabethan arrogance." After America's formal entry into the war following the Japanese attack on Pearl Harbor, Churchill addressed a joint session of Congress. On 26 December, in a speech shortwaved as far as Singapore, Churchill proclaimed, "The U.S. has drawn the sword for freedom and cast away the scabbard . . . the subjugated peoples of Europe [can] lift up their heads again in hope."

Churchill remained optimistic when, in the spring of 1942, the Japanese overran British possessions in the Far East and

Winston Churchill, London, 28 March 1940
Courtesy AP/Wide World Photos

sank the battleships *Prince of Wales* and *Repulse.* In a 15 February 1942 broadcast, Churchill talked of "drawing from the heart of misfortune the vital impulses of victory," and on 10 May he reviewed Allied successes during the two years since he had become prime minister.

When the tide of war shifted decisively in Britain's favor after El Alamein and the American invasion of French North Africa in late 1942, Churchill was euphoric. On 11 November he told listeners: "This is not the end. It is not even the beginning of the end. But it is perhaps the end of the beginning." To those who feared that the war would hasten the disintegration of the Empire, he reassured Britons that he had not "become the King's First Minister in order to preside over [its] liquidation."

The majority of Churchill's broadcasts after 1943 emphasized the certainty of victory and the necessity of continued Allied cooperation in the postwar era. On 26 January 1943 he reported on the Casablanca conference and the "no compromise peace" formula decided there. In a broadcast from London's Guildhall in June 1943, Churchill remarked on the success of the Allied bombing offensive against Germany. "Those who sowed the wind are now reaping the whirlwind." The year 1944 was, in his words, the "year of almost unbroken success." On 26 March 1944 he discussed Allied preparations for the much-anticipated second front in France. In his 9 November broadcast, he announced the end of the U-boat threat, listed the European capitals recently liberated by Allied forces, and declared that the enemy had been "beaten back into his own lair." On 23 November 1944 he gave a Thanksgiving Day speech from the Royal Albert Hall in honor of the United States' massive war-winning contribution.

On 8 May 1945 Churchill delivered the news of Germany's unconditional surrender, and on 13 May he made a 40-minute victory broadcast to the Empire and Commonwealth. In the general election that followed the end of the war in Europe, radio was Churchill's principal campaign instrument. Unfortunately, his vicious attacks on Clement Atlee and his 4 June 1945 broadcast comparing the Labour Party to the Nazi Gestapo were highly unpopular and helped to ensure his eventual defeat at the polls.

Final Years

Despite being out of the Cabinet between 1945 and 1951, Churchill continued to make notable broadcast speeches. His most famous is certainly the one delivered in Fulton, Missouri, in March 1946, when his stark depiction of the communist takeover in Eastern Europe and his call for Anglo-American vigilance against Soviet expansionism were instrumental in escalating Cold War tension. "From Stettin in the Baltic to Trieste in the Adriatic," he declared, "an iron curtain has descended across the Continent." On 19 September 1946 another significant postwar speech was made in Zurich, where Churchill strongly advocated European integration.

In 1951 Churchill absorbed the lessons of his 1945 election defeat and adeptly used the airwaves to secure the Conservatives' return to power. On 7 February 1952, as prime minister again, he delivered a magnificent eulogy of George VI over the BBC. His last major public address was made on 30 November 1954, on the occasion of his 80th birthday.

On 30 January 1965, Churchill's elaborate London funeral was comprehensively covered by the BBC and American radio networks. For three and a half hours, CBS's Alan Jackson and Morley Safer described the procession from Westminster Hall to St. Paul's Cathedral, Robert Trout reflected on Churchill's oratorical achievements, and ex-President Eisenhower delivered a moving tribute to his wartime friend.

ROBERT J. BROWN

See also British Broadcasting Corporation; Propaganda by Radio; Reith, John; Shortwave Radio

Winston Leonard Spencer Churchill. Born at Blenheim Palace, England, 30 November 1874. Eldest son of Lord Randolph Churchill and descendant of John Churchill, first Duke of Marlborough. Educated at Harrow School, 1888–92; cadet at Royal Military College, Sandhurst, 1893–94; Married Clementine Ogilvy Hozier, second daughter of Sir Henry and Lady Blanche Hozier, and grand-daughter of the Earl of Airlie, 1908; joined Queen's Own Fourth Hussars, 1895; service in India and Sudan, 1896–98; war correspondent in South Africa during Boer War, 1899–1900; Conservative member of Parliament for Oldham, 1900–1906; joined Liberal Party, 1904; under-secretary of state for the colonies, 1905–08; member of Parliament for Northwest Manchester, 1906–08; president of the Board of Trade, 1908–10; member of Parliament for Dundee, 1908–22; home secretary, 1910–11; first lord of the admiralty, 1911–15; chancellor of the Duchy of Lancaster, 1915; war service in France, 1915–16; minister of munitions, 1917–18; secretary of state for war and air, 1919–21; secretary of state for colonies, 1921–22; not in Parliament, 1922–24; chancellor of the exchequer, 1924–29; member of Parliament for Epping, 1924–45; reentered Conservative Party, 1925; first lord of admiralty, 1939–40; prime minister and minister of defense, 1940–45; opposition leader, 1945–51; member of Parliament for Woodford, 1945–64; prime minister, 1951–55. Recipient of Order of Merit, 1946; Knight of the Garter, 1953; Nobel Prize for Literature, 1953; honorary U.S. citizenship, 1963. Died in London, 24 January 1965.

Selected Publications

Winston S. Churchill: His Complete Speeches 1897–1963, edited by Robert Rhodes James, 8 vols., 1974

Further Reading

Briggs, Asa, *The War of Words,* volume 3 of *The History of Broadcasting in the United Kingdom,* London and New York: Oxford University Press, 1970

Dimbleby, Richard, "Churchill the Broadcaster," in *Churchill By His Contemporaries,* edited by Charles Eade, London: Hutchinson, 1953; New York: Simon and Schuster, 1954

Gilbert, Martin, *Churchill: A Life,* New York: Holt, and London: Heinemann, 1991

Humes, James C., *Churchill: Speaker of the Century,* New York: Stein and Day, 1980

Wenden, D.J., "Churchill, Radio, and Cinema," in *Churchill,* edited by Robert Blake and William Roger Louis, New York: Norton, 1993

Citizens Band Radio

Private Two-Way Radio Service

The Federal Communications Commission (FCC) defines "The Citizens Band Radio Service" (CB) as "a private, two-way, short-distance voice communications service for personal or business activities. The CB Radio Service may also be used for voice paging." In the United States and several other countries, a license is not required. In other countries, a license must be obtained or a fee paid, but no examination is required. Other countries with similar services include Canada, Australia, New Zealand, Japan, the United Kingdom, France, Germany, Italy, and Russia. While not broadcasting, CB radio is one of the most widely recognized uses of wireless technology.

Stations in the CB service are limited to a power output of 4 watts, with additional restrictions on antenna height. Most stations use AM, but a single sideband (SSB) is allowed with a peak effective power of 12 watts output. In some countries, FM is used. CB radios are used for both fixed and mobile communication over relatively short distances. In the United States, communication at ranges greater than 250 kilometers is prohibited by law.

Prior to the rapid growth of the CB service in the United States during the early 1970s, a Class D license was required, and a small fee had to be paid. A federal court decision made the fees illegal. Faced with no revenue to support the administration of issuing licenses, the FCC issued a blanket authorization allowing the CB service to operate within the constraints of 28 simple rules.

In the early days of the CB service, the FCC was divided into 18 regions. So, for example, call signs beginning with 18W indicated a station was somewhere in region 18 (region 18's central office was in Chicago, for example). The number of applications for CB licenses soon exhausted that call sign format. New calls signs were issued that followed an alphabetical-numerical mix and were issued sequentially. As the service became even more popular, the FCC stopped issuing official call signs completely, but encourages users of the service to identify using the call sign form of *K* followed by the operator's initials and zip code. This can lead to duplication of call signs, but the FCC doesn't seem to be concerned about that.

The Original UHF Service

Citizens Radio is a family of services, not just the one that rose to great heights of popularity during the 1970s. As originally conceived of and defined by the FCC—long before cellular mobile telephony—hobby-type conversations were explicitly forbidden in the service. Small businesses, many of which provided a service of some type—locksmiths, delivery services, and plumbers, for example—used the Citizens Radio Service to expedite and run their businesses efficiently. No one "owned" or was assigned to a particular channel. All users shared the channels available.

The FCC established the first Citizens Radio service in 1947. A group of frequencies in the 460- to 470-megahertz range was assigned. This service still exists and is properly called the General Mobile Radio Service (GMRS). A license is required. Recently, the microwave band of 31.0–31.3 gigahertz was also opened for GMRS operation.

Although UHF frequencies were widely used in military equipment during World War II, the near-microwave nature of the technology involved made production of commercially manufactured equipment expensive. Hobbyists were capable of building the required equipment but faced stiff certification requirements in order to legally use their creations.

Even so, the new service was appealing to some early pioneers. John M. Mulligan, who was employed as a radio engineer, became the first recorded CB licensee in 1958. Mulligan, who had ties to industry, built his own equipment. By year's end, 40 citizens in the United States held FCC licenses in the new service.

The same year, a single 3-watt experimental station heralded what would become Class D service, operating in the 27-megahertz range, often referred to by its wavelength of 11 meters. A total of 23 discrete channels were originally assigned to the Class D service. In 1977, to help relieve the congestion that had developed, The FCC increased the number of channels to 40. Several additional attempts to add more channels have not been successful.

For a few years in the 1950s, the 11-meter band was assigned to the Amateur Radio Service as compensation for other spectrum reassigned from the Amateur Service. Amateurs retained the nearby 10-meter frequency range.

By the 1960s, hobby magazines were filled with articles on building radio transceivers for the service as well as advertisements for kits that could be assembled. Heathkit, EICO, and Allied made many of the kits available. The EICO transceiver lives on today in reruns of television's *Andy Griffith Show*—it's one of the units seen in the Mayberry courthouse.

Gas Shortage Fuels Popularity

The 1972 gas shortage in the United States played a major role in making the CB service popular. Originally, truck drivers relied on their CB radios to help each other locate fuel. In a short time, the general public caught on and began to purchase CB radios as well.

Seven million units were sold in one year during the peak years of the 1970s. Even though CB has returned to relative obscurity in recent years, an estimated 3.5 million units are still sold each year. Today, the service's value as a traveler's aid and means of emergency communication has largely been supplanted by cellular telephones. CB sales in the trucking and freight industry, however, have never slowed.

Popular culture embraced CB radio, including the jargon developed by truck drivers. Even the FCC has its own "handle"—Uncle Charlie. The song "Convoy," written and performed by C.W. McCall, a marketing executive, helped to fuel sales of CB units. A movie by the same name; another titled simply *Citizens Band;* and the popular movie series *Smokey and the Bandit,* starring Burt Reynolds, Jackie Gleason, Sally Fields, and Jerry Reed, are further evidence of the impact of CB during the period.

Public Service

Long before the advent of 911 emergency telephone systems and cellular phones, CB radio provided an effective emergency communication system. There are numerous examples of how CB radio has been used for public service.

On 23 January 1962, long before CB's rise to popular icon status, Henry B. "Pete" Kreer, a CB enthusiast, recruited the Hallicrafters Company (a manufacturer of radio equipment) to sponsor the REACT program. Kreer got the idea during a Chicago snowstorm after using his CB radio to help a family stuck on an expressway with a very sick child and a disabled car. The REACT concept was simple. Initially, a team consisting of three members agreed to monitor for CB emergencies. Today, there are thousands of teams, with teams in nearly every community, to monitor for emergencies around the clock.

In 1964, with 800 teams in place, it became apparent that trying to monitor all 23 channels was a difficult if not impossible task. REACT called for the establishment of a voluntary National CB Emergency Channel. Thanks to their efforts, in 1967 the FCC designated channel 9 as the CB emergency channel, restricting communication on the channel to that associated with emergency operations.

Although 9 is the only channel on which use is legally restricted, other channels have become de facto standards. Throughout California and western states, channel 17 is the unofficial "trucker's" channel. Nationwide, channel 19 is used by truckers and other motorists, especially for speed-trap advisories. Channels from 30 to 40 are used for SSB communication. Although most CB activity uses AM modulation, SSB is authorized with the advantage that all radiated power is concentrated on the information being transmitted, rather than having a large portion consumed by a carrier wave.

Over the years, some have attempted to make the CB service into an unlicensed version of the Amateur Radio Service by modifying equipment to operate on frequencies not officially assigned, boosting power beyond what is allowed, and erecting antennae at variance with the law. CB operators and amateur radio enthusiasts or "hams" are often indistinguishable in the public mind. Operators in both services assist in natural disasters and provide communication for public events. Although both serve as valuable communication assets, they remain distinct services with different primary purposes.

JIM GRUBBS

See also Ham Radio

Further Reading

Brown, Robert M., *CB Radio Operator's Guide,* Blue Ridge Summit, Pennsylvania: G/L Tab Books, 1969; 2nd edition, by Brown and Paul Doreweiler, 1975

Buckwalter, Len, *ABC's of Citizens Band Radio,* Indianapolis, Indiana: Sams, 1962; 4th edition, 1976

Harwood, Don, *Everything You Always Wanted to Know about Citizens Band Radio,* New York: VTR, 1976

Herbert, Evan, *The Best Book on CB,* Tucson, Arizona: H.P. Books, 1976

Lieberman, Jethro Koller, and Neil S. Rhodes, *The Complete CB Handbook,* New York: Avon, 1976; as *The Complete 1980 CB Handbook,* New York: Avon, 1980

Long, Mark, Jeffrey Keating, and Albert Houston, *The Big Dummy's Guide to British C.B. Radio,* Summertown, Tennessee: Book, 1981; revised edition, as *The World of CB Radio,* edited by Long, Keating, and Bonnie Crystal, 1988

Perkowski, Robert L., and Lee Philip Stral, *The Joy of CB,* Matteson, Illinois: Greatlakes Living Press, 1976

CKAC

Montreal, Quebec French-Language Station

CKAC is the oldest French-language radio station in Montreal and remains one of the most popular in the region.

Origins

On 3 May 1922, a Montreal newspaper, *La Presse,* the largest French-language daily in North America, announced the creation of its radio station, CKAC. *La Presse* had just obtained a broadcasting license from the Canadian government, as had 22 other private corporations. It was the dawn of commercial broadcasting in Canada, and CKAC entered the business with a substantial financial investment and an exceptional founder and director, Jacques-Narcisse Cartier.

Cartier was doubly qualified for the post. He was both an expert technician who had traveled around the world and a journalist who had written for Montreal, British, and U.S. daily newspapers. He had been a longtime close collaborator with Guglielmo Marconi in his Canadian and U.S. East Coast enterprises. He had also developed a close friendship with David Sarnoff of the Radio Corporation of America (RCA), whom he had introduced to the business in 1909. Cartier had served in World War I as an expert in communications for the Canadian Armed Forces and the British Royal Air Force and had worked in New York at American Marconi.

From 1922 to 1937, CKAC and a few other private Quebec radio stations took on all of the responsibilities of a public radio service in the absence of state-run radio, which was not introduced until 1932. CKAC, because it was owned by a major newspaper, soon developed a very diverse content directed toward a variety of social groups and interests, unlike Montreal's English-language radio station, CFCF, which was created at the same time (as an extension of Marconi Wireless). CKAC broadcast news programs, reports on North American stock exchanges, and concerts as early as 1922–23.

Before the 1930s, CKAC was still in a developmental stage and was creating its own audience as well as its own market. In 1929 popular enthusiasm was so strong and stable that *La Presse* built a new transmitter just outside of Montreal in Saint-Hyacinthe, increasing the station's power from 5,000 to 50,000 watts and making it the most powerful station in eastern Canada. An extensive programming schedule then became possible. The news service was consolidated and broadened. CKAC's second director, J.-Arthur Dupont, created a remote studio at a stock brokerage firm and broadcast financial bulletins twice daily. CKAC also became an associate of the Columbia Broadcasting System (CBS) in the United States in order to have the opportunity of exchanging CKAC symphony orchestra concerts for concerts by U.S. orchestras broadcast by CBS. Owing to its affiliation with CBS, CKAC could also broadcast translated versions of news programs, which had been difficult previously because of limits set by the Canadian Press Agency with regard to radio transmission of its news briefs. In 1930 the transatlantic flight of the famous British airship R-100 was a major media event marked by live daily radio coverage from the airport. Even more significantly, when the R-100 made its return crossing to Great Britain, a single francophone journalist was on board, CKAC's J.-N. Cartier, who transmitted live daily reports from the dirigible. Thus CKAC was an enterprise bursting with potential when it was taken over by Louis-Philippe Lalonde in 1933. Lalonde would turn it into a financial success story for the next 30 years.

In 1938, CKAC became the news station of first choice for its program *Les nouvelles de chez-nous.* In 1939 the station created a complete news service.

During the same period, CKAC developed and implemented its educational mission, following the British and Western model. When the Quebec government decided, in 1929, to launch educational programming via radio, CKAC negotiated an agreement through which an important magazine-format program aimed at adult education was created and broadcast twice a week. The show was called *L'heure provinciale* and was produced by a renowned economist from the Université de Montréal, Édouard Montpetit. Its star was artist-musician-writer Henri Letondal. At about the same time, CKAC launched two other specialized educational programs, *L'heure universitaire* (The University Hour) and *L'heure catholique,* the latter being developed following the model of the U.S. program *The Catholic Hour.*

Development of Modern CKAC

In the early 1940s, cultural programs on CKAC were numerous and well structured, influenced by the arrival of Ferdinand Biondi who would be the pillar of the station's programming in music, literature, education, and news for 25 years. In the literary field, CKAC played an important role, complemented by state radio, in supporting the creation of a body of theatrical works and other dramatic forms. In 1938, the station launched *Théâtre de chez-nous,* which would present Quebec creations every week until 1947. During that same period, many authors whose work is now considered important were writing for the radio, including Robert Choquette, Gratien Gélinas, Jovette Bernier, Claude-Henri Grignon, and Ovila Légaré.

In the musical field CKAC evolved with changes in the collective culture of Quebec as well as with more generalized changes in musical genres. From orchestral pieces to evenings of dance music, from religious concerts to the development of the Quebec popular song, all musical expressions typical of Quebec were integrated into CKAC's schedule, with the objective of highlighting the Quebec identity as a counterweight to the cultural influence of the United States. Thus CKAC, in collaboration with *La Presse,* created a weekly chart of Québécois songs in 1959. In addition, the station continued to broadcast popular U.S. songs during appropriate time periods.

In the 1970s, CKAC continued its role as an essential public service by broadcasting live reports from the scenes of major events, by being on the air around the clock, and by maintaining an important team of journalists, among whom was the famous reporter Pierre Pascau.

CKAC remained the property of *La Presse* until 1969, when it was sold to Philippe de Gaspé Beaubien, owner of other stations and creator of the Télémédia network, a subsidiary of Power Corporation. This network, whose flagship was now CKAC, signed an agreement in 1988 that enabled it to rebroadcast programs from Radio France Internationale in Quebec.

CKAC has been broadcasting without interruption since 1922. It has always maintained a type of general programming that appeals to various strata of the population, with programs ranging from family listening to health care information to topics of benefit to underprivileged segments of society. It is still the most listened to station in the Montreal region as well as in Western Quebec. Audiences in Quebec still listen to radio regularly; their average is 23 hours per week per listener. And the station also continues to advance into the future, as demonstrated by its change to digital technology broadcasts in 1999.

PIERRE-C. PAGÉ

See also Canadian Radio and Multiculturalism

Further Reading

Vipond, Mary, *Listening In: The First Decade of Canadian Broadcasting, 1922–1932,* Montreal and Buffalo, New York: McGill-Queen's University Press, 1992

CKLW

Windsor, Ontario Station

With 50,000 watts of power, CKLW (AM) was a ratings winner in several U.S. markets during its heyday, broadcasting from the small town of Windsor, Ontario, across the Detroit River from Detroit, Michigan. Despite content restrictions by the Canadian Radio-Television and Telecommunications Commission (CRTC), CKLW programmed primarily a mixture of U.S. and British rock music to overpower even the major Detroit AM rocker, WKNR, in the late 1960s and early 1970s. Like many other AM outlets, however, CKLW later struggled for a new identity to hold its own with the increasingly popular FM band stations.

Origins

CKLW began its life as CKOK on 2 June 1932 with 5,000 watts of power at 540 kilohertz. CKOK was owned by Western Ontario Broadcasting Company, whose president and general manager until 1947 was M.G. Campbell. In 1933 the signal changed to 840 kilohertz and the call letters became CKLW. (The "LW" stands for London-Windsor, the Canadian towns it was licensed to serve.) The following year the signal moved to 1030 and finally came to rest at 800 kilohertz in 1941. In 1949 CKLW's power was raised to 50,000 watts. At night, because of the nature of the sky wave propagation of AM signals, CKLW could often be heard in 28 U.S. states and four Canadian provinces.

The station began as a network affiliate of the Columbia Broadcasting System (CBS). In 1935 CBS was dropped and an affiliation began with the Mutual Broadcasting System that lasted until 1960. A dual affiliation with the Canadian Broadcasting Corporation (CBC), a government-supported network, began in 1936 and lasted until 1950. Local programming during the 1930s and 1940s included *Vignettes of Melody* (light classical pieces from the station's staff) and the big band sound of *Make Believe Ballroom.* At one time in the 1950s different dayparts were dedicated to diverse genres of music, including classical, rock and roll, and country.

In the early 1960s program director John Gordon allowed his on-air talent considerable freedom in operating their shows. For instance, disc jockey Tom Shannon played and introduced many rhythm-and-blues records. The hodgepodge of music ended, however, with Radio-Keith-Orpheum (RKO) general ownership in 1963 and subsequent use of the Drake consultancy (Drake-Chenault was a major industry consulting firm). The use of consultants coincided with the hiring of Paul Drew as program director in 1967. Drew is credited for establishing the tight Top 40 format that led CKLW to dominate regional ratings. CKLW became known as the "Big 8" for its booming sound at 800 kilohertz. Popular disc jockeys Tom Shannon and Dave Shafer hosted neighborhood record hops at which local celebrities such as Stevie Wonder or Bob Seger occasionally would appear to lip-sync one of their current hit songs.

Changing Roles

In 1970 a 20 percent foreign ownership interest limit established by the CRTC in the late 1960s forced RKO to sell the station to Baton Broadcasting, owned by John Bassette, who already was a force in publishing and television in Canada. The popularity of the station continued to rise with its contemporary sound, however, and CKLW made its own playlists available at record stores as a clever marketing ploy to instill listener loyalty.

CKLW was well known for its news, but its reputation for newscasts fluctuated between critical acclaim and disapproval. In 1967 it won the Radio Television News Directors Association international award for coverage of the race riots in Detroit. Also in the late 1960s, the Drake consultancy established "20-20 News" at CKLW and other RKO stations. This resulted in a sensationalistic presentation of news 20 minutes before and after each hour—a format that had its critics. Newscaster Byron McGregor read everything intensely and quickly, emphasizing the tragic and sensational. Weekly cash prizes were given to listeners who called in stories and a $1,000 prize was awarded for the story of the year. In addition, Jo Jo Shutty-McGregor reported on traffic conditions from a helicopter in the busy Detroit area during the newscasts.

In the 1970s, in an attempt to appeal to an older demographic (ages 25 to 49 instead of its previous focus on the 12 to 35 age group), CKLW would drop the 20-20 format and deliver the news in a more conversational style. With FM taking away much of the younger audience from AM stations, program director Bill Gable created a type of adult contemporary/Top 40 hybrid. Dick Purtan from WXYZ-AM in Detroit joined CKLW in 1978 as their drive-time personality. He was famous for his comedy, including (with sidekick Tom Ryan) "put-on" calls to unsuspecting listeners. In 1979 Purtan won *Billboard*'s disc jockey of the year award.

In 1980 Gable left and disc jockey Pat Holiday became program director, taking a Top 40 approach. CKLW changed ownership a number of times throughout the 1980s as AM stations continued to seek their niche in the FM-dominated radio business. By the time a cable company in Toronto was ready to purchase CKLW, the station had changed to a big band format.

After 1970 CKLW and all Canadian outlets had to contend with the content regulations of the CRTC, which undertook to ensure a place for Canadian artists and culture in media increasingly dominated by the very popular U.S. music and influence. Initially the CRTC established that at least 30 percent of broadcast music must be by Canadians. CKLW music director Rosalie Trombley became known for her ability to pick both U.S. and Canadian hits in the 1970s (Bob Seger wrote about her in his song, "Rosalie"), as it was important for CKLW to be able to compete with the unrestricted U.S. stations across the river. By the late 1990s the percentage of required Canadian music was 35 percent with a 40 percent requirement looming on the horizon.

CHUM Limited, a Toronto-based company with multiple radio and television stations throughout Canada, bought CKLW in 1993. Its ownership of another AM radio station in the Windsor area made it a test case—the first duopoly in Canada. With CKLW's new ownership came a common format for AM stations: news/talk. Because of the Detroit market's importance to their survival, at the end of the 20th century the CHUM stations in Windsor (two FM and two AM, including CKLW) were required to play only a minimum of 20 percent Canadian music between 6:00 A.M. and 12:00 midnight.

Co-owned FM radio and television stations in Windsor have also used the call letters CKLW. CHUM owns the former CKLW-FM, now broadcasting as CIDR at 93.9 megahertz. CKLW television (Channel 9) was purchased by CBC in the 1970s. But the story of the original CKLW could be the story of the rise and fall of many AM stations, with its own subplot as a successful Canadian station in a major U.S. market.

LYNN SPANGLER

See also Canada; Canadian Radio and the Music Industry; Drake, Bill

Further Reading

CKLW website, <www.am800cklw.com>

Clandestine Radio

Illegal Radio Transmitters Supporting Propaganda or Political Movements

Clandestine radio stations are unlicensed transmitters that advocate civil war, revolution, or rebellion; they usually operate in secret. Because they desire to keep their actual identity unknown, they provide misleading information about their sponsorship, transmitter location, and goals. Sponsorship may be from one country that is funding broadcasting aimed at another country but concealing its activities. These stations represent political movements, guerrilla organizations, or one country at war with another that attempts to broadcast into the enemy nation under false pretenses. This type of radio is often confused with pirate stations, because neither uses a frequency assigned to them by a national authority or operates with authorization from the target country. International agreements consider both illegal. Clandestine broadcasts involve more than simple piracy, however: they are aimed at the overthrow of a government by revolutionary forces or by another state seeking to subvert an adversary without armed intervention. The Soviet Union considered the activities of United States–sponsored Radio Free Liberty and Radio Free Europe to be clandestine, especially during the 1956 Hungarian and 1968 Czechoslovak uprisings. Clandestine radio stations still operate in large numbers around the world, including stations in Asia, the Middle East, and Central America.

There are three categories of radio propaganda stations; two are considered clandestine (Soley, 1993). "Dark" (clandestine) and "white" are categories used by U.S. intelligence agencies. White radio stations are truthful about their location and purpose; examples are Radio Free Europe and Radio Martí (Voice of America–sponsored radio into Cuba). This category also includes foreign service and armed forces radio operations. Dark stations may be either "gray" or "black." Gray stations are operated by or attributed to local dissident groups but are often sponsored by foreign governments. The Free Voice of Iran is an example. Black radio stations are operated by an enemy of the target country or by guerrilla groups, and they keep their location and operation secret. Radio Venceremos, which operated for 12 years in El Salvador as the underground voice of the antigovernment guerrillas, is an example of a black clandestine station.

Operations before and during World War II

Clandestine radio operations began before World War II in Europe on stations such as Radio Free Spain during the Spanish Civil War and in such broadcasts as those of the Irish Republican Army in Northern Ireland, secret Communist broadcasts in Czechoslovakia, and broadcasts by anti-Nazi Germans. Other countries involved in clandestine radio operations before World War II included Italy, Great Britain, and France. Beginning in September 1934, Rudolf Wormys ran the first anti-Nazi clandestine stations in Germany.

The Soviet Union, France, Austria, and Spain all operated clandestine stations during World War II. The Dutch operated a clandestine station, known as "The Bugbomb," from England but claimed to be operating from within the Netherlands. The British began clandestine operations aimed at Germany in May 1940, which claimed untruthfully to be broadcasting from Germany. During World War II, Radio 1212 was operated by the psychological warfare branch of the U.S. 12th Army Group. It claimed to be operating in Germany but actually broadcast from Luxembourg. Truthful news and information were mixed with rumors and lies in an attempt to undermine German morale.

The Germans sponsored a gray clandestine operation named "New British Broadcasting," which began in February 1940. The station claimed to be operated by a British peace organization, but it was actually located in Germany and used scripts written by William Joyce ("Lord Haw Haw"). There is not much evidence that it was listened to by the British. The Germans also directed clandestine radio at the Soviet Union and at the United States through Radio Debunk, which claimed to be broadcasting from the Midwest but actually operated from Bremen, Germany. There is no evidence of any effect of these operations.

Operations during the Cold War, 1945–89

Cold War demands led to clandestine radio, sponsored by both the U.S. and Soviet governments, that aimed at overthrowing unfriendly governments. United States operations included the Voice of Liberation in Guatemala, which called for the democratically elected president to resign because he was trying to make Guatemala a communist dictatorship. In the summer of 1954, the president did resign and fled the country. The actual location of the transmitters was on Swan Island, off Honduras and in Nicaragua. The station was operated by U.S. Central Intelligence Agency (CIA) agents and by Guatemalans trained at a CIA base in Miami. Another clandestine CIA-operated station located on Swan Island was Radio Swan, which claimed that it was privately owned. In 1960 Radio Swan demanded the resignation of Fidel Castro.

Other operations during the Cold War included The Voice of the National United Front of Kampuchea, which claimed to be operating in Cambodia on behalf of the deposed leader, Prince Sihanouk, but which was actually broadcasting from Laos and was run by the CIA. Other stations operating from the same complex in Asia broadcast to North Vietnam, South Vietnam, Thailand, and Laos. In 1980 The Free Voice of Iran demanded that the Ayatollah Khomeini resign or face civil war. At first, it was reported that the station was operated by a general loyal to the deposed Shah. It was soon disclosed that the station's transmitters were in Egypt and were supported by the CIA.

Radio Quince de Septiembre (active from April 1981 to 1987) broadcast into Nicaragua and demanded the resignation of the newly installed Sandinista government. The station was run by contra guerillas based in Honduras and backed by money from the CIA. The funding was covert, because the U.S. Congress prohibited the use of funds to assist rebels. Covert radio operations were also conducted by other nations into Afghanistan, Angola, Cambodia, Ethiopia, Libya, Nicaragua, Poland, and Surinam.

During the Cold War, the Soviet Union also sponsored clandestine radio broadcasts. The Voice of the Turkish Communist Party and Our Radio (begun in 1958) broadcast from East Germany and Rumania to Turkey. The National Voice of Iran broadcast from the Soviet Union to Iran. In the 1970s, Radio Free Portugal transmitted from Romania and Hungary to Portugal against the latter's right-wing dictatorship. Other Soviet-sponsored clandestine radio stations included Ce Soir en France; Oggi in Italia; Radio España Independiente; and German Freedom Station 904, which broadcast from East to West Germany. None of these broadcasts ever disclosed that they transmitted from within the Soviet Union. Many, such as Greece's Voice of Truth (1975) and Radio Free Portugal (1974), went off the air during the 1970s. Radio España Independiente ended operations in 1977 after 36 years of operation and declared its mission accomplished. Those stations that survived into the 1980s were shut down by Soviet Premier Mikhail Gorbachev. The last clandestine broadcast from Soviet-backed stations to Europe ended in 1988. There have been no clandestine operations broadcast from Russia since the dissolution of the Soviet Union.

Operations after the Cold War

According to Soley (1993), the ending of the Cold War has led to *more,* not fewer, clandestine radio stations, as regional and ethnic conflicts have increased. Although Soviet-backed clandestine radio has largely disappeared, the United States continues to support clandestine stations in a number of countries, most notably those broadcasting into Panama and Iraq.

Clandestine radio operations are often conducted by indigenous groups fighting in a civil war. The Eritrean People's Liberation Front (EPLF) keeps people aware of progress in the war against Ethiopia using a clandestine radio service that began in May 1991. Radio has played a significant role in the movement's success. The broadcasts encouraged the voluntary participation of guerilla soldiers and enabled the movement to fight successfully against Ethiopia. The EPLF used its own broadcasting station, called the Voice of the Broad Masses of Eritrea, to declare their ideas of self-determination. Two Marxist clandestine stations are The Radio of the Sudanese People's Liberation Army (Radio SPLA), broadcasting against the Islamic fundamentalist regime in Khartoum, and the Voice of the Sarbedaran, an antigovernment station broadcasting to Iran on behalf of the Union of Iranian Communists.

In 1980 Radio Venceremos (literally, "We shall overcome") was begun by the Farabundo Martí National Liberation Front (FMLN) guerrillas in the mountains of El Salvador. During the civil war, broadcasts included war reports, messages for fighters, and political propaganda. The station called itself "the signal of freedom." With the acceptance of a peace plan in 1992, Radio Venceremos began legal operations and now reports everyday events and social and economic issues affecting the poor. One of the station's biggest problems is finding commercial support. During the war, the FMLN guerillas kidnapped many businessmen for ransom, and now they need the business community's support for advertising.

Other clandestine operations broadcast to Angola, Mozambique, Rwanda, and Afghanistan. Although these stations have roots in the Cold War, their continued operation may indicate ethnic rather than ideological war. For example, in Haiti, pro-Aristide stations continue to operate; in Algeria, La Radio de la Fidelitié began after the government canceled elections that fundamentalists were predicted to win; and in Burma, a clandestine radio began broadcasting after the military leadership refused to hold elections.

Despite the end of the Cold War, clandestine radio stations will continue operations in societies where there is a lack of openness, whether the government is military, socialist, or democratic. Their use will probably continue, because authoritarian regimes, political conflict, and civil war continue to threaten world peace. Radio is an inexpensive and technologically easy way to spread propaganda. Radio can reach everyone and is an appropriate medium for expression of suppressed views or to foment revolution, liberation, and rebellion.

MARY E. BEADLE

See also Propaganda; Shortwave Radio

Further Reading

Berg, Jerome, *On the Short Waves, 1923–45: Broadcast Listening in the Pioneer Days of Radio,* Jefferson, North Carolina: McFarland, 1999

Downer, Monica, "Clandestine Radio in African Revolutionary Movements: A Study of the Eritrean Struggle for Self-Determination," *Journal of Communication Inquiry* 17, no. 2 (1993)

Fortner, Robert S., *International Communication,* Belmont, California: Wadsworth, 1993

López Vigil, José Ignacio, *Mil y una historias de radio venceremos,* San Salvador, El Salvador: UCA, 1991; as *Rebel Radio: The Story of El Salvador's Radio Venceremos,* translated by Mark Fried, Willimantic, Connecticut: Curbstone Press, 1994

Nelson, Michael, *War of the Black Heavens: The Battles of Western Broadcasting in the Cold War,* Syracuse, New York: Syracuse University Press, and London: Brassey's, 1997

Soley, Lawrence C., "Radio: Clandestine Broadcasting, 1948–1967," *Journal of Communication* 32 (1982)

Soley, Lawrence C., "Clandestine Radio and the End of the Cold War," *Media Studies Journal* 7, no. 3 (1993)

Soley, Lawrence C., and John Spicer Nichols, *Clandestine Radio Broadcasting: A Study of Revolutionary and Counterrevolutionary Electronic Communication,* New York: Praeger, 1987

Street, Nancy Lynch, and Marilyn Matelski, *Messages from the Underground: Transnational Radio in Resistance and Solidarity,* Westport, Connecticut: Praeger, 1997

World Radio TV Handbook (annual) (provides a listing of clandestine radio frequencies)

Clark, Dick 1929–

U.S. Disc Jockey

During the middle and late 1950s, Dick Clark was one of the major figures pioneering rock music. An early career in radio enabled Clark to attain his position as disc jockey and host of television's *American Bandstand,* which propelled him to national fame and influence. The ever-boyish Clark—surely more than almost any other figure—symbolized the inescapable intersection between popular music styles, radio airplay, and television exposure used to create a new style of music and its new stars.

If rock music seemed to combine the big beat of rhythm and blues with the hick aspects of hillbilly music (indeed, Elvis Presley was originally known as the "Hillbilly Cat"), it was Clark who cleaned rock up, urbanizing it with clean-cut male talents who always appeared in suits and who kept themselves far from the scandal pages. In the 1950s, Clark signed talent to record labels that he owned and then promoted his stars on *American Bandstand*—Frankie Avalon, Fabian, and Bobby Rydell. He carefully watched over the careers of "his boys." Indeed, Clark marketed the clean-cut image throughout his career, as a glance at his published books reveals; he continually sold good behavior as well as music with a beat.

Through the late 1950s into the 1960s, Clark was a hot property. *American Bandstand* seemed to define what good young teens should listen to as acceptable rock music, certainly before the British invasion by the Beatles. His half-hour-long *Saturday Night Beechnut Show* defined what "good" teenaged baby boomers should be listening to—from Annette Funicello to Connie Francis, from Fabian to Frankie Avalon, all lip-syncing their records.

Thus, within the larger context of radio broadcasting, Clark boosted Top 40 as a radio format exclusively through his work on television. Radio broadcasting also proved to be Clark's means of becoming a TV star and later a producer. His radio work during the expanding postwar years in small-town and small-city radio in upstate New York prepared him to move to a major market, Philadelphia. But once the American Broadcasting Companies (ABC) picked up *American Bandstand* in 1957, it was television that served as Clark's major vehicle, with radio falling into the category of "other interests."

Clark's other significant link to radio and music recording came as a result of his ownership of several minor music labels in the late 1950s, whose artists he shamelessly plugged on *American Bandstand.* He was then accused of paying radio stations to play his star's records, and in 1959 the U.S. government began to scrutinize this high-profile figure for payola, the practice of bribing disc jockeys and later radio program directors to play certain songs. Pressured by Leonard Goldenson, head of ABC-TV, to choose between *American Bandstand* and his music recording business, Clark went the way of television after he was forced to humiliate himself by testifying in front of a Congressional committee. He moved to Hollywood and became a mainstay on television and behind its cameras through the last third of the 20th century.

DOUGLAS GOMERY

Dick Clark
Courtesy of Dick Clark Productions

See also Classic Rock Format; Contemporary Hit Radio/Top 40 Format; Disk Jockeys; Oldies; Payola; Rock and Roll Format

Dick Clark. Born Richard Wagstaff Clark, 30 November 1929, in Mt. Vernon, New York. Graduated from Syracuse University, 1951. Began radio career while a student at Syracuse on campus radio station as well as on WRUN, Utica, and WOLF, Syracuse; television newscaster at WKTV in Utica, New York, 1950–52; disc jockey, WFIL (AM), Philadelphia, 1952; moved to television for *American Bandstand,* 1956; after investigation for payola by U.S. Congress, sold interests in music publishing and recording companies and formed television production company, Dick Clark Productions, 1960; continued producing and hosting nostalgia weekly syndicated shows such as *Rock, Roll, and Remember.* Inducted into Rock and Roll Hall of Fame, 1993.

Radio Series

Dick Clark's Caravan of Music
Dick Clark's Music Machine
Dick Clark's National Music Survey
Dick Clark's Music Survey
Dick Clark's Rock, Roll, and Remember

Television Series

The Dick Clark Saturday Night Beechnut Show, 1958–60; *Dick Clark's World of Talent,* 1959; *Missing Links,* host, 1964; *In Concert,* executive producer, 1973–75; *$10,000 Pyramid; $20,000 Pyramid; $25,000 Pyramid; $50,000 Pyramid; $100,000 Pyramid,* host, 1973–89; *TV's Bloopers and Practical Jokes,* executive producer, cohost, 1984–86, 1988; *Puttin' on the Hits,* executive producer, 1985–89; *Live! Dick Clark Presents,* host, executive producer, 1988; *The Challengers,* host, executive producer, 1990–91; *Greed,* producer, 1999– ; *Winning Lines,* host, 2000–

Selected Publications

Your Happiest Years, 1959
To Goof or Not to Goof, 1963
Rock, Roll, and Remember (with Richard Robinson), 1976
Dick Clark's Program for Success in Your Business and Personal Life, 1980
Looking Great, Staying Young (with Bill Libby), 1980
Dick Clark's The First 25 Years of Rock and Roll (with Michael Uslan), 1981
The History of American Bandstand: It's Got a Great Beat and You Can Dance to It (with Michael Shore), 1985

Further Reading

Jackson, John A., *Big Beat Heat,* New York: Schirmer Books and Macmillan, 1991
Jackson, John A., *American Bandstand: Dick Clark and the Making of a Rock 'n' Roll Empire,* New York: Oxford University Press, 1997

Classical Music Format

Once a radio program staple, classical music has in recent years been relegated to a relative handful of stations, most of them noncommercial FM outlets. Sometimes called "fine-arts" programming, the format combines the playing of classical music with interviews, cultural programs, and news commentary.

Origins

At the inception of radio, classical music was omnipresent. Even before there were formal programs, stations would broadcast singers or orchestras performing familiar classics. Most early studios had a piano, and a pianist on call, useful for last-minute fill-in performances when a scheduled program for some reason could not be broadcast. Although popular music was also played, the classics were relied upon to fill airtime.

Many stations in larger markets retained full orchestras and featured them regularly. Somewhat ironically in light of the situation decades later—when broadcasts of the classics have become increasingly rare—classical music was the first musical style to achieve large-scale exposure on radio. Many people were exposed to classical music for the first time on the radio, because few people could afford live concert hall performances in the few cities where they were available. The provision of such music over the air was also a way of gaining radio respectability among upper-class listeners.

Although a rather extreme example, when station KYW first broadcast in mid-1921 in Chicago, it offered almost nothing but opera from the Chicago Civic Opera. Unfortunately, KYW soon discovered that the audience that wanted to hear opera all the time was relatively small.

Network Classical Music

Radio's golden age was certainly golden for classical music listeners as the Columbia Broadcasting System (CBS) and the National Broadcasting Company (NBC) vied to present prestigious orchestras in live performances from the late 1920s into the 1950s. A typical offering on a Sunday, such programs were also heard on other evenings, and virtually all of the broadcasts were live. Among the better known and longer-running program series were the following:

Voice of Firestone (1927–54, NBC; 1954–57, American Broadcasting Companies [ABC]) offered a mixture of popular and more serious music, becoming more focused on the classics after 1932. The program's theme music was composed by the sponsor's wife. The radio audience declined after the broadcasts were simulcast on television beginning in 1949.

Cities Service Concerts (1927–56, NBC) offered an hour-long program until 1940, then a half hour. The title varied, but for a seven-year period (1930–37) its top star did not. Young Jessica Dragonette (1910–80) became hugely popular with her renditions of classical solo works and developed a devoted following.

New York Philharmonic (1927–53, CBS) was the major offering of the second network. With the Philharmonic conducted by the noted Arturo Toscanini until 1936, the broadcasts from Carnegie Hall were useful exposure for the orchestra, helping to sell its growing number of recordings.

Philadelphia Symphony Orchestra (1931–57, largely on CBS) helped to popularize the lush "Philadelphia sound" developed first by Leopold Stokowski and after 1938 by longtime conductor Eugene Ormandy.

Music Appreciation Hour (1928–42, NBC Blue) featured conductor Walter Damrosch (1862–1950) providing and explaining classical music to a youthful audience. Broadcast on Fridays for an hour (the program ended when Damrosch refused to cut it to a half hour), the program was widely used in schools across the country.

Sinfonietta (1935–45, Mutual) was one of the few forays into classical music programming by the cooperative network. The orchestra was conducted by Alfred Wallenstein, and the program filled various half-hour time slots.

NBC Symphony Orchestra (1937–54, NBC) marked the epitome of network classical music presentations. Radio Corporation of America's (RCA) David Sarnoff paid dearly to coax Arturo Toscanini (1867–1957) out of retirement by letting him establish his own orchestra and paying a handsome salary. Also featuring conductors Artur Rodzinski and Leopold Stokowski, some recordings of these broadcasts are still available. Regular broadcasts ceased when Toscanini finally retired in April 1954.

The *Bell Telephone Hour* (1940–58, NBC) was actually a half hour long and melded light classics and sometimes popular orchestral music selected by conductor Donald Vorhees. Sponsored by American Telephone and Telegraph (AT&T), the program helped promote the dignified view the company had of itself and its role in society.

The Longines Symphonette (1943–49, Mutual; 1949–57, CBS) was unusual among these programs in featuring prerecorded programs.

Classics on Local Stations

For most of radio, however, classical music from the 1930s into the 1950s was at best an occasional offering, usually from a network broadcast. A few university-owned stations (e.g., the University of Wisconsin's WHA in Madison) provided classical music among other educational and cultural programs. But, otherwise, classical music largely disappeared from AM stations.

The development of FM radio from the 1940s into the 1950s, however, offered a new lease on life for classical music. By offering the classics (usually with recordings plus some live performances), FM outlets could differentiate themselves from the more common AM stations. Well into the 1960s, to both broadcasters and listeners, classical music meant FM radio—and vice versa. FM stations often published program guides to help listeners (and to sell advertising, though few made money), and they thrived on audiences developed by the high-fidelity craze that began in the 1950s. Many offered musical performances uninterrupted by announcements or commercials (again distinguishing themselves from AM) for those who wished to tape broadcasts off the air.

Major cities soon enjoyed one or more classical music FM stations, including New York's WQXR (1939); Chicago's WEFM (1940), initially owned by Zenith with call letters featuring the chairman's initials), and WFMT (1951), which by 1958 became the first FM outlet reflected in local market audience ratings; WWDC (1947) and WGMS (1948) in Washington, D.C., WFLN (1949) in Philadelphia; WCRB (1954) just outside of Boston; and WFMR (1956) in Milwaukee, to name only a few examples.

With the appearance of the first AM-FM nonduplication rules from the Federal Communications Commission in the mid-1960s, however, FM's days as a bastion of classical music were numbered. As FM frequencies became more valuable, thanks to the larger audiences (and thus greater advertising income), they employed a wider variety of more widely popular musical formats. Small-audience classical music stations often either were sold or changed their formats to something more lucrative. Classics once again became relatively elusive.

By the turn of the century, classical music programs appealed to only a small portion of the audience and were predominant on only a few dozen commercial and several hundred public radio stations, all catering to audiences with more education and higher income than the norm (and to the advertisers that want to reach them). A typical classical music station in the early 21st century provided not only music but also a blend of other cultural features designed to appeal to its audience.

For nearly three decades, Robert Lurtsema (1931–2000) hosted *Morning Pro Musica* each morning on Boston's WHDH, carried on public radio stations in New York and New England (and later by satellite to stations across the country). The program was broadcast five hours each day of the week (until 1993, when it shifted to weekends only) with musical selections introduced in Lurtsema's widely recognized slow and easy delivery. Programs often focused on a specific composer or theme, but music predominated. Along with Karl Haas' *Adventures in Good Music,* Lurtsema's program made classics readily available to sizeable audiences.

CHRISTOPHER H. STERLING

See also Adventures in Good Music; BBC Orchestras; Desert Island Disks; FM Radio; Metropolitan Opera; Promenade Concerts; WQXR

Further Reading

Chase, Gilbert, editor, *Music in Radio Broadcasting,* New York: McGraw-Hill, 1946

DeLong, Thomas A., "The Maestros," in *The Mighty Music Box: The Golden Age of Musical Radio,* by DeLong, Los Angeles: Amber Crest Books, 1980

Dunning, John, "Concert Broadcasts," in *On the Air: The Encyclopedia of Old-Time Radio,* by Dunning, New York: Oxford University Press, 1998

LaPrade, Ernest, *Broadcasting Music,* New York: Rinehart, 1947

Morton, David, "High Culture, High Fidelity, and the Making of Recordings in the American Record Industry," in *Off the Record: The Technology and Culture of Sound Recording in America,* by Morton, New Brunswick, New Jersey: Rutgers University Press, 2000

Stokowski, Leopold, "New Vistas in Radio," *Atlantic* Monthly (January 1935)

"Toscanini on the Air," *Fortune* (January 1938)

Classic Rock Format

Classic Rock is a music radio format that focuses on harder rock music from the late 1960s through the 1980s. It is a derivation of the Album-Oriented Rock (AOR) format that incorporates rock music from the same period along with current selections.

Classic Rock evolved from several earlier radio programming formulas that were attempts to provide alternatives to tightly formatted Top 40 radio stations of the 1950s and 1960s. One of the first was freeform radio, in which disk jockeys were given total control over the music played. Although music genres varied, freeform stations tended to feature music selections generally not heard on Top 40 stations. Freeform radio was also referred to as underground, progressive, or alternative radio.

The Progressive radio format emerged from freeform radio, but it had some structure designed by a music programmer. The Progressive format incorporated a rotation system for music categories. Disk jockeys followed the category rotation system but still had considerable latitude in the selection of specific songs. Progressive served as a bridge between freeform radio and the more mainstream AOR format that developed in the 1970s.

Like its predecessors, AOR was a rebellion against Top 40 in that it avoided chart hits in favor of longer album cuts by popular artists. It brought with it *music sweeps*—uninterrupted series of songs—and a less-structured, more laid-back announcer delivery style. Initially, AOR appealed to a young-adult, primarily male audience. During the 1970s AOR grew in popularity as Top 40 waned.

During the 1980s, however, AOR began to lose younger listeners as Top 40 regained popularity with the emergence of MTV. Younger listeners could no longer relate as well to standard AOR artists such as the Doors, the Grateful Dead, and the Moody Blues. As a result, a splinter of AOR, called classic rock, emerged to appeal to adult listeners. The format features hits of the past but with a harder musical edge than other popular music formats for adults. Typical artists in the Classic Rock format include Bob Seger, ZZ Top, Bruce Springsteen, and the Rolling Stones.

The Classic Rock format first appeared in 1983 in Dallas on WFAA-FM. The format often is classified as part of the vintage rock category that also include the Oldies format. Classic Rock is different from Oldies in that it features rock hits with a harder edge and generally does not include music from the 1950s and early 1960s. Arbitron, a radio audience research firm, includes Classic Rock as one of the 15 formats it recognizes nationally and defines Classic Rock's content as "the same universe of music as Album Rock, but without much current rock."

Some radio programming analysts have predicted that the Classic Rock format will disappear within a decade. They see adult listeners shifting to modern rock or alternative formats in the future, but as the 1990s drew to a close Classic Rock was gaining in popularity among "baby boomer" listeners. According to Arbitron, the format's share of the national radio audience has continued to increase over the past few years, with a peak of 4.9 percent in 2002.

FREDERIC A. LEIGH

See also Album-Oriented Rock Format; Oldies Format; Progressive Rock Format; Rock and Roll Format

Further Reading

Arbitron, *Radio Today,* New York: Arbitron, 1995
Keith, Michael C., *Voices in the Purple Haze: Underground Radio and the Sixites,* Westport, Connecticut: Praeger, 1997
Ladd, Jim, *Radio Waves: Life and Revolution on the FM Dial,* New York: St. Martin's Press, 1991
Neer, Richard, *FM: The Rise and Fall of Rock Radio,* New York: Villard, 2001
Stuessy, Joe, *Rock and Roll: Its History and Stylistic Development,* Englewood Cliffs, New Jersey: Prentice Hall, 1990; 3rd edition, by Stuessy and Scott David Lipscomb, Upper Saddle River, New Jersey: Prentice Hall, 1999

Clear Channel Communications Inc.

Texas-based Clear Channel Communications Inc. had by the late 1990s become, through mergers and acquisitions, the largest single owner of radio stations in the world with nearly 1,000.

The company began with the 1972 purchase of a struggling San Antonio, Texas, FM station for $130,000 by L. Lowry Mays and B.J. "Red" McCombs. Station KEEZ (later KAJA) operated for a number of years under San Antonio Broadcasting, the original company name. Three years later, Mays purchased WOAI-AM, a pioneering operation that had first gone on the air in 1922. (McCombs retains a 2.5% ownership in Clear Channel Communications. As of early 2003 he owned a number of car dealerships as well as the NFL Minnesota Vikings.)

Clear Channel Communications was incorporated in 1974 and grew quite slowly at first, becoming a publicly traded firm in 1984 and owning a dozen stations in several markets a year later. Clear Channel used a simple formula: buy low-priced stations, build up their revenues while controlling costs, and operate conservatively. By 1990, the company had expanded into television station ownership as well, but it was still just one of a host of group owners of broadcast stations.

Federal Communications Commission (FCC) deregulation of limits on radio station ownership after 1993 fueled the first burst of Clear Channel station purchases, but the 1996 Telecommunications Act provided the key for the huge expansion of Clear Channel. By June of that year, it became the first company to own more than 100 stations. Many of its takeovers involved one or two radio stations at a time; bigger multi-station deals would come late in the decade. By 1997, CEO Mays was about halfway up *Forbes* magazine's list of the 400 richest Americans.

The October 1998 takeover of Jacor Communications (then the second-largest owner of radio stations with 230 outlets) in a $2.8 billion stock deal moved Clear Channel toward the front of the radio owner pack. The transaction made Clear Channel the country's second-largest owner in number of stations and the third in total radio revenues. By 1999, radio provided 53 percent of total company revenue; billboards brought in 47 percent. Clear Channel also held equity interests in about 240 foreign radio stations, including outlets in Australia, Denmark, Mexico, and New Zealand.

In an agreement announced in October 1999, Clear Channel paid $23.5 billion to take over AMFM Inc. which owned 444 stations—320 FM and 124 AM stations. Combined with what Clear Channel already owned, this deal made it the largest group owner of stations in terms of numbers of outlets and revenues. But the deal also required the sale of about 110 stations collectively worth $4.3 billion to meet government limits on station ownership in individual markets. Early in 2000 the first 88 stations were sold to 17 companies, seven of which were minority-controlled. Early in 2000, Clear Channel Communications also purchased SFX Entertainment (a concert and sports producer and owner of a number of arenas) for $3.3 billion. The 19 March 2000 issue of *The New York Times* noted that "the company will have operations in 32 countries, [including] . . . 550,000 billboards and 110 entertainment venues. It will also own all or part of 1,100 radio stations, though some are being sold to satisfy regulators."

Clear Channel's brand of advertising synergy—selling combined advertising packages across radio, television, and billboards, especially in markets where it owns stations and billboards (virtually all of the cities where it also owns theaters and arenas)—has clearly played a major factor in its success. Company business strategy, as stated in its 1999 10-K filing with the Securities and Exchange Commission, makes clear the value of growth through acquisition and ownership of multiple stations in the same market:

> We believe that clustering broadcasting assets together in markets leads to substantial operating advantages. We attempt to cluster radio stations in each of our principal markets because we believe that we can offer advertisers more attractive packages of advertising options if we control a larger share of the total advertising inventory in a particular market. We also believe that by clustering we can operate our stations with more highly skilled local management teams and eliminate duplicative operating and overhead expenses. We believe that owning multiple broadcasting stations in a market allows us to provide a more diverse programming selection for our listeners.

While the company's very size (1,214 stations, of which 485 are in the 100 largest markets, with a total weekly audience of nearly 105 million as of mid-2002, generating $8.4 billion in annual revenue) attracted criticism and not a little carping from industry competitors, Clear Channel attracted further attention early in 2003 when it co-sponsored more than a dozen political rallies supporting the American incursion into Iraq. Stories criticizing these "Rally for America!" events first appeared on the internet and soon broke into general press reports. They argued the company was supporting the Bush administration just as the FCC was considering changes in radio ownership rules. They also felt Clear Channel stations were programming in similar fashion—a criticism the company firmly denied.

Alan B. Albarran

See also Hicks, Tom; Ownership, Mergers, and Acquisitions

Further Reading

"The Biggest Li'l Broadcaster in Texas," *Fortune* (19 August 1996)
Bryce, Robert, "What? A Quiet Texas Billionaire?" *New York Times* (19 March 2000)
Clear Channel Communications website, <www.clearchannel.com>
Elliott, Stuart, "Clear Channel in $3 Billion Deal to Acquire SFX Entertainment," *New York Times* (1 March 2000)
Forest, Stephanie, and Richard Siklos, "The Biggest Media Mogul You Never Heard Of," *Business Week* (18 October 1999)
Hagan, Joe, "Radio's 1,200-Station Gorilla," *Inside* (3 April 2001)
Rathbun, Elizabeth A., "Texas Size: Clear Channel Builds a Broadcast Dynasty," *Broadcasting and Cable* (7 October 1996)
Rathbun, Elizabeth A., "Clear Channel's Under-Fire Sale, "*Broadcasting and Cable* (13 March 2000)
Viles, Peter, and Geoffrey Foisie, "Clear Channel: Sunbelt Success Story," *Broadcasting and Cable* (23 July 1993)

Clear Channel Stations

Powerful Major-Market Radio Stations

Clear channels refers to a class of high-powered AM radio stations that from 1928 into the early 1980s operated with no (or only one or two) interfering stations broadcasting on their channels during evening hours. In other words, their operating channel was "cleared" of other outlets. The role and status of such stations was a matter of major radio industry controversy for decades. (Clear channel stations should not be confused with the Clear Channel Communications company, the owner of a large number of radio stations in the early 21st century.)

Origins

With its General Order No. 40 issued in August 1928, the Federal Radio Commission (FRC) reserved 40 of the then 96 AM channels to ensure listening options in rural areas. As the FRC put it: "On these 40 channels only one station will be permitted to operate any time during night hours, thus insuring clear reception of the station's program up to the extreme limit of its service range." First referred by the FRC as cleared or "clear channels" in a November 1931 order, these outlets came to represent the pinnacle of radio broadcasting.

All clear channels were located between 640 and 1200 kHz, and most were on or near the East and West coasts, with several in major Midwestern cities (four were located in Chicago). They were given the privilege of operating without other stations on the same channel in order to provide night-time service to so-called "white areas"—more than half the U.S. land mass—that could not receive a ground-wave primary local radio signal and thus depended on night-time sky wave transmission to receive distant higher-powered outlets.

From the beginning, these elite stations were the focus of controversy. On the one hand they provided service in rural areas that could often receive no other radio signal at night. On the other, they strongly resisted the formation of new local stations to serve such areas. Yet in an indication of things to come, the first clear station was "broken" just six months after the 1928 order when the FRC allowed stations WGY and KGO (both owned by General Electric) to share the same frequency, partially because they were on nearly opposite coasts. Two other clear channels were assigned for use by Canadian stations. With other decisions, only 32 stations remained truly "clear" by the time of the Federal Communication Commission's (FCC) formation in 1934.

At the same time, clear channel stations became identified with requests for higher or "super" power above the 50 kw limit. Cincinnati clear-channel station WLW (700 kHz) was given an experimental authorization in 1934 to use 500 kw of power—ten times that of any other station. Using its experimental W8XO, at first only in evening hours, then full-time, the outlet quickly became first choice of listeners in 13 states and second in six others. Under pressure from competitors in the U.S. and from Canada (unhappy with the station's reach into that country), WLW's daytime use of super power ended in March 1939, although occasional night-time use continued until late 1942.

Soon other clear channels petitioned the FCC for "super power," arguing that they could thus better serve rural areas.

Clear channel station managers formed the Clear Channel Group (CCG) in 1934 to put forth the views of independent (not network-owned) stations. The CCG lobbied hard for the use of superpower, as well as to protect existing clear channels.

The CCG testified at extensive FCC allocation hearings from 1936 to 1938. The commission's resulting engineering report generally supported the retention of clear channels for their evening rural service. In the late 1930s negotiations that led to the North American Regional Broadcasting Agreement (NARBA), the U.S. was given priority use of 32 of a total of 59 clear channels, while Cuba received one, and Mexico and Canada each got six. By this time clear channels were dubbed class I-A, while clears that were duplicated by at least one station at night on the same channel were dubbed I-B outlets.

Needing a still stronger lobbying voice, the CCG was largely converted to the Clear Channel Broadcasting Service (CCBS) beginning in 1941. With a larger budget and full-time staff, it became more focused on lobbying and public relations efforts as well as representation of clear channel station owner views before the FCC and Congress. But it faced a growing split between network-owned I-A stations and independent I-A outlets. The CCBS also focused on building alliances with farmer groups to create a stronger lobbying front in favor of retaining the clear channel classification.

Breaking Down the Clears

With the end of World War II, the FCC was able to turn its attention to pressing domestic matters, among them what to do about the simmering clear channel controversy. In early 1945 the commission initiated Docket 6741 to focus discussion on the policy tradeoffs (a few national vs. many local services) raised by the continued operation of clear channel stations. This proceeding became the primary arena for continued debate on the future of the I-A stations, including their service role and how much power they would be allowed to use. Some 40 days of hearings were held from January 1946 to January 1948, allowing a host of witnesses on all sides of the controversy to be heard. Many critiqued the clear channel stations for their relative lack of farm and agricultural programming (located in major cities, the clear channel stations programmed chiefly for urban audiences). The FCC briefly considered plans to combine the use of more and smaller AM and FM stations to meet the "white area" unserved audience problem. The CCBS proposed that 20 stations (not surprisingly all CCBS members) be allowed to use up to 750 kw of power. Throughout the hearings, the radio networks argued for retention of the status quo, which had served their interests well.

When the hearings adjourned, the FCC considered two plans that were variations on that proposed by the CCBS plus a third that would break down the clears to allow many other stations on the same channels. Senate hearings on these questions largely repeated the same arguments but also delayed any definitive FCC decision. At that point the commission turned to television allocations and essentially ignored clear channel issues for a decade. Only at the urging of many affected parties did the FCC reopen the Docket in 1958. At this point virtually all the clears shared their frequencies in daytime hours; this final debate concerned only their retention of cleared status in evening hours when radio signals traveled much further.

Finally in mid-1961 came resolution of Docket 6741 with the FCC decision that 11 of the 25 Class I-A stations would now be required to share their frequency with at least one unlimited time regional (class II) station. This "breaking" of the clears brought an attempt by the CCBS stations to roll back the commission action in Congress, and long hearings on several bills to do just that resulted. FCC and other radio station opposition killed those potential laws, and, upheld on court appeal, the FCC moved to break down the clear channels over the next several years. In the meantime, the commission considered what to do with the remaining dozen clear channel outlets, including continuing proposals to allow them to operate with super power up to 750 kw. Faced with a continued lack of progress on the issue and continuation of the status quo, in 1968 CCBS closed down its Washington office.

Two decades later the FCC voted to end the long-lasting controversy by allowing up to 125 unlimited time AM stations to use either the remaining clear channels or adjacent channels during evening hours, while protecting a 750-mile coverage radius for the original clear channel outlets. Attempts to roll back this final assault also came to nothing. Clear channel stations, although often still called that and remaining among the most powerful AM outlets, no longer operate as the sole occupants of their frequencies in evening hours. Service to rural "white" areas is now largely provided by a host of regional and local stations.

CHRISTOPHER H. STERLING

See also, in addition to individual stations mentioned, Farm/Agricultural Radio; Federal Communications Commission; Federal Radio Commission; Frequency Allocation; Licensing; North American Regional Broadcasting Agreement

Further Reading

Brown, Eric R., "Nighttime Radio for the Nation: A History of the Clear-Channel Proceeding, 1945–1972," Ph.D. diss., Ohio University, 1975

Caldwell, Louis G., "The Case for Clear Channels and High Power," Statement on Behalf of Certain Clear Channel Station Licensees before the Federal Communications Commission, 6 September 1938

"The FCC and the Clears," *Broadcast Engineering* (November 1961)

"FCC Cuts Back the Clears," *Broadcasting* (2 June 1980)

Foust, James C., *Big Voices of the Air: The Battle over Clear Channel Radio,* Ames: Iowa State University Press, 2000
Rogers, George Harry, "The History of the Clear Channel and Super Power Controversy in the Management of the Standard Broadcast Allocation Plan," Ph.D. diss., University of Utah, 1972
Smulyan, Jeffrey, "Power to Some People: The FCC's Clear Channel Allocation Policy," *Southern California Law Review* 44 (1971)
U.S. House of Representatives, Committee on Interstate and Foreign Commerce, *Clear Channel Broadcasting Stations: Hearings,* 87th Congress, 2nd Session, 1962

Cold War Radio

For four decades during the Cold War, international shortwave channels were filled with dueling radio broadcasts from "the East" and "the West." The broadcasts both reflected and projected prevailing government viewpoints in the multi-stage conflict.

Origins

Most historians date the beginnings of the Cold War to early 1946. It is not easy to say exactly when it began, or even to obtain clear agreement on its causes, but factors contributing to its initiation included the following: disagreements between the Soviet Union and its wartime Western partners over the dismemberment of Germany to break its political and military power; the disposition of Polish borders and the creation of a new Polish state; and a variety of other Soviet moves in establishing new governments in Eastern and Central Europe, where it was the occupying force at the conclusion of World War II. In February 1946 Stalin openly attacked the Western powers in a strident speech, and in March 1946 British Prime Minister Winston Churchill declared in Fulton, Missouri, that an "Iron Curtain" had descended to divide Eastern and Western Europe. In June 1948 Soviet troops set up a blockade around the western sectors of Berlin; this act led to the Berlin airlift to keep the few thousand American, British, and French troops there supplied and to provide food and fuel to the blockaded civilians of West Berlin. The blockade lasted 324 days before being lifted in 1949. In August 1961 the Soviets began erecting the Berlin Wall, using barbed wire at first, but gradually reinforcing this most potent symbol of the Cold War until it was finally breached in 1989.

Throughout this period, the radio services of various states in what was known as the bipolar world attempted to influence one another. In the Eastern Bloc, radio services such as Radio Moscow (USSR), Radio Berlin (East Germany), Radio Bucharest (Rumania), Radio Budapest (Hungary), Radio Prague (Czechoslovakia), Radio Sofia (Bulgaria), and Radio Tirana (Albania) broadcast the communist version of events, commentaries, features, and cultural programs to both the Western Bloc and nonaligned countries, attempting to influence their citizens' opinions and ideological commitments. Likewise, radio services such as the British Broadcasting Corporation (BBC) World Service (United Kingdom), Radio France International, Deutsche Welle (West Germany), the Voice of America (VOA; United States), Radio Netherlands, and Radio Belgium all broadcast in a similar vein, though they used the news values of the West—such as objectivity and separation of news and editorial comment—in their programs. They broadcast into both Eastern and Central Europe and to the Soviet Union, as well as to the nonaligned countries, in an effort to win "hearts and minds." In addition, the U.S. Central Intelligence Agency secretly funded the creation of two surrogate radio stations, Radio Free Europe and Radio Liberty, which were programmed by émigrés from the Eastern Bloc in an effort to provide citizens of the Soviet Union and the Warsaw Pact countries with news and commentary as they would have heard it if their media had been free to operate without ideological constraints. Still another service begun by the United States was called Radio in the American Sector (RIAS); this service ostensibly broadcast to the Allied occupation forces in Berlin, although its signal could easily be heard in much of East Germany, which surrounded the western enclave in the city.

American Radio Services

In the United States, at the end of World War II, the budget of the international wartime service, the VOA, was cut back significantly, just as all wartime budgets were. But the increasingly confrontational nature of the relations between the "superpowers" and their allies in NATO (North Atlantic Treaty Organization) and the Warsaw Pact led Congress to pass the Smith–Mundt Act in 1948, creating a permanent government information agency and providing increased funding for the VOA. With the outbreak of the Korean conflict in 1950, President Harry Truman initiated what he called a "Campaign of Truth" and called on the media to combat communist distortions of American actions and values by exposing them as lies and telling the truth about America. Various boards and agencies were created over the next couple of years to implement that request, and in 1953 Congress created the United States Information Agency as a permanent coordinating agency for all American information activities abroad. The VOA became the official voice of the United States. Three basic principles were adopted to govern its activities. First, the VOA was to become a consistently reliable and authoritative source of accurate, objective, and comprehensive news. Second, the VOA was to represent all aspects of American society and present a balanced and comprehensive view of significant American thought and institutions. It was not, in other words, to be merely a news service but was also to present programs about the arts, culture, science, and everyday life. Third, it would present the official policies of the U.S. government clearly and effectively and provide responsible discussion of and a venue for the expression of opinion about these policies.

Radio Free Europe and Radio Liberty were not to represent the United States; instead, they were to represent those in their target audiences. The opinions of those with "free" opinions—that is, opinions not under the control of the communist governments in power—were broadcast, along with news about the internal events of the target countries themselves. These two stations were engaged in psychological warfare and sought to stop the spread of communism in Europe and to destabilize the Soviet Union. RIAS was a similar operation concentrating on East Germany and broadcasting the opinions, news, and successes of the West (and particularly West Germany) to the people of West Berlin. It broadcast using both medium wave and shortwave at first (neither of which was needed to reach West Berlin) and eventually began broadcasting in FM. The East Germans said that the service's initials stood for "Revanchism, Intervention, Anti-Bolshevism, and Sabotage."

Soviet and Related Services

On the Soviet side, in addition to Radio Moscow, the Soviet Union funded Radio Peace and Progress and Radio Kiev; all of them broadcast the Soviet version of history, reported the progress of socialism, and sought to influence opinion in both the West and the nonaligned world, particularly in Europe, Africa, the Middle East, and the Indian subcontinent. In addition to those in the capitals of the Eastern European countries that were part of the Warsaw Pact, there were such services in Cuba (Radio Havana), China (Radio Beijing), and North Korea (Radio Pyongyang).

ROBERT S. FORTNER

See also BBC World Service; Board for International Broadcasting; International Radio Broadcasting; Jamming; Propaganda by Radio; Radio Free Asia; Radio Free Europe/Radio Liberty; Radio Martí; Radio Moscow; Shortwave Radio; Voice of America

Further Reading

Alexandre, Laurien, *The Voice of America: From Detente to the Reagan Doctrine,* Norwood, New Jersey: Ablex, 1988
Holt, Robert T., *Radio Free Europe,* Minneapolis: University of Minnesota Press, 1958
Lisann, Maury, *Broadcasting to the Soviet Union: International Politics and Radio,* New York: Praeger, 1975
Nelson, Michael, *War of the Black Heavens: The Battles of Western Broadcasting in the Cold War,* Syracuse, New York: Syracuse University Press, and London: Brassey's, 1997
Panfilov, Artem Flegontovich, *Broadcasting Pirates or Abuse of the Microphone: An Outline of External Political Radio Propaganda by the USA, Britain, and the FRG,* translated by Nicholas Bobrov, Moscow: Progress, 1981
Short, Kenneth R.M., editor, *Western Broadcasting over the Iron Curtain,* New York: St. Martin's Press, 1986
Soley, Lawrence C., *Radio Warfare: OSS and CIA Subversive Propaganda,* New York: Praeger, 1989
Wood, James, *History of International Broadcasting,* vol. 1, London: Peregrinus, 1992; vol. 2, London: Institute of Electrical Engineers, 2000

College Radio

College radio has a long history. A significant number of the earliest radio stations in the United States, such as WHA (then 9XM) in Madison, Wisconsin, were college stations. Many evolved into large professional enterprises, exemplified by those that are members of National Public Radio. However, "college radio" today generally conjures up a different picture. It might be an image of committed volunteer student disc jockeys playing obscure but artistically valuable music for a small but loyal audience, or it might be of a ragtag bunch of kids playing songs that nobody outside their small circle of friends wants to hear. Regardless, it is safe to say that college radio stations play a significant role in many communities and within the music industry, while differing in numerous ways from their more visible professional counterparts and also among themselves.

Despite the great variety, there are some generalizations that can be made about the underlying purposes of most college stations. For many, the primary focus is educational. Colleges may see this role as including educational and informational programming for the community, but it nearly always means that these stations serve academic departments whose scope includes broadcasting or journalism. College radio provides a training laboratory for students in those disciplines, as well as those in business, marketing, and other fields. Some college broadcasters see their mission as providing an entertainment or information service to the listening public, but they usually define themselves as an alternative to professional, tightly formatted stations. Still others exist primarily as a student activity alongside the myriad other extracurricular clubs on campus. These stations generally have a faculty or professional staff adviser, but they are operated as a hobby by and for students. Whatever the station's foremost reason for being, nearly all college stations serve multiple purposes, a fact also reflected in the unique programming and structure of many stations.

College radio is often associated with programs that do not adhere to the rigid niche format structure of professional, commercial radio. Some stations adopt a free-form approach in which almost anything goes, from classical music to poetry to punk rock, at the discretion of the person on the air. Another popular option is block programming, airing shows in many different styles but at specified times. One might hear a three-hour heavy metal show, followed by two hours of blues, which then leads into a two-hour mainstream jazz program, and then a half-hour news magazine, followed by a 90-minute sports talk show, two hours of hip-hop, an hour of contemporary jazz, and so on. Limited only by the number of hours in the day and the availability of qualified and interested students, block programming offers the advantage of allowing a station to serve many different constituencies both within the university and among the listening public. Even at stations that do program a single music format, there are often additional programs on the schedule. For example, college radio is often the outlet for play-by-play coverage of a school's athletic teams, particularly at smaller schools or for minor sports at large institutions. Many stations also make a significant commitment to local public-affairs and news programming.

The most popular single format in college radio is alternative rock. Approximately 70 percent of all college radio stations reportedly program the format; however, the specific execution can vary considerably from one station to another. Some stations concentrate on music far outside the mainstream, deliberately ignoring any release that gets played on MTV or professional radio stations, whereas others sound very much like typical commercial alternative rock stations.

However individual stations define *alternative,* college radio has a reputation for playing an important role in nurturing the careers of many top music stars by providing important early exposure. U2, R.E.M., and the Red Hot Chili Peppers are just a few of the many staples of commercial rock radio that first received attention via college radio. It was probably in the mid-to-late 1970s that the recording industry began to take college radio seriously. Record companies developed college radio marketing strategies and resources, including full-time college radio representatives. Radio and record industry trade press, such as *Gavin,* began to report college radio airplay, and college programming was of sufficient importance to attract its own trade journal. *CMJ New Music Report*, first published in 1979, is devoted exclusively to college programmers and the record promoters who target that market.

College radio has more in common with community radio than with professional, commercial broadcasting. However, finance, staffing, and the means of transmission also differ markedly from station to station. Money is a major concern for college radio, because inconsistent funding creates problems for all areas of the station's operation, from programming to engineering.

Most, but not all, college stations are noncommercial. In some cases, the station is licensed by the Federal Communications Commission (FCC) as noncommercial in the part of the FM band below 92 megahertz that is specifically set aside for that purpose. Other stations adopt a noncommercial policy by their own choice. Noncommercial stations rely on various combinations of student activity fees, state or college support, listener contributions, and underwriting donations from local business to finance operation. College stations that sell commercial time may also rely in part on these other funding mechanisms for a portion of their budget.

Depending on the station's purpose(s) and budgetary support, a college radio station staff may be all or mostly volunteer; they may have one or more professionals, sometimes a faculty member, involved in a management or advisory capacity; or there may be paid student or professional staff handling day-to-day operations. Students also staff some stations in part or in whole as a requirement in specific classes.

Acquiring an FCC license and following all the rules that apply to broadcasting are beyond the reach of many schools and student organizations. Therefore, many schools have chosen to take advantage of more affordable, accessible, and flexible unlicensed alternatives. Derisively dubbed "radiator radio," these are not broadcast stations but facilities that use campus electric or power lines to distribute their signal. In most cases, these permit either commercial or noncommercial operation and also free the school from the record-keeping and public-interest programming obligations imposed on all licensed radio stations.

The FCC's rules explicitly allow some kinds of very low power broadcasting without a license. These include AM carrier current using the electrical system of campus buildings as the antenna; micropowered AM transmitters; or "leaky coax," an FM alternative utilizing coaxial cable throughout a building or campus as the transmitting antenna. Acceptable unlicensed signals cover an area measured in yards rather than miles, broadcasting to only a single block or even just a single building. Cable television provides another unlicensed alternative, via cable FM or audio on regular cable television channels (perhaps as the audio background on a college or public access "bulletin board" channel). Some schools provide a signal through a public address system to reach audiences in a building's public spaces, a service dubbed "cafeteria" radio. Finally, the emergence of the internet as a means of transmitting programming presents colleges with an additional unlicensed radio outlet.

GREGORY D. NEWTON

See also Alternative Rock Format; Community Radio; Free Form Format; Low-Power Radio/Microradio; National Association of Educational Broadcasters; National Public Radio; Ten-Watt Stations

Further Reading

Bloch, Louis M., Jr., *The Gas Pipe Networks: A History of College Radio, 1936–1946,* Cleveland, Ohio: Bloch, 1980

Brant, Billy G., *The College Radio Handbook,* Blue Ridge Summit, Pennsylvania: Tab Books, 1981

CMJ New Music Report, <www.cmj.com>

College Broadcaster website, <www.rice.edu/cb>

Sauls, Samuel J., "Alternative Programming in College Radio," *Studies in Popular Culture* 21 (October 1998)

Sauls, Samuel J., *The Culture of American College Radio,* Ames: Iowa State University Press, 2000

Collingwood, Charles 1917–1985

U.S. Radio and Television Correspondent

Charles Collingwood was the youngest of "Murrow's Boys," a group of reporters hired by Edward R. Murrow to cover World War II in Europe for the Columbia Broadcasting System (CBS). Collingwood made a name for himself as a war correspondent by scooping his newspaper colleagues. By his 26th birthday, he was a top radio reporter with a great news sense and natural radio voice.

Collingwood was born in Three Rivers, Michigan, in 1917. Shortly thereafter, his father was appointed to a professorship of forestry at Cornell University in Ithaca, New York. This was one of many moves during his childhood. Despite this lack of stability, he was an outstanding student. In high school, he was president of the Student Council and a member of the National Honor Society. Upon graduation, he received a scholarship to Deep Springs School in California, a ranch school that consisted of 20 students taught by five professors under the Oxford tutorial system.

In 1937 Collingwood attended Cornell University and majored in pre-law. While in school, during the summers, he traveled extensively, primarily in the United States and in southern Canada. However, one of his trips took him through the Panama Canal. He excelled in his studies, in athletics, and in leadership positions. He was made a member of the Telluride Association, a special foundation that gave scholarships to outstanding students. He received a B.A. degree *cum laude* in 1939 and was a Rhodes Scholar.

While in England attending Oxford University, Collingwood worked for United Press in London. When the Rhodes

Charles Collingwood during 1952 election coverage
Courtesy CBS Photo Archive

committee discovered this dual career, they informed Collingwood that he must choose between scholarship and journalism. He chose scholarship, at least temporarily. The mounting tensions in Europe and the need to be a part of the war effort finally resulted in Collingwood's leaving Oxford in June 1940. In March 1941 he was hired by Edward R. Murrow, the head of CBS correspondents in Europe. His job was to broadcast news analyses, first from London and then from North Africa. His reports from North Africa made him famous in the United States, because his voice conveyed a sense of urgency. He was the first to report on the assassination of an Algerian leader and often scooped his colleagues.

In 1943 Collingwood was awarded the George Foster Peabody Award for the outstanding reporting of news. The Peabody Award Committee selected him unanimously, remarking that he "with the tools of inference, and fact, has conveyed to us, through the screen of censorship, an understanding of the troublesome situation in North Africa." During his stay in North Africa, he broadcast for two-and-a-half minutes twice a day from Radio Algiers on the CBS network. In October 1943, after four years abroad, he returned to the United States and reported to government officials on the importance of the "invasion of ideas" as well as invasions by armies; he also emphasized the importance of British and American cooperation. He then made a two-month lecture tour of the United States.

After the war, Collingwood began working in television news; during the 1950s he was the first United Nations correspondent for CBS News, the CBS White House correspondent, and the chief of the CBS London Bureau. In 1948 he covered the Republican National Convention for CBS radio and television. In 1959 he succeeded Murrow as host of *Person to Person;* for this show, he often left the studio and conducted interviews on location. He hosted a number of television specials, including "A Tour of the White House with Mrs. John F. Kennedy" in February 1962. From 1964 to 1975 he was the chief foreign correspondent for CBS News; in this capacity, Collingwood covered the war in Indochina and went to South Vietnam in 1965. In 1968 he was the first American network correspondent to be admitted to North Vietnam, and he appeared in broadcasts from Vietnam in late 1972 and early 1973.

Collingwood's work included a variety of other assignments that suggest his qualities as an educated and well-respected member of the CBS news team. He hosted *Adventure* (1953), a wildlife program. Following Charles Kuralt and Walter Cronkite in 1962–63, he anchored *Eyewitness to History,* which presented the cultural side of contemporary society. Other programs were *Chronicle,* a news program that alternated with *CBS Reports* from 1963 to 1964, and *Portrait* (1963), an interview show with guests as varied as Peter Sellers and General Curtis LeMay. At the same time that he was White House correspondent (1948–51), he hosted *The Big Question,* a live news discussion program. In the early 1950s, he appeared as a guest on *Youth Takes a Stand,* a discussion of current events with invited high school students and CBS correspondents, and he was a regular guest on *The Morning Show,* CBS's version of the National Broadcasting Company's (NBC) *Today* show.

Collingwood retired from CBS News in 1982 but served as special correspondent until his death. His many awards include the Peabody, the National Headliners Club award, and the Overseas Press Club award. In 1975 he was appointed a commander of the Order of the British Empire by Queen Elizabeth in recognition of his contribution to British–American friendship and understanding. He was also a chevalier of the French Legion of Honor.

He died of cancer on 4 October 1985 in New York City. William Paley, the founder of CBS, said at the time of Collingwood's death that he "represented . . . the highest standards of accuracy, honesty and integrity, leavened with humanity and sensitivity."

MARY E. BEADLE

See also Columbia Broadcasting System; Commentators; Murrow, Edward R.; News; Peabody Awards

Charles Cummings Collingwood. Born in Three Rivers, Michigan, 4 June 1917. United Press reporter, London, 1939–41; World War II correspondent for CBS in North Africa and Europe, 1941–46; first United Nations correspondent for CBS News, 1946–47; White House correspondent, 1948–51; chief of CBS London Bureau, 1957–59; correspondent, *CBS Reports,* 1961–71; chief foreign correspondent, CBS News, 1964–75; retired in 1982. Received Peabody Award; Overseas Press Club award; National Headliners Club Award; commander, Order of the British Empire; chevalier of French Legion of Honor. Died in New York City, 3 October 1985.

Television Series

Adventure, host, 1953–56; *The Big Question,* moderator, 1951–52; *CBS Reports,* correspondent, 1961–71; *Chronicle,* anchor, 1963–64; *Eyewitness to History,* anchor, 1962–63; *Person to Person,* anchor, 1959–61; *Portrait,* anchor, 1963; *A Tour of the White House with Mrs. John F. Kennedy,* 1962; anchor, Princess Anne's Wedding, 1973; anchor, President Nixon's Visit to China, 1972

Further Reading

Bliss, Edward, *Now the News: The Story of Broadcast Journalism,* New York: Columbia University Press, 1991

Cloud, Stanley, and Lynne Olson, *The Murrow Boys: Pioneers on the Front Lines of Broadcast Journalism,* Boston: Houghton Mifflin, 1996

McKerns, Joseph, editor, *Biographical Dictionary of American Journalism,* New York: Greenwood Press, 1989

Murray, Michael D., editor, *Encyclopedia of Television News,* Phoenix, Arizona: Oryx Press, 1999

Schoenbrun, David, *On and Off the Air: An Informal History of CBS News,* New York: Dutton, 1989

Columbia Broadcasting System

U.S. National Radio Network

The Columbia Broadcasting System (CBS), America's second radio network, grew out of the United Independent Broadcasters (UIB) network, which was incorporated on 27 January 1927. It became CBS after it was purchased by William S. Paley in 1928, and in the decades that followed CBS played a leading role in the development of network radio and in the evolution of radio broadcasting following the establishment of television as a primary entertainment medium. Today, CBS Radio is part of the media conglomerate Viacom Inc. and serves nearly 1500 radio stations nationwide with a variety of news, public affairs, information, and sports programs through Westwood One, a program syndication company. Through its radio subsidiary, Infinity Broadcasting, Viacom owns some 180 radio stations in 40 of the nation's largest markets. Infinity manages and holds an equity position in Westwood One.

Origins

The UIB network went on the air on 18 September 1927 with a string of 16 radio stations in 11 states. The network was not well financed, however, losing more than $200,000 in its first year of operation. In order to survive, UIB arranged for backing by the financially strong Columbia Phonograph Company, a leader in the record-pressing business. Columbia bought into UIB for $163,000. UIB, in turn, changed the name of its broadcast arm to the Columbia Phonograph Broadcasting Company. Later, when the network continued to sustain substantial losses, Columbia Phonograph withdrew from the network and took free broadcast time as payment for what it was owed.

One of the early advertisers on UIB was the Congress Cigar Company, which produced and aired *The La Palina Smoker* on the network. The musical program, put together by William Paley (the son of the company's owner) and named after one of its cigars, proved quite a successful advertising tool for the company, doubling sales of the brand in less than six months. Paley was delighted with the results of the program and became fascinated with the radio medium. He spent a great deal of time on his project and learned all that he could about radio and the UIB network. Although UIB was losing money, Paley felt the network had a future, and eventually he decided to buy it. On 25 September 1928, the 27-year-old Paley made it official, purchasing the UIB network for a reported $500,000 of his own money. His father soon bought into the network for $100,000 as a show of support for his son's undertaking.

Paley saw that expansion was a must for his fledgling network and quickly renegotiated the contracts UIB had with its affiliates to achieve three goals: (1) to lower the amount that the network paid stations for the broadcast time they provided; (2) to ensure a long-term association with the stations; and (3) to make sure that UIB was the only network carried by each affiliate. The stations were happy with the arrangement, because UIB was able to hire talent not available to them and to provide better and more programs than the stations could produce themselves locally.

With his existing affiliates taken care of, Paley invited other stations, mainly in the South, to join his new network with contracts similar to those he had just renegotiated. Twelve new stations joined. He later gained a few more affiliates in the Midwest, bringing the UIB network to 48 stations in 42 cities, but none on the West Coast.

By December 1928 UIB was broadcasting 21 hours a week from leased facilities at WABC in New York City and WOR in Newark, Delaware, and it desperately needed a station of its own from which to produce programs. For that purpose, Paley bought WABC in New York for $390,000, after selling shares in the network and investing another $200,000 of his own. WABC (which became WCBS in 1946) thus became the network's first company-owned station.

When Paley had taken over UIB three months earlier, it consisted of three companies: UIB, which supplied the airtime; the Columbia Broadcasting System (the old phonograph company unit), which sold the time to sponsors; and a unit that supplied programs. When Columbia Phonograph left UIB, it insisted that the word *phonograph* be removed from the name of the broadcast arm of the network but allowed UIB to keep the *Columbia* portion. As the on-air part of UIB, the Columbia Broadcasting System was what listeners were familiar with. To preserve this name recognition with radio audiences, Paley reorganized UIB, doing away with the broadcasting unit as a separate entity and merging all three UIB companies into one, named the Columbia Broadcasting System, Incorporated (CBS).

In the first six months of 1929 advertising sales picked up at CBS, and the movie studios began to take an interest in the new network. Just as the movie industry was to fear the impact television would have on theatergoing decades later, the industry was leery of radio broadcasting and decided that a link with the growing medium would be a good financial move. After lengthy negotiations, Paramount paid $5 million for half of CBS in June 1929. As part of the sales contract, the studio agreed to buy back the stock it transferred to CBS to make its 50 percent purchase if CBS earned $2 million within the next two years. Incredibly, CBS met the goal and bought out Paramount, even though the country was then in the depths of the Great Depression.

In 1929 CBS signed the Don Lee group of stations as network affiliates, giving CBS a West Coast link and making it a truly nationwide network. This was also the year it began its first daily news program and its first regular program of political analysis. Late in the year, the network moved into its newly completed headquarters on Madison Avenue in New York with 60 affiliates under contract and annual advertising sales of $4 million.

Development

During the 1930s and 1940s, CBS radio grew from infancy to maturity through a process of trial and error. Programs, largely music, variety, and comedy at first, increased in variety to include drama, soap opera, audience participation and quiz shows and, by the late 1930s, fledgling news efforts.

Early in the decade, with 400 employees in his employ at CBS, William Paley hired a new assistant, Edward Klauber, a former *New York Times* editor who was to become the number-two man at CBS. Another addition to the CBS staff was Paul White, a wire-service reporter who established strong journalistic standards and ethics for the new CBS news organization.

Paley was quick to recognize that growth and revenue for CBS could only come by obtaining new talent and programs to offer to sponsors, and he became adept at finding and signing performers for radio shows. His first big talent coup was to get Will Rogers, America's most popular philosopher–comedian, to agree to do a 13-week series for CBS in the spring of 1930. With Rogers aboard, Paley was soon able to woo comedians Fred Allen, George Burns, and Jack Benny to CBS radio, as well as Morton Downey, Bing Crosby, Kate Smith, and the Mills Brothers, all of whom went on to great success and fame.

CBS acquired its second station, WBBM in Chicago, in 1931 and began laying claim to being the number-one news network by virtue of the number of news bulletins it was airing. The network also began airing the *March of Time*, a weekly dramatization of the major news events of the previous week that was sponsored by *Time* magazine. Although considered melodramatic by some, the program became very popular and remained on CBS until 1937, when it moved to NBC.

Classical music programs were also quite popular in the early 1930s, and CBS signed the New York Philharmonic for Sunday afternoon broadcasts. In addition, the network formed its own Columbia Symphony Orchestra and presented thousands of programs of classical music in the years that followed.

Another popular type of program, which emerged on CBS and NBC in the early 1930s and would enjoy loyal audiences for nearly three decades, was the daily romance serial. Sponsored by the giant soap firms of the day, such as Procter and Gamble, Colgate–Palmolive, and Lever Brothers, the shows became a part of the daily lives of housewives across the country.

In 1933, just as CBS became the largest network with 91 affiliates, a high-stakes battle broke out between the radio and newspaper industries that threatened the network's news function. The conflict grew out of radio's steady rise in popularity, which caused newspaper publishers to fear that the new medium was siphoning off advertising revenue and news audiences. The American Newspaper Publishers Association voted not to print the radio industry's daily program schedules in their papers except as paid advertising. The publishers next pressured the newswire services to stop serving radio stations and networks.

Left without wire-service news, CBS, with the help of sponsorship from General Mills, formed its own news-gathering organization, the Columbia News Service, and placed bureaus in New York, Washington, Chicago, and Los Angeles. It also lined up correspondents as "stringers" in nearly every major American city and negotiated exchange agreements with a number of overseas news agencies in an effort to keep news flowing to the radio audiences. Paul White and his staff prepared three news programs each day at CBS, many times with stories the papers did not yet have. By the time the press–radio war ended some time later, CBS had established a strong commitment to providing news and information to America, a commitment that remains at the core of its modern-day radio offerings.

CBS's *American School of the Air* was a noncommercial supplement to regular classroom instruction, complete with a teacher's manual. The program featured geography, history, English, music, and drama for young people. It was regularly heard by 6 million children, but it was not able to make CBS the most popular network. During the 1934–35 broadcast season, radio's top five programs were all on NBC.

With 91 stations, CBS had more affiliates than NBC, but it continued to trail NBC in popularity. New programs were produced, and a number of policy changes were made during 1935 in an effort to move the network into the top spot. CBS established standards for the amount of advertising time it would permit per program and for the type of products that it would and would not advertise. Standards were also set that dealt with "fairness and balance" in all news and public information programs.

The extra effort seemed to pay off when, by 1936–37, radio's top five programs were on CBS. This was due largely, however, to the fact that the network had enticed three of NBC's most popular entertainers to the network: variety show host Major Bowes, singer Al Jolson, and comedian Eddie Cantor. *Major Bowes' Original Amateur Hour* was the most popular program of the day. CBS also soon took the *Lux Radio Theater* from NBC.

In an effort to serve as many audiences as possible, CBS also began to present the *Church of the Air* on Sundays and formed an advisory board to set policy for its educational programs and to choose shows suitable for children. In addition, it formed the Columbia Workshop in 1936 as an experimental theater of the air. CBS kept this program unsponsored to give it freedom and a chance to pioneer new radio techniques, especially in sound, electronic effects, and music, and many of the ideas the Workshop perfected later became broadcast industry standards. With many of its scripts written by the best-known writers of the day, such as Dorothy Parker, Irwin Shaw, and William Saroyan, the program achieved great critical acclaim.

Late in 1936, CBS established a base of operations in Hollywood in order to be able to originate radio shows from the West Coast. The move allowed the network to better serve that region and its time zone and gave CBS more access to Hollywood stars.

As the signs of war grew in Europe in 1937, Edward Klauber decided that CBS needed a European director, and the job went to Edward R. Murrow. Once overseas, Murrow hired journalist William L. Shirer, and the pair set about lining up cultural events, concerts, and other programs for CBS. When German troops entered Austria in 1938, Shirer, then in Vienna, flew to London to get the story out, and Murrow went to Austria. Shirer went on the air for CBS from the British Broadcasting Corporation (BBC) studios in London, and Murrow sent a shortwave broadcast from Vienna reporting the German takeover of Austria. The effort was the beginning of CBS's exemplary war coverage. Continuing to report from Europe, Murrow and Shirer put together the *CBS World News Roundup,* the first round-robin international radio news broadcast, with Murrow reporting from Vienna; Shirer from London; and other newsmen in Paris, Berlin, and Rome.

In the late 1930s CBS bought Columbia Records (the company from which it got its name) and opened its new $1.75 million Columbia Square studio/office complex on Sunset Boulevard in Hollywood, California. The network also continued to give its growing audience exciting entertainment programs while providing increasing amounts of news and information.

The War Years

The CBS foreign news staff grew from 3, including Murrow and Shirer, to 14, and then to 60 in 1941 as the United States entered the war. Of the war years, CBS President William Paley, who became the Deputy Chief of the Office of Psychological Warfare under General Dwight Eisenhower, later noted,

> [W]e adopted war themes on many of our programs. In dramatic shows, characters met wartime problems; the *American School of the Air* brought war news, information, and instruction to children; *Country Journal* gave farmers help in solving wartime agricultural problems; the *Garden Gate* promoted Victory gardens; *Church of the Air* broadcast talks by chaplains. There were also many new series exclusively about the war: *They Live Forever, The Man Behind the Gun, Our Secret Weapon* (Paley, 1979).

Kate Smith also conducted hugely successful war bond drives, and some CBS company-owned stations began a 24-hour-a-day schedule, serving as part of an air raid defense system and providing entertainment for defense workers on the overnight shift. CBS's foreign correspondents were its stars of the air. Edward R. Murrow became a hero, even before the United States entered the war, through a series of "rooftop" broadcasts during the 1940 blitz in London; William L. Shirer covered the surrender of France to Germany at Compiègne; Larry Lesueur provided regular shortwave reports from Moscow; and Howard K. Smith provided coverage from Berlin. Others were stationed throughout Europe and in North Africa and Asia.

Paul White oversaw the international news organization on a daily basis with the help of a news team of some 50 members, including a staff of shortwave listeners who kept him abreast of what was happening around the world. Many of the team's members went on to achieve individual fame as writers and commentators: Eric Sevareid, Robert Trout, Charles Collingwood, John Daly, Howard K. Smith, and of course Edward R. Murrow all became well known and gave CBS News great credibility.

In 1943 CBS acquired WCBS-FM in New York, its first FM radio station, and WBBM-FM in Chicago. It also lost its number-two man when CBS Vice President Ed Klauber suffered a heart attack and resigned.

On 6 June 1944, CBS went on the air at 12:30 A.M. to begin special coverage of the D-Day invasion, utilizing several of its commentators in New York, print reports from Washington and overseas, and live transmissions from London. Additional coverage was provided by correspondents in other European capitals, who kept the listeners updated on the progress of the invasion, and CBS's Charles Collingwood, who crossed the English Channel in an LST to report on the invasion from the beach at Normandy.

Between 7 December 1941 and 2 September 1945, the day the Japanese surrendered, CBS broadcast 35,700 wartime news and entertainment programs, including Norman Corwin's commemorative show *On a Note of Triumph,* which aired at the end of hostilities in Europe. A similar program, entitled *14 August,* was aired following the Japanese surrender.

Postwar Transitions

As the war ended, Frank Stanton, who had been hired in the 1930s as a research specialist to determine CBS listenership, became the network's president; William Paley moved up to become chairman of the board; and Edward R. Murrow was promoted to vice president and director of news and public affairs. By this time, CBS had once again fallen behind in the ratings battle, as 12 of the top 15 radio shows were on NBC.

By the end of 1947, CBS had put together 36 radio programs, but few were sponsored. It also established a news documentary unit to look at the subjects that were most affecting Americans at the time. In addition, Murrow resigned as a CBS vice president and returned to the air with *Hear It Now,* a talking history of World War II that evolved into the later television news-documentary program *See It Now.* As it gained popularity, the documentary became a mainstay of CBS programming. CBS also joined the other networks in providing live broadcasts of hearings of the House Un-American Activities Committee, which looked into the alleged presence of communist sympathizers in the motion picture industry.

The following year, CBS increased sponsorship for its own shows to 29, and 2 of the programs became among the nation's 10 most popular: *My Friend Irma* and *Arthur Godfrey's Talent Scouts.*

Near the end of the decade, CBS tax attorneys discovered a way for radio stars to save tax money by selling their programs as "properties," and CBS was able to "raid" the most popular NBC programs, including *Amos 'n' Andy* and Jack Benny. In January 1949 CBS signed other NBC stars: Bing Crosby; comedians Red Skelton, Edgar Bergen, George Burns and Gracie Allen, Ed Wynn, and Groucho Marx; singers Al Jolson and Frank Sinatra; and band leader Fred Warning. Soon, CBS had 12 of the Hooperating's "First 15," 16 of Nielsen's "Top 20," and an average audience rating that was 12 percent larger than that of any other network. CBS was definitely number one in American broadcasting. By the late 1940s, CBS and its rivals were able to use money made in radio to fund progress in television, with CBS allocating $60 million to that cause.

Decline of CBS Radio Network

By the end of 1955, television's ability to attract radio's evening audience became clear when the Nielsen ratings listed no night-time programs among radio's top 10. Searching for a way to keep radio audiences, CBS beefed up its news offerings and premiered the *CBS Radio Workshop* as a revival of the earlier Columbia Workshop. It showcased some of the best talent of the day and used exceptional imagination and creativity in providing critically acclaimed but unsponsored radio drama.

Between 1957 and 1960, Jack Benny, Bing Crosby, and *Amos 'n' Andy* left the air; CBS radio shortened its schedule and turned over more time to the affiliates; and radio stations across the country began to offer more music and less network programming.

By 1960 all three radio networks hit bottom financially, losing 75 percent of the sales they had had in 1948. CBS began to offer even more news, sports, and information programming, and in November 1960 it canceled its last surviving soap operas, putting an end to a chapter of radio history and relinquishing the genre to television.

As the 1960s progressed, CBS made new arrangements with its affiliates that allowed them to put "packages" of network programs together to meet their needs rather than having to take all the network offerings. With the move, profitability returned, and CBS radio changed fundamentally. No longer would the network be able to *tell* its affiliates what it would offer them; CBS would instead have to *ask* the stations what they needed and try to provide it. In 1974 CBS decided to hire E.G. Marshall to host the *CBS Radio Mystery Theater* in an attempt to reintroduce radio drama and the feel of programs from the golden age of radio. The hour-long show, which was run seven days a week using new scripts and some old production formulas, received mixed reviews, and many affiliates declined to carry it or aired it outside of prime time. It was clear that news, sports, and information programs were all stations wanted from the networks, and CBS vowed to provide it through its strong network news division. The decision has remained in place for nearly three decades.

Today CBS News serves both radio and television station affiliates. On the radio side, there are two entities, CBS Radio News and CBS Radio Sports, that produce news, information, and sports programming for distribution to more than 1500 stations through Westwood One, a program syndication company.

CBS Radio News provides stations in nearly every major market with hourly newscasts, instant coverage of breaking stories, special reports, updates, features, customized reports, and newsfeed material that alerts the stations receiving CBS material to what will be available to them in the following hours. Among the CBS Radio News productions is the World News Roundup, first broadcast in 1940 and said to be the longest-running newscast in America. CBS Radio Sports provides the affiliated stations with regular sportscasts, customized reports, features, and sporting events coverage.

Early in 2000, CBS merged with Viacom Inc., a global media company with interests in broadcast and cable television, radio, outdoor advertising, online entities, and other media-related fields. Viacom's holdings include MTV, Nickelodeon, VH1, BET, Paramount Pictures, Viacom Outdoor, UPN, TV Land, The New TNN, CMT: Country Music Television, Showtime, Blockbuster, and Simon & Schuster.

As part of the merger, Viacom also became the parent company of the CBS-owned Infinity Broadcasting Corporation, which operates more than 180 radio stations, the majority of which are in the nation's largest markets. Infinity manages and holds an equity position in the syndicator Westwood One, which is the major CBS radio program distributor.

JACK HOLGATE

See also American School of the Air; Don Lee Network; Infinity Broadcasting; Karmazin, Mel; KCBS; Kesten, Paul; Klauber, Ed; March of Time; Murrow, Edward R.; Network Monopoly Probe; News; Paley, William S.; Press-Radio War; Shirer, William L.; Stanton, Frank; Talent Raids; WCBS; White, Paul

Further Reading

Archer, Gleason L., "Travails of a Rival Radio Network," in *Big Business and Radio,* New York: American Historical Company, 1939; reprint, New York: Arno Press, 1971

Bergreen, Laurence, Look Now, Pay Later: The Rise of Network Broadcasting, Garden City, New York: Doubleday, 1980

Buxton, Frank, and Bill Owen, *Radio's Golden Age: The Programs and the Personalities,* New York: Easton Valley Press, 1966; revised edition, as *The Big Broadcast, 1920–1950,* New York: Viking Press, 1972; 2nd edition, Metuchen, New Jersey: Scarecrow Press, 1997

"CBS at 60," *Television/Radio Age* (28 September 1987)

"CBS: Documenting 38 Years of Exciting History," *Sponsor* (13 September 1965)

"CBS: The First Five Decades," *Broadcasting* (19 September 1977)

"CBS: The First 60 Years," *Broadcasting* (14 September 1987)

Metz, Robert, *CBS: Reflections in a Bloodshot Eye,* Chicago: Playboy Press, 1975

Paley, William S., *As It Happened: A Memoir,* Garden City, New York: Doubleday, 1979

Paper, Lewis J., *Empire: William S. Paley and the Making of CBS,* New York: St. Martin's Press, 1987

Slater, Robert, *This is CBS: A Chronicle of 60 Years,* Englewood Cliffs, New Jersey: Prentice-Hall, 1988

Sloan, William David, James Glen Stovall, and James D. Startt, editors, *The Media in America: A History,* Worthington, Ohio: Publishing Horizons, 1989; 2nd edition, Scottsdale, Arizona: Publishing Horizons, 1993

Smith, Sally Bedell, *In All His Glory: The Life of William S. Paley,* New York: Simon and Schuster, 1990

Columnists

Writing About Radio

Many listeners first learned about radio's performers and the broadcasting industry through columns appearing in daily newspapers and popular magazines. At first largely technical, these columns soon melded comment, criticism, and interviews to feed the public fascination with radio.

Newspapers

As one read a typical 1920–21 newspaper, it was often difficult to tell that radio existed. Because the new medium was capable of taking a person to an event as it was happening, newspapers saw radio as competition, and most decided to ignore it. A few newspapers offered an amateur radio column once or twice a week, because ham radio had become a popular hobby, but these columns were mainly about technical matters (such as how to build the latest equipment) or news from local ham radio clubs. Events that today might seem worthy of a bold headline (such as KDKA's first broadcast) were either mentioned in a paid advertisement or relegated to a short article somewhere inside the newspaper.

Among the first big-city newspapers to have a regular radio column was the *Boston Traveler;* the column's author was Guy Entwistle, an executive with the American Radio Relay League, an important ham radio organization. As a student at Tufts College, he had earlier worked at pioneer station 1XE. His column appeared three times a week beginning in February 1921. But the *Traveler* was the exception, and none of the other Boston newspapers covered radio with any degree of thoroughness until early 1922, by which time most of America's newspapers finally accepted that radio was more than a mere fad. Newspapers realized that the public was interested not just in how to build a radio (because various brands were now sold retail) but in learning more about the voices they heard on the air.

The *Detroit News* was probably the first newspaper to own a radio station (WWJ), and the newspaper clearly saw radio's promotional value. Owning the station was a fine reason to mention radio, because it was not competition in this case. Other newspapers soon followed suit, accepting the fact that radio was here to stay. Some began offering a full page of radio news by 1922, although at first it was mainly news of their own stations' programming. But even newspapers lacking a license began to offer radio coverage. For example, the *New York Tribune*'s radio page was edited by ham radio expert (and maritime disaster hero) Jack Binns. The *Boston Globe*'s radio editor, Lloyd Greene, also came from ham radio. Some of the first radio columns at the *Los Angeles Times* were written by John S. Daggett, who would soon become station manager of KHJ and who, as "Uncle John," would host a popular children's show. For those papers unable to afford a dedicated radio editor, several nationally known radio experts offered syndicated columns, usually of a technical nature—the best-known of these writers was Sidney Gernsback.

About this time, some newspapers decided to do more than operate and publicize their own stations. They began to allow their best columnists to read news (or to read from their own columns) on those stations. The famous editor of the *Brooklyn Daily Eagle,* H.V. Kaltenborn, was among the first journalists to not only read the news but also provide editorial commentary when he broadcast on the Signal Corps' New York station WVP (and later on WJZ) beginning in early April 1922.

As radio's popularity increased, the job of reporting about it became more essential. The development of radio networks certainly contributed to this increased interest: when the National Broadcasting Company (NBC; 1926) and the Columbia Broadcasting System (CBS; 1927) came along, they enabled hit songs and hit performers to be enjoyed across the entire country and even into Canada (before the Canadian Broadcasting Corporation [CBC] was created, Canadian stations often affiliated with U.S. networks). Wherever there was a station, the public was eager to learn about the celebrities and the announcers, some of whom had begun their careers in local markets before passing the audition for the network. Fans sent letters and telegrams to their favorite shows, and they also contacted radio reporters, asking for more in-depth information about the most popular stars. Even those newspapers that had been hesitant to have a radio page jumped in enthusiastically by the mid- to late 1920s—and readers welcomed the thorough radio coverage of such prestigious newspapers as the *New York Times* (where radio editor Orrin E. Dunlap, Jr., kept readers up on the latest trends from 1922 through 1940).

The job of newspaper radio editor became a very stable one, as well as one with some excellent "perks"—the radio editor not only attended many live performances but could conduct interviews with the stars and take part in the annual vote for the best programs and personalities of the year. A number of the radio editors who began writing their columns in the 1920s were still doing so several decades later: Alton Cook of the *New York World-Telegram,* Larry Wolters of the *Chicago Tribune,* and Howard Fitzpatrick of the *Boston Post,* just to name a few. In fact, during the late 1920s and early 1930s, some newspapers had two radio columnists, one specializing in musicians and the other writing mainly about air personalities and special events. A few key announcers wrote newspaper columns in the 1920s and later, usually promoting the relationship between the newspaper and their station: Boston's "Big Brother" Bob Emery and Kansas City's "Merry Old Chief" Leo Fitzpatrick were only two of the celebrity columnists. Bertha Brainard, who began her media career as a theater critic and then was hired by WJZ in New York to do theater reviews on radio, moved back and forth between radio and print, writing guest articles for both newspapers and magazines.

Unfortunately, it is difficult to acknowledge most radio columnists through the 1930s, because it was then customary to provide bylines to only a select few reporters, and these seldom included the radio columnist. There were exceptions, of course—some large newspapers, such as the *Washington Post, Los Angeles Times,* and *Boston Globe,* did give credit to those who wrote about radio: at the *Post,* for example, we know that Elizabeth Poe (who also wrote about music and did some announcing on the *Post*'s program on WRC) did some radio columns in the mid-1920s, and that the column known as "Dial Flashes" was written by Robert D. Heinl into the early 1930s. The *Los Angeles Times* gave a byline to its succession of radio reporters, from 1920s writers such as John Daggett and Dr. Ralph Power to William Hamilton Cline, Doug Douglas, and Carroll Nye in the 1930s. At the *Boston Globe,* Lloyd Greene gave way to Willard De Lue in the late 1920s, and then to Elizabeth Sullivan in the 1930s. But more often than not, newspapers that had radio listings and even some commentary about radio shows did not identify who wrote the articles.

Early Radio Magazines

Magazines reacted to the growing radio craze in a way similar to newspapers: some saw radio as a threat and ignored it, and others were quick to embrace it. When *Time* first appeared in 1923, it lacked a radio page and only reported on trends in print journalism. Only 15 years later would this policy change when in late 1938 a radio page was finally added. On the other hand, in spring 1922, *Literary Digest* expanded its science and technology section by adding a radio page, with stories about the owners, announcers, and performers on the air in various cities.

Variety was one of the periodicals that was initially very negative about radio. This is understandable because *Variety* was known for its coverage of—and identification with—movies and vaudeville, both of which feared radio's competition. Yet even by *Variety*'s admission, an increasing number of famous singers were already appearing on radio; one—Vaughn DeLeath, known as the "Original Radio Girl"—even became program manager of New York's WDT in 1923. By the mid-1920s, it was impossible to ignore the obvious fact that vaudeville was dying, but radio was not. At the decade's end, *Variety,* too, would have a radio page, with columnists who mainly covered New York, but by the early 1930s, *Variety* began gathering reports about radio in other major cities. Music reviewer Abel Green was one of the first writers at *Variety* to include radio coverage in his columns.

Initially, the most authoritative place to read about radio was in magazines dedicated exclusively to the new medium. One of the earliest and best-known was *QST,* the ham magazine that is still published. In 1919 *Radio Amateur News* first appeared, and although it mainly covered the technology and people of amateur radio, by 1921 it had expanded to include the latest happenings in commercial broadcasting. It was founded by Hugo Gernsback, an immigrant from Luxembourg who became a successful inventor, entrepreneur, and publisher of numerous radio and science fiction magazines. As commercial radio's popularity grew, the magazine changed its name to *Radio News;* among those who wrote articles for it were Lee de Forest and Reginald Fessenden. In 1925, Gernsback put New York radio station WRNY on the air, and he often wrote about the station in his magazine.

1922 was a big year for radio magazines—*Popular Radio, Radio Digest, Radio Broadcast, Radio* (formerly *Pacific Radio News*), *Radio in the Home,* and *Radio World* were all available, and other magazines that previously had a technology focus (such as *Science and Invention* and *Wireless Age*) added radio pages. Most had pictures of announcers and performers, as well as interviews. Another new element of radio magazines was that a few had women columnists. Jennie Irene Mix, a published author and music critic, wrote for *Radio Broadcast* in 1924–25; former publicist and actress Nellie Revell not only wrote a column for *Variety* in the late 1920s but also wrote for *Radio Digest* in the early 1930s.

Billboard was available, but it was not a music industry magazine yet; back then, it was known as *The Billboard* and was devoted to coverage of county fairs, circuses, and expositions. It had little reason to mention radio and seldom did. But in the early 1920s, *Billboard* did have a unique column about black actors, actresses, and musicians that mentioned certain black performers who appeared on radio, such as Eubie Blake and Noble Sissle. However, *The Billboard* would not pay much attention to radio until the late 1930s, when the magazine gradually changed its focus toward more music industry reporting. By the early 1940s, radio coverage and articles about popular music occupied more of a place in the magazine.

A second wave of radio magazines appeared in the 1930s—most in the "fan" genre with a focus on programs and stars—with *Song Hits, Radio Stars, Radioland, Radio Guide,* and *Radio Mirror,* to name a few. Also, some local radio magazines began to appear, such as *Radiolog,* which mainly covered New England, and *Microphone,* which began in Boston but had regional editions in other parts of the country. National magazines such as *Radio Guide* also produced locally written columns for each of the major regions they covered; some of these were written by the program directors of local stations whose personalities were being featured that month. (On the other hand, some magazines offered articles ostensibly written by the radio stars themselves; many of these were ghostwritten by publicists.)

Other Radio Publications

As radio stations began to achieve success and stability in radio's golden age, some began to put out their own publications. This enabled stations to reach out to their audience, offering photos of the announcers, columns written by various staff members, and even some contests; in cities where radio coverage in the newspaper was minimal, such magazines were especially welcome, but even in cities with thorough radio coverage, such as Chicago, stations used their magazines as vehicles for increased publicity. Beginning in the mid-1930s, for example, WLS radio published a weekly magazine called *Stand By.* Editor Julian Bentley interviewed station performers as well as national celebrities who had come to Chicago to perform. Station magazines were usually written by someone in the program department, with help from the networks, which provided photos and biographies of nationally known stars. Well into the 1940s, many stations put out their own "yearbooks," with pictures of the biggest station events of the previous year, stories about what the station had done in news and public service, and friendly messages from the station's performers and announcers. And beginning in 1938, there was a reference book that covered all of broadcasting: *Radio Annual* (later *Radio Annual/Television Yearbook*), edited by Jack Alicoate and Don Carle Gillette. Published by the staff of *Radio Daily,* it offered not only profiles of every station but afforded radio executives (the first edition featured columns by NBC's David Sarnoff, Professor of Education and CBS Adult Education Board member Lyman Bryson, and respected engineer Alfred N. Goldsmith, among others) to give their assessments of the state of the industry.

There were also many nonfiction (and nontechnical) books written by media critics or columnists of the 1930s and 1940s, offering their own view of radio's achievements and problems. These included Ruth Brindze's controversial *Not to Be Broadcast: The Truth about the Radio* (1937); Robert J. Landry's *Who, What, Why Is Radio* (1942); and Albert N. Williams' book of essays *Listening: A Collection of Critical Articles on Radio* (1948). In addition to his book, *Variety*'s radio editor, Robert J. Landry, also wrote several critical essays about radio for *Public Opinion Quarterly* in the early 1940s.

The 1933 college debate topic, "Resolved: That the United States should adopt the essential features of the British system of radio control and operation," gave rise to a host of critical (and supportive) columns about American radio. Several serious journals, such as the *Annals of the American Academy of Political and Social Science,* devoted issues to the medium or featured regular columns and commentary. And writers especially concerned about radio's role in education produced a host of booklets, talks, columns, and articles, which usually lamented the relative lack of serious programming, too much advertising, or other complaints. These were often reprinted and widely disseminated by various interest groups—including the radio industry itself.

Changes in Radio Reporting

With television's arrival in the late 1940s, many critics predicted the rapid demise of radio. Radio coverage in newspapers had definitely diminished during World War II, with coverage of movie stars getting most of the attention; some radio magazines changed their name to "Radio/Movie" magazines to reflect the public's strong interest in Hollywood. But writing about radio never disappeared: it could even be found in such top-drawer periodicals as *Atlantic Monthly, Harper's,* and the *Saturday Review of Literature,* among others, and some newspapers continued to focus attention on local radio stars. In the early 1950s, as television became more available, major newspapers from Los Angeles to Boston began putting out a weekly *TV/Radio* magazine, which included at least one page about local stations and the people who worked there.

In the 1950s and 1960s, a new group of columnists appeared. Now columnists were not just responsible for writing about radio, of course, but were expected to cover television and also popular culture. Some of the veteran radio columnists were still writing in the 1950s, and as music changed, it was quite a problem for many of them to objectively cover the new Top 40 radio stations, because they did not understand or like rock music. The *Boston Record*'s highly respected Bill Buchanan, who much preferred jazz and big bands, interviewed Dick Clark and pronounced *American Bandstand* a total waste of time. In the 1950s radio columnists were of two camps—the veterans, who still recalled the golden age and lamented what sounded like noise to them, and the new columnists, who thought the changes were exciting and appreciated the high energy of Top 40 announcers. If nothing else, the arrival of Top 40 got radio back into the newspaper again, even if it was only so that certain columnists could criticize it. Younger writers, such as Gary Deeb (who wrote the radio and television column for the *Chicago Tribune* during the 1970s and early 1980s) had a much easier time writing about Top 40. And Jane Scott, who began her radio writing in the mid-1960s after first being the "Teen Editor" at the *Cleveland Plain Dealer,* would go on to write about radio and music in Cleveland well into the 1990s. As other formats, such as progressive rock, emerged in the late 1960s and into the 1970s, some new newspapers emerged too, so-called "alternative" weeklies, which adopted album rock as the music of rebellion and wrote about the stations that played it. Radio columns were not as easy to find as they had once been, but in cities where there was a dominant radio station (such as Cleveland, with first WIXY in the Top 40 days and then WMMS-FM in the album rock period), there was always a newspaper or magazine writing about it. And for some radio fans, especially those who hoped to go into the industry, the 1970s were the time when radio editor Claude Hall's column in *Billboard* magazine was eagerly read.

Today

By 2000 few magazines are writing about disc jockeys, and those newspapers that have a radio/television editor also send him or her out to cover concerts and write about the club scene. The internet has taken up the slack, and many radio stations have developed their own webpages, where disc jockeys can write their own columns. Some stations still put out their own newsletters or publications, and popular industry publications such as *Billboard* and *Radio and Records* have columnists who write exclusively about radio; weekly newspapers such as Boston's *Phoenix* or New York's *Village Voice* have occasional articles about what's on the air these days. Media critics still write magazine articles about such issues as radio consolidation, and controversial personalities such as Rush Limbaugh and Howard Stern have been the subject of numerous essays. And although it is more difficult to find radio columns in the newspaper (many newspapers have even stopped publishing radio listings, a trend that started in the late 1980s), certain stations still know how to get publicity, as do certain personalities, and certain reporters still enjoy writing about the achievements of local broadcasters. As the industry continues to change, it is safe to assume radio will keep being discussed and analyzed in the print media.

DONNA L. HALPER

See also Critics; Dunlap, Orrin E., Jr.; Fan Magazines; Kaltenborn, H.V.; Taishoff, Sol; Trade Publications

Further Reading

Many early radio magazines (and newspapers with radio columns) are available on microfilm or in actual copies at public and university libraries. The Library of American Broadcasting (University of Maryland) is one archive that holds an extensive pamphlet file including many reprinted critical articles and columns.

Aly, Bower, and Gerald D. Shively, editors, *A Debate Handbook on Radio Control and Operation,* Norman: University of Oklahoma Bulletin, 1933
Brindze, Ruth, *Not to Be Broadcast: The Truth about Radio,* New York: Vanguard Press, 1937
Bryson, Lyman, *Time for Reason about Radio,* New York: Stewart, 1948
Frost, S.E., Jr., *Is American Radio Democratic?* Chicago: University of Chicago Press, 1937
Hettinger, Herman S., editor, "Radio: The Fifth Estate," *Annals of the American Academy of Political and Social Science* 177 (January 1935)
Hettinger, Herman S., editor, "New Horizons in Radio," *Annals of the American Academy of Political and Social Science* 213 (January 1941)
Landry, Robert John, *Who, What, Why Is Radio?* New York: Stewart, 1942

Combo

Announcer-Engineer Combination

The term *combo* is short for combination. In the radio industry, the term refers to a combo announcer, one who combines announcing with engineering duties such as playing recorded music and announcements.

Like all businesses, the radio industry experienced growing pains brought by technological advances. For radio stations, one of the growing pains was a labor cost growing out of the need to hire several employees for a disc jockey program. In the early days of radio, three employees were often needed to broadcast a program: two engineers (one to operate the audio console and to play transcribed materials and another to operate the station's transmitter) and an announcer to present spoken materials. As the industry grew, and as the control room and transmitter operating equipment became more sophisticated, station managers concluded they could save money by using "combo" announcers who could also perform the functions of engineers.

In the infancy of radio, the control room of a radio station operated as follows: the announcer was positioned by a microphone to read, or possibly to ad-lib, material that went on the air. The written material was called continuity and consisted of a daily file of all commercials and public service announcements, in chronological order, to be broadcast by the station. The station log provided a schedule of the announcements and programs, notifying both the announcer and the engineer of what should be read when.

So that the announcer could read a given announcement on the air, the engineer operating the audio console would turn on the microphone using the proper switch and volume control. The volume control, more commonly known as a "fader," "pot," or "mixer," was used to control the volume of audio current. The console contained a number of these faders, located in parallel series near the bottom of the unit. Each microphone, turntable, tape recorder, and network input had its corresponding fader. Another fader was used to control the input of a network into the console. The control room engineer had the responsibility of turning on the correct microphones or turntables and then using the fader to "ride gain," or maintain the appropriate volume for each microphone, each turntable, and so on.

The engineer at the audio control console was responsible for regulating the volume during a specific program or through a series of them. This was especially complex when radio stations broadcast live orchestras or bands, live vocalists, and live announcers. The engineer's responsibility was to regulate the volume so that the quality of transmission would not vary and so that distortion or inaudibility would not distract listeners.

In a relatively simple program, such as one in which an announcer hosted transcribed music, the engineer would turn on the switch for the correct microphone and then cue the announcer by pointing at him to begin reading or talking. During the message, the engineer would make certain the volume level was correct and would prepare to turn on the next microphone, turntable, and so on. Once the message had been read, the announcer would indicate completion of the message by pointing back at the engineer. This would be the signal for the engineer to turn off the microphone and to activate other switches for the next source of sound.

As broadcast equipment improved, stations adopted combo operations, and by the 1950s most small-market stations were combo. By "going combo," one person could operate the control room console, turntables, and tape recorders while also announcing live copy. A combo announcer had to combine several traits: an adequate voice to perform announcing duties and sufficient manual dexterity to simultaneously operate the equipment. Not all people could fulfill both roles.

Station managers also had the combo announcer read and record meters on the station transmitter and make necessary adjustments. Because many AM stations were required to sign off at night or were required to prevent interference with other stations by using directional transmission patterns to control the station's signal, correct transmitter operation was essential.

To perform transmitter adjustments at stations transmitting a directional pattern, a Federal Communications Commission (FCC) First Class radiotelephone operator's license, known as "First Phone," was needed until 1981. Many announcers enrolled in schools that taught them the basic knowledge needed to obtain a First Phone. A person who acquired a First Phone could announce, operate control room equipment, and make transmitter adjustments. First Phone announcers were often paid more than announcers who did not have the First Class FCC license, but the financial savings were important to smaller stations. Paying one combo announcer somewhat more than an announcer without a First Phone was financially preferable to paying several staff members.

Not all AM (and eventually FM stations) employed combo announcers. Stations with union agreements generally continued to subdivide the announcing, engineering, and control room operations. However, the majority of nonunion AM and FM radio stations now use combo announcers.

The introduction of digital technology has further altered combo operations. At many stations, recorded material, including music, commercials, and station promotional items, are placed on the hard drive of a computer. The announcer operates the audio console, but the computer controls the programming of the other items, that is, music, recorded commercials, and so on. The announcer only stops the computer for live inserts and then restarts it once the live insertion is complete. The rest of the time, the computer plays recorded music, recorded commercials, and so forth on the air in the correct order. In other cases, the announcer can also prerecord the verbal inserts he or she will include between commercials or recordings, and the computer can present the entire recorded program.

MIKE MEESKE

See also Automation; Control Board; Recording and Studio Equipment

Further Reading

Chester, Giraud, and Garnet R. Garrison, *Radio and Television: An Introduction,* New York: Appleton, 1950; 5th edition, as *Television and Radio,* by Chester, Garrison, and Edgar E. Willis, Englewood Cliffs, New Jersey: Prentice Hall, 1978

Ditingo, Vincent M., *The Remaking of Radio,* Boston: Focal Press, 1995

Halper, Donna L., *Full-Service Radio: Programming for the Community,* Boston: Focal Press, 1991

Keith, Michael C., *The Radio Station,* 5th edition, Boston: Focal Press, 2000

Comedy

Comedy on radio was a slow starter. Until the mid-1920s, music and various forms of talk provided most of the infant medium's programming. It is probably no coincidence that the most fertile period for radio comedy—and movie comedy, for that matter—was when times were hardest: the Great Depression and World War II. Americans needed the healthy release of laughter, and the young electronic medium was eager to oblige.

Vaudeville on the Air

Just as movies had first borrowed from the format of the proscenium stage—and later, television borrowed from radio—radio itself also initially borrowed from a preceding medium, vaudeville. As the Depression deepened in the early 1930s, people had less money for live entertainment, and vaudeville performers found themselves increasingly out of work. Fortunately for them, radio was proving to have a voracious appetite for talent, and although it was a major contributor to vaudeville's demise—again, along with movies—it was also something of a savior for many of its performers. Nearly all of radio's first stars came from vaudeville: Ed Wynn, Eddie Cantor, Burns and Allen, Jack Benny, Fred Allen, and many more.

Probably radio's earliest paid entertainers were Billy Jones and Ernie Hare, a song-and-comedy-patter duo. First appearing in 1921, they were known by various names depending on their sponsors: The Happiness Boys (a candy company), The

Interwoven Pair (socks), The Best Food Boys (mayonnaise), or The Taystee Loafers (bread).

Not only did radio comedy get its performers from vaudeville, radio adopted vaudeville's form as well. A missing component was the audience itself. Initially, broadcasting executives thought that the sound of laughter might be a distraction to listeners, so members of the technical crew or other visitors to the studio were under strict orders to remain absolutely silent during the performance. This practice didn't last very long: comedians gauge their timing and modulate their acts based on audience reaction. Eddie Cantor was the first to insist that audience members not only be allowed to laugh, but *encouraged* to do so. Although there was some criticism thereafter that occasionally comedians played too much to the studio audience at the expense of listeners at home, for the most part the radio audience accepted and even came to expect a live audience's reactions.

Another missing element was, of course, sight. Whereas a comedian on stage could engage in all manner of leers, sight gags, takes and double takes, even dropping his pants if things got really desperate, all this was lost on radio. Ed Wynn, "The Perfect Fool," would dress up in costume for his radio shows, saying he thought if he looked and felt funny, he'd sound funny. Yet much of his appeal depended on the broad, physical comedy of the stage, and his radio career was only moderately successful. But once comedians and comedy writers adjusted to this limitation, they learned to exploit it, frequently using it as a magician uses misdirection. For example, in a scene from *The Jack Benny Program*, a nervous Jack is riding in his vintage Maxwell auto, nagging Rochester to watch where he's going. "But Boss," protests Rochester, "*you're* driving!"

Because it soon became apparent that lengthy monologs grew tiresome to home listeners, a second voice in the form of a foil or "stooge" came into vogue. Frequently it was the announcer who, after introducing the star, would stick around for a few minutes to engage in comedic dialog, usually as the straight man. Graham McNamee bantered with Ed Wynn, Jimmy Wallington with Eddie Cantor, Harry Von Zell and later Kenny Delmar with Fred Allen, and, for more than 30 years, Don Wilson sparred gently with Jack Benny. Sometimes other characters filled this role, often in dialect. Eddie Cantor played straight man to Bert Gordon's "the Mad Russian" whose frenzied opening line, "How do you doooo," never failed to get a laugh. Several comedians called upon their wives. Fred Allen's wife, Portland Hoffa, always entered off-mike screeching, "Mister Aaaallen! Mister Aaaallen!," before launching into a description of Momma's latest letter from home. Mary Livingston was always around to puncture husband Jack Benny's latest pomposity. And George Burns was the quintessential straight man to wife Gracie Allen's scatter-brained humor.

J. Fred MacDonald (1979) writes that this device allowed the comedian to better delineate his own personality. Without Mary, Jack Benny's foibles were less "real" and therefore less funny. Fred Allen—one of radio's all time great wits—needed someone to react to, establishing a kind of almost detached bemusement that was the basis for much of his observational humor.

During the 1930s, the big, expensive, star-driven comedy-variety shows were the most popular form of entertainment on the air. All had several elements in common: they usually opened with a musical number, followed by a monolog (or dialog), then more music, one or more comedy skits, usually featuring guest stars from other shows or the movies, still more music, and a short closing bit with the guest star before saying good night. This formula, with nominal variations, satisfied listeners for more than twenty years.

Ethnicity and Race

And what did audiences laugh about? Frequently they laughed at ethnicity. To the modern ear, much of the humor of that era can seem insensitive, sometimes even bordering on cruel. But this was an America still in the process of digesting the second great wave of immigration, predominantly from Southern and Eastern Europe. While immigrants themselves often listened to the radio to discover their place in the new culture, native-born Americans were tuning in to hear caricatures and stereotypes of the recent arrivals. The Irish were usually portrayed as a police officers, if not as drunks. Asians—usually Chinese—were either obsequious launderers or mysterious and inscrutable villains. Mexicans were lazy, the French were great lovers, the British insufferable prigs. These and other stereotypes were commonly understood by audiences and formed the basis for numerous jokes and comedic situations.

For example, Minerva Pious portrayed the "typical" urban Jewish housewife, Mrs. Nussbaum, who was constantly "Yiddishizing" recognizable names, such as Emperor Shapiro-Hito (for Hirohito), Cecil B. Schlemiel (DeMille), Weinstein Churchill, and Heimie Wadsworth Longfellow. Other ethnic characters who would pop up on various shows were Jack Pearl's German Baron von Munchausen, Harry Einstein's Greek Parkyakarkas, and Mel Blanc's lazy Mexican known only as Si (pronounced sigh).

But in many ways, the ultimate ethnic stereotype was reserved for African-Americans. Just as movie audiences were accustomed to shiftless, superstitious, and subservient black characters like Stepin Fetchit, so were radio audiences offered a succession of black maids, handymen, and janitors whose foibles and frailties were often played for laughs.

But while movies at least provided employment for black actors, radio usually did not. The popular character Beulah, of *The Beulah Show*, was portrayed by a white man, Marlin

Hurt. Part of the studio presentation involved Hurt's standing among other actors with his back to the audience, turning around only to bellow his opening line in falsetto "colored" dialect, "Somebody bawl fo' Beulah?" Radio listeners could only wonder at the studio audience's astonished reaction.

The most popular, and longest running, black-impersonation act was the phenomenally successful *Amos 'n' Andy.* Freeman Gosden and Charles Correll, both white men, had come out of the minstrel tradition and they teamed up to create two black characters whose adventures spanned the entire life of radio's so-called "Golden Era." Their format eventually spawned the soap opera and the situation comedy. Another—and perhaps more revolutionary—innovation of *Amos 'n' Andy* was to create in listeners' minds a rich and varied black subculture filled with bankers, lawyers, doctors, and other professionals along with the more stereotypical scoundrels, braggarts, and ne'er-do-wells—all played by Gosden and Correll. In fact, take away the dialect, and one would be hard pressed to identify much that was particularly "black" about any of the program's plot lines or characterizations. Indeed, were it merely a minstrel show on radio, *Amos 'n' Andy* could hardly have riveted the nation's attention as it did. Listeners may have tuned in for the laughs, but they returned because of the fully developed characters and stories.

Amos 'n' Andy also influenced the creation of other programs, similar in form, if not in content. *Lum 'n' Abner* was a variation on the ethnic comedy known as the "rube" show. It featured two bumpkins who presided over the Jot 'em Down Store in the then-fictitious town of Pine Ridge, Arkansas. (In 1936, the town of Waters changed its name to Pine Ridge.) Creators Chester Lauck and Norris Goff played the title characters and everyone else who happened to come in to the store, such as Grandpappy Peabody, Snake Hogan, Doc Miller, and Squire Skimp. (Laureen Tuttle added female voices in 1937.) Sometimes the stories were complete in a single episode. Sometimes they could extend for weeks. When a woman asked Lum to watch her baby for a few moments, then disappeared, the story went on for 40 episodes.

The town of Cooper, Illinois, "40 miles from Peoria," was the setting for *Vic and Sade.* The Gooks were a so-called typical American couple who lived with their adopted son Rush "in a little house halfway up the next block." John Dunning calls the program an American original, in a category of its own making. Though it was a daily, daytime show, it was in no way a soap opera. In fact, it was not even a serial, but rather presented 10-minute sketches that individually stood on their own. In one episode, for example, Uncle Fletcher drops by the Gooks' house to make a long distance call to a family relative. But the then-complex process of getting a long distance line, coupled with the rest of the family's disputes on the proper telephone protocol, finally sends Fletcher home without ever making the call. Its creator, Paul Rhymer, populated the series with such goofy characters as Dottie Brainfeeble, Smelly Clark, Ruthie Stembottom, and Vic's cousin Ishigan Fishigan who hailed from Sishigan, Michigan, most of whom were only referred to but never heard. It was an understated show that eschewed big laughs in favor of smiles punctuated by occasional chuckles.

Ethnic humor became considerably toned down once World War II was underway and Hitler's racist policy of Jewish extermination became more widely understood. Suddenly it was no longer quite as funny to single out a person's racial or national origins as the basis for laughs. For example, one notices a distinct difference between the prewar and postwar portrayals of Jack Benny's black valet, Rochester (Eddie Anderson). Before the war, Rochester was a razor-carrying, craps-shooting womanizer. After the war, those attributes had all but disappeared. When, in 1945, *The Abbott and Costello Show* aired a sketch involving a Jewish loan shark who wanted two quarts of Lou's blood for collateral, the public criticism was immediate and emphatic.

Character and Cliché

Early radio comedy had been based—as in vaudeville—on jokes or gags. From *The Joe Penner Show* came this exchange. Penner: "Waiter, I must say, this is not very good goulash." Waiter: "I can't understand it. I used a pair of your best goulashes." But within a very few years, radio writers' extensively cross-indexed joke reference files had been exhausted. In 1934 Eddie Cantor called for an end to gag-style comedy, saying the public was no longer fooled by dressing up the old jokes and calling them new. Eventually the gags were subordinated to comedy based on characterization. And no one was more adept at that than Jack Benny.

Benny's on-air personality developed slowly over the years. In his radio debut in 1932, he is a suave, somewhat self-deprecating host, serving up jokes and quips between musical numbers. By 1940 his character is fully realized: stingy and vain, he supposedly plays the violin badly and never admits to being older than 39. One of the most celebrated episodes of the series is particularly instructive if one listens to the audience reaction. Jack is being held up, and the pistol-wielding thief growls the immortal line, "Your money or your life." Jack's cheapness is so well understood by this time that the studio audience begins to laugh immediately, even *before* he can deliver the intended laugh line, "I'm thinking it over!"—and that only after a very long pause allowing the laughter to build.

Another comedy program that depended heavily on characterization was *The Edgar Bergen/Charlie McCarthy Show.* But the character in question wasn't even really a person—except in the minds of audiences—but rather a ventriloquist's dummy. Charlie McCarthy was depicted as a mischievous and sometimes lascivious 10-12 year old boy, with Edgar Bergen playing a sort of ambiguous parent figure. Probably because he *was* a dummy, audiences accepted Charlie's sometimes lecherous come-ons to glamorous female guest stars. Had he actually been a child, this could have been highly objectionable.

Related to comedy based on character were the running gags or comedic clichés. These were situations or routines that became funnier by the very fact of their repetition. Audiences came to welcome each new variation on the familiar theme. Two of the most famous were the Benny-Allen feud and Fibber McGee's closet.

On one episode of his program in 1937, Fred Allen, following a dazzling guest performance by a ten-year-old violinist, ad-libbed, "Jack Benny should be ashamed of himself." Fortunately, Benny was listening and thought it was funny, so on his next program he reacted by defending his own prowess on the violin, making some disparaging remarks about Allen in the process, and the "feud" was on. The two programs played the supposed conflict for laughs until Allen finally left the air in 1949.

Fibber McGee and Molly was one of radio's longest running situation comedies. As played by real-life married couple Marian and Jim Jordan, Fibber was a lovable windbag and Molly his patient wife. This program may have had more running gags than any other, the most famous being a hall closet so stuffed full of junk that every time the door was opened everything would come crashing down in a nearly epic cacophony of sound. Listeners at home could either laugh at the closet of their imaginations or at the sound effects wizardry that went into its creation. At the end of the last clink, Fibber would inevitably say, "I've gotta clean out that closet one of these days."

War and Controversy

During World War II, radio comedy played its part in keeping homefront morale high. Most programs integrated war-related themes into their plotlines or sketches. The 1944 New Year's *Jack Benny Program* contains a sketch in which a metaphorical World Series baseball game is played between the Axis Polecats and the Allied All-Stars. Various military campaigns are transformed into hits, sacrifice flies, and walks. As the program ends, General Eisenhower is about to come to bat. Later in 1944, on *Fibber McGee and Molly,* Fibber thinks he has a brilliant idea that will revolutionize postwar travel, but he must travel from his home in Wistful Vista in order to pitch it to some government official. The trains are filled with servicemen either returning from or going on leave. No matter how hard he tries, Fibber can't get a ticket and is berated by everyone he meets for trying to take up valuable space that could be used by a soldier to get home. At the end of the program, the Jordans step out of character and appeal directly to the audience not to travel unless absolutely necessary. Comedy shows also addressed other topics like scrap drives, War Bonds, victory gardens, the rubber shortage and anything else that helped out "our boys."

The war had another effect on radio comedy. During the 1930s, comedians and writers had avoided potentially controversial topics such as politics in their plots or sketches. The only notable exception to this rule was humorist Will Rogers, whose rural-flavored, good-natured ribbing made his jibes palatable. ("I don't belong to an organized political party," he would say. "I'm a Democrat.") But when Rogers was killed in a plane crash in 1935, radio comedy became essentially a controversy-free zone. For example, at one point in 1940, Fibber McGee apologized for inadvertently saying "china" on the air when he meant dishes, acknowledging that "we can't say anything controversial."

Once America was in the war in late 1941, however, radio comedy took a turn for the political: references to national and world events, governmental leaders, and current issues were woven into scripts. Among the most bitingly satirical of the newer generation of comedians was Henry Morgan, whose program, *Here's Morgan,* began on a local station in New York before getting a spot on the Mutual Network. He once "interviewed" a businessman in a mythical southern state who said the new governor of Georgia—formerly associated with the KKK—was great for his business, manufacturing bed sheets. On another show, in the postwar era when housing was tight, he presented a dialog between two landlords, one of whom expressed dismay that the eighth floor of his tenement had caved in. When asked if anyone was hurt, the landlord replied, "No, just tenants." He wasn't particularly kind to business institutions, either. "You know," he said, "most people think of banks as cold, heartless, large institutions. And they're wrong. There are small ones, too." This kind of humor on radio would have been almost unthinkable only a few years earlier.

Situation Comedy

The postwar era also saw the rise of the situation comedy. Aside from its pictures, the format of the modern-day TV sitcom is virtually indistinguishable from that of its radio progenitor of the 1940s. The American family was the central location for many of them—*The Great Gildersleeve, The Aldrich Family, Father Knows Best, Blondie*—but sitcoms also found comedy in high school (*Our Miss Brooks),* in the blue collar workplace (*The Life of Riley),* a restaurant (*Meet Me at Parky's),* and even a bar (*Duffy's Tavern).*

One of the most popular was *The Phil Harris-Alice Faye Show,* in which Jack Benny's band leader and his wife, a popular singer and film actress, played fictionalized versions of themselves. The versatile actor-director Elliot Lewis played Frankie Remley, an actual member of Harris' band. The program grew out of the many wisecracks from the Benny show about the band's supposed incompetence (though it was obviously first rate), Harris' presumed inability to read (words *or* music), and Remley's purported drinking. These characteristics were extended and enlarged in the sitcom, and placed within the context of the zany Harris-Faye home life and Phil and Frankie's misadventures.

MacDonald calls these shows middle-class morality tales with the family portrayed as *the* vital American institution. Plots tended to revolve around insignificant misunderstandings that were resolved by show's end. Such core values as trust, love, honesty, and tolerance always triumphed. Of the top 10 programs in the 1947 season chosen by Protestant churches as those most faithfully portraying American life, five were situation comedies.

Television Takes Charge

But by the end of the 1940s, the end of an era was drawing near. After having been postponed first by war, and then by technical problems, television was now ready to take its place as the center of family home entertainment. Radio fought its upstart competitor with, among other things, a weekly, 90-minute comedy-variety extravaganza on NBC called *The Big Show,* hosted by the Broadway and film star Tallulah Bankhead and featuring numerous guest stars from all points on the entertainment compass. But it was too *much* and too late. Though lavish and expensive, it only lasted two seasons.

Other comedians and sitcoms were rapidly jumping ship to try out the new medium. Most of the old line vaudevillians were unable to make the transition, except for occasional guest appearances on TV variety shows like *The Colgate Comedy Hour.* A few did well, however. Jack Benny first appeared on the small screen in 1950 but continued to do the radio program concurrently with television until 1955. Bob Hope also ended his radio series in 1955 and continued to perform on television for more than 30 years. Red Skelton was even more successful on television than on radio because so much of his humor was visual. His weekly television series ran from 1951 to 1971 and was usually among the highest-rated shows on the air. But the most spectacular transition from radio to television was made by Milton Berle. His radio career had been indifferent at best, but his broad, visual form of comedy was perfect for the tube. What *Amos 'n' Andy* had done for radio 20 years earlier, Berle did for television: create excitement about the new medium and sell receiving sets.

On the other hand, Fred Allen had retired from radio in 1949, a victim of falling ratings and his own poor health. He did guest spots on television but never seemed really comfortable there. Allen was a "word" man in a visual medium. His last job on the air was as a panelist on the TV game show, *What's My Line?*

One of radio's lasting legacies was the situation comedy format. Although the sitcom found success on radio, it has flourished on television for even longer. Making their way to television from radio were, among others, *The Life of Riley, Father Knows Best, Burns and Allen, The Goldbergs, December Bride,* and a reworked form of Lucille Ball's radio series *My Favorite Husband,* retitled *I Love Lucy.*

Wit, Satire, and Shock

Although radio comedy on a national scale dwindled during the 1950s, replaced by local disc jockeys and personalities, there was still room for innovative young comics with a satirical edge to their humor. Bob Elliot and Ray Goulding, more familiarly known as *Bob and Ray,* had started their radio career in Boston, joining the NBC network for a daily 15-minute slot in 1951. Eventually they were heard on all the commercial networks at various times until 1960 and even did several limited series on National Public Radio in the 1980s. Their straight-faced, understated routines were frequently hilarious. They generally used no script, sometimes improvising absurd mock interviews as conducted by ace reporter Wally Ballou ("winner of seven international diction awards"), other times spoofing soap operas with scenes from *One Feller's Family* (a dig at the long-running *One Man's Family)* or *Mary Backstayge, Noble Wife.* The detective series, *Mr. Keen, Tracer of Lost Persons* became *Mr. Trace, Keener Than Most Persons.* They offered numerous ersatz premiums, such as the Bob and Ray Home Surgery Kit or membership in Heightwatcher's International ("six ample servings of low vitamins and nutrients in artificial colorings"). With parody, verbal nonsense, non sequitur, and wit they created what has been described as a surrealistic Dickensian repertory company, all of it clean, subtle and gentle.

Jean Shepherd's rambling, discursive, free-form style was a lineal audio descendant of that of Henry Morgan. Shepherd's program, *Night People,* was broadcast on New York's WOR from 1956 to 1977 and heard in 27 states, parts of Canada and as far south as Bermuda. He was a comic anthropologist, offering mock commentary on social and cultural trends and behavior. A radio raconteur, he would launch into a rambling chat with a central story in mind, often digressing wildly, sometimes playing "The Sheik of Araby" on the kazoo while rhythmically thumping his knuckles on his head, usually wandering back to his main point just as time was running out on his show. His extemporaneous storytelling has been compared to making pizza in the window of a restaurant.

When broadcasting was largely deregulated in the 1980s, standards for acceptable content were liberalized. This made way for so-called "shock jocks" like Morton Downey, Jr., Don Imus, and Andrew Dice Clay, radio personalities whose routine references to sex and use of crude language resulted in endless controversy, occasional fines from the FCC, and laments that the end of civilization was at hand. None have generated more notoriety than Howard Stern, self-styled King of All Media (AKA Fartman), who parlayed a local show in New York into one of the highest-rated programs in national syndication. Stern's clownish, flamboyant brand of humor is the lowest of low brow. He frequently describes his own sexual fantasies, engages in personal attacks, and serves up his own bizarre take on current events (he once wondered how necrophiliac Jeffrey Dahmer could get a fair trial unless there were more guys on the jury who wanted to have sex with dead men). He is rude, crude, and, to fans, often very funny.

At the other end of the spectrum, both figuratively and literally (being at the bottom of the FM dial) is humor served up by public radio. *Car Talk* features Click and Clack, the Tappet Brothers (Tom and Ray Magliozzi), dispensing car advice between self-deprecating jokes, funny letters from listeners, puzzlers, and features like "Stump the Chumps," in which callers are asked if advice they got from Click and Clack some time previous was any good (frequently it wasn't, but nobody really seems to mind). Michael Feldman's *Whad'Ya Know?* is a two-hour comedy/quiz on which audience members and callers compete for whimsical prizes. *Wait Wait... Don't Tell Me* plays the week's news for laughs and offers callers who correctly answer questions the highly coveted prize of veteran newscaster Carl Castle's voice on their answering machine. *Rewind* also lampoons the news through comic skits and extemporaneous commentary from guest comedians.

The one real throwback to an earlier era is Garrison Keillor's *Prairie Home Companion,* which, ironically, has been on the air longer than any of the original comedy-variety shows. Broadcast live before a large theater audience, its form—if not its content—is somewhat reminiscent of *The Fred Allen Show,* circa 1940. Host Keillor banters with guests, introduces musical acts (and often sings himself), and performs with his troupe in various comedy sketches and fake commercials for "sponsors" like Powdermilk Bisquits and the Catsup Advisory Board. The centerpiece is a weekly 20-minute monolog, "News from Lake Wobegon," in which Keillor tells stories and ruminates on life in his mythical Minnesota home town.

Although radio has certainly not abandoned comedy, it has yielded to television its place as America's primary purveyor of laughter. Mostly gone, then, is a form of humor that depends on listeners' active participation through imagination. Susan Douglas calls this "dimensional listening." For example, Jack Benny's money vault was never as funny on television as it had been on radio, when listeners conjured up their own visions of moats, chains, gates, and a bearded guard who had not seen the light of day since the Civil War. This was radio's contribution to comedy and has since passed into aural history.

ALAN BELL

See also, in addition to performers and programs mentioned in this essay, British Radio Programming; Canadian Radio Satire; The Goon Show; Shock Jocks; Situation Comedy; Stereotypes on Radio; Variety Shows; Vaudeville

Further Reading

Buxton, Frank, and Bill Owen, *Radio's Golden Age: The Programs and the Personalities,* New York: Easton Valley Press, 1966; revised edition, as *The Big Broadcast: 1920–1950,* New York: Viking Press, 1972

Douglas, Susan J., "Radio Comedy and Lingusitic Slapstick," in *Listening In: Radio and the American Imagination,* by Douglas, New York: Times Books, 1999

Dunning, John, *Tune in Yesterday: The Ultimate Encyclopedia of Old-Time Radio, 1925–1976,* Englewood Cliffs, New Jersey: Prentice-Hall, 1976; revised edition, as *On The Air: The Encyclopedia of Old-Time Radio,* New York: Oxford University Press, 1998

Firestone, Ross, editor, *The Big Radio Comedy Program,* Chicago: Contemporary Books, 1978

Flick Lives: "A Salute to Jean Shepherd," <www.flicklives.com/index.html>

Gaver, Jack, and Dave Stanley, *There's Laughter in the Air! Radio's Top Comedians and Their Best Shows,* New York: Greenberg, 1945

Harmon, Jim, *The Great Radio Comedians,* Garden City, New York: Doubleday, 1970

Havig, Alan R., *Fred Allen's Radio Comedy,* Philadelphia, Pennsylvania: Temple University Press, 1990

Hilmes, Michele, *Radio Voices: American Broadcasting, 1922–1952,* Minneapolis: University of Minnesota Press, 1997

Lackmann, Ronald W., *Same Time, Same Station: An A-Z Guide to Radio from Jack Benny to Howard Stern,* New York: Facts on File, 1996; revised edition, as *The Encyclopedia of American Radio: An A-Z Guide to Radio from Jack Benny to Howard Stern,* 2000

MacDonald, J. Fred, *Don't Touch That Dial! Radio Programming in American Life, 1920–1960,* Chicago: Nelson-Hall, 1979

Nachman, Gerald, *Raised on Radio,* New York: Pantheon Books, 1998

Poole, Gary, *Radio Comedy Diary: A Researcher's Guide to the Actual Jokes and Quotes of the Top Comedy Programs of 1947–1950,* Jefferson, North Carolina: McFarland, 2001

Radio Days: "Comedy Central," <www.otr.com/comedy.html>

Wertheim, Arthur Frank, *Radio Comedy,* New York: Oxford University Press, 1979

Commentators

Expressing Opinions on News Events

During the golden age of network radio news commentary, from the late 1930s into the 1950s, some of the best known broadcast journalists regularly presented their views on domestic and world events. The term *commentator* usually refers to a journalist who provides insight, comment, or opinion about current news events or trends. Such comment may be interspersed with the news itself or, more commonly, may be presented as a separately identified program or segment. It is generally understood that such comment represents the commentator's own ideas rather than the editorial views of station or network management. In recent years, though, because of the popularity of talk radio hosts who express strong political views, the definition has become blurred.

Origins

Commentary occurred before regular radio news reports had become established. H.V. Kaltenborn was offering opinion (just as he did in his newspaper column) on New York stations in the early 1920s—as, more occasionally, did other reporters in other cities. Many were purely local in their coverage and appeal; others became nationally known names.

The agreement signed in 1933 between the radio networks and press associations to end the "Press-Radio War" served to limit the number of daily newscasts but did not affect daily commentaries "devoted to a generalization and general news situations, so long as the commentators do not report spot news." Furthermore, unlike network newscasts, commentators could be sponsored. Many radio journalists were quickly reclassified as news commentators, and commentary got a new lease on life just as the world political situation cried for informed analysis.

Americans soon paid close attention to radio commentators, in part because they could identify more readily with the informed voice they could hear than with disembodied words on a newspaper or magazine page. Radio's commentators seemed to be reasoning with their listeners, not lecturing to them as they often seemed to be doing in print.

Determining Commentary's Place

By its nature, however, commentary deals with matters of controversy. Controversy creates diverse points of view and thus disagreements. The broadcasting industry has never been a particularly active presenter of diverse viewpoints: although they please some listeners, they will by their very nature anger others. So from the beginning, commentators faced unique pressures.

Early on, broadcasters sought a neutral or objective role in providing news and commentary. The National Association of Broadcasters added to its own code of good programming practices in 1939 a recommendation that

> News shall not be selected for the purpose of furthering or hindering either side of any controversial public issue nor shall it be colored by the opinions or desires of the station or the network management, the editor or others engaged in its preparation, or the person actually delivering it over the air, or, in the case of sponsored news broadcasts, the advertiser. . . . News commentators as well as other newscasters shall be governed by those provisions.

Mutual took the loosest approach to the code, often providing commentators thought too harsh or one-sided by the other networks. National Broadcasting Company (NBC) News called for coverage devoid of all personal feeling, thought, or opinion. The Columbia Broadcasting System (CBS) became the toughest of all when news head Paul White declared in 1943 that "the public interest cannot be served in radio by giving selected news analysts a preferred and one-sided position." He went on to conclude that the news analyst's job is "to marshal the facts on any specific subject and out of his common or special knowledge to present those facts so as to inform his listeners rather than to persuade them. . . . Ideally in the case of controversial issues, the audience should be left with no impression as to which side the analyst himself actually favors."

Kaltenborn's response to the CBS directive was that "no news analyst worth his salt could or would be completely neutral or objective. He shows his editorial bias by every selection or rejection from the vast mass of news material placed before him. Opinion is often expressed by the mere shading and emphasis." Walter Winchell on NBC Blue huffed, "Aren't we lucky that Patrick Henry's message didn't have to be reported by the Columbia Broadcasting System."

Pressures on commentators to tone down their message came from management (which always had a fear of alienating advertisers), from listeners unhappy with the views they heard, and often from politicians or others who were critiqued. Because advertisers paid the bills, they always had a strong say. The networks took different paths to solve the problem of perceived commentator bias. CBS and NBC attempted to stop any political slanting. On the other hand, Mutual and the American Broadcasting Companies (ABC) tried to offset bias (and, as the weak networks, gather both listeners and sponsors) by adding more commentators with differing points of view. Reviewing commentator careers, one notes that many began on CBS or NBC and ended up on ABC or Mutual.

In 1942 an Association of Radio News Analysts (ARNA) was formed by a group of 31 New York–based commentators as a craft guild. The membership of the organization included most of the leading figures in the business. Kaltenborn was president, Elmer Davis and Raymond Gram Swing were vice presidents, and Quincy Howe became secretary-treasurer. A number of potential members were excluded because of some doubts about whether they were really offering commentary. ARNA fought for such things as a proposal by Swing to eliminate the "middle commercial" as an interruption of the analysis or commentary. On the other hand, some commentators read their own commercials and (it was reported) did so with some enthusiasm. Gabriel Heatter switched from the war against Germany "to the great war against gingivitis—gingivitis, that creeps in like a saboteur." But ARNA's primary concern focused on the right to offer commentary in the first place.

On behalf of his fellow commentators, H.V. Kaltenborn advised network and station owners to "hire the best men you can get with the money you can pay" and then

> Tell them exactly what you expect, what you are trying to do with your station or network. Then give them their heads. If they get out of line, correct them. If they continually violate what you deem to be an essential policy, fire them. But don't pretend you are going to be able to prevent a commentator worth his salt from expressing his opinion.

Fang (1977) reports that the number of network news commentators rose from 6 in 1931 to about 20 when World War II began, whereas perhaps as many as 600 commentators were reporting news and analyzing events for networks and larger stations shortly after the war ended in 1945.

Examples of Radio Commentators

To supplement those radio commentators discussed in their own entries in this encyclopedia, here are brief summaries of the professional lives of a few more who were active from the 1930s to the 1960s.

Hilmar Robert Baukhage (1889–1976)

Beginning in 1932 on NBC Blue with a five-minute daily commentary, he identified himself simply as "Baukhage talking." By the 1940s, Baukhage was providing a 15-minute daily program on ABC, sometimes dealing with several topics but often focusing an extended essay on a single subject. He added part-time broadcasts on Mutual beginning in 1948 and joined that network full-time from 1951 to 1953. He provided a no-nonsense tone to his broadcasts that many listeners appreciated.

Cecil Brown (1907–87)

Brown was one of those reporters who seemed at his best when in a struggle with someone else. Hired by Murrow in 1940 to cover Italy, he had a short fuse and thus a short network career. Paul White had trouble with Brown, who in 1943 had just returned from a national speaking tour and had commented on the air that the American people seemed to have lost interest in the war. White questioned why Brown had not qualified his statement with a comment such as "From information I received in those interviews, I gathered that Americans are losing interest in the war." The network was already concerned about Brown, who was losing his sponsors. Brown's response was to resign and to call a news conference to declare himself a victim of censorship. Quite liberal in his views (e.g., he favored racial integration earlier than most), Brown moved to ABC and then NBC and left journalism in 1967.

Boake Carter (1903–44)

In the mid-1930s, Boake (Harold Thomas Henry) Carter was for a period the most popular radio news commentator in the United States. Carter's first chance at broadcasting came in 1930, when WCAU (the CBS affiliate in Philadelphia) wanted someone to broadcast a rugby match. His big broadcast break came with the Lindbergh baby kidnapping in 1932. National exposure placed Carter on CBS radio in direct competition with other major commentators. He was soon a favorite with listeners. Carter spoke very fast and used metaphors and clichés to create images in his listeners' minds. He also liked to put himself into his stories. Despite his English background, he was a strong isolationist, developed a bitter dislike for British foreign policy, and accused the British of trying to drag the United States into the war. He increasingly attacked the Roosevelt administration. In April 1938 CBS took Carter off the air as his ratings were declining and his attacks on others had grown too harsh. He continued as a newspaper columnist until his death.

Upton Close (1894–1960)

Born Joseph Washington Hall, Close became a prominent right-wing commentator on network radio, but that is far from how he began. After early experience in and writing books about China, Close entered radio as a result of extensive public lecturing (much like Lowell Thomas at about the same time). He began offering occasional broadcasts on NBC as a Far East authority. In 1942 he began a weekly Sunday political commentary, also on NBC, called *Close-Ups of the News*. But he became more conservative and even shrill as the war went on, and NBC, concerned with low ratings, took him off the air at the end of 1944. For another year he broadcast increasingly right-wing commentary for the Mutual network before retiring in 1946 and moving to Mexico.

Floyd Gibbons (1887–1939)
A traveler and a dashing foreign and war correspondent, Gibbons covered World War I for the *Chicago Tribune* and lost his left eye reporting on American forces fighting in Belleau Wood. For the rest of his life he wore an eye patch, which added to his adventurous image. He broadcast on the *Tribune*'s WGN in 1926 and then on NBC as the *Headline Hunter* in 1929, but that program provided more entertainment than hard news. Gibbons became the first network daily newscaster in 1930 (still on NBC), sponsored by the *Literary Digest* for six months, but he was soon replaced by Lowell Thomas.

Gabriel Heatter (1890–1972)
Heatter had worked in print journalism for a number of years and had done occasional news broadcasts. As with Boake Carter, his big break came in 1933, when New York station WOR assigned him to both report and comment on the Lindbergh kidnapping murder trial on which public attention was focused at the time. Heatter continued to report and comment on world events as war broke out in 1939. His familiar catch phrase first appeared early in the war when things were not going well for the Allies. After American naval forces sank a Japanese destroyer, Heatter began his evening broadcast saying "there is good news tonight." He continued to use the phrase throughout his career on the Mutual network, becoming known as a morale booster.

Edwin C. Hill (1884–1957)
Hill was another newspaperman (a feature writer for the *New York Sun* and King Features Syndicate) who later turned to motion pictures (his was the voice on Fox newsreels beginning in 1923) and in 1932 to radio. His *Human Side of the News* offered often sentimental features focused on people in the news. He became more politically conservative over time.

Don Hollenbeck (1905–54)
Hollenbeck worked as a print journalist, radio reporter, and photojournalist before he began broadcasting reports of World War II from London, North Africa, and Italy for the Office of War Information. He was sent to Algiers in time to join the Allied troops for the invasion of Salerno (1943). In Italy he was one of the first correspondents to begin broadcasting from Naples. On one occasion in 1946, Hollenbeck began a newscast following a singing commercial by saying, "The atrocity you have just heard is not part of this program." Not surprisingly, by noon he was on the street looking for another job. Some years later, he lost another job for criticizing Senator Joseph McCarthy. He joined CBS in 1948, but in 1954, ailing and depressed by attacks from McCarthy supporters, especially the Hearst newspapers, Hollenbeck took his own life.

Fulton Lewis Jr. (1903–66)
Lewis offered conservative commentary on the Mutual network for three decades. He worked for various Washington, D.C., newspapers in the 1920s and 1930s and became a substitute newscaster on Mutual's WOL in Washington in 1937. He was soon hotly popular on some 500 stations with his evening program touting his conservative and isolationist views, and radio became his primary medium. He led the successful battle to get broadcast reporters admitted to the congressional press galleries. He loved to feature the latest government boondoggle. He also loved to double up on adjectives to describe those he disagreed with (e.g., "an inexperienced, impractical, theoretical college professor"). He backed McCarthy in the early 1950s. He briefly tried television but was not successful. His audience began to diminish as his politics remained stuck in the far right.

Edward P. Morgan (1910–93)
After an early career in print media (newspapers and magazines), Morgan worked with CBS from 1951 to 1954 as both a broadcaster and producer, then with ABC (1955–75) where he offered commentary on radio and television. Sponsored by the American Federation of Labor and Congress of Industrial Organizations in what was radio's last regular network evening news program, he began with news and then concluded with what he called "the shape of one man's opinion," commenting on one or more issues in a script he prepared himself. Strongly incisive even when commenting on the media (as few in radio did), he often offered forthright, stinging critiques when he felt strongly about an event.

Drew Pearson (1897–1969)
Pearson began his journalism career as the diplomatic correspondent for both a magazine and a newspaper, where he gained extensive international experience. He was the coauthor (with several others, for the longest period with Jack Anderson) of the "Washington Merry-Go-Round," a widely syndicated newspaper column, beginning in 1932, generally taking on most conservatives and supporting liberals, all of this fed by good reporting and many leaks. In 1940 the success of the column led to his being offered a weekly NBC Blue half-hour Sunday commentary program, which carried the same kind of "hit-em-hard" investigative reporting, often about government mistakes or malfeasance. He fascinated listeners and readers alike with reports based on leaks or informants. Pearson had brief television stints on ABC and DuMont television in the early 1950s.

Howard K. Smith (1914–2002)
Smith was a Rhodes Scholar (1937–39) and worked for United Press International (1939–41). He was one of "Murrow's Boys" when he joined CBS in Berlin in 1941 and was on the "last train" out in December 1941. Smith was the chief European correspondent for CBS after Murrow. After the war, he provided commentaries on Douglas Edwards' television news in the 1950s on CBS. He resigned from CBS in 1961 over the degree of freedom to comment. ABC gave him a program and the freedom to comment, but the show lost its sponsor and Smith his program; he resigned from ABC in 1979.

Dorothy Thompson (1894–1961)
Dorothy Thompson began her career as a publicity writer for the women's suffrage movement as well as for advertising agencies. She went to Europe, where she interviewed many world leaders and sold articles to the Philadelphia *Public Ledger* and the Chicago *Daily News,* becoming a permanent foreign correspondent for both—one of the earliest female foreign correspondents. She was outspoken opponent of Nazism, which led to her expulsion from Germany in 1934. She began as a radio news commentator with NBC during the 1936 political conventions and had her own weekly commentary program a year later, which lasted to early 1945. She spoke rapidly—her listeners got used to a torrent of words in her allotted 15-minute program. She was generally liberal but could be independent and sometimes confusing. She continued her newspaper column and speech-giving until the late 1950s.

Decline

With the decline of radio networks in the 1950s, commentary began to disappear. It had been a rarity on local stations even in the 1930s and 1940s. There was little room for a 15-minute news program (let alone commentary) on tightly formatted stations that tried to retain the same sound appeal at all times. Advertisers, always uncomfortable with riling up listeners with controversy, increasingly turned away from supporting such programs. A few television journalists, including Walter Cronkite, avoided commentary on their primary news medium but did offer it over radio.

What radio news survived by the 1980s and 1990s was more of a quick "rip and read" nature, "plus traffic and sports," than anything providing in-depth reporting, let alone analysis. Talk shows with opinionated hosts (none of them journalists) replaced reasoned and thoughtful commentary, which could still be found in scattered moments on television but more consistently in the press. The quest for profit, which became stronger in the 1990s with industry consolidation, wiped out any chance that commentary might return.

Public radio offered an alternative that appealed to small but influential audiences. Daniel Schorr, for example, joined National Public Radio in 1985 after a career on CBS and Cable News Network (CNN) and provided reasoned political and foreign affairs commentary on *All Things Considered*—thus providing a tiny remnant of what had once been a radio staple.

CHRISTOPHER H. STERLING

See also All Things Considered; Collingwood, Charles; Davis, Elmer; Frederick, Pauline; Harvey, Paul; Hottelet, Richard C.; Howe, Quincy; Kaltenborn, H.V.; Kuralt, Charles; Murrow, Edward R.; Osgood, Charles; Press-Radio War; Sevareid, Eric; Shirer, William L.; Swing, Raymond Gram; Thomas, Lowell; Totenberg, Nina; Trout, Robert; White, Paul

Further Reading

Bliss, Edward, Jr., "The Oracles," in *Now the News: The Story of Broadcast Journalism,* by Bliss, New York: Columbia University Press, 1991
Brown, Cecil B., *Suez to Singapore,* New York: Random House, 1942
Bulman, David, editor, *Molders of Opinion,* Milwaukee, Wisconsin: Bruce, 1945
Culbert, David Holbrook, *News for Everyman: Radio and Foreign Affairs in Thirties America,* Westport, Connecticut: Greenwood Press, 1976
Fang, Irving, E., *Those Radio Commentators!* Ames: Iowa State University Press, 1977
Gibbons, Edward, *Floyd Gibbons, Your Headline Hunter,* New York: Exposition Press, 1953
Heatter, Gabriel, *There's Good News Tonight,* Garden City, New York: Doubleday, 1960
Herndon, Booton, *Praised and Damned: The Story of Fulton Lewis Jr.,* edited by Gordon Carroll, New York: Duell Sloan and Pearce, 1954
Hosley, David H., *As Good As Any: Foreign Correspondence on American Radio, 1930–1940,* Westport, Connecticut: Greenwood Press, 1984
Howe, Quincy, "Policing the Commentator," *Atlantic Monthly* (November 1943)
Knoll, Steve, "Demise of the Radio Commentator: An Irreparable Loss to Broadcast Journalism," *Journal of Radio Studies* 6, no. 2 (1999)
Morgan, Edward P., *Clearing the Air,* Washington, D.C.: Luce, 1963
Pilat, Oliver Ramsay, *Drew Pearson: An Unauthorized Biography,* New York: Harper's Magazine Press, 1973
Sanders, Marion K., *Dorothy Thompson: A Legend in Her Time,* Boston: Houghton Mifflin, 1973
Smith, Howard K., *Events Leading Up to My Death: The Life of a Twentieth-Century Reporter,* New York: St. Martin's Press, 1996
Wecter, Dixon, "Hearing Is Believing," *Atlantic Monthly* (August 1945)

Commercial Load

Amount of Advertising Carried on Radio

Commercial load refers to the total amount of time commercials are broadcast on radio during an hour or some other specific time period. Radio stations, unlike the print media, have a limited commercial inventory, a finite amount of time available for advertising "spots." A broadcast hour cannot be longer than 60 minutes, and the broadcast day cannot be longer than 24 hours, whereas newspapers and magazines can add as many pages as necessary. Further, only so many commercials can be packed into each hour's programming without losing a significant part of the audience.

At one time, the National Association of Broadcasters (NAB) code recommended a limit of 18 minutes of commercials per hour on radio. But the NAB discarded the code when in 1984 the U.S. Justice Department alleged that the standards, although voluntary, violated antitrust laws by promoting limits that discouraged competition.

Historically, astute station management has carefully limited commercial load. When the legendary Bill Drake reinvented top 40 radio at KHJ in Los Angeles in the 1960s, he maintained an "iron-clad" hourly limit on commercials. Drake ordered that commercials should not exceed 13 minutes, 40 seconds per hour, nearly one-third less than the U.S. average at the time. When FM finally became successful in the late 1960s and early 1970s, listeners perceived it as the "less-commercials band," and operators wanting to maintain their stations' success instituted firm policies limiting the number of spot announcements per hour.

Since the top 40 hit-music format emerged in the 1950s, radio stations have grouped commercials in clusters called "spot sets" (or "stop sets"). A common approach has been to promote longer "sweeps" of uninterrupted music, a strategy that requires fewer but longer commercial breaks. However, an Arbitron/Edison Media Research study found strong support for more frequent, and shorter, spot sets. Fifty-two percent of those surveyed preferred more frequent stops with shorter blocks of commercials, while 39 percent prefer longer programming blocks and longer blocks of commercials. The findings led the report's authors to recommend that radio stations consider changes in their spot-clustering paradigms, but only after conducting research of their own audience's listening habits.

The number of commercial minutes each hour is entirely up to the management of each individual station. Increasingly, writers on the topic are concerned that a trend toward increasing the number of commercials per hour is having a negative effect on radio listenership. The Radio Advertising Bureau reported that radio-advertising revenue exceeded $17 billion in 1999, up 15 percent from the previous year. However, radio listenership had declined 12 percent over the past decade, according to the consulting firm Duncan's American Radio, with only 15.4 percent of the national population age 12 and over listening in any quarter hour, 6 A.M. to midnight, down from 17.5 percent in 1989. One of the reasons for the decline, according to Duncan's, is the trend toward higher spot loads. Some sources report stations airing up to 22 commercial minutes per hour.

A 2000 study by Empower MediaMarketing of Cincinnati found that the number of paid advertisements on radio stations grew by about 6 percent in the previous year. The greatest increase was in the San Francisco-Oakland-San Jose, California, market, where the number of 10-, 30-, and 60-second spots increased by 20 percent (see Kranhold, 2000). A 2001 Arbitron study found that advertisers perceived radio as the most "cluttered" mass communications medium. Although clutter (or the absence of clutter) was not considered a key criterion by most advertisers, the report recommended reducing spot loads.

Another Arbitron report suggests that higher commercial loads are turning off audiences. In the report titled "Will Your Audience Be Right Back After These Messages?" Arbitron and Edison found that 42 percent of radio listeners had noticed that stations are airing more commercials, although, interestingly, listeners are not as likely to believe that their own favorite station is playing more spots. Young listeners seem to be the most annoyed by the trend. The report found 31 percent of listeners ages 12–24 said they were listening to radio less, while 17 percent of listeners in the 25–54 demographic, and 11 percent of those 55 and older, said they were listening to less radio. The report suggests that the greater number of commercials is a major reason for the decline in time-spent-listening. Advertising agencies, on the other hand, suggest the results show the need for more entertaining commercials.

A Washington, D.C., station, WWVZ, seemingly took to heart Arbitron's advice to reduce commercial spot loads. The station implemented a format with only two three-minute commercial breaks per hour. An advertising executive with Hill, Holliday in Boston, Karen Agresti, hailed the station's decision. "Clutter is one of the biggest problems in radio. For a station to take a lower load is great, and I hope more will do it," she said. WBLI-FM (Long Island, New York), a contemporary hits station, cut its commercial load from 16–17 minutes per hour to 10–11 minutes. The station gained 3.5 ratings points among its core audience, women 18–34.

In an attempt to make room for more commercials, some radio (and television) stations have begun using a device called "Cash," which uses audio delay and "intelligent microediting" to create up to six minutes of additional commercial time per hour. The use of Cash drew criticism from the president of the American Association for Agencies, O. Burtch Drake. Drake said radio will not benefit in the long run from creating more clutter. "You can shoehorn more commercials in, but it hurts both the station and the advertiser," Drake said. "That's why we are taking a very strong stand against this kind of technology."

As a way to cut through the clutter, and as an alternative to hiring high-profile celebrity endorsers, some advertising agencies are advising clients to find popular local radio personalities to endorse their products.

J.M. DEMPSEY

See also Advertising on Radio; Arbitron; Drake, Bill; National Association of Broadcasters

Further Reading

Bachman, Katy, "Listeners and Advertisers Cheer Spot-less Loads," *Mediaweek* (14 December 1998)

Bachman, Katy, "Listeners Turning Off Radio," *Mediaweek* (21 June 1999)

Eberly, Philip K., *Music in the Air: America's Changing Tastes in Popular Music*, New York: Hastings House, 1982

"Executioner" (article on Bill Drake), *Time* (23 August 1968)

Kiesewetter, John, "Big Radio Airs the Sound of Sameness," *Cincinnati Enquirer* (19 March 2000)

Kranhold, Kathryn, "Advertising on Radio Increases 6%; San Francisco Area Sees 20% Rise," *Wall Street Journal* (12 April 2000)

McWhorter, Ben, "Sales Insights! Radio's Biggest Spenders Speak Up!" Arbitron study, 2001, <arbitron.com/downloads/radiosbiggestpres.pdf>

Moran, Susan, "Radio Slips . . . ," *American Demographics* (May 1998)

Rathbun, Elizabeth A., "Clutter's in the Air," *Broadcasting & Cable* (17 April 2000)

Steinberg, Howard, "Radio as Volume Builder, Not Commodity," *Brandweek* (18 June 2001)

Stine, Randy J., "'Cash' Stokes Advertiser Concerns," *Radio World Newspaper* (1 March 2002)

Taylor, Tom, "Major Warning Signs for Radio," *M Street Daily* (8 December 1999)

"Top 40 Story: Bill Drake," *Radio and Records* (special supplement; September 1977)

Toroian, Diane, "Are Radio Stations Approaching Commercial Overload?" *St. Louis Post-Dispatch* (4 May 2000)

"Will Your Audience Be Right Back After These Messages? The Edison Media Research/Arbitron Spot Load Study" (sponsored by *Radio and Records*), <www.edisonresearch.com/SpotLoadSum.htm>

Commercials. *See* Advertising

Commercial Tests

Determining Audience Preferences

Radio advertisers have been interested in documenting the effects of their commercials since the 1930s. And since the 1980s, when TV commercials promoting radio stations became a major advertising category for local television stations, radio broadcasters have been interested in testing the efficacy of their TV ads.

Commercials (on radio or television) may be tested at any stage in the process of developing a campaign. For example, concept testing is conducted during the planning and writing stages. The campaign's appeal or its basic assumptions may be studied in focus group discussions. *Focus groups* are groups of survey participants who are chosen for their relevance to the research topic and guided through group discussions of that topic; for example, an advertiser wishing to test the potential effectiveness of a campaign to promote diapers would probably choose young mothers for participants in its focus group discussion.

Each version of an ad to be studied in commercial testing is referred to as an *execution*. If more than one execution has been created for a campaign, the object of commercial testing is to determine which execution will be more effective in producing the desired results for the advertiser. However, only relatively large advertisers produce more than one execution for a campaign. In terms of the number of commercials submitted to testing, the most common situation is a test of a single execution. In this case, the aim of commercial testing is to determine how well the commercial performs with each of its potential target audiences. In addition, testing may suggest the kind of media purchase justified by the effectiveness of the commercial. A poor commercial may not justify heavy spending on media.

What is measured in commercial testing? The most popular measures are called *scorecard measures*. They include recall, copy point recall, affinity toward brand or toward product and/or service, intent to purchase, and comparative brand preference.

Advertising strategists assume that *recall* (remembering) is produced by attention, so measurement of a subject's recall of advertisements is actually a measure of his/her attention to those ads at the time they were presented. Decades of research into advertising indicates that, by itself, consumer recall of a brand or product name is not a powerful inducement to purchase. Because of the relatively simple process for measuring recall and the straightforward analysis of data collected, however, the measure continues to be popular. A typical study to assess recall involves recruiting (by telephone) a sample of adults who watch television during known hours and who are interested in the kind of radio station portrayed in the TV commercial purchased to promote the radio station. Each recruited respondent is sent a videocassette containing a television program in which the test commercial and others are embedded. The morning after viewing the video, an interviewer calls to ask each study participant which products and services appeared in the video and what companies were represented. If 80 percent of respondents recall station KATT, then the recall score is 80 for KATT.

When using *copy point recall*, interviewers ask specific questions about features of the targeted product or features of the commercial. If 60 percent of all respondents can recall the key points from the commercial, the copy point recall is 60.

Affinity is a measure of what a person likes. It is assumed that when a consumer likes (has an affinity for) a product or sponsor, then he/she is more likely to purchase the product. Advertising research confirms a positive correlation between liking and purchase, but the relationship is not strong. Researchers often suspect that liking comes from previous exposures to the product. If a listener tunes to only one call-in show host, even if that consumer rarely listens to that host, the listener may report liking the host out of proportion to the amount of actual listening that takes place. Affinity can also be measured by scales that reflect several dimensions of liking. There is a growing preference for this kind of measure, as it may explain what considerations affect the magnitude of affinity in general. In the case of a call-in show host, listeners may like the fairness of the host and his/her treatment of callers but dislike the topics chosen for discussion on the show. This dislike for topics is likely to account for a lower-than-expected general affinity. Also, liking a product is often quite different from liking its manufacturer or dealer. The owner of a particular brand of car may dislike the quality of service provided by the dealer, so on the general affinity measure, the consumer reflects dislike although he/she retains a strong affinity for the car brand.

The *intent to purchase* question asked of a respondent may be as simple as, "Are you more likely to listen to station KATT after hearing this promotion?" Or the question may be embedded in a scenario such as "Suppose that you go to the store because you have run out of milk. You are making a special trip just for this product. When you get to the store, your favorite brand is sold out. What are the odds that you will go to another store rather than try the brand in this commercial, which is available in your store?" Advertising research shows that intent to purchase is a complicated mental process for the consumer. If a young man has been wearing the same brand of jeans for a number of years and has been pleased with that brand, a long period of time will be required to effect a change of preference to another brand (assuming the brands are similar). So exposure to a test of a radio commercial for the new brand may produce very little change in the intent measure, but that small change may be significant because the consumer previously had never considered a change in brands. This is an especially important point when the products are radio stations, as a consumer's attachment to a radio station is rarely a rational process subject to logical argument. Transfer of emotional allegiances to radio stations may initially occur very slowly, then accelerate at surprising speed.

The commercial test measure of *comparative brand preference* has considerable face validity (that is, it appears to be quite useful and reliable) for advertisers. But it is sometimes complicated to incorporate into a commercial test, and advertising research firms have risen or fallen in the past based on their handling of this measure. A typical comparative preference item might be, "If the brand you currently use was priced at $1.00 and you considered that price fair, what price would be fair for the product you heard about in the commercial?"

Tests of commercials can be performed by nearly all local market research firms. They can also be contracted for by large national firms known for commercial testing, such as Gallup and Robinson or Mapes and Ross.

JAMES E. FLETCHER

See also Audience Research Methods

Further Reading

Fletcher, James E., and Ernest Martin, Jr., "Message and Program Testing," in *Handbook of Radio and TV Broadcasting: Research Procedures in Audience, Program, and Revenues,* edited by Fletcher, New York: Van Nostrand Reinhold, 1981

Hartshorn, Gerald Gregory, *Audience Research Sourcebook,* Washington, D.C.: National Association of Broadcasters, 1991

National Association of Broadcasters, Research Committee, *Standard Definitions of Broadcast Research Terms,* New York: National Association of Broadcasters, 1967; 3rd edition, as *Broadcast Research Definitions,* edited by James E. Fletcher, Washington, D.C.: National Association of Broadcasters, 1988

Communications Act of 1934

Since 19 June 1934, the often-amended Communications Act of 1934 has served as the basic federal statute governing most forms of interstate and foreign wireless and wired electronic communications originating in the United States. Currently codified in Title 47 of the *United States Code,* the Act created the Federal Communications Commission (FCC) as the expert administrator of the statute. The act sets basic standards for radio station ownership, licensing, and operation in the public interest in the United States and its possessions. Congress' authority to legislate in this area is based on the Commerce Clause of the U.S. Constitution (Article I, Sec. 8). Congress posits that all uses of the electromagnetic spectrum are inherently interstate in nature.

Statutory History

The Communications Act of 1934 repealed and replaced the earlier Radio Act of 1927, itself the first federal statute dealing with broadcasting. The 1934 Act, a quintessential example of "New Deal" legislation, grew from a 1933 Department of Commerce study aimed at assessing the adequacy of federal regulation of electronic media. Decrying the division of regulatory powers among various agencies, the study recommended that Congress consolidate authority over almost all forms of interstate electronic media in a single regulatory agency. The resulting act abolished the Federal Radio Commission, whose authority had been limited to users of the electromagnetic spectrum (including radio stations) and transferred authority to a reconstructed and enlarged entity, the Federal Communications Commission. It shifted responsibility for interstate wired telephony and telegraphy from the Interstate Commerce Commission to the FCC. Portions of the 1934 Act dealing with broadcasting were, for the most part, unchanged from the earlier Radio Act of 1927. The primary purpose of the new law was to strengthen federal oversight of the telephone and telegraph industries and, by placing authority over radio, telephony, and telegraphy in a single agency, to recognize that the industries overlapped somewhat.

The act has been frequently amended since 1934. Most revisions modify just a few sections of the law. Congress, for example, has repeatedly changed parts of the act regulating how broadcasters treat candidates for public office. But Congress has also found it necessary to sometimes adapt the law to large changes in the field of telecommunications that were unanticipated in 1934. Major electronic media revisions have dealt with communications satellites (1962, 1999), public broadcasting (1968), and cable television (1984, 1992). Substantial revision with the Telecommunications Act of 1996 reflected congressional recognition that previously distinct parts of the electronic media were converging and sought to enhance competition between and within segments of the electronic media and, through reliance on marketplace-induced discipline, chipped away at the New Deal philosophy that the FCC's notion of what was in the public interest was inherently preferable to relying on what industry players would do in response to consumer demand. Although the act was written when television was in its infancy, it proved unnecessary to substantially amend it when television emerged in the 1940s and 1950s. For the most part, the radio provisions of the act were simply applied to television.

The Act, the FCC, and Related Agencies

In some respects, the act functions as a bare-bones framework for federal control of electronic media. The agency it created, the FCC, is frequently relied upon to fill in details through enactment of rules and regulations that must be consistent with the act. Radio broadcasters must comply with these FCC rules and regulations as well as with the language of the statute. Other federal laws, dealing with matters such as antitrust law, copyright law, and advertising law, also apply to radio, although they are not administered by the FCC. The federal statute preempts most state or local regulation of broadcasting, although general business, taxation, zoning, equal employment opportunity, and labor laws at the state level apply as long as they do not conflict with federal law.

Under the act, appeals of FCC decisions and actions are usually brought to the U.S. Court of Appeals for the D.C. Circuit, although appeals may sometimes begin in other circuits. Appeals of most FCC enforcement actions not involving licensing go to the U.S. District Courts. The U.S. Supreme Court has occasionally issued significant interpretations of the act. Only twice, however, has the Court found any part of the act unconstitutional: once, in 1996, when Congress tried to regulate indecent internet content, and earlier, in 1984, after Congress prohibited noncommercial educational broadcasters from supporting or opposing candidates for public office.

Major Provisions

Like the Radio Act of 1927, the 1934 Act mandates that the FCC regulate broadcasting in the "public interest, convenience or necessity" (Sec. 307[a]).

The act reenacted the parts of the Radio Act of 1927 that made the FCC a "technical traffic cop" of the air, so it authorizes the commission to set technical standards for radio. The FCC allocates spectrum space to all users except the federal government (whose spectrum use is overseen by the National Telecommunications and Information Administration, a part of the Department of Commerce). In times of national emergency, the act authorizes the president to assume control over all spectrum users, although that has never happened.

Under the act, radio station licenses can be granted for up to eight years. Since radio in the United States is a mature industry, with most licenses granted years ago, broadcasters rarely enter the industry by starting a new station. Rather, most enter the field by purchasing existing stations, a process that requires FCC approval. The act does not give the FCC power to directly regulate radio networks or program suppliers except insofar as those networks are also licensees of stations.

Prior to amendments in 1996, the act allowed for another party to file a competing application against a renewal applicant, and this often led to hearings in which the FCC compared the incumbent to the challenger. In 1996, however, Congress amended the act and eliminated such comparative hearings. Now, the FCC cannot entertain competing applications unless it first finds the incumbent unqualified for renewal. Under the act, broadcasters must be renewed if the station has "served the public interest, convenience, and necessity," if the station has not committed "serious violations . . . of [the] Act or the rules and regulations of the Commission; and . . . there have been no other violations by the licensee of [the] Act or the rules and regulations of the Commission which, taken together, would constitute a pattern of abuse" (Sec. 309[k][1]). It remains possible for outsiders to intervene in the licensing process, however, because the statute still allows anyone to file a Petition to Deny with the FCC, arguing that the incumbent's application for renewal should not be granted. Absent grievous misbehavior, however, incumbents are nearly automatically renewed.

The licensing standards are a mixture of statutory requirements and regulatory requirements created by the FCC. Licensees must be legally, financially, and technically qualified. Under the statute, the ownership of radio licenses by foreigners remains strictly limited to no more than 20 percent of total stock.

The act has been amended to require that most users of the electromagnetic spectrum (e.g., cellular phone systems and common-carrier satellite services) pay spectrum use fees, usually set through spectrum auctions. Congress, however, generally prohibits the FCC from charging broadcasters for spectrum. The theory is that, in exchange for free use of the spectrum, broadcasters provide free over-the-air broadcast services that promote the public interest. Broadcasters do pay small regulatory fees for such things as the processing of license applications by the FCC. Congress expects the FCC, through such fees, to recover annually an amount equal to its own cost of operation. In an economic sense, the FCC is expected to be minimally self-sustaining and, through spectrum auctions where they do apply, to generate substantial surplus revenue for the U.S. treasury.

In 1996 Congress amended the act and greatly liberalized radio ownership. It prohibited the FCC from setting any national limit on the number of stations owned and directed the commission to study (and presumably relax) within-market radio ownership limits.

Regulating Content

The act has long been schizophrenic about the regulation of radio content. Concerned about how they were treated by radio broadcasters, Congress directed in the Radio Act of 1927 that broadcasters provide equal opportunities for opposing candidates for public office to use stations. These provisions, sometimes erroneously called "equal-time" laws, were reenacted in the 1934 Act and, with some modifications, continue today. The provisions stipulate that radio (and television) broadcasters must treat legally qualified opposing candidates for all elected political offices alike. If a broadcaster, for example, sells advertising time to one candidate, the radio station must, within certain time limits, be prepared to sell equal amounts of time, with comparable audience potential, to opposing candidates at the same rate charged the first candidate. In 1959, however, Congress amended Sec. 315 of the act to exempt most news-related programming from these requirements. In 1971, Congress mandated in a new Sec. 312 [a] [7] that radio stations must provide for "reasonable access" to their stations by legally qualified candidates for federal elective office only—state and local offices such as governor or mayor are excluded—but commercial radio stations can fulfill this requirement exclusively through paid advertising time (Sec. 312[a][7]). During the 45 days before a primary election and the 60 days before a general election, the act specifies that candidates for any office cannot be charged more than the "lowest unit charge" for advertising on stations, and they can never be charged more than other commercial advertisers are charged for comparable uses. When candidates make use of stations under these sections of the act, broadcasters are powerless to censor what candidates say, even if their uses may be libelous, obscene, or offensive to viewers—and thus cannot be held legally liable for what a candidate says.

Despite this regulation of political content, the act's Sec. 326, in language from the Radio Act of 1927, prohibits the FCC from exercising "censorship over the radio communications or signals transmitted by any radio station" and states that "no regulation or condition shall be promulgated or fixed by the commission which shall interfere with the right of free speech by means of radio communication." But, in possibly contradictory terms, it also requires the FCC to regulate radio in the public interest. Between 1934 and 1984, the FCC—relying on the generic public-interest standard—exercised broad, categorical regulation of radio content. Congress has not interfered in the commission's modern pursuit of marketplace-based deregulation and has agreed with the FCC that relying on the marketplace is consistent with the public-interest standard of the act.

The Radio Act of 1927 and the Communications Act of 1934 as originally enacted by Congress forbade the broadcast of "obscene, indecent or profane utterances" by radio. In 1948, however, these sections were moved by Congress from the Communications Act to the United States Criminal Code (18 U.S.C. sec. 1464). Acting under the public-interest portions of the act, however, the FCC continues to enforce regulations that prohibit obscene radio broadcasts and that attempt to channel indecent broadcasts to times of day when few children are listening.

The statutory framework created by Congress in 1927 and 1934 has proven to be durable and flexible. Without major modification, it accommodated the displacement of AM radio by FM and the creation of satellite-delivered digital radio services. In the early 21st century, it appears that analog terrestrial radio broadcasting, a technology of the 1920s, can be replaced with DAB without major changes to this long-lived statute.

HERBERT A. TERRY

See also Equal Time Rule; Federal Communications Commission; Federal Radio Commission; Licensing; Obscenity/Indecency on Radio; Public Broadcasting Act; Regulation; Telecommunications Act of 1996; United States Congress and Radio; Wireless Acts of 1910 and 1912/Radio Acts of 1912 and 1927

Further Reading

Aufderheide, Patricia, *Communications Policy and the Public Interest: The Telecommunications Act of 1996,* New York: Guilford Press, 1999

Carter, T. Barton, Marc A. Franklin, and Jay B. Wright, *The First Amendment and the Fifth Estate: Regulation of Electronic Mass Media,* Mineola, New York: Foundation Press, 1986; 5th edition, 1999

The Communications Act of 1934, 47 USC 609 (1934) (Title 47 of the *United States Code*)

Emery, Walter Byron, *Broadcasting and Government: Responsibilities and Regulations,* East Lansing: Michigan State University Press, 1961; revised edition, 1971

McChesney, Robert Waterman, *Telecommunications, Mass Media, and Democracy: The Battle for the Control of U.S. Broadcasting, 1928–1935,* New York and Oxford: Oxford University Press, 1993

Olufs, Dick W., III, *The Making of Telecommunications Policy,* Boulder, Colorado: Rienner, 1999

Paglin, Max D., editor, *A Legislative History of the Communications Act of 1934,* New York and Oxford: Oxford University Press, 1989

Paglin, Max D., Joel Rosenbloom, and James R. Hobson, editors, *The Communications Act: A Legislative History of the Major Amendments, 1934–96,* Silver Spring, Maryland: Pike and Fischer, 1999

Community Radio

Small FM Noncommercial Stations

More than any other broadcast medium, community radio reflects the cultural diversity of a region. In the United States, for example, KILI in Porcupine, South Dakota, airs a morning drive program in the Native American Lakota language; Monterey, Virginia's WVLS broadcasts volunteer-produced community-affairs programs across the Shenandoah Valley. KRZA in Alamosa, New Mexico, offers bilingual programming to southern Colorado and northern New Mexico, and WWOZ fills the New Orleans airwaves with early and modern jazz, blues, gospel, and funk.

Community radio stations may be found in isolated hamlets and major cities. They may feature highly eclectic programming or be geared to serving one community exclusively. In spite of their differences, community stations have several qualities in common: they are governed by the communities they serve; they provide a sounding board for local politics and culture; and they are committed to reaching groups, particularly women and minorities, overlooked by other broadcasters. Community radio stations follow the Pacifica Foundation's practices of volunteer programming and listener sponsorship. Like the Pacifica stations, community stations often feature eclectic music and politically activist news and public-affairs programming. However, these stations tend to be smaller and less structured than Pacifica's high-profile stations. In addition, community stations are locally governed, whereas the licenses of Pacifica stations are held by a central board of directors.

Community radio's history began in 1962 when a former Pacifica KPFA station volunteer, Lorenzo Milam, founded KRAB-FM in Seattle, Washington. Whereas Pacifica was somewhat staid, with a more or less paternalistic approach to programming at the time, Milam embraced the then unheard of notion that radio stations should be run by the listeners themselves. Milam was a man of some financial means, and he eagerly committed his resources to his vision of a truly "public" radio system. In 1968 Milam and his partner, Jeremy Lansman, founded KBOO in Portland, Oregon, and KDNA in St. Louis, Missouri. Following a series of conflicts between the station's primarily white management and the African-American community, Milam and Lansman sold KDNA to a commercial firm in 1973 for more than $1 million. They used the sale's proceeds to fund 14 community stations around the country in the early 1970s.

The whimsy of Milam's and Lansman's intentions is reflected in the call letters for stations in what they termed the "KRAB Nebula": WORT in Madison, Wisconsin; WDNA in Miami; KOTO in Telluride, Colorado; WAIF in Cincinnati, Ohio; and KCHU ("the wettest spot on the dial") in Dallas, Texas. During the mid-1970s, a community station's typical broadcast day might consist of

> . . . music from India blended with readings from esoteric magazines, blues and jazz from very old or very new recordings and the '50s rock and roll antics of Screamin' Jay Hawkins' "I Put a Spell On You," followed by a rare classical recording by Enrico Caruso. Later in the week [listeners] may have turned to a feminist talk program, a program for the gay community . . . a 12-tone music program, poetry, a noon-hour interview with a flamenco guitarist, music of the Caribbean, news from the Reuters wire service and tapes of speeches by political activists of the '60s. Programs wouldn't necessarily appear at the right times, some announcers had difficulty pronouncing the titles of the works they were introducing, microphones wouldn't always be opened in time to allow a speaker to be heard, and much laughter was heard (Routt, et al., 1978).

Community conflicts at KCHU in Dallas led to Milam's withdrawal from the community radio movement. KCHU signed off the air on 1 September 1977, the first community station to cease broadcasting since KPFA's temporary sign-off in 1950.

Nevertheless, the community radio movement continued to grow. In 1975 the National Federation of Community Broadcasters (NFCB) was founded in Washington, D.C., as a professional support and advocacy group by 25 community stations. The community radio movement gradually moved beyond its counter-cultural past to embrace an array of minority-controlled stations serving Native American, Hispanic, and African-American communities. By the late 1990s the NFCB counted 140 member stations around the country. Of these stations, 46 percent are minority operated, and 41 percent serve rural communities. At the same time, many community radio stations in major metropolitan areas face the "mission versus audience" dilemma that plagues the public radio system. Do these stations stay true to their original mission of serving a variety of small audiences, or do they focus on capturing a single, larger, and more affluent audience? To ensure their financial survival, some large-market community radio stations have followed their Pacifica counterparts in abandoning their traditional, freewheeling eclecticism in favor of more homogeneous programming designed to attract "marketable" audiences.

At these and other community radio stations, debates over policy, programming, and funding are commonplace. Because of the strong ideological commitment of their participants, relations between volunteers (as well as between volunteers and staffers) may be emotionally charged and highly fractious. Democracy has never been noted for its efficiency, and, at times, dominant factions within community radio stations have adopted authoritarian models of leadership that are the antithesis of community broadcasting. Yet, despite a chronic lack of funds and occasional internecine conflicts, community radio stations continue to erase the line between broadcasters and listeners. Their accessibility, as well as the range of their programming, makes community radio in the eyes of many people the closest approximation to the ideal of "public" broadcasting in the United States.

TOM MCCOURT

See also Alternative Format; Australian Aboriginal Radio; Canadian Radio and Multiculturalism; College Radio; Localism in Radio; Low-Power Radio/Microradio; Milam, Lorenzo; National Federation of Community Broadcasters; Native American Radio; Pacifica Foundation; Ten-Watt Stations

Further Reading

Armstrong, David, *A Trumpet to Arms: Alternative Media in America,* Los Angeles: Tarcher, 1981
Engelman, Ralph, *Public Radio and Television in America: A Political History,* Thousand Oaks, California: Sage, 1996
Lewis, Peter M., and Jerry Booth, *The Invisible Medium: Public, Commercial, and Community Radio,* Washington, D.C.: Howard University Press, 1990
Milam, Lorenzo W., *Sex and Broadcasting: A Handbook on Starting a Radio Station for the Community,* 2nd edition, Saratoga, California: Dildo Press, 1972; 4th edition, as *The Original Sex and Broadcasting,* San Diego, California: MHO and MHO, 1988
Milam, Lorenzo W., *The Radio Papers: From KRAB to KCHU: Essays on the Act and Practice of Radio Transmission,* San Diego, California: MHO and MHO, 1986
Routt, Edd, James B. McGrath, and Fredric Weiss, *The Radio Format Conundrum,* New York: Hastings House, 1978

CONELRAD

Emergency Warning System

Instituted in 1951, CONELRAD served as America's first mandated nationwide emergency broadcast notification program. It was a direct result of official fears that Russian planes might try striking the United States with atomic bombs.

Only a decade earlier, Japanese aircraft had devastated Pearl Harbor, Hawaii, thus pulling the United States into World War II. Later, members of the Japanese attack force admitted that they had easily navigated to their target by simply homing in on the AM radio signal of Honolulu station KGMB. American military leaders and civil defense planners would not soon forget such a modus operandi, and so they sought to develop a way to keep local broadcast communication flowing without providing a beacon for an enemy.

Soviet Russia's 1949 acquisition of nuclear weaponry reminded nervous U.S. officials that bombers poised to deliver nuclear warheads could adroitly locate any of the several thousand American communities that had an AM radio outlet. On any given day or night in the New York metropolitan area, for example, each Russian flyer in the attacking squadron would have his choice of any one of over a dozen strong, standard broadcast stations. And to make matters worse, maps that ordinary folk, as well as spies, could buy from the federal government for a couple of dollars pinpointed the exact whereabouts of every significant AM transmitter tower.

In 1951 President Harry Truman approved a plan to control all domestic radio waves so that navigators in enemy aircraft could not be aided by listening to an American broadcast station. The plan for *con*trol of *el*ectromagnetic *rad*iation, which was simplified into the acronym "CONELRAD," was in practice a complex scheme of transmitter sign-offs and sign-ons, power reductions, and frequency shifts designed to confound hostile bomber crews. It had the potential to confuse loyal Americans, too.

A civil defense pamphlet printed shortly after CONELRAD's implementation explained that "at the first indication of enemy bombers approaching the United States, [the Commanding Officer of the Air Division Defense or higher military authority will instruct] all television and FM radio stations to go off the air." In the days before portable, battery-powered TVs or FM personal or automobile radios, no one considered either service a reliable means of conveying emergency information. Typically, television and FM stations received their cue to sign off through a silence-sensor device that detected the sudden absence of key AM outlets, which had also been ordered to be quiet. In daisy-chain fashion, all television, FM, and AM stations would go silent. Along with the TV and FM facilities, many of these AM stations were required to stay dark in order to make way for certain designated CONELRAD AM stations that, during the brief shutdown, had quickly switched their transmitter frequency to either 640 or 1240 kilohertz (whichever was closest to each particular station's regular Federal Communications Commission (FCC)-assigned dial position) and then returned to the air with less than normal output power. Understandably, antenna systems customized for, say, 1600 kilohertz, suffered efficiency loss when coupled to a jury-rigged 1240-kilohertz transmitter. Officials admitted that "the changeover to CONELRAD [frequency and power level] takes a few minutes" and suggested that the understandably anxious public "not be alarmed by the radio silence in the meanwhile."

Once the participating CONELRAD stations resumed broadcasting on their new (640-kilohertz or 1240-kilohertz) wavelength, they were all required to air the same emergency programming instructing the citizenry what to do next. During this information transmission, the CONELRAD outlets would sequentially shut down momentarily. The idea was to have, at any given time during the crisis, ample operating CONELRAD stations to reach the public while making normal radio station frequency, city-of-origin guides, and transmitter tower maps completely useless from an air-navigational standpoint.

In theory, attempting to decipher the true identity of a CONELRAD station would be like trying to identify which person, in an auditorium filled with whisperers, was intermittently whispering. In practice, though, not all of these elaborately cloaked CONELRAD facilities were effective conduits for vital communication. This was especially true at 1240 kilohertz, to which many of the participating stations were switched. In CONELRAD test runs, suburbanites near New Brunswick, New Jersey, tuning to the 1240 spot occupied by local WCTC and not-too-distant WNEW (now WBBR) New York heard little there but unintelligible cross-talk interference. The 640-kilohertz CONELRAD setup was the better bet. On that less crowded lower dial position, the result included noticeable station overlapping and some heterodyne whistling, but it delivered readable signals to much of the country.

A young broadcast buff, Donald Browne, recalled rushing home in late April of 1961 from his Bridgeport, Connecticut, high school to catch a CONELRAD dress rehearsal. He described this final CONELRAD system-wide test as sounding "real spooky," like something from *The Twilight Zone* television show. Browne noted that "several primary stations could be heard simultaneously on 640, all with the same program, each slightly delayed or out of phase with the others, like one weird echo effect . . . and probably scaring more listeners than they informed."

Most Cold War–era radio audiences took CONELRAD quite seriously. The government asked broadcasters to tout the warning system by ubiquitously airing public service announcements capped with a tiny jingle that went, "Six-forty, twelve-forty . . . Con-el-rad." Then people were urged to "mark those numbers on [their] radio set, now!" Starting in 1953, though, every AM radio sold in America was required to have a civil defense logo triangle factory-printed on its dial at 640 and at 1240 kilohertz.

In addition, CONELRAD regulations touched the amateur or "ham" radio community. As with commercial broadcast outlets, amateur stations were required to cease transmitting at the first sign of a CONELRAD activation. The consumer electronics maker *Heathkit* offered an inexpensive automatic alarm unit that would ring a bell and immediately cut off one's ham transmitter if any local broadcast station being monitored suddenly left the air. CONELRAD architects could take no chances with some unwitting 25-watt radio hobbyist who might innocently mention his backyard antenna's whereabouts during an atomic enemy sortie. Hams as well as staff at non-participating CONELRAD radio and TV stations knew to listen closely to the official 640/1240 facilities for the "Radio All Clear." Initiated by the Air Defense Commander (or higher military official), this relief meant that the CONELRAD emergency test had ended and heralded the resumption of normal transmissions over regular AM, FM, TV, amateur, and other FCC-licensed frequencies.

By the early 1960s, Soviet missiles, including those they briefly positioned in Cuba, made up a nuclear weapon delivery system far more sophisticated than an airplane navigated via some unsuspecting pop music radio station. Therefore, in 1963 CONELRAD was scrapped as obsolete. Its cumbersome 640/1240 frequency shifting, power reducing, and on/off sequencing went the way of the wind, but positive aspects of CONELRAD's warning scheme (such as employing a series of primary, participating stations to reach the public) were revamped into the Emergency Broadcast System, which stayed in effect through 1996, when it, in turn, was superseded by the Emergency Alert System.

PETER E. HUNN

See also Emergency Broadcasting System

Further Reading

Beeman, Bess, *A Resolution,* Women's Advisory Council for Defense and Disaster Relief, June 30, 1961 [pamphlet]

"CONELRAD," in *Federal Communications Commission Rules, Regulations, and Standards,* Paragraphs 4.51 through 4.57 (1952)
Kobb, Bennett, "The Last Radio Network," *Northern Observer Newsletter* (November 1990)
United States Federal Civil Defense Administration, *Six Steps to Survival: If An Enemy Attacked Today Would You Know What to Do?* Washington, D.C.: GPO, 1956

Conrad, Frank 1874–1941

U.S. Engineer and Pioneer Broadcaster

Frank Conrad was an engineer working for the Westinghouse Company when he began experimental broadcasts that led to the development of historic radio station KDKA in Pittsburgh, Pennsylvania, often described as the birthplace of broadcasting.

Conrad had dropped out of school in the seventh grade, but his curiosity and determination led to employment as a bench assistant in the Pittsburgh laboratory of Westinghouse in 1890. Because of his unusual motivation and mathematical ability, Conrad was able to work his way up in the company's engineering department and was assigned to the testing of electronic equipment. Along with this technical work at Westinghouse, Conrad devoted much of his time at his home in Wilkinsburg, near Pittsburgh, to electronic experimentation. Based on a wager concerning the accuracy of his watch, he constructed a wireless system in 1915 to receive time signals from the Naval Observatory station in Virginia. Becoming increasingly immersed in his radio interest, Conrad expanded the amateur wireless station located in his garage, licensing it as 8XK in 1916.

On 7 April 1917, following the entry of the United States into World War I, the government canceled the licenses of amateur stations, including that of 8XK, although Conrad did receive authorization to use the facility for periodic testing of Westinghouse military wireless equipment during the war. After the ban on amateur wireless was lifted in early October 1919, Conrad reestablished 8XK and began regular broadcasting. Much of the content of these early on-air efforts involved descriptions of the equipment being used by the station.

By mid-October 1919, Conrad was tiring of reading newspapers into the microphone and offering on-air recitations concerning the 8XK equipment, but he continued to be fascinated with broadcasting, especially with responses from listeners who described the strength of the 8XK signals. To minimize the need for constant talking and to further entice listeners, Conrad placed a phonograph in front of his microphone and began broadcasting recorded music. Listener response was positive, and growing requests for specific selections led Conrad to turn to the Hamilton Music Store in Wilkinsburg for additional recordings. The store complied, with the proviso that Conrad announce the availability of this music at the store. For two hours each Wednesday and Saturday, Conrad played his recordings, and the store reported increased sales of the music heard on 8XK. Thus radio, music, promotion, and advertising came together in a manner very foretelling of the future. This content was also supplemented with live music, both vocal and instrumental, and Conrad's two sons occasionally spelled their father as on-air hosts for this pioneer variety programming.

By September 1920, the station was drawing so many listeners that the Joseph Horne Department Store placed an announcement in the Pittsburgh *Sun* describing the Conrad broadcasts and advertising the availability of wireless sets for sale at the store, with prices starting at $10.

Harry P. Davis, a Westinghouse vice president, took note of the announcement and conceived the idea that the future profitability of Westinghouse lay in the manufacture of receiving sets, the sales of which could be greatly stimulated if listeners were provided with reliable signals and programming designed to capture their interest and attention. Davis convinced other company executives of the value of this opportunity for a profitable Westinghouse role in electronic communication, and the company submitted an application for a license on 16 October 1920. The license, specifying the call letters KDKA, arrived on 27 October, and Westinghouse hastily constructed a station, locating a studio in a tent on the roof of a building near the KDKA transmitter. KDKA was on the air in time for the station to begin broadcasting on election night, 2 November 1920, a date used by some historians to mark the actual inauguration of broadcasting in the United States. Frank Conrad, however, was not on hand at KDKA for this momentous event. He was standing by, back in Wilkinsburg at 8XK, ready to fill in on the air if trouble occurred at the Pittsburgh station.

Frank Conrad
Courtesy Library of American Broadcasting

In 1921 Frank Conrad became assistant chief engineer at the Westinghouse Company and continued with experiments and improvements in radio, including work with signals of different frequencies and with the reflection of sky waves from an ionized layer above the Earth. At a conference in London in 1924, Conrad demonstrated the use of shortwaves for long-distance broadcasting. He also directed Westinghouse efforts in the improvement of designs for transmitting and receiving equipment during the 1930s, earning 178 patents during his career.

B.R. SMITH

See also KDKA; Westinghouse

Frank Conrad. Born in Pittsburgh, Pennsylvania, 4 May 1874. Attended Starrett Grammar School through seventh grade; hired by Westinghouse in 1890; appointed general engineer at Westinghouse, 1904; received license to put experimental station 8XK on the air, 1916; KDKA licensed 27 October 1920 and began broadcasting 2 November 1920. Received Morris Liebmann Prize, Institute of Radio Engineers, 1926; honorary degree, University of Pittsburgh, 1928; Edison Medal, American Institute of Electrical Engineers, 1931; John Scott Medal, Institute of Philadelphia, 1933; Lamme Medal, American Institute of Electrical Engineers, 1936; Gold Medal, American Institute of the City of New York, 1940. Died in Miami, Florida, 10 December 1941.

Further Reading

Archer, Gleason Leonard, *History of Radio to 1926,* New York: American Historical Society, 1938

Barnouw, Erik, *A History of Broadcasting in the United States,* 3 vols., New York: Oxford University Press, 1966–70; see especially vol. 1, *A Tower in Babel: To 1933,* 1966

Douglas, George H., *The Early Days of Radio Broadcasting,* Jefferson, North Carolina: McFarland, 1987

Dunlap, Orrin Elmer, *Radio's 100 Men of Science: Biographical Narratives of Pathfinders in Electronics and Television,* New York and London: Harper, 1944

Kraeuter, David W., "Frank Conrad" in *Radio and Television Pioneers: A Patent Bibliography,* by Kraeuter, Metuchen, New Jersey: Scarecrow Press, 1992

Westinghouse Public Relations Department, "History of Broadcasting and KDKA Radio," in *American Broadcasting: A Source Book on the History of Radio and Television,* compiled by Lawrence W. Lichty and Malachi C. Topping, New York: Hastings House, 1975

Conrad, William 1920–1994

U.S. Radio, Television, and Film Performer and Director

William Conrad's vocal ability and full bass voice were the key to his early and long-lasting success in radio. Unlike many other radio actors, however, Conrad also became a successful director and producer for film and television and had a long and varied career.

Conrad was born to a theater-owning family in Kentucky who moved to Southern California when he was still a small boy. His initial radio experience came at age 17 when he took on announcing and later writing and directing roles at Los Angeles (Beverly Hills) station KMPC. His bass voice was already very expressive, especially for a teenager. He attended but appears not to have graduated from Fullerton Junior College (now California State University, Fullerton). During World War II, Conrad served initially as a fighter pilot with the U.S. Army Air Force until he was grounded because of night blindness. He finished his military obligation as a producer and director for the Armed Forces Radio Service.

Radio Work

Conrad is best known in radio history for his appearance as U.S. Marshall Matt Dillon in 480 episodes of the Columbia Broadcasting System (CBS) series *Gunsmoke,* broadcast from 1952 to 1961. His opening statement (recorded in 1952 and reused for years) remains in the minds of listeners:

> I'm that man, Matt Dillon, United States Marshall. The first man they look for and the last they want to meet . . . it's a chancy job, and it makes a man watchful, and a little lonely.

Before the program first appeared on television in 1955, the radio cast was given perfunctory auditions, but none of the radio performers transferred to the visual medium. Conrad had by then become a rotund man far different from the image he had created with his magnificent voice. He remained bitter for many years about the loss of the television role.

During this period, Conrad also played a variety of characters on many other radio series, including *The Whistler, Romance, The Lux Radio Theatre, Suspense, The Screen Guild Players,* and *The Philip Morris Playhouse.* His voice opened alternate weekly episodes of *Escape* (1947–54). For a period, he was appearing in 10 to 15 radio programs every week. Conrad later estimated he had appeared in no fewer than 7,500 radio broadcasts.

Film and Television Work

Beginning in the late 1940s, Conrad focused on film appearances, beginning with his film debut in *The Killers* in 1946. His success in performing the villain and other character roles obtained him roles in *Body and Soul* (1947), *Sorry, Wrong Number* (1948), and *East Side, West Side* (1949). His supporting roles continued more sporadically in the 1950s and included *The Naked Jungle* (1954) and *The Conqueror* (1956), but were too few and far between. Conrad thus began a related but more successful career as a producer or director of both films and television series. He was under contract to Warner Brothers as a producer and director for 15 years.

Conrad reached the peak of his career not in radio or films, but in television. He played the eponymous role of detective *Cannon* in the 1970s, appeared more briefly on *Nero Wolfe* in 1981, and finally played the older and more experienced man in the aptly named *Jake and the Fat Man* from 1987 to 1992. In the background Conrad undertook many television tasks, as narrator of various series and as the producer or director of others. He died on 11 February 1994 in North Hollywood, California.

EDD APPLEGATE AND CHRISTOPHER H. STERLING

William Conrad. Born in Louisville, Kentucky, 27 September 1920. Attended Fullerton (California) Junior College. Served as a fighter pilot in the U. S. Army Air Force, 1943–45. Worked as an announcer at radio station, KMPC, Los Angeles, California; actor-producer for radio series, *The Hermit's Cave,* 1935–mid 40s; film debut, *The Killers,* 1946; starred in radio series, *Gunsmoke,* 1952–61; directed television episodes of *Gunsmoke,* 1955; supporting actor in radio and television; directed and produced for radio, motion picture, and television; starred in several television series; well known narrator, including *The Bullwinkle Show* and *Buck Rogers in the 25th Century.* Died in Hollywood, California, 11 February 1994.

Radio Series

1935–mid 40s *The Hermit's Cave*
1946–52 *The Count of Monte Cristo*
1947 *Johnny Modero: Pier 23*
1947–54 *Escape*
1948 *The Front Page*
1948–50 *The Damon Runyon Theater*

1948–53	*This is Your FBI*
1949–51	*The Adventures of Sam Spade, Detective*
1949–60	*Yours Truly, Johnny Dollar*
1950	*Romance*
1951–52	*The Silent Men*
1952–61	*Gunsmoke*
1952–53	*Jason and the Golden Fleece*
1956–57	*The CBS Radio Workshop*

Films

The Killers, 1946; *Body and Soul,* 1947; *Arc of Triumph,* 1948; *To The Victor,* 1948; *Joan of Arc,* 1948; *Sorry, Wrong Number,* 1948; *East Side, West Side,* 1949; *Cry Danger,* 1951; *The Naked Jungle,* 1954; *The Conqueror,* 1956; *—30—,* 1959; *The Man from Galveston* (director), 1964; *Two on a Guillotine* (producer/director), 1965; *Brainstorm* (producer/director), 1965; *An American Dream* (producer), 1966; *A Covenant with Death,* 1967; *First to Fight,* 1967; *The Cool Ones,* 1967; *Countdown* (producer), 1968; *Chubasco* (producer), 1968; *Assignment to Kill* (producer), 1969; *Moonshine County Express,* 1977

Television

Escape, 1950; *Gunsmoke* (director), 1955; *The Rifleman* (director), 1958; *Klondike* (producer/director), 1960–61; *The Bullwinkle Show* (actor/narrator), 1961–73; *General Electric True* (director), 1962–63; *77 Sunset Strip* (director), 1963–64; *The Fugitive* (narrator), 1963–67; *George of the Jungle,* 1967; *D.A.: Conspiracy to Kill,* 1970; *O'Hara, U. S. Treasury,* 1971; *Cannon,* 1971–76; *The Wild, Wild World of Animals* (narrator), 1973–78; *Buck Rogers in the 25th Century* (actor/narrator), 1979–80; *The Return of Frank Cannon,* 1980; *Turnover Smith,* 1980; *Nero Wolfe,* 1981; *Jake and the Fat Man,* 1987–92

Further Reading

Corwin, Miles, "William Conrad; Star of 'Cannon,' 'Fatman,'" *Los Angeles Times* (12 February 1994)

DeLong, Thomas A., *Radio Stars: An Illustrated Biographical Dictionary of 953 Performers, 1920 through 1960,* Jefferson, North Carolina: McFarland, 1996

Falke, Ben, "The Harder He Works, the Fatter He Gets," *Biography News* (May 1974)

Lackmann, Ron, *Same Time, Same Station: An A–Z Guide to Radio from Jack Benny to Howard Stern,* New York: Facts on File, 1996; revised edition, as *Encyclopedia of American Radio: An A–Z Guide to Radio from Jack Benny to Howard Stern,* 2000

Raddatz, Leslie, "It Belonged to William Conrad, Radio's Matt Dillon," *TV Guide* (14 February 1976)

Stephens, Michael L., *Gangster Films: A Comprehensive Illustrated Reference to People, Films, and Terms,* Jefferson, North Carolina: McFarland, 1996

Weil, Martin, "Actor William Conrad Dies; Was 'Cannon' and 'Fatman,'" *Washington Post* (12 February 1994)

Consultants

Radio consultants, also known as "radio doctors" or "hired guns," advise stations on how best to increase listenership and thereby strengthen ratings. Consultants focus on improving a station's image or "sound," refining music playlists, and conducting audience research, all with the end goal of bringing success to a given station.

Development of Radio Consultants

Radio consulting as a profession had its beginnings at the end of the 1950s. By that time, commercial radio had evolved to the point of encompassing distinct programming formats. Rock and roll music changed the radio landscape: formats aimed at specific audiences came into being, notably Top 40, middle of the road, country, and beautiful music. According to Michael C. Keith (1987), radio consulting began in the U.S. Midwest. The first radio consultant, Mike Joseph, decided to start his own business after achieving success as a radio station program director. WMAX-AM in Grand Rapids, Michigan, was Joseph's first client; it found ratings success thanks to Joseph's expertise. More clients soon followed, including WROK in Rockford, Illinois; WKZO in Kalamazoo, Michigan; KDAL in Duluth, Minnesota; and WKBW in Buffalo, New York. Joseph's successive clients each achieved larger audiences, and the business of radio consulting took hold.

As a new subpart of the radio business, consultants initially faced a limited market. However, prospects for those who went into the radio consulting business, usually former station program directors, increased dramatically during the 1960s. During that decade, genre programming blossomed as the number and styles of music programming expanded to meet the needs of specific audiences. Consequently, the resulting fragmentation of formats increased the need for consultants.

Consultants worked at both the individual and agency level; one could work freelance as a one-person operation or at a consulting firm. In addition to these "hired guns" stations brought in to improve their ratings and on-air presentation, program syndicators and station rep companies started to enter the consulting side of the business. Indeed, for the next several decades, many program syndicators would provide their client stations with both the advice and the programming to increase ratings in one convenient package.

The 1960s saw station management in larger markets searching for even larger audiences to attract big advertisers and thus big profits. "Numbers became the name of the game," with a station's goal in any given market to increase listenership: "To be number one was to be king of the hill" (Keith, 1987). With the number-one rating status serving as "the holy grail" of radio stations during that period of "frag-out" in programming formats, stations experiencing poor numbers for several ratings periods sought the advice of consultants. The importance of consultants in maintaining a modicum of ratings success became apparent. As Keith notes, "'Call a consultant' became a cry commonly heard when a station stood at the edge of the abyss" (1987).

Although the growth in radio as big business had expanded in the 1960s, the industry saw even more expansion with the rise in popularity of FM during the 1970s. FM contributed to the doubling in the number of stations and the tripling of programming formats—and, of course, to the expansion of consulting opportunities. One particular new format served as a notable example of the significance of the need for and power of the radio consultant: album-oriented rock (AOR). Prominent radio consultant Donna Halper, in *Radio Music Directing* (1991), relates the power given to consultants by station owners trying out the new format. She quotes Kent Burkhart, one of the industry's best-known consultants and the one-time partner of Lee Abrams, an AOR expert:

> Back in the 1970s, AOR was still a fairly new format, so owners wanted us to have total control. They didn't want to leave anything to chance. . . . In fact, in those early days of Album Rock, many of the stations didn't even have a music director. They just had a PD [program director] who often guessed what music should be played.

With clients nationwide, Burkhart and Abrams' partnership, Burkhart/Abrams and Associates, became the most powerful consulting firm of the decade. Burkhart and Abrams provided their AOR clients with a playlist based on research, adding new albums each week. Their influence became such that record promoters could count on sales of albums they approved because as many as 100 stations could potentially play them.

Whereas the ascent of FM and the proliferation of format and music genres during the 1970s provided increased employment for consultants, another development—audience research—provided consultants with more complicated tasks. Keith (1987) points out that stations began to rely more on results of surveys, notably those conducted by the Arbitron Company. As FM began luring away AM listeners, smaller AM stations trying to increase listener shares posed a major challenge to consultants, a "Herculean" task that "only a few master consultants were up to."

Ironically, whereas stations had called on consultants to help them during FM's "infancy," the AM market was fertile for the consulting industry in the 1980s. With FM's dominance firmly established by then, the field expanded now to AM: "Radio consultants, who found themselves an integral part of FM's bid for prominence in the 1960s and 1970s, worked on the AM side with as much fervor in the 1980s in an attempt to reverse the misfortunes that befell the one-time ratings leader" (Keith, 1987).

By the mid-1980s, some three decades after Mike Joseph's initial foray into the consulting business, about a third of all radio stations used consultants. There were about 50 individual consultants in 1986; though that number held constant, the consulting business involves a degree of turnaround in that consultants must show positive results in order to remain in business. This number remained constant into the 1990s; of the more than 100 broadcasting consultants listed in various media directories in the United States, more than half specialized in radio (Keith, 1987).

Just as stations had become specialists in certain program formats, with fragmentation and "narrowcasting" helping to stimulate the consulting industry in the 1970s, consultants themselves began to cater to stations' particular program needs in the mid-1990s. Perhaps as an indication of the successes some consultants had achieved by that time, some stations brought in "niche consultants" to help boost ratings during specific dayparts as well as in specific formats. Morning shows in particular served as prime targets for consultants' services (Keith, 1997).

Although Keith (1987) had predicted that increased competition resulting from deregulation all but guaranteed the future of radio consulting in the 1980s, *Billboard* reported a trend toward consolidation of consulting agencies beginning in 1995. For instance, Stark (1995) found that radio consultants started teaming up for long-term joint ventures and referred their clients to rival agencies. Consultant alliances became the product of individual agencies' desire to do whatever it took for a client to succeed. As stations downsized, the demand increased for highly specialized people from "the outside"—consultants—that stations could rent rather than hire full-time.

Consolidation among consultants continued into 1999, and increased competition for work resulted in a shakeout in the business, with some individual consultants being forced to join companies or take other jobs in programming. Media groups—broadcast companies owning several stations—increasingly relied on their own in-house programmers, who effectively took the place of consultants. Additionally, those working for large companies specializing in one format have better long-term prospects than do individual consultants, who must cover several formats. Industry experts predict that those consultants who stay independent will have to provide their clients with expanded services.

Consultant Services

Although radio consultants ultimately aim to improve station ratings, they also advise management on ways to implement a change in format, to gain higher visibility, and to achieve a higher quality of on-air presentation. As their nickname "radio doctors" implies, consultants "diagnose the problems that impair a station's growth and then prescribe a plan of action designed to remedy the ills" (Keith, 1997). Initially, consultants treated ailments in programming, especially those involving playlists. As the industry increased in complexity, so, too, have the services offered by consultants. These range from making specific observations and suggestions regarding the performance of on-air talent to audience research. The array of services offered depends on the type of consultant.

There are two basic kinds of radio consultant: programming and full-service. Programming consultants focus primarily on the on-air aspect of a station's product—such as playlists and execution of format. Traditionally, they come from the programming side of the business; program directors get into consulting when they have achieved a record of ratings success. One can find consultants who specialize in particular formats, such as country and adult contemporary.

Full-service consultants, in addition to providing clients with programming expertise, offer a "package" of services that covers virtually every aspect of radio station operations: staff training and motivation, music, audience and market trend research, drug and alcohol counseling, sales and management consulting, union and syndication, music suppliers, and record company negotiating. The range of full-service offerings also extends to the use of engineering consultants and advice regarding business operations. Some consulting companies provide clients with programming (program syndicators). Clients can use all the programming services, or just part, either as recorded material or via satellite in conjunction with live announcers (Keith, 1987).

In some cases, station managers give consultants total control, such as when a station changes format or ownership. In other cases, consultants simply give objective advice regarding a station's performance. They also examine the competition and determine what other stations with the same format in a market present the best execution. Consultants also determine what call letters best reflect a station's desired image.

With the consulting industry becoming more specialized, some agencies focus solely on research. Research consultants go through survey data, study a station's market and target audience in terms of socio-economic and financial statistics, and may conduct music research to determine what most appeals to an audience.

Consultants also conduct research to discover what factors about a radio station the listening public likes and dislikes. To this end, consultants use three basic approaches: focus groups, callout research, and music testing. In focus group research, small groups of listeners or potential listeners serve as "sounding boards" regarding certain elements of programming. Researchers document the group's attitudes and emotions concerning a station's music, disc jockeys, news, and contests. Callout research refers to telephone surveys that measure respondents' opinions in empirical form. Music testing involves paying participants to listen to and evaluate songs, usually in an auditorium-like setting. A station then creates its playlist based on the results. Based on these types of research, consultants can make recommendations that have the greatest potential for success in a given market.

Consultants' duties include making in-house visits and examining a station's physical plant. This includes technical assessment of the station's signal strength and clarity. If needed, engineering consultants are brought in to make recommendations regarding the station's equipment.

Consultants usually research their client stations' performance and competition by listening to the station, either live or on tape. When a consultant arrives in the client's market city or town, he or she monitors the station, usually from a hotel room. This leaves the consultant free from distractions in order to assess the client's on-air presentation. Typically, a consultant takes notes on all aspects of a station's "sound." As described by Donna Halper, these include the following: music mix, "listenability," announcer effectiveness, the match between proclaimed format and music played, technical quality, station image, times songs are played, front or back sells, and use of call letters. Consultant Jim Smith looks at certain other basics of on-air execution, including production values, stop sets, newscasts, features, and promotions (Keith, 1987).

In addition to assessing competing stations' products, consultants compare what is aired during specific dayparts and even hours to what their client offers at those times. For example, a consultant would compare station A to station B in terms of songs played, times and lengths of commercial breaks, and the like. The consultant then compiles the information and submits a comprehensive report to the station. As with any type of evaluation, consultants' reports not only include constructive criticism of their client, but also should provide some positive feedback as well.

Consultant Characteristics

Consultant companies can range in size from 2 or 3 people to 50, with fees ranging from $500 to more than $1,200 a day. Consultants base their fees on the services the client wants and the size of the station. Most consultants have backgrounds in broadcasting, usually as station program directors. Those who have broadcasting experience hold a considerable advantage over those who do not. Most also have a thorough knowledge and understanding of radio broadcasting at all levels, including programming, sales, marketing, and promotion. Some obtain formal training in college, notably through research methods and broadcast management courses.

The consulting business as a whole does face obstacles in the radio industry—notably, that of gaining acceptance among broadcasters, who consider consultants a "necessary evil." Industry insiders cite negative perceptions held by some station managers and program directors regarding consultants; despite these perceptions, consultants are not all "charlatans intent on cleaning house and selling fad formats" (Keith, 1987). A good consultant's effectiveness requires, first, that station management make clear to staff the reasons why it is bringing in a consultant. Station managers also need to implement the consultant's recommendations effectively. Those who do often benefit from their investment. As Keith contends, "Statistically, those stations that use programming consultants more often than not experience ratings success" (1997).

ERIKA ENGSTROM

See also Drake, Bill; Programming Strategies and Processes; Trade Organizations

Further Reading

Halper, Donna L., *Radio Music Directing,* Boston: Focal Press, 1991

Keith, Michael C., *Radio Programming: Consultancy and Formatics,* Boston and London: Focal Press, 1987

Keith, Michael C., and Joseph M. Krause, *The Radio Station,* Boston: Focal Press, 1986; 4th edition, by Michael C. Keith, 1997; 5th edition, Boston and Oxford: Focal Press, 2000

"Radio Consultants Still in Shakeout," *Billboard* (11 September 1999)

Stark, Phyllis, "Consultancy Alliances Prosper," *Billboard* (10 June 1995)

Contemporary Christian Music Format

Part of a growing U.S. trend of religious formats on the air, this development of the past few decades combines the basic tenants of Christianity with popular music approaches that appeal to a broader audience.

Origins

Contemporary Christian Music (CCM) has grown over a long period of time. Its foundations are evident in the early hymns of various Protestant faiths. Overtures of intimacy and sentimentality were mixed with Christian music in the 19th century as the feminine ideal of piety combined with the temperance crusade emerged. Even militaristic themes characterized hymns in the early 20th century as the world and the United States fought several major wars.

By the 1960s, however, Evangelicals began to realize that "Bringing in the Sheaves" on Sundays couldn't begin to compete with weekday broadcasts of "Hey Jude" and "I Can't Get No Satisfaction," especially among younger listeners. Composer Ralph Carmichael began the CCM renaissance with pieces such as "Pass It On" and "He's Everything to Me." Musical creativity burst onto the Christian music scene as the younger generation brought its hippie culture, with music largely devoid of theological divisions, into various churches. Musicians such as Larry Norman, Andrae Crouch, Keith Green, Chuck Girard, and Randy Stonehill added the 1960s flavor of rock and roll to Christian music, thus creating the concept of Contemporary Christian Music.

Variations and Controversy

As with most musical formats, however, the genre of contemporary Christian music has splintered into many different kinds (15 different CCM types are listed in the 2001 *Directory of Religious Media*), and as a result a controversy arose within the evangelical community that continues today. How, some ask, can Christian music be used to evangelize if it sounds just like secular music? One argument holds that there is nothing in music that makes it Christian. A related controversy is the ability of some Christian musicians to cross over to the secular music world. Radio stations with commercial formats and their audiences either accepted Christian music or Christian artists attempted to create music designed for them. Christian recording artist Amy Grant and the group Sixpence None the Richer both had hit singles in the 1990s that prompted the Gospel Music Association (GMA) to redefine Christian music in association with the annual Dove Awards, which honor Christian recording artists. The GMA's criteria for defining music as Christian is that in any style whose lyrics are

> Substantially based upon historically orthodox Christian truth contained in or derived from the Holy Bible; and/or - An expression of worship of God or praise for His works; and/or - Testimony of relationship with God through Christ; and/or Obviously prompted and informed by a Christian world view.

GMA president Frank Breeden commented that "this statement is not intended to be the definition of gospel music for all time, nor is it meant to characterize music made by Christians that may not fit the criteria" (Grubbs, 1998).

These definitional controversies didn't prevent CCM from becoming a multi-million dollar industry by the mid-1990s. CCM's share of 1998 recording industry revenue exceeded the shares of jazz, classical, New Age, and soundtracks according to *Billboard* magazine. CCM record labels did not, however, escape the consolidation fever of the 1990s. By early 1997 three companies controlled all labels that produced CCM: Zomba Group (parent company of Benson Music Group), EMI (parent company of Sparrow, Star Song, ForeFront, and GospoCentric), and Gaylord Entertainment (parent company of Word Music). Other industries were also beneficiaries of CCM.

The 2001 *Directory of Religious Media* listed nearly 2,500 radio stations that provided some form of religious programming. Sixty-six percent of those were formatted with one of 15 different varieties of Christian music. Those formats included Adult Contemporary, Alternative, Christian Hit Radio, Contemporary Christian, Country, Gospel, Hispanic, Inspirational, Instrumental, Middle of the Road, Praise and Worship, Sacred, Southern Gospel, Specialty, and Urban/R&B. The other religious stations' formats focused on Christian news, talk, or preaching.

Aspiring Christian musicians can even major in CCM at Greenville College, a Christian liberal arts college in south-central Illinois. Music Department Chair Ralph Montgomery began the program in 1987; the most notable alumni are members of the recording group Jars of Clay.

The internet is as pervasive a presence in CCM as it is elsewhere. Numerous sites deal with every aspect of CCM, from the controversies noted earlier to sites developed by fans of various artists. CCM fans can listen to their favorite artists through internet-only audio streaming sites or through radio stations streaming their signals on their websites.

LINWOOD A. HAGIN

See also Evangelists/Evangelical Radio; Gospel Music Format; Religion on Radio

Further Reading

NRB Directory of Religious Media (annual)
Balmer, Randall, "Hymns on MTV," *Christianity Today* (15 November 1999)
Grubbs, Deanna, "Gospel Music Association Sets New Criteria for Dove Awards Eligibility," *Gospel Music Association News Release* (30 July 1998)
Howard, Jay R., "Contemporary Christian Music: Where Rock Meets Religion," *Journal of Popular Culture* 26, no. 1 (1992)
Howard, Jay R. and John M. Streck, "The Splintered World of Contemporary Christian Music," *Popular Music* 15, no. 1 (1996)
Martin, Yvi, "An Education with a Backbeat," *Christianity Today* (15 November 1999)
Olsen, Ted, "Will Christian Music Boom for New Owners?" *Christianity Today* (28 April 1997)
Romanowski, William D., "Roll Over Beethoven, Tell Martin Luther the News: American Evangelicals and Rock Music," *Journal of American Culture* 15, no. 3 (1992)
Schultze, Quentin, "The Crossover Music Question," *Moody Magazine* 93, no. 2 (October 1992)
Solomon, Jerry, "Music and the Christian," <www.probe.org/docs/music.html>

Contemporary Hit Radio/Top 40 Format

Contemporary hit radio (CHR) is a rock music format that plays the current best-selling records. The music is characterized as lively, upbeat rock or soft rock hits. The playlist generally consists of 20 to 40 songs played continuously throughout the day. Disc jockeys are often upbeat "personalities," and the format emphasizes contests and promotions. CHR stations tend to target a young demographic of both men and women, aged 18–34, with listenership extending into the 35-to-44 demographic cell.

CHR grew out of Top 40, which was developed in the late 1950s by Todd Storz and Gordon McLendon, who found success in playing the 40 most popular records. By the mid-1960s, the rise of rock music and FM led to audience fragmentation and a revitalized, tighter format with less chatter, refined by programmer Bill Drake. The format was successful but was also criticized for being too slick and dehumanized. The move to FM was initially met with resistance, because FM was regarded as an alternative listening medium. As a result, Top 40 underwent another face-lift and became known as contemporary hit radio.

The trade periodical *Radio & Records* (*R&R*) began using the term *contemporary hit radio* in 1980. The retitling of the format was orchestrated by consulting pioneer Mike Joseph. Joseph's CHR format featured a tight playlist of about 30 records with up-tempo sounds, fast rotations, limited recurrence, chart hit countdowns, and no more oldies and declining records. At that time, CHR songs were by such artists as Blondie, Billy Joel, Christopher Cross, Queen, Dan Fogelberg, and Pink Floyd. The format moved to FM and became virtually nonexistent on AM as most radio listening shifted to the higher-fidelity broadcast system. Soon many broadcasters abandoned their soft rock and album-oriented rock (AOR) formats in favor of CHR.

By the mid-1980s, CHR became the highest-rated format. There were two or more CHR stations in many medium to large markets. Close to 800 CHR stations were on the air in 1984, and this number increased to nearly 900 in 1985. In 1987 and 1988 CHR was the number-one format in both New York City and Los Angeles, according to Arbitron market reports.

As more stations flocked to the popular format, some turned to format segmentation as a way to broaden their core audience targets and to counterprogram against similar formats in their markets. In the mid-1980s CHR split into two directions. The basic CHR format became a mass-appeal, 12-plus format. A variation on that theme became "adult CHR," which went with softer announcing and added some oldies songs to the musical mix, attempting to appeal to the 25-to-34 demographic and divert audiences from adult contemporary (AC) stations. Some stations also went with hybrid formats, such as a Top 40/AOR format.

CHR has undergone even more fragmentation in recent years. Today the most common variations include CHR/pop, CHR/rhythmic, and adult CHR/hot AC. CHR/pop most closely resembles the original Top 40 format. It is the most current-based format, playing the hottest-selling popular songs of the day. As a result, the music may vary from rock and pop to dance and alternative, depending on what is most popular at the moment. The style is fast paced, with lots of audience interaction with on-air personalities. Examples of CHR/pop artists include Madonna, Sheryl Crow, Hootie and the Blowfish, and Red Hot Chili Peppers.

CHR/rhythmic is similarly fast paced and personality driven, but it is more dance oriented than CHR/pop is. More dance and urban hits are mixed into the format. Artists include Puff Daddy, En Vogue, Toni Braxton, and Baby Face.

Adult CHR/hot AC focuses on a slightly older demographic of 25- to 34-year-olds and is dominated more by female artists. The format includes a fair amount of pop alternative. The format includes pop-rock artists such as Alanis Morissette and Natalie Merchant, as well as such traditional hot AC artists as Phil Collins and Gloria Estefan.

Over the years, Top 40/CHR disc jockey announcing styles have changed. The early Top 40 jocks were heavy-voiced, shouting and cajoling their audiences. The mid-1960s change saw a reduction in disc jockey presence, with less chatter and more music. The 1970s saw even less aggressive, more mellow announcers. But with the reformation to CHR in the 1980s, the energetic, big-voiced personality reasserted itself. Irreverent morning shows grew in popularity. Still, audience loyalty is generally to the music and not to the disc jockey.

CHR has replaced and generally become synonymous with the term *Top 40,* although many still refer to hits-oriented music stations as Top 40, and CHR is sometimes distinguished as using a larger playlist than Top 40. CHR stations feature little, if any, news and public-affairs programming. Syndicated features that reflect the all-hit nature of the format, such as *American Top 40,* are typically aired to help attract listeners. Nonmusic features such as sporting events are rarely programmed, however. Contests and promotions are an integral element of programming at CHR stations. CHR audiences are perhaps more receptive than those of any other format to imaginative and entertaining promotions.

Its "more hits, more often" image led CHR stations to cluster commercials in spot sets after music sweeps. Commercials on CHR stations are designed to sound as slick and entertaining as the music they interrupt. Since CHR is production-intensive, liners and catch phrases are vital to the format.

Competition to CHR comes primarily from other CHR stations, which fragment the audience. Formats that share the highest percentages of CHR audiences include Spanish, alternative/modern rock, urban, and AC. AC attracts older demographics and women; AOR draws younger listeners and men. The prospects for CHR are good, however, because analysts believe radio listeners will always be interested in the hot new songs and artists of the day.

After being dismissed in the mid-1990s by some critics and advertisers as too teen oriented, CHR experienced a resurgence. An Interep Research study attributed its success to an increase in the median age of listeners and to a wide range of music available and suitable to the format. A crossover of playlists became a boon to CHR, as CHR stations were able to play many of the same hits that get airtime on other young adult formats, such as alternative, modern AC, and adult album alternative. Today there is a variation on the format called rhythmic or "Churban" (a blend of CHR and urban).

At the turn of the 21st century, CHR reached 14 million adults weekly in the 18-to-34 demographic. About 20 percent of the overall audience extends into the 35–44 range, and 8 percent are 45–54 years old. Most CHR listeners are female (56 percent). The CHR format audience is characterized as being active consumers of alcoholic beverages; restaurants; and all entertainment categories, especially movies; as well as of computers and electronic equipment. CHR has a fairly low cost per thousand (CPM), because its audience generally does not have much money, although the young demographic does have very active spending habits.

In 1998 CHR was the fourth most popular format, behind news/talk, country, and AC. Fall 1998 Arbitron ratings showed Top 40 to be rebounding in all dayparts except midday. By 2002 there were 646 CHR/Top 40 stations in the United States. The format was the 11th most popular for stations to carry.

LAURIE THOMAS LEE

See also Adult Contemporary Format; Urban Contemporary Format

Further Reading

"CHR Format Grows Up with Former Teens," *Broadcasting* (12 October 1998)

Keith, Michael C., *Radio Programming: Consultancy and Formatics,* Boston and London: Focal Press, 1987

Lynch, Joanna R., and Greg Gillispie, *Process and Practice of Radio Programming,* Lanham, Maryland: University Press of America, 1998

MacFarland, David T., *Contemporary Radio Programming Strategies,* Hillsdale, New Jersey: Erlbaum, 1990; 2nd edition, as *Future Radio Programming Strategies: Cultivating Listenership in the Digital Age,* Mahwah, New Jersey: Erlbaum, 1997

O'Donnell, Lewis B., Philip Benoit, and Carl Hausman, *Modern Radio Production,* Belmont, California: Wadsworth, 1986; 5th edition, Belmont, California, and London: Wadsworth, 2000

"Shakeout after the CHR Goldrush," *Broadcasting* (22 July 1985)

Sklar, Rick, *Rocking America: How the All-Hit Stations Took Over,* New York: St. Martin's Press, 1984

Control Board/Audio Mixer

Device to Manipulate Audio Signals

A radio station's control board (or audio mixer or console, or simply a "board") is the primary piece of studio equipment. It allows for the use of multiple audio signals, such as from a microphone or compact disc (CD) player; allows an operator to manipulate those signals, such as controlling the volume or combining two or more together; and allows signals to be recorded or broadcast. During any of these processes, audio signals can be monitored through meters and speakers.

Functions

Any control board or console serves five basic functions: to select, mix, amplify, monitor, and route an audio signal.

An operator can select (input) various sounds at the same time. Most typical of the radio work accomplished with a board is to mix voice and music (or sound effects), as in production of a commercial.

Control Board
Courtesy of Logitek Electronic Systems

A board can also amplify any sound source. This allows an operator to properly balance sound levels, as when an announcer talks over music (where microphone volume must exceed music volume so the voice is clearly heard).

Monitoring an audio signal can be either visual (by watching volume unit [VU] meters) or aural (by listening to speakers or through headphones).

Finally, the control board is used to route (output) signals to a recorder, another studio, or the transmitter.

The easiest way to understand operation of any control board is to look at one of its individual sound channels. Such a channel includes a group of switches, faders, and knobs in vertical alignment; each group controls one or two sound sources. The number of channels (boards typically have between 12 and 36 or even more) defines its capacity to handle multiple signals. Most boards allow more than one input (microphones, etc.) to be assigned to a each channel, though only one can be selected at a time. Regardless of configuration, the first two channels (from the left) of any board are usually designed for microphones, which always need special amplification. CD players, audio recorders, and other equipment can be patched into the remaining channels.

For any channel, signals can be sent ("output") to one of three destinations: program, audition, or auxiliary. When "program" is selected, a signal is directed to a recorder or transmitter. In the "audition" position, a signal can be previewed off-air. For example, a DJ may play a CD through channel 3 in the "program" position (in other words, on the air) while at the same time previewing another disc or tape through channel 4 in the "audition" position. The "auxiliary" (aux) or "utility" (utl) are often used in production, such as to send signals to another studio.

Volume Control and Monitoring

The volume or gain control is called a slider or fader. Such controls are variable resistors—much like water faucets in function. Raising the fader (pushing it away from you) increases the volume. Some older boards have rotary knobs called potentiometers ("pots") which fill the same function. Faders are easier to work with, as they provide a quick visual check of which channels are in use and at what level.

One way to judge volume is simply to listen, but this is a relative measure, and what one operator deems loud may seem quieter to another. To more objectively indicate volume, control boards include a volume unit indicator (VU meter). Most use a moving needle on a graduated scale, ideally registering between 80 and 100 percent. Above 100 percent the signal is peaking "in the red" (because that portion of the VU meter scale is usually indicated by a red line) and may distort. On the other hand, a signal consistently below 20 percent—and thus too quiet—is said to be "in the mud." Most newer boards offer VU meters with digital lights (LEDs) to indicate volume.

Sound can be monitored in different ways as it passes through a control board. A common mistake is to run studio monitors quite loud and think all is well, when in reality the program signal going through the audio board may be at too low a level. Most boards also have provision for monitoring their output through headphones. When microphones are on ("live"), monitor speakers in the same control or studio space are automatically muted to avoid feedback howls or squeals.

Another way to monitor (and preview) a sound source is to use a board's "cue" function, which allows any input to be previewed. Shifting a volume control into the cue position, usually marked on the face of the console, routes the audio signal to a cue speaker rather than on the air.

Many control boards have additional features which make them more flexible. For example, some boards will automatically turn a channel on when its fader is moved upward. Others include built-in clocks and timers. Many boards have simple equalizer (EQ) controls that increase or decrease certain frequencies, thus altering the sound of the voice or music by changing the tonal quality. These most often affect a range of frequencies—high, midrange, and low.

Digital Future

Like other radio equipment, the control board is rapidly progressing from analog to digital mode. Incoming audio signals, if not already digital, are converted, and they remain in digital form while being manipulated through the mixer and ultimately output. Such digital boards begin to add new features and capabilities, such as hard disk audio storage. The most striking feature of the digital board is often the addition of an LCD display screen that provides status information for each channel.

Although a digital board offers all the traditional functions of an analog board, it is usually more flexible. For example, instead of just two inputs per channel, a digital board may allow any channel to be assigned to any input. Such user-defined functions allow a board to be custom-designed for a particular use. Another form of digital control board is the virtual audio console. Instead of a physical piece of equipment in the studio, an operator manipulates an image of a control board on a computer screen. A virtual fader or other console control can be managed with simple point-and-click or drag-and-drop mouse commands.

Whether digital or analog, any control board is part mechanical contrivance and part creative component. While learning its technical operation is fairly easily accomplished, effective utilization of a board takes time and experience.

DAVID E. REESE

See also Audio Processing; Production for Radio; Recording and Studio Equipment

Further Reading

Alten, Stanley R., *Audio in Media: The Recording Studio,* Belmont, California: Wadsworth, 1996

Gross, Lynne S., and David E. Reese, *Radio Production Worktext: Studio and Equipment,* Boston: Focal Press, 1990; 3rd edition, 1998

Keith, Michael C., *Radio Production: Art and Science,* Boston: Focal Press, 1990

McNary, James C., *Engineering Handbook of the National Association of Broadcasters,* Washington, D.C.: National Association of Broadcasters, 1935; 9th edition, as *Engineering Handbook,* edited by Jerry Whitaker, 1999

Nardantonio, Dennis N., *Sound Studio Production Techniques,* Blue Ridge Summit, Pennsylvania: Tab Books, 1990

Nelson, Mico, *The Cutting Edge of Audio Production and Audio Post-Production: Theory, Equipment, and Techniques,* White Plains, New York: Knowledge Industry, 1995

O'Donnell, Lewis B., Philip Benoit, and Carl Hausman, *Modern Radio Production,* Belmont, California: Wadsworth, 1986; 5th edition, Belmont, California, and London: Wadsworth, 2000

Oringel, Robert S., *Audio Control Handbook: For Radio and Television Broadcasting,* New York: Hastings House, 1956; 6th edition, Boston: Focal Press, 1989

Siegel, Bruce H., *Creative Radio Production,* Boston: Focal Press, 1992

Sterling, Christopher H., *Focal Encyclopedia of Electronic Media* (CD Rom), Boston: Focal Press, 1998

Thom, Randy, *Audio Craft: An Introduction to the Tools and Techniques of Audio Production,* Washington, D.C.: National Federation of Community Broadcasters, 1982; 2nd edition, 1989

Controversial Issues, Broadcasting of

Broadcasting programs concerning controversial issues of public importance has been a subject of continuing U.S. public policy debate for nearly as long as radio broadcasting has existed. The basic conflict has been between broadcasters, who are concerned about not offending their audiences and advertisers, and the Federal Communications Commission (FCC), who argue—and are often upheld in court decisions—that provision of time for such content is a vital part of the public-interest standard by which broadcast stations are licensed. In recent years much of the controversy has evaporated thanks to deregulation.

This entry *excludes* most discussion of commentators, political candidates, the fairness doctrine, or station editorializing—all directly related, but treated separately in the encyclopedia.

Basis for Concern

Consideration of broadcasts about controversial issues begins with an understanding of three related matters: the First Amendment, censorship, and access. The First Amendment (1791) makes clear that "Congress shall make no law" affecting freedom of speech, of the press, or of religion. Countless statements of political theory and policy as well as numberless court decisions (most having nothing to do with broadcasting) have made clear for decades that in order to be effective citizens and voters, the public needs to be informed about public issues and the various points of view concerning them. A number of Supreme Court decisions have held that robust public debate is a central component of effective freedom of speech and of the democratic system itself.

Strictly defined, censorship in the American context means *prior restraint* of publication, broadcast, or speech by an act of government. It does *not* usually include private actions (such as those by broadcasters or advertisers) that might limit speech or access to a microphone by others. The term is usually applied far more generally and is often applied to corporate actions to restrict access or debate.

Media access divides into two concerns: media access *to* places or people in order to report news or public (sometimes seemingly private) affairs, and access *by* people (other than a broadcaster or his or her staff) or their ideas to broadcast facilities. Discussion of controversial issues on radio almost always involves the latter.

Shaping a Policy

As the potential value of radio as a means of shedding light on public controversies first became clear in the 1920s, policy makers and broadcasters alike began to focus on just what radio stations should or could do in support of public-affairs communication. Yet neither the Radio Act of 1927 nor the Communications Act of 1934 (until the latter was amended in 1959) said anything about radio coverage of controversial issues or fairness in doing so. Both acts *did* make clear that government had no right of censorship over radio content. The combination of having no clear statutory requirement to deal fairly with controversial issues on the one hand, with a very clear and firm statement of no censorship on the other hand, has made defining government policy-making in this field difficult.

The first important relevant policy statement—one often still referenced in modern decisions—is found in an early Federal Radio Commission licensing case in which the commission concluded, "In so far as a program consists of discussion of public questions, public interest requires ample play for the free and fair competition of opposing views, and the commission believes that the principle applies . . . to discussions of issues of importance to the public" (*Great Lakes Broadcasting Co.*, 1928; see Kahn, 1984).

In the late 1920s and early 1930s, a number of legal cases concerned broadcasters who sought either to obtain or to retain stations as personal mouthpieces (e.g., John Brinkley) or whose programs espoused strong political views with little or no chance for rebuttal by others (e.g., Father Charles Coughlin). In a few short years, most had been forced to share time with others of different views or to leave the air entirely.

Later cases and policy statements echoed the need to cover controversial issues, but to do so fairly. In March 1939, in an FCC statement on objectionable programming practices, one item listed was "refusal to give equal opportunity for the discussion of controversial subjects"; this statement underlines the twofold nature of the concern. First, stations should provide discussion of controversial subjects, and second, they should provide a fair balance of views on those subjects. Paralleling the government concern was the 1939 version of the National Association of Broadcasters' (NAB) *Standards of Practice,* which held that "as part of their public service, networks and stations shall provide time for the discussion of public questions including those of a controversial nature."

This relatively early version of the NAB radio self-regulatory code also made clear a long-standing industry practice of not *selling* airtime for the discussion of controversial issues. The code claimed that this was because broadcasters did not want a situation in which only those able to afford the time could be heard. But such a policy also avoided offending either audience members or advertisers with too much controversy. The no-sell provision was largely followed until relatively recently. The downside of all this was that such discussions were nearly always provided as sustaining programs—meaning at the broadcaster's expense. The code provisions remained unchanged until 1948, when the restriction on sponsorship was dropped.

In the meantime, other FCC decisions helped to pin down policy still further. At the end of World War II, the commission held that a station could not establish a blanket policy of not providing any time for discussion of controversial issues (*United Broadcasting Company [WHKC]*, 10 FCC 515, 1945). A year later, three radio licenses in California were renewed despite the stations' refusal to allow a noted atheist to offer his views on the air. The commission reiterated that although "the criterion of the public interest in the field of broadcasting clearly precludes a policy of making radio wholly unavailable as a medium for the expression of any view which falls within the scope of the constitutional guarantee of freedom of speech," there was no obligation on the part of a station to grant the request of any specific person for time to state his or her views (*Robert H. Scott*, 11 FCC 372, 1946).

The FCC *Blue Book* issued in 1946 devoted several pages to the discussion of public issues, going so far as to raise 19 questions about such broadcasts—but answering none of them. The section concluded that in its decisions on whether a licensee had served in the public interest, the FCC "would take into consideration the amount of time which has been or will be devoted to the discussion of public issues" *(Blue Book*, 1946, 40). Clearly, trying to avoid such programming was not going to please the licensing authority. Thus, the licensee had to make judgments about what issues to cover and which points of view to present.

The seeming hole in the Communications Act of 1934 was finally filled when a 1959 amendment to Section 315 made clear that licensees had an affirmative obligation "to afford a reasonable opportunity for the discussion of conflicting views on issues of public importance." Nearly four decades after radio broadcasting began, the country's basic communications statute finally and specifically included coverage of controversial issues as being a part of the public interest stations were licensed to serve.

Modern Era: Selling Controversy

The dominance of television in American life by the 1950s naturally shifted regulatory attention to that medium. Most cases concerning controversial issues focused on television programs, though the concerns raised paralleled those first evident with radio. And as discussed elsewhere, most controversial issue program questions were now dealt with in the context of the FCC's fairness doctrine, issued in 1949 and rescinded in 1987. The demise of the fairness doctrine, however, was but one of several factors that changed the face of controversial issue programming.

The end of the doctrine in 1987 made possible substantial expansion of political and other controversial talk programs on radio, because they no longer faced private or government fairness doctrine–based requests for response time. Rush Limbaugh, Oliver North, and many others with decided (usually conservative) political views would have had a difficult time maintaining their controversial programming in the face of a constant barrage of requests from audience members to respond to what they had heard.

Nor did the FCC any longer seem concerned that many stations no longer provided time for discussion of controversial issues. Detailed license renewal forms that asked about station policies concerning the amount of time provided for controversial issue programming disappeared in the 1980s, to be replaced by simple postcard forms with no program-relevant queries whatever. The growing number of stations in most markets made regulation of individual outlets seem less relevant. Therefore, neither FCC commissioners nor their staff any longer felt that each station in a market had to provide such programs—as long as at least *some* stations did. Any nearby public radio station was often the selected "mark" to pick up the slack.

More important, the economic basis of radio time devoted to discussion of controversial issues has changed radically. Once shunned by broadcasters, as noted above, the selling of time for expression of points of view, whether in short spots ("editorial advertising" or "advertorials") or in programs, had by the 1990s become accepted practice. No longer did broadcasters have to pick and choose among the minefield of potential controversial issues without even the saving grace of selling time to support programs dealing with such topics. By the early 2000s, controversial issue programs almost always meant time *sold* for that purpose, usually to one or more syndicated talkers with an axe (or several) to grind or an audience large enough to attract advertisers.

Some have argued that radio is thus no longer providing a minimum of public-affairs service to its listeners—that points of view have simply become another commodity for sale to the highest bidder. Others hold that radio is but one information conduit to the modern household and that audiences can obtain as much controversy as they desire from a combination of radio, television, periodicals, and the internet, to name only a few key media. In any case, far more people agree that government supervision or regulation of radio content is not the most effective means of creating an informed electorate.

CHRISTOPHER H. STERLING

See also Blue Book; Brinkley, John R.; Censorship; Commentators; Coughlin, Father Charles; Critics; Editorializing; Equal Time Rule; Fairness Doctrine; First Amendment and Radio; Limbaugh, Rush; Politics and Radio

Further Reading

Benjamin, Louise M., *Freedom of the Air and the Public Interest: First Amendment Rights in Broadcasting to 1935,* Carbondale: Southern Illinois University Press, 2001

Bensman, Marvin R., editor, *Broadcast Regulation: Selected Cases and Decisions,* Lanham, Maryland: University Press of America, 1983; 3rd edition, 1990

Brindze, Ruth, *Not to Be Broadcast: The Truth about the Radio,* New York: Vanguard Press, 1937

Kahn, Frank J., editor, *Documents of American Broadcasting,* New York: Appleton-Century-Croft, 1968; 4th edition, Englewood Cliffs, New Jersey: Prentice-Hall, 1984 (contains "The Great Lakes Statement")

Ripley, Joseph Marion, Jr., *The Practices and Policies Regarding Broadcasts of Opinions about Controversial Issues by Radio and Television Stations in the United States,* New York: Arno Press, 1979

Smead, Elmer E., *Freedom of Speech by Radio and Television,* Washington, D.C.: Public Affairs Press, 1959

Summers, Harrison Boyd, editor, *Radio Censorship,* New York: Wilson, 1939

Cooke, Alistair 1908–

American (British-Born) Journalist, Host, and Commentator

This British-born journalist is well known for his weekly radio talk series *Letter from America,* which at the time of writing has been broadcast continually by the British Broadcasting Company (BBC) since 1946, when it was originally titled *American Letter.* The weekly 15-minute journalistic talk on life in the United States became a unique institution in the history of radio journalism; it has been heard in 52 countries, and there have been more than 2,600 programs. Cooke's elegant style of writing and presentation gained huge popularity with listeners throughout the English-speaking world. Various critics have praised his ability to capture the human atmosphere of significant news events and, through anecdotes and concrete storytelling, to explain the importance of social, cultural, technological, and political changes. Essayist Harold Nicolson described him as "the best broadcaster on five continents." He has maintained a consistency that has ensured an almost uninterrupted delivery of his talks even when rare illness has meant confinement in the hospital or at his home. The first time the BBC had to use an old edition of the show because Cooke was incapacitated by illness was 17 December 1999.

Letter from America is very much in the genre of the commentaries of American radio network broadcasters of the 1930s and 1940s. In fact, Cooke's first journalistic success was with the National Broadcasting Company (NBC) network in 1936, when in live transmissions from London he provided daily radio reports on the abdication crisis involving Edward VIII and American divorcée Wallis Simpson. He continued to work as the network's London commentator until 1937, presenting a weekly broadcast called *London Letter* that had many similarities to the style and content of the later *Letter from America* BBC series. His journalism has spanned most of the significant events of the 20th century and early 21st century, including World War II, the assassination of President John F. Kennedy, the Vietnam War, the fall of the Berlin Wall, President Bill Clinton's impeachment hearings, and the suicide attacks on the World Trade Center in New York.

Cooke's success can be traced back through his scholarship at Cambridge to his relatively humble origins in Blackpool, Lancashire, in northwest England. He left behind a brother who worked as an assistant in a butcher's shop. The "Oxbridge" environment enabled Cooke to establish contacts and use opportunities that took him to Yale, Harvard, and even to a collaboration in Hollywood with Charlie Chaplin. His career has been marked by the ability to reinvent himself, to transcend the British class system, and to eventually change nationality (he became an American citizen in 1941). Cooke had no formal training as a journalist and entered the trade via review writing for *Granta,* the *Nation and Athenaeum,* and the *Manchester Guardian* rather than by honing his reporting skills on local and regional newspapers and small town/university radio stations. He was fortunate that the insecurity of freelance life was subsidized by the wealth of his first wife's family.

Once in front of a microphone, his broadcasting talent and professionalism became self-evident. A similar blossoming of ability occurred in print journalism when he began writing articles for the London *Times, Daily Herald,* and the *Manchester Guardian.* Although Kenneth Tynan described him as "one of the great reporters," there is no evidence that

Alistair Cooke
Courtesy AP/Wide World Photos

his journalism had any scoop value or political and social impact, because he was more of a chronicler and storyteller. It could be argued that Cooke became the most intelligent interpreter of American society and history for English-speaking people abroad. This was especially true of his memorable BBC television series *America—A Personal History,* first broadcast 1972–73, which has been compared to Kenneth Clark's *Civilisation.* During the 1950s, Cooke's popularity in America was also elevated through his hosting of the pioneer cultural television program *Omnibus,* which paralleled his later role on *Masterpiece Theatre* in the 1970s and 1980s on the Public Broadcasting Service (PBS).

He changed his Christian name from Alfred to Alistair in 1930, and he abandoned British citizenship in 1941 when he was young enough to enlist for Britain's war effort. His biographer argues that he had applied for American citizenship some years before, and the eventual granting of citizenship in 1941 was the result of a delay and not an unwillingness to enlist in the British forces. His collaboration with Charlie Chaplin on a film project about Napoléon was not a success, although Chaplin did invite Cooke to be assistant director on *Modern Times.* Cooke turned down the opportunity and instead returned to Britain as the BBC's film critic. It has been suggested that he obtained his first radio job as the BBC's film critic in 1934 after he had spotted on a newspaper billboard in Boston that this predecessor, the then prime minister's son Nigel Baldwin, had been fired. In fact Cooke started lobbying the BBC for the job two months before Baldwin had somewhat courageously launched a public attack on BBC bureaucracy and philistinism. He had gone so far as to allege "Prussianism in the BBC." Cooke's self-confident assertion that cinema was in urgent need of "serious, unsolemn propaganda rather than analysis" was looked at rather favorably after Baldwin's attack. BBC historian Asa Briggs unearthed a telegram Cooke had sent the BBC in 1936 that read, "Script today would have to be about two good books and game of ice hockey, for in 23 general releases and 6 new films nowhere to go for a laugh or a cry."

Cooke's radio debut was at Broadcasting House at 6:45 P.M. on 8 October 1934; his sympathetic (and official) biographer states that the surviving script of his first cinema talk suggests he had arrived "almost fully formed." After settling in America in 1937, he began producing short series of talks such as *Mainly about Manhattan* until a substantial investment by the BBC in wartime broadcasting to and from the United States led to a more regular commission for *American Commentary.*

Cooke has made a major contribution to musical and cultural programming in radio. His first series for BBC Radio was *The American Half Hour,* which ran for 13 episodes in 1935 and sought to dramatize various aspects of American life using actors and music. Cooke researched, wrote, and presented a 1938 series for the BBC titled *I Hear America Listening,* in which he traced the history of American folk music. It was the first serious attempt to mark the contribution of African Americans in the foundation of jazz music and involved acquiring rare recordings of workers on cotton fields and citrus plantations. In 1938 he was responsible for producing and commenting on a live performance of a jazz jam session from the roof garden of the St. Regis Hotel in Manhattan, New York, to BBC listeners in Britain.

His creative collaboration with the BBC producer Alan Owen between 1974 and 1987 resulted in several series investigating the origins and development of American popular music. A series for BBC Radio in 1986 and 1987 celebrated five significant American composers. *The Life and Music of George Gershwin* achieved considerable critical acclaim. In the six programs, Cooke argued that Gershwin deserved greater prominence as an original and important composer of American music.

By 2003 Cooke, at age 94, was still broadcasting with assurance. He took part in an end-of-millennium reflection on his series with his biographer Nick Clarke on BBC Radio 4.

TIM CROOK

See also British Broadcasting Corporation

Alistair Cooke. Born Alfred Cooke in Salford, near Manchester, 20 November 1908; second son of Samuel and Mary Cooke. Attended Blackpool Secondary School; Jesus College, Cambridge, Upper Second degree, 1930; studied at Yale and Harvard, 1932–34; film critic for the BBC, 1934–37; London correspondent for NBC, 1936–38; special correspondent on American Affairs for *The Times,* 1938–41; American feature writer for *The Daily Herald,* 1941–42; United Nations correspondent for *The Manchester Guardian,* 1945–48; chief correspondent for *The Guardian,* 1948–72; writer and presenter of BBC Radio's *Letter from America,* 1946–present; writer and presenter of BBC television's *America,* 1972–73. Received Peabody Award for Radio's Outstanding Contribution to International Understanding, 1952; honorary KBE for broadcasting and outstanding contribution to Anglo-American mutual understanding, 1973; Ellis Island Medal of Honor for special contribution to the culture and ethnic mix of New York, 1986; BAFTA lifetime award for contribution to television, 1991; Benjamin Franklin Medal, RSA; Howland Medal, Yale University.

Radio Series

1935	*American Half-Hour,* BBC
1935	*English on Both Sides of the Atlantic,* BBC and NBC
1936	*New York City to the Golden Gate,* BBC
1936–37	*London Letter,* NBC
1938	*I Hear America Singing,* BBC; *The Day and the Tune,* BBC
1938–39	*Mainly about Manhattan,* BBC
1940–45	*American Commentary,* BBC
1946–50	*American Letter,* BBC
1950–present	*Letter from America,* BBC
1958	*Letter from England,* BBC
1984	*The First Fifty Years: A Personal View of Social Life in Britain and the USA from 1900–1950,* BBC
1985	*Alistair Cooke's American Collection,* BBC
1986, 1987	*Life and Times of George Gershwin,* BBC

Television Series

Omnibus, ABC, CBS, and NBC, 1952 –61; *America,* BBC, 1972; *Masterpiece Theatre,* PBS, 1971–92

Selected Publications

Garbo and the Night Watchmen, 1937
Douglas Fairbanks, 1940
A Generation on Trial: U.S.A. v. Alger Hiss, 1950
Letters from America, 1951
Christmas Eve, 1952
A Commencement Address, 1954
Around the World in Fifty Years, 1966
Talk about America, 1968
General Eisenhower on the Military Churchill, 1970
America, 1973
Six Men, 1977
The Americans, 1979
Above London, 1980
The Patient Has the Floor, 1986
America Observed, 1988
Fun and Games, 1994

Masterpiece Theatre, 1995
Memories of the Great and the Good, 1999

Further Reading

Briggs, Asa, *The History of Broadcasting in the United Kingdom,* 5 vols., Oxford and New York: Oxford University Press, 1961–95 (see especially vols. 2–5)
Clarke, Nick, *Alistair Cooke: The Biography,* London: Weidenfeld and Nicolson, and New York: Arcade, 1999
Donovan, Paul, *The Radio Companion,* London: HarperCollins, 1991
Miall, Leonard, *Inside The BBC: British Broadcasting Characters,* London: Weidenfeld and Nicholson, 1994

Cooper, Giles 1918–1966

British Playwright

Giles Cooper was an Irish-born dramatist whose radio plays have been celebrated by the British Broadcasting Corporation (BBC) as innovative and experimental writing eminently suited to the sound medium. Such was his influence that for many years the BBC ran an annual radio playwriting awards program in his memory. His first broadcast play was *Thieves Rush In* on the Home Service in 1950. This was followed by more than 60 scripted programs of adaptations and original plays on the old Light, Home, and Third national radio networks. It is claimed that the creative use of sound in the play *The Disagreeable Oyster* on the Third Program in 1957 inspired the formation of the BBC Radiophonic Workshop. *Under the Loofah Tree* was the first play officially commissioned for it.

The Disagreeable Oyster is seen as the first of his radio plays to demonstrate the unusual quality of Cooper's imagination, fully utilizing the medium's capacity of facilitating leaps from objectivity to subjectivity. Cooper developed his voice and vision during a period when the Theatre of the Absurd and the phenomenology of existentialism were influential. This is reflected in the disquieting and pessimistic themes of horror expressed through hilarity in plays such as *Unman, Wittering, and Zigo; Pig In The Middle;* and *Without the Grail. Unman, Wittering, and Zigo* was later made into a film with David Hemmings and the other two were adapted for television.

Much of Cooper's work could be described as the investigation of meanings and motives beneath everyday normalities. He characterized a world or consciousness in which the sinister and the terrifying hid behind smooth polished surfaces of action and witty, hilarious dialogue. His *London Times* obituary writer stated that he showed "an almost frightening apprehension of the modern world and its ailments." The *Times* also claimed that he "was the most prolific and arguably the most original dramatist of our mass communications." As a stage dramatist, his first success *Never Get Out!* and other plays, including in 1950, *Everything in the Garden, Out of the Crocodile, The Spies Are Singing,* and *Happy Family,* have not stood the test of time. Unlike younger contemporaries such as Harold Pinter, Tom Stoppard, and Joe Orton, Giles Cooper's literary dramatic reputation has not gathered the necessary momentum to achieve powerful cultural resonance. Despite attempts by BBC Radio Drama to keep his reputation alive, his work is virtually unknown outside the United Kingdom and has not enjoyed any renaissance through association with contemporary styles of production. However, it is worth noting that the British Chichester Literary Festival held a reading of his radio plays *Unman, Wittering, and Zigo* and *Under the Loofah Tree* in July 2000. *Unman, Wittering, and Zigo,* which is understood to have been based on memories of his former school, Lancing College, with its nearby racetracks and clifftops, was also produced by Charterhouse independent school students in 1999. His son Ric Cooper said at the time of writing that amateur royalties for *Everything in the Garden, Happy Family,* and *Out of the Crocodile* continue to be paid for performances around the world. The Edward Albee version of *Everything in the Garden* sells in the U.S. and the school version of *Unman, Wittering, and Zigo* generates several hundred sales a year.

Cooper's radio plays have also received limited academic attention, although what has appeared has been of high quality. It is somewhat ironic that his name is remembered primarily for the BBC awards given in his honor. On 9 December 1966, the *Times* published a tribute emphasizing that he was "kind and helpful to new writers learning their trade, he was a most sympathetic listener and conversationalist and he had one of the most generous laughs in London." It stated that "a half hour's chat with Giles was a wonderful tonic because of his infectious cheerfulness and good humour."

The radio drama producer Donald McWinnie wrote in his introduction to the collection of six Giles Cooper radio plays, published in 1966, that he was introduced to his work by the editor of BBC Radio Features, Lance Sieveking. McWinnie said Cooper was "usually several steps ahead of current fashion without being sensationally avant-garde." McWinnie observed that Cooper "deals with inadequate human beings—or at least with people who by trying to resolve their problems create further problems for themselves."

Frances Gray defines Cooper's unique contribution to radio drama as creating a language "capable of depicting the twilight zones between illusion and disillusion." She also said that Cooper could "move from the real to the unreal and back again; he can leave us unsure whether we are hearing illusion or reality, he can even, in seconds, change our perception of what has already happened." She concluded that listening to a Giles Cooper radio play was akin to listening to a medium and that he demonstrated a "moral view in perfect harmony." Radio drama enabled Cooper to exploit the vision of large scale nudity in *The Disagreeable Oyster* that would have been impossible on television at the time and to use the process of having a bath as an entire moral universe in *Under the Loofah Tree*.

Louise Cleveland compares Cooper's radio plays with the works of Louis MacNeice and Samuel Beckett. Her analysis draws on interviews with BBC radio drama editors and directors Martin Esslin, John Tydeman, Richard Imison, and Donald MacWhinnie, and is based on reading and listening to hundreds of scripts and archive recordings. She also invests Cooper's work with significance in the development of radio drama, proposing that his work led to the medium's acquisition of styles and techniques that separate writing and production practice from the theatrical origins of the genre.

Gray points out some of the intertextual themes of Cooper's writing across radio, television, and stage theater. Cooper's adaptability was a key factor in BBC television's successful introduction of Georges Simenon's character "Maigret" to mainstream British television audiences as one of their *Sunday-Night Theatre* presentations—"Maigret and the Lost Life," which debuted on 6 December 1959. Cooper adapted the first radio script from Simenon's 1954 novel; the drama was produced/directed by Campbell Logan. The actual *Maigret* series began in October 1960, the first of what would become four series totaling 51 episodes, each 45 to 55 minutes in length. The BBC had acquired the rights, with Simenon's blessing, against worldwide competition, making it their most ambitious series production to that date. Cooper's work on the series (he wrote most of the television scripts) was recognized by a Guild of Television Producers' and Directors' Award in 1961. He also adapted Evelyn Waugh's *Sword of Honour* trilogy of war novels and Victor Hugo's *Les Miserables* for television, both of which were broadcast posthumously.

On 2 December 1966, Giles Cooper fell out of the open doorway of a fast-moving railway car near Surbiton. He had been returning home to Sussex from a writers' dinner in Central London. Upon investigating the bizarre circumstances of his death, the inquest jury returned a verdict of misadventure after a pathologist revealed that Cooper had consumed the equivalent of half a bottle of whiskey that evening.

TIM CROOK

See also BBC Radio Programming; Playwrights on Radio

Giles Cooper. Born in Carrickmines, County Dublin, Ireland, 9 August 1918. Educated at Lancing College and Grenoble University; trained as actor at Webber Douglas School, London; seven years in the British Army as infantry officer, including 1939–45; stationed in Burma; actor and writer until 1966. OBE and Guild of Television Producers and Directors Award 1961. Died in London, 2 December 1966.

Radio Series

1950	*Thieves Rush In*
1950s	Adaptations of *Lord of the Flies, The Day of the Triffids*
1956	*Mathry Beacon*
1957	*The Disagreeable Oyster*
1958	*Without the Grail; Under the Loofah Tree; Unman, Wittering, and Zigo; Dangerous Word; Before the Monday; Pig in the Middle*
1961	*The Return of General Forefinger*

Stage

Never Get Out! 1950; *Everything in the Garden,* 1962; *Out of the Crocodile,* 1963;*The Spies Are Singing,* 1966; *Happy Family,* 1966

Film

Unman, Wittering, and Zigo, 1971

Television

Maigret and the Lost Life, 1959; *Maigret the Series:* Episode 1 *Murder in Montmarte,* 1960; *Pig in the Middle,* 1963; *Without the Grail,* 1962; *Loop,* 1963; *Carried by Storm,* 1964; *Kittens Are Grave,* 1967; *Sword of Honour,* 1967; *Les Miserables,* 1967; *To the Frontier,* 1968

Further Reading

Cleveland, Louise, "Trials In The Soundscape: Achievements of the Experimental British Radio Play," unpublished doctoral dissertation, University of Wisconsin, 1973

Cooper, Giles, "Everything In The Garden," *New English Dramatists* 7, Harmondsworth, Middlesex: Penguin, 1963

Cooper, Giles, *Six Plays for Radio,* London: British Broadcasting Corporation, 1966
"Giles Cooper," obituary, London Times (5 December 1966)
Gray, Frances, "Giles Cooper: The Medium as Moralist," in *British Radio Drama,* John Drakakis, editor, Cambridge, New York: Cambridge University Press, 1981
Gray, Frances, "The Impacted Man: Giles Cooper And The Twentieth Century," in *Radio Literature Conference 1977 Conference Papers,* Durham: Durham University, 1979
Lewis, Peter Elfed, editor, *Radio Drama,* London, New York: Longman, 1981

Cooper, Jack L. 1888–1970

U.S. Disc Jockey and Radio Entrepreneur

William Barlow has called Jack L. Cooper the "undisputed patriarch of black radio in the United States." Cooper, a Chicago-based radio entrepreneur and personality, debuted on the medium in 1925 and within five years had become the most influential black man in the radio industry. He would prove that blacks could succeed as radio personalities, programmers, and entrepreneurs, thereby helping to ignite the accelerated growth of black-appeal radio in the post–World War II era.

When Cooper entered radio in 1925 at the age of 36, he had played semi-professional baseball, managed boxers, sang and danced in black minstrel shows, and toured the nation in his own vaudeville troupe. In 1924 he was covering the theater scene for the *Chicago Defender* when the paper transferred him to Washington, D.C. It was in the nation's capital that Cooper heard a black singing group on radio station WCAP and realized that the only time he ever heard blacks on radio was when they were singing, never speaking. Hoping to correct this omission, he approached WCAP about performing comedy on the air. A producer there hired him, but racial attitudes of the day dictated that he would perform only in stereotypical black dialect. The confinement to dialect and other indignities he faced at the station frustrated Cooper so much that he returned to Chicago in 1926. But WCAP had marked a turning point in his career; he saw opportunity in radio, opportunity that would do much to break the color line in broadcasting. He returned to Chicago determined to produce radio programming by black people for black people.

While Cooper continued to write for the *Defender,* he found an announcing position with radio station WWAE. But he failed to find an outlet in Chicago that would allow him to broadcast black-appeal programming until he met Joseph Silverstein, who owned WSBC, a small station that featured various ethnic programs. Silverstein gave Cooper airtime and paid him with proceeds from advertising that Cooper himself sold. On 3 November 1929 at 5:00 P.M., Cooper launched *The All-Negro Hour,* which featured vaudeville-like entertainment for the black audience. In producing and presenting *The All-Negro Hour,* Cooper pioneered black-appeal programming, beginning a trend that would spread with some vigor after World War II.

Within a year of *The All-Negro Hour*'s premiere, Cooper introduced a number of new programs for the black audience, most of them religious in nature. The Great Depression of the 1930s limited the amount of advertising income Cooper generated from his programs, but by the late 1930s his income and influence grew when he began brokering time on Chicago radio. Cooper bought airtime, initially from WSBC only, and then resold it at a large profit to individuals and groups who used the time for various purposes. In addition, as he produced more programs through his Jack L. Cooper Presentations and as the Depression's grip loosened, black and white businesses, both local and national, began buying advertising from Cooper with increasing regularity.

By 1947 Jack L. Cooper Presentations was producing programs on Chicago's WSBC, WHFC, WBEE, and WAAF. He controlled some 40 hours of airtime on the Chicago stations. The shows reflected Cooper's panoply of interests: in addition to religious and variety programs, he produced public-affairs and public service shows, live broadcasts of sporting events, dramas, and comedies.

As Cooper's presence on Chicago radio expanded, so did the trails he was blazing for the black-appeal format. It was unheard of in the 1930s and 1940s for a black man to control any programming, but Cooper's production company (which was based in his home) at its peak produced some 50 programs primarily for black audiences and at the same time made a handsome profit. He would serve as a model for the white owners and program directors who would haltingly begin to woo the black audience in the 1940s. WDIA in Memphis, the first all-black-appeal station in America (1949) would be the culmination of what Cooper had begun in the late 1920s.

Just as Cooper was making a way for the black format, he was also making a way for black employment in radio. The success of Cooper's programs hinged on the efforts of black employees: Cooper featured black actors and musicians and hired black writers and disc jockeys. In addition, blacks found employment on the operations side of his business. Barlow noted in 1999 that many of Cooper's "employee-trainees went on to successful careers in broadcasting; among the best known are Oliver Edwards, Eddie Plique, Manny Mauldin, William Kinnison, and Gertrude Roberts Cooper—the boss' third wife."

Among the other firsts in Jack L. Cooper's pioneering career was his status as America's first black disc jockey. One Sunday evening in 1932, *The All-Negro Hour*'s pianist walked out on the show at the last minute to protest Cooper's refusal to give her a pay raise. In her absence, the boss improvised by putting a microphone by a small record player and spinning records. Over the years, Cooper would continue to rely on records for much of his musical programming, although, according to many of his contemporaries, he often eschewed the blues music so popular among blacks in favor of jazz and white dance music.

Jack L. Cooper built a small empire in black radio, controlling significant periods of airtime in one of America's largest markets and hiring many blacks to work in the industry during a time when virtually no other opportunities were available to blacks in radio. He demonstrated that blacks could succeed in radio and that black-appeal programming could be profitable. In the final analysis, Cooper's work proved to be an important factor leading to the substantial growth of black-appeal programming and black employment in radio during the late 1940s and 1950s.

Cooper continued to broadcast into the 1960s, when failing eyesight forced him to retire. He died in 1970.

MICHAEL STREISSGUTH

See also African-Americans in Radio; Black-Oriented Radio; Stereotypes on Radio

Jack L. Cooper. Born in Memphis, Tennessee, 18 September 1888. Born the last of ten children and raised by foster family in Cincinnati, Ohio. Boxed in 160 amateur bouts; from early 1900s, played semiprofessional baseball and performed as dancer, singer, actor, and comedian in traveling vaudeville shows; wrote for *Chicago Defender*, mid-1920s; producer, announcer, actor, comedian, disc jockey, 1925–1960s, for WCAP (Washington, D.C.), WWAE (Chicago), WSBC (Chicago), WHFC (Chicago), WBEE (Chicago), WAAF (Chicago); established radio production company (Jack L. Cooper Presentations), 1932; established advertising firm (Jack L. Cooper Advertising Company), 1937. Died in Chicago, 12 January 1970.

Further Reading

Barlow, William, *Voice Over: The Making of Black Radio*, Philadelphia, Pennsylvania: Temple University Press, 1999

Dates, Jannette L., and William Barlow, editors, *Split Image: African Americans in the Mass Media*, Washington, D.C.: Howard University Press, 1990; 2nd edition, 1993

Newman, Mark, *Entrepreneurs of Profit and Pride: From Black-Appeal to Radio Soul*, New York: Praeger, 1988; London: Praeger, 1989

Passman, Arnold, "Jack L. Cooper: Did the First Black Disc Jockey Play the Blues?" *Living Blues* 65 (May–June 1985)

Passman, Arnold, "Jack L. Cooper: The First Black Disc Jockey," *Popular Music and Society* 16, no. 2 (Summer 1992)

Spaulding, Norman W., "History of Black Oriented Radio in Chicago, 1929–1963," Ph.D. diss., University of Illinois, 1981

Williams, Gilbert Anthony, *Legendary Pioneers of Black Radio*, Westport, Connecticut: Praeger, 1998

Cooperative Analysis of Broadcasting

First U.S. Radio Ratings Service

The Cooperative Analysis of Broadcasting (CAB), a nonprofit organization, was the first company to provide regular studies of the radio audience on a continuing basis. From 1929 to 1933 the service belonged to Crossley, Inc., headed by Archibald M. Crossley. With the emergence of the American Association of Advertising Agencies (AAAA) in 1934, Crossley turned over his service to a jointly financed cooperative venture consisting of national advertisers and agencies. By 1936

CAB was supported by the National Association of Broadcasters (NAB) and by 1945 the four major networks.

CAB was a marketing research organization. Its interest was not just in radio listening but in what advertisers in general got out of advertising. Radio program ratings were but one means of answering a number of questions and concerns that CAB addressed. Radio program ratings provided an answer to the question of how large an audience was. During this period, radio advertisers were also program producers, purchasing time periods from the radio networks and filling these periods with their own programs. Sponsors used CAB information about radio's listening audience to build programs that attracted the type of listeners most likely to buy their products. Because radio audiences were broadly based and were not already predetermined as magazines were by editorial policy, most advertisers concentrated merely on attracting large numbers of listeners.

Origins

Because national advertisers were some of the first companies to explore the use of radio as a means of advertising, it was perhaps not surprising that the Association of National Advertisers (ANA), through its radio commission (or committee), made the first attempt to answer basic questions regarding radio's anonymous audience and the national advertisers' possible customers. Because advertisers created programs and purchased time from networks, the first question that served as the focus for the first generation of audience research was that of network programs' relative popularity: what network programs were the most popular; that is, which programs drew the largest audiences relative to other programs? This question of program ratings served as the focus for CAB.

Crossley had started Crossley Inc. in 1918; it was one of three major organizations (the others being Roper and Gallup) that engaged in political polling. Political polling results were ranked by percentage points, much like ratings. When the ANA returned to Crossley to repeat an earlier audit, Crossley suggested that he should instead study radio's listening audience and listeners' program preferences. Although the radio committee of the ANA decided not to finance the study, the ANA did agree to endorse it if Crossley would underwrite it. By 1929 Crossley had gained endorsement from 30 sponsors, and by 1930 he began field work. Crossley's service functioned primarily as a network program rating service. Crossley, in fact, coined the term *rating,* although initially he used only the "identified listening audience," or what is now called the share, as the base for his ratings. Because national advertisers and agencies had developed Crossley's service, it initially served the 33 cities where the radio networks had outlets.

CAB's Method: The Telephone Recall

Although Crossley used a variety of techniques for different clients, including printed roster, mechanical recorder, personal interview, and coincidental, he selected a simple "next-day recall" telephone method to provide the first regular measure of network program audiences. The coincidental method employs an interview method in which the respondent is asked to state his or her listening to radio at the precise time ("coincidental" with) the interview. By comparison, next-day recall required respondents to recall behavior for the previous 24 hours. The recall method raised questions about respondent memory. Crossley's technique was to dial a telephone-based list and interview respondents about their previous day's listening.

Crossley chose the telephone recall method for four primary reasons. First, radio and telephone ownership had originally exhibited a high degree of congruence when he began his radio work in 1929—though radio homes would soon outstrip those with telephones. Second, telephones covered a wide area quickly. Third, recall meant that a great deal of information about listeners could be collected at little expense. Finally, sponsor identification was an important concern during the period. As Daniel Yankelovich noted in 1938, for many years advertisers held the belief that recall or registration was the most useful index of advertising effectiveness. Because most sponsors or agencies typically developed their own programs during this period or at least sponsored an entire program, advertisers sought to know the degree of registration between the programs and their product. Recall measured conscious impression. The chief drawback, of course, was that not all homes had telephones. By the late 1930s, twice as many homes had radios as owned telephones, meaning a large proportion of the radio audience was not reached by CAB researchers. Still, over the years Crossley's sample size grew four-fold from its starting point of about 100,000 households.

Enter Hooperatings

Crossley's service, under stiff competition after 1934 with the up-and-coming Hooperatings service, gradually lost ground. The foundation of Crossley's survey technique was originally quota sampling. This was a type of nonprobability sampling, meaning that the degree of sampling error could not be calculated (sampling error is the difference between the sample's results and the results that would be obtained if the whole population were surveyed). Quotas were set as to the number of respondents of varying demographics, such as geographical areas, age, gender, economic levels, and so forth that interviewers were to obtain. The goal in setting quotas was to ensure that the sample was distributed with respect to these characteristics in proportion to presumably known population totals. In response to attacks from C.E. Hooper of Hooperatings, Crossley changed to random sampling, a type of probability sample in which sampling error could be calculated. A random sample ensured that each unit used in the sample had an equal chance or probability of being selected. He also shifted to the coincidental method, or having interviewers call while the program audience was still tuned to their radios. Hooper also questioned CAB's tabulation procedures, the phrasing of questions asked, and distribution methods.

As the authority of CAB was whittled away by the compiler of a more convincing set of statistics, backers of CAB became alarmed. By January 1946 Hooper had substantial subscribers, and industry backers saw CAB's switch from telephone recall to Hooper's method of telephone coincidental as rendering CAB superfluous. The final straw for CAB was the financial withdrawal of ABC, CBS, and NBC, which left MBS as the only radio network member. The four networks had provided 40 percent of CAB revenues. The three advertising association backers, ANA, AAAA, and NAB, were left holding the bag. An attempt was made to cover this large operating cost gap through increasing dues and assessment. However, CAB had gradually lost more and more ground to Hooper, and in June 1946 CAB suspended its 17-year-old service.

Run as a cooperative membership organization, CAB was operated by a board of governors consisting of advertisers, agencies, and broadcasters. This cooperative structure led to a crucial difficulty that impaired CAB. This structure had hindered its efficient operation, making it much less responsive to the marketplace than Hooper. The committee's divergent idea and politics led to a bureaucratic situation that Crossley called "too many chiefs." Its committee structure meant that decisions were long in the making; decisions were typically compromises and were tied to interest rather than economic considerations; furthermore, results were not measured in terms of profits or loss in the marketplace. Hooper's private enterprise, on the other hand, was conscious of cost, was aware of the degree of acceptance from its clients, and was more responsive in general to the marketplace.

Both CAB's and Hooper's rating services were rating indices limited to a handful of urban cities. Neither provided national size estimates of the number of listeners to a given program, but rather were limited to comparative figures. Neither was designed to measure a true national program audience but was instead limited to urban telephone homes.

KAREN S. BUZZARD

See also Advertising Agencies; Audience Research; Consultants; Hooperatings; Programming Strategies and Processes

Further Reading

Beville, Hugh Malcolm, Jr., "Radio Services—Pre TV (1930–1946)," chapter 1 of *Audience Ratings: Radio, Television, and Cable,* Hillsdale, New Jersey: Erlbaum, 1985; revised edition, 1988

Buzzard, Karen S., *Chains of Gold: Marketing the Ratings and Rating the Markets,* Metuchen, New Jersey: Scarecrow Press, 1990

Buzzard, Karen S., "Radio Ratings Pioneers: The Development of a Standardized Ratings Vocabulary," *Journal of Radio Studies* 6, no. 2 (November 1999)

Hooper, C.E., "CAB Suspends Ratings Service," *Broadcasting* (24 June 1946)

Womer, Stanley, "What They Say of Two Leading Methods of Measuring Radio," *Printer's Ink* (7 February 1941)

Copyright

Protecting Intellectual Property

Copyright is the legal principle that protects the intellectual property rights of the author of a work. Any original literary, audio, or video material can be copyrighted by the author. The author has the exclusive right to control the reproduction, distribution, performance, or adaptation of that work. In order for an author to claim copyright, the work must have been produced in a fixed medium (written on paper, tape-recorded, or even typed onto a computer hard drive). Extemporaneous speeches or performances are not copyrighted.

There is a basic philosophy behind protecting copyright. Those who hope to make a profit from their creations are much more likely to be successful if others are prohibited from using the creations without paying for them. Copyright laws are created to encourage authors to write more with an assurance that their material cannot be illegally reproduced or altered. The U.S. Constitution states that Congress has the authority to "promote the progress of science and useful arts, by securing for limited times to authors . . . the exclusive right to their respective writings."

Most countries have some form of copyright law. In the United States, the 1976 Copyright Act is the law that provides much of the detail. There are a number of international copyright treaties and conventions, but there is not a single worldwide "copyright registration" that will protect an author's work globally. The United States is a member of the Berne Convention and the Universal Copyright Convention, which recognize copyrights of residents of member nations. Some countries (most notably China) have been criticized for their unwillingness to enforce copyright claims in their country, allowing unlimited copying without compensating the author.

Copyright exists the moment an author creates a work in a tangible form, but without providing notice and registering the work, the author stands little chance of enforcing the copyright. Notice requires that the author provide the word *copyright* or the international symbol ©, the name of the copyright holder, and the date of the copyright. For audio recordings (whether CDs, cassettes, or vinyl), the © symbol is replaced with a ℗, for "phonorecord." Without a notice, a copyright infringer may be able to claim an "innocent infringement." The infringer may successfully claim to have been unaware that the work was copyrighted or may claim not to have known whom to approach to request copyright permission.

Registration with the Copyright Office is a relatively simple procedure. Although registration is not required to have copyright, trying to make a claim in court that copyright has been violated is impossible without registration. In order to register copyright, an author submits a form (available on-line), a $30 registration fee, and two copies of the work. Although registration can be done months or years after a work is created, there are distinct legal advantages to registering the work within three months of its creation. Only those promptly registering copyright are entitled to recover attorney's fees and statutory damages in a lawsuit.

Radio and Copyright

Radio stations use copyrighted material that they do not own every day. Almost all of the music they play was written by someone other than a station employee. Radio stations are required to pay royalties to the authors of the music they play. To simplify the process, performing rights organizations such as the American Society of Composers, Authors, and Publishers (ASCAP) and Broadcast Music Incorporated (BMI) were created to collect royalties from stations and to distribute the royalties equitably among authors. Stations pay annual blanket licensing fees, which cover the cost of playing any songs from those libraries. The fees are based on the benefit each station derives from the copyrighted music. Highly profitable stations that play a lot of music in large markets are charged much more than are all-talk stations in small markets. Those fees are then equitably divided by the rights organizations based on their own determination of which authors' works were most used.

Not everything a radio station does is covered by music licensing fees. The annual fees paid to performing rights organizations only cover *performances* of the music. Those organizations do not collect fees for other uses of copyrighted material, such as reproducing the work or creating some kind of adaptation. If a radio station uses music in producing a commercial, that is not a performance but is instead the creation of a new, derivative work. Using a piece of copyrighted music in this way without permission of the author is a copyright violation. In order to comply with copyright law, stations or advertisers that use copyrighted music in their commercials need to seek the permission of the copyright holder, usually the author. From that point, it is all a matter of contract negotiation. The author can allow the use for a limited time, charge any fee for the use, or prohibit the use altogether.

Just as a radio station has to pay copyright fees for the music it plays, others might be obligated to pay copyright fees for playing music, even if the music comes from a radio station that has already paid copyright fees. Because the premise of copyright is to reward authors when others benefit from their work, retail establishments, restaurants, and other venues must pay for the use of copyrighted music. If a restaurant plays music that enhances the atmosphere, thereby contributing to its profitability, the restaurant might be responsible for paying a fee for the use of the music. In 1998 the U.S. Congress amended the law to exempt smaller establishments. Retail establishments under 2,000 gross square feet and restaurants under 3,750 gross square feet are exempt from having to pay copyright fees. Larger establishments are expected to pay fees much as a radio station would.

Limitations on Copyright

Copyright does not last forever. Under earlier copyright laws, rights were protected for 28 years and could be renewed for an additional 28 years. In 1978 the law was modified to protect copyright for the life of the author plus 50 years. A more recent modification extended the protection. For works created since 1978, copyright lasts for the life of the author plus 70 years. Because the legislation had to account for changes in existing copyright, duration can be somewhat more complicated for pre-1978 works. Works created before 1978 are generally protected for 95 years, with some rare exceptions (e.g., if a work was created in 1940 and not reregistered 28 years later). Because copyright endures beyond the life of the author, it can be willed just like any other piece of real property that the author owns. In fact, authors can sell or give away their copyrights before death, but the duration is still based on the author's life plus 70 years.

There are instances in which copyright is held not by an individual but rather by a corporation. It would be impossible to calculate the life of the author plus 70 years, because the corporation might go on indefinitely. For these *works for hire,* copyright lasts for 95 years from publication or 120 years from creation, whichever is greater.

There are times when copyright owners have no control over the works they've created. The doctrine of *fair use* allows for certain types of uses without permission or payment to the copyright holder. In determining whether a use is fair, four elements need to be considered: (1) the nature of the use, (2) the nature of the copyrighted work, (3) the amount and substantiality of the original that is used, and (4) the impact of the use on the potential market for the original.

Nature of the Use

Section 107 of the Copyright Act of 1976 states that the fair use of copyrighted works "for purposes such as criticism, comment, news reporting, teaching (including multiple copies for classroom use), scholarship, or research, is not an infringement of copyright." A 1994 U.S. Supreme Court decision determined that parody is also an acceptable purpose for a fair use. The nature of the use can be the sole factor distinguishing a fair use from a copyright infringement. For example, a radio station using a piece of music as the sound bed for a commercial would be infringing copyright. The same piece of music used by the same station as a sound bed for a news story would be a fair use, because news reporting is considered a fair use and so the use of the song as a news story element would be protected. Ironically, the same song used by the same station, if used as production music (e.g., as the theme song for the newscast) would be a copyright infringement. The nature of the use is an important consideration.

Nature of the Copyrighted Work

Authors who don't publish their works have a greater interest in keeping their works out of public sight. A poet or composer who creates work but prefers not to share it has greater protection from a claim of fair use than does a poet or composer who constantly tries to reach the largest possible audience. It would be difficult to claim a fair use of an unpublished work, even if the nature of the use were acceptable.

Amount and Substantiality Used

In order for use of copyrighted material to be considered fair, the amount of the original used should not be excessive. Unfortunately, no legally defined line separates the quantity considered fair from the amount that would not be considered fair. What we do know is that the judgment is made based both on qualitative and quantitative information. Although copying an entire work is not likely to be a fair use, use of just a small percentage of an original might be determined unfair if the use contains the essential part of the copyrighted work. The U.S. Supreme Court ruled that the *Nation* magazine infringed the copyright of Gerald Ford's memoirs when it published "only" a few hundred words from the original: the portion published was at the heart of what people most wanted to know (regarding Ford's dealings with Nixon) and was therefore significant.

Impact on the Potential Market

This is probably the most important consideration in determining whether a use is fair. Uses that harm authors by denying them profits are not likely to be considered fair. The issue is not whether the user profits, but rather whether the copyright owner's profit is reduced. Not-for-profit educational radio stations still must pay to use copyrighted music on the air, even though no one profits from airing the music. Musical parodies are protected in part because they do not harm the market for the original copyrighted work. The U.S. Supreme Court ruled that a 2 Live Crew parody of Roy Orbison's "Pretty Woman" was a fair use, stating in its reasoning that it could not imagine a potential purchaser trying to choose between the two versions.

An area of fair use where there is still some uncertainty is in home recording. A U.S. Supreme Court decision in 1984 held that video recording of television programs for noncommercial, in-home viewing was a fair use. That decision, however, essentially authorized "time shifting" of TV programs, that is, the taping of a program in order to watch it later. The Court never really addressed the issue of whether an individual would be allowed to amass a collection of videotaped programs as a fair use. The issue will undoubtedly be revisited as advancing digital technology makes high-quality reproductions easier to obtain. The recording industry already fears the possible explosion of MP3 digital audio recordings, many of which are made without compensation to the author. Their reproduction and distribution is made more rapid by the expansion of the internet and the increase of more advanced home computers. If in-home recording is a fair use, and individuals can access thousands of audio files, the recording industry could be severely affected.

Enforcement

In terms of enforcement, the government has no agency charged with seeking out copyright infringements. It is the responsibility of individual authors to protect their own copyrights, and for this reason many copyright infringements are never punished. In the case of music, performing rights organizations seek out copyright violators and take legal action through the courts, if necessary. Commercial studios, publishing houses, and other industries with a vested interest in protecting their copyrights are active in seeking out violators. Major violations are easily caught: ASCAP and BMI know which stations pay their fees and can monitor those that don't and charge them with a violation. Disney Studios is likely to find out if someone releases a film that is substantially similar to one of their own. But there are thousands of small venues where music is performed, hundreds of thousands of photocopiers, and millions of tape recorders. It is impossible for copyright owners to be able to enforce their rights in every possible arena. This does not mean, however, that the laws are any less real. Violations, no matter how small, can still be legally prosecuted. An appropriate analogy might be a traffic light in a small town at 3:00 A.M. If the light is red, it is still illegal to go. Anyone who does go through the red light is not likely to be caught. Nonetheless, the law still exists. The same is true of copyright. A number of copyright violations are unlikely to be discovered, but they are still violations and can be punished if discovered.

DOM CARISTI

See also American Society of Composers, Authors, and Publishers; Broadcast Music Incorporated

Further Reading

Middleton, Kent, and Bill F. Chamberlin, *The Law of Public Communication,* White Plains, New York: Longman, 1988; 5th edition, by Kent Middleton, Chamberlin, and Robert Trager, 2000

Stanford University Libraries: Copyright and Fair Use, <fairuse.stanford.edu>

Strong, William S., *The Copyright Book: A Practical Guide,* Cambridge, Massachusetts: MIT Press, 1981; 5th edition, 1999

United States Congress Senate Committee on the Judiciary, *The Copyright Office Report on Compulsory Licensing of Broadcast Signals: Hearing before the Committee on the Judiciary, One Hundred Fifth Congress, First Session, November 12, 1997,* Washington, D.C.: GPO, 1998

United States Copyright Office, <www.lcweb.loc.gov/copyright>

Corporation for Public Broadcasting

The Corporation for Public Broadcasting (CPB) was created by the passage of the Public Broadcasting Act of 1967. Conceived as a nonpartisan entity established to promote and protect public radio and television, the Corporation has been embroiled in controversies and conflicts with the very public broadcasting organizations it brought into existence. Despite the best intentions of Congress, the CPB has not been able to sustain a nonpartisan posture through most of its existence.

Origins

When Lyndon B. Johnson signed the Public Broadcasting Act into law on 7 November 1967, the Corporation for Public Broadcasting was created. This landmark legislation became Section 396 of the Communication Act of 1934. In his public remarks at the signing ceremony, President Johnson stated that the Corporation "will get part of its support from our Government. But it will be carefully guarded from Government or from party control. It will be free, and it will be independent—and it will belong to all our people."

Contained in the Corporation's charter is a Congressional declaration of policy that provides, in part, that it is in the public interest to encourage the growth and development of educational radio and television broadcasting; that freedom, imagination, and initiative at both the local and national levels are necessary for high-quality, diverse programming for public broadcasting; that federal support for public broadcasting is appropriate; and that a private corporation should facilitate system development and afford maximum protection from extraneous interference and control. The Corporation's board of directors consists of 15 members appointed by the U.S. President, with the advice and consent of the Senate. No more than eight members may be of the same political party, and all members must be United States citizens. Board members are appointed for a six-year term of office. All officers of the Corporation serve at the pleasure of the board of directors. Each year the CPB must submit an annual report to the President for transmittal to Congress that contains a detailed statement regarding its operations, activities, accomplishments, and financial condition.

CPB and Public Radio

The Corporation for Public Broadcasting was incorporated in 1968 and began the task of staff appointments in 1969. Although the Office of Radio Activities was organized in June 1969 with Albert Hulsen as its first director, the top priority for CPB was television and the formation of a TV interconnection system, the Public Broadcasting Service (PBS). Hulsen, and his successor Thomas Warnock, used this early period to gather information about the performance of public radio in the United States and to begin planning for what would become radio's interconnection system, National Public Radio (NPR), launched in 1971.

CPB and PBS began feuding soon after PBS was created. Lyndon Johnson had since left the White House and President Richard M. Nixon did not like much of what he saw on public television. The Nixon White House started applying pressure on CPB, and the board in turn started applying pressure on PBS. The resulting conflicts between CPB and PBS nearly destroyed the very public broadcasting system that the Corporation had been created to protect. Eventually a reorganization at PBS and a partnership agreement between CPB and PBS stabilized the system and left television much stronger than its radio counterpart. Public radio would need to engage in major reorganization itself if it hoped to get its share of the funding pie during joint negotiations between CPB, PBS, and NPR. That reorganization came as NPR merged with the Association of Public Radio Stations (APRS) to form a new NPR in May of 1977. This created a new political equation in public broadcasting. No longer could CPB and PBS make unilateral decisions about pubic broadcasting—whether funding allocations or system development—without considering public radio.

The overall prominence of the Corporation for Public Broadcasting that was witnessed during the first decade following the passage of the Public Broadcasting Act had clearly declined by the beginning of the 21st century. Revisions in Congressional appropriation procedures had substituted funding formulas for board room negotiations. Repeated calls for CPB to be a pass-through agent for the distribution of public monies rather than a policy-making organization for public broadcasting had stripped the Corporation of much of its public stature, and with its loss of image came a drain in talent. Partisan politics had inadvertently been built into the fabric of CPB regardless of the safeguards that had been written into the articles of incorporation. When the Republicans rule, the board is more likely to find liberal political bias in the system's public affairs programming. When the Democrats are in charge, the system's critics decry insufficient minority programming and barriers to access by independent producers.

CPB was created to help put into practice the visions of public service broadcasting that existed only in the form of idealistic rhetoric. Congress was willing to craft lofty language that gave the Corporation its mandate, but lawmakers were never willing to grant the system the kind of fiscal independence that the original system framers envisioned for the fulfillment of the dream for an alternative system of public radio and television stations that were above partisan politics and commercial marketplace imperatives. Given the political, social, and economic environment in which CPB has been forced to function, it is not surprising that its performance record as a catalyst for the development and preservation of public broadcasting has been no more impressive. Indeed, when all of the political and economic handicaps placed on CPB are factored into the performance equation, one might wonder how the Corporation has been able to sustain the level of effectiveness that it has achieved. Whether the Corporation for Public Broadcasting will be able to fulfill its true potential as a positive agent on behalf of public radio and television in the 21st century will largely depend on whether Congress will at long last create the insulated funding mechanism that was envisioned as vital to the Corporation's functioning some 35 years ago.

ROBERT K. AVERY

See also National Public Radio; Public Broadcasting Act; Public Radio Since 1967; Public Service Radio

Further Reading

Avery, Robert K., and Robert Pepper, "Balancing the Equation: Public Radio Comes of Age," *Public Telecommunications Review* 7, no. 2 (March/April 1979)

Burke, John Edward, *An Historical-Analytical Study of the Legislative and Political Origins of the Public Broadcasting Act of 1967,* New York: Arno Press, 1979

Carnegie Commission on Educational Television, *Public Television: A Program for Action,* New York: Bantam Books, 1967

Carnegie Commission on the Future of Public Broadcasting, *A Public Trust,* New York: Bantam Books, 1979

Engelman, Ralph, *Public Radio and Television in America: A Political History,* Thousand Oaks, California: Sage, 1996

Witherspoon, John, Roselle Kovitz, Robert K. Avery, and Alan G. Stavitsky, *A History of Public Broadcasting,* Washington, D.C.: Current, 2000

Corwin, Norman 1910–

U.S. Radio Playwright and Producer

Norman Corwin is widely considered radio's most revered and celebrated dramatist and poet. His career began during the medium's late infancy at the Columbia Broadcasting System (CBS) and extended into the 21st century with critically acclaimed work for public radio. Corwin wrote, directed, and produced more than 200 original radio plays that featured many of the most prominent actors of the day.

For his extraordinary output and talents, spirit and sensibilities, Corwin has been honored and admired by heads-of-state as well as by the literary and broadcast communities worldwide. Among his most noted radio works are *We Hold These Truths, On A Note of Triumph, 14 August, The Undecided Molecule, The Plot to Overthrow Christmas,* and *The Odyssey of Runyon Jones.* Along with countless awards, Corwin has the distinction of being the first writer inducted into the Radio Hall of Fame. In his landmark history of American broadcasting published several decades ago, Erik Barnouw proclaimed Corwin the "unofficial" poet laureate of radio. Today this designation has rightfully attained "official" status.

Words in the Air

Corwin was born to emigrant parents in Boston in 1910. Unlike his siblings, who pursued college studies after graduating from high school, Corwin chose to go to work. Enamored of the power and beauty of words since early childhood, he first sought and gained employment as a writer and reporter for newspapers. He landed his first job in 1927 at *The Recorder* in Greenfield, Massachusetts, and not long after he was able to transfer his growing skills to the pages of Springfield's *The Daily Republican.*

While at *The Daily Republican*, Corwin edited the news and produced a popular poetry feature for a local radio station, but he had not seriously considered a career in radio. Following a dispute with the station's management concerning its unwillingness to cover strike-related stories, Corwin headed for New York. After a brief stint as a film publicist for 20th Century Fox, he convinced experimental station W2XR (later WQXR) to let him offer a show called "Poetic License." Not long after the show's debut, Corwin was doing clever and witty recitations on NBC, and in 1937 these would capture the attention of CBS's William Lewis, who would hire him to direct the network's just formed *Columbia Workshop.*

Between 1938 and 1949, Corwin wrote, directed, and produced dozens of highly acclaimed radio plays as the CBS network provided him uncommon freedom to innovate during sustained (unsponsored) segments of its broadcast schedule. Later Corwin would attribute his great success to CBS's willingness to allow him to work without interference and strictures. Through evocative and imaginative scripts and technological mastery of the medium wherein the use of sound effects and music were redefined, Corwin took the radio drama in new and extraordinary directions. In his hands it became a legitimate and respected art form. Not only were his works rich in meaning and scale, but his output was amazing as well. In order to meet airtime demands, he often turned out a new play every week for months on end. His great energy and creativity brought added prestige to CBS.

Among Corwin's early network successes were *The Plot to Overthrow Christmas* (1938), which became a holiday standard in both broadcast and published form, and his dramatic disquisition against fascism, *They Fly Through the Air with the Greatest of Ease* (1939). Corwin's vast range became immediately apparent, and he soon was the driving force behind the medium's most auspicious series of programs offered under the rubric *The Columbia Workshop.* This distinguished effort provided a venue for the works of prominent literary figures of the day, such as Archibald MacLeish and Stephen Vincent Benet, as well as for Corwin himself. Many critics view CBS's *Workshop* as radio's high water mark.

Themes and Works

Corwin wrote programs that challenged the listener's emotions and intellect. His were dramas that dealt with the principle issues and concerns of the common man. Hatred, prejudice, deceit, and injustice were familiar themes in his plays. Given the era in which he worked, war became his foremost subject. For some critics, Corwin proved to be too much of a flag waver. The propagandistic and nationalistic nature and quality of many of his war-related scripts may have concerned some people but it won the praise and admiration of many more. Corwin was unabashedly patriotic, this stemming from his deep-seated passion for fairness and justice for all people. His dramas served to bookend the World War II era. Shortly after the bombing of Pearl Harbor and President Roosevelt's declaration of war, his play *We Hold These Truths*—celebrating the 150th anniversary of the Bill of Rights—was broadcast to the nation and commanded the largest audience for a play in the medium's history.

During this dark period in world affairs, Corwin contributed extensively to series like *This Is War* in 1942 and a host of other network programs that zealously promoted the Allied cause. To mark the war's end, Corwin penned and presented

Norman Corwin (center) directing
Courtesy of Norman Corwin

what many consider his masterwork, *On a Note of Triumph* in 1945. So poignant was its affect and impact that many who heard its broadcast were able to recite lines and even whole passages from it decades later. His war dramas had a galvanizing effect on the American public during a time when it was most needed.

Global conflict was not the only recurring or prominent theme in his protean repertoire during this period. Corwin excelled as much when writing about other aspects of the human condition. The more mundane experiences and actions of mankind intrigued him as did the curious machinations of the biological, mythical, and spiritual worlds.

His plays, although nearly always intended to convey a serious or thought-provoking message, were not unrelentingly somber in nature. Corwin possessed a wonderful comedic sense and loved to offer up delicious satires, fantasies, and parodies on politics, business, and human behavior. On the lighter side of Corwin's canon are *My Client Curley, A Soliloquy to Balance the Budget, Good Heavens, The Plot to Overthrow Christmas, The Cliche Expert,* and *The Undecided Molecule.* His keen humor and playful imagination are evident in dozens of other radio plays as well. Corwin also applied his pen to the making of operas, documentaries, and essays for consumption by network audiences. Versatility became one of his legendary hallmarks.

Perhaps the only person to have entered big time national radio through the portal of poetry, Corwin enjoyed writing in verse more than prose, and because of this most of his plays assume a distinct poetic life of their own. Among those who most greatly influenced him as a writer were poets like Walt Whitman and Carl Sandburg. The latter greatly admired Corwin's work and considered him one of the country's best practitioners of the rhyme form.

As the bard of radio's golden age, Corwin helped to lift the medium from its adolescence to its adulthood. His high ideals and moral vision demanded that radio become more socially and culturally astute. His symphonic (music was an integral ingredient of his plays) and altruistic inscriptions raised the stakes for practiced and aspiring radio dramatists alike and provoked and schooled its audience. The poet Carl Van Doren wrote that "Corwin was to radio what Marlowe was to the Elizabethan stage." Indeed, Corwin's work held its listeners in thrall as it conveyed the enduring truths—albeit both sad and delightful—about the human condition in unforgettable words and sounds.

Global Reach

Corwin's acclaim as a radio dramatist spread to the international arena and ultimately made him the most produced radio writer worldwide. Presaging this distinction, he set out on a global expedition to record the perspectives of people in many nations for use in a CBS documentary. Subsidized by the bestowal of the first ever Wendell Wilkie Memorial Award in 1946, this became a 13 part series called *One World Flight,* wherein many of the most compelling personalities of the time were interviewed on a host of propitious questions about the status of mankind. During Corwin's four month fact-finding journey he met and recorded the thoughts of people from all walks of life—heads-of-state, generals, waiters, actors, farmers, writers, composers, artists, orphans, and scientists. Corwin narrated the series, which particularly focused on the war's tragic aftermath and hopes for the future.

In one program he interviewed a widowed Italian woman and observed: "This voice and the echo of guns only lately stilled, and the silence of the cemeteries . . . the begging of alms, and the whimper of hungry children; this voice, and the mute rubble of wasted towns and cities—these were the sounds of need: need for the hope and for the reality of a united world."

Corwin has been cited as the first radioman to circle the globe as a journalist, and his passion for world justice and peace have informed his creative efforts throughout the last half of his exceptional career. Since the fade of radio's much heralded golden age following the arrival of television in the late 1940s, he has contributed to the efforts of the Voice of America, U.S. State Department, and several other world organizations by speaking on behalf of the poor and disenfranchised. In the early 1950s Corwin served as Chief of Special Projects for United Nations Radio. Over the decades many of his radio plays have been rebroadcast to underprivileged areas of the world, and he has been invited to conduct seminars on the medium in Europe, Africa, and Asia.

Golden Age to Graylist

In the 1950s, Corwin's star faded as television took over the American living room. The market for radio drama all but vanished as the networks pursued television's more lucrative bottom line. Within a few short years, Corwin's name was no longer a household word. Forced to seek work elsewhere, he wrote scripts for television and movies, where recognition for his writing gifts was soon forthcoming. In 1956, Corwin penned the screenplay *Lust for Life* and received an Academy Award nomination. In the years to follow, despite his appearance on the McCarthy era graylist of those suspected of associating with communists, he would be honored with several prominent awards and citations.

Corwin continued to earn the public's affection and attention with five stage plays and over a dozen books. Yet despite his foray into other writing venues, radio would still command his greatest attention. There would be a steady but modest flow of writing assignments for the medium even as it became dominated by pop music and deejays.

In the late 1990s, public radio would prompt a rediscovery of Corwin's ethereal artistry by digitally remastering and airing *13 by Corwin* and commissioning him to undertake a series of original works, culminating in his rousing salvo for New Years Eve 2000, *Memos to a New Millennium,* narrated by Walter Cronkite. At 90 years of age, the medium's grand "radiowright" (a term coined by Corwin) was still an undisputed master of the airwaves.

MICHAEL C. KEITH

See also Drama, U.S.; Playwrights on Radio; Poetry and Radio

Norman Lewis Corwin. Born in Boston, Massachusetts, 3 May 1910, one of four children of emigrant parents. Newspaper journalist, 1927–36; writer, producer, director at WBZ-WBZA, 1934; WQXR, 1937–38; and CBS, 1938–49; chief of special projects for United Nations Radio, 1950–55; freelance writer, producer, and director for radio, television, stage, and film, 1955–93; book author; magazine columnist, 1973–80; college lecturer, University of Southern California, 1980–present. Received American Academy of Arts and Letters Citation, 1942; Golden Globe Award, 1958; Academy Award nomination, 1958; Peabody Award, 1941 and 1945; DuPont Award, 1996; member of the Board of Governors and first Vice President of the Motion Picture Academy of Arts and Sciences. Inducted into the Radio Hall of Fame, 1993.

Radio Programs

1938 *Words without Music* (one play)
1939 *Words without Music* (one play); *Columbia Workshop* (one play); *So This Is Radio* (six plays)
1940 *Columbia Workshop* (one play); *Pursuit of Happiness* (one play); *Forecast* (one play); *Cavalcade of America* (one play)
1941 *26 by Corwin* (26 plays); *We Hold These Truths* (one play)
1942 *This Is War!* (six plays); *An American in England* (ten plays)
1943 *Cresta Blanca Carnival* (one play); *America Salutes the President's Birthday Party* (one play); *Transatlantic Call* (three plays); *Passport for Adam* (four plays)
1944 *Columbia Presents Corwin* (20 plays); *Election Eve Special* (one play)
1945 *U.N. San Francisco Conference Special* (one play); *On a Note of Triumph* (one play); *Columbia Presents Corwin* (seven plays); *VJ-Day Special* (one play); *Day of Prayer Special* (one play); *CBS Promotion Special* (one play); *Special: Radio's 25th Anniversary* (one play)
1947 *One World Flight* (13 plays); *Special: Committee for the First Amendment* (one play)
1949 *CBS Documentary Unit* (one play); *Un Radio* (one play)
1950 *The Pursuit of Peace* (two plays)
1951 *United Nations Radio* (one play))
1955 *United Nations Radio* (one play)
1990s *With Corwin* (six plays for public radio)

Selected Publications

They Fly through the Air with the Greatest of Ease, 1939
The Plot to Overthrow Christmas, 1940
Thirteen by Corwin, 1942
This Is War!, 1942
More by Corwin: 16 Radio Dramas, 1944
On a Note of Triumph, 1945
Untitled, and Other Radio Dramas, 1947

Further Reading

Bannerman, R. Leroy, *Norman Corwin and Radio: The Golden Years*, University: University of Alabama Press, 1986
Bell, Douglas, *Years of the Electric Ear: Norman Corwin*, Metuchen, New Jersey: Scarecrow Press, 1994
Keith, Michael C., compiler, *Talking Radio: An Oral History of American Radio in the Television Age*, Armonk, New York: M.E. Sharpe, 2000

Coughlin, Father Charles 1891–1979

U.S. Roman Catholic Priest and Radio Commentator

In the 1930s, Father Charles E. Coughlin, the "Radio Priest" from Royal Oak, Michigan, used radio broadcasts to assemble what was then the largest congregation in the history of Christianity. He also became the first Roman Catholic priest to make a serious impact on the U.S. political scene.

Early Years

Born of a Canadian mother and an Irish-American father in Ontario in 1891, Coughlin was ordained a Roman Catholic priest in 1916 as a member of the Basilian order. He assisted in several Michigan parishes, becoming a diocesan priest in Detroit in 1923. Already enjoying a reputation as a pulpit orator, his masses at the churches to which he was temporarily assigned regularly attracted overflow crowds.

In 1926 Coughlin became pastor of the just-dedicated Shrine of the Little Flower parish in Royal Oak, four miles up Woodward Avenue from Detroit's northern city limits. Although the parish had only 25 families at its inception, the enterprising young priest built a church to hold 600 people. The building process was anything but trouble-free. Raising funds for the new parish proved difficult, and the Ku Klux Klan, fearful of an increasing Roman Catholic populace in the area, burned a cross on the church lawn. Fortunately, Coughlin was introduced to Leo Fitzpatrick, station manager of powerful WJR radio, who was taken by the young pastor's plight. Fitzpatrick suggested that Coughlin employ his oratorical skills over WJR in order to create a more sympathetic climate for the Shrine parish and to appeal for financial support.

Father Charles Coughlin, 4 November 1938
Courtesy AP/Wide World Photos

Persuasion by Radio

Originally entitled the *Golden Hour of the Little Flower,* Coughlin's first broadcast was relayed from the Shrine on 3 October 1926. Initially the program was intended for children but gradually shifted to adult topics on general economic and political perils facing the country. For Coughlin soon discovered that such subjects struck a responsive chord with listeners, resulting in correspondence that often contained financial contributions. He organized the Radio League of the Little Flower (annual membership fee: $1) to stimulate donations that allowed him to purchase more radio exposure. In 1929 he bought time on Cincinnati's powerful clear-channel station WLW and began negotiations to add WMAQ (Chicago) as his enterprise's third station. Because WMAQ was a CBS-owned outlet, the matter was referred to network headquarters in New York. As a result, Coughlin was sold time on the CBS network. National visibility was at hand.

The cost of airtime soon was dwarfed by the rising tide of contributions that his widely distributed program elicited. As the Great Depression set in, Coughlin's offensive against the twin evils of communism and international banking resonated with many and further increased the popularity of his broadcasts. But when his attacks became more specific and mentioned President Herbert Hoover by name, CBS became nervous. Edward Klauber, the network's executive vice president, requested that the priest submit scripts for advance clearance. Coughlin's response came in his 4 January 1931 broadcast when he asked his listeners whether or not CBS should be allowed to censor him. CBS was inundated with 1,250,000 letters of protest, and Coughlin's messages were never prescreened.

CBS still eased him off its network the following April and NBC was not interested in being Coughlin's replacement chain. So WJR's Fitzpatrick contacted Alfred McCosker, his counterpart at WOR, New York. Together they set up a telephone-linked group of 11 stations that expanded to 26 outlets from Maine to Colorado by the autumn of 1932. Weekly cost for the landlines and airtime was $14,000.

Coughlin's program now openly laid blame for the Depression on President Hoover. Over a million letters of support poured in; Royal Oak's first post office was established to cope with his correspondence, and Father Coughlin became the subject of feature articles in radio fan magazines and newspapers across the country. His religious superior, Bishop Michael Gallagher, was a firm supporter of Coughlin's social justice agenda, so grumblings from prominent East Coast cardinals were of no concern. Thousands of visitors, Catholic and non-Catholic alike, made pilgrimages to the Shrine church, now graced by an imposing 150-foot stone tower upon which was carved a crucifix illuminated by floodlights. Coughlin's radio speeches, composed in a small study at the top of the tower, increased in both vehemence and popularity. A poll conducted by WOR named him the nation's most useful citizen of 1933, and when WCAU asked its Philadelphia listeners whether they wanted Father Coughlin or the New York Philharmonic on Sunday afternoon, there were 187,000 votes for the cleric and only 12,000 for the orchestra.

Controversy

Coughlin cheered the election of Franklin D. Roosevelt and remained a staunch New Deal supporter throughout 1933, FDR's first year in office. The priest's resonant brogue and passionate oratory advocated the nationalization of gold and revaluation of the dollar, policies that were originally favored by the Roosevelt administration as well. But Roosevelt never warmed to the "Radio Priest" and never extended to him the counselor status that Coughlin thought he deserved. So on 11 November 1934, Coughlin announced the formation of a National Union for Social Justice to lobby independently for his social and economic proposals. The break with Roosevelt became complete when the administration proposed joining the World Court, an entity Coughlin considered a tool of international bankers. His 27 January 1935 broadcast was a blistering attack on the proposal, resulting in 200,000 protest letters to Congress, a key factor in the government's abandonment of the plan.

Coughlin's social justice movement now converged with the Share-Our-Wealth platform of Louisiana's bombastic Senator Huey Long. But in September 1935, Long was assassinated. The news reached President Roosevelt during a meeting with Coughlin and Joseph P. Kennedy (father of the future president), a conference Kennedy had arranged in an unsuccessful attempt to reconcile the priest and the president. The next year, Coughlin's National Union for Social Justice joined with Long's Share-Our-Wealth backers to create the Union Party and to endorse the presidential bid of North Dakota Congressman William Lemke. The priest also founded his own newspaper, *Social Justice Weekly,* which soon achieved a circulation of 1 million copies. Such success emboldened Coughlin to promise that he would leave radio if he could not deliver 9 million votes to Lemke. When Lemke garnered less than 1 million ballots, Coughlin honored his pledge and took leave of his broadcast audience on 7 November 1936.

Three months later, however, he was back on the air, rationalizing that this turnabout occurred because it was the dying wish of his supportive superior, Bishop Gallagher. For the next two years, Coughlin continued his broadcast attacks on Roosevelt's New Deal and its failure to adopt the monetary reforms that the priest advocated. Beginning in mid-1938, with European war clouds gathering, Coughlin began to focus more on international affairs. In his November 20th radio program, he excused German Nazism as a necessary defense mechanism against communism and supported the Nazi theory that Jewish bankers were behind the Russian Revolution. Over the next year his broadcasts took a more and more anti-Semitic tone.

In October 1939, fearing government retaliation for the strident broadcast oratory of Coughlin and other radical political voices, the National Association of Broadcasters' Code Committee placed strict limitations on the sale of radio time to "spokesmen of controversial public issues." As the priest's airtime was all purchased at commercial rates, this new self-policing edict gradually eroded his network as his contracts with stations expired. He canceled his 1940–41 season (the program had always taken a summer hiatus anyway) and never returned to the airwaves. On 1 May 1942, Coughlin's banishment from the public stage was complete when his new superior, Archbishop Francis Mooney, ordered him to cease all writings and nonreligious activities for the duration of the war. Acting upon a request relayed by Roosevelt emissary Leo Crowley and expressions of concern from the Vatican, Mooney threatened to revoke Father Coughlin's priestly authority if he did not comply.

Always the obedient priest, Coughlin immediately abandoned publication of *Social Justice,* allowed the government to revoke its second-class mailing privilege, and retreated to the role of Shrine pastor. He served in that capacity until his retirement in 1966.

At the height of his prominence, Father Coughlin had a listenership of more than 30 million, received 400,000 letters per week, and was featured twice on the cover of *Newsweek.* In stark contrast, from 1966 until his death a decade later the "Radio Priest" lived unobtrusively, first in a small apartment behind his beloved Shrine of the Little Flower and then in a home he purchased in nearby Birmingham.

PETER B. ORLIK

See also Controversial Issues; Politics and Radio; Religion on Radio

Charles Edward Coughlin. Born in Hamilton, Ontario, Canada, 25 October 1891. Only child of Thomas Coughlin and Amelia Mahoney. Honors degree in philosophy, University of Toronto's St. Michael's College, 1911; St. Basil's Seminary, 1911–16; ordained priest, 1916; taught in Canadian Basilian schools, then served in several Michigan parishes; accepted into Detroit diocese, 1923; pastor, Shrine of the Little Flower parish, Royal Oak, Michigan, 1926–66; Sunday afternoon radio broadcast, *Golden Hour of the Little Flower,* 1926–40; founded National Union for Social Justice, 1934; ordered off air, 1942; retired, 1966. Died in Birmingham, Michigan, 27 October 1979.

Radio Series

1926–40 *Golden Hour of the Little Flower* (carried on CBS, 1929–31)

Publication

Social Justice Weekly, 1936–42

Further Reading

Carpenter, Ronald H., *Father Charles E. Coughlin: Surrogate Spokesman for the Disaffected,* Westport, Connecticut: Greenwood Press, 1998

Lee, Alfred McClung, and Elizabeth Briant Lee, editors, *The Fine Art of Propaganda: A Study of Father Coughlin's Speeches,* New York: Harcourt Brace, 1939

Marcus, Sheldon, *Father Coughlin: The Tumultuous Life of the Priest of the Little Flower,* Boston: Little Brown, 1971

Schlesinger, Arthur Meier, *The Politics of Upheaval,* Boston: Houghton Mifflin, 1960

Spivak, John Louis, *Shrine of the Silver Dollar,* New York: Modern Age, 1940

Tull, Charles J., *Father Coughlin and the New Deal,* Syracuse, New York: Syracuse University Press, 1965

Warren, Donald I., *Radio Priest: Charles Coughlin, the Father of Hate Radio,* New York: Free Press, 1996

Country Music Format

At the beginning of the 20th century, "country music" was a version of folk music. With field recordings of the late 1920s, it became categorized as "hillbilly" music, and then after World War II, entrepreneurs renamed it "country and western," a designation used throughout the 1950s and into the 1960s. With the rise of Nashville as the recording center, however, the "western" was dropped, and by the time it became an important radio format, the name "country" was widely accepted. Whatever the name, until the 1960s country music tended to be songs of poor white folk that were passed down generation to generation as the South and then the West were settled.

Early Country Radio

In 1927 Ralph Peer of Radio Corporation of America (RCA) Victor record company began to record country music performers, most notably Jimmie Rogers and the Carter family, and a commercial industry was born. The western side of country music was popularized in cowboy films of the 1930s and 1940s by such stars as Roy Rogers and Gene Autry. On the radio, western stars had regular programs: there were country music performances on live barn dances such as the *National Barn Dance* from Chicago; *Town and Country Time* from Washington, D.C.; and the *Grand Ole Opry* from Nashville.

During World War II, soldiers from the South and West took their music all around the world. In the postwar era, Hank Williams made country songs popular, and he was followed by Jim Reeves and Patsy Cline. As Top 40 took over radio airplay, country music—in the 1960s—emerged as an alternative genre centered in Nashville, with stars such as Johnny Cash, Jimmy Dean, Loretta Lynn, and Dolly Parton.

On radio, country music had long been confined to network programs (such as the *National Barn Dance* from Chicago and the *Grand Ole Opry* from Nashville), small-town stations, and the border radio stations in Mexico. With the decline of network programming and the rise of radio formats, non–network-affiliated stations were playing a substantial amount of country music as early as the late 1940s. Top 40 led the way in terms of playing a selected playlist from one genre of music. Country—from Nashville—did the same as it evolved during the 1960s.

The country music business realized the threat of rock and roll, and in the late 1950s the business reorganized what had been the annual country disc jockeys' convention into the Country Music Association to promote more country music on radio. Although the identity of the first country-formatted station will forever be debated, stations converted from the programming techniques of the network era to those of the format era. Stations such as WARL-AM (Arlington, Virginia); KXLA-AM (Pasadena, California); and KDAV-TV (Lubbock, Texas) played live and recorded country music almost all day by 1950.

A generation later, more than 300 radio stations broadcast recorded country music on a full-time basis, and over 2,000 more programmed country for part of the day. The Country Music Association deserves much of the credit for promoting country as an alternative format, but it was certainly helped by the fracturing of rock music during the 1960s and the alienation of its older audience. During the 1950s and 1960s, country format radio moved from its small-town base in the South (as well as in cities such as Los Angeles, where thousands of southerners had moved during the Great Depression) to cities all across the United States.

Country as a Format

By the mid-1960s, advertisers no longer thought of country radio stations as only being listened to by country folk. During the 1950s, it looked as though country would not survive the popularity of rock and Top 40 formatting, but the introduction of the "Nashville Sound"—typified by the now-classic recordings of Patsy Cline—proved that crossover hit making was possible. By the mid-1970s, country had its place in radio, with more than 1,000 stations playing country format. Country had become suburban—it had given a voice to adult problems, such as infidelity, boss hating, and the like, whereas pop music seemed stuck in teenaged concerns. Country radio listeners were therefore older and were nearly always white.

By the 1990s, one survey determined that country stations were number one in 57 of the top 100 radio markets in the United States. Many surveys found country the most popular format on radio, with such megastars as Garth Brooks, Reba McEntire, Alan Jackson, and Shania Twain. The country music format had surely reached a high point as the most popular radio music in the country.

Even in the mid-1990s, however, some argued that radio was becoming too formulaic. Country radio aimed programming at adults aged 18 to 34 who listened on their way to and from work, were made up of more women than men, and lived in the suburbs rather than the cities. Artists who did not appeal to these listeners, including Dolly Parton and Willie Nelson, were simply ignored. Pressed by advertisers seeking younger buyers, country radio ironically abandoned listeners over 49, the very fans who had helped build it into the nation's most popular format.

During the 1990s, grown-up baby boomers embraced country, and so advertisers willingly anted up millions of dollars in advertising spending to reach them. Using Donnelley's Cluster

The Grand Ole Opry celebrates its first new stage set in 22 years, 10 June 2000
Courtesy AP/Wide World Photos

Plus system, in 1990 the Arbitron ratings service found that 40 percent of all country fans fit into the system's most affluent groupings, compared with fewer than one-quarter of all Americans aged 12 and older.

The boom in country radio (and television) is well reflected in the career of Garth Brooks, a star who did not sell his first compact disc until 1989 and who by the close of the century was the best-selling popular music artist in history. American music has never seen a phenomenon like Brooks, who in 1996 at age 34 had reached number two—in just seven years. In the process, he eclipsed Elvis Presley, Michael Jackson, and the Beatles. During the early 1990s he sold an average of 8 million "units" per year. Radio stations featured Brooks' latest releases and captured millions of new listeners.

New ways of determining hits helped as well. In 1992 Brooks became the year's top-selling artist based on *Ropin' in the Wind,* released late in 1991, because computers were used to determine what was sold in stores rather than relying on telephone surveys. Brooks became, because of SoundScan computer counting, the first country artist to top *Billboard's* charts. *Ropin' in the Wind* became the first country album ever to top *Billboard's* year-end pop album chart. Country radio programmers used SoundScan data (a music sales reporting service for subscribers that integrates weekly retail store reports on how many CDs have been sold, providing results for individual markets, regions, or the nation) to determine their playlists.

The 25 May 1991 *Billboard* chart was the first done on SoundScan, and suddenly 15 more country albums showed up in the Top 200 than had been there a week before. In 1984 the country category showed only 8 gold (500,000 sales), 4 platinum (1,000,000-plus sales), and 7 multiplatinum albums. By 1991 the numbers were 24 gold, 21 platinum, and 8 multiplatinum.

At the beginning of the 21st century, the future of the country music format looked bright. The number of young people listening to country music had increased almost 70 percent during the 1990s, and although its popularity was leveling off as the decade ended, no one predicted that country's core popularity would decline anytime soon. According to the *Simmons*

Study of Media and Markets, country music was the choice of one-fifth of the 18-to-24 population, with growth among those aged 25 to 34, 35 to 44, and 45 to 54, most of these being just the listeners most desired by radio advertisers.

DOUGLAS GOMERY

See also Border Radio; Contemporary Hit Radio Format/Top 40; Gospel Music Format; Grand Ole Opry; Music; National Barn Dance; Oldies Format

Further Reading

Carney, George O., editor, *The Sounds of People and Places: Readings in the Geography of Music,* Washington, D.C.: University Press of America, 1978; 3rd edition, as *The Sounds of People and Places: A Geography of American Folk and Popular Music,* Lanham, Maryland: Rowman and Littlefield, 1994

Keith, Michael C., *Radio Programming: Consultancy and Formatics,* Boston: Focal Press, 1987

Kingsbury, Paul, editor, *The Encyclopedia of Country Music,* New York: Oxford University Press, 1998

Lewis, George H., editor, *All That Glitters: Country Music in America,* Bowling Green, Ohio: Bowling Green State University Popular Press, 1993

MacFarland, David T., *Contemporary Radio Programming Strategies,* Hillsdale, New Jersey: Erlbaum; 2nd edition, as *Future Radio Programming Strategies: Cultivating Listenership in The Digital Age,* Mahwah, New Jersey: Erlbaum, 1997

Malone, Bill C., *Country Music U.S.A.,* Austin: University of Texas Press, 1968; revised edition, 1985

Routt, Edd, James B. McGrath, and Fredric A. Weiss, *The Radio Format Conundrum,* New York: Hastings House, 1978

"Symposium: Country Music Radio," *Journal of Radio Studies* 1 (1992)

Cowan, Louis 1909–1976

U.S. Radio Producer and Executive

Louis Cowan is associated mainly with the creation of various quiz shows, both for radio and later for television. However, Cowan himself was most interested in creating shows with an intellectual slant for popular audiences. Along those lines, Cowan was influential during World War II in helping to promote positive portrayals of African-Americans and military personnel on radio programs.

Louis Cowan was born Louis Cohen in Chicago in 1909. While a student at the University of Chicago, Cowan was influenced by the noted communications researcher Harold Lasswell and gained an interest in how communication could shape opinion. After graduating in 1931, he changed his name to Cowan and began his own public relations firm. One account he gained was publicity for the radio program *Kay Kyser's College of Musical Knowledge.* Cowan later formed his own independent production company, Louis G. Cowan Productions, and in 1940 he created his first radio hit, *The Quiz Kids,* which ran on the National Broadcasting Company (NBC) Blue network. Cowan's concept for this program was to feature a panel of bright children, none older than 16, fielding questions sent in by listeners. The program was an early example of Cowan's attempt to blend his intellectual interests into a program aimed to appeal to a mass audience.

During World War II, Cowan moved to New York and volunteered his services to the Office of War Information (OWI). Cowan became a consultant and director of domestic affairs, working under Edward Kirby, who had transferred to the public relations division of the OWI from the National Association of Broadcasters. One assignment Kirby gave to Cowan was to encourage radio producers to develop positive portrayals of African-Americans; it was hoped that this would help decrease agitation and tension at a time when all branches of the U.S. armed forces were segregated. Cowan decided that the greatest prejudice and resistance to African-Americans was often found among lower-middle-class Americans, and he knew that these people could be reached effectively through daytime soap operas. Cowan therefore convinced Frank and Anne Hummert, the creators of many serials, to incorporate some African-American characters in their soap opera story lines—a very unusual occurrence at that point in radio history. In one case, the Hummerts introduced into *Our Gal Sunday* a young African-American in military training, Franklin Brown, who appeared intermittently during furloughs. In an even more elaborate effort, the Hummerts had the heroine in *The Romance of Helen Trent* rescued by an African-American doctor after she fell into an abyss; afterwards, she found a job for this doctor as a staff physician in a wartime factory.

While working at the OWI, Cowan also sought to promote the general image of all military personnel. He convinced the Hummerts to create a new serial around a comforting and problem-solving army chaplain. This concept became *Chaplain Jim,* which started on the NBC Blue network in 1942. Cowan had one other interesting assignment relating to this serial. He was given the task of briefing the newly appointed psychological warfare officer, Columbia Broadcasting System (CBS) president William Paley, on the purpose of this program. This would not be the only time the two men's paths crossed.

After the war, Cowan went back to independent production of radio programs. Acting on his interest in intellectual and issue-oriented programming, Cowan in 1946 created *The Fighting Senator,* a show that featured a principled state senator battling corruption around him. Around the same time, Cowan's wife, Polly, was producing a talk show called *Conversation,* which brought together writers, college professors, and other celebrities to discuss various topics. Though winning accolades, neither program attracted large audiences. Cowan did have one more radio hit, again with the quiz show format. *Stop the Music!* premiered on the American Broadcasting Companies (ABC) network in 1948 and quickly attracted a large following. The premise of the show was simple: an orchestra played popular songs and was periodically interrupted by the program's host, Bert Parks, who shouted, "Stop The Music!" A phone call was then made randomly somewhere in the United States, and if the person who answered could identify the song that had just been interrupted, he or she would receive a number of expensive gifts. *Stop The Music!* was initially scheduled to air opposite Fred Allen's popular comedy show. Allen's program ratings fell so drastically as a result that his show was cancelled in 1950.

During the 1950s, Cowan turned his attention to television and created what became the first large-prize TV quiz show, *The $64,000 Question.* This program was purchased by CBS in 1955. A few weeks after it premiered, Cowan sold his production company and accepted a job at CBS as vice president in charge of creative services. Two years later, in March 1958, he was promoted to president of CBS-TV. When it was discovered at the end of the decade that many of the big TV quiz shows, including *The $64,000 Question,* had been rigged, Cowan was dismissed by CBS. This was done despite the fact that Cowan had no association with *The $64,000 Question* after he sold his production company, and there was no evidence that Cowan ever knew about the rigging.

In his latter years, Cowan taught at Brandeis University and in Columbia University's Graduate School of Journalism. He also started his own publishing company and was cofounder of the American Jewish Committee's William E. Weiner Oral History Library. In 1976 Cowan and his wife died in an accidental fire in their apartment. Ironically, it was suspected that the fire was caused by faulty wiring in their television set.

RANDALL VOGT

See also Kyser, Kay; Office of War Information; Quiz and Audience Participation Programs

Louis George Cowan. Born Louis George Cohen in Chicago, Illinois, 16 December 1909. Graduated from University of Chicago, 1931; formed own public relations firm; entered independent production and created radio and television shows such as *The Quiz Kids, Stop the Music!* and *The $64,000 Question;* vice president of creative services, CBS, 1955; served as president of CBS television from March 1958 to December 1959. Died in New York City, 18 November 1976.

Further Reading

Cowan, Paul, *An Orphan in History: Retrieving a Jewish Legacy,* Garden City, New York: Doubleday, 1982

Critics

Like critics of theater, music, and art, radio critics have reviewed as well as helped audiences to interpret a host of radio programs. Radio, however, presented unique challenges to the critic. Since radio was a new and strictly aural and fleeting medium, professional critics struggled to find an appropriate way to critique the material presented. When radio first emerged as a mass communication tool, it and its content were dismissed as popular—even vulgar—"art." Intelligent writing about it was not taken as seriously as writing about theater, books, or film, thus making it difficult for radio critics to gain credibility. There also did not seem to be a purpose to reviewing programs that were played only once. Such difficulties, among others, left a scarcity of radio criticism in the early years of the medium, and what criticism did exist failed to assume much importance. With the advent of television, radio became further buried in the press. However, the emergence of television criticism furthered the cause of radio criticism. Many critics were, as they are today, labeled by their publishers as radio/television critics.

Challenge of Criticism

Radio columnists did not face the same hurdles as critics because columnists served to inform readers about coming attractions and gossip, with perhaps some superficial appraisal. The critic—who, like the columnist, also amused and informed—did so in a broader, deeper context of constant evaluation. The critic judged the significance of a broadcasting event, considered its impact, or related it to past events in broadcasting or other areas. Critics added artistic, philosophical, and sociological dimensions to program reviews and commented on the industry and government or public actions.

Unlike theater, film, and music, which offered discrete presentation formats, radio was on the air continuously, and the quantity and variety of programs were a burden to critics, who often were called upon to treat programs ranging from education and politics to commerce and entertainment. A wide variety of assignments were therefore available, and those critics who had come from the world of newspaper reporting (as many of them did) often had an edge over other critics, as they were equipped with the reportorial skills of speed and judgment.

The difficulties of critiquing something heard and not seen played out most noticeably in drama criticism. Radio drama was given an ungenerous report overall by some because it lacked a visual element and therefore could not, many contended, hold the audience's attention. Many early critics insisted that all elements of a play be identifiable; a production was praised if it was clear who was speaking or if a synopsis of the story was given before the start of the action. Critics often treated radio dramas as adaptations instead of creating new ways to critique.

It was a common habit of newspaper critics to mention and discuss actors and their performances but not the programs themselves, as most program titles would require plugging the product or sponsor and therefore result in free advertising, which would compromise the critic's integrity. In addition, the lack of credence given to radio criticism meant that radio sponsors became protected and pampered, and they rarely heard criticism. In general, radio criticism was thought of as outside the industry's needs because station leaders and sponsors made a point of claiming to be businessmen, not showmen; therefore, in the name of advertising, they were immune from theatrical standards.

Rise of Criticism

Many critics, most notably multimedia critic Gilbert Seldes, believed that the duty of broadcast critics was to propose change, but change that was workable within the advertising-supported commercial radio system. In this line of thinking, critics must understand and explain (perhaps today more than ever) the environmental constraints within which any electronic media organization must operate.

Several critics in the early days of radio issued calls for a responsible corps of radio critics, and many intellectuals acknowledged the importance of broadcast criticism in the name of preserving democracy.

Much early criticism was of a technical nature, wherein critics discussed such issues as transmission quality. Perce Collison, writing in the early 1920s, for example, critiqued sound quality as opposed to content. When critics wrote about broadcast news, which was a growing aspect of programming by the 1930s, they tended to comment on such aspects as sound effects and newscasters' voices.

Ralph Lewis Smith (1959), in his analysis of the U.S. broadcast system, lamented critics' coverage and opinions of the electronic functioning of a program, as if they were "scientific journalist[s]." This type of criticism only furthered the notion that broadcasting was not to be considered an art. The nuts and bolts discussion of radio subsided when people could buy ready-built radios as opposed to hobbyist-assembled kits. As radio's commercial potential became more obvious, "circuit talk" was replaced with gossip about radio stars. As an interest in radio personalities increased interest in radio overall, critical articles and commentary on individual shows and series as well as various personalities began to emerge.

Because there were no standards or precedents, broadcast critics experimented with column formats. In 1926 John Wallace presented his criticism in *Radio Broadcast* in the form of one long essay, two or three short reviews, and a few bright tidbits. In October 1942 in *Woman's Day*, Raymond Knight set up his material in newspaper form and called it "The Radioville Chronicle," which contained program reviews, a local gossip column, a classified section, and a notice of new shows.

Major Critics

During the 1920s and 1930s, most writing about programs was descriptive rather than critical, although there were exceptions. Volney Hurd and Leslie Allen, for instance, were among the earliest who produced insightful criticism in their 1930s columns in the *Christian Science Monitor*.

Radio producer Darwin Teilhet began writing a monthly broadcasting feature in *Forum* magazine from 1932–34 while he was in charge of radio production for the N.W. Ayer and Son advertising agency. When Teilhet began writing critical articles, his employer questioned their propriety; as a result Teilhet continued writing for several months under the name Cyrus Fisher—a prime example of the obstacles critics encountered within the radio industry.

A thoughtful critical approach to broadcasting was rare before World War II. Because radio proved to be an essential communication tool during the war, more attention was given to broadcast criticism after the war, not from scholars or journalists, but from the federal government. In 1946 the Federal Communications Commission (FCC) released its "Blue Book," officially known as "The Report on Public Service Responsibility of Broadcast Licensees." It examined overcommercialization and the lack of local and public affairs programming. One of its principal authors was Charles Siepmann, a former British Broadcasting Corporation (BBC) employee. The revelations in the Blue Book raised a storm of controversy about U.S. broadcasting that generated interest in more serious, professional critiquing of the medium.

Two months after the "Blue Book" was issued, critic John Crosby wrote his first daily radio column for the *New Herald Tribune*. Born in Milwaukee, Wisconsin in 1912, Crosby went to Yale for two years and then began a newspaper career, first on the *Milwaukee Sentinel* covering courts and police headquarters and then on the *New York Herald Tribune*. After serving in World War II for five years, Crosby returned to the *Tribune*. Reporters back from the war were so plentiful that the editors, unsure how to use Crosby, stuck him with writing a radio column. Having never owned a radio and barely having listened to one, he took the job expecting to wait for something better to materialize. Crosby eventually settled into his role and won fans with his wry, cynical wit. His column "Radio and Television" was syndicated from coast to coast to an audience of more than 18 million. Crosby's columns were so influential that some thought they helped to raise radio into the realm of legitimacy with music and theater.

Despite the fact that radio criticism had to share page space with the gossip column, it was the data and conclusions in the "Blue Book," Crosby's column, and the burgeoning of television simultaneously that paved the way for more and better broadcast criticism in the 1950s and beyond. From the 1950s to the present, critics in both the media and academia have contributed, albeit in small volume, to radio criticism.

Producer-turned-critic Albert N. Williams was associated with the National Broadcasting Company (NBC) from 1937–41 and wrote occasional articles for the *Saturday Review of Literature,* which became a monthly series in 1946–47. A published collection of his columns entitled *Listening: A Collection of Critical Articles on Radio,* one of the first such books on radio, appeared in 1948 with columns divided into categories of networks, programs, artisans, and advertising.

Lyman Bryson, a Columbia Broadcast System (CBS) consultant on public affairs, held a Sunday afternoon series on CBS radio, *Time for Reason—About Radio,* in 1946. One example of the type of program in the series was "Documentary and Actuality Programs." Guests, including Charles Siepmann and John Crosby, were part of the program to provide a broader point of view. It was the first time a major network had used its own facilities to tell listeners, in an extended series of talks and discussions, about the problems and possibilities of radio in the United Stats from the broadcasters' point of view. The idea was originally proposed by William S. Paley, CBS chairman of the board, in an address to the National Association of Broadcasters. Paley had asked for more intelligent criticism of the industry and for more activity by the industry in helping to provide background information for it. The idea was well received by critics, and the public and the series lasted until June 1947. A selection of the program material, "written from the broadcaster's point of view," was published a year later.

Previously a long-time drama critic, John Hutchens joined *The New York Times* in 1941 as a radio editor and columnist. When he left in 1944 for book reviewing, he was replaced by Jack Gould, who had been a *New York Herald Tribune* reporter in the mid-1930s and in the drama department of the *Times* from 1937–42. He was part of the *Times* radio department until 1944.

Robert Lewis Shayon was a producer-director for the Mutual Broadcasting System from 1938–42 and executive producer for CBS from 1942–49. Probably best known for the *You Are There* historical programs, he wrote for the *Saturday Review of Literature* and in 1950 became the *Christian Science Monitor*'s first television/radio critic.

Saul Carson was a critic for *The New Republic* from 1947–52. He started as an assistant radio editor under George Rosen at *Variety* in the early 1940s. He was a regular contributor to *Radio and TV Best* magazine, *The Nation,* and others.

Though not solely a radio critic, Ring Lardner, a comic *New Yorker* columnist, was a well-known sportswriter during World War I and also achieved literary success as a short story writer. When he was hospitalized with an illness in 1932, he spent a lot of time listening to the radio just as it was becoming a medium of mass entertainment, and he shared his observations in the *New Yorker.*

Other important critics and columnists from the 1920s through the 1950s included John Wallace, a prominent 1920s critic who wrote for *Radio Broadcast;* Ben Gross, broadcasting critic and columnist for the *New York Daily News;* George Rosen, radio and TV editor for *Variety;* Alton Cook and Harriet van Horne for the *New York World-Telegram;* Paul Cotton and Mary Little for the *Des Moines Register;* Stanley Anderson of the *Cleveland Plain Dealer;* Leonard Carlton of the *New*

York Post; and Edith Isaacs of *Theater Arts Monthly.* B.H. Haggin was a music critic who wrote for *The Nation, Hudson Review, The Dial,* and *Vanity Fair.*

Many cultural critics paid particular attention to radio, among them Bernard DeVoto, Frederick Lewis Allen, Harry Skornia, Wilbur Schramm, Paul Lazarsfeld, and Llewellyn White. These academic critics tended to critique the medium itself (as opposed to specific programs) and paved the way for radio studies by experts in communications, psychology, sociology, literature, and linguistics.

The best known of the cultural critics was Gilbert Seldes (1893–1969). Seldes did much for radio and television, increasing the interest in and serious attention paid to the popular arts. Seldes, unlike Shayon, Williams, and Teilhet (who were producers-turned-critics), started out as a critic then subsequently worked within the industry. He was perhaps the first American critic who devoted most of his career to examining popular as opposed to fine arts. Seldes was managing editor of one of the most famous "little magazines" of the 1920s, *The Dial,* which was modernist in its literary outlook but also enthusiastic about popular entertainment. His career also included stints as a theatrical producer, a radio writer and producer, the first director of programming for CBS television, and founding dean of the Annenberg School of Communications.

Seldes' most famous work, *The Seven Lively Arts* (1924, which included essays on theater as well as film and comic strips) helped to establish his reputation as an important critic. Seldes became a regular film critic for *The New Republic* and a columnist for the *New York Evening Journal* and *The Saturday Evening Post;* he also wrote a monthly broadcast column for the *Saturday Review of Literature,* authored books of criticism on American history and current events, and contributed articles to nearly every high- and middle-brow magazine of the time. Seldes' attention as a media critic eventually moved from film and radio to television, a shift exemplified in *The Public Arts,* his last major work, published in 1956.

Format and Content of Criticism

Radio program criticism, as differentiated from columns and general program information, was often found in such trade magazines as *Variety, Billboard,* and *Radio Daily.* Because the circulation of these publications was usually limited to show business professionals, the general listening audience did not benefit from such writing, to the chagrin of many critics. Leading critics also wrote for *Life, Collier's, Atlantic, Harpers,* and *The Quarterly Review of Film, Radio and Television.* Today, however, readers are likely to be more familiar with their local newspaper critic.

In some ways, critical radio coverage finds itself at the beginning of the 21st century as it was in the 1920s and 1930s, with program listings and perhaps a review here and there. Few, if any, professional critics now label themselves solely radio critics. Most cover television as well, along with a host of other electronic media. Television and radio columnist Robert Feder of the *Chicago Sun Times* and David Hinckley, critic-at-large for the (New York) *Daily News,* are two of the very few who semi-regularly still include radio criticism in their columns. Cultural criticism of radio in the late 20th century tends toward lamentations over the phenomenon of talk radio, including figures such as Howard Stern, Rush Limbaugh, and Dr. Laura Schlessinger.

KATHLEEN COLLINS

See also Blue Book; Columnists; Peabody Awards; Siepmann, Charles; Trade Press

Further Reading

Bryson, Lyman, *Time for Reason about Radio: From a Series of Broadcasts on CBS,* edited by William Cooper Ackerman, New York: Stewart, 1948

Crosby, John, *Out of the Blue: A Book about Radio and Television,* New York: Simon and Schuster, 1952

Kammen, Michael G., *The Lively Arts: Gilbert Seldes and the Transformation of Cultural Criticism in the United States,* New York: Oxford University Press, 1996

Landry, Robert J., "Wanted: Radio Critics," *Public Opinion Quarterly* 4 (December 1940)

Landry, Robert J., *Magazines and Radio Criticism,* Washington, D.C.: National Association of Broadcasters, 1942

Landry, Robert J., "Critics and Criticism in Radio," in *Selected Radio and Television Criticism,* edited by Anthony Slide, Metuchen, New Jersey: Scarecrow Press, 1987

Orlik, Peter B., *Critiquing Radio and Television Content,* Boston: Allyn and Bacon, 1988

Seldes, Gilbert, *The Great Audience,* New York: Viking Press, 1950

Seldes, Gilbert, *The Public Arts,* New York: Simon and Schuster, 1956

Shayon, Robert Lewis, *Open to Criticism,* Boston: Beacon Press, 1971

Siepmann, Charles Arthur, "Further Thoughts on Radio Criticism," *Public Opinion Quarterly* 5 (June 1941)

Siepmann, Charles Arthur, *Radio's Second Chance,* Boston: Little Brown, 1946

Siepmann, Charles Arthur, *Radio, Television, and Society,* New York: Oxford University Press, 1950

Smith, Ralph Lewis, "A Study of the Professional Criticism of Broadcasting in the United States, 1920–1955," Ph.D. diss., University of Wisconsin, 1959

Williams, Albert Nathaniel, *Listening: A Collection of Critical Articles on Radio,* Denver, Colorado: University of Denver Press, 1948

Crosby, Bing 1903–1977

U.S. Radio Personality and Entertainer

Often referred to as "America's Crooner," Bing Crosby was the defining male singer of his time and one of the most popular and successful stars of the 20th century. He dominated the recording, film, and radio industries for 30 years. Crosby's soft, conversational singing style, which critic Will Friedwald has described as a "warm B-flat baritone with a little hair on it" (1990), appealed to a broad audience of Americans and helped establish American popular song as both a legitimate art form and an extremely profitable mass media industry. Crosby's relaxed, modest star persona, folksy charm, and quick wit also contributed greatly to his success. The informality of his radio variety show and his easy banter with guests made listeners feel especially comfortable with him and ensured Crosby a prominent place in the hearts of millions.

Harry Lillis Crosby was born in Tacoma, Washington. His birth date is a matter of dispute, but the most recent research puts it at 3 May 1903. His father worked as a bookkeeper, while his Irish-Catholic mother raised the brood of seven Crosby children. Strong-willed, practical, and religiously devout, his mother was the strongest influence on Crosby's life, and he was her favorite child. Both parents were amateur musicians and encouraged a sense of popular music appreciation in their children. A bright but unmotivated student, young Bing was best known for his charm and his habit of whistling or humming while he walked. He earned his nickname because of his attachment to a humor feature called the "Bingville Bugle" in Spokane's Sunday paper.

In 1920 Crosby entered Gonzaga College and joined an amateur band as a drummer and singer (using a megaphone). His life changed course in 1923 when he met Al Rinker, another amateur bandleader and the brother of blues singer Mildred Bailey. Al persuaded Bing to join his band; after the band dissolved in 1925, the two men left Spokane for Los Angeles to try to make it in the big time. Mildred provided them with connections, and their act, "Two Boys and a Piano," proved so successful that they were hired by bandleader Paul Whiteman in late 1926. Whiteman paired the boys with Harry Barris, an up-and-coming musician and songwriter, and the trio became famous as the Rhythm Boys, recording and singing Barris songs such as "Mississippi Mud." They sang in a modern, jazz-influenced, intimate style that was new to most audiences. In 1930 the trio left Whiteman and signed to play with Gus Arnheim's Orchestra at the Coconut Grove in Los Angeles. Crosby frequently sang solos with the band, but he eventually left the trio and became well known as a solo performer to radio and nightclub audiences in the Los Angeles area. In September 1930 he married Fox starlet Dixie Lee, who was a much bigger star at the time.

Crosby emerged as a star himself in 1931 with the release of the short film *I Surrender Dear* in the summer (made in part to promote his new hit single) and the debut of his nightly 15-minute radio program for the Columbia Broadcasting System (CBS) in the fall. CBS built him up as a crooner to rival the National Broadcasting Company's (NBC) original crooner, Rudy Vallee, and by December, Crosby, Russ Columbo (also at NBC), and Vallee were the most popular singers in the nation.

Crosby's public image, however, was more controversial. He had missed the debut of his New York radio show because of his drinking, and he was widely regarded in the industry as unstable. In response to criticism, Crosby worked hard to change his playboy image; fan magazines helped by promoting Crosby as a devoted husband and father (son Gary was born in 1933) whose wild days were behind him. This image better suited the more socially conservative Depression years and helped ensure Crosby a broad audience. In addition, Crosby enlarged his repertoire of songs beyond romantic crooning ballads. His music producer at Decca Records, Jack Kapp, believed that Crosby could become a type of "musical everyman" by singing a variety of songs, including cowboy songs, Hawaiian songs, hymns, and holiday songs. His film career for Paramount Studios, which signed him in 1932, followed a similar formula, presenting Crosby in a variety of roles that underlined his cool, relaxed persona and his comic as well as his singing talents.

In 1936 his radio career hit a new high when he took over as host of the popular NBC variety show *The Kraft Music Hall* and remained there for ten years. This hour-long program starred Crosby as the host and primary vocalist and featured a number of comic players and star musicians. Crosby's comic sidekick, Bob Burns, known as the "Arkansas Traveler," amused audiences with his rube humor and remained with the program until 1941. His bandleader, John Scott Trotter, replaced Jimmy Dorsey in 1937 and remained with Crosby until the end of his radio career. As John Dunning (1998) has noted, Burns' humor helped balance the more serious musicians who appeared on the program, including a number of accomplished classical and jazz artists such as Jose Iturbi, Duke Ellington, and Jack Teagarden. Broadway star Mary Martin spent a year on the show in 1942, as did Victor Borge, a Danish concert pianist who served as an additional comic foil. The show also created new stars, such as Spike Jones and Jerry Colonna, both of whom were originally members of Trotter's band.

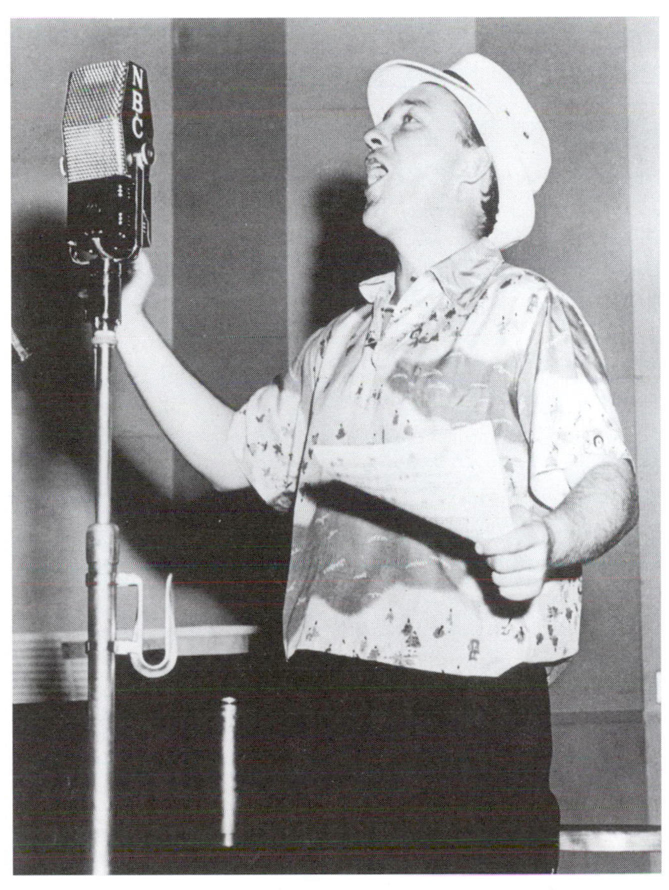

Bing Crosby
Courtesy Radio Hall of Fame

Crosby's personality held the show together, and he shaped it to suit his easygoing personality. Unlike other variety shows of the time, for example, Crosby wanted no applause after numbers, because he thought it slowed the pace of his program. Guest stars were treated with an informality that belied their importance and made them seem like members of the gang, and Crosby's own self-deprecating humor underlined the "everyman" quality so essential to his image. His witty delivery seemed so natural and effortless that fans did not believe the program was scripted. At the same time, however, Crosby was clearly in charge of his program and would walk out of rehearsals if he became annoyed. In 1945 he left both NBC and his sponsor when they refused to consider the possibility of taping his programs rather than performing them live. The technology for taping had been developed during the war, and Crosby wanted the greater scheduling flexibility it could offer him. NBC thought audiences would not tune in to canned entertainment, so Crosby walked out and moved his variety program to the American Broadcasting Companies (ABC). His *Philco Radio Time* show ran for three years very successfully, and Crosby's program is credited as initiating the tape era in radio.

Crosby's radio career continued through the mid-1950s with various variety shows, although none was as successful as those of the 1930s and 1940s. He appeared on various television specials and occasionally in films until his death on a Spanish golf course in 1977. His reputation has taken a beating since then: biographers have portrayed Crosby as an emotionally distant man, cold to his first wife and children, inflexible, and miserly. For better or worse, however, Crosby remains what *Life* magazine called him in 1945: "an American institution."

ALLISON MCCRACKEN

See also Audiotape; Film and Radio; Hollywood and Radio; Hope, Bob; Syndication; Variety Shows

Bing Crosby. Born Harry Lillis Crosby in Tacoma, Washington, 3 May 1903. Attended Gonzaga University School of Law, 1920–24; toured with Al Rinker as "Two Boys and a Piano," 1925–26; member of the Rhythm Boys trio in Paul Whiteman's band, 1926–30; radio debut with Paul Whiteman, WNBC, New York, *Old Gold Cigarettes*, 1928; performed at Cocoanut Grove with Gus Arnheim's Orchestra, 1930; solo radio debut on CBS, New York, 1931; host of variety show *The Kraft Music Hall*, 1936–46, and *Philco Radio Time*, 1946–49; first film appearance (with the Rhythm Boys), *The King of Jazz*, 1930; first starring role in feature film, *The Big Broadcast*, 1932; notable film roles in *Going Hollywood*, 1933, *Waikiki Wedding*, 1937, *Road to Singapore*, 1940, *Road to Morocco*, 1942, *Holiday Inn*, 1942, *White Christmas*, 1954, and *The Country Girl*, 1954. Received Academy Award for Best Actor, 1944, for *Going My Way*. Died outside Madrid, Spain, 14 October 1977.

Radio Series

1930	*Bing Crosby with the Gus Arnheim Orchestra*
1931–32	*Fifteen Minutes with Bing Crosby* (later *The Cremo Singer* for Cremo Cigars)
1933	*The Music That Satisfies*
1933–35	*Bing Crosby for Woodbury Soap*
1936–46	*The Kraft Music Hall Starring Bing Crosby*
1946–49	*Philco Radio Time Starring Bing Crosby*
1949–52	*The Bing Crosby Chesterfield Show*
1952–54	*The General Electric Show*
1954–56	*Bing Crosby*

Films

The King of Jazz, 1930; *Ripstitch the Tailor* (short), 1930; *Two Plus Fours* (short), 1930; *Check and Double Check* (cameo), 1930; *Reaching for the Moon* (unbilled cameo), 1931; *I Surrender Dear* (short), 1931; *Confessions of a Coed* (unbilled cameo), 1931; *One More Chance* (short), 1931; *Dream House* (short), 1931; *Billboard Girl* (short), 1931; *Blue of the Night* (short), 1931; *The Big Broadcast,* 1932; *Sing, Bing, Sing* (short), 1933; *College Humor,* 1933; *Star Night at the Cocoanut Grove* (short), 1933; *Too Much Harmony,* 1933; *Going Hollywood,* 1933; *Please* (short), 1933; *Bring on Bing* (short), 1933; *Hollywood on Parade* (short), 1933; *We're Not Dressing,* 1934; *Just an Echo* (short), 1934; *She Loves Me Not,* 1934; *Here Is My Heart,* 1934; *Mississippi,* 1935; *Two for Tonight,* 1935; *The Big Broadcast of 1936* (cameo), 1935; *Anything Goes,* 1936; *Rhythm on the Range,* 1936; *Pennies from Heaven,* 1936; *Waikiki Wedding,* 1937; *Double or Nothing,* 1937; *Swing with Bing* (short), 1937; *Doctor Rhythm,* 1938; *Don't Hook Now* (short), 1938; *Sing You Sinners,* 1938; *Paris Honeymoon,* 1939; *East Side of Heaven,* 1939; *The Star Maker,* 1939; *Road to Singapore,* 1940; *If I Had My Way,* 1940; *Rhythm on the River,* 1940; *Road to Zanzibar,* 1941; *Birth of the Blues,* 1941; *Holiday Inn,* 1942; *Angels of Mercy* (short), 1942; *Road to Morocco,* 1942; *Star Spangled Rhythm,* 1942; *My Favorite Blonde* (cameo), 1942; *Dixie,* 1943; *Going My Way,* 1944; *The Road to Victory* (short), 1944; *The Princess and the Pirate,* 1944; *Here Come the Waves,* 1944; *Out of This World,* 1945; *All-Star Bond Rally* (short), 1945; *Hollywood Victory Caravan* (short), 1945; *Duffy's Tavern,* 1945; *The Bells of St. Mary's,* 1945; *Road to Utopia,* 1946; *Road to Hollywood,* 1946; *Blue Skies,* 1946; *Variety Girl,* 1947; *My Favorite Brunette* (cameo), 1947; *Welcome Stranger,* 1947; *Road to Rio,* 1947; *The Emperor Waltz,* 1948; *A Connecticut Yankee in King Arthur's Court,* 1949; *Top o' the Morning,* 1949; *The Adventures of Ichabod and Mr. Toad,* 1949; *Riding High,* 1950; *Mr. Music,* 1950; *Angels in the Outfield* (cameo), 1951; *Here Comes the Groom,* 1951; *The Greatest Show on Earth* (cameo), 1952; *Son of Paleface,* 1952; *Just for You,* 1952; *Road to Bali,* 1952; *Scared Stiff* (cameo), 1953; *Little Boy Lost,* 1953; *White Christmas,* 1954; *The Country Girl,* 1954; *Anything Goes,* 1956; *Bing Presents Oreste* (short), 1956; *High Society,* 1956; *Man on Fire,* 1957; *Say One for Me,* 1959; *Alias Jesse James,* 1959; *Let's Make Love* (cameo), 1960; *Pepe* (cameo), 1960; *High Time,* 1960; *The Road to Hong Kong,* 1962; *Robin and the Seven Hoods,* 1964; *Stagecoach,* 1966; *Cinerama's Russian Adventure* (narrator), 1966; *Bing Crosby's Washington State* (narrator), 1968; *Cancel My Reservation* (cameo), 1972

Stage

Various vaudeville appearances as "Two Boys and a Piano," 1925–26

Various vaudeville and nightclub appearances in the Rhythm Boys, part of Paul Whiteman's Orchestra, 1927–30

The Cocoanut Grove with Gus Arnheim's Orchestra, 1930

Publication

Call Me Lucky (with Pete Martin), 1953

Further Reading

Bauer, Barbara, *Bing Crosby,* New York: Pyramid, 1977

Brackett, David, *Interpreting Popular Music,* New York: Cambridge University Press, 1995

Crosby, Edward J., *The Story of Bing Crosby,* New York and Cleveland, Ohio: World, 1946

Dunning, John, *On The Air: The Encyclopedia of Old-Time Radio,* New York: Oxford University Press, 1998

Friedwald, Will, *Jazz Singing: America's Great Voices from Bessie Smith to Be-Bop and Beyond,* New York: Scribner, 1990

Giddins, Gary, *Riding on a Blue Note: Jazz and American Pop,* New York: Oxford University Press, 1981

Giddins, Gary, *Bing Crosby, a Pocketful of Dreams: The Early Years, 1903–40,* Boston: Little Brown, 2001

Morgereth, Timothy A., *Bing Crosby: A Discography, Radio Program List, and Filmography,* Jefferson, North Carolina: McFarland, 1987

Pitts, Michael, and Frank Hoffmann, *The Rise of the Crooners,* Lanham, Maryland, Scarecrow Press, 2002

Shepherd, Donald, and Robert F. Slatzer, *Bing Crosby: The Hollow Man,* New York: St. Martin's Press, 1981

Crosley, Powel 1886–1961

U.S. Radio Inventor, Manufacturer, and Broadcaster

Powel Crosley, Jr., the founder of a radio company that became the world's largest manufacturer of receivers within two years, was known as "the Henry Ford of Radio" for his pioneering work in the development and manufacture of inexpensive radio sets that made radio affordable to millions. Crosley's sets, some of which sold for under $10, boosted radio's popularity, but because they were equipped with less sophisticated circuits, good reception necessitated stations of greater power. Crosley, the owner of WLW, the most powerful AM radio station ever to operate in the United States, became a promoter of increased power for all stations.

Origins

Crosley was born in Cincinnati, Ohio, on 18 September 1886, and as a child he was fascinated by automobiles. While not yet in his teens, to win a bet offered by his father he built his own small vehicle powered by a washing machine motor and got it to run at least around the block. But he would be frustrated all of his life by his inability to build a commercially successful automobile.

After two years in law school in Cincinnati, he quit to become a chauffeur and work around cars. He later worked for a number of automobile firms in Ohio and Indiana, which, in the early years of the automobile industry, were competing with Michigan in motorcar production. Crosley found success in selling many auto accessories and gadgets. For example, Crosley made a radiator cap to hold an American flag that was a big seller during World War I and a device to keep the front tires of a Model T Ford headed straight before the development of rack-and-pinion steering. Crosley also owned a small manufacturing plant making a number of wood products, including cabinets for phonographs. When he noticed that one of his employees had made a small wooden three-wheeled riding bike for his son from scrap wood, Crosley began making and selling the "Taylor Tot," named for that employee. He pioneered maximizing advertising value by means of market research by carefully tracking response to magazine advertisements and then reducing the size of the display to get the biggest response from the smallest expenditure.

Powel Crosley
Courtesy AP/Wide World Photos

Receiver Manufacturer

In February 1921 Crosley's nine-year-old son asked for a "radio toy" as a birthday present. By then radio was the rage among boys who liked to build gadgets. Juvenile books and magazines often featured radio stories and radio heroes, along with instructions for making a simple set at home. The story goes that Crosley's son had listened to a radio at a friend's house and wanted one of his own. Crosley went to a store, discovered that the least expensive retail set cost about $130, and instead bought a 25-cent instruction booklet and the parts.

Having learned that a homemade set could be assembled for about $25, Crosley hired two engineering students from the University of Cincinnati (one of whom, Dorman Israel, was later chairman of Emerson Radio and Phonograph). Crosley and the students designed a set they could manufacture on an assembly line in his phonograph cabinet plant and called it the "Harko." The Harko was introduced at about $20, but later the price was reduced to $9, plus batteries. Crosley liked to call this radio's successor, Harko, Jr., the "Model T of radio."

The first large sales campaign for the inexpensive sets was during the Christmas shopping season of December 1921: "It will tune from 200 to 600 meters, bringing in spark, voice and music, with an average amateur aerial." By July 1922, just a year after introducing the Harko, Crosley was producing 500 receivers a day and had become the world's largest manufacturer of radio sets and parts. During the 1930s the Crosley sets were extremely popular, especially the smaller kitchen models—many of which looked like the front grille of a car—made of a pre-plastic material called Bakelite in an art deco style. He purchased other, smaller radio companies, and the Crosley Radio Corporation (the word *Radio* was later dropped as other appliances were added) was a major business for 30 years.

Crosley remained interested in ideas for new products and improvements on existing appliances; his was the first company to make refrigerators with shelves in the doors—the Shelvador—controlling a patent that made millions. He purchased the Cincinnati Reds baseball team in 1934 and installed them in the renamed Crosley Field, and founded a professional football team in Cincinnati in the 1930s.

Broadcaster and Manufacturer

As a radio hobbyist, and to provide programming for purchasers of his sets (and advertising to gain further customers), he started a radio station in his home. This station evolved into WLW, "the Nation's Station." In 1934 the Cincinatti Reds were the first major league team to play a night game under lights, arranged by Crosley from his newly-renamed Crosley Field, so the play-by-play reports could be carried on his station during more popular listening hours. WLW was also a founding station of the Mutual Broadcasting System (MBS), established in 1934. Crosley was also a co-founder of the National Association of Broadcasters (NAB), founded in 1923.

The Crosley Corporation was an early television manufacturer and owned TV stations in Cincinnati, Dayton, Columbus, Indianapolis, and Atlanta. In 1945 the company was purchased by the Aviation Corporation (later Avco), and in 1954 was still the fifth-largest manufacturer of radio and TV sets. The Crosley line of household appliances, including broadcast receivers, was discontinued in 1956. (While the name is still used by a manufacturing company of appliances, there is no connection to the original firm or family.)

Small Cars

Crosley had long been fascinated by automobiles, and yearned to produce a popular economy model. In 1939 he introduced a small car intended to sell for about $300 and able to run 50 miles per gallon of gas. Opposition from the big car makers in Detroit, however, kept him from signing up dealers, so he tried to sell the car through department stores, such as Macy's in New York. World War II delayed manufacturing until 1946. In 1947 Crosley sold about 17,000 sedans, station wagons, delivery vans, and roadsters. The company's auto sales peaked the following year, at nearly 47,000. After years of "doing without" during the Depression and the war, Americans wanted bigger, roomier cars, and the appeal of smaller models declined. Because many of the parts for his small cars came from other manufacturers, such as the Willys Jeep, and were produced for bigger vehicles, it was hard to make an inexpensive car with the power Americans wanted for the open road. The Crosley auto plant shut down in July 1952.

When Crosley died in 1961, he left an impressive legacy that included pioneering developments in popularizing radio and television, leadership in the fledgling broadcasting industry, efforts toward production of economy cars, and, not the least of his contributions, nighttime baseball.

LAWRENCE W. LICHTY

See also Mutual Broadcasting System; Receivers; WLW

Powel Crosley, Jr. Born in Cincinnati, Ohio, 18 September 1886. Son of an attorney. Studied both engineering and law at the University of Cincinnati; developed a short-lived automobile company, then another in 1912; manufactured motorcycles, 1914–1917; owned American Automobile Accessories auto parts mail order business by 1920; created in 1921 what became (in 1924) Crosley Radio Corp.; began station WLW in 1921; purchased Cincinnati Reds baseball team and Redlands Field (renaming it Crosley Field) in 1934; introduced small Crosley auto in 1939 (sold to General Tire and Rubber in 1952); sold all radio interests to the Aviation Corporation (Avco) in 1945. Died in Cincinnati, 28 March 1961.

Further Reading

Douglas, Alan, "Crosley," in *Radio Manufacturers of the 1920s*, vol. 1, Vestal, New York: Vestal Press, 1988

Lichty, Lawrence W., "'The Nation's Station': A History of Radio Station WLW," Ph.D. diss., The Ohio State University, 1964

"Making Five Thousand Radio Sets A Day," *Popular Mechanics* (January 1925)

Perry, Dick, *Not Just a Sound: The Story of WLW*, Englewood Cliffs, New Jersey: Prentice Hall, 1971

Pyle, Howard S., "From Binding Post to Varnish," *Radio Broadcast* (February 1924)

Cross, Milton J. 1897–1975

U.S. Radio Announcer

Milton John Cross was a pioneer radio announcer who became famous as the voice of the Metropolitan Opera broadcasts and as a popular authority on serious music. Cross enjoyed a distinguished announcing career that lasted for five decades on the National Broadcasting Company (NBC) and on the American Broadcasting Companies (ABC), where his most prominent role was hosting the weekly nationwide broadcasts of the Metropolitan Opera. A one-time ballad singer on early wireless radio, Cross began his announcing career during broadcasting's pioneer days at WJZ in Newark, New Jersey. His tenure as a network announcer became one of the longest among all radio personalities. With his rare combination of announcing ability and knowledge of serious music, Cross occupies a unique place in radio history.

Born in New York City, Milton Cross grew up in the rough Hell's Kitchen section and dreamed of a career in music. After high school, he enrolled at the Institute of Musical Art (now part of the Juilliard School of Music) to study under the direction of Dr. Frank Damrosch. At the institute, he earned a certificate to serve as a music supervisor for public schools. Although he frequently sang as a tenor for several excellent church choirs in New York, he never applied for a school position. Instead, he became interested in radio shortly after the new medium began.

In 1921 Cross became aware of the new Westinghouse station, WJZ, in Newark and sought a position with the station as a singer. On 15 September of that year, while still attending the institute, he was given a chance to sing without pay on the station. The management liked his voice and quickly offered Cross its second announcing position, largely because of his excellent diction and his familiarity with foreign names and musical terms, and because he could fill in with a song whenever the need arose. Although he felt his future was in the music field, Cross accepted the position. His on-air duties included introducing speakers and performers, singing songs to fill time whenever speakers and other performers failed to show, delivering commercials for household supplies, reciting the Sunday funnies, reading children's stories, and other assorted announcing duties. He supplemented his small income from WJZ by singing in a Presbyterian church (his own faith), in Jewish synagogues, and in Catholic churches. Soon he was earning about $40 a week for his singing and announcing.

Cross joined the NBC announcing staff when the network was formed in 1926. Station WJZ became the flagship for NBC's Blue network in New York, and many of NBC's programs originated at WJZ, where Cross had begun his professional career. As a staff announcer, Cross was called upon to announce many different programs for the NBC network. However, his most notable career achievement came with his assignment to serve as announcer and host commentator for the radio broadcasts of the Metropolitan Opera.

In 1931 NBC decided to broadcast the Saturday afternoon performances of the Metropolitan Opera on a regular basis, and the network selected Milton Cross to serve as announcer-commentator. On Christmas Day 1931, Cross inaugurated NBC's Metropolitan Opera radio broadcasts with a performance of the opera *Hansel and Gretel*. From that initial network broadcast until his death in 1975, Cross missed only two Saturday afternoon Met broadcasts, and those were because of his wife's death. In his 43 years behind the microphone, Cross provided commentary on every opera in the Met's extensive repertoire. He often attended rehearsals and was noted for careful preparation of his opera commentaries, as well as his exquisite descriptions of the color, costumes, and scenery on the opera stage. In January 1975, at age 77, Cross was working on his material for an upcoming opera broadcast when he suffered a fatal heart attack, ending his remarkable 43-year career as the voice of the Metropolitan Opera. Cross has been credited with doing more than any other individual to make the Met a national institution. An estimated 12–15 million listeners tuned in from coast to coast each Saturday afternoon to hear the Metropolitan Opera's performances as explained by Milton Cross. To millions of opera fans, the resonant and cultured bass-viol voice of Milton Cross was said to have been more widely known than that of any other American, with the possible exceptions of President Franklin Roosevelt and Charlie McCarthy.

Through the opera broadcasts, Cross was instrumental in introducing opera to millions of Americans and in educating them about the operatic form of music. During the intermissions, he developed such popular features as "Opera Quiz," presented members of the cast, and tried to explain often-convoluted opera plots. In addition, Cross contributed to the public's knowledge of opera through well-known books such as *Complete Stories of the Great Operas* and *Favorite Arias from the Great Operas*.

Cross received numerous honors for his excellence as a radio announcer. He was acclaimed for his pure diction, his mellifluous voice quality, and his fine delivery. In 1929 the American Academy of Arts and Letters conferred upon Cross a gold medal with the highest honors possible for a radio announcer. His distinctive voice brought instant recognition wherever he appeared, and his great ability to pronounce the foreign words so commonly associated with opera was frequently praised.

This ability resulted from four years of studying German in high school and years of study of Italian at the Institute of Musical Art. He also studied French with tutors at NBC and diction at Columbia University. At the time of his death at age 77, Cross left a legacy of having introduced millions of Americans to live opera through the radio medium.

HERBERT H. HOWARD

See also Blue Network; Classical Music; Metropolitan Opera

Milton John Cross. Born in New York City, 16 April 1897. Studied at Institute of Musical Art (now Julliard School) and Columbia University; sang in various churches in New York City; worked as announcer at WJZ, Newark, 1921–26, NBC, 1926–43, and ABC, 1943–75; announced NBC programs, 1925–42; host and commentator for Saturday afternoon Metropolitan Opera performances on NBC, ABC, and *Texaco Metropolitan Opera Network,* 1931–75. Received American Academy of Arts and Letters gold medal, 1929; Handel Medallion by City of New York, 1969. Died in New York City, 3 January 1975.

Radio Series

1925–36 *The A&P Gypsies*
1930–32 *Slumber Hour*
1931–75 *Metropolitan Opera*
1934–37 *General Motors Concert*
1938–40 *Information Please*
1939–42 *Metropolitan Opera Auditions of the Air*

Selected Publications

Complete Stories of the Great Operas, 1947; revised edition *The New Complete Stories of the Great Operas,* 1957
Favorite Arias from the Great Operas, 1958
More Stories of the Great Operas (with Karl Kohrs), 1971; revised edition *The New Milton Cross' More Stories of the Great Operas,* 1980

Further Reading

DeLong, Thomas A., *Radio Stars: An Illustrated Biographical Dictionary of 953 Performers, 1920 through 1960,* Jefferson, North Carolina: McFarland, 1996
Truitt, Evelyn Mack, *Who Was Who on Screen,* New York: Bowker, 1974; 3rd edition, 1983

Crutchfield, Charles H. 1912–1998

U.S. Radio Executive

Beginning in the 1930s, Charles H. Crutchfield initiated numerous radio programming improvements and became a pioneer in the U.S. system of commercial broadcasting.

Crutchfield was born in 1912 in Hope, Arkansas. In 1920 the family moved to Spartanburg, South Carolina, where Crutchfield graduated from Spartanburg High School and enrolled in Wofford College. As he was walking home from classes one night in 1930, he decided to visit the studios of WSPA. When he got inside, the telephone rang. Since the announcer and the engineer were the only people in the studio, Crutchfield answered the phone and took a request from a woman who wanted a certain song played on the air. He wrote down the request and passed it on to the announcer. Immediately the station began getting more calls from listeners who wanted their requests and names on the air. Although Crutchfield did not realize it at the time, this was the first radio request program in the country.

It was not long before the owner of the station called, wanting to know what was going on at the station and if the young man with the deep, rich voice who answered the phone would like a part-time announcing job at the station. Crutchfield accepted immediately and thus began his broadcasting career.

After working at five more stations in North and South Carolina, in 1933 Crutchfield was hired as a staff announcer at one of the oldest commercially licensed radio stations in the country, WBT in Charlotte, North Carolina. At the time, WBT was owned by the Columbia Broadcasting System (CBS) and was a non-directional, 50,000-watt, clear channel station that reached all of the United States except the West Coast.

Crutchfield is probably best known for his role in a program called *The Briarhoppers.* In 1934 a Chicago entrepreneur wanted to sell patent medicines, such as Peruna iron tonic, Kolor-Back hair dye, and Radio Girl perfume, on the radio. He called WBT's station manager to see if the products could be showcased on a hillbilly music program. The station manager asked Crutchfield if the station had such a band, and Crutchfield replied that they did, knowing full well that no such band existed at WBT. Seizing the opportunity, Crutchfield got some

Charles H. Crutchfield
Courtesy WBT/WBTV Archives

hillbilly musicians together, named the band "The Briarhoppers," and went on the air with Crutchfield himself—as "Charlie Briarhopper"—acting as the emcee.

The Briarhoppers was immediately popular, garnering a huge and loyal audience. One reason for its success lay in its down-to-earth quality at a time when radio programs were decidedly highbrow and radio announcers spoke in a haltingly stilted and affected manner. But Crutchfield and the Briarhoppers would play upbeat music, tell corny jokes, and talk *with* the listeners rather than *at* them. Crutchfield's ad-libbed commercials often poked fun at the products, and he used a just-plain-folks delivery. Former *New York Times* critic John Crosby recalls that Crutchfield was the first man he knew of "to sit a listener down with a microphone across from him in a studio and tell the man about the product." The audience believed and bought, and thus was personal salesmanship born on radio. According to Crosby, it was Crutchfield and Arthur Godfrey who first did away with the stilted delivery of that era and popularized the one-to-one pitch so characteristic of the medium today.

The popularity of *The Briarhoppers* is difficult to imagine today. One of Crutchfield's promotions promised listeners a black-and-white photograph of the Briarhoppers in return for a box top from Peruna, and the station consequently received more than 18,000 box tops each week. When requests slowed down, a color photo of the Briarhoppers was offered, and later a color picture of the Last Supper kept bringing in floods of responses. *The Briarhoppers* did as much as any program to convince newspaper advertisers and American businesses of the power of radio as a medium for advertising.

Crutchfield was named WBT's program director in 1935, and he continued to introduce innovative programming. In 1936 he persuaded the Southern Conference to allow the first play-by-play radio coverage of its football games, and Crutchfield became the conference's first play-by-play announcer. Other notable programs included reenactments of baseball games (including improvisation when the wire service went out), the broadcast of an egg frying on the sidewalk during the summer, Rebel yells by old Confederate soldiers, the broadcast of the wedding of two nonagenarian former slaves, and a program targeted specifically to a black audience. Crutchfield's decision to air a local evangelist's revival meeting led to the launching of Billy Graham's broadcast career. Crutchfield was also responsible for the airing of station editorials long before it became a general practice among broadcasters.

Under Crutchfield's direction from 1937 to 1945, WBT won seven coveted *Variety* awards. In 1942 the station became the first ever to garner two Variety awards in one year—one for its contributions to the war effort and another for fostering racial goodwill and understanding through a program series that broke precedent with Southern tradition. In 1945 Crutchfield was named general manager of WBT and became the youngest chief executive officer of a 50,000-watt radio station in the nation. He became a director of the broadcasting subsidiary the following year, vice president in 1947, executive vice president in 1952, and president in 1963.

Crutchfield's influence in radio has been felt in other regions of the world. In 1951 the U.S. State Department sent him to Greece with the mission of setting up a nationwide radio network that would counter the barrage of communist propaganda flooding the country. Crutchfield represented the broadcasting industry when he and other American businessmen went to Russia in 1956 on a special mission. Upon his return he launched *Radio Moscow.* The program, designed to refute Soviet propaganda, won several awards and was syndicated nationally in 1960.

In 1949 Crutchfield signed North Carolina's first television station on the air, WBTV. Eventually he was elected president of WBT's and WBTV's parent company, Jefferson-Pilot Broadcasting Company, which owned five other radio stations; two television stations; a company that provided computer services for broadcasters; and two companies engaged in audio, film, and video commercial work. Although Crutchfield retired from Jefferson-Pilot in 1977, he was active in the media and public service until his death in 1998.

PATTON B. REIGHARD

See also Country Music Format; WBT

Charles H. Crutchfield. Born in Hope, Arkansas, 27 July 1912. Attended Wofford College, 1929–30; announcer, various stations, 1930–45; host of *The Briarhoppers,* 1934–45; program director, WBT, 1935–45; general manager, Jefferson-Pilot Broadcasting Company, 1945–63, director, 1946–80; vice president, 1947–52, executive vice president, 1952–63, president, 1963–77; established national radio network in Greece while on special mission for U.S. State Department, 1951; special mission to Soviet Union as broadcaster, 1956; produced *Radio Moscow* series, 1959–62; founded American Values Center, 1965; board of directors, Jefferson-Pilot Corporation, 1970–78; member National Commission on World Population, 1974–75; U.S. Chamber of Commerce Board of Directors, 1974–79; member, board of directors, Corporation for Public Broadcasting, 1976–82; president, Media Communications, 1978–98; founding chairman, North Carolina Board of Telecommunications Commissioners, 1979–86. Received Broadcast Preceptor Award from San Francisco State University, 1967; International Revenue Service Award, 1967; charter member North Carolina Association of Broadcasters Hall of Fame, 1970; honorary degree of Doctor of Humane Letters from Appalachian State University, 1973; first recipient of Abe Lincoln Railsplitter Award for Pioneering in Broadcasting, 1975; North Carolina Distinguished Citizen Award, 1977; honorary degree of Doctor of Humane Letters from Belmont Abbey College, 1979. Died in Charlotte, North Carolina, 19 August 1998.

Radio Series

1934–45 *The Briarhoppers*
1959–62 *Radio Moscow*

Further Reading

Fishman, Karen, compiler, *The Papers of Charles H. Crutchfield,* Broadcast Pioneers, Library of American Broadcasting, University of Maryland Libraries, <http://www.lib.umd.edu/lab/collections/crutchfield.html>

Crystal Receivers

Simple Early Radio Sets

The crystal receiver, popularly known as the "crystal set," was the device used by most people to listen to radio between 1906 and the early 1920s. The heart of the crystal receiver was the crystal itself, a small piece of silicon or galena, natural elements with the ability to detect radio frequency waves and to rectify or convert them into audio frequency signals. The crystal's ability to detect radio signals was discovered in 1906 by General Henry H.C. Dunwoody and G.W. Pickard. The crystal receiver was inexpensive and easy to construct, making it possible for even a young child to build a radio. Even today, the construction of a crystal receiver from a kit is often a young person's first introduction to radio technology.

As early as 1910, the three basic components and the instructions needed to construct a crystal receiver were available from mail-order electrical supply houses. All one needed was a spool of wire, a crystal detector, and earphones. The crystal detector consisted of a small piece of galena mounted in a lead base, approximately 1/4 inch in diameter, and electrically connected to a terminal. A tiny wire, called a "cat's whisker," made contact with the exposed top of the galena, and its small moveable handle allowed the listener to find the spot on the galena where the radio signal was the loudest. The cat's whisker was connected to a second terminal. One of these terminals was connected to the tuning coil.

A tuning coil was made by winding several hundred turns of thin, insulated wire around an empty oatmeal box or similar cylindrical object. A sliding piece of metal was positioned to move across the exposed coil windings for precise tuning. To the other terminal of the tuning coil a long wire, called the antenna, was connected and strung to a tree in the back yard, the goal being to get it as high as possible. The second terminal of the galena crystal/cat's whisker combination was connected to one wire of a headset or single earphone. The other earphone wire was connected to a ground, usually a metal stake driven into the earth. Sometimes a fourth component, a fixed or variable capacitor, was added to the circuit.

To understand what the crystal receiver meant to the early science of radio, it is necessary to look at the available wire-

Crystal set
Courtesy of Michael H. Adams

less detector technology in the early years of the 20th century. In Marconi's 1900 wireless, the receiving device used to translate the dots and dashes of his spark transmitter was a coherer, a small tube containing iron filings that closed like a switch when receiving the electromagnetic pulses of the Morse code. Each time the filings cohered, or caused the circuit to close, current from an in-series battery flowed; then, either a buzzer sounded, a telephone receiver clicked, or an inking device recorded a coded symbolic component of the message, a dot or dash. Then a small hammer would tap the filings apart, and the entire process began again to detect the next dot or dash. The coherer could only indicate to a radio operator if a spark signal was present. Such a system might receive five or ten words per minute and was unreliable. And although the coherer was a satisfactory receiver as long as the transmitter was of the spark-gap type, it would not work with the continuous-wave and voice-transmitting systems that quickly replaced the spark. The mechanical coherer did not allow a receiver to "hear" audio, obviously a serious technical impediment to the development of wireless telephone and radio broadcasting.

Between the crude mechanical coherer and the discovery of the detecting properties of the crystal, several intermediate systems of detecting were invented and used by two of the leading early radiotelephone inventors. Between 1900 and 1905, Reginald Fessenden's Liquid Barretter and Lee de Forest's similar Electrolytic Detector were able to detect both continuous-wave code transmission and audio. These were less reliable than the crystal detector that followed, but they did allow radio operators to hear the human voice. By 1906 de Forest was advertising a radiotelephone system with his vacuum tube, the Audion, as the detector. Whether liquid, crystal, or vacuum tube, this new generation of non-mechanical detectors that converted or rectified radio frequency into audio frequency really opened the door for the development of the radiotelephone.

When licensed radio for the public was introduced in 1920, it was believed that the financial basis for the new commercial radio service would derive from sales of manufactured radio sets. Large companies such as Westinghouse and General Electric introduced home radios, the most popular of which was a crystal receiver in a wood box with earphones and instructions, called the Radiola 1 and the Aeriola Jr. The vacuum tube detector, as pioneered and used by de Forest 15 years earlier, was still too expensive for most families, but there were higher-priced radios available that used the crystal as a detector but added a vacuum tube as an amplifier to increase the volume. By the mid-1920s, better programming caused a demand for radios that would play loud enough to drive a horn speaker, and manufacturers introduced radios that used vacuum tubes for both detector and amplifier.

The vacuum tube remained the technology of choice for detecting radio signals until the transistor finally replaced it in the 1960s. And what happened to the crystal receiver? It is still the entry-level radio technology of choice. Its components, in the form of kits, are readily available today. It is almost a rite of passage for a young boy or girl to build a crystal set, and technical museums still offer Saturday morning classes where parents and their children can learn to construct a crystal receiver. There is still a thrill from building your own radio, one that seemingly works by magic, using no batteries, no electricity, one that pulls faint programs from a local AM station, experiencing what your great-grandparents did almost a century ago.

MICHAEL H. ADAMS

See also De Forest, Lee; Fessenden, Reginald; Receivers

Further Reading

Aitken, Hugh G.J., *Syntony and Spark: The Origins of Radio*, New York: Wiley, 1976

Greenwood, Harold S., *A Pictorial Album of Wireless and Radio, 1905–1928*, Los Angeles: Clymer, 1961

Kendall, Lewis F., and Robert Philip Koehler, *Radio Simplified: What It Is—How to Build and Operate the Apparatus*, Philadelphia, Pennsylvania: Winston, 1922; revised edition, 1923

Sanders, Ian L., *Tickling the Crystal: Domestic British Crystal Sets of the 1920s*, Tunbridge Wells, England: Bentomel Publications, 2001

Sievers, Maurice L., *Crystal Clear: Vintage American Crystal Sets, Crystal Detectors, and Crystals*, Vestal, New York: Vestal Press, 1991

Verrill, A. Hyatt, *The Home Radio: How to Make and Use It*, New York: Harper, 1922

Cuba

Cuban radio has been linked to developments in the United States since its inception in 1922, mirroring Cuba's economic and political dependence on its northern neighbor. During its first four decades, Cuban radio followed the U.S. network system and broadcasting style. After the 1959 Cuban Revolution, the media was nationalized and relations with the United States were broken, but relations between the countries continued to frame Cuba's broadcasts, both domestic and international.

Pre-Revolutionary Radio: 1922 to 1958

On 10 October 1922 Cuba became the first country in Latin America to broadcast radio. Perhaps the 200 Cubans who owned radio sets (and until then had listened to U.S. stations) tuned in to that first broadcast—a speech in English by Cuban president Alfredo Zayas, made possible by the local phone company. Four days later, a second broadcast featured a speech by the president of the American Club of Havana, encouraging U.S. citizens to visit Cuba. Through 1934 the phone company station (PWX) broadcast music, vignettes about Cuba's natural and historical attractions, and news in English and Spanish to the United States.

A regulatory radio commission was created in 1929, but the government's role was limited to establishing power standards, assigning frequencies, and awarding licenses—which it often did in exchange for fees or favors. By 1930 there were a half-million radios and 61 stations on the island. Radio was quickly popularized as an entertainment medium featuring talk shows and live popular music (orchestras would play free in exchange for promotion).

Stations were private and were financed mostly by advertisers. They initially leased airtime to private announcers, who in turn solicited advertisers to finance their programs. In the 1930s these freelance announcers were replaced by advertising agencies, and soon two large soap manufacturers were in fierce competition for domination of the airwaves through their advertising departments.

In the 1940s, the two largest radio networks became associated with these manufacturers: Crusellas, with station CMQ, formerly PWX, became a subsidiary of Colgate-Palmolive; and Sabates, with Radio Habana Cuba–Cadena Azul, became a subsidiary of Procter & Gamble. It was not until the mid 1950s that the Cuban company Gravi shifted the balance by creating a third network.

In 1942 industrialists Abel and Goar Mestre bought 50 percent of station CMQ. After visiting the National Broadcasting Company (NBC) studios in New York, Goar Mestre decided to introduce rational planning in programs and schedules by organizing commercial spots into regular rotating blocks. Mestre also experimented with new formats such as specialized programming. Affiliate station CMBF played classical music only while Radio Reloj, founded in 1947, became the first 24-hour news station in the world. At this time, with a total audience of 85 percent of the population listening to radio daily, sophisticated survey research was introduced to serve commercial interests. By the end of the decade, CMQ was ahead of Radio Habana Cuba (RHC) in the ratings. Radio Progreso, with music and soap operas, rated third, and Mil Diez, an entertainment station run by the Popular Socialist Party, was fourth.

In 1947 CMQ's director of programming, Gaspar Pumarejo, started his own radio network, Union Radio, which became CMQ's main competitor. Two years later he inaugurated the first Cuban TV station. The introduction of the new medium caused a slight decline in radio's popularity. Resources were diverted to television, and with them went both advertisers and audiences, especially during the evening hours. Political battles, however, continued to be waged over the radio. As a result of frequent heated arguments, in 1950 the government passed a "right to reply" law that gave citizens the right to reply on the air to any accusations made against them. One very controversial program was that of Orthodox Party senator Eduardo Chibás on CMQ, which ended in 1951 when Chibás killed himself on the air. Radio Reloj, in turn, became the object of such battles in more dramatic ways—rebel forces occupied its facilities twice—once in 1952 by pro-Batista militias, and again in 1957 by revolutionaries.

At the end of the decade, the Batista government exerted tight control over broadcasts via financial subsidies as well as through direct censorship. In 1957, for example, rock and roll music was banned for allegedly promoting immorality. At the time there were 32 local commercial stations and five national networks—Union Radio, Radio Progreso, Circuito Nacional Cubano (founded in 1954 and using the former RHC infrastructure), Cadena Oriental de Radio, and CMQ. CMQ's signal was loud and clear thanks to high-power transmitters. However, the other stations' use of phone lines provided better coverage of the island at the expense of sound quality.

Radio and Revolution

On 1 January 1959, CMQ was taken over by employees involved in the underground struggle against the Batista regime; they announced the news of Fidel Castro's victory to the Cuban people. The new government immediately moved to control the media, which it had used for ideological purposes since Che Guevara founded Radio Rebelde in the Oriente Mountains in February 1958. The revolutionaries immediately eliminated all state subsidies, and over the course of two years, all stations were nationalized and placed under the administration of the Independent Federation of Free Radios (FIEL)—beginning with those such as Circuito Nacional Cubano that were associated with the former regime. As both foreign companies and advertising agencies were also nationalized, commercial advertising disappeared. By March 1961, public service announcements and political propaganda spots had replaced advertisements. Propaganda campaigns such as the 1962 Literacy Campaign and the 1970 Ten-Million-Ton Sugar Harvest Campaign were carried out to a great extent over radio.

To oversee operations and broadcasts, the Instituto Cubano de Radio (ICR; Cuban Institute of Radio) was founded on 24 May 1962. (It was renamed Cuban Institute of Radio and Television in 1976.) Revolutionary army commander and state prosecutor Jorge "Papito" Serguera was ICR's director between 1967 and 1974, when he was replaced by former urban guerrilla Nibaldo Herrera, a member of the council of ministers. A vice president for radio oversaw radio operations in coordination with the Partido Comunista de Cuba (PCC; Cuban Communist Party). The PCC Central Committee nominated all ICR officers, including the president, vice presidents, and station directors. The Communist Party Department of Revolutionary Orientation was in charge of media policies and exerted control over broadcast content.

Both the 1971 National Congress of Culture and Education and the 1975 First Cuban Communist Party Congress cemented radio's role in education and propaganda. The 1976 Communist Party Theses on Mass Media established policy guidelines in program design and production, which directed the media to "educate, inform, orient, organize and mobilize the population by appealing to reason and consciousness." The National Culture Council had an advisory role in educational programming. In addition, the armed forces and mass organizations (such as the Federation of Cuban Women and the Union of Communist Youth) used radio to disseminate their agendas and promote organizational membership.

Throughout the mid 1980s, Cuba's national stations were Radio Rebelde, based in Havana's former Circuito Nacional

Cubano studios and focusing on educational programming, including Russian lessons; Radio Liberación, formerly CMQ; CMBF, still an all-classical-music station; Radio Enciclopedia, dedicated to culture and the arts; and Radio Progreso, which featured variety programming. Because all stations shared the goal of ideological education, there was collaboration rather than competition between them. Most programming was locally produced, except for shows acquired by ICR through international exchanges, usually with other communist countries. Government control over content was absolute and extended to music programming, which had to be more than two-thirds Cuban, with all music by exiled artists excluded. Between 1973 and 1975, there was a ban on British and U.S. pop music that paralleled the repression of hippie aesthetics during that period.

Ideological Confrontation and International Broadcasts

The use of radio to disseminate ideology was not unique to the Cuban government. Since World War II, the United States used radio to combat communist ideology in Eastern Europe and, after Castro's victory, in Cuba as well. The United States and Cuba have a long history of disputes over frequencies and broadcasting power. In 1937, Cuba caused interference on 60 U.S. stations in retaliation for U.S. stations' intrusion into Cuban AM frequencies. In February 1960 both Cuba and the United States signed the updated North American Regional Broadcasting Agreement (NARBA). One month later the Eisenhower administration approved anti-communist broadcasts to Cuba, thereby subverting the agreement. Between 1960 and 1969, Radio Swan (later known as Radio Américas) broadcast to Cuba from Honduras. Swan featured old CMQ shows and exiled announcers whose voices were familiar to the Cuban public. As the U.S. administration initiated plans for the 1961 Bay of Pigs invasion, Swan's programming extended to 24 hours per day. At the same time, the U.S. Central Intelligence Agency (CIA) initiated clandestine broadcasts, and the Voice of America increased its Spanish programming from 30 minutes to 22 hours daily in both shortwave and AM (on Radio Marathon). Cuba jammed Radio Swan with CMBN's La Voz del INRA ("The Voice of the National Institute for Agrarian Reform") on the same frequency and with higher power. CMBN had a commercial format and broadcast mostly music and entertainment shows interspersed with propaganda messages and Castro's speeches. Russian transmitters and jamming devices allowed Cuban stations to broadcast at high power over several frequencies, to avoid U.S. penetration.

In May 1961 Radio Habana Cuba began shortwave broadcasts to disseminate the official views of the Cuban government and provide an alternative to the Voice of America, which was becoming its biggest competition. Throughout the 1960s and 1970s, RHC also offered a voice to representatives of leftist movements, from Colombian guerrillas to U.S. black nationalists such as Bobby Seale and Stokely Carmichael. Also in 1961, Cuba launched CMCA, "the friendly voice of Cuba," to the United States. CMCA was on the AM band 17 hours per day through 1967, with mostly Cuban music and cultural programming in English. Mainly staffed with North Americans, one of its typical features was a biweekly program about the plight of African-Americans, entitled *Negroes in Today's World*. During the same years (1962–67), Radio Free Dixie was recorded at Radio Progreso studios and broadcast in English three times per week to the southern United States, where it was widely heard. The brainchild of black nationalist Robert F. Williams Jr. (who lived in Havana at the time), the show featured African-American music and news about the Civil Rights struggle. In the 1970s, Cuban broadcasts to the United States became more sporadic, except for rebroadcasts of the Voice of Vietnam from RHC (in English) until 1976.

Responding to Radio Martí

Improved U.S./Cuban relations during the Carter administration (1976–1980) brought about a radio truce, but Ronald Reagan's election as U.S. president in 1980 refueled the confrontation. In 1980, Cuba resumed English-language international broadcasts that interfered with U.S. stations, and often rebroadcast Moscow Radio in English to the United States on several frequencies. On 27 August 1982, Reagan announced plans to launch Radio Free Cuba (later known as Radio Martí for Cuba's independence hero, Jose Martí). Although the bill would not be passed until more than two years later, Cuba immediately suspended a bilateral migration agreement and announced plans to upgrade its transmitting equipment and increase broadcasting power. In 1982 it backed out of the NARBA agreement, and on the night of 30 August 1982, it broadcast English-language programming for four hours, blacking out stations on five frequencies across much of the United States. North American newspapers reported a Cuban radio announcer saying in English, "We are bringing you the world news and some good Cuban music for your enjoyment." This was the strongest signal ever broadcast from Cuba, according to the Federal Communications Commission, and it moved the National Association of Broadcasters to oppose passage of the Radio Martí bill in order to avoid further Cuban interference. Cuban plans to launch the Voice of Cuba and Radio Lincoln in English to the United States were not carried out for financial reasons. Instead, the Cuban Institute of Radio and Television assessed Radio Martí's possible level of intrusion throughout the island and measured the power increases necessary for local stations to neutralize Martí's signal. Cuba also filled most available frequencies with domestic broadcasts.

United States Public Law 98-111 was passed on 4 October 1983, approving Radio Martí as a division of the Voice of America. Broadcasts began on 20 May 1985, Cuba's independence day. Cuba protested before the United Nations and responded by jamming Martí with noise. A center was established in Havana for the purpose of monitoring and transcribing Radio Martí's broadcasts, but no direct response was broadcast. However, when Cuban exile station Radio Mambí reached the Havana airwaves five months later, the Cuban government retaliated and immediately launched Radio Taíno toward Florida. Taíno began broadcasting in November 1985 on the same frequency as Radio Martí, in both English and Spanish. Like PWX in the 1920s and CMCA in the 1950s, Radio Taíno was a friendly voice, featuring cultural programming that included vignettes about Cuba's natural wonders and tourist attractions. Radio Taíno appealed to the nostalgia of older Cuban expatriates. The station's identification announcement was identical to that of 1950s CMQ, and popular announcers from pre-revolutionary radio days presented Cuban music from the 1940s and 1950s.

The "Rectification" Process

By 1983 there were 54 radio stations in Cuba—five national, 14 provincial, and 35 local, covering 99 percent of the island. At that time, an administrative redesign of the provinces required an accompanying readjustment of the network system in terms of frequencies, power, and schedules. As a result, Radio Liberación was merged with Rebelde in 1984, and all national stations improved their coverage. Both national and provincial stations extended their broadcasting day to 24 hours. Stereo was introduced in 1984 and FM in 1986.

During Radio Martí's first two years (1985–87), audience ratings for all Cuban stations declined. The ICRT decided to face up to the competition by improving the quality of programming and by appealing to young people, who preferred foreign stations' music and entertainment news. At the same time, Russian premier Gorbachev's *perestroika* translated in Cuba into a "Process of Rectification of Errors" that called for journalistic "transparency" and attention to young people's needs. To lead this renovation, Carlos Aldana was nominated head of the Communist Party's Department of Revolutionary Orientation, and Ismael González, a social psychologist, became head of the ICRT. Both men had a history as youth leaders in communist mass organizations. Following guidelines issued at the 1986 Third Communist Party Congress, González promoted audience research and led a generational renewal, at a time when the average age of media producers was 59 years.

Radio Rebelde became a symbol of this renovation. In May 1984 it appeared on the air with a completely new sound. Live news and sports broadcasts were interspersed with commercial-sounding propaganda spots that used catchy slogans and jingles. Announcers followed no written scripts, addressed the audience informally, and took live phone calls for the very first time. Also for the first time, drama shows presented the hardships of every day life. Beginning in 1987, Rebelde's top show was *Haciendo Radio* (Doing Radio), whose title recalled the new transparency line. It was a news program featuring live news broadcasts and investigative reporting of controversial issues.

More controversial was a local Havana station, Radio Ciudad de La Habana (Havana City Radio), that garnered top ratings with political humor and rock music shows. With a young and talented staff, Radio Ciudad emerged as the voice of Havana's youth. From 1986 to 1991, El Programa de Ramón (Ramon's Show) was its number one program. The first radio show named for an individual, it included biting political humor and featured the music of local underground rock bands. It was broadcast live every evening, reaching an audience of 300,000 in the city of Havana alone, with pirate tapes of the show circulating throughout the island. Next in popularity among young people was the classic rock show *Melomanía*, featuring the best of North American and British rock. But in 1990 Ismael González was deposed, and shortly thereafter these two shows were eliminated. The Rectification Period ended. Enrique Román, a former vice director of the Communist Party's Department of Revolutionary Orientation and director of the Communist Party's daily newspaper *Granma,* became the new ICRT president. A few months later, Carlos Aldana was also deposed on corruption charges and replaced as head of the Department of Revolutionary Orientation by former ambassador to Moscow and army colonel Ramón Balaguer.

Crisis and Reform

In the 1990s, following the collapse of the Soviet bloc, Cuba suffered a severe economic crisis. The ICRT was in desperate need of financial support—radio equipment was of Eastern European manufacture and fell into disrepair due to lack of spare parts. Advertising was now considered a viable solution to the problem, and incoming foreign investors were eager to promote their products in the emerging Cuban market. In 1994, Radio Taíno, a station born out of ideological confrontation with the United States (but with poor audience ratings), was revamped and became the first advertising broadcast outlet on the island since 1961. It was assigned an FM frequency for better sound quality, and new transmitters soon sent its signal to the island's urban centers and tourist resorts. While other stations curtailed their airtime in the face of crisis, Taíno expanded to 18 hours a day. Its new commercial format included dance music and entertainment news in both Spanish and English. Former Radio Ciudad de La Habana producers were hired by Taíno to appeal to urban youth. In its first year alone, Radio Taíno's hard currency revenues amounted to $250,000.

Due to pressures by new stakeholders such as advertisers and leaders in the music industry, the nature of radio production and broadcasting changed. Radio Taíno's premier program was *De 5 a 7*—the most popular music radio show during its six-year run. Although overt social critique was absent from *De 5 a 7*, stylistically it represented a revolution in Cuban radio broadcasting with its use of audio effects such as record scratching and a variety of special sound effects. The music played included the latest U.S. salsa hits, peppered with light show business news and cheerful concert announcements. The role of mass media was transformed. Rather than serving primarily as an instrument for ideological education, it became a tool for marketing products, services, and popular culture.

The need for economic recovery made Cuba's radio war with the United States a secondary concern. Cuba jammed U.S. stations with special programming only occasionally, as in retaliation for TV- Martí broadcasts (1989–90), to protest U.S. military intervention in Iraq and the former Yugoslavia, and during the custody battle over six-year-old Elián González (1999–2000). In 1999 ICRT vice president Ernesto López replaced Enrique Román to preside over the new commercial era and manage a $6 million budget for the years from 2000 to 2002, the highest ever.

In 2000 there were 55 radio stations in Cuba, all controlled by the state: six were national (Reloj, Rebelde, Progreso, CMBF, Taíno, and Enciclopedia), 16 provincial, 32 local, and one in shortwave (Radio Habana Cuba). There were 2.12 million radio sets, according to CIA estimates. The shows with highest audience ratings were Radio Reloj's morning news, Radio Progreso's late morning soap opera, and Radio Taíno's Latin dance music shows.

Cuba is a member of the International Telecommunication Union and is therefore bound by the Administrative Radio Regulations adopted by the 1979 World Administrative Radio Conference. Cuban stations are members of the Latin American and Caribbean Union of Radio Broadcasters, an association of public radio stations founded in 1985.

ARIANA HERNANDEZ-REGUANT

See also Cold War Radio; International Radio Broadcasting; Jamming; North American Regional Broadcasting Agreement; Radio Martí; Shortwave Radio

Further Reading

Lent, John A., *Mass Communications in the Caribbean,* Ames: Iowa State University Press, 1990

Salwen, Michael Brian, *Radio and Television in Cuba: The Pre-Castro Era,* Ames: Iowa State University Press, 1994

Smith, Wayne S., and Esteban Morales Dominguez, editors, *Subject to Solution: Problems in Cuban-U.S. Relations,* Boulder, Colorado, and London: Rienner, 1988

D

Daly, John Charles 1914–1991

U.S. Radio Journalist and Program Moderator

On 7 December 1941 John Charles Daly, Jr., made an indelible mark upon U.S. radio audiences with the first bulletin of the bombing of Pearl Harbor. Working for the Columbia Broadcasting System (CBS), Daly broke into the network at 2:25 P.M. Eastern time: "The Japanese have attacked Pearl Harbor, Hawaii, by air, President Roosevelt has just announced."

Although he was ridiculed for his English accent in prep school, Daly possessed one of the richest speaking voices on the air. He was born in Johannesburg, South Africa on 20 February 1914. He was the youngest son of John Charles Daly, Sr., a Boston geologist, and Helene Grant Tennant, an Englishwoman.

After his father died of complications from yellow fever in the Belgian Congo, young John Charles, his older brother John Grant, and their mother moved to Boston in 1923. Daly graduated from the Tilton School in New Hampshire and was offered a scholarship to attend Boston College in 1930. He worked as a telephone switchboard operator to make ends meet during the Depression, but his salary was later reduced because of the worsening economic times. Two years later, when the financial burden of supporting his ailing mother became too great, Daly was forced to drop out of college. He first took a job at a wool factory in New England and later moved to Washington, D.C., where he worked for a transit company.

Seeking to join the new and exciting mass medium of radio, Daly was hired as a Washington reporter with the National Broadcasting Company (NBC). He became known for his ability to speak extemporaneously with great ease under the time constraints of broadcast airtime. However, his trademark diction and eloquence on the air sometimes upset NBC's listeners; a few wrote letters to complain that Daly's English accent sounded too foreign and pretentious.

In 1936, at the age of 22, Daly jumped from NBC to CBS just as the latter network was expanding its news service. Daly soon became an original member of the legendary CBS News broadcasting team that worked under Edward R. Murrow.

Daly succeeded Robert Trout as the CBS White House correspondent. He had traveled more than 150,000 miles with Roosevelt throughout the United States, Europe, and South America by 1941. After the United States' entry into the war, he broadcast accounts of military operations in North Africa and Italy. Daly broadcast the first bulletins of Roosevelt's death from Warm Springs, Georgia on 12 April 1945.

As the new medium of television loomed, Daly was persuaded to host *What's My Line?* in 1950, in which a celebrity panel would guess the occupation of invited guests. The CBS network program eventually brought Daly the most fame and fortune of his broadcasting career.

In 1953 Daly was offered an unprecedented opportunity to become the vice president of news, special events, public affairs, religious programs, and sports with the rival American Broadcasting Company (ABC) network while continuing to host *What's My Line?* During the period 1953–60, Daly was ABC's only on-air news anchor on television. He was reported to have a quick temper and often clashed with management over news policy matters. Although Daly broke new ground by hiring a then largely unknown Howard Cosell to broadcast the first nightly sports report on national television, ABC executives said that Daly was a weak administrator who failed to hire other on-air talent to build the news division. In 1957 Daly publicly objected to the management hire of Mike Wallace to host a weekly personality interview program, *Night Beat*, saying that Wallace lacked hard news credentials. A decade later, Wallace would become the leading correspondent for the CBS newsmagazine, *60 Minutes*.

Daly resigned from ABC in November 1960 over a much-publicized dispute about the network's purchase of outside news programming. He was appointed director of the Voice of America (VOA) in 1967. Succeeding NBC journalist John

John Daly
Courtesy CBS Photo Archive

Chancellor, Daly was expected to enhance the agency's image in its broadcasts around the world. He resigned the following year, complaining about a lack of management autonomy from VOA's parent agency, the United States Information Agency. Leaving VOA a year later, Daly said he was entering a "state of semi-retirement."

DENNIS W. MAZZOCCO

See also News; Voice of America; World War II and U.S. Radio

John Charles Daly. Born in Johannesburg, South Africa, 20 February 1914. Attended the Tilton School, 1923–30; attended Boston College, 1930–32; radio news reporter and correspondent for NBC, Washington, D.C., 1936; CBS News radio correspondent, 1936–49; CBS White House correspondent, 1937–45; ABC News senior correspondent and news analyst, 1949–60; host, CBS weekly quiz show, *What's My Line?* 1950–67; ABC vice-president of news, special events, publicity, religious programs, sports, 1953–60; president, Tilton School Board of Trustees, 1963–1984; director, Voice of America, 1967–68; freelance writer for national publications, 1968–84; member of Artists and Writers Association, Association of Radio Analysts; former president, International Radio and Television Society, Overseas Press Club of America; honorary doctorates, St. Bonaventure University, American International College, honorary law degree from Norwich University. Died in Chevy Chase, Maryland, 24 February 1991.

Radio Series

1952–53 *What's My Line?*

Television Series

What's My Line (1950–67)

Further Reading

"Fluent Broadcaster: John Charles Daly (Man in the News)," *New York Times* (17 November 1960)

"His Line Is U.S. News: John Charles Daly, Jr. (Man in the News)," *New York Times* (30 May 1967)

Mazzocco, Dennis W., *Networks of Power: Corporate TV's Threat to Democracy,* Boston: South End Press, 1994

Metz, Robert, *CBS: Reflections in a Blood Shot Eye,* Chicago: Playboy Press, 1975

Persico, Joseph E., *Edward R. Murrow: An American Original,* New York: McGraw-Hill, 1988

Quinlan, Sterling, *Inside ABC: American Broadcasting Company's Rise to Power,* New York: Hastings House, 1979

Tomasson, Robert E., "John Daly, Newsman, Dies at 77; Host of TV's 'What's My Line,'" *New York Times* (26 February 1991)

Davis, Elmer 1890–1958

U.S. Radio Commentator

From 1939 to 1941, Americans listened to Elmer Davis describe the war in Europe and analyze its impact on their lives. When the United States entered World War II, Davis became head of the government agency charged with coordinating the release of all information about the war effort.

Davis was born in Aurora, Indiana. His father was a bank clerk and his mother a school teacher, later a principal. He attended Franklin College, near Indianapolis, and graduated in 1910 with a B.A. He taught Latin for a year in a high school, then, as a Rhodes Scholar, he studied classics at Oxford University. When he graduated, he traveled across Europe and met the woman who later became his wife, Florence MacMillan of Boston. He settled in London to do postgraduate work and hoped for a career as a teacher of ancient history.

But upon his father's death in 1913, Davis returned to the United States to take care of his mother. The two settled in New York City, and he began a career as a writer by working for *Adventure Magazine.* A year later, he became a reporter for the *New York Times,* where he worked his way up to foreign correspondent and later to editorial writer. He also wrote light fiction, including novels and short stories. In 1924 he became a full-time freelancer, writing both fiction and nonfiction. His journalistic articles appeared in many of the major magazines.

War in Europe seemed inevitable by 1939. Davis, who had previously substituted briefly for the Columbia Broadcasting System's (CBS) top commentator, H.V. Kaltenborn, was asked to become a full-time news analyst for CBS. Two weeks later, World War II broke out. Soon, Davis was the third most

Elmer Davis
Courtesy CBS Photo Archive

popular commentator in the country, following Kaltenborn and Lowell Thomas. At the height of his popularity, his listeners numbered 12.5 million.

When preparing his commentaries, Davis would look at the teletype and read the wire service news, but only take notes. He would then rewrite the material in his own style—concise and easy to understand. And he did some of his own reporting, interviewing political leaders and traveling to the scene. For example, he joined Edward R. Murrow, the CBS newsman, in England in 1941 for a month.

Davis liked to think of his audience as being rational and gave them the relevant facts behind the story. His analysis was insightful and frequently sounded like plain old common sense, although he did have a tendency to refer to the ancient Greeks and Romans that he had studied in his youth. His most memorable characteristic was to ask a question, followed by a pause, then "Well—" and the answer. His Midwestern twang, dry sense of humor, and flat delivery seemed to reflect the average American voice and was reassuring to many.

An admitted liberal, Davis was nevertheless not reluctant to criticize President Franklin Roosevelt. Davis favored the United States' remaining neutral during the war, but he had strong sympathies for the British. When Pearl Harbor brought America into the war in 1941, Davis became an ardent supporter of the war effort.

In one of his commentaries, he complained that too many government agencies were turning out information about the war. Reporters didn't know where to go for facts; important information was withheld from the public; and confusing, conflicting stories were the result. Davis recommended that the president create an office to oversee all of these efforts. In 1942 Roosevelt offered him the job, and Davis accepted.

The Office of War Information had the task of coordinating all national and international propaganda and the release of all news about the war to journalists. Davis favored getting as much information—including the negative—to the public as possible without violating security. When the military was reluctant to cooperate, President Roosevelt backed Davis up. Davis' lack of administrative experience led to some problems, but most journalists respected the efforts of one of their own to keep the news flowing during a difficult time.

His approach to drumming up domestic support for the war was to ask the networks for voluntary cooperation. He preferred that propaganda be integrated into regular programming, rather than being broadcast through special programming or didactic speeches. And broadcasters agreed.

When the war ended in 1945, Davis went to work for the new radio network, the American Broadcasting Companies (ABC), where he had a 15-minute program three times a week. He may have been motivated to move to ABC by a higher salary, but he also disliked CBS's policy that news analysts had to stick to the facts and not air their own opinions.

Davis was one of the first people in the industry to attack Senator Joseph McCarthy, the powerful Republican from Wisconsin who had little tolerance for criticism. McCarthy believed that the country was threatened by communism, but Davis felt the dangers of McCarthyism to freedom of speech and freedom of thought were even greater. He wrote a best-selling book of essays on the topic, *But We Were Born Free.* During the 1952 elections, some stations dropped Davis' program, but at his peak he had 150 stations carrying his commentaries.

In 1953 Davis had to retire from ABC because of his declining health. A year later, he tried a weekly commentary for ABC TV but again had to stop because of poor health. He suffered a stroke and died in 1958.

BARBARA MOORE

See also Commentators; Office of War Information; Propaganda; World War II and U.S. Radio

Elmer Davis. Born in Aurora, Indiana, 13 January 1890. Only child of Elam Holmes Davis and Louise Severin. Attended Franklin College, B.A., 1910; attended Oxford University (Rhodes Scholar), B.A. 1912; editorial staff, *New York Times,* 1914–24; freelance writer, 1924–39; CBS commentator, 1939–41; head of the Office of War Information, 1942–45; ABC commentator, 1945–54. Died in Washington, D.C., 18 May 1958.

Selected Publications

The Princess Cecilia, 1915
History of The New York Times, 1851–1921, 1921
Times Have Changed, 1923
I'll Show You the Town, 1924
Friends of Mr. Sweeney, 1925
The Keys of the City, 1925
Show Window, 1927
Strange Woman, 1927
Giant Killer, 1928
Morals for Moderns, 1930
White Pants Willie, 1932
Bare Living (with Guy Holt), 1933
Love Among the Ruins: Little Novels of Hard Times, 1935
We'll Never Be Any Younger, 1935
Not to Mention the War, 1940
Some Aspects of the Economics of Authorship, 1940
"War Information," in *War Information and Censorship,* 1943
But We Were Born Free, 1954
Two Minutes till Midnight, 1955
By Elmer Davis, edited by Robert Lloyd Davis, 1964

Further Reading

Bliss, Edward, *Now the News: The Story of Broadcast Journalism,* New York: Columbia University Press, 1991

Burlingame, Roger, *Don't Let Them Scare You: The Life and Times of Elmer Davis,* Philadelphia, Pennsylvania: Lippincott, 1961

Culbert, David Holbrook, *News for Everyman: Radio and Foreign Affairs in Thirties America,* Westport, Connecticut: Greenwood Press, 1976

Fang, Irving E., *Those Radio Commentators!* Ames: Iowa State University Press, 1977

Winkler, Allan M., *The Politics of Propaganda: The Office of War Information, 1942–1945,* New Haven, Connecticut: Yale University Press, 1978

Dees, Rick 1951–

U.S. Radio Personality

Rigdon Osmond Dees III (Rick Dees) grew up in Greensboro, North Carolina, where he began his radio career at age 17 upon being dared by another student to audition for a local radio station. Dees took the challenge, landed the job at WGBG, and has worked in radio ever since. He was known first as Rig Dees and then adopted the name Rick. Dees has combined his love for comedy and the absurd to create his own wacky style, which is performed for listeners by his "Cast of Idiots," a myriad of voices both self-created and done by his wife, Julie. Dees' unique radio style has inspired audiences to tune in and to make him number one in the ratings in every market in which he has worked. From WGBG he moved to WCOG in Greensboro in 1969. In 1971 Dees split time between WKIX in Raleigh and WTOB in Winston-Salem while working on his degree at the University of North Carolina.

Dees left North Carolina in 1973 to join WSGN in Birmingham, Alabama. He moved to Memphis, Tennessee, in 1976, where he worked for a short time at WMPS before moving over to WHBQ that same year. As an air personality in Memphis, Dees became known nationally in 1976 with his disco music parody "Disco Duck," which reached multi-platinum status by selling more than 4 million copies. The song made it to number one on the charts and landed Dees a People's Choice Award. In addition to "Disco Duck," Dees produced and wrote several comedy albums, including *I'm Not Crazy, Rick Dees' Greatest Hit (The White Album),* and *Put it Where the Moon Don't Shine.* His album *Hurt Me Baby, Make Me Write Bad Checks* was nominated for a Grammy Award.

In 1979 Dees moved to Los Angeles station KHJ-AM, where he was not particularly successful. But upon moving to Los Angeles' KIIS-FM in 1981, where he remained as of early 2003, Dees began a streak of winning *Billboard*'s Radio Personality of the Year award 15 consecutive times. Since 1982, he has hosted American Broadcasting Companies (ABC) Radio's internationally syndicated *The Rick Dees Weekly Top 40,* which airs to 30 million people via more than 400 stations in the United States and 70 other countries.

Dees was part of a 1984 Top 40 revival radio now known as "contemporary hit radio." The format relies on personalities or "superjocks" who are as important to the format as the music is. The format has been likened to that employed by 1960s air personalities such as Cousin Brucie Morrow, Murray the K, Alan Freed, Wolfman Jack, and others. Dees' fans are attracted by loony gags and bits. One of his staples is "Spousal Arousal," in which a spouse offers his or her partner romantic enticements. Another is "Battle of the Sexes," where men and women compete in trivia contests. Dees attributes his longevity in radio to his ability to stand for things that are different from the Howard Sterns of the world. On his own show, Stern frequently insults Dees.

Dees hosted the syndicated TV show *Solid Gold* (1983–84) and ABC-TV's largely unsuccessful late-night program *Into the Night Starring Rick Dees* (1990–91), a midnight talk show targeted at a younger viewing audience. His house band was Billy Vera and the Beaters, and the announcer was Lisa Canning. Greg Binkley and Bob Perlow provided comedy bits, along with "The Committee," four female senior citizens interacting with the host about a variety of topics. Dees said the show did not work because he was up against Johnny Carson's last year on the National Broadcasting Company's (NBC) *The Tonight Show.* Also, it was on during the Gulf War and was frequently delayed when Ted Koppel's ABC *Nightline* went overtime.

Other Dees television credits include appearances on *Roseanne, Married with Children,* and *Burke's Law* and voiceovers for animated children's programs such as *The Flintstones,* in which he was "Rock Dees." In the film *La Bamba,* the story of

Rick Dees
Courtesy CBS Photo Archive

rock and roll artist Richie Valens, Dees played Ted Quillen, the person who helped start Valens' career. In the animated full-length motion picture *The Jetsons: The Movie,* he played Rocket Rick. Dees has received numerous honors and awards for his accomplishments, including placement of his own star on the prestigious Hollywood Walk of Fame.

W.A. KELLY HUFF

See also Contemporary Hit Radio Format/Top 40; Disk Jockeys

Rick Dees. Born Rigdon Osmond Dees III in Jacksonville, Florida, 14 March 1951. Attended University of North Carolina; Worked at various radio stations including WGBG, Greensboro, North Carolina, 1968; WCOG, Greensboro, 1969; split time between WKIX in Raleigh, North Carolina, and WTOB in Winston-Salem, North Carolina, 1971; joined WSGN, Birmingham, Alabama, 1973; WMPS, Memphis, Tennessee, 1976, WHBQ, Memphis, 1976–79; KHJ-AM, Los Angeles, 1979–81; moved to KIIS-FM, Los Angeles, 1981–present. Recorded multi-platinum *Disco Duck,* 1976; first film appearance, *Record City,* 1977; appeared in *La Bamba,* 1987; host of *The Rick Dees Weekly Top 40,* 1982–present; first album, Grammy Award-nominated *Hurt Me Baby, Make Me Write Bad Checks*; other albums *I'm Not Crazy, Rick Dees' Greatest Hit* (*The White Album*), and *Put It Where the Moon Don't Shine.* Inducted into the Museum of Broadcast Communications Radio Hall of Fame, 1999.

Radio Series

1981–present *The Rick Dees Morning Show*
1982–present *The Rick Dees Weekly Top 40*

Television Series

Host, *Solid Gold,* 1983–84; host, *Into the Night Starring Rick Dees,* 1991; voiceovers, *Casper and the Angels,* 1979; guest starred in series including *The Brady Bunch Hour,* 1977; *The Love Boat,* 1977; *The Greatest American Hero,* 1981; *Married with Children,* 1987; *Roseanne,* 1988; *Diagnosis Murder,* 1993; *Burke's Law,* 1994; *Baywatch Nights,* 1995

Films

Record City, 1977; *Best Defense,* 1984; *The Gladiator,* 1986 (TV movie); *La Bamba,* 1987; *Jetsons: The Movie,* 1990

Further Reading

Hilliard, Robert L, and Michael C. Keith, *The Broadcast Century: A Biography of American Broadcasting,* Boston: Focal Press, 1992
Rick Dees website, <www.rick.com>

De Forest, Lee 1873–1961

U.S. Radio Inventor

A formally educated scientist whose inventions in some way have affected nearly every human life, Lee de Forest was one of the most important of the early inventors of radio and electronic technology. He is most known for his pioneering work with the vacuum tube—first as a detector of radio waves, then as an amplifier for long-distance telephone calls, and finally as the major technology of the radio transmitter, one still in use today. Although de Forest was responsible for some of the more significant radio technical accomplishments of the early 20th century, his career was one of continuing controversy: he was accused of stealing inventions from Reginald Fessenden and Edwin Howard Armstrong, he was accused but not convicted of business fraud, and his continual exaggeration of the facts surrounding his life and career caused him to become estranged from the radio engineering establishment. Even though he wrote an autobiography proclaiming himself "the Father of Radio," he never received the respect he actively sought his entire life.

Origins and Early Work

Lee de Forest was born in the Midwest but grew up in the Old South. Shortly after his birth in Council Bluffs, Iowa, in 1873, his father accepted a position as the president of Talledega College, a small, historically black school in Alabama. But although de Forest grew up in a rural environment, his education was formal, upper class, and thorough. He attended a private boys' school in Massachusetts, preparatory to his entrance into Yale University's Sheffield Scientific School, where he received the degree of Doctor of Philosophy, with an 1899 dissertation entitled "The Reflection of Hertzian Waves at the End of Parallel Wires." After graduation, he worked briefly for several Chicago companies, Western Electric among them.

But young de Forest wanted to start his own wireless business. In the beginning, he followed the work of Marconi, attempting to develop better communication between ships and shore stations. After several attempts to demonstrate that his version of spark transmitter and coherer receiver technology was superior to that of Marconi and others, de Forest finally received the noticed of a Wall Street promoter named Abraham White. In 1902 de Forest joined White in forming the De Forest Wireless Telegraph Company. Among their early customers were the War Department and the U.S. Navy. Under the guidance of White, a public offering of stock was made, public demonstrations were held, and radio equipment was sold. But characteristic of de Forest's entire career, the hyperbole surrounding the company was greater than its actual value, and although de Forest continued to invent, he was apparently unaware that White may have been engaging in less than ethical business practices.

Apparently, de Forest tired of telegraph-based wireless. A person of culture and lover of the opera, de Forest believed early on that the radiotelephone, or talking wireless, was going to be a way to send highbrow musical entertainment into homes. It was one thing to have to earn a living selling communications equipment to the navy, but his real passion was voice and music by wireless. In 1907 he formed the De Forest Radio Telephone Company—merely one of what would become a steady stream of companies with various backers. For the transmitting part of his radiotelephone, de Forest used a version of a Poulsen direct current arc, and the historical record shows that de Forest did attempt on several occasions to use this device to send the voices of opera singers to members of the press stationed at receiving sets. Even when testing the radiotelephone for the navy, he usually played some sort of phonograph music as the ships entered the harbor. De Forest was a showman, and he was one of the early pioneers in what would become radio broadcasting to an audience.

Audion and Later Inventions

Lee de Forest is best known for his improvements to the basic invention behind all radio and television, the vacuum tube. Earlier, Thomas Edison's electric lamp had been modified by the Englishman Ambrose Fleming, who added a second element, called a plate, and named the new invention the Fleming Valve. By 1906 de Forest had modified Fleming's valve by adding a grid to control and amplify signals; he called his device the Audion. As became apparent over the next few years, the inventor did not fully understand his own creation. Little did he realize then that the simple Audion was going to bring him fame, fortune, heartbreak, and high legal bills for most of his life.

One of de Forest's first major brushes with the legal system did not concern the Audion but happened as a result of fraud in his radiotelephone company, and in 1913 he and business partners James Smith and Elmer Burlingame went to trial for misleading stock offerings. Smith and Burlingame were found by a jury to be guilty, but de Forest was declared innocent. Shortly thereafter, he began a decades-long court battle with Edwin Armstrong over the invention of the regenerative circuit based on the Audion tube. Regeneration is like feedback: a

Lee de Forest (right) and Reginald Hawkins of the New York Public Library, 1952
Courtesy AP/Wide World Photos

small signal from the output of a vacuum tube is fed back into the input, thus making weak signals very strong. Both de Forest and Armstrong claimed discovery of this principle; the litigation lasted from 1914 to 1934, and although the courts would finally side with de Forest, the technical community did not. It was a hollow victory, which nearly destroyed both claimants.

Personally, de Forest suffered a series of failed marriages. The first of these, in 1906, was to a Lucille Sheardown, a marriage that ended in divorce the same year. The second, in 1907, was to Nora Blatch, who bore him a child, but Nora, a liberated woman with an engineering background, soon realized that she did not want to live under the shadow of de Forest. By 1911 the marriage had ended in divorce. By 1912 de Forest had remarried, this time singer Mary Mayo. Several children resulted, but by 1926, while in Europe, de Forest had married his fourth wife, the actress Marie Mosquini. Even though he had failed to divorce Mary Mayo, de Forest managed with legal help to marry Mosquini and remained married to her, apparently happily, for the rest of his life.

The Audion would continue to dominate de Forest's life. Moving to California in 1910, he worked for the Federal Telegraph Company at Palo Alto. While there, de Forest finally made his Audion tube perform as an amplifier and sold partial rights to American Telephone and Telegraph (ATT) as an amplifier for transcontinental wired phone calls. For this innovation he received $50,000, whereupon he returned to New York and started the Radio, Telegraph and Telephone Company. By the beginning of 1916, he had finally perfected his Audion for its most important task, that of an oscillator for the radiotelephone transmitter. By late 1916 de Forest had begun a series of experimental broadcasts from the Columbia Phonograph Laboratories on 38th Street, using his Audion as a transmitter of radio for one of the very first times. According to de Forest in a newspaper article published in late 1916, "The radio telephone equipment consists of two large Oscillion tubes, used as generators of the high frequency current" ("Air Will Be Full of Music Tonight," *New York Sun*, 6 November 1916).

Early Broadcasts

A few months later, de Forest moved his tube transmitter to High Bridge, New York, where one of the most publicized pre–World War I broadcasting events took place. Just as Pittsburgh's KDKA would attempt to broadcast an election exactly four years later, in 1920, de Forest used the most public of events, the Hughes–Wilson presidential election of November 1916, for his broadcast. The *New York American* installed a private wire, and bulletins were sent out every hour. The listener reports in the press were positive: "Seven thousand wireless telephone operators within a radius of 200 miles of New York City received election returns from the *New York American*. They heard not only election returns, but music as well." Because it happened in New York, was heard by a large audience, and received so much press attention, it was one of the single most important pre–World War I events in radio broadcasting. Beginning with his arc telephone experiments for the navy and his transmissions of opera music, and ending with his radio station at High Bridge in 1916, the evidence strongly suggests that Lee de Forest, more than any other individual, saw a potential for voice transmission beyond just a wireless replacement for two-way communication.

Lee de Forest's accomplishments in radio technology were both huge and unrewarded. His vacuum tube innovations between 1906 and 1916, although clouded by court battles, were nevertheless significant and long-lasting. In his later years he lived in Hollywood and worked on a variety of non-radio technical devices such as guidance systems for bombs. Most notable was his Phonofilm process, a way to make the movies talk by adding a synchronized optical soundtrack to the film. For that invention, he received an Oscar in 1959 from the Academy of Motion Picture Arts and Sciences. He continued to promote his legacy as the "Father of Radio," but his most important non-technical contributions to radio, his publicized pre-1920 broadcasts, were far in the past. He became increasingly paranoid, believing that his failure to achieve recognition was because of his "enemies." Following a long illness, he died in Los Angeles in 1961.

MICHAEL H. ADAMS

See also Armstrong, Edwin Howard; Early Wireless; Fessenden, Reginald; Fleming, John Ambrose

Lee de Forest. Born in Council Bluffs, Iowa, 26 August 1873. Educated at Mt. Hermon School for Boys; attended Yale University, Ph.D. in Physics, 1899; wireless and radio technology inventor and early broadcaster; major invention was the Audion (1906), a vacuum tube he developed as a detector, amplifier, and oscillator of radio waves. Broadcast Enrico Caruso from Metropolitan Opera (New York), 1910. Developed regenerative circuit, 1912; worked on sound motion picture system, 1920s; developed diathermy machines for medical use, and did some work with television, 1930s; during World War II conducted research with Bell Laboratories. Received honorary Oscar for his contributions to sound film, 1959. Died in Los Angeles, California, 30 June 1961.

Selected Publications

Father of Radio: The Autobiography of Lee De Forest, 1950

Further Reading

Aitken, Hugh G.J., *The Continuous Wave: Technology and American Radio, 1900–1932*, Princeton, New Jersey: Princeton University Press, 1985

Barnouw, Erik, *A History of Broadcasting in the United States*, 3 vols., New York: Oxford University Press, 1966–70; see especially vol. 1, *A Tower in Babel: To 1933*, 1966

Carneal, Georgette, *Conqueror of Space: An Authorized Biography of the Life and Work of Lee de Forest*, New York: Horace Liveright, 1930

The Complete Lee de Forest, <www.leedeforest.org>

Hijiya, James A., *Lee de Forest and the Fatherhood of Radio*, Bethlehem, Pennsylvania: Lehigh University Press, 1992

Lewis, Tom, *Empire of the Air: The Men Who Made Radio*, New York: HarperPerennial, 1991

Lubell, Samuel, "Magnificent Failure," *Saturday Evening Post* (17, 24, 31 January 1942)

Maclaurin, W. Rupert, "Lee de Forest," in *Invention and Innovation in the Radio Industry*, New York: Macmillan, 1949; reprint, New York: Arno Press, 1971

Mayes, Thorn L., "DeForest Companies," in *Wireless Communication in the United States: The Early Development of American Radio Operating Companies*, East Greenwich, Rhode Island: New England Wireless and Steam Museum, 1989

Delmar, Kenneth 1910–1984

U.S. Radio Actor and Announcer

Kenneth (Kenny) Delmar's career spanned the history of 20th-century media and popular entertainment. A versatile character actor, Delmar appeared in vaudeville, television, and motion picture productions, but he is most remembered for his work in radio. As Senator Beauregard Claghorn on *The Fred Allen Show* in the late 1940s, Delmar entertained millions each Sunday with his blustery rhetoric, his puns and malapropisms satirizing Southern culture, and his oft-imitated tag line, "that's a joke, son."

Kenny Delmar was born in Boston, Massachusetts, on 5 September 1910. His parents separated when he was a young boy, and he was raised in New York by his mother. A child of show business, he toured the country as early as 1918 with his mother and aunt, a vaudeville duo booked as the Delmar Sisters.

Early Radio Work

Delmar broke into radio in the 1930s as an announcer, working on such major network programs as the musical countdown show *Your Hit Parade*. He began doing more acting work throughout the decade, appearing as the vain Police Commissioner Weston on *The Shadow*, and in various roles on the newsreel dramatization program *The March of Time* and on Orson Welles' dramatic series *The Mercury Theater of the Air*. By the end of the decade he had established himself as one of radio's elite character actors.

Delmar's biggest role during this period was on the "War of the Worlds" broadcast of 30 October 1938. Before the broadcast, CBS censors had demanded that Orson Welles fictionalize the names of some real places and characters in the script, including changing Delmar's President Franklin Roosevelt character to the nameless "Secretary of the Interior." Delmar's impression of Roosevelt—a voice that by the late 1930s was intimately familiar to American radio listeners—was so realistic that it was a major inspiration for the panic felt by millions of listeners, despite the fact that the character was clearly identified as the "Secretary of the Interior" and not the actual President.

Senator Claghorn

Delmar's big career break came in 1946 with his acting work on *The Fred Allen Show* as Senator Beauregard Claghorn, the loudmouthed Southern politician who showed his sectional loyalty by drinking only from Dixie Cups, refusing to drive through the Lincoln Tunnel, and claiming that he was so Southern, "where I live we call the people from Alabama Yankees." Senator Claghorn was one of the residents of *Allen's Alley*, the imaginary street Fred Allen strolled down each week on his program beginning in the early 1940s, talking current events and sharing jokes with a geographically and culturally disparate quartet of recurring characters who occupied this same fantastical radio space. As the five-minute sketch developed over time and the Alley's characters solidified, Senator Claghorn shared the street with the Jewish housewife Mrs. Nussbaum, the feisty Irishman Ajax Cassidy, and the farmer Titus Moody. Delmar landed the Claghorn role after Minerva Pious, the actress who played Mrs. Nussbaum, heard Kenny Delmar do a hilarious southern impression and recommended him to Allen. Though Fred Allen wrote most of the "Allen's Alley" dialogue himself, Delmar contributed a great deal to Claghorn's character, claiming that he modeled Claghorn after a Texas rancher who picked him up while hitchhiking in the late 1920s and barely stopped talking.

Delmar's Senator Claghorn first appeared on 5 October 1945 and became an overnight sensation, the subject of millions of amateur impersonations throughout America. Claghorn offered hilarious tidbits on Southern life, such as his statement that his Thanksgiving feast always began with a "Memphis Martini . . . a tall glass of pure corn likker with a wad of cotton in it." Senator Claghorn's "that's a joke, son" became a national catchphrase. The new Allen's Alley quartet was a certifiable hit, and *The Fred Allen Show* became one of the most popular shows on radio, drawing some 20 million listeners on Sunday evenings. Consumers ate up Claghorn memorabilia in the form of shirts and compasses that only pointed south, and the character became so popular that Delmar parlayed the role into two comedy records, a feature film entitled *It's a Joke, Son*, and radio and television commercial work well into the 1960s.

Allen's Alley became popular to an American population emerging from the Depression and World War II, moving to the suburbs, and coming to grips with the tumultuous decades of the 1930s and '40s. The very idea of the nation was under intense revision by many Americans, especially as older conceptions of ethnicity and regionalism had been subjected to the crucible of war. As the most popular character on the most popular radio program of the immediate postwar years, Kenny Delmar's Senator Claghorn character played a considerable role in the shaping of national ideas about the South during the postwar period. Allen's Alley provided for a suburbanizing population a reminder of the old urban ethnic neighborhoods and people, and may have served for its millions of listeners as a sounding board for their own new postwar Americanness, as this anonymous "white" population increasingly left behind its older urban ethnic ways and mores for the mass media society of the suburbs. Kenny Delmar's character had an ambiguous role in this process, but the degree to which Senator Claghorn shaped national ideas about Southerners should not be overlooked. Some Southern newspapers printed anti-Claghorn editorials, and the real Senator Theodore Bilbo of Mississippi was an outspoken critic of what he saw as Delmar's insulting caricature of Southern politicians. Senator Claghorn was invoked on the Senate floor several times in the 1950s, both as an example and counterexample of the nature of Southern politics and politicians, and he was for many regular Americans of the World War II generation a national reference point on Southern culture.

Kenny Delmar as an actor never achieved the notoriety of his Senator Claghorn character, and he had wisely continued his character and journeyman work in radio throughout Claghorn's popularity. NBC considered giving Senator Claghorn his own show in 1949, but the plans never came to fruition. Delmar continued to work in radio after *The Fred Allen Show* went off the air in 1949, and he branched out into theater and television work as well. He also provided vocal work on cartoon shows into the 1970s, including work on *The Adventures of Hoppity Hooper* and *Underdog*. The cartoon character Foghorn Leghorn was lifted almost directly from Delmar's Claghorn character, though he received no royalties or official credit. He died on 14 July 1984 in Stamford, Connecticut.

MICHAEL STAMM

See also Allen, Fred; Comedy; War of the Worlds

Kenneth Delmar. Born in Boston, Massachusetts, 5 September 1910. Began radio work in the 1930s as an announcer; early dramatic roles on *The Shadow*, *The March of Time*, and *The Mercury Theater of the Air*. Played the F.D.R.-esque "Secretary of the Interior" on "The War of the Worlds" in 1938. Best known as Senator Beauregard Claghorn on *The Fred Allen Show*, 1945–49. Appeared in several prominent radio and television series as a supporting actor; starred in the feature film *It's a Joke, Son* in 1947. Provided voice work for animated cartoons such as *Underdog*, 1964–73. Died in Stamford, Connecticut, 14 July 1984.

Further Reading

Harmon, Jim, *The Great Radio Comedians*, New York: Doubleday, 1970

Havig, Alan, *Fred Allen's Radio Comedy*, Philadelphia: Temple University Press, 1990

Taylor, Robert, *Fred Allen: His Life and Wit*, Boston: Little Brown, 1989

Wertheim, Arthur Frank, *Radio Comedy*, New York: Oxford University Press, 1979

Demento. *See* Dr. Demento

Demographics

Defining the Radio Audience

Broadcast Research Definitions describes demographics as

> a system of categories by which a population is subdivided according to characteristics of the people who comprise it. The same term also describes audience reports which present audiences according to this system of categories. In broadcast ratings demographics most often refer to age and sex categories such as "Men 18–34" or "Women 25–49." The terms may also be used to describe categories based on marital status, education, etc. (Fletcher, 1988).

Demographics are important in the radio industry in two ways. First, advertisers use demographics to describe their customers and to buy audiences for commercials. This practice is the basic revenue transaction supporting commercial radio. Second, radio stations are programmed to produce audiences with the demographics their advertisers seek.

Core consumers of any product or service account for the day-to-day, year-to-year success of a consumer company. The core consumer spends more or purchases more units in the product/service category in question and is particularly important to the survival of companies providing such goods/services. For instance, Campbell's Soup identifies its bedrock offering as its condensed soups sold in red-and-white cans. The core consumer of this product purchases a case or more of these soups every month. The demographics of Campbell's core consumers are important to the company, and its advertising must reach them as well as additional consumers who use their condensed soup less often or seasonally.

These additional consumers are an example of special advertising opportunities for radio. For many years Campbell's has observed that more soup is purchased during the cold seasons and in areas most affected by cold weather. They also know that the typical seasonal purchaser of their soups is a mother with children at home who feels that adding soup to a child's diet in cold weather will improve resistance to colds. As a consequence, Campbell's regularly purchases additional advertising time on radio stations in geographical areas most affected by cold weather and on stations high in the demographic group comprised of women aged 30 and over, which presumably includes mothers with children at home.

Origins

In the 1930s, the early years of radio audience reports, the principal demographic of interest to advertisers was the number of households in the audience of particular programs. It was assumed that each household would own one radio receiver and that every member of the household would hear the programs broadcast through the household receiver. The result was that information about the programs tuned in by the household set was the prime information sought by researchers, and advertisers based their strategies on appealing to the product-purchasing decision makers in the household.

Demographics were so unimportant during radio's Golden Age that the Nielsen Radio Index (NRI) became the principal national radio ratings service in 1950 (right at the end of the period) without having the capability of reporting demographics. The NRI used audimeters attached to respondent receivers at home (and eventually in cars). These devices could only report which programs were accessed by which radios; they could not determine who was listening or whether anyone at all was listening.

The rise of television in the late 1940s marked the rise of demographic information's importance in selling radio time. By 1975—the year before its demise—The Pulse included these demographic categories in its reports: (1) Gender: male, female, total; (2) Age: teens, 18–24, 25–34, 35–49, 50–64, 65 and over, total; and (3) Ethnicity: Black, Hispanic, other, total. In the same year, Arbitron Radio reports included these demographics: (1) Gender: male, female, adults (male and female) and (2) Age: teens, 18–24, 25–34, 35–49, 50–64, 65+. Special weighting and interviewing procedures were in place for African-American and Latino listeners.

In the mid-1980s Tapscan had become an important selling tool for radio. Tapscan of Birmingham, Alabama, was a provider of radio sales software. The service was available only to full-service clients (stations that purchased all of the regularly scheduled Arbitron surveys of the market in question) of a radio rating service. One of the displays produced by Tapscan analysis was a ranking report that showed which stations in a market had the highest numbers of listeners in each of the demographic categories included in the rating report. If an advertiser wished to sell jeans in a given market, in-house research would reveal that the core jeans consumer was a person in the 18–24 age group, that males were more brand loyal, and that females made more purchases and were more sensitive to price. A Tapscan analysis of the stations in each market would reveal which stations could provide the most young male and female listeners and thus should be included in the campaign buy for a jeans campaign.

The greater availability of radio listener demographics and increased use of sales analysis software made possible an important media buying strategy—Optimum Effective Scheduling (OES) or "optimizers." The idea behind OES is that advertisers are best advised to purchase a combination of advertising opportunities from whatever stations in the market are necessary to deliver every listener in the demographic category sought by the advertiser. The software's calculations during data analysis take into account two factors called *recycling* and *sharing*.

Recycling refers to the tendency of some proportion of an audience to listen during more than one daypart or to more than one program on a given station or network. For example, 25 percent of a radio station's morning drive time audience is recycled to evening drive time. This means that only one out of four listeners in the morning will be present in the evening; hence, an advertiser determined to reach the optimum number of target listeners should buy commercials in both morning and evening drive time.

Sharing refers to listeners of more than one station, which may also be referred to as their having a "duplicated audience." Sharing also implies that there will be some members of the advertising target demographic groups who listen to one station only, so those members' favorite stations must be included in the OES calculation for advertisers who wish to reach them. OES software routines determine how many spots on which stations must be purchased to reach a designated percent of all radio listeners in a particular demographic group.

Optimum effective scheduling has become very important in national and regional media buying. It is also important in planning programming for a radio station. If an important audience listens exclusively to a particular radio station, then advertising from that station must be included in the advertising media plans of any business aiming for the demographic audience delivered by that station.

The need for OES presents some dilemmas to radio programmers. Should a station design its programs for homogeneous audiences from a relatively narrow set of demographic groups, increasing the likelihood that exclusive audiences will be included in buys made using the OES rationale? Or should the station attempt to appeal to a wider range of demographic groups, hoping that each group delivered will be useful to the campaigns of different advertisers? Contemporary radio stations programmed under either of these philosophies are usually prosperous.

One of the most important radio program consultants since the 1980s, Ed Shane, wrote the following while discussing station audiences that are not core to the station:

> An essential, of course, is to be able to judge whether there are others in the general audience. A jazz or new age station, for example, may have not only a loyal core, but also may have all the jazz or new age devotees in the market. More than one station has been bitterly disappointed to discover that there was no growth beyond their initial impact (Shane, 1991).

The situation is even more complicated for station clusters, which are now the norm in the major markets of the United States. Large radio groups today may own stations providing every major radio format, perhaps with five or six or more formats in the same market. The issue of which demographic groups a particular station should target with its programming must be answered by program executives and consultants who are responsible for stations with the same format in many cities, even when that format is failing to deliver its assigned demographics in one market served by the group. At the same time, a cluster of stations in any market attempts to use OES strategies to sell combinations of its stations whenever possible. Sales are often the responsibility of an integrated sales management team charged with selling all of the stations in the cluster.

Demographic Factors

Programmers must look at the demographic characteristics of the market as a whole in light of station formats and audiences already present in the market. The economy of the market is another important consideration. For instance, if the market is characterized by rapid home construction and high levels of home buying, the program planner will recognize that sales of products related to home improvement, major appliances, lawn and garden items, furniture, and interior design are likely to be important sources of revenue in the market. If the community hosts a rapidly growing number of small businesses, business-to-business advertising and ads that direct the attention of business managers to office supply websites will be potentially rewarding.

Various radio program materials produce different demographic patterns among listeners. Most music listeners, for example, gain their music listening preferences during the years in which they are becoming aware of popular music, their teens and young adult years. Although their tolerance for other music may increase over the years, they tend to retain favorites acquired in those formative years. As a consequence, oldies (music no longer in current release but popular with demographic groups socialized to it at a young age) are a major component of popular radio programming. The latest hits tend to be preferred by young audiences, other types of music appeal more consistently to women, and still others are the choice of blue-collar workers.

A programmer's experience will suggest one or more formats that will attract the desired demographic groups in the market. The station will then conduct studies, perhaps by focus group, to determine listener attitudes about these formats and about the lack of certain music or other radio programming in the market. The focus group discussion may explore the preferences of potential demographics for different parts of the day. Listeners to some formats, although enthusiastic about hearing more of their favorite music, may also wish to hear traffic and weather reports at some time of the day. If talk formats are being studied, popular topics for discussion and the behavior of talk show hosts may be explored in focus groups. Increasingly there will also be discussions about the role of commercials in these formats. Should the commercials be concentrated in a small number of breaks in programming? How many commercials in a row are tolerable?

Format Approaches

The radio station program format will consist of a set of principles or rules for assembling programming. Sometimes these rules are relatively simple: (1) three musical selections should be played in uninterrupted succession; (2) a hit should be heard in this musical style four times per hour; (3) an oldie harking back to the birth of this style should be played at least once per hour; (4) titles and performers should be announced both before and after clusters of musical selections; (5) during early morning hours the time should be announced four times per hour; (6) during commuting hours traffic highlights should be presented twice per hour; and so on.

When the programmer has defined the format, a model tape of the format will be produced in which the music to be played is represented by brief excerpts. This makes it possible for potential listeners to hear all elements of a format in a short space of time (perhaps five minutes) while getting a feel for the music to be included. These model tapes are then played to potential listeners in focus group or auditorium settings. The researcher will also play excerpts of programming from competing stations to measure the competitive appeal of the proposed radio format. The same session may also include examples of promotional materials for the station so that those attractive to the station's target demographic groups can be identified.

When a new station format is broadcast, station management will pay close attention to the demographic groups delivered by the station. An important analysis from audience research reports will be the number of minutes an average listener in a target group listens to the station. This statistic is called "time spent listening" (TSL). The programmer hopes that the station's target demographic groups' TSLs are relatively large and that they will grow after a change in format. It is also significant that exclusive audiences (audiences that tune to only one station) will be apparent in the station's target demographic groups. The exclusive audiences will guarantee the station a place in OES advertising plans. Station promotion will be designed to increase the number of listeners in the target groups, and improvements in programming will be intended to increase time spent listening for these demographics.

Many stations use satellite or other syndicated sources for their programming. There is some cost associated with acquiring programming in this way. For a syndicator or satellite music service to remain profitable, a similar sort of demographic research is essential.

Although demographic research was not important in the early days of radio, the role of demographics in both contemporary radio programming and advertising sales has become crucial. Commercial radio stations must command audiences in the demographic groups important to their advertisers in order to ensure their own survival.

JAMES E. FLETCHER

See also A.C. Nielsen Company; Arbitron; Audience Research Methods; Commercial Testing; Programming Research; Programming Strategies and Processes; Pulse Inc.

Further Reading

Beville, Hugh Malcolm, *Audience Ratings: Radio, Television, and Cable,* Hillsdale, New Jersey: Erlbaum, 1985; revised edition, 1988

Fletcher, James E., *Music and Program Research,* Washington, D.C.: National Association of Broadcasters, 1987

Fletcher, James E., *Broadcast Research Definitions,* Washington, D.C.: National Association of Broadcasters, 1988

Fletcher, James E., *Profiting from Radio Ratings: A Manual for Radio Managers, Sales Managers, and Programmers,* Washington, D.C.: National Association of Broadcasters, 1989

Shane, Ed, *Cutting Through: Strategies and Tactics for Radio,* Houston, Texas: Shane Media Services, 1991

Webster, James G., and Lawrence W. Lichty, *Ratings Analysis: Theory and Practice,* Hillsdale, New Jersey: Erlbaum, 1991; 2nd edition, as *Ratings Analysis: The Theory and Practice of Audience Research,* by Webster, Lichty, and Patricia F. Phalen, Mahwah, New Jersey: Erlbaum, 2000

Deregulation of Radio

Eliminating Old Rules

Radio deregulation refers both to a specific Federal Communications Commission (FCC) proceeding (1978–81) and to a more general—and continuing—trend of dropping or modifying existing laws, rules, and regulations. Briefly, *deregulation* means to remove or significantly modify existing regulation, either through FCC administrative action or by congressional legislation.

Origins

Contrary to popular opinion, deregulation is not new. Indeed, the basic concept that less government is best is an old shibboleth evident in most aspects of American life. There is a deep-seated feeling that cuts across political lines (or most of them) that competition, rather than regulation, will lead to lower prices and higher-quality products or services. Nowhere is this truer than in an expanding industry with new players clamoring to enter the marketplace.

Combining this background with the economics of government in the late 20th century created the seedbed for radio deregulation. On all levels, government was operating at a deficit for much of the period after World War II. Federal deficits mounted annually, forcing Congress and the executive branch to consider ways of cutting costs—or at least of carefully assessing the benefits of new, let alone existing, rules and regulations. Paperwork reduction became a byword after the 1970s as government sought to root out rules that were no longer needed but expensive to maintain.

When deregulatory consideration was applied to broadcasting, it involved a basic review of the practical meaning of the Communication Act's concept of "public interest, convenience or necessity," which had guided FCC decisions since 1934. As society changed, so did at least some views of what government could or should accomplish. To varying degrees an ideological battle, there was at least broad agreement that government could no longer do everything.

In 1972 FCC Chairman Richard Wiley initiated a search for "regulatory underbrush" that could safely be eliminated without harm to broadcasters or their audiences—and within six years the commission had dropped or modified some 800 mostly minor rules. Many concerned small technical changes, and others reduced reporting requirements. This exercise set a larger process in motion.

Radio Deregulation Proceeding: 1978–81

In 1978 the National Association of Broadcasters (NAB), always seeking ways of reducing the burden of government on its member stations, petitioned the FCC to consider dropping four requirements that affected radio. These included processing guidelines (used by the commission staff to decide on the granting of new and renewal license applications) that were designed to limit on-air advertising and promote non-entertainment programming. In addition, the NAB wanted the FCC to drop formal program log requirements as well as the complex process known as "ascertainment," which required licensees to learn more about their community of operation (and reflect that knowledge in their programming).

The petition was a good example of perfect timing, for it paralleled the thinking of many staffers in the FCC's Broadcast Bureau. Prompted in part by the NAB petition for rulemaking, the staff undertook its own studies of the radio business, noting especially how many stations had taken to the air since many of the rules had been established. They also developed a somewhat complex economic policy model that served to question the continuation of the rules the NAB targeted.

In September 1979 the commission issued a *Notice of Inquiry and Proposed Rulemaking* concerning the deregulation of radio. Pragmatic in tone, the long notice (80 pages in tiny *Federal Register* type) conceded that with the removal of the processing guidelines on programming there "will be a tendency toward program duplication and imitation" (Paragraph 144). But it also asked whether this would matter, given how many stations were now on the air, including multiple signals in all but the tiniest towns.

Release of the notice brought forth a torrent of public reaction. In the several months allowed for public comments, some 20,000 were filed, filling shelves of notebooks in the Broadcast Bureau's public file room. They were often emotional—arguing that dropping program guidelines would lead to the elimination of religious or some other kind of minority-interest programming or that letting go of the advertising guideline would lead to a flood of commercials on the air. Some claimed that elimination of the logging requirement would remove a useful tool for those who watched closely how well stations performed. In addition, critics held that elimination of the ascertainment rules would lead to even more "plain vanilla" radio, which would sound the same no matter where a facility was located.

The sheer amount of filed comments took the staff nearly a year to process and consider. Early in 1981, after concerted internal debate, the FCC released a *Report and Order* dropping the four radio rules that the NAB had originally proposed be dropped. On the very day of the order's release, the activist office of communications of the United Church of Christ filed a court appeal and requested that the rule change be stayed pending a final decision. Not surprisingly, the broadcasting industry cheered the FCC decision, restating that the rules being dropped had little to do with program quality or service to listeners.

The U.S. Court of Appeals largely upheld the FCC rulemaking. It remanded for further action one piece of the decision—that concerning the dropping of program logs. The FCC had replaced the logging rule with a requirement that stations develop a list of community problems and programs aired that addressed those problems. The original order called for this to be done annually, and with the court's remand, the FCC made this a quarterly process, with station reports going into each station's own public files rather than being sent to Washington. The change passed muster with the appeals court.

Looking back two decades later, it is difficult to understand the emotions this proceeding created at the time. The issues now seem small and marginal, though at the time many critics saw the FCC decision as a watershed. For if the commission no longer concerned itself with the content that radio stations provided, what was the point of regulation, and what would happen to the industry? How could licenses be issued in the public interest if there was no longer any effective measure of what the public interest was? How could people complain about and seek to improve local station practices if the prime tools they had used previously (the station's local market ascertainment study or composite logs showing a typical week's programming and advertising) were eliminated? These critics argued that the new "problems and programs" listing would not be much use, certainly not in the way the old rules had.

Continuing Deregulation: 1980–2000

The 1981 rulemaking applied only to commercial radio. However, in parallel rulemakings commenced in 1981, the FCC eventually dropped the same four rules for noncommercial radio in 1984, and for television stations a year later. As many had predicted and others had feared, the radio deregulation proceeding paved the way for more substantial actions in the years that followed.

At virtually the same time, the FCC dropped its long and often complex license renewal form, replacing it with a mere postcard with a handful of easy-to-answer questions. Where stations had previously often filed a box of material, the simple postcard itself would often suffice now. Congress joined in the process, lengthening radio station licenses from three years to seven in 1981, and to eight in the 1996 Telecommunications Act. And beginning in 1982, licensees could buy and sell stations like any other property when the FCC dropped its "anti-trafficking" rule, which had required that licenses be held for at least three years before they could be sold.

Many rules were not as emotionally charged. Beginning in 1981 the FCC steadily reduced its former requirements on how much engineering expertise a station needed to maintain. Reversing its traditional approach, the commission argued that as long as a station was not creating interference to others, how good or how poor its own signal was would be better regulated by marketplace competition than by stiff rules. A station with poor-quality signals would rapidly lose audience and advertisers. A boon especially to smaller stations, the relaxed rules allowed them to share engineers or merely to have one on call rather than on the premises.

Higher on the emotional scale was the 1987 elimination of the FCC fairness doctrine. The decision made clear that licensees were the absolute authority on what issues and points of view they aired. Another shibboleth collapsed in 1992 when the FCC first allowed an owner to control more than a single station of each type (AM and FM) in a given market. This spelled the end of the long-established "duopoly" rule, created in the 1940s when there were fewer than 1,000 stations on the air.

The 1996 Telecommunications Act included provisions making it very difficult to challenge a broadcast license (such a process had briefly become a sport in the late 1960s and early 1970s, though few stations actually lost their licenses). The FCC would now have to find the incumbent licensee undeserving of a continued license before it could even consider a possible challenger.

The same act greatly expanded the number of stations anyone could own. In the 1940s the FCC had created de facto rules allowing ownership of no more than seven AM and seven FM stations nationally. As the industry expanded, so did pressure to raise those admittedly arbitrary limits. Finally, in 1985, the FCC increased them by five stations each—to 12 and 12. The limits rose again in 1992, to 18 and 18, and to 20 and 20 by 1994. With the 1996 Telecommunications Act, Congress eliminated any national cap on radio station ownership. By the turn of the century, the largest radio owners controlled nearly 1,200 stations.

Did all of this deregulation change the face of radio broadcasting? Certainly the economic and structural changes concerning ownership have considerably modified what was once a business of many small groups or individual owners. But the evidence remains inconclusive that elimination of the FCC's radio license processing and content rules two decades ago made any difference that competitive pressures would not have brought about anyway.

CHRISTOPHER H. STERLING

See also Fairness Doctrine; Federal Communications Commission; First Amendment and Radio; Licensing; Localism in Radio; Ownership, Mergers, and Acquisition;

Public Interest, Convenience or Necessity; Regulation; Telecommunications Act of 1996

Further Reading

Broadcast Deregulation, New York: Station Representatives Association, 1979
Federal Communications Commission, "Ascertainment of Community Problems by Broadcast Applicants: Primer," *Federal Register* 41 (7 January 1976)
Federal Communications Commission, *Inquiry and Proposed Rulemaking: Deregulation of Radio,* 73 FCC 2d 457 (6 September 1979)
Federal Communications Commission, *In the Matter of: Deregulation of Radio—Report and Order in BC Docket* 79-219, 84 FCC 2d 968 (14 January 1981)
Office of Communication of the United Church of Christ v FCC, 707 F2d 1413 (D.C. Circuit, 1983)

Desert Island Discs

BBC Music Program

The *Guinness Book of Records* states that *Desert Island Discs* is the longest-running music program in the history of radio. The BBC Radio 4 program is the third longest-running radio program in the world, after *The Daily Service* (1928) and *A Week in Westminster* (1929). The program was devised and copyrighted by Roy Plomley, who began his broadcasting career in 1930s commercial radio on the station Radio Normandie. The original program was expected to be a series of six episodes with the first being transmitted on 29th January 1942. Since that first episode, thousands of celebrities—from members of the British royal family to prime ministers and stars of stage, screen, and television—have taken part in this long-running hit program that takes the form of a sort of parlor game. In the game, guests must talk about themselves, select eight pieces of music, and imagine that they are to be stranded on a desert island Robinson-Crusoe style, taking with them only a book, a record, and some luxurious inanimate object.

Culturally, *Desert Island Discs* represents a history of British social convention and the British perception of "celebrity." It is also a symbol of British public radio's function throughout the second half of the 20th century and continuing into the 21st. Certainly its longevity has been used by the British Broadcasting Corporation (BBC) to symbolize the BBC's role in British popular culture and social life. A BBC Television 2 program produced by Alan Yentob in 1984 marked the *Desert Island Discs'* 40-year anniversary, and Monica Sims, then BBC Radio 4 Controller, was filmed at a reception for members of the U.K. arts community proclaiming: "All of us, whether we are eminent or not, all feel we are potential castaways. We all have eight favorite records. . . Roy had always tried to get the best out of people." Ten years later the same television program was repackaged with film of the third host of the radio show Sue Lawley with the then Prime Minister John Major as castaway. This served to symbolize the BBC's importance in public life as well as ingratiate the Prime Minister by offering a soft and personal side to his character. As his castaway luxury he selected the Oval Cricket ground with a bowling machine. From 1988 Sue Lawley and her producer Olivia Seligman introduced a political and current affairs dimension to the program's content. During Roy Plomley's direction politicians were a rarity largely because "There is a drawback in inviting politicians to broadcast. . . . Unless it is in a news or political program, one is not allowed to discuss politics, and there doesn't seem much point in interviewing a politician if you can't discuss the subject dearest to his heart."

The change in political edge after Plomley's death was also reflected in the program's editorial migration from light entertainment to news and current affairs. The emphasis on news-related interviews served the BBC's objectives at that time to achieve greater public relations with its broadcasts and to increase audiences. It could also be argued that this trend reflected the developing notion of politician as celebrity. Politicians or individuals with controversial political views would inevitably prefer the *Desert Island Disc's* format because its structure of music selection was more entertainment orientated than politically focused.

In his book on the program, Roy Plomley theorized about the reasons for the program's success: "I believe *Desert Island Discs* adds a dimension to a listener's mental picture of a well-known person, giving the same insight he would receive from visiting the celebrity's home and seeing the books, pictures and furniture with which he surrounds himself." Sir Paul McCartney said the program "conjures up traditional British pleasures like the Great British Breakfast, Billy Cotton's Band Show—

very downbeat, very relaxed. I love its homeliness." The program's first producer Lesley Perowne said, "The reason people liked it so much is curiosity. I think everybody wants to know the private tastes of public people and this was a very good way of doing it." The host since 1988, Sue Lawley, believes the program's success is based on "marrying music with conversation and thereby creating life." A cost-conscious BBC executive who had to concede the copyrighting of the format to Roy Plomley asked "Why didn't we think of it before?" Comedian Arthur Askey, who holds the record for being the only four-time castaway, stated, "It's such a simple idea. That's part of its success. It's a wonder that somebody hadn't beaten Roy to it." Conceptually, radio throughout the world has established similar programs because of the popular appeal of the structure and the method of satisfying listener curiosity. But the endurance of the program on the BBC Home Service and then BBC 4 is attributable to the social stability of the audience, the continuity of license fee funding, and the need to maintain the audience-drawing component in cultural speech programming. The program is aired in the U.S. on some public radio stations. The BBC has a longstanding policy of syndicating its programming in partnership relays with public networks such as PRI and NPR.

Despite its light atmosphere, which was always encouraged by Roy Plomley's ritual of lunching guests at the Garrick Club before their interviews, there have been moments of poignancy and profound revelation of character. The concert pianist Artur Rubinstein said that despite playing in practically all the countries of the world, he refused to play in Germany: "I don't go to Germany out of respect for the dead. Unfortunately among the dead is my whole family." This contrasts with the bitter controversy of Sue Lawley interviewing the widow of British anti-Semite and Fascist leader Oswold Mosley in 1989; Diana Mosley used the program to express disbelief that the Nazis had murdered 6 million people.

The program has also produced unforgettable moments of drama and wit. When Tallulah Bankhead was asked how good a Robinson Crusoe she would be, she replied, "I can't even put a key in the door darling. I can't do a thing for myself. I never stand up if I can sit down and I never sit down if I can lie down." When British comedian Frankie Howard was asked how he would endure loneliness he replied, "It's better than the alternative"; when Plomley asked, "What's that?" Howard replied, "Dead."

John Kenneth Galbraith responded to the same question with "The whole idea doesn't appeal to me at all. I'm not especially gregarious. I can get along with my own dismal personality for a little while. I would hate to endure it for any length of time."

Other castaways had a pessimistic view of the isolated life on a desert island. Artur Rubinstein selected as his inanimate luxurious object a loaded revolver, Maureen O'Sullivan asked for tranquillizers, and Jonathan Miller wanted to take with him a cutthroat razor.

In March 2002 the BBC sought to celebrate 60 years of the program with an anniversary celebration gala of music at the Royal Festival Hall in London. The event was recorded for radio and television transmission. Presenter Sue Lawley introduced a selection of music chosen by castaways down the years before an audience of 2,500 people. However, the event did not turn out to be a public relations success. *The Sunday Telegraph* called it "The Gala from Hell" and the reviewer for the *London Times* pointed out that the complex pageant defied the secret of the program's success, which was simplicity. Despite these negative reviews, the BBC could argue that celebrating 60 years of *Desert Island Discs* in this way generated media coverage that has helped embed the program into popular cultural folklore.

TIM CROOK

Presenters
Roy Plomley, 1942–85; Michael Parkinson, 1986–88; Sue Lawley, 1988–

Producer/Creator
Roy Plomley

Producer
Lesley Perowne, Olivia Seligman, Angie Nehring

Programming History
BBC Radio 4 29 January 1942–present

Further Reading

Briggs, Asa, *The History of Broadcasting in the United Kingdom,* 5 vols., London and New York: Oxford University Press, 1961–95; see especially vol. 3, *The War of Words,* 1970

Donovan, Paul, *The Radio Companion: The A–Z Guide to Radio from Its Inception to the Present Day,* London: Grafton, 1991

Plomley, Roy, *Desert Island Discs,* London: Kimber, 1975

Plomley, Roy, *Days Seemed Longer: Early Years of a Broadcaster,* London: Eyre Methuen, 1980

Plomley, Roy, *Plomley's Pick of Desert Island Discs*, London: Weidenfeld and Nicolson, 1982

Snagge, J., and Barsley, M., *Those Vintage Years of Radio,* London: Pitman Publishing, 1972

Developing Nations

Radio's Role in the World's Poorer Regions

Even in these days of instantaneous satellite communications and expanding cyberspace connections within and outside of the world's poor and industrializing nations, radio remains the medium with the largest global audience because of its ability to reach diverse and remote rural populations, the illiterate, those with little access to education, and those who have no electricity. Cheap radio sets are within the reach of even the poorest communities, if only at a local gathering place. As a result, radio has been and continues to be used extensively for development purposes, clandestinely by dissidents seeking to circumvent or overthrow governments, and more popularly as a vehicle for the dissemination of cultural and informational programming.

Historical Underpinnings

Development Radio began in most Third World countries while they were still colonies of one of the European powers. Colonial systems were operated largely for colonists, and closely followed the pattern and programs of radio "at home." Such facilities became the initial basis of national radio systems as colonies became independent nations.

In the decades following World War II, the major industrial powers divested themselves of most of their colonies. Radio has been used in these new nations in both crude and sophisticated ways to educate people, to propagandize them, and/or to involve them in grassroots development projects. In the 1960s the traditional practices of peasants in the developing world were targeted for change by Western development experts interested in the diffusion of innovations in agriculture, in education, and in industrialization. Their assumption was that use of radio and other media would assist and speed up the broader development process.

Persuasive communications were seen as key to the adoption of technological innovations, which were to lead directly to individual and national development. The introduction in the late 1950s of transistorized radio—cheap, portable, and not requiring electricity—made it possible to broadcast programs about what were seen as modern practices that would help the masses break free of stagnant traditionalism.

Since then, however, much research has refined and changed this simplistic and value-biased dominant paradigm of what development is and how it is to be achieved; research has also indicated the limitations to what were previously seen as direct, powerful, and uniform media effects on individuals. In the 1970s, for example, Third World scholars, with what was to be termed a "dependency" view, took issue with the style and manner of Western-dominated development programming. External variables such as neocolonialism, top-down decision making, and lack of basic needs were taken into account, as were internal structural and political constraints such as land tenure systems and inefficient or corrupt government bureaucracies. Consequently, by the 1980s much radio programming addressed basic development projects designed to extend scarce resources and educational opportunities to remote and rural areas. Other factors in this change included the active participation of people at the grass roots; attempts at equity in the distribution of information; and the meeting of basic needs such as food, clean water, and education.

At the root of these newer perspectives was the recognition that the mass media in general, though still an important component in development, was not a magic means for implementing change. In addition, with illiteracy still rampant, print media were limited in their usefulness. Television was expensive to produce and disseminate. Computers with internet connection are for the privileged few. So radio, with its portability, low cost, and its accessibility for the illiterate and uneducated, remains the medium of choice for developing countries. Although much work has gone into using television, traditional media, and newspapers, development programming has focused on radio. This inexpensive and flexible medium continues to be the most frequently used of all the mass media in development projects. Radio is by far the most diversified and dispersed of the media, and its programming is the least expensive to produce.

Government Initiatives

The first act of government in new nations liberated from colonialism often was to set up national radio stations, which were seen as an integral part of propaganda and education efforts. These stations, typically in the national capital, are still seen as so important that they are often the first facilities to be occupied by rebels in any attempts to overthrow governments.

In addition to exploiting radio's potential as a conduit for propaganda, Third World governments have used radio to stimulate national integration in former colonies that have artificial borders encompassing many ethnic and language groups. National networks can simultaneously relay national news and programming in different languages across vast distances. Programming of national sporting events, or of traditional and indigenous music, is seen as a powerful integrator and reinforcer of national identity.

Radio broadcasting also has been used at both national and regional levels to disseminate information about government development projects. Official messages are transmitted countrywide in many local languages. Programming can run the gamut from official news and policies, to information about agricultural innovations, to music programs.

Often, educational programming on radio is supplemented by other media; for example, a program about family planning can be followed up with printed matter and audiocassettes and with visits by government workers to reinforce the initial message and to organize community discussions to build consensus. In addition, social marketing organizations can broadcast entertaining forms of programming. Mini soap operas, for example, have been used to promote nutritional beverages that counter dehydration in infants with diarrhea.

Considerable programming has been produced for rural farmers in an effort to motivate them to adopt specific agricultural innovations, such as the use of improved corn seed and fertilizers.

In recent years, however, as world lending institutions have insisted on open markets and democratization, many of these Third World government radio systems have been opened to commercial interests, which has resulted in less programming for development purposes.

Private Initiatives

Although governments have generally used radio to inform their populations about development projects, by contrast, development media planners in recent years have turned to community media rather than national systems. Programs are designed and carried out by members of the community. In the late 1970s, for example, project planners turned to farmers to produce programs for other farmers. In one such project, tape recorders and blank tapes were provided to volunteer workers, who in turn helped local people produce items for weekly broadcast. Some of these initiatives, however, had political ramifications in authoritarian countries and were therefore short-lived or heavily censored.

At the turn of the millennium, radio initiatives funded by international aid organizations such as UNESCO predominate. The creation of local rural radio stations coincided with the deregulation of national telecommunication monopolies, especially in Africa. Some current examples include the following. In Kenya, an "English in Action" program provides quality programming for secondary schools with few resources. In Papua New Guinea, a radio science project helps teachers improve basic education. In Mongolia, an informal distance education system provides learning opportunities to the nomadic women of the Gobi desert. In Suriname, community stations encourage dialogue among women and link communities in the interior through a network of interactive telecenters. In the Philippines, local radio allows people to express themselves on political, economic, and cultural subjects. In El Salvador, women tune in for a broadcast that informs them about their rights and encourages their involvement in the community. And in Somalia, where most media operations were destroyed during the civil war and those that remained were clan owned, a nonpartisan media organization produces programming on peace issues.

Radio as Popular or Dissident Medium

Audiences living under authoritarian rule continue to tune in to clandestine radio broadcasts by dissident or revolutionary groups seeking to overthrow a government. The aim is to persuade citizens and elicit their participation in the cause, as well as to damage the legitimacy of a regime. However, pirate stations usually have very little reach and must constantly move operations to avoid imprisonment or even execution.

Listeners who live in countries without a free press depend on external broadcasting to get their news. In authoritarian regimes in Africa, for example, listeners with shortwave radios still turn to the British Broadcasting Corporation (BBC) for reliable and accurate news about internal unrest. Before the end of apartheid in South Africa, broadcasts from the bordering nations of Zimbabwe, Namibia, and Botswana kept the oppressed South African majority informed of world condemnation of apartheid and of dissident activities.

Those Third World countries with a free press often also have lively programming on both national and private radio stations. Call-in shows are a particular favorite, although participants are limited to those living in or near cities. Jamaica's talk shows are one lively example.

Finally, as new computer technologies allow radio broadcasting over the internet, new concepts of radio programming are being formulated. For example, the Feminist International Radio Endeavour, based in Costa Rica, has for the past decade produced daily live shortwave broadcasts for and about women in the developing world. In 1999 Realaudio programs in English, French, and Spanish were made available on the internet. In addition to these non-profit programs, private commercial initiatives such as WorldSpace are sending digital audio programs via satellite.

MELINDA B. ROBINS

See also Africa; Arab World Radio; Asia; Brazil; Digital Audio Broadcasting; India; Mexico; Shortwave Radio; South America; South Pacific Islands

Further Reading

Casmir, Fred L., editor, *Communication in Development,* Norwood, New Jersey: Ablex, 1991
Hornik, Robert C., *Development Communication: Information, Agriculture, and Nutrition in the Third World,* New York: Longman, 1988
Jamison, Dean T., and Emile G. McAnany, *Radio for Education and Development,* Beverly Hills, California: Sage, 1978
Katz, Elihu, and George Wedell, *Broadcasting in the Third World: Promise and Performance,* Cambridge, Massachusetts: Harvard University Press, 1977
Maherzi, Lofti, "Radio" in *World Communication Report,* 2nd edition, by Maherzi, Paris: UNESCO, 1997
Melkote, Srinivas R., *Communication for Development in the Third World: Theory and Practice,* New Delhi and Newbury Park, California: Sage, 1991

Diary

Method of Audience Research

As the name suggests, a diary is a paper booklet in which a person is asked to record his or her listening to radio or television programs, noting when listening started and stopped, which station the set was tuned to, and other comments. A diary is typically used to record one week of listening, then returned for tabulation. Radio audience estimates, usually called ratings, are based on the tabulation of information obtained from these diaries.

Origins

In the medium's early days, many radio set builders and listeners were not interested in hearing specific programs so much as listening to as many different stations as possible. This "channel-surfing" was called DXing—an abbreviation for distance. DXers kept track of the stations they heard in log books that recorded when they heard a station, the frequency and/or call sign, slogans, programs, and the city of origin. Some computed the distance to the stations they heard, and there were contests sponsored by radio clubs and radio magazines with prizes given to those who compiled the most stations and greatest total distance. Although the diary method existed in the early days of radio, until the rise of television diary-keeping was not the mainstay of radio audience measurement.

The first systematic audience research using diaries was done by Professor Garnet Garrison at Wayne (later Wayne State) University in 1937, though he called it a "listening table." Garrison, who later taught for many years at the University of Michigan, was working on an "experiment developing a radio research technique for measurement of listening habits which would be inexpensive and yet fairly reliable." He noted that other methods most widely used at the time were the telephone survey, either coincidental or unaided recall, personal interviews, mail—sometimes called fan mail—analysis, surveys, and "the youngster automatic recording." He said that he had borrowed something from each method; since the listening table could be sent and retrieved by mail, it included a program roster, and was thought to be objective. The form he used was a grid from 6 A.M. to midnight, divided into 15-minute segments, that asked respondents to list stations, programs, and the number of listeners. Garrison concluded that "With careful attention to correct sampling, distribution of listening tables, and tabulation of the raw data, the technique . . . should assist materially in obtaining at small cost quite detailed information about radio listening." While his methodology was not adopted for about a decade, Garrison's "listening table" is essentially the way radio audience estimates are obtained to this day.

The Columbia Broadcasting System experimented with diaries in the 1940s but apparently thought the data was most applicable for programming research, for which it also used the program (or Lazarsfeld-Stanton) analyzer. CBS used information from diaries primarily to track such things as audience composition, listening to lead-in and lead-out programs, and charting audience flow and turnover. In the late 1940s, C.E. Hooper also added diaries to his telephone sample in areas that could not easily be reached by telephone. But this mixture of diary and coincidental data was never completely satisfactory. Indeed, one of the reasons for Nielsen's Audimeter winning out over Hooperatings was that the telephone method was confined to larger metropolitan areas, where TV first began to erode the radio audience. Hence, Hooper (unlike Nielsen) tended to understate the radio audience and therefore quickly lost the support of radio stations.

Arbitron Diaries

It was not until the end of the 1940s that diaries were introduced on a large-scale basis for providing syndicated audience research. James Seiler, director of research for the National Broadcasting Company's station in Washington, D.C., had for several years proposed using diaries to measure radio. NBC finally agreed to try a survey, not for radio, but for its new TV station in the market, agreeing to help pay for several tests. Seiler set up his own ratings service company in Washington and called it the American Research Bureau. He thought the name sounded very official, even patriotic. Later the name was shortened to ARB, and then to Arbitron when instant television ratings in larger cities were gathered electronically.

The American Research Bureau's first report was based on a week-long diary that covered 11–18 May 1949. By that fall, the company was also measuring TV viewing in Baltimore, Philadelphia, and New York. Chicago and Cleveland were added the next year. In spite of covering more markets the company grew slowly as both TV and the new diary method gained acceptance. Diaries were placed with TV viewers identified by random phone calls. From the beginning, Seiler was careful to list the number of diaries placed and those "recovered and usable." Also, "breakdowns of numbers of men, women, and children per set for specific programs [could] be furnished by extra tabulation."

Tele-Que, another research company, began diary-based television ratings in Los Angeles in 1947. In 1951 Tele-Que merged with ARB, thus adding reports for Los Angeles, San Francisco, and San Diego. During the 1950s ARB emerged as the prime rival to Nielsen's local TV audience measurement, especially after 1955 when it took over the local Hooper TV rating business. By 1961, ARB was measuring virtually every TV market twice a year and larger markets more often.

Local radio reports using the diary method were begun by Arbitron in 1965, nearly a quarter of a century after Garnet Garrison had recommended the method for the audio medium. Arbitron quit the TV measurement business in 1993 and now confines itself to radio.

The diary used by Arbitron today is not much different from the one used more than 50 years ago for television stations. Arbitron now measures more than 100 radio markets continuously and provides monthly ratings for about 150 areas for eight or more weeks each year. Diaries are still placed by phone call, then sent and retrieved by mail. A diary is sent to each member in the household who is 12 or older. The diary format asks the respondent to indicate each time she starts and stops listening to the radio, the call letters or station name, and suggests that if she is not sure of the call letters or station name, she should write in program name or dial setting. Respondents must also indicate whether the station is AM or FM and whether they are listening at home, in a car, or some other place. At the back of the diary are questions about age and gender for audience composition tabulations and other questions, typically on product usage.

While most diaries are still placed by telephone calls, special care is taken to place diaries personally in Spanish-speaking homes and residences in high-density ethnic areas. After careful editing of the diaries' listening reports, audience estimates are published for each market in a "ratings book" and are available online to subscribers for other detailed analysis. In transferring diary entries into computer data, the operators have a number of aids and checks in the computer program that allow them to check the accuracy of call letters in each market and other information. Some radio programmers like to go to Arbitron offices near Washington, D.C., to study the diaries. Images of all entries are available and can be sorted to observe them in many different categories. By examining actual diaries, programmers or a consultant hired by a station can determine whether people remember call letters or station slogans correctly. Often diary keepers write other comments that might be helpful. More detailed statistical analysis is possible by consulting the ratings book online, which allows subscribers to tabulate persons in the sample by any or all demographic categories. Such manipulations of data allow computation showing favorite station, sharing listeners with other stations, audience in zip code areas, the time spent listening to one station, and other categories, to name just a few.

Advantages

As envisioned by Garnet Garrison, diaries offer some significant advantages that account for their popularity. They offer a relatively inexpensive method for gathering a lot of information over the weekly period. But there are problems associated with the method. Responses—the rate is reported in each rating book—are often from only half of the sample. Younger males, for example, have a low return rate. Since the listeners who are more likely to keep a diary and provide accurate information are also likely to listen to some formats more than others, there is continuing controversy about rating results. Recently the growing use of telephone answering machines, cell phones used out of the home, and other factors make it harder to obtain diary keepers. Nonetheless, millions of diaries recording radio listening and TV watching are processed each year, and the broadcasting industry, advertisers, and advertising agencies depend on (and pay a high price for) the information obtained from a very simple little book that has been around for quite a long time.

How much longer the diary method will be used is not clear. Arbitron, in cooperation with Nielsen Media Research, is testing a small personal recorder that people might carry to keep tabs on all the wearer's electronic media use.

LAWRENCE W. LICHTY

See also A.C. Nielsen Company; Arbitron; Audience; Audience Research Methods; DXers/DXing; Hooperatings; Lazarsfeld, Paul F.

Further Reading

Beville, Hugh Malcom, Jr., *Audience Ratings: Radio, Television, and Cable,* Hillsdale, New Jersey: Erlbaum, 1985; 2nd edition, 1988
Chappell, Matthew Napoleon, and Claude Ernest Hooper, *Radio Audience Measurement,* New York: Daye, 1944
Lumley, Frederick Hillis, *Measurement in Radio,* Columbus: Ohio State University Press, 1934
Webster, James G., and Lawrence W. Lichty, *Ratings Analysis: Theory and Practice,* Hillsdale, New Jersey: Erlbaum, 1991; 2nd edition, as *Ratings Analysis: The Theory and Practice of Audience Research,* by Webster, Lichty, and Patricia F. Phalen, Mahwah, New Jersey, and London: Erlbaum, 2000

Digital Audio Broadcasting

Replacing Analog Radio Stations

Several different digital radio standards or systems are being operated in different parts of the world. The system that has been in development the longest and that is in full-time operation in the most countries is called EUREKA 147. In the United States, which only selected a national digital standard in 2002, the phrase "high-definition radio" was coming into use by 2003 to suggest a parallel with developments in digital television. This entry surveys digital audio broadcasting (DAB) developments in Britain, Scandinavia, and the United States, with reference to other regions of the world as well.

DAB Basics

Digital audio broadcasting may be seen as the third stage in the use of the electromagnetic spectrum to transmit radio broadcasting services after analog AM and FM, both of which are prone to interference. In AM's case this is caused by static and other unwanted signals, by sky waves reflected from the ionosphere, and by other stations on the same or nearby frequencies. FM's main problems stem from unwanted reflections from high-rise buildings and other objects that cause what engineers term multipath distortion.

There are probably about 2.5 billion receivers with AM or medium wave reception capability in the world. Despite the expansion of FM, AM remains the most widely available means of reception. Both national and international radio broadcasters continue to rely on AM for much of their transmission requirements. The capacity of AM's medium waves to reach beyond horizons and thus much further than FM transmitters makes it essential and irreplaceable. For many countries, DAB is too expensive because a complex and extensive transmitter network is required.

DAB represents a major break with this analog technology. Like all digital systems, it converts the original material into streams of "zeroes" and "ones", which are then reconverted to recreate the original information. DAB, and specifically the EUREKA 147 system, differs from earlier digital systems in being capable of transmitting over a number of different "platforms," including both terrestrial (land) and satellite (delivering services either separately or jointly) and over a large section of the electromagnetic spectrum: from 30 MHz to 3 GHz for mobile reception, and higher for fixed reception.

DAB can be accomplished in any of three ways: (1) in-band, on-channel (IBOC) or in band, adjacent channel (IBAC) using existing AM and FM terrestrial frequencies (actually blank spaces between frequencies); (2) terrestrially over another broadcast band (S-band in the United States and L-band elsewhere); and (3) by satellite digital audio radio services (SDARS), which bypasses terrestrial broadcasters by sending signals directly to consumers. (*See separate entry,* Digital Satellite Radio.)

The main advantages of DAB, compared with analog radio broadcasting, are many. First, DAB produces a much closer (technically accurate) replication of the original sound reproduced by the receiver, together with easier/automatic tuning than analog techniques.

Second, DAB is more efficient in its use of the radio spectrum and has a lower power requirement. Using the "multiplex" system, a number of radio transmitters carrying multiple signals can "overlap" on the same frequency—if broadcasting the same material. A "single frequency network" means that only one frequency needs to be used to cover a wide area—including a whole country. This compares well with the amount of spectrum needed for analog services. The overall digital signal is produced by 1,536 "carrier frequencies" that are distributed over a 1.5-megahertz band. The majority of these carriers are noise-free, and this, coupled with error-correction techniques, means there is no interference to any of the services.

Third, there is at least the potential for many more services in the same spectrum, compared with analog transmissions. This is because the total digital "bit rate" available on the single frequency—1.2 megabits—can be "sliced" in an almost limitless number of configurations, so more services can be "squeezed" into the same part of the spectrum. However, there is a widely accepted minimum bit rate per second (bp/s) thought to be necessary to provide acceptable quality of either mono or stereo transmissions, and a lower rate to be used for text, graphics, and pictures without causing distortion or interference.

Finally, material other than sound can also be transmitted with the "radio" signal, because many other types of media can also be converted and then reconverted in roughly the same way. Thus digital methods can transform sound broadcasting into a multimedia system.

The main *disadvantages* of digital radio broadcasting are the following: first, a variety of different systems are being developed (unlike AM or FM, there is as yet no agreement on a worldwide standard); second, one transmitter serving a number of different radio services on one multiplex means that less well established, independent organizations are likely to be excluded; finally, listeners will have to buy new radio receivers that are at least initially significantly more expensive than those for analog systems. Indeed, the total consumer expense for this technology will be vastly larger than what stations have to pay out.

EUREKA 147

The earliest digital radio system to reach adoption was the European EUREKA standard. The main reason for its development was to provide a new, distinct outlet for the European consumer electronics industry, which had been overshadowed by those in Japan and other "tiger" economies of the Far East. The name "EUREKA 147" derives simply from the prosaic fact that it was the 147th system to be developed under the umbrella of the EUREKA project, which was launched by 17 countries and the European Commission in 1985.

Work on a possible DAB system began in 1981 at a research institute, the Institut für Rundfunktechnik. The original base of the EUREKA 147 project and its main participant was a (West) German research institute—the DLR Projekttraeger Informationstechnik in Cologne. Both Germany and France contributed 36 percent of the costs of the project, the United Kingdom just 6 percent.

By 2003 the World DAB Forum indicated that some countries around the world were operating a DAB system using the EUREKA standard. Outside Europe, the system is particularly well developed in former British colonies, notably Singapore, Hong Kong, Australia, Canada, and India. China is also operating the system, and thus the two most populous countries in the world have committed themselves to the "European" standard. One of the main advantages claimed for EUREKA is that it can be transmitted on all the main broadcasting platforms, terrestrial and satellite—the latter clearly important for covering large and sparsely populated areas, such as Canada and Australia, and over a large number of frequency bands.

Digital Radio Mondiale

Although EUREKA is undoubtedly the most developed and operationally established DAB system, several others are either in operation or are close to being so. A system developed by a private United States–based company, WorldSpace, designed for international broadcasting to developing countries by satellites, became fully operational in 1999, but the system requires new transmitters and, most significantly, new and completely different radio receivers. There is also a system for transmitting shortwave in digital form, called Telekom-TELEFUNKEN-Multicast, which was developed in Germany. These systems can be transmitted alongside current analog services on the same parts of the radio spectrum and may be converted by receivers requiring relatively little modification.

Another system comes from a consortium called Digital Radio Mondiale (DRM) and is designed primarily to complement AM broadcasting; this system was originally also developed as part of the EUREKA projects. DRM is a worldwide consortium of broadcasters, receiver and transmitter manufacturers, transmission companies, regional broadcasting unions, and research institutes. It is coordinating the development, standardization, and market roll-out of the system from 2003 onward. While the original drive came mainly from international shortwave broadcasters, the consortium now involves both domestic and international radio companies. Beginning in 1998, work proceeded very quickly and system development was completed during early 2001. In April 2001 the International Telecommunication Union (ITU) approved the DRM system for analog AM broadcasting. The DRM system overcomes the many defects of analog broadcasting—especially its often poor sound quality, fading, and interference. It is capable of greatly increasing the audio and reception quality, making it similar to good FM.

As with other digital technologies like DAB and the WorldSpace system, provision of supplemental program information or other information independent from the program is possible. It will also be possible to manufacture receivers in such a way that there will be no more need to search for alternative frequencies. The name of a station would be sufficient to receive the wanted program. The receiver will automatically retune and change the frequency to the best one for the respective station chosen. This will be especially important for the many stations in Africa, Europe, and Asia that operate on several different AM frequencies.

DRM was established as a worldwide consortium to develop a single system for digital broadcasting in the frequency bands below 30 MHz. The formal inauguration of the group took place in Guangzhou, China, in March 1998. It is a not-for-profit organization registered in Switzerland with a single goal: to bring affordable, digital sound and services to the world radio market. By mid-2001, DRM had 69 members from 27 different countries. A project office is located in Geneva.

A second proposal, from the United States, was also designed specifically for medium wave broadcasting. Developed by USA Digital Radio (which later merged with Lucent to form iBiquity), it takes a different approach from DRM, requiring 30 kHz high frequency channels. DRM and USA Digital Radio were cooperating so that listeners could receive both systems with the same receiver. However, since the merger with Lucent this cooperation has come to a halt, and it will be up to the receiver manufacturers to decide which system will be implemented. Only the DRM system fulfils all the ITU requirements for long-, medium- and shortwave, and will therefore be the replacement of the analog broadcasting system below 30 MHz in the future.

DAB in the United Kingdom

In the United Kingdom, the development of DAB using the EUREKA 147 standard has been slow—certainly in the commercial sector. The BBC began full test transmissions in September 1995. Within the frequency spectrum allocated by the U.K. government to DAB (217.5–230 MHz), there is room for seven frequency blocks; two of these have been allocated for national radio coverage, one for Independent National Radio services and one for BBC national services. The remaining five multiplexes have been allocated for local/regional radio services, to be awarded by the Radio Authority (which regulates and licenses U.K. commercial radio).

The national commercial multiplex, run by a consortium called Digital One, carried up to ten services—simulcasts of the three national commercial stations, plus a classic rock station, a spoken word service featuring readings of books and plays, and a middle-of-the-road music station. The roll-out of Digital One's network of transmitters happened faster than those for the BBC's national DAB multiplex—quickly reaching 85% of the U.K. population. However, there were setbacks, as the suppliers of a radio version of the Independent Television News (ITN) rolling news service, and of finance and business channel Bloomberg, pulled out of the multiplex, which meant that by the end of 2002 two channels were broadcasting nothing more interesting than "tone." In addition to the national multiplexes, over 45 licences for local and regional services were expected to be issued by the beginning of 2004—each offering a "bouquet" of services, including simulcasts of FM and AM services. Of the "digital only" services, typically about three of these would be stations broadcasting in other parts of the country as well as, usually, two or three which have been created specifically for the DAB system in that area. In all, over 300 digital services were being offered on U.K. DAB transmitters.

There was a huge incentive for the commercial services to pay for their analogue services to be available on the DAB "platform": if they did so, the 1996 Broadcasting Act compelled the sector's licensing body and regulator, the Radio Authority, to automatically renew their analogue licenses, which would otherwise be re-advertised every eight years. A Communications Bill—scheduled to become law by fall 2003—extended a further renewal period to 12 years, meaning that those investing in DAB transmission would have their analogue licenses on a 20 year lease. With FM licenses at a premium and otherwise liable to intense competition, the cost of simulcasting on DAB seemed a relatively small price to pay for the knowledge that their main services were secure.

An amendment to the 1996 Act of Parliament allowed the commercial operators to double the percentage of the multiplex "cake" used for non-audio services. This opened up increased possibilities for the development of commercial text-based services. Unlike the internet, the number of consumers downloading such services on DAB receivers is unlimited and does not increase the cost to the supplier, so this seemed likely to be an attractive additional source of revenue for the commercial operators.

Although no such financial incentive was on offer to the BBC, the Corporation's enthusiasm and commitment to DAB remained undimmed—although some senior managers were on record stating that, with hindsight, they had probably invested in it too quickly and too heavily. By the end of 2002 the BBC's national multiplex was offering simulcasts of its five national networks; plus the BBC World Service; an extra sports channel; a new music channel—6 Music—featuring rock sessions from the BBC's archives and pitched at the "young middle aged" audience; BBC 7 playing archive comedy and drama programs plus original, new programming, for children; and the BBC Asian Network. All of these include scrolling text showing program information and news and sports headlines as appropriate.

Despite the proliferation of program services, however, the take-up of digital radio receivers was slow and patchy, with industry estimates that only about 60,000 DAB sets had been sold by the summer of 2002. The main reason for the apparently unenthusiastic response from the listening public was price—manufacturers and campaigners for DAB were caught in a classic economic dilemma: where only a relatively few consumers were prepared to pay several hundred pounds for a DAB receiver, large-scale, and therefore cheaper, production of the electronic chips and sets required mass sales. The early models for home use were designed for the high end of the consumer hi-fi market and therefore appealed to a relatively small number of enthusiasts with high disposable incomes.

The introduction of the first sub-£100 portable (designed only for plug-in use) receiver in the middle of that year led to a more than doubling of DAB radio purchases in the crucial run-up to Christmas. Indeed, consumer demand far outstripped supply with customers being put on waiting lists. At least one manufacturer brought out a portable combined DAB radio and CD player, and DAB radios available for automobiles expanded and became cheaper, lighter, and easier to fit.

By the spring of 2003 at least two small, battery-powered sets using third generation DAB chips were on the market. The integration of DAB receivers into cell-phones seemed imminent (several models already incorporated an FM receiver, with 2 percent of adults listening to radio via their cell-phone) and key manufacturers were promoting DAB receivers. Crucially, the BBC began to heavily promote its DAB (only) services, with program listings included in the best-selling magazine the *Radio Times*—published by the Corporation's commercial arm—and a major on-air promotion on both radio and TV in the summer of 2003. Continued support for DAB from national government was made explicit when the Secretary of State for Culture, Media and Sport (at a meeting of the Social Market Foundation in March 2003) praised the BBC and commercial sectors for putting aside their normal rivalries for a common development and promotion of DAB and said this had helped ensure that digital radio was not only "a great British success story" but that the U.K. was the most advanced country in Europe in its development. (In fact, Sweden could probably claim that title, certainly in terms of population coverage.) The Minister also opened the possibility of a date for switching off analogue radio transmitters.

One rather unexpected phenomenon was a considerable increase in radio being consumed through television. All the digital TV platforms—terrestrial, satellite, and (most) cable—allowed a range of radio services, usually including all the national DAB services. Most significant of these was probably the re-launched national TV digital terrestrial service—with financial backing by the BBC—which allowed consumers to watch and listen to a range of free-to-air services received through the "normal" TV aerial. This meant that hundreds of thousands of consumers, who were reluctant to pay for satellite dishes and subscription fees and were certainly unlikely to be "early adopters" of DAB radio sets, were now exposed to the DAB stations. By the end of 2002, 16 percent of U.K. adults—nearly 8 million—were listening to radio services via their TV.

All the national DAB services—BBC and commercial—were "live streamed" on the internet. Although reliable figures for individual services were hard to come by, it was estimated that around 12 percent of adults listened to "streaming" radio via the internet in a typical week, although much of this was to services originally created for analogue radio. Several DAB receivers were developed for plugging into computers. Indeed, some of the DAB non-audio services were developed specifically for this sort of integrated use. As well as seeing the scrolling text and program information available on "normal" DAB radios, a range of other information and web links were also presented on screen.

The link with the internet and the increased possibilities of interactivity between broadcaster and listener, as well as the integration between the different media, seemed likely to be a trend that was both increasing and intriguing. In early 2003 the MXR consortia—a joint venture between Chrysalis Radio, Capital Radio, Jazz FM, and the Guardian Media Group—announced an agreement with U.S.-based technology firm Command Audio to launch "on-demand audio." The facility will allow listeners to their DAB multiplexes to listen to local news, sports, travel, and business bulletins whenever they want. The ability for individual DAB listeners to "rewind," to "time-shift" programming elements, had for some time been one of the promised benefits of the technology.

All of this raised fascinating questions about the nature of radio as a medium and its relationship to the audience. Broadcasters had to consider the different ways that DAB was being consumed (through DAB radio receivers, TV sets, online, and over cell-phones), as well as the implications for this in their programming and marketing.

The United Kingdom's 1996 Broadcasting Act and the Radio Authority's interpretation of it has resulted in the geographical and population areas of local digital licences broadly matching the existing coverage areas of analog broadcasters. The effect of this is to greatly restrict the flexibility of the system: as with analog broadcasts, separate services cannot be broadcast on the same frequency, and there has to be sufficient distance between transmitters broadcasting different material on the same frequency to avoid mutual interference. At the local level, DAB is no more efficient than analog in the use of the frequency spectrum. This inefficiency and the fact that having at least two frequencies to carry national networks means that most areas will (at least until 2007, when other frequencies may be available) have a maximum of two "local" multiplexes.

A unique feature of the transmission arrangements of DAB in the United Kingdom is that BBC local/regional analog services have a "must carry" requirement on commercial multiplexes, but there is no automatic right of access for local/regional analog commercial stations, nor is there any limit to the number of services also broadcast—simulcasted—on analog.

A study of the license applications for commercial multiplexes reveals how the perceived "early adopters" market—the people who are most likely to be the first to invest in digital radio receivers—is linked with particular types of programming designed to appeal to them. One of the key demographic groups is clearly affluent males aged 25 to 45.

It also seems clear that EUREKA 147 is, to some extent, "technologically deterministic" in that it "naturally" favors large rather than small geographic areas of broadcasting, even though there is no overriding technical reason why national, and especially regional, services should be favored over local. The pattern adopted in the United Kingdom greatly favors large area services and, almost inevitably, large commercial interests.

DAB in Scandinavia

The development of EUREKA 147 DAB is on a fast track in the Nordic countries. Leaping into DAB before there is a robust popular market eager to receive digital signals is an unattractive commercial proposition. Thus, government initiatives are driving the transformation process, and Nordic public broadcasting companies bear much of the responsibility for building the DAB infrastructure necessary to expand this market.

Three factors underlie rapid Scandinavian DAB development. First, DAB significantly increases the possibilities for transmitting radio signals. There is no longer room for any development of AM or FM analog signals given the rapid growth of private and public channels since the mid-1980s (the bandwidth required for one FM channel is sufficient for six DAB channels). Second, DAB significantly increases production efficiency in radio programming. Public service companies in northern Europe are adding new radio channels to compete more effectively with the private sector via audience segmentation. The digital platform offers synergistic possibilities at a higher speed and efficiency. "Versioning" content, a process whereby the same "raw content" can be reconfigured in a variety of ways appropriate to the program format and audience interests of respective channels, is increasingly important in this context. Finally, DAB significantly increases integration possibilities across and within media industries. DAB is the only viable opportunity for radio broadcasting to maintain and advance its position in an increasingly competitive market place.

Although private broadcasting companies in Scandinavia have not yet been deeply involved with DAB development, that will soon change. The cost for a car DAB receiver was nearly $1,000 at the turn of the century. Seven manufacturers are competing in the DAB receiver retail market, and prices are dropping. Four of the best-known companies are Clarion, Sony, Pioneer, and Grundig. It is expected that the cost of a combined receiver and boot box will decline by half in the early 2000s. As signal coverage increases and receiver purchase prices decrease, the open market for DAB broadcasting will surge.

The transition to digital systems challenges the culture of public service radio. As the number of channels multiply and the scope of the private sector involvement increases, it is increasingly difficult for any public service company to provide a comprehensive and competitive range of radio program services. Achieving that nonetheless remains the heart of democratic principles that characterize and legitimate the approach.

Despite the certainty of DAB development in the Nordic countries, the digital future is laden with risks. A key factor hinges on when FM signals will be eliminated. That is at least a decade away from happening, although the pace will ultimately depend on consumer receiver purchasing decisions. Still, DAB developments are taking place more rapidly in the Nordic region than in the United States because there are fewer vested interests fighting the transition, stronger government involvement, and a common EUREKA 147 standard. Further, there are experienced and capable public broadcasting companies managing much of the risk as a tax-supported initiative.

DAB in the United States: IBOC

The United States has taken a different approach to digital radio. On 1 August 1990, the Federal Communications Commission (FCC) initiated a *Notice of Inquiry* into DAB development and implementation. The FCC and broadcasting business became interested in DAB after learning that Europe was developing its EUREKA 147 system. By 1991 the American radio industry responded with USA Digital Radio (USADR), a partnership of CBS, Westinghouse Electric Corporation, and Gannett to develop an in-band, on-channel (IBOC) system of DAB to eventually replace AM and FM. USADR preferred IBOC over other methods because it allowed stations to continue existing analog AM and FM service as they developed new digital signals that eliminate multipath and noise and reduce interference.

IBOC DAB has been called the "Holy Grail solution" because broadcasters can convert from analog to digital without service disruption and with low start-up costs while maintaining their heavily promoted dial positions. Initially the National Association of Broadcasters (NAB) supported DAB implementation using the EUREKA 147 system on L-band (500 to 1500 MHz), which it believed would give AM and FM equal footing. Incensed broadcasters caused NAB to change their position and support IBOC.

On 26 August 1992, USADR successfully delivered IBOC DAB on the expanded AM band at 1660 kHz in Cincinnati. At the September NAB radio show in New Orleans, USADR demonstrated its system using WNOE-AM and NPR affiliate

WWNO-FM. In 1993 NAB's DAB task force officially endorsed IBOC because it believed the FCC would never allocate alternative additional spectrum for DAB. The Electronic Industries Association (EIA), which had held that no system should be selected until all types were tested, struck a compromise with the NAB, agreeing that other systems would not be considered unless IBOC systems were shown not to meet terrestrial DAB requirements.

While much of the world appears convinced L-band is the best DAB spectrum, the FCC supported use of the S-band (2310 to 2360 MHz), in part because of the difficulty in shifting existing L-band spectrum users in the United States. Most non-FCC experts agreed S-band will be more expensive and less effective than L-band. As a result of the S-band decision, U.S. DAB will be IBOC, causing global incompatibility. At NAB's 1993 Las Vegas convention, USADR introduced broadcasters to its IBOC system, demonstrating that its IBOC system was more fully developed than any system other than EUREKA 147.

USADR and Lucent Digital Radio (LDR) agreed to work together on IBOC in May 1997, making broadcasters more optimistic about DAB's future. USADR and LDR worked jointly for about 10 months but ended their alliance early in 1998. Digital Radio Express (DRE), another IBOC developer, allied with USADR in late 1999. In October 1998 USADR petitioned the FCC to open a rulemaking proceeding to make its system the DAB standard. In 1999 a number of the larger U.S. radio groups invested in USADR. USADR's new corporate status was important because it demonstrated that much of the radio business believed in IBOC. Other broadcasters and electronics manufacturers, including receiver makers, soon fell into line. At NAB's 1999 meeting in Las Vegas, some broadcasters and manufacturers called for a "Grand Alliance" like the one struck with digital television (DTV). Robert Struble, USADR president, accurately stated that as a coalition from the beginning, USADR already was the Grand Alliance.

The FCC issued a DAB *Notice of Proposed Rulemaking* on 1 November 1999, more than nine years after its first *Notice of Inquiry.* The commission believed it was time

> to determine whether an IBOC model and/or a model utilizing new radio spectrum would be the best means of promptly introducing DAB service in the United States. By initiating this proceeding now, we can foster the further development of IBOC systems, as well as new-spectrum DAB alternatives, help DAB system proponents identify design issues of public interest dimension and, where possible, encourage modifications that advance these policy objectives.

In October 2002, the FCC provisionally approved the technical standard offered by iBiquity digital, the company controlled by the 15 largest radio broadcasters. The system allows AM and FM broadcasters to begin transmitting digital signals while continuing to offer their analog service. Initial broadcast equipment began to reach the market in late 2002 while the first consumer receivers became available early in 2003.

Fears remained, however, that using more of a station's frequency assignment (as the digitized signal does) might threaten some sub-carrier services such as reading for the blind, carried by many noncommercial stations. The FCC order allowed temporary authority for digital operation until such problems could be resolved. FM stations may offer digital signals at all hours, but AM stations are at least temporarily limited to daytime hours only because of their more complex evening signal propagation.

DAB's Future

Despite the impressive development of DAB in many parts of the world, its detractors had not been fully disarmed by the early 2000s. Many smaller and community radio operators had not been able to afford to pay for the still scarce capacity on multiplexes. The U.K.'s Community Media Association (CMA) which, in the early days of the development of DAB, had been led to believe that the new system would allow small-scale not-for-profit stations to (at last) find a secure and widespread outlet on a mass media, was bitterly disappointed when it became clear that both the transmission pattern and licensing structure would largely exclude their model of radio stations. The Association tended to view the licensing and regulatory structure of DAB in the U.K. as re-enforcing the dominance of a few major institutional operators—the larger commercial radio groups and the BBC—whereas the CMA favored a greater plurality of ownership and control, as well as of the "voices" and perspectives able to gain access to the public airwaves. The increase in frequency allocation in DAB may help to increase the range and type of operators using DAB, but these arguments highlight tensions on the more fundamental political and economic questions of access and regulation of broadcasting.

Other DAB detractors continued to question whether the public would embrace the new system *en masse*, in particular whether DAB receivers would ever fall in cost to the level where most listeners would be prepared to replace their many analogue radios and even whether there was sufficient interest in the increased choice of services to motivate listeners to make the digital switch. Many asked whether the text and other non-audio services were merely gimmicks, leading such critics to dub the new system "TV without the pictures." Whereas multichannel digital TV had been embraced by the British consumer, it was often argued that radio was not perceived as providing the same level of sophisticated entertainment, information, and education, with a consequence that many citizens were reluctant to pay a premium for increased radio services or the claimed increase in audio quality. Certainly, the early and sometimes persisting claims that DAB provided CD-quality sound had rebounded, amid complaints that the sound quality on many of the individual stations on the multiplexes had been degraded in order to accommodate more services. Broadcasters admitted they had to make compromises between program choice and sound quality.

There was also the continuing, nagging question as to whether the U.K. had backed the wrong system. Putting aside the argument as to whether radio needed to go digital at all, were the alternative and fast-developing standards such as IBOC likely, in the long term at least, to provide a more cost-effective, easily received, radio system? Furthermore, the integration and fast development of mobile (cell) phone technology and the internet, especially with broadband access, confuses and even threatens the concept of "radio" as a separate and distinct medium.

It may well be that the EUREKA 147 system, although undoubtedly a proven, sophisticated, and robust transmission standard, will prove too complex for sufficiently cheap receivers to be produced for the mass market, thus making the system commercially viable. The refusal of governments and broadcasters in the United States and Japan—the world's two largest economies—to adopt the EUREKA system, and the subsequent reluctance of some of the major manufacturers to develop inexpensive receivers, has prevented EUREKA from becoming the world standard. In the long term, the answer may be for receivers and transmitters to operate a "pick and mix" system: terrestrial IBOC for existing broadcasters and EUREKA for new ones. The development of the WorldSpace, Sirius, and XM Radio satellite-transmitted systems will also be followed with great interest to see if they are successful in developing their respective markets.

RICHARD RUDIN, W.A. KELLY HUFF,
GREGORY FERREL LOWE, GRAHAM MYTTON

See also AM Radio; Audio Streaming; Digital Recording; Digital Satellite Radio; FM Radio; Internet Radio; Virtual Radio

Further Reading

BBC: Research and Develoment: DAB <www.bbc.co.uk/rd/projects/dab/>

Barboutis, Chris, "Digital Audio Broadcasting: The Tangled Webs of Technological Warfare," *Media, Culture, and Society* 19 (1997)

Digital Radio Development Bureau (DRDB) website, <www.drdb.org>

Digital Radio Mondiale website, <www.drm.org/indexdeuz.htm>

Eureka 147 Consortium website, <www.eurekadab.org/eureka_147_consortium.htm>

Feder, Barnaby J., "F.C.C. Approves a Digital Radio Technology," *New York Times* (11 October 2002)

Federal Communications Commission, *Amendment of the Commission's Rules with Regard to the Establishment and Regulations of New Digital Audio Radio Services,* Docket MM 90-357, FCC Rcd 5237 (1990)

Federal Communications Commission, *Digital Audio Broadcasting Systems and Their Impact on the Terrestrial Radio Broadcast Service Notice of Proposed Rule Making,* Docket MM 99-325 (1999); available online at <www.fcc.gov/Bureaus/Mass_Media/Notices/1999/fcc99327.txt>

"Heavenly Music: Digital Radio Finally Arrives," *The Economist* (16 March 2002)

Henry, Shannon, "Clearing a Path for Digital Radio," *Washington Post* (9 October 2002)

Hoeg, Wolfgang, and Thomas Lauterbach, editors, *Digital Audio Broadcasting: Principles and Applications,* New York: Wiley, 2002

Huff, W.A. Kelly, *Regulating the Future: Broadcasting Technology and Governmental Control,* Westport, Connecticut: Greenwood Press, 2001

Lax, Stephen, *Beyond the Horizon: Communications Technologies: Past, Present, and Future,* Luton: University of Luton Press, 1997

Mirabito, Michael M., and Barbara L. Morgenstern, *The New Communications Technologies,* Boston: Focal Press, 1990; 4th edition, 2000

Radio Advertising Bureau, Radio listening via new technologies, <www.rab.co.uk/news/html/NewTechnologiesQ402.htm>

World DAB Forum: Frequently Asked Questions, <www.worlddab.org/dab/whatis.htm>

WorldSpace website, <www.worldspace.com>

"USA Digital Radio and Digital Radio Mondiale To Collaborate On a Worldwide Standard for Digital Radio," *Car Sound & Performance* (22 May 2001)

Digital Recording

Radio's transition from an analog to a digital medium began with the arrival of the compact disc (CD) in the early 1980s. Since then, radio has embraced digital audio technologies ranging from the first CD players, digital tape recorders, digital effects processors, and digital audio workstations to the more recent introduction of hard disk recorders, digital exciters for transmitters, digital audio consoles, and digital audio file transfer and streaming on the internet.

Digital audio provides superior reproduction of sound and additional benefits useful in a radio station's operation. Except for a station's microphones (which may become the sole analog source at a radio station), it is possible for a radio station's audio chain to be completely digital, from production and storage to playback and processing, before being sent to a digital exciter and on to the transmitter. In the many countries where digital audio broadcasting standards are now in place, transmitters and receivers are now also digital, completing the final links to make radio a totally digital medium. Many radio stations also distribute digital transmissions of their programming on the internet, making a digital version of radio's signal available to listeners with a computer, audio card, and internet connection.

Digital audio has dramatically improved the quality of radio's on-air sound and has also brought many operational enhancements to the production process used to create radio programs. The clear benefits of digital audio have motivated nearly universal adoption among radio stations of some type of digital audio recording and playback equipment. A list of digital audio equipment found in radio stations today includes CD players, CD recorders, open-reel stationary-head digital recorders, rotating-head digital audio tape recorders, mini disc recorders, and hard-disk recording systems. Computers with specialized software and audio cards with inputs and outputs to interface with the other audio equipment in the station provide digital audio replacements for tape recorders, the splicing block, and other production and processing equipment. Digital versions of other equipment, such as audio consoles, telephone hybrids, effects processors, compressors, limiters, microphone processors, studio-transmitter links, on-air audio processing equipment, exciters, and transmitters, are rapidly becoming the standards as aging analog audio equipment is replaced.

Digital Audio Basics

A sound itself is not digital. Sound is created when an object vibrates and causes the molecules of the medium surrounding it (usually air) to vibrate. These vibrations or sound waves are transferred through the air until they reach someone's ear or a microphone. At the microphone, sound is transduced (converted) into electrical energy and becomes analog audio. The characteristics of this electrical energy are analogous to the original sound energy. This electrical energy can be amplified, manipulated, stored, or transmitted as analog audio; however, it can also be digitized and then amplified, manipulated, stored, or transmitted as digital audio.

Digital audio is created by converting analog audio into a stream of binary code, a series of ones and zeroes, representing the measurements of the characteristics of the original sound. This binary code represents measurements made of samples of the original audio representing the sound energy. The binary code can be recorded and stored on any device capable of reading and storing digital data. Magnetic tape, computer floppy disks, hard disks, and optical disks can store the digital information. These data can also be transmitted as pulses through copper wire, fiber-optic cable, or as radio frequency energy through the air. An exact, full-fidelity reproduction of the original audio can be created from the stored or transmitted digital code, copied without generation loss, and easily processed for creative and technical purposes. Unlike analog audio, the digital signal is not as subject to the limitations imposed by the storage medium or the electronics of the equipment. In the analog world, the tape itself adds noise, copying adds more noise, the amplifier adds noise, and so on. The dynamic range and frequency response of the original sound are also reduced as analog audio, because the analog system has inherent limitations in reproducing sound faithful to the original. The methods used to record and process digital audio minimize these limitations.

The digitization of analog audio involves four stages: filtering, sampling, quantizing, and coding. First, the audio is sent through a low-pass filter to prevent unwanted higher frequencies from becoming audible. This process is called anti-aliasing. Then the analog signal is divided, which determines the sampling rate. The more often the signal is sampled and measured, the more accurate the recreation of sound will be. Sampling rates are typically 32 kHz, 44.1 kHz, or 48 kHz. This means the signal is sampled either 32,000, 44,100, or 48,000 times every second. A measurement is made during every sampling period using a multidigit binary number. This binary number is called a word. The number of bits in a word is word length. A 1-bit measurement, for example, would only be able to discriminate between presence and absence of voltage. If n is the number of bits in the word length, the number of levels of measurement is 2^n. An 8-bit system provides 256 levels of voltage measurement. A 16-bit system has 65,536 possible levels, and a 20-bit system provides 1,048,576 levels. Systems with more quantizing levels have more accuracy and wider signal-to-noise ratios. The last stage of the digitizing process is the coding stage, in which the bits are placed in a precise order for recording or output to another digital device. During this coding stage, each word is identified in the bit stream. Error correction minimizes the impact of storage defects. The binary code is then distributed or recorded as pulses of magnetic energy.

Moving audio to the digital domain for recording and reproduction purposes provides a number of advantages. Compared to analog audio, digital audio has an improved frequency response, wider dynamic range, immeasurable noise and distortion, and no degradation or generation loss in multiple digital recordings. Some audiophiles have been critical of the digital audio recording process, suggesting that when sound is digitized, it loses its warmth and can sound too sterile and even harsh. Radio has generally rejected those concerns and has continued to replace analog audio equipment with digital equivalents.

The Compact Disc

In 1980 the Philips and Sony corporations joined forces to create an optical disc for digital audio. The two companies agreed on a CD standard, a 12-cm optical disc using 16-bit/44.1-kHz sampling. The CD player and disc were introduced in Europe in the fall of 1982 and in the United States in the spring of 1983. As record and production library companies began to release their catalogs on CD, radio began using CD players in their production and air studios. Manufacturers developed CD players with features such as a shuttle control as well as a model that played CDs inserted in a special protective case, creating a process similar to the use of a broadcast cartridge machine. The CD changer, capable of handling multiple CDs, was also found useful at many radio stations. Many broadcasters used consumer models because the audio quality was the same for both. Not only did the CDs sound better than the vinyl long-playing and 45-rpm records, but the CD format was also much more efficient to use. CDs could be cued and started faster and, with care to protect the disc from scratches, would allow endless replays without degradation of sound quality. Additional data encoded in the compact disc provided precise track timings, indexing, and continuous monitoring of playing time of tracks and programs.

The CD player uses a laser to read the data encoded in the microscopic circular pits on the disc. The binary code is stored in a series of pits and lands in the disc. A pit is an indentation in the groove; a land is a flat area with no indentation. A photoelectric cell reads the amount of light reflected from the pits and lands and emits voltage to recreate the digital code representing the audio waves. As needed, the digital output of the CD can be converted back to analog audio or sent as digital output to a digital recorder or console. The only problem with the CD was that its content was limited to prepackaged material offered by the manufacturer. However, the recordable CD was soon on the way.

The recordable CD (CD-R) was launched in 1988 but initially was not widely adopted as a production tool by radio. The first CD recorders were relatively expensive, and the disc recording was permanent: the disc could not be erased and recorded on again. Recently, with the introduction of the rewritable CD (CD-RW), lower-cost CD-R recorders, and CD-R drives installed in computers, the recordable CD has attracted more attention from broadcasters, for use as an archival and production tool and as a component in digital automation systems.

The digital versatile disc (DVD) is not yet a factor in the radio environment, but it most likely will be. The DVD is the same diameter and thickness as a CD, but a difference in design and manufacturing provides eight times the capacity of a CD by creating a dual layer. There are currently five recordable DVD formats. DVD-R and DVD+R discs can be recorded only once. DVD-RW and DVD+RW can be rewritten. The DVD-RAM is used for recording computer data only.

Digital Audio Tape and Mini Disc Recording

Although the CD was quickly adopted by most radio stations shortly after its introduction, digital audio recording had a more difficult time gaining a foothold in radio. Commercial digital audio recorders have been available since the 1970s and early 1980s. Sony and Denon introduced adapters that made it possible to record digital audio on videotape recorders. Open-reel two-track and multitrack digital recorders (Digital Audio Stationary Head) were employed in recording studios but were not widely used in radio. By the early 1990s, however, the rotary-head digital audio tape recorder (R-DAT) was finding a place in radio production.

R-DAT machines use essentially the same digitizing scheme as the CD, but they use a rotating helical-scan tape head to record and read the large quantity of information representing the audio signal on the small cassette tape. The result is audio recordings with characteristics similar to the CD with the additional flexibility of being able to record and rerecord. R-DAT recorders were adopted as a cost-effective, high-quality production and on-air playback tool, especially in automation systems. The use of R-DAT has been supplanted somewhat in recent years by other digital formats, including the mini disc.

Although originally intended as a consumer product, the mini disc recorder is finding a niche in broadcasting. The mini disc offers a more portable, less expensive alternative to hard disk recording, and the portable mini disc units provide a digital alternative for field recording. The mini disc offers nonlinear access, track identification, and a recording time of 74 minutes. It has low noise, low distortion, and a wide dynamic range, but its use of data compression limits its use in critical recording.

Computer-Based Recording, Editing, and Digital Distribution

Open-reel analog recordings have at least one advantage over open-reel digital tapes: analog recordings are easier to edit than a digital tape. Because a digital tape has to be running at speed in order to decode the data to recreate the digital audio, digital open-reel recorders record and play on an analog head for editing and cueing purposes. The digital tape can then be marked, cut, and spliced like an analog tape. Physical editing is not possible with the DAT cassette tape, and electronic editing on a DAT recorder requires some finesse. Moving the recorded digital information to a computer hard disk opened the way for the rapid deployment of computer-based digital audio recording and editing systems, which have revolutionized audio production for radio.

By the early 1990s, audio could be recorded to media other than tape. Increased hard disk capacity, faster computer processing speeds, and new compression methods combined to make recording directly to a computer's hard disk a viable alternative to recording on analog tape. Software programs and audio input/output cards were developed to be used on inexpensive personal computers to create digital audio recordings that sounded better than the recordings created on professional analog equipment. Even consumer products could create professional-sounding results in radio production studios. Editing software was introduced that would allow nondestructive editing of the audio material. These programs typically provide a visual representation of the audio waveform, which can be marked, highlighted, cut, copied, pasted, and moved within and between sound files. Precise, noise-free edits are performed that can be readjusted and fine-tuned as needed without destruction of the original sound file.

There are numerous multitrack recording and editing programs used by radio stations, which, when combined with compatible high-quality computer audio cards, allow desktop computers to perform the same functions as multichannel recorders, production consoles, and effects processors—which cost thousands of dollars more—all in one computer.

Once the digital audio exists as a file in a networked computer, local area networks, wide area networks, and the internet allow these sound files to be distributed and shared internally or externally. An increasingly common distribution approach is the use of MP-3 files. The MPEG-1 layer 3 recording technology (commonly known as MP-3) is a digital audio file compression method increasingly used by radio stations to send and receive programs and programming elements through the internet. This form of distribution becomes cost-effective and important as advertising agencies and production companies start to distribute commercials, programming, and other information digitally. As radio groups consolidate and combine station operations and look for economies of scale, digital distribution of content will become even more important. After a commercial is created in the production studio of one of the stations in the group, it can be distributed instantly to all the other stations on the computer network.

Digital Audio Processing

After audio has been converted to digital form, it can be manipulated or processed for creative and technical reasons. Modern radio production studios often have at least one digital effects processor, which efficiently creates various combinations of digital effects, such as echo, reverb, pitch changing, phasing, flanging, and many others. Computer software-based recording and editing programs also have digital audio processing and effects as part of the package. Digital audio processing is also used in the station's air chain, running microphones through processors that convert the signal to digital before processing to strengthen and improve the sound quality of the announcer's voice. Digital processing of the audio signal before it is sent to the transmitter provides one last measure of limiting, compression, and other subtle adjustments to give the station's audio a distinctive, full sound.

JEFFREY D. HARMAN

See also Audio Processing; Audio Streaming; Digital Audio Broadcasting; Recording and Studio Equipment

Further Reading

Alten, Stanley R., *Audio in Media,* Belmont, California: Wadsworth, 1981; 6th edition, 2002

Daniel, Eric D., C. Denis Mee, and Mark H. Clark, editors, *Magnetic Recording: The First 100 Years,* New York: IEEE Press, 1999

Gross, Lynne S., and David E. Reese, *Radio Production Worktext: Studio and Equipment,* London and Boston: Focal Press, 1990; 3rd edition, Oxford and Boston: Focal Press, 1998

Huber, David Miles, and Robert E. Runstein, *Modern Recording Techniques,* 5th edition, Boston: Focal Press, 2001

Kefauver, Alan P., *Fundamentals of Digital Audio,* Madison, Wisconsin: A-R Editions, 1998

O'Donnell, Lewis B., Philip Benoit, and Carl Hausman, *Modern Radio Production,* Belmont, California: Wadsworth, 1986; 5th edition, as *Modern Radio Production: Production, Programming, and Performance,* 2000

Pohlmann, Ken C., *Principles of Digital Audio,* 4th edition, New York: McGraw-Hill, 2000

Talbot-Smith, Michael, *Broadcast Sound Technology,* London and Boston: Butterworths, 1990; 2nd edition, Oxford and Boston: Focal Press, 1995
Talbot-Smith, Michael, editor, *Audio Engineer's Reference Book,* Oxford and Boston: Focal Press, 1994; 2nd edition, 1999
Watkinson, John, *An Introduction to Digital Audio,* Oxford and Boston: Focal Press, 1994
Watkinson, John, *The Art of Sound Reproduction,* Oxford and Woburn, Massachusetts: Focal Press, 1998; 2nd edition, 2002

Digital Satellite Radio

For more than 100 years, radio has been transmitted by electronic analog waves modulated by voice or frequency variance. At the beginning of the 21st century, digital signals beamed from communications satellites could change American radio from a medium with thousands of local stations into a national radio service with only a few content providers. In the new system, a listener could drive from coast to coast and remain tuned to the same CD-quality signal all the way. In the United States, two corporations and hundreds of investors are betting billions of dollars that Americans will embrace the new digital system of satellite radio.

Sirius Corporation vs. XM Corporation

Two corporations are at the front of the race to bring satellite radio to the American consumer: Sirius, headquartered at the Rockefeller Center in New York City, and XM, which has its offices on New York Avenue in Washington, D.C. Both companies take a similar approach to satellite broadcasting. Each is beaming a digital signal from a satellite to antennas the size of a playing card. The antennas, mounted in the consumers' cars or homes by suction cups, feed the signal into digital radio receivers that produce CD-quality audio with at least 100 different format selections. As of early 2003, XM offered 70 music channels and 31 talk channels while Sirius advertised 60 music channels and 40 talk channels. Both companies have signed well-known stars to provide special programming for subscribers. Both companies—and here is the big gamble—are charging for their audio services: initially $9.95 (XM) and $13.95 (Sirius, which offers more channels without advertising).

Investors at XM and Sirius are gambling that enough listeners are dissatisfied with the current fare on AM and FM radio stations that they will be willing to pay a small monthly fee to receive programming unavailable on analog terrestrial stations. Executives at both corporations note that almost 30 percent of all recordings sold at music stores come from artists receiving little or no radio airplay. The reasoning goes like this: There may not be enough fans for alternative country acts such as Lyle Lovett or Steve Earl or for new-age performers such as Yanni to support a local radio station format. There are, however, enough of them scattered across the country to make a nationwide satellite feed economically feasible.

Given the sheer cost of both projects, niche programming alone will not offer the kind of return on investment stockholders of either company are looking for. Thus, the need for big-name performers. XM gave a channel to Grammy-winning producer and composer Quincy Jones. Former Yes member Jon Anderson is using the same approach, while Ted Nugent gets his own talk show.

Lee Abrams, creator of the album-oriented rock (AOR) sound of the 1970s, provides the consulting for all 50 music formats for XM. Abrams claims that XM will not emulate traditional radio. He understands subscribers are paying for the audio services and expect something different for their money. Abrams delivers expanded selections of classic rock artists such as Bob Dylan, The Beatles, or Led Zeppelin to counter complaints from traditional radio listeners that playlists have become repetitive. Abrams also delivers channels for contemporary alternative rockers as well as multiple jazz, country, and blues formats.

Sirius has given a channel to rock superstar Sting, who produces a daily live show with original and recorded music. With the main studios in New York City, Sirius plans to offer a number of live, in-studio concerts from artists who pass through the city while on tour. National Public Radio (NPR) provides two channels of talk and information, including an original morning program for satellite listeners only. Programmers point out that Sirius, with multiple rock, jazz, country, blues, and talk formats, plans no commercial advertising at this time, while some of XM's channels have up to six minutes of commercials per hour.

Sirius Corporation used to be known as CD Satellite Radio Service. Focus groups and marketing studies found consumers were confusing the name with the audio CDs available in music stores. Furthermore, company executives felt the name CD no longer implied cutting edge technology. So, a name change to Sirius was ordered early in 2000.

Making Satellite Technology Pay Off

Several companies launch commercial communication satellites. XM hired Sea Launch to place its two Hughes Corporation satellites (named *Rock* and *Roll)* into orbit from an ocean platform located 4,600 miles west of South America. Sirius hired Space System/Loral to build and launch three satellites into orbit from the former Soviet republic of Kazakhstan. The enormous cost of the high orbit satellites for both corporations may be partially offset by leasing unused space on the various transponders to other companies with communications needs.

In order to maintain a constant signal, or footprint, over a specific region of the planet, satellites must remain in roughly the same position relative to the earth. Sirius and XM are using two different systems to meet that goal. XM uses two geostationary satellites positioned at 22,300 miles above the Earth. At that height, the speed of the satellite's orbit matches the speed of the rotation of the earth. Therefore, the satellites appear to be stationary in the sky.

Sirius uses three satellites in an inclined elliptical constellation. Elliptical orbit means the satellites are in a lower orbit moving across the sky. Each satellite spends at least 16 hours a day over the United States and at least one satellite is placing a footprint over the continental U.S. at all times. Both companies have a spare satellite on the ground in case of a catastrophic failure.

What kind of return will the two American corporations need to stay afloat? Wall Street analysts predict that each company will have to attract a minimum of 4 million subscribers within five years to break even. Such numbers are possible, as has been proven by successful satellite radio ventures in other countries.

Satellite Radio Worldwide

A privately owned American corporation with immense international ties, WorldSpace is the current leader in digital satellite technology. WorldSpace claims a potential audience of 4.6 billion people on five continents. Launched in October of 1998, the geostationary satellite Afri-Star offers three overlapping signals to the continent of Africa with 50 audio channels and multi-media programming available on each signal beam. Asia-Star was successfully placed into orbit with a similar programming array over the Asian continent in March of 2000.

A third satellite, Ameri-Star, will service South and Central America as well as Mexico. This satellite has no current plans to broadcast to the United States or Canada. However, Worldstar has signed a cooperative agreement with XM to share technological innovations.

Delivering 24 digital radio signals, Orbit Satellite TV and Radio Network serves the Middle East, North Africa, and parts of Asia with "socially responsible" broadcasts in Arabic and other languages. Originally chartered with the Italian Ministry of Post and Telecommunications, Orbit signed with Telespazio to provide space segment services and launched programming in 1994 with 16 TV and four radio channels. Orbit now boasts business offices and uplink centers throughout the Mediterranean, Middle East, and the Indian Subcontinent.

Orbit first used transponder space on InTelSat satellites, the international telecommunications satellite consortium established by the United Nations. In 1999, Orbit expanded its coverage area by 22 million households when it began to also broadcast on Arabsat, a satellite placed into space by a cooperative of Middle Eastern nations. Orbit contracts with numerous international providers for programming content including CNN and ESPN Radio.

Several other satellites carry digital audio and analog radio programming. The Eutelsat array and the Astra satellites provide a radio footprint over Europe. Panamsat and Brasilsat provide programming for Central and South America. Other countries may well adapt to satellite radio faster than America does. Much of Europe and Asia are already used to the concept of a national radio service. Britain's BBC or Germany's Deutsche Welle have for years broadcast a national signal through the use of relay transmitters. A satellite service is simply a logical extension of that system.

Broadcast Opposition

As early as 1982, the Federal Communications Commission (FCC) began to develop regulations for direct broadcast satellites, or DBS. Signals would be provided to consumers via a three-meter dish antenna. Local broadcasters immediately attacked the proposed service with charges that a national TV service would undermine the localism provided by traditional television broadcasting.

In October 1992, the FCC again acted on an industry DBS proposal, this time for radio. The original proposal called for a Satellite Digital Audio Broadcasting System or DAB to be located in 50 MHz of the S band (2310–2360 MHz) with the intent to create a system that would provide a national service.

Again, local broadcasters rolled out the same arguments used against the DBS system: "The current number of FM and AM stations serving the United States represents the highest level of audio diversity available in the world." In comments submitted to the FCC the association added, "A competing satellite service presents a potential danger to the United States' universal, free, local radio service and, thus, to the public interest it serves."

A consortium of radio group owners also weighed in with the following comments to the FCC: "National radio stations raise a troubling question of undue concentration of control of the media, an issue that has been consistently a concern to the commission." The joint comments added, "erosion of audiences and advertising revenues caused by satellite radio would inevitably destroy the ability of many [existing] stations to offer these services."

Broadcasters opposed only the national delivery of a digital satellite signal and not the use of digital technology. In fact, the National Association of Broadcasters (NAB) began proposing a system referred to as in-band, on channel (IBOC). This system would replace terrestrial analog transmitters with digital transmitters for better signal quality and reception. Local broadcasters would then repeat satellite feeds of the various program services with inserts for local advertising and announcements.

However, satellite programmers counter that the broadcasters are overemphasizing the continued importance of localism. Listeners, they say, do not tune exclusively to one station. While local news, weather, and information are important, research shows many listeners will tune to a station specifically for music or entertainment programming. Satellite programmers believe listeners will tune to their service for specific formats and return to local stations when they need local information. As one programmer stated, "Program directors who think localism is fundamental to successful radio should look at the success of *USA Today.*"

The Box in the Car or Home

Ultimately, the success of satellite radio in the United States depends upon whether consumers will buy the new digital receiver necessary to pick up the satellite signal. Sirius and XM settled a patent lawsuit in March of 2000 by agreeing that all digital radios will eventually have the capacity to receive both audio feeds. Deals have now been made with Sony, Alpine, Pioneer, Clarion, and other audio manufacturers to produce the receivers. Major retailers have agreed to market and sell the receivers for $200.00 and up, depending on features and installation.

Some of the retailers, including consumer electronic shops, car sound shops, and automobile dealers, are offering installation for car receivers. Distributors are also marketing a portable device designed to translate a digital satellite signal for existing radios. In January of 2003, Delphi introduced the SkyFi radio, advertised as the first plug and play portable digital satellite radio.

However, the real key to success is cooperation by major automobile manufacturers in offering digital radios as available equipment in new cars. XM has agreements with General Motors (GM) and Honda. Sirius has a deal with Ford and tentative agreements with Chrysler and five other automakers. The deal with Ford is particularly beneficial because the buyer gets a satellite radio and a two-year subscription to Sirius audio services.

The drawback, if the current marketing arrangement stays in place, is that owners of GM cars will have to take XM audio services while Ford owners will receive Sirius. Will listeners continue to pay for audio services when the initial deals run out with the car companies? As one industry observer put it, "Twenty-five years ago, TV viewers would have thought it odd to pay for watching television." Today, a majority of American homes have cable television and pay a monthly fee for expanded video service.

The Impact on Local Radio

If satellite radio succeeds, there will be some impact on local radio stations.

First, three critical audience-rating categories could be affected: time spent listening, or TSL; average quarter hour listening, or AVQ; and cumulative audience numbers, or cume. If enough listeners tune out local stations to listen to satellite radio, the local stations will report reduced numbers in those categories and will have to lower the rates charged for advertising.

Second, the concept of localism in radio in recent years has really come to mean the ability to generate local ad revenue, rather than the desire to program for regional interests of the community of license. Radio stations are no longer compelled by the FCC to provide specific programming to meet the needs of the local community. Music playlists are almost identical by format in every market. A listener can drive across the country today and find virtually the same 30 or 40 songs being rotated in each format by stations across the dial. If the majority of stations are playing similar music and running many of the same syndicated programs, then the broadcasters' "localism" argument is greatly diminished.

Despite their more than 100 channels of specialized programming, both Sirius and XM radio added subscribers more slowly that expected. Toward the end of 2002, Sirius was reaching only about 30,000 paying customers while XM (which had aired earlier) was reaching better than ten times as many. But each needed more than a million subscribers in order to survive, and in 2002 it appeared such levels would take at least a couple more years to achieve. One result of the slow growth was that their stock prices declined to record low levels as many observers concluded there was room for only one such service, not two. Both continued to seek additional investors and claimed they were in the competition for the long haul. At the same time, both trimmed their expenses, closing down unused studios and cutting back staff. Talk of consolidation lingered in the air.

Nonetheless, the investors for both XM and Sirius believe that the public will eventually buy into the idea of satellite radio. They believe enough of the public has grown tired of traditional radio's repetitive playlists and commercial saturation to give satellite radio a try at $10.00 per month. Even radio giant Clear Channel Communications hedged its bet by becoming a major investor in XM. It is not clear that satellite radio will succeed as numerous challenges remain. However, executives at XM and Sirius are literally betting millions of dollars that the time is right for a new national radio service.

CORLEY DENNISON

See also Clear Channel Communications; Digital Audio Broadcasting; Virtual Radio

Further Reading

Aherns, Frank, "Radio's Race to Space," *Washington Post* (10 September 2001)
Irwin, Neal, "XM Puts Satellite into Space," *Washington Post* (19 March 2001)
Malaysia HITZ Satellite Radio Network, <www.astro.com.my/radio/hitz.htm>
Markels, Alex, "100 Channels, But Where Are the Subscribers?" *New York Times* (3 November 2002)
RUsirius (The Internet Guide to Satellite Radio), <www.rusirius.com>
Sirius Satellite Radio, <www.siriusradio.com>
WorldSpace Satellite Radio, <www.worldspace.com>
XM Satellite Radio, <www.xmradio.com>

Dill, Clarence Cleveland 1884–1978

U.S. Communications Policy Maker

Clarence C. Dill was one of the key co-authors of the 1927 Radio Act and the 1934 Communications Act. In the formulation of the law, Dill helped set the traditional concepts that still govern electronic media. It was Dill who proposed a commission for the regulation of radio. The concept passed in 1927, was reinstated in 1934, and continues today as the regulatory authority. Believing the commission would control and regulate the continuing issues of monopoly, censorship, and spectrum utilization, Dill took those issues into the Senate and provided leadership for passage of the law.

Dill's role in the earliest legislative history of broadcasting was limited only by his late entry into the congressional arena. Secretary of Commerce Herbert Hoover and Representative Wallace H. White, Jr. had worked to develop radio law since the early 1920s, but they were unable to interest the Senate in their proposals. Dill provided that Senate leadership. His first bills mirrored the work of Representative White. However, as Dill educated the Senate, radio entered a critical period in its history, often called the period of chaos. Following the first session of the 69th Congress, the Zenith case (*United State v Zenith Radio Corporation,* 16 April 1926) and the Attorney General's Opinion (*Opinions* of the Attorney General, 8 July 1926) left the Secretary of Commerce with no regulatory authority, and Dill emerged as a significant leader.

During the conference committee negotiations between the Senate and the House over the pending radio bills, White was deeply involved in his own reelection campaign, which left Dill as the primary author in charge of bringing the House and Senate versions of the bill together. Dill convinced Representative White that a commission should be established to function for one year. This would theoretically provide Congress with enough time to review the law before regulatory authority reverted back to the Secretary of Commerce. White accepted this as a temporary solution, but Dill openly declared that this commission was merely the beginning of a lasting communications commission. In this Conference Committee, Dill not only achieved this important directional victory, but out of the committee came a unified proposal with emphasis on the phrase "public interest, convenience or necessity." These key words provided the overall regulatory authority and were used rhetorically by both Dill and White to answer all major objections to the passage of the bill. By the time the bill came out of conference committee, the public pressure for passage was intense. Dill obtained quick passage of the bill that he felt protected the public interest. Dill's comments before the Senate during its deliberations on radio law reflect his attitude, the political atmosphere, and his own feelings toward the radio industry: "large corporations have invested large sums of money with little return on their investments. They hope for bigger returns in the future. I am not sure that it would be wise . . . to put too many legislative shackles around the industry at this state of its development" (*Congressional Record,* 69th Cong., 1927, 68, pt. 3:3027). Shifting with the political current of a demanding electorate, Dill achieved passage for the bill that set the foundation for electronic media legislation.

Representative White wanted control of radio in the hands of the Secretary of Commerce. However, by 1934 he was supportive of the commission, and the Democrats were in charge. Dill was now the chair of the Senate Commerce Committee, and he chaired the subcommittee that was holding hearings on the proposed 1934 legislation. The key issue was placing the telephone and telegraph under control of the commission, a provision suggested by both Presidents Hoover and Roosevelt. The 1934 Communications Act passed with the basic provisions of the Radio Act still intact. It added only commission jurisdiction over telephone and telegraph, changed the name of the commission from the Federal Radio Commission to the Federal Communications Commission, and added commissioners to help with the increased workload.

Dill was a product of the 1920s era, when business and businessmen were held in high esteem by society. The business ethic permeated all areas of American life. According to Frederick Lewis Allen (1931) radio was the "youngest rider" on this "prosperity bandwagon." Dill fought in the Senate for financial assistance for the fledgling radio industry. Besides being an advocate of business interests, Dill was a conservationist. The Teapot Dome and Elk Hills oil scandals had left the United States outraged by the exploitation of public resources. Senator Dill had been a principal in the Teapot Dome investigations as a member of the Public Lands Committee. The scandals and resulting investigations left a strong impression on Dill, and as a result he became an ardent supporter of the public ownership of the airwaves and an antagonist of radio monopolies.

Dill was first elected to the House of Representatives in 1915, but his career was cut short because of his vote against the United States' entry into World War I. After the war, he was reelected, this time to the Senate. Tucker and Barkley, describing Dill's ability as an astute politician, wrote, "he can smell the change of public opinion . . . a month or six weeks before anyone else in the chamber" (1932). Dill was a progressive politician, proudly associating himself with William E. Borah (R-Idaho), Robert La Follette (R-Wisconsin), Thomas Walsh (D-Montana), Burton K. Wheeler (D-Montana), and James E. Watson (R-Indiana). Dill described his own role as a leader of the Senate debate over radio legislation as "a one-eyed man among the blind."

Dill served in the House of Representatives from 1915 to 1919, and he served as the Senator from Washington State from 1923 to 1935. On 12 July 1934, after 12 years in the Senate, he announced his resignation, stating, "The most common fault of public men is that they do not know when to quit . . . I do not want to make that same mistake" (*Tacoma Tribune*, 12 July 1943). Upon leaving the Senate, he returned to Spokane, Washington, where he had a successful law practice. He ran for governor in 1940 and made an unsuccessful second run for the House in 1942. During his tenure in Congress, he worked primarily on radio, the Grand Coulee Dam, and hydropower legislation. This work set the foundation for a legal practice that he continued until his death. He died in Spokane, Washington, in 1978.

DONALD G. GODFREY

See also Communications Act of 1934; Public Interest, Convenience, or Necessity; United States Congress and Radio; White, Wallace H.; Wireless Acts of 1910 and 1912/Radio Acts of 1912 and 1927

Clarence Cleveland Dill. Born in Fredericktown, Ohio, 21 September 1884. Attended Ohio Wesleyan University, graduated 1907; taught high school in Dubuque, Iowa, and Spokane, Washington, 1907–10; deputy prosecuting attorney and secretary to governor of the state, Spokane, 1911–13; elected to House of Representatives, 1915; lost House seat in 1919, after which he resumed his law practice in Spokane; served 12 years, U.S. Senate, 1922–35; after leaving Senate practiced law in both Washington, D.C., and Spokane; member, Washington state Columbia Basin Commission and special assistant to the U.S. Attorney General, 1946–53. Died in Spokane, Washington, 1978.

Selected Publications

How Congress Makes Laws, 1936
Radio Law: Practice and Procedure, 1938
Where Water Falls, 1970

Further Reading

Allen, Frederick Lewis, *Only Yesterday*, New York: Harper and Row, 1931

Biographical Directory of the United States Congress <bioguide.congress.gov/scripts/bibdisplay.pl?index=D000345>

Godfrey, Donald G., "Senator Dill and the 1927 Radio Act," *Journal of Broadcasting* 23, no. 4 (Fall 1979)

Godfrey, Donald G., "Senator Clarence C. Dill: Father of Grand Coulee Dam and Early Congressional Radio Authority," *The Ohio Wesleyan Magazine* 54 (May 1979)

Irish, Kerry E., *Clarence Dill: The Life of a Western Politician*, Pullman: Washington State University Press, 2000

McChesney, Robert W., *Telecommunications, Mass Media, and Democracy: The Battle for Control of U.S. Broadcasting, 1928–1935*, New York: Oxford University Press, 1993

Paglin, Max D., editor, *A Legislative History of the Communications Act of 1934*, New York: Oxford University Press, 1989

Tucker, Ray Thomas, and Frederick R. Barkley, *Sons of the Wild Jackass*, Seattle: University of Washington Press, 1932

Disc Jockeys (DJs or Deejays)

What announcers were to listeners during the golden age of network radio—the voice or image of a program, network, or individual station—so disc jockeys (or deejays or simply DJs) became beginning in the mid-1950s. The style of the person playing records on the air determined, to a considerable extent, the identity of that radio station. Many DJs became regional stars, and a few became nationally known. By the early 21st century, however, the heyday of radio's popular DJ was long over.

Origins

A number of radio pioneers played records over the air in the medium's experimental days—Lee de Forest on several occasions, Reginald Fessenden, "Doc" Herrold in San Jose, and Westinghouse engineer Frank Conrad—in the years before World War I. So did many other known or nameless announcers (at the medium's inception, announcers were allowed to use only their initials on the air) at numerous stations during the 1920s and 1930s, though playing records over the air was frowned upon as a poor use of a scarce medium. Musicians' unions and record manufacturers also encouraged the use of live rather than "canned" music. Years later, stations located along the Mexican border aimed their formats of country music and conservative Christian programs toward American listeners; these programs were hosted by people with warm voices who seemed like comfortable neighbors and were soon widely popular.

The decline of network radio, beginning in the late 1940s, paved the way for the rise of the DJ. For now stations were forced to be free agents, not mere local conduits of national programs. Having to fend for themselves, stations turned to the easiest and least expensive way of filling time—playing recorded music.

1930 saw the introduction of one of the first "musical clock" programs (playing a set format of musical types, time notices, advertisements and news within a one-hour period), on KYW in Chicago, hosted by Ms. Halloween Martin. Certainly one of the earliest DJs and one of the first women in such a role, Martin intermixed recorded music with regular time cues and other chatter. The two-hour program moved to WBBM in 1934 when KYW relocated to Philadelphia. There it lasted for a decade, then enjoyed two more years on WCFL. The "musical clock" format remained a radio mainstay well into the 1950s.

Often cited as the first real "disc jockey," Martin Block first hosted New York station WNEW's *Make-Believe Ballroom* in 1935. Using a format borrowed from stations on the Pacific Coast, the program was hugely successful, entering syndication in 1940 (by 1946 it was carried on some 30 stations, making Block the most highly paid radio performer for a brief time). Another New York innovation, WNEW's listener-request-based *Milkman's Matinee* gained success as an overnight/early morning program hosted by Stan Shaw.

The term "disc jockey" seems to have come into use about 1940; it had its basis in such earlier terms as music jockey or record jockey. The "jockey" part may have been based on "riding" a record to fame (or riding gain on an individual record).

Several historians point to March 1946 and Al Jarvis at Los Angeles station KLAC as being "the first all disc-jockey station with identifiable personalities" (see Passman, 1971). Gene Norman, Dick Haynes, Alex Cooper, and Bob McLaughlin led the Los Angeles market's radio ratings into the 1950s. One popular feature was a daily evening "top 10" countdown program. With the KLAC model, the role of the DJ, as opposed to the earlier announcer, became clearer. The DJ combined the playing of music with chatter and intense station and record promotion. The DJ was also pushing his personality, not merely announcing what was being aired.

As other stations added DJs to their rosters, a National Association of Disc Jockeys, based in New York, was formed in the late 1940s. The need to cut back on station costs helped promote still wider use of DJs to perform some of their own technical work (some had broadcast engineer licenses) as well as to talk around the music, thus creating the "combo" role.

But external factors were also increasing the DJ's importance. The development of the 45 rpm record, soon used for pop singles, and the expansion of the market for portable radios, sharply increased the number of young people who listened to radio. This audience grew even faster after 1954 with the introduction of the first transistor radios and the proliferation of car radios. Thanks to these trends, by the 1950s DJs had become central to the popular music record industry—with both good (increased record sales) and bad (a bribery scandal in the late 1950s) results.

Rise of Top-40 DJ

The "golden age" of radio DJs came with the inception of Top-40 radio in the mid-1950s. Todd Storz (at KOWH in Omaha and WTIX in New Orleans) in 1953, and Gordon McLendon (at KLIF in Dallas) paved the way for countless others. They created a formula of tight formats, constant repeating of top popular songs, and flamboyant DJs. The DJs generated listener interest through promotional stunts (flag-pole sitting, treasure hunts) local events (record hops, remote broadcasts), extensive advertising, and other audience-building events. DJs also played an increasingly central role in the marketing of popular records. The record industry and radio had forged one of the most profitable alliances in the history of electronic media—a relationship that survives into the 21st century.

Two examples of the rising DJ illustrate the trend. Alan Freed became one of the first nationally known radio DJs, first at WJW in Cleveland and by 1955 on WINS in New York. He played what was then called rhythm-and-blues or "race music"—an important forerunner to rock 'n' roll. Tensions existed between largely black rhythm-and-blues musicians (and the "cover music" recorded by white groups to make it acceptably "white" in sound) and rising rock 'n' roll, and Freed was one of the first to bridge the gap. Freed became the most important victim of the late 1950s payola scandal. On the other hand, Dick Clark's *Bandstand* on Philadelphia's WFIL beginning in 1952 (it became an afternoon television staple in 1957) gave strong impetus to the Top-40 approach and served as a template for radio DJs working in this format. But Clark also pushed a squeaky clean image for rock with dress codes for the dancers, as well as hosts, although he, too, would be implicated in the payola scandal in 1959.

Yet the trend toward Top-40 DJs was anything but smooth. Even KLAC in Los Angeles, the first all-DJ station, announced as early as 1958 that it would revert to mere announcers' voices in pre-produced music formatting, arguing it would gain 13 percent more music by not using DJs. More music (and commercials) would result, with less focus on personality. The change would also provide for greater management control over the music played. A few years later, programmer Bill Drake would all but remove the DJ from the on-air mix at KHJ by imposing major strictures on non-music elements. DJs quickly dropped from high profile personalities to little more than record "liner note" readers at many stations. In the industry parlance of the time, they had been "Draked."

Perhaps the highlight of DJ excess—at least in the public eye—were two national DJ conventions hosted by Todd Stortz. The first, in Kansas City in 1958, was a relatively staid affair. Not so the second and much bigger one (with about 2,500 attendees) in Miami Beach in 1959 that brought considerable bad press—"booze, broads, and bribes" read the newspaper headlines. The seamy underside of popular music radio was out there for all to see. What was happening, of course, was the public revelation of "payola" (record makers or producers paying DJs to play specific records), which had long been a part of the music industry. In part created by independent record makers trying to break into markets largely controlled by major distributors, the payola scandal became headline news at the end of the 1950s. Nor was the scandal limited to big cities: to the record makers, "every DJ was important in his own town." The DJs tried to respond with the formation of a National Disc Jockey Association in Milwaukee in mid-1959. At its first convention in Minneapolis the NDJA developed a code of ethics to show listeners that the business was trying to regulate itself.

In 1965 "Boss Radio" was introduced on KHJ in Los Angeles by Bill Drake and Gene Chenualt. This newest twist to Top 40 involved "bigger" everything: huge contests and promotions, greatest events, hottest play lists (super-targeted and narrowed), and the coolest DJs (Ron Jacobs, Robert W. Morgan, The Real Don Steele), who always said just the "right" thing to enhance their larger-than-life images. After its success in Los Angeles, Drake and Chenault implemented the "Boss" approach at other RKO General stations around the country and influenced the sound of Top-40 radio personalities from then on.

Format and DJ Variety

A major change in the presentation style of contemporary music radio DJs occurred in 1966 with the introduction at KMPX-FM in San Francisco of commercial Underground Radio. A year later, Tom ("Big Daddy") Donahue, after tuning in the eclectic programming of all night radio deejay Larry Miller on the little known KMPX, decided to execute his plan for commercial underground radio at that same station. He is regarded by many as the father of that format. Deejays at FM stations employing this sound (characterized by long album cuts and counterculture ruminations) assumed personas that were in stark contrast to their rock 'n' roll AM radio counterparts. Underground DJs did not scream and shout at their listeners; rather, they spoke softly and conveyed a "naturalness" uncommon for programming targeted at the under thirty-year-old listeners. Furthermore, the format made room for more female DJs, something in rare supply on the airwaves up to this time. Underground radio was ultimately co-opted in the early 1970s by the corporations that owned them. Management imposed tighter format criteria and sought to eliminate the rambling, sometimes incoherent, monologues and diatribes for which DJs using this format were famous.

Meanwhile, Beautiful Music stations—initially AM and ultimately FM stereo (mostly automated) outlets—fostered the "deep" voice announcer style that was felt to resonate with their mostly lush instrumental music playlists and mature and temperate image. As with other formats of the period, but especially this format, women announcers were unwelcome as their voices were perceived as too high. In the 1970s and later, when the format adjusted its image (renaming itself Easy Listening) to appeal to a more youthful audience, styles of announcing became less voice-centric and stilted and somewhat more natural and conversational.

At the same time, the hyper-specialization that occurred in radio programming in the 1970s and 1980s inspired a number of variations in DJ styles. Foremost among them was the brazen, the bold, and the brash as exemplified by the on-air performances of morning DJs throughout the country who emulated the scatological rants of Don Imus and Howard Stern and who worked to create zany "zoo"-like atmospheres, high on talk and offering less music—a format highly popular with some audiences.

Decline

Many once-popular DJs, such as Dick Biondi, are now heard on "golden oldie" stations playing the same music for the same listeners—both now decades older. Because of the migration of music programming to the FM band in the 1980s, talk became the standard fare at most AM stations and in many ways its savior. Despite the harsh criticism leveled against right-wing broadcasters such as Rush Limbaugh and G. Gordon Liddy, such figures are in great part responsible for AM's continued survival. But these are, of course, political commentators rather than true DJs.

By 2000 many industry observers sensed a growing dearth of future opportunities for DJs in radio. This was the result of many factors, chief among them being the greater use of satellite syndication (and thus a need to separately program each local market), ever-tighter playlist standardization, music format fragmentation, and ownership consolidation with its usual resulting lay-offs of personnel. The increasing presence of "voice tracking" and "cyber jocking" (the use of a few DJs in many different markets thanks to pre-recording and re-use) is, among other things, killing off the role of smaller markets as radio training grounds for a new generation of DJs.

All of this has led to more of a coast-to-coast sameness in what remains of radio's DJ sound—what might be termed the "malling" (or mauling) of assembly-line radio. Indeed for many formats, including those of classical or jazz music, people can more readily listen to CDs or tapes (or, increasingly, the internet) for all the luck they are likely to have in finding their favorite music on the air. And this lack of offering special and interesting radio personalities, combined with insufficient musical variety, may prove fatal determinants of radio's future as a venue for music.

So the story of the radio DJ may have come full circle—from invention in the late 1930s and early 1940s, to the glory days of the 1950s and 1960s high personality DJ, to a slow decline in the decades since then, to the "plain vanilla" sound of much of music radio today. Radio is once again the home for a more amorphous radio voice, not unlike those early (and unknown) announcers.

MICHAEL C. KEITH AND CHRISTOPHER H. STERLING

See also Adult Contemporary Format; Album-Oriented Rock Format; American Top-40; Automation; Biondi, Dick; Block, Martin; British Disk Jockeys; Clark, Dick; Classic Rock Format; Contemporary Hit Radio Format/Top 40; Dees, Rick; Donahue, Tom; Drake, Bill; Drew, Paul; Dr. Demento; Dunbar, Jim; Everett, Kenny; Female Radio Personalities and Disk Jockeys; Freed, Alan; Freed, Paul; Gabel, Martin; Herrold, Charles; Hulbert, Maurice "Hot Rod"; Imus, Don; Joyner, Tom; Kasem, Casey; Morrow, "Cousin Brucie"; Murray the K; Murray, Lyn; Music; Oldies Format; Payola; Recordings and the Radio Industry; Rock and Roll Format; Shaw, Allen; Shepherd, Jean; Shock Jocks; Stern, Howard; Talk Radio; Tracht, Doug "Greaseman"; Transistor Radios; Urban Contemporary Format; Williams, Bruce; Williams, Jerry; Williams, Nat D.; Wolfman Jack; Wright, Early

Further Reading

Belz, Carl, *The Story of Rock,* New York: Oxford University Press, 1969; 2nd edition, London: Oxford University Press, 1972

DeLong, Thomas A., *The Mighty Music Box: The Golden Age of Musical Radio,* Los Angeles: Amber Crest Books, 1980

Holson, Laura M., "With By-the-numbers Radio, Requests Are a Dying Breed," *New York Times* (11 July 2002)

Keith, Michael C., *Voices in the Purple Haze: Underground Radio and the Sixties,* Westport, Connecticut: Praeger, 1997

MacFarland, David, *The Development of the Top 40 Radio Format,* New York: Arno Press, 1979

Mainelli, John, "Who Killed the DJs?" *The New York Post* (20 March 2001)

Passman, Arnold, *The Deejays,* New York: Macmillan, 1971

Segrave, Kerry, *Payola in the Music Industry: A History, 1880–1991,* Jefferson, North Carolina: McFarland, 1994

Smith, Wes, *The Pied Pipers of Rock 'n' Roll: Radio Deejays of the 50s and 60s,* Marietta, Georgia: Longstreet Press, 1989

Disney. *See* Radio Disney

Documentary Programs on U.S. Radio

Documentary programs did not play a large role in the history of American radio broadcasting. For three decades from the late 1920s, typically only one or two hours of documentary programs were presented on the national networks during the evening hours. In the 1930s and 1940s there were two to three hours of documentaries each week during the daytime as well, most of them produced and/or presented in association with educational institutions and intended for students listening in schools. By the late 1940s there were also several hours of documentary and other factual dramas on weekends, especially on Sunday afternoons (a time period that later would become the "intellectual ghetto" for television).

The documentary programs that were presented, although small in number, often were inventive in their use of voices, music, and sound effects: they created great prestige for the networks and helped stations satisfy requirements for public service programming. Some received much critical acclaim and are still remembered as the pinnacle of radio writing and production.

"Drama documentaries" broadcast during World War II were among the most exciting and accomplished examples of the art of radio and are still studied and enjoyed by students of radio history. Many American documentaries were produced at this time dealing with U.S. history and patriotism. These programs told war stories about America's fighting men and about its allies—especially the British—or encouraged civilians to conserve resources and support the war effort (and reduce inflation), especially by buying savings bonds. The forms and techniques of such documentaries, combined with the traditions of the film documentary, were precursors to the television documentaries that began in the 1950s.

Origins

The earliest American radio dramas included documentary-type programs such as *Biblical Dramas* and *Great Moments in History,* both on NBC during the season of 1927–28. For several seasons following, at least one radio series presented historical stories or biographies in a dramatic form.

In 1931 the Columbia Broadcasting System (CBS) began *American School of the Air,* which was intended for students. Although this was more specifically an "educational program," it did have documentary elements—and an eavesdropping audience of adults listening at home. The landmark program ran for nearly two decades.

On 6 March 1931, *The March of Time* began airing on CBS and was carried by 20 of the approximately 80 CBS affiliates. Each program dramatized several important news stories. Initially sponsored *Time* magazine, *The March of Time* had been developed by Fred Smith, who earlier had dramatized news at WLW-Cincinnati and for syndication. Although the news stories were "re-created"—actors portrayed the characters of important events—this was the first important documentary program on radio. Under various sponsors, *The March of Time* was broadcast for 13 seasons and was partly responsible for Time, Inc. developing a monthly newsreel. By the late 1930s *The March of Time* on film had evolved from a collection of several short reports—the format of the typical newsreel at the time—to a documentary on a single topic. At the end of the 1930s the radio program was off the air for two seasons, but it returned during World War II, and increasingly it made use of the actual persons featured in the news stories.

During the 1930s and World War II, the high cost of using telephone lines for the transmission of remote stories and the lack of portable, high-quality recording equipment that allowed for easy editing made studio re-creation of events easier and less expensive than coverage of the actual event.

In 1936, to create a better image for itself, the DuPont chemical company began sponsorship of *Cavalcade of America.* A broadcast of 2 September 1936 told "the story of rayon," which was said to "rank with the automobile and radio in its speed of development and in the way it has opened a wide new field of employment." The program dramatized events in American history, although it carefully avoided any mention of gunpowder or dynamite, which were also manufactured by the company. Commercials always spoke of the progress and benefits of chemistry. This program introduced many young listeners to American history; it continued for nearly two decades until the end of most network dramatic programs.

Also in 1936, CBS introduced *We The People.* This program dramatized stories of generally well-known people, using music and a narrator, and was said to be based on actual documents.

In the late 1930s some factual programs were produced with or by the U.S. Office of Education, the Smithsonian Institution, the Library of Congress, the Rockefeller Foundation, and a number of universities. Some, such as *The Ballad Hunter,* produced by John Lomax, about folklore and folk songs, used recordings made on location. Alan Lomax, John's son, also produced a 1941 program with the Library of Congress about people displaced by the Tennessee Valley Authority. Alan Lomax would eventually collect thousands of hours of sound and film recordings, and later develop television documentary series, about folklore.

William Lewis, as CBS vice president of programming, had been given carte blanche by the network for experimental dramas; he assembled the group that created Columbia

Workshop. From 1936 until 1941, Columbia Workshop produced experimental dramas and labeled some of these programs "documentaries," especially when they dealt with social problems and issues. In February 1939, *Words Without Music* presented "They Fly Through the Air with the Greatest of Ease," written and directed by Norman Corwin: the program described airmen bombing homes and then strafing the people who fled them. Accompanied by narration, dialogue, and sound effects, the story of the pilots involved their completing their missions of destruction but then being themselves shot down. A pilot describes the sight as one of the bombs hits its target: "Gee, that's fascinating! What a spread! Looks just like a budding rose unfolding!" Although it was not noted during the radio program, the speaker was Vittorio Mussolini, a pilot and the son of Il Duce, the Italian dictator. But Corwin did make this clear in his introduction to the published version of the radio play: "One group of horsemen gave the impression of a budding rose unfolding, as the bombs fell in their midst and blew them up. It was exceptionally good fun." —Vittorio Mussolini. (The younger Mussolini had written his impressions of dropping bombs on Ethiopian cavalry and watching the horses and riders as they were blown to bits.)

The program excoriated "all aviators who have bombed civilian populations and machine-gunned refugees." While the U.S. was still neutral, and public opinion was evenly divided on the coming war in Europe, CBS was brave not to censor this program or other similar dramas that were based on fact. "They Fly Through the Air with the Greatest of Ease" received an Ohio State Institute for Education by Radio award in May 1939, as the program "best demonstrating the cultural, artistic, and social uses of radio." It was a great boost for Corwin's career, and it was the first of his many dramatized documentaries about the events leading up to World War II and the war itself.

Lewis now told Corwin that he wanted a new series that would build pride among Americans and promote self-awareness of the American heritage. The result was the series *Pursuit of Happiness,* which CBS publicity described as "dedicated to the brighter side of the American scene," bringing us "reminders that today, with thankfulness and humility, we Americans still enjoy our constitutional rights to life, liberty, and the pursuit of happiness." Begun in October 1939, the program presented a spectrum of Americana. It was not an immediate success. But the third program, "Ballad for Americans," a musical written by Earl Robinson and John Latouche and sung by Paul Robeson, received much critical praise. By the time the series ended in May 1940—although there had been many arguments along the way—Corwin had successfully explored his idea of presenting "American sound patterns and phenomena." This series would set the pattern for Corwin's and many other CBS programs throughout the coming time of war.

World War II

On 15 December 1941, all four networks presented the Corwin production "We Hold These Truths," honoring the 150th anniversary of the Bill of Rights: it presented short dramatized stories illustrating the importance of, and conflicts inherent in, each of the rights. Although the program had been in preparation for several months, its broadcast just one week after the attack on Pearl Harbor gave it special emotional appeal. In 1942 Corwin produced a series called *An American in England,* which sought to show how our allies lived, worked, and fought. As previous documentaries had been, these programs were mostly studio-produced dramatizations, but this series distilled research on actual people.

During World War II, the major emphasis of the networks was on reporting the news of the war. Most of that reporting was by newscasters in studios, mainly in New York. There was, of course, much news from London, and some from Asia, transmitted via shortwave—especially at the time of very big stories such as the June 1944 D-Day invasion. The vast majority of such reporting was broadcast live. Recordings were used only sparingly because available equipment was heavy, fragile, unreliable, and required too much battery power for use over long periods.

During the seasons of 1942–43 to 1944–45, radio saw a three-fold increase in the amount of documentary programming. Virtually all of the new programs were war related, often produced with the Office of War Information, and included salutes of each branch of the armed services. Because of a special wartime excess profits tax, many companies preferred to buy sponsorship of programs with money that would otherwise go for taxes, and some companies were willing to pay for documentary and information programs that in the past would have had to be self-sustaining. This additional funding meant that broadcasters could produce more elaborate and expensive documentaries.

At the end of the war in Europe, all networks again carried a Corwin-produced dramatized documentary called "On A Note of Triumph." The program was well received and was subsequently released as a phonograph album. On that same night of 8 May 1945, however, the real future of documentary was revealed—although there seems to be no evidence that this change was understood at the time. An NBC program, arranged primarily by Prof. Garnet Garrison of Wayne State University in Detroit, presented a history of the war through phonograph recordings of the most important speeches and events of the war. There was a narrator as well as musical bridges, and some speeches had to be re-created because recordings were not available. This was the first use (on such a scale and for such an important occasion) of a substantial compilation of the actual voices and sounds recorded over a period of years.

Postwar Developments

In June 1946, Norman Corwin left New York for a four-month trip around the world "in a search for common ties and yearnings for world unity." He returned with hundreds of hours of interviews recorded on magnetic wire and acetate discs. By this time plastic magnetic audio tape, to which all of the material was transferred, made editing much easier. The twelve programs produced from this material aired on CBS from 14 January to 1 April 1947 under the series title *One World Flight.*

Although the word "documentary" was now regularly used to describe programs (CBS had formed a documentary unit in 1946), studio re-creation was still the preferred means of presenting non-fiction material. Robert Lewis Shayon created *CBS Is There,* which began 7 July 1947: it dramatized historical events as if they had been covered by CBS correspondents at the time that they happened. For example, on 7 December of that first season, the program was "The Exile of Napoleon," with CBS reporters "covering" the events that marked "the end of the Napoleonic era." The next season the title was changed to *You Are There,* and the program continued on radio until 1950. It later had two runs on CBS television narrated by Walter Cronkite.

Many stations began using audio tape as an aid in gathering news and covering actualities. Yet it was another medium, the phonograph record, that helped show the way to documentary based on audio compilation of bits of recorded sound. In 1948 Fred W. Friendly and Edward R. Murrow compiled a history of the 20th century to that time; marketed as "I Can Hear It Now", it was based on fragments from newsreel sound tracks and on radio recordings. But many items were also re-created, or edited drastically for a more dramatic effect. The success of the record (first released on 78 rpm records, and then re-released in the new LP format), and the several that followed, showed that there was a viable market for audio compilation history.

Working again with Friendly as producer, Murrow made the next significant breakthrough in documentary with his reporting of the 1950 election. "A Report to the Nation" was broadcast just 48 hours after election day; it was a compilation of the voices and sounds of the campaign—speeches, commercials, song, rallies, winners' declarations, and losers' laments. Within a few weeks this new format became the basis of a weekly program called *Hear It Now,* which the announcer introduced "as a document for ear." A segment on 9 February 1951 called "Biography of a Pint of Blood" was not only one of the most dramatic of the series but also showed how audiotape recording would be used in future documentaries. With Murrow narrating, the program offered listeners the story (in interviews, reportage, and the sounds of the events) of one pint of blood—its being donated in the U.S., its transportation to the war front in Korea, and its use in saving a wounded soldier. The Murrow commentary is spare, limited mostly to essentials needed for transition. Now the real reporter was the tape recorder gathering reality sound, to which narration as needed could be added. But with the decline of radio network audiences because of television competition, in April 1951 CBS cut radio advertising rates for the first time ever. The future of broadcast documentary programming, combining with the traditions of the film documentary, was now on television. *Hear It Now* lasted only that first season; it returned to television in the fall as *See It Now.*

Documentary After 1950

During the next half century there were many fine radio and audio documentary programs and series. But none would attract the audience, achieve the critical praise, or reach such large audiences as their counterparts on television.

For four seasons beginning in 1956, NBC Radio presented *Biography in Sound;* often it was the story of a performer and was based on radio, phonograph record, and movie sound clips, as well as recorded interviews with colleagues and historians. Also in 1956 NBC produced 33 half-hour summaries of the history of radio called *Recollections at 30.* Often the phonograph records of radio performers had to be substituted for actual recordings of the radio programs because none could be found—an unfortunate result of the networks' earlier ban on using recordings of programs or of their carelessness: many of the recordings that were made were either lost or destroyed.

From the middle 1950s, most radio stations developed new music formats; there was little radio documentary. Sometimes audio-only reporting was used, especially if it was much less expensive and more convenient than bulky and expensive film equipment. In 1965 there were two notable series about poverty in America. *This Little Light* was a 10-program series produced in Mississippi by Chris Koch and Dale Minor for Pacifica Radio. Westinghouse Broadcasting also produced for the stations it owned a series on poverty called *Outskirts of Hope.*

CBS radio sent its own reporters to cover the Vietnam war, but from the beginning news president Fred Friendly insisted that they also provide TV coverage. ABC produced a weekly radio documentary about the Vietnam War for several years. NBC and CBS produced a very few radio documentaries each year, usually including one that was a summary of news for that year. These documentaries, as well as similar collections from Associate Press Radio, were often also made available as phonograph albums.

Since the 1970s, even smaller and more portable audio-recording equipment has made advances in the documentary art possible, but there are now few stations where radio documentaries can still be heard. Currently, National Public Radio in the U.S., the CBC in Canada and the BBC in Britain each typically broadcast several hours of documentaries each week. Producer

David Isay has won numerous awards for his sound documentaries, many of which have aired on National Public Radio.

In December 2000 ABC News produced what it called the first "Webumentary," a biography and recollection of John Lennon on the 20th anniversary of his death.

LAWRENCE W. LICHTY

See also Cavalcade of America; Corwin, Norman; Friendly, Fred W.; Hear it Now; Isay, David; Lewis, William B.; March of Time; Recreations of Events

Further Reading

Arnheim, Rudolf, *Radio,* London: Faber, 1936
Bannerman, R. Leroy, *Norman Corwin and Radio: The Golden Years,* University: University of Alabama Press, 1986
Barnouw, Erik, editor, *Radio Drama in Action: Twenty-Five Plays of a Changing World,* New York and Toronto: Farrar and Rinehart, 1945
Hesse, Jurgen, *The Radio Documentary Handbook,* New York: Self Counsel Press, 1987
Lichty, Lawrence W., and Malachi C. Topping, compilers, *American Broadcasting: A Source Book on the History of Radio and Television,* New York: Hastings House, 1975 (see especially the tables on the types and number of quarter-hours of programming on the networks)
Paget, Derek, *True Stories: Documentary Drama on Radio, Stage and Screen,* Manchester, England: Manchester University Press, 1990
Radio Documentaries site, <www.transom.org>
Ryan, Milo, editor, *History in Sound,* Seattle: University of Washington Press, 1963

Dolby Noise Reduction

Reducing Unwanted Noise in Broadcasts and Recording

Dolby Noise Reduction (NR) makes audio signals clearer by reducing noise in the signal. There are a number of forms of Dolby NR, but in its simplest form, as applied in Dolby B NR, high-frequency hiss is reduced by about 10 dB using a process called *companding:* on low-level signals, the high frequencies are boosted during recording (or transmission) and then cut by a correspondingly appropriate amount during playback (or reception). Other noise reduction techniques have been applied with Dolby A, C, SR, and S.

Origins

Ray Dolby founded Dolby Laboratories in London in 1965. Dolby's career began in high school when he worked for Ampex Corporation in Redwood City, California. While still in college he worked with a team of Ampex engineers to invent the first practical video tape recorder, which was introduced to the broadcast industry in 1956 (Dolby was largely responsible for the machine's electronics). Dolby graduated from Stanford University in 1957 and received his Ph.D. in physics from Cambridge University in 1961. Dolby Laboratories moved its headquarters to San Francisco in 1976.

The first commercial product from the company was Dolby A, a noise reduction system using audio compression and expansion to reduce background hiss without the discernible side effects of conventional wide-band companders. More than any other feature, the system's freedom from side effects is what differentiated Dolby NR from previous attempts at audio noise reduction and ultimately helped earn its reputation. Also incorporated in this new noise reduction system was the ability to treat only soft signals, leaving unprocessed the loud signals that naturally mask noise.

A consumer version of the noise reduction system, called Dolby B, was released in 1968. Instead of dividing the signal into fixed multiple bands (as did Dolby A), Dolby B NR used a single, less-costly sliding band of compression that reduced noise in the higher frequency hiss region where most of the noise of consumer tape recording occurs. But the system needs the reference levels to be precisely set. Dolby NR is perhaps best known for its use with magnetic tape. The first product to use Dolby B was an open reel tape recorder made by KLH in 1968, but it made its biggest impact in the compact cassette format; with its inherent tape hiss problem and slow speed, cassette tape was a natural for Dolby B noise reduction.

Applications to Radio

The first experimental FM broadcasts using Dolby NR were made by WFMT in Chicago in June 1971 using the model 320

B noise reduction system. In March 1972 WQXR in New York City began full-time FM broadcasting using Dolby B.

In June 1973 it was proposed that Dolby B be joined with 25 microseconds pre-emphasis for FM broadcasting. In February 1974 this became the Dolby FM system demonstrated to the Federal Communications Commission (FCC). In June the FCC approved Dolby FM broadcasts. By August 1975 about 100 U.S. stations were broadcasting in Dolby FM. The use of Dolby FM moved outside the United States in October when the Canadian Department of Communications approved its use for broadcast.

Reducing Noise in Digital Signals

Concurrent with the development of these analog systems, Dolby Labs began research into digital audio in 1982. The primary goal was to find ways to reduce the amount of data required to transmit and store high-quality digital audio. Dolby Labs moved into the world of digital encoding schemes with AC-1, a refined form of Adaptive Delta Modulation (ADM). AC-1 was introduced in 1984, when bit rate reduction was in its infancy, and adopted the next year for use in a number of direct satellite broadcast and cable distribution systems. The Australian Broadcasting Corporation adopted it for direct broadcast satellite in its AUSSAT 1 in October 1985.

Dolby began its work in digital radio with the Digital Studio to Transmitter Line (DSTL) system, which was demonstrated for the first time in September 1991 in San Francisco at the National Association of Broadcasting's "Radio '91" conference. DSTL is a design featuring low time-delay implementation of Dolby AC-2, its second-generation digital encoding scheme using ADM. It features two high-quality channels and two auxiliary channels in the same spectrum space used for narrowband FM composite signals. Dolby DSTL was first installed in May 1992 at WWKX FM in Providence, Rhode Island.

Dolby Fax, for linking worldwide facilities with digital audio, began U.S. sales in March 1994. Dolby Fax uses Dolby AC-2 digital audio coding over two ISDN lines for high-quality transmission worldwide. In May 1994 DMX for Business began as the first direct broadcast satellite service, with Dolby AC-3 digital audio.

Dolby Surround is a matrix process that enables any stereo (two channel) medium, analog or digital, to carry four-channel audio. Encoded program material is fully compatible with mono and stereo playback, and listeners with playback systems incorporating Dolby Surround Pro Logic decoding receive four-channel surround sound. In February 1995 the BBC broadcast the first radio production in Dolby surround, "Bomber," on BBC Radio 4, and "Batman," "The Adventures of Superman," and "The Amazing Spiderman" on BBC Radio 1. Italy's Radio 101 has used Dolby Surround for all its evening entertainment programming. In March 1995 "West Coast Live" became the first U.S. radio show to be regularly broadcast in Dolby surround. In January 1997 the first Dolby Surround Pro Logic system for in-car use was launched at the Detroit Motor Show.

Dolby moved into the world of personal computer (PC) Surround Sound in April 1996 when Dolby and Microsoft signed a letter of intent to jointly develop PC surround technologies and specifications supporting the use of Dolby Digital AC-3 and Dolby Surround Pro Logic. Dolby Digital (sometimes known as AC-3 for the technology on which it is based) is a perceptual coding for consumer applications that enables storing and transmitting between one and 5.1 audio channels at a low data rate. Dolby Digital makes it possible to store five audio channels using less data than is needed to store just one channel on a compact disc. This was followed by Dolby Net, a low-bit-rate version of Dolby Digital introduced for low-bandwidth applications such as real time streaming internet audio, in November 1996.

In January 1997 Virtual Dolby Surround and Virtual Dolby Digital were introduced at the huge Consumer Electronics Show in Las Vegas. They enable surround sound effects from desktop computers with just two speakers. In October 1999 real time multichannel audio was streamed successfully over the internet using 5.1-channel Dolby Digital.

In 1999 Dolby Laboratories introduced Dolby E, a professional multichannel audio coding technology that allows a single AES/EBU (digital audio) pair to carry up to eight channels of audio, as well as digital metadata, through broadcast facilities. It is a convenient, simple, cost-effective conversion of two-channel broadcast facilities to multichannel audio.

In the world of radio, Dolby Lab's impact has been most felt with Dolby NR, as well as several other programs such as Dolby Surround, DSTL, Dolby Digital, Dolby Fax, and Dolby E. Although no one can predict the future of radio and the interaction with new technologies, given its track record, Dolby Labs will likely play a role in radio's future—on the airwaves, on the internet, or wherever it goes in the new millennium.

BRAD MCCOY

See also Audio Processing; Audiotape; Digital Recording; High Fidelity; Stereo

Further Reading

Dolby Labs, <www.dolby.com>

Eargle, John, *Handbook of Recording Engineering,* New York: Van Nostrand Reinhold, 1986

Elen, Richard, "Noise Reduction Systems" in *Sound Recording Practice,* edited by John Borwick, London and New York: Oxford University Press, 1976; 3rd edition, Oxford and New York: Oxford University Press, 1987

Honoré, Paul M., *A Handbook of Sound Recording: A Text for Motion Picture and General Sound Recording,* South Brunswick, New Jersey: Barnes, 1980
Marco, Guy A., and Frank Andrews, editors, *Encyclopedia of Recorded Sound in the United States,* New York: Garland, 1993
Runstein, Robert E., *Modern Recording Techniques,* Indianapolis, Indiana: Sams, 1974; 5th edition, by David Miles Huber and Runstein, Boston: Focal Press, 2001
White, Glenn D., *The Audio Dictionary,* Seattle: University of Washington Press, 1987; 2nd edition, 1991
Woram, John M., *Sound Recording Handbook,* Indianapolis, Indiana: Sams, 1989

Donahue, Tom 1928–1975

U.S. Disc Jockey and Radio Station Executive

In the late 1960s and early 1970s, Tom Donahue helped to revolutionize radio broadcasting in the United States. He is regarded as the "Godfather," if not the "father," of "free-form" rock and roll radio. In 1967 at KMPX-FM in San Francisco, he introduced the commercial underground radio format, which eventually became Progressive and later Album-Oriented Rock radio. Known as "Big Daddy," Donahue was an imposing figure. He weighed over 350 pounds; sported a thick, dark beard; and spoke with a deep bass voice. He began his radio programs with the line, "I'm here to clean up your face and mess up your mind."

Donahue got his first disc jockey job in 1949, at age 21, at WTIP in Charleston, West Virginia. He moved to radio stations in Maryland and Pennsylvania and eventually was hired in 1951 at WIBG in Philadelphia, where, in addition to the early 1950s popular tunes commonly heard on the air, he played rhythm and blues and rock and roll music.

In 1961 he moved to San Francisco, becoming a disc jockey at KYA, which had a Top 40 musical format. Working with music programmer Bill Drake, Donahue became a "kingpin" disc jockey at the station after helping to make it one of the top-rated operations in its market. The success of Donahue and the station was attributed in part to the fact that he and other disc jockeys heavily involved themselves with the music they played, often holding meetings to discuss the merits of records heard on the air. This distinguished the KYA disc jockeys from others in the business, who saw their jobs as simply springboards to careers in television or movies and who cared little about Top 40 music.

Donahue's interests grew beyond his air shift and music programming jobs. Together with his partner, fellow KYA radio disc jockey Bobby Mitchell, he operated a small record label, Autumn Records. He employed a 19-year-old disc jockey, Sylvester Stewart (Sly Stone), at soul music station KDIA to supervise the production of records by artists Bobby Freeman, the Beau Brummels, and The Great Society, with lead singer Grace Slick.

In the spring of 1965 Donahue left KYA to devote more time to his entrepreneurial interests. He operated a radio consultant service and published a music tipsheet, *Tempo,* which kept track of record sales. He also owned racehorses and produced music concerts, including a 1966 performance by the Beatles at Candlestick Park.

By the mid 1960s the Vietnam War and the civil rights movement began to dominate the news, and attitudes of young people on those issues, as well as on sex, drugs, and fashion, began to change. In response, Donahue and Mitchell opened Mother's, San Francisco's first psychedelic nightclub, which featured bands such as The Byrds and Lovin' Spoonful.

Although Donahue recognized that Top 40 music had been dominated by rock and roll in the 1950s and had developed into a formidable economic industry, he believed it had now stagnated, failing to keep pace with changing tastes in American popular culture. For example, in order to play music by an emerging popular group, The Doors, stations often shortened the 6:50 "Light My Fire" to three minutes, and they virtually ignored "The End," which ran 11:35. To be sure, there were in the mid- to late-1960s stations that experimented by playing longer musical pieces; the Pacifica Foundation's WBAI-FM in New York City and Massachusetts Institute of Technology's WTBS in Cambridge were two such stations. In 1966 an alternative format was presented on commercial station WOR-FM in New York City. However, the "underground" radio format was extremely rare.

Donahue longed to return to radio to program a station reflecting America's burgeoning new musical and artistic landscape. He envisioned a "free-form" format, where disc jockeys were not constrained to be "boss jocks," constantly upbeat and energetic as they chatted in between two-minute pop songs with silly lyrics and commercials for products promising to control dandruff and acne.

In the spring of 1967, Donahue began contacting FM stations, hoping he could convince one to try the proposed format. He felt it was possible, because in the 1960s FM radio still had small audiences and formats mostly of classical or beautiful music, or else they were simulcasts of an AM station's signal. He discovered KMPX in San Francisco, a poor station that made ends meet by selling airtime to anyone who had money to pay for it. Most of the time the station provided foreign language programming.

Donahue convinced the station's owners to let him on the air as a disc jockey and music programmer, and on 7 April 1967 he went to work with the format he would later call "freak freely" radio. There would be no playing of pop singles. Instead, he encouraged the station's disk jockeys to play music from a variety of formats, ranging from rock to blues, to folk, to folk rock. Cuts could be as long as the disc jockey desired, and talk by the announcers was conversational and unhurried. There would be no announcing of the current time and temperature. And there would be no musical jingles heard, as was the case on Top 40 stations.

KMPX's commercial "underground rock" became an instant success in the San Francisco area. In June of 1967 Chinese language programming was dropped completely, and the "underground radio" format was extended to 24 hours a day. Donahue was soon invited to employ the same format at Los Angeles FM station KPPC.

In November of 1967 Donahue was interviewed in the second issue of the magazine *Rolling Stone,* where he touted the free-form format and derided Top 40 music as "dead, and its rotting corpse is stinking up the airways." Though Donahue overstated his case against AM radio, the changes he brought to the radio music business did force Top 40 stations to rethink their programming, and in time they added album-oriented groups to their programming lists.

In March of 1968 Donahue and his air staff had a falling out with KMPX management, and they went on strike. They moved to San Francisco station KSAN, where Donahue was hired as program director and later became vice president and general manager. In May of 1972 he participated in a radio special honoring the fifth anniversary of free-form music radio. He began the program with Chinese music, to represent the foreign language show his disc jockeys replaced.

In the early 1970s the popularity of the free-form radio approach waned at commercial stations as managers began to reestablish stricter control over disc jockeys and their playlists.

Donahue was still working as KSAN general manager when he died of a heart attack on 28 April 1975. He was inducted posthumously into the Rock and Roll Hall of Fame in Cleveland, Ohio, in 1996.

ROBERT C. FORDAN

See also Album-Oriented Rock Format; Consultants; Free Form Format; Underground Radio

Tom Donahue. Born in 1928. Disc jockey, WTIP, Charleston, South Carolina, WIBG, Philadelphia, Pennsylvania, and KYA, KMPX, and KSAN, San Francisco, California. Posthumously inducted into Rock and Roll Hall of Fame, 1996. Died in San Francisco, 28 April 1975.

Further Reading

Fong-Torres, Ben, *The Hits Just Keep on Coming: The History of Top 40 Radio,* San Francisco: Miller Freeman Books, 1998

Hall, Claude, and Barbara Hall, *This Business of Radio Programming: A Comprehensive Look at Modern Programming Techniques Used throughout the World,* New York: Billboard, 1977

Keith, Michael C., *Voices in the Purple Haze: Underground Radio and the Sixties,* Westport, Connecticut: Praeger, 1997

Ladd, Jim, *Radio Waves: Life and Revolution on the FM Dial,* New York: St. Martin's Press, 1991

Post, Steve, *Playing in the FM Band: A Personal Account of Free Radio,* New York: Viking Press, 1974

Selvin, Joel, *Summer of Love: The Inside Story of LSD, Rock and Roll, Free Love, and High Times in the Wild West,* New York: Dutton, 1994

"Top 40 Music Is Dead, and Its Rotting Corpse Is Stinking Up the Airways," *Rolling Stone* 1 (November 1967)

Don Lee Broadcasting System

The California-based Don Lee network was one of the longer lasting and more influential of the regional radio networks that first appeared in the 1930s. Before its demise in 1950, the network also pioneered local television broadcasting.

Origins

In 1926 Don Lee (a Los Angeles resident whose wealth had been built on his exclusive distribution rights for Cadillac automobiles in California) purchased KFRC in San Francisco. In 1927 he acquired KHJ in Los Angeles to promote auto sales. He then joined the McClatchy radio stations in 1928 and formed a western network that became affiliated with the Columbia Broadcasting System (CBS).

Don Lee died in 1934, and his son Thomas took over what continued to be known as the Don Lee Broadcasting System. Thomas Lee broke with McClatchy, and in 1936 signed with the Mutual network. Through the remainder of the 1930s and during the 1940s, the Don Lee Broadcasting System served as Mutual's West Coast nerve center. Operations were centered at KHJ in Los Angeles, with KFRC in San Francisco and KGB in San Diego. By the late 1940s the Don Lee Broadcasting System also controlled 17 other stations in smaller California towns. KALE in Portland, Oregon, anchored that state's ten owned-and-operated stations located in smaller Oregon towns. KVI in Seattle and KNEW in Spokane led Washington's nine stations. The properties of the Don Lee Broadcasting System also included three stations in Idaho; KATO in Reno, Nevada; KOOL in Phoenix, Arizona; KCNA in Tucson, Arizona; and KHON in Honolulu, Hawaii.

Pioneering TV

For all its regional radio success, the Don Lee Broadcasting System's truly pioneering work was in television broadcasting. Although RCA is usually given most of the credit for pioneering American television, the Don Lee Broadcasting System deserves at least equal billing. Its radio profits funded television experiments that began as a simple television laboratory in 1930. On 14 November 1930, auto-dealer-turned-broadcaster Don Lee hired Philo Farnsworth protégé Harry Lubcke as his director of television. By 10 May 1931 Lubcke and his staff had built television equipment of sufficient quality to convince the Federal Radio Commission to give the Don Lee Broadcasting System permission go on the air experimentally. On the last day of 1931, its experimental TV station W6XAO went on the air on a UHF frequency for one hour per day, a schedule maintained through the 1930s.

With a transmitter and studios located in the Lee building's second floor in downtown Los Angeles, TV broadcasting experiments commenced on a slow, steady basis through the 1930s. Highlights included 1933 news footage of a Southern California earthquake as well as USC football games. In October 1937 the Don Lee Broadcasting System presented a gala high-profile premiere, broadcasting both on radio and television the opening of the 27th annual Los Angeles Motor Car Dealers' Automobile Show from the Pan Pacific Auditorium. (This event represented for Don Lee Broadcasting System the equivalent of the more fabled 1939 RCA demonstration at the World's Fair.) Every year the Don Lee Broadcasting System television station broadcast the Pasadena Tournament of Roses Parade, and in 1940 moved its transmitter to the top of Mount Lee, above the fabled "Hollywood" sign. The Don Lee company was apparently on the verge of initiating mass market TV broadcasting—in Hollywood's back yard—but the bombing of Pearl Harbor on 7 December 1941 put further experimentation on hold. Throughout World War II, the Don Lee Broadcasting System concentrated on radio.

Decline

After the war, Thomas Lee continued to concentrate on his highly profitable radio business and thus did not push to convert experimental W6XAO into formally licensed KTSL-TV (KTSL stood for "Thomas S. Lee") until May 1948. This came after Lee opened a $2.5 million radio and television complex at Hollywood and Vine with 18 studios covering 112,000 square feet, for television, AM, and FM broadcasting. Within the studio, Lee management sought to link radio with TV programming, not convinced that television would prove to be a stand-alone medium. Thomas Lee reasoned that simulcasting radio and TV programs would be the wave of the future. The Don Lee Broadcasting System's most famous show, *Queen for a Day,* was an example of this thinking.

Although radio broadcasts of *Queen for a Day* began on 29 April 1945, it would be two years before Lee gave the go-ahead to simulcast the program on the still-experimental TV station W6XAO. But when the simulcast finally occurred, its success led Lee to commit to television. Broadcasting live from a Hollywood theater- restaurant called the Moulin Rouge, host Jack Bailey interviewed four or five women who poured out tales of woe. The one receiving the most audience applause became "Queen for a Day" and was showered with gifts, her problems momentarily solved, at least in a material sense. James Cagney's sister Jeanne offered fashion shows in breaks between sob stories. *Queen for a Day* became a Los Angeles sensation through the early 1950s, and later ran on the

National Broadcasting Company (NBC) and American Broadcasting Company (ABC) networks.

Thomas Lee's death on 13 January 1950 was the beginning of the end of the Don Lee Broadcasting System. (He had been in poor health for years and either fell or jumped from the 12th floor of his broadcasting center.) Operations continued under long-time Vice President Lewis Allen Weiss until May, when, under the terms of Lee's will, the broadcast properties were put up for sale. In November, General Tire and Rubber Company's bid of $12.3 million was accepted. General Tire folded the Don Lee Broadcasting System into its broadcast empire, which then included the Mutual radio network, the Yankee radio network, and TV/radio station WOR of New York. By the end of the 1950s, nothing remained of the previous Don Lee Broadcasting System as a regional network or a group of influential radio stations.

DOUGLAS GOMERY AND CHUCK HOWELL

See also Mutual Broadcasting System

Further Reading

Murray, Michael D., and Donald G. Godfrey, editors, *Television in America: Local Station History from Across the Nation,* Ames: Iowa State University Press, 1997

Torrence, Bruce T., *Hollywood: The First Hundred Years,* New York: Zoetrope, 1979

Writers' Program, *Los Angeles: A Guide to the City and Its Environs,* New York: Hastings House, 1941

Dr. Demento 1941–

U.S. Radio Show Host

Since 1971 Dr. Demento has been the host of a weekly program of novelty songs and comedy skits, syndicated to more than 100 stations across the United States. Mixing the works of such legends as Spike Jones, Tom Lehrer, Monty Python, Stan Freberg, and Frank Zappa with contemporary comedians, comedy troupes, and musicians, the show has filled a unique niche in contemporary radio programming.

Barret Hansen, the future Dr. Demento, was born in Minneapolis in 1941. From his early years, it was clear that music would be a focus in his life. Prompted by his father, an amateur pianist, Hansen began taking piano lessons at the age of six. He also started a record collection, purchased mostly from a local thrift shop offering thousands of old 78-rpm disks.

The spread of rock and roll in the 1950s captivated Hansen, and he was particularly interested in rock's roots in rhythm and blues and country. After dabbling in disc jockey work in high school, Hansen attended Reed College in Portland, Oregon, where he worked as student manager of the campus FM station and graduated in 1963 with a degree in classical music. He then moved on to UCLA, where he wrote a master's thesis in 1967 on the evolution of rhythm and blues in the 1940s and early 1950s.

After graduate school in 1968, Hansen worked in Los Angeles for Specialty Records, a rhythm and blues label, and helped release a series of classic reissues. He also began hosting radio programs. This was during the famed era (late 1960s and early 1970s) of free-form FM radio, also frequently referred to as "progressive" or "underground" FM radio. Emerging from both coasts and a number of college radio stations, the relatively short-lived free-form era was a period of personal rather than corporate radio. Stations professed a concern with playing good music rather than salable product, and they allowed disc jockeys the freedom to construct their own playlists rather than follow a preconceived format. In 1970 Hansen made a series of guest appearances on KPPC, the legendary free-form radio station, joining a show hosted by Steven Siegal, who was later known as "The Obscene Steven Clean." Siegel coined the name "Dr. Demento" for Hansen, and the guest spots were such a success that in 1971 Los Angeles station KPPC gave Hansen his own weekly program of rock rarities. Dr. Demento expanded the playlist to include such classics as "The Purple People Eater," "Transfusion," and "The Monster Mash," and listeners quickly demanded more of such fare.

In 1972 KMET-FM in Los Angeles became the new home of *The Dr. Demento Show.* The show first captivated the local arena, becoming the most listened-to Sunday evening radio program in Los Angeles, then the nation, when it went into syndication in 1974. At this point, Dr. Demento became a well-known figure with his own unique look, including a top hat he first wore on an album cover in 1975, and a "Rare Records" t-shirt from a local collectors record store.

The Dr. Demento Show's success has been built on a foundation of both classic novelty hits and submissions from amateur artists. Of the former, Demento's resurrection of Spike

Barry Hansen, aka Dr. Demento, 24 February 2000
Courtesy AP/Wide World Photos

Jones is a good example. Jones and his band, The City Slickers, specialized in zany melodies and lyrics, with a wide variety of unusual noises and sound effects thrown in to produce a uniquely whimsical sound. Jones met with great success on radio and in records beginning in the World War II era, producing such hits as "Der Fuehrer's Face" and "You Always Hurt the One You Love." However, his popularity faded by the mid-1950s, and he died in 1965. His recorded works became a cornerstone of *The Dr. Demento Show*'s playlist, and Dr. Demento referred to Jones as "the King of Dementia." This exposure, plus the addition of a number of Jones' classics to Dr. Demento's various novelty collection vinyl (now CD) releases, introduced a new generation to the mad music of Spike Jones.

Although such classic recordings have continued to be a key part of the show, Dr. Demento has also received submissions of amateur works. Perhaps his most significant find through this process was famed parodist "Weird Al" Yankovic. When only in his teens and a fan of *The Dr. Demento Show*, Yankovic mailed in a home-recorded parody. After a number of years of such submissions, Yankovic finally reached a more mainstream audience with the single "Another One Rides the Bus," which he performed live on *The Dr. Demento Show* in 1980. Thus, *The Dr. Demento Show* helped spawn Yankovic's enormous popularity as rock music's best-known parodist.

Such songs as Yankovic's parodies are the backbone of Demento's weekly show. A typical broadcast opens with an hour of a general playlist: skits and songs from the Kids in the Hall, Tom Lehrer, Monty Python, Allan Sherman, Julie Brown, and Cheech and Chong; songs such as "They're Coming to Take Me Away," "Fish Heads," and "Dead Puppies"; novelty cuts from otherwise straight rock acts such as The Police and AC/DC; and outright strange experiments, such as songs sung by Leonard Nimoy and William Shatner. The subsequent half hour is oriented around a single theme, such as songs about streaking, baseball, the internet, and so on. The remaining airtime consists of the "Funny Five," the top five most requested songs of the week. Notably, virtually all of the two-hour playlist comes from Dr. Demento's own record collection, built up since his days of buying records at the thrift shop.

Outside of his radio work, Dr. Demento also works on research projects, helps to compile various comedy collections, and continues to build his archival collection. He also writes on music history, penning numerous magazine articles, liner notes, and encyclopedia entries.

Dr. Demento emerged during a unique period in FM radio history, and his subsequent success owes a great deal to his love for and knowledge of all kinds of music and his ability to convey his enthusiasm to listeners. When asked what he saw as his place in radio history, Dr. Demento replied: "I try to open people's ears to lots of different kinds of music, old and new, that they might not otherwise be exposed to. I've entertained some people and opened their ears to musical discoveries, and helped the careers of some great artists."

CHRISTINE BECKER

See also Disk Jockeys

Dr. Demento. Born Barret Hansen in Minneapolis, Minnesota, 2 April 1941. Attended Reed College, Portland, Oregon, B.A. in classical music, 1963; University of California, Los Angeles, M.A. in music, 1967; worked as researcher for Specialty Records, 1968–71, and for Warner Bros. Records, 1972–79; hosted *The Dr. Demento Show*, KPPC, 1971–72; moved to KMET, 1972; national syndication, 1974.

Radio Series

1971–present *The Dr. Demento Show*

Publications

"Summertime Blues from Percy Faith to Alice Cooper: A History of the Summer Song," *Waxpaper* (2 June 1978)

Rhino's Cruise Through the Blues (as Barry Hansen), 2000

Recordings

Demento's Mementos, 1982; *Dr. Demento Presents the Greatest Novelty Records of All Time*, 1985; *Dr. Demento Presents the Greatest Christmas Novelty CD of All Time*, 1989; *Dr. Demento's 20th Anniversary Collection*, 1991; *Spooky Tunes and Scary Melodies*, 1994; *Dr. Demento's 25th Anniversary Collection*, 1995; *Dr. Demento's 30th Anniversary Collection*, 2000; *The Very Best of Dr. Demento*, 2001

Further Reading

Fitzgerald, Terence J., "'Weird Al' Yankovic," *Current Biography*, (February 1999)

Ladd, Jim, *Radio Waves: Life and Revolution on the FM Dial*, New York: St. Martin's Press, 1991

Post, Steve, *Playing in the FM Band: A Personal Account of Free Radio*, New York: Viking Press, 1974

Young, Jordan R., *Spike Jones Off the Record: The Man Who Murdered Music*, Beverly Hills, California: Past Times, 1994

Drake, Bill 1938–

U.S. Radio Station Consultant

Bill Drake changed the landscape of Top 40 radio in the 1960s and 1970s. His fast-paced, forward-moving "more hits/less talk" approach to contemporary music formatting generated attractive ratings, but by 1970 had diminished the traditional influence of loquacious disc jockeys and left dozens of cookie-cutter stations that sounded alike. Drake accomplished this synthesis by compacting hit-music radio into a precise system that could be easily staffed and centrally controlled.

Disc Jockey Years

It is said that a knee injury nudged Bill Drake into radio programming. The Georgia Teacher's College student had originally hoped that his basketball scholarship would someday net him a contract with a professional team. When that was no longer possible, Drake (who was then known by his given name, Philip Yarbrough) decided to pursue a broadcasting career. The Donalsonville, Georgia native had started announcing while a teenager at WMGR in nearby Bainbridge. By 1957 he carried that experience to college and was hired by WWNS in Statesboro, Georgia. Four years later, Drake took to the microphone on Atlanta's WAKE. Pronounced "wake," the pioneer Top 40 music station had a manager with a penchant for rhyme; he suggested that Yarbrough go on the air as "Bill Blake on WAKE." Yarbrough countered with "Drake" (his mother's maiden name). That day, listeners to the pint-sized station (1,000 watts during the day and 250 watts at night) heard the name that would eventually be on the lips of radio people worldwide.

Bartell Broadcasting sold WAKE in 1961. As Drake was showing promise as program director in addition to performing as a disc jockey, Bartell moved him to their San Francisco outlet, KYA. This was another traditional Top 40 music station; programming focused on disc jockeys spinning records between frequent barrages of half-minute weather jingles, chatter, commercials of sundry lengths and tempo, sound effects, and time tones. Drake quickly concluded that listeners considered most of this as noise in the way of hearing more songs. Eliminating unwanted talk could convert the station into a real people-pleaser.

Armed with notebooks containing short station promotion announcements, as well as a plotted format clock showing precisely where these (and the music) should run, Drake started streamlining KYA. All rearranging was designed to keep listeners from tuning elsewhere. He concluded that few disc jockeys conveyed sufficient personality and charisma to be preferred over music. Although endowed with a resonant "radio voice" himself, Drake used the microphone only long enough to say the minimum about a song or station promotion, being very careful never to give the music short shrift. From his example, his on-air staff got the message not to ad-lib. Contrary to his critics, however, Drake welcomed a DJ air style where a personality could be perceived, as long as it did not noticeably impede the music flow. He found few who could master his ideal combination of great voice, impeccable pacing, friendly and enthusiastic tone, and compact presence.

The Fresno Fight

KYNO owner Gene Chenault needed programming advice for his Fresno, California station. His spunky Top 40 outlet practically owned the audience, but had started getting heavy local competition during 1962. Chenault's competition was KMAK, which was tightly formatted by Ron Jacobs. By the summer of 1963, Jacobs had conquered San Bernardino with KPEN and was busy toppling Chenault's Fresno station when Drake arrived.

The resulting radio war was fought with constant format tightening, increasing amounts of hit music, exciting station promotional stunts, and continual "ante upping." Each station monitored the other and had staff enthusiastically dabbling in a fraternal radio version of industrial espionage. When KMAK, for example, offered listeners a $1,000 prize, KYNO people would quickly counter with a $2,000 jackpot. Marathon remote broadcasts, where fans could see their favorite radio personality attempt to transmit solo all weekend, or sweepstakes in which listeners were invited to search for a single (cleverly hidden) "lucky key" to some station treasure, were thwarted by sleeping pills mysteriously ending up in a tired marathon disc jockey's coffee and by the strange appearance of decoy keys throughout Fresno.

Throughout the intrigue, Drake crystallized his belief that a music station should sound like it is always in motion. Every bit of announcing was positioned to scream, "This station has everything you like, and our supply is endless!" By simply blasting a short station identification jingle immediately prior to a hit song, he gave KYNO a bit more edge (by implying, "even our jingles move quickly out of the way to keep your favorite hits coming") than the competition. When the smoke cleared, Chenault's better-funded KYNO nosed past its rival. Drake had defeated Jacobs, but their Fresno clash set the stage for the rise of new radio formatting.

Formation of Drake-Chenault

KYNO's recaptured dominance gave Chenault the idea of exploring possibilities beyond Fresno. He envisioned a firm from which he could market radio-program doctoring to numerous ailing stations with Drake as the surgeon. The two associates had already achieved victories at KYNO and at the Stockton facility owned by Chenault's acquaintances. By 1964 the partners were ready for their Drake-Chenault Enterprises (also known as American Independent Radio and Drake-Chenault Productions) to spread the small-city successes into larger media markets. San Diego would serve as their first proving ground.

Station KGB offered the new consulting firm an opportunity to establish a miracle-worker reputation. The San Diego AM station would have to fight its way past two relatively sophisticated Top 40 adversaries. Drake's installation of his tight, uncluttered, more-music format soon made the competition sound outdated. The tremendous resulting KGB audience jump proved to be Drake-Chenault's ticket to what would be its most memorable radio foray.

For group-owner RKO-General was searching for ways to shore up its radio division. Although WOR (New York) stayed nicely afloat, other RKO outlets required continuous bailing out. Chief among the problems was KHJ (Los Angeles), a 5,000-watt AM station that seemed hopelessly bypassed in the ratings. Drake-Chenault signed a consulting agreement in 1965 for what they saw as a make-or-break debut. Radio managers would be monitoring their Los Angeles efforts closely as few felt the market needed another rock and roll station. Competitors included KFWB "Color Radio Channel 98," and the 50-kilowatt KRLA. Both used talkative air personalities known by most southern California listeners.

Drake hoped to use a KHJ win as a door into dozens of wounded stations, but he felt that three active assignments (Fresno, San Diego, and Los Angeles) was, for the time being, a prudent cap. He wanted unfettered time (away from corporate distractions) to ponder how best to keep his format concepts fresh, so he only promised RKO that he would be present in the KHJ offices about one day per week. That necessitated the hiring of a KHJ-based program director who clearly understood how best to achieve Drake-Chenault goals.

"Boss Radio"

Ron Jacobs, onetime Fresno competitor, filled the bill as program director of the anticipated Los Angeles operation. Chenault was especially pleased to have Jacobs and Drake on the same programming team and was said to have compared the amalgamation to going into battle (against the other area stations) with both Generals Grant and Lee on one side. A "more hit music" format, very similar to Drake's Fresno and San Diego output, was readied for KHJ's conversion. RKO's promotion director, Clancy Imislund, suggested to Drake and Jacobs that KHJ be dubbed "Boss Radio." The programmers judged "boss" (teenage slang for "cool") to be old hat, but reluctantly agreed to adopt it.

Boss Radio KHJ hit the airwaves early in May 1965. By fall the once-comatose station had rocketed to the top of Los Angeles' radio ratings and was the talk of the business. Its tight playlist (*never* any dead air), with 30 current hit songs (and with older favorites mixed in) sounded sleeker than rivals still working at getting through their top 40 or more records. Drake reasoned that he would rather have his outlets airing, for example, the 29th most popular tune, as opposed to the 40th or 58th. Also different from the competition was Drake's strictly mandated 12-minutes-per-hour commercial maximum (a third less than the industry standard). This gave KHJ time to air at least two more songs each hour than KFWB or KRLA.

Formats were typically devised on paper by Drake, then implemented by Jacobs. Though Drake-Chenault was clearly responsible for putting the KHJ conversion in motion, insiders (such as music radio pundit Bill Gavin) credit Jacobs with the milestone contemporary music station's day-to-day success. His connection to what flowed from KHJ kept the talented air staff, cache of station promotions, and every detail of the Boss Radio image highly polished. Not long after the format rollout, Jacobs heard a brief jingle burst, "KHJ-Los Angeles," immediately followed by the disc jockey exclaiming, "It's three o'clock in Los Angeles." Instantly hit with the redundancy (and possible negative effects on listeners), he commanded the air staff to announce Los Angeles as Boss Angeles. The memorable, "classic Drake" nuance serves as one example of how Jacobs constantly honed the programming concepts typically associated with Drake-Chenault. He stayed with KHJ until spring 1969.

National Expansion

RKO officials now wanted Drake to duplicate the KHJ success elsewhere in their chain. Consequently, in 1966 Boss Radio cloning took place at KFRC (San Francisco), WRKO (Boston, as "Now Radio"), CKLW (Windsor, Ontario/Detroit, Michigan), and WHBQ (Memphis). Each installation worked wonders, solidifying Drake's miracle-worker reputation. His modification of WOR-FM New York (which received a hybrid of oldies, album cuts, and touches of the Top 30) didn't match KHJ's impact. In terms of FM receiver penetration of well under 50 percent in 1968–69, however, it generated respectable listenership.

The Drake-Chenault agreement with RKO allowed the programmers to work with stations in media markets not served by RKO. During the early days of the pact, critics predicted that the Drake format's tightly controlled, high-energy sound was purely a province of the trendy West Coast and would never be accepted outside California. Prior to the Boston assignment, Drake-Chenault had been hired to try out Boss Radio in Oklahoma. Tulsa's mid-American KAKC listeners embraced what was by then (1966) dubbed "The Drake Sound," though it was also the brainchild of others—such as Jacobs.

Brainstorming kept the format fresh as Drake understood the value of being linked with associates able to translate ideas about mainstream radio listeners into concrete programming policy. In the days before digital satellite transmission or web streaming, Drake monitored client stations via telephone. Any disc jockey not conforming to the spirit of Boss Radio might receive a corrective call from THE boss. Except for fielding those dreaded calls, few directly heard from (or saw) Drake. He seemed to be both everywhere and nowhere. A chain of command—from him to the national program director to the local program director then down to the announcer—seemed to make edicts more powerful and enhanced the Drake mystique.

In 1970 *Newsweek* called Drake "the most powerful force in broadcast rock . . . with complete control of programming, commercial time, and even the hiring and firing of local disc jockeys." Technically, such a stranglehold was against Federal Communications Commission (FCC) rules requiring station licensees to be in total control of their stations, but the magazine had a point to make on behalf of recording companies. They complained that, unless Drake put a particular record on his well-researched 30-song playlist, it had little chance of being spun by other hit music broadcasters. Songs over three minutes (eating up precious airtime in which another tune could be started) had difficulty making the "Boss 30" grade. A promising four-minute song was often abridged in a Drake-Chenault–sanctioned production room.

By 1970 the firm had 40 client stations paying upwards of $10,000 monthly for its services. Reportedly, some felt cornered into such an association. Outwardly secure broadcasters, hearing rumors that the station across town was talking to Drake-Chenault, were known to panic, then hire the consultants first as an insurance measure. The term *Fake Drake* surfaced as a euphemism for outlets, often in smaller markets, choosing to save the consulting fee by imitating the "factory-authorized" format. The imitation typically involved a set of short Drake-like jingles; a copy of the weekly KHJ, CKLW, or WRKO music survey; cassette aircheck tapes of those stations; and an erstwhile Drake client station's disc jockey bent on becoming a noted programmer.

Changing Opportunities

The FCC's mid-1960s directive that most FM stations could no longer simulcast programming of their commonly owned AM sisters provided a new marketplace for Drake's output. RKO asked him to develop a format that could be inexpensively executed via automation on its FM properties. In late 1968 a middle-of-the-road (softer rock hits) offering, called "Hit Parade," was installed in Boston, Los Angeles, and San Francisco. Its tempo soon increased, as did interest in the canned fare on the part of other budget-minded FM operators. By the mid-1970s, "Hit Parade" and several other ready-to-run formats (such as "Solid Gold," "Classic Gold," "XT-40," "Super Soul," and "Great American Country") brought the taped Drake repertoire to about 200 stations in markets of various sizes. Associates tested a progressive rock offering but indicated that Drake never related to (for example) Grateful Dead or Janis Joplin album cuts, and they backed away from the genre.

The advent of increasing competition through separate FM formats, plus 18- to 34-year-olds' changing taste in music, rendered the original Drake sound somewhat shopworn by 1973. Consequently, the RKO AM stations showed signs of ratings erosion. The hit single, pop music radio's 45-rpm traditional stock in trade, was exhibiting a sales slide. Instead, young people started directing their attention toward the long-playing 33 1/3–rpm albums, which were primarily heard (without frequent jingles, promotion, commercialization, or staged forward momentum) on the emerging FM rock outlets. The 1960s generation, which Drake had cleverly schooled to demand a "more music/less talk" approach, discovered that even the most unsophisticated FM rock station (typically staffed by taciturn announcers who just wanted to let the latest album track all the way through) delivered more music than Boss Radio ever did. RKO officials insisted that Drake personally spend more time immersed in solving its stations' audience exodus. Agreement, however, could not be reached, and RKO ended the relationship in 1973.

Following the RKO termination, Drake-Chenault sought another Los Angeles programming platform. It came in the form of struggling KIQQ (FM), dubbed K-100 to denote a 100.3-megahertz dial position. The duo was granted stock ownership in the venture and five years to make K-100 a profit center. Also at stake was a chance to sink KHJ and embarrass RKO. Although much effort went into making K-100 into a ratings giant, the operation was filled with tension (compared to the sense of collegial fun that had driven the 1965 KHJ debut), and it never made a similarly significant impact on southern Californians. K-100 was sold in 1978.

By the late 1970s, the 24-hour syndicated format genre that Drake's firm had helped pioneer found itself swimming in a sea of competitors. Improved automation gear and, eventually, satellite delivery made the program distribution process easier for newcomers to enter. Even though from about 1980 Drake's programming had largely disappeared from major-market stations, his name still commanded respect in broadcasting circles. He and Chenault continued syndicating their turnkey formats through about 1980, when Drake-Chenault Enterprises was quietly sold (and eventually folded into Jones Satellite Radio).

In 1990 Drake actively reentered programming to help fine-tune KRTH (FM), KHJ's former sister. To his credit, Bill Drake found contemporary music radio mired in discordant presentation and intuitively built it a concise programming formula followed in principle by scores of broadcasters today.

PETER E. HUNN

See also Automation; Chenault, Gene; Consultants; Contemporary Hit Radio Format/Top 40

Bill Drake. Born Philip Yarbrough (sometimes documented as Yarborough) in Waycross, Georgia, 1938. Attended Georgia Teachers College (now Georgia Southern University); announcer, WMGR, Bainbridge, Georgia, 1955–57; announcer, WWNS, Statesboro, Georgia, 1957–61; announcer/program director, WAKE, Atlanta, Georgia, 1961; program director/announcer, KYA, San Francisco, California, 1961–62; program director, KYNO, Fresno, California, 1962–64; programming consultant, KSTN, Stockton, California, 1963–65; cofounder/president, Drake-Chenault Enterprises, 1963–ca. 1980; programming consultant, KGB, San Diego, California, 1964–72; programming consultant, KHJ, Los Angeles, California, 1965–73; programming consultant/co-owner, KIQQ FM, Los Angeles, 1973–78; programming consultant, KRTH FM, Los Angeles, early 1990s.

Further Reading

Eberly, Philip K., *Music in the Air: America's Changing Tastes in Popular Music, 1920–1980,* New York: Hastings House, 1982

Fong-Torres, Ben, *The Hits Just Keep on Coming: The History of Top 40 Radio,* San Francisco: Miller Freeman Books, 1998

Fornatale, Peter, and Joshua E. Mills, *Radio in the Television Age,* Woodstock, New York: Overlook Press, 1980

Goulart, Elwood F., "The Mystique and Mass Persuasion: A Rhetorical Analysis of the Drake-Chenault Radio Programming, 1965–1976," Master's thesis, Humboldt State University, 1976
Passman, Arnold, *The DeeJays*, New York: Macmillan, 1971
"Rock and Roll Muzak," *Newsweek* (9 March 1970)
Sklar, Rick, *Rocking America: An Insider's Story: How the All-Hit Radio Stations Took Over*, New York: St. Martin's Press, 1984

Drake, Galen 1907–1989

U.S. Radio Announcer and Program Host

Known for his homespun advice and storytelling, radio host Galen Drake was a consistent audience draw on local and national airwaves during the 1940s and 1950s. Once dubbed "The Most Convincing Microphone Voice" by *Radio Life* magazine, Drake pitched thousands of products over radio during the course of his career and hosted various programs, all of which revolved around his just-plain-folks commentary and philosophy. Drake reached the pinnacle of his radio career during the 1950s, when he simultaneously hosted a weekly variety show from New York City over the Columbia Broadcasting System (CBS) and two daily programs over radio station WCBS in New York.

Although Drake never acquired the popularity of fellow CBS star Arthur Godfrey, the two shared a somewhat similar style. Many listeners appreciated their advertisements because they seemed genuine, not overly fawning toward the products. Drake, like Godfrey, delivered his product appeals in a conversational, spontaneous manner, which he altered depending on his mood and experiences with the products; he rarely scripted his pitches, using only a few hastily sketched notes.

When Drake wasn't selling products, he hosted radio shows that were essentially forums for his advice, statements of philosophy, and stories; the conversational style that made his product pitches so convincing also characterized his performance on radio shows. Drake liked to sound spontaneous, a quality he achieved by striding into the radio studio less than five minutes before airtime, clutching only the sparsest of notes. Although he planned topics for discussion, he claimed to have no set outline to address them. "How am I going to know what I'll say until I start talking?" he once exclaimed. "Do you draw up diagrams of your conversation when you go visiting friends?"

The raconteur's "conversations" with his radio audience could cover multiple subjects in the course of one 45-minute program. For example, on one broadcast in 1947 he dealt with night vision, the effect of emotions on driving, mountain climbing, inventors, and juvenile delinquency. He constantly peppered his shows with commonsense observations: "There are two times to keep your mouth shut: when you're swimmin' and when you're angry" or "What a man must do he can do. When he says he cannot, he means he will not."

Drake told interviewers that he based his advice and truisms on the words of philosophers and psychiatrists he studied as well as on the homegrown ideals that his father (who was in the furniture business) articulated. Drake's knack for delivering his advice and truisms grew from conversing, while still young, with many adults (three half siblings were grown at the time of Drake's birth) as well as from his acting, which began at Long Beach (California) Polytechnic High School during the 1920s.

Drake's high school acting coincided with his employment at radio station KFOX in Long Beach, where he sang and acted in radio plays. After graduating from high school in 1926, he remained at KFOX and also directed plays at the Long Beach Community Playhouse. In 1940 he debuted on the airwaves of CBS-affiliated KSFO in San Francisco, California, and two years later he landed at the CBS-owned KNX in Los Angeles, California. Drake's association with CBS radio would be his avenue to national fame.

Over KSFO and KNX, Drake hosted the *Sunrise Salute* and *Housewives Protective League* programs, which were created by Fletcher Wiley and produced locally at CBS radio stations around the country. The programs touted products approved by a panel of housewives and featured advice and stories for their female audience. Drake perfected his intimate radio delivery (which, however, some critics derided as smug and arrogant) as a voice of the Housewives Protective League (HPL). It was through his product marketing on the Fletcher Wiley programs that he came to define himself as a product pitchman rather than an entertainer. "I'm not a radio star," he once said. "I'm a partner in a client's business when I get his product to sell."

Galen Drake
Courtesy CBS Photo Archive

In 1944 the radio pundit moved to the East Coast to broadcast for the HPL over radio station WJZ in New York City. Three years later Drake went to WCBS in New York, which in 1954 would become the base of his first CBS network program, the *Galen Drake Show.* The show aired Saturday mornings at 10:15 EST and offered a mixture of music and Drake's wisdom and interviews. *Billboard* magazine reviewed the program in 1954 shortly after its debut: "Drake chats away, tells a few stories and interviews members of the studio audience who have the most unusual hobbies, are young grandmothers and are from out-of-town. There is no pretentiousness about the stanza or its entertainment, and yet it's most effective" (see Morse, 1954).

Drake's CBS program performed so well that in 1957 ABC-TV offered him a television show. However, Drake's appeal failed to translate to the new medium, leaving him to cling to his radio base. He continued on WCBS until 1959, when he moved to radio station WOR in New York. His network program was picked up by the Mutual Broadcasting System, but Drake was nearing the end of his radio career. By 1963 he was off the air, a victim, perhaps, of diminishing radio audiences and the explosion of disc jockey programs. The personality's last notable show had been his nightly readings of the Bible over WOR, which began in 1960. In the mid-1960s, Drake returned to Long Beach, where he did occasional advertisements and acted in the Salvation Army's *Heartbeat Theater,* a syndicated radio drama. Drake died of lung cancer in 1989.

MICHAEL STREISSGUTH

Galen Drake. Born Foster Purcell Rucker in Kokomo, Indiana, 26 July 1907. Youngest of two children born to Theodore and

Flora Rucker; attended University of California, Los Angeles, 1926; wrote, directed, sang, and acted in radio plays, KFOX, Long Beach, California, 1926–40; host and announcer, KSFO, San Francisco, 1940–42; KNX, Los Angeles, 1942–44; WJZ, New York City, 1944–47; WCBS, New York City, 1947–59; WOR, New York City, 1959–early 1960s. Died in Long Beach, California, 30 June 1989.

Radio Series

1945–58 *Galen Drake Show* (also heard as *This Is Galen Drake*)

Selected Publications

This Is Galen Drake, 1949
What You Can Do Today, 1960

Further Reading

Bigsby, Evelyn, "Sunrise Salute," *Radio Life* (5 March 1944)
David, Miles, "Galen Drake: Radio's Highest-Paid Copywriter," *Sponsor* (21 September 1953)
Morse, Leon, "Galen Drake Show," *Billboard* (13 February 1954)

Drama on U.S. Radio

The American radio industry is and has been the largest in the world. Before the rise of television, U.S. radio drama was creative as well as commercially popular. Radio also provided a marketing dimension for the Hollywood film industry. The political need to gain government support for the network monopoly was one factor behind the substantial network investment in quality sustaining (non-advertiser supported) programs.

As radio reinvented itself as a music medium in the 1950s, dramatic productions quickly fell out of favor. The competition from popular dramatic story telling on television was too intense. While the counter-culture movement of youth protest in the 1960s generated a demand for radio that offered a space for intelligent speech and alternative music, it did not support storytelling. Unlike the radio situation in Europe, Canada, and Australia, for example, American radio drama funding was not centralized into one service. Radio drama projects relied on entrepreneurial projects to raise funds from sponsors, private foundations, station budgets, and such public resources for cultural projects as the National Endowment for the Arts and the National Endowment for the Humanities. The developing publishing market for spoken-word cassettes and CDs in the 1980s and 1990s provided further commercial collateral for production budgets.

Origins

Early U.S. drama developed in individual stations across the country. Even before World War I, Charles "Doc" Herrold organized schedules for a station in San Jose, California, which in 1914 included transmitting a live play. A drama series was broadcast by station WGY in Schenectady, New York, in 1922. The WGY players began with a full-length production of Eugene Walter's *The Wolf* and soon established a regular Friday night schedule of two-and-a-half-hour performances of anything from *The Garden of Allah* to Ibsen's *The Wild Duck*. For three years not a single playwright asked for payment. An orchestra provided music in the silences between scenes and acts. By 1923 the WGY players had launched a $500 radio drama prize competition to encourage scripts specifically tailored to the medium. The rules stipulated plays that were "clean," avoided "sex dramas," and employed small casts of five to six characters. One hundred scripts were submitted, but the play selected did not result in a successful broadcast.

As WGY productions began to be shared with other stations by early 1924, some of the actors were paid and other stations started radio drama centers of production. WLW in Cincinnati broadcast *A Fan and Two Candlesticks* by Mary MacMillan, which was followed a week later by the balcony scene from *Romeo and Juliet.* Drama became a weekly event. The transmission of *When Love Wakes*, an original play written by program manager Fred Smith, may have been the first play written especially for radio when it aired 3 April 1923, nine months before the transmission of Richard Hughes' *Danger* on the BBC in January 1924.

During these early years, radio stations realized that it was easier to control sound levels within a studio than to depend upon a theater stage as an arena for drama performance. In WGY's case, engineers designed microphones hidden in lampshades in case actors became nervous at the sight of a bare microphone. Chicago's KYW took to the air specializing in broadcasting operas. *Radio Digest* observed in October 1923 that the radio play was increasing rapidly in popularity and that many eastern stations had their own theatrical groups.

That same year, as a publicity device, WLW in Cincinnati provided airtime for the Shuster-Martin School to perform drama readings.

One of the first sponsored dramas was probably included in the *Eveready Hour,* which was launched in December 1923 on WEAF, New York City, to sell Eveready batteries. By 1927 the *Eveready House* program was producing a prestigious drama on a monthly basis, with each production auditioned before the sponsor three weeks before airing. Actors were now being paid $75 to $125 per performance. Rosaline Greene was praised for her portrayal of Joan of Arc.

Golden Age of Popular Radio Drama

The golden age of U.S. radio drama really began in the late 1920s when first the National Broadcasting Company (NBC) and later the Columbia Broadcasting System (CBS) began distributing networked programs. Drama was included along with dance and jazz concerts, because drama had "an aura of respectability." The titles included *Great Moments in History, Biblical Dramas, Real Folks, Main Street,* and *True Story.* For the next three decades, virtually all American network drama and comedy programs were in the format of series in that they used a continuing cast of characters and provided programs on a regular (usually weekly) schedule. Individual episodes usually stood alone, each one a complete story, but such fare was built around continuing characters who would be around the following week. The growing number of network series were supplemented with anthology (sometimes dubbed "prestige") drama programs, whose characters and stories both varied (such as the long-running *Lux Radio Theater*), with no continuing elements save perhaps for a host and regular scheduled air time (and probably a sponsor).

One kind of radio series—the daytime soap opera—offered something additional; their continuing plot lines further focused the series to become a serial that combined continuing characters with stories that often lasted for years, with subplots melding into one another in a never-ending fashion. Each episode depended on the episodes that had gone before and led directly to the following episodes, though stories moved very slowly. Serial announcers usually began each episode with a paragraph or two describing what was happening for listeners who had missed any segments. The plots were purposely designed to hook listeners into regular attendance.

African-Americans and Radio Drama

The cultural gulf separating the white majority from the African-American minority (until the civil rights movement of the 1960s) tended to distort and sometimes censor the presentation of the black identity and story-telling culture. African-American performers had negotiated roles in white-interpreted and -mediated arenas for popular story telling. Although African-Americans performed some parts in *Amos 'n' Andy* (1928–60), "blacked up" whites Freeman Gosden and Charles Correll played the central characters; Madaline Lee played the duo's secretary Miss Blue; and Eddie Green performed the role of Stonewall, the lawyer. To put it mildly, this was an unsatisfactory context for the expression of African-American identity. Eddie Anderson's role as Rochester, Jack Benny's black valet, represented mainstream participation in radio's popular story-telling culture, but was also controversial in the perpetuation of the "Jim Crow" syndrome of racial stereotyping. Another distorting and comic exploitation of African-American women as maids was the character and series *Beulah* (1945–54). The first actor to play *Beulah* was William Hurt, a white man, who coined the character's famous catchphrases "Love dat man!" and "Somebody bawl for Beulah?" The paradoxical ambiguity of radio's representation of stereotypes became evident in 1947 after Hurt died from a heart attack and the Academy Award–winning African-American actress Hatie McDaniel took over the part.

World War II saw the introduction of positively drawn African-American soldier characters in daytime serials as the U.S. government sought to promote the contribution of black servicemen and reduce racial tension within the armed forces.

The problem of negative racial stereotypes and chronic discrimination against African-Americans' participation was challenged by the work of Richard Durham at WMAQ in Chicago between 1948 and 1950. Durham originated, wrote, and directed a series called *Destination Freedom* that dramatized the stories of black achievers such as Benjamin Banneker, Sojourner Truth, Brooker T. Washington, Marian Anderson, and Joe Louis. Durham had been editor of the black newspaper the *Chicago Defender.* The scripting of 91 half hour episodes and their production remains a significant event in American radio drama history. The most notable and widely praised episode was the dramatization of the accomplishments of heart surgeons Dr. Daniel Hale Williams and Dr. Ulysses Grant Dailey in "The Heart of George Cotton," originally aired in Chicago on 8 August 1948 and restaged in 1957 on the networked CBS Radio Workshop.

In 1944 Langston Hughes collaborated with the British radio drama producer D.G. Bridson to create a ballad-opera exploring the friendship between black Americans going to war with the people of Britain. The cast included Ethel Waters, Canada Lee, Josh White, and Paul Robeson; *The Man Who Went To War* was produced and performed in New York but only heard via shortwave by 10 million BBC listeners in Britain. Langston Hughes also wrote *Booker T. Washington In Atlanta,* commissioned by the Tuskegee Institute and CBS. Despite joining The Writers' War Board after Pearl Harbor in December 1941, he faced blacklisting pressures from the House of Un-American Activities from October 1944. Erik

Barnouw wrote that even when American networks had commissioned and produced work by black writers, affiliate stations in the Southern states would often block the broadcast by substituting a local program of musical records.

Sustaining and Prestige Drama

Experimental radio drama of exceptional quality was produced in the United States during the golden age as a by-product of the commercial success of the networks. Advertising profits financed such programming in unsold time. A "marquee status" in radio culture sparked by competition between CBS and NBC helped to generate such programs. Even the poor relation of U.S. radio networks, Mutual (MBS), would commission 21-year-old Orson Welles and his Mercury Theatre to produce a six-part dramatization of Victor Hugo's *Les Miserables* in 1937—ambitious storytelling and dramatic performance on a grand scale. Through a series called *Columbia Workshop* (1936–47), CBS was the first network to experiment with using sound effects for creative and cultural storytelling. Starting in July 1936 the programmers advanced radiophonic techniques for sparking the imagination. The discovery of sound filters that evoked ghostly phenomena gave birth to *The Ghost of Benjamin Street.*

The production of the script *The Fall of the City* by poet Archibald MacLeish presented a social and political attack on totalitarianism and ambition in production. The large "crowd" cast, the special location of performance at New York's Seventh Regiment Armory, and the quality of the cast, which included Burgess Meredith and the young Orson Welles, combined to establish a radio drama broadcast on 4 March 1937 that defined the potential of the medium. *The Fall of the City* invested production confidence in the idea of a drama written in verse for radio. Barnouw (1945) described the resulting competition for prestigious drama projects between the networks. CBS contributed a series of Shakespeare productions featuring John Barrymore. NBC recruited Arch Oboler, who started with a production of his own play *Futuristics* and then persuaded the network to support a series of experimental horror stories, *Lights Out.* Oboler founded a tradition of science fiction horror and melodrama that continued in a series bearing his name.

CBS also signed up Orson Welles and the *Mercury Theatre of the Air* (1938). A formidable production, performance, writing, musical composition, sound design, and directing team of Orson Welles, John Houseman, Howard Koch, and Bernard Herrmann fashioned classic and contemporary novels and plays into highly charged hour-long sequences of live radio entertainment. The subtle sound-design creativity of Ora Nichols advanced the interface of sound and imagination for listeners. The adaptation of the H.G. Wells novelette *The War of the Worlds* at the end of October 1938 would write radio drama into social and cultural history and send Orson Welles and his troubadours to Hollywood.

It was also in 1938 that CBS vice president William B. Lewis hired Norman Corwin to make a series of half-hour programs on Sundays to experiment with poetry. Corwin would become a tour de force in writing and radio drama. The series *Words without Music, Pursuit of Happiness,* and *Twenty-Six by Corwin* established his reputation. From his verse play for the festive season of that year, *The Plot to Overthrow Christmas,* to constitutional and historical pageants such as *We Hold These Truths,* Corwin contributed a body of literature and direction for radio that resonated in and had considerable influence on the English-speaking world. The poetics of his writing was also embedded with political poignancy; *They Fly through the Air* was an audio equivalent of Picasso's famous painting on the bombing of Guernica, and his *Seems Radio Is Here to Stay* is a verse essay on the beauty and potentialities of the medium.

An analysis of Corwin's verse play *The Undecided Molecule,* aired only weeks before the detonation of the atomic bombs in Japan in 1945, reveals writing, directing, and performance in advance of its time. Groucho Marks played the role of a judge metaphorically trying the idea of the atom bomb and mankind's use of it in a surreal courtroom. It was a culturally and politically subversive weave of irony, spiced with postmodernist and existential wit and a tour de force of production and performance that served to define Corwin's power and achievement in the history of world radio drama. It was also an elegant and powerful demonstration of radio drama's literary credentials.

Detective Drama

The golden age produced a genre of audio-noir detectives both male and female. *The Adventures of Maisie,* featuring a character who globe-trotted the high seas from one exotic port to the next, began with a man being slapped when asking for a light and the catchphrase "Does that answer your question, Buddy?" *The Adventures of Nero Wolfe,* based on the popular novels by Rex Stout, proved that you could be a good private detective and so fat that your assistant Archie would have to do all your legwork for you. *The Fat Man* (1946–51) had the central character Brad Runyon starting each episode as the announcer spoke these words: "There he goes into the drugstore. He steps on the scale. Weight 237 pounds! Fortune: Danger!" *The Adventures of Philip Marlowe* (1947–51), *The Adventures of Sam Spade Detective* (1946–51), and *The Adventures of the Thin Man* (1941–50) arose out of the successful novels and films featuring the characters so named. The character Nora Charles, first played by Claudia Morgan in the radio series, amounted to a curious blend of femme fatale and positive gender representation. Nora, with a voice that purred with sexuality, was a supersleuth. *Martin Kane, Private Eye* (1949–53) was an example of a radio detective series that made the successful transformation to television. The highly successful *Sherlock Holmes* (1930–36; 1939–46) culminated with British actors Basil Rathbone and Nigel Bruce in the roles of Holmes and Watson.

The invisibility of radio extended the boundaries of imaginative devices in story telling. *The Shadow* (1930–54) worked best on radio because the character had the power to be invisible. *The Shadow* and other series, such as *The Green Hornet* (1936–52) and *Nick Carter—Master Detective* (1943–55), transplanted the myth of the Western into the urban environment. The Western was also present in U.S. popular drama through, among others, the series *The Lone Ranger* (1934–56). *The Adventures of Dick Tracy* (1935–48) was based on the comic-strip detective created by Chester Gould and is another example of the radio detective genre that cross-pollinated newspaper/magazine comic strips and films.

The police radio series was a cultural mechanism for the mythologizing of Edgar Hoover's FBI or "G-men" and the large-city police departments. Notable series included *Dragnet, Call the Police, Calling All Cars, Crime Does Not Pay, Gangbusters, Famous Jury Trials, Official Detective, Renfrew of the Mounted Police, Silver Eagle Mountie, This Is Your FBI, The FBI in Peace and War, True Detective Mysteries,* and *Under Arrest.*

The detective genres generated controversy because of their stereotypically negative representations of Asian-Americans. *Charlie Chan* (1932–48) was built around an Asian-Hawaiian private eye who was always played by white American actors. The series *Fu Manchu* (1929–33), arising out of the Collier magazine stories, offered another example of the demonizing and typecasting of Chinese or Asians as "untrustworthy, inscrutable" villains.

"Soap Opera" Drama

Conditions for developing a thriving market of radio serials were ripe in 1930s America. National networks were expanding because they provided "free" entertainment once one owned a receiver. Further, and despite the Depression, there was a continuing market for products that could improve the quality of life in the home, including soap products and washing machines. Radio provided an excellent means of reaching out to the growing market of women concerned with such purchases, and seeking entertainment. Radio could meet the demand for entertainment about family and identity, about fantasy and the idea of home, about struggle against adversity and achievement, and about people who could be admired and respected. From these circumstances, the soap opera was born.

In Los Angeles Carlton Morse began writing episodes for the series *One Man's Family* in 1932; it would last for 28 years. The story of the Barbour family depended on its addictive narrative drive for its success. Radio had become the stage for its own popular American version of Galsworthy's *Forsyte Saga.* In 1939 Morse created *I Love a Mystery,* featuring the exotic adventures of three global adventurers from the A-1 Detective Agency. He wrote every word and directed every script; his craft had him up at 4 A.M. and kept him busy seven days a week. Barnouw (1945) wrote that radio was becoming a mecca for acting talent that could no longer find work in the theater. Radio was helping listeners—and more than a few creative people—buck the Depression.

Serials that charted social mobility and advancement secured lucrative sponsorship because the audience could identify with the reality and aspirations such programs embodied. *The Goldbergs* (1929–50) performed the ritual of a social journey from the Lower East Side of New York City into the middle-class suburbs; Pepsodent was a product that helped enhance the smile on their faces. Writer and actor Gertrude Berg created *The Goldbergs* in her quest to dramatize Jewish family life. Susan J. Douglas (1999) has observed the value of the research by Herta Herzog into the relationship between women listeners and soap operas: "The melodramatic narratives and strong female characters of daytime serials—coupled with the intimacy of the medium—provided powerful points of identification." Herzog's research in the 1930s showed that, to one woman listener, soap opera "teaches me as a parent how to bring up my child." Popular radio drama during the golden age was an opportunity for women's self-empowerment, and as Barnouw observes, "Almost one-third [of listeners] spoke of planning the day around serials." *Little Orphan Annie* (1930–42) could be comfortably associated with the coziness and nourishment of the Ovaltine bedtime drink. *Myrt and Marge* had the attitude that came with chewing gum made by Wrigley. *Buck Rogers 25th Century* (1932–36, 1939–40, 1946–47) was the kind of dream that listeners could think about when crunching on their breakfast cereals produced by Kellogg.

Many of the key producers and writers who controlled the form were highly educated and independent women. Irna Phillips probably wrote more words and made more money in the field of soap opera than anyone else. Her prodigious industry was founded in Chicago, and once she hit her stride no one could match her ability to invent story lines or dictate six scripts a day and write 3 million words a year. She was the creator of *Today's Children, Woman in White, Right to Happiness, The Guiding Light, Road of Life,* and *Lonely Women.* Her gift to the history of gender representation is that she invented and sustained women characters who were role models to American women listeners because they had strength and dignity and could hold their social position equally with men.

Another key center for popular series and serial production was founded and developed by advertising executive Frank Hummert and his assistant Anne S. Ashenhurst, who later became his wife. They established a team of writers distinguished by the talented *Chicago Daily News* reporter Robert D. Andrews, who probably generated more than 30 million words of storytelling for radio. Andrews' first serial story was *Three Days Lost.* Within a year of being hired by the Hummerts he was turning out five radio scripts a day. The author of a book on *How to Write for Radio* was convinced that Andrews was really three or four writers and that his name was the brand title of a writing syndicate. The Hummerts generated legends in the tradition of soap opera entertainment, some of which transferred to television.

In some respects the soap opera boom of the golden age could be described as the "Wild West of Writing." It was a Klondike for authors, advertisers, and networks because 80 percent of the programming of a network station in a big city market in 1939 was made up of wall-to-wall daytime soap episodes, most of them merely 15 minutes long. Elaine Carrington was an example of a short story writer who found that her ability to produce story outlines and scripts for daytime serials could make her a rich author. She conceived and wrote *Red Adams,* which became *Red Davis* and then *Pepper Young's Family* as the sponsors changed.

In the daytime soap by Jane Crusinberry, *The Story of Mary Marlin* (1935–45, 1951–52), Crusinberry dramatized a character who became a female U.S. senator. *Mary Marlin* was one of the highest-rated daytime serials after 1937, and because Crusinberry retained control of all the writing, the series was able to transcend the political and social compromises that arose from sponsor-controlled "factory writing." During World War II Crusinberry originated and wrote a series called *A Woman of America* (1943–46), which starred Anne Seymour and dramatized the history of achieving women in the United States.

The Legacy of World War II

The political struggle between communism and fascism through the 1930s and the years of World War II coincided with the most intense period of the "Golden Age" of radio. Howard Blue (2002) examines the work of 17 radio dramatists and writers who deployed the radio drama arts in their battle against fascism. Between 1941 and 1945 they were allied with commercial radio networks, private agencies, and U.S. government propaganda organizations that wished to rally the listening imagination in the fight against German Nazism and Japanese militarism.

Blue documents how the chill wind of the Cold War made casualties of left of center writers, directors, and actors who had been politically motivated in their creative engagement with radio drama during wartime. The personal memoir by the radio actor Joseph Julian, published in 1975 as *This Was Radio,* provides a compelling and agonizing account of how Senator Joe McCarthy's witch-hunting could snuff out a career virtually overnight.

The compendium of 25 radio plays edited by Erik Barnouw and published in 1945 as *Radio Drama in Action* represents a significant body of literature from this period of broadcasting. Morton Wishengrad's *The Battle of the Warsaw Ghetto,* broadcast by NBC on the eve of the Day of Atonement in 1943, demonstrated how radio dramatization of actuality could succeed where radio journalism had failed. There was no effective contemporary reporting of the extraordinary rebellion by young Jewish fighters in the Warsaw Ghetto in April 1942. Wishengrad's research and literary imagination combined with the direction of Frank Papp, a music score by Morris Mamorsky, and acting performance by Arnold Moss to represent a vital moment in history.

Radio Drama in the Shadow of Television

With the development of U.S. network television beginning in 1948, radio drama's days were numbered. Within just a few years, audiences and advertisers had begun the rapid migration to the video medium, and network schedules grew sparse. The McCarthy witch hunts and Cold War paranoia also damaged American radio drama at the same time. Opportunities for significant network projects and corporate sponsorship were not coming to anyone who was perceived to be left of center. Orson Welles, Norman Corwin, and Paul Robeson had characteristics and track records that could be regarded as left wing. Along with thousands of talented writers, actors, and directors, they could be perceived as politically subversive. Blacklisting generated self-censorship and drove a community of artists into exile. Others were silenced, their credentials ruined. Perhaps one of the more absurd manifestations of this cultural anomie was the 1952–54 syndicated series *I Was a Communist for the FBI,* in which film actor Dana Andrews infiltrated organizations as a double agent and week by week roamed episodes with titles such as *The Red Among Us, The Red Waves,* and *The Red Ladies.*

Money and talent became concentrated in television. There were courageous and worthy projects, such as *CBS Mystery Theater* between 1974 and 1977 produced by Himan Brown, which tried to turn back the clock. Such ambition and concentration of resources, however, was not sustained by audience figures and the interest of sponsors.

Drama on Public Radio

With the end of commercial radio's golden age in the early 1950s, radio drama was for many years an art hidden behind the cornucopia of television programming. Still, the mutually advantageous relationship between the Hollywood film industry and radio networks during the 1940s and 1950s exemplified by *Lux Radio Theatre, Hollywood Hotel/Premiere,* and *Hollywood Star Preview/Star Playhouse, Star Theater* left seeds for future development and opportunity. In the 1970s Himan Brown produced original radio dramas for CBS radio. Still later, George Lukas donated the sound rights of his *Star Wars* stories to his former university radio station. National Public Radio (NPR) invested more than $200,000 in a 13-part dramatization including members of the film's cast, music composer, and sound designer.

Satellite distribution generated a renaissance in interest in audio drama and created a new audience among young people. In some respects interest in the U.S. *Star Wars* project was similar to the interest shown by young audiences in Britain in the radio series *The Hitchhiker's Guide To the Galaxy.* Filmic music and multi-track sound design techniques combined with the cult of science fiction to produce success.

The period 1971 to 1981 witnessed the development of *Earplay,* a National Public Radio Drama Production Unit based in Madison, Wisconsin, under the artistic direction of Karl Schmidt. The project was substantially funded from federal sources and generated radio drama script competitions for new writers. It eventually developed large-scale collaborations with the BBC in England and commissioned well-known established writers such as Edward Albee, David Mamet, and Arthur Kopit.

National Public Radio's rival American Public Radio (APR, which later became Public Radio International) also generated interest in original spoken word story telling through the work of Garrison Keillor in Minnesota. NPR, in Washington, D.C., continued to courageously distribute *NPR Playhouse,* but, despite the enthusiasm of producer Andy Trudeau, radio drama became a cultural artifact in the tapestry of U.S. radio. Trudeau even commissioned an original series of new Sherlock Holmes dramatizations starring Edward Petherbridge, and NEH and NEA funding supported worthy cultural drama projects such as Samuel Beckett and German *Horspiel* seasons, which were the initiative of Everett Frost. Unfortunately, the poor take-up by NPR affiliate stations of *NPR Playhouse* programs resulted in its demise in September 2002.

Independent Producers and Radio Drama by Artisans

American radio drama has moved from a mass-appeal service based on daily or weekly series to a far narrower format aimed at small but elite audiences. There would appear to be no shortage of ambition and commitment from small independent producers all over the United States who use efficient modern digital technology to produce original plays and dramatizations that are crafted for a connoisseur audience mainly in public radio. New York–based independent producer Charles Potter and Random House have established a niche interest in the radio Western, with audiobooks that sell well in the retail market and are also carried by some radio stations over the air. The internet, the audio drama cassette/CD/minidisc market, and digital radio offer accessible, low-cost networks of distribution. Furthermore, it is a low-risk genre for ideas and counter-culture. Companies such as Ziggurat, ZBS Foundation, Atlanta Radio Theatre, The Radio Repertory Company of America, Hollywood Theater of the Ear, The Radio Play—The Public Media Foundation, Midwest Radio Theatre Workshop, LA Theatre Works, Shoestring Radio Theatre, and many others have established significant output of original productions. Notable directors/dramatists include Yuri Rasovsky, Eric Bauersfeld, Joe Frank, and David Ossman.

During the 1980s and 1990s, WBAI, the Pacifica radio station in New York, was the arena for an interesting development in the art of the live community radio play. The station's arts director Anthony J. Sloan catalyzed much of this work. Sloan observed that "Most people did taped drama because it's safer. BBC does radio drama every day, but it's canned. I like live radio drama because the adrenaline flows for the actors. They know that not only is this live, but, guess what, it's only one-time. You get some incredible performances." Sloan orchestrated a series of media pageants that have occupied the streets of New York, the studios of WBAI, the satellite frequency of Pacifica Network programming, and the worldwide web with orchestras of musical, dramatic, and acoustic artistic expression fused by captivating, bold narratives. The productions were not short half-hour or one-hour sequences. They spanned five-and-a-half hours of airtime. Philosophically challenging, politically controversial, intellectually stimulating, emotionally invigorating dimensions of communication combined with complex sound production techniques and live performances on the sidewalks of the Lower East Side and various landmarks in the urban geography of New York City. The grassroots dimension of this work was an indicator of how radio drama could strengthen its identity and cultural value with its audiences. The *Leaving(s) Project,* transmitted on the night of 26 January 1996, comprised two live story-telling events over five-and-a-half hours. Larry Neal's play *The Glorious Monster in the Bell of the Horn* was presented before a live audience at the New Knitting Factory in the Tribeca section of Manhattan. The play was structured in the style of the epic opera based on the Brothers Grimm's *Peter and the Wolf,* wherein characters are identified by musical instruments. Then the multimedia event blossomed into "a journey piece" from different locations of the New York metropolitan area. There were six different groups of characters leaving New York for various reasons who were forced to deal with personal crises on their way to an Amtrak train at New York's Penn Station. Their interweaving storylines highlighted current social, political, spiritual, and artistic issues. All the disparate journeys were acted out live with moving microphones on location and culminated in a dramatic finale at Penn station. The realism of the event is indicated by the fact that the fictional characters intended to board the 3:45 A.M. Amtrak red-eye service leaving New York, which was actually waiting to leave one of the platforms at the end of the broadcast. The event began at 10 P.M. on Friday night and continued until 3:45 the following morning. It could be heard in stereo on WBAI 99.5 FM, received by satellite on 360 community radio stations, and heard nationally and internationally on the worldwide web.

Early 21st century U.S. radio drama can be described as "the age of the artisan," whereas the period before the 1950s could be described as "the age of the Network." Audio dramatic techniques are also widely used in advertising and public information spots, so narrative creativity in radio is not a totally lost art.

TIM CROOK

See also, in addition to individual shows and people mentioned in this essay, Blacklisting; Hollywood and Radio; National Public Radio; Pacifica Foundation; Playwrights on Radio; Poetry and Radio; Public Radio International; Science Fiction; Soap Opera; WBAI; Westerns

Further Reading

Augaitis, Daina, and Dan Lander, editors, *Radio Rethink: Art, Sound, and Transmission,* Banff, Alberta: Walter Phillips Gallery, 1994

Barnouw, Erik, editor, *Radio Drama in Action: Twenty-Five Plays of a Changing World,* New York and Toronto, Ontario: Farrar and Rinehart, 1945

Blue, Howard, *Words At War: World War II Radio Drama and the Postwar Broadcasting Industry Blacklist,* Lanham, Maryland: Scarecrow Press, 2002

Brooke, Pamela, *Communicating through Story Characters: Radio Social Drama,* Lanham, Maryland: University of America Press, 1995

Burnham, Scott G., "Soundscapes: The Rise of Audio Drama in America," *AudioFile* (July 1996)

Crook, Tim, *Radio Drama: Theory and Practice,* London and New York: Routledge, 1999

Daley, Brian, and George Lucas, *Star Wars: The National Public Radio Dramatization,* New York: Ballantine Books, 1994

Douglas, Susan J., *Listening In: Radio and the American Imagination: From Amos 'n' Andy and Edward R. Murrow to Wolfman Jack and Howard Stern,* New York: Times Books, 1999

Fink, Howard, "The Sponsor's v the Nation's Choice: North American Radio Drama," in *Radio Drama,* edited by Peter Elfed Lewis, London and New York: Longman, 1981

Fink, Howard, "On the Trail of Radio Drama: Organizing a Study of North American and European Practices," *Journal of Radio Studies* 6, no. 1 (1999)

Grams, Martin, Jr., *Radio Drama: A Comprehensive Chronicle of American Network Programs, 1932–1963,* Jefferson, North Carolina: McFarland, 2000

Guralnick, Elissa S., *Sight Unseen: Beckett, Pinter, Stoppard, and Other Contemporary Dramatists on Radio,* Athens: Ohio University Press, 1996

Hilmes, Michelle, *Radio Voices: American Broadcasting, 1922–1952,* Minneapolis: University of Minnesota Press, 1997

Julian, Joseph, *This War Radio: A Personal Memoir,* New York: Viking, 1975

Kahn, Douglas, and Gregory Whitehead, editors, *Wireless Imagination: Sound, Radio, and the Avant-Garde,* Cambridge, Massachusetts: MIT Press, 1992

Shingler, Martin, and Cindy Wieringa, *On Air: Methods and Meanings of Radio,* London and New York: Arnold, 1998

Worlds without End: The Art and History of the Soap Opera (exhibition catalogue), New York: Abrams, 1997

Drama Worldwide

Varied Traditions of Radio Narrative

Radio has been and continues to be a substantial venue for serious and popular storytelling throughout the world. As English has become the global media language, it is possible to identify plays that have been the seed of stories in film, books, stage theater, television, and the internet. Arthur Miller and David Mamet are two leading U.S. playwrights who have acknowledged the debt they owe to radio drama for influencing and developing their writing abilities. The politically controversial Italian playwright Dario Fo excelled in the radio medium. Wolfgang Borchert and Peter Handke are literary giants in post-war German culture, and their literary reputations stem from their radio output. The director and sound play artist Klaus Schöning has articulated a distinct and original movement in the radio drama genre. The foundations of Orson Welles' film directing genius may well lie in his radio experience as much as in his theater work. His experience in writing, performing, and directing in radio is greater in volume and range than for any other medium in which he worked. Archibald MacLeish, one of the United States' leading 20th-century poets, wrote radio plays of exceptional literary quality. Norman Corwin developed a contemporary form for radio verse drama and fused it with the contemporary resonance of world events and U.S. history as it was happening. While the non-English speaking world has also created galaxies of storytelling cultures in radio drama, our knowledge of them has been compromised by the limited amount of translation and critical writing available.

Functions

Radio drama as storytelling tends to serve a number of purposes. It can define national cultural identity through original writing and performance. Dylan Thomas' *Under Milk Wood*, described as "a play for voices," is central to the consideration of Welsh literary achievement in the post-war period and is respected internationally as an example of radio drama literature par excellence.

Radio drama can be a location for the exploration of social, cultural, and sometimes political anxieties through popular soaps and long-running series. The popular soap has been a staple in the history of radio drama in many Latin American countries. Public and state radio in Poland has a long-standing tradition of supporting "high cultural" writing as well as the popular in plays and serials.

Through social action dramas, radio can warn, educate, and improve society through a blend of information, education, and entertainment. The British Broadcasting Corporation's (BBC) long-running radio soap opera *The Archers* was conceived as a method of improving the efficiency of farming techniques when there was still rationing in the aftermath of World War II. Social action dramas are prevalent in the developing world, and radio offers an opportunity to engage in the oral culture traditions of African and Asian countries.

Radio dramas also offer an arena for showcasing and adapting novels and short stories as well as stage literature from live theater. This has been the case in most Scandinavian and European countries. Hungary, France, Germany and Italy have extensive archives of scripts and recordings demonstrating radio drama's commitment to producing their countries' leading writers and dramatists.

Finally, radio drama has been used as propaganda and in support of war. Conventional plays and series are constructed to influence listeners ideologically, and the techniques of dramatization have been harnessed to fake enemy broadcasts, to deceive military forces, and to weaken the morale of the enemy's home population. In the early 1950s the CIA used audio drama techniques to fake the sound of a non-existent army and airforce to overthrow a left wing government in Guatemala.

Political Economy of Radio Drama

The success of soap operas in the United States and Australia during the 1930s and 1940s represented mass communication and popular expression for new writing on an enormous scale. Dramatizations of literature and significant original plays had their place when the profit-led radio corporations saw an advantage in prestigious productions impressing commercial sponsors. Sometimes a minority cultural program such as Orson Welles' *Mercury Theatre on the Air* pulled off a publicity stunt, such as the 1938 *War of the Worlds*, and this popularized radio drama in terms of entertainment.

But spending profits for prestige was not a recipe for stability and growth in the arts. If the capitalist, corporate moguls of the 20th century became the equivalent of patron princes from the Renaissance, they were adept in applying the ruthlessness of those princes when cutbacks were needed. Publicly funded national radio networks have been the cultural umbrellas for most of the original radio drama produced in the world and the sound dramatization of prose and poetry. However, the financial relationship between radio drama production centers and their state-funded national radio networks is often clouded by the potential for political compromise and economic expediency. Radio drama has been at the mercy of economic instability and the political pressure to reduce public expenditure. The prerequisite for public funding is sometimes predicated on how well state-funded radio drama performs in comparison with the audience surveys of its commercial counterparts. As a result, radio drama in some countries such as Australia now concentrates on high culture and experimentation rather than the maintenance and production of popular series and serials. In France, radio drama has undergone a painful reappraisal through rationalization and cutbacks.

United Kingdom

Leading playwrights such as Samuel Beckett and Tom Stoppard have written extensively for the radio medium. Stoppard's stage play *Indian Ink* was conceived and first produced as the radio play *In the Native State*. Britain's playwright Caryl Churchill had nine of her radio plays produced by the BBC up until 1973, when her stage work began to be recognized at the Royal Court Theatre. Hanif Kureishi, regarded as one of Britain's leading Asian writers, famous for his film *My Beautiful Launderette* and his novel *The Buddha of Suburbia*, was first produced in radio. Sue Townsend, Harold Pinter,

Alan Ayckbourn, Alan Plater, Anthony Minghella, Angela Carter, Alan Bleasdale, Willy Russell, and Louis MacNeice are a few other literary luminaries whose writing roots were planted in radio drama.

One of the paradoxes of radio drama is that highly accomplished and revered writers who have chosen to specialize in this field remain locked in a cabinet of obscurity. Rhys Adrian, who died in 1990 after having written 32 plays broadcast by BBC radio, is an example of such a writer little known today.

Giles Cooper cultivated the art of dramatically counterpointing the exterior and the interior of characters who felt themselves "trapped in the contemporary machinery of modern life and who were unable to escape." Cooper wrote over 60 scripts for BBC radio. His 1957 play *The Disagreeable Oyster,* along with the production of Samuel Beckett's *All That Fall,* was fundamental in creating the need for a permanent sound workshop to create aural images based on effects and abstract musical rhythms.

In 1939/1940 BBC Radio Drama commissioned and produced *In the Shadow of the Swastika,* which offered a humanist challenge to the anti-Semitic prejudice engendered by Nazi ideology. The occasion of war stimulated a production that may have commanded the highest audience for any play broadcast in the history of radio. Norman Corwin's *We Hold These Truths* was commissioned to commemorate the 150th anniversary of the United States Bill of Rights. This coincided with the Japanese attack at Pearl Harbor, and eight days after the attack, Corwin assembled a cast of the country's leading actors including James Stewart and Orson Welles and broadcast to an audience of 60 million listeners across all the U.S. networks. Many of the dramas and drama-documentaries produced by Corwin throughout the war were preoccupied with themes that promoted the Allied cause and challenged the morality of the Axis powers.

Germany

Germany possesses a rich and diverse critical tradition of the radio drama form, but the language barrier means that its wealth of texts is inaccessible to the radio drama communities of the English-speaking world. It is also ironic that one of the most prolific, versatile, and widely published critical analysts of British radio drama is the German academic Horst Priessnitz. Radio drama has also been frequently used as propaganda. Nazi Germany used skilful mixtures of popular music and drama to psychologically intimidate Allied troops and civilian populations. They were sometimes aided by United States and British fascists. The U.S. academic Frederick Wilhelm Kaltenbach used dramatic scripts in overseas English broadcasts to attack the British position in the war. He translated a radio play by Erwin Barth von Wehrenalp called *Lightning Action* to celebrate the German victory in Norway. Twelve scenes were recorded on 5 April 1941, and the cast included the British film actor Jack Trevor and other ex-patriots. He also satirized Roosevelt's Lease-Lend Bill with a series of dramatic talks called *British Disregard for American Rights.* In May 1944 U.S.-German academic Otto Koischwitz wrote a doomsday radio play for the D-Day invasion forces and their families at home that was broadcast by shortwave to the United States. The actress Mildred Gillars ("Axis Sally") played the part of a GI's mother who in a tear-stained monologue predicted disaster and grief.

An argument could be made that German radio drama is distinctive for the greater importance it has had within the cultural traditions of German drama and literature. There is evidence that *Hörspiel* has generated a cornucopia of inspiration and originality in other storytelling media. The Weimar Republic era is distinguished by the work of playwright Bertolt Brecht, composer Paul Hindemith, critic Walter Benjamin, and composer Kurt Weill. They and many other pioneers understood the relationship between sound drama and radio reception and interrogated and explored modernist ideas of stream of consciousness found in symbolist novels and short stories, imagistic poetry, and developments in modern theater.

Whereas German Radio Drama has mirrored the BBC in generating a powerful canon of popular drama, dramatized literature, soap opera, light comedy, and detective and mystery series, German *Hörspiel* has become established as a literary art and drama form equal to stage theater, film, television, and literature. Alfred Döblin's novel *Berlin Alexanderplatz* reached a popular audience through German radio in 1929 before it was produced for television by Rainer Werner Fassbinder. The nihilistic torpor of Nazi propaganda radio suffocated expression in the art form, and after the war the cultural role that radio played offered a rich arena for diversity and quality. The shortage of printing paper and the destruction of theater and cinema meant that radio drama contributed to the recovery and development of German storytelling. Decentralization of the structure of public stations meant that several centers of radio drama production flourished simultaneously.

The positive public response to the transmission of Wolfgang Borchert's *The Outsider* in 1947 illustrated the centrality of *Hörspiel* in the country's social, psychological, and cultural psyche. Bochert, Ingeborg Bachmann, and Gunter Eich illustrate the force of *literarische Hörspiel,* and Peter Handke, Jurgen Becker, and Reinhard Lettau are auteurs who represent that development of *Neue Hörspiel* that sought to explore new ideas and philosophies about the prosody of sound and thought in modern and postmodern society.

Canada

Original radio drama thrived in Canada because the government had decided to support public radio through the Canadian Broadcasting Corporation (CBC). Canadian radio drama experienced a golden age of literary and dramatic expression from the1940s to the 1960s. Canadian radio drama has been re-inventing itself in the public sector. CBC has maintained an imaginative and flexible approach to international co-production. Director James Roy helped originate the 1996 series *Searching Paradise,* co-produced between CBC in Toronto, ABC in Perth, and BBC Wales in Cardiff. He also pioneered the introduction of *The Diamond Lane*, a lively modern soap drama sitting comfortably with CBC's peak time morning format for 2 years during the late 1990s. The series featured live performances of actors as commuters traveling to work using a freeway lane reserved for vehicles containing three or more passengers and interacting with the on air presenters in the studio. The development of the English Speaking Radio Drama Association has facilitated the exchange of publicly funded radio plays such as CBC's *Mourning Dove* and work done in New Zealand, Hong Kong, Wales, Los Angeles, Australia, and South Africa.

Sweden

Swedish public radio has discovered new storytelling forms to find a new generation of radio drama listeners. The dramaturges at Swedish Radio have cleverly propelled radio drama into the mainstream of artistic and cultural debate by creating a five minute soap opera that is broadcast within the peak listening morning slot between 7:50 and 8:00 A.M. Eva Stenman-Rotstein at the publicly funded Swedish Radio Broadcasting Corporation has steered a series of evolutionary changes to young people's radio drama that has captured a new generation of listeners. Storytelling for children and young people has also attracted sizable audiences.

Asia and Africa

The literary and dramatic traditions of African, Arabic, and Asian countries are virtually unknown in Europe and North America and other English-speaking countries. Many western societies are struggling to reconcile themselves to a past history of racist structure of education, imperialist history, and negative stereotyping of other countries' political, industrial, social, and cultural values. Furthermore, as these societies seek to realize their own multi-cultural status, interesting examples of creative reception in radio drama is emerging. British Asians and Afro-Caribbeans have found a confident voice in writing and production.

The satirical comedy drama series *Goodness Gracious Me* originated on BBC Radio 4 and has successfully transferred to television. Whereas the BBC of 1929 would produce *Shakuntala* or *The Lost Ring* by Kalidasa with a translation by a European academic, direction and performance by an all white European cast, and orientalist attitudes to promotion and representation, the BBC of 2003 would mediate a classic work from an Asian by transferring the interpretation to Asian writers, directors, and performers. The Africa Service of the BBC has been a substantial patron of writers, dramatists, and poets from Africa, and producers have sought to create a co-production dynamic between London's Bush House and the writers, directors, actors, and audiences in African countries. An annual competition produces a series of radio plays called African Performance.

Japan

In the early 1990s the Television and Radio Writers' Association of Japan set up and ran an international award called the Morishige Audio Drama Contest. Over a period of four years, productions were entered from all parts of the world and were given equal treatment. The representation in this competition revealed a thriving and comprehensive infrastructure of radio drama practice and tradition from African and Asian countries that had not been as well represented in European dominated international competitions and festivals.

The Television and Radio Writers' Association had about 900 freelance writer/members and set up the award to stimulate the support of radio drama within the domestic and international radio industry. It also sought to make "a contribution to cultural exchange." The Morishige Award succeeded in its second objective. The first objective foundered on the lack of funds to continue the contest. The Asia-Pacific Broadcasting Union had also been seeking to encourage cultural celebration and exchange. In 1993 the Morishige selection committee were in the position to consider three plays entered by the state broadcasting system NHK, based in Tokyo and Nagoya, and further entries from Kyushi Asahi Broadcasting Company, Tokyo-FM Broadcasting Company, the Tokyo Broadcasting system, and Nippon Cultural Broadcasting. These plays demonstrated a fertile and competitive field of audio play production in both publicly funded and commercial sectors of the industry.

An entertaining representation of Japanese radio drama can be found in the film by Koki Mitani, *Welcome Back Mr. McDonald* (1998). This is a backstage farce that uses a live radio drama to send up Japanese society, American blockbuster mentality, and the prima donna values of show business.

Korea

The Korean Broadcasting System's (KBS) entry for the 1993 Morishige Prize was *The Angel's Curse* written by Choi Jae-Do, directed by Cho Won-Suk—KBS's Chief Producer of radio drama. At this time Korea, like Japan, had a thriving broadcast dramatists' association with 500 members who were also established poets, novelists, and stage playwrights. Korean radio drama could draw upon a profession of 400 actors. KBS was producing 200 or more single plays every year. New writers for radio were continually being brought on through special competitions, and the commercial broadcasting corporations produced 30 new radio dramatists every year in this way. There was also evidence of radio drama production and broadcasting in China, Mongolia, Uganda, Egypt, Hong Kong, Malawi, and South Africa.

India

In 1993 plays by the director Kamal Dutt had been entered in the Morishige award from All India Radio, which at the time of writing presented a considerable range of Indian radio drama on the internet. In 1991 P.C Chatterji observed that the powerful theater movement in several parts of India had only "marginally affected the field of radio drama." He said that the ordinary run of radio play was "not of a high standard" because the rate of payment was poor and there was little hope of their utilization elsewhere. He cited *Tumhare Ghum Mere Hain* (Your Woes Are Mine) by Delhi playwright Reotic Sharan Sharma and *Harud* (Autumn) by Shankar Raina as examples of radio texts that had successfully transferred to theater and film and been recognized internationally.

Radio Drama Futures

From 1990 there has been a maturing of a global spoken word market so that the talking book, sound drama, or sound dramatization have been fighting for equal space on the shelves with traditional books. Radio drama's ephemeral status as an art form could be at an end. The performance of a dramatic script no longer exists just in the fleeting moment of a live stage event. It is being captured on cassette, compact disc, mini-disc, computer file, and other means of electronic storage for replay. Multimedia and the internet offer exciting dimensions to sound drama production and storytelling. The radio dramatist has been liberated from the dimension of short-lived terrestrial sound broadcasts.

Erik Ohls and the Swedish Radio Theatre in Finland have been pioneering the use of the internet for the promotion and more meaningful distribution of radio drama as an art form. Distribution of linear sound narrative can interact with lateral channels of sound, text, animation, and photographic and video images on the worldwide web. The web is also liberating from the point of view of control and means of production. The British website www.irdp.co.uk is an example of the internet being used as an independent space for new writing in radio outside the territories of state-funded broadcasting. Sound communication on the internet is not subject to government licensing and censorship. Transmission is instantly stable in the international dimension and the technology is affordable.

In 2001 BBC radio commissioned a project called "the Lab" to explore the futures of audio drama in terms of interactive communication technology. The internet, with its hypertextual lateral routes of structured narrative and digitalization, seemed able to expand the potentiality of sound storytelling. Similar experimentation has been undertaken by the drama department of Austria's public radio service.

The challenge facing radio drama producers of the 21st century seems to center on how the radio play can attract the younger generation when the form is not relevant or central to their media consumption. The BBC in Britain and public radio broadcasters in Europe and elsewhere have been trying without much success to establish thriving audio drama projects with young people and ethnic communities using new forms of interactivity via the internet. The fact remains that the most lucrative resources of radio drama funding in European public networks are controlled largely by middle-aged and middle-class people.

The development of digital radio in Britain has resulted in the establishment of two national radio channels dedicated to audio dramatic genres, *BBC 7* and *Oneword* (commercial). The BBC has the advantage of a huge back catalogue and the cushion of guaranteed funding from the license fee. Diversification of access to the broadcasting spectrum and an equality of opportunity in public funding may be a potential solution to continuing decline in radio drama activity in Britain and the rest of the world.

TIM CROOK

See also, in addition to individual shows and people mentioned in this essay, Canadian Radio Drama; Drama, U.S.; Playwrights on Radio; Poetry and Radio; Science Fiction; Soap Opera; Westerns

Further Reading

Adams, Douglas, *The Hitch-Hiker's Guide to the Galaxy: The Original Radio Scripts,* edited by Geoffrey Perkins, London: Pan, 1985

Allan, Andrew, *All the Bright Company: Radio Drama Produced by Andrew Allan,* edited by Howard Fink and John Jackson, Kingston, Ontario: Quarry Press, and Toronto, Ontario: CBC, 1987

Arnheim, Rudolf, *Radio,* London: Faber, 1936

Ash, William, *The Way to Write Radio Drama,* London: Elm Tree, 1985

Augaitis, Daina, and Dan Lander, editors, *Radio Rethink: Art, Sound, and Transmission,* Banff, Alberta: Walter Phillips Gallery, 1994

Beck, Alan, *Radio Acting,* London: Black, 1997

Beck, Alan, *The Invisible Play: The History of British Radio Drama, 1922–1928* (CD-ROM), Canterbury, England: Kent University, 2000
British Broadcasting Corporation, *Author and Title Catalogues of Transmitted Drama, Poetry, and Features, 1923–1975,* Cambridge: Chadwyck-Healey, and Teaneck, New Jersey: Somerset House, 1977
Brooke, Pamela, *Communicating through Story Characters: Radio Social Drama,* Lanham, Maryland: University Press of America, 1995
Callow, Simon, *Orson Welles: The Road to Xanadu,* London: Cape, 1995; New York: Viking, 1996
Cantril, Hadley, Hazel Gaudet, and Herta Herzog, *The Invasion from Mars: A Study in the Psychology of Panic with the Complete Script of the Famous Orson Welles Broadcast,* Princeton, New Jersey: Princeton University Press, 1940
Chatterji, P.C., *Broadcasting in India,* New Delhi and Newbury Park, California: Sage, 1987; revised edition, 1991
Crisell, Andrew, *Understanding Radio,* London and New York: Methuen, 1986; 2nd edition, London and New York: Routledge, 1994
Crook, Tim, *Radio Drama: Theory and Practice,* London and New York: Routledge, 1999
Cusy, Pierre, and Gabriel Germinet, *Théâtre radiophonique: Mode nouveau d'expression artistique,* Paris: Chiron, 1926
Daley, Brian, and George Lucas, *Star Wars: The National Public Radio Dramatization,* New York: Ballantine, 1994
Döhl, Reinhard, *Das neue Hörspiel,* Darmstadt, Germany: Wissenschaftliche Buchgesellschaft, 1988
Döhl, Reinhard, *Das Hörspiel zur NS-Zeit,* Darmstadt, Germany: Wissenschaftliche Buchgesellschaft, 1992
Drakakis, John, editor, *British Radio Drama,* Cambridge and New York: Cambridge University Press, 1981
Durham, Richard, *Richard Durham's Destination Freedom: Scripts from Radio's Black Legacy, 1948–50,* edited by J. Fred MacDonald, New York: Praeger, 1989
Fink, Howard, "The Sponsor's v the Nation's Choice: North American Radio Drama," in *Radio Drama,* edited by Peter Elfed Lewis, London and New York: Longman, 1981
Fink, Howard, "On the Trail of Radio Drama: Organizing a Study of North American and European Practices," *Journal of Radio Studies* 6, no. 1 (1999)
Gielgud, Val Henry, *British Radio Drama, 1922–1956: A Survey,* London: Harrap, 1957
Guralnick, Elissa S., *Sight Unseen: Beckett, Pinter, Stoppard, and Other Contemporary Dramatists on Radio,* Athens: Ohio University Press, 1996
Independent Radio Drama Productions website, <www.irdp.co.uk>
Kahn, Douglas, and Gregory Whitehead, editors, *Wireless Imagination: Sound, Radio, and the Avant-Garde,* Cambridge, Massachusetts: MIT Press, 1992
Lewis, Peter Elfed, editor, *Radio Drama,* London and New York: Longman, 1981
Lewis, P.M., "Referable Words in Radio Drama," in *Broadcast Talk,* edited by Paddy Scannell, London and Newbury Park, California: Sage, 1991
McWhinnie, Donald, *The Art of Radio,* London: Faber, 1959
Richards, Kieth, *Writing Radio Drama,* Sydney: Currency Press, 1991
Schöning, Klaus, editor, *Hörspielmacher: Autorenporträts und Essays,* Königstein im Taunus, West Germany: Athenäum, 1983
Schöning, Klaus, editor, *Neues Hörspiel: Essay, Analysen, Gespräche,* Frankfurt: Suhrkamp, 1970
Schöning, Klaus, editor, *Spuren des neuen Hörspiels,* Frankfurt: Suhrkamp, 1982
Shingler, Martin, and Cindy Wieringa, *On Air: Methods and Meanings of Radio,* London and New York: Arnold, 1998
Stoppard, Tom, *Stoppard: The Plays for Radio, 1964–1991,* London: Faber, 1994

Drew, Paul 1935–

U.S. Radio Personality and Executive

There are four phases to the career of broadcaster and entrepreneur Paul Drew: early in his career, Drew influenced teenagers in Atlanta and the South as a disc jockey introducing the new sounds of Top 40 radio and traveling with the Beatles. Later, his influence extended to the broadcast industry in the United States and Canada as a vice president of RKO General Radio. He formed several companies that linked the United States and Japan in entertainment ventures. And he was the first director of Radio Martí, establishing Voice of America broadcasts to Cuba.

On the Air

As an Atlanta disc jockey, Drew avoided the "hyped" sound fashionable at Top 40 radio, opting instead for a soft, conversational approach that won him high ratings among teenagers on three stations. After stints at stations in his home state of Michigan, Drew moved to Atlanta in 1957 to join WGST, where he was advertised as "Atlanta's most music-wise DJ."

In 1961 Drew joined Atlanta's WAKE at the invitation of his neighbor, WAKE Program Director Bill Drake. His show was a combination of tidbits about the artists whose records he played and descriptions of mythical "submarine races in Piedmont Park," an excuse for teenage sweethearts to sit in their cars by the park's lake.

It was his move to WQXI in 1964 that gave Drew his national reputation for picking hits. First as music director and later as program director, Drew attracted the attention of both the record industry and the radio industry. Because of that early visibility, he called his ten years in Atlanta radio his most important, because "I came in as nobody and left as somebody."

During the 1960s, Drew was dubbed "the Fifth Beatle" (and occasionally "the bald-headed Beatle") because he was the only broadcaster to travel with the Beatles on all their American tours. His daily reports from the tours were heard by millions of radio listeners. On Christmas Eve 1964, Drew produced a worldwide special with the Beatles from London.

Executive and Entrepreneur

For five years during the 1970s, Drew was vice president of programming for the radio division of RKO General Corporation, supervising formats as diverse as talk, Top 40, oldies, adult contemporary, and classical. At the time, RKO owned stations in Boston, New York, Memphis, Miami, Washington, San Francisco, Los Angeles, and Chicago.

The connections Drew made with recording artists throughout his years in radio resulted in a series of live concerts and artist specials produced for both radio and television. Featured were Neil Diamond, Chicago, Elton John, Olivia Newton-John, Cher, Frank Sinatra, and a long list of others.

He introduced the Japanese singing group Pink Lady to the United States and established them as the only Japanese artists to have both an American hit single and their own prime-time network TV show (on the National Broadcasting Company [NBC]).

His business interests expanded beyond broadcasting and music to specialize in Japan with a variety of companies, including Paul Drew Enterprises, The USA Japan Company, the Mobotron Corporation, and the 2151 Corporation (the last two in partnership with Sony Corporation and the family of Sony founder Akio Morita).

Concurrently, Drew was commissioner to the California Motion Picture Council and served under two California governors. He was also a White House adviser for President Jimmy Carter's energy program and served on the Commission of Ceremonies for the 1984 Olympic Games in Los Angeles.

In 1984 Drew received a nonpolitical appointment from President Ronald Reagan to be the first director of Radio Martí, which established broadcasts of the Voice of America aimed at Cuba. When he received the call from a friend of President Reagan, Drew had just returned from one of many trips to Japan, where he had accompanied California Governor Jerry Brown and introduced Brown to Japanese business opportunities for the state of California. Drew described himself as a "life-long Democrat" but accepted the appointment from a Republican administration because, as he stated, "I was doing something for my country." After the Radio Martí experience, Drew confessed: "The radio part was easy. The political part was not."

During his years in broadcasting, Drew guided and developed many well-known air talents and programmers. In *Network 40* magazine, Gerry Cagle noted, "Perhaps Paul's greatest legacy lies in the success of those he hired." The list of broadcasters who worked for Drew includes consultants Jerry Clifton, Don Kelly, and Guy Zapoleon; industry writers Gerry Cagle, Jerry Del Colliano, Walt "Baby" Love, and Dave Sholin; and air personalities Rick Dees, Charlie Van Dyke, Jay Thomas, and Dr. Don Rose.

In 1999 Drew and his wife Ann moved from their longtime home, Los Angeles, to Forsyth, Georgia, where he claimed he "was not officially retired."

ED SHANE

See also Radio Martí

Paul Drew. Born in Detroit, Michigan, 10 March 1935. Attended Wayne State University; began career at WDET-FM, Detroit, 1954; air personality at WHLS-AM, Port Huron, Michigan, 1955; night-time host, WGST-AM, Atlanta, Georgia, 1957; moved to WAKE, Atlanta, 1961; air personality, WQXI, Atlanta, 1964; music and program director at WQXI; program director at WIBG, Philadelphia, Pennsylvania, CKLW, Windsor, Ontario, Canada, KFRC, San Francisco, California, WGMS, Washington, DC, and KHJ, Los Angeles, California, 1968–73; while at KHJ, named vice president of programming for parent company, RKO General Broadcasting; director, Radio Martí, 1984–85; formed several companies, including Paul Drew Enterprises, Mobotron Company, and 2151 Corporation (a partnership with RayKaySony).

Further Reading

Cagle, Gerry, "Paul Drew: Still 'Cagey' after All These Years," *Network* 40 (17 February 1995)

Farber, Erica, "Publisher's Profile—Paul Drew, Legendary Programmer," *Radio and Records* (15 October 1999)

Fitzherbert, Tony, "A Profile of Legendary CKLW," *Monitoring Times* (April 1991)

Fong-Torres, Ben, *The Hits Just Keep on Coming: The History of Top 40 Radio,* San Francisco: Miller Freeman Books, 1998

"In Retrospect—Paul Drew, Master Programmer," *Radio Ink* (7 September 1992)

McCoy, Quincy, *No Static: A Guide to Creative Radio Programming,* San Francisco: Miller Freeman Books, 1999

Duffy's Tavern

Comedy Program

—*"Duffy's Tavern, where the elite meet to eat, Archie the manager speaking, Duffy ain't here. Oh, hello, Duffy . . ."*

Every week, a ringing phone and Archie's nasal New York accent invited listeners into *Duffy's Tavern,* a weekly situation comedy set in a dilapidated pub in the heart of Manhattan's east side. Running the place on behalf of the ever-absent Duffy, Archie the manager and his cohorts—Eddie, Finnegan, Clancy the Cop, Miss Duffy (Duffy's daughter), and others—welcomed a new guest star or guest character every week into a defiantly low-class atmosphere of barbed but friendly give and take.

Duffy's was famous for its play with (and mistreatment of) language, especially Archie's constant malapropisms. "Leave me dub you welcome to this distinctured establishment," Archie said to guest star Vincent Price, "and leave me further say, Mr. Price, that seldom have we behooved such an august presentiment to these confines. . . . And feel assured, Mr. Price, that your visit is a bereavement from which we will not soon recover." As comedian Georgie Jessel once chided Fred Allen on *The Texaco Star Theatre:* "Fred! Two split infinitives and a dangling metaphor—people will think this is Duffy's Tavern!"

The pilot for *Duffy's Tavern* aired on 29 July 1940 on the Columbia Broadcasting System (CBS) radio program *Forecast,* which aired previews and pilots of proposed CBS shows in order to gauge audience reaction. The reaction in this case was enthusiastic, and CBS picked *Duffy's* up as a weekly half-hour program beginning in March 1941, running on Saturday nights at 8:30. It moved to two more time slots over the next year, until October 1942, when it switched to National Broadcasting Company's (NBC) Blue network, running at 8:30 on Tuesdays. When NBC Blue became the separate network American Broadcasting Companies (ABC), *Duffy's Tavern* moved to the NBC network proper, where it ran for the next seven years.

Duffy's Tavern was largely the brainchild of its star, Ed Gardner (born Ed Poggenburg), who first created the character of Archie for the CBS program *This Is New York* in 1939. Archie's right-hand man was "Eddie the Waiter," played by Eddie Green, an African-American comedian with a dry wit, who often took the wind out of Archie's sails. Green's part was notable at the time for its lack of stereotype or inferiority: the writers for *Duffy's Tavern* received an award during "Negro History Week" in 1946 for providing Green with such positive, racially inoffensive material.

The most faithful patron of the tavern was the monumentally stupid "Clifton Finnegan," played by veteran radio comic Charlie Cantor (no relation to Eddie Cantor). Cantor had originated the slow-talking, even slower-witted stooge character Socrates Mulligan for the "Allen's Alley" segment on the *Fred Allen Show,* and he eventually transplanted a renamed Mulligan into *Duffy's Tavern.* Cantor was an enormously experienced radio talent, "the Great Mr. Anonymous of radio," appearing on programs such as *The Shadow, Abie's Irish Rose, Dick Tracy, The Life of Riley,* and *Baby Snooks.* At one point he was in such demand that he performed in 26 programs in one week, and for several months he appeared in the same time slot on three different networks in two recordings and one live show.

Ed Gardner's wife Shirley Booth originally played "Miss Duffy," a young woman on the lookout for marriageable men, pursuing them almost as energetically as they fled from her. After Gardner and Booth divorced in 1942, at least 12 different actresses essayed the role of Miss Duffy, most notably Florence Halop and Sandra Gould, whose combined tenure lasted approximately six years.

In 1944 production moved from Manhattan to Hollywood (though of course the tavern remained eternally in New York), and in 1945, Paramount released a *Duffy's Tavern* feature film, starring the central cast of the radio show as their tavern characters (Gardner, Green, and Cantor, with Ann Thomas as the Miss Duffy du jour), surrounded by literally dozens of Paramount contract players appearing as themselves (among them Bing Crosby, Dorothy Lamour, Alan Ladd, and Paulette Goddard). The film received tepid reviews—critics appreciated the star-studded stage show within the movie much more than the framing story featuring the radio characters—but the radio show itself kept going strong.

An NBC report in March 1949 identified *Duffy's Tavern* as one of the network's top four programs, "vital to the maintenance of a strong position in the industry," alongside Fred Allen, Fibber McGee, and Bob Hope. That year there were rumblings within NBC and rumors in the newspapers that Gardner planned to take the show to CBS; in fact, Gardner went so far as to obtain a release from the program's contract with longtime sponsor Bristol-Myers to free up his options to court another network. However, NBC found *Duffy's* important enough to renegotiate. During the 1949 summer hiatus, Gardner and NBC moved the program in its entirety—including equipment, staff, and performers—to Puerto Rico, to take advantage of a 12-year tax holiday intended to attract new industry.

When *Duffy's* returned to the airwaves in the fall of 1949 with recordings sent in from Puerto Rico, the show had a new sponsor, Blatz Brewing Company, which eventually caused some trouble for NBC. Some stations could not or would not allow beer advertising, and some that allowed beer would not accept the accompanying wine trailer ads, causing a number of stations to drop the program entirely. During its last year, *Duffy's Tavern* relied on multiple sponsors, including Radio Corporation of America (RCA) Victor and Anacin.

At the very end of 1951, despite a temporary rise in ratings after the Puerto Rico move, *Duffy's Tavern* was cancelled. In 1954, a syndicated *Duffy's Tavern* television show appeared, starring Gardner as Archie and Alan Reed (the radio show's Clancy the Cop) as Finnegan, but reviews were strongly negative, and the program did not last. By that time, the tavern's style of humor was considered old-fashioned and long past its prime. During the 1940s, however, *Duffy's Tavern* had been one of the mainstays of radio comedy, and long before television's *Cheers* came along, Duffy's embodied "the place where everybody knows your name."

DORINDA HARTMANN

Cast

Archie the manager	Ed Gardner
Eddie the waiter	Eddie Green
Clifton Finnegan	Charlie Cantor
Miss Duffy	Shirley Booth (1940–43), Florence Halop (1943–44, 1948–49), Sandra Gould (1944–48), Sara Berner, Pauline Drake, Helen Eley, Gloria Erlanger, Margie Liszt, Helen Lynd, Connie Manning, Florence Robinson, Hazel Shermet, Doris Singleton
Clancy the cop	Alan Reed
Wilfred, Finnegan's little brother	Dickie Van Patten
Colonel Stoopnagle	F. Chase Taylor
Dolly Snaffle	Lurene Tuttle

Producer/Creator
Ed Gardner

Producers
Mitchell Benson, Rupert Lucas, Jack Roche, and Tony Sanford

Programming History

CBS	29 July 1940 (pilot aired on Forecast)–June 1942
NBC Blue	October 1942–June 1944
NBC	1944–52

Further Reading

Beatty, Jerome, "What's His Name?" *American Magazine* 136 (July 1943)
Hutchens, John K., "A Very Fine Joint," *New York Times* (23 November 1941)
Mooney, George A., "Talk of the Tavern," *New York Times* (6 April 1941)
"Negro History Week Citations Go to 18 for Aid to Race Relations," *New York Times* (11 February 1946)
"New York Hick," *Time* (21 June 1943)
Schumach, Murray, "Regarding Archie," *New York Times* (2 February 1947)
Williams, Richard L., "Duffy's Latin Tavern," *Life* (13 February 1950)

Duhamel, Helen 1904–1991

U.S. Broadcaster and Executive

In the sparsely populated areas of the Midwest, some women were able to advance professionally in radio broadcasting as regional leaders in the industry. Such was the case with Helen S. Duhamel.

During the Depression of 1929, Helen Duhamel, at that time a mother of two small children, found herself in a unique position. Her father-in-law, Alex Duhamel, the owner of the Duhamel Trading Post in Rapid City, South Dakota, was facing near-bankruptcy and turned to her for advice. She encouraged her father-in-law to hold on to his business. During the next two years, Alex Duhamel managed to stay afloat, but he finally asked Helen Duhamel to take over. Using her background as a bookkeeper and seeking advice from the best businessmen in Minneapolis, she devised an innovative plan. By dividing the Trading Post building into smaller sections, she was able to sell some of the units and rent out others. By 1937 she had paid off the mortgage, and the Duhamel Company was out of debt in only five years. By 1940, Helen Duhamel had bought out all the other Duhamel heirs and owned the Duhamel Trading Post free and clear. Located near Ellsworth Air Force Base and Mount Rushmore, the Trading Post prospered during the war years of the 1940s.

From the inception of the first radio station in Rapid City in 1936, Duhamel recognized its potential as a profitable business. As a retail merchant, she had invested in radio advertising and found it an important factor in the success of her business enterprises. In 1943 she became a stockholder in Black Hills Broadcasting, which included KOBH Radio.

Observing the growing interest in radio, Duhamel began purchasing more and more stock in KOBH Radio, which became KOTA in 1945. By 1954 she had bought out all the other stockholders. When attending early broadcasting conferences, she was the only female owner of a radio station present.

Both KOTA and KOBH Radio had original offices in the Alex Johnson Hotel across the street from the Duhamel Trading Post in downtown Rapid City. The station was set up on the 10th floor, with its offices in the solarium on the 11th floor. After Duhamel's acquisition of the radio station, the operation was eventually moved to new headquarters above the Trading Post.

Near the end of World War II, the radio station applied to the Federal Communications Commission (FCC) for permission to expand from a tiny 150-watt station to 5,000 watts. U.S. Representative Francis Case of South Dakota assisted in getting the needed approval. He discovered that the military used the local radio station signals as a homing device for a guidance system at night. By approaching the War Department, Case was able to ensure FCC acceptance of the KOTA application. On 1 January 1945, the application was accepted. The call letters were changed to KOTA, and the station logo was changed to "KOTA—Chief Signal Station in the Old Sioux Nation."

KOTA Radio became a Columbia Broadcasting System (CBS) affiliate, and the stockholders purchased a used transmitter and built three directional towers south of town on Highway 79.

Duhamel expanded her radio operations to include television. In order to bring live network TV to South Dakota, she installed the world's longest privately owned microwave system. In 1966, as an equal partner in South Dakota Cable, she brought cable television to rural South Dakota.

In 1976 Duhamel was elected to the South Dakota Broadcasters Association Hall of Fame. In 1991 she was also selected to be included in the Nebraska Broadcasters' Hall of Fame. Helen S. Duhamel died in 1991 at the age of 87. Bill Duhamel, her son, continues to head the highly successful, privately owned Duhamel Broadcasting Enterprises.

MARY KAY SWITZER

See also Women in Radio

Helen S. Duhamel. Born in Windsor, Missouri, 26 November 1904. Bought an interest in KOBH Radio, Rapid City, South Dakota, 1943; after the station became KOTA, she took over the operation and eventually developed Duhamel Broadcasting Enterprises, 1954; spent 22 days behind Iron Curtain as representative from National Association of Broadcasters, 1969; recognized as first woman president of state broadcasting association, 1961; received Jaycees' "Boss of the Year Award," 1965; McCall's "Golden Mike Award"; Alfred P. Sloan Radio-Television Award for Distinguished Public Service; special letter of commendation from the President of the United States for her stations' public service during the devastating Black Hills flood,1972; elected to South Dakota Broadcasters' Association Hall of Fame, 1976; Nebraska Broadcasters' Association Hall of Fame, 1992. Died in Rapid City, 8 November 1991.

Further Reading

Lewis, Dale, *Duhamel: From Ox Cart . . . to Television*, Chamberlain, South Dakota: Register-Lakota, 1993

"Millennium All-Stars," *Rapid City Daily Journal* (27 September 1999)

Dunbar, Jim 1932–

U.S. Talk-Show Host

Jim Dunbar was shot at five times by a disturbed listener. He may have been the only radio air personality to interview the "Zodiac" serial killer. As one of the pioneers of the news-talk radio format, Dunbar's long broadcast career has been marked by peculiar fate and fortuitous innovation, culminating in his election to the Radio Hall of Fame in 1999. "I'm still a little flabbergasted being up there on a wall," Dunbar said of his plaque in the Chicago-based Hall of Fame. "It's still in a way kind of surreal, seeing your name up there between Tommy Dorsey and Don Dunphy, the old fight announcer" (all Dunbar quotations are from an interview with the author).

Dunbar's career began in 1952 and included stints as a disc jockey and newscaster in Manhattan, Kansas, and Detroit, Michigan, before he became program director and morning disc jockey at WDSU in New Orleans in 1957. Four years later, he moved to WLS in Chicago as assistant program director and disc jockey.

WLS was "a very special station that owned the Midwest," Dunbar said. "We had the Midwest by the ears." He spent three years there but yearned to relinquish on-air duties and be a program director again. "I turned 30 and faced the fear that I might be playing Patti Page and Pat Boone the rest of my career," Dunbar said. "I hated that music. I'm a jazz fan."

Dunbar was hired in 1963 as program director for KGO in San Francisco, a station that consistently ranked last in a 12-station market despite adopting numerous formats. "They tried everything from German bund music to bird whistles, but nothing seemed to work," Dunbar said. "Yellow-cab dispatchers had a bigger audience."

As program director, Dunbar was charged with the responsibility of finding a format to pull KGO out of the ratings basement. He chose a new concept, news-talk, which he helped to shape. "I don't want to take credit for being the inventor of news-talk radio," Dunbar said. "A lot of people shared the credit. I just put together some things that had worked elsewhere, and that I thought would work here"—notably news, talk, and humor. "I had nothing to lose, so I thought, 'Let's give them something so compelling that they would hang on through the next commercial break,'" Dunbar recalled. "Radio had always been background. What we did was we made it foreground."

As Dunbar defined it, the news-talk format is "fundamentally a bulletin service with a heavy emphasis on traffic, along with a series of talk-interview programs that would hold people," Dunbar said. "It was bubble gum of the mind."

Dunbar introduced a duo of consummate practical jokers to the KGO mix. Jim Coyle and Mal Sharpe were masters of street pranks, devising an outlandish premise and then taping ad-lib interviews with unsuspecting passersby. They attempted to persuade people to graft chicken wings to their foreheads to enable them to fly, they tried to persuade a grocer to stock pre-bitten fruit, they sought recruits for a private army of San Franciscans to invade Los Angeles and endeavored to rent pigeons in Golden Gate Park for $1.50 per hour. The well-dressed, straight-faced duo recorded these exchanges while "pushing our victims as far as they'll go before they take a poke at us," Coyle said ("Rent-A-Pigeon," *Newsweek*, 13 January 1964). After they produced an album of their gags in 1963, Dunbar gave Coyle and Sharpe their own three-hour nightly show on KGO. "That's what I feel proudest of," Dunbar said of his career accomplishments. "They were so different from anything else on the air. They were funny and unusual. They helped establish the difference between us and other stations."

Despite Dunbar's innovations, the new format was not an immediate success. After one year, KGO's station manager wanted to abolish the news-talk format and switch to rock and roll. "I told him, 'We are about to turn the corner, and you are making a big mistake,'" Dunbar said. The station manager relented, and one year later KGO was one of the top-ranked stations in the market.

Shortly after taking the reins as program director, Dunbar reversed his decision to stay off the air and made himself an afternoon talk-show host. "Vietnam and conservation were pretty much all anyone wanted to talk about" during the 1960s and 1970s, Dunbar recalled. In September 1973 a psychologically disturbed listener, recent immigrant Lawrence Kwong, believed he heard Dunbar's voice inside his head, threatening him. "He thought I was going to kill him, so he decided to kill me first," Dunbar said. Kwong stood on the other side of the studio window during Dunbar's show and fired five shots. The station had recently installed bulletproof glass, so Dunbar was spared. The enraged Kwong was undeterred, though, and headed for the studio door. When station advertising salesman Ben Munson tried to intervene, Kwong killed him and then committed suicide.

Another unbalanced listener called in to Dunbar's show, claiming to be the notorious Zodiac killer who had committed a string of unsolved murders in California in the 1970s. He promised to give himself up if attorney Melvin Belli agreed on air to represent him. Dunbar invited Belli to the studio, where "Belli so dominated the conversation the guy hung up and called back 54 times in an hour and a half," Dunbar said. The caller made arrangements to surrender to police with Dunbar

Jim Dunbar
Courtesy Radio Hall of Fame

and Belli present, but he never showed. Dunbar is confident that the caller was the real Zodiac killer because he knew details of murders that had not been publicly revealed. (The killer has never been caught.)

In 1975, Dunbar moved to the morning drive-time shift on KGO, a position he retained until his retirement in July 2000. Dismayed by the plethora of shock-jocks and rude talk-show hosts on the airwaves today, Dunbar said he got out at the right time. "We are now imposing on the listener so much, and picking at scabs, that it has gone from intrusive to invasive," he said. Instead he wishes that other show hosts would emulate his approach. "You reveal what your heart is telling you and you let people respond to that, one way or another," Dunbar counseled. "That's how to do a talk show."

RALPH FRASCA

See also Disk Jockeys; KGO; Radio Hall of Fame

Jim Dunbar. Born 9 October 1932. Started radio career at WKAR in East Lansing, Michigan, 1952; served in U.S. Army, 1953–55; disk jockey and newscaster, WHDH, Manhattan, Kansas, 1953–55; disk jockey, WXYZ, Detroit, Michigan, 1955–56; program director and disk jockey, WDSU, New Orleans, Louisiana, 1956–60; assistant program director and on-air talent, WLS, Chicago, 1960–63; program director, KGO, San Francisco, 1963–65; afternoon talk show host, KGO, 1965–74; co-anchor of KGO Radio Morning News, 1974; hosted KGO-TV's morning talk show and anchored 5 P.M. news, 1965–79. Received Best Anchor Team award, Associated Press Television and Radio Association of California-Nevada, 1994; Lifetime Achievement Award from Northwestern University, Medill School of Journalism, 1999; Radio Hall of Fame, 1999.

Further Reading

Graysmith, Robert, *Zodiac,* New York: St. Martin's Press, 1986

Kava, Brad, "KGO's Soothing Voice," *San Jose Mercury News* (26 November 1999)

Marine, Craig, "Radio Stalwart Dunbar Is Stepping down," *San Francisco Examiner* (17 March 2000)

"Rent-A-Pigeon," *Newsweek* (13 January 1964)

Whiting, Sam, "The Man with the Mike: Coyle and Sharpe's Street Ambush Comedy Finds a New Audience Almost 40 Years Later," *San Francisco Chronicle* (19 January 2000)

Dunlap, Orrin E. 1896–1970

U.S. Writer, Editor, and Radio Publicist

Orrin Dunlap was an important newspaper radio editor, a prolific author of books on radio and television, and an important corporate radio publicist working in and reporting about radio's first several decades.

Dunlap's radio experience began in 1912 with a home-built amateur radio transmitter in the attic of his Niagara Falls, New York, home. He gained commercial experience on the Great Lakes as a Marconi Wireless Telegraph operator. During World War I, he served as a U.S. Navy coast station radio operator in Maine. After receiving his B.S. from Colgate in 1920, Dunlap did some graduate coursework at the Harvard Business School before finding employment with the New York–based Hanff-Metzger advertising agency.

He became the *New York Times'* first radio editor in 1922, just as the national craze for the new medium was reaching its peak. Carr Van Anda, the *Times'* managing editor, asked Dunlap to develop a regular radio section for the paper. Dunlap would hold the post for 18 years, becoming a widely read critic and columnist and one of the most influential commentators about radio.

At the same time, Dunlap began writing books about radio; the first one, *The Radio Manual,* grew out of the many technical information requests from *Times* readers. Dunlap was able to write about technical matters in a clear fashion for those with little or no background. Eventually, he would publish 13 volumes on radio and related topics. Among them were two volumes on radio advertising (among the first on the subject to appear); a history of radio; a biography of Marconi in which the inventor cooperated; several books on television (the first being drawn from his columns in the *Times,* his only book so based); a reference book on radio inventors; a chronology of radio and television; and an overall history of telecommunications, the last edition of which appeared after his death. Though all long out of print, several of Dunlap's books remain standard reference works today.

In 1940 Dunlap joined the Radio Corporation of America (RCA) as manager of the company's information department, rising to become a vice president (in 1947) of advertising and publicity before retiring. For most of this period he wrote RCA's annual report.

CHRISTOPHER H. STERLING

See also Columnists; Radio Corporation of America

Orrin Elmer Dunlap, Jr. Born in Niagara Falls, New York, 23 August 1896. Served in U.S. Navy as radio operator during World War I; attended Colgate University, B.S., 1920; employed by Hanff-Metzger Agency, 1920–22; radio editor, *New York Times*, 1922–40; manager, information department, Radio Corporation of America (RCA), 1940–47; vice president of advertising and publicity, RCA, 1947–61. Received Marconi Medal of History from Veterans Wireless Operators Association, 1945. Died in New York City, 1 February 1970.

Selected Publications

The Radio Manual, 1924
The Story of Radio, 1927; revised edition, 1935
Advertising by Radio, 1929
Radio in Advertising, 1931
The Outlook for Television, 1932
Talking on the Radio: A Practical Guide for Writing and Broadcasting a Speech, 1936; revised edition, 1948
Marconi: The Man and His Wireless, 1937; revised edition, 1938
The Future of Television, 1942; revised edition, 1947
Radio's 100 Men of Science: Biographical Narratives of Pathfinders in Electronics and Television, 1944
Radar: What Radar Is and How It Works, 1946
Understanding Television: What It Is and How It Works, 1948; revised edition, 1951
Radio and Television Almanac: Men, Events, Inventions, and Dates That Made History in Electronics from the Dawn of Electricity to Radar and Television, 1951
Communications in Space: From Wireless to Satellite Relay, 1962; revised editions, 1964, 1970

Durante, Jimmy 1893–1980

U.S. Radio, Film, and Stage Performer

When Jimmy Durante performed, he appeared to the world like a king penguin in basic black evening wear and a shapeless black fedora. Strong in his stride and tireless while performing, he was known for his rapid speech, gravelly voice, and cathedral nose. His legacy continues every Christmas season when he is heard as the narrator of *Frosty, the Snowman*.

Origins

Durante was born to Italian immigrant parents in New York City on the kitchen table in his parents' apartment. His parents gave him piano lessons as he grew up. From 1910 until 1914 Durante worked at Diamond Tony's on Coney Island. Billed as "Ragtime Jimmy," he became a talented piano player and learned the ins and outs of handling a crowd.

Durante worked his way up to the Alamo Club in Harlem, where he was bandleader and talent booker. He worked there until 1921 and moonlighted at other clubs. He also made a series of recordings with the Original New Orleans Jazz Band. During his time at the Alamo, Durante met his wife, Jeanne Olson. They were married in 1921.

During the 1920s, Durante gravitated toward vaudeville, where he became part of a comedy music team with Lou Clayton and Eddie Jackson, known as the "Three Sawdust Bums." He opened his own place, The Club Durant, which, although it did not last long, made Jimmy Durante a star. After the club closed in 1925, the three men continued performing music and comedy at various theaters and finally at the Palace Theater.

Durante developed much of his characteristic style during his vaudeville days. He purposefully misused the English language and enthusiastically delivered his signature "Hotcha-cha." He also became known for tunes such as "I Ups to Him," "I'm Jimmy, That Well-Dressed Man," and his theme song, "Inka-Dinka Doo."

Clayton, Jackson, and Durante made their Broadway debut in Ziegfelds *Show Girl* in 1929. Durante had a chance to do a solo and let his personality shine through. The "Bums" also made their film appearance in *Roadhouse Nights,* filmed on Long Island in the Paramount studio. Durante played Helen Morgan's piano accompanist. He knew that he was not the matinee idol type, and he found his popularity increasing as he aged and became loved by another generation.

The next Broadway appearance for the trio was *The New Yorker* (1930–1931), playing speakeasy gangsters. At that time Durante was offered a five-year contract by Metro Goldwyn Mayer (MGM), breaking up the trio. Lou Clayton became Jimmy's manager and confidant until Clayton's death in 1950.

Durante appeared in 17 films over the next four years. In his first film he supported William Haines in *The New Adventures of Get-Rich-Quick Wallingford* (1931). During his early film career, there were few memorable moments. He took time off to reteam with Clayton and Jackson in the Broadway show *Strike Me Pink*. The play got modest reviews. Although his film career was not spectacular, Durante appeared in a number of films from 1936 to 1941. He also appeared on Broadway in the Cole Porter revue *Red, Hot & Blue* (1936–37, with Ethel Merman and Bob Hope) and then in *Stars in Your Eyes* (1939, again with Merman).

Radio

Durante was a guest on many radio shows in the 1930s and 1940s. He appeared with Eddie Cantor, Rudy Vallee, Fred Allen, and others, making appearances as well on such programs as *Fibber McGee and Molly* and *Duffy's Tavern*. In 1943 Durante began appearing on CBS's *Camel Comedy Caravan* with Garry Moore. In 1945 the program was renamed *The Jimmy Durante-Garry Moore Show.* Moore left in 1947, but Durante continued with the show through 1949. While appearing on radio, Durante also appeared in six films at MGM. He was a great comic foil for Van Johnson and June Allyson. In the 1950s he made only three films, starring with Donald O'Connor in *The Milkman, The Great Rupert,* and *Beau James.*

Film and Television

During the late 1940s and the 1950s, some of Durante's greatest success came from television. He co-hosted NBC's *Four Star Revue* in 1950, as well as the *All Star Revue* (1951–52) and the *Colgate Comedy Hour* (NBC, 1953–54). In 1954 he starred in his own show, which ran through 1956. Though his show went off the air, Durante continued to guest-star on television shows and in specials. Baby boomers knew him in the early 1960s from his various appearances on the *Ed Sullivan Show.* In 1969 he supplied the narration for the CBS cartoon *Frosty, the Snowman*, which still runs every year at Christmas. His last television appearance was on the *Sonny and Cher Comedy Hour* in 1971.

Unlike most of his contemporaries, Durante seemed ageless from the late 1940s throughout the early 1970s. In 1972, he suffered a stroke and his health declined. Wheelchair bound, he appeared on various award shows, making his last public appearance at his 83rd birthday bash in 1976. His fame was

Jimmy Durante
Courtesy Library of American Broadcasting

rekindled when his recordings of "As Time Goes By" and "Make Someone Happy" were used on the soundtrack of the movie *Sleepless in Seattle* (1993), putting Durante on the charts in the 1990s.

ANNE SANDERLIN

See also Comedy; Vaudeville

James Francis Durante. Born in New York City, 10 February 1893. Began his music career playing piano in beer gardens in Coney Island; in one club, he played piano for then-waiter Eddie Cantor; organized a jazz/Dixieland band in Harlem, New York; opened the Club Durant with Eddie Jackson and Lou Clayton, 1923; club closed down by Prohibition officers, 1924; with Jackson and Clayton played Parody Club, 1924–26, then entered vaudeville; film debut with Jackson and Clayton, *Roadhouse Nights*, 1930; wrote songs, including "Inka Dinka Doo;" moved to Hollywood after the trio broke up and made numerous motion pictures while also appearing on Broadway; began doing radio with Garry Moore in 1933; starred on *The Jimmy Durante Show*, NBC-TV, with former partners Jackson and Clayton, 1954–56. Recipient: George Foster Peabody Award, 1950; Emmy Award, Best Comedian, 1953. Died in Santa Monica, California, 29 January 1980.

Radio Series

1933–34 *The Chase and Sanborn Hour*
1935–36 *The Jumbo Fire Chief Program*
1943–45 *The Camel Comedy Caravan*
1943–47 *The Jimmy Durante-Garry Moore Show*
1947–50 *The Jimmy Durante Show*
1951–52 *The Big Show*

Television

NBC's Four Star Revue, 1950; *The All Star Revue,* 1951–52; *The Colgate Comedy Hour,* 1953–54; *The Jimmy Durante Show,* 1954–56; *Alice Through the Looking Glass,* 1966; *Jimmy Durante Presents The Lennon Sisters,* 1969

Films

Roadhouse Nights, 1930; *The New Adventures of Get-Rich-Quick Wallingford,* 1931; *Cuban Love Song,* 1931; *The Christmas Party,* 1931; *Wet Parade,* 1932; *Speak Easily,* 1932; *The Phantom President,* 1932; *The Passionate Plumber,* 1932; *Blondie of the Follies,* 1932; *What! No Beer?,* 1933; *Meet The Barron,* 1933; *Hell Below,* 1933; *Broadway to Hollywood,* 1933; *Student Tour,* 1934; *Strictly Dynamite,* 1934; *Palooka,* 1934; *George White's Scandals,* 1934; *Hollywood Party,* 1934; *Carnival,* 1935; *Land Without Music,* 1936; *Little Miss Broadway,* 1938; *Start Cheering,* 1938; *Sally, Irene and Mary,* 1938; *Melody Ranch,* 1940; *You're in the Army Now,* 1941; *The Man Who Came to Dinner,* 1941; *Two Girls and a Sailor,* 1944; *Music For Millions,* 1944; *Two Sisters From Boston,* 1946; *It Happened in Brooklyn,* 1947; *This Time for Keeps,* 1947; *On an Island with You,* 1948; *The Milkman,* 1950; *The Great Rupert,* 1950; *Beau James,* 1957; *Pepe,* 1960; *Il Giudizio Universale* aka *The Last Judgment,* 1962; *Billy Rose's Jumbo,* 1962; *It's a Mad Mad Mad Mad World,* 1963

Stage

Show Girl, 1929; *The New Yorkers,* 1930; *Strike Me Pink,* 1933; *Jumbo,* 1935–36; *Red, Hot and Blue,* 1936–37; *Stars in Your Eyes,* 1939; *Keep off the Grass,* 1940

Selected Publications

I've Got My Habits On (with Chris Smith and Bob Schafer), 1921
Night Clubs (with John Christian Kofoed), 1931

Further Reading

Cahn, William, *Good Night, Mrs. Calabash: The Secret of Jimmy Durante,* New York: Duell Sloan and Pearce, 1963
Fowler, Gene, *Schnozzola: The Story of Jimmy Durante,* New York: Viking Press, 1951
Robbins, Jhan, *Inka Dinka Doo: The Life of Jimmy Durante,* New York: Paragon House, 1991

Durham, Richard 1917–1984

U.S. Writer and Radio Dramatist

From June 1948 to August 1950, Richard Durham was the force behind Chicago radio's *Destination Freedom,* a lyrical, politically outspoken weekly half-hour series of programs that dramatized the lives and accomplishments of various contemporary and historical black leaders. The series was an uncompromising, well-written presentation of dignified African-American images during a time when few such images existed in American media. Durham's scripts creatively and consistently rallied against racial, social, and economic injustice.

Early Years

Richard Durham was one of eight children born to a father who was a farmer and a mother who was a schoolteacher in rural Raymond, Mississippi. In 1923, when Durham was six years old, his family moved to Chicago, joining the large migration of African Americans from the agricultural South who sought better employment and education opportunities in the industrial North. Durham attended Chicago public schools, became an avid reader, briefly enrolled in Northwestern University, and wrote poetry in his spare time.

During the Depression, Durham landed a job with the Illinois Writers Project, a state program that was an outgrowth of the federal government's Works Progress Administration. As a member of the project's radio division, he wrote for several shows that aired on Chicago stations during the early 1940s.

Ever the versatile writer, Durham also worked as an editor and print journalist for the Chicago-based and black owned *Chicago Defender* newspaper and *Ebony* magazine during World War II. After the war he was hired to write for a 15-minute weekly drama series entitled *Democracy USA,* which aired on WBBM, Chicago's Columbia Broadcasting System (CBS) affiliate station. Durham continued to hone his writing skills by creating a soap opera that dealt with the trials and triumphs of a fictitious black family. As one of the first soap operas of its day concentrating on African American characters, *Here Comes Tomorrow* aired on Chicago radio's WJJD from 1947 to 1948.

Richard Durham
Courtesy of Clarice Durham

But Durham also longed to counter the stereotypical portrayal of African Americans on shows such as *Amos 'n' Andy* and *Beulah* by dramatically highlighting the accomplishments of black historical, cultural, educational, and political leaders—from figures such as slavery abolitionist Harriet Tubman and Haitian revolutionary Toussaint L'Ouverture, or cultural icons John Henry and Stackalee, to more contemporary movers and shakers such as poet Gwendolyn Brooks and statesman Ralph Bunche. *Destination Freedom* became Durham's brilliant means of expression.

Destination Freedom

Durham fortuitously approached Chicago's National Broadcasting Company (NBC) affiliate WMAQ with his series proposal during a time when there was a somewhat more liberal atmosphere in the post–World War II media. In fact, the director of WMAQ's public affairs and education division was Judith Waller, a woman who was passionate about and committed to public-service programming.

Durham was granted a weekly half-hour spot, and *Destination Freedom* debuted at 10 A.M. on Sunday morning, 27 June 1948. Each week, the series opened with an a cappella rendering of the African-American spiritual "Oh Freedom." An announcer briefly introduced the episode's specific focus, and then Durham began weaving his story—cleverly using actors, sound effects, and music to bring his scripts to life.

Durham was an inventive writer who used innovative storytelling methods. Often, objects or concepts became personalities. For example, in "Anatomy of an Ordinance," urban slums were personified by an arrogant character who was proud of his discriminatory origins, and Louis Armstrong's trumpet verbally guided listeners through the great musician's life in "The Trumpet Speaks." Additionally, in "The Rime of the Ancient Dodger," humorously rhymed verse helped to dramatize Jackie Robinson's integration of major-league baseball in 1947.

In one of the series' strongest productions, the award-winning "The Heart of George Cotton," Durham cast a human heart as a narrator. Accompanied by the sound of fluctuating heartbeats, Durham's heart character intimately involves the audience in an open-heart operation. This episode paid tribute to African-American surgeons Ulysses Grant Dailey and the legendary Daniel Hale Williams—the first doctor to successfully suture a human heart in 1893.

One of Durham's recurring themes maintained that until all people enjoyed social and economic freedom, the fight against oppression would continue. In fact, *Destination Freedom* characters such as slavery revolt leader Denmark Vesey and reconstruction senator Charles Caldwell actually verbalized this sentiment—one rarely heard on radio at that time. Also, Durham championed women's rights in his characterizations of such African-American pioneers as abolitionist Sojourner Truth, educator Mary McLeod Bethune, choreographer Katherine Dunham, and journalist/activist Ida B. Wells.

Durham spent hours in the library poring over historical documents to find material for his scripts. He attempted to avoid overtly didactic scriptwriting approaches, opting to tell stories that might emotionally capture listeners and encourage them to draw their own conclusions.

Apparently, his approach worked. The show was enthusiastically received by many Chicago listeners—despite an airtime that Durham disliked because of its proximity to Sunday morning worship services. By the end of its first year, *Destination Freedom* had garnered several awards, along with praise from citizens' groups, the Chicago Board of Education, and Illinois Governor Adlai Stevenson. But there were also complaints. Some considered the series too radical, and groups such as the American Legion and the Knights of Columbus protested certain episodes. But because of the largely positive attention that *Destination Freedom* brought to the station, WMAQ continued the series. However, the station never sought to broadcast the series to a national audience, in part because it was believed that Southern stations would refuse to air it.

Inside WMAQ, which financially supported *Destination Freedom* (the *Chicago Defender* and the Urban League also briefly sponsored the series), Durham regularly fought with station censors. There were attempts to soften Durham's characterization of Revolutionary War hero Crispus Attucks, and proposed programs on such legendary figures as Paul Robeson and Nat Turner were rejected because they were considered too controversial. One show on abolitionist and statesman Frederick Douglass was so heavily edited that it would have been only 20 minutes long had it been produced with the proposed cuts. Durham and his predominantly African-American cast of socially conscious actors protested, and much of the edited material was restored.

WMAQ ended its support of Richard Durham's *Destination Freedom* during the summer of 1950, sparked by the rising conservatism of the budding anti-Communist period. After Durham's departure, the series lasted for a short period featuring traditional white heroes. But Durham sued WMAQ, preventing the station from continuing to use the *Destination Freedom* name.

Richard Durham sustained his politically astute creativity through work as a Chicago-based television scriptwriter, newspaper editor, author, and political speechwriter until his death in 1984. But he is perhaps best remembered for his brilliantly written and historically significant *Destination Freedom* radio series.

SONJA WILLIAMS

See also Black-Oriented Radio; Playwrights on Radio; WMAQ

Richard Durham. Born in Raymond, Mississippi, 6 September 1917. Briefly attended Northwestern University; wrote radio scripts, Illinois Writer's Project of the Works Progress Administration, 1939–43; feature writer, editor, and freelancer for *The Chicago Defender* newspaper and *Ebony* magazine, 1943–46; writer for WBBM's weekly drama, *Democracy U.S.A.,* 1946–48; creator and writer of the African-American soap opera *Here Comes Tomorrow* on WJJD, Chicago, 1947–48; creator and writer of radio series *Destination Freedom,* 1948–50; freelance writer for soap operas and science fiction television shows, 1950–52; education director for United Packinghouse Workers Union, 1952–57; freelance political speechwriter, 1957–62; editor of newspaper *Muhammad Speaks,* 1963–70; creator and head writer for WTTW public television series *Bird of the Iron Feather,* 1970; author (with Muhammad Ali) of *The Greatest: My Own Story,* 1975; speechwriter for Congressman Harold Washington, 1982–83. Received Poetry Award, Northwestern University, 1937; Page One Award, Chicago Newspaper Guild, 1945; Wendell L. Wilkie Award *(Democracy USA),* 1946; Citation from President Harry S. Truman *(Democracy USA),* 1946; Citation from Illinois Governor Adlai Stevenson (*Destination Freedom*), 1949; Chicago Commission on Human Relations, 1949; National Conference for Christians and Jews, 1949; South Central Association of Chicago, 1949; Institute for Education by Radio, Ohio State University, 1949; Emmy Award *(Bird of an Iron Feather),* 1970; Literary Hall of Fame, Chicago State University, 1999. Died in New York City, 27 April 1984.

Radio Series

1940–43	*At the Foot of Adams Street/Legends of Illinois/Great Artists*
1946–48	*Democracy U.S.A.*
1947–48	*Here Comes Tomorrow*
1948–50	*Destination Freedom*
1957	*The Heart of George Cotton* and *Denmark Vesey*

Television Series

Bird of the Iron Feather, 1970

Selected Publications

The Greatest: My Own Story (with Muhammad Ali), 1975
Richard Durham's Destination Freedom: Scripts from Radio's Black Legacy, 1948–50, edited by J. Fred Macdonald, 1989

Further Reading

Barlow, William, *Voice Over: The Making of Black Radio,* Philadelphia, Pennsylvania: Temple University Press, 1999
"Black Radio: Telling It Like It Was," Washington, D.C.: Radio Smithsonian, Smithsonian Productions, 1996 (a 13-part radio documentary series with Lou Rawls)
Dates, Jannette L., and William Barlow, editors, *Split Image: African Americans in the Mass Media,* Washington, D.C.: Howard University Press, 1990; 2nd edition, 1993
MacDonald, J. Fred, "Radio's Black Heritage: Destination Freedom, 1948–50," *Phylon* 39, no. 1 (March 1978)
Savage, Barbara Dianne, *Broadcasting Freedom: Radio, War, and the Politics of Race, 1938–1948,* Chapel Hill: University of North Carolina Press, 1999

DXers/DXing

Tuning Distant Stations

DX is the telegrapher's abbreviation for "distance." It came into common use among early amateur radio operators to refer to those who concentrated on working with other operators at great distances. DXing continues as a focus of many modern-day amateurs.

The terms "DX" and "DXing" were also used among radio listeners. In broadcasting's earliest days, radio listeners were known as BCLs, for "broadcast listeners," and those BCLs who were interested in listening not for program content but for the thrill of hearing distant stations were called DXers. Their hobby was (and still is) known as DXing, and their goal was to hear as many stations, from the farthest locations, as possible.

For a time in the early 1920s, long-distance broadcast listening was popular with a large portion of the population. Radio—especially radio from distant places—was a new experience, and everyone wanted to see how far they could "get" with their equipment. However, as the novelty of radio wore off and network broadcasting started, DXing became the preserve mainly of the technically inclined and of hard-core distance aficionados.

Shortwave broadcasting would not come to the attention of most listening hobbyists until around 1924, and so in the United States most DXing before then was done on the standard broadcast (medium wave) band, with domestic stations the targets. The relatively small number of stations and the resulting absence of the channel blocking that is common today, coupled with the prevalence of daytime-only operations, made it possible for a conscientious, well-equipped night hound to hear a large number of the stations that were operating.

DXing's attraction was captured by journalist Charlotte Geer in her 1927 poem "Another One":

You may pick out an average young man
Who has nothing especial "agin it,"
And draw up a comfortable chair
And settle him in it.

Then you mention the call of the West
Thus tempting his spirit to roam
That he early may tire of nooks by the fire
The tame little voices of home.

When the chimes of the clock tinkle ten
The rest of the folks go to bed,
You must then take the youngster in hand
And fasten the phones on his head.

You must lead him and prod him by turns
He'll yawn and seem bored and forlorn
Till he hears that first call from the coast—
And behold a DXer is born.

The "call from the coast" refers to the ability of east-coast listeners, late at night and under the right conditions, to hear stations in California as outlets farther east signed off at local sundown.

Although pioneer radio fans built their own sets, as the number of DXers increased a distinct market was recognized by equipment manufacturers, and soon radios with special features for long-distance reception were being produced. The most important elements to good reception were sensitivity (the ability to pick up weak signals), selectivity (the ability to separate adjacent signals), and frequency readout (the ability to know what frequency the equipment is tuned to). To meet these needs and the like needs of the amateur radio operators, the communications receiver was developed. These receivers, which appeared beginning in 1933 and continue on the market today, emphasize technical capabilities rather than appearance or simplicity of operation. Among the principal early manufacturers of such sets were Hallicrafters, Hammarlund, and National. Today the major producers of high-quality, semi-professional receivers include Drake, Icom, AOL, and Japan Radio Company.

With the discovery that shortwave signals could propagate around the globe, what had been, in the United States, largely a search for U.S., Canadian, and Caribbean stations on the standard broadcast band took on a worldwide flavor as DXers tried their hand at the shortwave frequencies. Many countries began international shortwave services, and domestic shortwave broadcasting became commonplace in many foreign countries as well, particularly those with large geographic areas to cover. These latter stations made particularly good DX targets. Some listeners also went outside the broadcasting frequencies, preferring to tune in to amateur radio operators; transmissions from ships, planes, and police; and other users of the shortwave spectrum.

The need for up-to-date station information led to the publication of magazines such as *Radio Index*, *All Wave Radio*, and *Official Short Wave Listener Magazine*. These publications were devoted either entirely or in part to DXing. Other magazines had special sections for long-distance radio enthusiasts. Clubs such as the Newark News Radio Club and the International DXers Alliance were formed, and periodic bulletins containing members' "loggings" and other DX information issued. Special broadcasts were scheduled over stations at times when they might not ordinarily be heard, and contests were held to compare DXing prowess.

Besides hearing the stations, many DXers collected QSLs. QSL is the telegraph abbreviation for the acknowledgment of receipt of a signal or a message. QSLs are cards or letters that stations issue to DXers, confirming that it was in fact their station that was heard, based on the listener's description of the programming. QSLs usually take the form of distinctive cards and letters containing information about the station and its location. Some DXers also make it a practice to make audio recordings of their DX catches and thus preserve their listening experiences.

Notwithstanding the common availability of local radio signals over even the simplest equipment, long-distance radio listening still has a devoted following. Although the pervasiveness of high-power, all-night broadcasting has made DXing on the broadcast band more difficult, on the shortwave bands the combination of improved receivers and higher-power transmitters has led to easier tuning and more reliable reception.

As a result, whereas DX was the main objective of long-distance radio enthusiasts during most of radio's developmental period, "shortwave listening"—listening for program content rather than distance—is often the purpose today. Some listeners follow particular specialties. On shortwave they may listen to clandestine radio stations; unlicensed pirate stations; or small stations from exotic parts of the world, such as Indonesia or the Andes. On the broadcast band, some listeners concentrate on domestic stations, while others focus on foreign broadcasters.

Although the number of DXers has declined over the years, there is still much truth in the observation made by radio pioneer Hugo Gernsback in 1926: "I can not imagine any greater thrill," he wrote, "than that which comes to me when I listen, as I often do, to a station thousands of miles away. It is the greatest triumph yet achieved by mind over matter."

JEROME S. BERG

See also Ham Radio; International Radio; Shortwave Radio

Further Reading

Berg, Jerome S., *On the Short Waves, 1923–1945: Broadcast Listening in the Pioneer Days of Radio,* Jefferson, North Carolina: McFarland, 1999

DeSoto, Clinton B., *Two Hundred Meters and Down: The Story of Amateur Radio,* West Hartford, Connecticut: American Radio Relay League, 1936

Magne, Lawrence, editor, *Passport to World Band Radio,* Penn's Park, Pennsylvania: International Broadcasting Services (annual; 1984–)

E

Early Wireless

Induction and Conduction before 1900

Although Marconi, Lodge, Popov, and their contemporaries were the first electricians to make practical use of electromagnetic waves for communication, varied forms of wireless telegraphy had existed for 50 years before their work at the end of the 19th century. Most of the wireless inventors of this early era were Americans.

Henry and Morse

Joseph Henry, at Princeton University, conducted numerous experiments with induction, or the tendency of a primary electric current in a conductor to stimulate a secondary current in another conductor nearby. During one group of experiments in 1842, which Henry called "induction at a distance," he measured secondary current as far as 200 feet away from the primary circuit. Although the effects were more pronounced with the instantaneous discharge of a Leyden jar (electrostatic induction) at the primary circuit, they were also present when a continuous current passed through a coil (electromagnetic induction). Henry also noted that the wireless electricity seemed to oscillate, but he had no instruments sensitive enough to study this phenomenon. Neither Henry nor his students used induction for telegraphy, but his methods and results were well known to scientists on both sides of the Atlantic.

In the same year, Samuel F.B. Morse, using different technology, devised a working wireless telegraph. Having failed at submarine telegraphy when a ship's anchor hooked and cut the cable, Morse came up with a novel solution to the problem. By then, electricians knew how to use a natural conductor, like earth or water, as the return link to complete an electrical circuit. Morse reasoned that the same natural conductor could replace the wire in the link between transmitter and receiver as well. By attaching the ends of the wires leading from the battery and telegraph key to metal plates, doing the same with wires from a galvanometer, and then submerging the sets of plates on opposite banks, Morse successfully sent a message across a canal some 80 feet wide. The next year, 1843, his assistants sent messages a mile across the mouth of the Susquehanna River.

Morse soon became preoccupied with building a wired telegraph system and never applied for a patent on the wireless scheme. About ten years later, however, James Bowman Lindsay, a Scottish schoolteacher, applied the same natural conduction technology to the task. Apparently Lindsay was unaware of Morse's prior work. By using larger metal plates spaced more widely apart, he increased transmission distance to two miles across the River Tay at Dundee. Lindsay received British patent 1,242 for this invention in 1854, but he failed to raise enough capital to build a prospective transatlantic installation.

Loomis and Ward

American dentist Mahlon Loomis, of Washington, D.C., devised a wireless telegraph system using the upper atmosphere as one conductor, the earth as the other, and the difference in electric potential between the two as the power supply. In 1866 he raised two kites, attached to grounded wires, from mountains in northern Virginia and transmitted a signal between them, a distance of 18 miles. Loomis later adapted this apparatus for telegraphy and, he claimed, telephony. The most interesting facet of this invention was that it worked best when the kites were at the same altitude, leading some historians to speculate that Loomis had hit upon the principle of syntony, or tuned resonance. Although he received U.S. patent 129,971 in 1872, Loomis failed to get a requested appropriation from Congress, and several potential groups of investors went broke in the numerous financial panics that followed the U.S. Civil War.

Another U.S. patent for wireless telegraphy, number 126,356, issued to William Henry Ward of Auburn, New

York, actually preceded the Loomis patent by three months. Ward was an author of religious tracts and a vigorous promoter of his own inventions, which included naval signal flags, bullet-making machines, bomb shell fuses, and pomade for the hair. Loomis had made Ward's acquaintance in the late 1850s when Ward agreed to take a set of Loomis-patented false teeth under consignment to a trade show in Europe. Because Loomis furnished no illustrations or model with his patent application, it is difficult to tell how much similarity existed between the two inventions. The Ward wireless telegraph, however, seems to be based on the absurd idea that convection currents in the atmosphere would carry the electric signal to the receiver. Neither system found a ready market.

The 1870s were a busy decade for electricians, whether they were scientists, engineers, or tinkering inventors. Laboratories large and small sprang up in the United States and Europe for both educational and commercial purposes. The new professional societies, as well as academic and trade publications, facilitated communication, so electricians knew what other electricians were doing. Although there was scant progress in wireless telegraphy, there were several interesting events.

Various experiments, including those by Elihu Thompson and Thomas Edison in the United States and by David Hughes in Great Britain, suggested the existence of electromagnetic waves, as predicted by James Clerk Maxwell. Edison called them the "etheric force." Hughes unwittingly built a complete radio system, with a sparking coil as a transmitter and a carbon microphone as a coherer, or receiver. He gave up these experiments, however, when directors of the Royal Society visited his lab to witness a demonstration and assured him that the effects were due to induction alone.

Wireless Telegraphy

In 1878 Alexander Graham Bell attempted to build a Morse-type wireless telephone using water as a natural conductor. He tested this device the next year with only limited success on the Potomac River near Washington, D.C. When he learned that John Trowbridge at Harvard was pursuing similar research, Bell dropped that idea in favor of one more promising—sending a voice signal on a beam of reflected light.

With a thin diaphragm of reflective mica at the mouthpiece, the device captured light waves from the sun, modulated the light as sound waves moved the diaphragm, and reflected the beam toward the receiver. The receiver was a parabolic reflector coated with silver and aimed at a piece of selenium, a natural photoelectric transducer. Once the selenium converted the light to electricity, the signal traveled to a regular telephone receiver. Bell called it the Photophone, received U.S. patent 235,199 in 1880, and exhibited the device widely in both the United States and Europe. Later experiments proved that the device worked with less expensive lampblack substituting for the selenium and with any form of radiant energy. The Photophone then became known as the Radiophone, the first application of the term *radio* to wireless communication.

Although European electricians were captivated by the device, and Bell himself thought it a greater invention than the telephone, American Telephone and Telegraph (AT&T) saw no commercial possibility and soon ceased its development. The company continued to exhibit the Photophone as a novelty through the St. Louis World's Fair of 1904, but Bell was so disgusted that he gave his original model to the Smithsonian Institution and halted all active involvement with the company that had once borne his name and was based on his patents.

Amos Dolbear, another early telephone pioneer, became interested in wireless telephony by accident. While working in his lab at Tufts College in 1882, he noticed that he was still hearing sounds from a receiver, even though the wire to the transmitter on the other side of the room was disconnected. Upon examination, he determined that the signal was traveling by electrostatic induction. Through experimentation, Dolbear built a wireless telephone system that employed grounded aerials at both transmitter and receiver and worked well at distances of up to a mile. He demonstrated the invention by transmitting both voice and music at scientific conferences in the United States, Canada, and Europe, receiving U.S. patent 350,299 in 1886. Dolbear's transmitter was capable of generating electromagnetic waves and was in many respects similar to Marconi's of a decade later, but the receiver lacked any device to detect radio waves. Dolbear never attempted to sell his wireless telephone.

In Great Britain during the 1880s, both William Preece and Willoughby Smith developed functional wireless telegraphs to transmit across relatively short distances. They used both natural conduction and induction technologies to solve practical problems such as how to communicate with offshore islands and lighthouses and with workers in a coal mine. Preece, an engineer at the British Post Office, had been interested in wireless since he witnessed Lindsay's work in 1854. Later he became Marconi's chief advocate with the British government.

Moving Telegraphy

The possibility of communicating instantly by telegraph with moving trains generated substantial interest in the United States. William Wiley Smith, a telephone office manager, devised an electrostatic wireless telegraph that used existing lines running beside the tracks and received U.S. patent 247,127 in 1881. Although Smith's invention did not work very well, his partner, Ezra Gilliland, was a childhood friend of Thomas Edison. Three years later, Gilliland convinced Edison to buy the patent, improve the technology, and promote it to the rapidly growing railroad industry as a safety and convenience appliance. Gilliland and Edison's interest in railroad telegraphy grew when they learned that another inventor, Lucius Phelps of New York, had just applied for a patent on a similar system. Phelps, moreover, was preparing to demonstrate a working model in New York City during February 1885. Edison quickly filed a new patent application based on improvements to the prior Smith patent. The Patent Office called the two applications into interference hearings. Meanwhile, Granville Woods, a black inventor from Cincinnati, filed still a third application for a railroad telegraph system. Once again, the Patent Office declared the Woods and Phelps applications to be in interference.

In a series of hearings that ended in 1887, Woods was the ultimate victor. He proved that as early as 1881 he had shown sketches and models of his invention to friends in his neighborhood. Shortly thereafter, he had lost his job and contracted smallpox, delaying his progress. After reading about the Phelps demonstration in *Scientific American*, he quickly gathered his old notes and models and went to his lawyer. In the meantime, the patent examiner determined that the Smith patent, controlled by Edison, had no priority because it used electrostatic induction, whereas the Phelps and Woods inventions used electromagnetic. After this ruling, Edison dropped out of the hearings and merged efforts with Phelps.

Woods received several U.S. patents for his invention, beginning with number 371,241, and Phelps did likewise, starting with number 334,186. But all of this activity was in vain, for the railroads had no interest in wireless telegraphy at that time. With no laws compelling safety or emergency communication improvements, they saw the technology as added expense without compensatory revenue. So none of the inventors profited. Eventually Edison modified his application and received U.S. patent 465,971 in 1891. He sold the patent to Marconi some years later for a nominal sum. After the failure to market the railroad telegraph, interest in wireless waned in the United States until after Marconi brought his system to the America's Cup races of 1899.

In a widely read article of 1891, John Trowbridge concluded that the technologies of Henry and Morse were inefficient for long-distance wireless communication and, with no tuning mechanism, were limited to a single message at a time. The last champion of natural conduction and induction wireless was probably Nathan Stubblefield, a farmer from Kentucky who learned about electricity by reading *Scientific American* and *Electrical World*. Stubblefield demonstrated an induction wireless telephone for neighbors as early as 1892. He generated considerable publicity in 1902 with a natural conduction wireless telephone that he displayed to enthusiastic crowds in Washington, D.C., and Philadelphia. One remarkable feature of this system was its ability to broadcast a signal to multiple receivers simultaneously. Stubblefield intended to use it to disseminate news and weather information, but he was the victim of a stock fraud scheme that left him destitute. By the time he received U.S. patent 887,357 in 1908 for his induction system, superior technology existed. But he became very paranoid that others were seeking to profit from his work, and he eventually died of starvation in 1928.

Landell and Tesla

Two other inventors had the opportunity to develop wireless communication by electromagnetic waves prior to Marconi but failed to do so. Roberto Landell de Moura, a Catholic priest from Brazil, studied physics at the Gregorian University in Rome while he prepared for the priesthood. When he returned to Brazil in 1886, Landell set up a laboratory and began electrical experiments. His interest turned to wireless. He built acoustic telephones, a model of Bell's Photophone, and, soon after he learned of Hertz's and Branly's work, his own electromagnetic wave transmitter and receiver. Then he combined the three into one multifunction wireless telephone system. By 1893 Landell was sending messages over distance of five miles. Then, two years later, a powerful bishop witnessed a demonstration. He was so unnerved by the voices coming from nowhere that he declared the apparatus the work of the devil and ordered Landell to stop his work.

Shortly thereafter, fanatics broke into Landell's laboratory, destroyed the apparatus, and set fire to the building. It took Landell five years to regroup, but he eventually received Brazilian patent 3,279 in 1900 and traveled to the United States to pursue patents and development. Hampered by poor legal advice, illness, and inadequate knowledge of the U.S. patent system, Landell nevertheless persisted and received three U.S. patents, beginning with number 771,917 in 1904. By then Marconi and others controlled the market for radio.

Like Landell, Nikola Tesla was trained in physics at a European university. He came to the United States in 1884 to work first for Edison and then for Westinghouse. In his work with high-frequency alternating currents, Tesla discovered that the alternators also generated continuous electromagnetic waves that could be used to transmit signals. He demonstrated this phenomenon as early as 1891 but made no practical application of it until he built a radio-controlled toy boat, for which he received U.S. patent 613,809 in 1898. The next year, he established a laboratory in Colorado Springs, Colorado, from which he intended to send wireless messages to an international exposition in Paris in early 1900. But in the midst of his radio experiments, Tesla became fascinated with the possibility for wireless distribution of electric power through the earth itself and devoted most of the rest of his life to devising methods to accomplish that goal.

ROBERT HENRY LOCHTE

See also Fessenden, Reginald; German Wireless Pioneers; Hertz, Heinrich; Landell de Moura, Father Roberto; Lodge, Oliver; Marconi, Guglielmo; Popov, Alexander; Tesla, Nikola

Further Reading

Appleby, Thomas, *Mahlon Loomis: Inventor of Radio,* Washington, D.C.: s.n., 1967

Blake, George G., *History of Radio Telegraphy and Telephony,* London: Radio Press, 1926; reprint, New York: Arno Press, 1974

Dunlap, Orrin Elmer, *Radio's 100 Men of Science: Biographical Narratives of Pathfinders in Electronics and Television,* New York and London: Harper, 1944

Fahie, J.J., *A History of Wireless Telegraphy,* London and Edinburgh: Blackwood, and New York: Dodd Mead, 1899; 2nd edition, 1901; reprint, New York: Arno Press and The New York Times, 1971

Hawks, Ellison, *Pioneers of Wireless,* London: Methuen, 1927; reprint, New York: Arno Press, 1974

Lochte, Robert, "Reducing the Risk: Woods, Phelps, Edison, and the Railway Telegraph," *Timeline* 16, no. 1 (January–February 1999)

Seifer, Marc J., *Wizard: The Life and Times of Nikola Tesla,* Secaucus, New Jersey: Carol, 1996

Sivowitch, E.N., "A Technological Survey of Broadcasting's 'Pre-History,' 1876–1920," *Journal of Broadcasting* 14 (Winter 1969–70)

Earplay

Public Radio Drama Series

Earplay was an anthology series created in an effort to produce a variety of U.S. radio dramas in the late 1970s and early 1980s. Centered at radio station WHA in Madison, Wisconsin, it involved numerous prominent playwrights and scores of actors more than two decades after the last major commercial radio dramas had left the air. *Earplay* was a leading source of drama for member stations of the burgeoning National Public Radio (NPR) network.

The series began in 1971 as a grant proposal submitted to the Corporation for Public Broadcasting by project director Karl Schmidt, a University of Wisconsin professor and station manager of WHA. Schmidt had begun his radio career as a juvenile actor in 1941 and was involved as an actor and director in commercial, public, and armed forces radio. In the 1960s he produced a series of stereo dramas under the auspices of the National Center for Audio Experimentation. In many ways, station WHA, with a long history of producing radio drama, was an ideal place for the development of new forms of that genre. For 30 years the station's School of the Air applied the techniques and forms of drama for instructional purposes. In the early 1970s, WHA had particularly strong ties to Canadian actors and writers, BBC writers and directors, and producers from Radio Nederland.

Under the terms of its grant, the primary purpose of *Earplay* was "to develop drama in audio forms which are intelligible, enjoyable, and useful to more, rather than fewer, people." A second purpose was "the establishment of a testing ground for playwrights and plays." (In the project's latter days, its producers would increasingly disagree as to whether intelligibility or experimentation should predominate.) The creators of *Earplay* proposed an emphasis on original dramas—works by new playwrights that did not demand long attention spans and had strong plot lines, a high degree of intelligibility, enjoyable listening potential, and relevance to life in the United States in the 1970s. There was a conscious effort to "compete with substantial ambient noise levels in the listening circumstance" by emphasizing short dramas (usually less than half an hour long) and avoiding reliance upon subtle sound effects.

Between 1975 and 1979 *Earplay* produced more than 150 radio dramas ranging in length from a few minutes to more than an hour. During those years the dramas were distributed to public radio stations on long-playing records. After 1981 the medium was reel-to-reel tape as part of *NPR Playhouse*. Later the plays were repackaged in half-hour installments under the title *Earplay Weekday Theatre*.

Earplay dramas sounded very different from radio plays of the 1930s, 1940s, and 1950s. Most of them were in stereo, and the acting had a closer, more intimate quality. In some cases the language also was very different and prompted advisories to subscribing stations. In the broadcasts of both Archibald MacLeish's *J.B.* and Edward Albee's *Listening,* for example, there were a half-dozen warnings about "sensitive material."

Another departure from U.S. radio tradition was the variable length of *Earplay* dramas, although they were often packaged in one- or two-hour blocks.

Like traditional radio dramas, *Earplay* productions typically involved six or fewer actors, and they often emphasized narration. In most cases the story line was straightforward, but occasionally the plays were more in the realm of the "sound collage."

In their preface to the 1979 *Earplay* program information, the producers noted that the 1979 season reflected a commitment to "give the most promising American playwrights an opportunity to speak to a national audience through the unique medium of radio." During that same year, *Earplay* received both the Peabody Award and the Armstrong Award. The dramas were produced in various locations, including New York, Los Angeles, and Chicago. Most of the postproduction work took place at WHA; the technical director was Marv Nonn.

Among the playwrights commissioned to create *Earplay* dramas were some of the most distinguished of the day: John Mortimer, Donald Barthelme, Larry Shue, Vincent Canby, Alan Ayckburn, Gamble Rogers, John Gardner, Anne Leaton, Athol Fugard, Tim O'Brien, and Archibald MacLeish. Robert Anderson's play *I Never Sang for My Father* was later adapted into a major motion picture. Other plays went on to success on Broadway: *The Water Engine* by David Mamet, *Lightning* by Edward Albee, and *Wings* by Arthur Kopit, which won the Prix Italia. In order to defray the expense of involving major playwrights, *Earplay* initiated the International Commissioning Group with drama producers in England, Ireland, West Germany, Norway, Sweden, and Denmark, who added rights fees for their countries to *Earplay*'s U.S. fees to provide payments attractive to major playwrights.

Although *Earplay* was carried by a significant number of public radio stations, many years after the series ended Karl Schmidt reflected that its demise came in part from its inclination to raise difficult issues without offering answers, so that the working person coming home after a tiring day would be inclined to pass over them in favor of more refreshing fare. During the late 1970s and early 1980s, Schmidt turned *Earplay* into a series, partly because series were more economical to produce than individual dramas, and partly because he found that many writers preferred structural guidelines to the indefinite length of the earlier format. In order to put *Earplay* resources to the most efficient use, Schmidt stipulated that the episodes in the series were to be approximately half narration and half dialogue. They were: *A Canticle for Leibowitz* (16 half-hours), based on the science fiction novel by Walter Miller, Jr.; *Happiness,* Anne Leaton's multipart radio drama based on the musings of a middle-aged Texas woman; and *Something Singing,* Christian Hamilton's play about abolitionist Amos Bronson Alcott—all of which were distributed through National Public Radio. Contracts allowed three years of use in any and all non-commercial stations in the U.S., with a three-year renewal option that was not exercised.

NORMAN GILLILAND

See also WHA and Wisconsin Public Radio

Actors
Jay Fitts, Pat King, Carol Cowan, Karl Schmidt, and Martha Van Cleef, Meryl Streep, Vincent Gardenia, Laurence Luckinbill, Brock Peters, Lurene Tuttle, Leon Ames, etc.

Producers/Directors
Karl Schmidt, Howard Gelman, Daniel Freudenberger

Programming History
NPR 1972–86

The Easy Aces

Comedy Program

The Easy Aces was a comedy written by Goodman Ace that first aired in 1930 and enjoyed a 15-year run on the Columbia Broadcasting System (CBS) and National Broadcasting Company (NBC) networks. The show was known for simple plot lines that allowed Jane Ace to contort the English language inexhaustibly while her usually bemused but sometimes horrified husband supplied witty commentary. Like *The Burns and Allen Show, The Easy Aces* relied on the device of a scattered, illogical wife flummoxing the logic and control of her husband. Despite the current anachronistic, even sexist ring of the stereotypical scatterbrained wife, the character had thrived in vaudeville and was still common in mainstream comedy programming well past the middle of the 20th century.

Born in Kansas City, Missouri on 15 January 1899, Goodman Ace's writing talents led him to the *Kansas City Journal–Post* after a stint at haberdashery and other odd jobs. Jane Epstein, also from Kansas City, was born to a local clothing merchant on 12 June 1905. By the time they married in 1922, Ace was regularly exercising his wit in his weekly column of drama and movie reviews for the *Journal–Post*. For instance, while reviewing a play billed as a scenic extravaganza, he offered the comment that the "sets were beautiful—both of them." Since 1922 the paper had sent news to local station KMBC via live feed from a cramped studio off the news room. By the late 1920s Ace's growing interest in radio caused him to approach KMBC's station manager Arthur Church with the proposal for a 15-minute program entitled *The Movie Man*. Church granted Ace the show at the rate of $10 a week, money that supplemented his income from the *Journal–Post*. As a critic in the newspaper and on the air, Ace met many of the vaudevillians and celebrities who performed in Kansas City. He counted George Burns, Gracie Allen, Groucho Marx, Fred Allen, and Jack Benny among his friends, and even contributed jokes to some of Benny's early radio shows.

Ultimately, however, it was chance, not personal acquaintance, that led to the birth of *The Easy Aces* and Ace's initial success on radio. One Friday in late 1930 his broadcast ended, but the following program (which was to feature Heywood Broun) failed to come through on the network feed. A technician signaled Ace to ad-lib. Jane happened to have been watching from the lobby, so Ace motioned her in, introduced her as his roommate, and the couple bantered fluidly for 15 minutes. They spoke of a local man recently murdered while playing a game of bridge, and during a break Ace instructed Jane as follows: "You be dumb; I'll explain the finer points of bridge, and why murder is sometimes justified." Once on the air she fell effortlessly into character, asking "Would you care to shoot a game of bridge, dear?" and later wondering why "Whenever I lose, you're always my partner."

The audience responded favorably to the broadcast, prompting a local drugstore chain to offer the Aces an initial 13-week contract for two weekly shows. After this contract expired, the advertising firm Blackett-Sample-Humert (BSH) shopped the show around, quickly selling it to the advertisers for Lavoris mouthwash. During this period, the program was broadcast on CBS from Chicago, where Ace negotiated a weekly salary of $500. Foreshadowing future clashes with sponsors, Ace made light of a Lavoris executive's contention that the program had aired five minutes late one evening. The sponsor responded by canceling the show. The Aces then moved to New York, where Frank Hummert of BSH secured the sponsorship that allowed the show to continue.

Easy Aces, Goodman and Jane Ace
Courtesy Radio Hall of Fame

The initial improvised broadcast established many of the enduring characteristics of *The Easy Aces*. While the bridge game was eventually abandoned, plots remained fairly uncomplicated. Largely character-driven, typical shows featured Jane visiting a psychiatrist, seeing an astrologer, or serving on a jury. She was the central character from the beginning and her relaxed delivery was the key to the program's continuing popularity. From the beginning, too, Ace's asides imbued the show with an air of sophistication and witty urbanity. The Aces' characters were clearly identified as upper-class and urban—he played a highly paid advertising executive. As Ace later phrased it, they lived "in the typical little eastern town, New York City: population eight million . . . give or take one."

Jane's misspoken words and phrases, referred to as "janeacesims," were responsible for most of the laughs on the program. Among her weekly linguistic slips were statements that she feared "casting asparagus" on a friend's character, had "worked her head to the bone," made "insufferable friends," and had relatives "too humorous to mention." More often than not, the malapropisms seemed to make a satiric point, such as: "Congress is still in season," or "I got up at the crank of dawn." The seeming deliberateness of her linguistic twists was enhanced by her sober, deadpan delivery. Context might also lead listeners to believe that her deranged phrasing was occasionally intentional. For instance, when Ace got her brother a menial job, she asked him if he would receive a "swindle chair" like the one Ace had. Sometimes the mood was darker, as in "We are all cremated equal."

Although the show appealed to radio insiders and sustained a loyal following, it did not receive high ratings. Ace was proud of the show's low ratings and even bragged about them in the print advertisements for the show. He was disturbed, however, by sponsor Anacin's attempt to encroach on the program's autonomy. When the drug's parent company complained about a change in the music on the show, Ace fired off a letter criticizing Anacin's flimsy packaging. As he later put it, "They thought up a clever answer to that, which was 'You're fired.'"

After losing its sponsorship, *The Easy Aces* spent two years in syndicated reruns until it was revived as *mr. ace and JANE,* a 30-minute format with a live audience, a full orchestra, and a larger cast. The show also had a brief run on television in the 1949–50 season, but it did not translate well to that medium. By this time Goodman Ace had started writing for Danny Kaye's television program. In later years he would continue working in television, most notably for Perry Como and Milton Berle. Jane Ace's activities in broadcasting tapered off, and after a stint as *Jane Ace, Disc Jockey* in 1952, she became largely inactive in the field. Jane Ace died in New York on 11 November 1974. Goodman died on 18 March 1982.

BRYAN CORNELL

See also Comedy; George Burns and Gracie Allen Show

Cast

Ace	Goodman Ace
Jane	Jane Ace
Paul Sherwood, Jane's brother	Leon Janey
Marge	Mary Hunter
Mrs. Benton	Peggy Allenby
Betty	Ethel Blume (1939)
Carl	Alfred Ryder (1939)
Neil Williams	Martin Gabel (1939)
Laura the maid	Helene Dumas
Miss Thomas	Ann Thomas
Announcers	Ford Bond (1930–45), Ken Roberts (1948)

Producer/Creator

Goodman Ace

Programming History

KMBC Kansas City	October 1930–1931
WGN Chicago	late 1931–February 1932
CBS	March 1932–January 1935
NBC	February 1935–October 1942
CBS	October 1942–January 1945
as *Mr. Ace and Jane*	February 1948–May 1949

Television

December 1949–June 1950

Futher Reading

Ace, Goodman, *Ladies and Gentlemen, Easy Aces,* Garden City, New York: Doubleday, 1970
Gaver, Jack, and Dave Stanley, *There's Laughter in the Air! Radio's Top Comedians and Their Best Shows,* New York: Greenberg, 1945
Singer, Mark, "Goody," *The New Yorker* (4 April 1977)
Wertheim, Arthur Frank, *Radio Comedy,* New York: Oxford University Press, 1979

Easy Listening/Beautiful Music Format

The term *easy listening* refers to a program format characterized by the presentation of orchestral and small-combo instrumental music intended to elicit moods of relaxation and tranquility among its target audience of older adult listeners. Vocal music is intermixed and may include solo artist and choral recordings made popular by recognized personalities as well as "cover" renditions of original performances. Easy listening achieved its greatest popularity in the 1970s and was broadcast predominantly by FM stations. It was common for listeners in major markets to have from three to five such stations from which to choose. In the early 1980s, stations began to defect from easy listening in favor of formats with the youthful orientation advertisers were seeking. The format's popularity declined throughout the 1990s, to the point that easy listening stations by 2000 attracted less than 1 percent of the radio audience.

Origins

The origin of the phrase *easy listening* is undetermined. Its predecessors, "good music" and "beautiful music," were format descriptors popularized in the 1950s and 1960s by FM stations seeking differentiation from the Top 40 and middle of the road formats that dominated AM radio. Stations initially offered instrumental-laden good music programming to retail business operators on a subscription basis as a means of inducing customer relaxation. Good music programming in the 1950s was transmitted via the FM subcarrier frequency, which precluded reception by the general public. Growing listener interest led broadcasters to shift the good music format to their FM main channels in the early 1960s. The good music format subsequently evolved into beautiful music.

Stations that pioneered the good music format in the late 1950s included several noted AM outlets. KIXL in Dallas; WOR in New York City; WPAT in Paterson, New Jersey; and KABL in Oakland-San Francisco, all defined the easy listening presentation and accelerated the format's popularity. KIXL programmer Lee Seagall emphasized the importance of matching music tempo with the hour of the day—upbeat during the morning but down-tempo, soft, and romantic in the evening. Gordon McLendon's KABL relied upon interstitial poetry selections to create distinction. Two evening specialty programs—WOR's *Music from Studio X* and the *Gaslight Revue* on WPAT—influenced WDVR Station Manager/Program Director Marlin Taylor in 1963 to extend the format on this stand-alone FM station to around-the-clock presentation for Philadelphia listeners.

Pairing good music with the FM medium was fortuitous for performers, broadcasters, and listeners. When the Federal Communications Commission directed FM broadcasters in the mid-1960s to curtail the practice of simulcasting the programming of their AM sister stations, this format emerged as the de facto FM format standard. FM broadcasting, in contrast with AM, exhibited the sonic advantages of high-fidelity reproduction and stereophonic sound. The subtle nuances of orchestral performances, diminished by the process of low-fidelity, monaural AM transmission, sprang from FM receivers with astonishing clarity and accuracy. After years of languishing in the shadow of AM radio, the FM medium began to assert its identity as a separate and technically superior mode of broadcasting.

What It Is

Industry followers generally regard the easy listening format as a beautiful music derivative. Beautiful music is a program format featuring soft instrumental and vocal recordings directed to a target audience of predominantly middle-aged female listeners. Lush, melodic, and subdued in its presentation, the format is carefully planned and executed to offer listeners a quiet musical refuge for escaping from everyday distractions. Programmers regard announcer chatter as intrusive and tend to limit the spoken word to brief news reports, time checks, and weather forecasts. Commercial interruptions are similarly minimized, and the construction of message content reflects the format's low-key delivery approach. The objective is to provide listeners with a background musical environment that complements their daily activities.

Beautiful music resides in the development of Muzak, a registered trademark denoting the mood-enhancing background music service pioneered in the 1920s by Brigadier General George Owen Squier. Muzak programming was piped via leased telephone circuits into America's factories to stimulate productivity and into retail stores and restaurants to elevate patrons' spirits.

Programmers for beautiful music and easy listening founded their presentations in the lush, layered string arrangements popularized by the Andre Kostelanetz, Percy Faith, and Mantovani orchestras. Trade publications began using the format labels interchangeably during the mid-1970s. But musicologists cite several distinguishing characteristics between the formats. Beautiful music delved more deeply into the repertoire of 20th-century popular music composers than did easy listening. Tunes by Cole Porter, Jerome Kern, and Hoagy Carmichael—pop music's "standards"—were common fixtures on beautiful music playlists. In contrast, easy listening stations adopted a more contemporary music viewpoint, favoring a greater infusion of fresher sounds. Tunes by Burt Bacharach, Henry Mancini, and John Lennon and Paul McCartney were commonly integrated into the presentation.

Greater distinction between the formats was evidenced by programmers' approaches to vocal music, in terms of both style and quantity. Performances by "traditional" vocalists, such as Perry Como, Andy Williams, and Frank Sinatra, as well as the cover arrangements produced by artists including the Ray Conniff and Anita Kerr choral ensembles, were sparingly interspersed into the presentation of beautiful music. Easy listening stations, which tended to blend instrumentals and vocals on a more proportionate basis, gravitated toward the pop sounds of AM Top 40. Some of the "softer"-sounding hits by Elton John, Billy Joel, the Carpenters, and others qualified for airplay. Up-tempo performances, particularly those punctuated by intrusive guitar licks or intensive percussion, were generally passed over by easy listening programmers.

Key Producers

James Schulke, a former advertising executive and FM radio proponent, established SRP in 1968 to syndicate an approach to the good music presentation that he termed *beautiful music.* Capitalizing on the good music format's mood-music heritage, SRP Vice President/Creative Director Phil Stout constructed each quarter-hour of programming on a foundation of lush orchestral and vocal arrangements of popular music standards. One nuance in execution—the segue—differentiated the beautiful music from the good music format. Stout insisted that transitions between recordings flow in such a manner as to preserve the emotions elicited in listeners by the tempo, rhythm, and sound texture of the performances.

Schulke marketed Stout's concept as the "matched flow" approach to format execution. It was not uncommon for Stout to expend up to two days' effort to assemble a single hour of SRP programming. Collectively, the SRP team's insistence on musical cohesiveness and technical integrity proved successful in attracting and holding listeners. SRP's beautiful music format typically surpassed other formats in Arbitron's "time-spent-listening" measurements, a desirable position for stations catering to advertisers who sought message frequency over reach.

Bonneville was founded in 1970 by Marlin R. Taylor, a veteran programmer who pioneered easy listening on Philadelphia's WDVR in 1963. Taylor meshed a keen music sensibility with an adroit understanding of listeners' tastes in transforming the low-rated WRFM into New York City's number-one FM outlet. Success with WRFM inspired Taylor to launch Bonneville as the vehicle for extending his programming expertise to a clientele that grew in the early 1980s to approximately 180 stations.

Beautiful music prospered in the politically conservative 1970s because its musical message resonated with a nation of silent-majority listeners who had outgrown Top 40 and had never connected with progressive rock, country and western, or ethnic formats. By the end of that decade, more than a dozen syndicators, including Peters Productions, Century 21, and KalaMusic, competed with Bonneville and SRP for affiliates and listeners. As a result, it was common for two or three stations in each of the top 25 markets to vie for a share of the audience.

Beautiful music stations were unable to sustain the momentum, however. "Light" and "soft" adult contemporary (AC) stations, which burgeoned in the latter half of the 1970s, steadily siphoned beautiful music's target listener, the middle-aged female, during the 1980s. In a sense, both beautiful music and light AC were easy listening formats (see Josephson, 1986). Unlike beautiful music, which cultivated reputations with listeners as a background music companion, light AC was vocal-intensive, personality driven, and foreground focused in its presentation.

Switching from beautiful music to light AC improved the revenues of many stations. Advertising revenue erosion for beautiful music stations was attributed to the fact that its aging audience was spending a disproportionately low percentage of its disposable income on consumer goods. Light AC stations, whose demographic profiles skewed toward younger female adults, delivered the audience that had become most desired by national advertisers.

During a 1980 industry conference, broadcasters reached general agreement about the perceptual distinctions listeners had drawn between stations that positioned themselves as either easy listening or beautiful music outlets. In an effort to shed negative images of beautiful music as passive, background radio, broadcasters confirmed easy listening as a more appropriate positioning descriptor. The phrase *easy listening*, they agreed, evoked positive and active feelings of involvement by listeners with stations.

Decline

It was a calculated decision made by an industry about to confront a period of dramatic change in listener preferences. Easy listening/beautiful music, which Arbitron reported as radio's number-one format in 1979, slipped into second position behind adult contemporary the following year. The remainder of the 1980s proved to be a period of redefinition for easy listening.

Two distinct waves of defection by instrumental mood-music stations occurred, and each was precipitated by advertising industry pressures for stations to deliver younger, more upscale listeners. The first wave, in 1982–83, swept through the major markets. Where multiple easy listening outlets had once competed, now the lower-rated easy listening stations moved in other programming directions. Some of these stations subtly shifted toward soft AC, excising the instrumental music in favor of full-time vocals. Others pursued ratings success with entirely different formats. A second wave of abandonment occurred in 1988–89, when most of the remaining stations vacated easy listening. Traditional, "standards"-inspired instrumental music virtually disappeared from the airwaves.

The descriptive phrase *easy listening* became more indefinite in the radio lexicon of the 1990s, subsuming not only soft AC but the new age and smooth jazz genres as well. An emphasis on announcer personality and other formatics (sports play-by-play, traffic reports, and promotions) aligned easy listening with other mainstream music formats more closely than ever. As format fragmentation increased during this decade, the phrase *easy listening* generally gave way to variations on the soft AC theme.

BRUCE MIMS

See also Adult Contemporary Format; Formats; Middle of the Road Format; Schulke, James; Soft Rock Format; Taylor, Marlin R.

Further Reading

"Beautiful Music, Beautiful Numbers," *Broadcasting* (21 July 1975)
"Beautiful Music's Tag May Change," *Billboard* (6 September 1980)
Borzillo, Carrie, "Beautiful Music Gets a Makeover," *Billboard* (27 March 1993)
Eberly, Philip K., *Music in the Air: America's Changing Tastes in Popular Music, 1920–1980*, New York: Hastings House, 1982

"Hard Times for Easy Listening: List of Stations Dropping Format Grows," *Broadcasting* (28 November 1988)
Harris, Lee, "A Beautiful Opportunity to Resurrect a Format," *Radio World* (9 July 1997)
Josephson, Sanford, "Easy Listening vs. 'Soft' Contemporary: Shades of Gray," *Television/Radio Age* (17 March 1986)
Knopper, Steve, "Beautiful Music Gone, Not Forgotten," *Billboard* (21 June 1997)
Lanza, Joseph, *Elevator Music: A Surreal History of Muzak, Easy-Listening, and Other Moodsong,* New York: St. Martin's Press, 1994; London: Quartet, 1995
"Marlin Taylor Gets Beautiful Ratings with 'Beautiful Music,'" *Television/Radio Age* (7 September 1970)
Opsitnik, James, "Monday Memo," *Broadcasting* (5 November 1990)
Patton, John E., "Easy Listening Shake Out," *Television/Radio Age* (14 March 1983)
Ross, Sean, "Radio Takes Hard Look at Easy Format," *Billboard* (3 December 1988)
Routt, Edd, James B. McGrath, and Frederic A.Weiss, *The Radio Format Conundrum,* New York: Hastings House, 1978
"A Sour Note for Beautiful Music," *Broadcasting* (23 August 1982)
"Tailored Radio from Marlin Taylor," *Broadcasting* (11 June 1979)

The Edgar Bergen and Charlie McCarthy Show

Comedy Variety Program

Although few would have guessed it when the program first appeared on the air, a ventriloquist act became one of radio's longest-running comedy shows. The creation of Edgar Bergen, the smart aleck Charlie McCarthy character soon had the country in stitches.

The hour-long variety broadcast show was a highly popular format in the mid-1930s, and ventriloquist Edgar Bergen knew that his 17 December 1936 guest spot on Rudy Vallee's *The Royal Gelatin Hour* was a chance to break away from the uncertainties of nightclub engagements, party entertainment jobs, and a declining vaudeville circuit. Newspaper radio pages puzzled over Vallee's decision to "waste" airtime on an act that seemed to require being *seen,* and program insiders had even stronger doubts. However, Vallee asked his audience to give the newcomer a fair hearing, and top-hatted Edgar Bergen and his dummy, Charlie McCarthy, were so successful that they stayed for 13 weeks.

On 9 May 1937, the shy ventriloquist and his brash alter ego began hosting the National Broadcasting Company's (NBC) *The Chase and Sanborn Hour* on Sunday evenings at 8:00, one of radio's most desirable time slots. The Bergen-McCarthy program would see changes of title, length, personnel, sponsorship, and emphasis over the next two decades, but it would remain among the highest-rated of all programs until the summer of 1956, when network radio was rapidly yielding audiences to television.

Chase and Sanborn's program budget could afford a parade of Hollywood guest stars and the weekly services of singers Dorothy Lamour and Nelson Eddy, conductor Werner Janssen, and emcee Don Ameche. In the second half of each program, mischief-making Charlie McCarthy mocked W.C. Fields for his drunkenness and crabbiness. Volleys from the McCarthy-Fields feud are among the best-remembered lines in radio comedy—for instance, Fields' threat to whittle Charlie into a set of venetian blinds and the dummy's punning response, "That makes me shutter." On 12 December 1937, Mae West sent a shiver through the sponsor and network ranks when her sultry reading of an Adam and Eve sketch on the show drew widespread protests. More happily, the program pioneered remote broadcasts from military installations, and in 1939 Bergen introduced his second radio dummy, cheerfully slow-witted Mortimer Snerd.

Becoming a briskly paced half-hour show in January 1940, the retitled *Chase and Sanborn Program* lost regulars Ameche and Lamour and placed renewed emphasis on Bergen's and his dummies' interplay with guests such as Charles Laughton, Carole Lombard, Clark Gable, and Errol Flynn. Ray Noble led the orchestra, and, in sketches exploiting the differences between British and American English, he often rivaled Mortimer Snerd in comic "dumbness"; Bud Abbott and Lou Costello offered variations on their "Who's on first?" routine. In 1944 Bergen added a third dummy, aging and man-hungry Effie Klinker, who, when asked if she had anything to say to the listeners, blurted out her telephone number to any interested male. The lineup of celebrity guests halted in 1948 when Chase and Sanborn prepared to drop its sponsorship. In a transitional phase,

W.C. Fields, Dorothy Lamour, Charlie McCarthy, and Edgar Bergen
Courtesy of Bergen Foundation

Don Ameche returned with Marsha Hunt for a recurring segment as the eternally at-odds couple, the Bickersons.

Moving to the Columbia Broadcasting System (CBS) in 1949 as a result of William S. Paley's "talent raid" on NBC, *The Charlie McCarthy Show* continued to adjust to rapid changes in radio and in the sociopolitical climate. Called *The Edgar Bergen and Charlie McCarthy Show* in 1954 and *The New Edgar Bergen Hour* in 1955, Bergen's program now emphasized citizenship in a larger world by inviting scientists, professors, as well as military, political, and diplomatic figures to discuss their careers. Bergen and his friends vacated their regular time slot on 1 July 1956, and except for appearances on his first sponsor's 100th and 101st anniversary programs in 1964 and 1965, Edgar Bergen's radio career had ended. He went on to host a television quiz show in 1956 and 1957. Bergen died in his sleep in 1978 after the third Las Vegas performance of a planned farewell nightclub tour. Charlie McCarthy, who once had his own room in Bergen's Beverly Hills home, is now housed in a Smithsonian Institution display case.

In its early years, the Bergen-McCarthy program prompted an avalanche of Charlie McCarthy books, dolls, spoons, radios, and other products. In most radio households, the show was the measure of Sunday evening family listening, yet Charlie McCarthy's leering attentions to female guests challenged propriety, and today the show's stereotypic treatment of W.C. Fields' drinking, Effie Klinker's "old maid" status, and Mortimer Snerd's good-natured rural stupidity would draw protests from many quarters. Still, the Bergen-McCarthy shows remain among the most popular in "old-time radio" circulation.

In retrospect, Bergen's success lay in his decision to build his act on a cluster of ironic impressions. Opposing the hard-times grain of the 1930s, Bergen and Charlie wore elegant evening clothes, and Charlie's monocle and upper-crust English accent gave an initial impression that was startlingly at odds with his earthy brashness. Charlie assumed the calculating speech rhythm of the schoolyard sharpie, and he wasted little respect on his elders, particularly the sometimes preachy Bergen: "I'll clip ya', Bergie, so help me, I'll mo-o-ow ya' down!" was Charlie's signature threat. Bergen sometimes flubbed his lines, but cocky Charlie rarely did. Charlie often ridiculed Bergen's lip movements and complained that his creator had grown wealthy by stinting on the boy's allowance. Thus Charlie seemed to have the upper hand and the last word, but all the while Bergen's hand and mouth animated the creature of wood and cloth. The two were sharp opposites, yet one was entirely the creation of the other.

RAY BARFIELD

See also Ameche, Don; Comedy; Vallee, Rudy; Variety Shows

Cast

Charlie McCarthy	Edgar Bergen
Mortimer Snerd	Edgar Bergen
Effie Klinker	Edgar Bergen
Series regulars (various periods)	Don Ameche, Dorothy Lamour, Nelson Eddy, W.C. Fields, Dale Evans, Bud Abbott and Lou Costello, Pat Patrick, Jack Kirkwood, others

Producers

Tony Stanford, Sam Pierce

Programming History

NBC	1937–48
CBS	1949–56

Further Reading

Bergen, Candice, *Knock Wood,* New York: Linden Press, and London: Hamilton, 1984
Grams, Martin, Jr., *The Edgar Bergen and Charlie McCarthy Broadcast Log,* Delta, Pennsylvania: Martin Grams, Jr., n.d.
Wertheim, Arthur Frank, *Radio Comedy,* New York: Oxford University Press, 1979

Editorializing

Expressing a Station's Point of View

The broadcast of editorials by radio stations has enjoyed an uneven history for both legal and economic reasons. Although legally allowed since 1949 and actively encouraged by the FCC in later years, most stations rarely editorialize on any issue.

An editorial is the expression of the point of view of the station (or network) owner or management. News commentary, in which a single newsperson expresses an opinion about one or more news events, is not an "editorial," because that individual is rarely understood to be speaking for management. The parallel to newspapers is apparent—the editorial page matches what is addressed here, whereas the "op ed" page offers other (and sometimes disagreeing) individual expressions of opinion.

Origins

Many stations took to the air in the 1920s and 1930s specifically so that their owners could express their points of view on one or more controversial issues. Indeed, some early stations that became subject to Federal Radio Commission (FRC) sanctions got into trouble because of their one-sided approach to religious or political issues. Early administrative FRC and court decisions made clear that stations—as part of their requirement to operate in "the public interest, convenience or necessity"—should provide a balanced program menu, allowing a variety of points of view to be expressed. Nothing specific was said about whether or not stations could editorialize in the first place.

At the network level, editorializing was frowned upon in the 1930s. Although "comment" was fairly common among newscasters, and some programs of news "commentary" were common by the late 1930s—indeed, the role of news commentator became more widely recognized—such programs did not express the editorial opinion of network or station management and were thus not looked at in the same way by the regulators.

The *Mayflower* Case

Any question about the legality of station editorials was removed by the *Mayflower* case in 1941.

The Yankee Network's WAAB in Boston had presented editorials in 1937–38. This became a matter of legal concern in 1939, when the Mayflower Broadcasting Corporation (one of whose owners was a disgruntled former WAAB employee who felt stations should not editorialize) applied for the same frequency, throwing the stations into a comparative hearing before the FCC as required by provisions of the 1934 Communications Act. When the commission initially dismissed the Mayflower application for unrelated reasons, the company asked the FCC to reconsider based on the editorializing question. Early in 1941 the FCC issued its final decision, which, while upholding the renewal of WAAB's license, made clear that "a truly free radio cannot be used to advocate the causes of the licensee. . . . the broadcaster cannot be an advocate." The commission argued that given the limited number of frequencies available (and thus the number of stations that could broadcast), each licensee had to remain impartial. In using a public resource (its frequency assignment), a station took on the responsibility of being open to expression of all points of view, not merely its own. Because WAAB had ceased editorializing in 1938 when the FCC questioned the practice, its license was renewed.

Perhaps surprisingly, given broadcasters' fierce defense of their First Amendment rights, there was little industry complaint about the decision. Indeed, many broadcasters were secretly relieved that the FCC had eliminated the editorial option, because they knew that no matter what was said, any editorial could make listeners and advertisers unhappy. As stations grew to depend more on commercial support, avoiding controversy became ever more important.

Encouraging Editorials

Only after World War II did the issue arise anew. Early in 1948 the FCC held eight days of hearings on the question of stations' editorializing and dealing with controversial issues. From those hearings came a mid-1949 decision allowing editorials, in effect reversing the *Mayflower* ban. As the commissioners put it, "We cannot see how the open espousal of one point of view by the licensee should necessarily prevent him from affording a fair opportunity for the presentation of contrary positions." Little did anyone see at the time how this decision—which allowed but did not actively encourage editorials—would lay the groundwork for the hugely controversial fairness doctrine in years to come.

Eleven years later, the commission became more positive, including "editorialization by licensees" as the 7th of 14 specific program types held to be in the public interest. More stations began to offer at least occasional editorials, with some larger outlets hiring dedicated staff for the purpose. But the majority of stations never editorialized, and many others did so only infrequently. For a number of years, *Broadcasting Yearbook* kept track of the number of stations providing editorials. Of AM stations reporting (including most but not all of those then on the air), 30 percent editorialized in 1959, and six years later, 61 percent did, though nearly half of those did so only occasionally.

One indicator of the (temporary, as it turned out) growth of station editorializing was the formation of the National Broadcast Editorial Association (NBEA) in the early 1970s. The association grew to more than 200 members (usually the editorial director of a station); issued a quarterly publication, the *NBEA Editorialist;* and held annual conventions. One of the better-known station editorial directors was Don Gale, of the KSL stations in Salt Lake City, who over two decades broadcast some 5,000 editorials before his 1999 retirement. Each was broadcast three times a day on radio and twice on television. Ed Hinshaw at Milwaukee's WTMJ and Phil Johnson at WWL in New Orleans wrote or broadcast editorials for both radio and television for more than a quarter century.

By 1977 fewer stations reported editorializing activity, and most of those were only doing occasional editorials. A 1982 study identified more than 1,200 stations editorializing, many of them AM-FM-TV combinations. But later surveys suggested that the number was both much smaller and in decline. In September 1991 the NBEA was absorbed by the National Conference of Editorial Writers, most of whose members worked for newspapers.

The decline in station editorials after about 1980 was driven largely by economics. Increasing competition for advertising dollars and the need not to make potential clients irate certainly contributed. The demise of the fairness doctrine in 1987 probably contributed as well. So did the general demise of radio news and public-affairs programming on many stations. Because most focus on music, radio stations no longer compare themselves to newspapers in their community role. With that change in identification came the demise of editorializing.

CHRISTOPHER H. STERLING

See also Commentators; Controversial Issues; Fairness Doctrine; Mayflower Decision; News; Public Affairs Programming

Further Reading

Federal Communications Commission, *In the Matter of the Mayflower Broadcasting Corp. and the Yankee Network, Inc. (WAAB),* 8 FCC 333 (16 January 1941)

Federal Communications Commission, *In the Matter of Editorializing by Broadcast Licensees,* 13 FCC 1246 (1 June 1949)

Federal Communications Commission, *Report and Statement of Policy re: Commission En Banc Programming Inquiry,* 44 FCC 2303 (29 July 1960)

Kahn, Frank J., editor, *Documents of American Broadcasting,* New York: Appelton-Century-Croft, 1968; 4th edition, Englewood Cliffs, New Jersey: Prentice-Hall, 1984

Sterling, Christopher, "Broadcast Station Editorials, 1959–1975," in *Electronic Media: A Guide to Trends in Broadcasting and Newer Technologies, 1920–1983,* by Sterling, New York: Praeger, 1984

Summers, Harrison Boyd, compiler, *Radio Censorship,* New York: Wilson, 1939

Education about Radio

Developing University Curricula and Degrees

One measure of a topic's social importance or role is whether colleges and universities conduct research and offer courses (or even degrees) concerning that subject matter. Such academic recognition becomes a touchy question when the topic is largely commercial and enjoys a popular following—as did radio broadcasting by the mid-1920s. The study of mass communication was initially shaped with the first academic programs in newspaper journalism in the early 20th century. Radio broadcasting, however, presented something quite different with its emphasis on popular entertainment.

Origins

Perhaps ironically, the first Ph.D. dissertation on radio broadcasting was published as a book long before college or university organized studies of radio existed. Hiram Jome's *Economics of the Radio Industry* (1925) was based on the author's economics doctorate earned at the University of Wisconsin. Two years later, Stephen Davis' *The Law of Radio Communication* (1927) inaugurated yet another field of serious study, again long before most law schools offered courses on the subject. A series of lectures by radio leaders was delivered as part of a business policy course at the Harvard School of Business in 1927–28 (the lectures appeared as a book), and the first regular course about radio was organized in 1929 at the University of Southern California.

Soon additional scholarly apparatus became evident. The first scholarly journal article concerning radio appeared on the pages of the *Quarterly Journal of Speech* in 1930, when Sherman Lawton discussed principles of effective radio speaking. Two years later he expanded that article into the first college-level textbook on radio, *Radio Speech*. Initial courses began to appear elsewhere, usually within English or speech departments. By 1933 an early survey showed that 16 colleges and universities offered at least one radio course. In 1937 the first comprehensive radio textbook appeared, which would run through four editions over the next two decades—Waldo Abbot's *Handbook of Broadcasting*.

The pace of development speeded up in the final years before World War II. By 1938 another survey showed that more than 300 institutions offered at least one course in radio. Furthermore, eight now offered a bachelor's degree and two offered a master's degree; in 1939 the Universities of Iowa and Wisconsin began offering a Ph.D. with an emphasis in the study of radio broadcasting. By 1939–40 some 360 schools offered about 1,000 courses in 14 different subject categories (including electrical engineering, which was the majority), of which the most common non-technical topics included radio speech, a survey course, scriptwriting, and program planning/production.

Postwar Expansion

Broadcast education entered a period of substantial growth, and there was a concerted movement to develop standards, if not actual accreditation, for degree programs in radio (and soon television). In 1945 one committee published a brief set of suggested standards for radio degree programs. By 1948 a government survey reported that more than 400 schools provided at least a single radio course, with 35 offering non-engineering degrees in radio broadcasting. However, not everything was in place as television began to make its appearance.

It is often said that a true academic field of study needs at least one national (or international) association of like-minded scholars and a research journal. In 1948 a step was taken toward the first of these with the creation of the University Association for Professional Radio Education (UAPRE), founded by about a dozen universities. The title reflected a tension evident in education for radio—were such courses and degrees designed primarily to turn out personnel for the industry, or was this media education to be more in the liberal arts tradition? UAPRE was set up specifically to accredit university and college degree programs, but it was unable to achieve that goal, though for reasons having nothing to do with radio but rather with the complex politics of establishing any national accrediting process.

By 1950 some 420 colleges and universities offered courses, and 54 provided non-engineering degrees: 30 bachelor's, 15 master's, and 3 Ph.D. Five years later, there were at least 81 course sequences leading to radio/broadcasting degrees. In 1955 the UAPRE gave way to the Association for Professional Broadcasting Education (APBE), which a year later began publishing *Journal of Broadcasting* as the first dedicated scholarly journal in the developing field. A year later a national survey reported 93 bachelor's degree programs (3,000 majors), 56 master's degree programs (over 400 students), and 15 Ph.D. programs of study (122 candidates). Broadcast educators now met annually with the National Association of Broadcasters, and many academics were becoming active in the more senior speech and journalism academic organizations. More scholarships, internships, and research opportunities were becoming available every year.

At that point, of course, all eyes were quite literally on television, and radio courses and research began to disappear. A handful of announcing courses survived, but most other broadcast programs focused very strongly on television—and given the costs of such education and the lack of student interest in radio, radio studies were fairly quickly abandoned. However, the field was now sufficiently established to move away from a focus on professional education alone. APBE became the Broadcast Education Association (BEA) in 1973, thereby better recognizing the many liberal arts broadcast programs that had developed.

Radio's Revival

In the 1980s radio began to reappear in college and university curricula. This was due to a combination of factors, first in the industry and then in education.

Certainly the growing number of radio stations (with their entry-level positions for college and university graduates) was a factor. So was the revival of educational and public radio beginning in the 1970s and the reappearance of radio drama documentary broadcasts. The growing complexity of radio formats added a degree of depth to the medium that it had not previously possessed: there was more to study and understand than before. As audience and market research became more widespread and important in radio, people had to be educated in these areas. The explosive popularity of talk radio in the 1980s put radio in the political limelight. Radio was increasingly in the news: with controversial disc jockeys and talk show hosts having impact on elections, its appeal to young people as a possible career option increased exponentially. At the other end of the spectrum, growing interest in the "golden age" of radio increased interest in the history and sociology of radio. As in other parts of American life, more women and minorities were being employed by radio, opening further paths to success in the medium. Technology played a growing role as well, as the radio industry became increasingly computerized and automated, made plans to begin digital operations, and became more widely available on the internet.

Just as radio managers seemed to place more emphasis on the educational credentials of those trying to enter the field, colleges and universities seemed to rediscover radio as well. There were several indicators, including the reappearance of radio-only comprehensive and production-oriented textbooks, the new *Journal of Radio Studies* founded as an annual in 1992 and expanding to twice a year by 1998 (as the first academic journal focused on radio), and the (perhaps belated) formation of a broadcast and internet radio division within BEA in 2000. Radio began reappearing in course titles in university programs across the country, coming full circle seven decades after the academic discovery of the medium.

CHRISTOPHER H. STERLING

See also Barnouw, Erik; Broadcast Education Association; College Radio; Educational Radio to 1967; Intercollegiate Broadcasting System; Lazarsfeld, Paul F.; Museums and Archives of Radio; National Association of Educational Broadcasters; Office of Radio Research; Siepmann, Charles A.

Further Reading

Abbot, Waldo, *Handbook of Broadcasting: How to Broadcast Effectively,* New York: McGraw-Hill, 1937

Davis, Stephen Brooks, *The Law of Radio Communication,* New York: McGraw-Hill, 1927; reprint, 1992

Directory of Colleges Offering Courses in Radio and Television (annual)

Federal Radio Education Committee, *Suggested Standards for College Courses in Radiobroadcasting,* Washington, D.C.: U.S. Office of Education, Federal Security Agency, 1945

Head, Sydney W., and Leo A. Martin, "Broadcasting and Higher Education: A New Era," *Journal of Broadcasting* 1 (Winter 1956–57)

Kittross, John Michael, "A History of the BEA," *Feedback* 40 (Spring 1999)

Lawton, Sherman Paxton, *Radio Speech,* Boston: Expression, 1932

Lichty, Lawrence W., "Who's Who on Firsts: A Search for Challengers," *Journal of Broadcasting* 10 (Winter 1965–66)

Niven, Harold, "The Radio-Television Curricula in American Colleges and Universities," *Journal of Broadcasting* 4 (Spring 1960)

Niven, Harold, "The Development of Broadcasting Education in Institutions of Higher Learning," *Journal of Broadcasting* 5 (Summer 1961)

The Radio Industry: The Story of Its Development, As Told by Leaders of the Industry to the Students of the Graduate School of Business Administration, George F. Baker Foundation, Harvard University, Chicago, New York, and London: Shaw, 1928; reprint, as *The Radio Industry: The Story of Its Development, As Told by Leaders of the Industry,* New York: Arno Press, 1974

Smith, Leslie, "Education for Broadcasting, 1929–1963," *Journal of Broadcasting* 8 (Fall 1964)

Educational Radio to 1967

Well before radio broadcasting became an entertainment and sales medium, it was used for education. When Charles D. Herrold started the first radio broadcasting station in the United States in 1909, it was largely intended as a laboratory for students of the Herrold School of Radio in San Jose, California. When Lee de Forest, also in the first decade of the 20th century, tested his firm's radio apparatus using recordings of opera and other classical music, it was partly to introduce others to the music he loved.

There has been conflict from the early 1920s until the present between those who believe there is a need for nonprofit educational broadcasting and those who do not. For example, many commercial broadcasters believe that the frequencies occupied by noncommercial educational radio might be better employed for advertising. On the other hand, although its potential has never been fully realized, educational radio—like the internet—has such high potential that its use for education attracted many enthusiastic supporters.

This conflict is pervasive. In the late 1920s, the Federal Radio Commission (FRC) encouraged shared-time broadcasters to persuade educational institutions to give up their licenses. In the early 1930s, the Wagner-Hatfield Amendment was proposed to reserve channels for nonprofits. In 1941 channels were first set aside or reserved for education on the new FM band. In 1946, when the Federal Communications Commission (FCC) issued its "Blue Book" (*Public Service Responsibility of Broadcast Licensees*), it called upon commercial stations to cater to the needs of nonprofit organizations. However, the storm of protest from commercial stations about the "Blue Book" buried this idea at the time.

But although the record shows that educational radio has often been impressively effective in schools and adult education, it frequently was underappreciated. Educational broadcasting was cavalierly dismissed for decades, even though it was originally touted as a great advance. For decades, arguments in favor of a separate educational radio service relied on an analogy with the public service philosophy of agricultural extension intrinsic to the land grant colleges.

Origins: 1920–33

Indeed, one land grant institution, the University of Wisconsin, owns WHA, one of the oldest radio broadcasting stations in the United States. Land grant colleges were interested in outreach—the extension of their expertise to isolated regions. As a result, they first used radio both to provide lectures on topics of more or less general interest and to give specific information on topics intended to serve citizens of their state, such as agriculture and home economics. A few tried to use radio for fund raising in the early 1920s, with very limited success.

A number of stations were merely the tangible results of experiments by engineering and physics faculty and students (such as Alfred Goldsmith, who operated 2XN at City College of New York, 1912–14) who wished to explore the phenomenon of wireless telephony. However, once they had tinkered to their heart's content, the daunting need to find content for their transmissions led, in many instances, to these stations' being turned over to departments of speech, English, music, and extension services. Another reason for this transfer was that the cost of programming, once the technical facilities met the FRC's standards, was a major expense.

Even before World War I, radio telegraph and some radio telephone experimentation had taken place at such colleges and universities as Arkansas, Cornell, Dartmouth, Iowa, Loyola, Nebraska, Ohio State, Penn State, Purdue, Tulane, and Villanova, in addition to Wisconsin. After stations were allowed back on the air in 1919 following the war, regular broadcasting service was started by many of these and others either for extension courses or strictly for publicity.

By 1 May 1922, when the number of radio stations was beginning a meteoric rise, the Department of Commerce's *Radio Service Bulletin* reported that more than 10 percent (23 of 218) of the broadcasting stations then on the air were licensed to colleges, universities, a few trade schools, high schools, religious organizations, and municipalities. This proportion soon climbed, with 72 stations (13 percent) licensed to educational institutions among the 556 on the air in 1923, and 128 (22 percent) of a similar total (571) in 1925.

Although early records are imprecise—noncommercial AM stations were issued the same type of license as commercial ones—many of these stations were short-lived. Not only did it appear that there was no pressing need for educational radio, but other factors pared the number on the air from 98 in 1927 (perhaps 13 percent of all stations) to half that (43, or 7 percent) in 1933. Although nearly 200 standard broadcast (AM) stations were licensed to educational institutions through 1929, almost three-quarters were gone as early as 1930. The 47 stations that remained, however, were tenacious—and roughly half of them continued to serve their audiences at the beginning of the 21st century.

The number of both commercial and noncommercial stations dropped in the late 1920s, largely because of the expensive technical requirements imposed by the FRC under the Radio Act of 1927. Additionally, the FRC's 1927–29 reallocations and elimination of marginal and portable stations often gave desirable channels to commercial operators at the expense of educational institutions. The costs of new interference-reducing transmitters and other facilities were imposed just prior to the economic dislocations of the Great Depression—which dried up funds for colleges and commercial companies alike.

In addition, many university administrators saw no value in radio. At best, running a radio station, except where it also served curricular needs or could be used effectively by agricultural extension services, seemed to be a lot of work and expense for limited reward during tough times. Many institutions appeared glad that the FRC's new rules gave them an excuse to drop this expensive toy. This lack of interest closed many educational stations.

Into the 1930s, some educational stations found themselves pressured by local shared-time broadcasters to give up their frequencies for full-time use by commercial stations. To sweeten the deal, commercial broadcasters often offered to air some educational programming. Because commercial stations had more lobbying clout and deeper pockets and, in many cases, appeared to be more stable, closing down educational stations sounded very attractive to policy makers. It also sounded good to timid college administrators who didn't recognize the benefits of using radio for teaching, tended to allocate funding in other directions, and found commercial offers a "win-win" choice—until the promised educational programming was crowded out by advertising-supported programs.

A hard core of educational broadcasters successfully claimed that their stations were serving both the public interest and the interests of their parent institutions. But where there were not enough trained and interested personnel to argue for retention and financial support, many college and university stations no longer had an educational mission—nor, usually, a license.

Concern for their dwindling numbers and interest in ways of using radio effectively led radio educators to band together. In mid-1929, the Advisory Committee on Education by Radio was formed with backing from the Payne Fund, the Carnegie Endowment, and J.C. Penney, but it died before 1930 without having had much effect. In 1930 two rival organizations that would represent educational radio for a decade were organized: the National Advisory Council on Radio in Education and the National Committee on Education by Radio. The council worked with grants from the Rockefeller Foundation and the Carnegie Endowment and called on commercial stations to meet education's needs. The committee, with support from the Payne Fund, asked that nonprofit educational entities be given 15 percent of all station assignments—which would double the number of educational stations; the committee also attacked "commercial monopolies" and disagreed with what it called the "halfway" measures of the council.

One result of the proselytizing by both organizations was a Senate-mandated 1932 FRC survey of educational programs on both commercial and noncommercial stations. Having carefully timed their survey for National Education Week, when most stations scheduled some educational programs, the FRC concluded that commercial stations were adequately filling educational needs. Congress was not completely convinced, and Senators Wagner and Hatfield sponsored an unsuccessful amendment to the 1934 Communications Act allocating 25 percent of broadcast facilities to nonprofit organizations. After this failed, another dozen of the remaining educational stations began to take advertising in 1933 to meet operating costs and to cover American Society of Composers, Authors, and Publishers (ASCAP) music licensing fees.

The campaign for a set-aside of channels for education continued, resulting in Section 307 (c) of the Communications Act of 1934, which told the new Federal Communications Commission to "study the proposal that Congress by statute allocate fixed percentages of radio broadcasting facilities to particular types or kinds of non-profit radio programs or to persons identified with particular types or kinds of non-profit activities." The FCC held extensive hearings—100 witnesses, 13,000 pages of transcript—in the fall of 1934 and (as might have been anticipated) recommended against reservation of frequencies, for educator cooperation with commercial stations and networks, and for the establishment of another committee.

The resulting Federal Radio Education Committee (FREC) was originally composed of 15 broadcasters; 15 educators or members of groups such as the Parent-Teacher Association; and 10 government officials, newspaper publishers, and others, and it was headed by U.S. Commissioner of Education John Studebaker. "Broadcasters" ranged from network heads to the National Broadcasting Company's (NBC) music and education directors. The committee's purposes were to "eliminate controversy and misunderstanding between groups of educators and between the industry and educators" and to "promote actual cooperative arrangements between educators and broadcasters on national, regional, and local bases." A "subcommittee on conflicts and cooperation" was to try to deal with any friction between "the commercial and the social or educational broadcasters."

Holding On: The 1930s

FREC operated a script exchange (used by 108 stations and a large number of local groups in its first full year) and planned 18 major research projects so it could make valid recommendations. Funding—roughly two-thirds from educational foundations and one-third from the broadcasting industry (through the National Association of Broadcasters)—did not live up to expectations, and roughly half of the projects had to be dropped. Although some useful publications, a supply of standardized classroom receivers, and a great deal of script distribution resulted, the FREC actually held only one full meeting. It disbanded after World War II. Nevertheless, for more than three decades, most educational broadcasters looked to the U.S. Office of Education, headed by Franklin Dunham and Gertrude Broderick, for coordination, sympathy, and low-cost practical assistance.

Because there were then no educational networks of any sort, the exchange of scripts was the best way to give educational stations and their audiences access to programming prepared by others. Beginning in early 1929, the Payne Fund supported daily *Ohio School of the Air* broadcasts on powerful commercial station WLW in Cincinnati. These programs, intended for in-school listening, were produced in the studios of WOSU, then and now licensed to the Ohio State University in Columbus. The Ohio state legislature appropriated money to partly cover production costs, teacher guides, and pupil materials. Later, WOSU, in the center of the state, became the primary transmitter of this program. Another early educational series for below-college-level classroom listening was the *Wisconsin School of the Air,* which began on university-owned WHA in the fall of 1931. WHA started a "College of the Air" two years later.

During radio's "golden age" from 1934 until the end of World War II, educational radio survived—but barely. The 43 educational AM stations on the air in 1933 dropped to no more than 35 by mid-1941. Roughly half had been on the air more than 15 years, 12 were commercially supported, and 7 of these were affiliated with a commercial network, airing educational programs only a few hours a day. One was operated by a high school, 2 were operated by church-affiliated educational groups, 9 by agricultural schools or state agricultural departments, and 11 by land grant universities, mostly in the Midwest. Only 11 stations were licensed for unlimited broadcast time, about half of them in the 250- to 5,000-watt category.

Commercial stations and national networks regularly scheduled some avowedly educational programming. In the 1920s, commercial broadcasters started "radio schools of the air," lectures, and even courses for credit. WJZ (New York) began such broadcasts in 1923, WEAF (New York) followed, and WLS (Chicago) started its *Little Red Schoolhouse* series in 1924. Such programs, often in association with school boards, were long-lived and useful to the stations' images. For example, WFIL (Philadelphia) aired the *WFIL Studio Schoolhouse,* a series of daily radio programs, complete with teachers' manuals, in the mid-1950s. The *Standard School Broadcasts* (Standard Oil Company) were broadcast over a number of California stations into the 1960s.

A 1928 FRC study of 100 stations in the western United States had found that radio was supplying more features and more plays but fewer children's programs and less educational material than three years earlier. The study also found that, of the 54 hours that the average station was on the air each week, 5 were devoted to education and lectures other than on farm subjects, and 3 hours were on farm reports and talks. FCC hearings in 1934 contained testimony that networks and larger stations were more cooperative with educators than were small and independent stations. Although the amount of such programming did not rise appreciably over the next two decades, neither did it fall until most commercial radio adopted all-music formats in reaction to the success of television in the 1950s.

Educational programming tended to be of three kinds: classroom instruction in English, history, social studies, and other disciplines, intended for classes from kindergarten to college; extension study, typically in fields that would be useful to the state or region being served, such as detailed agricultural practices and marketing news; and general cultural programs, such as classical music, which were inexpensive to produce.

Few stations of any sort produced dramatic programs, because the need for proven stories, talented performers, and special skills (such as sound effects) tended to require a major investment in time as well as money. College football and baseball games were often carried on college stations—at least until they became popular enough that more-powerful commercial stations offered to carry them. One potential audience given special attention by nonprofit stations was children, with programs of all types from stories and games to health instruction.

But the failure to provide an adequate number of channels on the AM band for education was not forgotten. National organizations continued to agitate for an adequate number of channels or to coordinate the efforts of existing stations. These groups included the FREC, the National Advisory Council on Radio in Education, the Institute for Education by Radio, the National Association of Educational Broadcasters (NAEB), and the National Committee on Education by Radio.

The Institute for Education by Radio was established at the Ohio State University in 1930 by I. Keith Tyler, and it hosted annual practitioner conferences on educational radio until 1960. Its published proceedings are a good source of contemporary thinking about what radio was doing for education—and might still do with sufficient support.

The NAEB was founded in 1934, but it traced its lineage to the Association of College and University Broadcasting Stations, established during the rush to get education on the air in 1925. NAEB was primarily a program idea exchange during the period up until World War II, but some of its 25 members also experimented with off-the-air rebroadcasting.

The National Committee on Education by Radio, mentioned earlier, worked hard for allocation of educational channels and sponsored annual conferences from 1931 to 1938, when its Rockefeller Foundation funding ended. One of the committee's more interesting initiatives was to help establish local listening councils—groups of critical listeners who would work with local broadcasters to improve existing programs and plan new ones.

In an attempt to secure educational channels without stepping on the toes of commercial broadcasters, the National

Committee on Education by Radio proposed that new educational stations be assigned to channels in the 1500–1600 kilohertz range, just above the standard (AM) broadcasting band of the time. These stations, however, typically would have had less range than those on lower frequencies and would have had to purchase new transmitters and antennas in many instances. Furthermore, there was no way of ensuring that receivers would be built that could pick up this band. As a result, most educators had little enthusiasm. (These frequencies eventually had limited use for high-fidelity AM experimental broadcasting, using 20-kilohertz-wide channels. In March 1941 they were added to the regular AM band.)

Early in 1938, the FCC reversed its earlier position and established the first specific spectrum reservations for noncommercial broadcast use, in the 41- to 42-megahertz band. After the FCC set aside 25 channels for in-school broadcasting, the Cleveland Board of Education was licensed (as WBOE) in November 1938. The next year, this allocation was moved to the 42- to 43-megahertz band, and the broadcasters using it were required to change from AM to the new frequency modulation (FM) mode of modulation. The first noncommercial educational FM stations were authorized in 1941. Because FM required a much wider bandwidth, this allocation provided only five channels, on which two fully licensed (and possibly five experimental) stations were transmitting to radio-equipped classrooms by late 1941. When commercial FM went into operation, education was assigned channels at the bottom of the overall band, which were fractionally easier for listeners to receive than higher ones. The interference-free reception and high audio fidelity of FM initially were very attractive, and half a million sets were sold before all civilian radio manufacturing was ended by the demands of World War II.

Postwar Rebirth: To 1967

This opportunity to establish new educational stations was welcomed. Although the wartime construction "freeze" of 1942 exempted educational broadcasting stations, scarcity of construction materials and broadcast equipment—together with the slow decision-making processes of educational institutions—meant that educational FM broadcasting at the end of the war in 1945 consisted of about 25 AM stations, plus 12 FM authorizations and 6 FM stations on the air. Nevertheless, many potential educational broadcasters used the war years for planning, realizing that finally, after more than a decade of talking about educational radio, they had the means to accomplish it.

But there were complications. In 1945 the FCC shifted the educational FM allotment to 88–92 megahertz, at the bottom of a new FM band. Because only a handful actually had to replace transmitters, this change was reluctantly accepted. The number of noncommercial educational FM stations grew substantially after World War II. Although these data are not completely reliable, it appears that there were 10 such stations on the air at the start of 1947 and 85 by 1952, 14 percent of all FM stations then on the air. The number kept rising and the proportion remained respectable: 98 in 1953, 141 in 1958, 209 in 1963, and 326 (16 percent) on 1 January 1968.

Perhaps chastened by earlier experiences with educational AM radio, institutions of higher education, school districts, and nonprofit community groups were initially hesitant to apply for FM licenses. Apart from past disappointments, high costs, and the need for colleges and universities to earmark resources to serve the millions of postwar students, potential operators were wary of the continuing paucity of homes with FM receivers and the possible effects of another new medium, television.

Late in 1948 the FCC, recognizing the burdens of high cost and the limited or campus-only uses planned by some colleges, approved a new class of 10-watt stations. These low-power FM stations were immediately popular and constituted more than one-third of the 92 educational FM stations on the air in 1952.

Audio tape recording's potential was first realized by Seymour Siegel of WNYC in 1946. Recording also made it possible to reuse programs; for example, one station was still rebroadcasting some 1945 children's programs in the 1990s—and they still were attracting youthful audiences. In 1951 the Kellogg Foundation provided funds to NAEB to establish a tape duplication operation at the University of Illinois, Urbana, to facilitate a non-interconnected "bicycle network" by which tape recordings of one station's programs were mailed to other stations in succession. Soon, more than 40 stations were participating. The NAEB Tape Network was made possible by the postwar introduction of high-fidelity magnetic tape recorders, which replaced clumsy and fragile 15- or 16-inch transcription discs.

However, some postwar regional or statewide FM networks were established, as in Alabama and Wisconsin. In the latter, eight stations provided a full day's programming to schools, colleges, and adults in most of the state. The National Educational and Radio Center, established in Ann Arbor, Michigan, in 1954 and later moved to New York, eventually started providing some taped radio programs through a subsidiary. At least one independent group of noncommercial stations, affiliated with the Pacifica Foundation, still operates stations across the country; another, the KRAB Nebula, organized by Lorenzo Milam, was more ephemeral.

Many educationally licensed stations programmed a great deal of classical or folk music and jazz, which were timeless and inexpensive. While education, music, and talk programs were generally inexpensive, news coverage was not, and educational radio therefore played a fairly minor role in the listening habits of most members of radio's audience into the 1960s.

During the period up to 1967, educational radio remained an orphan in many ways. Although given general support in regulatory matters by the FCC and Congress, funding was a perennial local problem. Studies of educational radio usually arrived at the obvious conclusions that money and a national rather than strictly local image were needed. On a local level, some educational radio stations served the public interest economically and well. Some acted both as training laboratories—for example, municipally owned WNYC trained interns from colleges across the country, and its sister station, WNYE, used high school students to create programs for schools throughout New York City—and as sources of a wider variety of programming for the many people who were not well served by the commercial radio broadcasting industry. Although not the first to lobby, the 1966 Wingspread Conference on Educational Radio as a National Resource came at the right time to argue for federal funding for radio, a proposal that succeeded in including radio in the 1967 Public Broadcasting Act. This provision of some facilities and program funding through the Corporation for Public Broadcasting and NPR after 1967 gave educational radio a very different look.

JOHN MICHAEL KITTROSS

See also American School of the Air; Blue Book; Community Radio; Corporation for Public Broadcasting; Federal Communications Commission; Goldsmith, Alfred; Low-Power Radio/Microradio; National Association of Educational Broadcasters; National Public Radio; Pacifica Foundation; Public Broadcasting Act; Public Radio Since 1967; United States; WHA and Wisconsin Public Radio

Further Reading

Atkinson, Carroll, *American Universities and Colleges That Have Held Broadcast License,* Boston: Meador, 1941
Atkinson, Carroll, *Broadcasting to the Classroom by Universities and Colleges,* Boston: Meador, 1942
Atkinson, Carroll, *Radio Network Contributions to Education,* Boston: Meador, 1942
Atkinson, Carroll, *Radio Programs Intended for Classroom Use,* Boston: Meador, 1942
Baudino, Joseph E., and John M. Kittross, "Broadcasting's Oldest Stations: An Examination of Four Claimants," *Journal of Broadcasting* 21 (Winter 1977)
Blakely, Robert J., *To Serve the Public Interest: Educational Broadcasting in the United States,* Syracuse, New York: Syracuse University Press, 1979
Darrow, Benjamin Harrison, *Radio: The Assistant Teacher,* Columbus, Ohio: Adams, 1932
Edelman, Murray J., *The Licensing of Radio Services in the United States, 1927–1947: A Study in Administrative Formulation of Policy,* Urbana: University of Illinois Press, 1950
Engelman, Ralph, *Public Radio and Television in America: A Political History,* Thousand Oaks, California: Sage, 1996
Frost, S.E., *Education's Own Stations,* Chicago: University of Chicago Press, 1937
Gibson, George H., *Public Broadcasting: The Role of the Federal Government, 1912–1976,* New York: Praeger, 1977
Herzberg, Max J., editor, *Radio and English Teaching: Experiences, Problems, and Procedures,* New York and London: Appleton-Century, 1941
Hill, Harold E., and W. Wayne Alford, *NAEB History,* 2 vols., Washington, D.C.: National Association of Educational Broadcasters, 1965–66
Levenson, William B., *Teaching through Radio,* New York: Farrar and Rinehart, 1945
Lingel, Robert J.C., *Educational Broadcasting: A Bibliography,* Chicago: University of Chicago Press, 1932
Maloney, Martin J., "A Philosophy of Educational Television," in *The Farther Vision: Educational Television Today,* edited by Allen E. Koenig and Ruane B. Hill, Madison: University of Wisconsin Press, 1967
Marsh, C.S., *Educational Broadcasting: Proceedings of the National Conference on Educational Broadcasting, 1936,* Chicago: University of Chicago Press, 1937
Marsh, C.S., *Educational Broadcasting: Proceedings of the National Conference on Educational Broadcasting, 1937,* Chicago: University of Chicago Press, 1938
McGill, William J., *A Public Trust: The Report of the Carnegie Commission on the Future of Public Broadcasting,* New York: Bantam Books, 1979
Perry, Armstrong, *Radio in Education: The Ohio School of the Air, and Other Experiments,* New York: The Payne Fund, 1929
Radio and Education (annual; 1931–34); as *Education on the Air* (annual; 1934–) (proceedings of annual conferences on educational broadcasting)
Robertson, Jim, *TeleVisionaries: In Their Own Words, Public Television's Founders Tell How It All Began,* Charlotte Harbor, Florida: Tabby House Books, 1993
Siepmann, Charles Arthur, *Radio's Second Chance,* Boston: Little Brown, 1946
Sterling, Christopher H., and John M. Kittross, *Stay Tuned: A History of American Broadcasting,* 3rd edition, Mahwah, New Jersey: Lawrence Erlbaum, 2002
Stewart, Irvin, editor, *Local Broadcasts to Schools,* Chicago: University of Chicago Press, 1939
United States Federal Communications Commission, *Public Service Responsibility of Broadcast Licenses,* Washington, D.C.: Government Printing Office, 1946

Edwards, Bob 1947–

U.S. Radio Host

As the solo host of *Morning Edition* since its first broadcast in 1979, Bob Edwards has shaped the sound of National Public Radio (NPR) in the morning. He has captured the morning news audience that was largely abandoned by commercial outlets and built a loyal weekly listenership of more than 12 million.

Edwards is a man of contradictions. He has won most of the major awards in broadcasting, but he likes to say that his job security lies in the fact that no one else wants to work his hours. He's an elected union official who insists on every nickel in a contract, yet he spends weeks each year on the road raising money for public radio stations.

Edwards came to NPR in 1974 as a newscaster. In a matter of months, he was chosen to cohost *All Things Considered* (*ATC*). Susan Stamberg, who shared the microphone with Edwards for five years, says he made up for his limited experience with clear writing and "that terrific voice."

Edwards actually came to the job with a background in commercial radio; he worked first at WHEL-AM in New Albany, Indiana, and then as an anchor and newscaster for Armed Forces Radio and Television when he was in the Army in Korea. After the army, Edwards moved to Washington, D.C., where he worked as a part-time anchor for WTOP-AM while earning a master's degree in communication from American University. At American University, he absorbed the gospel of Edward R. Murrow–style journalism from Ed Bliss, Jr., a former newswriter for Murrow at the Columbia Broadcasting System (CBS). Bliss passed on Murrow's insistence on active, uncluttered language and his love of sound in telling the story.

Susan Stamberg says that Edwards had a very old-fashioned news style when he came to NPR but that he made her "understand the greatness of Edward R. Murrow." She says Edwards' understated style provided a good balance for her exuberance: "His coolness permitted me to be hotter." Edwards says he learned to be like the straight man in a comedy team, a skill that served him well later in his 12-year broadcast relationship with Red Barber.

Edwards' tenure on *ATC* came during the years that Frank Mankiewicz ran NPR and began seriously marketing it. "There was a lot of buzz," Edwards says. "That was the point at which people started taking us seriously as a news source."

Edwards and Stamberg felt mildly threatened in 1978 when the network began planning for a morning program. NPR's news resources were stretched thin already, and the *ATC* hosts weren't anxious to share reporters and production help with a rival that got on the air before they did. They were relieved when, after a year of planning, the first *Morning Edition* production team came up with a pilot that was laughably bad. It was chatty and sounded, says Edwards, like "a bad talk show in a small market." The network had already promoted the new show heavily to program directors at member stations, so it had to come up with something. Management fired most of the first *ME* team, recruited producers from other NPR shows, and then called on Bob Edwards to fill in as host for 30 days, until someone new could be hired. "Thirty days didn't sound so long," he says, "although it meant getting up each day at 1:00 A.M." By the time the 30 days had passed, it was apparent that Edwards' calm and reassuring presence was a perfect fit to the mood of a morning audience.

Edwards credits producer Jay Kernis with getting him through the early days of the program by crafting a system that could support a single anchor through two hours of news, interviews, and features each day. The system meant scripting as much of the anchor talk as possible and preparing thorough questions for every interview.

Edwards' own interests helped set the show's agenda. He was a sports fan, a relative rarity among NPR staffers, and he warmed to the show's daily sports segments. The need for daily sports features helped bring the show commentators such as Bill Littlefield and Frank DeFord. Most significantly, it brought Red Barber out of retirement to do a four-minute commentary every Friday at 7:35 EDT. As a veteran of more than 50 years in radio, the "Ol' Redhead" had no interest in cranking out a taped segment each week. He wanted a live conversation with the host, a prospect that Edwards found daunting, because the great sportscaster was as apt to talk about philosophy, religion, or raising camellias as about sports. Edwards eventually wrote a book about his on-air friendship with Barber, *Fridays with Red*, saying that the relationship helped him grow in confidence and maturity as a broadcaster.

Edwards' favorite awards show the range of concerns he brings to *Morning Edition*. He won his first Gabriel Award in 1987 from the National Catholic Association of Broadcasters for a story called "Bill of Sale: A Black Heritage," about a Maryland man who assembled a small museum of artifacts from his own family's enslavement, including the record of the transaction in which one of his ancestors was sold as chattel. Edwards won a second Gabriel in 1990 for "Born Drunk," a five-part series about fetal alcohol syndrome. In 1995 he won the Alfred I. duPont–Columbia University Award for "The Changing of the Guard: The Republican Revolution." A highlight of that series was Edwards' interview with Representative Dick Armey of Texas, in which the House majority leader averred that voters would be so happy with Republican leader-

Bob Edwards, host of *Morning Edition*
Courtesy National Public Radio

ship that the GOP's promise to limit lawmakers' terms wouldn't mean much. The comment caused an uproar among Republicans, who insisted on the promises in the GOP's "Contract with America," and Armey was forced to backpedal. In 1999 *Morning Edition* received a George Foster Peabody award lauding Edwards as "a man who embodies the essence of excellence in radio."

Edwards was on the air the morning of 11 September 2001 when terrorists attacked New York and Washington, D.C. NPR's coverage of events that day earned a George Foster Peabody Award and a duPont-Columbia Award.

Edwards still gets up at one o'clock in the morning. He has to. As Susan Stamberg says, "He's hitched to the nation's alarm clock, and he's the first voice you hear. He's reassuring. I honestly believe that people's days are different when Bob's not there."

COREY FLINTOFF

See also All Things Considered; Barber, Red; Mankiewicz, Frank; Morning Edition; National Public Radio; Stamberg, Susan; WTOP

Bob Edwards. Born in Louisville, Kentucky, 16 May 1947. B.S. in commerce, University of Louisville, 1969; M.A. in communication, American University, 1972; served in U.S. Army, television news anchor, American Forces Korea Network, Seoul, Korea; attained rank of specialist fifth class, 1969–71; news and program director, WHEL-AM, New Albany, Indiana, 1968–69; news anchor, WTOP-AM, Washington D.C., 1972; correspondent and night editor, Mutual Broadcasting System, Washington D.C., 1972–73; began career at National Public Radio as associate producer, 1974. Unity Award in Media, 1983; Corporation for Public Broadcasting's Edward R. Murrow Award, 1984; honorary Doctor of Public Service from University of Louisville, 1985; Gabriel award, 1987, 1990; honorary LHD from Grinnell College, 1991; Alfred I. duPont-Columbia University Award, 1995; George Foster Peabody Award, 1999. National vice president of the American Federation of Television and Radio Artists, from 1988.

Radio Series

1974–79 *All Things Considered*
1979– *Morning Edition*

Publications

Fridays with Red, 1993

Further Reading

Collins, Mary, *National Public Radio: The Cast of Characters*, Washington, D.C.: Seven Locks Press, 1993
Looker, Tom, *The Sound and the Story: NPR and the Art of Radio*, Boston: Houghton Mifflin, 1995
Siegel, Robert, editor, *The NPR Interviews, 1994*, Boston: Houghton Mifflin, 1994
Stamberg, Susan, *Every Night at Five: Susan Stamberg's All Things Considered Book*, New York: Pantheon, 1982
Stamberg, Susan, *Talk: NPR's Susan Stamberg Considers All Things*, New York: Turtle Bay Press, 1993
Wertheimer, Linda, editor, *Listening to America: Twenty-five Years in the Life of a Nation, As Heard on National Public Radio*, Boston, Houghton Mifflin, 1995

Edwards, Ralph 1913–

U.S. Radio Host and Producer

While Ralph Edwards is perhaps best known today for his work in television, his involvement with radio spanned several decades dating back to 1929. He is one of a handful of radio stars that caused a town to change its name.

Early Years

Edwards spent his formative years on the family farm in Colorado. The family moved to Oakland, California when he was 16 and he completed high school there in 1931. He began his broadcast career in 1929, while still in high school, at station KROW in Oakland. The station manager had been impressed with a high school play Edwards wrote and hired him to write scripts for radio—at the munificent pay of $1.00 per script. He soon took on general announcing (and some acting) duties as well.

Edwards entered the University of California, Berkeley in 1931, and had earned a BA in English by 1935, intent on becoming an English teacher. While in school, however, he kept a hand in radio, working at KTAB in Oakland. Upon graduation, with few jobs available to him in education at that time, Edwards decided to work in radio.

Radio Years

Not finding much in the way of full-time employment in California, Edwards hitch-hiked to New York to try his luck there. He took on various part-time announcing duties in 1936 while living hand-to-mouth. Slowly things picked up. He was soon performing actor duties in several network plays, and in 1938 won a coveted full-time staff announcer's position at CBS, beating nearly 70 rivals. Within a year he was announcing some 45 network programs, mostly daytime offerings, every week, and was said to be one of the busiest announcers in the business.

Over the next several years Edwards would serve as host or announcer on dozens of programs, among them *Vic and Sade, The Phil Baker Show, Major Bowes' Original Amateur Hour, Life Can Be Beautiful, The Quiz Show,* and *The Horn and Hardart Children's Hour.* Despite his success as a network announcer and emcee, he was not content (the work was monotonous and boring) and sought greater opportunities.

This move involved his recollecting a childhood game, which in turn led to his creation in 1940 of what would become one of radio's pioneer game shows, *Truth or Consequences.* Edwards sold the idea to the Compton Agency and Procter & Gamble and that package was sold to CBS in March, setting up his own production firm. Edwards was then 26. The program was the number one audience participation program for the next three years (though now on NBC). Prizes were minor ($10 or $15); everyone listened for the silly consequences of the (usually) wrong answers to the questions posed. *Truth or Consequences* was broadcast live until 1948, repeated twice on its Saturday schedule, at 8 P.M. for the Eastern and Central Time Zones, and three hours later for broadcast on the West Coast. Edwards later remembered they used the same scripts but new contestants for the rebroadcast. The program continued to enjoy enormous ratings success for years, becoming a network staple. As Dunning relates, many of the stunts (the "consequences") were highly elaborate, some stretching out for weeks at a time, often including audience participation.

In 1950 the New Mexico town of Hot Springs changed its name to Truth or Consequences in honor of the top-rated radio show. To mark the program's approaching tenth anniversary, Edwards had offered to host an annual celebration in and nationwide broadcast from any town willing to change its name to Truth or Consequences. The Hot Springs Chamber of Commerce spread the news of the offer to advertise the city free of charge. No longer would they be confused with other towns named Hot Springs across the country. A special city election voted by a margin of ten to one to change the name to Truth or Consequences. After a protest by nearly 300 area residents another election saw the votes in favor win by an even greater margin. Edwards kept his promise and aired the first live, coast-to-coast broadcast of *Truth or Consequences* from the newly named New Mexico town. Until the late 1990s, Edwards returned annually to the town of 7,000 people for the name-change celebration Fiesta. (In two later referenda, in January 1964, and again in August 1967, townspeople voted to retain the name. There is even a Ralph Edwards Park.)

In 1948, based on a minor "consequence" of his famous program, Edwards would create *This Is Your Life,* debuting first on radio and eventually enjoying a longer run (1952–61) on television, making Edwards a household name. During his long career, he also made a name for himself through his significant efforts for charity. During World War II, Edwards' formidable work for the War Bond drives won him wide praise, and his later involvement with the American Heart Association and the March of Dimes brought him further accolades.

Later Career

Edwards would enjoy considerable success in the television medium as both a producer and host. Ralph Edwards

Productions became part of Aquarius Productions in 1957 after 17 years as a sole proprietorship operation. After their long network runs, with several different hosts, both *This is Your Life* (1971–72, 1983) and *Truth or Consequences* (1966–74,1977–78, 1987) were revived for first-run syndication. In 1978 all the Edwards companies (there were a number of them) and programs were merged into Aquarius and renamed Ralph Edwards Productions. That company began to syndicate *People's Court* in 1981.

CHRISTOPHER H. STERLING

See also Quiz and Audience Participation Programs

Ralph Livingstone Edwards. Born in Merino, Colorado, 13 June 1913. Moved to Oakland, California, 1926. Graduated high school, 1931. University of California, Berkeley, BA in English, 1935. Began broadcast career in 1929 at KROW, Oakland; full-time employment with CBS in New York City, 1938–40. Began *Truth or Consequences* on NBC (and Ralph Edwards Productions), 1940.

Radio Series

1940–56 *Truth or Consequences*
1948–50 *This Is Your Life*

Further Reading

Current Biography (July 1943)
DeLong, Thomas A., "Aren't We Devils?" in *Quiz Craze: America's Infatuation with Game Shows,* by DeLong, New York: Praeger, 1991
Dunning, John, *Tune in Yesterday: The Ultimate Encyclopedia of Old-Time Radio, 1925–1976,* Englewood Cliffs, New Jersey: Prentice-Hall, 1976; revised edition, as *On the Air: The Encyclopedia of Old-Time Radio,* New York: Oxford University Press, 1998
Nachman, Gerald, *Raised on Radio,* New York: Pantheon Books, 1998

Election Coverage

Radio Reports of National Campaigns

Elections provided some of the earliest content for the new medium of radio. In 1920, as Westinghouse prepared to issue KDKA's first radio broadcast, they chose to debut with reports of national election results. With a small audience, to whom "simple receiving sets" had been distributed, and advertising placed in the local papers to spark the interest of amateur radio enthusiasts, on the evening of 2 November 1920, the national election returns were broadcast to an audience of perhaps 500 to 1,000 listeners.

The early attraction of radio to politics was not immediately returned in kind by politics, as was evidenced in a comment regarding the 1924 election: "The effect of the election on radio was more important than the effect of radio on the election result" (Chester). However, a dependence on the new medium was soon to follow. Through a historical review of the election coverage on radio, this essay traces the role that radio played in the coverage of political campaigns and the early public dependence on the radio for election coverage.

Early Election Coverage

The first political campaign in which radio played a major role was the 1924 presidential contest. Strategists for President Calvin Coolidge's reelection effort saw the immediate advantages of the medium for "silent Cal," who found radio's requirement of short and simple language more compatible with his own style than the traditional round of presidential candidate stump speeches. One campaign advisor noted, "Speeches must be short. Ten minutes is a limit and five minutes is better" (Chester).

Campaign speeches and addresses by the candidates and other political advocates made up radio's major election coverage in these early days. The Republican Party even purchased their own radio station, on which they broadcast campaign addresses at all available hours in the day. This saturation sparked the Radio Corporation of America (RCA) to suggest that a limit of one hour per day be reserved for political address broadcasting and that such addresses be limited to 15 minutes each. The 1924 election coverage also included the first coverage on radio of the national party conventions. Both the Republican and Democratic conventions were carried to listeners across the nation. Although some political observers were sure that this marriage of radio and elections was a fad, those who predicted that political campaigns would never be the same were closer to the mark.

By the 1928 presidential campaign, 40 million people were able to follow the election campaigns via the radio, which had become an integral campaign tool, so much so that a candidate's "radio personality" was a part of the election dialogue. Republican Herbert Hoover was not an exciting radio personality, but he was credited for not offending his audiences. Democratic opponent Alfred Smith, however, had an accent that tended to alienate voters from the Southern states that he desperately needed to carry. The 1928 campaign also saw the rise in coverage of radio campaign addresses by people other than the presidential candidates themselves. Republicans continued to build on the perceived advantages of brevity in the new medium by creating 30 short speeches that covered the main points of Hoover's campaign. Lasting five minutes each, these were delivered by well-known local people over 174 local stations.

By 1932, with radios in more than 12 million homes, voters witnessed new radio strategies in the presidential campaigns. Franklin Roosevelt established a new precedent in radio address, flying to Chicago to accept his party's nomination prior to the adjournment of the national convention and thus broadcasting his acceptance address by radio from the convention. Other changes in radio coverage evolved during this campaign cycle, with the Columbia Broadcasting System (CBS) and the National Broadcasting Company (NBC) each allotting three periods a week for political addresses and refusing to sell airtime to either party before the parties had officially selected their presidential candidates. During the 1936 campaign, both candidates, Roosevelt and Kansas Governor Alfred Landon, toured the country with many of their speeches broadcast over radio. A unique variation of campaign speeches in the 1936 campaign focused on foreign language broadcasts. Both parties produced radio messages in a variety of languages, with the Republicans using as many as 29 languages in numerous messages varying in length from 100-word spots to 30-minute talks. Although Landon enjoyed many powerful newspaper endorsements, many believe that Roosevelt's victory can be at least partially credited to his effective use of radio during the campaign.

In Roosevelt's last campaign, in 1944, the press noted that Republican nominee Thomas Dewey, then-governor of New York, might be the first real competition Roosevelt had faced in radio performance. It was an unfulfilled prediction; a poll conducted by the American Institute of Public Opinion found that the attitude Dewey conveyed while speaking on the radio hindered audience approval, further underscoring the need for candidates to understand and develop successful radio communication strategies.

During the 1948 election, Democratic incumbent Truman successfully developed his own approach to radio address. As opposed to Roosevelt's manuscript style of speech preparation, Truman found both success and effectiveness in extemporaneous speaking. Truman embarked on a successful "whistle-stop" tour during the campaign, and by simply and consistently repeating his message throughout the tour, he inadvertently capitalized on the fundamental nature of radio by reaching masses of voters with the same message.

Throughout the 1948 campaign, a larger percentage of President Truman's speeches were covered by local radio stations than were those of Dewey, again the Republican nominee. Whereas about 50 of the incumbent president's addresses were covered by radio, only about 34 of Dewey's were broadcast to voters. Also significant was the fact that most of Truman's speeches carried on radio were broadcast from his whistle-stop tour, giving them a background of excitement and timeliness. The coverage of Truman's speeches was particularly important for him, because the Democratic Party had limited funds to pay for broadcasts. In addition, many of the "whistle stops" where the actual speeches were given did not have local radio stations, and national coverage was essential for his message to reach large numbers of voters. Truman's upset victory was due in part to radio.

In the 1952 election, radio found a mass media partner as television made its first significant appearance in a political election. Although Eisenhower is credited for the first use of presidential televised spot advertising, surveys conducted after the election found that he made his greatest gains on radio (Chester). Nonetheless, the years when radio coverage would dominate elections were over.

Election Advertising

Early in radio coverage of elections, the blending of straightforward coverage and party and candidate advertising in election campaigns was established. By the time of the 1928 election, candidates and parties were routinely purchasing time for the airing of their speeches and messages. Advertising expenditures in 1928 reached $650,000 for the Democrats and $435,000 for the Republicans, as radio stations began charging presidential candidates for airtime after they had given their nomination acceptance speeches. However, during the 1936 campaign, Republicans began to run shorter spot advertisements on behalf of Landon, including language-specific spots directed to the minority and immigrant votes.

In the 1944 presidential election, both parties employed short, one-minute radio spots. The Democrats sought to remind voters that they should blame the Republicans for the Depression with quips such as "Hoover depression," and the Republicans sought to tie the Democrats to the war with slogans such as "End the war quicker with Dewey and Bricker." Testimonial five-minute spots were also employed by the Democrats, with speakers such as Vice President Henry A. Wallace and vice presidential candidate Truman.

During the 1948 presidential election, Dewey declined the use of five-minute spot announcements, feeling confident that he was ahead in the polls. The Republican national committee instead made use of 30- and 60-second spot ads to urge the people to vote Republican; the Democrats made very little use of spot announcements. Instead, Truman concentrated on reaching the voters through political speeches broadcast from his whistle-stop tour.

With Eisenhower's introduction of televised political spot ads in 1952, radio advertising began to take a back seat to the advertising dollars spent on television. However, as radio stations grew in popularity and number, candidates began to use the target marketing of certain radio stations to reach particular types of voters. The increasing numbers of radio advertising dollars in presidential candidate budgets in the seven decades from 1928 to 2000 is noteworthy. Radio advertising expenditures topped $650,000 in 1928, and by 2000 both parties were spending several million dollars each on radio advertising. A major reason for the resurgence of interest in radio advertising in the 1980s and 1990s has been the ability of radio to target increasingly diverse and segmented audiences.

Election Debates

An idea that was ahead of its time, a proposal for the first presidential debate to be broadcast on radio was actually put forth as early as the 1924 election campaign. Two decades later, radio hosted a 1948 Republican primary debate held in Portland, Oregon; the debate was a forerunner to the famous Nixon-Kennedy debates of 1960. The choice of the issue about which both the leading Republican candidates, Thomas Dewey, governor of New York, and Harold Stassen, former governor of Minnesota, were in complete disagreement set the stage for a traditional, non-moderated debate. The debate topic focused on whether the Communist Party should be outlawed in the United States. Stassen took the affirmative, and Dewey supported the negative. Although Stassen entered the debate with a slim lead in the state of Oregon, shortly thereafter (and strongly linked to his performance during the radio broadcast debate), he ultimately lost both his lead in Oregon and the Republican nomination.

The first Nixon-Kennedy debate in 1960 has been considered an important turning point in the presidential election. The debate aired on both television and radio. Interestingly, although the television appearance both hindered Nixon and fundamentally affected his future campaign strategies, his radio ratings from the debate were notably positive. Not only did Nixon seem to come across far better than Kennedy, but those who *heard*, rather than viewed, the debate chose him as the winner by a substantial margin. Debates before and for many years after 1960 were scarce because of the FCC's regulatory view that required *all* candidates to be included—or given equal time. This gradually changed in the 1970s and 1980s, leading to almost regular presidential candidate debates.

To this day debates make up a significant amount of the election coverage available to voters, and many debates are covered simultaneously by radio and television stations. However, so much more attention goes to the television medium that radio coverage of debates does not attract much attention in the modern political system.

Assessment of Modern Election Coverage on Radio and Its Impact

From its beginnings as the medium that brought election returns into the homes of Americans in 1920, radio coverage has evolved into a medium whose significance in American political coverage is quite limited. As television supplanted radio in the second half of the 20th century, radio's role in election coverage maintained some significance only in lower-level elections, in state and local contests where candidates were unable to expend major resources on television coverage.

Exceptions to the general trend of reduced radio coverage of elections can be found at the national level in the aggressive coverage of the National Public Radio (NPR) system and state public radio outlets. The growth in the 1970s of all-news radio stations in some communities also brought a renewed interest in covering election matters in state and local communities. The all-talk radio formats of the 1980s and 1990s also heightened radio coverage of and concentration on political and election matters.

The impact of radio election coverage is not easy to measure. However, by 1944 a Roper survey reported that 56 percent of those interviewed felt they received the "most accurate news" about the presidential campaign from the radio. Although Roper surveys from 1959 to 1980 indicate that the American people's reliance on radio as a news source decreased from 34 percent in 1959 to 18 percent in 1980, American radio listeners' use of this medium to acquaint themselves with local candidates ranked even lower from 1971 to 1984. In 1971 6 percent of the respondents to a Roper survey indicated that they became acquainted with local candidates through the radio. Although the percentage rose to 10 percent in 1976, from 1980 to 1984 the percentage remained stable at

6 percent. During the presidential election years of 1973, 1976, and 1984, 6 percent, 4 percent, and 4 percent, respectively, indicated radio as a source through which listeners became acquainted with candidates for public office. However, in 1987 46 percent of the respondents to a Roper survey indicated that the radio was a "somewhat important" source of information for learning about presidential candidates, and 19 percent indicated radio as a "very important" source. This increase probably stemmed in part from the rise in popularity of political talk radio.

More recent polls from the 1990s have reflected the increased attention radio election coverage has received from voters. A Roper survey conducted from May to June 1992 suggested that 18 percent of respondents gathered information about the 1992 presidential election campaign from the radio, with a decrease to 12 percent by November 1992. However, a November 1996 Roper survey found that 19 percent of the respondents indicated that they gathered information about the 1996 presidential election campaign from the radio. Most recently, a November 1998 Roper survey reported that 23 percent of the respondents indicated that they gathered information about the 1998 off-year election campaigns in their state and district from the radio.

Although radio began its political coverage by broadcasting presidential election results, recognition of the potential impact of radio on the election process came quickly. Whether through radio coverage of candidate addresses, the development of political advertising, or the coverage of major campaign events such as conventions and debates, strategists and ultimately candidates recognized that radio could reach millions of voters. That accessibility thus influenced the need to effectively present the candidate and the candidate's message in this medium.

Undoubtedly, radio's election coverage brought the political process much closer to the American people. Not only could those with a radio in their living room suddenly sit in on their party's national convention, but they could also hear campaign addresses delivered by presidential candidates in distant states. Even more important, the American people could listen to a candidate deliver an address on specific issues in real time, and they could judge for themselves the information as well as the candidate's delivery style and personality. In essence, radio, through its election coverage, handed the American people the opportunity to participate in and make decisions about candidates firsthand without relying solely on press interpretations.

LYNDA LEE KAID AND MARY CHRISTINE BANWART

Further Reading

"... And As for Radio," *Broadcasting* (5 September 1988)
Archer, Gleason Leonard, *History of Radio to 1926,* New York: The American Historical Society, 1938
Berkman, D., "Politics and Radio in the 1924 Campaign," *Journalism Quarterly* 64 (Summer-Autumn 1987)
Carcasson, Martin, "Herbert Hoover and the Presidential Campaign of 1932: The Failure of Apologia," *Presidential Studies Quarterly* 28 (Spring 1998)
Carroll, Raymond L., "Harry S. Truman's 1948 Election: The Inadvertent Broadcast Campaign," *Journal of Broadcasting and Electronic Media* 31 (Spring 1987)
Chester, Edward W., *Radio, Television, and American Politics,* New York: Sheed and Ward, 1969
Cornwell, Elmer E., *Presidential Leadership of Public Opinion,* Bloomington: Indiana University Press, 1965
Dobrez, Tom, "Radio: The Secret Weapon," *Campaigns and Elections* 17 (August 1996)
Dreyer, Edward C., "Political Party Use of Radio and Television in the 1960 Campaign," *Journal of Broadcasting* 8 (Summer 1964)
Gest, Ted, "Clinton: Tune-in, It's Party Time," *U.S. News and World Report* (16 September 1996)
Roper, Burns W., *Public Attitudes toward Television and Other Media in a Time of Change,* New York: Television Information Office, 1985
Summers, Harrison B., "Radio in the 1948 Campaign," *The Quarterly Journal of Speech* 34 (1948)
Swafford, Tom, "The Last Real Presidential Debate," *American Heritage* 37 (February–March 1986)
Willis, Edgar E., "Radio and Presidential Campaigning," *Central States Speech Journal* 20 (Fall 1969)
Wolfe, G. Joseph, "Some Reactions to the Advent of Campaigning by Radio," *Journal of Broadcasting* 13 (Summer 1969)

Electronic Media Rating Council. *See* Media Rating Council

Elliott, Bob. *See* Bob and Ray

Ellis, Elmo 1918–

U.S. Radio Executive

For more than 40 years, Elmo Ellis was a fixture in Atlanta radio. Dubbed "The Dean of WSB Radio," Ellis served as a national radio guru in the 1950s at the time the medium was falling behind television. A prolific writer and producer, Ellis brought innovations to Atlanta radio that are still being applied by stations across the country.

Origins

Ellis was born in 1918, in Alabama. He attended West Blocton High School (where he played quarterback on his high school football team) and was a Phi Beta Kappa graduate of The University of Alabama in 1940. A cousin of radio great Mel Allen, Ellis studied journalism at Alabama, but he did not take radio courses until his senior year. He applied for a job at WSB in Atlanta after his graduation and was hired as the station's promotions and publicity director. His first award came in 1940, from *Variety,* for best local station programming.

During World War II, Ellis joined the Army Air Corps and worked as a military broadcaster, writing and producing programs for every network. He also worked with Glenn Miller. Ellis told *The Atlanta Constitution* he was in Fort Worth, Texas, and had collaborated with Miller on a 1942 Army Air Corps radio show. Ellis said the bandleader had liked his work and wanted Ellis to join his radio production unit in New York, but the commanding officer rejected the transfer request. Ellis was still stateside when Major Miller, his band, and staff crashed into a foggy English Channel in December 1944.

In 1944, Ellis married his wife, Ruth, whom he had met in Texas. When he left the Army, Ellis worked in New York before returning to WSB and Atlanta in 1947. The station was about to launch its television efforts, and Ellis' former bosses wanted him back for the transition.

Postwar Activities

WSB-TV went on the air 29 September 1948, with Ellis producing previews in Atlanta. Planning had been underway for more than one year, and Ellis was at the heart of the operation. He introduced the station's first game shows, do-it-yourself shows, and talk shows. Elllis and WSB-TV were a huge success in those early years.

In late 1951, WSB Radio and TV officials asked Ellis to return to the radio side as program director. Radio as an industry was dying, since most of the night-time stars and revenue had migrated to the younger medium. "I hadn't thought much about radio in four years," Ellis told *The Atlanta Constitution.* "But I have always been a loyal type person, and I felt my bosses really needed me over there." He said in a 1994 interview that he did not want to return to radio, but as the radio side of WSB was languishing, Ellis obliged. WSB AM became known as "Atlanta's Radio-Active Station," and Ellis became its program director. He went on to write a list of 100 pointers that were published by Broadcast Music Incorporated, then he penned a groundbreaking article for *Broadcasting* magazine called "Removing the Rust from Radio." In the article—and subsequent speaking engagements across the country—Ellis simply told station managers what had worked for him at WSB radio. According to *The Atlanta Constitution,* Ellis suggested radio "get off its podium, drop its pomposity and put on its roller skates. Mingle with the citizens, collecting and reflecting actions and opinions." He suggested theme days (Old Timers Day, Sweet Music Day, etc.). Congratulate newly elected civic club leaders, he said. Promote radio's flexibility with station breaks such as, "You can take a bath and still listen to WSB Radio." Ellis wrote, "We must change from an entertainment-dominated medium to a locally oriented service. Remember, radio brings the event to the audience. This is different from television, which takes the audience to the event."

During his tenure, Ellis contributed a number of well-respected programs to the station. The musical format was middle of the road, and the programming covered everything from news to sports to entertainment. One innovation, "sky-copter traffic," is still being used today. The Shining Light award, which he introduced, is still being awarded on a regular basis after a hiatus.

Ellis also contributed to the scholarly world of radio, collaborating with Cox Broadcasting Corporation president J. Leonard Reinsch on the textbook *Radio Station Management* and writing *Opportunities in Broadcast Careers*. The latter is still in print, most recently updated in 1999.

In late 1981, after close to 30 years on the radio side of WSB, Ellis retired at age 63 as vice president and general manager of WSB AM-FM. He had held the general manager position for 17 years and had been a vice president of Cox Broadcasting for 12 years.

But Ellis did not leave the field of journalism. He began writing weekly opinion columns for the *Marietta Daily Journal* and *Neighbor* newspapers, and he continued to be on the air with "Life Management" spots aired weekly on smaller radio stations in Georgia. He is still an on-demand speaker and volunteer in his Sandy Springs community.

Ellis has won almost every award a Georgia broadcast journalist can win. In addition to the Peabody he won in 1966 for his "Viewpoint" editorials and "Pro and Con," Ellis is a Hall of Fame member of the University of Georgia's Di Gamma Kappa Broadcasting Honor society. In 1985, Ellis was the third person to be elected to the Georgia Association of Broadcaster's Hall of Fame. In September 1995, he won the Georgia Music Hall of Fame's Mary Tallent Pioneer Award. When he won the Ralph McGill Award from the Society of Professional Journalists at its annual Green Eyeshade banquet, he told *The Atlanta Constitution,* "Having been a member of the organization for more than 50 years, I am wondering if this award isn't just for longevity. I accept it as a tribute to the memory of Ralph McGill."

In 1995, Ellis was one of three Atlanta broadcasters named to *Radio Ink* magazine's list of 75 people who "made a distinctive and major impact on the radio industry." He joined, among others, Ronald Reagan and Rush Limbaugh on that list.

GINGER RUDESEAL CARTER

Elmo Ellis. Born in Alabama, 11 November 1918. Graduated Phi Betta Kappa, University of Alabama, 1940. Joined the Army Air Corp and worked as a military broadcaster during World War II. Became promotions and publicity director, WSB, Atlanta, Georgia, 1940; wrote and produced network radio programs in New York; returned to WSB, Atlanta, Georgia, 1947; helped to develop WSB-TV, the first television station in the South, 1948–51; WBS officials requested his return to radio as program director, 1951; developed live coverage radio reporting, 24-hour news and weather format, and other innovative radio programming techniques; general manager, WSB radio, 1964–81; vice president, WSB radio, 1969–81. Recipient: George Foster Peabody Award, 1966; elected to Georgia Association of Broadcasters' Hall of Fame, 1985; Ralph McGill Award, Society of Professional Journalists, 1993; Outstanding Alumnus, University of Alabama, 1993; member, Hall of Fame, University of Georgia's Di Gamma Kappa Broadcasting Honor Society; Mary Tallent Pioneer Award, Georgia Music Hall of Fame, 1995; inducted into University of Alabama's Communication Hall of Fame, 1999; Hugo Black Award, University of Alabama, 2000.

Selected Publications

Radio Station Management (with J. Leonard Reinsch), 2nd edition, 1960

Opportunities in Broadcasting Careers, 1981; revised edition, 1999

Emergencies, Radio's Role in

Radio's news and entertainment roles have long been taken for granted. In times of emergency, however, the medium often rises to the occasion to play a unique public service role. In times of natural or man-made disasters, radio becomes a prime means of social surveillance, a link with the outside world, and a source of information. Radio can mitigate problems by promoting disaster preparedness, keeping people out of harm's way, assisting in rescue coordination and relief efforts, and facilitating rehabilitation and reconstruction efforts. Radio often has the first reports of impending natural disasters, be they tornadoes, hurricanes, or volcanic eruptions.

Radio's potential was demonstrated even before the inception of broadcasting. While wireless aided in rescuing people from several maritime disasters early in the 20th century, notably during the loss of the liner *Republic* in 1909 when more than 1500 people were saved thanks to a distress call, the 1912 *Titanic* disaster focused public attention on what the medium could do. On her maiden voyage in mid-April of 1912, the huge passenger liner struck an iceberg and began to sink. Her two wireless operators stayed at their posts almost to the end, sending both "CDQ" and the newer "SOS" emergency signals to both nearby ships and the distant shore. The sole operator on the Cunard liner *Carpathia* heard the signals and the ship steamed 55 miles to rescue the 700 survivors in boats several hours after *Titanic* went down. For weeks thereafter, the wireless operators (one of whom perished) were newspaper heroes.

Early Emergency Broadcasting

The new role of radio became clear with two events in the spring of 1937. Massive snow melt flooding of the Ohio and Mississippi Rivers inundated towns and countryside alike, and local stations often were the only link with the outside world for days at a time. Stations that were themselves flooded out loaned their personnel to those still on the air. Regular program schedules were replaced with day and night reporting (sometimes around the clock) and radio broadcasters directed rescue teams where they were most needed. Some stations became arms of official state or federal agencies and provided a personal message service that might normally have been an illegal point-to-point use of radio stations. Radio's immediacy and portability were well demonstrated.

The most spectacular disaster covered on radio was the burning of the German passenger airship *Hindenburg* as it attempted to land in Lakehurst, New Jersey, near New York City on 6 May 1937. Although 36 people were killed in the fiery crash, 62 survived the disaster, which was hard to believe as the nation watched the subsequent newsreel coverage. But people first heard about the crash as Chicago station WLS reporter Herb Morrison reported what was presumed to be a routine landing. Morrison's gripping eye-witness account, in which he cried with a broken voice, "This is one of the worst catastrophes in the world," was aired on the networks the next day, all of them suspending their usual rule against use of recorded programs.

In all too many later natural disasters—floods, tornadoes, and hurricanes—radio stations provided the crucial warnings of impending trouble and then the critical links with rescue help and the outside world. People soon learned to turn to radio if concerned about the weather or some unusual event. Radio's growing transistor-driven portability in the 1950s continued the medium's unique role even in an age of television. Most radio stations have emergency generators and portable studio-transmitter links that allow reporters to get close to the scene of disasters or emergencies and provide on-the-spot reporting.

One sure indicator of radio's central role in emergencies was its use in government emergency communications schemes. Beginning with the CONELRAD system (1952–63), which can still be seen in old radios with tuners marked for the two frequencies (640 kilohertz and 1240 kilohertz) to be used in national or regional emergencies, and progressing to the Emergency Broadcast System (1963–97), radio was to play a key role in Cold War civil defense planning and emergency warning schemes.

Radio's Role in More Recent Disasters

In the 1960s, radio's emergency role showed in two man-made emergencies. When President John F. Kennedy was shot in Dallas, Texas, on 22 November 1963, radio was often the first medium most people tuned to; television was far less common in schools and in the workplace then. As that Friday afternoon wore on, one could see people clustered around portable or car radios trying to learn the latest from Dallas, including the swearing in of a new president. During the massive overnight electric power failure in most of northeastern U.S. in late 1965, WBZ's Bob Kennedy became famous for his reassuring radio coverage during the many dark hours in the Boston area. New York DJs and news reporters filled much the same role there in the almost total absence of television reporting. Those who owned battery-powered transistor radios could tune in local radio personalities who did their best to communicate what was going on, how widespread it was, and when the lights began to come back on (the next morning). Many argued later that radio's collective voice helped to avert a widespread panic in the darkness.

In early 1989 stations in the San Francisco Bay area were on top of the Loma Prieta earthquake and provided the first reports of downed bridges and collapsed and burning buildings, and thus assisted in crowd control and channeled rescue workers to where they were most needed. Later that same year, stations in the Caribbean and along the U.S. coast warned of the looming hurricane Hugo, one of the most powerful storms in years. Station WSTA in the Virgin Islands assisted the Federal Emergency Management Agency (FEMA) and for several days was the only source of news and information for and about the people on the devastated island. The station served as a government communications center, emergency police dispatcher, and chief contact for emergency medical personnel. Stations in southern Florida played similar roles when hurricane Andrew struck in 1992 and wiped out many communities south of Miami.

In January 1998 a huge ice storm struck Maine and Eastern Canada, cutting electricity for at least hours, often days, and up to two weeks in some isolated communities. Thousands also lost their telephone connections, and the state's emergency broadcast system was knocked out as well. Hundreds had to move to central shelters. Throughout the storm, two radio stations managed to stay on the air and launched call-in shows. WWBX-FM in Bangor began "Storm Watch" around the clock, combining an aural bulletin board with message relay, town meetings, and a sharing and coordination center. The station linked people with other people, announced the location of shelters, and passed on cold-weather survival tips. WVOM-FM, also in Bangor, stayed on the air thanks to propane gas carried up to its transmitter. It, too, provided a makeshift command center, often a voice in the dark for thousands with portable radios.

A year later, flooding from heavy rains ravaged eastern North Carolina. Stations again scrapped entertainment programming and went into a 24-hour emergency mode, reporting what was happening, linking people with safety spots on high ground, and helping to funnel rescue workers and food supplies where they were most needed.

When terrorists struck New York's World Trade Center (WTC) and Washington's Pentagon on 11 September 2001, radio again came to the fore as a primary means of media communication, especially to those in or near the attacked areas. With the loss of the multi-station antenna atop the WTC's North Tower, television reception was lost for much of the metropolitan New York region, save for cable subscribers. While across the nation many tuned to cable news services or the internet, thanks to battery-powered portables, radio was again the prime means of initial news reports and guidance for many listeners.

While television stations play similar roles, not everyone can receive signals if power is lost. Radio's portability and pervasiveness in cars, offices, schools, and homes makes it the medium of first resort in disasters and emergency conditions.

LYOMBE EKO AND JOANNE GULA

See also CONELRAD; Emergency Broadcast System; Hindenburg Disaster; News; World War II and U.S. Radio

Further Reading

Benthall, Jonathan, *Disasters, Relief, and the Media,* London and New York: I.B. Tauris, 1993

Deppa, Joan, *The Media and Disasters: Pan Am 103,* London: Fulton, 1993; New York: New York University Press, 1994

Disasters and the Mass Media: Proceedings of the Committee on Disasters and the Mass Media Workshop, February 1979, Washington, D.C.: National Academy of Sciences, 1980

Flint, J., "FCC Wants to Overhall EBS," *Broadcasting* 122 (1992)

Fortner, Robert S., *International Communication: History, Conflict, and Control of the Global Metropolis,* Belmont, California: Wadsworth, 1993

Garner, Joe, *We Interrupt This Broadcast: Relive the Events That Stopped Our Lives—from the Hindenburg to the Death of Diana,* Naperville, Illinois: Sourcebooks, 1998

Harrison, Shirley, editor, *Disasters and the Media: Managing Crisis Communications,* New York: Palgrave, 1999

Heyer, Paul, *Titanic Legacy: Disaster As Media Event and Myth,* Westport, Connecticut: Praeger, 1995

Hogan, Warren L., editor, *Hurricane Carla: A Tribute to the News Media, Newspaper, Radio, Television,* Houston, Texas: Leaman-Hogan Company, 1961

Mooney, Michael MacDonald, *The Hindenburg,* New York: Dodd Mead, and London: Hart-Davis, MacGibbon, 1972

Straubhaar, Joseph D., and Robert LaRose, *Communications Media in the Information Society,* Belmont, California: Wadsworth, 1996; 2nd edition, as *Media Now: Communications Media in the Information Age,* 2000

"Virgin Islands AM Beats Hugo," *Broadcasting* (30 October 1989)

Emergency Broadcast System

Warning Listeners of Disaster

The Federal Communications Commission (FCC) established the Emergency Broadcast System (EBS) in 1963 to provide the government with a means of quickly contacting U.S. citizens in the event of an emergency. Originally conceived for national defense purposes, its mission was expanded to include natural as well as man-made disasters. The Emergency Alert System (EAS) replaced the EBS system in 1997. Although the technology and specific rules differ, the intent of the EAS remains the same as EBS.

Origins

In the 1950s the United States was increasingly concerned about the possibility of Soviet aggression. During the Cold War, the United States instituted a number of actions to protect itself from a Soviet attack. One fear expressed by the military was that the many American radio transmissions could serve as navigational aids for enemy aircraft (as they had for Japan's 1941 attack on Pearl Harbor). In 1951 the United States instituted a program called Control of Electromagnetic Radiation (CONELRAD). Under this plan, all nonmilitary radio transmissions except for those on two frequencies—640 kilohertz and 1240 kilohertz—would cease. The military decided that limiting all transmissions to only those two frequencies would provide the best system for informing the public while limiting the usefulness of U.S. transmitters for locating military targets. Radio receivers sold after 1953 were required to mark the two designated frequencies with triangles to indicate their civil defense role. This requirement remained in effect until the EBS replaced CONELRAD in 1963.

The fear of Soviet aggression continued in the 1960s, but technology had advanced to the point that radio navigation for foreign bombers was no longer the anticipated threat. Instead, the United States feared a long-distance missile attack, which would not rely on radio navigation, so CONELRAD was replaced with the EBS in 1963. Although still serving the purpose of alerting the nation in the event of foreign aggression, there was no longer a need to order all transmitters off the air. Instead, only those lower-powered stations, usually without the staff or financing to keep a 24-hour emergency operation running, would be ordered off the air during an emergency.

EBS Operations

Unfortunately, the original system was susceptible to frequent false alarms. The early EBS warnings occurred when an EBS station stopped transmitting (known as dropping carrier). Other radio stations, required by law to monitor the EBS station, would then be alerted that the EBS had been activated. The problem was that EBS stations might stop transmitting as the result of a loss of power or because an electrical storm had caused a power surge. In 1973 the dual tone alert system was devised, which dramatically reduced the number of false alarms. EBS activation then required a purposeful triggering of the alert signal by a station employee rather than a station's passive dropping carrier. In the 1970s the EBS was seen as a valuable resource that could be used for more than just defense purposes. Added to the system's mission was notification of such natural disasters as earthquakes and tornadoes. Americans quickly became familiar with the EBS from the weekly tests that were required of all broadcast stations.

EBS was always taken seriously by the FCC, which has disciplined stations that failed to comply with the rules. Some stations ignored weekly tests. Others did not maintain the necessary equipment for monitoring and transmitting a signal. Although those transgressions occurred relatively frequently, it was the occasional willful violation of EBS rules that received the most attention and the most severe penalties. In 1991, during the Persian Gulf War, St. Louis disk jockey John Ulett announced that the United States was under nuclear attack and used the EBS tone to validate the hoax. Two hours later the station apologized for the prank, which management claimed it was unaware of until it actually aired. The FCC was not persuaded by either the apology or the station's lack of complicity and fined KSHE-FM $25,000 for the infraction.

Creating EAS

The need to revise the EBS became increasingly apparent. The two-toned signal was considered to be an annoyance by both broadcasters and listeners. The EBS's success depended on a "daisy chain" of alerts, requiring emergency personnel to contact a station, that station to broadcast a tone, and other stations to receive the tone and retransmit the information. Any breaks in the chain resulted in emergency information not being relayed. In 1989 one station did not rebroadcast an alert about the San Francisco earthquake because the operator on duty did not know how to use the equipment and all the station's engineering staff was attending the World Series baseball game. There have been numerous reports of stations that did not retransmit important weather information because of equipment or operator failure. The overall reliability of the system was called into question.

In 1994 the FCC acted to phase in a replacement of the EBS with the newer EAS by 1997. One change particularly appreciated by broadcasters was the shortening of the length of the alert signal. Instead of the two-tone transmission of nearly 30 seconds, the tone was reduced to only eight seconds. Although weekly testing continues, the alert tone is only required as part of monthly tests, thereby making the system much less intrusive. Weekly tests can actually be conducted that are nearly imperceptible to the audience.

For emergency personnel, EAS is an improvement over EBS because the digital signal can be triggered remotely without the involvement of station employees, thus decreasing the likelihood that messages will go unannounced. What's more, because EAS uses a digital signal, it is compatible with cellular phones, pagers, and other devices. The new law required cable systems to participate and made alerts available in Spanish and in visual forms for the deaf. If used at the national level, only the president or his representative can activate the EAS. Local activation can come from several sources, including National Weather Service and Federal Emergency Management Agency offices. In all the years of CONELRAD, EBS, and EAS, there has never been a national activation of an emergency alert.

DOM CARISTI

See also CONELRAD; Emergencies, Radio's Role in

Further Reading

Chartrand, Sabra, "Warning of Disasters, Digitally," *New York Times* (7 November 1993)

Federal Communications Commission, *Emergency Broadcast System: Rules and Regulations,* Washington, D.C.: Government Printing Office, 1977

Lambrecht, Bill, "KSHE Is Fined $25,000 for Fake Warning of Nuclear Attack," *St. Louis Post-Dispatch* (25 April 1991)

"Shrill Emergency Broadcast Test Soon to Be a Cold War Relic," *New York Times* (17 November 1996)

Emerson Radio

Pioneer of Small Radio Receivers

From the early 1930s into the 1950s, Emerson Radio (not to be confused with the older and larger Emerson Electric Co.) was one of the larger manufacturers of radio receivers in the United States. The company's story parallels the decline of American manufacturing in the late 20th century.

Origins

Victor Hugo Emerson, a former Columbia Phonograph company manager, created the Emerson Phonograph Company in 1915–16. He was already well known in audio circles for his 14 patents in sound recording and reproduction granted from 1893 to 1905, and he continued to receive patents until 1922. His firm manufactured both phonographs and records (including a talking books line for children), riding the early mechanical era recording boom of the World War I years. When business fell off sharply after the war, the company found itself overextended and went into receivership in December 1920.

Benjamin Abrams (1893–1967), his younger brothers Max and Lewis, and Rudolph Kamarak purchased the remaining assets of Emerson Phonograph in 1922 and formed the Emerson Radio and Phonograph Corporation. The senior Abrams would run the operation for the next four decades. After selling off the record business, the new firm entered the radio receiver manufacturing business in 1924 as a small player among several giants and created and marketed some of the first radio-phonograph combination devices. With the advent of the Great Depression and the failure of many other radio manufacturing businesses, Emerson took a new direction.

Emerson and Small Radios

The Depression forced radio manufacturers to provide smaller and cheaper sets, a trend first pursued by several smaller California–based companies. But many of the initial "midget" sets were poorly made and unsuccessful. The image of small radios began to change with a 1932 Emerson product.

The "Model 25" was ten inches wide, six and a half inches high and four inches deep. It was a four-tube receiver with a six-inch speaker, weighing about six pounds. At a then low cost of $25.00, the Model 25 sold about a quarter of a million units from late 1932 into the first half of 1933 and helped place Emerson on the map. Demand was so strong that for several months the manufacturing line had trouble keeping up.

Building on this breakthrough success, Emerson quickly focused its attention on other small radios. Some were novelty items built around popular movies (*Snow White and the Seven Dwarfs*), movie or radio stars *(Mickey Mouse)*, and even people in the news (the Dionne quintuplets). All of them (as well as similar products from other firms) sold well, as did a $9.95 compact radio of 1937. The "Little Miracle" set of 1938 was a five-tube superheterodyne receiver offered in a variety of styles and colors as combination and plastic cabinetry began to take over more expensive all-wood cabinets. The 1939 two-tube "Emersonette" took things further—a tiny six-by-five-inch receiver selling for $6.95.

Emerson was now making more than 1 million sets a year. Perhaps the peak of Emerson's prewar success came with the 1940 "Patriot" model, with a design by Norman Bel Geddes based on the American flag (it came in red, white, or blue cabinets with the other two colors as trim).

By 1941, about 80 percent of all radio sales in the U.S. were of compact models, though Emerson was among the first to offer FM receivers when that service began commercial operation that year. But it was on the strength of its small radios that Emerson's portion of the total American radio market rose from a mere 1.5 percent to 17.5 percent between 1932 and 1942. Riding this success, in 1943 the company went public, selling 40 percent of its stock.

Expansion

Emerson produced its first television sets in 1947, although it continued to offer many radio models. In 1951–52 Emerson first offered a new "pocket" portable radio with subminiature tubes designed by Raytheon. This led to the Model 747 in 1953, a tiny and eminently portable table radio that weighed only 22 ounces. One of the last miniature pretransistor radios, the Model 747 sold for $40 even though it received poor review notices in *Consumer's Research Bulletin.*

But change was in the air. By 1954 radio made up only 15 percent of company revenues. Emerson offered its first audio tape recorders in 1955 and by the late 1950s was producing combination tiny tube and transistor radios and a growing variety of television models. In 1958, Emerson bought the DuMont consumer electronics manufacturing operation and began to use that brand name along with its own.

Later Developments

Emerson's last full year of independent operation was 1964. The company was sold to National Union Electric in 1965 and absorbed Pilot radio the same year. By the end of the year, the combined firm owned 20 subsidiaries that continued the sale of consumer electronics under both the Emerson and DuMont names. After several years in the red, owing in part to rising consumer electronic imports, the company began to shift away from this focus in 1972.

Major Electronics Corporation of Brooklyn bought the Emerson name in 1973 and four years later renamed itself after the earlier company. It dropped the last U.S. manufactured product (fittingly, a phonograph) in 1980. By 1983–84, imported televisions and videocassette recorders made up two thirds of company sales. Most were manufactured in Korea but sold under the Emerson name. The once-revered H.H. Scott brand was taken over in 1985, and the company moved to New Jersey. The Scott line was discontinued in 1991.

In October the ever-smaller firm declared bankruptcy, and 60 percent of its stock was taken over by Fidenas Investments, a Swiss firm. Emerson began to retail car audio systems in 1995, licensing its name to several important Korean and Chinese consumer electronic products. But the turn of the 21st century, the company was down to about 100 employees dealing with product import and distribution. Emerson no longer manufactured anything. About half of its output was sold each year to Wal-Mart and a quarter to the Target discount chain.

CHRISTOPHER H. STERLING

See also Receivers; Transistor Radios

Further Reading

"Emerson Phonograph Company," *Victor and 78 Journal* (Winter 1997–98)

Emerson Radio and Phonograph Corporation, *Small Radio: Yesterday and in the World of Tomorrow,* New York: Emerson Radio and Phonograph Corporation, 1943

Halasz, Robert, "Emerson," in volume 30 of *International Directory of Company Histories,* edited by Jay P. Pederson, Detroit, Michigan: St. James Press, 2000

Schiffer, Michael Brian, *The Portable Radio in American Life,* Tucson: University of Arizona Press, 1991

"Equal Time" Rule

Political Broadcasting Regulations

Sections of the Communications Act of 1934 and related rules and regulations of the Federal Communications Commission (FCC) require that candidates for political office in the United States be treated equitably in their purchase or other use of broadcast time. These so-called equal time provisions are controversial and have been considerably modified over the years.

The Law

At first—in American elections between 1920 and 1926—there was no regulation of political broadcasting, and some stations did not allow candidates on the air with political appeals. As Congress moved to pass the Radio Act of 1927, the initial bill contained no provisions concerning political candidates' use of radio. Only a Senate amendment that grew out of concern that radio might exert too much control over the political process led to the inclusion of Section 18, which required that

> If any licensee shall permit any person who is a legally qualified candidate for any public office to use a broadcasting station, he shall afford equal opportunities to all other such candidates for that office in the use of such broadcasting station, and the [Federal Radio Commis-

sion] shall make rules and regulations to carry this provision into effect.

The provision did not *require* that political candidates be granted access to airtime, for the section continued, "No obligation is imposed upon any licensee to allow the use of its station by any such candidate." Section 18, without change, became Section 315 of the Communications Act of 1934.

A remarkably small number of amendments have only slightly altered the meaning of these words in the intervening years. The first, in 1952, added a provision prohibiting broadcasters from charging political candidates higher rates than those charged other advertisers for "comparable use." A more fundamental amendment added in 1959 exempted from Section 315 requirements any appearance by a candidate in a "bona fide" (meaning controlled by the broadcaster, not the candidate) newscast, news interview, news documentary, or on-the-spot news coverage if the appearance of the candidate was incidental to the program. Finally, in 1971 Congress narrowed the ability of broadcasters to avoid political advertisements or broadcasts when it modified another section of the act (Section 312[a][7]) to state that not allowing candidates for federal office (candidates for president, vice president, or for seats in the House and Senate) access to the air might be grounds for revocation of a station's license. Stations could still avoid dealing with candidates for state and local offices.

For decades, broadcasters were caught in a legal bind—Section 315 specifically enjoined them from censoring any remarks made by political candidates, yet stations could still be held liable for any defamatory comments candidates might make. The Supreme Court finally removed this danger by holding in the 1959 *WDAY-TV* decision that, given the no-censorship requirement in the law, stations could not be held responsible for whatever candidates might say.

FCC Rules and Regulations

As required by changing circumstances, over the years the FCC has defined and refined what the relatively few words in Section 315 are to mean in practice. Such definition has focused on phrases such as "legally qualified candidate"; "equal opportunities"; "use" of a station; how to determine rates charged; and, after 1959, which programs were entirely exempt from the provisions. Indeed, the rules vary for primary and general elections as to who is covered and for what period of time (requiring, for example, an even more stringent "lowest unit charge" price requirement for candidates' advertising 45 days before a primary and 60 days before a general election). Stations must maintain detailed and up-to-date records of all requests for political time and of all actual sales and/or uses of airtime for a period of two years.

These many complications grew steadily more involved after 1960 with an accumulation of numerous FCC decisions and court cases; the confusion finally led to publication by both the FCC and the National Association of Broadcasters of regularly revised booklet-length "primers" or "catechisms," usually in question-and-answer format, of the latest rule interpretations. Despite their complexity—or perhaps because of it—broadcasters, candidates, and the public refer to these rules by the shorthand term of *equal time,* even though far more is involved than merely an equitable provision of time.

As but one example—prior to 1960, debates between or among candidates counted as a "use" requiring Section 315 treatment under FCC rules. For the 1960 election, Congress temporarily suspended Section 315 (leading to the so-called Great Debates between Kennedy and Nixon, which were carried on both radio and television). Effective in 1975, the FCC allowed debates to occur without triggering Section 315 requirements if the debates were sponsored by a disinterested third party—initially, the League of Women Voters. Only in 1983 were broadcasters themselves allowed to sponsor debates as one of the "bona fide" news exemptions in Section 315. National and local debates have increasingly become a staple of American elections since then.

The broadcast industry has attempted repeatedly to have the "equal time" requirements dropped—at least for radio (because of the large number of stations) if not for television as well. Indeed, in the 1980s, a deregulation-minded FCC made the same recommendation to Congress. But all such efforts have generally fallen on deaf ears, given that those who must act to make the change—representatives and senators—are the very people who depend on the access the law provides.

CHRISTOPHER H. STERLING

See also Communications Act of 1934; Fairness Doctrine; Federal Communications Commission; Politics and Radio; Wireless Acts of 1910 and 1912/Radio Acts of 1912 and 1927

Further Reading

Barrow, Roscow L., "The Equal Opportunities and Fairness Doctrines in Broadcasting: Pillars in the Forum of Democracy," *University of Cincinnati Law Review* 37 (Summer 1968)

Farmer's Educational and Cooperative Union v WDAY Inc., 360 US 525 (1959)

Friedenthal, Jack H., and Richard J. Medalie, "The Impact of Federal Regulation on Political Broadcasting: Section 315 of the Communications Act," *Harvard Law Review* 72 (January 1959)

Ostroff, David H., "Equal Time: Origins of Section 18 of the Radio Act of 1927," *Journal of Broadcasting* 24 (Summer 1980)

A Political Broadcast Catechism, 3rd edition, Washington, D.C., National Association of Broadcasters, 1956; 15th edition 2000

United States Federal Communications Commission, *The Law of Political Broadcasting and Cablecasting: A Political Primer*, Washington, D.C.: GPO, 1978; new edition, 1984

European Broadcasting Union

International Association of Public Service Broadcasters

The European Broadcasting Union (EBU) calls itself "the largest professional association of national broadcasters in the world." It is the umbrella organization of public service broadcasting organizations in Europe and beyond. The EBU is a nonprofit organization that is not affiliated with any national government or transnational political institution such as the European Union. Its headquarters are located in Geneva, Switzerland. The EBU's radio collaboration is called Euroradio. The organization facilitates program exchange, develops and provides technological support and legal advice, and lobbies for the continued existence of public broadcasting. In February 2000, the organization celebrated its 50th anniversary.

History

The EBU was founded in February 1950 during the European Broadcasting Conference in Torquay, England, as an international organization for public service broadcasting institutions. This was at a time when many Western European countries, still recovering from World War II, were rebuilding their radio networks and beginning to develop television facilities. Its early task was to help members exchange their programs, as well as to lend technological assistance when needed. Today, the EBU has moved far beyond its original scope of Western European countries, counting 69 active members in 50 countries and 49 associate members in 30 countries. Reflecting the end of the Cold War, in 1993 the EBU merged with the Organisation Internationale de Radiodiffusion et Télévision (OIRT), the former organization of Eastern European Broadcasters.

The EBU is a major international broadcasting institution that negotiates broadcasting rights for its members, coordinates co-productions, operates satellites, consults on legal issues, and stimulates European cultural life. Over the last decade, it has become more aggressively involved in media policy. The EBU is now a major player, lobbying European and international institutions defending and guaranteeing the survival of public broadcasting despite the growing commercial competition in this area. The EBU has a staff of more than 250 people, including a dozen in Moscow and another dozen in the U.S.

Members and Finances

The EBU allows as active members only radio and television broadcasters committed to the public service idea of broadcasting, highlighting information, education, and entertainment. Public service broadcasters tend to be financed mostly but not exclusively through viewers' fees, membership, or taxes on televisions or radios. The EBU statute officially asks for broadcasting organizations that fulfill a public service mission and achieve 98 percent national penetration. Its 69 active members are from 50 countries in Europe and adjacent areas in North Africa and the Middle East. Member organizations are in countries such as Austria, Bosnia, France, Hungary, Ireland, Israel, Portugal, Russia, Tunisia, and the United Kingdom. Some of the more influential organizations relating to radio are the financially strong British Broadcasting Corporation (BBC) and the Association of German Broadcasters (ARD). Moreover, the EBU also includes 49 associate members in 30 countries in the Americas, Africa, Asia, and Australia, utilizing and feeding into the EBU network.

The EBU is financed through annual membership fees, which are determined by the number of radio and television households each member reaches. In addition, members also must pay for the technical transmission costs of EBU news items and programs that they air. In the last decade, the EBU commercialized some of its services, offering them on a fee basis to nonmembers. It also sublicenses rights to broadcast sporting events. In 1999 the EBU's annual revenue was 407 million Swiss francs (about U.S.$255 million), including 189 million Swiss francs for rights to broadcast sport events (about U.S.$118 million) and 117 million Swiss francs (about U.S.$75 million) for network transmission charges. Euroradio has a potential reach of 400 million listeners. It transmits approximately 2,000 concerts and operas, 400 sports events, and 120 major news events per year.

Organization

After the inauguration of the EBU in 1950, the first major event for the organization was the 1953 coronation of Queen Elizabeth II, which became the world's first live multinational television transmission. The Eurovision program exchange office was originally opened in Brussels in 1955, but a year later it moved most of its functions with the EBU to Geneva, where the operations were fully centralized by 1993. Since 1961 the EBU coordinates and transmits a daily television news feed (EVN) to all members. The late 1960s brought the change of international transmission to satellite. In 1970 the EBU opened a bureau in New York and another in Washington in 1987, followed by an office in Moscow.

The increasing deregulation of national media systems changed the media landscape in Europe tremendously. Facing increasing competition and membership interest from commercial broadcasters in most European countries, in 1990 the EBU underscored its status as an organization of public service broadcasters in its Marino Charter. Although its television transmission had formerly been based on free exchange systems between active members, the EBU commercialized its operations (Eurovision Network Services) and began selling transmission rights to non-members in 1994. In the last decade of the 20th century, the organization moved further away from its original mission by launching several television and radio outlets. Since 1998, for example, the EBU has offered its members a free classic radio program throughout the night *(Euroclassic-Notturno)*.

In the last few years, both television and radio operations have been shaped by the change from analog to digital technology. The EBU and its technical departments developed and facilitated for its members many new technological advancements into high definition television (HDTV), digital transmission, and digital audio broadcasting (DAB). Recently the EBU has ventured into collaborations and sub-licensing with commercial broadcasters, to avoid investigations by the EU commission for anticompetitive practices. (Several commercial broadcasters have challenged the EBU's exclusive agreements in court and in political committees.)

EBU Radio Department

Although the EBU radio department tends to be in the shadow of its more visible Eurovision branch, it is nevertheless an important international cooperation for radio program exchange and technology transfer. Radio has been part of the EBU since its launch in 1950. Over the years its expansion has reflected the major technical transformations in this field, with the EBU engineers often in the forefront of developing new technologies. In 1989 the EBU radio department launched Euroradio to provide "international exchange of high quality digital sound programs," as its mission statement declares. It began digital transmission of its programs in 1994 via the Euroradio Control Centre (ERC) in Geneva. The Euroradio network is based on 40 satellite-to-earth stations utilizing a digital transmission system. The network currently operates on a digital system incorporating an audio base-band of 20 kilohertz and auxiliary data channels. It transmits its programs via two carriers on the Eutelsat II F4 satellite.

The EBU radio department serves under the direction of the EBU radio committee, an elected body of member representatives who provide guidelines for programming strategies. The radio department organizes radio program exchanges between all or several members. The programs are mostly music concerts, sports events, and news or current affairs. The department also arranges conferences and events for members and holds workshops for foreign journalists and radio producers. Members often contact the department when they need professional, technological, or legal advice. EBU also publishes the "EBU Radio News Fax Letter."

In addition, the EBU supports the development of specialized programming such as radio features and documentaries, radio dramas, educational and other programs for young people. With these niche and even experimental radio programs, the EBU supports the public service mission of its members, who often are required by law to inform, educate, and entertain a variety of societal groups, including minorities. For instance, EBU organizes the International Radio Feature Conference, in which feature producers can present and compare their work. Another example is the Radio Drama Project Group that organizes workshops, initiates radio dramas (often from smaller, less affluent members), and stimulates collaboration between radio drama producers in all member countries. The Project Group biennially commissions a major radio play. In the past, well-known writers such as Mario Vargas Llosa and Anthony Burgess have been among those whose work was aired.

To celebrate the new millennium, the EBU initiated a number of radio projects supplied by its members. These projects included, among others, an experimental audio art collage with sounds of the century, a collection of commissioned compositions, and a series of panels and lectures on human rights.

Music

The music department initiates new music projects, coordinates music transmission between member stations, and negotiates music broadcasting rights for its members. Annually it organizes and offers more than 2,000 live and deferred musical events such as operas, classical concerts, jazz, and rock via the Euroradio transmission system. One hundred of those are within regularly scheduled program slots; the others can be ordered on demand. EBU members can select and request all events via the internet through EBU's MUS internet software. Members can download programs directly from the internet through SATMUSIC, "a program which synchronizes the automatic recording of high quality digital sound programme exchanges for EBU members," according to its web site. The Euroradio Control Centre also has a direct satellite connection to the Metropolitan Opera in New York to distribute performances to its members. The music department also coordinates the *Euroclassic-Notturno* produced by the BBC. It provides an all-night program with light classical concerts by member stations; these include breaks to allow localized announcements by the members' hosts.

The music department has recently ventured into compact disc production by publishing a collection of traditional music on the Ocora label. It sponsors the Euroradio Big Band Concert and the international Forum of Young Performers. This music sponsorship becomes more important as national public service stations face political pressure to tighten their budgets. Euroradio's sponsored musical events may not be as well known as the European Song Contest organized by Eurovision, but they have become an important stimulus of music culture in Europe.

News

The news division offers local feed to other members, sets up broadcasting centers on location as important events occur, and supports foreign correspondents' efforts to broadcast radio reports to their listeners as fast as possible. The provision of timely radio news transmissions is very important in many European countries, where a majority of people turn to radio for breaking national and international news during the day. Supported by the EBU's foreign bureaus, the news division also coordinates coverage of international political events such as elections, conferences, and political conventions.

Other tasks of the news department include the negotiation of access rates to other transmission networks and development and implementation of new technology, such as digital satellite telephone stations and lightweight news-gathering equipment. The division also organizes a joint EBU–NABA (North America Broadcasters Association) conference on radio news and current affairs. In special meetings, division staff discuss issues such as international news flow and the ethics of international news reporting.

Sports

In its first years of existence, the EBU was typically guaranteed the broadcast rights to major sport events such as the Olympics and soccer championships, as there were no commercial counterparts with similar coverage that were able to bid for the rights. In countries with more than one EBU member, these public stations would often share or take turns in covering major events. But the rise of commercial television and radio in the early 1980s changed the situation tremendously when new media moguls began to bid for rights to popular sports events. As a result the cost of broadcasting rights has skyrocketed.

The EBU captured the rights to broadcast the 2000 Olympic Games despite being outbid by Murdoch's FOX network. The International Olympic Committee decided on EBU because of its terrestrial penetration that no satellite provider could match at the time. Similarly, the EBU won the rights to air track-and-field competitions but lost the rights for the next two soccer World cups to the commercial Kirch Group from Germany. Now the EBU must sometimes rely on sublicense agreements and the support of national parliaments to keep such events on terrestrial channels.

The radio department typically negotiates jointly with the EBU television department. On location, it supports national broadcasters in their work. Besides top sports events, the EBU, following its public service mission, is also interested in broadcasting minor sports that commercial stations do not find appealing. In this role, the EBU radio covers and assists at about 150 sporting events each year.

Technology

The radio division is part of a broader EBU project that attempts to develop a more efficient and automated system of transmitting traffic and travel information to viewers, licensers, and agent systems. The goal is to keep public service stations updated with the latest technology, enabling them to counter commercial competition in this area. Moreover, the EBU has initiated joint experiments with web-radio and interactive internet radio. For more than a decade it has initiated the development of digital studio and transmission equipment.

Collaborations

The EBU works in partnership with all major international broadcasting organizations such as Asia Pacific Broadcasting Union (ABU), North American Broadcasters Association (NABA), the Union of National Radio and Television Organizations in Africa (URTNA), the Arab States Broadcasting Union (ASBU) and the Organizations de la Television Iberoamericana (OTI).

Future

As an increasing number of commercial stations build their market share by offering entertainment programs with mass appeal and outbidding public stations for rights to popular sports events, national political bodies ask public broadcasters to be more fiscally responsible and approve increases in viewer fees or licenses very reluctantly. As a result European national and regional public service broadcasters face mounting pressure from both sides. The EBU is more important then ever in helping the public broadcasting concept to survive. Members can synchronize their efforts, more efficiently share their resources, and use the institution as a powerful lobbying organization on an international level. In addition, at a time when commercial format radio is gaining a strong hold in many European countries, the EBU radio department can help to maintain their members' mission of broadcasting programs with less mass appeal, such as minority programs, radio drama, or cultural events. The EBU's support for public service radio is a way of ensuring that radio remains an information and education medium in Europe and beyond.

ELFRIEDE FÜRSICH

See also International Radio Broadcasting; Public Service Broadcasting

Further Reading

Aldridge, Meryl, and Nicholas Hewitt, editors, *Controlling Broadcasting: Access Policy and Practice in North America and Europe,* Manchester and New York: Manchester University Press, 1994

Avery, Robert K., editor, *Public Service Broadcasting in a Multichannel Environment: The History and Survival of an Ideal,* New York: Longman, 1993

Brack, Hans, *The Evolution of the EBU Through Its Statutes from 1950 to 1976,* Geneva: European Broadcasting Union, 1976

Browne, Donald R., *Electronic Media and Industrialized Nations: A Comparative Study,* Ames: Iowa State University Press, 1999

EBU: European Broadcasting Union, 25 Years, Geneva: EBU, 1974

European Broadcasting Union website, <www.ebu.ch>

Noam, Eli M., *Television in Europe,* New York: Oxford University Press, 1991

Paulu, Burton, *Broadcasting on the European Continent,* Minneapolis: University of Minnesota Press, 1967

Evangelists/Evangelical Radio

Conservative Protestant Religious Stations and Programs

Evangelical radio forms a distinct subgenre within religious radio, referring primarily to programs with a teaching/preaching format, often incorporating hymns or other kinds of sacred music. The intent of evangelical radio programming is to convert unbelievers or to reconvert lapsed Christians by stressing the Bible's call to repent and accept Christ as a personal savior.

It is difficult to be precise about what comprises "evangelical radio" as distinct from other forms of religious radio, because in some sense every effort at putting religion on the air constitutes an invitation to learn more about the principles being presented—and because radio itself, as an advertising-saturated medium, is built around an evangelistic pattern. Nearly everything on radio, from ubiquitous 60-second car sales spots to public radio fund drives, has something of an evangelical ring to it.

In addition, the term *evangelical religion* means different things to different groups of people, and the meaning has changed over the course of the 20th century. An *evangelist* can mean any person who seeks to convert another to his or her own religious beliefs. In the popular mind, many tend to link the terms *evangelical* and *fundamentalist,* because both offer a Bible-centered worldview with an emphasis on personal conversion. However, the two are not coterminous. The modern evangelical movement had its roots in the fundamentalist movement of the 1910s and 1920s and maintained ties to fundamentalism through Bible colleges, summer camps, publishing, and radio, but a group of so-called neo-evangelicals decisively broke with fundamentalism by the late 1950s under the leadership of Billy Graham. Contemporary evangelicals occupy a middle ground between liberal religion and fundamentalism. Today, *evangelicalism* refers to a loose coalition of conservative Protestant groups in North America, including Baptists, Holiness-Pentecostalists, nondenominational evangelists, and charismatic Protestants.

Although not all of these traditions have been equally involved in media evangelism, the media have been an important tool of American evangelicals in their quest to fulfill the so-called Great Commission, the instruction of Jesus Christ to his followers: "Go ye into all the world, and preach the gospel to every creature" (Mark 16:15). The preaching of the Christian message in or to every country is considered by many evangelists to be a necessary precondition for the second coming of Christ (e.g., Matt. 24:14), and radio was touted as a providential means to accomplish this important end.

Origins

Revivalism—that is, religious meetings that work through music, word, and emotional appeal to encourage conversion in those attending—lent itself naturally to the new medium of radio in the 1920s. Revivalism grew out of the highly successful late 19th- and early 20th-century mass-audience revival campaigns of Dwight Moody and Billy Sunday and traveling revival movements such as Chautauqua. As promoters of radio evangelism were fond of reminding potential donors, a single broadcast could reach more people than even Dwight Moody had been able to reach in a lifetime. Because folk and campmeeting revivalism was centrally an aural experience—the spoken and heard word being experientially more powerful than the written and read word—the format of revival sermons and meetings made it onto radio with little adaptation.

In fact, radio evangelism is one of the medium's oldest program genres. In the largely unregulated early years of radio broadcasting in the United States, municipal and private stations alike were in search of material to fill time. Evangelists, ever on the lookout for ways to speak to larger and larger audiences, stepped in to fill the need and never left the airwaves, though their presence was not always so sought after by station owners and broadcasters. In addition, some evangelistic denominations and religious organizations developed their own stations to promote gospel on the air: WMBI Chicago (owned by the Moody Bible Institute), KFUO St. Louis (Lutheran Church, Missouri Synod), and KFSG Los Angeles (International Church of the Foursquare Gospel) were three of the earliest, all coming on the air within two years of each other in the mid-1920s; in 2003, all three were still broadcasting as Christian radio stations.

However, once network radio was firmly entrenched, evangelicals found it harder to access airtime, even with donations from loyal listeners and supporters. Their often strident "hellfire and damnation" message worked against both networks' desire for mass audiences and advertisers' appeals for consumer spending, making evangelical radio a risk for network broadcasters. Additionally, airtime became more expensive, and evangelical radio was largely dependent on listener donations for the funds to purchase airtime. The National Broadcasting Company (NBC) and the Columbia Broadcasting System (CBS) early on developed a policy of donating a block of airtime for religious broadcasting to representatives of the major religious groups in America. As a fragmented and largely grassroots movement from the 1920s through the 1940s, American evangelicalism found itself unable to obtain this donated airtime from the networks.

Instead, evangelical radio focused on buying time on individual stations or developing its own small-scale independent networks. Although Christian benevolent organizations (e.g., the Gideons and the Christian Business Men's Association) made regular and sometimes substantial donations to evangelical radio efforts, most broadcasters relied heavily on individual donations to pay for airtime. Listeners sometimes proved creative and resourceful in scraping together small amounts of money to send to support their favorite broadcasts.

Mutual Broadcasting System—the only national network that then sold time for religion—collected over $2.1 million for its religious broadcasts in 1942. In 1944 this amount had jumped to $3.5 million, a full quarter of the network's income. By 1943, reported *Variety* magazine, an estimated $200 million was "rolling into church coffers each year from radio listeners," allowing racketeers and "religious pirates" to get rich quick without adhering to standards of accounting. Broadcasters' appeals for funds proved increasingly controversial; Mutual eventually prohibited on-air appeals for funding. In the late 1940s, for example, the vigilant director of religious activities at Mutual Broadcasting, Elsie Dick, insisted that Walter Maier (speaker of the *Lutheran Hour* on her network) refrain from using even relatively vague statements, such as "if you want these broadcasts to continue, write to us to assure us of your interest." Broadcasters on Mutual could not follow a request to "pray for the work of this broadcast" with the program's mailing address, as this would violate Mutual's policy against the solicitation of funds. Individual stations that accepted religious broadcasts sometimes also established similar policies. Broadcasters sometimes circumvented these restrictions by offering "free" merchandise such as Bibles, tracts, calendars, or commemorative pins. Through the written requests of listeners for promotional items, broadcasters could build a mailing list for direct-mail appeals instead of using airtime for financial appeals.

Examples of Evangelical Radio

Although it is impossible to be exhaustive in listing all evangelists and their programs throughout the years, a few examples illustrate the genre and its approach to missionizing America and the world: the *Back to the Bible Hour,* the *Radio Bible Hour,* the *Lutheran Hour,* and the *Old Fashioned Revival Hour.*

Started in 1939 by Theodore Epp, the *Back to the Bible Hour* was a daily gospel broadcast and Bible-study program originating from Lincoln, Nebraska. By the mid-1950s, Epp could claim that the "sun never set" on *Back to the Bible,* which was heard somewhere in the world at any given minute through AM, FM, or shortwave. The ministry had its own large two-story building in downtown Lincoln, where half a million letters were received annually and where some 300,000 copies of the *Good News Broadcaster* and the *Young Ambassador* (the latter aimed at teenagers) were printed and mailed each month. A staff of over 150 workers and volunteers provided music and choir direction for the broadcast, sorted and answered mail, taped and shipped out recordings to stations, staffed a round-the-clock prayer room, and coordinated appearances of Epp and his field evangelists at rallies and meetings. Epp authored 70 books, started a Bible correspondence school, and founded a Back to the Bible Missionary Agency (now International Ministries). At his death in 1985, the program was continued with speakers Warren Wiersbe (1985–92) and Woodrow Kroll (1992–present); it is currently syndicated on over 385 stations and on-line.

More controversial and colorful was the stridently fundamentalist *Radio Bible Hour* of the Reverend J. Harold Smith, which began in 1935 in South Carolina but for some years was forced off American stations after its pugnacious attacks on the Federal Council of Churches. In 1953 Smith moved the *Radio Bible Hour* to a Mexican border station, XERF, in Ciudad Acuna, just south of the Texas border, where he continued to broadcast until Mexico banned English language religious broadcasts from superpower border stations. The program can still be heard over 50 stations and on-line.

The *Lutheran Hour,* sponsored by the International Lutheran Laymen's League of the conservative Lutheran Church, Missouri Synod, is the most widely syndicated evangelical radio program. Still heard on 1,200 stations worldwide and on-line, the *Lutheran Hour* was hosted from 1930 to 1950 by Dr. Walter A. Maier, a professor at Concordia Theological Seminary in St. Louis. His successors continued Maier's sermon-and-song format and its nondenominational approach to Christian outreach.

Finally, perhaps the best-known evangelical radio program was the *Old Fashioned Revival Hour,* the creation of southern California evangelist Charles Fuller in 1934. Recorded for many years in front of a live audience at the Long Beach Auditorium, Fuller's *Revival Hour* combined lively choral and barbershop-style revival hymns with energetic preaching; the program pulled audiences of 20 million listeners weekly by the mid-1940s. Fuller was also active behind the microphone in bringing together evangelical broadcasters in the 1940s and 1950s to advocate for paid-time programming through the National Association of Evangelicals and the National Religious Broadcasters.

One important thrust of radio evangelism—and a strategy employed by each of the media evangelists noted above—has been to extend religious broadcasting worldwide in the various languages of each nation. Clarence Jones, who worked in the late 1920s on broadcasts from the Chicago Gospel Tabernacle, was one of the pioneers of long-range overseas religious radio. He founded radio station HCJB in Quito, Ecuador. Many North American evangelists, while securing time and broadcast airspace on the AM spectrum, also quietly used shortwave or mailed transcription discs to overseas stations in order to be heard by as wide as possible a swath of the globe. The *Lutheran Hour,* for example, broadcasts in over 130 countries in a multitude of languages, from French to Quechua to Zulu.

Although reaching the far corners of the globe was one goal of radio evangelists, reaching the hearts of listening individuals was the other and related goal. In other words, massive broadcast coverage mattered only as far as that coverage would convert people one at a time.

Trends since the 1960s

Billy Graham's evangelistic mass-media campaigns beginning in 1957 helped catapult evangelicalism back to a position of cultural influence. His *Hour of Decision* program was the first religious broadcast to be carried as a paid-time evangelistic broadcast on the American Broadcasting Company (ABC) radio network. And since the 1960s, the Christian media industry has literally exploded in growth, with television, publishing, music, and internet being added to radio. In 1971 there were 400 stations airing religious programming; in 1999 there were over 1,730 such stations, with 1,400 of those considered "full-time" religious stations airing 15 hours or more of religious programming per week.

Despite the dynamic growth, the format of evangelistic programs on radio has changed remarkably little. In stark contrast to the secular end of the contemporary radio industry, most evangelical radio programs are sponsored by a single organization or ministry. Although some are widely syndicated, nearly all evangelical programs are confined to Christian-format radio stations, in keeping with the radio industry's trend toward niche marketing. Gospel-oriented preaching programs continue to thrive within the world of Christian media, but since the 1960s, some religious broadcasters have expressed concern that evangelical radio is a religious ghetto, serving its own rather than reaching new converts.

As early as 1961, liberal Protestant Charles Brackbill, member of the Broadcasting and Film Commission of the National Council of Churches of Christ in America, criticized what he described as radio evangelism's tired format: "the loud and the sad and the intense voices pouring out on the faithful with their 'heartfelt' pleas for 'letters,' pictures, or books or blessed handkerchiefs." Evangelism, Brackbill argued, should seek to reach the unchurched through proven commercial broadcasting techniques: a quick first impression, a catchy "hook," and lots of repetition. In 1964 the Mennonites tried 30- and 60-second ad spots for the gospel message, criticizing traditional radio programming for attracting "an audience which already has some tendency toward spiritual orientation," in the words of Dr. Henry Weaver, the developer of the series. HCJB founder and director Clarence Jones lamented in 1970 that missionary stations tended to drift in the direction of speaking to believers—who, after all, were the source of any station's continuing funds (incidentally, many of these same concerns would surface over televangelism, where the audience numbers, production costs, and cultural stakes were even higher).

Some scholars expressed concern that evangelism as a type of radio program would be replaced by magazine or talk-format programs or by Christian music programming. In the late 1980s, for example, one study suggested that only 37 percent of all programs on religious stations focused on preaching or teaching. However, according to the *2000 Directory of Religious Media,* "teaching/preaching" is still the largest category among religious radio station formats, followed by "inspirational," "gospel," and "Southern gospel." This same directory lists hundreds of individual evangelistic programs, some syndicated or beamed by satellite to over 1,000 radio stations. Many of the radio programs are also broadcast on-line through the radio stations' websites, expanding the notion of "evangelical radio" far beyond the physical reaches of a radio signal. Perhaps internet broadcasting will prove another fruitful growth area for religious broadcasting. How successful evangelical radio is in reaching unchurched people, for whom the "good news" is "news" indeed, remains an open question. But there is no doubt that radio evangelism is, and has been throughout the century, a popular and profitable genre that is in no danger of vanishing from the American scene.

TONA J. HANGEN

See also Far East Broadcasting Company; Gospel Music Format; McPherson, Aimee Semple; Religion on Radio

Further Reading

Carpenter, Joel A., *Revive Us Again: The Reawakening of American Fundamentalism,* New York: Oxford University Press, 1997

Dick, Donald, "Religious Broadcasting, 1920–1965: A Bibliography," *Journal of Broadcasting* 9 (Summer 1965)

Directory of Religious Media (annual; 1994–)

Dorgan, Howard, *The Airwaves of Zion: Radio and Religion in Appalachia,* Knoxville: University of Tennessee Press, 1993

Erickson, Hal, *Religious Radio and Television in the United States, 1921–1991: The Programs and Personalities,* Jefferson, North Carolina: McFarland, 1992

Martin, William, "The God-Hucksters of Radio," *Atlantic Monthly* 225 (June 1970)

Neely, Lois, *Come Up to This Mountain: The Miracle of Clarence W. Jones and HCJB,* Opa Locka, Florida: World Radio Missionary Fellowship, and Wheaton, Illinois: Tyndale House, 1980

NRB: National Religious Broadcasters (1998–)

Schultze, Quentin J., editor, *American Evangelicals and the Mass Media,* Grand Rapids, Michigan: Academie Books/ Zondervan, 1990

Stout, Daniel A., and Judith Mitchell Buddenbaum, editors, *Religion and Mass Media: Audiences and Adaptations,* Thousand Oaks, California: Sage, 1996

Sweet, Leonard I., editor, *Communication and Change in American Religious History,* Grand Rapids, Michigan: Eerdmans, 1993

Ward, Mark, *Air of Salvation: The Story of Christian Broadcasting,* Manassas, Virginia: National Religious Broadcasters, and Grand Rapids, Michigan: Baker Books, 1994

Everett, Kenny 1944–1995

British Disc Jockey

Kenny Everett (who changed his name from Maurice Cole in the mid-1960s) was fascinated by radio from childhood; with money gained from newspaper delivery rounds, he bought reel-to-reel tape recorders on which he recorded inventive mini-programs: music interspersed with comedy clips and speeches by politicians edited to humorous effect.

In 1964 he sent one of these programs, "The Maurice Cole Quarter of an Hour Show" (which, typically, actually ran for some 20 minutes), to the British Broadcasting Corporation (BBC) in London. To his amazement and delight he was invited onto a morning magazine program on the Light Programme (the Corporation's light music and speech network), on which the tape was played in full and Everett was interviewed. Although he impressed the BBC producers sufficiently to be invited back for a formal audition, he was told there were no vacancies at the corporation. However, one of the producers was covertly acting as a "talent spotter" for a new "pirate" radio station, the United States–backed Radio London, which was to become one of the most successful of the unauthorized stations broadcasting from outside the United Kingdom's territorial limits. Soon after boarding the converted U.S. minesweeper that served as Radio London's studios and transmitter, Everett met Canadian Dave Wish—who changed his name to Dave Cash. The two became firm friends and, after listening to tapes of a double-headed disc jockey show on KLIF in Texas, the *Charlie and Harrigan Show,* which featured comic characters wisecracking with a generally irreverent style, the pair began the *Kenny and Cash Show,* which became hugely popular. Everett would often stay up all night recording what were quickly to become his trademark comedy pieces for his own shows and "zany" promotions for other disc jockeys.

However, Everett's unpredictable comments and outspoken views—especially on religion—led to his being dismissed by the station. He had taken exception to the pronouncements from U.S. evangelist Garner Ted Armstrong, who paid for a half-hour daily program that was scheduled in the middle of Everett's. The disc jockey frequently edited Armstrong's tapes to make it appear as if the evangelist were promoting violent crime, and Armstrong was generally lampooned by Everett, who had once been to a college for training Catholic priests. Armstrong happened to be visiting England during one of Everett's outbursts, and the evangelist demanded Everett be removed from the station or he would remove his program—one of Radio London's most lucrative contracts.

For a time Everett recorded a program for Radio Luxembourg, but he was again dismissed, this time for admitting in a newspaper interview that he had smoked cannabis. Forgiven by Radio London, he returned to the station in 1966—this time at ten times his original salary, a then enormous sum of £150 a week—in time to be chosen by the Beatles to accompany them to the United States to be the official disc jockey on what was to be their last tour. During this tour he forged a close relationship with his fellow Liverpudlians—then the biggest show business phenomenon in the world—and this was to pay off in subsequent exclusive interviews, previews of the group's albums, and the Fab Four's recording special jingles for Everett on both BBC and commercial radio.

When it became clear that the government was intent on forcing the pirates off the air, Everett jumped ship and this time was accepted by the BBC, which was about to launch the pop network Radio 1. At first, though, Everett was restricted to a weekly lunchtime show and recording jingles and promotions. His patience was rewarded with a weekly Sunday morning program and, eventually, a daily show. However, Everett's outspokenness and refusal to obey instructions from the "suits" led to his being sacked by the BBC in April 1970. Officially this was because of a potentially libelous remark—although clearly intended humorously—about the wife of a government minister. In reality, though, this incident provided the excuse the BBC mandarins had been looking for since Everett began attacking the Musicians' Union, whose leaders were in delicate discussions with the corporation over the amount of airtime given to playing music off of records (the Union thought that unlimited use of such recordings would deprive its members of their livelihood).

Then began a fallow period for the disc jockey, who was by this time widely recognized as one of the greatest on-air talents in the United Kingdom. Everett recorded programs for a few BBC local radio stations and did some television work. Then in early 1973 the BBC relented, and he was allowed back on Radio 1, but only if he would record his shows to allow producers to "vet" them before broadcast. Their judicious editing left some programs running considerably shorter than their allotted transmission time. However, salvation from this unsatisfactory existence came with the start of legal commercial radio in October 1973. Everett became one of the first—and most important—hires of the greater London service, Capital Radio. By Christmas of that year, Everett had teamed up again with Dave Cash to revive the *Kenny and Cash Show* at the vital breakfast period—again this was a huge success. However, now Everett's personal life was to have a dramatic impact on his career. Although he married in 1969, Everett had struggled throughout his adult life with latent homosexuality and had had a series of crushes on heterosexual men. When one of these—an engineer at Capital—rebuffed him in 1975, Everett made a serious attempt at suicide, leaving a note explaining his anguish at the unrequited love. However, much to Everett's evident frustration, he was found just before the sleeping tablets had a lethal effect.

After a period of recuperation, Everett returned to a weekly program at Capital. This was one of the most creative periods of his career, during which he created a comic space serial, Captain Kremmen and the Krells. Kremmen—forever saving the world from the evil Krells—was Everett's alter ego, inspired by 1950s BBC radio serials such as *Journey into Space.*

In 1978 Thames Television persuaded him to do a highly successful and inventive television show for the Independent Television (ITV) network. In 1982 the show switched to the BBC and was to run for five years at prime time on the mainstream BBC-1 channel. In October 1981 he also switched to BBC for his radio work—this time on the supposedly easy listening network, Radio 2. A risqué joke about Prime Minister Margaret Thatcher on the last program under his contract led to his final departure from BBC radio. In 1988 Everett returned to Capital Radio for a daily program on the golden oldies AM service, "Capital Gold," which he continued until July 1994, when, increasingly ill from AIDS-related illnesses, he felt obliged to give up work. He died in April 1995. His funeral service a few days later was attended by many of the biggest names in British show business. Everett's "Wireless Workshop" studio was donated by his sister Kate to the Paul McCartney–sponsored Liverpool Institute for the Performing Arts.

RICHARD RUDIN

See also British Broadcasting Corporation; British Commercial Radio; British Disk Jockeys; British Pirate Radio; Capital Radio

Kenny Everett. Born Maurice Cole in Seaforth, Liverpool, England, 25 December 1944. Attended St. Bede's Secondary Modern and Peter Claver College; disc jockey, Radio Luxembourg, 1964–65, Radio London, 1966–67; BBC Light Programme/Radio 1, 1967–70; host, *Nice Time*, Granada Television for ITV network, 1968; disc jockey, BBC, 1970–73, Capital Radio, London, 1973–80, various independent local radio stations, 1975–78; weekly program, BBC Radio 2, 1981–83; host, *The Kenny Everett Television Show*, 1982–1987; Capital Gold, 1988–1993; team leader, *That's Showbusiness*, BBC-TV, 1989–92. Received numerous awards for television and radio work, including Sony Gold award, 1994. Died in London, England, 4 April 1995.

Publication

The Custard Stops at Hatfield, 1982

Further Reading

Barnard, Stephen, *On the Radio: Music Radio in Britain*, Milton Keynes, Buckinghamshire, and Philadelphia, Pennsylvania: Open University Press, 1989
Chapman, Robert, *Selling the Sixties: The Pirates and Pop Music Radio*, London: Routledge, 1992
Lister, David, *In the Best Possible Taste: The Crazy Life of Kenny Everett*, London: Bloomsbury, 1996
Rocos, Cleo, and Richard Topping, *Bananas Forever: Kenny Everett and Me*, London: Virgin, 1998